Water, Life and Civilisation

Climate, Environment and Society in the Jordan Valley

Water, Life and Civilisation provides a unique interdisciplinary study of the relationships between climate, hydrology and human society from 20,000 years ago to 100 years into the future. At the heart of the book is a series of case studies that integrate climate and hydrological modelling with palaeoenvironmental and archaeological evidence to generate new insights into the Neolithic, Bronze Age and Classical periods in the Jordan Valley. The volume not only develops our understanding of this most critical region, but provides a new approach and new methods that can be utilised for exploring the relationships between climate, hydrology and human society in arid and semi-arid regions throughout the world.

This volume describes how state-of-the-art models can simulate the past, present and future climates of the Near East, reviews and provides new evidence for environmental change from geological deposits, builds hydrological models for the River Jordan and associated wadis and explains how present-day urban and rural communities manage their water supply. It demonstrates how the theories and methods of meteorology, hydrology, geology, human geography and archaeology can be integrated to generate new insights into not only the past, from the hunter-gatherers of the Pleistocene to classical civilisation, but also the present and future. As such, it is an invaluable reference for researchers and advanced students concerned with the impacts of climate change and hydrology on human society, especially in the Near East.

STEVEN MITHEN is Professor of Early Prehistory and Pro-Vice Chancellor for International and External Engagement at the University of Reading. Having originally studied at the Slade School for Fine Art, he has a BA degree in Archaeology (Sheffield University), an MSc in Biological Computation (York University) and a PhD in Archaeology (Cambridge University). He was appointed a lecturer at the University of Reading in 1992, where he has since served as Head of the School of Human and Environmental Sciences (2003–2008) and Dean of the Faculty of Science (2008–2010) prior to his present appointment as a Pro-Vice Chancellor. He directs archaeological fieldwork projects in western Scotland, where he is attempting to reconstruct Mesolithic settlement patterns, and in southern Jordan where he is excavating the early Neolithic village of WF16 in Wadi Faynan. In addition to such archaeological research, he has sought to develop interdisciplinary approaches to the past by integrating archaeology with theories and methods from the environmental and cognitive sciences. He is the author of several books including *The Prehistory of the Mind* (1996), *After the Ice* (2003), *The Singing Neanderthals* (2005) and *To the Islands* (2010), and editor of *The Early Prehistory of Wadi Faynan* (2007, with Bill Finlayson) and *Hunter-Gatherer Landscape Archaeology* (2000). Steven Mithen was elected as a Fellow of the British Academy in 2003.

EMILY BLACK is a senior research fellow at the University of Reading. After completing a BA in Natural Sciences at the University of Cambridge and a DPhil in Andean tectonics at the University of Oxford, in 2000 she was appointed a post-doctoral research fellow at the Climate Division of the National Centre for Atmospheric Science. In 2005, she took up the post of project manager of the Water, Life and Civilisation project. She has published widely in the scientific literature on a variety of topics, including Middle East climate change, African rainfall variability and seasonal forecasting.

INTERNATIONAL HYDROLOGY SERIES

The **International Hydrological Programme** (IHP) was established by the United Nations Educational, Scientific and Cultural Organization (UNESCO) in 1975 as the successor to the International Hydrological Decade. The long-term goal of the IHP is to advance our understanding of processes occurring in the water cycle and to integrate this knowledge into water resources management. The IHP is the only UN science and educational programme in the field of water resources, and one of its outputs has been a steady stream of technical and information documents aimed at water specialists and decision-makers.

The **International Hydrology Series** has been developed by the IHP in collaboration with Cambridge University Press as a major collection of research monographs, synthesis volumes, and graduate texts on the subject of water. Authoritative and international in scope, the various books within the series all contribute to the aims of the IHP in improving scientific and technical knowledge of freshwater processes, in providing research know-how and in stimulating the responsible management of water resources.

EDITORIAL ADVISORY BOARD
Secretary to the Advisory Board
Dr Michael Bonell *Division of Water Science, UNESCO, I rue Miollis, Paris 75732, France*

Members of the Advisory Board
Professor B. P. F. Braga Jr *Centro Technológica de Hidráulica, São Paulo, Brazil*
Professor G. Dagan *Faculty of Engineering, Tel, Aviv University, Israel*
Dr J. Khouri *Water Resources Division, Arab Centre for Studies of Arid Zones and Dry Lands, Damascus, Syria*
Dr G. Leavesley *US Geological Survey, Water Resources Division, Denver Federal Center, Colorado, USA*
Dr E. Morris *Scott Polar Research Institute, Cambridge, UK*
Professor L. Oyebande *Department of Geography and Planning, University of Lagos, Nigeria*
Professor S. Sorooshian *Department of Civil and Environmental Engineering, University of California, Irvine, California, USA*
Professor K. Takeuchi *Department of Civil and Environmental Engineering, Yamanashi University, Japan*
Professor D. E. Walling *Department of Geography, University of Exeter, UK*
Professor I. White *Centre for Resource and Environmental Studies, Australian National University, Canberra, Australia*

TITLES IN PRINT IN THIS SERIES
M. Bonell, M. M. Hufschmidt and J. S Gladwell *Hydrology and Water Management in the Humid Tropics: Hydrological Research Issues and Strategies for Water Management*
Z. W. Kundzewicz *New Uncertainty Concepts in Hydrology and Water Resources*
R. A. Feddes *Space and Time Scale Variability and Interdependencies in Hydrological Processes*
J. Gibert, J. Mathieu and F. Fournier *Groundwater/Surface Water Ecotones: Biological and Hydrological Interactions and Management Options*
G. Dagan and S. Neuman *Subsurface Flow and Transport: A Stochastic Approach*
J. C. van Dam *Impacts of Climate Change and Climate Variability on Hydrological Regimes*
D. P. Loucks and J. S. Gladwell *Sustainability Criteria for Water Resource Systems*
J. J. Bogardi and Z. W. Kundzewicz *Risk, Reliability, Uncertainty, and Robustness of Water Resource Systems*
G. Kaser and H. Osmaston *Tropical Glaciers*
I. A. Shiklomanov and J. C. Rodda *World Water Resources at the Beginning of the Twenty-First Century*
A. S. Issar *Climate Changes during the Holocene and their Impact on Hydrological Systems*
M. Bonell and L. A. Bruijnzeel *Forests, Water and People in the Humid Tropics: Past, Present and Future Hydrological Research for Integrated Land and Water Management*
F. Ghassemi and I. White *Inter-Basin Water Transfer: Case Studies from Australia, United States, Canada, China and India*
K. D. W. Nandalal and J. J. Bogardi *Dynamic Programming Based Operation of Reservoirs: Applicability and Limits*
H. S. Wheater, S. Sorooshian and K.D. Sharma *Hydrological Modelling in Arid and Semi-Arid Areas*
J. Delli Priscoli and A. T. Wolf *Managing and Transforming Water Conflicts*
H. S. Wheater, S. A. Mathias and X. Li *Groundwater Modelling in Arid and Semi-Arid Areas*
L. A. Bruijnzeel, F. N. Scatena and L. S. Hamilton *Tropical Montane Cloud Forests*
S. Mithen and E. Black *Water, Life and Civilisation: Climate, Environment and Society in the Jordan Valley*

Water, Life and Civilisation

Climate, Environment and Society in the Jordan Valley

Edited by Steven Mithen and Emily Black
University of Reading

CAMBRIDGE UNIVERSITY PRESS
Cambridge, New York, Melbourne, Madrid, Cape Town, Singapore,
São Paulo, Delhi, Dubai, Tokyo, Mexico City

Cambridge University Press
The Edinburgh Building, Cambridge CB2 8RU, UK

Published in the United States of America by Cambridge University Press, New York

www.cambridge.org
Information on this title: www.cambridge.org/9780521769570

© Steven Mithen and Emily Black 2011

This publication is in copyright. Subject to statutory exception
and to the provisions of relevant collective licensing agreements,
no reproduction of any part may take place without
the written permission of Cambridge University Press.

First published 2011

Printed in the United Kingdom at the University Press, Cambridge

A catalogue record for this publication is available from the British Library

Library of Congress Cataloging-in-Publication Data

Water, life & civilisation : climate, environment, and society in the Jordan Valley / edited by Steven Mithen and Emily Black.
    p. cm. – (International hydrology series)
Includes bibliographical references and index.
ISBN 978-0-521-76957-0 (Hardback)
  1. Hydrology–Jordan River Watershed. 2. Water-supply–Jordan River Watershed–History. 3. Jordan River Watershed–Antiquities.
4. Water and civilization. 5. Climate and civilization. I. Mithen, Steven J. II. Black, Emily. III. Title: Water, life, and civilisation.
GB791.W384 2011
551.48095694–dc22   2010028679

ISBN 978-0-521-76957-0 Hardback

Cambridge University Press has no responsibility for the persistence or
accuracy of URLs for external or third-party internet websites referred to
in this publication, and does not guarantee that any content on such
websites is, or will remain, accurate or appropriate.

Bruce Sellwood recording a section of the Lisan Marl for the Water, Life and Civilisation project, 2006.

This volume is dedicated to Professor Bruce Sellwood (1947–2007).

Bruce was a pioneer of integrating palaeoclimatic modelling and geological research. He was an inspirational figure within the Water, Life and Civilisation Project and has been sorely missed by his colleagues for both his academic contributions and bonhomie.

Contents

List of figures	*page* ix
List of tables	xxii
List of contributors	xxiv
Acknowledgements	xxvii

1 Introduction: an interdisciplinary approach to Water, Life and Civilisation   1
   Steven Mithen and Emily Black

Part I Past, present and future climate   11

2 The present-day climate of the Middle East   13
   Emily Black, Brian Hoskins, Julia Slingo and David Brayshaw

3 Past climates of the Middle East   25
   David Brayshaw, Emily Black, Brian Hoskins and Julia Slingo

4 Future climate of the Middle East   51
   Emily Black, David Brayshaw, Julia Slingo and Brian Hoskins

5 Connecting climate and hydrological models for impacts studies   63
   Emily Black

Part II The palaeoenvironmental record   69

6 A review of palaeoclimates and palaeoenvironments in the Levant and Eastern Mediterranean from 25,000 to 5,000 years BP: setting the environmental background for the evolution of human civilisation   71
   Stuart Robinson, Stuart Black, Bruce Sellwood and Paul J. Valdes

7 Palaeoenvironments of the southern Levant 5,000 BP to present: linking the geological and archaeological records   94
   Claire Rambeau and Stuart Black

8 Using proxy data, historical climate data and climate models to investigate aridification during the Holocene   105
   Emily Black, David Brayshaw, Stuart Black and Claire Rambeau

9 Palaeoenvironmental and limnological reconstruction of Lake Lisan and the Dead Sea   113
   Stuart Black, Stuart Robinson, Richard Fitton, Rachel Goodship, Claire Rambeau and Bruce Sellwood

Part III Hydrological studies of the Jordan Valley   129

10 The impacts of climate change on rainfall-runoff in the upper River Jordan: methodology and first projections   131
    Andrew Wade, Emily Black, Nicola Flynn and Paul Whitehead

11 Modelling Dead Sea levels and rainfall: past, present and future   147
    Paul Whitehead, Dan Butterfield, Emily Black and David Plinston

12 The hydrology of the Wadi Faynan   157
    Andrew Wade, Paul Holmes, Mohammed El Bastawesy, Sam Smith, Emily Black and Steven Mithen

13 Future projections of water availability in a semi-arid region of the eastern Mediterranean: a case study of Wadi Hasa, Jordan   175
    Andrew Wade, Ron Manley, Emily Black, Joshua Guest, Sameeh Al Nuimat and Khalil Jamjoum

Part IV Human settlement, climate change, hydrology and water management   189

14 The archaeology of water management in the Jordan Valley from the Epipalaeolithic to the Nabataean, 21,000 BP (19,000 BC) to AD 106   191
    Bill Finlayson, Jaimie Lovell, Sam Smith and Steven Mithen

| 15 | From global climate change to local impact in Wadi Faynan, southern Jordan: ten millennia of human settlement in its hydrological context
Sam Smith, Andrew Wade, Emily Black, David Brayshaw, Claire Rambeau and Steven Mithen | 218 |

| 16 | Palaeoenvironmental reconstruction at Beidha, southern Jordan (*c.* 18,000–8,500 BP): Implications for human occupation during the Natufian and Pre-Pottery Neolithic
Claire Rambeau, Bill Finlayson, Sam Smith, Stuart Black, Robyn Inglis and Stuart Robinson | 245 |

| 17 | The influence of water on Chalcolithic and Early Bronze Age settlement patterns in the southern Levant
Jaimie Lovell and Andrew Bradley | 269 |

| 18 | Modelling water resources and climate change at the Bronze Age site of Jawa in northern Jordan: a new approach utilising stochastic simulation techniques
Paul Whitehead, Sam Smith and Andrew Wade | 289 |

| 19 | A millennium of rainfall, settlement and water management at Humayma, southern Jordan, *c.* 2,050–1,150 BP (100 BC to AD 800)
Rebecca Foote, Andrew Wade, Mohammed El Bastawesy, John Peter Oleson and Steven Mithen | 302 |

Part V Palaeoeconomies and developing archaeological methodologies 335

| 20 | The reconstruction of diet and environment in ancient Jordan by carbon and nitrogen stable isotope analysis of human and animal remains
Michela Sandias | 337 |

| 21 | Irrigation and phytolith formation: an experimental study
Emma Jenkins, Khalil Jamjoum and Sameeh Al Nuimat | 347 |

| 22 | An investigation into the archaeological application of carbon stable isotope analysis used to establish crop water availability: solutions and ways forward
Helen Stokes, Gundula Müldner and Emma Jenkins | 373 |

| 23 | Past plant use in Jordan as revealed by archaeological and ethnoarchaeological phytolith signatures
Emma Jenkins, Ambroise Baker and Sarah Elliott | 381 |

Part VI Society, economy and water today 401

| 24 | Current water demands and future strategies under changing climatic conditions
Stephen Nortcliff, Emily Black and Robert Potter | 403 |

| 25 | Water reuse for irrigated agriculture in Jordan: soil sustainability, perceptions and management
Gemma Carr | 415 |

| 26 | Social equity issues and water supply under conditions of 'water stress': a study of low- and high-income households in Greater Amman, Jordan
Khadija Darmame and Robert Potter | 429 |

| 27 | The role of water and land management policies in contemporary socio-economic development in Wadi Faynan
Khadija Darmame, Stephen Nortcliff and Robert Potter | 442 |

| 28 | Political discourses and public narratives on water supply issues in Amman, Jordan
Khadija Darmame and Robert Potter | 455 |

Part VII Conclusions 467

| 29 | Overview and reflections: 20,000 years of water and human settlement in the southern Levant
Steven Mithen and Emily Black | 469 |

Index 481

Colour plates appear between pages 196 and 197.

Figures

1.1 Disciplinary aims and interdisciplinary interactions of the Water, Life and Civilisation Project. *page* 4
1.2 Hierarchical modelling from global circulation models to socio-economic impacts (courtesy of David Viner). See colour plate section. 5
1.3 The geographical scope of the climate modelling within the Water, Life and Civilisation project and the case study region, indicating the key research localities. 6
1.4 Water, Life and Civilisation team members during an orientation visit to Jordan in October 2004, here seen at the Iron Age tell of Deir 'Alla. See colour plate section. 7
2.1 Location of rain gauges. Top: Global Historical Climate Network (GHCN) gauges within Europe, Middle East and North Africa. Bottom: gauge data within the Middle East. Circles indicate GHCN monthly data; diamonds are gauges from the World Meteorological Organisation Global Summary of the Day (GSOD; daily data of very variable quality); stars are stations with daily data, provided by the Israeli Meteorological Service. 15
2.2 Mean climate over the Mediterranean. From top to bottom: December–February total precipitation; December–February SLP; December–February track density. See colour plate section. 17
2.3 Seasonal cycle in various rainfall statistics for the stations shown in the map to the right. The x-axis gives the month and the y-axis the statistic in question. The error bars represent the inter-annual standard deviation from one of the stations. All rainfall units are millimetres. From top to bottom, the statistics are: total monthly rainfall; mean number of rainy days in the month; mean rain per rainy day; mean maximum daily rainfall in the month; probability of rain given rain the day before (upper group of curves) and probability of rain given no rain the day before (lower group of curves). 18
2.4 Annual total rainfall in Jordan and Israel superposed on the orography. The contours are based on the data from the gauges shown in Figure 2.1. The dashed contours are sketched from published sources (US Geological Survey, 2006) because we were unable to obtain suitable quality data in eastern Jordan. See colour plate section. 19
2.5 Mean correlation versus mean distance apart for rainfall stations within Jordan and Israel. The solid line is cross-correlations between all stations; the dotted line is cross-correlations between grid squares of the same latitude, and the dashed line represents cross-correlations between grid squares of the same longitude. 19
2.6 Composite daily anomalies during the four GWL regimes that favour rainfall most strongly (WA, SWA, SWZ and NWZ – abbreviations defined in Table 2.1). Left set: daily rainfall anomaly composites over the Mediterranean (box shown on the top right plot); right set: daily SLP anomaly composites over the Mediterranean and Atlantic. See colour plate section. 20
2.7 Composites of precipitation, track density and SLP during January based on the five wettest and driest Januaries in a box with minimum longitude 34°, maximum longitude 36°, minimum latitude 31°, maximum latitude 33°. See colour plate section. 21
2.8 Histograms of rainfall total for the box defined in Figure 2.7 for positive and negative phases of the NAO, EAWR, East Atlantic pattern and for warm and cold Niño sea surface temperature (SST) anomalies. Negative phases or cold SSTs are shown by no shading and positive phases or warm SSTs by grey shading. 21
3.1 The forcings used to drive the global and regional models. (a) Greenhouse gas concentrations. (b) The annual cycle of insolation at the top of the atmosphere in experiment PREIND (units W m^{-2}). (c) The anomaly in the annual cycle of top of atmosphere (TOA) insolation applied to experiment 6kaBP (units W m^{-2}). (d) Annual mean insolation anomalies at the top of the atmosphere in each of the time-slice experiments (units W m^{-2}). 27
3.2 (a) The area of land surface modifications over North Africa and the Arabian Peninsula in the 'Wet Sahara' (WS) experiments (+ shows grid points that are converted from mostly desert to uniform savannah/shrubland and × are converted to open water).

(b) Imposed land ice-sheet changes between experiments 8kaBP and PREIND (shading shows the change in surface height, in metres). (c) Ocean heat flux convergence in experiments PRESDAY – 6kaBP (W m^{-2}). (d) Ocean heat flux convergence anomaly applied to experiment 8kaBP (W m^{-2}). (e) Sea surface temperature (SST) difference between experiments 8kaBP and 8kaBPNOICE (shading, °C) and sea ice difference (contours at 5% and 30%) for June–August. (f) As (e), but for December–February.   28

3.3 The seasonal distribution of precipitation over the Mediterranean. (a) GPCC dataset (June–September). (b) GPCC dataset (December–February). (c) As (a) but from the regional model in experiment PRESDAY. (d) As (b) but from the regional model in experiment PRESDAY. (e) As (a) but using the ERA-40 dataset. (f) As (b) but using the ERA-40 dataset. Units mm day^{-1}. In (a) and (b) missing data areas are blacked out. In (c) to (f), black squares mark the regions where GPCC data are missing.   31

3.4 Annual mean SAT and precipitation during the pre-industrial period. The top row panels show (a) SAT (°C) in experiment PRESDAY, and (b) the difference (°C) found in experiment PREIND (i.e. PREIND – PRESDAY). The middle row panels show results from the global model where (c) is the precipitation in experiment PRESDAY (mm day^{-1}) and (d) is the fractional difference (%) found in experiment PREIND (i.e. [PREIND – PRESDAY] × 100/ PRESDAY). The bottom row (e, f) is identical to the middle row but uses downscaled data from the regional model. For the difference plots (b, d, f), areas where the differences are statistically significant at (b) 99%, (d) 90% and (f) 70% confidence are indicated by black crosses. Areas of extremely low precipitation (less than 0.2 mm day^{-1}) in experiment PRESDAY are blacked out in the difference plots. See colour plate section.   32

3.5 Hemisphere average SAT differences. (a) Northern Hemisphere average SAT change relative to experiment PREIND (°C). Data points from the time-slice experiments are marked by crosses, and data points from experiment 8kaBPNOICE are marked by triangles (experiment PRESDAY is shown at time = −0.2 kaBP). (b) As (a) but for Southern Hemisphere.   33

3.6 Changes in SAT across the Holocene time-slice integrations. (a) Annual mean SAT change (experiment 6kaBP – PREIND, °C). Panels (b) and (c) are as (a), but for boreal summer and winter seasons, respectively. (d) The change in the strength of the seasonal cycle of SAT between experiment 6kaBP and PREIND (the strength of the cycle is defined as the maximum monthly mean SAT minus the minimum monthly mean SAT, units °C). (e) Boreal winter SAT change (6kaBP – PREIND, colours, °C) in the regional model and lower tropospheric winds (850 hPa) in experiment PREIND (arrows, units m s^{-1}). (f) As (e) but for boreal summer. In panels (b) to (d), areas where the changes are statistically significant at the 90% level are marked with black crosses. See colour plate section.   34

3.7 The annual cycle of zonal mean SAT anomalies in experiment 6kaBP relative to experiment PREIND. (a–c) Zonal mean SAT anomaly including (a) both ocean and land points, (b) land points only, and (c) ocean points only (units °C). (d) Outgoing longwave radiation anomalies at the top of the atmosphere (boreal summer, for experiment 6kaBP – PREIND, units W m^{-2}). For (a) to (c) contours are at ±0.25, 0.5, 1, 2 °C.   36

3.8 The lower tropospheric circulation, as given by the 850 hPa streamfunction. (a) Experiment PREIND during December–February. (b) Difference between experiments 6kaBP and PREIND during December–February; shaded areas indicate negative values. (c) As (a) but for June–September. (d) As (b) but for June–September. The circulation is along streamlines and is cyclonic (anticlockwise) around negative values. The contour interval is the same in (a) and (c), and is four times greater than that in (b) and (d).   38

3.9 Differences in boreal summer precipitation across the Holocene time-slice integrations. The top row shows the precipitation in experiment PREIND (units mm day^{-1}) using data from (a) the global model and (b) the regional model. The middle row shows the fractional change in precipitation (units %) in experiment 6kaBP relative to experiment PREIND (i.e. [6kaBP – PREIND] × 100/ PREIND), using data from (c) the global model and (d) the regional model. Panel (e) is similar to (c) but for experiment 8kaBP-WS. Panel (f) shows the fractional precipitation changes averaged over the SAHEL box (in the global model, as marked in panels (a) and (c)) and the CAUCUS box (in the regional model, as marked in panels (b) and (d)) in the time-slice experiments. Data points from the time-slice experiments are marked by × symbols whereas the + symbols mark data points from experiments 6kaBP-WS and 8kaBP-WS and triangles mark data points from experiment 8kaBPNOICE (experiment PRESDAY is shown at time = −0.2 kaBP). In panels (c) to (e), areas where the changes are statistically significant at the 90% level are marked with black crosses. Areas of extremely low precipitation (less than 0.2 mm day^{-1} for the global model and

0.05 mm day^{-1} in the regional model) in experiment PREIND are blacked out in panels (b) to (e). See colour plate section.   39

3.10 Differences in boreal winter precipitation across the Holocene time-slice integrations using the global model. (a) Experiment PREIND (units mm day^{-1}). (b) The fractional change in precipitation (units %) in experiment 6kaBP relative to experiment PREIND (i.e. [6kaBP − PREIND] × 100/PREIND). (c) As (b) but for experiment 8kaBP. (d) As (b) but for experiment 8kaBPNOICE. (e) As (b) but for Early Holocene experiments (8kaBP + 10kaBP + 12kaBP) minus the Late Holocene experiments (2kaBP + 4kaBP + 6kaBP). Panel (f) shows the fractional precipitation changes averaged over the boxes marked in panels (a) to (e). Data points from the time-slice experiments are marked by × symbols whereas triangles mark data points from experiment 8kaBPNOICE (experiment PRESDAY is shown at time = −0.2 kaBP). In panels (b) to (e), areas where the changes are statistically significant at the 90% level are marked with black crosses. Areas of extremely low precipitation (less than 0.2 mm day^{-1}) in experiment PREIND are blacked out in panels (c) and (d). See colour plate section.   41

3.11 Differences in boreal winter storm tracks across the Holocene time-slice integrations using the global model. (a) The storm track in experiment PREIND (units of storms per month passing through a 5° spherical cap). (b) Storm track difference (expt 6kaBP − PREIND). (c) As (b) but for experiment 8kaBP. (d) As (b) but for experiment 8kaBPNOICE. (e) As (b) but for Early Holocene experiments (8kaBP + 10kaBP + 12kaBP) minus the Late Holocene experiments (2kaBP + 4kaBP + 6kaBP). (f) The fractional storm track changes averaged over the boxes marked in panels (a) to (e). Data points from the time-slice experiments are marked by × symbols (experiment PRESDAY is not shown). In panels (b) to (e), areas where the differences are statistically significant at the 90% level are marked with black crosses, and areas of high orography (in excess of 1,200 m) are blacked out. Thick black contours in (b) to (e) show the 10 storms per month contour from experiment PREIND. Prior to display, the storm track diagnostics have been smoothed to improve readability.   42

3.12 The upper and lower tropospheric σ (the susceptibility of the mean state to weather system growth, as described in Section 3.2.7). (a) Values of σ in experiment PREIND (units s^{-1}) at 925 hPa. (b) As (a) but for 400 hPa. (c) Difference in σ between experiments 6kaBP and PREIND at 925 hPa. (c) As (b) but at 400 hPa. (e) As (c) but for experiment 8kaBP. (f) As (d) but for experiment 8kaBP. Data are only shown for the Northern Hemisphere extratropics and the contours have been lightly smoothed to improve readability.   43

3.13 Differences in boreal winter precipitation across the Holocene time-slice integrations from the regional model. (a) Experiment PREIND (units mm day^{-1}). (b) The fractional change in precipitation (units %) in experiment 6kaBP relative to experiment PREIND (i.e. [6kaBP − PREIND] × 100/PREIND). (c) As (b) but for experiment 8kaBP. (d) As (b) but for experiment 8kaBPNOICE. (e) as (b) but for early Holocene experiments (8kaBP + 10kaBP + 12kaBP) minus the late Holocene experiments (2kaBP + 4kaBP + 6kaBP). (f) The fractional precipitation changes averaged over the boxes marked in panels (a) to (e). Data points from the time-slice experiments are marked by × symbols, whereas triangles mark data points from experiment 8kaBPNOICE (experiment PRESDAY is shown at time −0.2 kaBP). In panels (b) to (e), areas where the differences are statistically significant at the 90% level are marked with black crosses. Areas of extremely low precipitation (less than 0.2 mm day^{-1}) in experiment PREIND are blacked out in panels (c) and (d). See colour plate section.   45

3.14 Differences in the boreal winter storm tracks across the Holocene time-slice integrations using the regional model. (a) The storm track in experiment PREIND (units of storms per month passing through a 5° spherical cap). (b) Storm track difference (expt 6kaBP − PREIND). (c) As (b) but for experiment 8kaBP. (d) As (b) but for experiment 8kaBPNOICE. (e) As (b) but for the average of the Early Holocene experiments (8kaBP + 10kaBP + 12kaBP) minus that for the Late Holocene experiments (2kaBP + 4kaBP + 6kaBP). (f) The fractional storm track differences averaged over the box marked in panels (a) to (e). Data points from the time-slice experiments are marked by × symbols (experiment PRESDAY is not shown). In panels (b) to (e), areas where the changes are statistically significant at the 90% level are marked with black crosses and areas of high orography (in excess of 1,200 m) are greyed out. Thick black contours in (b) to (e) show the 10 storms per month contour from experiment PREIND. The storm tracking analysis is performed on a coarse grid (hence the extremely coarse orographic features shown), and prior to display here the storm track diagnostics have been further smoothed to improve readability.   46

4.1 Model domains and modelled and observed topography. Top: model grid points (dots) on the model topography for the large domain. Bottom left: small domain (Middle East only) used for the ensembles. Bottom right: observed

4.2 Left set of panels: seasonal cycles in various statistics of the weather for eight stations (black lines), and the regional climate model baseline ensemble (grey shaded area indicates the ensemble range). The x-axis gives the month and the y-axis gives the mean rainfall statistic during that month. From top to bottom the statistics are: mean total monthly rainfall, mean total number of rainy days, mean rain per rainy day, mean maximum monthly rainfall, monthly mean probability of rain given no rain the day before, monthly mean probability of rain given rain the day before. Rainfall is given in millimetres for all the plots. Right panel: the location of the stations and the RCM time-series box. The crosses indicate RCM grid points.   54

   topography for the eastern Mediterranean and Middle East. Topography is given in metres above sea level for all the plots. See colour plate section.   53

4.3 Seasonal cycle in precipitation change under an A2 scenario by 2070–2100. Significance at the 95% level is shown by a dot in the grid square. Top set: monthly mean absolute change in rainfall (mm) over the whole of the Mediterranean under an A2 scenario. Bottom set: monthly mean percentage change in rainfall (%) over the East Mediterranean only. See colour plate section.   55

4.4 Change in the January climate (temperature, precipitation, sea-level pressure and 850 mb track density) over the Mediterranean under an A2 scenario by 2070–2100. See colour plate section.   56

4.5 Change in daily rainfall probabilities. Significance at the 95% level is indicated by a dot within the grid square. Top row, left panel: absolute change in the probability of rain given no rain the day before; right: absolute change in the probability of rain given rain the day before. Bottom row: as above but for percentage changes for the southeast part of the region only. See colour plate section.   57

4.6 Left set of panels: seasonal cycles in various statistics of the weather for the baseline ensemble (light grey polygon); the A2 ensemble (dark grey polygon) and the B2 integration (dashed line) for the box shown in Figure 4.2. The x-axis gives the month and the y-axis gives the mean rainfall statistic during that month. From top to bottom the statistics are: mean total monthly rainfall, mean total number of rainy days, mean rain per rainy day, mean maximum monthly rainfall, monthly mean probability of rain given no rain the day before, monthly mean probability of rain given rain the day before. Rainfall is given in millimetres for all the plots. Right set of panels: difference in ensemble means between the A2 scenario integration and the baseline integration for the statistics shown on the left. Filled bars indicate significance at the 95% level.   58

4.7 Percentage change in January precipitation under an A2 scenario by 2070–2100 for eight IPCC models. The model name abbreviations on the plots are: CSIRO Mark 3.0 (csmk3); GFDL CM 2.0 AOGCM (gfcm20); HadCM3 (hadcm3); IPSL CM4 (ipcm4); MRI-CGCM2.3.2 (mrcgcm);NCAR CCSM3 (nccsm); GFDL CM 2.1 AOGCM (gfcm21); MIMR MIROC3.2 (medium resolution). See colour plate section.   59

4.8 Top: mean percentage change in January precipitation predicted under an A2 scenario for 2070–2100 for the IPCC models shown in Figure 4.7; middle: mean percentage change in January precipitation predicted under a B1 scenario for 2070–2100 for the IPCC models shown in Figure 4.7. Bottom: difference in the mean percentage change between the A2 and B1 (B1 – A2). See colour plate section.   60

5.1 Location of gauges referred to in the chapter superposed on the topography. The inset map shows the location of the main map. Crosses are the locations of the stations provided by the Israeli Meteorological Service. The star is Tafilah and the circles are monthly data used in Chapters 12 and 13 and referred to here.   65

5.2 Seasonal cycles of rainfall statistics for the observations (black line), the weather generator based on observed statistics (dashed line) and the weather generator based on the predicted statistics (grey line). Top left: rainfall probabilities (upper lines are PRR and lower lines are PDR); bottom left: rainfall amount fractional frequency histogram; right: rainfall totals (mm day^{-1}).   66

5.3 Quantile–quantile plot of observed versus simulated rainfall amounts for a single gamma distribution (filled circles) and for spliced gamma/extreme value distribution (open circles). The line represents a $y = x$ function on which the circles would lie if the theoretical distribution perfectly matched the observations.   66

6.1 Map of the eastern Mediterranean and Levant region showing the locality of major features and sites discussed in the text. 1: Ghab Valley; 2: Hula Basin; 3: Peqiin Cave; 4: Israeli coastal plain; 5: Ma'ale Efrayim Cave; 6: Soreq Cave; 7: Jerusalem West Cave; 8: Wadi Faynan; 9: Ocean Drilling Program (ODP) Site 967; 10: site of core M44-1-KL83; 11: site of core GeoB5804–4; 12: site of core GeoB5844–2. Map produced with GMT (http://gmt.soest.hawaii.edu/).   72

6.2 Reconstructed air temperatures from the GISP 2 ice core in Greenland (after Alley, 2000). The timing and duration of the Last Glacial Maximum (LGM) is the

LIST OF FIGURES   xiii

        same as the 'LGM Chronozone Level 1' as defined by Mix *et al.* (2001).   72

6.3   Climate model outputs for the LGM and the present day. (A) Present-day winter (DJF) precipitation (precipitation in mm per day); (B) LGM winter (DJF) precipitation; (C) present-day summer (JJA) precipitation; (D) LGM summer (JJA) precipitation; (E) present-day annual precipitation; (F) LGM annual precipitation; (G) LGM winter (DJF) snowfall (snowfall in mm per day); (H) LGM summer (JJA) snowfall. Note that panel H is blank because there is no snowfall in summer.   74

6.4   Climate model outputs for the LGM and present day. (A) Present-day winter (DJF) temperature (°C); (B) LGM winter (DJF) temperature; (C) present-day summer (JJA) temperature; (D) LGM summer (JJA) temperature; (E) present-day average annual temperature; (F) LGM average annual temperature; (G) LGM annual average precipitation minus evaporation in mm day^{-1}; (H) LGM annual average wind strength (in m s^{-1}) and vectors. See colour plate section.   74

6.5   Compilation of lake level curves for Lake Lisan/the Dead Sea and Lake Tiberias. (A) Frumkin *et al.* (1994); (B) Neev and Emery (1995); (C) Landman *et al.* (2000); (D) timing of massive salt deposition (Yechieli *et al.*, (1993); Neev and Emery (1967); there is some uncertainty regarding the exact age of the sediments, hence the dashed line. Shading of the YD here indicates the range of the two best dates); (E) Bartov *et al.* (2002, 2003); (F) Hazan *et al.* (2004). EHWP = Early Holocene Wet Phase; YD = Younger Dryas; H1 = Heinrich Event 1; LGM = Last Glacial Maximum; H2 = Heinrich Event 2.   75

6.6   An integrated, schematic lake level curve (solid black line) for the Lake Lisan/Dead Sea based upon various studies. This curve is designed primarily to illustrate lake level trends over time for ease of comparison with other proxy data. For the period 25 to 13 cal. ka BP the integrated curve is an approximate average of Neev and Emery (1995), Bartov *et al.* (2002, 2003) and Landmann *et al.* (2002). From 13 to 9 cal. ka BP we have used the data from Neev and Emery (1967), Begin *et al.* (1985), Yechieli *et al.* (1993) and Stein (2001) which suggest a major lake level fall between 13 and 11 cal. ka BP. From 9 cal. ka BP onwards we have followed the study of Frumkin *et al.* (1994).   76

6.7   Palaeoclimate of the Israeli coastal plain, as interpreted from palaeosols (Gvirtzman and Wieder, 2001). Black dots show position of age model tiepoints. S1 = Sapropel 1, B-A = Bølling–Allerød; other abbreviations as in Figure 6.5.   77

6.8   Palynology of the Hula Basin (Baruch and Bottema, 1991) and the Ghab Valley (Niklewski and van Zeist, 1970) with the proposed chronostratigraphy of Rossignol-Strick (1995). Horizontal axes in %; in the left-hand figure, the percentage for each taxon refers to concentration of that pollen taxon with respect to total 'Arboreal pollen+non-arboreal pollen'. In the right-hand figure, the percentage scale refers to the relative proportions of 'trees + shrubs' and 'Chenopodiaceae+*Artemisia*'.   78

6.9   Speleothem stable-isotope data (Bar-Matthews *et al.*, 2003; Vaks *et al.*, 2003) and reconstructed air temperatures (McGarry *et al.*, 2004).   80

6.10   Gastropod oxygen isotope data from the Negev Desert (Goodfriend, 1991).   82

6.11   (A) Foraminiferal LGM annual, summer and winter SST reconstructions (Hayes *et al.*, 2005), calculated using an artificial neural network (ANN). (B) Temperature anomalies for annual, summer and winter SSTs during the LGM, compared with modern-day values (Hayes *et al.*, 2005). Anomaly values were calculated by subtracting modern-day SSTs from the glacial values. The black dots represent the sites of the cores from which the LGM data were obtained. This figure is a reproduction of part of Figure 9 in Hayes *et al.* (2005). See colour plate section.   83

6.12   Compilation of eastern Mediterranean Sea palaeoclimatic records from Site 967 (Emeis *et al.*, 1998, 2000, 2003) and MD84–461 (Fontugne and Calvert, 1992). The 'δ^{18}O/p.s.u. value' is a coefficient used in the calculation of SSS that relates salinity and δ^{18}O$_{seawater}$ (see Emeis *et al.*, 2000 for more details).   84

6.13   Compilation of northern Red Sea palaeoclimatic records (Arz *et al.*, 2003a, b; core names defined therein).   84

6.14   Compilation of terrestrial and marine palaeoclimatic proxy data for the Levant and eastern Mediterranean. Also shown is the ice-core record from GISP2 (Greenland). References: 1: Alley (2000); 2: see Figure 6.6; 3: Bar-Matthews *et al.* (2003); 4: Arz *et al.* (2003a); 5: Emeis *et al.* (2000, 2003); 6: Gvirtzman and Wieder (2001); 7: Rossignol-Strick (1995); 8: Magaritz (1986), Goodfriend and Magaritz (1988); 9: Magaritz and Heller (1980); Goodfriend (1990, 1991, 1999); 10: Reeder *et al.* (2002).   86

6.15   Summary of climatic conditions at the LGM (A), peak of the Bølling–Allerød warm phase (B), the Younger Dryas (C) and during the early Holocene/S1 (D). Turbidite data from Reeder *et al.* (2002); alkenone SSTs and SSS data from Emeis *et al.* (2000, 2003) and Arz *et al.* (2003a, b); speleothem data from Bar-Matthews *et al.* (1997, 1999, 2000, 2003), Frumkin *et al.* (1999b, 2000), Vaks *et al.* (2003) and McGarry *et al.* (2004); pollen data from

Niklewski and van Zeist (1970), Baruch and Bottema (1991) and Rossignol-Strick (1995); lake levels from Figure 6.6 (this study); LGM annual SSTs calculated from foraminiferal assemblages taken from Hayes *et al.* (2005); early Holocene SSS values are from Kallel *et al.* (1997a); Israeli coastal plain palaeosol data from Gvirtzman and Wieder (2001); Negev data from Magaritz (1986), Goodfriend and Magaritz (1988); Goodfriend, (1999); southern Jordan fluvial data from McLaren *et al.* (2004). T = temperature (all in °C), S = salinity, P = precipitation, NAP = non-arboreal pollen, AP = arboreal pollen, Maps drawn with GMT (http://gmt.soest.hawaii.edu/). Coastlines do not account for any changes in sea level or sedimentation. 87

7.1 Location of the different localities and regions presented in the text. (A) Map of the southern Levant and mean annual rainfall (from EXACT 1998). 1 – Hula Basin (Baruch and Bottema, 1999; Cappers *et al.*, 1998; Rosen, 2007). 2 – Upper Galilee caves (Issar, 2003). 3 – Ma'ale Efrayim Cave (Vaks *et al.*, 2003). 4 – Soreq Cave (Bar-Matthews and Ayalon, 2004; Bar-Matthews *et al.*, 1998, 1999, 2003). 5 – Israeli coastal plain (Gvirtzman & Wieder, 2001). 6 – Wadi Faynan (Hunt *et al.*, 2004, 2007; McLaren *et al.*, 2004; Grattan *et al.*, 2007). 7 – Elat shorelines (Shaked *et al.*, 2002). 8 – Wadi Muqat (Abboud, 2000). 9 – Cores GA 112–110 (Schilman *et al.*, 2001a,b).
10 – Jordan Valley (Hourani and Courty, 1997).
11 – Northern Negev Desert (Goodfriend, 1991, 1999).
12 – Southern Negev (Amit *et al.*, 2007). 13 – Qa'el-Jafr Basin (Davies, 2005). 14 – Central Negev Highlands (Rosen *et al.*, 2005; Avni *et al.*, 2006). 15 – Birkat Ram Lake, Golan Heights (Schwab *et al.*, 2004). 16 – Wadi ash-Shallalah (Cordova, 2008). 17 – Wadi al-Wala and the Madaba-Dhiban plateau (Cordova *et al.*, 2005; Cordova, 2008). 18 – Tel Lachish (Rosen, 1986).
19 – Nahal Qanah Cave (Frumkin *et al.*, 1999a). 20 – Nahal Zin, Negev (Greenbaum *et al.*, 2000). (B) Map of the Dead Sea area. 96

7.2 Compilation of several proxies for the middle to late Holocene in the southern Levant. Red and blue bars (see colour plate section) show interpreted climate fluctuations (wetter/drier conditions). Archaeological periods from Rosen (2007). Dead Sea levels: 1 – Frumkin and Elitzur (2002). 2 – Klinger *et al.* (2003). 3 – Enzel *et al.* (2003). 4 – Bookman *et al.* (2004). 5 – Migowski *et al.* (2006). (A) Dead Sea levels in 1997 (Migowski *et al.*, 2006). Lake Kinneret levels: 6 – Hazan *et al.* (2005). Calculated rainfall: 7 – From the Soreq cave record; Bar-Matthews and Ayalon (2004). (B) Present-day mean annual rainfall in Soreq area. 8 – From tamarisk wood, Mount Sedom cave; Frumkin *et al.* (2009). (C) Present-day mean annual rainfall at Mount Sedom. 9 – Climatic change from pollen indicators according to Neumann *et al.* (2007). Our synthesis is presented at the bottom of the figure. See colour plate section. 97

8.1 Summary of rainfall signal from the proxy data for the Middle East and Europe described in the text. Pluses indicate higher rainfall and minuses lower rainfall during the early Holocene as compared with the early/mid-Holocene. 108

8.2 Comparison between the observed and modelled climate. Top set, left: GPCC precipitation for the whole Mediterranean (top) and for the Middle East only (bottom); right: RCM precipitation for the whole Mediterranean (top) and for the Middle East only (bottom). Bottom set, left: NCEP reanalysis temperature for the whole Mediterranean (top) and for the Middle East only (bottom); right: RCM temperature for the whole Mediterranean (top) and for the Middle East only (bottom). See colour plate section. 109

8.3 Comparison of modelled and observed track densities (in number of tracks per month per 5 degree spherical cap). Left: mean January track density in the reanalysis. Right: mean January track density in the RCM large-domain baseline scenario. Both figures are based on tracking of features in the 850 mb vorticity field. See colour plate section. 110

8.4 Modelled changes in October–March precipitation (top, in mm) and December–February track density (bottom, in number of tracks per month per 5 degree spherical cap for (from left to right) late Holocene minus early Holocene; future (2070–2100) – present (1961–1990); and driest years – wettest years from 1948–1999. See colour plate section. 110

9.1 Map showing the initial extent of Lake Lisan and present-day Dead Sea (after Stein *et al.*, 2009). Inset shows the structural setting for the region. 113

9.2 Map showing the location of sites 1–4 sampled within this study. 114

9.3 Photograph showing the Lake Lisan Grey Unit and White Unit on the East side of the Jordan Valley. 116

9.4 The U-decay series chain. 117

9.5 The Th-decay series chain. 117

9.6 Compilation figure of previously published lake level data from Lisan sediments, predominately from the west side of the Jordan Valley together with information from this study coming from the east side. 118

9.7 Site 4 used in this study. Inset shows stromatolite in cross-section, with intercalated gravels above and fine-grained sediments below. 123

LIST OF FIGURES

9.8 Site 1. Inset shows laminated bands of aragonite (white) and grey (silicate-rich) units. 123
9.9 Site 3. Inset shows detailed layers of the sediments with annual bands. 124
9.10 The data presented in this chapter (black circles) together with the elevation/age data from Enzel *et al.* (2003) (grey squares). 124
10.1 A schematic map of the upper River Jordan. 133
10.2 An overview of the modelling framework. 137
10.3 Modelled and observed daily mean flows in the Jordan river at Obstacle Bridge from 1 October 1988 to 30 September 1993. 138
10.4 The relationship between monthly mean flow and monthly mean rainfall for different values of PDR. Top: monthly mean flow plotted against monthly mean rainfall. High PDR are filled circles and low PDR are unfilled circles. Bottom: histograms of flow for different ranges of monthly total rainfall for high PDR (filled bars) and low PDR (unfilled bars). The ranges are given on the figure. 140
10.5 The relationship between monthly mean flow and monthly mean rainfall for different values of PRR. Top: monthly mean flow plotted against monthly mean rainfall. High PRR are shaded circles and low PRR are unfilled circles. Bottom: histograms of flow for different ranges of monthly total rainfall for high PRR (shaded bars) and low PRR (unfilled bars). The ranges are given on the figure. 141
10.6 Flow duration curves for each of the sensitivity studies compared with the flow duration curve for the generated time-series based on the observed statistics of the weather (sensitivity studies labelled on the figure). 142
10.7 Comparison between the flow duration curves for halving PRR and halving PDR. The right-hand figure is a zoom of the high flow region of the left-hand figure, which shows all the data. 142
10.8 Projected changes in the monthly rainfall totals at Degania Bet, Israel, from the HadRM3 and weather generator models for 2070–2100 under the SRES A2 scenario. 143
10.9 Modelled daily mean flows in the Jordan river at Obstacle Bridge for control (1961–1990) and scenario (2071–2100) periods. 144
11.1 Map Showing the Dead Sea catchment area, with Lake Kinneret in the north. 148
11.2 A GIS representation of the digital terrain of the Jordan Valley and the Dead Sea. 148
11.3 Estimates of changing Dead Sea levels over the past 25,000 years (Enzel *et al.*, 2003 – upper graph; Black *et al.*, Chapter 9 of this volume – lower graph). 148
11.4 Dead Sea hypsometric curves showing relationships between sea elevation, surface area and volume. 149
11.5 The shoreline of the Dead Sea reconstructed for four different depths. The 170 m depth equates to that at the Last Glacial Maximum, which occurred at approximately 20 cal. ka BP based on the palaeoenvironmental evidence summarised in Robinson *et al.* (2006). 150
11.6 Dead Sea elevations from 1860 to 2009, showing rapid decline since the 1960s. 150
11.7 Jerusalem rainfall from 1846 to 1996. 152
11.8 Regression of Dead Sea level change against Jerusalem rainfall for the period 1860–1960 using decadal averages. 152
11.9 Observed and modelled Dead Sea levels 1860–1960. 152
11.10 Extension of the observed and modelled data to include the recent period of abstraction and sea level decline. 152
11.11 Predicted simulation of the levels to 2050 assuming continued abstraction from the River Jordan. 153
11.12 Predicted sea levels in the future assuming climate change for eight future realisations without the effects of abstraction, and two climate change scenarios that do include the abstraction. 154
11.13 The effects on Dead Sea elevations assuming a major water transfer from either the Red Sea or the Mediterranean Sea into the Dead Sea. Shown are three water transfer rates of 1,690, 1,900 and 2,150 million m^3 per year for the years 2020 to 2040, and then transfer rates falling to match the water abstraction rate of 800 million m^3 per year. 154
11.14 Estimated rainfall over the past 9,000 years based on the Enzel *et al.* (2003) Dead Sea elevations. 155
11.15 Estimated rainfall over the period 8–250 ka BP based on the Black *et al.* (Chapter 9, this volume) Dead Sea elevations. 155
12.1 A schematic map of the Wadi Faynan, its major tributaries and settlements. PPNB, Pre-Pottery Neolithic B. 158
12.2 The geology of the Wadi Faynan area. Source Geological Map of Jordan 1:250,000, prepared by F. Bender, Bundesanstalt für Geowissenschaften und Rohstoffe, Hannover 1968 [Sheet: Aqaba-Ma'an and Amman]. Reproduced with permission. Not to scale. © Bundesanstalt für Geowissenschaften und Rohstoffe. See colour plate section. 161
12.3 A geological cross-section made 5 km to the north of the Wadi Faynan. Source: Geological Map of Jordan 1:250,000, prepared by F. Bender, Bundesanstalt für Geowissenschaften und Rohstoffe, Hannover 1968 [Sheet: Aqaba-Ma'an and Amman]. Reproduced with permission. Not to scale. © Bundesanstalt für Geowissenschaften und Rohstoffe. See colour plate section. 162

12.4 A subset of the Landsat image of the Wadi Faynan area acquired on 08 March 2002. 162

12.5 This picture was taken just to the south of the Jebel Hamrat al Fidan and shows how a farmer has tapped the groundwater held close to the surface by the granitic barrier by digging a network of trenches to expose the water. The water is pumped from the trench and used to irrigate fields of watermelon. 162

12.6 Rainfall patterns (isohyets) in the region of the Wadi Faynan (marked by the circle). Source: Department of Civil Aviation, Jerusalem, 1937–38. 165

12.7 Sample site locations in the Wadi Faynan from the 2006, 2007 and 2008 field seasons. 167

12.8 A conceptual model of the key water stores and pathways in the Wadi Faynan. Precipitation on the limestone and plateau soils in the upper reaches is likely to be the key aquifer recharge mechanism. The water will flow laterally through the limestone and sandstones until it emerges at a contact point between the two or at a contact between the sandstone and the aplite-granite. Re-infiltration (transmission loss) occurs as the water flows along the channel network. The Jebel Hamrat al Fidan forces the water to return close to the surface. Base map source: Geological Map of Jordan 1:250,000, prepared by Geological Survey of Germany, Hannover 1968 (Sheet: Aqaba-Ma'an). Reproduced with permission. Not to scale. See colour plate section. 168

12.9 The flow duration curve based on the flows simulated in the Wadi Faynan over the period 1937 to 1973. The Pitman model was applied to simulate flows in the Wadi Faynan to a point on the channel network adjacent to the ancient field-system immediately downstream of the Ghuwayr–Dana confluence. The highest simulated flow of 98 $m^3 s^{-1}$ occurred in response to a mean daily rainfall-event of 90 mm. 170

13.1 A schematic map of the southern Ghors which includes the Wadi Hasa. 176

13.2 The modelling framework used to simulate the rainfall-runoff response in the Wadi Hasa catchment. 180

13.3 Observed and simulated mean monthly flows in the Wadi Hasa at (a) Tannur and (b) Safi. Simulated flows are shown for the calibration and scenario climate conditions. 183

13.4 Observed versus simulated (calibration period) mean monthly flows in the Wadi Hasa at (a) Tannur and (b) Safi. 183

13.5 Observed and simulated flow-duration curves for the Wadi Hasa at (a) Tannur and (b) Safi. Simulated flows are shown for the calibration (1923–2002) and scenario climate conditions. 184

13.6 Observed and simulated mean monthly flows, averaged over the 80-year simulation period, in the Wadi Hasa at (a) Tannur and (b) Safi. Simulated flows are shown for the calibration (1923–2002) and scenario climate conditions where both precipitation and PET are adjusted and where precipitation alone is adjusted (Scenario ΔP) for the model simulations of flow at Safi. 186

14.1 Map of the study region showing Epipalaeolithic and Pre-Pottery Neolithic A sites referred to in the text. 193

14.2 Map of the study region showing Pre-Pottery Neolithic B and Pottery Neolithic sites referred to in the text. 194

14.3 Map of the study region showing Bronze Age, Iron Age and Nabataean sites referred to in the text. 194

14.4 Massive concentration of chipped stone at Kharaneh IV, Wadi Jilat (© B. Finlayson), marked by the roughly oval, darkened area immediately in front of and next to the two figures. 194

14.5 Remnants of a brushwood hut at Ohalo II, showing location adjacent to Lake Tiberias (Kinneret) (© S. Mithen). 195

14.6 Extensive use of pisé and mud plaster at the Pre-Pottery Neolithic A settlement of WF16, Wadi Faynan, showing excavations of April 2009 (© S. Mithen). 196

14.7 Experimental PPNB buildings at Beidha (© B. Finlayson). 196

14.8 Archaeological remains of the Pre-Pottery Neolithic B 'mega-site' of Basta (© S. Mithen). 197

14.9 Archaeological remains of the Pre-Pottery Neolithic B site of Ba'ja, located within a steep-sided siq (© S. Mithen). 198

14.10 Excavation at the Pottery Neolithic settlement Sha'ar Hagolan (© Y. Garfinkel). 199

14.11 Cross-section of the Pottery Neolithic well at Sha'ar Hagolan (© Y. Garfinkel). 200

14.12 Plan of the Pottery Neolithic B settlement at Wadi Abu Tulayha, showing the relationship between the settlement structures, the barrage and the proposed cistern (Str M, W-III) (© S. Fujii). 202

14.13 The Pre-Pottery B 'outpost' settlement at Wadi Abu Tulayha (© S. Fujii). 203

14.14 The proposed Pre-Pottery Neolithic B barrage in Wadi Abu Tulayha (© S. Fujii). 203

14.15 The proposed Pre-Pottery Neolithic B barrage in Wadi Ruweishid (© S. Fujii). 204

14.16 Interior of the proposed Pre-Pottery Neolithic B cistern in Wadi Abu Tulayha (© S. Fujii). 204

14.17 Excavation of a remnant of a terrace wall in the vicinity of the Pottery Neolithic settlement of and projected extent of wall 'Dhra (© B. Finlayson). 205

14.18 Reconstruction of Pottery Neolithic cultivation plots supported by terrace walls at 'Dhra (© B. Finlayson). 206
14.19 Khirbet Zeraqoun, in the northern highlands of Jordan (© S. Mithen). 207
14.20 Tell Handaquq and the Wadi Sarar (© S. Mithen). 208
14.21 Iron Age Tell Deir Allah, looking across the Jordan Valley where water canals were built to supply the settlement (© S. Mithen). 211
14.22 Aqueduct at Humayma (© R. Foote). 212
14.23 Cistern at Humayma with an arched roof (scale provided by Dr Claire Rambeau) (© R. Foote). 213
14.24 Nabataean water channel in the siq at Petra (© S. Mithen). 213
15.1 Location of Wadi Faynan catchment in relation to Khirbet Faynan and present-day precipitation isohyets (mm yr^{-1}). 220
15.2 Contemporary Bedouin dam in Wadi Ghuwayr, used to help channel water into plastic pipes for transport to fields to west. 220
15.3 Contemporary Bedouin reservoir in Wadi Faynan. The water here is drawn from groundwater flow in the Wadi Ghuwayr using plastic pipes (as shown in Figure 15.2). 220
15.4 Schematic representation of present-day vegetation cover in Wadi Faynan catchment (after Palmer *et al.*, 2007, figure 2.11). 221
15.5 Map of main archaeological sites discussed in this chapter in relation to modern settlements and principal wadi systems. 222
15.6 PPN sites in Wadi Faynan. (a) PPNA WF16; (b) PPNB Ghuwayr 1. 223
15.7 Schematic representation of EBA water harvesting system in Wadi Faynan (WF 1628) showing several cross-wadi walls and check dams built to deflect surface runoff onto surrounding landscape (after Barker *et al.*, 2007b, figure 8.24). 223
15.8 WF4 field-system looking northwest. 223
15.9 Mean annual rainfall at Tafilah for each time slice against 0 ka (control) mean. 225
15.10 Mean monthly rainfall amounts for each time slice and control experiments. 226
15.11 Comparison of palaeo-rainfall estimates used in this study with those derived from Soreq Cave sequence. 227
15.12 Summary of mean daily flow rates for each month of time-slice and control experiments. Infiltration rate is 8 mm per day for all time slices. 231
15.13 Schematic diagram of Wadi Faynan showing the potential impact of changes in vegetation on hydrological processes. (a) Present day. Infiltration is limited to areas of colluvium, small areas of vegetation and soil cover in upper catchment (coloured grey), and the (saturated) wadi channel. The majority of the wadi comprises bare, rocky slopes and gravel terraces with very low infiltration rates, which induce high runoff. (b) Early Holocene. Dense vegetation cover and increased soil cover in the wadi system increases areas of infiltration (coloured grey). This increases potential percolation in upper catchment and reduces surface runoff generation. 231
15.14 Comparison of monthly flows for 12 ka BP time slice under different infiltration scenarios. 233
15.15 Comparison of flow duration curves for 12 ka BP simulation under high and low infiltration scenarios. 233
15.16 Comparison of monthly flows for 6 ka BP time slice under different infiltration scenarios. 233
15.17 Comparison of monthly flows for 2 ka BP time slice under different infiltration scenarios. 234
15.18 Comparison of monthly flows for each time slice and 0 ka BP (control) simulations under infiltration scenarios proposed for the Holocene. The key indicates time; infiltration rate. 234
15.19 Palaeo-rainfall (bars) and palaeo-flows (line) for Holocene scenarios and 0 ka (control) simulations for Wadi Faynan. 235
15.20 Summary of monthly flows for proposed early Holocene (12–8 ka BP) and 0 ka BP (control) simulations. The key indicates time; infiltration rate. 236
15.21 Summary of monthly flows for proposed mid-Holocene (6–4 ka BP) and 0 ka BP (control) simulations. The key indicates time; infiltration rate. 236
15.22 Summary of monthly flows for proposed late Holocene (4, 2 and 0 ka BP (control)) simulations. The key indicates time; infiltration rate. 237
15.23 Present-day cultivation in the WF4 field-system, irrigated using groundwater flow captured by plastic pipes in Wadi Ghuwayr (see Figures 15.2 and 15.3). 240
16.1 Map of the southern Levant showing average annual rainfall (modified from EXACT, 1998) and the location of the study area. 246
16.2 Schematic map showing the surroundings of the archaeological site of Beidha. The letters A to E refer to the locations from which the panoramic pictures presented in Figure 16.3 were taken. 247
16.3 (A) Panoramic view from the site, looking southwest. (B) View of ancient gravel terraces, looking south-southwest. Robyn Inglis for scale. (C) View from the site on the tufa section and the dunes, looking southwest. (D) View from a sandstone outcrop on the

site and the Seyl Aqlat, looking north-northwest. (E) View from a sandstone outcrop on the alluvial plain and the present-day bed of the Wadi el-Ghurab. Looking southeast. Picture credits: C. Rambeau and R. Inglis, May 2006. 248

16.4 Chronological framework at Beidha. Dates marked by triangles: this study. Black lines: data from Byrd (1989, 2005), recalibrated. 249

16.5 Schematic succession of sedimentary layers at the site section, as illustrated in Figures 16.2 and 16.3, and location of the levels from which new radiocarbon dates have been obtained. 250

16.6 Spring carbonate (tufa) sequence, U-series dates and average sedimentation rates. Italic numbers refer to dates obtained on one sub-sample only; other dates were obtained with the isochron technique on multi sub-samples. Numbers 1–6 relate to sedimentation rates as calculated in Table 16.3. 256

16.7 Granulometry analysis from the site section and other sediments from the valley. 259

16.8 Oxygen stable isotope curves for the tufa (left) and site (right) sections. Grey highlights refer to times of occupation at the archaeological site (Natufian and Neolithic). White dots in the tufa section correspond to isotopic compositions potentially influenced by disequilibrium effects (e.g. evaporation). 260

16.9 Oxygen and carbon isotopic composition and covariation (trend lines) for the site and tufa sections at Beidha. Evaporation/disequilibrium trend from Andrews (2006). The black box contains the samples that may be considered independent from disequilibrium effects. 261

16.10 Comparison between the Soreq isotopic record (left; modified from Bar-Matthews *et al*., 1999) and the Beidha spring-carbonate record (right; this study). Open dots correspond to samples from the Beidha record that show potential disequilibrium effects (e.g. evaporation). Light grey highlights indicate the probable time of the Younger Dryas (YD). 264

17.1 Southern Levant showing drainage basins and survey coverage, with the North Rift Basin and North Dead Sea Basin indicated. oPt refers to occupied Palestine territory. 272

17.2 Cost distance maps and site locations for the three time periods: (a) Chalcolithic, (b) EBI, (c) EBA. Top left, to permanent sites; top right to routes; bottom left, to springs; bottom right, to wadi. 274

17.3 Cost distance against altitude for wadi, site, route and spring for the three altitudinal sectors and three time periods. (a) North Rift Basin; (b) North Dead Sea Basin. Best fit lines represent the general trends in each sector. Altitudinal sectors, <−200 m, 200 m to 300 m and >300 m. CHL, Chalcolithic. 279

17.4 Cost distance pairs, site v. route, site v. spring, route v. spring and wadi v. spring regression plots for the three altitudinal sectors and three time periods. The first variable is on the horizontal axis, second on the vertical, dotted line shows line of equal value. (a) North Rift Basin; (b) North Dead Sea Basin. 282

18.1 Location map of Jawa showing catchment of Wadi Rajil with rainfall isohyets (rainfall data represent average annual rainfall 1931–1960). Source: NRA Jordan. Rajil catchment after Helms (1981). 290

18.2 Detail of the Wadi Rajil, storage ponds and Jawa. (After Helms, 1981 p. 157.) 291

18.3 Long-term changes in rainfall and temperature from the HadSM3 GCM, compared with Dead Sea levels (after Frumkin *et al*., 2001) and palaeo-rainfall estimates derived from analyses of Soreq isotope sequence (after Bar-Matthews *et al*., 2003). 297

18.4 Sustainable population levels at Jawa with varying water storage volumes for (a) high, (b) medium and (c) low rainfall scenarios. 298

18.5 Sustainable population levels as a function of pond storage under a range of rainfall conditions. 299

18.6 Percentage population failure at Jawa as a function of rainfall. 299

19.1 Location of Humayma in the hyper-arid zone of southern Jordan. 303

19.2 Site plan of the Humayma settlement centre (S. Fraser, courtesy J. P. Oleson) and aerial photograph of the site, looking north (courtesy D. Kennedy). 304

19.3 Humayma environs showing locations of pertinent hydraulic structures surveyed in the Humayma Hydraulic Survey. See colour plate section. 306

19.4 Reservoir at Humayma. (after modern restoration, originally covered; Courtesy E. De Bruijn). 306

19.5 Covered reservoir at Humayma (R. Foote). 306

19.6 Section of the aqueduct at Humayma (J. P. Oleson). 307

19.7 Terraced hillside at Humayma (J. P. Oleson). 308

19.8 Aerial photograph of the area north and east of the Humayma Settlement Centre (coustesy D. Kennedy archive, 26.002). 308

19.9 The Wadi Yitm catchment in southern Jordan (M. El Bastawesy). 314

19.10 The badlands at Humayma (R. Foote). 315

19.11 WorldView 1 Panchromatic satellite image from Eurimage, Jordania WV01 taken in 2008, showing the Wadi Amghar, the palaeo northern Wadi Qalkha tributary head waters, the badlands, the present-day northern Wadi Qalkha catchment headwaters and northern part of the Humayma settlement centre. 316

19.12 The interpolated path of the aqueduct between the surveyed points (turquoise circles) (M. El Bastawesy and R. Foote). See colour plate section. 317

19.13 Digital elevation models, comparing palaeo- and present-day northern Qalkha sub-catchment elevations and wadi pathways. Present-day, left; palaeo, right panel (M. El Bastawesy). 318

19.14 Details of 1926 (left) and 2008 (right) photos, comparing part of the badlands (after M. El Bastawesy). 318

19.15 Details of 1926 and 2008 photos, comparing a section of the Wadi Amghar. The fine lines sketch as it was in 1926 (after M. El Bastawesy). 319

19.16 Schematic diagram of the Humayma water balance model (A. Wade). 320

20.1 Map of Jordan indicating the sites mentioned in the text. 338

20.2 Mean isotopic values (±1 standard deviation) for humans and domestic animals from Pella. Figures in parentheses indicate number of individuals. Abbreviations: MB/LB = Middle Bronze/Late Bronze; LR/Byz = Late Roman/Byzantine. 342

20.3 Mean isotopic values (±1 standard deviation) for humans and domestic animals from Yaʻamūn. Figures in parentheses indicate number of individuals. Abbreviations: MB/LB = Middle Bronze and Late Bronze; LR/Byz = Late Roman/Byzantine. 342

20.4 Mean isotopic values (±1 standard deviation) for humans from Gerasa (seventh century AD), Pella (third–fourth centuries AD), Yaʻamūn (Late Roman/Byzantine); Yajūz (Byzantine) and Saʻad (Late Roman/Byzantine) in the north of Jordan. 343

21.1 Map showing location of crop growing sites. 351

21.2 Irrigation and evaporation (both in mm) by crop development stage. Crop development stages (given on the x axis) follow the Food and Agricultural Organisation convention of Initial (Init.), Crop development (Dev.), Mid-Season (Mid.) and Late. Late barley and late wheat are shown separately, reflecting differences in their development. DA, Deir 'Alla; KS, Khirbet as Samra; RA, Ramtha. 353

21.3 Harvesting barley at Khirbet as Samra after the third growing season. 355

21.4 (A) Well silicified conjoined phytolith. (B) Poorly silicified conjoined phytolith. 357

21.5 Extractable silicon from soil samples taken before and after experimentation. 358

21.6 Crop yield for wheat. 359

21.7 Crop yield for barley. 360

21.8 Weight percent of phytoliths to original plant matter processed for wheat. 361

21.9 Weight percent of phytoliths to original plant matter processed for barley. 362

21.10 Correlation between non-irrigated barley and rainfall. 363

21.11 Correlation between mean weight percent of phytoliths and levels of extractable silicon. 363

21.12 (A) Single dendritic long cell. (B) Single cork-silica cell. 363

21.13 Comparison of the percent of cork-silica cells and dendritic long cells for the wheat samples. 363

21.14 Comparison of cork-silica cells and dendritic long cells for the barley samples. 364

21.15 Comparison of well silicified to poorly silicified conjoined forms from the wheat samples. 364

21.16 Comparison of well silicified to poorly silicified conjoined forms from the barley samples. 366

21.17 Percent of conjoined cells in the wheat samples for all sites, years and irrigation regimes. 368

21.18 Percent of conjoined cells in the barley samples for all sites, years and irrigation regimes. 369

22.1 Flowchart illustrating the different environmental parameters determining the carbon stable isotope composition ($\delta^{13}C$) of C_3 plants (for references, see text). 375

22.2 Location of crop growing stations in Jordan used in the WLC experiments (map reproduced from Fig. 1 of Mithen et al., 2008). Khirbet as Samra N 32° 08′, E 36° 09′, 564 m asl, Ramtha N 32° 33′, E 32° 36′, 510 m asl and Deir 'Alla N 32° 13′, E 35° 37′, 234 m bsl. Altitudes were obtained from the Google Earth database and were checked against OS maps (Ordnance Survey 1949a and b and 1950). 377

22.3 Graph showing mean Δ (‰) values for each irrigation regime at the three sites with trendlines fitted by polynomial regression. 377

23.1 Map of Jordan showing location of sites. 381

23.2 (A) Juma in front of his winter tent. (B) Inside the public part of the tent. (C) Inside the domestic area of the tent. (D) Um Ibrahim (Juma's wife) cooking bread on the domestic hearth. 388

23.3 Images of the main single-celled phytolith morphotypes identified: (A) irregular shaped dicotyledon (platey) from Tell esh-Shuna; (B) polyhedral shaped psilate from Tell Wadi Feinan; (C) trapeziform psilate from Tell esh-Shuna; (D) bilobate from WF16; (E) rondel from Tell esh-Shuna; (F) siliceous aggregate (usually bright orange) from Tell esh-Shuna; (G) globular echinate (date palm) from Tell esh-Shuna; (H) elongate dendriform from Tell esh-Shuna; (I) trapeziform crenate from Tell esh-Shuna; (J) globular echinate (date palm) from Tell Wadi Feinan; (K) elongate psilate from WF16. 391

23.4 Images showing conjoined phytoliths. (A) Elongate dendriform and short cells formed in a wheat husk from Tell esh-Shuna; (B) conjoined parallepipedal bulliforms from Tell-esh Shuna; (C) conjoined elongate sinouate from a reed stem from Juma's summer tent; (D) elongate dendriform and short cells formed in a barley husk from Humayma; (E) elongate dendriform and short cells formed in a barley husk from Tell esh-Shuna; (F) conjoined elongate sinouate from a reed stem from Ghuwayr 1; (G) bilobates from Tell Wadi Feinan; (H) sinuous long cells from Ghuwayr 1; (I) elongate dendriform and short cells formed in a wheat husk from Juma's summer tent; (J) jigsaw puzzle from Tell esh-Shuna. 392

23.5 Weight percent of phytoliths by context category and chronological period. 393

23.6 Water availability index showing percentage of long-cell phytoliths relative to the sum of short and long cells by context category. 394

23.7 Woody material index showing percent of siliceous aggregates by context category. 394

23.8 Plot of two first axes of the PCA showing the taxonomic origin of the phytolith types. The individual vectors (phytolith types) are unlabelled for clarity reasons. 394

23.9 Plot of two first axes of the PCA showing the archaeological samples (used for the analysis) and the ethnoarchaeological samples (passively plotted) (JT, Juma's tent). 394

23.10 Plot of two first axes of the PCA highlighting the samples from fields and channel. 395

23.11 Plot of two first axes of the PCA highlighting the affinities of hearth and other fire-related contexts. 395

23.12 Plot of two first axes of the PCA showing the chronological patterns. 397

24.1 The general geography of Jordan. 404

24.2 Rainfall distribution in Jordan (Source: Ministry of Environment, 2006). 405

24.3 Jordan's surface water resources (Source: Ministry of Environment, 2006). 408

24.4 Jordan's groundwater resources (Source: Ministry of Environment, 2006). 409

24.5 Top left: percentage change in annual total rainfall predicted for 2070–2100 under an A2 emissions scenario. Top right: seasonal cycle in total rainfall in the box shown on the top left figure, for a group of model integrations for a present-day climate (dark grey polygon); a 2070–2100 A2 scenario climate (pale grey polygon) and a single B2 scenario model integration (line). Bottom left: observed annual total precipitation (based on the publicly available GPCC gridded dataset). Bottom right: percentage changes in annual total rainfall under an A2 scenario projected on to the observed present-day totals. See colour plate section. 412

25.1 Schematic diagram and corresponding map of the pathway of water reuse in northwest Jordan. 417

25.2 Map of northwest Jordan showing the locations of soil sampling and farmer interviews. 418

25.3 Nutrient inputs and nutrient requirements of barley at the research sites (nutrient uptake of barley is based on data from Cooke, 1982). 420

25.4 Salinity of the saturation extract (ECe) from Ramtha after one and two years of irrigation with reclaimed water (error bars give standard error from the mean). 421

25.5 Salinity of the saturation extract (ECe) from Khirbet as Samra after one and two years of irrigation with reclaimed water (error bars give standard error from the mean). 421

25.6 Salinity of the saturation extract (ECe) from sites irrigated with reclaimed water for extensive periods of time (error bars give standard error from the mean). 422

25.7 Leaching methods and timings of leaching described by farmers irrigating with reclaimed water in the Jordan Valley. 423

25.8 Modelled scenario of leaching timings on the Cl concentration in the soil solution at Deir Alla. 424

26.1 Water charges in Amman, Jordan and a selection of other cities in the MENA region. Exchange rates in 2004 (from Magiera et al., 2006). 430

26.2 The Greater Amman urban region: general location map (revised and adapted from Lavergne, 2004). 431

26.3 The social areas of Greater Amman and the residential areas sampled for the household interviews. 433

26.4 Typical roof-top storage tanks on a property in the northwestern suburbs of Amman. 435

26.5 Income-related variations in water storage capacity among the interview respondents. Each of the 25 high-income and low-income respondents is represented by a vertical bar. 436

26.6 Typical water delivery tanker in Greater Amman, Jordan. 437

27.1 Rainfall distribution and water resources in Jordan and location of the study area. Adapted from Darmame (2006). 444

27.2 The regional and national context of the Faynan and Qurayqira villages. 444

27.3 The typical small dwelling provided by the government in Faynan village. 444

27.4 The tent: place of traditional hospitality and shrubs on the 1 dunum plot. 445

27.5 The everyday challenges of water access for the Al-Azazma tribe. 445

27.6 A young girl bringing back water from Wadi Ghuwayr for cooking. 445

27.7 Land use in Faynan and Qurayqira villages. 445

LIST OF FIGURES

27.8 Rudimentary dam in Wadi Ghuwayr to collect water prior to transfer through pipes for irrigation. 447

27.9 Water storage basin near the farm for irrigation. 448

27.10 Watermelons immediately prior to harvest in the Faynan area. 450

27.11 Watermelons being packed on to trucks in Faynan for transport north to Amman and for export. 450

28.1 In a country area of the southern Amman Governorate, people use water directly from the piped network. 457

28.2 An illegal connection from the principal pipe of the network, southern Amman Governorate. 457

28.3 The service area of LEMA. From Darmame (2006). 459

28.4 The total supply duration (hours per week) in Greater Amman, summer 2006. Source: LEMA Company, personal communication, Anmar 2007. See colour plate section. 460

28.5 Supply duration (hours per week) in Greater Amman by geographical sector, January 2003 to August 2006. 461

Tables

1.1 Water, Life and Civilisation: project members. *page* 5
2.1 The 15 most common GWL regimes and their relationship to Jordan Valley rainfall and large-scale modes of variability. 20
2.2 Distribution of rainfall quintiles for NAO-positive and NAO-negative years. 22
3.1 The experimental configuration. 29
3.2 Summary of the changes in the time-slice experiments (expressed relative to the pre-industrial control run). 48
9.1 Uranium series ages for samples analysed as part of this study. 122
10.1 A summary of rainfall and flow data collated for the application of the hydrological models to simulate the flows in the upper River Jordan. 135
10.2 The mean monthly minimum and maximum near-surface air temperatures observed at Ramtha, Jordan and those simulated by the HadRM3P regional climate model during a control run (1961–1990) and for the A2 scenario (2071–2100). 143
11.1 Mean rainfall and Dead Sea levels for the period 1860–1960. 150
12.1 Baseflow measurements in the Wadi Ghuwayr made during the 2006, 2007 and 2008 field seasons. 166
12.2 Peak floods in the Wadi Ghuwayr and the Wadi Dana estimated using open-channel techniques. 171
12.3 Surface and groundwater flow statistics for the Wadi Faynan generated by running the Pitman hydrological model with modified rainfall inputs and/or a modified infiltration rate parameter. 171
13.1 Meteorological data collated for the hydrological study of the Wadi Hasa. 179
13.2 Delta change values calculated at four precipitation and two near-surface air temperature measurement stations to compare daily mean precipitation ratios and temperature differences between the HadRM3P modelled control period (1961–1990) and the A2 scenario period (2071–2100). 185
15.1 Summary of synthetic Tafilah rainfall series for time slices (results of 100-year simulation). 225
15.2 Synthetic monthly mean rainfall (mm per month) at Tafileh for each time slice (mean of 100-year total). 226
15.3 Comparison of palaeo-rainfall estimates. 227
15.4 Summary of results from Hunt *et al.* (2007) (table 12) showing changing patterns of Holocene vegetation in Wadi Faynan. 229
15.5 Summaries of simulated wadi flows under changing precipitation scenarios. 231
15.6 Summaries of simulated wadi flows under changing precipitation and infiltration scenarios. 232
15.7 Comparison of simulated median groundwater flows between first and last decade of simulations, for 12 ka scenarios (8 mm hr^{-1}, 100 mm hr^{-1}) and 0 ka (control) period. 232
15.8 Infiltration rates and rainfall values used in proposed simulation of Holocene evolution of Wadi Faynan. 234
15.9 Summary of proposed wadi flows for Holocene Wadi Faynan. 234
16.1 All radiocarbon dates from Beidha. 249
16.2 Uranium-series dates from the tufa series. 257
16.3 Sedimentation rates calculated for various intervals of the tufa section. 260
17.1 Number of sites and average altitude for the North Dead Sea Basin for all sectors and time periods (includes all survey data, sites may be listed in more than one column). 277
17.2 Number of sites and average altitude for the North Rift Basin for all sectors and time periods (includes all survey data, sites may be listed in more than one column). 277
17.3 Statistics of cost distance values for the North Rift Basin for all sectors and time periods (single time period data). 278
17.4 Statistics of cost distance values for the North Dead Sea Basin for all sectors and time periods (single time period data). 278

LIST OF TABLES

18.1 Distribution types and characteristics for key parameters used in the Monte Carlo Analysis for the high rainfall condition. 296
19.1 Humayma water storage features, divided by sub-catchment. 309
19.2 Catchment characteristics of the four sub-catchments used in the model-based water balance assessment rendered in present-day and proposed historic areas. 320
19.3 Summary of the parameter distributions used in the model simulations. 321
19.4 Estimates of the storage capacity available in the cisterns, reservoirs and created pools throughout the Humayma catchment. 323
19.5 Estimates of the surface area of the uncovered cisterns, reservoirs and created pools throughout the Humayma catchment. 324
19.6 The modelled mean and median populations at Humayma and modelled runoff for a succession of rainfall and water management scenarios. 327
21.1 Description of crop growing sites. 352
21.2 Amount of applied irrigation by crop, year and growing site. 352
21.3 Selected water quality parameters for reclaimed water from the crop growing stations based on samples collected in the field and available data. 354
21.4 Record of plots covered with mesh. 355
21.5 Soil physical and chemical properties at the crop growing stations. 356
21.6 Methodology used for analysis of extractable silicon. 357
21.7 Dry ashing methodology employed for extracting phytoliths from modern plants. 357
21.8 Absolute numbers of well silicified and poorly silicified conjoined phytoliths counted with standard deviations from the wheat samples (all standard deviations were calculated using the STDEV function in Excel 2007). 365
21.9 Absolute numbers of well silicified and poorly silicified conjoined phytoliths counted with standard deviations from the barley samples. 367
22.1 Table showing mean Δ values in ‰. 378
23.1 Chronology of the sites studied. 386
23.2 Taxa present by chronological period. 393
23.3 Single-cell phytoliths by family and subfamily for the various context categories. 396
24.1 Climatic classification according to rainfall distribution. 405
24.2 Sources of water used by sectors in Jordan in 2000 (million cubic metres, MCM). 406
24.3 Jordan water use (MCM) by sector according to the Ministry of Water and Irrigation. 407
25.1 Water quality parameters at Khirbet as Samra, Ramtha and Deir Alla based on data determined through water sampling and published data. 420
25.2 Comparison between the average ECe in the root zone (10–40 cm depth) of soil irrigated to 100% and 120% of the crop water demand for two years. 422
26.1 Socio-demographic profile of the respondent households. 433
26.2 Distribution of respondent households by income group. 433
26.3 Occupational categories of the heads of households included in the sample. 434
26.4 Residential profiles of the respondent households. 434
26.5 Distribution of respondent households by size of dwelling. 434
26.6 Household water supply from the public network. 434
26.7 Average household water storage capacity by income group. 435
26.8 Respondent households by water storage capacity. 435
26.9 Household water consumption and cost levels. 436
26.10 Method of payment of water bill. 437
26.11 Details of the use of networked water by households. 437
26.12 Gendered aspects of the management of water within households. 438
26.13 The use made of collected rainwater by households. 438
26.14 Awareness of selected water tariff issues. 439
26.15 Variations in household satisfaction levels with the water sector as a whole. 439
26.16 Overall household satisfaction score with the water sector by income group. 439
26.17 Satisfaction with different aspects of the water supply system. 440
26.18 Perceived improvement in the water supply system since privatisation in 1999. 440
26.19 Priority accorded to water supply issues by households. 440
28.1 Selected performance indicators concerning urban water and sanitation utilities in selected cities of the MENA region, 2004. 458
28.2 Targets set in the LEMA contract for the reduction of unaccounted for water (UFW) and water supply improvements. 460
28.3 Proportion of water subscribers in Greater Amman by total duration of water supply per week during summer 2006. 461

Contributors

Ambroise Baker
Long-Term Ecology Laboratory,
School of Geography and the Environment,
University of Oxford,
Oxford, OX1 3QY, UK

Emily Black
Department of Meteorology,
University of Reading,
Earley Gate, PO Box 243,
Reading, RG6 6BB, UK

Stuart Black
Department of Archaeology,
School of Human and Environmental Sciences,
University of Reading,
Whiteknights, PO Box 227,
Reading, RG6 6AB, UK

Andrew V. Bradley
Department of Geography,
University of Leicester,
University Road,
Leicester, LE1 7RH, UK

David Brayshaw
Department of Meteorology,
University of Reading,
Earley Gate, PO Box 243,
Reading, RG6 6BB, UK

Dan Butterfield
Water Resource Associates,
PO Box 838, Wallingford OX10 9XA, UK

Gemma Carr
Centre for Water Resource Systems,
Vienna University of Technology,
Karlsplatz 13/222, A-1040 Vienna,
Austria

Khadija Darmame
The French Institute of the Near East,
BP 830413,
Amman, 11183,
Jordan

Mohammed El Bastawesy
National Authority for Remote Sensing and Space Sciences,
Cairo, Egypt

Sarah Elliott
School of Human and Environmental Sciences,
University of Reading,
Whiteknights, PO Box 227,
Reading, RG6 6AB, UK

Bill Finlayson
Council for British Research in the Levant (CBRL),
10 Carlton House Terrace,
London,
SW1Y 5AH, UK

Richard Fitton
Talisman Energy Inc.,
Suite 3400, 888 – 3rd St Southwest,
Calgary,
Alberta, Canada

Nicola Flynn
School of Human and Environmental Sciences,
University of Reading,
Whiteknights, PO Box 227,
Reading, RG6 6AB, UK

Rebecca Foote
c/o School of Human and Environmental Sciences,
The University of Reading,
Whiteknights, PO Box 227,
Reading, RG6 6AB, UK

LIST OF CONTRIBUTORS

Rachel Goodship
School of Human and Environmental Sciences,
The University of Reading,
Whiteknights, PO Box 227,
Reading, RG6 6AB, UK

Joshua Guest
School of Human and Environmental Sciences,
The University of Reading,
Whiteknights, PO Box 227,
Reading, RG6 6AB, UK

Paul Holmes
School of Human and Environmental Sciences,
The University of Reading,
Whiteknights, PO Box 227,
Reading, RG6 6AB, UK

Brian Hoskins
Department of Meteorology,
University of Reading,
Earley Gate, PO Box 243,
Reading, RG6 6BB, UK

Robyn Inglis
Department of Archaeology,
University of Cambridge,
Downing Street,
Cambridge, CB2 3DZ, UK

Khalil Jamjoum
National Centre for Agricultural Research and Extension,
PO Box 639,
Baq'a 19381, Jordan

Emma Jenkins
School of Applied Sciences,
Bournemouth University,
Christchurch House,
Talbot Campus, Poole,
Dorset, BH12 5BB, UK

Jaimie L. Lovell
Council for British Research in the Levant (CBRL),
10 Carlton House Terrace,
London, SW1Y 5AH, UK

Ron Manley
Water Resource Associates Limited,
PO Box 838,
Wallingford, OX10 9XA, UK

Steven Mithen
University of Reading,
Whiteknights, PO Box 227,
Reading, RG6 6AB UK

Gundula Müldner
School of Human and Environmental Sciences,
The University of Reading,
Whiteknights, PO Box 227,
Reading, RG6 6AB, UK

Stephen Nortcliff
School of Human and Environmental Sciences,
The University of Reading,
Whiteknights, PO Box 227,
Reading, RG6 6DW, UK

Sameeh Al Nuimat
Permaculture Project,
CARE International – Jordan,
PO Box 950793 – Amman 11195,
Jordan

John Peter Oleson
Department of Greek and Roman Studies,
Box 3045, University of Victoria,
Victoria BC, V8W 2P3, Canada

David Plinston
Water Resource Associates,
PO Box 838,
Wallingford, OX10 9XA, UK

Robert B. Potter
School of Human and Environmental Sciences,
The University of Reading,
Whiteknights, PO Box 227,
Reading, RG6 6AB, UK

Claire Rambeau
c/o School of Human and Environmental Sciences,
The University of Reading,
Whiteknights, PO Box 227,
Reading, RG6 6AB, UK

Stuart Robinson
Department of Earth Sciences,
University College London,
Gower Street,
London, WC1E 6BT, UK

Michela Sandias
c/o School of Human and Environmental Sciences,
The University of Reading,
Whiteknights, PO Box 227,
Reading, RG6 6AB, UK

Bruce Sellwood
c/o School of Human and Environmental Sciences,
The University of Reading,
Whiteknights, PO Box 227,
Reading, RG6 6AB, UK

Julia Slingo
Department of Meteorology,
University of Reading,
Earley Gate, PO Box 243,
Reading, RG6 6BB, UK

Sam Smith
Department of Anthropology and Geography,
Oxford Brookes University,
Headington Campus,
Gipsy Lane,
Oxford, OX3 OBP, UK

Helen R. Stokes
Department of Archaeology,
University of York,
King's Manor,
York, Y01 7EP, UK

Paul J. Valdes
School of Geographical Sciences,
University of Bristol, University Road,
Bristol, BS8 1SS, UK

Andrew Wade
School of Human and Environmental Sciences,
The University of Reading,
Whiteknights, PO Box 227,
Reading, RG6 6AB, UK

Paul Whitehead
School of Human and Environmental Sciences,
The University of Reading,
Whiteknights, PO Box 227,
Reading, RG6 6AB, UK

Acknowledgements

The authors and editors are grateful to the Leverhulme Trust for generously funding the Water, Life and Civilisation (WLC) project (Grant no. F/00239/R), and thus enabling them to carry out the research described in this volume.

The majority of this research has been undertaken in Jordan and we are immensely grateful to the following for their support: Princess Sumaya bint Hassan, patron of the archaeological field work in Wadi Faynan undertaken by the Council for British Research in the Levant (CBRL) and its collaborators; Dr Abed al-Nabi Fardour (Director of National Centre for Agricultural Research and Technology Transfer, NCARTT), Khalil Jamjoum (NCARTT) and Sameeh Nuimat (Ministry of Agriculture), with regard to Chapters 12, 21, 22 and 24); Dr Fawwaz al-Khraysheh (Director General of the Department of Antiquities) for permission to undertake fieldwork in Wadi Faynan and at Jawa; Nasr Khasawneh and Mohammed al Zahran, who were the Department of Antiquities Representatives for fieldwork in Jawa and Wadi Faynan, respectively; Professor Dawud Al-Eisawi of the University of Jordan, for its ongoing collaboration regarding pollen analyses of the Dead Sea peaty deposits (sampled during the WLC project and still being analysed), and Sagida Abu-Seir (University of Jordan) for sample preparation and identification of the pollen species; Dr Mohammed Najjar (Council for British Research in the Levant) for advice and information regarding Chapters 12 and 13). Professor Zaid al-Sa'ad, Professor M. El-Najjar and Dr A. Al-Shorman (Institute of Archaeology and Anthropology of Yarmouk University Jordan), with regard to Chapter 20; and Dr L. Khalil (Department of Archaeology of the University of Jordan), for permission to sample skeletal remains for Chapter 20. With regard to the research concerning water issues in Jordan and especially with regard to present-day Amman (Chapters 24–28), we are grateful to: Professor Nasim Barham (University of Jordan); Dr Philipp Magiera (Team Leader, German Technical Cooperation/Ministry of Water and Irrigation, Improvement of the Steering Competence in the Water Sector Project) for consultations concerning the National Water Master Plan for Jordan; Dr Max Bobillier (Operations Director, LEMA/Ministry of Water) for discussions concerning upgrading the water network of Greater Amman; and staff of the Greater Amman Municipality GIS and Planning Departments for background information on urban planning and residential land issues in Greater Amman.

Equally valued have been the many residents of Jordan who generously gave up time to help with our research. We are especially grateful to those households in Greater Amman, and farmers in the Jordan Valley and Wadi Faynan who provided interviews as cited in Chapters 24, 26, 27 and 28 of this volume. The farmer interviews would not have been possible without the translation assistance from a number of individuals to whom we are grateful. The WLC archaeological and hydrological work in Wadi Faynan (Chapters 12, 14, 15) was only possible through the help and support provided by our friends in the local Bedouin communities, notably Abu Fawwaz, Abu Sael and Juma Ali Zanoon. We would also like to thank Juma Ali Zanoon for generously allowing the authors of Chapter 23 to take sediment samples from his tents, and to Haroun al-Amarat who provided translation. We are also grateful to the Maayah family from Madaba for their warm welcome to Claire Rambeau and support of her work in the vicinity of the Dead Sea region and Beidha.

We are grateful to the CBRL for providing administrative and logistical support, especially with regard to fieldwork in Jordan. We would particularly like to thank Professor Bill Finlayson, in his role as Director of the CBRL, for facilitating permits for archaeological and geological sampling of sites, and the Department of Antiquities of Jordan, for having awarded those permits.

We have benefited greatly from the support and advice of numerous other academics and bodies outside Jordan and our own University, to whom we are grateful: the Israeli Meteorological service for providing the daily data for rain stations along the Jordan River referred to in Chapters 2, 4, 5, 10 and 13; David Hassell, David Hein and Simon Wilson (Hadley Centre) for their invaluable assistance and advice on regional modelling, and for providing lateral boundary conditions as used in Chapter 4; Michael Vellinga (Hadley Centre) for providing the sea surface temperature data used to calibrate the slab ocean model in Chapter 3; Yuval Bartov (Colorado School of Mines), Stephen Calvert (University of British Columbia). Kay-Christian Emeis

(Institute for Biogeochemistry and Marine Chemistry), Mebus Geyh (Nidersächsisches Landesamt für Bodenforschung) and Angela Hayes (University of Limerick) for contributing data to Chapter 6; Professor Fabrice Monna (University of Burgundy) for his help regarding the ongoing analyses of the Dead Sea peaty sediments; Frank Farquharson and Helen Houghton-Carr (Centre for Ecology and Hydrology, Wallingford) for allowing the authors of Chapter 12 access to hydrological records in their overseas archive; Steve Savage (Arizona State University) for allowing the authors of Chapter 17 access to the raw data of the JADIS database; Yehuda Dagan and the staff at the Israeli Antiquities Authority for discussing survey methodology in Israel with regard to Chapter 17; Paul Valdes (Bristol University) for access to the Bristol University BRIDGE Climate Change data for Chapter 18 and advice regarding palaeoclimate modelling; Professor J. Rose (University of Arkansas) for enabling access to skeletal remains for Chapter 20; Dr Stephen Bourke and Dr I. Kehrberg (University of Sydney) for advice regarding the analysis of skeletal remains from Jordan for Chapter 20; Dr Carol Palmer (CBRL) and Dr Mark Nesbitt (Kew Gardens) for advice on setting up the crop growing experiments described in Chapter 21; Dr Mohammed al-Najjar, Dr Douglas Baird (University of Liverpool), Graham Philip (University of Durham) and Alan Simmons (University of Nevada) for allowing the authors of Chapter 23 permission to sample their archaeological sites; Anson MacKay (University College London) for advice on using Canoco for Chapter 23.

We are grateful to Elsevier for allowing permission to republish a WLC paper from the *Quaternary Science Reviews* (2006, **25**: 1517–1541) as Chapter 6 of this volume. We acknowledge the support of Stuart Robinson by a Royal Society University Research Fellowship. With regard to our academic colleagues at the University of Reading, we would especially like to thank those who have advised on the climate modelling: Lois Steenman-Clark and Jeff Cole of NCAS-Climate for advice on high performance computing, and Kevin Hodges of ESSC and the University of Reading for giving permission for the authors of Chapters 2, 3 and 4 to use his storm tracking software, and providing support and advice; Charles Williams for useful discussion throughout the project and help with setting up the regional climate model; Tim Woollings and Paul Berrisford for their contributions to the analysis during the early part of WLC.

The project has been fortunate in having the support of clerical and technical staff from the University of Reading. We would especially like to thank Jane Burrell, the WLC project administrator, Chris Jones and Cheryl Foote (University of Reading), and Penny Wiggins and Nadja Qaisi (Council for British Research in the Levant) for administrative support. Tina Moriarty, John Jack and Emilie Grand-Clément prepared samples for the Beidha study (Chapter 16); Ian Thomas carried out the spatial analysis of water quality data described in Chapter 12, while Anne Dudley and Dave Thornley carried out the water sample analysis described in that chapter; Ambroise Baker, Sarah Elliott, Kim Carter and Geoff Warren processed the soil and water samples described in Chapter 21; Tina Moriarty and Paul Chatfield advised on the sample processing and statistics described in Chapter 22; Bruce Main set up the mathematical model described in Chapter 24; We are especially grateful to Sophie Lamb who helped to prepare all of the illustrations for this volume, and to our copy-editor Lindsay Nightingale.

Finally, we would like to thank the four academics who were our 'critical friends' throughout the project, attending our annual meetings and providing immensely constructive advice: Professor Richard Bradley (Archaeology, University of Reading), Professor Robert Gurney (Meteorology, University of Reading), Professor Neil Roberts (Physical Geography, Plymouth University) and Professor Tony Wilkinson (Archaeology, Durham University).

1 Introduction: an interdisciplinary approach to Water, Life and Civilisation

Steven Mithen and Emily Black

This volume is an outcome of a five-year research project (2005–2009) based at the University of Reading, UK, entitled Water, Life and Civilisation. This project's aim was:

> to assess the changes in the hydrological climate of the Middle East and North Africa (MENA) region and their impact on human communities between 20,000 BP and AD 2100, with a case study of the Jordan Valley.

The project arose from a decision by the Leverhulme Trust to fund one or more projects under the heading 'Water, Life and Civilisation', each funded by an award of up to £1.25 million, advertising for applications in October 2003. Quite why the Leverhulme Trust selected this theme is unknown, but it was one that provided an ideal fit to research interests within the School of Human and Environmental Sciences at the University of Reading. The School had been formed in August 2003 from the previous Departments of Archaeology, Geography, and Soil Science, and the Postgraduate Research Institute for Sedimentology, with the avowed aim of developing interdisciplinary research. There was already research collaboration between Archaeology and the Department of Meteorology, exploring the impact of Pleistocene climate change on hominin dispersals from Africa (Hughes and Smith, 2008; Smith *et al.*, in press). In light of expertise within Geography regarding hydrology and development studies, and within Archaeology regarding the emergence of complex society, the Leverhulme Trust's request provided an excellent opportunity to realise the potential for interdisciplinary research provided by the new School; moreover it would be able to do so by addressing a research theme of global significance.

1.1 WATER, LIFE AND CIVILISATION

The planet is facing a water crisis: one billion people do not have access to safe drinking water. Two billion people have inadequate sanitation. By 2025 almost one-fifth of the global population is likely to be living in countries or regions with absolute water scarcity while two-thirds of the population will most probably live under conditions of water stress (UNWater, 2007). Excessive extractions of ground- and surface-water are causing rivers to run dry and wetlands to shrink; freshwater supplies are becoming polluted. Throughout the world, political tension is rife within and between countries over access to precious and dwindling water supplies, tensions that hinder peace processes and threaten to erupt into armed conflict (Gleick, 2006; Barlow, 2007). Population growth, economic development and urbanisation place unrelenting pressure on the planet's water resources. Ongoing, human-induced, climatic change is likely to have a further detrimental impact on water supplies to large sectors of the global population: precipitation is forecast to decrease and evaporation to increase in precisely those areas that are already suffering from water stress (Parry *et al.*, 2007; UNEP, 2007).

So if the term 'civilisation' is taken to mean a society in which people live and act in a civil manner to each other, 'Water, Life and Civilisation' is a phrase that encapsulates the immense challenge we face in securing a future for the planet in which all life, human and otherwise, has an adequate and uncontested water supply – or at least one in which drought-induced starvation, inadequate sanitation and water-based conflict are minimised. Research contributing towards that end must necessarily come from a wide array of disciplines ranging from climate science to development studies, achieving an integration that goes beyond mere multi-disciplinary or even interdisciplinary approaches: the challenge of 'Water, Life and Civilisation' epitomises the post-disciplinary academic world towards which we must strive.

One mark of a civilised society is knowledge about the past, both for its practical value – the challenges of the future cannot be resolved unless one understands how they have arisen – and for its own inherent value. The term 'civilisation' can also be used in a historical sense to refer to those ancient societies that

Water, Life and Civilisation: Climate, Environment and Society in the Jordan Valley, ed. Steven Mithen and Emily Black. Published by Cambridge University Press.
© Steven Mithen and Emily Black 2011.

reached high levels of social complexity, economic development and/or cultural attainment, as in the 'Mayan', 'Sumerian', Egyptian', 'Indus' and 'Inca' civilisations. Key attributes are high degrees of centralised authority and social stratification, extensive craft specialisation and trade networks, writing, art and monumental architecture (Trigger, 2003). While such attributes are shared, there is nevertheless immense social, economic and cultural variability, and quite different historical trajectories of development for such 'civilisations', making the term of limited academic value.

One further attribute that appears to have been shared by all of those ancient societies designated as 'civilisations' is sophisticated systems of water management to provide water for drinking, irrigation and, in some cases, social display. There has been a long history of archaeological study exploring and debating the relationship between water availability, water management and the 'rise of civilisation', or more generally the emergence of social complexity. The 'oasis theory' originally proposed by Pumpelly in 1908 and developed by Childe (1928) suggested that during a period of aridity immediately after the end of the ice age in the Near East, people, animals and plants were forced to cluster around oases, which led to domestication. While this particular proposition is no longer supported, changes in the overall quantity and annual distribution of precipitation between the Late Pleistocene and early Holocene epochs remain central to debates about the origin of farming, not only in the Near East but also in other regions of the Old and New Worlds (see the review in Mithen, 2003). The role of water management has been a key issue in theories about how small-scale farming communities grew into towns and ultimately state societies. Karl Wittfogel's monumental 1957 volume entitled *Oriental Despotism: A Comparative Study of Total Power* proposed that state societies in Asia were dependent upon the building of large-scale irrigation works. Wittfogel's 'hydraulic hypothesis' argued that such works required forced labour and a large, complex bureaucracy, both of which provided the basis for what he termed 'despotic' rule (Wittfogel, 1957). The anthropologist Julian Steward made a similar argument in his 1955 volume *Irrigation Civilization: A Comparative Study*, claiming that irrigation was the catalyst for state formation (Steward, 1955).

Robert McCormick Adams attempted to test the hydraulic hypothesis with regard to the rise of Mesopotamian civilisation (Adams, 1966; Adams, 1978) and found that complex systems of canals and irrigation came after the appearance of cities and the indicators of bureaucratic statehood, rather than before as Wittfogel had proposed. The same was found for the emergence of the Mesoamerican archaic state (Scarborough, 2003). Moreover, it soon became evident that societies throughout history have developed sophisticated techniques of water management but did not evolve into states, and that some of the most complex water management systems concerned reservoir management rather than irrigation. Indeed, what has emerged during the past few decades of archaeological research is a far more complex and diverse association between water and society than Wittfogel, Steward or anyone else had ever imagined, one that defeats the construction of a single grand theory such as the hydraulic hypothesis. Cross-cultural studies now emphasise the astonishing variety of methods of water management, and the even greater variety of relationships to land, labour and power (Scarborough, 2003). So we can use the phrase 'Water, Life and Civilisation' to encapsulate one of the key themes of archaeological research: understanding the complex relationships between water availability, water management and the emergence of social complexity.

1.2 THE IMPACT OF CLIMATE CHANGE: PAST, PRESENT AND FUTURE

There is an overlap between our concerns for the future and studies of the past: the extent to which securing a water supply has been in the past, and will be in the future, a driver for social and economic change. Another overlap is with the impact of climate change and how this might be assessed. The Intergovernmental Panel on Climate Change (IPCC Core Writing Team *et al.*, 2007) predicted a rise of global temperatures of 2.0–5.4 K by 2100 under a 'business as usual' (A2) scenario, one in which greenhouse gas emissions continue at their ongoing rate to the end of the century; even with a substantial reduction in emissions (for example under a B1 scenario), temperatures are predicted to rise by 1.1–2.9 K by 2100. The impact of such climate change on human communities will primarily be reductions in available water in some vulnerable regions, caused by changing patterns of precipitation allied with increased evaporation, which in turn influence food security. Furthermore, some climate models project increases in the frequencies of extreme precipitation and temperature events (e.g. Gao *et al.*, 2006), leading to more floods and droughts, with potentially catastrophic consequences. As such, there is a need to refine the spatial resolution of climate models to make them relevant to those charged with devising policy and plans to both mitigate and adapt to climate change. This has motivated the development of global climate models (GCMs), which have resolutions in the order of 20 km; Kitoh *et al.* (2008) describes the application of such a model to the Middle East. An alternative approach, which avoids the computational expense of running global high-resolution models, is to implement a Regional Climate Model (RCM) for the area of interest. Even when high resolution, state-of-the-art climate models are used, some form of statistical bias correction and downscaling is usually required, if the output is to be used to drive quantitative impacts models.

Studies of the past and current climate can inform projections for the future. At one level, comparison between the output of model integrations and observations of the recent and distant past is a test of how well a climate model can capture both large-scale climate processes and finer-scale variability. In particular, comparison with proxy observations of the past (e.g. the past 12,000 years) is an opportunity to assess how well models can simulate the response of the regional climate to substantially perturbed global forcings through time. This is highly relevant to future scenarios, in which the global forcings, such as net radiation, are substantially different from the present day and recent past. At another level, when used in conjunction, climate models and proxy data can deepen our understanding of large- and small-scale responses to strongly perturbed forcings, enabling us to interpret future climate projections. For example, the aridification that occurred in the Middle East during the Holocene has some parallels with the decrease in rainfall projected for the region over the next century under some future scenarios.

While archaeologists have always had an interest in the potential impact of climate change on the development of human society (Lubbock, 1865; Childe, 1928), this new level of understanding has allowed questions to be asked about the relationship between specific climatic events and cultural change; geologists and hydrologists are also increasingly associating changes in the palaeoclimatic record to changes in the nature of human society (e.g. Cullen *et al.*, 2000; deMenocal, 2001; Issar and Zohar, 2004). For instance, the onset of markedly more humid conditions in Africa 70 kyr ago (ka) has been proposed as the reason for the dispersal of modern humans (Scholz *et al.*, 2007), while the impacts of the 8.2 ka and 4.2 ka events have been proposed as playing a causal role in cultural change of society in the Near East (Cullen *et al.*, 2000; deMenocal, 2001; Rosen, 2007). Perhaps one of the key findings of recent years leading to this concern with the impact of climate change on human communities is the rapidity with which such change can occur: on some occasions, abrupt and dramatic changes in temperature and rainfall happened within the span of human memory and potentially individual human lifetimes (Alley *et al.*, 2003).

While it is appropriate to ask about the impact of these 'events', archaeologists and others have rarely gone beyond subjectively identifying a supposed chronological match between a particular spike or trough in a climate curve and a particular cultural event, followed by speculative claims that the former caused the latter. Insufficient attention has been paid to how the supposed global climatic event might have been environmentally manifest at the regional or local level and how that in turn affected the relevant economy and society. Just as is the case for the future, the past impacts would have been mediated by changing levels of precipitation, affecting the water supply and ultimately the provision of food. And just as we need Regional Climate Models to predict what may happen in the future, we need the same type of models to explore how the global climatic events detected in the marine and ice cores might have been manifested at a spatial and temporal scale relevant for past communities. Moreover, we also need an understanding of the society, economy and technology of those communities to assess how resilient and/or adaptable they might have been to such environmental change.

Archaeologists today are fortunate to be working at a time when the threat of future climate change has generated an investment in climate modelling research which they can exploit for their own interests in the past. Climate modellers are also fortunate to be working at a time when archaeologists and geologists have made significant progress in reconstructing past environments and relating changes in the terrestrial record to the climatic events detected in the more continuous ice core and marine records: how else can climate modellers verify their models other than by testing their ability to predict not future but past climate change, in other words against what actually happened in the past? The archaeological and geological records now provide a multitude of proxies for changing temperature and rainfall and their environmental consequences, such as changing patterns of vegetation. So the question arises whether the RCMs are able to draw on the known levels of greenhouse gases and patterns of solar radiation in the past to predict those changes in temperature and precipitation that are known to have happened: if they cannot, how can we have confidence that those same models provide valid predictions of future climate change?

1.3 THE WORLD, MIDDLE EAST AND NORTH AFRICA AND THE JORDAN VALLEY

The issues concerning water availability in the past, present and future are of global significance: they are relevant to all geographical areas of the planet and all human communities. There are, however, regions where the potential impacts of climate change are more pressing than others, partly because of existing problems of population pressure, environmental degradation and/or conflict: sub-Saharan Africa, Bangladesh and Mexico City are notable examples. But the Middle East is perhaps the pre-eminent region where the availability and management of water is most inextricably linked to society and economy.

Ownership and access to water sources has been central to the ongoing Arab–Israeli dispute, both as a cause of conflict and as one of the areas in which cooperation has been achieved (Allan, 2001). The water sources within this region have no respect for political boundaries, the watersheds and aquifers being transnational in their extent – although it might be argued that through its land annexations and construction of its wall, Israel has been attempting to adjust its national boundaries to encompass the extent of aquifers. Population growth, agricultural development, industrialisation and tourism are placing ever greater demands on water availability, at a time when rainfall is projected to decrease and evaporation to increase during the next century (Mariotti *et al.*, 2008; Evans, 2009). Ambitious plans involving vast financial investment are under way to transport water over long distances, epitomised by the Red Sea–Dead Sea pipeline (Glausiusz, 2010).

While the future, present and recent past of the Middle East cannot be understood without reference to the role of water, neither can its distant past. This is the region where both the earliest farming communities and ancient civilisations emerged; it has been the focus for archaeological theories relating water availability and management to major social and economic change. The access that archaeologists now have to climate, hydrological and palaeoenvironmental research should enable significant progress to be made on why farming, urbanisation and civilisation emerged in this region of the world.

In October 2003, when the Leverhulme Trust requested applications for projects on the theme of 'Water, Life and Civilisation', the Department of Archaeology at the University of Reading was already engaged in a long-term project exploring the origin of farming within Jordan, undertaking excavations at the early Neolithic site of WF16 in Wadi Faynan in collaboration with the Council for British Research in the Levant (CBRL), specifically between Steven Mithen at the University of Reading and Bill Finlayson, the Director of the CBRL (Finlayson and Mithen, 2007). That wadi, semi-arid today, also provided dramatic archaeological evidence for the role of water management in the Roman/Byzantine period in the form of a ruined aqueduct, reservoir and field system, and in rural economic development in the expansion of tomato and melon farming using irrigation by recently settled Bedouin following the formation of a cooperative in 2003 (Barker *et al.*, 2007).

1.4 THE SCOPE OF THE WATER, LIFE AND CIVILISATION (WLC) PROJECT

In light of the ongoing archaeological research interests in the Jordan Valley, collaboration with the CBRL, the overall significance of the Middle East, and the potential for interdisciplinary research provided by the mix of disciplines represented in the School of Human and Environmental Sciences and Department of Meteorology, a proposal was submitted to the Leverhulme Trust to fund a University of Reading project on Water, Life and Civilisation. The aim outlined at the outset of this chapter was, of course, enormously broad. It has the potential to encompass a vast range of studies covering not only climate modelling, hydrology and archaeological studies, but also history, politics and international relations. Even with the relatively large sum of

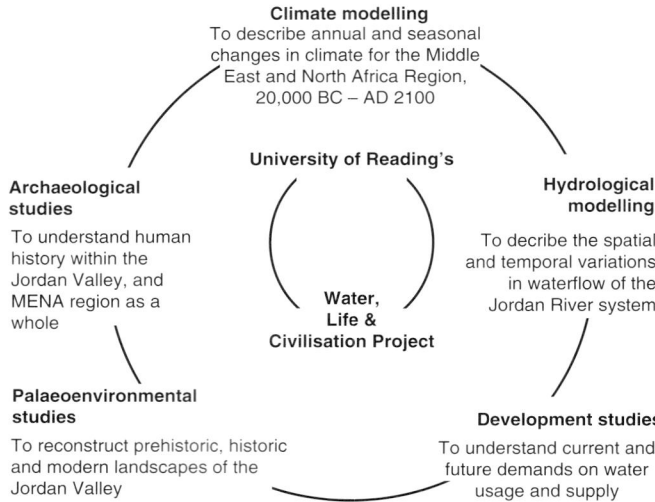

Figure 1.1. Disciplinary aims and interdisciplinary interactions of the Water, Life and Civilisation Project.

£1.25m that was available, the extent of research that could be undertaken was constrained, partly by resources and partly by the range of expertise available at the University of Reading. Some of these limits were identified at the start of the project – such as a decision not to address the recent history and politics of water. Others arose during the course of the research as time and resources became gradually exhausted or initial research opportunities were unable to be developed. For instance, there were plans to undertake a study of water usage within a Roman city, such as Jerash, and within the Islamic sugar industry of the Jordan Valley. Neither proved possible with the available time and funding.

One of the over-riding intentions of the project was to develop the interdisciplinarity that is widely recognised as essential for the study of water. This is difficult to realise in practice because of the manner in which academic institutions are structured by departments and faculties, often competing with each other for resource from the same institution. External pressures also serve to maintain disciplinary-based research, such as the Research Assessment Exercise in the United Kingdom. While the development of interdisciplinarity was a key objective, the WLC project also recognised that it needed to make substantive contributions to the core disciplines around which it was structured if the interdisciplinarity was to avoid becoming a superficial sharing of ideas and data. In this light, and to make the best of the available funding and expertise, the WLC project was structured around five sub-projects, each with their own Principal Investigators, aims and project-funded staff: Meteorology, Palaeoenvironments, Archaeology, Hydrology and Development Studies (Figure 1.1, Table 1.1).

The aim was to achieve the interdisciplinarity both informally, simply by regular discussions and seminars between the

Table 1.1 *Water, Life and Civilisation: project members*

Sub-project	Principal investigators	Post-doctoral research assistants	PhD students	MSc/BSc students	Other collaborators
Meteorology	Brian Hoskins Julia Slingo	Emily Black David Brayshaw			
Palaeoenvironmental studies	Stuart Black Bruce Sellwood	Stuart Robinson Claire Rambeau	Richard Fitton	Robyn Inglis Rachel Goodship	Paul Valdes
Archaeology	Steven Mithen Bill Finlayson	Emma Jenkins Sam Smith Andrew Bradley Rebecca Foote	Michela Sandias	Ambroise Baker Sarah Elliott Helen Stokes	Jaimie Lovell Gundula Mueldner Sameer Numait Khalil Jamjoum John Oleson
Hydrology	Paul Whitehead	Andrew Wade Nicola Flynn Dan Butterfield		Joshua Guest	Paul Holmes Ron Manley David Plinston Mohammed El Bastawesy
Development studies	Rob Potter Stephen Nortcliff	Khadija Darmame	Gemma Carr		Sameer Numait Khalil Jamjoum Nasim Barham

participants to make each other aware of the issues, language, methods and data sources of each discipline, and by formal integration of research. Figure 1.2 illustrates how this was envisaged in terms of a chain of models: a global circulation model would inform a regional circulation model for the study region; the RCM would generate input into a hydrological model for a localised area that would predict water availability, whether for a specified time period in the past or the future. This would, in turn, inform a study of either archaeological remains or a contemporary community. In this regard, the hydrology was identified as the key link between changing patterns of climate and human society – the link that has been most evidently missing from previous research.

Further forms of interdisciplinary engagement were envisaged. We aimed to undertake one or more direct comparisons of the environmental projections from an RCM for a specified region at a specified date in the past and the reconstruction of those environments by using a variety of geochemical proxies to evaluate the veracity of the RCM. We also aimed to make use of palaeoenvironmental reconstructions to inform the interpretation of archaeological sites within the study region and the study of contemporary water use, in terms of day-to-day household strategies to cope with limited water supply, to help formulate proposals for how water may have been managed at this level in the past.

The geographical scope of the climate modelling was chosen to include the whole Mediterranean and some of the eastern Atlantic, so as to allow the RCM to capture the eastward passage

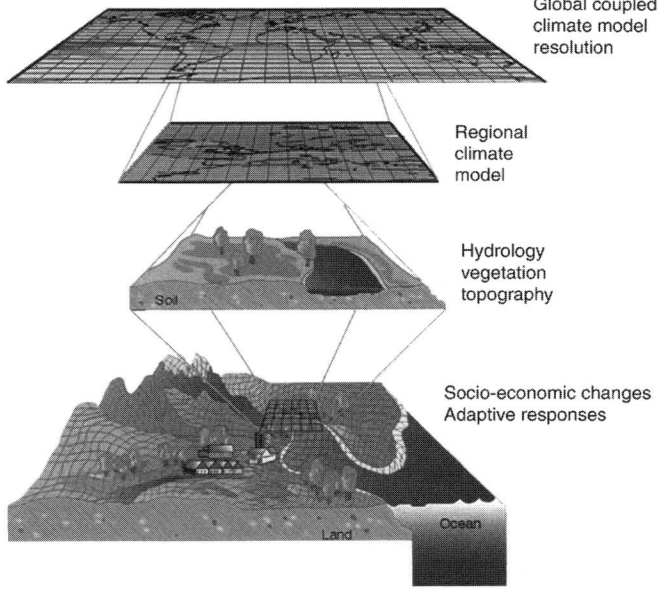

Figure 1.2. Hierarchical modelling from global circulation models to socio-economic impacts (courtesy of David Viner). See colour plate section.

of cyclones over the Mediterranean. The region had an eastward coverage from Portugal to Iraq and a southward coverage from southern Europe to North Africa (Figure 1.3). The model used was HadRM3, a derivative of HadAM3, an atmosphere-only GCM, which is widely used by the climate research community.

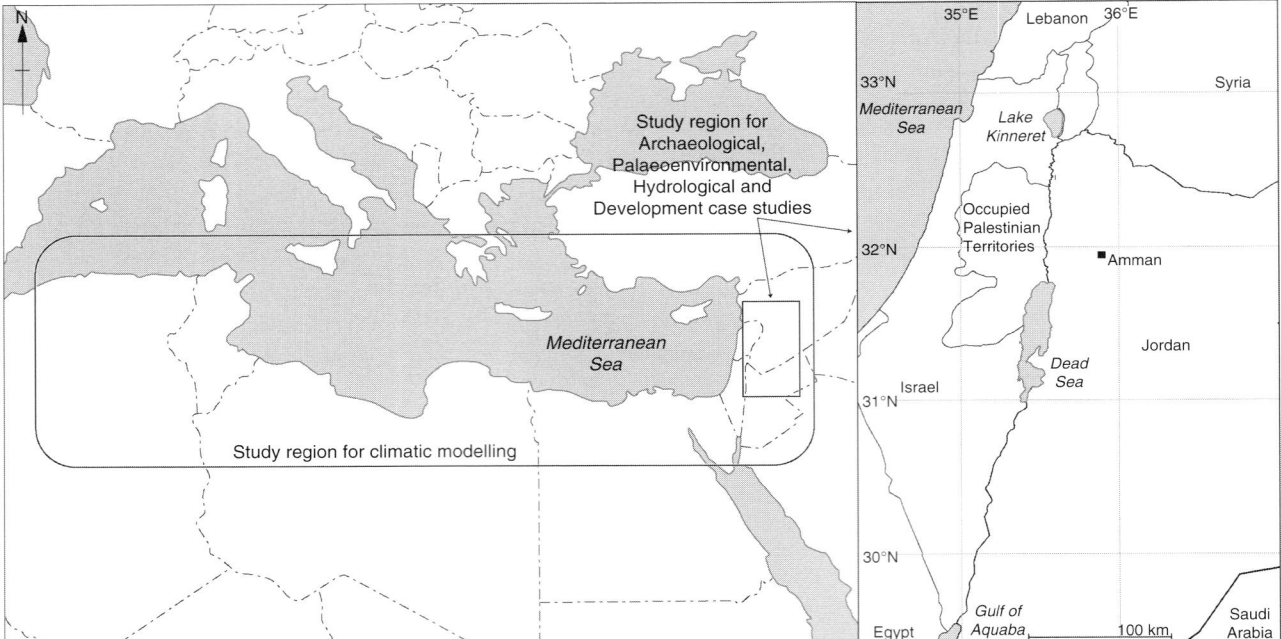

Figure 1.3. The geographical scope of the climate modelling within the Water, Life and Civilisation project and the case study region, indicating the key research localities.

The version of HadRM3 used in these studies has a resolution of $0.44° \times 0.44°$. The model was informed by lateral boundary conditions derived by the fully coupled model HadCM3 for the future climate integrations, and the slab model HadSM3 for the palaeoclimate integrations. While the Palaeoenvironmental, Hydrological, Archaeological and Developmental studies could feasibly have been undertaken anywhere within this region, the decision was taken to concentrate these in Jordan, and more specifically (but not exclusively) in the Jordan Valley and Wadi Araba regions, broadly between Amman and Aqaba – the southern reaches of the southern Levant and the western arm of the 'fertile crescent' (Figure 1.3). This was partly for pragmatic reasons in terms of the existing research activity in this region, the support of the Council for British Research in the Levant (CBRL) with regard to securing research permits and logistical help in the field, and political stability. More importantly, however, these were identified as the regions with the greatest potential for making substantive contributions towards our disciplinary and interdisciplinary research goals.

1.5 PROJECT DEVELOPMENT

Some elements of the research to be undertaken by each of the sub-projects were clearly defined from the start of the project as a whole. The initial work of the Meteorology sub-project was to provide a study of the present, past and future climates of the Middle East and North Africa region (Chapters 2, 3 and 4). Such work would provide the foundation for the interdisciplinary studies, the specific nature of which had to develop during the course of the project. Similarly, it was clear from the start of the project that the Palaeoenvironmental sub-project had to undertake a review of the evidence for past climates and environments in the study region to provide a framework for any case studies that would follow (Chapters 6 and 7). Those reviews involved some comparison between the output of GCMs/RCMs and the palaeoenvironmental data, but a more explicit comparison became necessary for which a study of changing precipitation in the eastern Mediterranean during the middle Holocene was developed (Chapter 8). It was also evident from the start of the project that further geochemical studies of the Lisan sediments from the former extent of the Dead Sea would be beneficial, especially because these would be on the eastern exposures (i.e. in Jordan), the majority of previous studies having been undertaken on the western bank (Chapter 9). During the course of the project, the potential for palaeoenvironmental reconstruction for the environs of the Neolithic site of Beidha by the analysis of carbonate deposits was identified and pursued (Chapter 16); following some success with that research, the same method was applied to carbonate deposits within Wadi Faynan. In addition, peaty deposits dating from at least 7,000 BP were identified within artesian springs in the vicinity of the Dead Sea, resulting in the extraction of cores for pollen and geochemical analysis.

The changing level of the Dead Sea, both of the past and projected into the future, was also an immediate target for the Hydrology sub-project, as this allowed an engagement between the hydrology and the palaeoenvironmental studies (Chapter 11), along with a broader understanding of the impact of climate change on the River Jordan (Chapter 10). This project then undertook a hydrological study of two of the major wadis in Jordan, Wadi Faynan (Chapter 12) and Wadi Hasa (Chapter 13), with the first of these providing the basis for an interdisciplinary study of long-term human settlement.

In the Archaeology sub-project, there was a clear need to review the existing evidence for water management especially prior to the historic periods (Chapter 14). Such evidence comes in the form of structures – dams, cisterns, aqueducts – and in the character of plant remains that could only have been grown under conditions of irrigation. The latter are critically important because structures such as irrigation channels may have been relatively ephemeral and left no archaeological trace. Rosen and Weiner (1994) and Araus *et al.* (1997) had claimed that the use of irrigation can be identified from the character of phytolith assemblages and the isotopic composition of charred grain, respectively. It was immediately apparent in the WLC project that the methodology they proposed required rigorous testing, and consequently a plant growing experiment was established to support the subsequent analysis of archaeological materials (Chapters 21, 22 and 23). The isotopic analysis of human and animal bone for the reconstruction of past diet and herd management strategies, and how these may have been influenced by changing water availability, was also identified as a priority, although a specific question to be addressed could not be defined until the availability of skeletal samples was determined (Chapter 20).

The Archaeology sub-project was also the main driver in identifying case studies for the interdisciplinary research in terms of developing and applying a chained set of climate and hydrological models to facilitate the interpretation of archaeological remains. During the development of the required methodology to link the climate and hydrological models, it became evident that an explicit description of the critical role of a 'weather generator' that provides the link from one to another was required (Chapter 5). Numerous potential case studies of archaeological sites were initially identified and evaluated, resulting in three being chosen: a multi-period study in Wadi Faynan from the Neolithic to the Byzantine period (Chapter 15), a study of the Bronze Age site of Jawa in northern Jordan (Chapter 18) and a study of the Nabataean–Islamic settlement of Humayma in southern Jordan (Chapter 19). Jawa and Humayma were selected because of pre-existing surveys and interpretations of their water management structures (Helms, 1981; Oleson, 1992; Oleson, in press) that were ripe for further interpretation by our interdisciplinary approach. For these studies a landscape analysis of Chalcolithic/Early Bronze Age settlement was undertaken, covering Jordan, Israel and the Occupied Palestinian Territories, exploring how the distribution of settlement may have been influenced by the availability of water (Chapter 17).

Figure 1.4. Water, Life and Civilisation team members during an orientation visit to Jordan in October 2004, here seen at the Iron Age tell of Deir 'Alla. See colour plate section.

Finally, the Development sub-project began with a clear need to review the overall situation regarding water usage in contemporary Jordan (Chapters 24 and 28) and the potential impact of future climate change. This formed the overarching background for a suite of studies concerning contemporary patterns and processes of water usage in urban and rural areas (Chapters 25, 26 and 27). These included detailed analysis of the use of treated waste-water in the agricultural realm (with a particular emphasis on both farmers and policy decision-makers), and the examination of social variations in the storage and use of water in the primate capital city of Greater Amman. A further line of research provided a detailed assessment of the development impacts of irrigated agriculture in Wadi Faynan.

The need to develop research within core disciplines prior to undertaking the interdisciplinary studies meant that during the first two years of the WLC project (2005, 2006), the extent of formal interaction between the sub-projects was relatively limited. Informally, however, this was achieved by a joint field trip to the Jordan Valley at the start of the project for group orientation (Figure 1.4), most critical for the meteorologists who would not be undertaking field work. It was subsequently met by means of an annual review meeting where each sub-project presented its ongoing work. To give this a self-critical edge we invited Professor Neil Roberts (Physical Geography, Plymouth University), Professor Richard Bradley (Archaeology, University of Reading), Professor Tony Wilkinson (Archaeology, Durham University) and Professor Robert Gurney (Meteorology, University of Reading) to attend these meetings and provide critical comment and advice regarding progress of each sub-project and the WLC project as a whole. Field work in Jordan was conducted by each of the sub-projects (except for Meteorology) combining field trips by staff from different sub-projects whenever possible. Small-scale excavations were undertaken in Wadi Faynan and at Humayma to secure sediment samples from Nabataean/Roman/Byzantine fields for phytolith analysis. As from the start of the third year of the project, the sub-project structure began to dissipate, as the boundaries between what was archaeology, hydrology, meteorology and so forth effectively became so blurred as to be of little value. As well as regular conference presentations by individual members of the project, an annual presentation of its ongoing research was made at the British Association for Near Eastern Archaeology. In November 2009 the WLC project organised a joint discussion meeting of the British Academy and Royal Society on the theme of Water and Society: Past, Present and Future, inviting participants from comparable interdisciplinary research projects based in Germany (GLOWA) and the United States (Mediterranean Landscapes Project).

1.6 THIS VOLUME

This volume provides reports on approximately 80% of the research that was undertaken between 2005 and 2010 by the Water, Life and Civilisation project. Some of this has already been published as journal articles but is brought together here to provide a near-comprehensive coverage of the research and its outcomes. The volume is broadly structured by the sub-projects, with the majority of the interdisciplinary studies being included in Part IV, the section entitled Human Settlement, Climate Change, Hydrology and Water Management. The WLC research not included within this volume is that which was started within the timeframe and with funding of the project, but which has extended beyond 2010. Most notable of this is the analysis of Dead Sea peat deposits and Wadi Faynan travertine deposits, both currently being undertaken by Dr Claire Rambeau, the multivariate analysis of data from the crop growing experiments by Dr Emma Jenkins and a review of water management technology of the Roman, Byzantine and Islamic periods by Dr Rebecca Foote. An attempt has been made to produce individual chapters and sections that can be read as stand-alone products but which also contribute to a single integrated volume. The final chapter (Chapter 29) of this volume seeks to draw the various research strands together to address the overall contribution of the project to our understanding of Water, Life and Civilisation.

REFERENCES

Adams, R. (1966) *The Evolution of Urban Society: Early Mesopotamia and Prehistoric Mexico*. Chicago: Aldine.

Adams, R. (1978) Strategies of maximization, stability and resilience in Mesopotamian society, settlement and agriculture. *Proceedings of the American Philosophical Society* **122**: 329–335.

Allan, J. A. (2001) *The Middle East Water Question: Hydro-Politics and the Global Economy*. London: I. B. Tauris.

Alley, R. B., J. Marotzke, W. D. Nordhaus et al. (2003) Abrupt climate change. *Science* **299**: 2005–2010.

Araus, J. L., A. Febrero, R. Buxo et al. (1997) Identification of ancient irrigation practices based on the carbon isotope discrimination of plant seeds: A case study from the south-east Iberian Peninsula. *Journal of Archaeological Science* **24**: 729–740.

Barker, G., D. Gilbertson and D. Mattingly, eds. (2007) *Archaeology and Desertification: The Wadi Faynan Landscape Survey, Southern Jordan*. Wadi Faynan Series Vol. 2, Levant Supplementary Series 6. Oxford, UK and Amman, Jordan: Council for British Research in the Levant in Association with Oxbow Books.

Barlow, M. (2007) *Blue Covenant: The Global Water Crisis and the Coming Battle for the Rights to Water*. New York and London: The New Press.

Childe, V. G. (1928) *The Most Ancient Near East*. London: Norton Company.

Cullen, H. M., P. B. deMenocal, S. Hemming et al. (2000) Climate change and the collapse of the Akkadian empire: Evidence from the deep sea. *Geology* **28**: 379–382.

deMenocal, P. B. (2001) Cultural responses to climate change during the late Holocene. *Science* **292**: 667–673.

Evans, J. P. (2009) 21st century climate change in the Middle East. *Climatic Change* **92**: 417–432.

Finlayson, B. L. and S. Mithen (2007) *The Early Prehistory of Wadi Faynan, Southern Jordan: Archaeological Survey of Wadis Faynan, Ghuwayr and Al Bustan and Evaluation of the Pre-Pottery Neolithic A Site of Wf16*. Wadi Faynan Series 1, Levant Supplementary Series 4, ed. B. Finlayson and S. Mithen. Oxford, UK and Amman, Jordan: Oxbow Books and the Council for British Research in the Levant.

Gao, X., J. Pal and F. Giorgi (2006) Projected changes in mean and extreme precipitation over the Mediterranean region from a high resolution double nested RCM simulation. *Geophysical Research Letters* **33**: DOI 1029/2005GL024954.

Glausiusz, J. (2010) New life for the Dead Sea? *Nature* **464**: 1118–1120.

Gleick, P. H. (2006) Water and terrorism. In *The World's Water 2006–2007: The Biennial Report on Freshwater Resources*, ed. P. H. Gleick. Washington DC: Island Press pp. 1–28.

Helms, S. (1981) *Jawa: Lost City of the Black Desert*. New York: Cornell University Press.

Hughes, J. K. and S. J. Smith (2008) Simulating global patterns of Pleistocene hominin morphology. *Journal of Archaeological Science* **35**: 2240–2249.

IPCC Core Writing Team, R. K. Pachauri and A. Reisinger (2007) Summary for Policymakers. In *IPCC, 2007: Climate Change 2007: Synthesis Report. Contribution of Working Groups I, II and III to the Fourth Assessment Report of the Intergovernmental Panel on Climate Change*. Geneva: IPCC.

Issar, A. S. and M. Zohar (2004) *Climate Change – Environment and Civilization in the Middle East*. Berlin: Springer-Verlag.

Kitoh, A., A. Yatagai and P. Alpert (2008) First super-high resolution model projection that the ancient 'Fertile Crescent' will disappear in this century. *Hydrological Research Letters* **2**: 1–4.

Lubbock, J. (1865) *Prehistoric Times, as Illustrated by Ancient Remains, and the Manners and Customs of Modern Savages*. London: Williams and Norgate.

Mariotti, A., N. Zeng, J.-H. Yoon et al. (2008) Mediterranean water cycle changes: Transition to drier 21st century conditions in observations and CMIP3 simulations. *Environmental Research Letters* **3**: 1–7.

Mithen, S. J. (2003) *After the Ice: A Global Human History, 20,000–5000 BC*. London: Weidenfeld and Nicolson.

Oleson, J. P. (1992) The water-supply system of ancient Auara: preliminary results of the Humeima Hydraulic Survey. In *Studies in the History and Archaeology of Jordan IV*, ed. M. Zaghoul. Amman: Department of Antiquities pp. 269–276.

Oleson, J. P. (in press) Humayma Excavation Project, Final Report Vol. I: The site and the water-supply system. *Annual of the American Schools of Oriental Research*.

Parry, M. L., O. F. Canziani, J. P. Palutikof, P. J. van der Linden and C. E. Hanson, eds. (2007) *Contribution of Working Group II to the Fourth Assessment Report of the Intergovernmental Panel on Climate Change*, 2007. Cambridge and New York: Cambridge University Press.

Rosen, A. (2007) *Civilizing Climate: Social Responses to Climate Change in the Ancient Near East*. Plymouth: AltaMira Press.

Rosen, A. M. and S. Weiner (1994) Identifying ancient irrigation: a new method using opaline phytoliths from emmer wheat. *Journal of Archaeological Science* **21**: 125–132.

Scarborough, V. L. (2003) *The Flow of Power: Ancient Water Systems and Landscapes*. Santa Fe NM: SAR Press.

Scholz, C. A., T. C. Johnson, A. S. Cohen *et al.* (2007) East African megadroughts between 135 and 75 thousand years ago and bearing on early-modern human origins. *Proceedings of the National Academy of Sciences of the United States of America* **104**: 16416–16421.

Smith, S. J., S. Mithen and J. K. Hughes (in press) Explaining global patterns in Lower Palaeolithic technology. Simulation of Hominin dispersal and cultural transmission using stepping out. In *Patterns and Process in Cultural Evolution*, ed. S. Shennan. Berkeley: University of California Press.

Steward, J. H. (1955) *Irrigation Civilization: A Comparative Study*. Washington DC: Pan American Union.

Trigger, B. (2003) *Understanding Early Civilizations*. Cambridge: Cambridge University Press.

UNEP (2007) *GE04 Global Environment Outlook (Environment for Development)*. New York: United Nations Environment Programme.

UNWater (2007) *Coping with water scarcity: Challenges of the twenty first century. Prepared for World Water Day*, 2007. Retrieved December, 2008 http://www.unwater.org/wwd07/download/documents/escarcity.pdf.

Wittfogel, K. A. (1957) *Oriental Despotism: A Comparative Study of Total Power*. New Haven CT: Yale University Press.

Part I
Past, present and future climate

2 The present-day climate of the Middle East

Emily Black, Brian Hoskins, Julia Slingo and David Brayshaw

ABSTRACT

The arid climate of the Middle East means that variations in rainfall on all timescales from days to years have an enormous impact on the people who live in the region. Understanding this variability is crucial if we are to interpret model simulations of the region's climate and make meaningful predictions of how the climate may change in the future and how it has changed in the past (Chapters 3 and 4). This study uses rain gauge measurements in conjunction with other meteorological data to address the following questions. How does rainfall vary from day to day and from year to year? How does rainfall vary spatially within Jordan and Israel? How does the atmospheric circulation over the Mediterranean region affect the daily probability of rain? What effect do large-scale modes of variability such as the North Atlantic Oscillation have on rainfall variability in the region?

2.1 INTRODUCTION

Variability in precipitation has posed a considerable challenge to the population of the Middle East throughout the Holocene, and continues to be a key issue today. Understanding this variability is crucial for the design and interpretation of climate model experiments that characterise how precipitation has changed in the past and predict how it will change in the future. This chapter focuses on the aspects of precipitation variability that are of greatest importance for the climate modelling and hydrological studies described in later chapters within this volume – namely the spatial variability of precipitation within Jordan and Israel, and the effect of circulation over the Mediterranean and Atlantic on daily to inter-annual precipitation variability. To this end, we address the following questions:

- How does annual total precipitation vary spatially in Jordan and Israel, and how spatially coherent is the inter-annual variability?
- What synoptic conditions over the Mediterranean are favourable for rainfall in Israel?
- How does the occurrence of these synoptic conditions relate to large-scale modes of variability such as the North Atlantic Oscillation (NAO) and the East Atlantic/West Russia pattern?

There is a large body of published work on Middle East climate variability, and several studies of larger-scale climatic variability of relevance to the region. The sparse number of rain gauges and inconsistent quality of the data (see for example US Geological Survey et al., 2006) has led most studies to focus on temporal rather than spatial variability of rainfall in Israel and Jordan. An exception to this is an analysis of annual rainfall over Israel, which used a statistical regression model to explore the influence of topography, the proximity of the sea, latitude and the differing trajectories of air masses arriving in the north and south of the country (Goldreich, 1994). That study concluded that the pronounced changes in rainfall totals over short distances are primarily related to the orography and the trajectory of air masses entering the region. A study of Jordanian precipitation variability identified three homogeneous rainfall zones: the highlands in the northwest, the surrounding lower elevation region, and the south and east. As in Israel, the topography was found to be of greatest importance in the north, where winter cyclones bring moist air from the Mediterranean rather than dry air from the Sinai Desert. Furthermore, precipitation in the north was found to be more spatially coherent on all temporal timescales. The coherence of precipitation in northern Israel and Jordan was also reported in Ziv *et al.* (2006), who found that the mean inter-annual correlation between groups of stations in the north and central Israel was 0.8, which is statistically significant at the 99% level and indicates a high degree of spatial coherence. Although the inter-annual variability in annual total rainfall exhibits high spatial coherence during all seasons and at all locations, there is some month to month variability, with precipitation relatively spotty in the early part of the season

Water, Life and Civilisation: Climate, Environment and Society in the Jordan Valley, ed. Steven Mithen and Emily Black. Published by Cambridge University Press.
© Steven Mithen and Emily Black 2011.

(September and October) and becoming more coherent as the season progresses (Kutiel and Paz, 1998).

The proximity of the Middle East to Europe and its mid-latitude winter climate raises the possibility that the Eurasian/Atlantic modes of variability that affect Europe in winter also affect the Middle East. In boreal winter, the NAO is a dominant pattern of circulation variability over the North Atlantic (see for example Jones et al., 2003). The NAO can be thought of as an oscillation in atmospheric mass, and hence sea level pressure, between the northern and subtropical Atlantic, with centres of action near Iceland and the Azores. During its positive phase, there is low pressure in the Icelandic region and high pressure in the Azores. This increased north–south pressure gradient causes storms to move across the Atlantic on a more northerly track, leading to wet winters in northern Europe and dry winters in southern Europe. In contrast, when the NAO is negative, pressure in the Icelandic region is high and pressure in the Azores is low. This reduced north–south pressure gradient causes storms to move across the Atlantic on a more west–east track, leading to dry winters in northern Europe and wet winters in southern Europe (see for example Wallace and Gutzler, 1981). Cullen et al. (2002) and Cullen and deMenocal (2000) showed that the influence of the NAO extends as far east as Turkey. Another prominent teleconnection pattern, which affects winter variability in the Mediterranean region, is the East Atlantic/West Russia mode (EAWR). The EAWR consists of two anomaly centres – one over the Caspian Sea and the other over Western Europe. In its positive phase, there is low pressure over southwestern Russia and western Europe, and high pressure over northwestern Europe, which is associated with dry conditions over Europe and the Mediterranean and wet conditions in the Middle East; the converse is true for the negative phase (Krichak et al., 2000).

Although there have been several studies of the effect of Eurasian/Atlantic teleconnections on Middle East rainfall there is little consensus. For example, Ben-Gai et al. (2001) and Ziv et al. (2006) report that Israel rainfall is poorly correlated with the NAO, while Eshel and Farrell (2000; 2001) suggest that east Mediterranean rainfall is modulated by the circulation of the Atlantic–Mediterranean region. Krichak et al. (2002) found a link between east Mediterranean rainfall and the NAO/EAWR pattern, and argued that variability in the region is primarily affected by the EAWR, with secondary influence from the NAO (Krichak et al., 2002; Krichak and Alpert, 2005a, 2005b). These inconsistencies probably reflect the relatively weak and indirect nature of the teleconnection. In any case, discrepancies between study regions, time periods considered and the rainfall datasets that were used make it difficult to compare results from different studies.

Although the previous discussion has focused on the lower atmosphere, variability within the upper atmosphere has also been found to affect Middle East precipitation – particularly for Israel. An upper-level trough extending from Eastern Europe towards the eastern Mediterranean affects rainfall in the southern Levant. Both the position and intensity of the trough vary, so that during rainy years, it is centred over Israel, and during dry years, it is centred further west (Ziv et al., 2006). Another example of the importance of upper-level processes is the teleconnection with the North Caspian Pattern (NCP). The NCP is a mode of upper-level variability with centres of action over eastern Europe/Asia and the English Channel. During NCP negative years there is an increased southwesterly circulation towards western Turkey. Conversely, during NCP positive years, there is an increased northwesterly circulation towards East Europe and increased northeasterly circulation towards the Black Sea (Barnston and Livezey, 1987). Although its influence on European rainfall is generally small in comparison with other modes of variability, such as the NAO, the NCP has been found to be significantly correlated with rainfall in Israel (Kutiel and Paz, 1998; Kutiel et al., 2002).

This chapter will focus on aspects of Middle East climate variability that are of particular relevance for the climate modelling and hydrological studies described in later chapters of this volume. The spatial variability and coherence of rainfall is a key issue for downscaling, so this is addressed in section 2.4 through a review of the published literature and a new study of coherence. As has been described above, the link between the Eurasian modes of variability and Middle East climate is unclear. Rather than attempting to disentangle the influences of the various teleconnection patterns on precipitation, section 2.5 first looks at the large-scale synoptic conditions that favour rainfall in Israel, and then relates these to the aforementioned indices. This way, we aim to address (i) to what degree the larger-scale circulation over the Mediterranean affects Middle East precipitation and (ii) how the circulation patterns that favour rainfall are related to European/Atlantic teleconnection patterns. Both of these questions are relevant to the issue of how long-term changes in global circulation (for example the thermohaline circulation) and more local circulation (for example, the Mediterranean storm track) may affect precipitation in the Middle East.

2.2 DATA AND METHODOLOGY

2.2.1 Rainfall data

The monthly precipitation data used in this study were based either on gauge records or on an interpolated 0.5° gridded dataset provided by the Global Precipitation Climatology Centre (GPCC). The choice depended on whether it was considered more important to have a long record of data from the same stations (for example for analysis of spatial coherence) or a clear

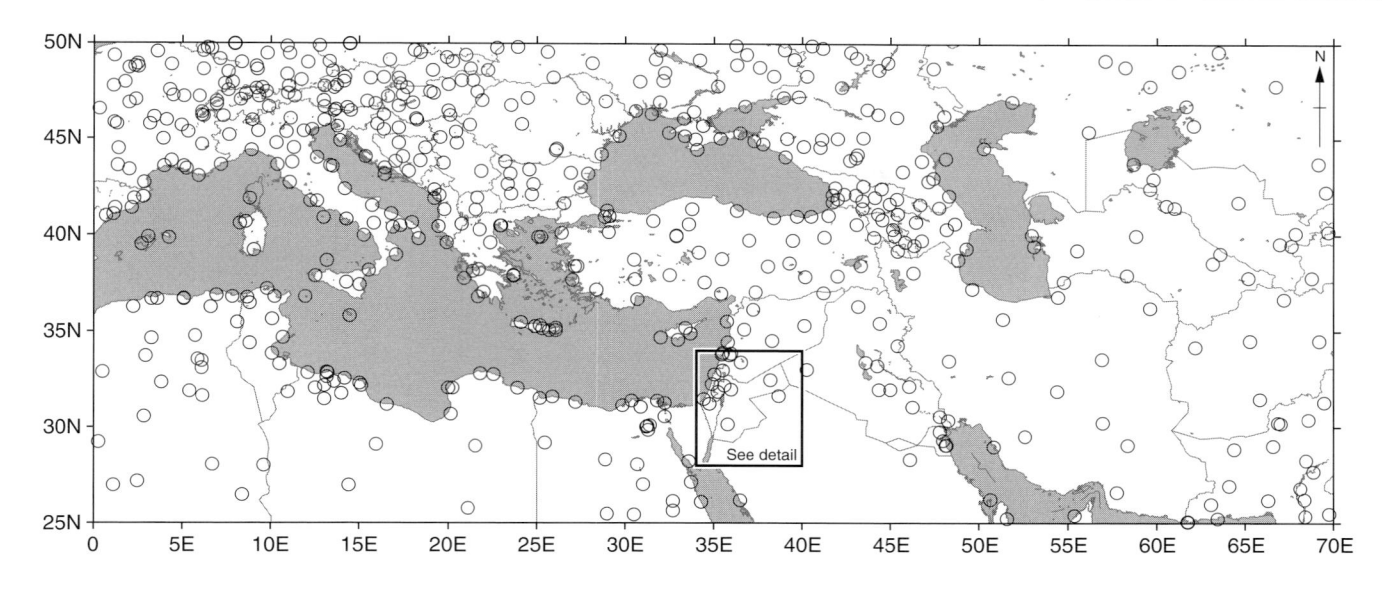

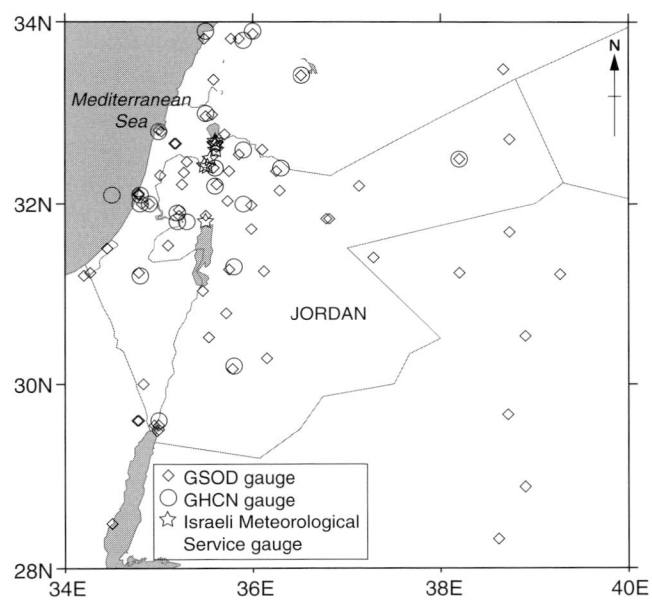

Figure 2.1. Location of rain gauges. Top: Global Historical Climate Network (GHCN) gauges within Europe, Middle East and North Africa. Bottom: gauge data within the Middle East. Circles indicate GHCN monthly data; diamonds are gauges from the World Meteorological Organisation Global Summary of the Day (GSOD; daily data of very variable quality); stars are stations with daily data, provided by the Israeli Meteorological Service.

idea of spatial pattern with no missing points (for example for the evaluation of the climate model or survey of the mean climate). The GPCC rainfall data is a 0.5° × 0.5° gridded product that is based on 9543 quality controlled stations for 1951–2000. The product used in this study was developed during the Variability Analysis of Surface Climate Observations (VASClimO) project (Schneider *et al.*, 2008).

Figure 2.1 shows the location of rain gauges within the study area. It can be seen that many parts of the region contain few gauges from the Global Historical Climate Network (GHCN), which provided much of the data to the GPCC. Because of the sparse number of observations included within the product for this region, many of the grid points would have been based on interpolations over large distances. Therefore, in order to avoid being misled by artefacts of the interpolation, analyses of monthly to inter-annual variability were carried out using either individual rain gauges or a gridded version of the available gauge data that was developed especially for this study. Away from the study area, only the GHCN data were included. In the study area, these data were supplemented by station data provided by the Israeli Meteorological Service, and data extracted from the World Meteorological Office Global Summary of the Day

(GSOD) archive. The gridding was carried out at monthly time intervals on a 1° × 1° grid for 1950–1999.

The gridding of the data was done in several stages in order to avoid biasing the results. First, stations were selected for input into the dataset. To be included the station had to have less than 10% missing data over the time period considered and pass a basic quality control test that checked the data for homogeneity, for repeated data (for example several years reporting exactly the same values) and for identical data from different stations. Many of the GSOD stations failed the quality control tests – primarily because of missing data. A rainfall climatology was calculated for all grid squares for which data were available. The data from each station were standardised and the mean standardised anomaly within each grid point calculated for every month. The standardised anomaly was then projected onto the climatology.

All daily analyses were carried out using individual stations. Daily rainfall data for 1985 onwards from nine stations along the River Jordan were provided by the Israeli Meteorological Service. Elsewhere in Jordan and Israel, data were taken from the World Meteorological Organisation GSOD archive. Like the monthly data, all daily data were subjected to a basic quality control process that involved checking the data for homogeneity, for repeated data and for identical data from different stations.

2.2.2 Other data and indices

The NAO index we used is based on station measurements of sea level pressure in Iceland and the Iberian Peninsula. Details of how the index was calculated can be found in Jones *et al.* (1997) and www.cru.uea.ac.uk/cru/data/nao.htm. Other indices were taken from the Climate Prediction Center (CPC) website (http://www.cpc.ncep.noaa.gov/data/teledoc/telecontents.shtml). Sea level pressure data were taken from the National Centers for Environmental Prediction (NCEP) reanalysis, which is a 2.5° × 2.5° product based on an optimal interpolation between observational and modelled data (Kalnay *et al.*, 1996). Although, as has been described above, we used gauge-based measurements for the main analysis of spatial and temporal precipitation variability, NCEP reanalysis precipitation data were used for the daily rainfall composites described in section 2.5.1. Although there are serious problems with the way that the reanalysis represents both spatial and temporal variability (see for example Diro *et al.*, 2009), in this case there was no alternative. It should be noted, however, that the composites based on rainfall reanalysis were constructed for qualitative survey only, and we made no attempt to analyse them quantitatively.

2.2.3 Tracking of weather systems

Low-level weather systems in the Mediterranean were identified and tracked using the TRACK software developed at the Natural Environment Research Council, Environmental Systems Science Centre by Kevin Hodges (see http://www.nerc-essc.ac.uk/~kih/TRACK/Track.html and Hodges, 1994). This method automatically and objectively identifies features in a meteorological field and tracks them in 6-hourly data. The data used here were 850 mb vorticity, derived from the ERA40 reanalysis, which were filtered to exclude very long- and short-wavelength features. The tracks were then collated and the track density calculated (the methodology is described fully in Hodges, 1994; Hodges, 1995).

2.2.4 Semi-objective Grosswetterlagen (GWL) catalogue

Synoptic variability over Europe can be characterised by considering the circulation as being in a particular regime. Regimes can be thought of as consecutive synoptic patterns that follow particular spatial developments. These regimes can last anything from a few days to a fortnight or more, and are associated with characteristic weather patterns. There are various methods of defining synoptic regimes, ranging from manual identification of a regime on a particular day to fully automated cluster techniques. Manual identification is time-intensive and subject to error, while the fully automated techniques often produce poor results. To overcome these problems, we used a semi-objective method that combines both the subjective definition of regimes based on the well-established GWL catalogue (a list of the subjectively determined synoptic regime over Europe for every day, developed by the German Weather Service), with an automatic way of characterising each day (James, 2007). In this method, composites of sea level pressure (SLP) for each regime, based on the subjective GWL catalogue, are derived. To determine what regime a given day belongs to, the pattern correlation between each regime SLP composite and the SLP for the day are calculated, and the day assigned to the regime with which its correlation is highest. An additional consideration factored into the analysis is that regimes must last for more than two days.

2.3 THE MEAN CLIMATE

Figure 2.2 shows the boreal winter total precipitation, SLP and track density (based on tracking of features in the 850 mb vorticity field) over the Mediterranean, southern Europe and North Africa. The mean path that the mid-latitude cyclones follow over the Mediterranean (the Mediterranean storm track) is shown as a low in the mean SLP and as a maximum in the track density. It can be seen that the Mediterranean storm track extends to the Middle East. The Middle East's position on the edge of the storm track, however, indicates that even small variations in the mean trajectory of cyclones during a season can have a large effect on the total rainfall.

influence of the Indian monsoon (Rodwell and Hoskins, 1996). Figure 2.3 shows the mean seasonal cycle for various rainfall statistics between 1985 and 1999 for several stations in the Jordan Valley. It can be seen that most precipitation is experienced during the Mediterranean cyclone season between November and March, and that the summer is completely dry. The large inter-annual standard deviation in the number of rainy days compared to the mean rain per rainy day suggests that the inter-annual variability in total rainfall is related largely to variability in the number of rainy days, with the variability in rainfall intensity having a lesser effect.

It can be seen that the maximum rainfall and mean rain per rainy day at the peak of the rainy season (DJF: December, January, February) remain steady, and decrease at the margins of the season, perhaps reflecting a reduction of the intensity of cyclones. The lower-most figure shows the probability of rain given rain the day before, and the probability of rain given no rain the day before, which are measures of the duration and frequency of rainy events. It can be seen that during the peak of the rainy season (DJF), both the duration and frequency of events stays fairly constant. In the early part of the rainy season (October and November), the frequency of rainy events increases gradually, although the duration of each event is the same as for the latter part of the rainy season (DJF). This supports the idea that most rainfall is delivered by large-scale, long-lasting cyclones throughout the season, but that the intensity and frequency of the cyclones at the peak of the rainy season is greater than at its margins.

2.4 THE SPATIAL PATTERN OF PRECIPITATION

Rainfall in Israel and Jordan ranges from over 700 mm/year to less than 50 mm/year. Figure 2.4 shows that there are large changes in annual rainfall amounts over relatively short distances. Previous studies have identified several controls on spatial variability in rainfall in Jordan and Israel, including the proximity of the Mediterranean, the trajectory of air masses entering the region, orography and urban development (see for example Goldreich, 1994 and Freiwan and Kadioglu, 2008). In the north of the region, annual precipitation is strongly influenced by orography, with precipitation in Galilee and Jerusalem (elevations ~800 m) exceeding 600 mm/year and precipitation in the Jordan Valley (elevations of −400 m to 0 m) less than 200 mm/year. In contrast, the south is very dry and precipitation is not influenced by orography. For example Ma'an has an elevation of 1069 m and annual total rainfall of less than 100 mm (Diskin, 1970; Goldreich, 1994). This north–south contrast arises from differences in the trajectories of the depressions that bring most of the winter precipitation to the Middle East. In the north,

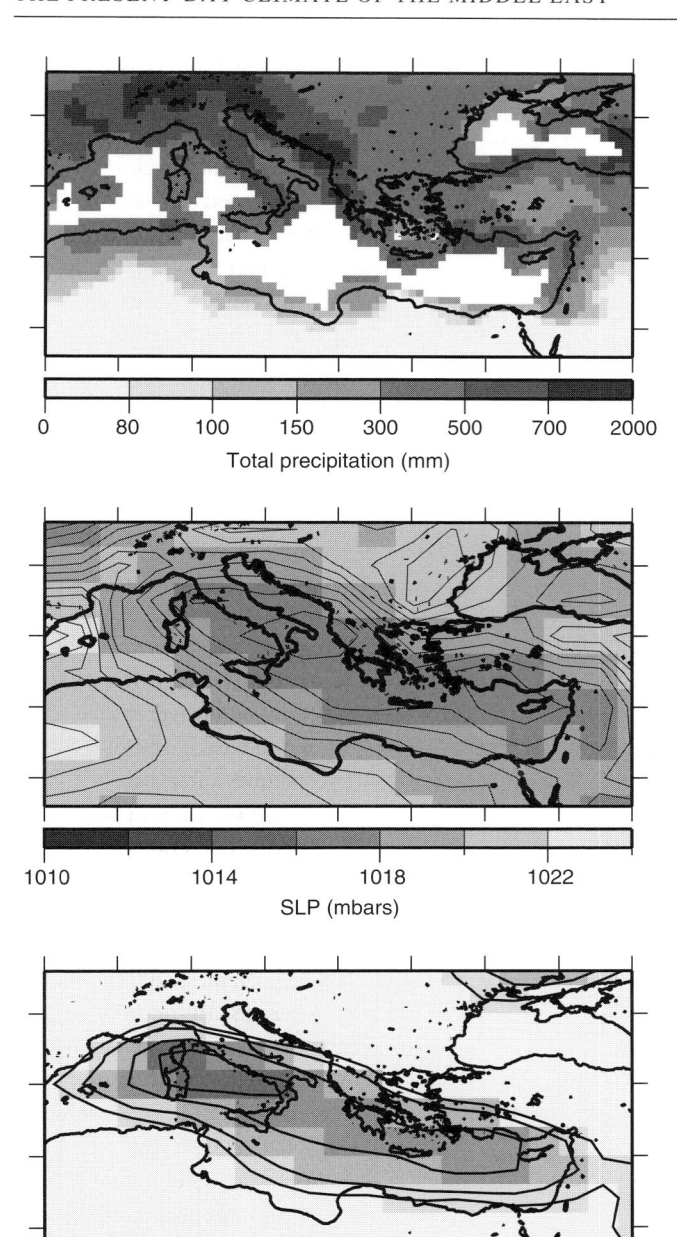

Figure 2.2. Mean climate over the Mediterranean. From top to bottom: December–February total precipitation; December–February SLP; December–February track density. See colour plate section.

The Middle East lies on the boundary between the tropics and extra-tropics, and thus has a climate that is influenced by both mid-latitude and tropical phenomena. This is exemplified by the seasonal cycle in precipitation. In boreal winter, as has been described above, the mid-latitude cyclones bring rain to the region; the summer, in contrast, is dry because of the remote

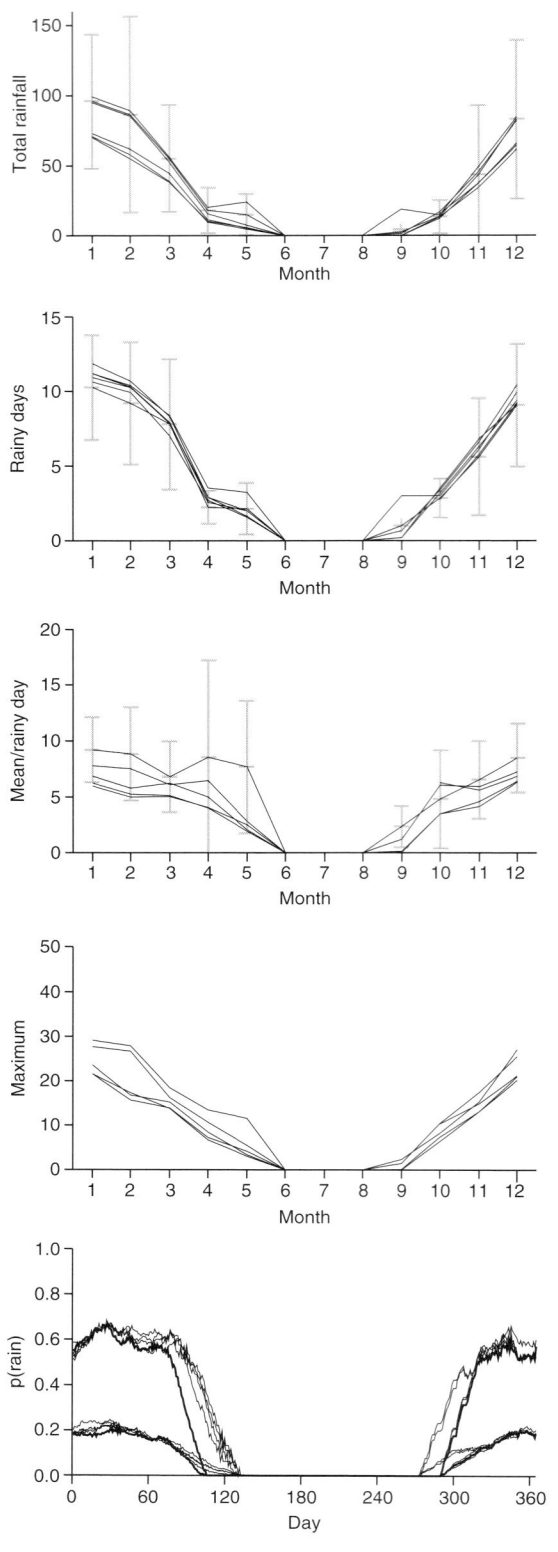

Figure 2.3. Seasonal cycle in various rainfall statistics for the stations shown in the map to the right. The *x*-axis gives the month and the *y*-axis the statistic in question. The error bars represent the inter-annual standard deviation from one of the stations. All rainfall units are millimetres. From top to bottom, the statistics are: total monthly rainfall; mean number of rainy days in the month; mean rain per rainy day; mean maximum daily rainfall in the month; probability of rain given rain the day before (upper group of curves) and probability of rain given no rain the day before (lower group of curves).

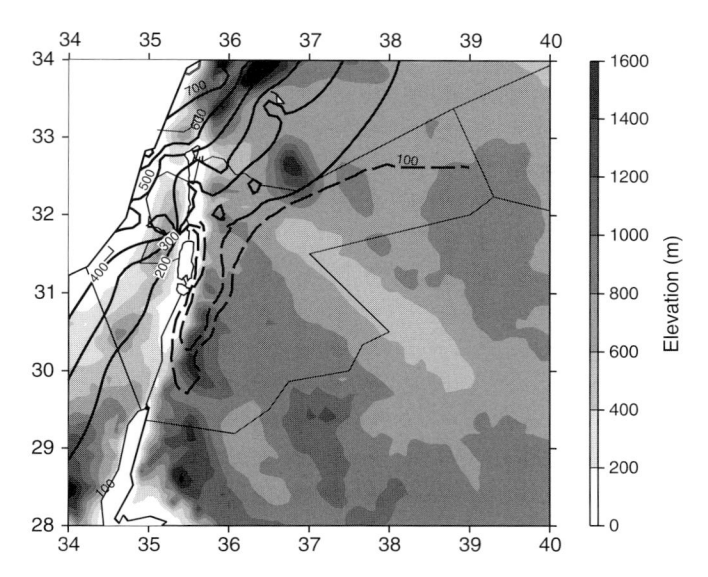

Figure 2.4. Annual total rainfall in Jordan and Israel superposed on the orography. The contours are based on the data from the gauges shown in Figure 2.1. The dashed contours are sketched from published sources (US Geological Survey, 2006) because we were unable to obtain suitable quality data in eastern Jordan. See colour plate section.

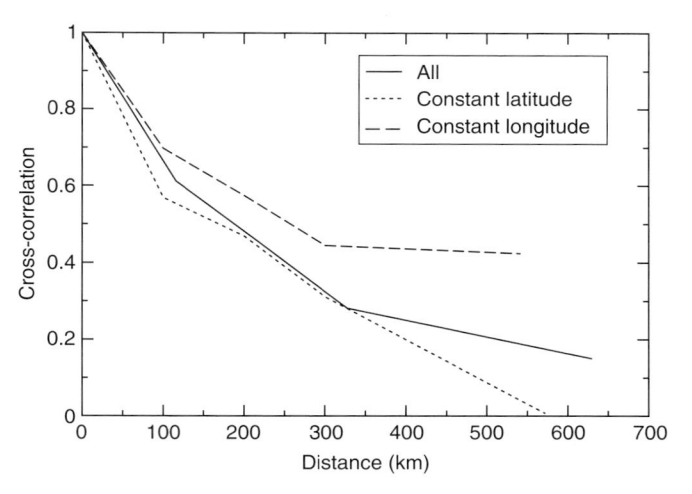

Figure 2.5. Mean correlation versus mean distance apart for rainfall stations within Jordan and Israel. The solid line is cross-correlations between all stations; the dotted line is cross-correlations between grid squares of the same latitude, and the dashed line represents cross-correlations between grid squares of the same longitude.

these systems bring moist Mediterranean air and hence cause precipitation; in the south, on the other hand, the same large-scale depressions bring dry Sinai desert air.

Despite the large spatial variability in annual totals, the inter-annual variability of precipitation in Jordan and Israel is fairly coherent. This is illustrated by Figure 2.5, showing the cross-correlation between annual rainfall at grid squares in Jordan and Israel versus the distance between the grid squares (using the uninterpolated gridded dataset developed for this study). It can be seen that at distances of up to 100 km, the mean cross-correlation is around 0.6 for all data, and around 0.7 if only grid squares with constant longitude are considered. It should be noted, however, that the density of stations in the north and west is far greater than that in the south and east. In the southern, drier parts of Jordan and Israel, it has been reported that the rainfall is far less coherent (Freiwan and Kadioglu, 2008). The rest of this chapter, and all of the subsequent hydrological case studies, which use climate model data, focus on Israel and west Jordan. This is where rainfall is coherent on inter-annual timescales and is clearly related to large-scale features such as the passage of cyclones through the region.

The variability in the daily statistics of the weather from north to south along the Jordan River is shown in Figure 2.3. Although the shapes of the seasonal cycles of total rainfall for the stations are evidently similar, the total amounts of rainfall change along the River Jordan. Most of the differences in rainfall totals result from differences in rainfall intensity, with the number of rainy days varying little from one station to another. This suggests that most precipitation is delivered by cyclones that cause precipitation over large areas, with the intensity of precipitation modulated by local variations in terrain.

2.5 DAILY TO INTER-ANNUAL VARIABILITY

2.5.1 Variability on daily to inter-annual timescales

The question of how circulation in the Mediterranean and Atlantic affects the daily probability of rainfall is key to understanding the large-scale controls on rainfall at all timescales. As was described in section 2.2, the synoptic weather patterns over the Atlantic can be characterised using an objective version of the Grosswetterlagen (GWL) system developed by the German Weather Service (James, 2007). In order to determine which synoptic weather patterns are associated with high rainfall in the Middle East, composites of SLP and rainfall for the regimes with the highest probability of rainfall were derived for an example station in the Jordan Valley (see Figure 2.6). The rainfall probabilities at the example station are given in Table 2.1, which summarises data for each of the GWLs considered. It should be noted that the same regimes would have been selected for any of the stations shown on Figure 2.3 although, as would be expected, the rainfall probabilities vary subtly. Although 29 objective GWLs can be defined, only the most common 15 were considered because of the short time-series. This approach is preferable to compositing SLP on rainy days, because it avoids the problem of superposing different synoptic regimes.

Table 2.1 *The 15 most common GWL regimes and their relationship to Jordan Valley rainfall and large-scale modes of variability*

Regime	Abbreviation	Percentage probability of the regime	Percentage probability of rain when the regime is active	Ratio of the probability of the regime during NAO+ and NAO− years	Ratio of the probability of the regime during EAWR+ and EAWR− years
Anticyclonic westerly	WA	11.1	37.4	4.0	2.4
Cyclonic westerly	WZ	12.8	13.7	2.7	0.4
South-shifted westerly	WS	3.1	30.8	0.6	0.2
Maritime westerly (Block E. Europe)	WW	7.8	18.9	2.2	0.7
Anticyclonic southwesterly	SWA	6.7	33.0	3.2	0.7
Cyclonic southwesterly	SWZ	3.8	39.6	1.3	2.3
Anticyclonic northwesterly	NWA	5.1	25.8	0.6	5.5
Cyclonic northwesterly	NWZ	4.6	32.0	1.7	0.5
High over Central Europe	HM	5.8	24.6	0.9	7.1
Zonal ridge over Central Europe	BM	4.5	10.0	2.3	0.9
Low over Central Europe	TM	1.4	5.6	2.3	0.5
Anticyclonic northerly	NA	1.2	22.2	0.2	0.2
Cyclonic northerly	NZ	1.2	16.7	1.1	1.0
Icelandic high, ridge Central Europe	HNA	4.4	18.9	0.4	3.0
Icelandic high, trough Central Europe	HNZ	1.5	26.3	0.1	1.9

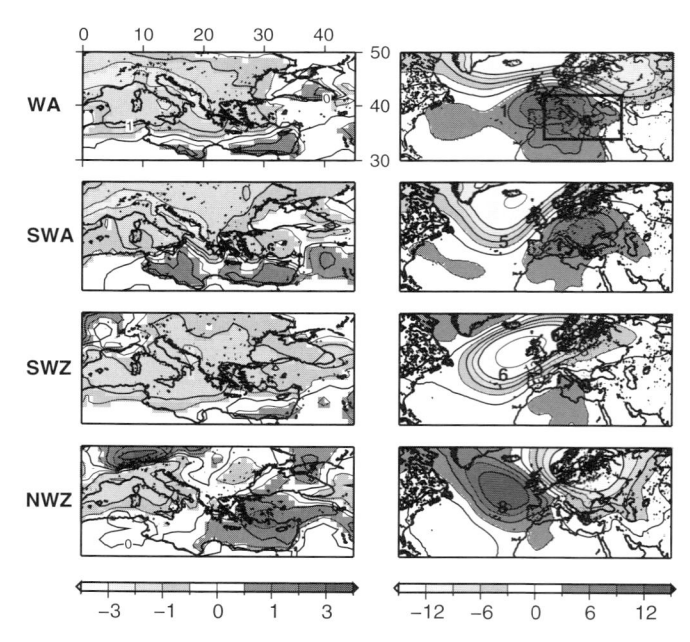

Figure 2.6. Composite daily anomalies during the four GWL regimes that favour rainfall most strongly (WA, SWA, SWZ and NWZ – abbreviations defined in Table 2.1). Left set: daily rainfall anomaly composites over the Mediterranean (box shown on the top right plot); right set: daily SLP anomaly composites over the Mediterranean and Atlantic. See colour plate section.

The four regimes associated with the largest probabilities of rainfall are: anticyclonic westerly, anticyclonic southwesterly, cyclonic southwesterly and cyclonic northwesterly. The first three of these are associated with high pressure and low rainfall in central and western Europe, and higher rainfall in the southeast Mediterranean. The fourth is associated with a large low and high rainfall over central and east Europe and the whole east Mediterranean.

There is considerable inter-annual variability of rainfall in the Middle East. For example, annual rainfall in Jerusalem in the twentieth century ranged from 218 mm to 1153 mm. In order to relate the daily patterns described above to the inter-annual variability, Figure 2.7 compares composites of various climate variables for the five rainiest and five driest Januaries for a box with minimum longitude 34°, maximum longitude 36°, minimum latitude 31° and maximum latitude 33°. The precipitation and track density are shown for the Mediterranean region, and the SLP is shown for a wider region. The resemblance between the differences between the wettest and driest years, and the first three synoptic regimes shown in Figure 2.6 are striking. Like the regimes associated with high rainfall probabilities described above, during wet years, rainfall is low in western Europe and high throughout the Middle East. The track density composites fit in with this picture, with high rainfall in the Middle East

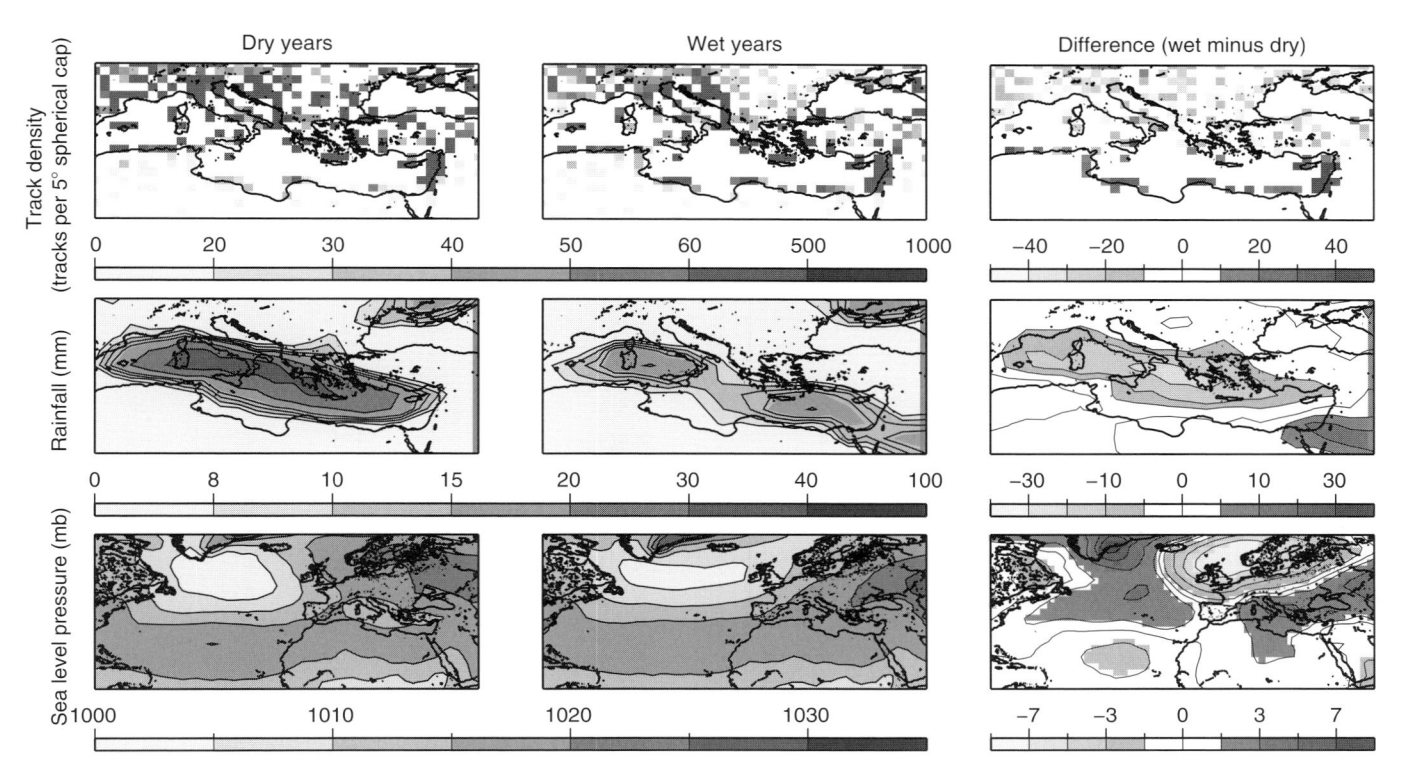

Figure 2.7. Composites of precipitation, track density and SLP during January based on the five wettest and driest Januaries in a box with minimum longitude 34°, maximum longitude 36°, minimum latitude 31°, maximum latitude 33°. See colour plate section.

associated with a weakening and southward shift of the storm track. This is consistent with the SLP patterns, which indicate that high rainfall in the study area is associated with high pressure over Europe and an intensified Cyprus low, resulting in a slackening of the west–east SLP gradient. On a larger scale, during rainy years, there is tendency to high pressure in the subtropical Atlantic and low pressure in the North Atlantic. The superposition of the SLP anomalies from the four rainy synoptic regimes is clear on this large scale, with the lows associated with the anticyclonic and cyclonic southwesterly regimes partially cancelling out the highs that occur during the westerly anticyclonic and cyclonic north-westerly regimes in the central Atlantic.

2.5.2 The impact of large-scale modes of variability on inter-annual precipitation variability

The inconsistencies between the published studies described at the beginning of this chapter reflect the fact that the link between Middle East precipitation and Eurasian modes of climate variability is not clear. Figure 2.8 shows the distribution of rainfall for positive and negative phases of four modes of variability: the North Atlantic Oscillation (NAO), El Niño Southern Oscillation (ENSO), and the East Atlantic (EA) and the East Atlantic/West Russia (EAWR) teleconnection patterns. It can be seen that there is no clear difference between the rainfall histograms for positive and negative phases of the East Atlantic pattern or ENSO. The

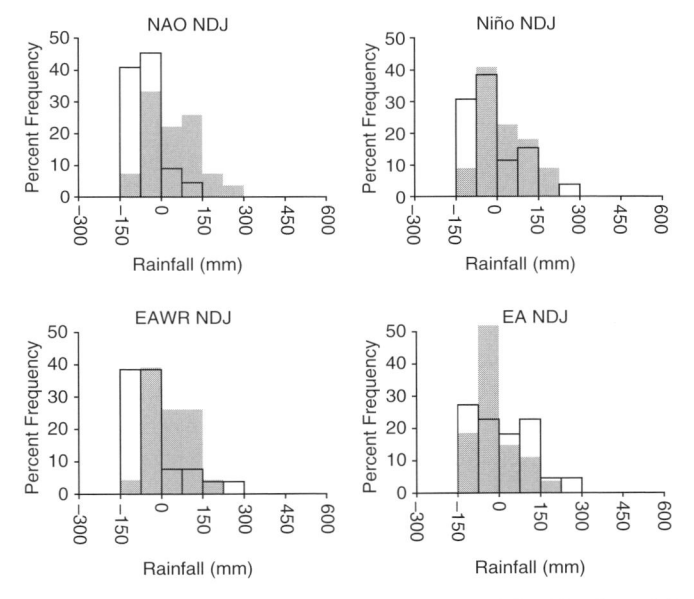

Figure 2.8. Histograms of rainfall total for the box defined in Figure 2.7 for positive and negative phases of the NAO, EAWR, East Atlantic pattern and for warm and cold Niño sea surface temperature (SST) anomalies. Negative phases or cold SSTs are shown by no shading and positive phases or warm SSTs by grey shading.

rainfall histograms for the positive and negative phases of the EAWR pattern hint at a link – with positive phases of the EAWR pattern associated with higher rainfall (consistent with Krichak

Table 2.2 *Distribution of rainfall quintiles for NAO-positive and NAO-negative years*
Bold numbers indicate that there are significantly more NAO-positive years than NAO-negative years in the rainiest quintile (assessed using the method described in Mason and Goddard, 2001).

October						November					
Quint	1	2	3	4	5	Quint	1	2	3	4	5
NAO-	5	4	4	6	4	NAO-	5	6	6	6	2
NAO+	5	6	6	4	6	NAO+	5	4	4	4	**8**
December						January					
Quint	1	2	3	4	5	Quint	1	2	3	4	5
NAO-	6	6	5	6	1	NAO-	4	5	6	4	1
NAO+	4	4	5	4	**9**	NAO+	5	5	4	6	**9**
February						March					
Quint	1	2	3	4	5	Quint	1	2	3	4	5
NAO-	7	4	4	4	1	NAO-	4	6	**1**	**3**	4
NAO+	2	6	6	6	**9**	NAO+	5	4	**9**	**7**	6

et al., 2002 and Krichak and Alpert, 2005a). However, the clearest link is with the NAO. During NAO positive years there is greater variability, which is reflected by a greater proportion of very rainy years. In fact, nine out of the ten rainiest Decembers occur when the NAO is in its positive phase (see Table 2.2). The relationship between the positive phase of the NAO and high rainfall is not universal: during some positive NAO years, rainfall is well below average. Moreover, there is no converse relationship between NAO-negative years and very low rainfall. In order to quantify these changes of distribution and estimate their significance, the rainfall was ranked and divided into five bins (quints). The numbers of NAO positive and NAO negative years falling into each quint is shown in Table 2.2. It can be seen that the strength of the relationship between the NAO and Middle Eastern rainfall varies within the rainy season, being strongest in November, December, January and February, and weakest in October and March.

The association between the NAO and Middle East rainfall on inter-annual timescales can be related to the likelihood of the synoptic regimes discussed in section 2.4. Table 2.1 shows that the four regimes most favourable for rainfall are all more likely when the NAO is in its positive phase. In other words, when the NAO is positive, circulation regimes over Europe that strongly favour rainfall in the Middle East are more likely to occur. This leads to more rainy days during NAO-positive years and hence to higher rainfall over the season. It should be noted that the probability of rain for a given synoptic regime does not depend on the state of the NAO or EAWR (data not shown here). This implies, perhaps not surprisingly, that the teleconnection between large-scale modes of variability and precipitation in the Middle East arises via perturbations to the circulation over the Mediterranean. This suggests that on centennial to millennial timescales, changes in the NAO or EAWR are likely to affect Middle East rainfall only insofar as they affect the circulation over the Mediterranean. Any change in the links between these modes of variability and the circulation patterns over the Mediterranean would be expected to change their statistical relationships with Middle East rainfall. The changing strength of the teleconnection between the European circulation and the NAO has previously been noted (e.g. Jones *et al.*, 2003), suggesting that the apparent teleconnection between the NAO and Middle East rainfall is unlikely to be constant in time.

The relationship between daily variability, synoptic regimes over western Europe and the Mediterranean, and large-scale modes of variability reported here is broadly consistent with the published literature. In particular, the dipole pattern of a wet Europe and western Mediterranean being associated with a dry southeast Mediterranean and Middle East has previously been recognised (e.g. Kutiel and Paz, 1998 and Krichak *et al.*, 2000). Consistent with this, several other studies have reported that the association between precipitation and large-scale modes of variability such as the NAO and EAWR pattern have opposite polarities in western Europe and the Middle East (for example Krichak and Alpert, 2005a and Ziv *et al.*, 2006).

2.6 DISCUSSION

This chapter set out to describe the aspects of Middle East precipitation and larger-scale Mediterranean climate that are most relevant for the design, interpretation and exploitation of climate model experiments for the region. The studies of water availability described in subsequent chapters of this volume rely on hydrological models that have been calibrated using station data. Even a 'perfect' climate model (or indeed a gridded observed rainfall dataset) can thus only be exploited if rainfall variability within its grid squares can be downscaled to a point. In other words, the potential usefulness of climate model output rests on how well precipitation within its grid squares can be downscaled to represent precipitation at a point within the square. The scale over which rainfall is coherent is therefore key to the design of climate model experiments that aim to provide data to hydrological models.

This study has shown that daily to inter-annual rainfall variability over Israel and West Jordan is primarily affected by the circulation over the Mediterranean. This would be expected to cause a degree of spatial coherence in the rainfall variability at these timescales. Direct analyses of inter-annual variability in the observed rainfall record demonstrated that Middle East rainfall is coherent at scales of less than 200 km – particularly within physiographic zones. At distances greater than 200 km, the degree of coherence reduces sharply, which suggests that standard-resolution climate models (horizontal resolution of the order of 250 km or greater) would have significant difficulty in representing precipitation variability at individual stations. A regional climate model (resolution of the order of 50 km), on the other hand, has the potential to provide useful information for hydrological modelling. Moreover, these results raise the possibility that increasing climate model resolution beyond 50 km would add little value, and that finer-scale variability is best represented using a statistical rainfall model. Such a method is described in Chapter 5 of this volume.

The hydrological studies reported here focus on changes at centennial to millennial timescales. Therefore, if we are to use climate model data to provide estimates of how rainfall at a particular station may change, we need to consider the spatial scales of climate change over long timescales. Proxy data are too sparse to examine this issue directly, so we need to look instead at how the large-scale controls on variability affect precipitation in the present day, and how these controls vary on centennial to millennial timescales. This study has demonstrated that the primary control on daily to inter-annual variability is the circulation in the Mediterranean basin and that variability in this circulation leads to coherence in rainfall over scales of at least 100 km. If precipitation change over centennial to millennial timescales results from changes in the Mediterranean circulation, it is likely that these changes are coherent within 50 km grid squares. Chapters 3 and 4 of this volume describe how large-scale controls on rainfall, such as the Mediterranean storm track, have changed in the past and are predicted to change in the future, and how these changes have affected or will affect precipitation.

The dependence of Middle East precipitation on the larger-scale circulation suggests that precipitation variability has the potential to change markedly. In particular, the link between inter-annual variability and the Mediterranean storm track is such that any change in the storm track would have a significant impact (see Chapters 3 and 4 in this volume). On even longer time scales, changes in the thermohaline circulation and solar insolation would be expected to have a strong impact on Middle East precipitation (see Chapter 3). Large changes in Dead Sea levels (Chapters 9 and 11) suggest that the rainfall climate of Israel has indeed changed markedly during the Holocene, with the mean rainfall being approximately twice today's values before 8 ka.

2.7 CONCLUSIONS

Much of the Middle East is arid, with annual rainfall less than 200 mm. The small amount of rain is crucial for the local population, with both spatial and temporal variability having a significant impact on human communities. This study has shown that inter-annual variability in rainfall is coherent on scales of 100 km and that daily to inter-annual variability is influenced by the large-scale circulation. The daily probability of rainfall is affected by the synoptic regime over Europe and the Atlantic, while the occurrence of these regimes is determined, in turn, by slowly varying large-scale modes of variability, such as the NAO and the EAWR pattern.

In summary:

- Middle East precipitation is highly variable spatially, with annual total rainfall exceeding 500 mm near the coast and less than 50 mm in the south and east.
- Precipitation exhibits a degree of spatial coherence at spatial scales of 100 km or less within the Middle East, particularly when longitude is held constant.
- The probability of precipitation is affected by circulation over the Mediterranean. This is reflected in SLP composites for wet and dry years and by certain circulation regimes favouring rainfall.
- The circulation over the Mediterranean is strongly influenced by the NAO and EAWR pattern.
- There is a link between large-scale Eurasian/Atlantic teleconnections (such as the NAO and EAWR), and Middle East rainfall. However, the relationship is non-linear and probably reflects the link between the large-scale state of the atmosphere and Mediterranean circulation.

REFERENCES

Barnston, A. G. and R. E. Livezey (1987) Classification, seasonality and persistence of low-frequency atmospheric circulation patterns. *Monthly Weather Review* **115**: 1083–1126.

Cullen, H. M. and P. B. deMenocal (2000) North Atlantic influence on Tigris–Euphrates streamflow. *International Journal of Climatology* **20**: 853–863.

Cullen, H. M., A. Kaplan, P. A. Arkin and P. B. deMenocal (2002) Impact of the North Atlantic Oscillation on Middle Eastern climate and streamflow. *Climatic Change* **55**: 315–338.

Diro, G. T., D. I. F. Grimes, E. Black, A. O'Neill and E. Pardo-Iguzquiza (2009) Evaluation of reanalysis rainfall estimates over Ethiopia. *International Journal of Climatology* **29**: 67–78.

Diskin, M. H. (1970) Factors affecting variations of the mean annual rainfall in Israel. *Bulletin of the International Association of Scientific Hydrology* **15**: 41–47.

Eshel, G. and B. F. Farrell (2000) Mechanisms of eastern Mediterranean rainfall variability. *Journal of the Atmospheric Sciences* **57**: 3219–3232.

Eshel, G. and B. F. Farrell (2001) Thermodynamics of Eastern Mediterranean rainfall variability. *Journal of the Atmospheric Sciences* **58**: 87–92.

Freiwan, M. and M. Kadioglu (2008) Spatial and temporal analysis of climatological data in Jordan. *International Journal of Climatology* **28**: 521–535.

Goldreich, Y. (1994) The spatial-distribution of annual rainfall in Israel – a review. *Theoretical and Applied Climatology* **50**: 45–59.

Hodges, K. I. (1994) A general method for tracking analysis and its application to meteorological data. *Monthly Weather Review* **122**: 2573–2586.

Hodges, K. I. (1995) Feature tracking on the unit-sphere. *Monthly Weather Review* **123**: 3458–3465.

James, P. M. (2007) An objective classification method for Hess and Brezowsky Grosswetterlagen over Europe. *Theoretical and Applied Climatology* **88**: 17–42.

Jones, P. D., T. Jonsson and D. Wheeler (1997) Extension to the North Atlantic oscillation using early instrumental pressure observations from Gibraltar and south-west Iceland. *International Journal of Climatology* **17**: 1433–1450.

Jones, P. D., T. J. Osborn and K. R. Briffa (2003) Pressure-based measures of the North Atlantic Oscillation (NAO): A comparison and an assessment of changes in the strength of the NAO and its influence on surface climate parameters. In *The North Atlantic Oscillation Climate Significance and Environmental Impacts*, ed. J. W. Hurrell, Y. Kushnir, G. Ottersen and M. Visbeck. Washington DC: American Geophysical Union pp. 51–62.

Kalnay, E., M. Kanamitsu, R. Kistler *et al.* (1996) The NCEP/NCAR 40-year reanalysis project. *Bulletin of the American Meteorological Society* **77**: 437–471.

Krichak, S. O. and P. Alpert *et al.* (2005a) Decadal trends in the east Atlantic-west Russia pattern and Mediterranean precipitation. *International Journal of Climatology* **25**: 183–192.

Krichak, S. O. and P. Alpert (2005b) Signatures of the NAO in the atmospheric circulation during wet winter months over the Mediterranean region *Theoretical and Applied Climatology* **82**: 27–39.

Krichak, S. O., P. Kishcha and P. Alpert (2002) Decadal trends of main Eurasian oscillations and the Eastern Mediterranean precipitation. *Theoretical and Applied Climatology* **72**: 209–220.

Krichak, S. O., M. Tsidulko and P. Alpert (2000) Monthly synoptic patterns associated with wet/dry conditions in the Eastern Mediterranean. *Theoretical and Applied Climatology* **65**: 215–229.

Kutiel, H., P. Maheras, M. Turkes and S. Paz (2002) North Sea Caspian Pattern (NCP) – an upper level atmospheric teleconnection affecting the eastern Mediterranean – implications on the regional climate. *Theoretical and Applied Climatology* **72**: 173–192.

Kutiel, H. and S. Paz (1998) Sea level pressure departures in the Mediterranean and their relationship with monthly rainfall conditions in Israel. *Theoretical and Applied Climatology* **60**: 93–109.

Mason, S. J. and L. Goddard (2001) Probabilistic precipitation anomalies associated with ENSO. *Bulletin of the American Meteorological Society* **82**: 619–638.

Rodwell, M. J. and B. J. Hoskins (1996) Monsoons and the dynamics of deserts. *Quarterly Journal of the Royal Meteorological Society* **122**: 1385–1404.

Schneider, U., T. Fuchs, A. Meyer-Christoffer and B. Rudolf (2008) Global precipitation analysis products of the GPCC. Global Precipitation Climatology Centre (GPCC) DWD Internet Publikation 1–12.

US Geological Survey, Israeli Hydrological Service, Israeli Meteorological Service, Israeli Soil Erosion Research Station, Jordanian Meteorological Department, Ministry of Water and Irrigation-Jordan, Palestinian Meteorological Office and P. W. Authority (2006) *Application of Methods for Analysis of Rainfall Intensity in Areas of Israeli, Jordanian and Palestinian Interest*. Reports of the Executive Action Team, Middle East Water Data Banks Project, http://exact-me.org/ri/rain2/index.htm.

Wallace, J. M. and D. S. Gutzler (1981) Teleconnections in the geopotential height field during the Northern Hemisphere winter. *Monthly Weather Review* **109**: 784–812.

Ziv, B., U. Dayan, Y. Kushnir, C. Roth and Y. Enzel (2006) Regional and global atmospheric patterns governing rainfall in the southern Levant. *International Journal of Climatology* **26**: 55–73.

3 Past climates of the Middle East

David Brayshaw, Emily Black, Brian Hoskins and Julia Slingo

ABSTRACT

In this chapter, we develop an improved understanding of the Mediterranean's past climate through a series of 'time-slice' climate integrations relating to the past 12,000 years, performed using a version of the Met Office Hadley Centre's global climate model (HadSM3). The output is dynamically downscaled using a regional version of the same model to offer unprecedented spatial detail over the Mediterranean. Changes in seasonal surface air temperatures and precipitation are discussed at both global and regional scales along with their underlying physical drivers.

In the experiments the Mediterranean experiences more precipitation in the early Holocene than the late Holocene, although the difference is not uniform across the eastern Mediterranean. The results suggest that there may have been a relatively strong reduction in precipitation over the eastern Mediterranean coast during the period around 6–10 thousand years before present (kaBP). The early Holocene also shows a stronger seasonal cycle of temperature throughout the Northern Hemisphere but, over the northeast Mediterranean, this is mitigated by the influence of milder maritime air carried inland from the coast.

3.1 INTRODUCTION

Understanding the changes in the Mediterranean climate during the Holocene period is a challenging problem, but one that is critical to interpreting long-term change in human settlement. The region at present displays marked seasonality with dry, hot summers and cool, wet winters. From a dynamical perspective, both tropical and extratropical processes are important: in the summer the region lies within the latitudes of the Northern Hemisphere subtropical anticyclones whereas in winter it is dominated by the extratropical cyclones of the Mediterranean storm track.

The region is also characterised by a complex coastline and orography. As such, it is marked by strong spatial and temporal contrasts. For example, the so-called Levant experiences relatively high levels of precipitation in winter, while the Syrian and Arabian Desert to the south and east receives almost none.

In order to assess the detailed regional patterns of change in this region through the Holocene period, a new series of 'time-slice' climate simulations have been performed to investigate the climate response to inter-millennial changes in solar radiation (insolation), greenhouse gas (GHG) concentrations and land ice-sheet changes over the past 12,000 years. Each time slice is separated by a period of 2,000 years. To help to provide the level of spatial detail sought by archaeological and palaeographic studies in the Mediterranean, the simulations are dynamically downscaled to approximately 50 km over the Mediterranean using a regional climate model. This regional model is physically and dynamically consistent with the global model and adds spatial detail associated with the coastlines and orography, but does not, in general, produce any significant modification of the broad patterns of climate. This is consistent with the one-way nesting of the regional model (the regional model is driven by the global model without the regional model 'feeding back' to the global model). Compared with previous modelling studies (e.g. the PMIP-2 project, Braconnot et al., 2007), the modelling configuration used in this study offers greatly increased spatial and temporal detail for the Mediterranean region.

This chapter focuses on a selection of climatological variables that are selected to be particularly relevant from an archaeological and palaeographic perspective: seasonal- and annual-mean changes in surface air temperatures (SATs) and precipitation. This chapter begins with an overview of the experimental configuration, its validation under modern-day climate conditions, and a description of the major climatological drivers

Water, Life and Civilisation: Climate, Environment and Society in the Jordan Valley, ed. Steven Mithen and Emily Black. Published by Cambridge University Press.
© Steven Mithen and Emily Black 2011.

affecting the Holocene period (section 3.2). Section 3.3 discusses the climatological changes seen in the recent past (since the Industrial Revolution). Sections 3.4 and 3.5 then discuss the changes in SAT and precipitation, respectively, across the Holocene simulations. Section 3.6 presents a summary of the results along with discussion and conclusions. Where appropriate, comparisons are made with previous relevant climate modelling studies and large-scale palaeoclimate records and syntheses. Detailed comparison against eastern Mediterranean palaeorecords is discussed in Chapter 8 of this volume.

3.2 EXPERIMENTAL CONFIGURATION AND MODEL VALIDATION

The 'time-slice' experiments are performed using the Hadley Centre HadSM3 global climate model. For each experiment, the climate 'forcings' (GHG levels, variations in the Earth's orbit, sea level and land ice-sheets) correspond to those estimated for a particular time period in the past 12,000 years (experiments are separated by 2,000-year intervals). After an initial spin-up period (5 model years), the model reaches a forced equilibrium and is then run for 20 model years. This was extended to 40 model years for the 'pre-industrial' and the 8 kaBP simulations, as these were initially targeted as being of particular interest. By performing such multi-year runs, it is possible to form an estimate of the interannual variability of the model, and thereby assess the statistical significance of any changes that are seen (the methods involved are discussed in section 3.2.7). The HadRM3 regional climate model, forced by boundary conditions from the global model, is used to dynamically downscale the output over the Mediterranean area (owing to the application of the AMIP-II method to interpolate the monthly mean sea-surface temperature (SST) fields recorded by the global model (Taylor *et al.*, 2000), the regional model runs are one to two years shorter than the global simulations).

Additional sensitivity experiments are performed to assess the atmospheric response to:

- changes in the land-surface properties over subtropical North Africa around 8 kaBP and 6 kaBP; and
- changes in the high-latitude land ice-sheet cover at 8 kaBP.

The 'time-slice' simulations should, therefore, be considered as a series of sensitivity tests, each indicating the atmospheric response to a particular set of forcings, rather than as a deterministic hind-cast of the climate at a given time period. Furthermore, the experiments represent the model's equilibrium response to a fixed forcing, rather than a 'transient' response to an evolving forcing. For the Holocene, where the forcings generally change slowly (on millennial timescales), this is unlikely to affect the model output because the slowly evolving components of the climate system which have memory over this timescale, such as land ice-sheets and ocean currents, are prescribed in these experiments. However, when examining the relatively rapid change in climate between pre-industrial conditions and the present day (timescales of decades), the equilibrium response in the model may be somewhat different from the real transient response. For example, the model's state that is in equilibrium with GHG concentrations fixed at 1960 levels will tend to be somewhat warmer than if the model was run forwards with the observed increasing levels of GHG concentrations from 1800 to 1960.

The model configuration and the forcings used in the experiments are discussed in the following sections and are shown in Figures 3.1 and 3.2. A summary of the experiments is provided in Table 3.1.

3.2.1 The global model

The HadSM3 GCM and its variants are widely used in the climate research community. The atmospheric component is described in detail by Pope *et al.* (2000) and is used here in its standard configuration with a resolution of N48L19 (2.5° latitude by 3.75° longitude with 19 vertical levels).

A well-mixed slab of 50 m depth with a simple sea-ice model (Hewitt *et al.*, 2001) is used to represent the ocean. The heat flux convergence in the slab (Figure 3.2c) is calibrated to reproduce the annual cycle of SSTs from a pre-industrial control run using the fully coupled HadCM3 model (which includes a dynamical ocean model). The assumption is made that the changes in the heat flux convergence in the ocean may be neglected. As a consequence the heat flux convergence in the slab ocean is held fixed throughout all the experiments, although in the early Holocene experiments (see Table 3.1) there is some modification to the heat flux convergence to account for ocean points that are lost because of lower sea levels, as discussed in section 3.2.5.

The use of a slab ocean model simplifies the modelling process considerably when compared with using a fully coupled dynamical ocean. In particular, dynamical ocean models require long spin-up periods to reach an equilibrium state, typically hundreds to thousands of model years for the type of experiments performed here. As such, they are very computationally expensive. Previous studies using such fully coupled models, such as the PMIP-2 project (Braconnot *et al.*, 2007) have therefore either focused on a limited set of time periods (e.g. pre-industrial, 6 kaBP, 21 kaBP) rather than a full span over the Holocene, or used simpler EMIC-type models (Claussen *et al.*, 1999) with very coarse spatial resolution and simplified atmospheric processes.

As noted above, in order to use the slab ocean model it is necessary to assume that the heat transport and currents within the ocean are constant across the time periods. In some regions

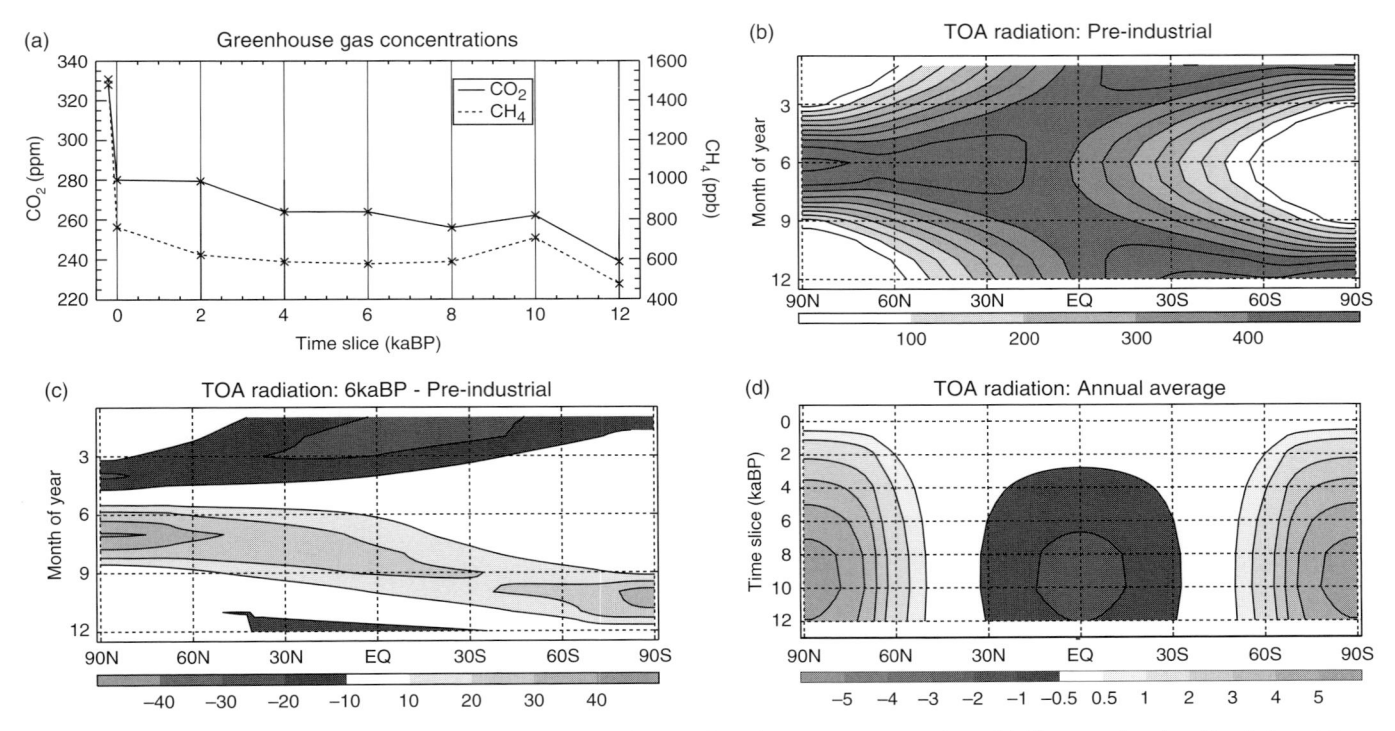

Figure 3.1. The forcings used to drive the global and regional models. (a) Greenhouse gas concentrations. (b) The annual cycle of insolation at the top of the atmosphere in experiment PREIND (units W m^{-2}). (c) The anomaly in the annual cycle of top of atmosphere (TOA) insolation applied to experiment 6kaBP (units W m^{-2}). (d) Annual mean insolation anomalies at the top of the atmosphere in each of the time-slice experiments (units W m^{-2}).

this may lead to a decoupling between the atmosphere and the ocean (in reality, the atmospheric circulation influences the ocean circulation and vice versa). Nonetheless, given the wide range of time periods under investigation and the need to simulate atmospheric processes at reasonably high resolution in order to represent the Mediterranean storm track, the slab ocean model offers a good compromise. Where possible, the slab model's performance has been compared to the performance of previous studies using coupled models and, in general, the slab model is found to agree reasonably well with the coupled ocean–atmosphere PMIP-2 simulations for 6 ka (Braconnot et al., 2007), as discussed in sections 3.4 and 3.5.

3.2.2 The regional model

The HadRM3 model is a high-resolution, limited-area version of the global model, driven by lateral boundary conditions provided by the global model. The horizontal resolution is $0.44° \times 0.44°$ with 19 vertical levels, and the regional domain extends eastward from Portugal to Iraq and southward from Southern Europe to North Africa. The HadRM3 model requires SSTs to be prescribed, which is achieved by using a spatially interpolated version of the time-varying monthly mean SST fields calculated by the global model. As such, the SSTs in the regional model are consistent with the lateral boundary conditions used to drive it.

3.2.3 The forcing of the climate during the Holocene period: time-slice experiments

The time-slice experiments are forced by changes in GHG concentrations (CO_2 and CH_4 only), the Earth's orbital parameters, land ice-sheets and sea level. These are discussed separately below.

The atmospheric concentrations of CH_4 and CO_2 used in the simulations are shown in Figure 3.1a. The values are consistent with Jansen et al. (2007) but with trace gases (e.g. N_2O and CFCs) held constant throughout all the simulations. The GHG concentrations used in the 'present-day' experiment ('PRESDAY', 328 ppm for CO_2 and 1,600 ppb for CH_4, corresponding roughly to the period 1960–1990) are substantially higher than those used in the 'pre-industrial' experiment ('PREIND', 280 ppm and 790 ppb, respectively, corresponding to the period around 1800). This change is consistent with an increase in the global-mean, annual-mean radiative forcing of about 1 to 1.5 W m^{-2} in experiment PRESDAY relative to experiment PREIND (Jansen et al., 2007), and can be expected to be associated with an increase in the equilibrium global-mean, annual-mean surface temperature of around 1 °C, which is similar to that seen in the observational record (Trenberth et al., 2007).

As Figure 3.1a shows, prior to industrial times the changes in atmospheric CO_2 and CH_4 were relatively small during the

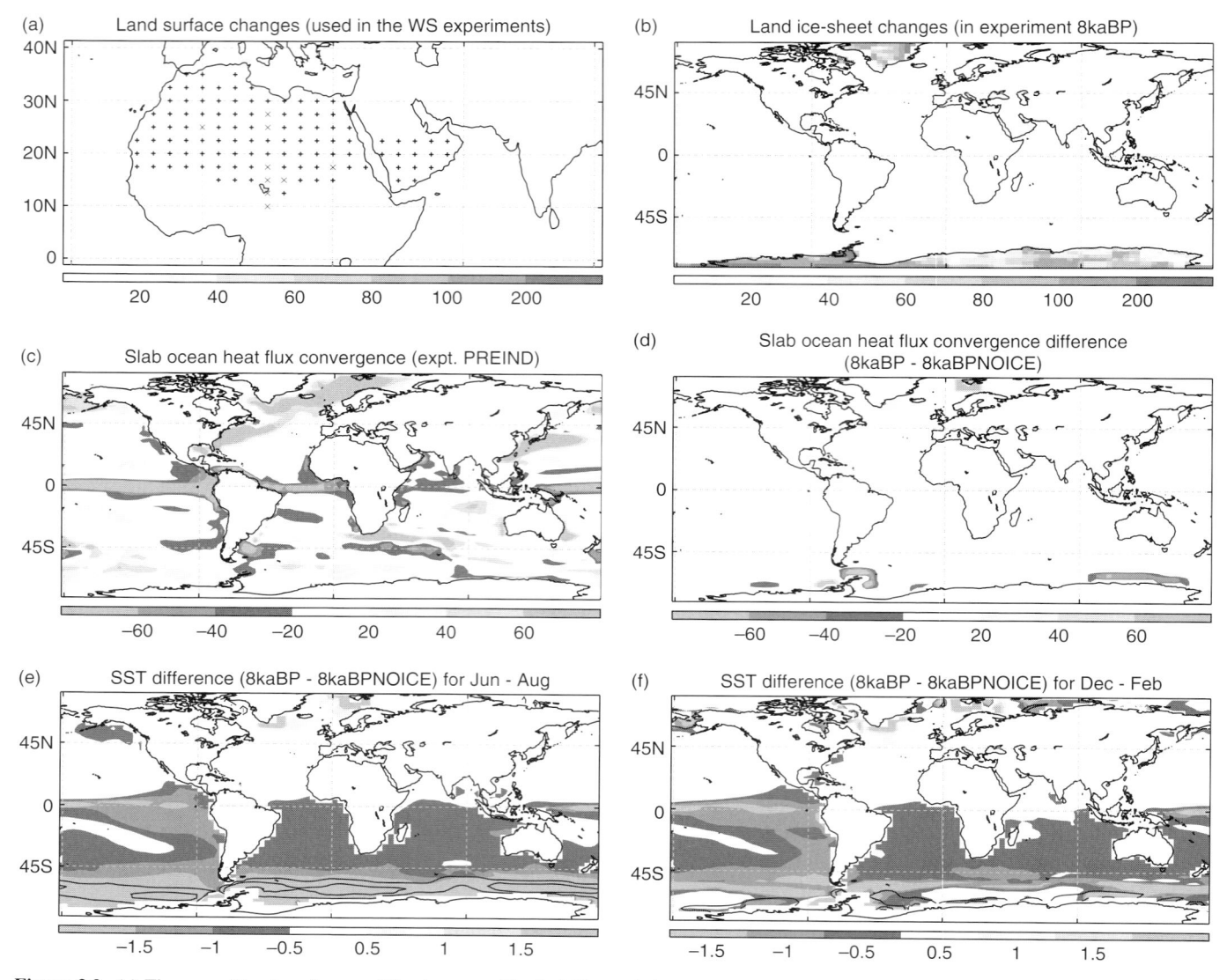

Figure 3.2. (a) The area of land surface modifications over North Africa and the Arabian Peninsula in the 'Wet Sahara' (WS) experiments (+ shows grid points that are converted from mostly desert to uniform savannah/shrubland and × are converted to open water). (b) Imposed land ice-sheet changes between experiments 8kaBP and PREIND (shading shows the change in surface height, in metres). (c) Ocean heat flux convergence in experiments PRESDAY – 6kaBP (W m^{-2}). (d) Ocean heat flux convergence anomaly applied to experiment 8kaBP (W m^{-2}). (e) Sea surface temperature (SST) difference between experiments 8kaBP and 8kaBPNOICE (shading, °C) and sea ice difference (contours at 5% and 30%) for June–August. (f) As (e), but for December–February.

Holocene. GHG levels have generally increased during the Holocene, with the combined radiative forcing in the early period (12 kaBP) being approximately 1 W m^{-2} lower than the pre-industrial level, and the radiative forcing due to changes in CO_2 being stronger than that due to those in CH_4. The changes in GHG concentrations can therefore be expected to contribute to a gradual warming of the planet over the Holocene.

Variations in the Earth's orbital parameters are calculated following Berger (1978). The impact these variations have on the Earth's radiation budget is profoundly different from that of the GHG concentrations, which is largely seasonally and latitudinally invariant. In particular, the differences in the orbital parameters produce very little change in the global- (or hemispheric-) mean annual-mean incident shortwave radiation (insolation) received at the top of the atmosphere (approx. 0.05 W m^{-2} at 10 kaBP) but they have a strong impact on the seasonal cycle of insolation and its latitudinal structure.

During the Early Holocene, differences in the Earth's orbital parameters led to stronger insolation at the top of the atmosphere in boreal summer and weaker during boreal winter when compared with modern levels (Figure 3.1). The early Holocene seasonal cycle of insolation in the Northern Hemisphere is

Table 3.1 *The experimental configuration*

The column marked GHG denotes the atmospheric concentrations of the major greenhouse gases (CO_2 and CH_4), with PI indicating pre-industrial conditions and PD indicating present-day (1960–1990) levels. 'Orbital' denotes the time period corresponding to the orbital parameters used in the integration. Sea level and land ice-sheets have changed relatively slowly over the past 6,000 years, so present-day conditions are used for the mid- to late Holocene integrations (6 kaBP and more recent), as shown in the columns 'Ice' and 'Sea level'. Experiments featuring the imposed land surface changes over North Africa are indicated by the column marked 'Wet Sahara'. Experiments marked with * include the ocean heat flux error described in section 3.2.5. (EH) and (LH) denote the experiments included in the 'Early Holocene' and 'Late Holocene' ensembles of experiments.

Experiment	GHG	Orbital	Sea level	Ice	Wet Sahara
PRESDAY	PD	PD	PD	PD	No
PREIND	PI	PD	PD	PD	No
2kaBP (LH)	2kaBP	2kaBP	PD	PD	No
4kaBP (LH)	4kaBP	4kaBP	PD	PD	No
6kaBP (LH)	6kaBP	6kaBP	PD	PD	No
8kaBP* (EH)	8kaBP	8kaBP	8kaBP	8kaBP	No
10kaBP* (EH)	10kaBP	10kaBP	10kaBP	10kaBP	No
12kaBP* (EH)	12kaBP	12kaBP	12kaBP	12kaBP	No
6kaBP-WS	6kaBP	6kaBP	PD	PD	Yes
8kaBP-WS*	8kaBP	8kaBP	8kaBP	8kaBP	Yes
8kaBPNOICE	8kaBP	8kaBP	PD	PD	No

therefore stronger while that in the Southern Hemisphere is weaker. For 30° N at 6 kaBP the seasonal cycle is stronger by approximately 30–40 $W\,m^{-2}$ or approximately 15% (Figure 3.1c). The latitudinal differences in insolation are shown in Figure 3.1d, which indicates that the orbital variations led to stronger annual-mean high latitude insolation in the early Holocene (approx. 5 $W\,m^{-2}$ at 10 kaBP) and weaker near the equator (approx -1 $W\,m^{-2}$ at 10 kaBP).

Land ice-sheet differences are generally small in the recent past although, for time-periods earlier than 6 kaBP, larger ice-sheets are imposed based on the reconstruction developed by Peltier (1994). The extent of the ice at 8 kaBP is shown in Figure 3.2b.

Differences in sea level are imposed following the synthesis presented in Jansen *et al*. (2007). Such differences are small in the recent past (i.e. 6 kaBP and later) and are therefore neglected, this being consistent with the approach adopted in the PMIP-2 project (Braconnot *et al*., 2007). For periods earlier than 6 kaBP, the sea level is lower than present day by 12 m (at 8 kaBP), 36 m (10 kaBP) or 56 m (12 kaBP). The redistribution of the ocean heat flux convergence from the 'lost' ocean grid points is discussed in section 3.2.5.

3.2.4 The forcing of the climate during the Holocene period: sensitivity experiments

Three sensitivity experiments are performed:

- Experiment 6 kaBP-WS ('WS' denotes 'Wet Sahara') is identical to experiment 6 kaBP but with the land surface over a wide area of subtropical North Africa and the Middle East prescribed to have enhanced vegetation and open water. This scenario is consistent with the palaeorecords presented by, for example, Jolly *et al*. (1998) and Drake and Bristow (2006). The extent of the vegetation difference is shown in Figure 3.2a.

- Experiment 8 kaBP-WS is conceptually similar to experiment 6 kaBP-WS, but relates to the conditions of the 8 kaBP time period (and therefore experiment 8 kaBP).

- Experiment 8 kaBPNOICE examines the impact made by the extent of the land ice-sheets and sea-level change under the conditions of the 8 kaBP time slice. In particular, experiment 8 kaBPNOICE is identical to experiment 8 kaBP but uses present-day land ice-sheets and sea level (see Table 3.1). The extent of the difference in ice cover is shown in Figure 3.2b.

3.2.5 Redistribution of 'lost' ocean heat flux convergence in the early Holocene

It was intended that the ocean heat flux convergence over the 'lost' ocean grid points (see section 3.2.1) in the Early Holocene experiments (8, 10 and 12 kaBP) would be distributed primarily within the area immediately adjacent to the lost grid point, with the remainder spread evenly across the global ocean. However, an error in our configuration led to the imposition of a seasonally strong negative heat flux convergence around the coast of Antarctica and a strong positive heat flux convergence around a small number of grid points at high northern latitudes (see Figure 3.2d). The error is largest in boreal summer but small in boreal winter (vanishingly small in November). The implications of this error are discussed below.

This negative heat flux anomaly near Antarctica leads to stronger sea-ice formation in the Southern Ocean, particularly during boreal summer (Figures 3.2e and 3.2f) and part of the ocean's negative heat flux convergence in these sea-ice covered grid points is then redistributed again across the Southern Hemisphere oceans to reduce model instability (Crossley and Roberts, 1995). This second redistribution leads to an annual-mean negative heat flux convergence over the Southern Hemisphere oceans of approximately 0.5 Wm^{-2} (similar for each of experiments 8 kaBP, 10 kaBP and 12 kaBP) and an annual-mean cooling of approximately 0.7–0.8 °C in the SSTs of the southern tropics (Figures 3.2e and 3.2f).

Throughout this chapter, the limitations and the impacts of the error are discussed wherever they are believed to significantly affect the interpretation of the results. However, this feature is not believed to have significantly affected the broad patterns of change in the tropics and the Northern Hemisphere during the Holocene, or the conclusions of this chapter beyond the limitations specifically referred to in the text.

3.2.6 Model validation under present climate

The modelling configuration is similar to that presented in Chapters 2 and 4 of this volume and is also described in Black (2009). As such, only a limited discussion of the model's validation under present-day climate is provided here.

Previous studies have found that the global atmospheric model is able to provide a good representation of many of the key features of the climate system. In particular, given the relatively coarse resolution of the model, the global distribution of precipitation compares favourably with reanalysis and observational datasets (which provide the best available reconstructions of the global climate over the past 50 years), such as ERA-40 (Uppala et al., 2005) and GPCC (Rudolf et al., 1994). The model also provides a good representation of the major atmospheric storm tracks (Stratton, 1994) that are crucial to understanding the distribution of precipitation in the winter mid-latitudes.

The detailed structure of the precipitation in the Mediterranean region is shown in Figure 3.3. The regional model clearly shows the characteristics of the seasonal cycle (dry summers, wet winters) and many of the spatial features in the observational datasets (e.g. stronger precipitation over orographic regions such as the Alps and to the east of the Black Sea; strong north–south and east–west precipitation gradients over the Jordan Valley). The model does, however, have a tendency to precipitate too much over the western Sahara in summer (compare Figures 3.3a, 3.3c and 3.3e) and the coastal precipitation over the Jordan Valley in winter does not extend as far inland as the GPCC dataset appears to suggest (compare Figures 3.3b and 3.3d).

Despite these limitations, the model provides a sufficiently accurate representation of the precipitation characteristics of the region under present-day conditions, suggesting that the main physical processes that influence the distribution of precipitation in the Mediterranean are being reasonably well captured.

3.2.7 Diagnostics

To understand many of the changes in precipitation described in this chapter it is necessary to examine changes in the Northern Hemisphere storm tracks. The storm tracks are diagnosed using 6-hourly model data by tracking cyclonic circulation maxima on the 850 hPa pressure surface after the large-scale background circulation has been removed. For a track to be recorded, a minimum lifetime of two days is required. This method is described in more detail by Hoskins and Hodges (2002).

Mid-latitude weather systems (or 'storms') tend to form in regions of strong horizontal temperature gradients which are also regions of strong vertical wind shear ('baroclinicity'). Following Hoskins and Valdes (1990), a commonly used measure of this is the maximum growth rate of weather-system-like disturbances in a simple westerly flow. This is referred to as the Eady growth rate, σ, and is proportional to the latitudinal temperature gradient. For more details, see Hoskins and Valdes (1990).

As the storm tracks respond to changes in both near-surface baroclinicity (typically caused by contrasts in the local surface temperatures) and upper tropospheric baroclinicity (which may, for example, be caused by tropically driven changes in the subtropical jet through the Hadley cell), the Eady growth rate is measured in both the lower (925 hPa) and upper (400 hPa) troposphere (see, e.g., Brayshaw et al., 2008).

The statistical significance of differences in the diagnostics between various integrations is assessed using the non-parametric Wilcoxon–Mann–Whitney (WMW) test (Wilks, 1995). It is assumed that the data from each model year are serially uncorrelated.

3.3 THE RECENT PAST, PRE-INDUSTRIAL TO PRESENT DAY

The model's equilibrium response to the changes in GHG concentrations over the industrial period is shown in Figure 3.4. In the global average, the annual-mean SAT in the PREIND experiment is approximately 1 °C cooler than in experiment PRESDAY, a difference that is comparable to that seen in the observational record (Trenberth et al., 2007). The spatial distribution of the signal is remarkably uniform although it is stronger over land and is particularly visible at high latitudes. This is in common with other studies that have examined the effects of GHG concentrations (Meehl et al., 2007). For the land areas of the eastern

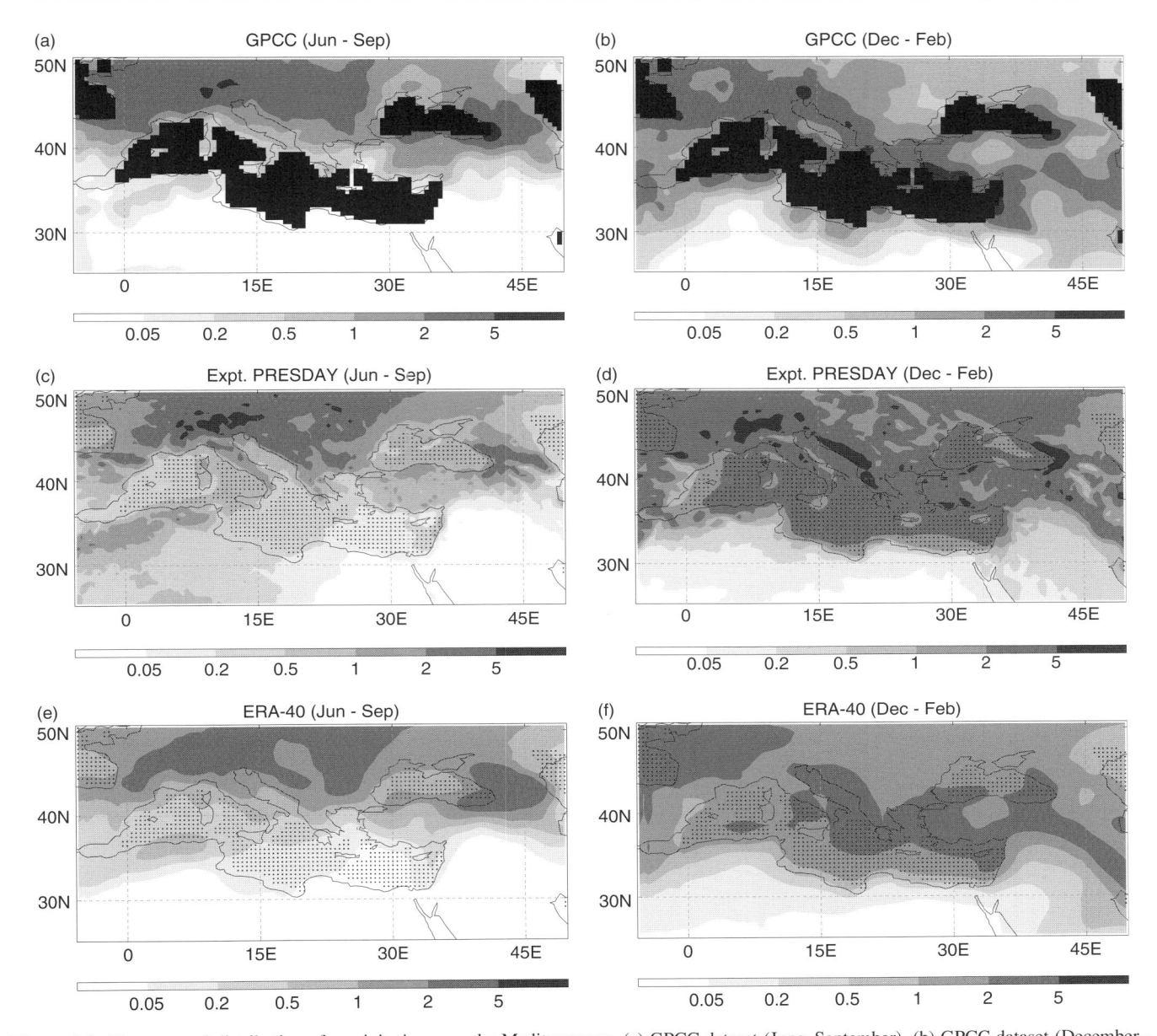

Figure 3.3. The seasonal distribution of precipitation over the Mediterranean. (a) GPCC dataset (June–September). (b) GPCC dataset (December–February). (c) As (a) but from the regional model in experiment PRESDAY. (d) As (b) but from the regional model in experiment PRESDAY. (e) As (a) but using the ERA-40 dataset. (f) As (b) but using the ERA-40 dataset. Units mm day^{-1}. In (a) and (b) missing data areas are blacked out. In (c) to (f), black squares mark the regions where GPCC data are missing.

Mediterranean, the temperature signal is slightly stronger in summer than winter (by about 0.5 °C, not shown).

In Figure 3.4b, the lower SAT in experiment PREIND appears to be approximately uniform across the North Atlantic basin. This stands in contrast to other simulations that have examined the SAT response to increasing GHG concentrations using fully coupled ocean–atmosphere models under modern-day conditions. These coupled simulations tend to show a minimum in the SAT response over the North Atlantic consistent with a weaker Meridional Overturning Circulation (MOC) in the North Atlantic Ocean at higher levels of atmospheric GHG concentrations (Wood et al., 1999; Dai et al., 2005; Meehl et al., 2007). The weaker MOC leads to less heat being transported polewards in the North Atlantic Ocean, locally opposing the global-scale atmospheric warming associated with increased GHG levels. This suggests that the negative SAT differences between experiment PREIND and experiment PRESDAY should be less strong over that region than those shown in Figure 3.4b (as seen in observations (Hegerl et al., 2007)).

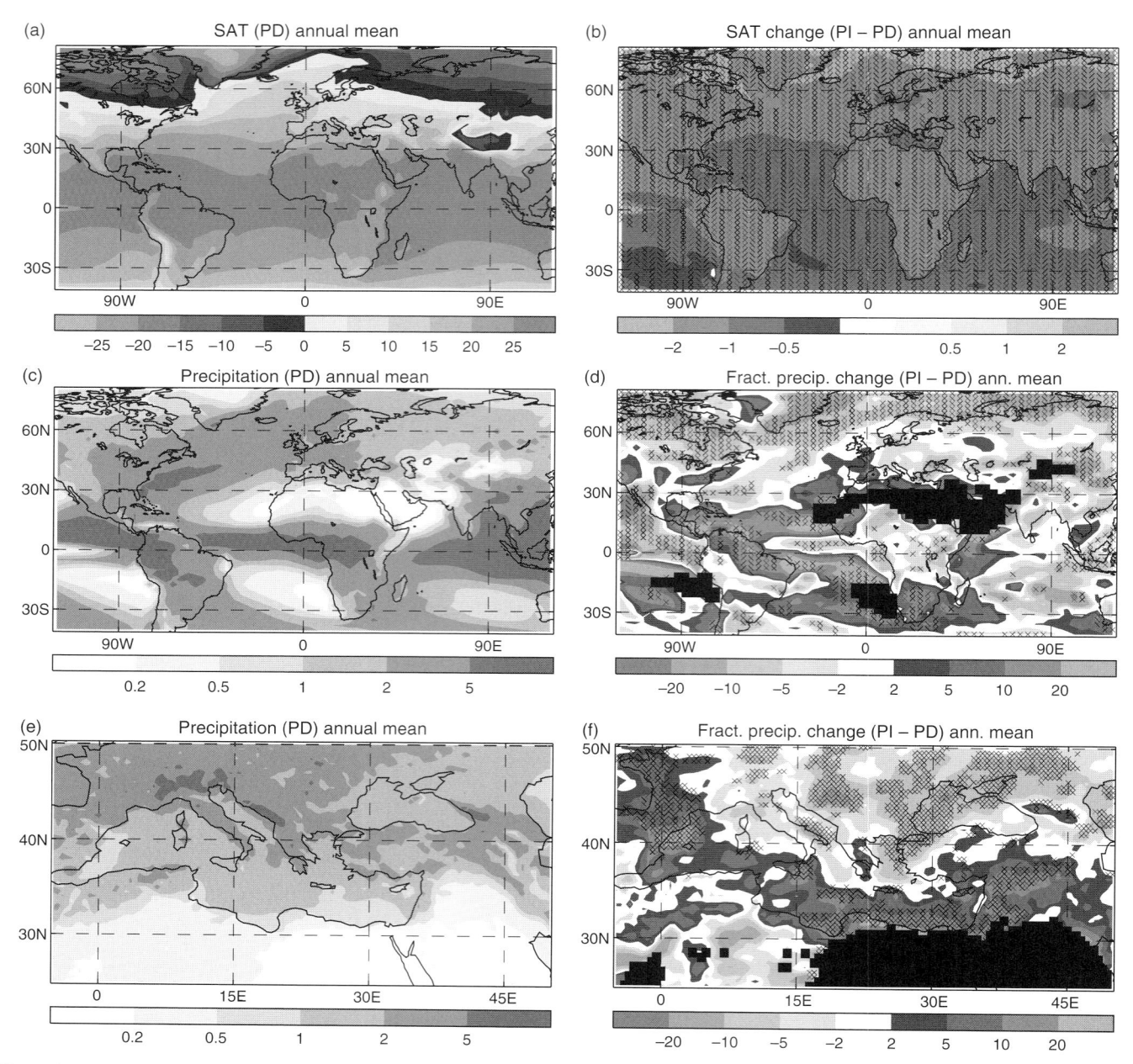

Figure 3.4. Annual mean SAT and precipitation during the pre-industrial period. The top row panels show (a) SAT (°C) in experiment PRESDAY, and (b) the difference (°C) found in experiment PREIND (i.e. PREIND – PRESDAY). The middle row panels show results from the global model where (c) is the precipitation in experiment PRESDAY (mm day^{-1}) and (d) is the fractional difference (%) found in experiment PREIND (i.e. [PREIND – PRESDAY] × 100/PRESDAY). The bottom row (e, f) is identical to the middle row but uses downscaled data from the regional model. For the difference plots (b, d, f), areas where the differences are statistically significant at (b) 99%, (d) 90% and (f) 70% confidence are indicated by black crosses. Areas of extremely low precipitation (less than 0.2 mm day^{-1}) in experiment PRESDAY are blacked out in the difference plots. See colour plate section.

The difference in planetary-scale annual-mean precipitation is shown in Figure 3.4d. Although the response is more complex than the SAT signal, there is a general weakening in the contrast between dry and wet latitudes. More specifically in experiment PREIND there is reduced precipitation in the rainy areas of the deep tropics and high latitudes but enhanced precipitation in the dry areas of the subtropics, as evident from a comparison of Figures 3.4c and 3.4d. The overall strength of the atmospheric hydrological cycle is weaker in experiment PREIND than in PRESDAY: the global-mean annual-mean precipitation is reduced by approximately 2%, which is consistent with previous climate modelling studies (summarised in Meehl *et al.*, 2007).

In the eastern Mediterranean, the regional model simulations show reduced precipitation over much of Europe (~5–10%), extending southwards to the northern part of the Mediterranean basin (Figure 3.4f), which is consistent with the signal seen in the global model (Figure 3.4d). Over the southern and eastern part of the Mediterranean basin, experiment PREIND gives more precipitation (5–20%), consistent with the generally wetter subtropics. It should, however, be noted that this difference represents a very small amount of actual precipitation, with the total precipitation in this area being typically less than 1 mm day^{-1}. As in experiment PRESDAY and modern observations (e.g. the GPCC dataset, Figure 3.3), most of the precipitation in experiment PREIND in the eastern Mediterranean region falls in winter.

The overall structure of the differences in precipitation and SAT differences is therefore broadly consistent with that which would be expected on the basis of both observations, as summarised in Trenberth et al., 2007 (see their Figure 3.13) and the IPCC ensemble of model responses to GHG forcings (Meehl et al., 2007).

A considerable body of literature exists on the response of the climate system to GHG changes over the industrial period, which is beyond the scope of this chapter to review in full. Instead, the focus here is to examine the changes in climate that have occurred on inter-millennial timescales during the Holocene. The remainder of this chapter will therefore use the pre-industrial experiment (PREIND) as the 'reference state', thereby removing the impact of the rapid increases in GHG concentrations during the past 200 years and allowing the response to the relatively slow changes in forcing during the Holocene to be seen more clearly. In order to compare the subsequent results against experiment PRESDAY (representing present-day conditions), it is necessary to recall that pre-industrial conditions were generally around 1 °C cooler than present day, with somewhat weaker precipitation in the northern Mediterranean and stronger in the south.

3.4 THE HOLOCENE: SURFACE AIR TEMPERATURE

The SAT response produced by the model in the time-slice experiments is shown in Figures 3.5 and 3.6 and is discussed in terms of an annual-mean change and the seasonal cycle below.

3.4.1 The annual-mean hemisphere-mean SAT response

Figure 3.5a shows that the time-slice experiments simulate a warming trend in the annual-mean SAT over the Northern Hemisphere across the Holocene, rising from approximately 1.7 °C below pre-industrial levels at 12 kaBP. If the annual-mean hemisphere average SAT response to GHG concentrations is assumed to be approximately linear, then much of the difference between experiment 12 kaBP and experiment PREIND can be attributed to differences in the chemical composition of the atmosphere. In particular, the GHG difference between experiment PREIND and experiment PRESDAY is similar to that between experiment 12 kaBP and experiment PREIND (CO_2 is lower in the latter experiment in each pair by approximately 40–50 ppm; see Figure 3.1a) and, in the former case (PREIND – PRESDAY), is associated with a difference in the annual mean Northern Hemisphere SAT of approximately 1–1.2 °C (Figure 3.5a). This suggests that approximately 50–70% of the difference in the Northern Hemisphere SAT between experiments 12 kaBP and PREIND may be attributable to a direct response to GHG concentrations. Similar arguments suggest that approximately 0.6–0.8 °C of the difference in the annual mean Northern Hemisphere SAT between experiments 8 kaBP and PREIND and approximately 0.3–0.5 °C of the difference between experiments 6 kaBP and PREIND may be associated with the lower GHG concentrations relative to pre-industrial times.

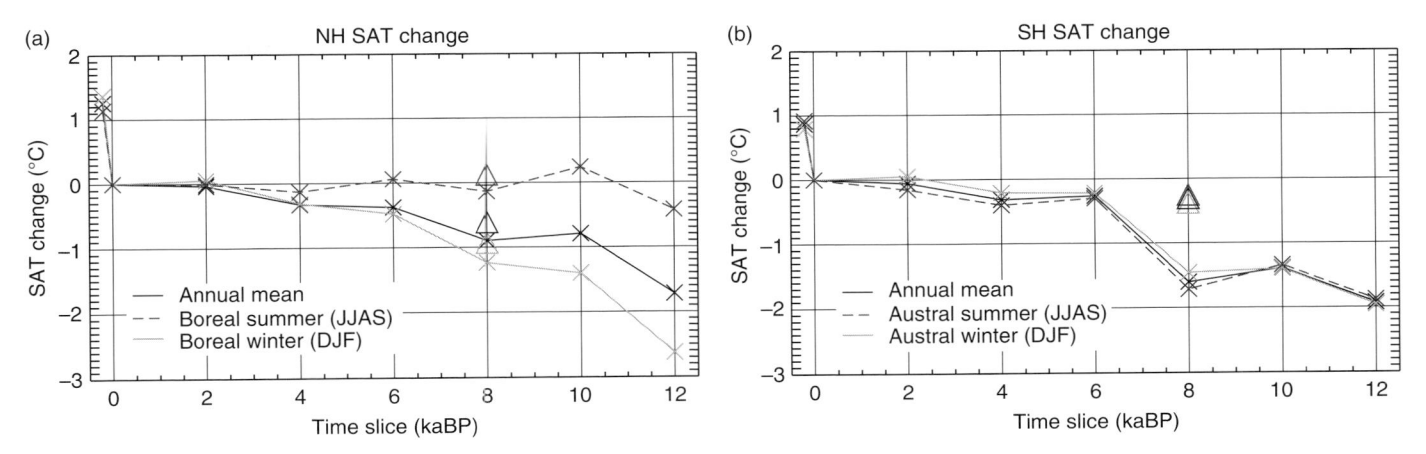

Figure 3.5. Hemisphere average SAT differences. (a) Northern Hemisphere average SAT change relative to experiment PREIND (°C). Data points from the time-slice experiments are marked by crosses, and data points from experiment 8kaBPNOICE are marked by triangles (experiment PRESDAY is shown at time = −0.2 kaBP). (b) As (a) but for Southern Hemisphere.

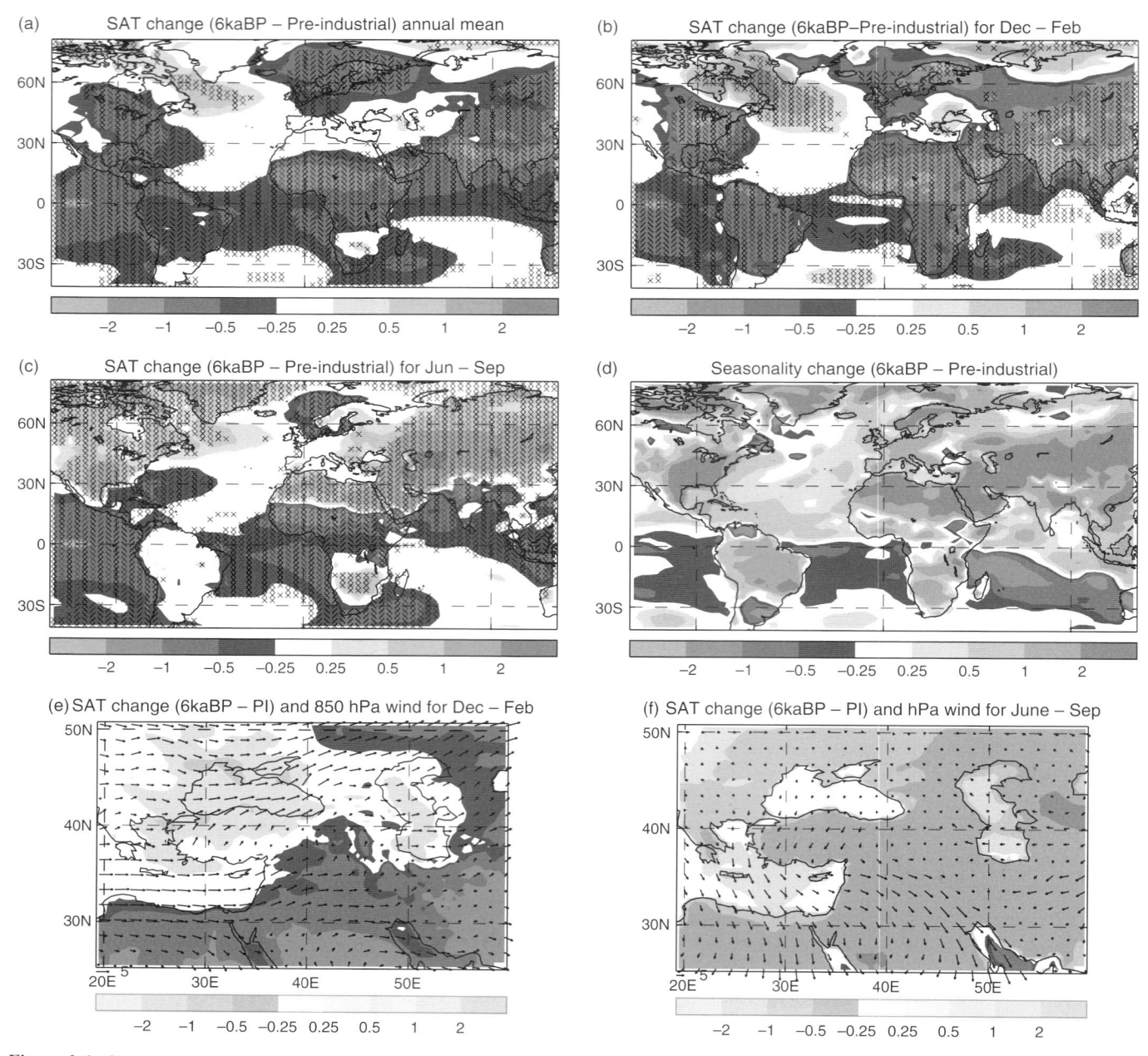

Figure 3.6. Changes in SAT across the Holocene time-slice integrations. (a) Annual mean SAT change (experiment 6kaBP − PREIND, °C). Panels (b) and (c) are as (a), but for boreal summer and winter seasons, respectively. (d) The change in the strength of the seasonal cycle of SAT between experiment 6kaBP and PREIND (the strength of the cycle is defined as the maximum monthly mean SAT minus the minimum monthly mean SAT, units °C). (e) Boreal winter SAT change (6kaBP − PREIND, colours, °C) in the regional model and lower tropospheric winds (850 hPa) in experiment PREIND (arrows, units m s^{-1}). (f) As (e) but for boreal summer. In panels (b) to (d), areas where the changes are statistically significant at the 90% level are marked with black crosses. See colour plate section.

The difference between experiments 8 kaBP and 8 kaBP-NOICE provides an indication of the role of the land ice-sheets on the Northern Hemisphere annual-mean SAT (Figure 3.5a). In particular, SATs are approximately 0.2–0.3 °C higher in experiment 8 kaBPNOICE, consistent with the increased surface height and lower surface albedo associated with the land ice-sheets in experiment 8 kaBP (the impact of the ice-sheet can be expected to be larger during earlier time periods). It therefore appears possible to explain much of the annual-mean Northern Hemisphere SAT signal in the time-slice simulations as a combination of the direct response to GHG concentrations and ice-sheet changes.

The situation in the Southern Hemisphere (Figure 3.5b) is more difficult to interpret in the early Holocene, as a consequence of the errors in the ocean heat flux configuration (described in section 3.2.5). In particular, the difference between experiment 8 kaBP and either of experiments 6 kaBP or 8 kaBP-NOICE (lower SATs by approximately 1–1.2 °C) is associated with a mixture of: increased surface elevations over Antarctica; spuriously strong sea-ice formation suppressing SATs over the Southern Ocean particularly in Austral winter (Figures 3.2e and 3.2f); and a spurious negative oceanic heat flux convergence over the whole of the Southern Hemisphere's oceans ($\sim 0.5\,\mathrm{W\,m^{-2}}$, see section 3.2.5).

These differences due to the ocean heat flux error are, however, strongest in the high southern latitudes and weaker in the tropics (approximately 0.7–0.8 °C in the tropics). Furthermore, the impact on the Northern Hemisphere average SAT appears to be small. The annual-mean SAT differences in the Northern Hemisphere between experiments 8 kaBP and 8 kaBPNOICE do not exceed 0.3 °C and at least some of this difference will be a 'real' response to the larger land ice-sheets in the Northern Hemisphere in experiment 8 kaBP. As such, the broad character of lower Northern Hemisphere annual-mean SATs during the early Holocene is thought to be robust and is consistent with the response expected from the changing level of GHG concentrations.

3.4.2 The annual-mean spatial pattern of SAT response

The spatial distribution of the differences (relative to experiment PREIND) in the annual mean SAT is shown for experiment 6 kaBP in Figure 3.6a. The tropical latitudes have generally lower SATs than in the pre-industrial simulation (approximately 0.5 °C averaged over 30° S to 30° N, not shown), consistent in sign with weaker annual mean insolation (Figure 3.1d, approximately $0.5-1\,\mathrm{W\,m^{-2}}$ averaged over the same region) and lower GHG levels (Figure 3.1a). This pattern is consistent with palaeorecords in the tropics that indicate lower SSTs in the mid- to early Holocene than in the late Holocene (e.g. Rimbu et al., 2004).

The high northern latitudes are warmer than experiment PREIND, consistent with stronger annual mean insolation, but the signal is perhaps rather weak given the strength of the insolation anomalies (in experiment 6 kaBP, the maximum SAT difference is 0.5 °C in the zonal average at high northern latitudes, not shown, whereas the annual-mean insolation difference is in excess of $4\,\mathrm{W\,m^{-2}}$). This demonstrates the important role played by latitudinal heat transport through the system, although it should be remembered that not all of the stronger 6 kaBP insolation will act to heat the surface (surface albedo is likely to be high given the presence of sea-ice) and the globally colder 6 kaBP SATs (associated with lower GHG concentrations) will also tend to oppose any localised warming signal.

The detailed spatial structure of the SAT change is, however, complex. Over the Labrador Sea experiment 6 kaBP is up to 1 °C warmer than experiment PREIND (Figure 3.6a), consistent with less sea-ice cover in experiment 6 kaBP (as described by Otto-Bliesner et al., 2006). A similar feature can be seen in some of the PMIP-2 ensemble members, although it is unclear from Braconnot et al. (2007) whether it occurs in the ensemble mean. In contrast, experiment 6 kaBP is colder than experiment PREIND over Scandinavia and the Norwegian Sea (Figure 3.6a), a feature that seems contrary to the alkenone-derived North Atlantic SST records of Rimbu et al. (2003). The colder 6 kaBP SATs over the Norwegian Sea also do not appear to feature in the PMIP-2 runs (Otto-Bliesner et al., 2006; Braconnot et al., 2007). It should, however, be noted that the PMIP-2 6 kaBP simulations are expected to be generally warmer with respect to their corresponding pre-industrial control simulation than the experiment set described here (consistent with the smaller difference in the imposed GHG concentrations between the two time periods in the PMIP-2 project). This difference in the experimental configurations will therefore tend to act to exaggerate the strength of the cold signal in Figure 3.6a relative to that seen in the PMIP-2 simulations.

The annual-mean SAT difference over the eastern Mediterranean region shown in Figure 3.6a is rather weak, particularly over Turkey and the Black and Caspian seas. This is discussed in more detail below in the context of the seasonal cycle.

3.4.3 The seasonal cycle of SAT

While much of the hemisphere-average annual-mean response can be understood in terms of changes in GHG concentrations, the seasonal cycle of the SAT anomalies in the Northern Hemisphere is strongly influenced by the changes in the orbital forcing and their impact on the insolation at the top of the atmosphere. This is clearly shown by the difference between the summer and winter averages of Northern Hemisphere SAT (Figure 3.5a).

The cool early Holocene winters are consistent with the relatively weak insolation during winter – the January insolation at 30° N in experiment 6 kaBP is around 5% lower than in experiment PREIND. The magnitude of this difference is amplified by the lower GHG concentrations, estimated to be associated with an annual-mean cooling of around 0.3–0.5 °C in experiment 6 kaBP. The combined signal is therefore consistent with the strong observed reduction in winter SAT (approx. 0.5 °C in experiment 6 kaBP, Figure 3.5a). In contrast, the summer season experiences increased solar radiation during the early Holocene: the July insolation at 30° N in experiment 6 kaBP is around 5% higher than in experiment PREIND, which acts to oppose the cooler temperatures that would be expected from the lower levels of GHG concentrations. The summer and winter SAT differences over the Northern Hemisphere are therefore consistent with the

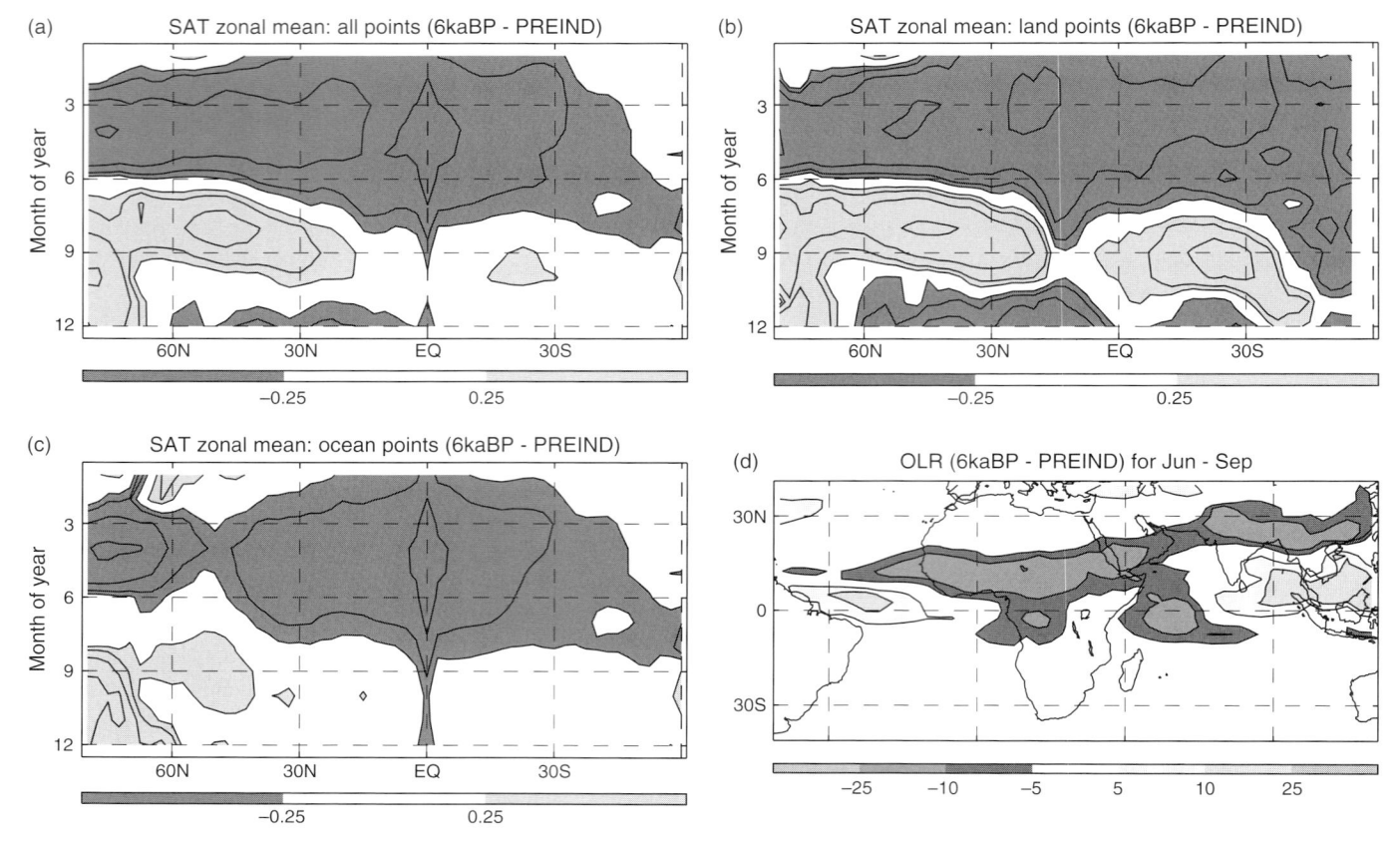

Figure 3.7. The annual cycle of zonal mean SAT anomalies in experiment 6kaBP relative to experiment PREIND. (a–c) Zonal mean SAT anomaly including (a) both ocean and land points, (b) land points only, and (c) ocean points only (units °C). (d) Outgoing longwave radiation anomalies at the top of the atmosphere (boreal summer, for experiment 6kaBP – PREIND, units W m^{-2}). For (a) to (c) contours are at $\pm 0.25, 0.5, 1, 2$ °C.

spread of the PMIP-2 coupled ocean–atmosphere model ensemble, as shown in Braconnot *et al.* (2007) (note that the GHG concentrations used in the PMIP-2 configuration do not change significantly between the pre-industrial and 6 kaBP experiments).

It is also clear from Figure 3.5a that the peak SAT anomalies do not always coincide with the standard meteorological 'seasons' (the annual average line lies closer to the winter average than the summer). Assuming that the climate system's response to the radiative forcing is approximately linear and ignoring any non-linear relationships such as may exist between, for example, cloud cover and temperature, this can be linked to two key factors.

Firstly, the insolation anomalies are not perfectly in phase with the pre-industrial insolation cycle (compare Figure 3.1b with 3.1c). The seasonal progression for the Northern Hemisphere in the mid-Holocene is slightly delayed with respect to the pre-industrial cycle; in other words the positive insolation anomalies lag behind the peak insolation in the Northern Hemisphere, although the situation is somewhat reversed over the Southern Hemisphere. As such, the strongest Northern Hemisphere positive insolation anomalies occur in June–September (after the peak insolation in late June) and the strongest negative insolation anomalies occur in December–April (after the minimum insolation in late December).

The second factor is the memory of the components of the climate system, that is, the rate at which the land and the ocean respond to changes in insolation (in general the ocean will respond more slowly than the land, as discussed below). Figure 3.7a shows the seasonal evolution of the latitudinal response of SAT in experiment 6 kaBP. Comparison between Figure 3.7a and Figure 3.1c shows a number of common features:

- A broad latitudinal range of lower SAT between January and June. This is consistent with the largest negative SAT anomalies lagging the largest negative insolation anomalies.
- A diagonal line of positive anomalies in SAT starting at 60° N in July and reaching 30° S in September/October. This is consistent with the largest positive SAT anomalies lagging the largest positive insolation anomalies. The negative values near the equator (and near 10° N in Figure 3.7b) are associated with increased soil moisture and cloud cover associated with enhanced tropical convection consistent with the enhanced radiative forcing during the boreal

summer period. This is clearly visible in the lower levels of outgoing longwave radiation over tropical Northern Africa (indicating more cloud) in Figure 3.7d and is discussed in more detail in Section 3.5.

The lag between the peak insolation and SAT anomalies is considerably greater over ocean areas than over land areas, consistent with the greater heat content of the ocean surface mixed layer, and is visible when the latitudinal structure of the SAT seasonal cycle is compared over land (Figure 3.7b) and ocean (Figure 3.7c). The large surface heat content of the ocean also ensures that the peak seasonal SAT anomalies over the ocean are typically weaker than those over land. This is a pattern widely reproduced in palaeoclimate models (e.g. Braconnot et al., 2007). This indicates that, although the insolation anomalies experienced over the Holocene are symmetric around a latitude band, the atmospheric circulation response pattern can be highly asymmetric, and the local response can be strongly dependent on the global spatial structure of the surface properties.

The spatial structure of the seasonal cycle of SAT anomalies in the 6 kaBP simulation is shown in Figures 3.6b and 3.6c. As expected from the hemisphere-average picture described above, there are more areas showing cooling than warming, particularly in the tropics where SATs are lower throughout the year. Almost all of the Northern Hemisphere displays a stronger seasonal cycle of SAT, particularly over continental areas, with increases of over 3 °C in the seasonal cycle over much of Asia, Figure 3.6d.

The changes in SAT are, however, smaller in coastal regions, particularly over the eastern side of the mid-latitude ocean basins (for both the seasonal cycle and the annual mean, Figures 3.6a–c). This is consistent with the eastward advection of milder maritime air by the time-mean westerly flow in these regions. Figure 3.6e shows the situation in the eastern Mediterranean region, where mean westerly and southwesterly flow advects relatively warm air from the oceans and seas into the land areas during winter. In summer, mean westerly and northwesterly flow could similarly be expected to advect relatively cool air inland, although this is not apparent in Figure 3.6f. The influence of local orography is visible in both seasons with the low-level flow deflected northwards to the west of the Anatolian Plateau in winter (near 30° E, 38° N) and around the southern flank of the Zagros Mountains (near 45° E 35° N) in both seasons. There is also evidence of a northward deflection as the westerly flow approaches the Jordanian Plateau and Syrian Desert during winter.

On a larger scale, the differences in the large-scale winter circulation patterns in the lower troposphere in experiment 6 kaBP (Figure 3.8b) are consistent with a stronger maritime influence across Southern Europe and the eastern Mediterranean coast. This pattern is consistent with that proposed in the palaeo-data–model comparison described in Bonfils et al. (2004).

However, it appears to be inconsistent with the conclusions of Rimbu et al. (2003), who suggest that there has been a continual weakening of a mean circulation pattern similar to that of the positive phase of the North Atlantic Oscillation (NAO) during the Holocene (i.e. the mean state was more NAO-positive like in the mid-Holocene). A more recent study by Gladstone et al. (2005) finds that the difference in the mean near-surface circulation patterns between the PMIP-2 6 kaBP and pre-industrial simulations is inconsistent between models, with some models showing an NAO-positive like signature at 6 kaBP (e.g. the FOAM, HadCM3 and MIROC models) and others with an NAO-negative like signature (e.g. the CCSM3 model).

Notwithstanding the uncertainties surrounding the model responses, the eastward advection of relatively mild maritime air (particularly during winter) in the eastern Mediterranean acts to mitigate the extremes of summer and winter temperatures over the coastal region. It therefore seems reasonable to suggest that, although the seasonal cycle of SAT throughout much of the Northern Hemisphere was markedly stronger during the early Holocene, it is likely that this enhanced seasonal cycle was somewhat less pronounced in the Levant region.

3.5 THE HOLOCENE: PRECIPITATION

The differences seen in the boreal summer and winter distributions of precipitation will now be discussed separately.

3.5.1 Summer precipitation

Figure 3.9 shows the distribution of summer precipitation simulated by the model in experiment PREIND and how it has changed across the time-slice experiments. Consider first the large-scale changes seen in experiment 6 kaBP relative to the pre-industrial control run (Figure 3.9c).

In experiment 6 kaBP, there is a very large change in the structure of the intertropical convergence zone (ITCZ) over the Atlantic and the West African monsoon system. In particular, the heavy precipitation associated with the ITCZ is shifted northwards in the Atlantic (by approximately 5–10° latitude), and the precipitation over the Sahel region is enhanced by approximately 20–50% (Figures 3.9c and 3.9f). A similar northward shift and intensification of monsoonal rain is also seen over southeast Asia.

These changes in the ITCZ are consistent with the changes in the boreal summer tropical insolation in two ways:

- The 6 kaBP insolation anomalies (relative to pre-industrial levels) are generally stronger in the northern tropics than the southern tropics (Figure 3.1c), consistent with shifting the thermal equator to the north.

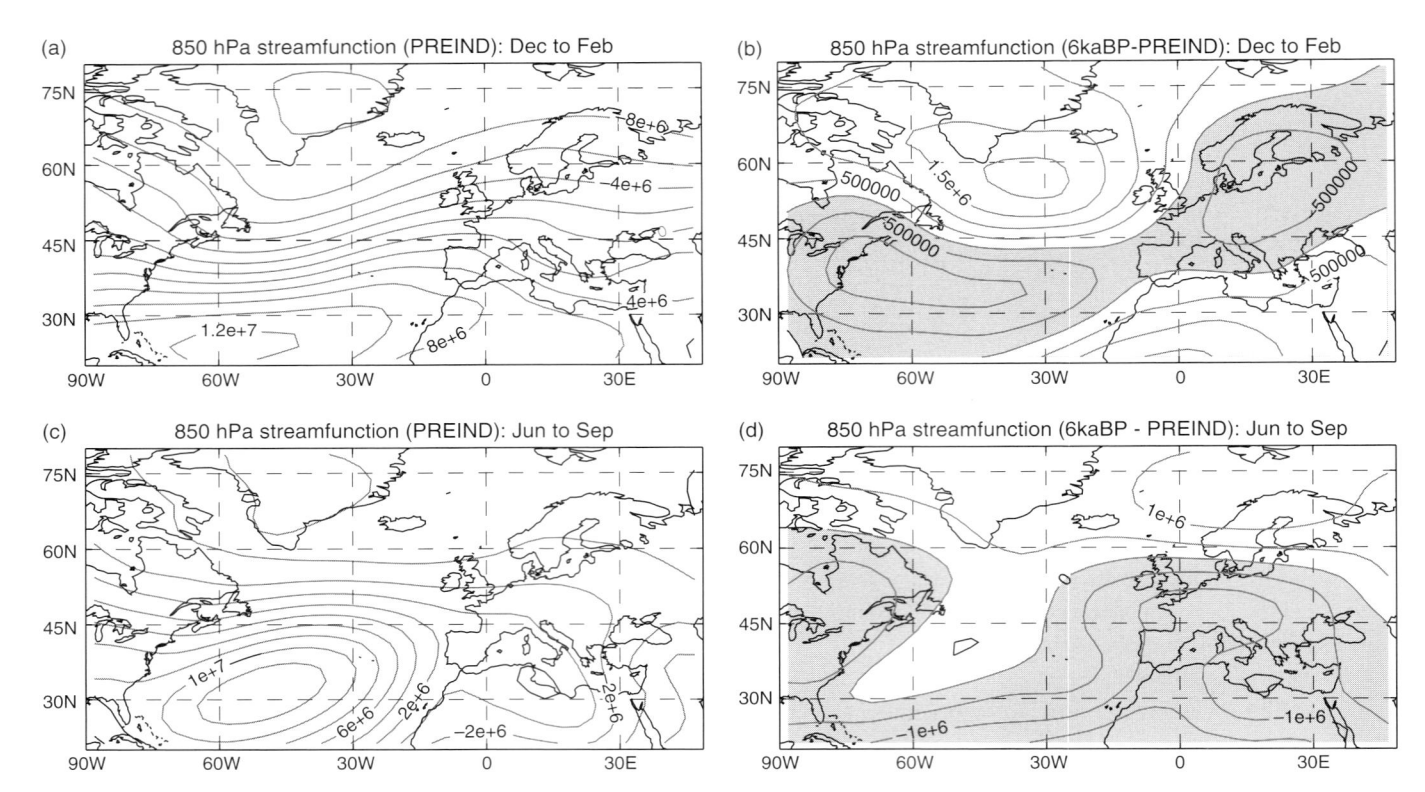

Figure 3.8. The lower tropospheric circulation, as given by the 850 hPa streamfunction. (a) Experiment PREIND during December–February. (b) Difference between experiments 6kaBP and PREIND during December–February; shaded areas indicate negative values. (c) As (a) but for June–September. (d) As (b) but for June–September. The circulation is along streamlines and is cyclonic (anticlockwise) around negative values. The contour interval is the same in (a) and (c), and is four times greater than that in (b) and (d).

- The stronger seasonal cycle of insolation in experiment 6 kaBP will, in the absence of clouds and soil moisture, lead to a stronger land–sea surface temperature contrast and therefore stronger ascent over the land. This is consistent with an enhancement of the climatological onshore flow in the West African monsoon and precipitation over the continent. It should be noted, however, that in the time mean the increased cloud cover and soil moisture associated with the precipitation is consistent with reduced summer SAT over the affected region, as shown in Figures 3.6c and 3.7b (also the reduced outgoing longwave radiation in Figure 3.7d).

The northward offset of the ITCZ and the extension of the West African monsoon are strongest in the Early Holocene experiments (appearing to peak in experiment 10 kaBP, Figure 3.9f). This is consistent with what would be expected from the insolation anomalies (which reach their peak relative to pre-industrial levels at around 8–10 kaBP, Figure 3.1d). It should, however, be noted that in the Early Holocene experiments (time periods 8–12 kaBP), the ocean heat flux error (described in Section 3.2.5) is likely to enhance this signal as it will project onto the Atlantic SST dipole pattern (cool south tropical Atlantic and a warm north tropical Atlantic) discussed by, for example, Braconnot *et al.* (2007).

The extension of the West African monsoon during the early to mid-Holocene has been examined in many climate modelling experiments (not least the participants in the PMIP-2 model intercomparison project; Braconnot *et al.*, 2007) and is seen in palaeo-observations documenting the extent of vegetation and lake-level changes over North Africa (e.g. Jolly *et al.*, 1998; Drake and Bristow, 2006; Wanner *et al.*, 2008).

Several studies (e.g. Claussen *et al.*, 1999; Texier *et al.*, 2000; Braconnot *et al.*, 2007) have examined the role of vegetative feedback processes on the monsoon extension. In this process, increased vegetation over North Africa leads to lower surface albedo and therefore stronger surface heating. This heating drives ascent and stronger onshore winds carrying moisture from the tropical Atlantic. The atmospheric moisture is precipitated out in the ascent region, providing the soil moisture to maintain the vegetation.

The impact of changes in the vegetation over North Africa is examined in experiments 6 kaBP-WS and 8 kaBP-WS (compare Figures 3.9c and 3.9e, also see Figure 3.9f). Experiment 6 kaBP-WS has markedly larger increases in precipitation over the Sahel region than experiment 6 kaBP (experiment 6 kaBP-WS shows a nearly 120% increase in experiment PREIND compared with a 40% increase in experiment 6 kaBP; Figure 3.9f). This increase

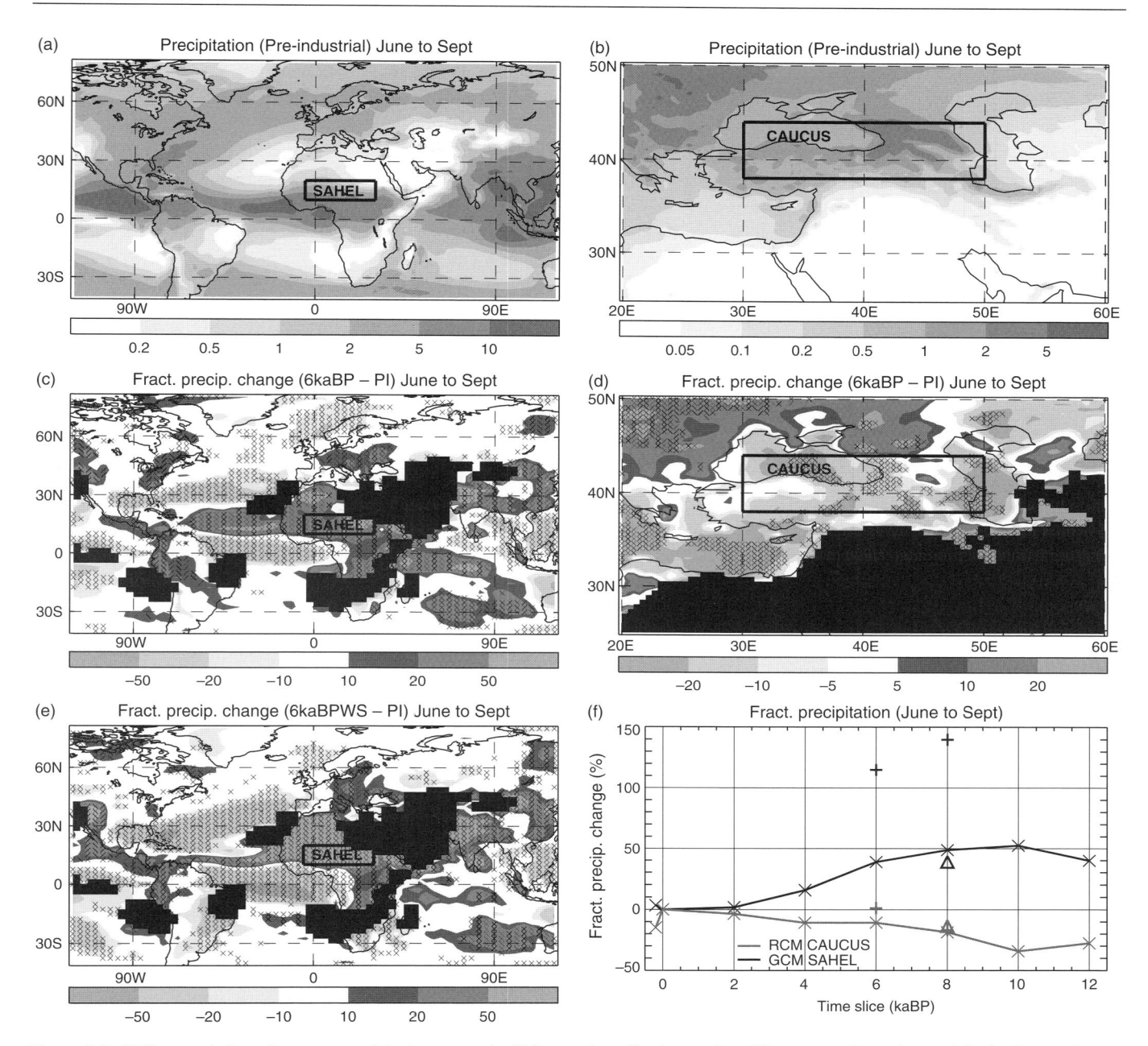

Figure 3.9. Differences in boreal summer precipitation across the Holocene time-slice integrations. The top row shows the precipitation in experiment PREIND (units mm day^{-1} using data from (a) the global model and (b) the regional model. The middle row shows the fractional change in precipitation (units %) in experiment 6kaBP relative to experiment PREIND (i.e. [6kaBP − PREIND] × 100/PREIND), using data from (c) the global model and (d) the regional model. Panel (e) is similar to (c) but for experiment 8kaBP-WS. Panel (f) shows the fractional precipitation changes averaged over the SAHEL box (in the global model, as marked in panels (a) and (c)) and the CAUCUS box (in the regional model, as marked in panels (b) and (d)) in the time-slice experiments. Data points from the time-slice experiments are marked by × symbols whereas the + symbols mark data points from experiments 6kaBP-WS and 8kaBP-WS and triangles mark data points from experiment 8kaBPNOICE (experiment PRESDAY is shown at time = −0.2 kaBP). In panels (c) to (e), areas where the changes are statistically significant at the 90% level are marked with black crosses. Areas of extremely low precipitation (less than 0.2 mm day^{-1} for the global model and 0.05 mm day^{-1} in the regional model) in experiment PREIND are blacked out in panels (b) to (e). See colour plate section.

is even more marked in the 8 kaBP experiments (a 140% increase in pre-industrial in experiment 8 kaBP-WS compared with a 50% increase in experiment 8 kaBP; Figure 3.9f), consistent with the stronger insolation anomalies in this earlier time period.

The north- and northeastward extent of the ITCZ and West African monsoon does not, however, extend into the Mediterranean region in any of the simulations performed here. To demonstrate this it is important to examine two key regions more closely: Northwest Africa (the area around Morocco and Algeria) and the eastern Mediterranean.

With regard to Northwest Africa, Figure 3.9e suggests that there is an increase in precipitation in experiment 6 kaBP-WS compared with experiment PREIND (over 50%). This should be treated with some caution as it occurs on a relatively low baseline amount in experiment PREIND (less than 0.5 mm day^{-1}; Figure 3.9a) and summer precipitation in that region is overestimated in experiment PRESDAY when compared with reanalysis or observational data (e.g. ERA-40 and GPCC, Figure 3.3). Furthermore, even if the precipitation increase is a realistic feature, it is unlikely that the precipitation has a tropical moisture source (as is the case in the ITCZ) as the time-mean low-level flow over this area is generally not from the south (Figure 3.8c). The precipitation over Northwest Africa is therefore not simply a direct northward extension of the ITCZ and West African monsoon system.

The regional downscaling over the eastern Mediterranean region is shown in Figure 3.9b and 3.9d. It is clear from these figures that much of the eastern Mediterranean and the Middle East experiences a decrease in precipitation (20–50% over the Caucus region marked in Figure 3.9d; see also Figure 3.9f), from a relatively low starting point (1–2 mm day^{-1} in experiment PREIND; Figure 3.9b).

The relatively low levels of precipitation over the eastern Mediterranean in experiment PREIND are consistent with both the latitudinal and longitudinal position of this region. Latitudinally, the eastern Mediterranean lies within the descending air associated with the subtropical high pressure belt (Figure 3.8c), which tends to inhibit the vertical motion required for convection and tropical precipitation. In terms of longitude, the ascent over Southeast Asia associated with the Indian monsoon is also consistent with enhanced descent over the eastern Mediterranean, further suppressing convection and precipitation, as per the monsoon–desert mechanism described by Rodwell and Hoskins (2001). Although the former of these two factors can perhaps be expected to be weaker in the early Holocene because the precipitation minima in the subtropical highs tend to receive slightly more rainfall under lower GHG concentrations (cf. experiment PREIND – PRESDAY in Figures 3.4d and 3.4f), the latter is intensified: precipitation over Southeast Asia is increased by 20–50% in experiment 6 kaBP relative to experiment PREIND (Figure 3.9c). This increase in the strength of the Indian monsoon is consistent with enhanced land–sea contrasts over Southeast Asia, driven by changes in insolation.

When considered as a whole, these simulations indicate that it is unlikely that the particular forcing mechanisms examined (orbital, land ice-sheets, sea-level change, North African vegetation change) are capable of producing significant summer precipitation over the Jordan Valley region during the early to mid-Holocene. This is consistent with the interpretation provided by Arz *et al.* (2003) and the results of the PMIP-2 modelling project (e.g. Braconnot *et al.*, 2007). The results therefore suggest that care should be taken to acknowledge the specific meteorological situation of the eastern Mediterranean when interpreting the northward extensions of the ITCZ, particularly in the context of schematic palaeosynthesis maps (such as Figure 18 of Wanner *et al.* (2008)), which sometimes indicate that the ITCZ reached 35° N during the mid-Holocene in the eastern Mediterranean.

3.5.2 Winter precipitation: large-scale patterns

Figure 3.10 shows the large-scale distribution of boreal winter precipitation simulated by the global model in experiment PREIND and the changes across the time-slice experiments. These changes are discussed first before examining the detailed regional structure over the Mediterranean. (It should be noted here that impact of the error in the ocean heat flux convergence is discussed separately in Section 3.5.4. This error is not, however, believed to affect the general patterns of change in the present discussion.)

A number of robust large-scale features are identifiable in these experiments (all anomalies are expressed relative to experiment PREIND):

- Precipitation over tropical continents is lower (up to 10–20% in experiment 6 kaBP; Figure 3.10b), while over the tropical oceans there are mixed signals (over the tropical Indian Ocean, precipitation is greater by around 20% in experiment 6 kaBP, Figure 3.10b).
- Precipitation averaged over the whole Mediterranean area is somewhat stronger (up to 10–15% in some areas in experiment 6 kaBP, Figure 3.10b, and in the area average by 5–10% in the Early Holocene experiments, Figure 3.10f) while precipitation over the high-latitude North Atlantic and Northern Europe is weaker (2–5% in the area average, Figure 3.10f). This pattern is consistent with the interpretation of palaeoreconstruction in Cheddadi *et al.* (1997) and the interpretation given by Bonfils *et al.* (2004).

The first of these – the relative weakening of tropical precipitation over land areas compared with over the ocean – can be understood as a direct response to the insolation anomalies in the tropics: negative insolation anomalies in the tropics during boreal winter (Figure 3.1) are consistent with a weakening of the land–sea contrast. In other words, the surface heating over the land areas is reduced relative to that over the ocean, leading to relatively less ascent and precipitation over land and more over the ocean.

Although this response occurs primarily in the deep tropics, it is capable of affecting remote regions. For example, increased precipitation over the tropical Indian Ocean in Figure 3.10 is

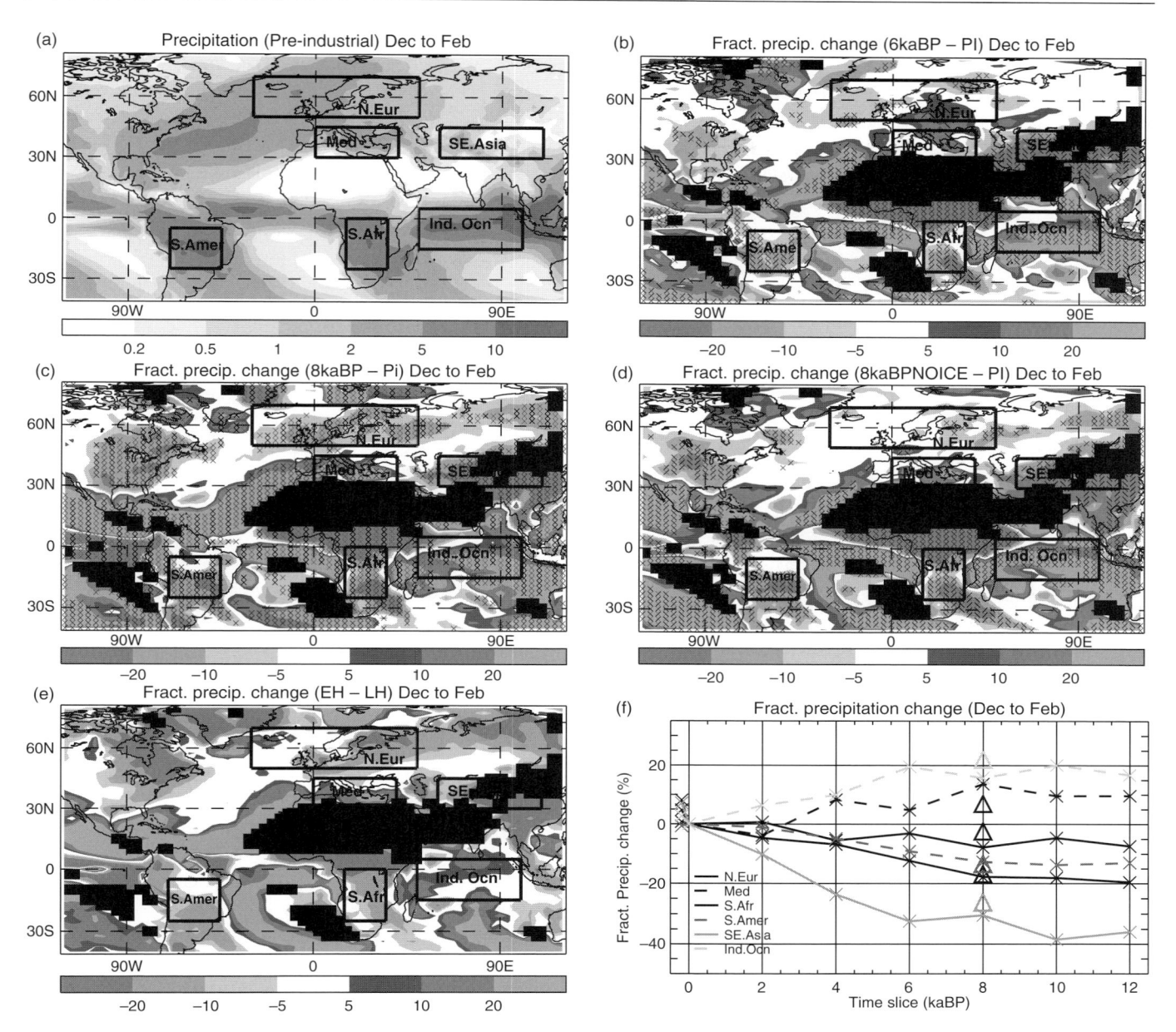

Figure 3.10. Differences in boreal winter precipitation across the Holocene time-slice integrations using the global model. (a) Experiment PREIND (units mm day^{-1}). (b) The fractional change in precipitation (units %) in experiment 6kaBP relative to experiment PREIND (i.e. [6kaBP − PREIND] × 100/PREIND). (c) As (b) but for experiment 8kaBP. (d) As (b) but for experiment 8kaBPNOICE. (e) As (b) but for Early Holocene experiments (8kaBP + 10kaBP + 12kaBP) minus the Late Holocene experiments (2kaBP + 4kaBP + 6kaBP). Panel (f) shows the fractional precipitation changes averaged over the boxes marked in panels (a) to (e). Data points from the time-slice experiments are marked by × symbols whereas triangles mark data points from experiment 8kaBPNOICE (experiment PRESDAY is shown at time = −0.2 kaBP). In panels (b) to (e), areas where the changes are statistically significant at the 90% level are marked with black crosses. Areas of extremely low precipitation (less than 0.2 mm day^{-1}) in experiment PREIND are blacked out in panels (c) and (d). See colour plate section.

associated with reduced precipitation over large areas of Central to Southeast Asia (Barlow et al., 2005). Inspection of Figure 3.10 suggests a similar relationship exists in the precipitation patterns in the time-slice experiments described here (e.g. see the 'Ind. Ocean' and 'SE. Asia' boxes in Figure 3.10), with the influence perhaps extending westward to the area of Iran.

The second feature (precipitation changes over the Mediterranean and Europe) is consistent with changes in the North Atlantic storm track, and its downstream extensions over Europe and the Mediterranean.

The characteristics of the North Atlantic storm track are illustrated in Figure 3.11. Although the patterns of the differences

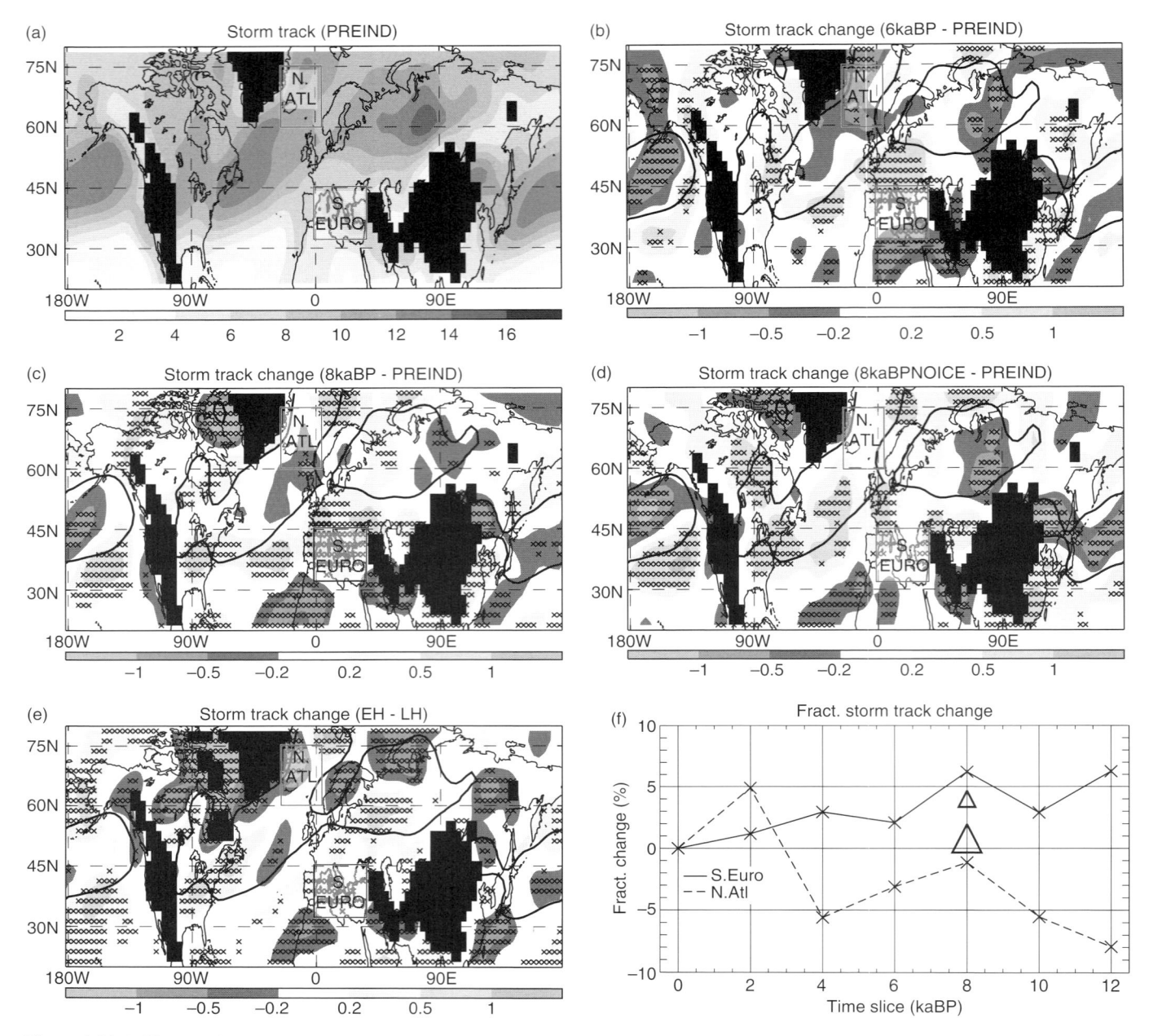

Figure 3.11. Differences in boreal winter storm tracks across the Holocene time-slice integrations using the global model. (a) The storm track in experiment PREIND (units of storms per month passing through a 5° spherical cap). (b) Storm track difference (expt 6kaBP − PREIND). (c) As (b) but for experiment 8kaBP. (d) As (b) but for experiment 8kaBPNOICE. (e) As (b) but for Early Holocene experiments (8kaBP + 10kaBP + 12kaBP) minus the Late Holocene experiments (2kaBP + 4kaBP + 6kaBP). (f) The fractional storm track changes averaged over the boxes marked in panels (a) to (e). Data points from the time-slice experiments are marked by × symbols (experiment PRESDAY is not shown). In panels (b) to (e), areas where the differences are statistically significant at the 90% level are marked with black crosses, and areas of high orography (in excess of 1,200 m) are blacked out. Thick black contours in (b) to (e) show the 10 storms per month contour from experiment PREIND. Prior to display, the storm track diagnostics have been smoothed to improve readability.

across the experiments are rather noisy, it is possible to see in the earlier time-slice experiments that the storm track is weaker in the northern and stronger in the southern part of the eastern North Atlantic relative to experiment PREIND. This shift in the position of the storm track is perhaps most clear in Figures 3.11b and 3.11c, but the general trend can also be seen in the difference between the North Atlantic (N. Atl) and Southern Europe (S. Euro) boxes in Figure 3.11f.

This is consistent with an intensification of storm activity in the Mediterranean area (up to 5–10% stronger in experiment

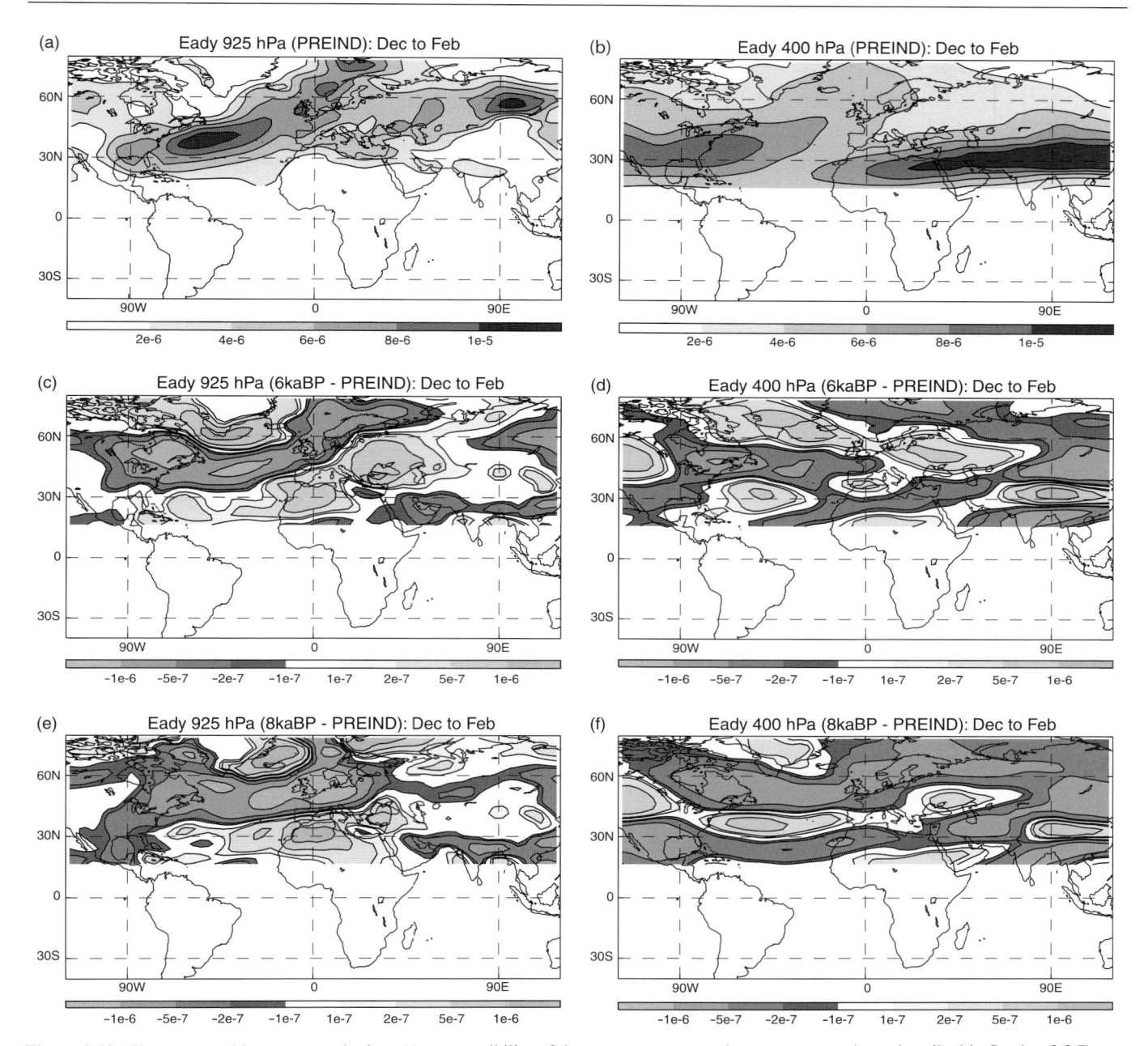

Figure 3.12. The upper and lower tropospheric σ (the susceptibility of the mean state to weather system growth, as described in Section 3.2.7). (a) Values of σ in experiment PREIND (units s^{-1}) at 925 hPa. (b) As (a) but for 400 hPa. (c) Difference in σ between experiments 6kaBP and PREIND at 925 hPa. (c) As (b) but at 400 hPa. (e) As (c) but for experiment 8kaBP. (f) As (d) but for experiment 8kaBP. Data are only shown for the Northern Hemisphere extratropics and the contours have been lightly smoothed to improve readability.

8 kaBP than experiment PREIND, Figure 3.11f). The spatial structure of this large-scale change in the track corresponds very well with the precipitation changes seen over the Atlantic, European and Mediterranean sector (compare Figures 3.10 and 3.11), consistent with the expectation that weather system/storm activity is the dominant source of precipitation for the mid-latitudes during the winter season. The drivers controlling the changes in the storm track are, however, rather complex (a fuller discussion can be found in Brayshaw et al., 2010).

Figure 3.12 shows the maximum Eady growth rate, σ, a measure of the susceptibility of the mean state to the growth of weather systems as described in Section 3.2.7. This is given in both the upper and lower troposphere for the Atlantic and European sector. Consider first the changes in the lower tropospheric σ in experiment 6 kaBP compared with experiment PREIND.

In the storm track region (approximately 40–60° N over the western North Atlantic Ocean), the lower tropospheric σ is reduced (Figure 3.12c). This is consistent with the changes in the SAT shown in Figure 3.6c (warm water to the north and cold to the south reduces the meridional gradient of surface temperature), which is in turn consistent with the latitudinal changes in insolation (i.e. stronger negative insolation anomalies at low latitudes than high latitudes in Figure 3.1c). In itself, this lower tropospheric signal could be expected to lead to a weaker storm track which is perhaps shifted slightly southwards as the storms became more closely associated with the upper tropospheric baroclinicity on the subtropical jet (Brayshaw et al., 2008).

However, σ is also modified in the upper troposphere (Figure 3.12d). In particular, the pattern over the North Atlantic in Figure 3.12d is such that σ is lower than that in experiment PREIND at 25° N and 45° N and greater at 30–35° N and 60° N. In isolation, this could perhaps be expected to lead to a weaker storm track which is more closely associated with the lower tropospheric baroclinicity caused by the mid-latitude surface temperature gradients (Brayshaw et al., 2008).

These two drivers (upper and lower tropospheric baroclinicity) therefore both act to give a weaker storm track. Examination of the storm track changes in the time-slice experiments (Figure 3.11) suggest that, over the eastern North Atlantic, the storm track response is a tripole pattern consistent with the storm track being generally weaker in the early Holocene but also broader on both the northern and southern flanks (e.g. Figures 3.11b and 3.11c). Similar patterns of change are also visible in the North Pacific, particularly in the eastern part of the basin (e.g. Figure 3.11b). This broadening of the North Atlantic storm track in the earlier time periods is perhaps further enhanced by stronger land–sea contrast across the east coast of North America during winter (Brayshaw et al., 2009; visible here in the southwest–northeast line of enhanced storm activity near 60° W, 40° N in Figures 3.11b to 3.11e).

The differences between the storm track in experiment 6 kaBP and that in experiment PREIND are consistent with weakened lower tropospheric westerly winds in the storm track region around 45° N and enhanced westerlies to the south (Edmon et al., 1980; Hoskins et al., 1983). This pattern is clearly visible in Figure 3.8b, as has already been discussed. It agrees well with the palaeodata interpretations of Bonfils et al. (2004) over Europe but tends to project onto a negative NAO-like state (in contrast to the positive NAO-like state described by Rimbu et al., 2003).

Over the northwest part of the North Atlantic there are also marked differences in the storm tracks between the mid- to late Holocene experiments (2 kaBP, 4 kaBP, 6 kaBP) and the Early Holocene experiments (8 kaBP, 10 kaBP, 12 kaBP), consistent with the storm track being disrupted by the greater extent of the land ice-sheets over Greenland and northeastern Canada in the early Holocene (Figure 3.11e). It is, however, difficult to completely separate the influence of the land ice-sheets from the affects of the model configuration error (as described in section 3.2.5).

3.5.3 Winter precipitation: the eastern Mediterranean

The patterns of precipitation and storm track changes in the Mediterranean obtained from the regional simulations are shown in Figures 3.13 and 3.14.

As discussed previously, the global simulations indicate that the Mediterranean region experienced a drying trend from the early to the late Holocene. Consistent with this, the Mediterranean storm track simulated by the regional model does indeed weaken between the Early Holocene experiments and experiment PREIND (Figure 3.14f). However, the detailed spatial distribution of precipitation changes is rather more complex. Figure 3.13 provides an indication of the simulated changes in precipitation changes across the time-slice experiments.

In the 4 kaBP and 6 kaBP experiments the differences in the eastern Mediterranean can be separated into two main areas:

- In the north and west (Greece, Turkey; the Anatolia box in Figure 3.13b) there is generally more precipitation than experiment PREIND, consistent with the more southward position of the storm track in the North Atlantic and European sectors in the earlier experiment (as described in section 3.5.2).
- In the south and east (Iraq, Iran, northern Saudi Arabia) there is less precipitation (shown for experiment 6 kaBP in Figure 3.13b), perhaps consistent with the influence of stronger precipitation over the Indian Ocean (as described in section 3.5.2).

In the earlier time-slice experiments (8 kaBP, 10 kaBP and 12 kaBP), the pattern of differences with respect to experiment PREIND is, however, somewhat changed (the impact of the error in the slab ocean model's heat flux convergence (section 3.2.5) on these results is discussed in section 3.5.4). In these experiments there is more precipitation over the Mediterranean (Figures 3.13c and 3.13e), extending northwards and eastwards into the continental areas, consistent with the inland advection of moisture by the seasonal-mean westerly winds in the lower troposphere (Figure 3.6e). As such, the detailed orography of the region plays a key role in determining the distribution of precipitation, with stronger precipitation anomalies over the upstream (southern and western) side of the Zagros mountains and Arabian Plateau clearly defining the Fertile Crescent region (Figures 3.13c and 3.13e).

Figure 3.13f shows how this transition between the two characteristic patterns of precipitation differences affects a variety of locations in the eastern Mediterranean across the Holocene. This figure suggests that:

- The eastern areas ('Saudi' and 'Highland') experience a gradual increase in precipitation across the Holocene period.
- The northern part of the eastern Mediterranean ('Anatolia') experiences a gradual fall in precipitation across the Holocene period.

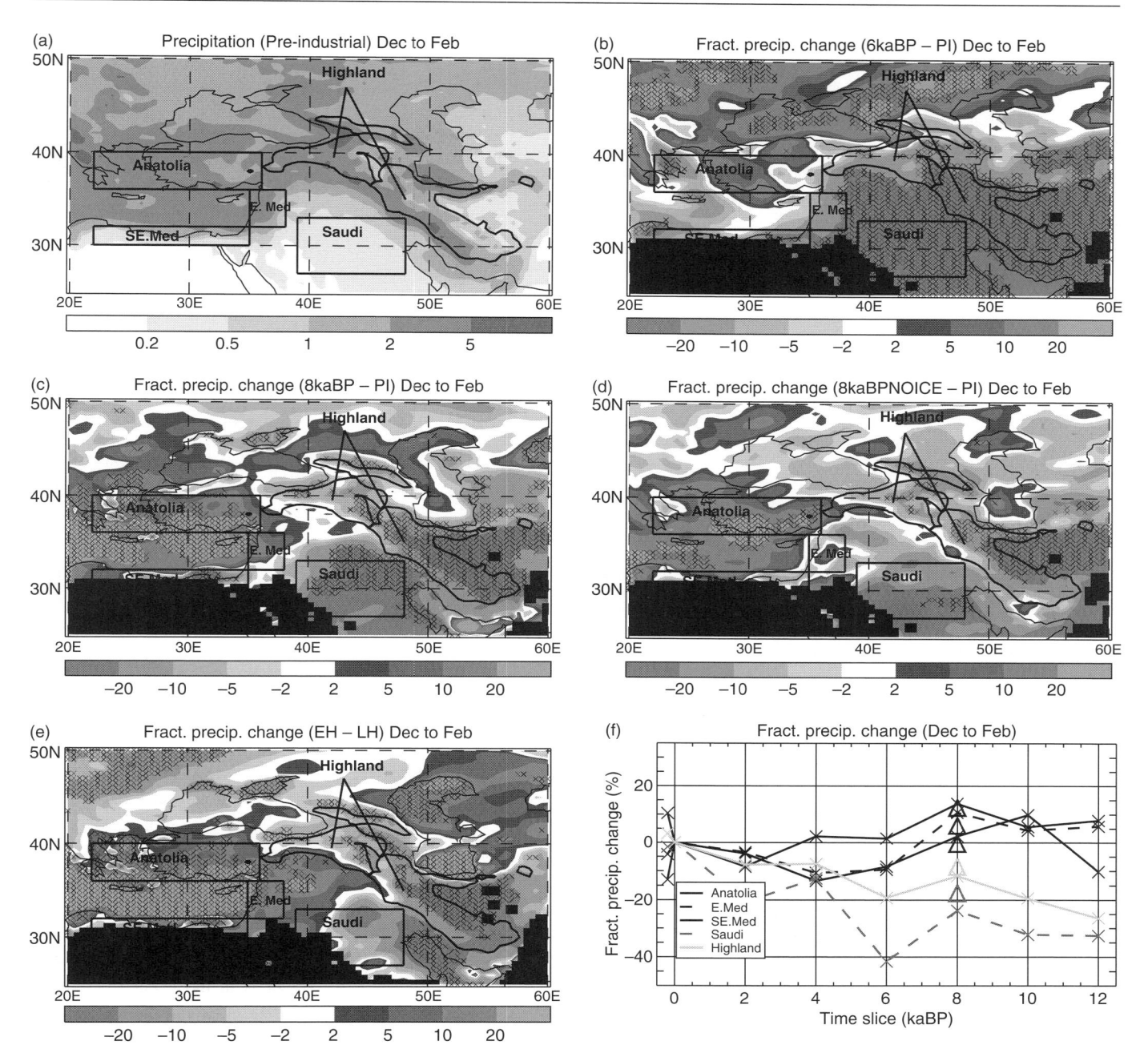

Figure 3.13. Differences in boreal winter precipitation across the Holocene time-slice integrations from the regional model. (a) Experiment PREIND (units mm day^{-1}). (b) The fractional change in precipitation (units %) in experiment 6kaBP relative to experiment PREIND (i.e. [6kaBP − PREIND] × 100/PREIND). (c) As (b) but for experiment 8kaBP. (d) As (b) but for experiment 8kaBPNOICE. (e) as (b) but for Early Holocene experiments (8kaBP + 10kaBP + 12kaBP) minus the Late Holocene experiments (2kaBP + 4kaBP + 6kaBP). (f) The fractional precipitation changes averaged over the boxes marked in panels (a) to (e). Data points from the time-slice experiments are marked by × symbols, whereas triangles mark data points from experiment 8kaBPNOICE (experiment PRESDAY is shown at time −0.2 kaBP). In panels (b) to (e), areas where the differences are statistically significant at the 90% level are marked with black crosses. Areas of extremely low precipitation (less than 0.2 mm day^{-1}) in experiment PREIND are blacked out in panels (c) and (d). See colour plate section.

- The eastern coastline of the Mediterranean ('E. Med', and possibly extending inland through the Levant and perhaps eastward into the Fertile Crescent) experiences high precipitation in the early Holocene, before dropping off rapidly to low values in the mid- to late Holocene and gradually recovering to moderate pre-industrial levels (this change is discussed in more detail in section 3.5.4).

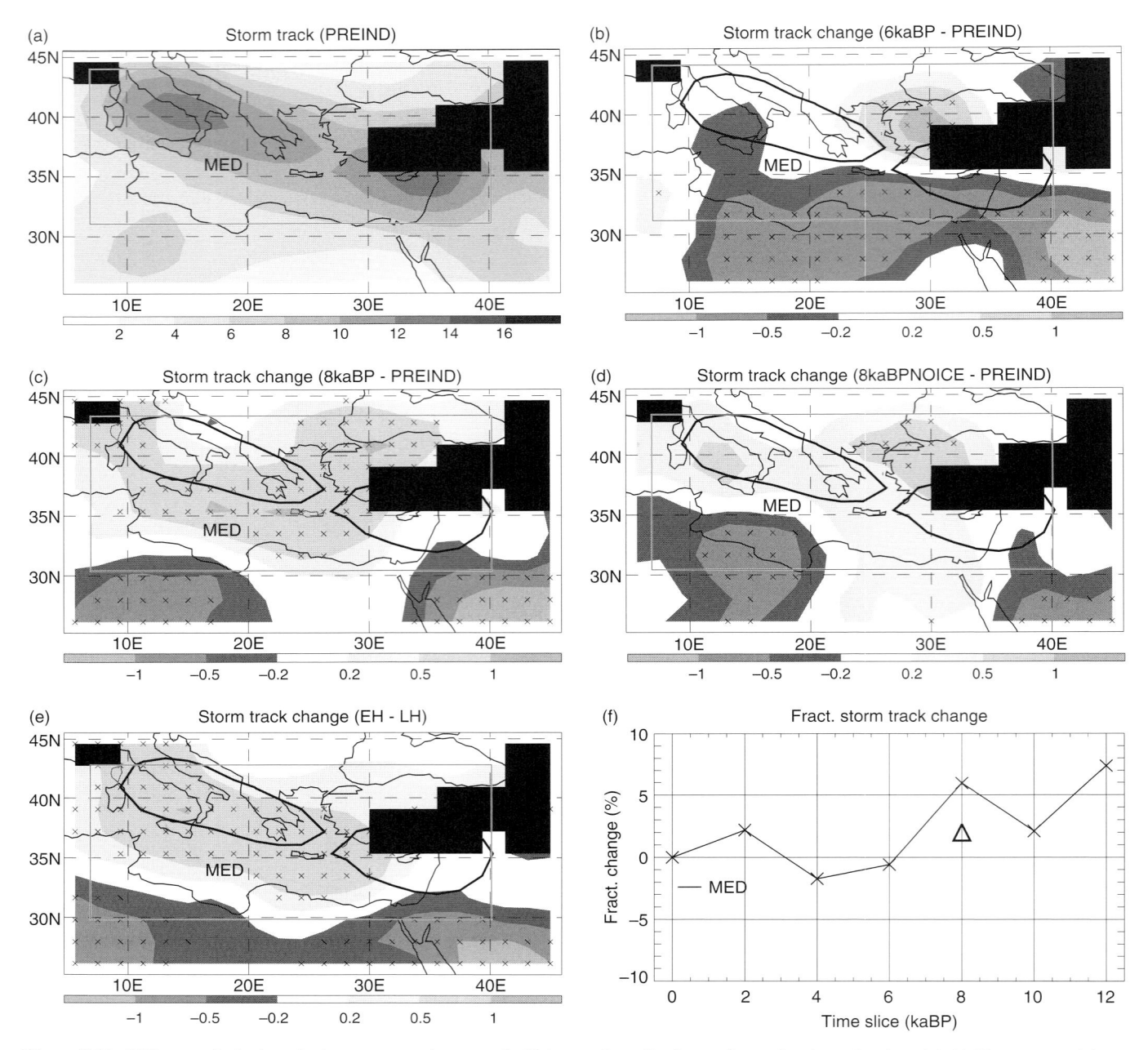

Figure 3.14. Differences in the boreal winter storm tracks across the Holocene time-slice integrations using the regional model. (a) The storm track in experiment PREIND (units of storms per month passing through a 5° spherical cap). (b) Storm track difference (expt 6kaBP − PREIND). (c) As (b) but for experiment 8kaBP. (d) As (b) but for experiment 8kaBPNOICE. (e) As (b) but for the average of the Early Holocene experiments (8kaBP + 10kaBP + 12kaBP) minus that for the Late Holocene experiments (2kaBP + 4kaBP + 6kaBP). (f) The fractional storm track differences averaged over the box marked in panels (a) to (e). Data points from the time-slice experiments are marked by × symbols (experiment PRESDAY is not shown). In panels (b) to (e), areas where the changes are statistically significant at the 90% level are marked with black crosses and areas of high orography (in excess of 1,200 m) are greyed out. Thick black contours in (b) to (e) show the 10 storms per month contour from experiment PREIND. The storm tracking analysis is performed on a coarse grid (hence the extremely coarse orographic features shown), and prior to display here the storm track diagnostics have been further smoothed to improve readability.

- The southeastern coastline ('S.E. Med') shows similar behaviour to the eastern coastline but the signal is less clear.

Figure 3.14 shows the differences in the Mediterranean storm track associated with these time-slice experiments using the regional model. In particular, it is possible to see a marked difference between the situation in the 6 kaBP and 8 kaBP experiments (Figures 3.14b and 3.14c). In experiment 6 kaBP the storm track is weakened in the southern part of the Mediterranean basin, particularly approaching the eastern coastline, and strengthened in the north over western Turkey (indeed in experiments 2 kaBP, 4 kaBP and 6 kaBP there is relatively little change in the storm track averaged over the whole Mediterranean basin, Figure 3.14f). Experiment 8 kaBP, by contrast, has a stronger storm track throughout the whole ocean basin. The difference between these two responses is clear between the average of the Early and Late Holocene experiments (Figure 3.14e), with the storm track averaged over the whole basin being generally stronger in the Early Holocene experiments than in the Late Holocene experiments (Figure 3.14f). The changes in the structure of the Mediterranean storm track in the regional model are therefore consistent with the precipitation signals seen in Figure 3.13.

3.5.4 Winter precipitation: the Jordan Valley during the mid-Holocene

This change in precipitation between the Early Holocene and Late Holocene experiments is potentially of considerable significance to the interpretation of archaeological and palaeological records in the Jordan Valley region. Although a full discussion of the relationship of this model signal to regional palaeorecords is reserved until Chapter 8, it is worth describing here a dynamical mechanism that could potentially account for the change.

As discussed in section 3.5.2, the major Northern Hemisphere winter storm tracks (Atlantic and Pacific) are generally weaker and shifted southwards in the early Holocene experiments relative to experiment PREIND. Furthermore, as one of the primary drivers of this is the weaker latitudinal gradient of boreal winter insolation during the early Holocene, it is reasonable to expect that the storm tracks may be at their weakest and with their most southerly extent in experiments 8 kaBP and 10 kaBP (there is some indication of this in Figures 3.11e and 3.11f). The relatively low levels of GHG concentrations in the early Holocene would also be expected to enhance this southward offset further (Yin, 2005). This southward shift could be expected to lead to greater feeding of synoptic disturbances into the Mediterranean.

In addition to this, the land–sea surface temperature contrast over the northern coast of the Mediterranean is stronger in the early Holocene experiments than in experiment PREIND (consistent with the greater thermal inertia of the ocean mixed layer; see Figure 3.6b and the discussion in section 3.4). This is consistent with enhanced near-surface σ in the Mediterranean region, as visible in Figures 3.12d and 3.12e. In the upper troposphere, there are also suggestions of increased σ over the northern Mediterranean in the early Holocene (Figures 3.12d and 3.12f) associated with changes in the tropical circulation (due to the weaker insolation during boreal winter in the tropics in the early Holocene).

Each of these three factors (feeding of disturbances, lower tropospheric σ, and upper tropospheric σ) could be expected to enhance storm activity over the Mediterranean area during the early Holocene. This, therefore, leads to the intriguing possibility that a combination of small, gradual changes in each of these three factors could have led to a relatively large, fast transition in the strength of the Mediterranean storm track around 6–8 kaBP.

It should be noted that the ocean heat flux error (described in section 3.2.5) does influence the change between experiments 6 kaBP and 8 kaBP. In particular, the cooling over the southern tropical oceans associated with the error in experiment 8 kaBP is consistent with exacerbating the changes in the tropical circulation relative to experiment 6 kaBP which can be expected to exaggerate the change in the storm tracks. It is therefore relevant to examine the responses in experiment 8 kaBPNOICE (which does not include the ocean heat flux error), and note that a similar but weaker change in the storm tracks and precipitation does occur between experiments 8 kaBPNOICE and 6 kaBP as between experiments 8 kaBP and 6 kaBP. In particular, Figure 3.13 shows that in both experiments 8 kaBP and 8 kaBPNOICE the precipitation is more concentrated over the Mediterranean basin than in experiment 6 kaBP in the regional model. Similarly, Figure 3.14c shows that the character of the anomalies in the Mediterranean storm track in experiment 8 kaBPNOICE lies somewhere between the responses seen in experiments 6 kaBP and 8 kaBP.

It is, however, more difficult to see a 'transition' of this type in the global model (Figures 3.10 and 3.11), although there is a suggestive eastward extension of the Mediterranean precipitation across its eastern coastline in the 8 kaBP and 8 kaBPNOICE experiments (Figures 3.10c and 3.10d).

3.6 DISCUSSION AND CONCLUSIONS

A series of time-slice sensitivity tests using a consistent set of global and regional atmospheric models has been performed. These experiments investigate the climate response in the Mediterranean region relating to a range of climatological 'forcing' conditions experienced over the Holocene period and up to present day.

The following general patterns have been observed (and are summarised in Table 3.2):

- GHG changes have led to a gradual rise in annual-mean SATs since the early Holocene, with a particularly rapid increase since the 'pre-industrial' period. The signal is relatively uniform spatially and is broadly consistent with a wide body of previous literature.
- The Northern Hemisphere seasonal cycle of SAT has weakened since the early Holocene, particularly over land areas, primarily associated with changes in insolation

Table 3.2 *Summary of the changes in the time-slice experiments (expressed relative to the pre-industrial control run)*
Key: EH = Early Holocene; LH = Late Holocene; Str = strength, Pos = position; NH = Northern Hemisphere, AP = Arabian Plateau; S = south; N = north; + = increase; − = decrease; 0 = no change. Multiple symbols indicate the relative strength of the change; * denotes that the change can be considered as a spreading out of the storm track rather than a simple southward shift.

	Storm tracks				SAT		Precipitation				
	North Atlantic		Mediterranean		Seasonal cycle		Summer	Winter			
Expts.	Str.	Pos.	Str.	Pos.	NH continents	East Med Coast	East Med	All Med	Anatolia	East Med Coast	AP
LH	−	S*	+	N	++	+	0	+	+	−	−
EH	−	SS*	++	S	+++	++	0	++	++	+	−

due to variations in the Earth's orbital parameters. The change in the seasonal cycle of SAT is somewhat weaker over coastal regions (particularly the west coast of the mid-latitude continents), consistent with the inland advection of milder air from the oceans. The meteorological proximity of the Levant region and the Jordan Valley to the eastern Mediterranean is therefore likely to have acted to ameliorate the harsh seasonal SAT cycle in the early Holocene.

- The West African monsoon and the ITCZ in boreal summer have reduced in intensity and shifted southwards between the early Holocene and the pre-industrial period (consistent with the changes in the top of the atmosphere insolation). This change is reinforced over the Sahel region by the die-back of green vegetation there, and is consistent with a wide variety of palaeo-observations and modelling studies. The West African monsoon and the ITCZ do not, however, extend into the eastern Mediterranean in any of the simulations performed here, consistent with the monsoon–desert mechanism proposed by Rodwell and Hoskins (2001).
- The boreal winter Northern Hemisphere storm tracks have intensified between the early Holocene and pre-industrial conditions and the North Atlantic storm track has shifted northwards, away from the Mediterranean region. This appears to be consistent with changes in both upper and lower tropospheric baroclinicity (σ) driven by insolation changes and the palaeodata interpretation of Bonfils et al. (2004), but contrasts with other studies such as Rimbu et al. (2003); Rimbu et al. (2004).
- The Mediterranean storm track has weakened from the early Holocene to pre-industrial times, which is physically consistent with the changes observed in the North Atlantic storm track and also reduced land–sea contrast at the Mediterranean's northern coastline.
- Precipitation over the Mediterranean during boreal winter (December to February) has generally reduced between the early Holocene and the pre-industrial period, consistent with the Mediterranean storm track changes.

Although the global model clearly simulates a reduction in precipitation over the Mediterranean through the Holocene (Figure 3.13), the detailed regional spatio-temporal patterns over the eastern Mediterranean are complex. In particular, there is some evidence to suggest that the precipitation may have reduced relatively swiftly in the period around 6–10 kaBP and that this may have been associated with a subtle shift in the North Atlantic storm track and with a rather larger change in the Mediterranean storm track. At the very least, there is a suggestion that the Mediterranean climate could be highly sensitive to relatively small changes in forcing around this period (which is also approximately coincident with the formation of the S1 sapropel layer in the Mediterranean; Rohling and De Rijk, 1999) and that the atmospheric dynamics of this period are worthy of further investigation. This should also include a fuller examination of the complex seasonal cycle of Mediterranean storm activity (Trigo et al., 1999) and its impact on precipitation, particularly in the late winter and into early spring.

It is important, however, to note that the model used in this study does not include a full three-dimensional dynamical ocean. As such, there is a range of ocean–atmosphere interactions that are not simulated here. In particular, it would be expected that the atmospheric changes described here would lead to changes in ocean circulations such as the Gulf Stream, equatorial upwelling, and the Antarctic Circumpolar Current. These oceanic feedbacks may lead the coupled response to be different to those discussed here. For example, a weaker, more southerly, North Atlantic storm track in the early Holocene may lead to a weaker Gulf Stream current and, consequently, changes in the near surface temperature gradients that are important for storm growth. Similarly, changes in the Atlantic Ocean's MOC may help to explain the discrepancy between the model's simulation of annual-mean SAT over Scandinavia; palaeo-observations (e.g. Davis et al., 2003); and fully coupled model simulations (e.g. Braconnot et al., 2007; Otto-Bliesner et al., 2006).

While these areas may appear geographically remote from the climate of the Jordan Valley, it is clear that an improved understanding of them is vital to understand the patterns of change over the Mediterranean. Indeed, it is not clear how to reconcile the interpretations of palaeo-observations provided in Bonfils *et al.* (2004) (suggestive of a mean state projecting onto NAO-negative like conditions and consistent with the patterns observed here) with the NAO-positive interpretation provided by Rimbu *et al.* (2003). It is therefore evident that further work is required to fully understand the storm track and large-scale flow patterns over the North Atlantic and Mediterranean during the mid-Holocene. The need for this dynamically based understanding is further emphasised when one considers the differing mid-Holocene flow patterns over the North Atlantic and Europe seen in the PMIP-2 simulations (Gladstone *et al.*, 2005).

This study has focused on the time-mean state of the climate (i.e. seasonal or annual means). However, human societies are often more vulnerable to extremes and variations, such as the frequency of droughts or floods on a range of timescales from days to decades. Assessing changes in such variability is a major challenge for climate modelling (and palaeoscience more generally), but will be necessary in order to better understand the full impact of climate on both past and future communities. Notwithstanding these limitations, the experiments described here provide useful insight into some of the dynamical mechanisms behind the inter-millennial scale climate change experienced in the eastern Mediterranean during the Holocene period. It is anticipated that future work will further examine the dynamical mechanisms involved, particularly in the context of the eastern Mediterranean precipitation change in the period around 6–10 kaBP.

REFERENCES

Arz, H. W., F. Lamy, J. Patzold, P. J. Muller and M. Prins (2003) Mediterranean moisture source for an Early-Holocene humid period in the Northern Red Sea. *Science* **300**: 118–121.

Barlow, M., M. Wheeler, B. Lyon and H. Cullen (2005) Modulation of daily precipitation over Southwest Asia by the Madden-Julian Oscillation. *Monthly Weather Review* **133**: 3579–3594.

Berger, A.L. (1978) Long term variations of daily insolation and Quaternary climatic changes. *Journal of the Atmospheric Sciences* **35**: 2362–2367.

Black, E. (2009) The impact of climate change on daily precipitation statistics for Jordan and Israel. *Atmospheric Science Letters* **10**: 192–200.

Bonfils, C., N. de Noblet-Ducoudre, J. Guiot and P. Bartlein (2004) Some mechanisms of Mid-Holocene climate change in Europe, inferred from comparing PMIP models to data. *Climate Dynamics* **23**: 79–98.

Braconnot, P., B. Otto-Bliesner, S. Harrison *et al.* (2007) Results of PMIP2 coupled simulations of the Mid-Holocene and Last Glacial Maximum – Part 1: experiments and large-scale features. *Climate of the Past* **3**: 261–277.

Brayshaw, D. J., B. Hoskins and M. Blackburn (2008) The storm track response to idealised SST perturbations in an aquaplanet GCM. *Journal of the Atmospheric Sciences* **65**: 2842–2860.

Brayshaw, D. J., B. Hoskins and M. Blackburn (2009) The basic ingredients of the North Atlantic storm track. Part I: Land–sea contrast and orography. *Journal of the Atmospheric Sciences* **66**: 2539–2558.

Brayshaw, D., B. Hoskins and E. Black (2010) Some physical drivers of changes in the winter storm tracks over the North Atlantic and Mediterranean during the Holocene. *Philosophical Transactions of the Royal Society A*, **368**: 5185–5223.

Cheddadi, R., G. Yu, J. Guiot, S. Harrison and I. C. Prentice (1997) The climate of Europe 6000 years ago. *Climate Dynamics* **13**: 1–9.

Claussen, M., C. Kubatzki, V. Brovkin *et al.* (1999) Simulation of an abrupt change in Saharan vegetation in the Mid-Holocene. *Geophysical Research Letters* **26**: 2037–2040.

Crossley, J. F. and D. L. Roberts (1995) *The Thermodynamic/Dynamic Sea Ice Model.* Unified Model Documentation Paper **45** Met Office.

Dai, A. G., A. Hu, G. A. Meehl, W. M. Washington and W. G. Strand (2005) Atlantic thermohaline circulation in a coupled general circulation model: Unforced variations versus forced changes. *Journal of Climate* **18**: 3270–3293.

Davis, B. A. S., S. Brewer, A. C. Stevenson *et al.* (2003) The temperature of Europe during the Holocene reconstructed from pollen data. *Quaternary Science Reviews* **22**: 1701–1716.

Drake, N. and C. Bristow (2006) Shorelines in the Sahara: geomorphological evidence for an enhanced monsoon from palaeolake Megachad. *The Holocene* **16**: 901–911.

Edmon, H. J., B. Hoskins and M. E. McIntyre (1980) Eliassen-Palm cross sections for the troposphere. *Journal of the Atmospheric Sciences* **37**: 2600–2616.

Gladstone, R. M., I. Ross, P. J. Valdes *et al.* (2005) Mid-Holocene NAO: A PMIP2 model intercomparison. *Geophysical Research Letters* **32**: L16707, DOI:10.1029/2005GL023596.

Hegerl, G. C., F. Zwiers, P. Braconnot *et al.* (2007) Understanding and attributing climate change. In *Climate Change 2007: The Physical Science Basis. Contribution of Working Group I to the Fourth Assessment Report of the Intergovernmental Panel on Climate Change*, ed. S. Solomon, D. Qin, M. Manning *et al.* Cambridge and New York: Cambridge University Press.

Hewitt, C. D., C. A. Senior and J. F. B. Mitchell (2001) The impact of dynamic sea-ice on the climatology and climate sensitivity of a GCM: a study of past, present, and future climates. *Climate Dynamics* **17**: 655–668.

Hoskins, B. J. and K. I. Hodges (2002) New perspectives on the Northern Hemisphere winter storm tracks. *Journal of the Atmospheric Sciences* **59**: 1041–1061.

Hoskins, B. J. and P. J. Valdes (1990) On the existence of storm-tracks. *Journal of the Atmospheric Sciences* **47**: 1854–1864.

Hoskins, B. J., I. N. James and G. H. White (1983) The shape, propagation and mean-flow interaction of large-scale weather systems. *Journal of the Atmospheric Sciences* **40**: 1595–1612.

Jansen, E., J. Overpeck, K. R. Briffa *et al.* (2007) Palaeoclimate. In *Climate Change 2007: The Physical Science Basis. Contribution of Working Group I to the Fourth Assessment Report of the Intergovernmental Panel on Climate Change*, ed. S. Solomon, D. Qin, M. Manning *et al.* Cambridge and New York: Cambridge University Press.

Jolly, D., I. C. Prentice, R. Bonnefille *et al.* (1998) Biome reconstruction from pollen and plant macrofossil data for Africa and the Arabian peninsula at 0 and 6000 years. *Journal of Biogeography* **25**: 1007–1027.

Meehl, G. A., T. F. Stocker, W. Collins *et al.* (2007) Global climate projections. In *Climate Change 2007: The Physical Science Basis. Contribution of Working Group I to the Fourth Assessment Report of the Intergovernmental Panel on Climate Change*, ed. S. Solomon, D. Qin, M. Manning *et al.* Cambridge and New York: Cambridge University Press.

Otto-Bliesner, B. L., E. C. Brady, G. Clauzet *et al.* (2006) Last Glacial Maximum and Holocene climate in CCSM3. *Journal of Climate* **19**: 2526–2544.

Peltier, W. R. (1994) Ice-age paleotopography. *Science* **265**: 195–201.

Pope, V. D., M. L. Gallani, P. R. Rowntree and R. A. Stratton (2000) The impact of new physical parametrizations in the Hadley Centre climate model: HadAM3. *Climate Dynamics* **16**: 123–146.

Rimbu, N., G. Lohmann, J. H. Kim, H. W. Arz and R. Schneider (2003) Arctic/North Atlantic Oscillation signature in Holocene sea surface temperature trends as obtained from alkenone data. *Geophysical Research Letters* **30**: DOI 10.1029/2002GL016570.

Rimbu, N., G. Lohmann, S. J. Lorenz, J. H. Kim and R. R. Schneider (2004) Holocene climate variability as derived from alkenone sea surface temperature and coupled ocean-atmosphere model experiments. *Climate Dynamics* **23**: 215–227.

Rodwell, M. J. and B. J. Hoskins (2001) Subtropical anticyclones and summer monsoons. *Journal of Climate* **14**: 3192–3211.

Rohling, E. J. and S. De Rijk (1999) Holocene Climate Optimum and Last Glacial Maximum in the Mediterranean: the marine oxygen isotope record. *Marine Geology* **153**: 57–75.

Rudolf, B., H. Hauschild, W. Rueth and U. Schneider (1994) Terrestrial precipitation analysis: Operational method and required density of point measurements. In *Global Precipitations and Climate Change*, ed. M. Desbois and F. Desalmond. Berlin: Springer-Verlag pp. 173–186.

Stratton, R. A. (1994) Report on aspects of variability in high-resolution versions of HadAM3. *Hadley Centre Technical Note* **53**.

Taylor, K. E., D. Williamson and F. Zwiers (2000) The sea surface temperature and sea-ice concentration boundary conditions for AMIP II simulations *PCMDI Report 60*. California: Lawrence Livermore National Laboratory Program for Climate Model Diagnosis and Intercomparison.

Texier, D., N. de Noblet and P. Braconnot (2000) Sensitivity of the African and Asian monsoons to Mid-Holocene insolation and data-inferred surface changes. *Journal of Climate* **13**: 164–181.

Trenberth, K. E., P. D. Jones, P. Ambenje *et al.* (2007) Observations: surface and atmospheric climate change. In *Climate Change 2007: The Physical Science Basis. Contribution of Working Group I to the Fourth Assessment Report of the Intergovernmental Panel on Climate Change*, ed. S. Solomon, D. Qin, M. Manning *et al.* Cambridge and New York: Cambridge University Press pp. 433–497.

Trigo, I. F., T. D. Davies and G. R. Bigg (1999) Objective climatology of cyclones in the Mediterranean region. *Journal of Climate* **12**: 1685–1696.

Uppala, S. M., P. W. Kallberg, A. J. Simmons *et al.* (2005) The ERA-40 re-analysis. *Quarterly Journal of the Royal Meteorological Society* **131**: 2961–3012.

Wanner, H., J. Beer, J. Butikofer *et al.* (2008) Mid- to late Holocene climate change: an overview. *Quaternary Science Reviews* **27**: 1791–1828.

Wilks, D. S. (1995) *Statistical Methods in the Atmospheric Sciences: An Introduction*. San Diego: Academic Press.

Wood, R. A., A. B. Keen, J. F. B. Mitchell and J. M. Gregory (1999) Changing spatial structure of the thermohaline circulation in response to atmospheric CO_2 forcing in a climate model. *Nature* **399**: 572–575.

Yin, J. H. (2005) A consistent poleward shift of the storm tracks in simulations of 21st century climate. *Geophysical Research Letters* **32**: DOI 10.1029/2005GL023684.

4 Future climate of the Middle East

Emily Black, David Brayshaw, Julia Slingo and Brian Hoskins

ABSTRACT

A survey of publicly available data from the Intergovernmental Panel on Climate Change (IPCC) suggests that the Middle East will become significantly drier as greenhouse gas levels rise – with potentially devastating consequences. Simulating the climate of the eastern Mediterranean and the Middle East is, however, a tough challenge for climate models and those results should be interpreted with caution. The cyclones which migrate from west to east across the Mediterranean in winter and early spring, and which deliver much of the annual precipitation to the Middle East, are not well resolved by global climate models of the type included in the IPCC archive. Furthermore, the local climate is modified by coastlines and mountains throughout the region. For these reasons we provide a supplement to the IPCC results with simulations from a regional climate model. As in the global models, the regional model projects that, under an A2 (business-as-usual) scenario, precipitation will decrease significantly in the Middle East. Further investigation of the daily statistics of the weather, along with tracking of weather systems in the present day and future climate scenarios, suggest that the dominant mechanism for these changes is a reduction in the strength of the Mediterranean storm track. The Mediterranean storm track is fairly well simulated by the regional climate model, increasing confidence in this projection. In contrast to a business-as-usual scenario, under scenarios in which emissions slow down or stabilise then reduce, the rainfall changes projected are small, and barely significant in the eastern Mediterranean. This suggests that the potentially devastating precipitation decreases projected under a business-as-usual scenario may be avoided if greenhouse gas emissions are mitigated.

4.1 INTRODUCTION

The IPCC fourth assessment report suggested that the eastern Mediterranean region will become significantly drier under a future climate scenario, with potentially devastating impact on the population (Christensen *et al.*, 2007). Here we use data from the IPCC archives along with regional model integrations to explore this projection in more detail. Our main focus is precipitation because of its key importance to hydrological modelling.

There have been several studies of observed climate change in the Middle East. An examination of changes in temperature and rainfall in Israel at stations with long records showed that boreal winter rainfall in Israel increased in the second half of the twentieth century (Ben-Gai *et al.*, 1998; 1999). A study of precipitation changes over Europe using observations and CMIP3 simulations suggested that an observed slight reduction in precipitation over much of continental Europe can be attributed to anthropogenic forcing (Mariotti *et al.*, 2008). However, a study of temperature and precipitation extreme indices suggested that while there have been significant trends in temperature extremes, any trends in precipitation extremes are weak and insignificant (Zhang *et al.*, 2005).

Other studies have used model projections to investigate Middle East climate change during the twenty-first century. A survey of precipitation projections from 18 global climate models revealed that rainfall is projected to decrease over the Middle East, with statistically significant decreases projected by the end of the twenty-first century (Evans, 2009). These results are consistent with a study of water cycle changes in the CMIP3 models, which show a widespread reduction in water availability over much of southern Europe and the Mediterranean because of decreased precipitation and increased evaporation (Mariotti *et al.*, 2008). A study of temperature and precipitation extremes in Northern Israel using a regional model suggested that under an A2 scenario, temperature will increase and precipitation will decrease over Israel during the twenty-first century. Under a B2

Water, Life and Civilisation: Climate, Environment and Society in the Jordan Valley, ed. Steven Mithen and Emily Black. Published by Cambridge University Press.
© Steven Mithen and Emily Black 2011.

scenario, although temperature was projected to increase significantly, no significant trend in precipitation was projected. These changes in the mean were projected to be accompanied by increases in the frequency of extreme temperature and precipitation events (Alpert et al., 2008). An investigation of precipitation and stream flow changes in the Middle East, using a global climate model with 20 km resolution, suggested that under an A1b scenario there will be a large enough decrease in precipitation and increase in evaporation to cause the fertile crescent to disappear (Kitoh et al., 2008).

The mechanisms behind these changes in the climate have been the subject of several studies. A statistical study of precipitation processes in the Middle East in a future and current climate suggested that the projected drying may be accompanied by a change in the dominant precipitation mechanism from being directly driven by storm tracks to having a greater dependence on the upslope flow of moist air masses (Evans, 2008). This would be consistent with the weakening of the storm track in the southern Mediterranean that has been projected by several studies (Bengtsson, 2006; Lionello and Giorgi, 2007; Pinto et al., 2007).

The questions this chapter aims to address are:

- How is precipitation projected to change on daily to seasonal timescales for the Middle East?
- What are the mechanisms for these changes?
- How much confidence can we have in these projections?

The climate of the Middle East is modulated by complex topography (see for example Goldreich, 1994). This means that the spatial variability in precipitation cannot be adequately simulated by standard-resolution global climate models, which typically have a horizontal resolution of 2.5° or coarser. For this reason, a regional climate model (RCM) was used for most of the analysis in this chapter, and for all the inputs into the hydrological models described in subsequent chapters within this volume.

Section 4.2 describes the methodology and data used. Section 4.3 describes how well the regional model simulates key aspects of the Mediterranean climate. Section 4.4 examines changes in the climate on daily to monthly to seasonal timescales and includes discussion of the possible mechanisms for these changes. Section 4.5 describes the key sources of uncertainty in the model projections. Section 4.6 discusses the reliability of the simulations and their usefulness for impacts studies.

4.2 DATA AND MODELLING METHODOLOGY

4.2.1 Observed data

All observations of rainfall referred to in this study are based on rain gauge measurements. Daily rainfall data for 1985 onwards from nine stations along the River Jordan were provided by the Israeli Meteorological Service (see Figure 4.2 for the locations of these stations). GPCC data were used for the larger-scale comparisons between the observed and modelled rainfall data. The GPCC rainfall data used in this study is a $0.5° \times 0.5°$ gridded product, which is based on rain gauge measurements (Schneider et al., 2008). The modelled temperature data were compared to NCEP reanalysis data (Kalnay et al., 1996). Topographic data were taken from United States Geological Survey, GTPOP30 digital elevation map dataset (http://topotools.cr.usgs.gov/). More details of the observed datasets are given in Chapter 2 of this volume.

4.2.2 IPCC model integrations

As part of the IPCC fourth assessment, 23 global GCMs were integrated under baseline and future scenarios (Solomon et al., 2007). For this study, eight models were chosen. For each model, a baseline integration and future scenarios were assessed. The criteria used to select the eight models were availability of precipitation data and a horizontal resolution of 3.75° or higher for the A2 and B1 scenarios. The A2 scenario is a 'business as usual' scenario, in which large increases in greenhouse gas concentrations are projected by the end of the twenty-first century. The B1 scenario projects lower emissions than the A2 scenario, with GHG emissions peaking in the middle of the twenty-first century and then declining. Both scenarios are described fully in the IPCC special report on emissions scenarios (Nakicenovic et al., 2001).

Monthly climatological data were extracted from the IPCC data portal for each scenario and each model (http://www.ipcc-data.org/). In order to compare the models, the data were re-gridded at 3.75° longitude and 2.5° latitude, which is equivalent to the coarsest-resolution model included.

4.2.3 Regional model setup

The RCM chosen for this study is based on HadAM3P, which is a global, atmosphere-only model developed at the Hadley Centre. The Hadley Centre regional models have been applied successfully in both the tropics and extratropics (see for example Hassell and Jones, 1999; Durman, 2001). The model was integrated for present-day conditions (1960–1989) and for 2070–2099 under A2 and B2 scenarios (described above). Present-day land-cover was assumed for all integrations. The future scenarios included changes in SST and CO_2.

RCMs are applied over a limited area and consequently require input both at the surface and lateral boundaries of the domain. The lateral boundary conditions were derived from integrations of HadAM3P forced with prescribed surface boundary conditions (SST and sea-ice fraction) that were based on observations, and projections generated by the global coupled model, HadCM3. The surface boundary conditions for the RCM

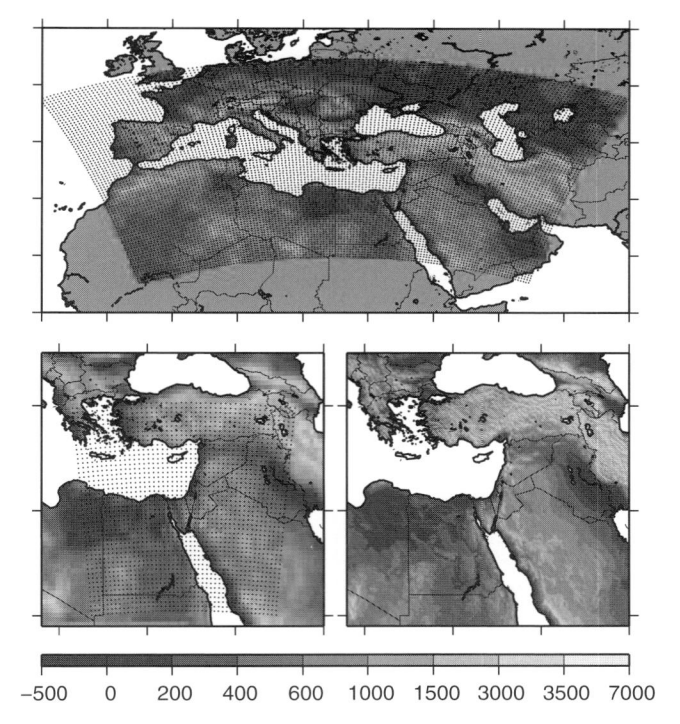

Figure 4.1. Model domains and modelled and observed topography. Top: model grid points (dots) on the model topography for the large domain. Bottom left: small domain (Middle East only) used for the ensembles. Bottom right: observed topography for the eastern Mediterranean and Middle East. Topography is given in metres above sea level for all the plots. See colour plate section.

(SST, sea-ice fraction and land cover) are similarly prescribed, based on HadCM3 projections and observations. For further background on the regional model, see http://precis.metoffice.com/docs/PRECIS_Handbook.pdf.

The RCM was run with a horizontal resolution of 0.44° (~44 km), which permitted a sufficiently large domain, while representing significantly more topographic variation than is possible in a global model. The choice of resolution is justified by the analysis of interannual rainfall coherence presented in Chapter 2 of this volume, which demonstrated that at scales of less than 100 km, the interannual variability is highly coherent, although the annual total rainfall is modulated by the orography. This means that it should be possible to downscale the rainfall to the scales needed for the hydrological modelling (approximately 10 km) statistically in the catchments of interest (see this volume, Chapters 10 and 13). Furthermore, the coherence was found to decline sharply at scales of more than 200 km, which suggests that the increase in resolution from around 250 km to 44 km gained by using a regional model adds value to the impacts studies described in subsequent chapters.

Two domains were used (see Figure 4.1). The large domain was chosen to include the whole Mediterranean and southern Europe. This size of domain was needed because the interannual rainfall variability in the Middle East is determined by the occurrence of circulation regimes over the Mediterranean, and the consequent perturbations in the storm track (see Chapter 2 of this volume). Although a large domain was needed for investigating the mechanisms of precipitation change, comparison between integrations carried out over the large and small domains (not shown here) revealed no significant differences in the simulated precipitation over Jordan and Israel – provided the boundaries of the small domain were sufficiently far from the regions of interest. For this reason, in addition to the single-realisation large domain integrations, three member ensembles for the baseline climate and A2 scenario were run for a smaller domain. Finally, a single B2 scenario integration was run over the small domain.

In summary, the RCM integrations carried out were:

- a baseline scenario (1961–1990) for the full region driven with lateral boundary conditions derived from a run of HadAM3P forced with observed climatological SST and land surface at the lower boundary;
- an A2 emission scenario (2070–2100) for the full region driven with lateral boundary conditions derived from HadAM3P forced with HadCM3 A2 scenario SST changes added to the observed SST;
- a three-member ensemble of a baseline scenario (1961–1990) for the eastern Mediterranean only, driven with lateral boundary conditions derived from an AMIP type run of HadAM3P and observed climatological SST and land surface at the lower boundary;
- a three-member ensemble of an A2 emission scenario (2070–2100) with lateral boundary conditions derived from HadAM3P forced with HadCM3 A2 scenario SST changes added to the observed SST;
- a single-member B2 scenario with lateral boundary conditions derived from HadAM3P forced with HadCM3 A2 scenario SST changes added to the observed SST.

4.3 HOW WELL DOES THE REGIONAL MODEL REPRESENT THE PRESENT-DAY CLIMATE?

This section considers the RCM's ability to represent the seasonal cycle in precipitation statistics in Israel. For more detailed discussion of the model's representation of the large-scale spatial patterns in temperature and precipitation, see Chapter 8.

Figure 4.2 compares the seasonal cycles of various rainfall statistics for the baseline scenario with data from nine stations along the River Jordan in Israel. Although the model captures the general pattern of a dry summer and wet winter, precipitation at the peak of the rainy season is grossly underestimated. This underestimation of rainfall in Jordan and Israel during DJF is a

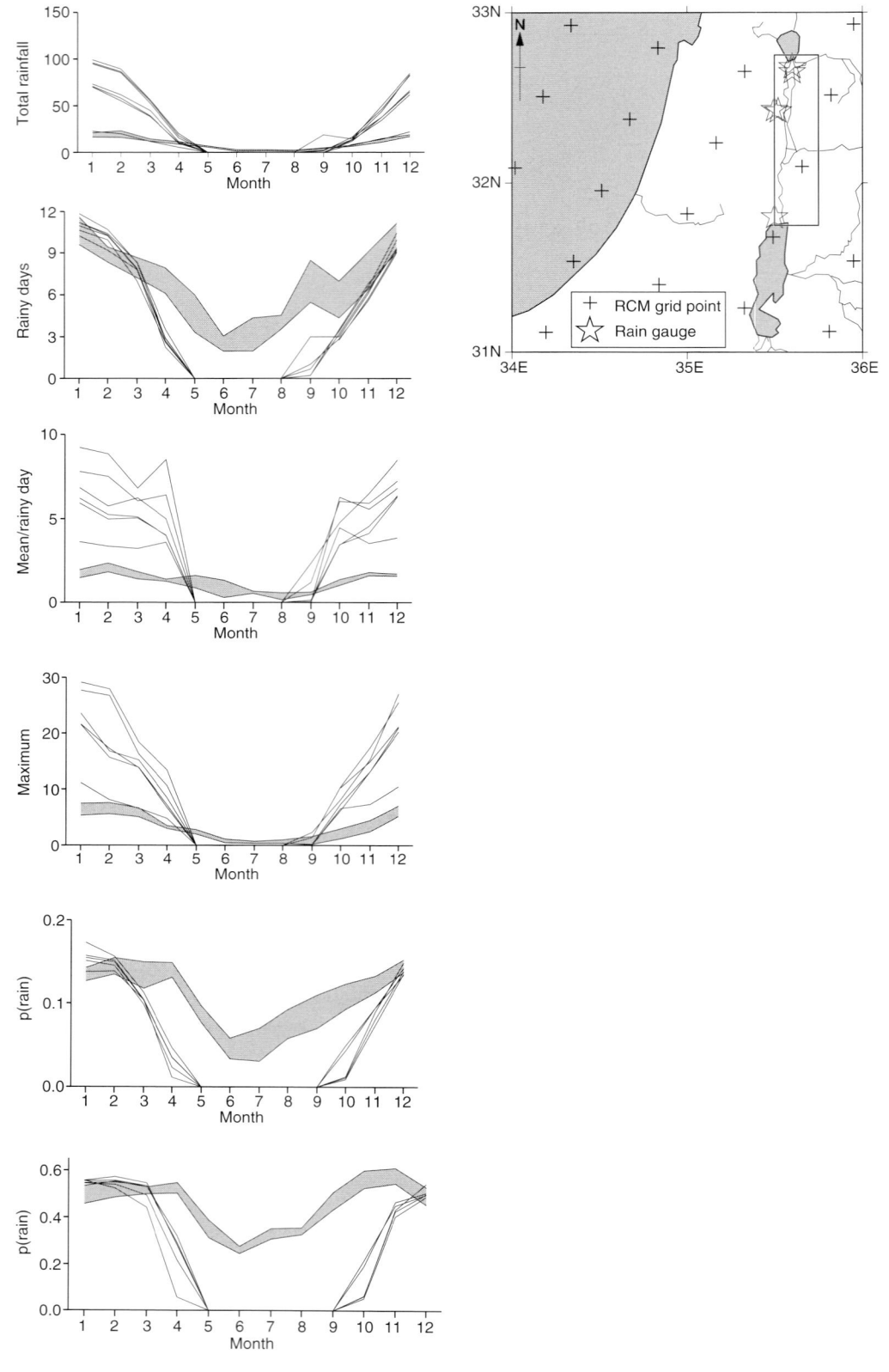

Figure 4.2. Left set of panels: seasonal cycles in various statistics of the weather for eight stations (black lines), and the regional climate model baseline ensemble (grey shaded area indicates the ensemble range). The *x*-axis gives the month and the *y*-axis gives the mean rainfall statistic during that month. From top to bottom the statistics are: mean total monthly rainfall, mean total number of rainy days, mean rain per rainy day, mean maximum monthly rainfall, monthly mean probability of rain given no rain the day before, monthly mean probability of rain given rain the day before. Rainfall is given in millimetres for all the plots. Right panel: the location of the stations and the RCM time-series box. The crosses indicate RCM grid points.

feature of other regional models, including RegCM2, RegCM3 and MM5. In a recent study using MM5, with a 27 km resolution the mean annual total rainfall near the Mediterranean coast was ~290 mm, compared with an observed value of ~550 mm (Evans, 2008). In another study where RegCM (50 km resolution) was driven by boundary conditions based on HadCM3, the winter precipitation in Israel and western Jordan ranged from less than 100 mm to ~250 mm, compared with the observed precipitation, which exceeds 400 mm near the Mediterranean coast (Lionello and Giorgi, 2007). Several simulations of RegCM3 driven with different boundary conditions exhibited similar biases, with the annual total rainfall along the Mediterranean coast being ~200–300 mm compared with more than 500 mm observed (Krichak et al., 2007). Even a state-of-the-art global model with 20 km resolution underestimated the precipitation in Jordan, although the amount of coastal rainfall was closer to that observed (Kitoh et al., 2008). In a study of precipitation processes using RegCM2 at 25 km resolution, the model's underestimation of rainfall at the peak of the rainy season was attributed to its failure to resolve the coastal mountains (Evans et al., 2004). The RCM used in this study had a lower resolution (~44 km) and was thus even less able to resolve the coastal mountains.

During the rainy season, there tend to be several days of rain as a cyclone passes, followed by a dry period. The frequency and duration of these rainy events are reflected, respectively, by the probabilities of rain given no rain the day before and rain given rain the day before. Like most climate models, the RCM generates small amounts of rainfall on almost every day. Therefore, in this study, a rainy day in both the observations and model is defined as one with at least 0.1 mm of rain. Figure 4.2 shows that the probability of rain given rain the day before and rain given no rain the day before, and hence the number of rainy days, are adequately simulated by the model for the winter season. However, the mean rain per rainy day and maximum rainfall are far lower in the model baseline scenario than in observations. This suggests that the underestimation of monthly rainfall totals by the RCM reflects an underestimation of rainfall intensity.

It can be seen that there are rainy days during the summer and hence the rainfall probabilities do not fall to zero. Rainfall on these days is, however, low and summer rainfall does not contribute significantly to the annual total. It is not possible to make a formal comparison between summer rainfall in the model and observations because the observed summer rainfall was not recorded. However, the summer season is known to be extremely dry in the Middle East, and rainy days are rare.

Overall, the regional model simulations presented in this study are of similar quality to other recent regional model simulations of the Middle East present-day climate. In particular, the frequency and duration of rainy events is captured, increasing confidence in projections of changes in these statistics. In contrast, because it is poorly simulated, projections of changes in rainfall intensity should be regarded with caution. The disparity in the quality of the model's simulation of different rainfall statistics is a strong argument for considering individual statistics separately – the approach taken in this study.

4.4 PROJECTED CHANGES IN THE CLIMATE ON MONTHLY TO SEASONAL TIMESCALES

4.4.1 Changes in climate on monthly to seasonal timescales

Figure 4.3 shows the change in the monthly mean rainfall over the Mediterranean and southern Europe. In October, the projected

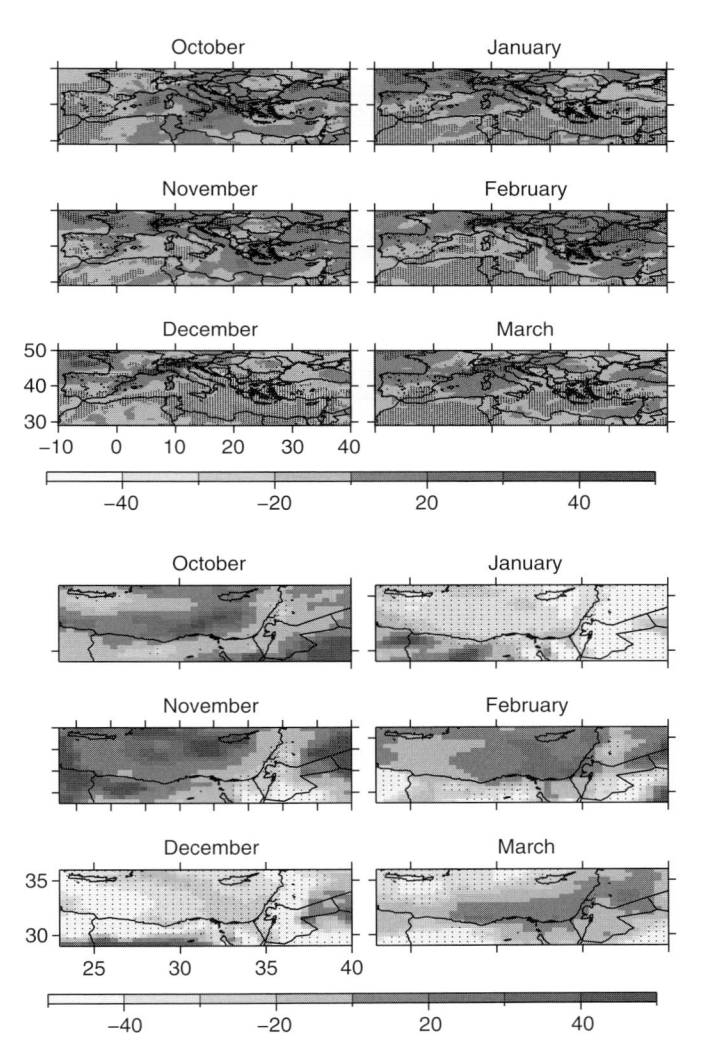

Figure 4.3. Seasonal cycle in precipitation change under an A2 scenario by 2070–2100. Significance at the 95% level is shown by a dot in the grid square. Top set: monthly mean absolute change in rainfall (mm) over the whole of the Mediterranean under an A2 scenario. Bottom set: monthly mean percentage change in rainfall (%) over the East Mediterranean only. See colour plate section.

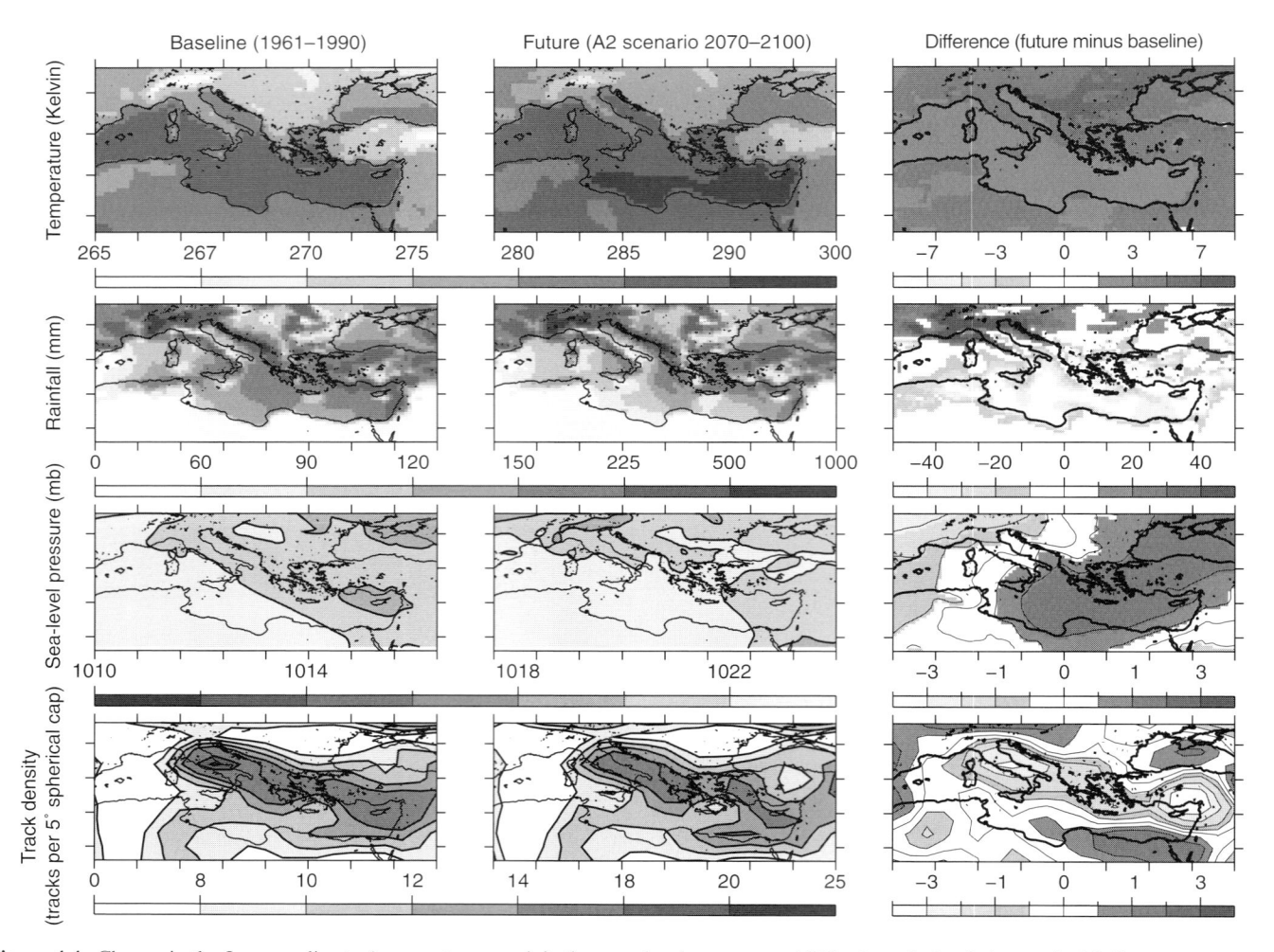

Figure 4.4. Change in the January climate (temperature, precipitation, sea-level pressure and 850 mb track density) over the Mediterranean under an A2 scenario by 2070–2100. See colour plate section.

changes in Middle East and eastern Mediterranean rainfall are small and insignificant. In November, a significant decrease in precipitation is projected in Jordan and Israel. The largest changes are projected at the peak of the rainy season, December and January, when significant reductions in rainfall (of the order of 40%) are projected over much of the eastern Mediterranean region. By February and March the rainfall changes projected for the Middle East are small both in magnitude and in percentage terms.

On a larger scale, in October and November, the eastern Mediterranean is projected to get wetter and the projected decrease in precipitation alluded to above is localised to the Middle East. In December and January, rainfall is projected to decrease significantly over the whole Mediterranean. The pattern is more complicated in February and March, with some parts of the Mediterranean projected to get wetter, and other parts to get drier.

In order to investigate these rainfall changes in more detail, the following section focuses on January. We chose to present results from a single month rather than a season because the large-scale patterns of changes in rainfall vary significantly over the course of boreal winter and autumn. Figure 4.4 shows how the January temperature, rainfall, sea-level pressure (SLP) and storm track are projected to change under an A2 scenario. It can be seen that the whole Mediterranean region is projected to get warmer, with the greatest warming (over 4 K) in southeastern Europe, and somewhat less warming over the Mediterranean Sea itself (1–2 K). The Middle East is projected to get approximately 3 K warmer under this scenario. This is accompanied by a reduction in rainfall over the Middle East, southeast Europe and the Mediterranean, with the largest reductions in Israel and western Jordan. Further north and west, rainfall is projected to increase. The projected change in SLP is consistent with the rainfall signal, with higher SLP projected in the east and lower pressure in the west. This reflects a slackening of the east–west gradient of SLP and a far weaker Cyprus low under an A2 scenario. The bottom plots show that the reduction in rainfall within the eastern Mediterranean and the change in the SLP pattern reflects a weakening of the Mediterranean storm track. Interestingly, the track density is projected to increase, or decrease only slightly, near the southeast Mediterranean coast. Most of these weather systems, however, originate in north Africa, and comparison with the projected changes in rainfall suggests that they do not bring significant amounts of rain to Israel or Jordan.

These results are consistent with those previously published. A study using the ECHAM5 model projected that, in a future climate, the northern hemisphere storm tracks will move polewards, resulting in a reduction in the strength of the Mediterranean storm track of the order projected here (Bengtsson *et al.*, 2006). This was supported by a regional model study, which projected a decrease in rainfall in the southern Mediterranean, largely caused by a reduction in the number of cyclones crossing the Mediterranean (Lionello and Giorgi, 2007).

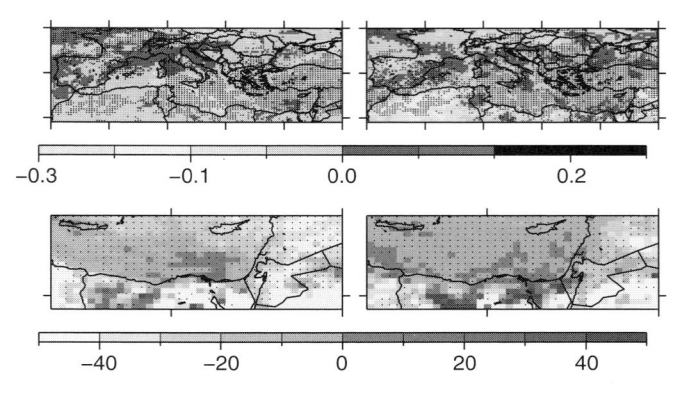

Figure 4.5. Change in daily rainfall probabilities. Significance at the 95% level is indicated by a dot within the grid square. Top row, left panel: absolute change in the probability of rain given no rain the day before; right: absolute change in the probability of rain given rain the day before. Bottom row: as above but for percentage changes for the southeast part of the region only. See colour plate section.

4.4.2 Changes on daily timescales

Figure 4.5 shows how the probability of rain has changed for January. It can be seen that, in Jordan, there is a small but significant reduction in the probability of rain, both given rain the day before, and given no rain the day before. In Israel, there is a significant reduction in the probability of rain given rain the day before, and a slight increase in probability of rain given no rain the day before. On a wider scale, over much of the Mediterranean, there is a significant reduction in the probabilities of rain both given rain and given no rain the day before, which means that both the duration and frequency of rainy events are projected to decrease in the Jordan River area – although these results do not apply to the whole regional model domain.

The changes in the statistics of the weather reported in this study are consistent with the projected changes in the Mediterranean storm track. Both the frequency and duration are projected to reduce, with the reduction in duration somewhat greater in absolute terms than the reduction in frequency. This is indicative of a reduction in the proportion of rainfall delivered by longer-lasting rainy events, such as would result from the passage of Mediterranean cyclones.

The seasonal cycles in several daily precipitation statistics for 2070–2100 and 1960–1990 near the Jordan River are compared in Figure 4.6. Both the A2 ensemble and the B2 realisation are compared with the baseline ensemble. It can be seen that, under an A2 scenario, precipitation is projected to decrease throughout the rainy season. The shape of the seasonal cycle in total rainfall is also projected to change markedly, with more rain at the beginning of the rainy season than the end and a less pronounced peak of the rainfall season. Under a B2 scenario the projected changes to the seasonal cycle are similar, although the magnitude is lower. This strong decrease in winter rainfall accompanied by a smaller or insignificant change in the spring is also seen in a study using MM5 (Evans, 2008).

These changes are associated with a reduction in rainy days and hence rainfall probabilities at the peak of the rainy season. The projected reduction in rainfall intensity is smaller than the projected reduction in the number of rainy days. Under both an A2 and a B2 scenario it can be seen that, although the maximum daily rainfall is similar to the baseline scenario, the rainfall intensity is on average a little lower at the peak of the rainy seasons. The changes in rain per rainy day are, however, not significant and should be regarded with caution because the regional model does a poor job of simulating rainfall intensity (see section 4.3).

4.5 KEY SOURCES OF UNCERTAINTY

4.5.1 Uncertainty due to model bias and discrepancies between models

Model bias and discrepancies between models are major causes of uncertainty. The comparisons with observed data described in section 4.3 showed that the model simulated some aspects of the Middle East and Mediterranean climate far better than others. Specifically, while the model could simulate the present-day probabilities of rain and the storm track, it grossly underestimated the rainfall intensity over the whole eastern Mediterranean.

It is interesting to note that the underestimation of rainfall reported here is a feature of several other regional climate

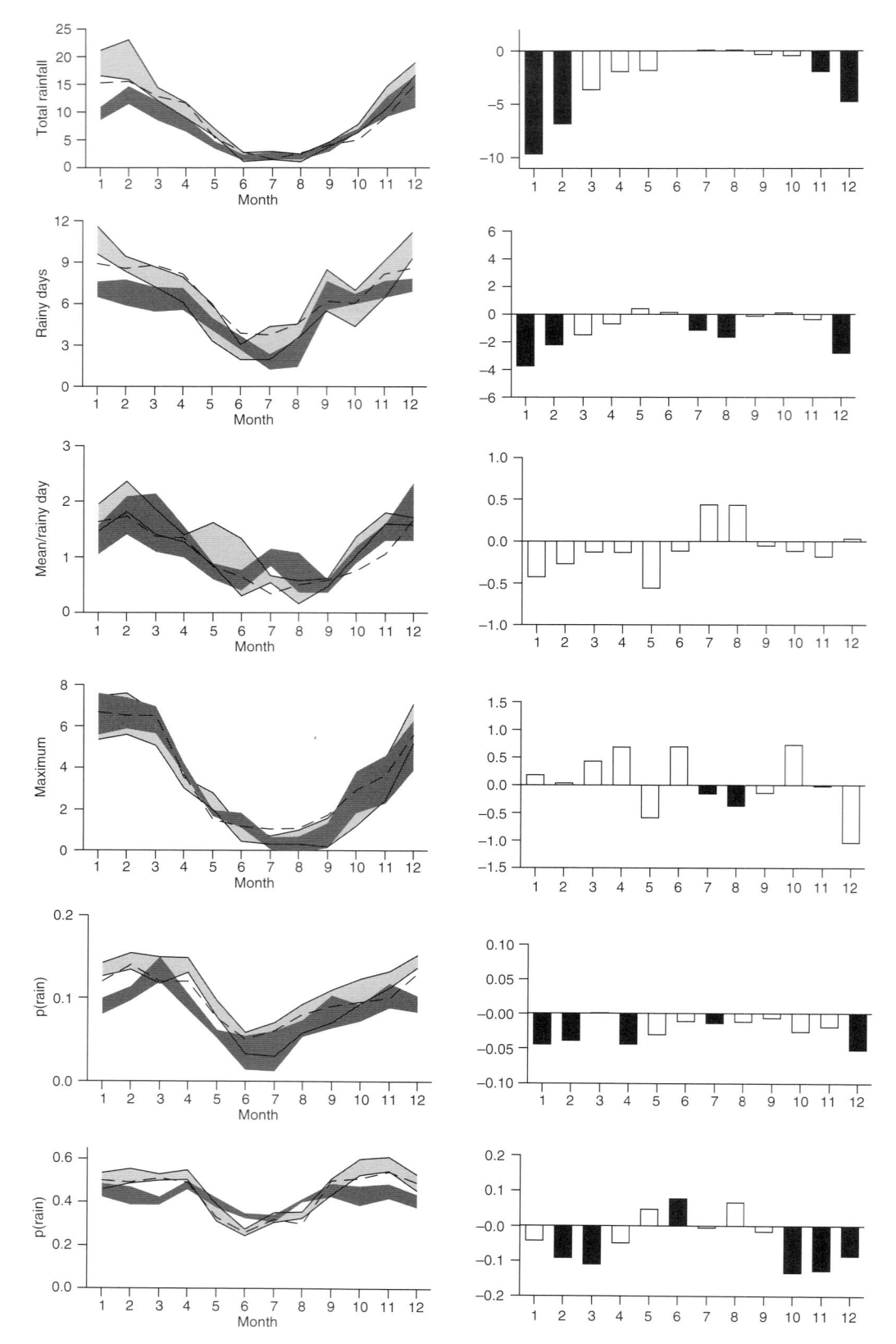

Figure 4.6. Left set of panels: seasonal cycles in various statistics of the weather for the baseline ensemble (light grey polygon); the A2 ensemble (dark grey polygon) and the B2 integration (dashed line) for the box shown in Figure 4.2. The *x*-axis gives the month and the *y*-axis gives the mean rainfall statistic during that month. From top to bottom the statistics are: mean total monthly rainfall, mean total number of rainy days, mean rain per rainy day, mean maximum monthly rainfall, monthly mean probability of rain given no rain the day before, monthly mean probability of rain given rain the day before. Rainfall is given in millimetres for all the plots. Right set of panels: difference in ensemble means between the A2 scenario integration and the baseline integration for the statistics shown on the left. Filled bars indicate significance at the 95% level.

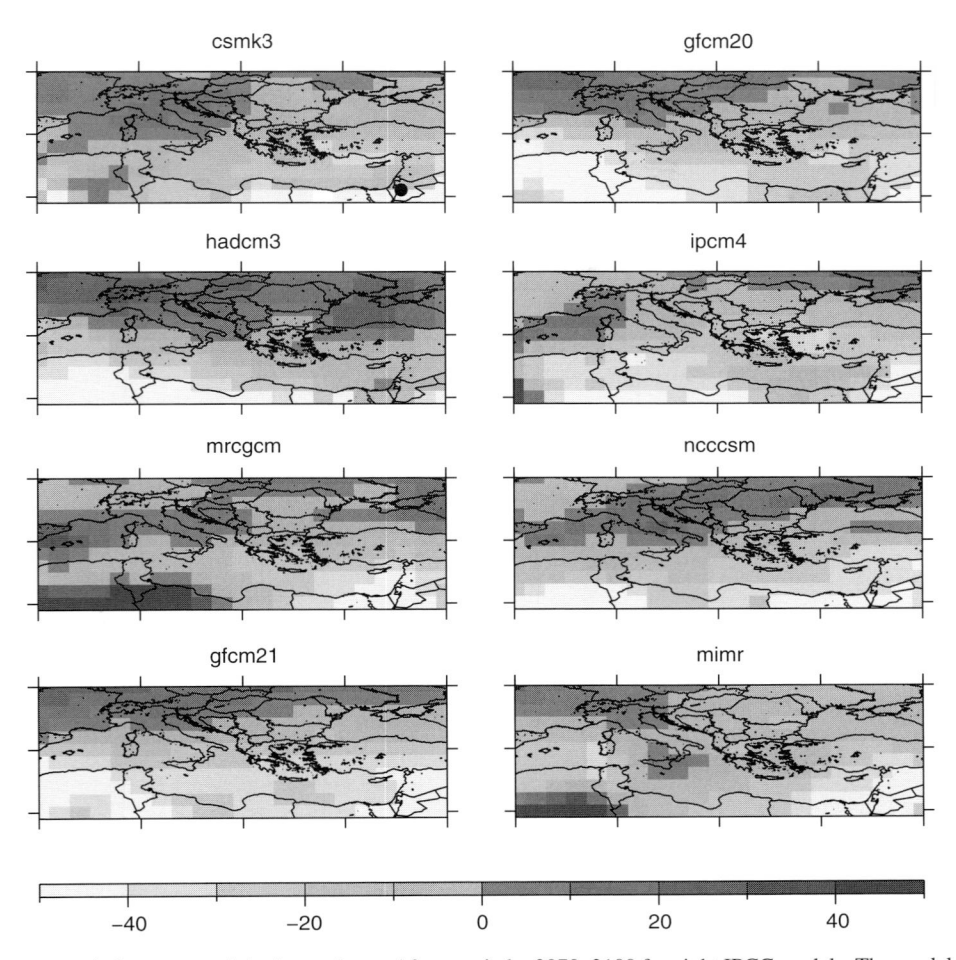

Figure 4.7. Percentage change in January precipitation under an A2 scenario by 2070–2100 for eight IPCC models. The model name abbreviations on the plots are: CSIRO Mark 3.0 (csmk3); GFDL CM 2.0 AOGCM (gfcm20); HadCM3 (hadcm3); IPSL CM4 (ipcm4); MRI-CGCM2.3.2 (mrcgcm); NCAR CCSM3 (nccsm); GFDL CM 2.1 AOGCM (gfcm21); MIMR MIROC3.2 (medium resolution). See colour plate section.

models including RegCM3, RegCM2 and MM5 (see section 4.3). This demonstrates that agreement between models does not necessarily mean that the underlying climate processes are being simulated correctly. Thus, while discrepancies between models undoubtedly add to the uncertainty, similarities do not guarantee reliable projections.

It is nevertheless worth noting that the regional modelling results presented here are broadly consistent with published wider surveys of global models included in the IPCC fourth assessment. Figure 4.7 shows the percentage change in January precipitation projected under an A2 scenario by eight of the models used during the IPCC assessment. January was chosen because it is the month for which the regional model projects the greatest change in precipitation. It can be seen that there is a degree of consistency between the models, although there is some variation in the patterns of projected changes, and considerable variation in the amount of change projected. In particular, all the models projected that the eastern Mediterranean and Middle East will get drier, while further north, the climate will get wetter.

The results from the modelling reported in this chapter agree with published surveys of the models included in the IPCC fourth assessment, which project a decrease in annual total rainfall in the Middle East by the end of the twenty-first century under an A2 scenario (Evans, 2009; Mariotti *et al.*, 2008). Several regional and high resolution global model studies also project a significant drying in the eastern Mediterranean by the end of the twenty-first century (Lionello and Giorgi, 2007; Evans, 2008; Kitoh *et al.*, 2008).

4.5.2 Uncertainty due to initial conditions and stochastic variability

As described previously, Figure 4.6 shows the mean seasonal cycle in various rainfall statistics for two sets of integrations: a four-member A2 scenario ensemble for 2070–2100 and a four-member baseline ensemble. The ensemble members have the same horizontal boundary conditions, but are driven by lateral boundary conditions derived from integrations of the global atmospheric model set off on successive days. Comparing model integrations that were started with subtly different states of the atmosphere enables us to assess the uncertainties that arise from our lack of knowledge of the state of the atmosphere at the outset of the climate model runs (the initial conditions).

Focusing on the total rainfall, it can be seen that, while there is some difference between ensemble members, during January and February, the two ensembles are well separated. In the early part of the rainy season (October–December), the ensembles are less clearly separated, both because the projected changes are smaller and the ensemble spread is greater. The sensitivity to initial conditions suggested by the large spread of ensemble members in the early part of the rainy season may reflect a high degree of stochastic variability. Further investigation with a larger ensemble would be required to establish the significance of the projected rainfall changes in November and December.

4.5.3 Uncertainty due to emissions scenario

As well as being a significant source of uncertainty, the sensitivity of Middle East precipitation change to emissions scenario is important sociologically and politically. The previous sections have shown that, under a B2 scenario, the projected reduction in rainfall is far smaller than under an A2 scenario. The A2 scenario is one of the more extreme, in which the present trends in GHG emissions continue, leading to CO_2 emissions doubling by the end of the century. In this section, we explore the importance of scenarios further by comparing the A2 scenario with the B1 scenario, in which emissions peak in the middle of the twenty-first century and then decrease to twentieth-century levels by 2100. The results are shown in Figure 4.8, which compares the changes in winter rainfall under the A2 and B1 scenarios for the eight IPCC models referred to in section 5.1. It can be seen that, even if emissions reduce to pre-twenty-first-century levels by the end of this century, a degree of drying is inevitable. However, there are large differences between the A2 and B1 scenarios. In particular, under the A2 scenario, the drying extends further north than in the B1 scenario, and, most relevant for the case studies presented in subsequent chapters of this volume, there is projected to be less drying over the eastern Mediterranean and Middle East under a B1 scenario.

These results are consistent with published work on the sensitivity of changes in precipitation and Mediterranean cyclones to emissions scenario. RegCM2 projects large decreases in rainfall under an A2 scenario in the Middle East. Under a B1 scenario, the changes are weak and insignificant. This was found to reflect a smaller reduction in the number of cyclones crossing the Mediterranean (Lionello and Giorgi, 2007). Another study, using RegCM, projected that under an A2 scenario, the changes in rainfall would

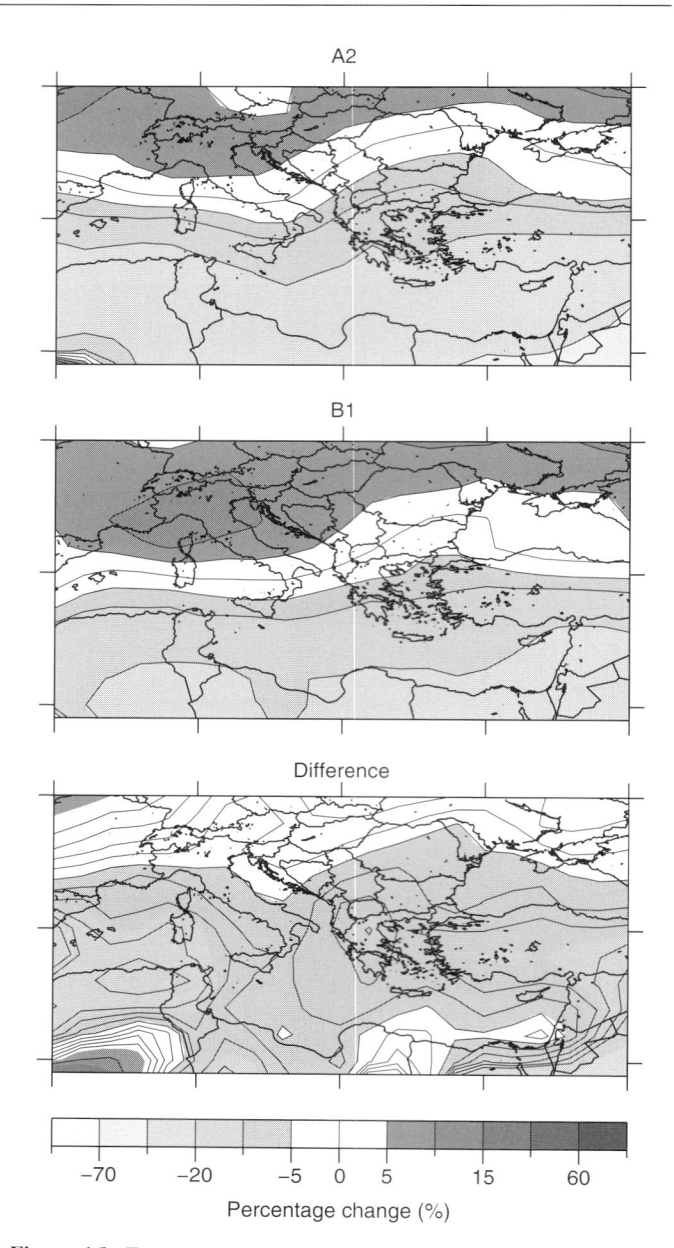

Figure 4.8. Top: mean percentage change in January precipitation predicted under an A2 scenario for 2070–2100 for the IPCC models shown in Figure 4.7; middle: mean percentage change in January precipitation predicted under a B1 scenario for 2070–2100 for the IPCC models shown in Figure 4.7. Bottom: difference in the mean percentage change between the A2 and B1 (B1 – A2). See colour plate section.

be large and significant by the end of the century, while under a B2 scenario any changes would be small and insignificant (Giorgi *et al.*, 2004). In a study of northern hemisphere storm tracks using the ECHAM5/MPI-OM1 coupled model, significant changes in storm tracks were projected under A1B, A2 and B1 scenarios with the changes under the B1 scenario weaker than under the A-family scenarios (Pinto *et al.*, 2007).

4.6 DISCUSSION

The model integrations presented in this chapter project that, under an A2 scenario, the Middle East will get significantly drier, with smaller changes projected under less extreme emissions scenarios. In subsequent chapters of this volume these projections will be used to drive hydrological models and hence to assess how water availability might change in Jordan and Israel. In this context, the usefulness of these results depends on the reliability of the model projections, and on how we account for the inevitable uncertainties.

For a given emissions scenario and set of initial conditions, the credibility of our projections of rainfall change rests on how well climate models can simulate the features underlying the projected change. The survey of changes in the daily statistics of the weather presented here suggests that the projected reduction in Middle East rainfall at the peak of the rainy season is caused by a reduction in the strength of the Mediterranean storm track, which causes a reduction in both the frequency and duration of rainy events. The model's capacity to simulate these features in the present day lends credence to the projections. Moreover, because the amount of rain per rainy day is not projected to change significantly in an A2 scenario, it is straightforward to account for the model bias when deriving time-series for input into hydrological models (see Chapter 5, this volume).

The uncertainties due to initial conditions are particularly significant in the early part of the rainy season (October–December), when the changes in rainfall are relatively small. This could reflect a degree of unpredictability – i.e. the dominant controls on rainfall being internal atmospheric variability rather than SST or greenhouse gas concentrations. It would be necessary to do a large ensemble of integrations to investigate this further. Without a large ensemble, it was not possible to take into account directly the effect of initial conditions when using this data to drive hydrological models. Instead, multiple realisations of a statistical rainfall model were used to simulate stochastic variability.

The uncertainty due to emissions scenario is critical from both a scientific and social point of view. It has been shown here, and in other published work, that under emissions scenarios in which CO_2 emissions increase slowly or stabilise, the projected reduction in precipitation in the Middle East is far smaller than it is under an A2 scenario, although some change is inevitable. In fact, under a B1 or B2 scenario, any decreases in rainfall are projected to be small and statistically insignificant.

From a socio-economic point of view, it should be noted that some of the projections of changing water availability presented in this volume and elsewhere, and arguably the future habitability of parts of Jordan and Israel, are subject to a business-as-usual emissions scenario. In other words, severe reductions in rainfall, and the consequent social hardship, might be avoided, if greenhouse gas emission can be mitigated.

4.7 CONCLUSIONS

1. Under an A2 scenario, Middle East rainfall is projected to reduce significantly.
2. The dominant cause of this reduction is a reduction in rainy events, probably caused by a reduction in the strength of the Mediterranean storm track.
3. The regional climate model simulates the Mediterranean storm track and hence the frequency and duration of rainy events in Jordan and Israel fairly well, lending credibility to the projections.
4. Under less extreme emissions scenarios the reductions in rainfall are far less severe, which means that climate change in the Middle East could be, to some degree, mitigated.
5. As well as emissions scenario, model bias and initial conditions cause significant uncertainty in the model projections.

REFERENCES

Alpert, P., S. Krichak, H. Shafir, D. Haim and I. Osentinsky (2008) Climatic trends to extremes employing regional modeling and statistical interpretation over the E. Mediterranean. *Global and Planetary Change* **63**: 163–170.

Ben-Gai, T., A. Bitan, A. Manes, P. Alpert and S. Rubin (1998) Spatial and temporal changes in rainfall frequency distribution patterns in Israel. *Theoretical and Applied Climatology* **61**: 177–190.

Ben-Gai, T., A. Bitan, A. Manes, P. Alpert and S. Rubin (1999) Temporal and spatial trends of temperature patterns in Israel. *Theoretical and Applied Climatology* **64**: 163–177.

Bengtsson, L., K. I. Hodges and Roeckner, E. (2006) Storm tracks and climate change. *Journal of Climate* **19**: 3518–3543.

Christensen, J. H., B. Hewitson, A. Busuioc *et al.* (2007) Regional Climate Projections. In *Climate Change 2007: The Physical Science Basis. Contribution of Working Group I to the Fourth Assessment Report of the Intergovernmental Panel on Climate Change*, ed. S. Solomon, D. Qin, M. Manning *et al.* Cambridge and New York: Cambridge University Press.

Durman, C. F., J. M. Gregory, D. C. Hassell, R. G. Jones and J. M. Murphy (2001) A comparison of extreme European daily precipitation simulated by a global and a regional climate model for present and future climates. *Quarterly Journal of the Royal Meteorological Society* **127**: 1005–1015.

Evans, J. P. (2008) Global warming impact on the dominant precipitation processes in the Middle East. *Theoretical and Applied Climatology* **99**: 389–402.

Evans, J. P. (2009) 21st century climate change in the Middle East. *Climatic Change* **92**: 417–432.

Evans, J. P., R. B. Smith and R. J. Oglesby (2004) Middle East climate simulation and dominant precipitation processes. *International Journal of Climatology* **24**: 1671–1694.

Giorgi, F., X. Bi and J. Pal (2004) Mean, interannual variability and trends in a regional climate change experiment over Europe. II: Climate change scenarios (2071–2100). *Climate Dynamics* **23**: 839–858.

Goldreich, Y. (1994) The spatial distribution of annual rainfall in Israel – a review. *Theoretical and Applied Climatology* **50**: 45–59.

Hassell, D. and R. Jones (1999) *Simulating Climatic Change of the Southern Asian Monsoon Using a Nested Regional Climate Model (HadRM2)*. HCTN 8: Hadley Centre for Climate Prediction and Research.

Kalnay, E., M. Kanamitsu, R. Kistler *et al.* (1996) The NCEP/NCAR 40-year reanalysis project. *Bulletin of the American Meteorological Society* **77**: 437–471.

Kitoh, A., A. Yatagai and P. Alpert (2008) First super-high-resolution model projection that the ancient 'Fertile Crescent' will disappear in this century. *Hydrological Research Letters* **2**: 1–4.

Krichak, S. O., P. Alpert and P. Basset (2007) The surface climatology of the eastern Mediterranean region obtained in a three-member ensemble climate change simulation experiment. *Advances in Geosciences* **12**: 67–80.

Lionello, P. and F. Giorgi (2007) Winter precipitation and cyclones in the Mediterranean region: future climate scenarios in a regional simulation. *Advances in Geosciences* **12**: 153–158.

Mariotti, A., N. Zeng, J.-H. Yoon et al. (2008) Mediterranean water cycle changes: transition to drier 21st century conditions in observations and CMIP3 simulations. *Environmental Research Letters* **3**: 1–7.

Nakicenovic, N., J. Alcamo, G. Davis et al., eds. (2001) *IPCC Special Report on Emissions Scenarios (SRES)*. Geneva: GRID-Arendal.

Pinto, J. G., U. Ulbrich, G. C. Leckebusch et al. (2007) Changes in storm track and cyclone activity in three SRES ensemble experiments with the ECHAM5/MPI-OM1 GCM. *Climate Dynamics* **29**: 195–210.

Schneider, U., T. Fuchs, A. Meyer-Christoffer and B. Rudolf (2008) Global precipitation analysis products of the GPCC. Global Precipitation Climatology Centre (GPCC) DWD Internet Publikation 1–12.

Solomon, S., D. Qin, M. Manning et al., eds. (2007) *Climate Change 2007: The Physical Science Basis. Contribution of Working Group I to the Fourth Assessment Report of the Intergovernmental Panel on Climate Change*. Cambridge and New York: Cambridge University Press.

Zhang, X. B., E. Aguilar, S. Sensoy et al. (2005) Trends in Middle East climate extreme indices from 1950 to 2003. *Journal of Geophysical Research-Atmospheres* **110**: DOI 10.1029/2005jd006181.

5 Connecting climate and hydrological models for impacts studies

Emily Black

ABSTRACT

Driving hydrological models with climate data is a tough challenge – whether the data are from the observational record or climate models. One reason for this is that many hydrological models require long daily time-series of precipitation and evaporation. The scarcity of appropriate observed data in many parts of the MENA region is therefore a potential constraint for the development of such models. Although climate models have the capacity to produce daily time-series for the whole region, the results of impacts studies driven directly by model output would be prejudiced by model error – particularly in precipitation, which is one of the most difficult variables to simulate. This chapter describes how these problems can be addressed by using a simple statistical rainfall model (weather generator) in conjunction with a regional climate model. This enables climate model bias to be corrected, observed monthly data to be disaggregated and the length of a precipitation time-series to be extended.

5.1 INTRODUCTION

Driving hydrological models with climate model data provides a key means to understand the hydrological systems of the MENA region, how water availability has changed in the past and how it is projected to change in the future. In cases when it is not appropriate or not possible to use raw model output to drive a hydrological model, a statistical rainfall model (also known as a stochastic weather generator) can be used as an intermediate step. This chapter describes the development of a simple weather generator and how it can be used, in conjunction with observed data and a climate model of present-day precipitation, to correct model bias, disaggregate monthly data, extend time-series and provide data for regions where there are no observations. This weather generator is then used in Chapters 10, 11, 12 and 13 of this volume.

Stochastic weather generators are statistical models that produce synthetic time-series of one or more variables, such as precipitation and temperature, based on the underlying characteristics of the weather. They vary from simple models that aim to represent rainfall at a single station (such as the one developed for this study), to complex systems that can simulate the spatial interrelationships between several variables (e.g. Semenov and Barrow, 1997) or capture climate trends (e.g. Samuels et al., 2009). The complexity of the weather generator needed for any particular study depends on the availability of data, the type of climate and the purpose for which it is being used.

There are several reasons why using a weather generator may be preferable to driving hydrological models directly with observed or climate model data. Climate models have significant difficulty representing spatial and temporal variability in precipitation (see for example Chapters 4 and 8 of this volume). As precipitation is the primary driver of hydrological models, these errors introduce large uncertainties into our assessments of climate impacts on the hydrological system. Weather generators provide a means of producing daily time-series of precipitation and other variables based on bias corrected climate model output (see for example Jones et al., 2009 and Chapters 10, 11, 12 and 13 of this volume).

Running climate models is computationally expensive, which means that models with high enough resolution to produce output relevant to impact studies can often only be run for relatively short periods, typically 30–200 years. Moreover, there are few stations in the world with more than 200 years of good quality daily data, and none in the Middle East. In these situations, weather generators that have been informed by the observed or modelled climate statistics can be used to generate the long time-series required to assess the risk of rare events. This assumes, of course, that the statistics of the weather can be accurately extrapolated over long time periods.

Water, Life and Civilisation: Climate, Environment and Society in the Jordan Valley, ed. Steven Mithen and Emily Black. Published by Cambridge University Press.
© Steven Mithen and Emily Black 2011.

Weather generators can compensate for a lack of observed data or for the insufficient resolution of climate models. This may mean disaggregating monthly data, if monthly but not daily data is available (e.g. Hansen and Ines, 2005), or, in data-sparse regions, interpolating data between stations (e.g. Hutchinson, 1995). In a similar vein, weather generators can be used to compensate for the relatively low resolution of many climate models by either interpolating the statistics across grid boxes (as in Jones et al., 2009), or projecting changes onto a point (see later sections of this chapter).

5.2 DESIGN AND DEVELOPMENT OF A WEATHER GENERATOR

Several studies in this volume concerning hydrology and settlement in the southern Levant required a weather generator capable of simulating daily precipitation at a point. A simple system was chosen over a more complex one for several reasons. There are few stations with more than 30 years of daily precipitation data in the study region, and none with this quantity of temperature and solar radiation data. This would make it difficult to train the weather generator to incorporate co-variability between different aspects of the weather. In any case, histograms of hydrological effective rainfall (HER) derived using daily varying and climatological potential evaporation (PE) are virtually indistinguishable (plots not shown). This probably reflects the fact that in a hot climate like that of the Jordan Valley, any moisture that does not infiltrate the surface is rapidly evaporated for the entire range of PE experienced, except in the circumstances of particularly intense rainfall. It can be surmised, therefore, that co-variability between precipitation and variables such as cloudiness and temperature has little impact on flow. As a consequence, generating time-series of such additional variables would add little value. Some weather generators are designed to simulate rainfall at several stations simultaneously to incorporate co-variance between rainfall stations. The weather generator being developed here is, however, intended primarily for generating input for hydrological models that have been calibrated using data from a single station. Therefore, even though significant coherence in daily rainfall variability exists on the catchment scale, there is no need to incorporate spatial coherence into the weather generator.

Precipitation weather generators have two parts: the occurrence of precipitation – whether a given day is rainy or not – and the amount of rain per rainy day. When simulating precipitation occurrence, it is necessary to account for persistence. In other words, precipitation events tend to happen in clumps: several dry days followed by several wet days. This is particularly true for boreal winter rainfall in the Middle East, which mainly results from the passage of cyclones over the region (see Chapter 2). Weather generators can take this feature into account either by modelling precipitation happening in spells of dry or wet days, with the length of a spell having a particular probability (spell-length models), or by considering the probability of rainfall on a given day, conditional on whether it rained on previous days (Markov models). First-order Markov models take into account the rainfall occurrence just one day previous to the one in question; higher-order Markov models take several days into account.

The choice of method depends on the availability of data and the type of climate. In practice, spell-length models require at least 25 years of data to produce good results and tend to perform badly in arid regions (Roldan and Woolhiser, 1982). Moreover, in one of the first descriptions of a rainfall occurrence model, it was demonstrated that a first-order Markov model can reproduce the lengths of rainy spells in Tel Aviv (Gabriel and Neumann, 1962). For these reasons, a first-order Markov chain method was chosen.

When estimating the rainfall probabilities from observed or model data, each day of the year was considered separately in order to account for the seasonal cycle. For the Jordan Valley such daily data were available for the boreal winter. The mean climatological precipitation probabilities for a given day during this period were based on data for the day in question, and two days either side (i.e. five days). This meant that for a 20-year time-series the mean climatological conditional rainfall probabilities for each day were based on 100 values (5 values for each year for 20 years). The process was then repeated for each day between September and May. In practice, even with this smoothing and averaging, the resulting seasonal cycle was noisy. The values were therefore smoothed with a 10-day running mean. For the boreal summer, because no observed data were available, both the rainfall probabilities given rain and given no rain the day before were set to 0.01.

Given the conditional rainfall probabilities, it is straightforward for the weather generator to determine whether or not rainfall should occur. A random number between 0 and 1 is generated for each day. If the random number is less than the relevant probability for that day, then it rains, and if it is greater than the probability, it is dry. So for a summer day, if the random number is greater than 0.99, there will be rain and if it is less than 0.99 there will be no rain.

There are various approaches to the second part of the weather generator: simulating rainfall amount. If the generated time-series are of approximately the same length or shorter than the data with which the weather generator was trained, it may be possible simply to interpolate the observed distribution of rainfall intensity (Semenov and Barrow, 1997). This approach has the advantage of avoiding the biases and assumptions that inevitably arise when trying to fit a theoretical distribution to observed rainfall. If the time-series being generated is longer than that with which the weather generator was trained, there is the potential for bias caused by a failure to sample the distribution adequately. For example, a 30-year time-series cannot be expected to pick up a 10,000-year event. If such an event did occur during the 30-year period, its probability would be greatly inflated and the distribution distorted. Of course, it is possible to filter out such outliers and avoid this type of bias. Nevertheless, such an approach cannot accurately estimate the probability of extremes.

The weather generator in this study was required to produce time-series of 100 years or more, based on at most 30 years of training data, and for this reason, a theoretical distribution was used. It was found that a gamma distribution was not able to simulate the extremes of the rainfall, so a theoretical extremes distribution was used to estimate values greater than the 95th percentile. The parameters of the distributions were estimated using the maximum likelihood method, as described in Wilks (1995). This method of using two distributions spliced together is described by Furrer and Katz (2007), who found this hybrid approach to be an improvement over using a single theoretical distribution to represent the entire range of rainfall.

5.3 HOW WELL CAN THE WEATHER GENERATOR REPRODUCE THE OBSERVED SEASONAL CYCLE?

The weather generator was trained with the same daily rainfall data that were used to calibrate the hydrological model described in Chapter 10 of this volume: station 320650 near Mount Keenan and Obstacle Bridge.

Figure 5.2 compares the seasonal cycles of rainfall and the rainfall amounts as produced by the weather generator against the observed data. It can be seen that while the weather generator does a good job of simulating rainfall amounts at the peak of the rainy season, and captures the onset of the rainy season during September and October, it overestimates rainfall during March. This overestimation is probably related to the smoothing carried out for the parameter estimation, which would have the effect of including features of the January and February climate in the estimation of March rainfall probabilities and intensity.

As well as the seasonal cycle in rainfall totals, Figure 5.2 shows the daily rainfall statistics. It can be seen that rainfall probabilities are accurately simulated throughout the year. Comparison between the observed and simulated rainfall intensity histograms shows that the weather generator does fairly well at capturing intensity, although there is some bias around 20 mm (approximately the 90th percentile): the weather generator tends to simulate a somewhat larger fraction of days with rainfall around this amount than is observed.

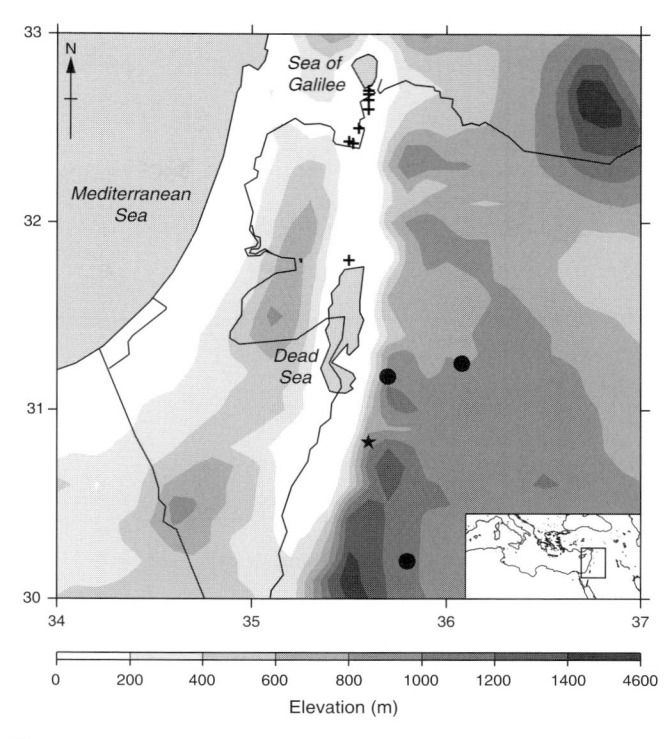

Figure 5.1. Location of gauges referred to in the chapter superposed on the topography. The inset map shows the location of the main map. Crosses are the locations of the stations provided by the Israeli Meteorological Service. The star is Tafilah and the circles are monthly data used in Chapters 12 and 13 and referred to here.

The methodology used to simulate rainfall amount was explored in more detail in Figure 5.3, which shows the ranked rainfall amount on each rainy day for the simulated and observed time-series. In other words, the rain on the nth rainiest day in the observed time-series is plotted against the rain on the nth rainiest day in the simulation. Both a single gamma and spliced gamma/extreme value distributions are shown. It can be seen that, while there is still some bias, the spliced distribution adopted for the WLC weather generator is a considerable improvement over the single gamma distribution.

5.4 USING THE WEATHER GENERATOR IN CONJUNCTION WITH A REGIONAL CLIMATE MODEL TO DISAGGREGATE MONTHLY RAINFALL DATA

Elsewhere in this volume, the weather generator is used in conjunction with the regional climate model to disaggregate monthly data (e.g. Chapter 13). We have developed a method of doing this that exploits the available observations of daily precipitation, and the regional model's ability to reproduce the main features of the spatial pattern of precipitation within Jordan and Israel (see Chapter 4). The method used has two steps.

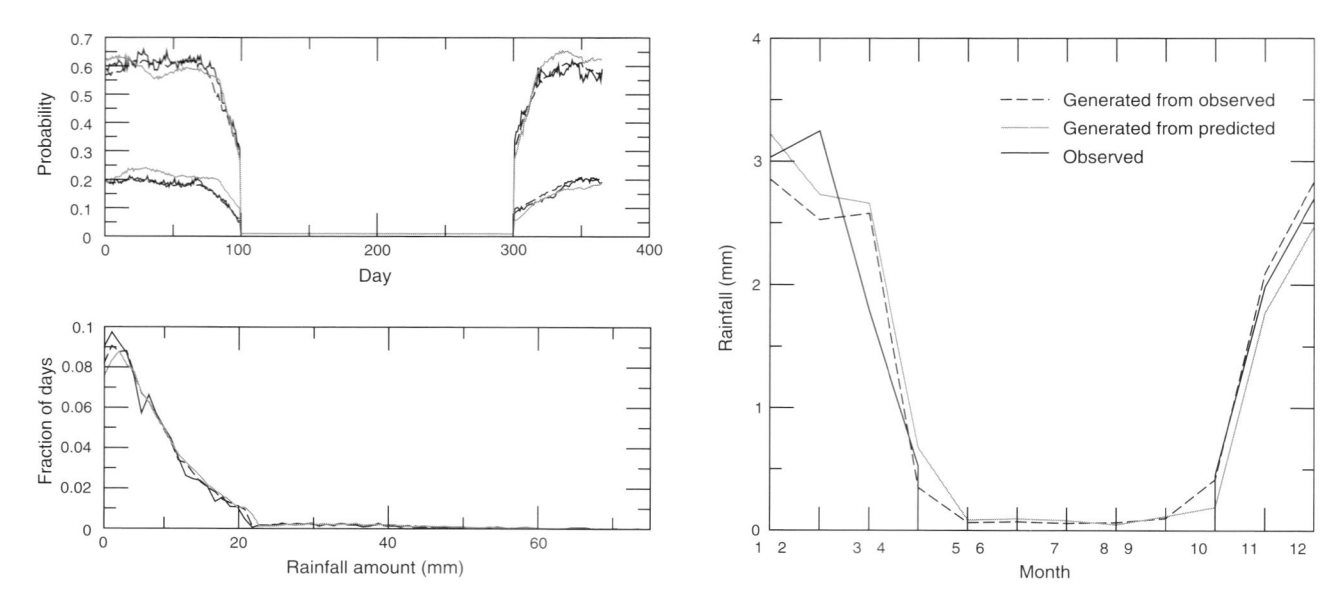

Figure 5.2. Seasonal cycles of rainfall statistics for the observations (black line), the weather generator based on observed statistics (dashed line) and the weather generator based on the predicted statistics (grey line). Top left: rainfall probabilities (upper lines are PRR and lower lines are PDR); bottom left: rainfall amount fractional frequency histogram; right: rainfall totals (mm day^{-1}).

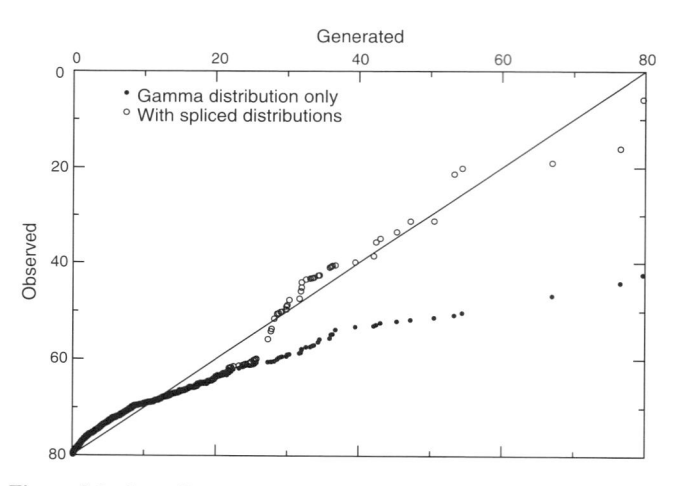

Figure 5.3. Quantile–quantile plot of observed versus simulated rainfall amounts for a single gamma distribution (filled circles) and for spliced gamma/extreme value distribution (open circles). The line represents a $y = x$ function on which the circles would lie if the theoretical distribution perfectly matched the observations.

First, we undertake a correction of the RCM bias in rainfall probabilities. The RCM bias in rainfall probabilities is estimated by comparing observations from the nearest station, for which daily data are available, with the probabilities derived from the RCM baseline integration for the appropriate grid point. The correction is then applied to the RCM probabilities for the grid point of the predicted time-series, producing a predicted rainfall probability seasonal cycle. This step assumes that the RCM bias is the same for both locations.

Second, we make an estimation of rainfall intensity. The rainfall intensity histogram is assumed to be a scaled version of the histogram at the nearest station for which daily data are available. In order to determine what factor the rainfall distribution should be scaled by, the statistical rainfall model is driven by the corrected baseline rainfall probabilities (as described above) and the rainfall distributions scaled by different factors. The annual total rainfall for each time-series is then calculated and compared to the observations. The scaling factor which best replicates the observed annual total rainfall from each location is selected.

In order to test the method, data from Tafilah (also spelt Tafileh) in western Jordan were used in conjunction with the regional climate model to predict the seasonal cycle in rainfall at station 320650 near Obstacle Bridge. The results are shown in Figure 5.2. It can be seen that the regional model is able to account for the differences in the rainfall probabilities and shape of the seasonal cycle between Tafilah and 320650. Moreover, it is clear that the scaled Tafilah intensity is a good approximation to the observed intensity at 320650. The weather generator's ability to predict the statistics of the rainfall at Tafilah with a reasonable level of accuracy reflects the regional model's skill in capturing the spatial variability in rainfall statistics. This lends credence to both the weather generator and the regional model.

5.5 DISCUSSION

This chapter described both the development and implementation of a simple weather generator for hydrological studies of water availability in Jordan and Israel. It has been shown that the weather generator can produce realistic daily time-series for the present day, when it is informed by the observed annual total and a regional model capable of capturing the shape of the seasonal cycle in monthly totals, the seasonal cycle in daily probabilities and the gross spatial pattern of precipitation. The weather generator can be used in this way to produce time-series at a point because the daily occurrence of rainfall is coherent (in other words, if it is raining at one place within a grid square, there is a significant probability that it is raining elsewhere).

Several studies in this volume use the weather generator, in conjunction with regional climate model integrations, to produce past and future daily time-series at points within Jordan and Israel. Of course, the weather generator cannot correct all climate model biases, nor can its success with present-day rainfall be taken as proof that it can simulate the past and future. In addition, using a weather generator may introduce additional bias, particularly when it is trained with short time-series, as is the case here. The largest error in the climate model simulations is the rainfall intensity, with the daily probabilities well represented in comparison. Estimates of rainfall intensity are therefore based entirely on observations (with percentage changes derived for the past and future from projected changes in the annual total), and the rainfall probabilities are based on climate model data, corrected using the observations from the nearest available station. Treating the data in this way involves two sets of assumptions: first, that the errors in the rainfall probabilities are sufficiently uniform that corrections from one station can be applied to another; second, that the changes in observed rainfall total and probabilities projected for the past and future are accurate.

The first assumption would seem reasonable, considering the climate model's ability to capture present-day spatial variability, and the weather generator's ability to predict station rainfall statistics when informed by the climate model and observed data. In itself, the second assumption appears the more dubious. However, the comparisons with the observed record reported in Chapters 3, 4 and 8 lend credence to the climate model, and suggest that, in certain circumstances, the assumption that the model is capturing changes in annual totals and rainfall probabilities is valid. The question of climate model credibility is discussed in depth in Chapters 3, 4 and 8 of this volume.

As was mentioned above, the weather generator itself introduces biases into the predictions. Being based on a first-order Markov model, it does not take into account changing rainfall probabilities within rainy or dry spells, which tends to mean that the length of rainy spells is underestimated (Wilks and Wilby, 1999). Moreover, the assumption that rainfall intensity does not change within the season biases the monthly precipitation totals. These compromises in the methodology are an inevitable consequence of the lack of observed daily rainfall data with which to train the model. In practice, however, the evaluations performed here show that these biases, while detectable, are small, and that the daily time-series generated by the weather generator are a vast improvement over the raw climate model data.

A more serious bias is in the treatment of extremes. Even though we have attempted to account for sampling bias by extrapolating the data, it is likely that significant error remains. Furthermore, it is not possible to evaluate the weather generator's treatment of extremes using the short time-series available. Therefore, the methodology described should be used only with the greatest of caution to investigate the effect of changes in climate extremes on the hydrological system.

5.6 SUMMARY AND CONCLUSIONS

- A weather generator has been developed as an intermediate step between the climate model output and the hydrological models described in subsequent chapters.
- The weather generator simulates the seasonal cycle in rainfall reasonably well when forced with observed data.
- When informed with a regional model simulation of the present-day climate in conjunction with observed annual totals, the weather generator is capable of accurately predicting the seasonal cycle in rainfall statistics for individual rain gauges. This lends credence both to the weather generator and to the climate model.

REFERENCES

Furrer, E. M. and R. W. Katz (2007) Generalized linear modeling approach to stochastic weather generators. *Climate Research* **34**: 129–144.

Gabriel, K. R. and J. Neumann (1962) A Markov chain model for daily rainfall occurrence at Tel Aviv. *Quarterly Journal of the Royal Meteorological Society* **88**: 90–95.

Hansen, J. W. and A. V. M. Ines (2005) Stochastic disaggregation of monthly rainfall data for crop simulation studies. *Agricultural and Forest Meteorology* **131**: 233–246.

Hutchinson, M. F. (1995) Stochastic space-time weather models from ground-based data. *Agricultural and Forest Meteorology* **73**: 237–264.

Jones, P. D., C. Harpham, C. G. Kilsby, B. Glenis and A. Burton (2009) Projections of future daily climate for the UK from the Weather Generator. UK Climate Projections Science Reports **3**. University of Newcastle.

Roldan, J. and D. A. Woolhiser (1982) Stochastic daily precipitation models: 1. A comparison of occurrence processes. *Water Resources Research* **18**: 1451–1459.

Samuels, R., A. Rimmer and P. Alpert (2009) Effect of extreme rainfall events on the water resources of the Jordan River. *Journal of Hydrology* **375**: 513–523.

Semenov, M. A. and E. M. Barrow (1997) Use of a stochastic weather generator in the development of climate change scenarios. *Climatic Change* **35**: 397–414.

Wilks, D. S. (1995) *Statistical Methods in the Atmospheric Sciences: An Introduction.* San Diego: Academic Press.

Wilks, D. S. and R. L. Wilby (1999) The weather generation game: a review of stochastic weather models. *Progress in Physical Geography* **23**: 329–357.

Part II
The palaeoenvironmental record

6 A review of palaeoclimates and palaeoenvironments in the Levant and Eastern Mediterranean from 25,000 to 5,000 years BP:[1] setting the environmental background for the evolution of human civilisation

Stuart Robinson, Stuart Black, Bruce Sellwood and Paul J. Valdes

ABSTRACT

The southern Levant has a long history of human habitation, and it has been previously suggested that climatic changes during the Late Pleistocene to Holocene stimulated changes in human behaviour and society. In order to evaluate such linkages it is necessary to have a detailed understanding of the climate record. We have conducted an extensive and up-to-date review of terrestrial and marine climatic conditions in the Levant and eastern Mediterranean during the past 25,000 years. We firstly present data from general circulation models (GCMs) simulating the climate for the Last Glacial Maximum (LGM), and evaluate the output of the model by reference to geological climate proxy data. We consider the types of climate data available from different environments and proxies and then present the spatial climatic 'picture' for key climatic events. This exercise suggests that the major Northern Hemisphere climatic fluctuations of the past 25,000 years are recorded in the eastern Mediterranean and Levantine region. However, this review also highlights problems and inadequacies with the existing data.

6.1 INTRODUCTION

The effects of global climate change on human society, on the environments we inhabit and on the sustainable development of our planet's poorest people are of prime concern to all. Predicting changes in water availability, local environments and climates will be the key to determining which areas of the world will require greatest assistance in dealing with increased global warmth and climate change in the coming century, and beyond. However, the detailed linkages between the development of human civilisation, anthropogenic activities, climate and environmental change remain poorly understood. Through a more detailed study of how climate and environmental changes influenced the history of human civilisation, it is hoped that a better understanding of the relationship between these key elements can be deduced. The Levantine region (encompassing the modern countries of Israel, Jordan, Lebanon, Syria and parts of Egypt and Saudi Arabia; Figure 6.1) has a long history of human habitation (e.g. Issar and Zohar, 2004), making it an ideal area to study linkages between climatic, environmental and societal change.

Palaeoclimatic records, such as those obtained from ice cores, demonstrate the generalised climatic evolution of the past 25,000 years (Figure 6.2). The long-term trend during this interval is from the cold, glacial conditions of the Late Pleistocene to the warm, interglacial conditions of the Holocene. This long-term trend is punctuated by several shorter climatic events (Heinrich events, the Younger Dryas and the '8.2 ka cold event') that seem to have had a significant effect on climatic conditions in much of the Northern Hemisphere. The origin of these events may be linked to the rapid input of cold freshwater into the North Atlantic causing a disturbance in oceanic circulation and local climatic regimes (e.g. Broecker and Denton, 1989; Bond et al., 1992, 1993; Alley, 2000; Alley and Ágústsdóttir, 2005; Rohling and Pälike, 2005). The Levantine region straddles the arid/semi-arid boundary and is thus highly sensitive to climatic changes. Therefore, rapid climatic changes in the past may have had sufficient effect on local environmental conditions to cause adaptation in human behaviour and society.

In the Levant region, where the evidence for human civilisation extends furthest back in time, it has been suggested that there are causal links between major Northern Hemisphere climatic events and the evolution of human civilisation (e.g. as reviewed

Water, Life and Civilisation: Climate, Environment and Society in the Jordan Valley, ed. Steven Mithen and Emily Black. Published by Cambridge University Press.
© Steven Mithen and Emily Black 2011.

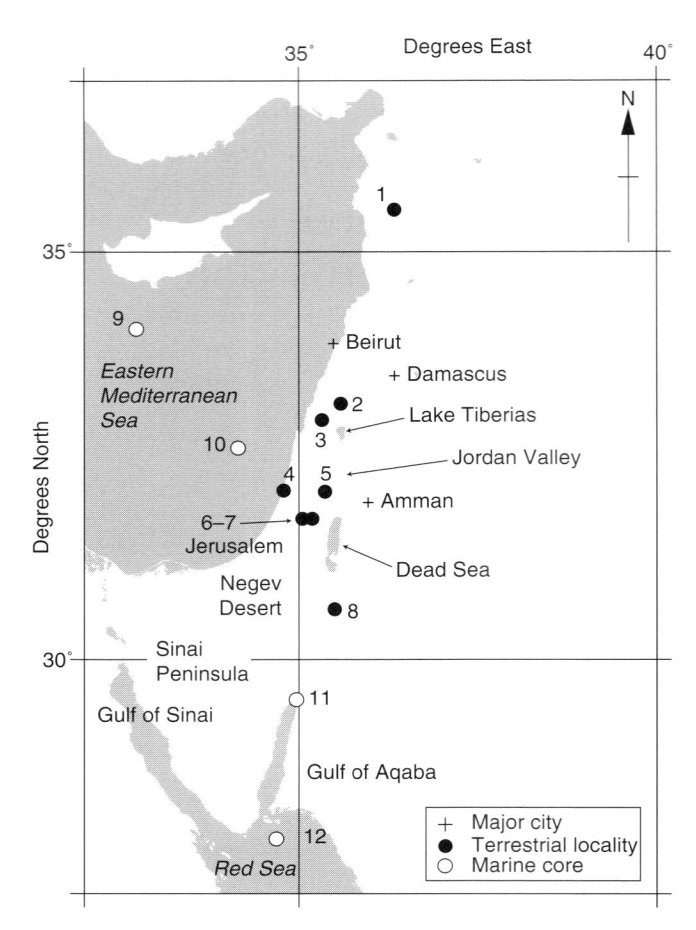

Figure 6.1. Map of the eastern Mediterranean and Levant region showing the locality of major features and sites discussed in the text. 1: Ghab Valley; 2: Hula Basin; 3: Peqiin Cave; 4: Israeli coastal plain; 5: Ma'ale Efrayim Cave; 6: Soreq Cave; 7: Jerusalem West Cave; 8: Wadi Faynan; 9: Ocean Drilling Program (ODP) Site 967; 10: site of core M44–1-KL83; 11: site of core GeoB5804–4; 12: site of core GeoB5844–2. Map produced with GMT (http://gmt.soest.hawaii.edu/).

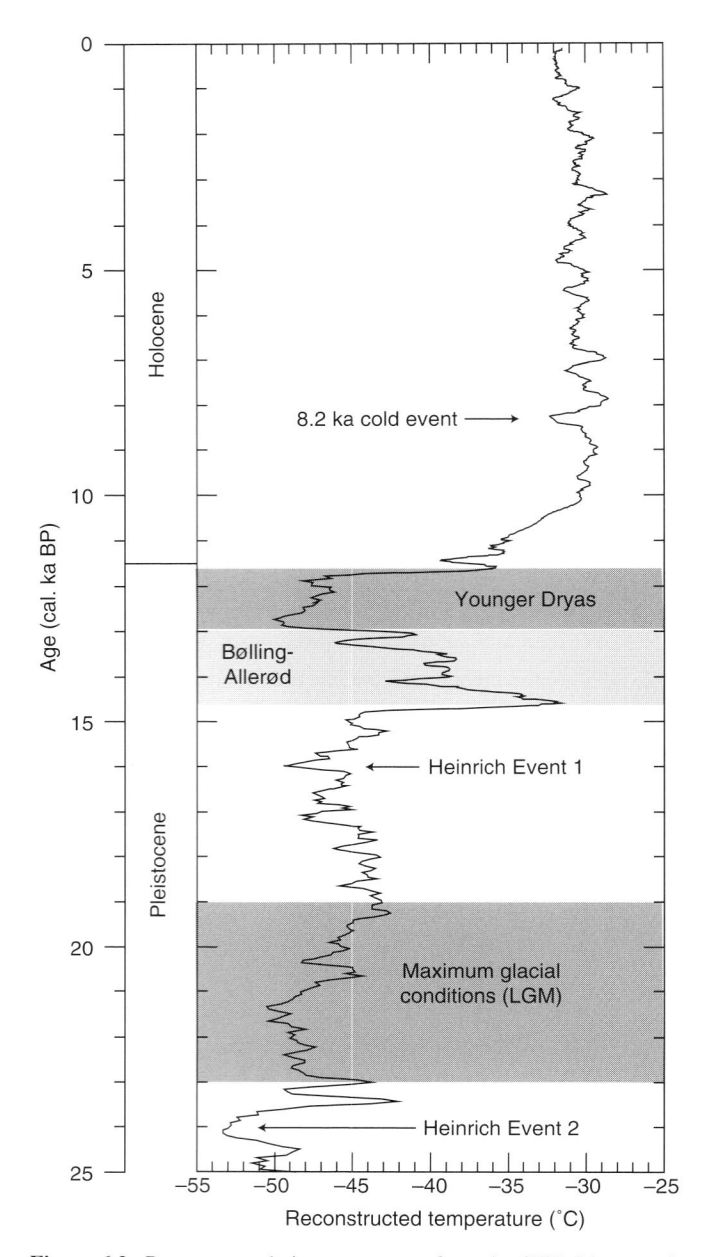

Figure 6.2. Reconstructed air temperatures from the GISP 2 ice core in Greenland (after Alley, 2000). The timing and duration of the Last Glacial Maximum (LGM) is the same as the 'LGM Chronozone Level 1' as defined by Mix *et al.* (2001).

in: Moore and Hillman, 1992; Harris, 1996; de Menocal, 2001; Mithen, 2003; Issar and Zohar, 2004). The development of hunter–gatherer sedentism and the earliest semi-permanent settlements in the Levant occurred during the late Glacial Interstadial, known as the Bølling–Allerød warm interval (e.g. Bar-Yosef and Belfer-Cohen, 1992; Moore and Hillman, 1992; Bar-Yosef, 2000). During the Younger Dryas human communities in the Levant generally returned to a hunter–gatherer existence (e.g. Bar-Yosef and Belfer-Cohen, 1992; Bar-Yosef, 2000), possibly as a response to colder, drier climatic conditions. The 'Neolithic revolution', including the advent of farming and the first villages, occurred after the Younger Dryas during the warm early Holocene. Unfavourable climatic conditions at about 8.2 and 4.2 cal. ka BP (calendar (calibrated) kiloyears before present) have been invoked as the cause of the abandonment of PPNB (Pre-Pottery Neolithic B) farming towns in the Jordan Valley and the collapse of the Akkadian civilisation, respectively (e.g. Cullen *et al.*, 2000; de Menocal, 2001; Mithen, 2003).

A number of authors have reviewed various aspects of Late Quaternary palaeoclimates and palaeoenvironments in the eastern Mediterranean region (e.g. Goodfriend, 1999; Issar, 2003; Issar and Zohar, 2004), but these reviews have often focused on specific time periods (i.e. in the studies cited above, the Holocene), environments or sets of techniques. The aim of this work is to provide a thorough, yet accessible, review of the current state of our knowledge of palaeoclimates, and palaeoenvironments in the context of palaeoclimatic changes, throughout the Levant and eastern Mediterranean Sea from 25 to ~5 cal. ka BP. We focus on this time period as it represents (i) the interval in which the magnitude of environmental and climatic change was largest (and therefore most easily observed in the geological record); (ii) the period during which human behaviour changed from a mobile to sedentary lifestyle and (iii) the period least subject to anthropogenic environmental change, thereby allowing a clearer examination of the 'natural' effects of climate change on human behaviour. After 5 cal. ka BP, the magnitude of climate changes is much smaller and it is deemed inappropriate to discuss these here. We note, however, the possibility of a smaller magnitude event at 4.2 cal. ka BP (e.g. Cullen et al., 2000; de Menocal, 2001), whose impact was significant for, by now, large, urbanised communities such as those in Egypt and Akkadia. Because of the unique climatic setting of the Levant, we have restricted our review to the region shown in Figure 6.1. Reviews of palaeoclimatic conditions in the adjoining areas (i.e. Turkey and North Africa) during the Late Pleistocene–Holocene have been published elsewhere (e.g. Roberts and Wright, 1983; Gasse and Fontes, 1992; Roberts et al., 2004; Verschuren et al., 2004). We review the different types of palaeoclimatic data available for both terrestrial and marine environments, and synthesise these data for key intervals. In the course of this exercise we pay particular attention to the quality, reliability and accuracy of the various proxies and their age models. This study provides an overview of climatic conditions in the region and allows the identification of key geographic and stratigraphic gaps.

6.2 DATING

A key aim of this review is to compare and contrast various proxies, which have been dated by a variety of techniques. All ages reported in this paper are in calendar (calibrated) years (or kiloyears, ka) before present (cal. years BP or cal. ka BP). Uncalibrated radiocarbon ages (^{14}C ages) have been recalibrated using OxCal v. 3.10 (Bronk Ramsey, 1995, using atmospheric data from Reimer et al., 2004), to allow comparison with calendar ages (such as those from ice cores) and uranium-series ages (principally from carbonates). It should be noted that ^{14}C calibration using tree rings extends currently to around 13 cal. ka BP; calibrated ages for the preceding Late Pleistocene period are much less secure, so that the possibility of mis-correlation increases.

6.3 CLIMATE MODELS

In recent years, general circulation models (GCMs) have been used to evaluate past climates. These models use the laws of physics and an understanding of past geography to simulate climatic responses. They are objective in character and it is now possible to compare results from different GCMs for a range of times and over a wide range of parameterisations for the past, present and future (e.g. in terms of predictions of surface air temperature, surface moisture and precipitation).

We present here some outputs on climate for the Last Glacial Maximum (LGM) from model simulations generated using the HadAM3 version of the UK Meteorological Office (UKMO) coupled atmosphere–ocean GCM (Pope et al., 2000). The model was developed at the Hadley Centre for Climate Prediction and Research, which is part of the UKMO. The LGM simulation follows the protocol of the Palaeoclimate Model Intercomparison Project (PMIP, see Joussaume and Taylor, 1995). The boundary conditions on the model include the CLIMAP sea surface temperature reconstructions (CLIMAP, 1981) and Peltier (1994) ice-sheet reconstructions. The climate model provides new insights for the formulation of working hypotheses, providing a range of parameterisations, some of which can be expected to leave geological proxy signatures (e.g. temperature, precipitation, salinity) and others which, though significant, may not (e.g. snowfall, cloud cover, wind velocities, surface pressure). Lakes (as closed systems) respond through time to changes in the balance between precipitation and evaporation (P–E). Similarly rainfall, combined with the overall temperature regime, will influence palaeosol generation. The geological record of such sediments can be compared directly with climate model outputs.

We have run simulations for the LGM, and it is instructive to compare the outputs from the model with modern observations (Figure 6.3 and Figure 6.4). Most precipitation today occurs in the eastern Mediterranean during the winter months (December, January and February, DJF; Figure 6.3A). In the far north of the area, precipitation, mostly from the Atlantic, falls as rain (and some snow) at between 2 and 4 mm per day (Anatolian Uplands northwards of 36°N). There is a rapid decline in the amount of precipitation southwards (see Figure 6.3E), falling to between 0.5 and 1.0 mm day^{-1} around 30°N. This marked contrast in rainfall, between north and south, has been noted by previous authors and, by reference to palaeoclimatic geological proxy data, is believed to have operated in past times too (e.g. Enzel et al., 2003 and references therein). Average winter (DJF) temperatures range between 8 and 12°C. During the summer months (June, July and August, JJA) virtually the whole region experiences high temperatures and almost no precipitation (Figure 6.4). In the south, and over the Dead Sea, there is a marked excess of evaporation over precipitation which, coupled with the major abstraction of water from the Jordan system, has resulted in Dead Sea levels dropping by around 0.5 m per annum in recent years. The annual average rainfall between about 31° and 36°N is around 180 mm yr^{-1}, rising along the coastal strip as far east as Jerusalem, to around 400–450 mm per annum (Figure 6.3E).

Model results for the LGM at first sight seem very similar to the observations for the present, with most precipitation

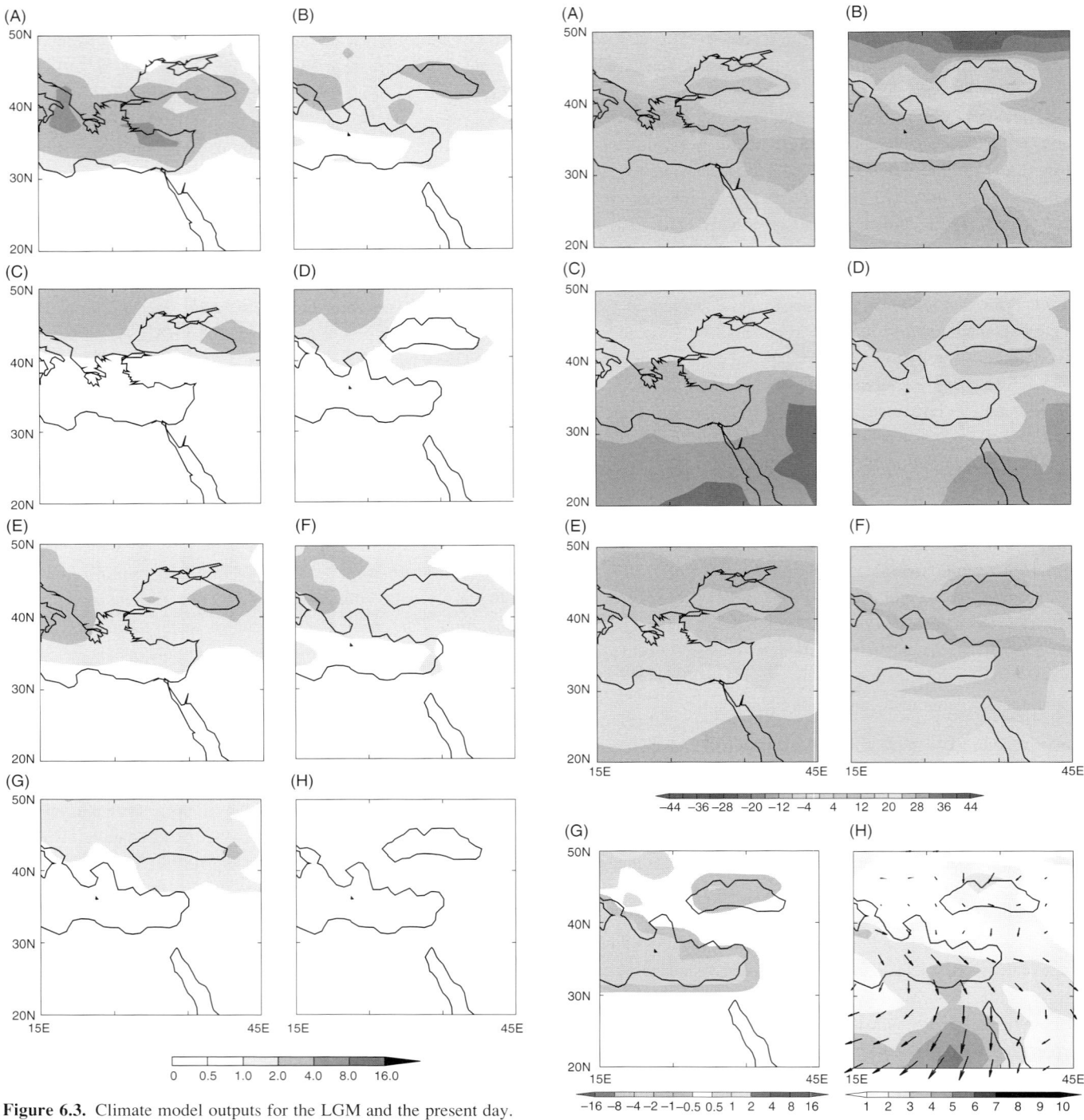

Figure 6.3. Climate model outputs for the LGM and the present day. (A) Present-day winter (DJF) precipitation (precipitation in mm per day); (B) LGM winter (DJF) precipitation; (C) present-day summer (JJA) precipitation; (D) LGM summer (JJA) precipitation; (E) present-day annual precipitation; (F) LGM annual precipitation; (G) LGM winter (DJF) snowfall (snowfall in mm per day); (H) LGM summer (JJA) snowfall. Note that panel H is blank because there is no snowfall in summer.

Figure 6.4. Climate model outputs for the LGM and present day. (A) Present-day winter (DJF) temperature (°C); (B) LGM winter (DJF) temperature; (C) present-day summer (JJA) temperature; (D) LGM summer (JJA) temperature; (E) present-day average annual temperature; (F) LGM average annual temperature; (G) LGM annual average precipitation minus evaporation in mm day^{-1}; (H) LGM annual average wind strength (in m s^{-1}) and vectors. See colour plate section.

modelled to occur during the winter (DJF; Figure 6.3B), but with 1–2 mm day^{-1} precipitation extending between 37 and 31°N, particularly adjacent to the coastal strip. Summers, like today, are modelled to be very dry throughout most of the area (Figure 6.3C). However, during the LGM, much of the winter precipitation over the Anatolian uplands falls as snow, released in a major spring thaw (Figure 6.3G and H). The resolution of the model (2.5° in latitude and 3.75° in longitude) precludes depiction of the Mt Hermon Jordan catchment as a discrete area, but it too was likely to have received very heavy snowfalls. Anatolia is modelled to receive some rain even during the summer months (Figure 6.3). The overall pattern is for a small but significant increase in precipitation across the northern parts of the area. Modelled temperatures are much lower than present (Figure 6.4), and account for less evaporation, but still evaporation exceeds precipitation (Figure 6.4G). The lower temperatures help to explain how lake levels (e.g. Lake Lisan) could have grown during the glacial phase, even though the Earth climate system was generally colder and drier. Winds during the LGM are modelled to be out of the north and northwest, generally strengthening southwards (Figure 6.4H). We will now examine the climate record from 25 cal. ka BP to 5 cal. ka BP, as recorded by a range of geological data, and consider model performance for the LGM in light of these, so that its possible application to intervening times might be evaluated.

6.4 DESCRIPTION OF THE CLIMATE RECORDS

6.4.1 Non-marine sedimentary records

LACUSTRINE SEDIMENTS AND LAKE LEVELS

The long-term rise and fall of lake levels is the result of the interplay between climate (controlling precipitation, runoff, evaporation and global sea level) and tectonics (controlling basin shape, catchment area and sill depths). Although the Jordan Rift Basin is tectonically active, and has been throughout the Pleistocene and Holocene (e.g. Marco et al., 1996; Enzel et al., 2000; Migowski et al., 2004), most authors have suggested that climate was the dominant control on lake levels in the rift valley (e.g. Begin et al., 1985; Frumkin et al., 1991, 1994, 2001; Yechieli et al., 1993; Neev and Emery, 1995; Stein, 2001; Bartov et al., 2002, 2003; Frumkin and Elitzur, 2002; Landmann et al., 2002; Enzel et al., 2003; Hazan et al., 2005). During the last glacial period, Lake Lisan covered much of the Jordan Valley (e.g. Neev and Emery, 1967) encompassing the present-day Dead Sea and Lake Tiberias (also known as Sea of Galilee or Lake Kinneret). Sometime in the late glacial (probably after the Younger Dryas), Lake Lisan shrank to a configuration similar to the present situation. Lake levels have been reconstructed by combined

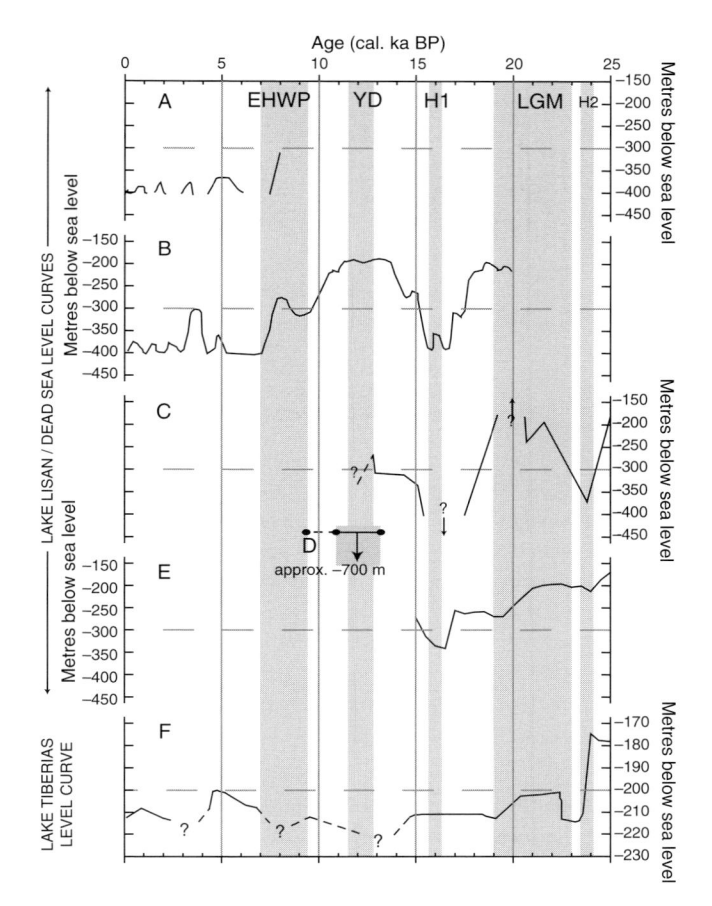

Figure 6.5. Compilation of lake level curves for Lake Lisan/the Dead Sea and Lake Tiberias. (A) Frumkin et al. (1994); (B) Neev and Emery (1995); (C) Landman et al. (2000); (D) timing of massive salt deposition (Yechieli et al. (1993); Neev and Emery (1967); there is some uncertainty regarding the exact age of the sediments, hence the dashed line. Shading of the YD here indicates the range of the two best dates); (E) Bartov et al. (2002, 2003); (F) Hazan et al. (2004). EHWP = Early Holocene Wet Phase; YD = Younger Dryas; H1 = Heinrich Event 1; LGM = Last Glacial Maximum; H2 = Heinrich Event 2.

chronostratigraphy and facies analysis (e.g. Begin et al., 1985; Yechieli et al., 1993; Neev and Emery, 1995; Bartov et al., 2002, 2003; Landmann et al., 2002; Hazan et al., 2005), and dating of flood sediments in elevated saltcaves and karstic topography (e.g. Frumkin et al., 1991, 1994, 2001; Frumkin and Elitzur, 2002). Figure 6.5 shows a compilation of different lake level curves for the interval from 25 cal. ka BP to the present for both Lake Lisan and Lake Tiberias. All the studies agree that lake levels in the Lake Lisan/Dead Sea system were relatively high during the last glacial and relatively low during much of the Holocene. However, on a shorter timescale (<5000 years) there are clearly a number of inconsistencies between the different studies, which may be critical for understanding the palaeoclimatic significance of lake level curves in the region. A major problem is the difficulty in constructing high-resolution age models for

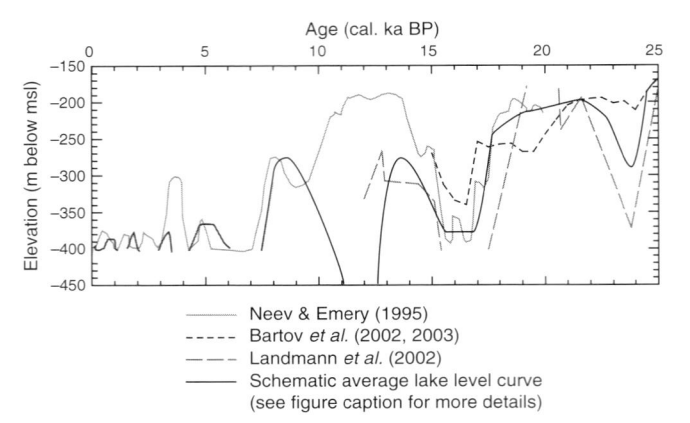

Figure 6.6. An integrated, schematic lake level curve (solid black line) for the Lake Lisan/Dead Sea based upon various studies. This curve is designed primarily to illustrate lake level trends over time for ease of comparison with other proxy data. For the period 25 to 13 cal. ka BP the integrated curve is an approximate average of Neev and Emery (1995), Bartov et al. (2002, 2003) and Landmann et al. (2002). From 13 to 9 cal. ka BP we have used the data from Neev and Emery (1967), Begin et al. (1985), Yechieli et al. (1993) and Stein (2001) which suggest a major lake level fall between 13 and 11 cal. ka BP. From 9 cal. ka BP onwards we have followed the study of Frumkin et al. (1994).

sequences that do not contain sufficient datable materials and do contain a significant number of hiatuses by virtue of their marginal-lake setting. Additionally, the deposition of thick evaporites has meant that some parts of the sections are not dated (e.g. Yechieli et al., 1993). Neev and Emery (1995) studied more distal lake-facies present in boreholes and constructed their lake level curve by using downhole gamma-ray logs to identify evaporitic (lowstand) and siliciclastic (highstand) sediments, but their age model is based upon a combination of ^{14}C dates that may be compromised by old carbon and climatic assumptions inferred from the archaeological record. By comparing and carefully considering the merits of the different datasets, we have produced an integrated reconstruction (solid black line in Figure 6.7) of Lake Lisan/Dead Sea levels to ease comparison between lake level data and other climatic proxies.

Maximum lake levels for the last glacial were achieved at about 25 cal. ka BP (Bartov et al., 2002, 2003; Landmann et al., 2002). Both Bartov et al. (2003) and Landmann et al., (2002) suggest that a major lowering in Lake Lisan occurred at about 24 cal. ka BP, approximately synchronous with Heinrich Event 2 (H2; 23.8 cal. ka BP) although the magnitude of lake level fall is unclear (compare Landmann et al., 2002; Bartov et al., 2003). At the LGM, lake levels were high (compared with H2) but may have been falling during the peak of LGM conditions, synchronous with increased local aridity proposed at this time (see discussion below), and as would be predicted from GCM results (Figure 6.4G). All the datasets seem to suggest the existence of a lowstand centred about 16 to 17 cal. ka BP, synchronous with

Heinrich Event 1 (H1) in the North Atlantic (Bartov et al., 2003). Bartov et al. (2002, 2003) suggest that lake levels peaked after H1 at 15 cal. ka BP. Both Neev and Emery (1995) and Landmann et al. (2002) suggest that a highstand was reached about 13 cal. ka BP (during the Bølling–Allerød warm interval), although the magnitude and duration is vastly different in both studies. Between 13.2 and 11.4 cal. ka BP a substantial halite layer was deposited in the current Dead Sea Basin (Neev and Emery, 1967; Yechieli et al., 1993), which Neev and Emery (1967) interpreted as the result of a lake level drop to about −700 m below present-day sea level. Yechieli et al. (1993) correlate this extremely arid event with the Younger Dryas of Northern Europe. In contrast, Neev and Emery (1995) found no evidence for salt deposition at this time in the southern basin and suggested that lake levels were high during this period of time. However, the bracketing of ages around the salt layer in the northern basin and other evidence from speleothems and palaeosols (discussed in more detail below) suggests that the Younger Dryas in the Levant was more likely to have been an arid period marked by extremely low lake levels.

Frumkin et al. (1991, 1994) dated Holocene wood fragments found in flood sediments in elevated salt caves to determine the timing of highstands. A first highstand is indicated as occurring ~8.5 cal. ka BP, roughly synchronous with a small highstand in the Neev and Emery (1995) curve. By 6.5 cal. ka BP, the level of the Dead Sea had fallen sufficiently to result in the desiccation of the southern basin (Frumkin et al. 1991, 1994, 2001; Neev and Emery, 1995). Frumkin et al. (1991, 1994) and Neev and Emery (1995) appear to be broadly in agreement over the timing of relatively higher lake levels at about 5 cal. ka BP.

Absolute lake levels in Lake Tiberias and Lake Lisan were similar in the last glacial (see figure 8 in Hazan et al. 2005) but started to diverge at ~19 cal. ka BP for reasons that are currently unclear but are probably a combination of tectonics, basin morphology and climatic change. However, both lakes generally display similar variations in lake-level throughout the past 25,000 years (Figure 6.5; Hazan et al. 2005). Maximum levels were attained in Lake Tiberias during the last glacial between 26 and 24 cal. ka BP. A major lowering of lake levels at 24 cal. ka BP may have been a consequence of Heinrich Event H2. The level of Lake Tiberias rose slightly after this event, reaching a highstand at about 22.5 cal. ka BP after which there was a gradual decline until about 15 cal. ka BP. At that point in time, lake levels fell, although the timing and magnitude is uncertain. During the Younger Dryas Lake Tiberias was probably very low before rising in the early Holocene. Hazan et al. (2005) report a highstand at ~5 cal. ka BP, synchronous with high lake levels in the Dead Sea.

PALAEOSOLS

Palaeosol sequences on the coastal plain of Israel have been constrained by luminescence and radiocarbon dating (Gvirtzman

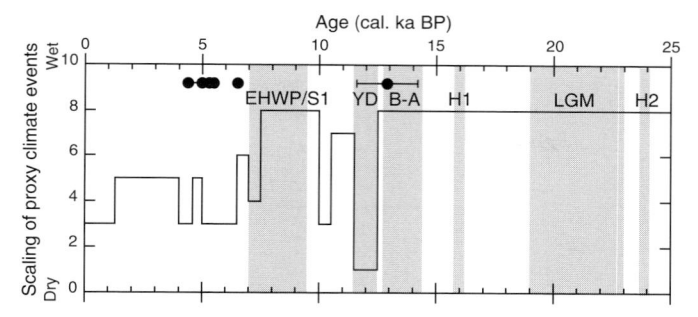

Figure 6.7. Palaeoclimate of the Israeli coastal plain, as interpreted from palaeosols (Gvirtzman and Wieder, 2001). Black dots show position of age model tiepoints. S1 = Sapropel 1, B-A = Bølling–Allerød; other abbreviations as in Figure 6.5.

and Wieder, 2001). Through examination of a number of soil characteristics (including magnetic susceptibility, particle-size distribution, clay mineralogy and soil micromorphology), Gvirtzman and Wieder (2001) classified sequences of palaeosols, aeolianites and dune sands for the past 53,000 years. The different soil-parent materials and soil types were rated on a semi-quantative 'wet to dry' scale. The temporal pattern (Figure 6.7) bears some similarities to other proxy data from speleothems and Lake Lisan/Dead Sea levels. However, because of the semi-quantitative nature of the palaeosol record (and hence step-like function), some of the nuances of palaeoclimatic changes are lost. Furthermore, the duration of some palaeosol events has been estimated on the basis of soil/sediment type, as no dates exist for those horizons (Figure 6.7). It is striking to note that the palaeosol record suggests relatively wet (or moist) conditions during the late Glacial (prior to 12.5 cal. ka BP) and from 10 to 7.5 cal. ka BP (Figure 6.7) that require an excess of precipitation over evaporation. Gvirtzman and Wieder (2001) suggest that rainfall must have been at least 350 mm year^{-1} but may have been as high as 800 mm year^{-1} (the GCM output for the LGM indicates 360–750 mm year^{-1} with both strong latitudinal zonation and seasonality, Figure 6.3F). However, these authors suggest that the cold, dry conditions during the peak of the LGM were not recorded by the palaeosol sequence. From 12.5 to 11.5 cal. ka BP there was a rapid accumulation (~700 mm year^{-1}) of atmospheric dust in the palaeosols (compared with 22–83 mm year^{-1} in present-day Israel). Gvirtzman and Wieder (2001) consider this aeolianite to be the local record of the Younger Dryas interval and representative of increased dust transport from the Saharan and Arabian deserts.

FLUVIAL SEDIMENTS

The study of fluvial environments has the potential to yield information about changes in hydrological regime and climate. However, hiatuses and a lack of datable materials can hamper such research. Although Late Pleistocene to Holocene fluvial sedimentation in the Levant has been discussed in terms of environmental change (e.g. Grossman and Gerson, 1987; Oguchi and Oguchi, 2004), the lack of absolute dates makes it difficult to compare these studies with other well-dated climate records. However, in the Wadi Faynan region of southern Jordan McLaren *et al.* (2004) were able to identify and date a succession of fluvial terraces, alluvial fans and aeolian sediments, allowing them to place some constraint on the possible palaeoclimatic evolution of this region. Through a limited number of ^{14}C and OSL (optically stimulated luminescence) dates they were able to construct a relative stratigraphy that recorded some of the facies changes in the area during the Quaternary. Aeolian sediments (in places interbedded with fluvial units), dated to about 13.7 cal. ka BP, suggest a dry climatic phase. Perennial meandering stream deposits, thought to be early Holocene in age, suggest that the climate was wetter at this time (possibly 9,500 to 8,000 cal. years BP; McLaren *et al.*, 2004; Hunt *et al.*, 2004). A wind-blown unit indicates that after 7,400 cal. years BP climatic conditions were increasingly arid. Unfortunately, the sequence, relative timing and absolute dating of facies changes in Wadi Faynan are extremely complicated and difficult to resolve, and thus McLaren *et al.* (2004) were only able to draw very broad conclusions about the impact of climatic changes on fluvial facies changes.

6.4.2 Terrestrial palaeobotanical records

Temporal variations in vegetation are important indicators of climatic and environmental change. The best preserved and most informative palynological records come from continuous sedimentary sections with good age control. Lake sediments are often ideal for this purpose, whereas fluvial sediments are less desirable because of possible hiatuses and the increased potential for oxidation of organic matter. Additionally, 'natural' sediment accumulations (i.e. lacustrine or fluvial sediments) are preferable to sediments that accumulate at archaeological sites, where the potential for an anthropogenically induced taphonomic bias is high. Because of this we decided only to include sites without human interference (i.e. natural sites) in this review and so have omitted much of the strictly archaeological record (e.g. Darmon, 1988; Miller, 1998).

High-resolution Late Pleistocene–Holocene continental palynological records from the Levant are available from the Ghab Valley in Syria (e.g. Niklewski and van Zeist, 1970; van Zeist and Woldring, 1980; van Zeist and Bottema, 1982; Yasuda *et al.*, 2000), the Hula Basin in northern Israel (e.g. Horowitz, 1971, 1989; Baruch and Bottema, 1991, 1999) and the Birkat Ram crater lake, also in northern Israel (Weinstein, 1976; Schwab *et al.*, 2004). Other more limited palynological records exist from fluvial (e.g. Hunt *et al.*, 2004) and coastal marsh (e.g. Kadosh *et al.*, 2004) sites, although these will not be discussed in detail in this review.

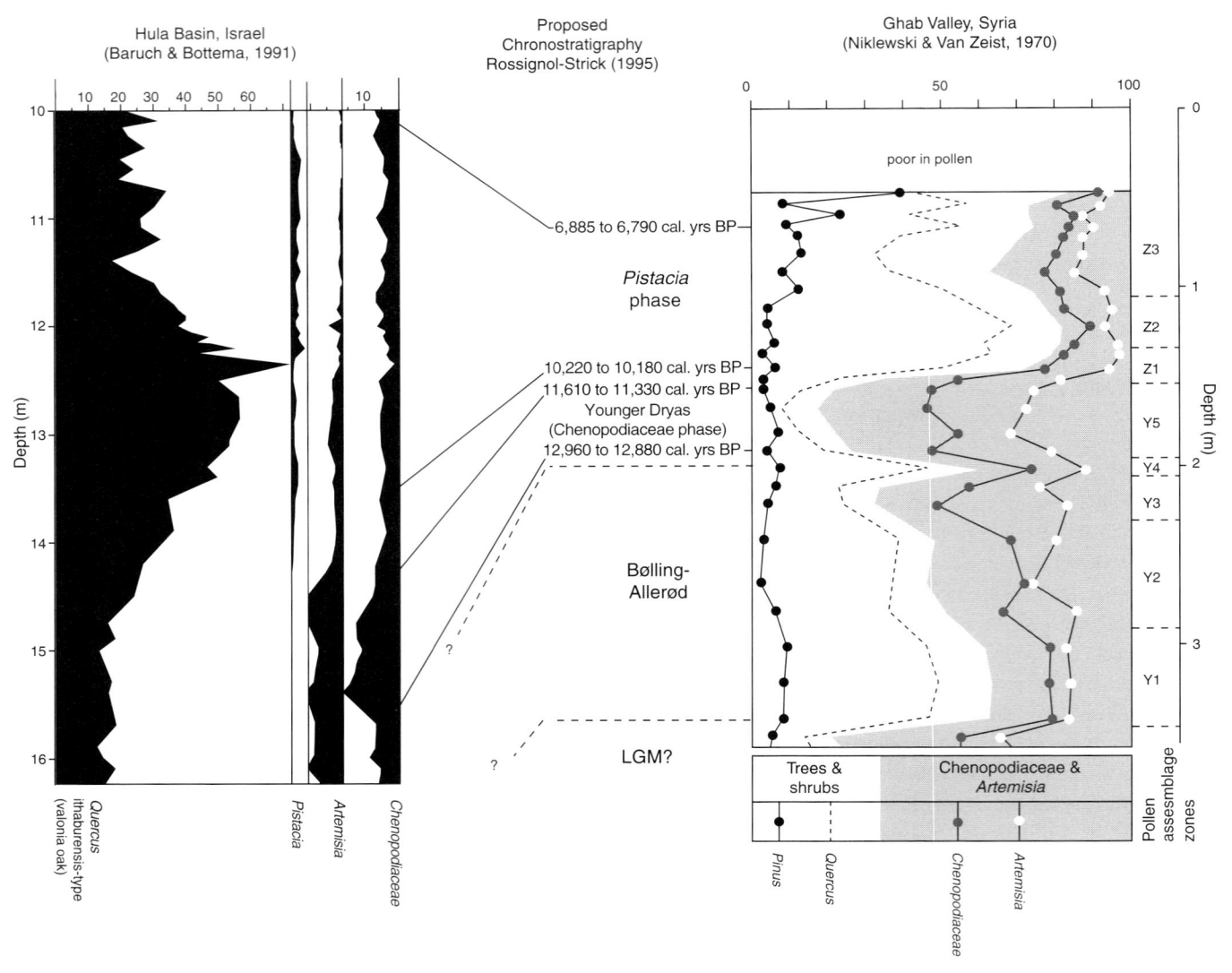

Figure 6.8. Palynology of the Hula Basin (Baruch and Bottema, 1991) and the Ghab Valley (Niklewski and van Zeist, 1970) with the proposed chronostratigraphy of Rossignol-Strick (1995). Horizontal axes in %; in the left-hand figure, the percentage for each taxon refers to concentration of that pollen taxon with respect to total 'Arboreal pollen+non-arboreal pollen'. In the right-hand figure, the percentage scale refers to the relative proportions of 'trees + shrubs' and 'Chenopodiaceae+*Artemisia*'.

The Ghab Valley and Hula Basin records (Figure 6.8) have formed the basis for most palaeovegetational histories and reconstructions in the Levant region (e.g. van Zeist and Bottema, 1991). However, controversy has existed for some time over the reliability of the age models for these records and the inferred sequences of vegetational changes in the Levant (e.g. Rossignol-Strick, 1995; Hillman, 1996; Meadows, 2005). The original published age models for the Ghab Valley (Niklewski and van Zeist, 1970; van Zeist and Woldring, 1980; van Zeist and Bottema, 1982) and Hula Basin (Tsukada, quoted in van Zeist and Bottema, 1982; Baruch and Bottema, 1991, 1999) suggest that during the Younger Dryas, forest cover and humidity (as defined by arboreal pollen values) increased in Syria, while in northern Israel forest cover contracted and aridity increased (see Baruch and Bottema, 1991; Bottema, 1995; Rossignol-Strick, 1995). Baruch and Bottema (1991) and Bottema (1995) took the view that during the Younger Dryas two climatic subregions existed, unlike the present-day uniform climate regime. In contrast, Rossignol-Strick (1995) suggested that the age models for both the Ghab Valley and the Hula Basin were suspect. Rossignol-Strick (1995) presented palynological data from marine cores in the Mediterranean and the Western Arabian Sea. Through a combination of oxygen-isotope stratigraphy, radiocarbon dating and palynology, Rossignol-Strick (1995) showed that in marine cores the Younger Dryas was marked everywhere by an increase in Chenopodiaceae. This plant is associated with saline soils, arid conditions and areas with less than 100 mm mean annual rainfall. The 'Chenopodiaceae phase' was found to be synchronous in marine cores from the Mediterranean and Arabian seas, suggesting that this was a regional phenomenon across the entire Levantine and Arabian area. Rossignol-Strick (1995) identified a regional 'Pistacia phase' in the early Holocene (10.2 to 6.7 cal. ka BP) suggesting mild winters and mean annual precipitation between 300 and 500 mm. By identifying the Chenopodiaceae and Pistacia phases in the Ghab and Hula pollen records, Rossignol-Strick (1995) was able to provide an alternative chronostratigraphy for these cores (Figure 6.8). This revised stratigraphy suggests that, in contrast to the conclusions of Baruch and Bottema (1991), the climatic evolution of the northern (Ghab) and southern (Hula) Levant was actually very similar during the Late Pleistocene to early Holocene. Attempts have been made to refine the radiocarbon age models and chronology of vegetational change of the Hula Basin by applying a correction based upon the $\delta^{13}C$ of the sample (e.g. Cappers et al., 1998, 2002). However, Meadows (2005) presented a review of the Cappers et al. (1998, 2002) method from which he concluded that the revised radiocarbon chronologies for the Hula Basin are still contradictory with other marine, terrestrial and archaeological records of palaeoclimate and palaeoenvironment. In common with Meadows (2005), we take the view that the Hula and Ghab pollen sequences record regional changes in pollen assemblages during the Late Glacial transition that can be correlated with the marine record (as in Rossignol-Strick, 1995), suggesting that during the LGM and Younger Dryas, conditions were dry and cold. In contrast, the Bølling–Allerød and early Holocene were warmer and wetter intervals. In particular the early Holocene may have been very wet with mild winters, as evidenced by the high abundances of oak and Pistacia (Rossignol-Strick, 1999). These inferences are in keeping with the wider regional marine records of palynological change (Rossignol-Strick, 1995).

The Late Pleistocene–Holocene pollen record from the volcanic crater lake of Birket Ram (Golan Heights, Israel) shows clear changes in the abundance of arboreal/non-arboreal pollen (Weinstein, 1976). Unfortunately, there is no independent age model for this record; Weinstein (1976) simply correlated the pollen record with the records from the Hula Basin that were available at the time (Horowitz, 1971). A more recent study by Schwab et al. (2004) described new cores from the lake, but unfortunately these only covered the past 6,500 years.

Less continuous palynological records are provided by coastal marsh and fluvial sediments. Kadosh et al. (2004) provided some pollen evidence from the Israeli coastal plain for a drier climate than present during the Younger Dryas and a wetter climate during the early Holocene. Hunt et al. (2004) provided palynological and palaeobotanical evidence from fluvial sediments in Wadi Faynan (southern Jordan) for a diverse forest assemblage in the early Holocene. On the basis of a pollen assemblage rich in tree pollen, Poaceae, Artemisia, Plantago and steppe herbs, Hunt et al. (2004) suggested that before 8.0 cal. ka BP Wadi Faynan was on the margin of a Mediterranean forest zone with a mean annual precipitation of about 200 mm. From ~8 cal. ka BP to 7.4 cal. ka BP, the proportion of steppeland vegetation increased, probably as a response to decreasing rainfall. Hunt et al. (2004) suggested that rainfall during this period was ~150 mm per year. After this period drought indicators decrease in abundance (suggesting a slight increase in rainfall) but trees fail to recover, which Hunt et al. (2004) attributed to increased grazing pressure and anthropogenic activities rather than climate change.

6.4.3 Terrestrial geochemical records

SPELEOTHEMS

Speleothems can provide high-resolution geochemical data ($\delta^{18}O_{spel}$, $\delta^{13}C_{spel}$ and $^{87}Sr/^{86}Sr_{spel}$) that can be dated by uranium-series dating. The oxygen-isotopic composition of speleothem carbonate is a consequence of the $\delta^{18}O$ of local precipitation, itself controlled by several factors including latitude, altitude and air temperature (McDermott, 2004). Evaporative processes in semi-arid/arid regions further modify groundwaters in the vadose zone, before reaching the point of speleothem formation. The interpretation of $\delta^{18}O_{spel}$ records also depends on the assumption that the carbonate has been precipitated in isotopic equilibrium with cave drip-water throughout the interval of study (e.g. Hendy, 1971; Schwarz, 1986; McDermott, 2004). The carbon-isotopic composition of speleothem carbonate is controlled by $\delta^{13}C$ of soil CO_2, temperature and, in closed systems, dissolution of bedrock carbonate. The $\delta^{13}C$ of soil CO_2 in modern soils is primarily controlled by the varying proportions of C_3 ($\delta^{13}C \approx -23$ to -34‰) and C_4 ($\delta^{13}C \approx -8$ to -16‰) plants and, thus, large shifts in $\delta^{13}C_{spel}$ have been attributed to major changes in vegetation. However, in some cases evaporative processes, degassing of CO_2 from groundwaters and precipitation of vadose zone carbonate may lead to anomalous $\delta^{13}C_{spel}$ values. As with carbon and oxygen, a number of sources contribute to the strontium-isotopic composition of speleothem carbonate ($^{87}Sr/^{86}Sr_{spel}$). Principal among these are sea water (directly and through sea spray), deposition of aeolian dust in the soil cover and dissolution of bedrock (e.g. Bar-Matthews et al., 1999).

Speleothem records covering the entire time interval of interest are rare in the Levant region. However, speleothems from the Soreq Cave (Bar-Matthews et al., 1996, 1997, 1999, 2000, 2003; Ayalon et al., 1999) and the Jerusalem West Cave (Frumkin et al., 1999a, 2000; Frumkin and Stein, 2004) provide continuous, high-resolution records of the palaeoclimate of central Israel during the past ~25 cal. ka BP. Other speleothem records in the region provide shorter records of various time slices since the LGM. The peak of the LGM (25 to 19 cal. ka BP) is missing in the Ma'ale Eryaim Cave, possibly owing to very low rainfall at this time east of the central mountain ridge in Israel

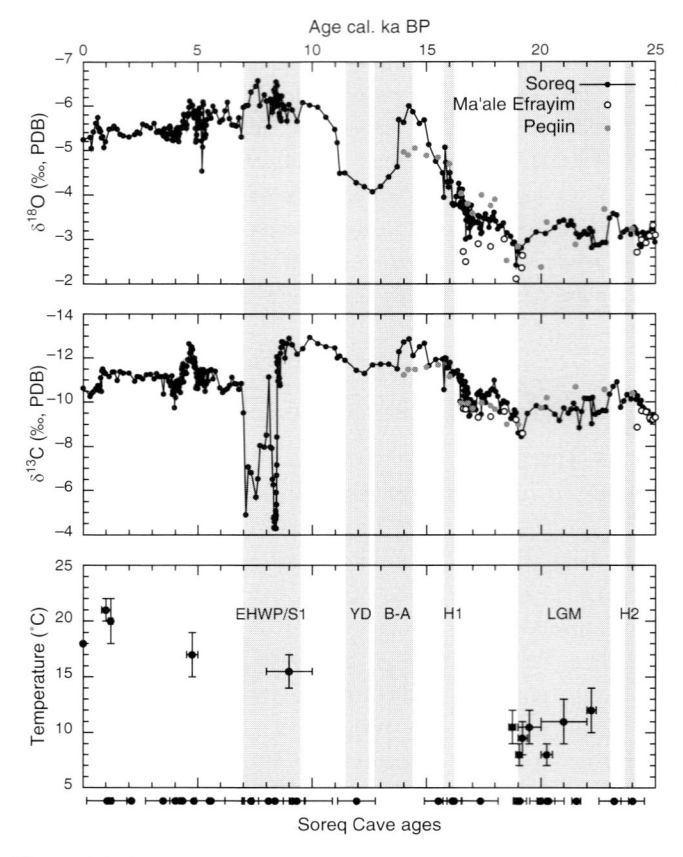

Figure 6.9. Speleothem stable-isotope data (Bar-Matthews *et al.*, 2003; Vaks *et al.*, 2003) and reconstructed air temperatures (McGarry *et al.*, 2004).

(Vaks *et al.*, 2003). However, the interval from 19 to 16 cal. ka BP is present, suggesting more precipitation at that time (Vaks *et al.*, 2003). In the very long (250 kyr) low-resolution record from Peqiin Cave (northern Israel), hiatuses exist from 15 to 7 cal. ka BP and from 5 to 0 cal. ka BP (Bar-Matthews *et al.*, 2000). Mid to late Holocene (~0 to 6 cal. ka BP) records exist from Nahal Qanah Cave, Israel (Frumkin *et al.*, 1999b), and caves in the Galilee region (Issar *et al.*, 1992), although these will not be discussed here.

The speleothem palaeoclimate record from Soreq Cave consists of a composite of several individual speleothems of varying ages, which overlap in time allowing correlation between them (Bar-Matthews *et al.*, 1999). The chronology of the composite record is provided by ^{230}Th–^{234}U (TIMS) ages. Against this time frame, Bar-Matthews *et al.* (1996, 1997, 1999, 2000, 2003) have generated high-resolution records of $\delta^{13}C_{spel}$, $\delta^{18}O_{spel}$ and $^{87}Sr/^{86}Sr_{spel}$ (Figure 6.9). The oxygen-isotope data clearly show peaks and troughs that are likely to be recording the major climatic events of the past 25,000 years including the LGM, Bølling–Allerød and the Younger Dryas. However, both the timing and duration of these last two events are not quite as would be predicted from ice-core data, but this could be due to the lack of U–Th ages during this interval (Figure 6.9). For the past 250 ka, Bar-Matthews *et al.* (2003) have demonstrated a close match, and almost constant offset, between speleothem $\delta^{18}O$ and a marine record based on planktonic foraminiferal $\delta^{18}O$ values in the eastern Mediterranean Sea. This suggests that climate events in the eastern Mediterranean Sea and on land were linked for at least the past 250 ka. Given this correlation, it is likely that changes in sea-surface temperatures (SSTs) were also registered in continental air temperatures, although it is not possible to use $\delta^{18}O_{spel}$ to calculate temperature directly (Bar-Matthews *et al.*, 2003). Bar-Matthews *et al.* (1999, 2000, 2003) highlight the occurrence of very low speleothem and foraminiferal $\delta^{18}O$ events during interglacials, including one in the Holocene at ~8 ka that they suggest is the result of increased rainfall. Bar-Matthews *et al.* (2003) used alkenone SST reconstructions (see below) as a proxy for land temperatures and were able to calculate from their speleothem record the $\delta^{18}O$ of rainwater ($\delta^{18}O_{rain}$) and the amount (in mm) of 'palaeo annual rainfall'. For the past 15 kyr, the $\delta^{18}O_{rain}$ record is most positive during the Younger Dryas, at about 5.1 ka and after 2.5 ka, indicating that these periods were intervals with low rainfall and increased aridity. Conversely the $\delta^{18}O_{rain}$ record is more negative prior to the Younger Dryas, during the early Holocene (11 to 7.5 ka) and at about 4.7 ka, indicating wetter, warmer conditions. According to the calculations of Bar-Matthews *et al.* (2003), annual rainfall was about 600 mm during the early Holocene (~7.5 to 8 ka), falling to a low of 220 mm at about 5.1 ka. There was a small peak in rainfall at about 4.7 ka which was followed by increasing aridity. They did not calculate the amount of rainfall for the Younger Dryas.

In order to gain a more quantitative estimate of terrestrial temperatures, McGarry *et al.* (2004) measured the hydrogen-isotopic composition (δD) of fluid inclusions within speleothems from the Soreq, Peqiin and Ma'ale Efrayim Caves. They used a modification of a method proposed by Matthews *et al.* (2000) which considers the relationship between δD and $\delta^{18}O$ of fluid inclusions. The $\delta^{18}O$ of the fluid inclusions are calculated from the $\delta^{18}O$ of the enclosing speleothem carbonate with a consideration of the range of possible temperatures at which the speleothem could have grown. This approach yielded average estimates of land temperatures that display similar trends to SSTs derived from marine alkenones in the eastern Mediterranean. The calculated temperatures from McGarry *et al.* (2004) for the LGM are in the range of ~7–14 °C (Figure 6.9). Holocene values are considerably warmer, varying from 14–17 °C between 8 and 10 cal. ka BP to a maximum estimate of 20–22 °C between 0.8 and 1.2 cal. ka BP.

The different $\delta^{13}C_{spel}$ records all display similar trends from the LGM to the start of the Holocene (e.g. Bar-Matthews *et al.*, 1999, 2000, 2003; Frumkin *et al.*, 2000). A general shift to more negative values is recorded from ~20 cal. ka BP to 15 cal. ka BP and a plateau from this point to ~10 cal. ka BP. These data are consistent with a shift from a C_4-rich flora to a mixed C_3/C_4 vegetation cover. Bar-Matthews et al. (1999, 2000, 2003) reported a large positive carbon-isotope excursion from ~8.5 to 7 cal. ka in their record from Soreq Cave in central Israel. They concluded that the positive excursion is related to an early Holocene rainfall 'deluge', associated with sapropel S1, which stripped soil cover, thereby reducing the influence of soil CO_2 and soil processes on the $\delta^{13}C$ of groundwaters. The Bar-Matthews et al. explanation is supported by other proxy data such as the abundance of detrital material incorporated into the speleothems and Sr-isotope evidence that suggests increased weathering during the positive carbon-isotope excursion. In contrast, Frumkin et al. (2000) found no evidence for this excursion in their record from Jerusalem West Cave. It is possible that the large $\delta^{13}C$ excursion observed in the Soreq Cave is not recognised elsewhere because of either the small number of speleothems available, or a unique, highly localised response to increased rainfall in the Soreq Cave region.

Strontium isotope data from the speleothems in Soreq Cave (Ayalon et al., 1999; Bar-Matthews et al., 1999) and the Jerusalem West Cave (Frumkin and Stein, 2004) display similar long-term trends associated with glacial–interglacial transitions. High $^{87}Sr/^{86}Sr$ ratios are observed during the last glacial, with a rapid fall to lower values during deglaciation. Both Bar-Matthews et al. (1999) and Frumkin and Stein (2004) conclude that during the last glacial the principal sources of strontium were sea spray and dust particles. The wetter and warmer conditions of the Holocene led to increased weathering of bedrock and thus a large decline in $^{87}Sr/^{86}Sr$ ratios towards bedrock values. The lesser influence of aeolian dust during the Holocene is also reflected by the lower Sr-concentrations in speleothem carbonate (Ayalon et al., 1999; Bar-Matthews et al., 1999; Frumkin and Stein, 2004).

LACUSTRINE SEDIMENTS

Stein et al. (1997) used Sr-isotope and elemental-ratio (Mg/Ca, Sr/Ca) evidence to demonstrate that during the Late Pleistocene, Lake Lisan operated in two modes that resulted in the precipitation of aragonite and gypsum. Aragonite formed when an increased supply of freshwater to the lake surface drove a density stratification of the lake. Gypsum was precipitated when this stratification was absent and the lake was able to mix and (possibly) overturn. Precipitation of gypsum is prevalent between 23–22 cal. ka BP and 16–15 cal. ka BP, suggesting dry climatic conditions during the LGM (e.g. Abed and Yaghan, 2000) and Heinrich Event H1.

Magaritz et al. (1991) measured the $\delta^{13}C$ of bulk organic matter from sediment cores collected from the Dead Sea shore (those used by Yechieli et al. (1993) to reconstruct lake levels, see discussion above). These data show a distinct shift from early Holocene values of −24 to −25‰ towards more positive values of between −21 and −23‰ at about ~9.4 cal. ka BP. At ~5.7 cal. ka BP, the data record a return to more negative values before returning to more positive values at, or after, about 2 cal. ka BP. Magaritz et al. (1991) interpreted the positive shifts as a change from a C_3 flora, to a mixed C_3/C_4 flora. In present-day Israel, the presence of C_3 and C_4 plants is strongly controlled by rainfall and C_4 plants are generally restricted to areas with less than 300 mm mean annual rainfall (Goodfriend, 1988). Thus, the more positive carbon-isotope data were interpreted by Magaritz et al. (1991) as representing drier climates and periods of lower mean annual rainfall. The data from Magaritz et al. (1991) would thus seem to indicate a relatively dry and arid Holocene after ~9.4 cal. ka BP, with a brief return to wetter conditions at ~5.7 cal. ka BP, at odds with other indicators of palaeoclimatic change (e.g. speleothems, marine proxies). It should be noted that the age constraints on this dataset are extremely limited and there has been no attempt to identify the origin of the organic matter analysed. There is considerable isotopic variability within individual terrestrial plants (e.g. +3‰ between leaves and wood; Leavitt and Long, 1982) and thus much of the signal described by Magaritz could simply be due to preservational differences. Furthermore, at low total organic carbon (TOC) values (<0.2 wt % in Magaritz et al., 1991) there is potential for unrepresentative values of 'terrestrial' $\delta^{13}C$ to be obtained.

MOLLUSCS

In the Levant a number of studies have used the stable-isotope geochemistry ($\delta^{13}C$, $\delta^{18}O$) of gastropods to detect changes in rainfall source areas (e.g. Magaritz and Heller, 1980; Goodfriend, 1991), and to investigate temporal vegetational changes related to the amount of precipitation (Goodfriend, 1988, 1990). Low-resolution data from Magaritz and Heller (1980) suggest that between ~12.9 and 12.5 cal. ka BP (during the Younger Dryas), the climate of Israel was more arid than at present. Goodfriend (1991) showed that during the early Holocene (~10.95 cal. ka BP) $\delta^{18}O$ values of gastropods from southern Israel were similar to modern values (Figure 6.10). However, at ~6.8 cal. ka BP $\delta^{18}O$ values were depleted by 2‰, which Goodfriend (1991, 1999) suggests is the result of more frequent storms reaching the region from northeast Africa.

The $\delta^{13}C$ of organic matter contained within gastropod shells has been used to track changes in the abundance of C_3/C_4 plants in the Negev Desert during the Holocene (Goodfriend, 1988, 1990, 1999). Based upon the $\delta^{13}C_{org}$ of gastropods, Goodfriend (1988, 1990, 1999) has shown that the C_3/C_4 transition zone was ~20 km further south in the middle Holocene (~7.4 to 3.2 cal. ka BP), compared with its present position, implying that there was considerably more rainfall in the region at this time. In the early Holocene (~11 to 7.8 cal. ka BP) a considerable area of the Negev Desert appears to have been covered by C_3 plants,

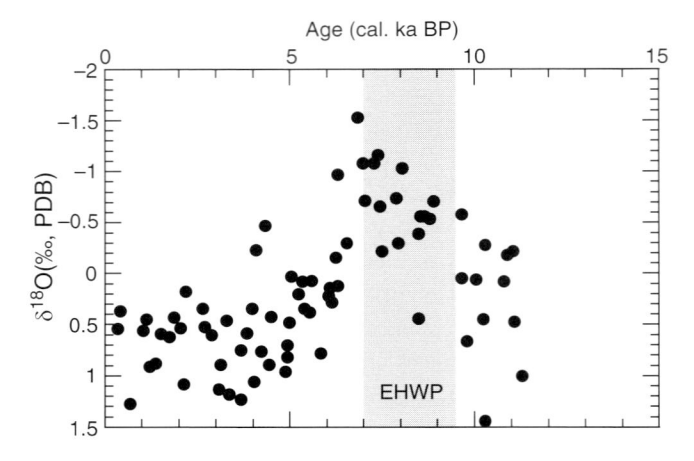

Figure 6.10. Gastropod oxygen isotope data from the Negev Desert (Goodfriend, 1991).

implying even higher rainfall. Goodfriend (1999) suggests that the annual rainfall at this time in the area north of the present-day 100 mm isohyet was more than 290 mm.

CALCRETES

In the Negev Desert periods of pedogenesis occurred during wetter climates, resulting in the development of palaeosols containing calcrete nodules (Goodfriend and Magaritz, 1988). Between the phases of pedogenesis drier conditions existed, particularly during the LGM, that resulted in extensive erosion. Stable isotope ($\delta^{13}C_{cal}$, $\delta^{18}O_{cal}$) data from Late Pleistocene (~16.7 to 12.8 cal. ka BP) soil carbonate nodules in the Negev Desert are enriched in ^{13}C and ^{18}O compared with those from older horizons (Magaritz, 1986). This has been interpreted as the result of a southward shift in the desert boundary, widespread pedogenesis and a possible change in the source of the precipitation in the region (Magaritz, 1986; Goodfriend and Magaritz, 1988). Additionally within the isotope data, there is a strong north–south carbon-isotope gradient, which suggests a decrease in plant productivity and an increase in C_4 plants towards the south (Goodfriend and Magaritz, 1988). These factors are consistent with a strong north–south rainfall gradient, suggesting that the major source area of precipitation was to the north or northwest, thereby excluding the possibility that monsoonal rains contributed to rainfall during the Late Pleistocene wet periods. Holocene oxygen-isotope values from the most southerly site (Ramat Hovav) are extremely positive ($\delta^{18}O = +2.5‰$), which Magaritz (1986) suggests is the result of extreme aridity and vigorous evaporative processes. Unfortunately, the lack of nodules in some horizons and changes in facies did not allow Magaritz (1986) to produce a high-resolution time-series of calcrete stable-isotope data, making it very difficult to fully understand palaeoclimatic changes from the calcrete data alone.

6.4.4 Marine records

DEEP-SEA CORES

A large number of deep-sea cores have been collected from the eastern Mediterranean and northern Red Sea, providing a good spatial record of Late Pleistocene–Holocene sedimentation in pelagic and hemipelagic settings. These records are generally continuous, yet are limited in their temporal resolution by typical pelagic sedimentation rates of 2 to 5 cm kyr^{-1}. In a few areas where turbidites are a significant feature of the sedimentary regime, extremely high sedimentation rates are attained (up to 2 m kyr^{-1}; Reeder et al. 2002). A number of palaeoceanographic proxies have been applied to these cores yielding information about SSTs, sea-surface salinity (SSS) and continental runoff.

Modern planktonic foraminiferal assemblages in core-top sediments vary with the temperature and salinity of the water in which they lived (see CLIMAP Project Members, 1976; Hayes et al., 2005). Since the 1970s transfer functions have been used to relate quantitatively foraminiferal assemblages with temperature and salinity, allowing the reconstruction of these parameters in the geological past. For the eastern Mediterranean various reconstructions exist (e.g. CLIMAP Project Members, 1976; Thiede, 1978; Thunell, 1979; Kallel et al., 1997a; Hayes et al., 2005) that use a number of different transfer functions. The recent study of Hayes et al. (2005) calculated LGM palaeotemperatures at 37 Mediterranean sites using a core-top calibration dataset from 274 sites in the Mediterranean and North Atlantic. Hayes et al. (2005) used a revised analogue technique and artificial neural networks to calculate SSTs. Their reconstructions suggest that mean annual temperatures in the eastern Mediterranean (Levantine Basin) during the LGM were approximately 2 °C colder than present. Much of this difference seems to have been manifested as colder summer temperatures, whilst winter temperatures show much less difference compared with present values (Figure 6.11). Model results (Figure 6.4) are broadly in agreement with such an interpretation. Kallel et al. (1997a) calculated SSTs for the Holocene using modern analogues of fossil foraminiferal assemblages. Their results suggest that during the early Holocene (~8 to 9 cal. ka BP) SSTs in the eastern Mediterranean were similar to present, although in the central and western areas they were somewhat cooler.

An alternative SST proxy is based upon the $U^{k'}_{37}$ index, a measure of alkenone unsaturation ratios that are dependent on SST (Prahl et al., 1988). This independent measure of SST, in conjunction with estimates of global ice volume, makes it possible to calculate SSS from the oxygen-isotopic composition of planktonic foraminifera ($\delta^{18}O_{foram}$). Using the alkenone method, Emeis et al. (1998, 2000, 2003) have derived estimates of SSTs and SSSs at three Mediterranean sites, including one in the Levantine Basin at ODP Site 967 (Figure 6.12). Their reconstruction suggests that

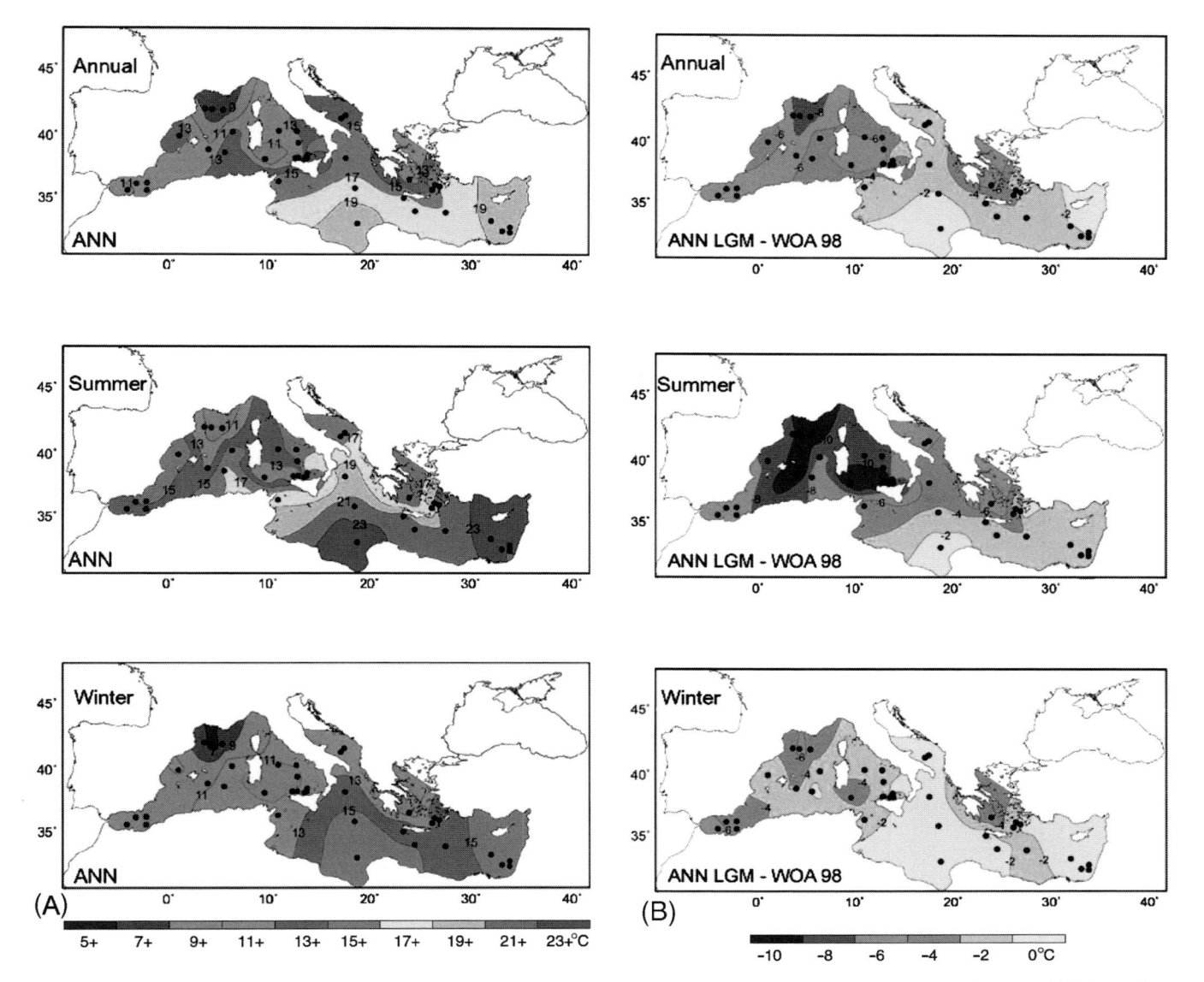

Figure 6.11. (A) Foraminiferal LGM annual, summer and winter SST reconstructions (Hayes *et al.*, 2005), calculated using an artificial neural network (ANN). (B) Temperature anomalies for annual, summer and winter SSTs during the LGM, compared with modern-day values (Hayes *et al.*, 2005). Anomaly values were calculated by subtracting modern-day SSTs from the glacial values. The black dots represent the sites of the cores from which the LGM data were obtained. This figure is a reproduction of part of Figure 9 in Hayes *et al.* (2005). See colour plate section.

during the LGM the eastern Mediterranean was at least 5 to 6 °C cooler than present (SST ≈ 12 °C). Between 18 and 9.5 cal. ka BP (tiepoints in the age model) there appears to have been an interval of much warmer temperatures, followed by much cooler temperatures. These intervals could correspond to a warm Bølling–Allerød and a cooler Younger Dryas, although the interpolated dates of the age model are not in agreement with the accepted ages of these events. This highlights the difficulty of sampling and dating short (centennial and millennial) events at low to moderate sedimentation rates, which may have been variable. The earliest Holocene is characterised by rising temperatures and lowering salinities. During deposition of sapropel S1 (an organic-rich unit dated between ~9.5 and 7.5 cal. ka BP), SSS was lower than present (owing to increased freshwater input) and SSTs were similar to present. A brief saline event during S1 is interpreted by Emeis *et al.* (2000) as a reduction in freshwater input during this time of otherwise high freshwater input.

Arz *et al.* (2003a, b) also used the alkenone method to produce a record of SSTs and SSSs in the Northern Red Sea spanning the Late Pleistocene to Holocene (Figure 6.13). This record suggests that during the Late Pleistocene the northern Red Sea was cooler by ~4 °C and more saline by ~10‰ than at present. From about 18 cal. ka BP to ~14 cal. ka BP, SSTs were even colder, ranging from ~16 to 21 °C. At about 16.5 cal. ka

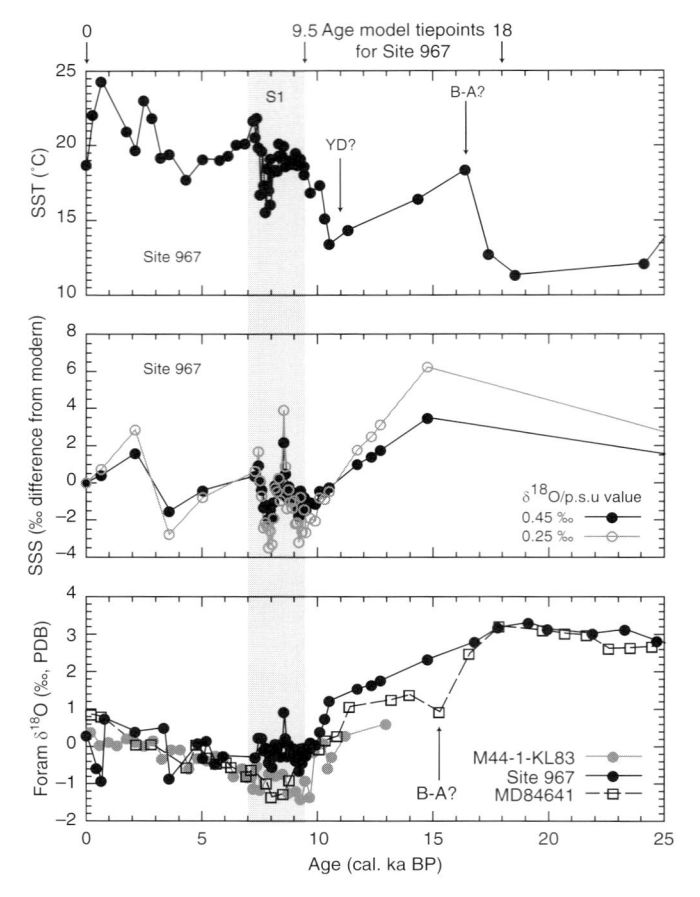

Figure 6.12. Compilation of eastern Mediterranean Sea palaeoclimatic records from Site 967 (Emeis et al., 1998, 2000, 2003) and MD84–461 (Fontugne and Calvert, 1992). The 'δ^{18}O/p.s.u. value' is a coefficient used in the calculation of SSS that relates salinity and $\delta^{18}O_{seawater}$ (see Emeis et al., 2000 for more details).

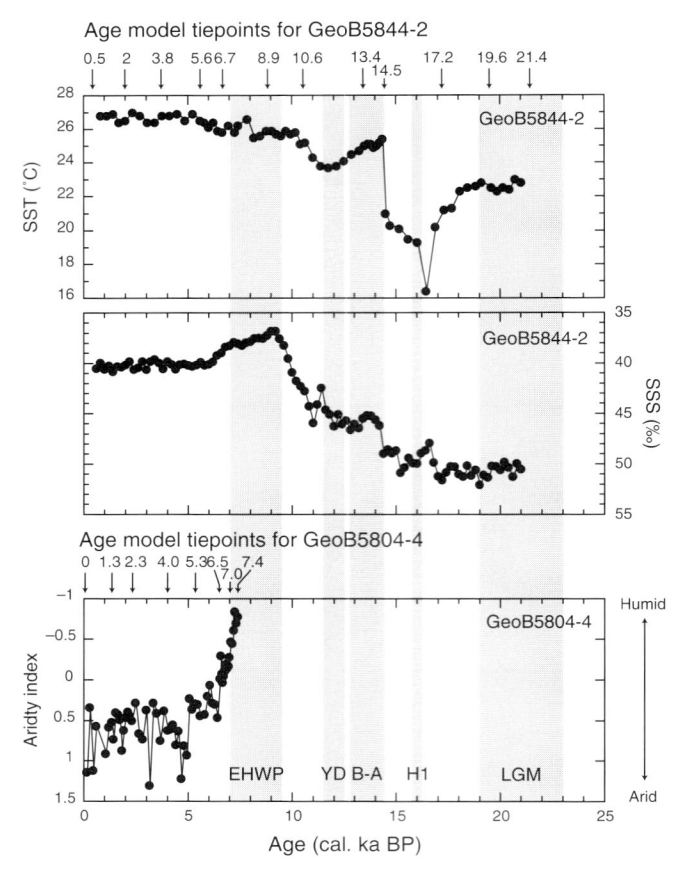

Figure 6.13. Compilation of northern Red Sea palaeoclimatic records (Arz et al., 2003a, b; core names defined therein).

BP, coincident with Heinrich Event H1, a minor freshening and a major cooling to ~16 °C is observed in the northern Red Sea. Arz et al. (2003a) suggest that the cooling was either a result of enhanced westerlies or an enhanced northeast monsoon. A rapid 4.5 °C increase in palaeotemperatures occurred at ~14.5 cal. ka BP marking the onset of the Bølling–Allerød warm period. SSTs during the Younger Dryas were cooler. The transition out of the Younger Dryas was not sharp in the Red Sea, suggesting a gradual improvement in climatic conditions into the early Holocene. Salinity during the Younger Dryas is slightly higher than during the Bølling–Allerød, suggesting increased aridity. After the Younger Dryas, SSTs gradually increased during the Holocene (Arz et al., 2003a, b). SSSs decreased after the Younger Dryas, reaching a minimum during the early Holocene 'humid period' (Arz et al., 2003a, b), an interval with low aridity index values (Figure 6.13). After ~6.5 cal. ka BP, the Red Sea became warmer, more arid and more saline, consistent with Mediterranean climatic records (Arz et al., 2003b).

Other attempts at calculating Mediterranean salinities from $\delta^{18}O_{foram}$ have been attempted using various different corrections for temperature. Thunell and Williams (1989) assumed that the difference in oxygen-isotope values between 8 cal. ka BP and the present is entirely due to salinity changes and not due to temperature or ice-volume (considered by Thunell and Williams to be identical to present at 8 cal. ka BP). Using this assumption and estimates of ice-volume at the LGM, Thunell and Williams (1989) estimated that the eastern Mediterranean was more saline at the LGM by 2.7‰. They estimated that at 8 cal. ka BP, SSSs were fresher by 2.9‰, coincident with the early Holocene humid period. Whilst this method may yield sensible indications of changes in salinity, it does not represent an overly rigorous estimate of SSSs, as the study was based largely on averages of many sites and does not account for local hydrographic and climatic variations. Kallel et al. (1997a) used their foraminiferal assemblage temperatures and $\delta^{18}O_{foram}$ records to produce a detailed picture of salinity variation during the early Holocene, suggesting that surface waters were fresher during this interval and that surface salinity was almost homogenous across the entire Mediterranean during the deposition of S1. However, Rohling and De Rijk (1999) delivered a cautionary warning, suggesting that salinities estimated from corrected $\delta^{18}O_{foram}$ values may be overestimated as these studies do not consider temporal changes in the oxygen-isotopic composition of rainfall and river waters.

In addition to reconstructing the physical (temperature (*T*), salinity (*S*)) marine environment, geochemical analysis of deep-sea cores can be used to reconstruct changes and sources of terrigenous input and, thus, be used to estimate changes in runoff. Calvert and Fontugne (2001) showed that quartz/clay, Si/Al and Zr/Al ratios were higher during the LGM than the Holocene, suggesting increased wind speeds and dust transport in glacial times. Such an interpretation is compatible with GCM output of increased windspeeds during the LGM (Figure 6.4H). Many studies have focused primarily on the early Holocene sapropel S1, which is thought to have resulted from increased freshwater supply into the Mediterranean that inhibited circulation and oxygenation of deep waters (e.g. Rossignol-Strick *et al.*, 1982; Rossignol-Strick, 1985). Krom *et al.* (1999) attempted to determine the provenance of sediment in the eastern Mediterranean during S1 by measuring major elements and strontium isotopes ($^{87}Sr/^{86}Sr$) in deep-sea cores, in a core in the Nile delta and in Saharan dust. They showed that there was a major change in sediment source concomitant with changes in surface salinity as determined by oxygen isotopes. Krom *et al.* (1999) suggested that a decrease in Saharan dust input coupled with an increase in Ionian and Levantine riverine input best explained their geochemical data in the western Levantine and Ionian basins. This conclusion suggests that the increased outflow from the River Nile did not influence sedimentation over the entire eastern Mediterranean (challenging the conclusions of Rossignol-Strick *et al.* (1982) and Rossignol-Strick (1985)) and that local increases in riverine input (particularly in the west) were also important for the development of S1. Freydier *et al.* (2001) and Scrivner *et al.* (2004) used Nd isotopic variations to demonstrate the importance of the Nile outflow for sites east of Crete (approximately 0.25° E) during S1. Scrivner *et al.* (2004) also suggest that the peak of Nile outflow may have been in the central part of the sapropel event, and that other freshwater discharges were more important than the Nile in the early and late stages of the sapropel.

An alternative method of investigating past changes in runoff is provided by deep-water turbidites. Reeder *et al.* (2002) have shown that Late Pleistocene–Holocene turbidites in the eastern Mediterranean (Herodotus Basin) were partially controlled by changes in sea level and climate. These turbidites were dated by radiocarbon and characterised as to their source (Nile Cone, Libyan/Eygptian shelf or Anatolian rise). During the LGM (up to ~17 cal. ka BP) no turbidites were derived from the Nile, attributed by Reeder *et al.* (2002) to the cooler and drier climate in the Nile hinterland at this time. In contrast, after the LGM, turbidites deposited in the Herodotus Basin were exclusively sourced from the Nile and contain an abundance of terrestrial organic matter. Reeder *et al.* (2002) suggested that these units correspond to increased Nile outflow associated with the early Holocene humid phase and sapropel S1. After ~6 cal. ka BP, only thin turbidites originating from the North African shelf are present, which is consistent with a decreased Nile outflow and an increase in aridity over the Nile region in the late Holocene.

CORAL RECORDS

At present the only coral records for the Levantine region come from the Red Sea. The coral record here is highly discontinuous, with reefs forming predominantly during interglacial times (e.g. Reyss *et al.*, 1993; Gvirtzman *et al.*, 1992). Accordingly, the only corals that were formed in the past 25,000 years are Holocene in age (e.g. Gvirtzman *et al.*, 1992; Gvirtzman, 1994; Moustafa *et al.*, 2000; Felis *et al.*, 2004), and even these only date from ~6.5 cal. ka BP onwards. The stable-isotope and elemental geochemistry of corals has been used extensively to reconstruct palaeoceanographic conditions at high resolutions, yet the Holocene Red Sea corals have received little attention. Moustafa *et al.* (2000) presented stable carbon and oxygen isotope data from *Porites* spp. corals from the northern Gulf of Aqaba that range in age from ~6.5 to 5.1 cal. ka BP. Some of their samples displayed seasonal variations in oxygen-isotope values. The amplitude of these changes was greater than present and annual growth rates were lower, suggesting to Moustafa *et al.* (2000) that the Mid-Holocene was characterised by a larger seasonal SST contrast, reduced SSS variations and increased terrigenous input. However, Arz *et al.* (2003b) suggest that in the Red Sea the major humid interval of the Holocene had terminated by ~7.25 cal. ka BP. Unfortunately, given the brevity of the Moustafa *et al.* (2000) coral record it is difficult to know whether the coral data are recording a transitional phase (between extremes of aridity/humidity) or a minor humid event unrecognised in the Arz *et al.* (2003b) aridity records.

6.5 DISCUSSION

The key marine and terrestrial palaeoclimatic records from the eastern Mediterranean and Levant are summarised in Figure 6.14. This figure provides a broad view of the climatic evolution of this region over the past 25,000 years at different localities and from different proxies. The following discussion focuses on the major climatic events during this interval and attempts to summarise the spatial picture of climatic conditions during these events. By viewing 'time slices' of climate (Figure 6.15) it is hoped that the reader can gain a more informed appreciation for where the data come from and, importantly, where gaps exist in our knowledge.

6.5.1 Heinrich Event 2, H2 (~23.8 cal. ka BP)

Heinrich Events were times of intense cooling resulting from the addition of freshwater to the North Atlantic inhibiting heat

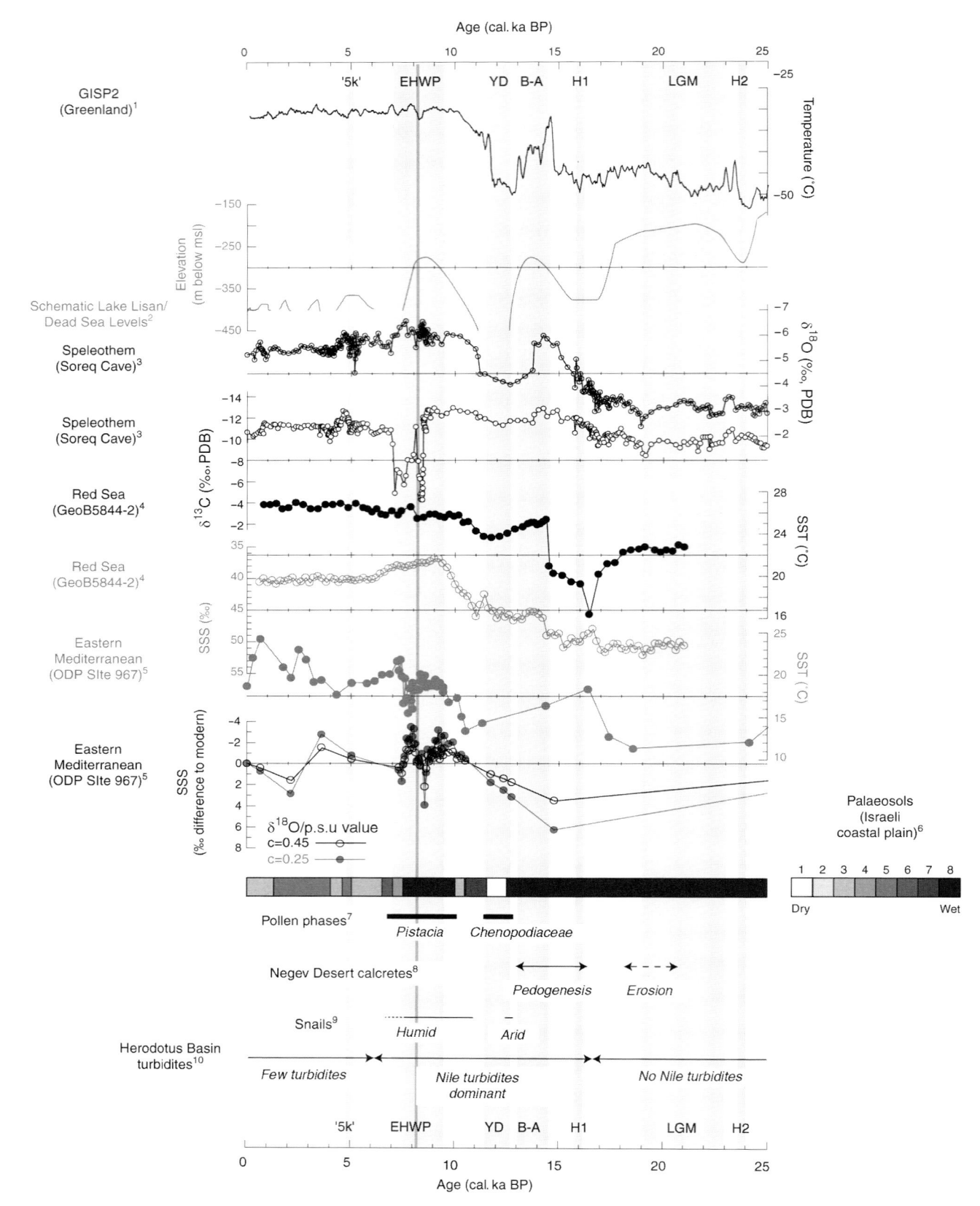

Figure 6.14. Compilation of terrestrial and marine palaeoclimatic proxy data for the Levant and eastern Mediterranean. Also shown is the ice-core record from GISP2 (Greenland). References: 1: Alley (2000); 2: see Figure 6.6; 3: Bar-Matthews *et al.* (2003); 4: Arz *et al.* (2003a); 5: Emeis *et al.* (2000, 2003); 6: Gvirtzman and Wieder (2001); 7: Rossignol-Strick (1995); 8: Magaritz (1986), Goodfriend and Magaritz (1988); 9: Magaritz and Heller (1980); Goodfriend (1990, 1991, 1999); 10: Reeder *et al.* (2002).

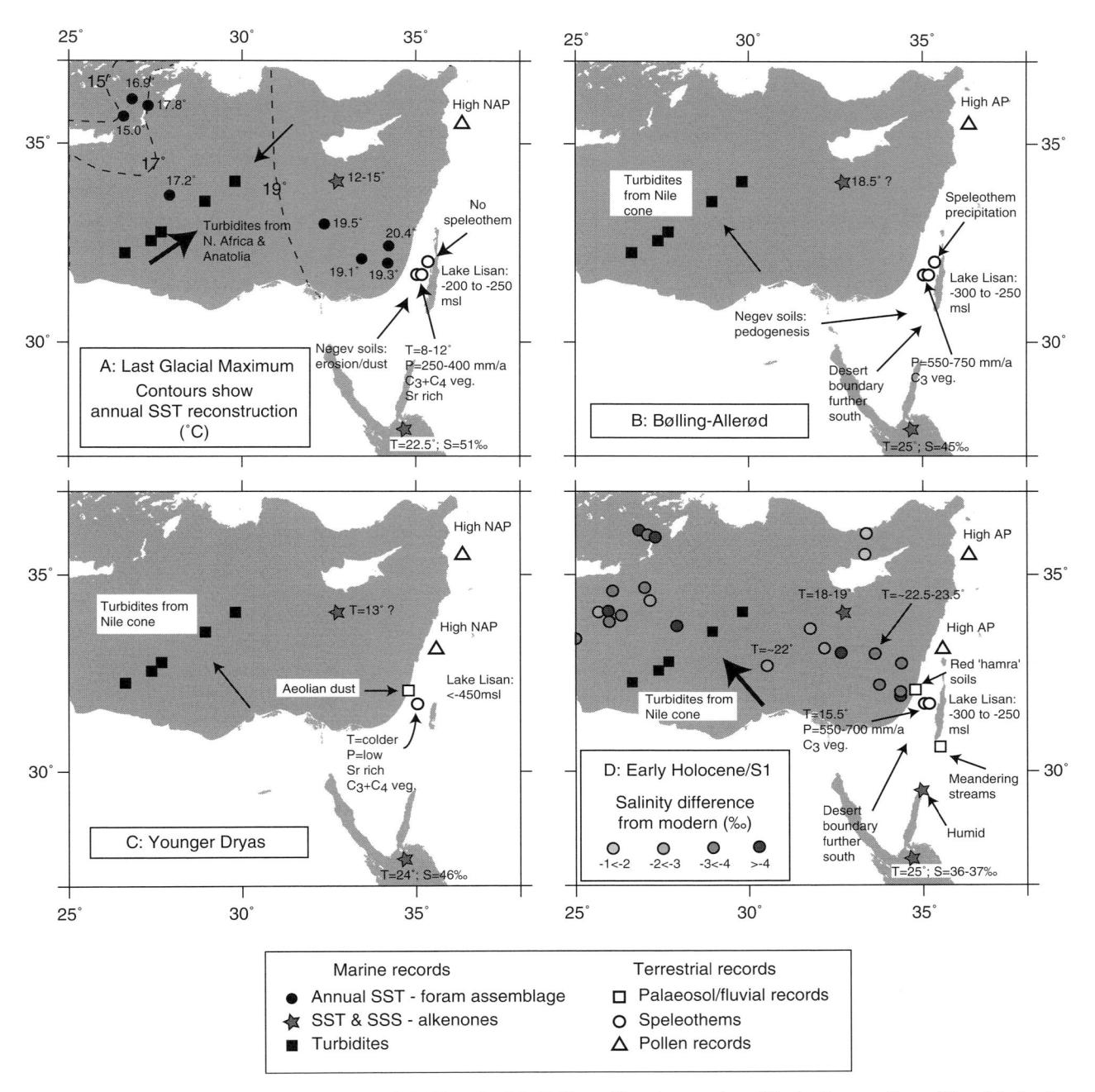

Figure 6.15. Summary of climatic conditions at the LGM (A), peak of the Bølling–Allerød warm phase (B), the Younger Dryas (C) and during the early Holocene/S1 (D). Turbidite data from Reeder *et al.* (2002); alkenone SSTs and SSS data from Emeis *et al.* (2000, 2003) and Arz *et al.* (2003a, b); speleothem data from Bar-Matthews *et al.* (1997, 1999, 2000, 2003), Frumkin *et al.* (1999b, 2000), Vaks *et al.* (2003) and McGarry *et al.* (2004); pollen data from Niklewski and van Zeist (1970), Baruch and Bottema (1991) and Rossignol-Strick (1995); lake levels from Figure 6.6 (this study); LGM annual SSTs calculated from foraminiferal assemblages taken from Hayes *et al.* (2005); early Holocene SSS values are from Kallel *et al.* (1997a); Israeli coastal plain palaeosol data from Gvirtzman and Wieder (2001); Negev data from Magaritz (1986), Goodfriend and Magaritz (1988); Goodfriend, (1999); southern Jordan fluvial data from McLaren *et al.* (2004). T = temperature (all in °C), S = salinity, P = precipitation, NAP = non-arboreal pollen, AP = arboreal pollen, Maps drawn with GMT (http://gmt.soest.hawaii.edu/). Coastlines do not account for any changes in sea level or sedimentation.

transport from the tropics to northern latitudes (e.g. Bond *et al.*, 1992, 1993). In the Levant and eastern Mediterranean, H2 appears to have been marked by a sharp lowering in lake levels (Bartov *et al.*, 2002, 2003; Landmann *et al.*, 2002; Hazan *et al.*, 2005) and a small positive excursion in speleothem oxygen isotope values (in the Soreq Cave), interpreted as a cooling event (dated at ~25 cal. ka BP by Bar-Matthews *et al.* (1999)). Bartov *et al.* (2003) suggests that cold-water input into the

Mediterranean during H2 caused a reduction of evaporation and precipitation in the eastern Mediterranean region, leading to transient cooling and an evaporation excess over the Levant region. Unfortunately the abundance of data for H2 in the eastern Mediterranean and Levantine region is low.

6.5.2 Last Glacial Maximum (~23 to 19 cal. ka BP)

The palaeoclimatic records of the LGM in the eastern Mediterranean and Levant suggest that the region was generally cooler and more arid than present, generally in good agreement with the predictions made by the climate models (Chapter 3, this volume). Mean annual SSTs predicted by foraminiferal assemblages (Hayes et al., 2005) vary from 15 °C north of Crete to about 19 °C in the far eastern Mediterranean (Figure 6.15A). However, alkenone temperature estimates from ODP Site 967 suggest lower SSTs of ~12 °C (Emeis et al., 2000, 2003), some 1 to 3 °C less than even the winter temperatures predicted by Hayes et al. (2005). This discrepancy could be due to the absence of foraminiferal assemblage temperature estimates in the region around Site 967 in Hayes et al. (2005). The Hayes et al. (2005) reconstruction is in good agreement with other alkenone temperature estimates further west (e.g. Cacho et al., 2000), suggesting that the techniques should give comparable results in the eastern Mediterranean. Salinity estimates from sites in the central and western Mediterranean are elevated at the LGM compared with present values, suggesting increased evaporation and/or decreased freshwater input (e.g. Kallel et al., 1997b; Emeis et al., 2000). Unfortunately, reliable SSS estimates for the eastern Mediterranean are currently unavailable, although Thunell and Williams (1989) suggest an average salinity 2.7‰ greater than present. The northern Red Sea also appears to have been marked by cooler temperatures and increased salinity at the LGM (Arz et al., 2003a), although at this time the reduced connection between the global ocean and the Red Sea would also have affected palaeoceanographic conditions (Siddall et al., 2003, 2004). The speleothem record from Soreq Cave displays a gradual increase in $\delta^{18}O$ values from ~21 to 19 cal. ka BP, which Bar-Matthews et al. (1997, 1999) interpret as a decrease in temperature. McGarry et al. (2004) calculate an air temperature range of ~8–12 °C above the Soreq Cave during the LGM. Bar-Matthews et al. (1997) estimated that mean annual rainfall above the Soreq Cave during the period spanning the LGM was between 250 and 400 mm, considerably less than the 500 mm that accumulates there at present. The absence of speleothem precipitation during the interval from 25 to 19 cal. ka BP in the Ma'ale Efrayim Cave (in the modern rain shadow of the Judean Mountains) is probably due to the unavailability of soil zone water at this time (Vaks et al., 2003) further suggesting low rainfall at this time in the Levant region. Dry, cool conditions during the LGM are also evidenced by erosion in the Negev Desert (Goodfriend and Magaritz, 1988), the precipitation of gypsum in the Dead Sea (Abed and Yaghan, 2000) and elevated non-arboreal pollen values in Syria (according to the revised Ghab stratigraphy of Rossignol-Strick, 1995). Speleothem carbon-isotope values also become more positive at this time, possibly suggesting an increase in the proportion of C_4 plants consistent with cooler and drier conditions (Bar-Matthews et al., 1997, 1999). The strontium isotopic composition of speleothem carbonate indicates that dust transport and sea spray were elevated compared with bedrock dissolution during the LGM (Bar-Matthews et al., 1999; Frumkin and Stein, 2004), implying lower amounts of precipitation. Elemental ratios in eastern Mediterranean sediments also suggest increased wind speeds and dust transport during the LGM (Calvert and Fontugne, 2001). Lake Lisan levels remained high during the LGM although they may have been falling. Clearly, despite the dry climate, evaporation in the Dead Sea Rift was sufficiently compensated by water supply to maintain lake levels or, at least, prevent an extremely large regression. In the absence of rainfall, riverine input from melting snow (as suggested by the GCM) may have been significant.

6.5.3 Heinrich Event 1, H1 (16.0 cal. ka BP)

As with H2, H1 in the Levant is marked by a lowering of lake levels (Bartov et al., 2002, 2003) and a small positive oxygen-isotope excursion in the Soreq Cave speleothem record (Bar-Matthews et al., 1999). Unfortunately the eastern Mediterranean marine records of SST and SSS are at insufficient resolution to detect whether H1 had a significant effect. Red Sea SSTs are generally low from ~17 to 15 cal. ka BP compared with the LGM. A sharp drop in SST and small decrease in salinity appears to have occurred at ~16 cal. ka BP which Arz et al. (2003a) postulate is a response to atmospheric cooling during H1, although strangely there is no oxygen-isotope response during this time period. This cooling is in keeping with the hypothesis of Bartov et al. (2003) that coldwater input into the Mediterranean during H2 caused a reduction of evaporation and precipitation in the eastern Mediterranean region, leading to transient atmospheric cooling and an evaporation excess over the Levant region. However, how the Red Sea became fresher during an interval of increased evaporation is unclear, although long-term sea-level rise and changes in Red Sea circulation may explain some of the salinity decrease.

6.5.4 Bølling–Allerød warm interval (~15 to 13 cal. ka BP)

Lake levels during the Bølling–Allerød (Figure 6.15B) appear to have been reasonably high, suggesting an increase in the precipitation/evaporation ratio from the conditions during H1. The revised pollen diagram from the Ghab Valley seems to suggest that oak forests were prevalent during the Bølling–Allerød, consistent with warmer and wetter conditions (see Rossignol-Strick, 1995). According to Rossignol-Strick (1995), deciduous oak requires at least 500 mm of rainfall per year, consistent with estimates from speleothems of 550 to 750 mm year^{-1} (Bar-Matthews et al., 1997). The speleothem stable-isotope data are also suggestive of an increase in C_3 plants and increased temperatures during this interval (Bar-Matthews et al., 1997, 1999, 2003). The eastern Mediterranean marine record of the Bølling–Allerød is unclear, probably as a result of age model problems and the low temporal resolution of the samples (a result of low sedimentation rates). However, warm temperatures (~18 °C) occur at Site 967 at ~16.5 cal. ka BP (Emeis et al., 2000, 2003), which could feasibly be part of a mis-dated Bølling–Allerød event, although this is impossible to determine at this time. A single data point at ~15 cal. ka BP in the MD84641 oxygen-isotope record (data from Fontugne and Clavert, 1992) may record a peak in warmth and/or a low in salinity during the glacial–interglacial transition. In the Red Sea, where sample temporal resolution is higher, SSTs rose steeply at ~14.5 cal. ka BP, concomitant with a small decrease in SSSs.

6.5.5 Younger Dryas (~12.7 to 11.5 cal. ka BP)

Although there has been some debate about climatic conditions during the Younger Dryas in the Levant, based mainly upon the poorly dated palynological records (e.g. Baruch and Bottema, 1991, 1999), it is clear from our review of multiple datasets that the period was extremely arid and, most likely, cold compared with the Bølling–Allerød and the Holocene (Figure 6.15C). Extreme aridity is evidenced by the sedimentary record in the form of massive salt deposition and a lowering of lake levels in Lake Lisan (e.g. Yechieli et al., 1993) and deposition of wind-blown sediments on the Israel coastal plain (Gvirtzman and Wieder, 2001). The high abundances of Chenopodiaceae and *Artemisia* suggest rainfall of <150 mm per year. Rossignol-Strick (1995) suggests that in the Levant the Younger Dryas was the most arid period of the Late Pleistocene. Speleothems record a return towards glacial oxygen and carbon isotope values suggesting cooling, aridity and an increase in C_4 vegetation. Bar-Matthews et al., (2003) calculated that $\delta^{18}O_{rain}$ was more positive during the Younger Dryas than in any interval of the past 15 cal. ka BP consistent with lower rainfall during this period.

The marine record of the Younger Dryas is less conclusive. Cool SSTs (~13 °C) are recorded at Site 967 at about 11 cal. ka BP (Emeis et al., 2000, 2003), which could be part of the Younger Dryas. Surprisingly, there does not appear to be any change in salinity at this time or a major shift in foraminiferal $\delta^{18}O$ values. However, further west several records show a return towards positive glacial oxygen isotope values, lower alkenone SSTs and higher salinities during the Younger Dryas (e.g. Kallel et al., 1997b; Cacho et al., 2001). In the Red Sea, there was a cooling spanning the Younger Dryas, although salinity does not appear to have changed greatly during this interval. Given the marked climatic changes recorded by terrestrial proxies it seems surprising that the eastern Mediterranean marine record does not show a clear climatic response during the Younger Dryas. Higher-resolution records from sites with high sedimentation rates are required to determine whether the impact of the Younger Dryas on marine environments in the eastern Mediterranean Sea was significant or not.

6.5.6 The early Holocene and sapropel S1 (~9.5 to 7 cal. ka BP)

The early Holocene appears to have been the wettest phase of the past 25,000 years across much of the Levant and eastern Mediterranean (Figure 6.15D). Elevated rainfall is evidenced on land by increases in *Pistacia* and oak in the Ghab and Hula pollen records (Rossignol-Strick, 1995, 1999), southward migration of the Negev Desert boundary (Goodfriend, 1999), meandering streams in southern Jordan (McLaren et al., 2004) and the deposition of 'red hamra' type palaeosols (indicating an excess of precipitation versus evaporation) on the Israeli coastal plain (Gvirtzman and Wieder, 2001). Lake levels also returned to relatively high levels following the massive drawdown during the Younger Dryas (e.g. Frumkin et al., 1994). The Soreq Cave speleothem oxygen-isotope record displays a shift towards more negative values, consistent with warmer, wetter conditions; Bar-Matthews et al. (2003) estimate rainfall amounts of between 550 and 700 mm year^{-1}. The carbon-isotope record from Soreq is remarkable, displaying a large (~8‰) excursion towards more positive values. This is interpreted by Bar-Matthews as resulting from greatly enhanced rainfall removing soil cover and increasing bedrock dissolution. The absence of this excursion in other speleothem records (e.g. Frumkin et al., 2000) suggests that soil removal may have been a highly localised phenomenon. The speleothem $\delta^{13}C$ record from the Jerusalem West Cave is relatively negative during the early Holocene, compared with the Late Pleistocene or the late Holocene, suggesting relatively more C_3 vegetation at this time, consistent with elevated rainfall (Frumkin et al., 2000). McGarry et al. (2004) calculated a terrestrial palaeotemperature of ~16 °C during the early Holocene.

Sediments rich in organic carbon (sapropel 1 or S1) were deposited in the eastern Mediterranean under anoxic conditions between ~9.5 and 7 cal. ka BP (e.g. Emeis et al., 2003). At Site 967, SSTs are estimated to have been ~20 to 22 °C whilst salinity was some 2 to 4‰ less than modern values at the beginning and end of sapropel deposition (Emeis et al., 2000, 2003). Foraminiferal oxygen-isotope values also suggest a freshening of surface waters during S1. Mediterranean sapropels have been linked to minima in the precession cycle, and it has been postulated that anoxic conditions were induced through stratification of the water column driven by a low-salinity surface layer (e.g. Rossignol-Strick *et al.*, 1982; Rossignol-Strick, 1985; Rohling and Hilgen, 1991; Rohling, 1994; Kallel *et al.*, 1997a). The source of this low-salinity water may have been from an enhanced Nile River outflow, driven by increased rainfall in the Sahara (e.g. Rossignol-Strick *et al.*, 1982; Rossignol-Strick, 1985). Evidence for this in the eastern Mediterranean comes from increased Nile-Cone-sourced turbidites in the Herodotus Basin (Reeder *et al.*, 2002) and more radiogenic Nd-isotopic values in planktonic foraminifera at ODP Site 967 (Scrivner *et al.*, 2004). Further west, however, trace element data and Nd-isotopic compositions suggest that influence of the Nile water there was minimal (Krom *et al.*, 1999; Freydier *et al.*, 2001). This is supported by the work of Kallell *et al.* (1997a) who argued that the uniform oxygen-isotopic values of surface waters in the eastern and central Mediterranean resulted from a uniformly distributed freshwater source across the entire region that most probably originated from enhanced precipitation over the entire basin. Increased rainfall in the Levant led to an increase in runoff into the Red Sea, as recorded by elevated amounts of terrigenous sediment and a minimum in salinity estimates.

Some records of S1 show a brief interruption of anoxic conditions, lasting approximately 200 years (e.g. Rohling *et al.*, 1997; Myers and Rohling, 2000). Myers and Rohling (2000) link this to a brief cooling event (evident in the Adriatic Sea) that induced some circulation of deep and intermediate waters in the Mediterranean. This event is dated by Rohling *et al.* (1997) to between 7.9 and 7.7 cal. ka BP, placing it very close in age to the Northern Hemisphere cooling that started at ~8.6 cal. ka BP, which may have lasted until 7.9 cal. ka BP (Rohling and Pälike, 2005). Rohling and Pälike (2005) have shown that the so-called 8.2 cal. ka BP cold event (recognised in the Greenland ice core) is superimposed upon a general background of climatic deterioration. Whether the interruption of S1 is related to the 8.2 cal. ka BP cold event or the more general climatic decline at this time is unclear. Bar-Matthews *et al.* (1999) claim that an excursion in speleothem $\delta^{13}C$ at about 8.2 cal. ka BP represents a sudden cooling and decrease in precipitation during the terrestrial 'deluge'. The SST data from Site 967 and the Red Sea are inconclusive (Figure 6.15), but it is interesting to note that at Site 967 there is a peak in salinity that is dated to ~8.2 cal. ka BP. Unfortunately, given the uncertainty in the age models for both speleothems and marine cores, it may prove very difficult to determine conclusively whether cooling and increased salinity in the eastern Mediterranean about 8,000 years ago was related to general cooling or the '8.2 ka cold event'.

6.5.7 Mid-Holocene wet event (~5 cal. ka BP)

There is some suggestion in the terrestrial data that there may have been a slightly wetter phase in the mid-Holocene (~5,000 years BP). Lake levels were higher at this time (Frumkin *et al.*, 1994), precipitation calculated from speleothems rose to >500 mm year^{-1} at ~4.8 cal. ka BP, and the palaeosol record suggests a slighty wetter climate at about 4.8 cal. ka BP. There is a suggestion of a brief wetter phase in the snail record at ~4.5 cal. ka BP. Marine records from the Mediterranean and Red Sea do not show any significant variation during the mid-Holocene, although coral records that date from 6.5 to 5.1 cal. ka BP suggest increased humidity (relative to present).

6.6 CONCLUSIONS

1. The major Northern Hemisphere climatic shifts of the past 25,000 years are present in the geological record of the eastern Mediterranean and Levant.
2. The marine and terrestrial palaeoclimatic records from the region are in general agreement over the timing of changes in climate. However, the records do not always agree as to the severity of some events (such as the Younger Dryas). This may be a problem with the sample resolution of marine sites with low to moderate sedimentation rates, or may be real, caused by differing rates of responses in different proxies and environments.
3. The LGM and early Holocene have received considerable attention, and this is reflected in the availability of data for these intervals. However, data coverage for the transition between these two extremes of climate is poorer. Further studies focused on the Bølling–Allerød and Younger Dryas are required in order to understand the climatic regime of the Levant during these intervals.
4. High-resolution age models are required to understand whether climate events in different regions and environments are synchronous or of the same duration. Where age models are good, there seems to be good agreement between the timing and duration of events, irrespective of environment or geographic position.
5. Output from GCMs for times such as the LGM are generally compatible with geological proxy data, giving confidence in the models themselves.
6. Model output throws light on processes operating within the climate system, and to possible responses within the environment. In particular, models alert us to seasonally dependent factors, which are often difficult to discern from the imperfectly preserved sedimentary record.

ENDNOTES

1 This chapter has been previously published as: Robinson, S. A., S. Black, B. Sellwood and P. J. Valdes (2006) A review of palaeoclimates and palaeoenvironments in the Levant and Eastern Mediterranean from 25,000 to 5000 years BP: setting the environmental background for the evolution of human civilisation. *Quaternary Science Reviews* **25**: 1517–1541. We are grateful to the editor of *QSR* and the authors for permission to republish in this volume.

REFERENCES

Abed, A. M. and R. Yaghan (2000) On the palaeoclimate of Jordan during the last glacial maximum. *Palaeogeography, Palaeoclimatology, Palaeoecology* **160**: 23–33.

Alley, R. B. (2000) The Younger Dryas cold interval as viewed from central Greenland. *Quaternary Science Reviews* **19**: 213–226.

Alley, R. B. and A. M. Ágústsdóttir (2005) The 8k event: cause and consequences of a major Holocene abrupt climate change. *Quaternary Science Reviews* **24**: 1123–1149.

Arz, H. W., J. Pätzold, P. J. Müller and M. O. Moammar (2003a) Influence of Northern Hemisphere climate and global sea level rise on the restricted Red Sea marine environments during termination I. *Paleoceanography* **18**, DOI 10.1029/2002PA000864.

Arz, H. W., F. Lamy, J. Pätzold, P. J. Müller and M. Prins (2003b) Mediterranean moisture source for an Early-Holocene humid period in the Northern Red Sea. *Science* **300**: 118–121.

Ayalon, A., M. Bar-Matthews and A. Kaufman (1999) Petrography, strontium, barium and uranium concentrations, and strontium and uranium isotope ratios in speleothems as palaeoclimatic proxies: Soreq Cave, Israel. *The Holocene* **9**: 715–722.

Bar-Matthews, M., A. Ayalon, A. Matthews, E. Sass and L. Halicz (1996) Carbon and oxygen isotope study of the active water-carbonate system in a karstic Mediterranean cave: implications for paleoclimate research in semiarid regions. *Geochimica et Cosmochimica Acta* **60**: 337–347.

Bar-Matthews, M., A. Ayalon and A. Kaufman (1997) Late Quaternary paleoclimate in the eastern Mediterranean region from stable isotope analysis of speleothems at Soreq Cave, Israel. *Quaternary Research* **47**: 155–168.

Bar-Matthews, M., A. Ayalon, A. Kaufman and G. J. Wasserburg (1999) The eastern Mediterranean paleoclimate as a reflection of regional events: Soreq cave, Israel. *Earth and Planetary Science Letters* **166**: 85–95.

Bar-Matthews, M., A. Ayalon and A. Kaufman (2000) Timing and hydrological conditions of Sapropel events in the eastern Mediterranean, as evident from speleothems, Soreq Cave, Israel. *Chemical Geology* **169**: 145–156.

Bar-Matthews, M., A. Ayalon, M. Gilmour, A. Matthews and C. J. Hawkesworth (2003) Sea–land oxygen isotopic relationships from planktonic foraminifera and speleothems in the eastern Mediterranean region and their implication for paleorainfall during interglacial intervals. *Geochimica et Cosmochimica Acta* **67**: 3181–3199.

Bartov, Y., M. Stein, Y. Enzel, A. Agnon and Z. Reches (2002) Lake levels and sequence stratigraphy of Lake Lisan, the late Pleistocene precursor to the Dead Sea. *Quaternary Research* **57**: 9–21.

Bartov, Y., S. L. Goldstein, M. Stein and Y. Enzel (2003) Catastrophic arid episodes in the Eastern Mediterranean linked with the North Atlantic Heinrich events. *Geology* **31**: 439–442.

Baruch, U. and S. Bottema (1991) Palynological evidence for climatic changes in the Levant ca. 17,000–9,000 B.P. In *The Natufian Culture in the Levant*, ed. O. Bar-Yosef and F. R. Valla. Ann Arbor, Michigan: International Monographs in Prehistory, Archaeological Series 1, pp. 11–20.

Baruch, U. and S. Bottema (1999) A new pollen diagram from Lake Hula: vegetational, climatic and anthropogenic implications. In *Ancient Lakes: Their Cultural and Biological Diversity*, ed. H. Kawanabe, G. W. Coulter and A. C. Roosevelt. Ghent: Kenobi Productions, pp. 75–86.

Bar-Yosef, O. (2000) The impact of radiocarbon dating on old world archaeology: past achievements and future expectations. *Radiocarbon* **42**: 23–39.

Bar-Yosef, O. and A. Belfer-Cohen (1992) From foraging to farming in the Mediterranean Levant. In *Transitions to Agriculture in Pre-History*, ed. A. B. Gebauer and T. D. Price. Madison: Prehistory Press, pp. 21–48.

Begin, Z. B., W. Broecker, B. Buchbinder *et al.* (1985) Dead Sea and Lake Lisan levels in the last 30,000 years: a preliminary report. *Israel Geological Survey Report* **29**/85: 1–18.

Bond, G., H. Heinrich, W. Broecker *et al.* (1992) Evidence for massive discharges of icebergs into the North Atlantic Ocean during the last glacial period. *Nature* **360**: 245–249.

Bond, G., W. Broecker, S. Johnsen *et al.* (1993) Correlations between climate records from North Atlantic sediments and Greenland ice. *Nature* **365**: 143–147.

Bottema, S. (1995) The Younger Dryas in the Eastern Mediterranean. *Quaternary Science Reviews* **14**: 883–891.

Broecker, W. S. and G. H. Denton (1989) The role of ocean–atmosphere reorganizations in glacial cycles. *Geochimica et Cosmochimica Acta* **53**: 2465–2501.

Bronk Ramsey, C. (1995) Radiocarbon calibration and analysis of stratigraphy: the OxCal program. *Radiocarbon* **37**: 425–430.

Cacho, I., J. O. Grimalt, F. J. Sierro, N. Shackleton and M. Canals (2000) Evidence for enhanced Mediterranean thermohaline circulation during rapid climate coolings. *Earth and Planetary Science Letters* **183**: 417–429.

Cacho, I., Grimalt, J. O., Canals, M. *et al.* (2001) Variability of the western Mediterranean Sea surface temperature during the last 25,000 years and its connection with Northern Hemisphere climatic changes. *Paleoceanography* **16**: 40–52.

Calvert, S. E. and M. R. Fontugne (2001) On the late Pleistocene–Holocene sapropel record of climatic and oceanographic variability in the Eastern Mediterranean. *Paleoceanography* **16**: 78–94.

Cappers, R. T. J., S. Bottema and H. Woldring (1998) Problems in correlating pollen diagrams of the Near East: a preliminary report. In *The Origins of Agriculture and Crop Domestication*, ed. A. B. Damania, J. Valkoun, G. Willcox and C. O. Qualset. Aleppo: International Center for Agricultural Research in Dry Areas, pp. 160–169.

Cappers, R. T. J., S. Bottema, H. Woldring, H. van der Plicht and H. J. Streurman (2002) Modelling the emergence of farming: implications of the vegetation development in the Near East during the Pleistocene–Holocene transition. In *The Dawn of Farming in the Near East*, ed. R. T. J. Cappers and S. Bottema. Berlin: *Ex Oriente*, Studies in Early Near Eastern Production, Subsistence and Environment 6, pp. 3–14.

CLIMAP Project Members (1976) The surface of the ice-age Earth. *Science* **191**: 1131–1137.

CLIMAP (1981) Seasonal reconstructions of the earth's surface at the last glacial maximum: Geological Society of America Map and Chart Series, v. MC-36.

Cullen, H. M., P. B. deMenocal, S. Hemming *et al.* (2000) Climate change and the collapse of the Akkadian empire: evidence from the deep sea. *Geology* **28**: 379–382.

Darmon, F. (1988) Essai de reconstitution climatique de l'Épipaléolithique au début du Néolithique ancient dans la region de Fazaël-Salibiya (Basse Vallée du Jourdain) d'après la palynologie. *Comptes Rendus Académie des Sciences, Paris* **307**: 677–682.

deMenocal, P. B. (2001) Cultural responses to climate change during the Late Holocene. *Science* **292**: 667–673.

Emeis, K.-C., H.-M. Schulz, U. Struck *et al.* (1998) Stable isotope and temperature records of sapropels from ODP Sites 964 and 967: constraining the physical environment of sapropel formation in the Eastern Mediterranean Sea. *Proceedings of the Ocean Drilling Program, Scientific Results* **160**: 309–331.

Emeis, K.-C., U. Struck, H.-M. Schulz *et al.* (2000) Temperature and salinity of Mediterranean Sea surface waters over the last 16,000 years: constraints on the physical environment of S1 sapropel formation based on stable oxygen isotopes and alkenone unsaturation ratios. *Palaeogeography, Palaeoclimatology, Palaeoecology* **158**: 259–280.

Emeis, K.-C., H. Schulz, U. Struck *et al.* (2003) Eastern Mediterranean surface water temperatures and $\delta^{18}O$ composition during deposition of sapropels in the late Quaternary. *Paleoceanography* **18**, DOI 10.1029/2000PA000617

Enzel, Y., G. Kadan and Y. Eyal (2000) Holocene earthquakes inferred from a fan-delta sequence in the Dead Sea graben. *Quaternary Research* **53**: 34–48.

Enzel, Y., R. Bookman, D. Sharon *et al.* (2003) Late Holocene climates of the Near East deduced from Dead Sea level variations and modern regional winter rainfall. *Quaternary Research* **60**: 263–273.

Felis, T., Lohmann, G., Kuhnert, H. *et al.* (2004) Increased seasonality in Middle East temperatures during the last interglacial period. *Nature* **429**: 164–168.

Fontugne, M. R. and S. E. Calvert (1992) Late Pleistocene variability of the carbon isotopic composition of organic matter in the Eastern Mediterranean: monitor of changes in carbon sources and atmospheric CO_2 concentrations. *Paleoceanography* **7**: 1–20.

Freydier, R., A. Michard, G. De Lange and J. Thomson (2001) Nd isotopic compositions of Eastern Mediterranean sediments: tracers of the Nile influence during sapropel S1 formation. *Marine Geology* **177**: 45–62.

Frumkin, A. and Y. Elitzur (2002) Historic Dead Sea level fluctuations calibrated with geological and archaeological evidence. *Quaternary Research* **57**: 334–342.

Frumkin, A. and M. Stein (2004) The Sahara–East Mediterranean dust and climate connection revealed by strontium and uranium isotopes in a Jerusalem speleothem. *Earth and Planetary Science Letters* **217**: 451–464.

Frumkin, A., M. Magaritz, I. Carmi and I. Zak (1991) The Holocene climatic record of the salt caves of Mount Sedom, Israel. *The Holocene* **1**: 191–200.

Frumkin, A., I. Carmi, I. Zak and M. Magaritz (1994) Middle Holocene environmental change determined from the salt caves of Mount Sedom, Israel. In *Late Quaternary Chronology and Paleoclimates of the Eastern Mediterranean*, ed. O. Bar-Yosef and R. S. Kra. Tucson: University of Arizona Press, pp. 315–322.

Frumkin, A., D. C. Ford and H. P. Schwarz (1999a) Continental oxygen isotopic record of the last 170,000 years in Jerusalem. *Quaternary Research* **51**: 317–327.

Frumkin, A., I. Carmi, A. Gopher *et al.* (1999b) A Holocene millennial-scale climatic cycle from a speleothem in Nahal Qanah Cave, Israel. *The Holocene* **9**: 677–682.

Frumkin, A., D. C. Ford and H. P. Schwarz (2000) Paleoclimate and vegetation of the last glacial cycles in Jerusalem from a speleothem record. *Global Biogeochemical Cycles* **14**: 863–870.

Frumkin, A., G. Kadan, Y. Enzel and Y. Eyal (2001) Radiocarbon chronology of the Holocene Dead Sea: attempting a regional correlation. In *Near East Chronology: Archaeology and Environment, Proceedings of the 17th International 14C Conference*, ed. H. J. Bruins, I. Carmi and E. Boaretto. *Radiocarbon* **43**, 1179–1189.

Gasse, F. and J.-Ch. Fontes (1992) Climate changes in northwest Africa during the last deglaciation (16–7 ka BP). In *The Last Deglaciation: Absolute and Radiocarbon Chronologies*, ed. E. Bard and W. S. Broecker. Berlin: Springer-Verlag, pp. 295–325.

Goodfriend, G. A. (1988) Mid-Holocene rainfall in the Negev Desert from ^{13}C of land snail shell organic matter. *Nature* **333**: 757–760.

Goodfriend, G. A. (1990) Rainfall in the Negev Desert during the middle Holocene, based on ^{13}C of organic matter in land snail shells, *Quaternary Research* **34**: 186–197.

Goodfriend, G. A. (1991) Holocene trends in ^{18}O in land snail shells from the Negev Desert and their implications for changes in rainfall source areas. *Quaternary Research* **35**: 417–426.

Goodfriend, G. A. (1999) Terrestrial stable isotope records of Late Quaternary paleoclimates in the Eastern Mediterranean region, *Quaternary Science Reviews* **18**: 501–513.

Goodfriend, G. A. and M. Magaritz (1988) Palaeosols and late Pleistocene rainfall fluctuations in the Negev Desert. *Nature* **332**: 144–146.

Grossman, S. and R. Gerson (1987) Fluviatile deposits and morphology of alluvial surfaces as indicators of Quaternary environmental changes in the southern Negev, Israel. In *Desert Sediments: Ancient and Modern*, ed. L. Frostick and I. Reid. Geological Society of London Special Publication **35**, pp. 17–29.

Gvirtzman, G. (1994) Fluctuations of sea level during the past 400,000 years: the record of Sinai, Eygpt (northern Red Sea). *Coral Reefs* **13**: 203–214.

Gvirtzman, G. and M. Wieder (2001) Climate of the last 53,000 years in the Eastern Mediterranean, based on soil-sequence stratigraphy in the coastal plain of Israel. *Quaternary Science Reviews* **20**: 1827–1849.

Gvirtzman, G., J. Kronfeld and B. Buchbinder (1992) Dated coral reefs of southern Sinai (Red Sea) and their implication to late Quaternary sea levels. *Marine Geology* **108**: 29–37.

Harris, D. R., ed. (1996) *The Origins and Spread of Agriculture and Pastoralism in Eurasia*. London: UCL Press.

Hayes, A., M. Kucera, N. Kallel, L. Sbeffi and E. J. Rohling (2005) Glacial Mediterranean sea surface temperatures based on planktonic foraminiferal assemblages. *Quaternary Science Reviews* **24**: 999–1016.

Hazan, N., M. Stein, A. Agnon *et al.* (2005) The late Quaternary liminological history of Lake Kinneret (Sea of Galilee), Israel. *Quaternary Research* **63**: 60–77.

Hendy, C. H. (1971) The isotopic geochemistry of speleothems I. The calculation of the effects of the different modes of formation on the isotopic composition of speleothems and their applicability as palaeoclimatic indicators. *Geochimica et Cosmochimica Acta* **35**: 801–824.

Hillman, G. (1996) Late Pleistocene changes in wild plant-foods available to hunter-gatherers of the northern Fertile Crescent: possible preludes to cereal cultivation. In *The Origins and Spread of Agriculture and Pastoralism in Eurasia*, ed. D. R. Harris. London: UCL Press, pp. 159–203.

Horowitz, A. (1971) Climatic and vegetational developments in northeastern Israel during Upper Pleistocene–Holocene times. *Pollen et Spores* **13**: 255–278.

Horowitz, A. (1989) Continuous pollen diagrams for the last 3.5 M.Y. from Israel: vegetation, climate and correlation with the oxygen isotope record, *Palaeogeography, Palaeoclimatology, Palaeoecology* **72**: 63–78.

Hunt, C. O., H. A. Elrishi, D. D. Gilbertson *et al.* (2004) Early-Holocene environments in the Wadi Faynan, Jordan. *The Holocene* **14**: 921–930.

Issar, A. S. (2003) *Climate Changes during the Holocene and their Impact on Hydrological Systems*. Cambridge: Cambridge University Press.

Issar, A. S. and M. Zohar (2004) *Climate Change: Environment and Civilization in the Middle East*. Berlin and Heidelberg: Springer-Verlag.

Issar, A. S., Y. Govrin, M. A. Geyh, E. Wakushal and M. Wolf (1992) Climate changes during the upper Holocene in Israel. *Israel Journal of Earth Sciences* **40**: 219–223.

Joussaume, S. and K. E. Taylor (1995) Status of the Paleoclimate Modeling Intercomparison Project. *Proceedings of the First International AMIP Scientific Conference: Monterey, USA, WCRP-92*, pp. 425–430.

Kadosh, D., D. Sivan, H. Kutiel and M. Weinstein-Evron (2004) A late Quaternary paleoenvironmental sequence from Dor, Carmel Coastal Plain, Israel. *Palynology* **28**: 143–157.

Kallel, N., M. Paterne, J.-C. Duplessy *et al.* (1997a) Enhanced rainfall in the Mediterranean region during the last sapropel event. *Oceanologica Acta* **20**: 697–712.

Kallel, N., M. Paterne, L. Labeyrie, J.-C. Duplessy and M. Arnold (1997b) Temperature and salinity records of the Tyrrhenian Sea during the last 18,000 years, *Palaeogeography, Palaeoclimatology, Palaeoecology* **135**: 97–108.

Krom, M. D., A. Michard, R. A. Cliff and K. Strohle (1999) Sources of sediment to the Ionian Sea and western Levantine basin of the Eastern Mediterranean during S-1 sapropel times. *Marine Geology* **160**: 45–61.

Landmann, G., G. M. Abu Qudaira, K. Shawabkeh, V. Wrede and S. Kempe (2002) Geochemistry of the Lisan and Damya Formations in Jordan, and implications for palaeoclimate. *Quaternary International* **89**: 45–57.

Leavitt, S. W. and A. Long (1982) Evidence for $^{13}C/^{12}C$ fractionation between tree leaves and wood. *Nature* **298**: 742–744.

Magaritz, M. (1986) Environmental changes recorded in the Upper Pleistocene along the desert boundary, Southern Israel. *Palaeogeography, Palaeoclimatology, Palaeoecology* **53**: 213–229.

Magaritz, M. and J. Heller (1980) A desert migration indicator – oxygen isotopic composition of land snail shells. *Palaeogeography, Palaeoclimatology, Palaeoecology* **32**: 153–162.

Magaritz, M., S. Rahner, Y. Yechieli and R. V. Krishnamurthy (1991) $^{13}C/^{12}C$ ratio in organic matter from the Dead Sea area: palaeoclimatic interpretation. *Naturwissenschaften* **78**: 453–455.

Marco, S., M. Stein, A. Agnon and H. Ron (1996) Long-term earthquake clustering: a 50,000-year paleoseismic record in the Dead Sea Graben. *Journal of Geophysical Research* **101**: 6179–6191.

Matthews, A., A. Ayalon and M. Bar-Matthews (2000) D/H ratios of fluid inclusions of Soreq cave (Israel) speleothems as a guide to the Eastern Mediterranean Meteoric Line relationships in the last 120 ky. *Chemical Geology* **166**: 183–191.

McDermott, F. (2004) Palaeo-climatic reconstruction from stable isotope variations in speleothems: a review. *Quaternary Science Reviews* **23**: 901–918.

McGarry, S., M. Bar-Matthews, A. Matthews *et al.* (2004) Constraints on hydrological and paleotemperature variations in the Eastern Mediterranean region in the last 140 ka given by the δD values of speleothem fluid inclusions. *Quaternary Science Reviews* **23**: 919–934.

McLaren, S. J., D. D. Gilbertson, J. P. Gratten *et al.* (2004) Quaternary palaeogeomorphologic evolution of the Wadi Faynan area, southern Jordan. *Palaeogeography, Palaeoclimatology, Palaeoecology* **205**: 131–154.

Meadows, J. (2005) The Younger Dryas episode and the radiocarbon chronologies of the Lake Huleh and Ghab Valley pollen diagrams, Israel and Syria. *The Holocene* **15**: 631–636.

Migowski, C., A. Agnon, R. Bookman, J. F. W. Negendank and M. Stein (2004) Recurrence pattern of Holocene earthquakes along the Dead Sea transform revealed by varve-counting and radiocarbon dating of lacustrine sediments. *Earth and Planetary Science Letters* **222**: 301–314.

Miller, N. (1998) The macrobotanical evidence for vegetation in the Near East c. 18000/16000 BC to 4000 BC. *Paléorient* **23/2**: 197–207.

Mithen, S. J. (2003) *After the Ice: A Global Human History, 20,000–5000 BC*. London: Weidenfeld and Nicolson.

Mix, A. C., E. Bard and R. Schneider (2001) Environmental process of the ice age: land, oceans, glaciers (EPILOG). *Quaternary Science Reviews* **20**: 627–657.

Moore, A. M. T. and G. C. Hillman (1992) The Pleistocene to Holocene transition and human economy in southwest Asia: the impact of the Younger Dryas. *American Antiquity* **57**: 482–494.

Moustafa, Y. A., J. Pätzold, Y. Loya and G. Wefer (2000) Mid-Holocene stable isotope record of corals from the northern Red Sea. *International Journal of Earth Sciences* **88**: 742–751.

Myers, P. G. and E. J. Rohling (2000) Modeling a 200-yr interruption of the Holocene sapropel S1. *Quaternary Research* **53**: 98–104.

Neev, D. and K. O. Emery (1967) The Dead Sea. *Geological Survey of Israel Bulletin* **41**: 1–147.

Neev, D. and K. O. Emery (1995) *The Destruction of Sodom, Gomorrah, and Jericho, Geological, Climatological and Archaeological Background*. Oxford: Oxford University Press.

Niklewski, J. and W. van Zeist (1970) A late Quaternary pollen diagram from northwestern Syria. *Acta Botanica Neerlandica* **19**: 737–754.

Oguchi, T. and C. T. Oguchi (2004) Late Quaternary rapid talus dissection and debris flow deposition on an alluvial fan in Syria. *Catena* **55**: 125–140.

Peltier, W. R. (1994) Ice-age paleotopography. *Science* **265**: 195–201.

Pope, V. D., M. L. Gallani, P. R. Rowntree and R. A. Stratton (2000) The impact of new physical parametrizations in the Hadley Centre climate model: HadAM3. *Climate Dynamics* **16**: 123–146.

Prahl, F. G., L. A. Muehlhausen and D. L. Zahnle (1988) Further evaluation of long-chain alkenones as indicators of paleoceanographic conditions. *Geochimica et Cosmochimica Acta* **52**: 2303–2310.

Reeder, M. S., D. A. V. Stow and R. G. Rothwell (2002) Late Quaternary turbidite input into the east Mediterranean basin: new radiocarbon constraints on climate and sea-level control. In *Sediment Flux to Basins: Causes, Controls, and Consequences*, ed. S. J. Jones and L. E. Frostick. Geological Society of London Special Publication **191**, pp. 267–278.

Reimer, P. J., M. G. L. Baillie, E. Bard et al. (2004) IntCal04 terrestrial radiocarbon age calibration, 0–26 Cal Kyr BP. *Radiocarbon* **46**: 1029–1058.

Reyss, J.-L., A. Choukri, J.-C. Plaziat and B. H. Purser (1993) Radiochemical dating of coral reefs from the west coast of the Red Sea: first stratigraphic and tectonic implications. *Comptes Rendus Académie des Sciences Paris* **317**: 487–492.

Roberts, N. and H. E. Wright Jr (1983) Vegetation, lake-level and climatic history of the Near East and Southwest Asia. In *Global Climates since the Last Glacial Maximum*, ed. H. E. Wright Jr, J. E. Kutzbach, T. Webb III et al. University of Minnesota Press, pp. 194–220.

Roberts, N., Stevenson, T., Davis, B. et al. (2004) Holocene climate, environment and cultural change in the circum-Mediterranean region. In *Past Climate Variability through Europe and Africa*, ed. R. W. Battarbee, F. Gasse and C. E. Stickley. Dordrecht: Springer, pp. 343–362.

Rohling, E. J. (1994) Review and new aspects concerning the formation of Eastern Mediterranean sapropels. *Marine Geology* **122**: 1–28.

Rohling, E. J. and S. De Rijk (1999) Holocene climate optimum and Last Glacial Maximum in the Mediterranean: the marine oxygen isotope record. *Marine Geology* **153**: 57–75.

Rohling, E. J. and Hilgen, F. J. (1991) The Eastern Mediterranean climate at times of sapropel formation: a review. *Geologie en Mijnbouw* **70**: 253–264.

Rohling, E. J. and H. Pälike (2005) Centennial-scale climate cooling with a sudden cold event around 8,200 years ago. *Nature* **434**: 975–979.

Rohling, E. J., F. J. Jorissen and H. C. De Stigter (1997) 200 year interruption of Holocene sapropel formation in the Adriatic Sea. *Journal of Micropalaeontology* **16**: 97–108.

Rossignol-Strick, M. (1985) Mediterranean Quaternary sapropels, an immediate response of the Africa monsoon to variation of insolation. *Palaeogeography, Palaeoclimatology, Palaeoecology* **49**: 237–263.

Rossignol-Strick, M. (1995) Sea-land correlation of pollen records in the Eastern Mediterranean for the glacial–interglacial transition: biostratigraphy versus radiometric time-scale. *Quaternary Science Reviews* **14**: 893–915.

Rossignol-Strick, M. (1999) The Holocene climatic optimum and pollen records of sapropel 1 in the Eastern Mediterranean, 9000–6000 BP. *Quaternary Science Reviews* **18**: 515–530.

Rossignol-Strick, M., W. Nesteroff, P. Olive and C. Vergnaud-Grazzini (1982) After the deluge: Mediterranean stagnation and sapropel formation. *Nature* **295**: 105–110.

Scrivner, A. E., D. Vance and E. J. Rohling (2004) New neodynmium isotope data quantify Nile involvement in Mediterranean anoxic episodes. *Geology* **32**: 565–568.

Schwab, M. J., F. Neumann, T. Litt, J. F. W. Negendank and M. Stein (2004) Holocene palaeoecology of the Golan Heights (Near East): investigation of lacustrine sediments from Birkat Ram crater lake. *Quaternary Science Reviews* **23**: 1723–1731.

Schwarz, H. P. (1986) Geochronology and isotopic geochemistry of speleothems. In *Handbook of Environmental Isotope Geochemistry*, ed. P. Fritz and J.-Ch. Fontes. Amsterdam: Elsevier, pp. 271–303.

Siddall, M., E. J. Rohling, A. Almogi-Labin et al. (2003) Sea-level fluctuations during the last glacial cycle. *Nature* **423**: 853–858.

Siddall, M., D. A. Smeed, Ch. Hemleben et al. (2004) Understanding the Red Sea response to sea level. *Earth and Planetary Science Letters* **225**: 421–434.

Stein, M. (2001) The sedimentary and geochemical record of Neogene–Quaternary water bodies in the Dead Sea Basin – inferences for the regional paleoclimate history. *Journal of Paleolimnology* **26**: 271–282.

Stein, M., A. Starinsky, A. Katz et al. (1997) Strontium isotopic, chemical, and sedimentological evidence for the evolution of Lake Lisan and the Dead Sea. *Geochimicia et Cosmochimica Acta* **61**: 3975–3992.

Thiede, J. (1978) A glacial Mediterranean. *Nature* **276**: 680–683.

Thunell, R. C. (1979) Eastern Mediterranean Sea during the Last Glacial Maximum: an 18000 years BP reconstruction. *Quaternary Research* **11**: 353–372.

Thunell, R. C. and D. F. Williams (1989) Glacial–Holocene salinity changes in the Mediterranean Sea: hydrographic and depositional effects. *Nature* **338**: 493–496.

Vaks, A., M. Bar-Matthews, A. Ayalon et al. (2003) Paleoclimate reconstruction based on the timing of speleothem growth and oxygen and carbon isotope composition of a cave located in the rain shadow in Israel. *Quaternary Research* **59**: 182–193.

van Zeist, W. and H. Woldring (1980) Holocene vegetation and climate of Northwestern Syria. *Palaeohistoria* **22**: 111–125.

van Zeist, W. and S. Bottema (1982) Vegetational history of the Eastern Mediterranean and the Near East during the last 20,000 years. In *Palaeoclimates, Palaeoenvironments and Human Communities in the Eastern Mediterranean Region in Later Prehistory*, ed. J. L. Bintliff and W. van Zeist. BAR International Series **133**, pp. 277–321.

van Zeist, W. and S. Bottema (1991) Late Quaternary Vegetation of the Near East. Beihefte zum Tübinger Atlas des vorderen orients, Reihe A (Naturwissenschaften), 18.

Verschuren, D., K. R. Briffa, P. Hoelzmann et al. (2004) Holocene climate variability in Europe and Africa: a PAGES-PEP III time stream 1 synthesis. In *Past Climate Variability through Europe and Africa*, ed. R. W. Battarbee, F. Gasse and C. E. Stickley. Dordrecht: Springer, pp. 567–582.

Weinstein, M. (1976) The late Quaternary vegetation of the Northern Golan. *Pollen et Spores* **18**: 553–562.

Yasuda, Y., H. Kitagawa and T. Nakagawa (2000) The earliest record of major anthropogenic deforestation in the Ghab Valley, Syria: a palynological study. *Quaternary International* **73/74**: 127–136.

Yechieli, Y., M. Magaritz, Y. Levy et al. (1993) Late Quaternary geological history of the Dead Sea area, Israel. *Quaternary Research* **39**: 59–67.

7 Palaeoenvironments of the southern Levant 5,000 BP to present: linking the geological and archaeological records

Claire Rambeau and Stuart Black

ABSTRACT

In this chapter we review climatic and environmental changes during the middle to late Holocene in the southern Levant, and their potential impact on human communities. The Holocene is characterised in the eastern Mediterranean region by a trend towards aridity. Several climatic fluctuations are superimposed on this general trend, but contradictory palaeoenvironmental evidence often renders these variations difficult to characterise and date accurately. Nevertheless, the cultural changes at the end of the Early Bronze Age and the Byzantine period appear to be clearly related to a shift towards aridity. Relatively wetter conditions seem to prevail during the Chalcolithic to Early Bronze Age, and during the Hellenistic to Byzantine period. The first of these humid periods may be related to a decline in settlements in the Israeli coastal plain during the Early Bronze Age, caused by increased flooding and the spread of diseases in a marshy environment. The second is associated with thriving agriculture. Arid conditions are prevalent during the Early Islamic period to modern times. More contradictory information characterises the Middle Bronze to Iron Age, but arid conditions are likely to have been dominant at the end of the Middle Bronze Age as well as during the first half of the Iron Age. Phases of climate instability occurred at $c.$ 5,200–5,000 BP and $c.$ 1,000 BP, the latter being coeval with the demise of the Decapolis society.

7.1 INTRODUCTION

The climatic and environmental changes of the Middle to late Holocene are generally less intense than the large fluctuations marking the end of the Pleistocene and the onset of the Holocene (Chapter 6, this volume). A series of studies in the eastern Mediterranean, however, have indicated that the middle to late Holocene was less 'uneventful' than previously thought (e.g. Issar, 1990; Rosen, 1995; Schilman et al., 2001b; Issar, 2003). Furthermore, as mentioned by Heim et al. (1997), 'in desert margin areas even minor climatic variations can result in major environmental changes'.

Middle to late Holocene climate changes in the eastern Mediterranean seem to consist of a long-term trend towards more aridity, punctuated by rather short-term episodes that may have had a marked impact on human communities. The most prominent of these events is probably the one occurring at $c.$ 4,200 BP, which has been held responsible for the collapse of the Akkadian civilisation (e.g. Cullen et al., 2000; deMenocal, 2001). A short-term wet period at around 5,000 BP has also been suggested (Chapter 6, this volume).

The middle to late Holocene in the eastern Mediterranean region is a time of increased human control and modification of the landscape, which renders palaeoenvironmental records difficult to interpret. The response of human communities to climate variations itself may have changed through time, as societies progressed from small settlements, only responsible for their own subsistence, towards bigger centres orientated towards more economically productive agriculture (Rosen, 1995) and buffered against small-scale climate change by, for instance, participation in extensive trade networks. When progressing from the early Holocene towards the modern time period, it becomes increasingly difficult to differentiate between the impacts of climate change and of human activity on environmental change (e.g. Heim et al., 1997; Neumann et al., 2007; Cordova, 2008; Frumkin et al., 2009), these sometimes working in contradictory directions to each other (Avni et al., 2006). Pollen sequences, in a context of increasing cultivation and vegetation-clearance, are probably the most sensitive type of record, and indicate principally the expansion of agricultural practices

Water, Life and Civilisation: Climate, Environment and Society in the Jordan Valley, ed. Steven Mithen and Emily Black. Published by Cambridge University Press. © Steven Mithen and Emily Black 2011.

(e.g. Baruch, 1986, 1990; Baruch and Bottema, 1991; Heim *et al.*, 1997; Baruch and Bottema, 1999; Schwab *et al.*, 2004; Neumann *et al.*, 2007). Increased anthropogenic impact on the environment, therefore, exacerbates the usual constraints on palaeoenvironmental reconstructions, which include uncertainties over dating, varying timescales, potentially discontinuous records, and the interpretation of multiple proxies in terms of climate variations.

In this chapter we summarise the environmental changes during the middle to late Holocene period in the southern Levant, and indicate their potential impact on human communities and cultures of the region. More detailed reviews can be found in Issar (2003) and Rosen (2007). The major types of palaeoenvironmental evidence available in the eastern Mediterranean are detailed in Chapter 6 of this volume. Palaeoenvironmental records are available from several localities of the southern Levant (Figure 7.1), although they tend to be scarcer in the more arid zones (Southern and Eastern regions, Figure 7.1).

It must be noted, however, that in light of the amount of contradictory information contained in the literature, collating all evidence into a single history of environmental change for this period and this region is a difficult task. As such, we wish to start with a caveat borrowed from Rosen (2007): 'like bathing in the Dead Sea, the results must be accepted with a sizeable portion of salt'.

7.2 THE TREND TOWARDS ARIDITY

Numerous studies attest to the presence of increasing trends towards aridity throughout the Holocene to the present day. The Dead Sea levels were much lower during the past 8,000 years than during the early Holocene (Chapter 9, this volume, Figure 9.7). In central Israel, the isotopic composition of speleothems shows a general evolution towards less rainfall from *c.* 7,500 BP onwards, consistent with a progressive increase in temperatures from around 7,000 to 500 BP (Bar-Matthews *et al.*, 2003; see also Figure 6.14, this volume, Chapter 6). Sediments in northeastern Jordan (Wadi Muqat; Abboud, 2000) indicate a shift to drier conditions at *c.* 5,000 BP, following more humid conditions. Long-term trends have been noted in the isotopic, chemical and sedimentological evolution of deposits from the southeastern Mediterranean, reflecting the aridification process that began *c.* 7,000 years ago in the Levantine region (Schilman *et al.*, 2001a; 2001b). The onset of aridification, following a wetter period, seems also to be recorded at Wadi Faynan at *c.* 6,500 BP (Hunt *et al.*, 2004; McLaren *et al.*, 2004; Grattan *et al.*, 2007; Hunt *et al.*, 2007).

In the Jordan Valley, the transition to a modern climate is marked at about 6,500–5,500 BP by climate instability leading to generalised erosion, high-energy flows and river incision (Hourani and Courty, 1997). The changes in shoreline progradation at Elat, which started *c.* 5,000 years ago, may also reflect a regional change towards aridity (Shaked *et al.*, 2002). The $\delta^{18}O$ composition of land snails in the Negev shows a progressive increase from *c.* 6,800 to 5,200 cal. years BP, when the more arid conditions of the late Holocene were established, followed by modern-like values from *c.* 5,200 to 500 cal. years BP (Goodfriend, 1991, 1999). Palaeosols from the Israeli coastal plain during the past 6,000 years reflect more arid conditions than during the early Holocene (Gvirtzman and Wieder, 2001).

The detailed pattern of the climatic fluctuations superimposed on this general trend is less clear, and varies between the different areas of the southern Levant. Whereas climate fluctuations are discernable in most parts of the region, a few zones, situated in the modern arid to hyper-arid regions, seem to have remained under the influence of climatic conditions similar to those of today from at least the start of the Holocene. This is the case in particular for the southern Negev Desert (e.g. Amit *et al.*, 2006, 2007), and the eastern Jordan plateau (Davies, 2005). The Negev Highlands seem to have been subjected to a prevalent erosion regime for at least the past *c.* 10,000 years (Rosen *et al.*, 2005; Avni *et al.*, 2006). Holocene climate fluctuations in the rainshadow region east of the central mountain ridge of Israel (Ma'ale Efrayim Cave) and in the northern Negev were insufficient to induce the precipitation of speleothems, which have deposited in these regions during earlier, wetter periods (Vaks *et al.*, 2003). These observations corroborate the analysis of Enzel *et al.* (2008), who estimated that north–south climatic gradients, similar to the present-day or even accentuated, between arid/hyper-arid and Mediterranean zones, were already established in the Late Pleistocene.

7.3 THE CHALCOLITHIC PERIOD (4500–3800 BC/*c.* 6,500–5,800 BP) AND THE EARLY/INTERMEDIATE BRONZE AGE (3800–2000 BC/*c.* 5,800–4,000 BP)

Pollen evidence from the northern part of the region (e.g. Hula Basin: Baruch and Bottema, 1999; Lake Kinneret: Baruch, 1990; Birkat Ram Lake: Schwab *et al.*, 2004), as well as the pattern in fluvial sedimentary sequences in central Israel and central and northern Jordan (Wadi ash-Shallalah: Cordova, 2008; Madaba-Dhiban plateau: Cordova *et al.*, 2005; Cordova, 2008; Tel Lachish: Rosen, 1986) suggests that most of the Chalcolithic and Early Bronze Age featured relatively humid conditions, with intensified cultivation of Mediterranean species, and the creation of alluvial plains. The Moringa Cave, on the western Dead Sea shore, was occupied during the Chalcolithic (at *c.* 6,500 BP), which may indicate increased humidity in this currently arid environment (Lisker *et al.*, 2007).

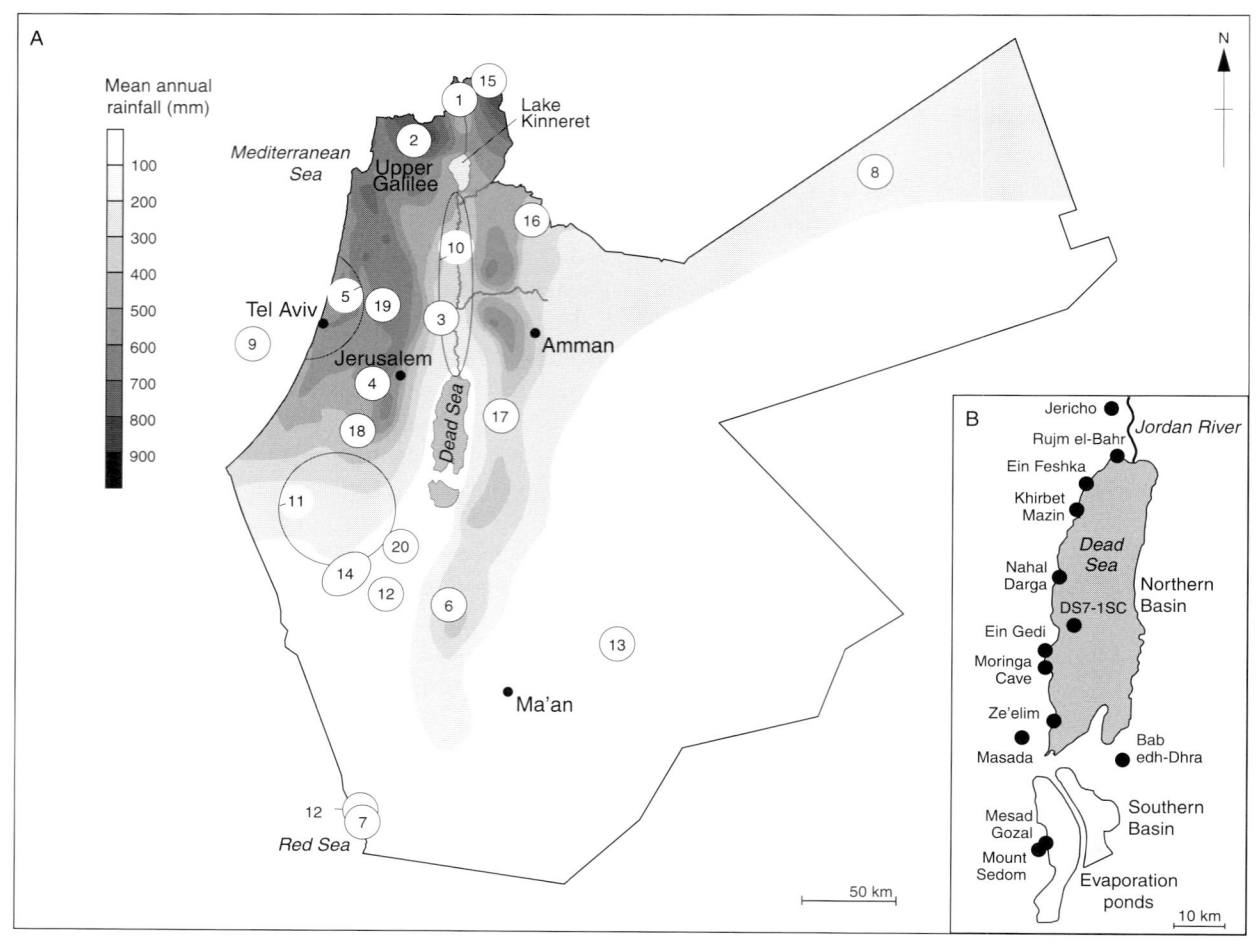

Figure 7.1. Location of the different localities and regions presented in the text. (A) Map of the southern Levant and mean annual rainfall (from EXACT 1998). 1 – Hula Basin (Baruch and Bottema, 1999; Cappers *et al.*, 1998; Rosen, 2007). 2 – Upper Galilee caves (Issar, 2003). 3 – Ma'ale Efrayim Cave (Vaks *et al.*, 2003). 4 – Soreq Cave (Bar-Matthews and Ayalon, 2004; Bar-Matthews *et al.*, 1998, 1999, 2003). 5 – Israeli coastal plain (Gvirtzman & Wieder, 2001). 6 – Wadi Faynan (Hunt *et al.*, 2004, 2007; McLaren *et al.*, 2004; Grattan *et al.*, 2007). 7 – Elat shorelines (Shaked *et al.*, 2002). 8 – Wadi Muqat (Abboud, 2000). 9 – Cores GA 112–110 (Schilman *et al.*, 2001a,b). 10 – Jordan Valley (Hourani and Courty, 1997). 11 – Northern Negev Desert (Goodfriend, 1991, 1999). 12 – Southern Negev (Amit *et al.*, 2007). 13 – Qa'el-Jafr Basin (Davies, 2005). 14 – Central Negev Highlands (Rosen *et al.*, 2005; Avni *et al.*, 2006). 15 – Birkat Ram Lake, Golan Heights (Schwab *et al.*, 2004). 16 – Wadi ash-Shallalah (Cordova, 2008). 17 – Wadi al-Wala and the Madaba-Dhiban plateau (Cordova *et al.*, 2005; Cordova, 2008). 18 – Tel Lachish (Rosen, 1986). 19 – Nahal Qanah Cave (Frumkin *et al.*, 1999a). 20 – Nahal Zin, Negev (Greenbaum *et al.*, 2000). (B) Map of the Dead Sea area.

The isotopic record from Soreq Cave and the related palaeo-rainfall calculations (Bar-Matthews and Ayalon, 2004; Figure 7.2; see also Bar-Matthews *et al.*, 1998, 1999, 2003) suggest more abundant rainfall for the period 6,500–4,150 BP than for the following periods. The Soreq record shows low oxygen isotopic ratios and the lowest carbon isotopic values for the past 6,500 years at *c.* 4,800–4,600 BP, which is interpreted as reflecting a particularly humid period of the record (Bar-Matthews *et al.*, 2003; Bar-Matthews and Ayalon, 2004; Figure 7.2). The isotopic composition of land snails in the Negev also hints at increased annual rainfall until *c.* 3,200 BP (Goodfriend, 1990, 1999).

High Dead Sea levels (above the sill that separates the northern and southern Dead Sea basins, see also Chapter 9, this volume; Figure 7.2) seem to prevail from *c.* 5,500 years and possibly before, and persist until at least *c.* 4,300 BP (Frumkin and Elitzur, 2002; Enzel *et al.*, 2003; Klinger *et al.*, 2003; Migowski *et al.*, 2006; see also Frumkin *et al.*, 1994; Neev and Emery, 1995; Frumkin, 1997; Frumkin *et al.*, 2001). However, they follow a period (*c.* 6,250–5,900 BP) of low Dead Sea levels (Figure 7.2), potentially featuring the drying-up of the shallow, southern basin (Frumkin and Elitzur, 2002); this may correlate with the arid period attested by the presence of a salt tongue in a core from the southern Dead Sea basin *c.* 7,500–5,000 BP (Neev and Emery, 1995; Migowski *et al.*, 2006).

Very high levels may have been attained in the Dead Sea at *c.* 5,000 BP (Frumkin *et al.*, 1994; Frumkin, 1997; Frumkin *et al.*,

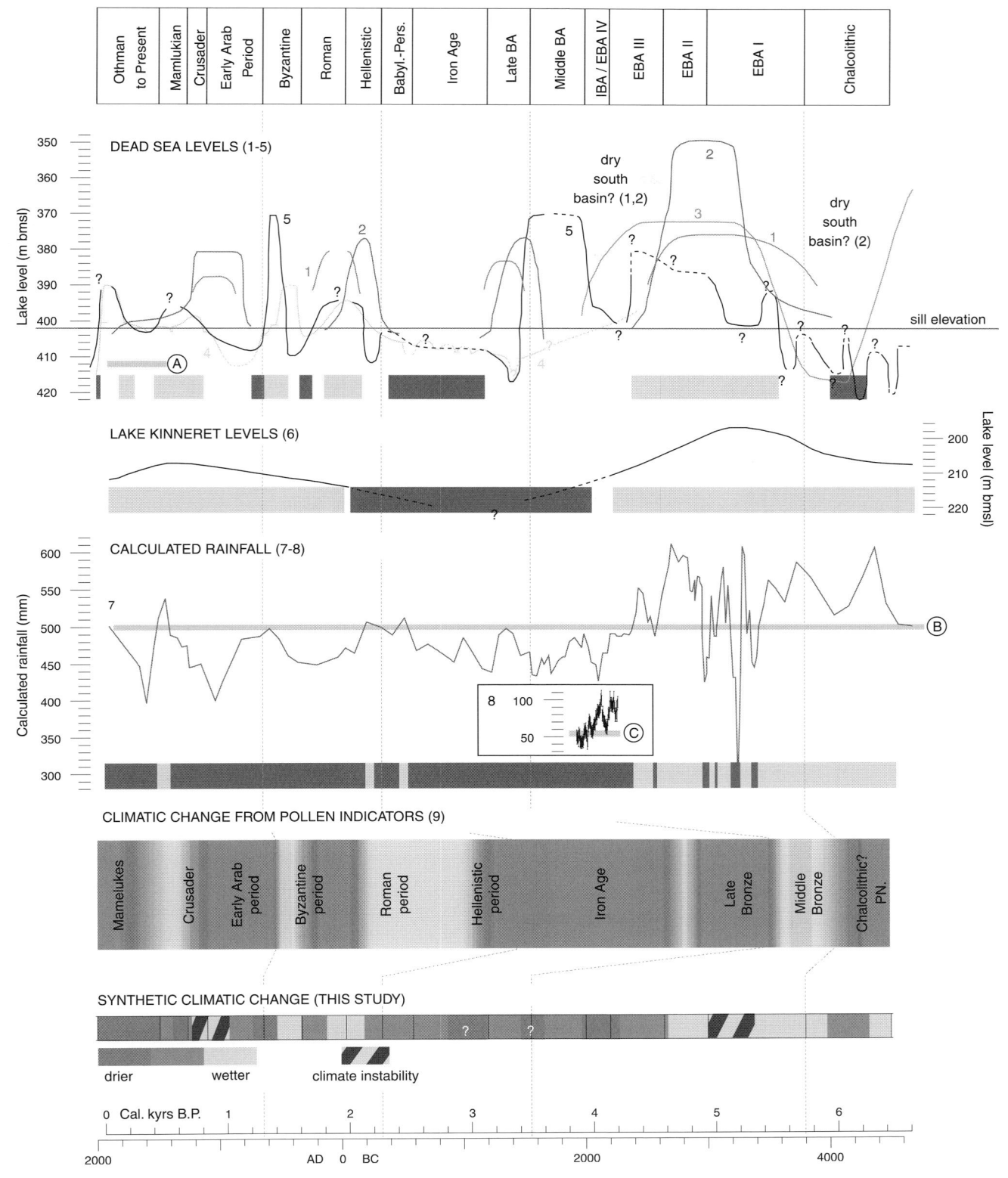

Figure 7.2. Compilation of several proxies for the middle to late Holocene in the southern Levant. Red and blue bars show interpreted climate fluctuations (wetter/drier conditions). Archaeological periods from Rosen (2007). Dead Sea levels: 1 – Frumkin and Elitzur (2002). 2 – Klinger *et al.* (2003). 3 – Enzel *et al.* (2003). 4 – Bookman *et al.* (2004). 5 – Migowski *et al.* (2006). (A) Dead Sea levels in 1997 (Migowski *et al.*, 2006). Lake Kinneret levels: 6 – Hazan *et al.* (2005). Calculated rainfall: 7 – From the Soreq cave record; Bar-Matthews and Ayalon (2004). (B) Present-day mean annual rainfall in Soreq area. 8 – From tamarisk wood, Mount Sedom cave; Frumkin *et al.* (2009). (C) Present-day mean annual rainfall at Mount Sedom. 9 – Climatic change from pollen indicators according to Neumann *et al.* (2007). Our synthesis is presented at the bottom of the figure. See colour plate section.

2001; Frumkin and Elitzur, 2002; Klinger *et al.*, 2003). A coeval mid-Holocene lake transgression at Lake Kinneret (low-resolution curve from Hazan *et al.*, 2005; Figure 7.2), culminating at *c*. 5,200 BP, deposited sediments on top of archaeological layers from the Early Bronze Age site of Tel Bet Yerach. In the Jordan Valley, the decrease in high-energy mudflows hints at less violent and less concentrated rain episodes at *c*. 5,500 BP (Hourani and Courty, 1997). Palaeosols from the Israeli coastal plain (Gvirtzman and Wieder, 2001) suggest a relatively dry period at *c*. 6,500–5,000 BP, but followed by a wetter period at 5,000–4,600 BP. Faust and Ashkenazy (2007) link the decline in settlements in the Israeli coastal plain during the Early Bronze Age (EBA) II–III (*c*. 5000–4300 BC) to increased precipitation inducing flooding, destruction of cultivated lands, and illnesses due to the development of marsh environments.

7.4 CLIMATE INSTABILITY AT *c*. 5,200 BP

Within the generally wetter period of the Early Bronze Age, a good deal of climate instability occurred at *c*. 5,300–5,000 BP. A dry event marked by the deposition of gypsum is recorded in the Dead Sea sediments at *c*. 5,200 BP (Migowski *et al.*, 2006). Frumkin *et al.* (1999a) note a prominent $\delta^{13}C$ negative peak and low $\delta^{18}O$ values in their isotopic record from Nahal Qanah Cave (central Israel) at *c*. 5,100 BP (but the chronology of this sequence is not very well constrained), which they interpret as reflecting wetter conditions. Large fluctuations are recorded in the Soreq record within the period 5,400–4,800 BP; alternating maxima and minima in the $\delta^{18}O_{speleothem}$ and the associated calculated rainfall (Bar-Matthews *et al.*, 1999; Bar-Matthews and Ayalon, 2004; Figure 7.2) indicate frequent climatic variations. Bar-Matthews *et al.* (2003) relate the sharp decrease in calculated rainfall at *c*. 5,150–5,200 BP (Bar-Matthews and Ayalon, 2004; Figure 7.2) to a historically recorded drought period (Bar-Matthews *et al.*, 2003 and references therein).

7.5 THE END OF THE EARLY BRONZE AGE AND THE '4.2' Ka EVENT

The end of the Early Bronze Age is marked by a decline in cultivation as seen in Lake Kinneret (Baruch, 1986). Pollen indicative of more arid conditions occurs in the Hula Basin (*c*., 3,900 BP, Baruch and Bottema, 1999, amended chronology from Cappers *et al.*, 1998; Rosen, 2007). The stable isotope composition of tamarisk wood from the Mount Sedom Cave (Frumkin *et al.*, 2009; Figure 7.2) also indicates a progressive drying for the period *c*. 4,200–3,900 BP, with rainfall decreasing by half. The record shows a succession of droughts, with a prominent but short-lived event at *c*. 4,000 BP, followed by a longer event at *c*. 3,900 BP that ultimately killed the tree. This last event also seems contemporaneous with the abandonment of Bab edh-Dhra (also known as Bab-adh-Dhra), on the southeastern shore of the Dead Sea (Frumkin *et al.*, 2009).

The isotopic record at Soreq indicates very dry conditions, starting around 4,300 BP and culminating 250 years later (Figure 7.2). Swamp conditions ceased to exist on the Israeli coastal plain during the Intermediate Bronze Age (Faust and Ashkenazy, 2007 and references therein). Sediments from the northern Red Sea indicate a major evaporation event at *c*. 4,200 BP (Arz *et al.*, 2006). Low or decreasing Dead Sea levels are recorded around this time (Frumkin *et al.*, 1994; Frumkin, 1997; Frumkin *et al.*, 2001; Frumkin and Elitzur, 2002; Klinger *et al.*, 2003; Enzel *et al.*, 2003; Migowski *et al.*, 2006), coeval with canyon downcutting on the eastern shore of the Dead Sea and the partial erosion of the Early Bronze Age city of Bab edh-Dhra (Frumkin and Elitzur, 2002 and references therein). A dry event seems to be marked at *c*. 4,200–4,000 BP within the Dead Sea sediments by the deposition of gypsum (Frumkin *et al.*, 2001; Migowski *et al.*, 2006) and salt (Neev, 1964; Frumkin and Elitzur, 2002). An episode of increased aeolian input at Wadi Faynan may also be associated with this episode of aridity (Hunt *et al.*, 2007). Floodplains are being incised at this time on the Madaba-Dhiban plateau (*c*. 4,500–4,000 BP, Cordova *et al.*, 2005). The palaeosols from the Israeli coastal plain record drier conditions for the period *c*. 4,600–4,000 BP (Gvirtzman and Wieder, 2001).

The termination of the prevalent humid conditions of the Early Bronze Age, and the transition to the more arid Middle Bronze Age, may have been responsible for the demise of several major Early Bronze Age settlement centres (e.g. Frumkin *et al.*, 1999a; Enzel *et al.*, 2003; Cordova *et al.*, 2005; Frumkin *et al.*, 2009) and the transition to a more nomadic Middle Bronze Age society. On the Madaba-Dhiban plateau, the site of Khirbet Iskander was abandoned after the EB IV period, possibly following the destruction of the fertile alluvial plain by stream incision at *c*. 4,000 BP (Cordova *et al.*, 2005; Cordova, 2008). A decline in cultivated species, however, seems to occur even before, sometime during EBA III to EBA IV (Cordova, 2008).

In light of the evidence listed above, the aridification event at *c*. 4,200 BP seems to be a progressive process, taking place slowly over several centuries. Archaeological sites situated near a perennial source of water, such as Bab edh-Dhra, may have managed to resist longer in a context of increasing aridity, whereas other sites were abandoned earlier (e.g. the urban centre of Arad in the northern Negev, at *c*. 4,600 BP (Issar, 2003; Frumkin *et al.*, 2009)).

7.6 THE MIDDLE BRONZE TO IRON AGE (2000–586 BC/c. 4,000–2,500 BP)

The pattern of climate evolution during the Middle Bronze to Iron Age is somewhat less clear than for the previous periods, with conflicting climatic evidence being more frequent. Some data suggest the return to wetter conditions after the dry event at c. 4,200 BP, whereas other studies infer arid conditions for the same period. The evidence from the Dead Sea is contradictory. Migowski et al. (2006) interpret the Dead Sea sequences as reflecting generally high levels until c. 3,500 BP (Figure 7.2). Other authors, however, have determined low lake levels marking increasingly arid conditions (Frumkin et al., 1991, 1994; Frumkin, 1997; Frumkin et al., 2001; Frumkin and Elitzur, 2002; Klinger et al., 2003) until c. 3,600–3,450 BP, possibly with a dry southern basin (Frumkin and Elitzur, 2002; Klinger et al., 2003). Migowski et al. (2006; see also Neumann et al., 2007) note an arid episode and a lowstand starting at c. 3,400 BP until c. 2,150 BP. However, the period c. 3,450–3,150 BP seems to be marked by relatively highstand conditions (Frumkin et al., 1991, 1994; Frumkin, 1997; Frumkin et al., 2001; Frumkin and Elitzur, 2002; Klinger et al., 2003). Low Dead Sea levels, however, are generally agreed upon for the period 3,100–2,300 BP (Figure 7.2). Between c. 3,100–2,600 BP, archaeological evidence furthermore indicates lake levels lower than 388 m below sea level, mbsl (buildings at Mesad Gozal, in the southern Dead Sea basin) and around 395–400 mbsl (docks at Rujm el-Bahr and Khirbet Mazin in the northern basin; Frumkin and Elitzur, 2002 and references therein).

The marine sediment record from the eastern Mediterranean (Schilman et al., 2001b) suggests a humid interval at 3,500–3,000 BP, possibly coeval with high Dead Sea levels, as well as with high lake levels in the Sahara and in Turkey (Schilman et al., 2001b and references therein). The isotopic composition of land snails in the Negev still indicates relatively humid conditions until c. 3,200 BP (Goodfriend, 1990, 1999).

Hazan et al. (2005) considered that the Lake Kinneret levels fell between c. 4,000 and 2,000 BP (Figure 7.2). The alluvial sequences from central Israel (Rosen, 1986) indicate prevailing dry conditions and wadi incision during the Middle Bronze to Iron Age. The palaeosols from the Israeli coastal plain seem to record slightly wetter conditions for the period c. 4,000–1,300 BP, but this may be in fact associated with the mostly wetter conditions of the following Hellenistic to Byzantine period (Gvirtzman and Wieder, 2001).

In the pollen records, this time period is characterised in the Birkat Ram Lake by forest regeneration, following the agricultural demise at the end of the Early Bronze Age (Golan Heights, Schwab et al., 2004). However, Baruch (1990) indicates widespread forest clearance and olive cultivation in the Dead Sea and Lake Kinneret area during the second and first millennium BC, and Neumann et al. (2007) show that Middle Bronze Age sediments from the Dead Sea suggest increasingly more humid conditions, which they relate to the widespread development of agriculture and the prosperity of Canaanite towns in the Judean Mountains (Neumann et al., 2007 and references therein).

A collapse of agriculture and the settlement of arid conditions was inferred at the end of the Middle Bronze Age (c. 3,500 BP; Neumann et al., 2007 and references therein), possibly coeval with a dramatic decrease in lake levels (Migowski et al., 2006). The following Late Bronze Age seems to be marked in the Dead Sea pollen record by low arboreal pollen and the dominance of species related to a more arid climate.

Abandonment of settlements linked with very arid climates seems to occur in the Middle East at c. 3,300–3,200 BP (Issar and Zohar, 2004). Arid conditions appear to prevail during the Iron Age in the Dead Sea region (Neumann et al., 2007). However, the Iron Age is marked in the Hula Basin region by an increase in *Olea* pollen percentages, coeval with the widespread cultivation and processing of olives in the southern Levant (Baruch and Bottema, 1999, amended chronology from Cappers et al., 1998; Rosen, 2007).

Arid conditions are also suggested by the eastern Mediterranean marine record for the period c. 3,000 to c. 2,000 BP (Schilman et al., 2001b), and by the sediment record from Lake Kinneret for the period c. 3,000–2,550 BP (Dubowski et al., 2003). The palaeo-rainfall curve from Bar-Matthews and Ayalon (2004) suggests generally drier (calculated rainfall below average modern-day values) and stable climatic conditions for the whole Middle Bronze to Iron Age period.

A wet event seems, however, to be recorded in the Nahal Qanah Cave record (central Israel) around 3,200 BP (although the chronology of this sequence is not very well defined; Frumkin et al., 1999b). A return towards wetter conditions may be witnessed by the short period of stability in the fluvial sequences of northern Jordan at c. 2,800 BP (Cordova, 2008), coeval with relatively high percentages of tree pollen and moderate olive cultivation, and the colluvial deposition in Central Israel during the Late Iron Age, following dry conditions (Rosen, 1986). In contrast, the positive excursion noticeable in the isotopic records from both Lake Kinneret and the upper Galilee caves at c. 2,800–2,600 BP (Issar, 2003) may indicate a drier climate. There are suggestions for a wetter climate starting during the Iron Age at Wadi Faynan (Hunt et al., 2007).

7.7 THE BABYLONIAN TO BYZANTINE PERIOD (586 BC TO AD 638/c. 2,500–1,300 BP)

The period c. 2,100–1,800 BP is generally considered to be one of high Dead Sea levels (Figure 7.2). Archaeological evidence includes mooring emplacements situated at a high elevation on the Dead Sea north shore during the second and first centuries BC

(Bar-Adon, 1989; Bookman et al., 2004; Rosen, 2007). At Ein Gedi, a bathhouse structure situated at ~390 mbsl was covered by shore deposits sometime before AD 300 (Bookman et al., 2004 and references therein). Historical records indicate a very rainy period in the first century BC that ended just before the start of the Christian calendar (Klein, 1986; Bookman et al., 2004).

High lake levels are followed by a drop at c. 1,700–1,600 BP and another highstand at c. 1,500–1,300 BP (Frumkin and Elitzur, 2002; Enzel et al., 2003; Klinger et al., 2003; Migowski et al., 2006; Figure 7.2; see also Frumkin et al., 1999a; Frumkin et al., 2001).

The period 2,550–1,600 BP in the Lake Kinneret sedimentological, chemical and isotopic record has been interpreted by Dubowski et al. (2003) to reflect more humid conditions, with a maximum at c. 2,000 BP. Hazan et al. (2005) indicate a rise in Lake Kinneret levels after 2,000 BP (Figure 7.2). The algae and diatom assemblages from Lake Kinneret at c. 2,500 BP were linked to an increase in the nutrient supply to the lake, connected with increased human activity (Pollinger et al., 1986). An increase in cultivated plants and the decline of the natural forest characterise the pollen record (Baruch, 1986; Heim et al., 1997; Baruch and Bottema, 1999, amended chronology from Cappers et al., 1998, Schwab et al., 2004; Neumann et al., 2007; Rosen, 2007; Cordova, 2008; Figure 7.2). Olive cultivation (and cultivation of other plants such as grape or walnut) was at its maximum during the Roman to Byzantine period, coeval with more humid conditions, favourable to agriculture (Neumann et al., 2007 and references therein). The stable isotopic composition of tamarisk wood fragments from the siege at Masada (AD 70–73) also indicates humid conditions, with around twice the present-day annual rainfall, and possibly associated with a slightly cooler climate (Yakir et al., 1994; Isaar and Yakir, 1997).

The Iron Age to Byzantine period was characterised in Wadi Faynan by wetter conditions (Hunt et al., 2007). Renewed alluvial sedimentation in Central Israel (Rosen, 1986) may indicate more suitable climate conditions for agriculture during the phase of Byzantine settlement. Cordova et al. (2005) also note a period of resumed fluvial accumulation, sometime between the Roman and the Early Islamic periods. Wetter conditions for the period c. 2,500–2,100 BP are also indicated by the restricted phase of travertine deposition in the Moringa Cave (Lisker et al., 2007). A humid interval at 1,700–1,000 BP has been identified in the marine sequences from the eastern Mediterranean (Schilman et al., 2001a, 2001b), coeval with enhanced Nile supplies in nutrient and particulate organic matter.

A short period of more arid conditions seems to be marked at Ze'elim (Dead Sea area, Neumann et al., 2007) between 1,800 and 1,600 BP, with a marked decrease of olive and other anthropogenic indicators percentages in the pollen record, coeval with the lake level drop (Frumkin et al., 1991; Issar and Zohar, 2004; Figure 7.2). More humid conditions resume at Ze'elim during the later Byzantine period. The more northern record at Ein Feshkha remained unaffected by this event (Neumann et al., 2007).

In the Negev area, fewer floods, and of smaller amplitude, were recorded during the time c. 1,700–1,400 BP (Greenbaum et al., 2000). Roman to Early Islamic water-harvesting farms were developed in the Negev Highlands and other areas of the Levant, and triggered the local re-deposition of sediments (Avni et al., 2006).

7.8 CLIMATE CHANGE AT c. 1,400 BP AND THE DEMISE OF THE BYZANTINE SOCIETY

A series of climatic events, showing increased aridity, seem to be recorded at the point between the Byzantine and Early Islamic cultures, and may have been responsible for this transition. At that time there was a generalised abandonment of important centres and a return to nomadic practices following the Arab invasion (e.g. Heim et al., 1997; Schwab et al., 2004).

The transition between the Byzantine and Early Islamic period is characterised by a drop in lake levels (Frumkin et al., 1991; Enzel et al., 2003; Bookman et al., 2004; Migowski et al., 2004) with salt deposition in the northern Dead Sea Basin (Bookman et al., 2004; Migowski et al., 2006; Heim et al., 1997).

This is coincident with an episode of aridity marked in Wadi Faynan at about 1,400 BP by increased aeolian deposition (Hunt et al., 2007). Dunes also formed in the Negev, at c. 1,400–1,100 BP (Tsoar and Goodfriend, 1994). Pollen records show a sharp decrease in cultivation at the end of the Byzantine period, coeval with reforestation (Baruch and Bottema, 1999; amended chronology from Cappers et al., 1998; Schwab et al., 2004; Neumann et al., 2007; Rosen, 2007) and increase in aridity indicators (Heim et al., 1997). Historical records mention severe droughts, hunger and plagues for the sixth to ninth centuries AD (Klein, 1986; Bookman et al., 2004). The Byzantine towns of the Negev were also abandoned during this period of increased aridity (Rubin, 1996; Neumann et al., 2007).

7.9 THE EARLY ISLAMIC PERIOD (AD 638–1099 / c. 1,300–850 BP)

Conditions at the start of the Early Islamic period seem to be mostly arid in nature. An episode of dune formation occurred on the Israeli coastal plain and covered Hellenistic to Byzantine artefacts (Issar, 1968; Issar et al., 1989; Neumann et al., 2007). The increased aridity that started at the end of the Byzantine period seems to have reached its maximum around the end of the first millennium AD (Issar and Zohar, 2004). Dry conditions were recorded in the Israeli coast palaeosols starting at 1,300 BP (Gvirtzman and Wieder, 2001). From Byzantine times, the wadi incision processes dominate in northern and central Jordan (Cordova, 2008). At Wadi Faynan, after c. 1,400 BP the environmental conditions progressively acquired their modern characteristics, with increasing aridity (Hunt et al., 2007).

The pollen sequences for this period show low counts of cultivated species and forest regeneration following the decline of agriculture at the end of the Byzantine period (Baruch, 1986; Baruch, 1990; Schwab et al., 2004; Neumann et al., 2007). Arid conditions seem to persist until the transition to the Crusader period (Neumann et al., 2007). High isotopic ratios for both the upper Galilee Cave record at 1,200–1,000 BP and the record from Lake Kinneret at c. 1,400–900 BP (Issar, 2003) are consistent with a drier climate. The Lake Kinneret isotopic record (Dubowski et al., 2003) also suggests drier conditions for the period c. 1,250–900 BP. Calculated rainfall values (Bar-Matthews and Ayalon, 2004) in the Soreq area are consistently lower than modern values (Figure 7.2).

Between 1,300 and 1,200 BP, Dead Sea levels are low (Frumkin et al., 1991; Enzel et al., 2003; Bookman et al., 2004; Migowski et al., 2006; Figure 7.2). However, rising Dead Sea levels (Bookman et al., 2004; Migowski et al., 2006), or even a highstand (Frumkin et al., 1991; Frumkin and Elitzur, 2002; Issar, 2003; Klinger et al., 2003), are inferred for the period c. 1,100–800 BP.

There are other indications that wetter conditions may have occurred within the Early Islamic period. Schilman et al. (2001b) suggest increased humidity at about 1,300 BP indicated by the negative excursion in $\delta^{18}O_{G.ruber}$ and the deposition of the organic-rich layers in the Mediterranean. A correlation between high productivity events and high Nile water levels is seen at c. 1,400 BP (Schilman et al., 2001a). A highstand, dated by archaeological artefacts, is recorded at Lake Kinneret at c. 1,200 BP (late seventh–eighth century AD; Hazan et al., 2005), and suggests a rapid and destructive rise of the lake during the Early Arabic period (Marco et al., 2003). This sudden rise and an increase in high-energy sediment flux have been inferred to explain the abandonment and degradation of Roman and Byzantine piers and jetties on the lake shores (Hazan et al., 2005 and references therein).

7.10 CLIMATE INSTABILITY AROUND 1,000 BP AND THE DEMISE OF THE DECAPOLIS SOCIETY

A period of climate instability seems to have occurred at c. 1,000 BP. A period of high-intensity runoff is recorded in the Mount Sedom record at c. 1,000–700 BP (Frumkin et al., 2001). This is also a period of increased runoff at Nahal Darga on the Israeli Dead Sea shore (Kadan, 1997; Frumkin et al., 2001) and increased flood frequency and magnitude at Nahal Zin in the Negev (1,400–900 BP, the highest for the past 2,000 years (Greenbaum et al., 2000). This event is not witnessed in the deep Dead Sea records. Frumkin et al. (2001) suggest that it may have been triggered by an increase in local storms in southern Israel, in connection to an active Red Sea trough (which today causes most of the intense rainfall and flooding in the southern Negev).

A short-lived episode of floodplain deposition occurred during the Early Islamic period (tenth to eleventh centuries AD) on the Madaba-Dhiban plateau, with pollen reflecting a wet environment (Cordova, 2008). Historical documents report high rainfall and prosperity for the tenth and eleventh centuries AD (Klein, 1986; Bookman et al., 2004).

A period of low surface-water palaeo-productivity, linked with low Nile levels, is recorded in the eastern Mediterranean at c. 1,000 BP (Schilman et al., 2001a). A high point in the oxygen isotopic record at c. 900 BP, interpreted as a drop in annual rainfall, is also recorded in the Soreq curve (Bar-Matthews et al., 1999; Bar-Matthews and Ayalon, 2004; Figure 7.2), indicating a dry event.

The demise of the Decapolis society (Lucke et al., 2005) occurred in the tenth century AD after a long period of prosperity. The Decapolis settlements relied mainly on rain-fed agriculture, and the shallow soils in this area are highly vulnerable to droughts (Lucke et al., 2005). This suggests that a decrease in rainfall over northern Jordan (Lucke et al., 2005), or increased climate instability, may hold some responsibility for the abandonment of settlements in the region.

7.11 THE PAST 850 YEARS

Relatively dry conditions seem to have prevailed during the past 850 years. Calculated rainfall values (Bar-Matthews and Ayalon, 2004) in the Soreq area are consistently lower than modern values, except during a short period at 450–550 BP (Figure 7.2). The Israeli coast palaeosols indicate more arid conditions (Gvirtzman and Wieder, 2001) and erosion prevails on the Madaba-Dhiban plateau (Cordova, 2008). It is considered that Dead Sea levels stayed slightly above the sill elevation from c. 800–700 BP until at least 400 BP, and possibly until about 50 years ago (Figure 7.2). Historically recorded maximum levels attain ~390 mbsl in the late nineteenth century (Frumkin et al., 2001; Enzel et al., 2003; Bookman et al., 2004 and references therein;). The artificial drop in the Dead Sea levels that occurred in the twentieth century is dramatic but not of exceptional magnitude for the late Holocene (Bookman et al., 2004).

The Medieval optimum (c. 1,100–600 BP) is recorded at Wadi Faynan (Hunt et al., 2007) and in the northern Sinai/northern Negev deserts (Tsoar, 1995) by relatively moister conditions. In the Dead Sea area, the end of the Crusader/early Mameluke period seems to be marked by conditions more favourable to agriculture, possibly linked to increased rainfall and higher lake levels (Neumann et al., 2007). At Lake Kinneret, a short very wet period, followed by cooler and drier conditions, seems to occur during the time interval 900–170 BP (Dubowski et al., 2003).

A wet event at c. 800 BP seems to be recorded in most of the eastern Mediterranean sediments (Schilman et al., 2001a, 2003). A transitional period featuring several very large floods in the Negev (c. 880–530 BP, Greenbaum et al., 2000) and pollen records characteristic of a dry climate (c. 800–600 BP, Neumann et al., 2007) was also evidenced in the palaeoenvironmental record.

A period of high productivity in the eastern Mediterranean occurred at about 650 BP, coeval with high Nile water levels (Schilman et al., 2001a). Increased humidity in the fifteenth century also seems coeval with a rise in Dead Sea levels (Neumann et al., 2007). The low point in the oxygen isotopic records of Soreq Cave at c. 500 BP may indicate a wet event, followed by a $\delta^{18}O$ maximum and dry event at c. 350 BP (Bar-Matthews et al., 1999; Figure 7.2).

Marked arid conditions seem to prevail at Wadi Faynan during the 'Little Ice Age' (Hunt et al., 2007) with fewer, and smaller, floods in the Negev indicating a drier climate (c. 530–60 BP, Greenbaum et al., 2000). Tsoar (1995) notes that the northern Sinai/northern Negev Desert was under more arid conditions during most of the seventeenth and eighteenth centuries. The positive excursion in $\delta^{18}O_{G.ruber}$ at c. 270 BP is interpreted by Schilman et al. (2001b) as reflecting the coldest time during the past 3,600 years.

7.12 CONCLUSIONS

A trend towards aridity is evident in the palaeoenvironmental records throughout the Holocene to the present day. Climatic fluctuations superimposed on this general trend (Figure 7.2, synthetic climatic change) are more difficult to determine, and the different records from the southern Levant often present contradictory information for the middle to late Holocene period.

It seems clear, however, that the cultural developments at the end of the Early Bronze Age and the Byzantine period, respectively, relate to a shift towards more arid conditions. Relatively humid conditions seem to prevail during the Chalcolithic to Early Bronze Age, and during the Hellenistic to Byzantine period. The decline in settlements in the Israeli coastal plain during the Early Bronze Age (EBA) II–III may have been linked to increased precipitation inducing flooding and illnesses related to marshy environments. The Hellenistic to Byzantine period is a time of thriving agriculture, marked by an increase of anthropogenic indicators in the pollen sequences.

Mostly arid conditions are recorded during the Early Islamic period to modern times. Information available for the Middle Bronze to Iron Age is more contradictory, but arid conditions may have prevailed at the end of the Middle Bronze Age as well as during the first half of the Iron Age (Figure 7.2, synthetic climatic change). Phases of climate instability occurred at c. 5,200–5,000 BP and c. 1,000 BP, the latter potentially holding some responsibility for the demise of the Decapolis society.

The high climatic gradients characteristic of the southern Levant may partly explain the contradictions shown in the various palaeoenvironmental records, with different regions reacting variously to small-amplitude climatic fluctuations. More local studies, based on well-dated sequences, may help to refine our understanding of middle to late Holocene climate change in the southern Levant.

REFERENCES

Abboud, I. A. (2000) Palaeoenvironment, palaeoclimate, and palaeohydrology of Burqu' Basin, Al-Badia, NE Jordan. Unpublished PhD thesis: Baghdad University.

Amit, R., Y. Enzel and D. Sharon (2006) Permanent Quaternary hyperaridity in the Negev, Israel, resulting from regional tectonics blocking Mediterranean frontal systems. *Geology* **34**: 509–512.

Amit, R., J. Lekach, A. Ayalon, N. Porat and T. Grodek (2007) New insight into pedogenic processes in extremely arid environments and their paleoclimatic implications – the Negev Desert, Israel. *Quaternary International* **162**: 61–75.

Arz, H. W., F. Lamy and J. Patzold (2006) A pronounced dry event recorded around 4.2 ka in brine sediments from the northern Red Sea. *Quaternary Research* **66**: 432–441.

Avni, Y., N. Porat, J. Plakht and G. Avni (2006) Geomorphic changes leading to natural desertification versus anthropogenic land conservation in an arid environment, the Negev Highlands, Israel. *Geomorphology* **82**: 177–200.

Bar-Adon, P. (1989) Excavations in the Judean Desert. In *Antiquities*, ed. A. Zusman and D. Straws. Jerusalem: The Antiquities Authorities pp. 3–14.

Bar-Matthews, M. and A. Ayalon (2004) Speleothems as palaeoclimate indicators, a case study from Soreq Cave located in the Eastern Mediterranean Region, Israel. In *Past Climate Variability through Europe and Africa*, ed. R. W. Battarbee, F. Gasse and C. E. Strickley. Dordrecht: Springer pp. 363–391.

Bar-Matthews, M., A. Ayalon and A. Kaufman (1998) Middle to Late Holocene (6,500 yr. period) paleoclimate in the eastern Mediterranean region from stable isotopic composition of speleothems from Soreq Cave, Israel. In *Water, Environment and Society in Times of Climatic Change*, ed. A. Issar and N. Brown. Dordrecht: Kluwer Academic Publishers pp. 203–214.

Bar-Matthews, M., A. Ayalon, A. Kaufman and G. J. Wasserburg (1999) The Eastern Mediterranean paleoclimate as a reflection of regional events: Soreq cave, Israel. *Earth and Planetary Science Letters* **166**: 85–95.

Bar-Matthews, M., A. Ayalon, M. Gilmour, A. Matthews and C. J. Hawkesworth (2003) Sea-land oxygen isotopic relationships from planktonic foraminifera and speleothems in the Eastern Mediterranean region and their implication for paleorainfall during interglacial intervals. *Geochimica et Cosmochimica Acta* **67**: 3181–3199.

Baruch, U. (1986) The late Holocene vegetational history of Lake Kinneret (Sea of Galilee), Israel. *Paleorient* **12**: 37–48.

Baruch, U. (1990) Palynological evidence of human impact on the vegetation as recorded in late Holocene lake sediments in Israel. In *Man's Role in the Shaping of the Eastern Mediterranean Landscape*, ed. S. Bottema, G. Enjes-Nieborg and W. van Zeist. Rotterdam: Balkema.

Baruch, U. and S. Bottema (1991) Palynological evidence for climatic changes in the Levant ca. 17,000–9,000 BP. In *The Natufian Culture in the Levant*, ed. O. Bar-Yosef and F. R. Valla. Ann Arbor: International Monographs in Prehistory pp. 11–20.

Baruch, U. and S. Bottema (1999) A new pollen diagram from Lake Hula. Vegetational, climatic, and anthropogenic implications. In *Ancient Lakes: Their Cultural and Biological Diversity*, ed. H. Kawanabe, G. W. Coulter and A. C. Roosevelt. Ghent: Kinobi Productions pp. 75–86.

Bookman, R., Y. Enzel, A. Agnon and M. Stein (2004) Late Holocene lake levels of the Dead Sea. *Geological Society of America Bulletin* **116**: 555–571.

Cappers, R. T. J., S. Bottema and H. Woldring (1998) Problems in correlating pollen diagrams of the Near East: a preliminary report. In *The Origins of Agriculture and Crop Domestication*, ed. A. B. Damania, J. Valkoun, G. Willcox and C. O. Qualset. Aleppo, Syria: ICARDA pp. 160–169.

Cordova, C. E. (2008) Floodplain degradation and settlement history in Wadi al-Wala and Wadi ash-Shallalah, Jordan. *Geomorphology* **101**: 443–457.

Cordova, C. E., C. Foley, A. Nowell and M. Bisson (2005) Landforms, sediments, soil development, and prehistoric site settings on the Madaba-Dhiban Plateau, Jordan. *Geoarchaeology – an International Journal* **20**: 29–56.

Cullen, H. M., P. B. deMenocal, S. Hemming *et al.* (2000) Climate change and the collapse of the Akkadian empire: evidence from the deep sea. *Geology* **28**: 379–382.

Davies, C. P. (2005) Quaternary paleoenvironments and potential for human exploitation of Jordan Plateau desert interior. *Geoarchaeology – an International Journal* **20**: 379–400.

deMenocal, P. B. (2001) Cultural responses to climate change during the late Holocene. *Science* **292**: 667–673.

Dubowski, Y., J. Erez and M. Stiller (2003) Isotopic paleolimnology of Lake Kinneret. *Limnology and Oceanography* **48**: 68–78.

Enzel, Y., R. Bookman, D. Sharon *et al.* (2003) Late Holocene climates of the Near East deduced from Dead Sea level variations and modern regional winter rainfall. *Quaternary Research* **60**: 263–273.

Enzel, Y., R. Arnit, U. Dayan *et al.* (2008) The climatic and physiographic controls of the eastern Mediterranean over the late Pleistocene climates in the southern Levant and its neighboring deserts. *Global and Planetary Change* **60**: 165–192.

EXACT (Executive Action Team Middle East Water Data Banks Project) (1998). Overview of Middle East Water Resources – Water resources of Palestinian, Jordanian, and Israeli interest http://water.usgs.gov/exact/overview/index.htm.

Faust, A. and Y. Ashkenazy (2007) Excess in precipitation as a cause for settlement decline along the Israeli coastal plain during the third millennium BC. *Quaternary Research* **68**: 37–44.

Frumkin, A. (1997) The Holocene history of the Dead Sea levels. In *The Dead Sea, the Lake and its Setting*, ed. T. M. Niemi, Z. Ben-Avraham and Y. Gat. Oxford: Oxford University Press pp. 237–248.

Frumkin, A. and Y. Elitzur (2002) Historic dead sea level fluctuations calibrated with geological and archaeological evidence. *Quaternary Research* **57**: 334–342.

Frumkin, A., M. Magaritz, I. Carmi and I. Zak (1991) The Holocene climatic record of the salt caves of Mount Sedom. *The Holocene* **1**: 191–200.

Frumkin, A., I. Carmi, I. Zak and M. Magaritz (1994) Middle Holocene environmental change determined from the salt caves of Mount Sedom, Israel. In *Late Quaternary Chronology of the Eastern Mediterranean*, ed. O. Bar-Yosef and R. S. Kra. Tucson: University of Arizona Press pp. 315–322.

Frumkin, A., I. Carmi, A. Gopher *et al.* (1999a) A Holocene millennial-scale climatic cycle from a speleothem in Nahal Qanah Cave, Israel. *The Holocene* **9**: 677–682.

Frumkin, A., D. C. Ford and H. P. Schwarcz (1999b) Continental oxygen isotopic record of the last 170,000 years in Jerusalem. *Quaternary Research* **51**: 317–327.

Frumkin, A., G. Kadan, Y. Enzel and Y. Eyal (2001) Radiocarbon chronology of the Holocene Dead Sea: attempting a regional correlation. *Radiocarbon* **43**: 1179–1189.

Frumkin, A., P. Karkanas, M. Bar-Matthews *et al.* (2009) Gravitational deformations and fillings of aging caves: the example of Qesem karst system, Israel. *Geomorphology* **106**: 154–164.

Goodfriend, G. A. (1990) Rainfall in the Negev Desert during the middle Holocene, based on ^{13}C of organic matter in land snail shells. *Quaternary Research* **34**: 186–197.

Goodfriend, G. A. (1991) Holocene trends in ^{18}O in land snail shells from the Negev Desert and their implications for changes in rainfall source areas. *Quaternary Research* **35**: 417–426.

Goodfriend, G. A. (1999) Terrestrial stable isotope records of Late Quaternary paleoclimates in the eastern Mediterranean region. *Quaternary Science Reviews* **18**: 501–513.

Grattan, J. P., D. D. Gilbertson and C. O. Hunt (2007) The local and global dimensions of metalliferous pollution derived from a reconstruction of an eight thousand year record of copper smelting and mining at a desert-mountain frontier in southern Jordan. *Journal of Archaeological Science* **34**: 83–110.

Greenbaum, N., A. P. Schick and V. R. Baker (2000) The palaeoflood record of a hyperarid catchment, Nahal Zin, Negev Desert, Israel. *Earth Surface Processes and Landforms* **25**: 951–971.

Gvirtzman, G. and M. Wieder (2001) Climate of the last 53,000 years in the eastern Mediterranean, based on soil-sequence stratigraphy in the coastal plain of Israel. *Quaternary Science Reviews* **20**: 1827–1849.

Hazan, N., M. Stein, A. Agnon *et al.* (2005) The late quaternary limnological history of Lake Kinneret (Sea of Galilee), Israel. *Quaternary Research* **63**: 60–77.

Heim, C., N. R. Nowaczyk, J. F. W. Negendank, S. A. G. Leroy and Z. BenAvraham (1997) Near East desertification: evidence from the Dead Sea. *Naturwissenschaften* **84**: 398–401.

Hourani, F. and M.-A. Courty (1997) L'évolution climatique de 10 500 à 5500 B.P. dans la vallée du Jourdain. *Paleorient* **23**: 95–105.

Hunt, C. O., H. A. Elrishi, D. D. Gilbertson *et al.* (2004) Early-Holocene environments in the Wadi Faynan, Jordan. *The Holocene* **14**: 921–930.

Hunt, C. O., D. D. Gilbertson and H. A. El-Rishi (2007) An 8000-year history of landscape, climate, and copper exploitation in the Middle East: the Wadi Faynan and the Wadi Dana National Reserve in southern Jordan. *Journal of Archaeological Science* **34**: 1306–1338.

Issar, A. (1968) Geology of the central coastal plain of Israel. *Israel Journal of Earth Sciences* **17**: 16–29.

Issar, A. S. (1990) *Water Shall Flow from the Rock: Hydrogeology and Climate in the Lands of the Bible*. Heidelberg: Springer.

Issar, A. S. (2003) *Climate Changes during the Holocene and their Impact on Hydrological Systems*. Cambridge: Cambridge University Press.

Issar, A. S. and D. Yakir (1997) Isotopes from wood buried in the Roman siege ramp of Masada: the Roman period colder climate. *Biblical Archaeologists* **60**: 101–106.

Issar, A. S. and M. Zohar (2004) *Climate Change – Environment and Civilization in the Middle East*. Berlin: Springer-Verlag.

Issar, A. S., H. Tsoar and D. Levin (1989) Climate changes in Israel during historical times and their impact on hydrological pedological and socio-economic systems. In *Paleoclimatology and Paleometeorology*, ed. M. Leinen and M. Sarthein. Dordrecht: Kluwer Academic Publishing pp. 525–542.

Kadan, G. (1997) Evidence for Dead Sea lake-level fluctuations and recent tectonism from the Holocene fan-delta of Nahal Darga. Unpublished MSc thesis: Ben-Gurion University of the Negev.

Klein, C. (1986) Fluctuations of the level of the Dead Sea and climatic fluctuations in Israel during historical times. Unpublished PhD thesis: Hebrew University, Jerusalem.

Klinger, Y., J. P. Avouac, D. Bourles and N. Tisnerat (2003) Alluvial deposition and lake-level fluctuations forced by Late Quaternary climate change: the Dead Sea case example. *Sedimentary Geology* **162**: 119–139.

Lisker, S., R. Porat, U. Davidovich *et al.* (2007) Late Quaternary environmental and human events at En Gedi, reflected by the geology and archaeology of the Moringa Cave (Dead Sea area, Israel). *Quaternary Research* **68**: 203–212.

Lucke, B., M. Schmidt, Z. al-Saad, O. Bens and R. F. Huttl (2005) The abandonment of the Decapolis region in Northern Jordan – forced by environmental change? *Quaternary International* **135**: 65–81.

Marco, S., M. Hartal, N. Hazan, L. Lev and M. Stein (2003) Archaeology, history, and geology of the A.D. 749 earthquake, Dead Sea transform. *Geology* **31**: 665–668.

McLaren, S. J., D. D. Gilbertson, J. P. Grattan *et al.* (2004) Quaternary palaeogeomorphologic evolution of the Wadi Faynan area, southern Jordan. *Palaeogeography Palaeoclimatology Palaeoecology* **205**: 131–154.

Migowski, C., A. Agnon, R. Bookman, J. F. W. Negendank and M. Stein (2004) Recurrence pattern of Holocene earthquakes along the Dead Sea transform revealed by varve-counting and radiocarbon dating of lacustrine sediments. *Earth and Planetary Science Letters* **222**: 301–314.

Migowski, C., M. Stein, S. Prasad, J. F. W. Negendank and A. Agnon (2006) Holocene climate variability and cultural evolution in the Near East from the Dead Sea sedimentary record. *Quaternary Research* **66**: 421–431.

Neev, D. (1964) *The Dead Sea*. Geological Survey of Israel Report Q/2/64.
Neev, D. and K. O. Emery (1995) *The Destruction of Sodom, Gomorrah, and Jericho*. New York: Oxford University Press.
Neumann, F. H., E. J. Kagan, M. J. Schwab and M. Stein (2007) Palynology, sedimentology and palaeoecology of the late Holocene Dead Sea. *Quaternary Science Reviews* **26**: 1476–1498.
Pollinger, U., A. Ehrlich and S. Serruya (1986) The planktonic diatoms of Lake Kinneret (Israel) during the last 5000 years—their contribution to the algal biomass. In *Proceedings of the 8th International Diatom Symposium*, ed. M. Richard. Koenigstein: Koeltz pp. 459–470.
Rosen, A. (1986) Environmental change and settlement at Tel Lachish, Israel. *Bulletin of the American Schools of Oriental Research* **263**: 55–60.
Rosen, A. (2007) *Civilizing Climate: Social Responses to Climate Change in the Ancient Near East*. Plymouth: AltaMira Press.
Rosen, A. M. (1995) The social response to environmental change in Early Bronze Age Canaan. *Journal of Anthropological Archaeology* **14**: 26–44.
Rosen, S. A., A. B. Savinetsky, Y. Plakht et al. (2005) Dung in the desert: preliminary results of the Negev Holocene Ecology Project. *Current Anthropology* **46**: 317–327.
Rubin, R. (1996) Urbanization and settlement in the Negev Desert in the Byzantine Period. In *The Mosaic of Israeli Geography*, ed. Y. Gradus and G. Lipshitz. Beersheva: Ben Gurion Press pp. 373–401.
Schilman, B., A. Almogi-Labin, M. Bar-Matthews et al. (2001a) Long- and short-term carbon fluctuations in the Eastern Mediterranean during the late Holocene. *Geology* **29**: 1099–1102.
Schilman, B., M. Bar-Matthews, A. Almogi-Labin and B. Luz (2001b) Global climate instability reflected by Eastern Mediterranean marine records during the late Holocene. *Palaeogeography Palaeoclimatology Palaeoecology* **176**: 157–176.
Schilman, B., A. Almogi-Labin, M. Bar-Matthews and B. Luz (2003) Late Holocene productivity and hydrographic variability in the eastern Mediterranean inferred from benthic foraminiferal stable isotopes. *Paleoceanography* **18**.
Schwab, M. J., F. Neumann, T. Litt, J. F. W. Negendank and M. Stein (2004) Holocene palaeoecology of the Golan Heights (Near East): investigation of lacustrine sediments from Birkat Ram crater lake. *Quaternary Science Reviews* **23**: 1723–1731.
Shaked, Y., S. Marco, B. Lazar et al. (2002) Late Holocene shorelines at the Gulf of Aqaba: migrating shorelines under conditions of tectonic and sea level stability. *EGU Stephan Mueller Special Publication Series* **2**: 105–111.
Tsoar, H. (1995) Desertification in Northern Sinai in the eighteenth century. *Climatic Change* **29**: 429–438.
Tsoar, H. and G. A. Goodfriend (1994) Chronology and palaeoenvironmental interpretation of Holocene aeolian sands at the inland edge of the Sinai-Negev erg. *The Holocene* **4**: 244–250.
Vaks, A., M. Bar-Matthews, A. Ayalon et al. (2003) Paleoclimate reconstruction based on the timing of speleothem growth and oxygen and carbon isotope composition in a cave located in the rain shadow in Israel. *Quaternary Research* **59**: 182–193.
Yakir, D., A. Issar, J. Gat et al. (1994) ^{13}C and ^{18}O of wood from the Roman siege rampart in Masada, Israel (AD 70–73): evidence for a less arid climate of the region. *Geochimica et Cosmochimica Acta* **58**: 3535–3539.

8 Using proxy data, historical climate data and climate models to investigate aridification during the Holocene

Emily Black, David Brayshaw, Stuart Black and Claire Rambeau

ABSTRACT

When used in conjunction, climate models and palaeoenvironmental data can lead to a more complete understanding of past climate than is possible using either method in isolation. Moreover, the veracity of climate models can be evaluated, which then lends credence to their use for predicting future climate change. In this study we investigate the transition to aridity in the eastern Mediterranean that occurred in the Holocene, a transition with marked consequences for settlement of the Middle East. We show that the general pattern of a transition during the Holocene to a wetter northern Europe and a drier Middle East is seen in both the palaeoenvironmental record and climate model simulations. The pattern of precipitation changes projected by the climate model for the past is similar to those projected for the end of the twenty-first century under GHG-driven climate change. The climate model's ability to represent the past – as tested against palaeoenvironmental observations – thus lends credence to the future projections.

8.1 INTRODUCTION

The Holocene climate of the Middle East can be investigated using climate proxies and geological evidence (Chapters 6 and 7, this volume) or through climate modelling (Chapter 3). In this chapter, we show how combining these two approaches has the potential to deepen our understanding of the past climate. We also describe how comparison between the climate model output, proxy data and historical observations provides a means of evaluating the climate models and assessing the credibility of future projections.

Climate models and palaeoclimate proxies can be used together to investigate the Earth system's sensitivity to climate forcings. For example, similiarities between the proxy observations and output from a climate model forced with freshwater influx into the Labrador Sea support the hypothesis that the widespread cold event at 8.2 ka BP was primarily caused by a weakening of the thermohaline circulation (THC), rather than directly by changes in the solar forcing (Renssen et al., 2001; Wiersma and Renssen, 2006). Detailed understanding of this event provides information about the response of the climate system to a weakening of the THC, which is particularly relevant because a disturbance of the THC is a possible consequence of anthropogenic climate change (see for example Cubasch et al., 2001).

Comparison between proxy data and climate model output can also be used to investigate the response of the climate system to forcings that change gradually through time, such as orbital parameters and greenhouse gas (GHG) concentrations. There have been few such studies of the evolution of Holocene climate in Europe and the Middle East, with most work focused on the Last Glacial Maximum or earlier periods (as described in Chapter 6 of this volume). In an early study, Huntley and Prentice (1993) compared pollen reconstructions for the present day and for 3 ka, 6 ka and 9 ka with the COHMAP (Cooperative Holocene Mapping Project) model output. Although some broad similarities between the observations and models are evident, the coarse resolution of the model (4° latitude by 7.5° longitude) precluded comparison of spatial variability at country scales. Other studies that have compared the PMIP (Palaeoclimate Model Intercomparison Project) Mid-Holocene (6 ka) runs with the European pollen reconstructions presented in Cheddadi et al. (1997) have had inconsistent results, with most models unable to replicate the comparatively small differences between 6 ka and the present day. Moreover, the coarse resolution of the PMIP models (typically 2.5° or coarser) made detailed comparisons difficult on sub-continental scales.

Water, Life and Civilisation: Climate, Environment and Society in the Jordan Valley, ed. Steven Mithen and Emily Black. Published by Cambridge University Press.
© Steven Mithen and Emily Black 2011.

In this study, we compare regional model simulations of European and Middle East precipitation for the early and late Holocene with published palaeoenvironmental data. We chose this as a case study because a survey of proxy evidence for the Middle East (Chapters 6 and 7 of this volume) suggested significant changes in the climate, and yet the period is not so far in the past as to be irrelevant to the archaeological studies presented elsewhere in this volume (as would be the case with, say, the Last Glacial Maximum). The proxy data considered in this chapter are drawn from a wider region than the data used in other chapters in this volume. This allows us to explore the hypothesis, proposed in Chapter 3, that large-scale changes in the storm tracks were closely associated with the aridification during the Holocene.

As well as providing the background for the studies of Holocene settlement and water availability described in later chapters of this volume, the palaeoclimate studies have ramifications for future projections for Europe and the Middle East. In particular, climate models respond to changes in the so-called physical 'forcings' that are used to drive them, and the change in these forcings are in some respects stronger across the Holocene period (measured in millennia) than over the historical record (approximately the past 100 years). As such, testing the ability of climate models to reproduce the gross patterns of changes across the Holocene period against proxy data provides some confidence that the same models can be used to reliably produce simulations of climate under twenty-first-century forcings which may be markedly different to those of the present day (although it should be noted that the anticipated changes in climate forcing in the twenty-first century are somewhat different in character from those experienced during the Holocene, as described in Chapter 3). These gross comparisons between model and proxy data across the Holocene period therefore complement model–data comparisons that use more recent historical data (i.e. direct meteorological observations over the past 100 years), which, rather than testing the climate model's gross response to large-scale changes in forcing, instead provide insight into the model's capacity to simulate daily or interannual variability and spatial patterns in precipitation. Thus, considered in conjunction, comparisons between historical data, climate proxies and climate model output may be a powerful test of the credibility of the climate projections for the next century.

In summary, the questions addressed by this chapter are as follows.

- Do proxy data for Europe and the Middle East support the hypothesis presented in Chapter 3 that the aridification of Jordan and Israel during the Holocene was part of a wider pattern of precipitation change, which resulted from changes in the storm tracks?
- How well does the regional climate model (RCM) represent the spatial patterns of present-day precipitation and temperature?
- To what extent is the RCM's simulation of the differences between late and early Holocene precipitation consistent with the proxy evidence?
- To what extent can the aridification of the Middle East that occurred in the past be considered an analogue for the drying projected for the future?

The remainder of this chapter is organized as follows. Section 8.2 describes the data and methodologies used. Section 8.3 describes the evolution of the climate from a wet early Holocene to aridity using palaeoclimate records that span the transition. Section 8.4 first considers how well the model can simulate present-day climate, and then compares the model simulations of the differences between early and late Holocene with proxy observations. Section 8.5 discusses the results, with particular focus on their relevance for future projections. Section 8.6 draws some brief conclusions from the study.

8.2 DATA AND METHODOLOGY

The sources of the palaeoenvironmental data referred to in this Chapter are described in section 8.3. The following section summarises the climate data and modelling methodologies.

8.2.1 Observed rainfall data

All observations of rainfall referred to in this study are based on rain gauge measurements. GPCC data were used for the larger scale comparisons between the observed and modelled rainfall data. The GPCC rainfall data used in this study is $0.5 \times 0.5°$ gridded product, based on rain gauge measurements (Schneider et al., 2008). The modelled temperature data were compared to NCEP reanalysis data (Kalnay et al., 1996). Topographic data were taken from the United States Geological Survey GTPOP30 digital elevation map dataset (http://topotools.cr.usgs.gov/). More details of the observed datasets are given in Chapter 2 of this volume.

8.2.2 Modelling methodology

This section summarises the palaeoclimate and future climate modelling methodologies described in detail in Chapters 3 and 4, respectively.

Holocene climate variability was investigated through a series of time-slice experiments for 12 ka, 10 ka, 8 ka, 6 ka, 2 ka, the pre-industrial period and the present day. For each time slice, the climate forcings were specified to correspond to those experienced at the time period in question. Specifically, the forcings included were changes in GHG concentrations (CO_2 and CH_4 only), the Earth's orbital parameters, land ice sheets and sea level. These integrations cannot be considered a true snapshot of the climate at a particular time because of the internal variability of the atmosphere, uncertainties in the initial conditions of the atmosphere and ocean, and omission of potentially important feedbacks and forcings, such as land surface change. However, comparison between time slices gives insight into how the climate has responded to changes in natural forcings during the Holocene.

Each integration lasted 20 years – a length chosen to give some idea of interannual variability without being prohibitively expensive. The global runs were carried out using the Hadley Centre HadSM3 global climate model. The HadRM3 regional climate model, forced by boundary conditions from the global model, was then used to dynamically downscale the output over the Mediterranean area.

The future climate integrations followed a generally similar method, with a global climate model (in this case HadCM3) providing lateral boundary forcing for the HadRM3 regional model. The only forcing applied was changes in GHG concentrations. As with the past time slices, the impact of land surface change was not considered. In this chapter, two simulations are considered:

- a baseline scenario (1961–90) driven with lateral boundary conditions derived from a run of HadAM3P forced with observed climatological sea-surface temperature (SST) and land surface at the lower boundary;
- an A2 emission scenario (2070–2100) driven with lateral boundary conditions derived from HadAM3P forced with HadCM3 A2 scenario SST changes added to the observed SST.

8.3 HOLOCENE CLIMATE IN THE MIDDLE EAST AND EUROPE

There is substantial evidence from the palaeoclimate record that during the early Holocene, the Levant was wetter than it is today. In particular, there was an increase in *Pistacia* and oak in the pollen records (Horowitz, 1971, 1989; Rossignol-Strick, 1995, 1999) and a southward migration of the Negev Desert boundary (Goodfriend, 1999). Further west, the deposition of red palaeosols on the Israeli coastal plain indicates an excess of precipitation versus evaporation (Gvirtzman and Wieder, 2001). The increase in annual rainfall suggested by all these records is consistent with higher Dead Sea levels (e.g. Neev and Emery, 1995; Klinger *et al.*, 2003; Robinson *et al.*, 2006). Conservative estimates of palaeo-rainfall from cave sequences at this time suggest values of 550 and 700 mm yr^{-1} in comparison with 300 mm yr^{-1} or less in that region now (Bar-Matthews *et al.*, 2003). A doubling of rainfall relative to the present day is supported by the Dead Sea modelling described in later chapters of this volume, which suggests rainfall of at least 1,200 mm yr^{-1} in Jerusalem in the early Holocene (compared with 600 mm yr^{-1} now).

Widening our perspective to southern Europe, there is evidence for erosive events caused by elevated rainfall in Crete (Maas *et al.*, 1998), Greece (Hamlin *et al.*, 2000), Italy (Rowan *et al.*, 2000) and southern Spain (Candy and Black, 2009) which resulted in highly localised soil removal (intensive soil erosion is also a proposed explanation for the large positive carbon isotope excursion in Soreq Cave speleothems; Bar-Matthews *et al.*, 1997). Lacustrine records across Europe are also responsive to increased rainfall (Roberts *et al.*, 2008) and show a significant shift in isotopic signals as a result of the increased rainfall.

The marine records for the eastern Mediterranean and Red Sea also suggest wetter and warmer conditions in the early Holocene than previously recorded. Inorganic carbon-rich deposits (sapropel 1 or S1) were deposited in the eastern Mediterranean under anoxic conditions between ~9.5 and 7 cal. ka BP over an extensive area (e.g. Emeis *et al.*, 2003). The Mediterranean sapropels formed as a result of anoxic conditions induced through stratification of the water column driven by a large freshwater surface layer (e.g. Rossignol-Strick *et al.*, 1982; Rossignol-Strick, 1985; Rohling and Hilgen, 1991; Rohling, 1994; Kallel *et al.*, 1997). Reeder *et al.* (2002) suggest that the source of this large freshwater surface layer may have been an enhanced Nile River outflow, driven by increased rainfall in the Sahara (as postulated by Rossignol-Strick *et al.*, 1982; Rossignol-Strick, 1985) and the Ethiopian highlands.

On a continental scale, there have been several studies of pollen assemblages during the Holocene for the whole of Europe. Huntley and Prentice (1993) investigated changes in vegetation for the whole Holocene, contrasting the present day with 3, 6 and 9 ka. The results suggest significant variation in available moisture (precipitation minus evaporation) during the Holocene. After 6 ka, to first order, the vegetation within Europe looks similar to the present day. The conclusions of this study are supported by two subsequent studies of pollen assemblages at 6 ka, which suggest that differences in winter precipitation between 6 ka and the present day were small and statistically insignificant. Nevertheless, some spatially coherent signals are evident, and there is broad agreement between the two studies. The first study, described in Wu *et al.* (2007), used observed pollen data in conjunction with a biome model (BIOME6000) to constrain differences in precipitation and temperature between 6 ka and the present day. The second study (Cheddadi *et al.*, 1997)) used lake level and pollen data to assess moisture availability within Europe. Both studies found there to be somewhat more moisture available in south and east Europe than today and also found wetter conditions in more northerly regions. However, the BIOME study (Wu *et al.*, 2007) suggested that the Alps were

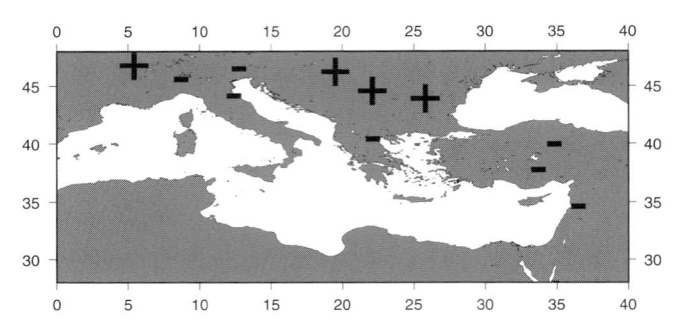

Figure 8.1. Summary of rainfall signal from the proxy data for the Middle East and Europe described in the text. Pluses indicate higher rainfall and minuses lower rainfall during the early Holocene as compared with the early/mid-Holocene.

wetter, while the pollen and lake level study (Cheddadi *et al.*, 1997) suggested that they were drier.

The most marked change for northern and central Europe occurred between 9 and 6 ka when colder winters, warmer summers and greater available moisture led to the replacement, in some regions, of the deciduous forests, which had dominated prior to 9 ka, by mixed and boreal forests (Landmann *et al.*, 2002). In contrast to the situation in northern Europe, geological and isotopic data from lakes in Israel, Turkey, Greece and Spain suggest more available moisture (greater precipitation) in the early Holocene in southern Europe (Landmann *et al.*, 1996; Giralt *et al.*, 1999; Roberts *et al.*, 2001). Changes in vegetation type and pollen assemblages for southern Europe are consistent with the hypothesis that, at this time, Mediterranean-type vegetation replaced the deciduous forests that dominated in the early Holocene (Huntley and Prentice, 1993). The contrast between northern and southern Europe suggested by the pollen and lake level studies is borne out by other geochemical data. For example, comparison of $\delta^{18}O$ time-series for three speleothems suggests that climate variability in southern Europe was in anti-phase with that in southwest Ireland throughout the Holocene (McDermott *et al.*, 1999). The differences between late and early Holocene climate for the Middle East and for Europe are sketched in Figure 8.1.

8.4 COMPARING CLIMATE MODELLING RESULTS WITH OBSERVATIONS

8.4.1 Comparison with historical data

The following section summarises the model evaluations described in Black (2009), which additionally includes a formal statistical comparison between observed and modelled precipitation and temperature.

It can be seen from Figure 8.2 that the RCM does a reasonable job of simulating the spatial pattern of rainfall over the Mediterranean and southern Europe. The observed rainfall maxima over the Alps, the shore of the Black Sea and the north Mediterranean coast are all evident in the RCM simulation, although the actual rainfall amounts are lower than those observed. The spatial variability of seasonal mean temperature is also well represented, with the model capturing the zonal transition from high temperatures in Italy and southern France to lower temperatures in Turkey.

Focusing on the eastern Mediterranean, it can be seen that the model correctly simulates the observed zonal and meridional temperature gradients. It also captures some aspects of the spatial variability in rainfall including the sharp west–east gradient from the relatively wet Mediterranean coastal region to the arid east, and the increase in rainfall from the arid south to a wetter climate in the Turkish highlands. However, although the model captures some aspects of the spatial variability of eastern Mediterranean rainfall, it underestimates the amount at the peak of the rainy season, in some areas by a factor of two or more.

The model's ability to capture the general structure of the Mediterranean storm track is highly relevant to this study because most rainfall at the peak of the rainy season is delivered by cyclones crossing the Mediterranean (Chapter 2), and because it is postulated that long timescale variability in precipitation is primarily driven by changes in the storm tracks (Chapters 3 and 4). Figure 8.3 compares the mean track densities during January for the regional model baseline scenario with reanalysis data (calculated following the methodology described in Bengtsson, 2006 and references therein). It can be seen that, although the storm track is somewhat more diffuse than that observed, the regional model captures its general orientation and location. This is consistent with the model's realistic simulation of the daily probabilities of rainfall, which suggests that the frequency and duration of cyclones entering the River Jordan region are captured (see Chapter 4).

8.4.2 Comparison with proxy data

The time-slice modelling results suggested that gradual changes in GHG concentrations and solar insolation resulted in a poleward shift of the Mediterranean storm track. The shift in the storm track led to a reduction in winter precipitation in the Middle East and southern Europe and an increase further north within Europe during the course of the Holocene. In particular, precipitation reduced over the Mediterranean, southernmost Europe and the Middle East (including Italy, Greece, Turkey, Jordan and Israel) and increased in central and eastern Europe (including Romania, Bulgaria, Hungary and Germany). These results are shown in Figure 8.4, which depicts the difference between the modelled mean October–March precipitation and December–February track density in the early Holocene (mean of the 12, 10 and 8 ka time slices) and late Holocene (mean of the 6, 4, 2 and pre-industrial time slices).

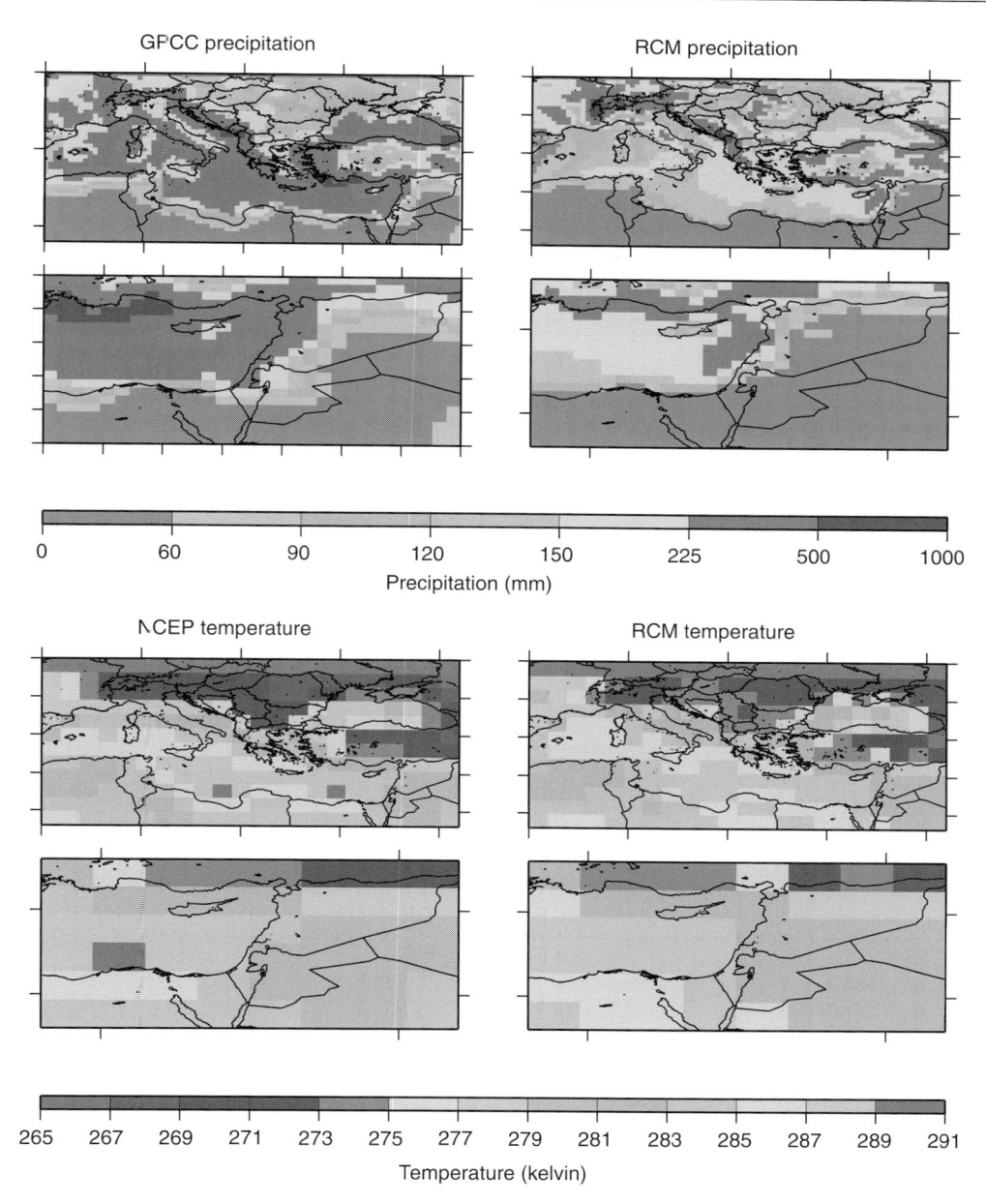

Figure 8.2. Comparison between the observed and modelled climate. Top set, left: GPCC precipitation for the whole Mediterranean (top) and for the Middle East only (bottom); right: RCM precipitation for the whole Mediterranean (top) and for the Middle East only (bottom). Bottom set, left: NCEP reanalysis temperature for the whole Mediterranean (top) and for the Middle East only (bottom); right: RCM temperature for the whole Mediterranean (top) and for the Middle East only (bottom). See colour plate section.

Comparison between the model (Figure 8.4) and proxy observations (Figure 8.1) shows clear similarities in the broad patterns of precipitation. In both cases, southernmost Europe and the Middle East become more arid during the Holocene, while eastern and northern Europe become wetter. Furthermore, the model captures the observed location of the boundary between a wetter and drier late Holocene in the eastern Mediterranean. The general pattern of increasing rainfall in northern Europe accompanied by a decrease in southern Europe indicated by the proxy data supports the hypothesis, suggested by the palaeoclimate modelling, that the aridification of the Middle East during the Holocene was caused by a weakening and poleward shift of the storm track.

Although the spatial pattern in rainfall changes during the Holocene is captured reasonably well, the ~25% increase in rainfall projected for Jerusalem would not be sufficient to cause

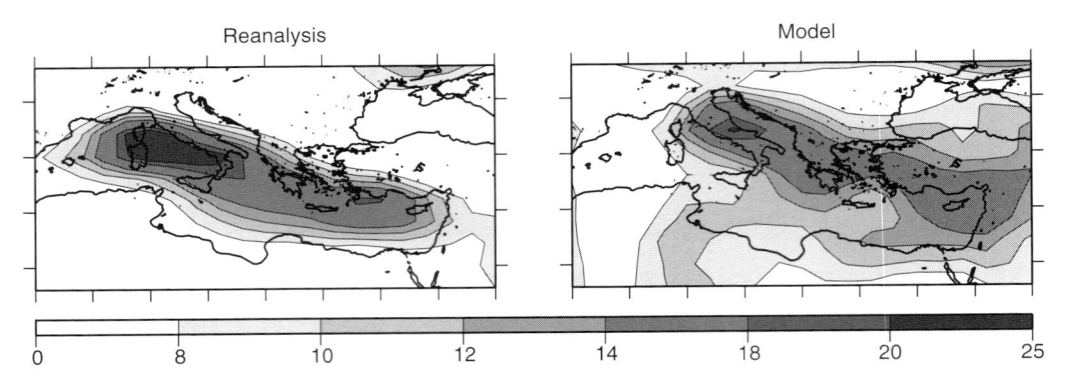

Figure 8.3. Comparison of modelled and observed track densities (in number of tracks per month per 5 degree spherical cap). Left: mean January track density in the reanalysis. Right: mean January track density in the RCM large-domain baseline scenario. Both figures are based on tracking of features in the 850 mb vorticity field. See colour plate section.

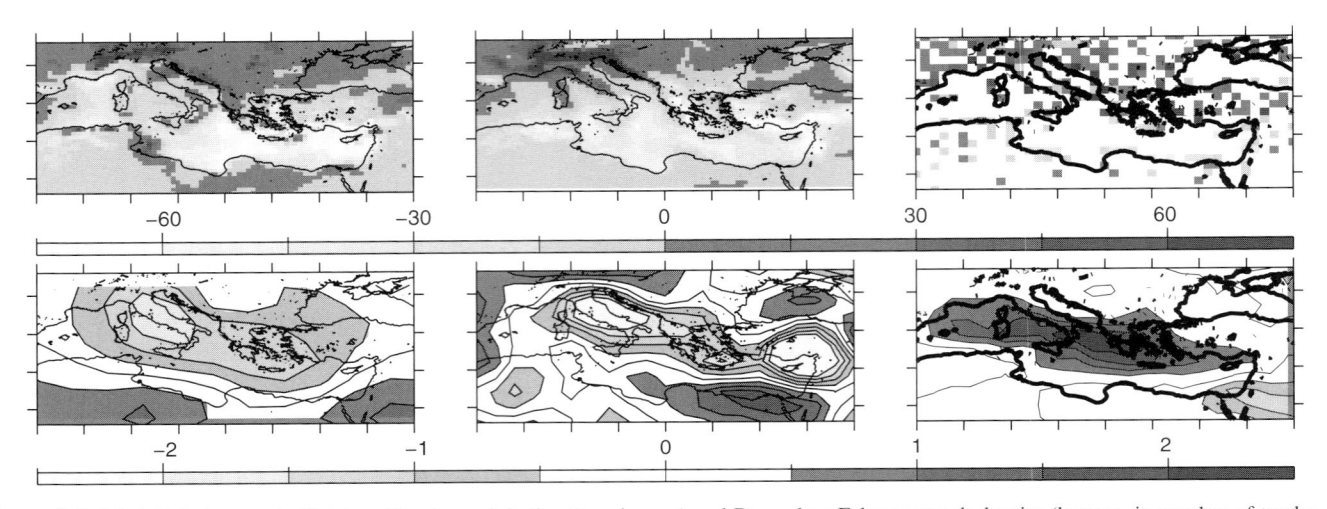

Figure 8.4. Modelled changes in October–March precipitation (top, in mm) and December–February track density (bottom, in number of tracks per month per 5 degree spherical cap for (from left to right) late Holocene minus early Holocene; future (2070–2100) – present (1961–1990); and driest years – wettest years from 1948–1999. See colour plate section.

the observed changes in Dead Sea level, which require a doubling in rainfall (Chapter 11 of this volume). This suggests either that the changes in the storm track cannot, in themselves, explain the higher precipitation during the Early Holocene, or that the model is biased. There is some evidence that feedbacks not included in the RCM time-slice integrations would generate extra rainfall. Sensitivity studies, in which a green Sahara was imposed on the model (described fully in Chapter 3), indicate that changes in the African land surface may have contributed some extra rainfall. Nevertheless, the biases in the model revealed by the comparison with historical data suggest that much of the deficit in rainfall is attributable to model bias.

8.5 RAMIFICATIONS OF THE RESULTS OF THIS STUDY FOR FUTURE PROJECTIONS

For a given emissions scenario and set of initial conditions, the credibility of our projections of rainfall change rests on how well climate models can represent the features underlying the projected change. This chapter and Black (2009) have demonstrated that the general spatial patterns and seasonal cycles of precipitation and temperature are well represented, although precipitation intensity is underestimated. Moreover, the study presented in Chapter 4 of this volume and alluded to in this chapter, along with several published studies using other regional and global models (for example Bengtsson, 2006; Lionello and Giorgi, 2007), shows that the Mediterranean storm track can be adequately simulated by climate models. These studies further suggest that the Mediterranean storm track will weaken in response to increasing GHG concentrations. Chapter 4 of this volume suggested that this weakening would be the primary cause of the projected reduction in precipitation. The fact that the dominant cause of the reduction in rainfall is consistently projected by several climate models lends credence to the projection. Confidence in the robustness of the projected gross patterns of precipitation change is further increased by comparison with proxy data, which indicates that the RCM is capable of representing some features of the precipitation response to changes in global forcings. However, as has been described earlier in this chapter and also in Chapters 3 and 4 (and in references therein), the RCM's representation of some aspects of precipitation remains poor, and significant uncertainty remains – particularly with regard to projections of rainfall intensity.

The previous discussion has proposed that, in a broad sense, the RCM's ability to represent both the present-day patterns in precipitation and the regional precipitation response to a strongly perturbed climate forcing lends a degree of credibility to the future projections. More specifically, the possibility exists that the aridification during the Holocene is an analogue for the projected drying in future. At one level, this would seem not to be the case because the changes in global forcing factors during the Holocene are different from those projected for the future. In the Holocene, the primary driver of climate change is variability in orbital forcing, which leads to changes in the seasonal cycle and latitudinal distribution of insolation. Over the next 100 years, in contrast, the primary driver of climate change would be a rise in GHG concentrations. However, in some respects, the global changes in winter climate simulated and observed in the past resemble those projected for the future. For example, in the boreal winter, both the orbitally forced insolation changes experienced in the Holocene, and the increases in GHG concentrations projected for the future, lead to a warming of the tropics. The response of the Mediterranean climate, and specifically precipitation in the Middle East, to these large-scale changes in the climate in the past is analogous to that projected for the future. Therefore, to some degree, the aridification of the Middle East in the Holocene can be considered an analogue for the aridification projected for the future – even though the ultimate forcing factors are different.

The striking similarity between the precipitation changes during the Holocene (both observed and simulated by the climate model) and those projected over the next 100 years adds weight to this argument (Figure 8.4). In particular, reduced precipitation over the Mediterranean and its northern and eastern margins, along with increased precipitation further north and west, is seen in both sets of simulations. Moreover, these results suggest that a weakening of the storm track underlies the aridification in both cases.

The Middle East experiences large interannual variability in precipitation. For example, the standard deviation of Jerusalem rainfall is approximately 30% of the mean. However, Figure 8.4 shows that although the magnitude of rainfall variability on interannual timescales is similar to the changes projected for the future and experienced during the Holocene, the underlying mechanisms are different: while long-term aridification results from a weakening of the Mediterranean storm track, unusually dry years result from a strengthening and focusing of the storm track in the north of the Mediterranean. Consistent with this, on interannual timescales dry conditions in the Middle East tend to be restricted to the eastern margin of the Mediterranean (including Jordan, Israel, Lebanon and some of Syria) with wet conditions prevailing further north and west (see Chapter 2 for more detail).

In summary, this brief comparison has shown that the changes projected for the future have more in common with those experienced during the Holocene than with variability during historical times. Perhaps this suggests that, in this context at least, the past may yet prove to be the key to the future.

8.6 CONCLUSIONS

1. Both climate and proxy observations suggest that the Middle East and southern Europe have become drier during the Holocene, and that northern Europe has become wetter.
2. The pattern of precipitation changes during the Holocene simulated by the RCM are similar to those projected by the end of the twenty-first century under an A2 scenario.
3. The RCM's capacity to replicate the general pattern of precipitation changes observed in the proxy and observed record lends credence to the future projections. Moreover, the fact that increases in GHG concentrations and solar insolation have caused a poleward shift in the storm track in the past adds credibility to the projection of such a change in the future.

REFERENCES

Bar-Matthews, M., A. Ayalon, M. Gilmour, A. Matthews and C. J. Hawkesworth (2003) Sea–land oxygen isotopic relationships from planktonic foraminifera and speleothems in the Eastern Mediterranean region and their implication for paleorainfall during interglacial intervals. *Geochimica et Cosmochimica Acta* **67**: 3181–3199.

Bar-Matthews, M., A. Ayalon and A. Kaufman (1997) Late quaternary paleoclimate in the eastern Mediterranean region from stable isotope analysis of speleothems at Soreq Cave, Israel. *Quaternary Research* **47**: 155–168.

Bengtsson, L. H., K. I. Hodges and E. Roeckner (2006) Storm tracks and climate change. *Journal of Climate* **19**: 3518–3543.

Black, E. (2009) The impact of climate change on daily precipitation statistics for Jordan and Israel. *Atmospheric Science Letters* **10**: 192–200.

Candy, I. and S. Black (2009) The timing of Quaternary calcrete development in semi-arid southeast Spain: investigating the role of climate on calcrete genesis. *Sedimentary Geology* **218**: 6–15.

Cheddadi, R., G. Yu, J. Guiot, S. Harrison and I. C. Prentice (1997) The climate of Europe 6000 years ago. *Climate Dynamics* **13**: 1–9.

Cubasch, U., G. A. Meehl, R. Boer et al. (2001) Projections of future climate change. In *Climate Change 2001: The Scientific Basis. Contribution of Working Group I to the Third Assessment Report of the Intergovernmental Panel on Climate Change*, ed. J. T. Houghton, Y. Ding, D. J. Griggs et al. Cambridge and New York: Cambridge University Press.

Emeis, K.-C., H. Schulz, U. Struck et al. (2003) Eastern Mediterranean surface water temperatures and $\delta^{18}O$ composition during deposition of sapropels in the late Quaternary. *Paleoceanography* **18**, DOI 10.1029/2000PA000617.

Giralt, S., F. Burjachs, J. R. Roca and R. Julia (1999) Late Glacial to Early Holocene environmental adjustment in the Mediterranean semi-arid zone of the Salines playa-lake (Alacante, Spain). *Journal of Paleolimnology* **21**: 449–460.

Goodfriend, G. A. (1999) Terrestrial stable isotope records of Late Quaternary paleoclimates in the eastern Mediterranean region. *Quaternary Science Reviews* **18**: 501–513.

Gvirtzman, G. and M. Wieder (2001) Climate of the last 53,000 years in the eastern Mediterranean, based on soil-sequence stratigraphy in the coastal plain of Israel. *Quaternary Science Reviews* **20**: 1827–1849.

Hamlin, R. H. B., J. C. Woodward, S. Black and M. G. Macklin (2000) Sediment fingerprinting as a tool for interpreting long-term river activity: the Voidomatis basin, north-west Greece. In *Tracers in Geomorphology*, ed. I. D. L. Foster. Chichester: John Wiley & Sons pp. 473–501.

Horowitz, A. (1971) Climatic and vegetational developments in northeastern Israel during Upper Pleistocene–Holocene times. *Pollen et Spores* **13**: 255–278.

Horowitz, A. (1989) Continuous pollen diagrams for the last 3.5 M.Y. from Israel: vegetation, climate and correlation with the oxygen isotope record. *Palaeogeography Palaeoclimatology Palaeoecology* **72**: 63.

Huntley, B. and I. C. Prentice (1993) Holocene vegetation and the climates of Europe. In *Global Climates since the Last Glacial Maximum*, ed. H. E. Wright Jr, J. E. Kutzbach, T. Webb III et al. Minneapolis: University of Minnesota Press pp. 136–168.

Kallel, N., M. Paterne, J. C. Duplessy et al. (1997) Enhanced rainfall in the Mediterranean region during the last sapropel event. *Oceanologica Acta* **20**: 697–712.

Kalnay, E., E. Kanamitsu, R. Kistler et al. (1996) The NCEP/NCAR 40-year reanalysis project. *Bulletin of the American Meteorological Society* **77**: 437–471.

Klinger, Y., J. P. Avouac, D. Bourles and N. Tisnerat (2003) Alluvial deposition and lake-level fluctuations forced by Late Quaternary climate change: the Dead Sea case example. *Sedimentary Geology* **162**: 119–139.

Landmann, G., G. M. Abu Qudaira, K. Shawabkeh, V. Wrede and S. Kempe (2002) Geochemistry of the Lisan and Damya Formations in Jordan, and implications for palaeoclimate. *Quaternary International* **89**: 45–57.

Landmann, G., A. Reimer, G. Lemcke and S. Kempe (1996) Dating Late Glacial abrupt climate changes in the 14,570 yr long continuous varve record of Lake Van, Turkey. *Palaeogeography Palaeoclimatology Palaeoecology* **122**: 107–118.

Lionello, P. and F. Giorgi (2007) Winter precipitation and cyclones in the Mediterranean region: future climate scenarios in a regional simulation. *Advances in Geosciences* **12**: 153–158.

Maas, G., M. G. Macklin and M. J. Kirkby (1998) Late Pleistocene and Holocene river development in Mediterranean steepland environments, southwest Crete. In *Palaeohydrology and Environmental Change*, ed. G. Benito, V. R. Baker and K. J. Gregory. Chichester: Wiley.

McDermott, F., S. Frisia, Y. M. Huang et al. (1999) Holocene climate variability in Europe: evidence from delta O-18, textural and extension-rate variations in three speleothems. *Quaternary Science Reviews* **18**: 1021–1038.

Neev, D. and K. O. Emery (1995) *The Destruction of Sodom, Gomorrah, and Jericho*. New York: Oxford University Press.

Reeder, M. S., D. A. V. Stow and R. G. Rothwell (2002) Late Quaternary turbidite input into the east Mediterranean basin: new radiocarbon constraints on climate and sea-level control. In *Sediment Flux to Basins: Causes, Control and Consequences*, ed. S. J. Jones and L. E. Frostick. Bath: Geological Society Publishing House, Bath (2002), pp. 267–278.

Renssen, H., H. Goosse, T. Fichefet and J. M. Campin (2001) The 8.2 kyr BP event simulated by a global atmosphere-sea-ice-ocean model. *Geophysical Research Letters* **28**: 1567–1570.

Roberts, N., M. D. Jones, A. Benkaddour et al. (2008) Stable isotope records of Late Quaternary climate and hydrology from Mediterranean lakes: the ISOMED synthesis. *Quaternary Science Reviews* **27**: 2426–2441.

Roberts, N., J. M. Reed, M. J. Leng et al. (2001) The tempo of Holocene climatic change in the eastern Mediterranean region: new high-resolution crater-lake sediment data from central Turkey. *The Holocene* **11**: 721–736.

Robinson, S. A., S. Black, B. Sellwood and P. J. Valdes (2006) A review of palaeoclimates and palaeoenvironments in the Levant and Eastern Mediterranean from 25,000 to 5000 years BP: setting the environmental background for the evolution of human civilisation. *Quaternary Science Reviews* **25**: 1517–1541.

Rohling, E. J. (1994) Review and new aspects concerning the formation of eastern Mediterranean sapropels. *Marine Geology* **122**: 1–28.

Rohling, E. J. and F. J. Hilgen (1991) The eastern Mediterranean climate at times of sapropel formation – a review. *Geologie en Mijnbouw* **70**: 253–264.

Rossignol-Strick, M. (1985) Mediterranean Quaternary sapropels, an immediate response of the African monsoon to variation of insolation. *Palaeogeography Palaeoclimatology Palaeoecology* **49**: 237–263.

Rossignol-Strick, M. (1995) Sea-land correlation of pollen records in the Eastern Mediterranean for the glacial–interglacial transition: biostratigraphy versus radiometric time-scale. *Quaternary Science Reviews* **14**: 893–915.

Rossignol-Strick, M. (1999) The Holocene climatic optimum and pollen records of sapropel 1 in the eastern Mediterranean, 9000–6000 BP. *Quaternary Science Reviews* **18**: 515–530.

Rossignol-Strick, M., W. Nesteroff, P. Olive and C. Vergnaud-Grazzini (1982) After the deluge: Mediterranean stagnation and sapropel formation. *Nature* **295**: 105–110.

Rowan, J. S., S. Black, M. G. Maclin, B. J. Tabner and J. Dore (2000) Quaternary environmental change in Cyrenaica evidenced by U–Th, ESR and OSL dating of coastal alluvial fan sequences. *Libyan Studies* **31**: 5–16.

Schneider, U., T. Fuchs, A. Meyer-Christoffer and B. Rudolf (2008) Global precipitation analysis products of the GPCC. Global Precipitation Climatology Centre (GPCC) DWD Internet Publikation 1–12.

Wiersma, A. P. and H. Renssen (2006) Model-data comparison for the 8.2 ka BP event: confirmation of a forcing mechanism by catastrophic drainage of Laurentide Lakes. *Quaternary Science Reviews* **25**: 63–88.

Wu, H. B., J. L. Guiot, S. Brewer and Z. T. Guo (2007) Climatic changes in Eurasia and Africa at the last glacial maximum and Mid-Holocene: reconstruction from pollen data using inverse vegetation modelling. *Climate Dynamics* **29**: 211–229.

9 Palaeoenvironmental and limnological reconstruction of Lake Lisan and the Dead Sea

Stuart Black, Stuart Robinson, Richard Fitton, Rachel Goodship, Claire Rambeau and Bruce Sellwood

ABSTRACT

Lake Lisan existed between 75 and 14 ka BP and provides a unique opportunity for investigating climate change for the southern Levant region. The exposures of Lake Lisan deposits on the eastern shores of the current Dead Sea have received limited study in comparison to the western shores. Here we present new U-series and elevation data for a sequence of Lisan deposits ranging between 25 and 8 ka BP. The data show an elevation–age record consistent with previously published data while providing new information about the extent of decrease in lake levels in the Late Pleistocene and early Holocene. All dates cited in this chapter are calibrated; those that were uncalibrated in their original publications were calibrated using the online OxCAL program.

9.1 INTRODUCTION

Lake Lisan existed from approximately 70 to 15 cal ka BP (see Kaufman, 1971, Kaufman *et al.*, 1992, Schramm *et al.*, 2000; Hasse-Schramm *et al.*, 2004 and references therein) and extended up to 200 km along the Dead Sea Transform at its highest levels (Hazan *et al.*, 2005; Chapter 6, this volume; Figures 9.1, 9.2). During its history, the lake fluctuated between highstands of 165–180 mbsl (metres below sea level) and lowstands of perhaps as much as 700 mbsl (Chapter 6, this volume; Begin *et al.*, 1985, Bartov *et al.*, 2002). During this time Lake Lisan deposited a series of laminated sediments consisting of aragonite, gypsum and clastic material within the lake and along its margins. In addition, stromatolites (solidified algal mats) and diatoms were deposited in the northern parts of the lake which formed the bulk of the biogenic components within Lake Lisan (Begin *et al.*, 1974, Begin *et al.*, 2004). During times of high

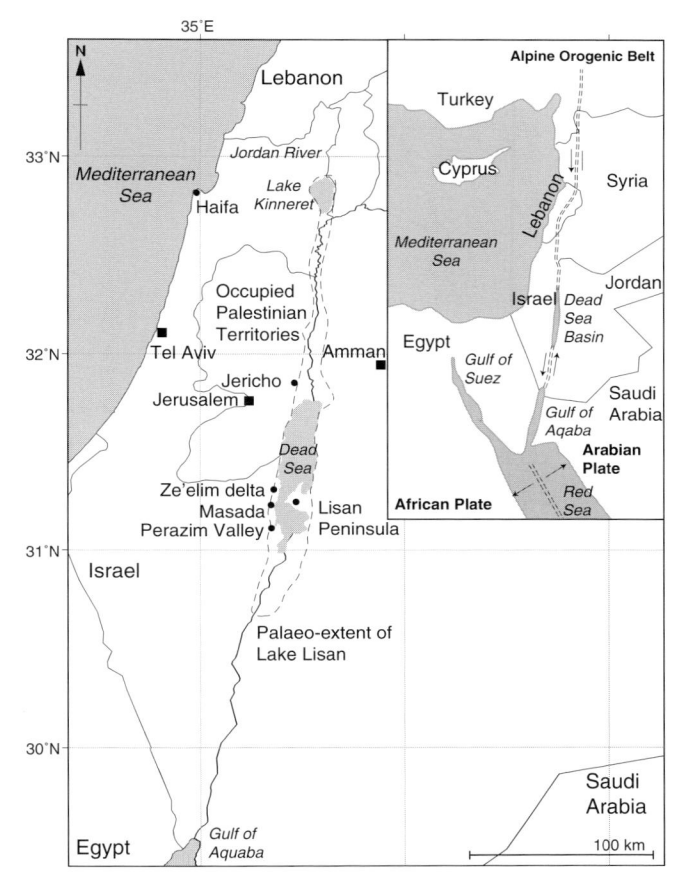

Figure 9.1. Map showing the initial extent of Lake Lisan and present-day Dead Sea (after Stein *et al.*, 2009). Inset shows the structural setting for the region.

freshwater input in the south, aragonite–detritus pairs were deposited, whereas gypsum and clastics were deposited when freshwater influx decreased and lake levels declined (Stein *et al.*, 1997).

Latterly the Lake Lisan system evolved into the present-day Dead Sea basin which is composed of two sub-basins (Figure 9.1)

Water, Life and Civilisation: Climate, Environment and Society in the Jordan Valley, ed. Steven Mithen and Emily Black. Published by Cambridge University Press.
© Steven Mithen and Emily Black 2011.

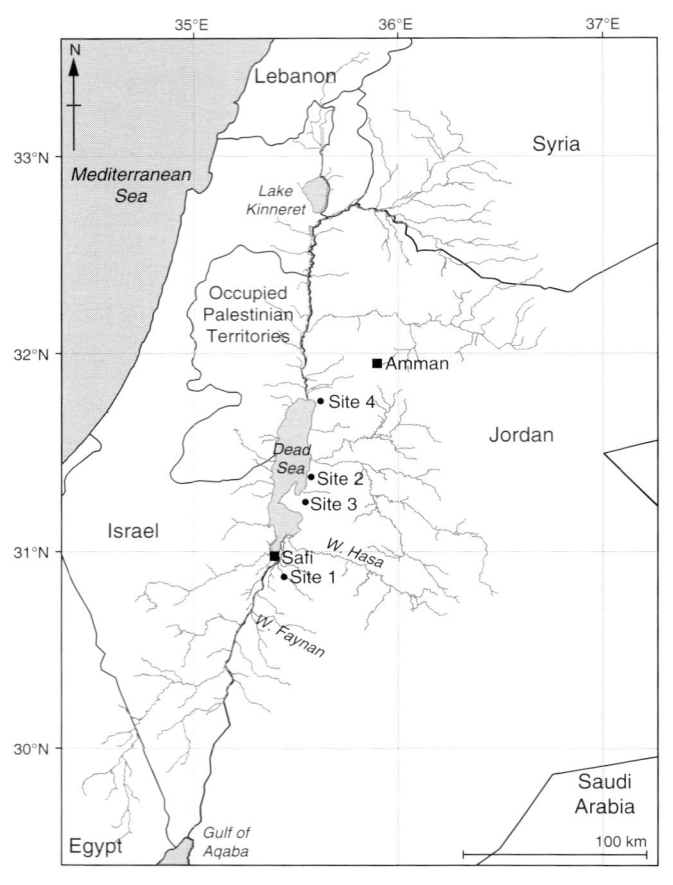

Figure 9.2. Map showing the location of sites 1–4 sampled within this study.

An important reason for studying Lake Lisan and the modern Dead Sea systems is that both are 'amplifier lakes', whereby the lake stratigraphy and water levels respond rapidly to climatic changes. These basins also have no natural outflows, so the records of these climate fluctuations are recorded within the basins and unlike ephemeral systems are not subject to removal of parts of the record. Thus, Lake Lisan and the Dead Sea make an ideal study for rapid climate change response over the Late Pleistocene and early Holocene periods. In this chapter, the context of the little-studied eastern shores will also be set within the larger body of data that exists from the western shores of Lake Lisan. Thus, the main goal of this chapter is to review the tectonics, stratigraphy, chronology, lake level fluctuations and chemical evolution of the Lake Lisan system and to present new data from the eastern shores of the Dead Sea collected during the Water, Life and Civilisation project which is set within this context.

9.2 SETTING AND EARLY HISTORY OF THE DEAD SEA

9.2.1 Tectonics and sedimentation

Lake Lisan was formed in a chain of morpho-tectonic basins (Dead Sea Basin, Sea of Galilee) within the larger Dead Sea Transform (Figure 9.1 and inset). This transform is approximately 1000 km in length and extends from the divergent plate boundary along the Red Sea to the Alpine Orogenic Belt of Turkey (Figure 9.1 inset). In the Dead Sea region the transform is mainly a sinistral strike-slip fault system, which was initiated 30–25 Ma ago as the African and Arabian plates began to separate (Garfunkel and BenAvraham, 1996). At this time, continental break-up created the Red Sea and some of the plate motion was taken up by the Dead Sea Transform (Figure 9.1 inset). The tectonic configuration of the transform changes to the north of Lebanon where it bends to the right and results in a compressional regime in that area (Garfunkel, 1981).

The Dead Sea Basin partly occupied by the present Dead Sea formed the deepest part of Lake Lisan. It is approximately 150 km in length and extends from the southern Jordan Valley near Jericho to the Arava Valley in the south (Figure 9.1 and inset). It is flanked on the east by a 1–1.4-km-high plateau and by a lower crest to the west. The Dead Sea Basin fundamentally consists of a rhomb-shaped pull-apart basin that is the result of the sinistral separation of a series of left-stepping faults within the transform valley (Figure 9.1 and inset). The valley walls that flank the strike-slip fault are delimited by normal faults that rise above the transform valley to the east and the west. Strike-slip motion in the valley totals approximately 105 km, as estimated from correlation of rocks on either side of the fault trace

separated by a sill (the Lisan Peninsula) at ~402 mbsl (e.g. Enzel et al., 2003, Bookman (Ken-Tor) et al., 2004, Migowski et al., 2004). The northern part of the basin consists of a deep (300 m), hypersaline lake receiving water from the Jordan River and runoff. During most of the late Holocene, Dead Sea levels were below the sill (402 mbsl) and therefore restricted to the northern basin (Enzel et al., 2003, Bookman (Ken-Tor) et al., 2004). The shallow southern basin – now covered by artificial evaporation ponds – offers a large surface to evaporation that acts as a buffer to lake level rises. A significant increase in water input is therefore necessary to raise lake levels well above the sill elevation (Chapter 6, this volume; Bookman (Ken-Tor) et al., 2004). Thus, significant drops in lake levels during arid periods are shown by low levels in the northern basin, deposition of salt in the middle of the lake, and accumulation of clastic sequences on the lateral margins (Bookman et al., 2004; Chapter 6, this volume). Large fluctuations in lake levels, together with accurate dating methods, allow the possibility of understanding the timing and extent of these variations, implying that they have the potential to be sensitive recorders of hydrological conditions and changes in the Dead Sea drainage basin over time (Enzel et al., 2003, Migowski et al., 2004).

(Garfunkel and BenAvraham, 1996). The basin architecture is largely asymmetric, with steeper slopes toward the eastern marginal faults (<2 km wide with 30° gradients), and shallower slopes toward the western marginal faults (4–5 km wide and 7° gradient). To the north, the portion occupied by the present Dead Sea is the deepest part of the basin, extending to more than 700 mbsl (Chapter 6, this volume). The north and south slopes of this deeper portion are thought to be either fault-induced or formed through flexure of underlying crust (Garfunkel and BenAvraham, 1996). The southern and northernmost portions of the basin are approximately 300–400 mbsl and are thus shallow extensions of the deeper part of the basin.

The Dead Sea Basin is believed to have been a site of sediment accumulation since the early Miocene (15 Ma), when a drainage system flowed across the transform area towards the Mediterranean Sea (Garfunkel and BenAvraham, 1996, and references therein). Evidence for such a system comes from the clastic deposits of the Hazeva Formation, which contain lithologies suggesting derivation from the southern Sinai and Saudi Arabia. Later, uplift of the bounding transform flanks sometime in the Late Miocene disrupted the drainage system and stopped the delivery of sediments to the area.

Following deposition of the Hazeva Formation, the Sedom Formation, a thick marine-derived evaporite sequence, formed in the Dead Sea Basin (Stein, 2001). At this time it is thought that a tongue of Mediterranean water from the north formed a lagoon in the Dead Sea Basin, depositing thick beds of halite interbedded with clastics, gypsum, anhydrite and dolomite (Stein, 2001). Because accumulation rates of evaporites can be rapid, the deposition of the Sedom Formation may indicate either the infilling of a pre-existing depression, or relatively quick subsidence and syn-deposition within the basin (Garfunkel and BenAvraham, 1996). Subsequent burial of the thick evaporite sequences eventually led to salt diapirism (e.g. the Sedom diaper, Lisan diaper), and deformation of the overlying sediments. These deposits are exposed in southern Israel in Mount Sedom.

Stein *et al.* (2000) used strontium and sulfur isotopes to investigate the chemical evolution of brines associated with the Mount Sedom evaporites. These authors found that Sr isotope ratios were consistently below those inferred for Pliocene sea water, while S isotopes suggested evolution of the brine from an evaporated seawater source. To explain this, Stein *et al.* (2000) proposed that some of the original brine had infiltrated into the surrounding Cretaceous bedrock, and through water–rock interaction had caused dolomitisation of wall-rocks and alteration of the original brines to Ca-chloridic brines. These altered brines were then later refluxed into the evaporating Sedom lagoon, changing the isotopic composition. More specifically, the Sr isotope values of the infiltrated brine were lowered toward Cretaceous values, which in turn lowered Sr values of the Sedom lagoon waters upon reflux (Stein *et al.*, 2000).

Klein-BenDavid *et al.* (2004) also studied brine evolution using ratios of sodium, bromine and chlorine from Ca-chloridic brines collected from various locations in the Dead Sea Transform. They found that the brines had Na, Br and Cl concentrations that deviated from an experimental seawater evaporation curve. To explain this they suggested that early marine brines evolved from two different lagoon configurations. In one mode the lagoon was connected to the sea, and the evaporation and influx of sea water were in steady state. At times when the connection to the sea was more restricted, the lagoon switched to a mode in which evaporation dominated, and inflow and reflux were minimal. The mixing of the two brine types explains the deviation in Na/Cl and Br/Cl from the experimental marine evaporation trajectory. The formation of this Sedom brine is important for the chemical composition and evolution of Lake Lisan. Interaction of brine and fresh water influenced the sedimentation in the lake and the chemical composition of lake waters (Katz and Kolodny, 1989, Stein *et al.*, 2000, Torfstein *et al.*, 2005).

During the Quaternary, the Dead Sea Basin accumulated lacustrine, fluvial and alluvial sediments in the main water bodies and along their margins (Stein, 2001). These include the dominantly clastic deposits of the Amora/Samra Formation, the laminated and clastic units of the overlying Lisan Formation, and the clastic units of the post-Lisan Damya Formation. During this time, sedimentation did not keep pace with basin subsidence such that a topographic depression, with no outflow, was continually in existence. The depression occupied by the modern Dead Sea, with elevations lower than 700 mbsl, is probably the area of fastest subsidence in the basin's recent past (Garfunkel and BenAvraham, 1996). Tectonic movement in the Late Pleistocene and Holocene is clearly evident from seismically induced deformation of laminated sediments (seismites) (Migowski *et al.*, 2004, Marco and Agnon, 2005).

9.2.2 Stratigraphy: the Lisan Formation

Begin *et al.* (1985) provided a good overview of the stratigraphy of the Lisan sediments, describing 13 stratigraphic sections along a transect from the inferred southern extent of Lake Lisan to Lake Kinneret in the north. The sections were chosen as close as possible to the centre of the lake in order to minimise sampling marginal facies. The Lisan Formation was defined as the beginning of fine varve-like laminations lying above the mostly clastic Samra Formation, although it was noted that the distinction between the Samra and Lisan was increasingly difficult toward the clastic palaeo-margin of Lake Lisan.

Begin *et al.* (1974) divided the Lisan sequence into two major units, a lower 'Laminated Member', and an upper 'White Cliff Member'. In the south, the Laminated Member consisted of varve-like laminations of aragonite and detrital material that

Figure 9.3. Photograph showing the Lake Lisan Grey Unit and White Unit on the East side of the Jordan Valley.

was composed chiefly of quartz, calcite, dolomite, clay and lithics. Near Masada (Figure 9.1 and inset) the Laminated Member was found to contain some gypsum intercalations, while to the north and south of this area gypsum beds were rare in the Laminated Member. Towards the north, the Laminated Member contained increasing amounts of detrital material and diatoms. This was interpreted as evidence of a salinity gradient from north to south, as exists in the region today. The upper White Cliff Member (seen also in Figure 9.3 north of the present Dead Sea), in contrast, contained very prominent beds of gypsum along with aragonite and detrital material, particularly in the southern sections. Again, towards the north the amount of gypsum decreased and was replaced by diatomite with some aragonite and clastic material. Within the gypsiferous sections in the southern parts of the basin the gypsum content increased up section and was the thickest near Masada, a trend similar to the lower Laminated Member. A pattern of increasing evaporite content from north to south has also been noted by other authors (Abed and Yaghan, 2000, Landmann et al., 2002) and is consistent with the modern climatic and limnological conditions in the Dead Sea (Chapter 6, this volume).

More recently, authors have divided the Lisan Formation into three units, aptly named the Lower, Middle and Upper Members (e.g. Abed and Yaghan, 2000, Machlus et al., 2000, Bartov et al., 2002, Landmann et al., 2002). In the south, along the Perazim Valley, in the offshore sections of the Masada plain and on the Lisan Peninsula (Figure 9.1 and inset), the Lower and Upper Members consist chiefly of aragonite and gypsum laminations interbedded with detrital material (Machlus et al., 2000, Bartov et al., 2002, Landmann et al., 2002). The Middle Member, by contrast, is dominated by detrital material composed of sand, silt and clay, and is believed to represent a period of lower water levels (Stein et al., 1997). Although gypsum beds are found in the Lower Member, they are generally thicker towards the top of the Lisan Formation, in the Upper Member. Indeed, the prominent cliff-forming gypsum beds at the top of the Lisan Formation are the White Cliff Member of Begin et al. (1985). In the north, along the Jordan River near Ghor el Katar, Abed and Yaghan (2000) describe a slightly different lithologic sequence. While the Upper and Lower units in this area consist of the familiar aragonite and detrital laminations, with increasing gypsum abundance towards the top of the Upper Member, the Middle unit is composed of organic-rich clays (illite, smectite, with minor kaolinite), with no evaporite deposition.

9.3 CHRONOLOGY

The detailed chronology of Lisan deposits has been obtained principally through U-series dating (e.g. Kaufman, 1971, Kaufman et al., 1992, Schramm et al., 2000, Klinger et al., 2003, Hasse-Schramm et al., 2004), although radiocarbon dating, varve-counting and lithostratigraphic correlation with dated sections have also been extensively employed (Kaufman, 1971, Abed and Yaghan, 2000, Machlus et al., 2000, Schramm et al., 2000, Bartov et al., 2002, Landmann et al., 2002, Prasad et al., 2004). Uranium-series dating offers the advantage of longer temporal resolution than radiocarbon, and does not need to be calibrated to calendar ages. In addition, because the Lisan Formation contains abundant carbonate, it is well suited for U-series dating.

9.3.1 U-series: theory

Dating of marine and lacustrine carbonates is mainly achieved through ^{230}Th dating, alternatively referred to as U/Th dating (Edwards et al., 2003). Theoretically, this technique covers from 3 years to 600,000 years and is based on the first portion of the ^{238}U decay chain (Figure 9.4), in which ^{238}U decays through a series of intermediate isotopes to ^{230}Th and eventually to ^{206}Pb (Figure 9.4). This dating technique is based on the principle of calculating the time it takes a system to return to equilibrium after it has been disrupted (Bourdon et al., 2003). Equilibrium occurs if a system has been closed to any perturbation for at least six half-lives of the longest-lived daughter in the series. After this amount of time the activities of parent–daughter isotope pairs (e.g. ^{234}U–^{230}Th) are equal, meaning that the number of decay events per unit of time for the parent and daughter is equal (Bourdon et al., 2003). In the case of lacustrine carbonates, perturbation of equilibrium conditions refers to the fractionation of U from Th during weathering and in the hydrologic cycle (Edwards et al., 2003). Such fractionation occurs because of the difference in solubility between U and Th, since U is readily dissolved into natural waters and then incorporated into

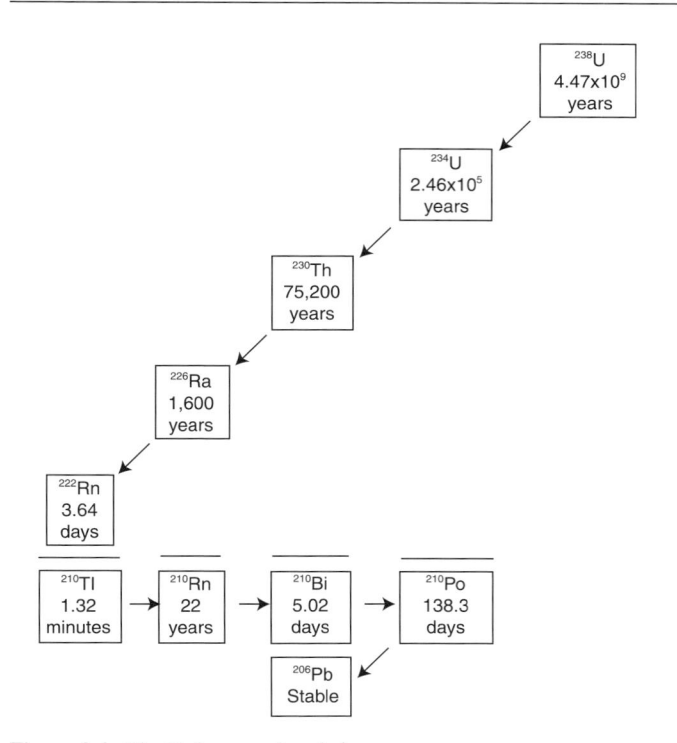

Figure 9.4. The U-decay series chain.

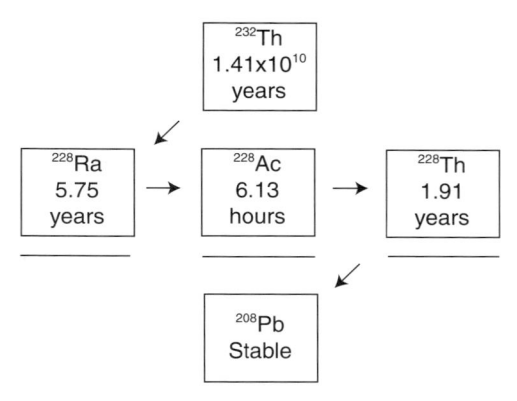

Figure 9.5. The Th-decay series chain.

authigenic minerals upon precipitation, whereas Th is relatively insoluble and remains undissolved or adsorbed to particles in suspension (Edwards *et al.*, 2003). Based on this process, if the isotopic ratios in a sample can be deduced, then the amount of time since the fractionation of U from Th can be calculated.

The calculation of U/Th ages requires some assumptions to be made. The first assumption is that the system has remained closed to chemical exchange since the time of fractionation and precipitation (Edwards *et al.*, 2003). For lacustrine carbonates, this necessitates that samples have not undergone any diagenetic alteration that would change U or Th concentrations since the time of precipitation. A second assumption is that no initial daughter isotopes were incorporated into the carbonate at the time of precipitation. If this second assumption is violated, however, it is possible to correct for this by additional measurements (Bischoff and Fitzpatrick, 1991, Luo and Ku, 1991, Edwards *et al.*, 2003). One method for correction involves estimating the amount of initial daughter present (usually by measurement of ^{232}Th in the same sample; Figure 9.5) or by constructing isochrons (Bischoff and Fitzpatrick, 1991, Luo and Ku, 1991, Edwards *et al.*, 2003). Isochron techniques involve constructing linear regressions through a series of data points taken from coeval samples (Edwards *et al.*, 2003). By plotting ratios of nuclide pairs (e.g. ratio of ^{230}Th/^{232}Th to ^{234}U/^{232}Th or ratio of ^{234}U/^{232}Th to ^{238}U/^{232}Th) on a Rosholt-type diagram one can estimate the initial values of ^{230}Th in the sample and correct for this in the age calculation. Isochron techniques require that the coeval samples have differing amounts of detritus and carbonate such that there is a spread in the isotope ratios in order to construct a linear regression (Candy and Black, 2009). A further assumption is that the (^{230}Th/^{232}Th) ratio is constant within coeval samples, i.e. that the contaminating detritus contained a constant ratio of (^{230}Th/^{232}Th). In Lisan aragonites the (^{230}Th/^{232}Th) ratio was similar to that in the detrital material (Kaufman, 1971, Hasse-Schramm *et al.*, 2004) and it has been suggested that the error in isochron ages was likely to be small. An additional complication for calculation of U/Th ages based on isochrons arises from the presence of hydrogenous Th in a sample. Previous work (Hasse-Schramm *et al.*, 2004) corrected for the hydrogenous Th component by estimating an average amount of hydrogenous Th in each sample with a (^{230}Th/^{232}Th) ratio, thus allowing the initial Th-free end member to be estimated from the linear regression. Hasse-Schramm *et al.* (2004) also used trace element data to infer that, in addition to detrital Th contamination, samples also contained a hydrogenous Th component.

9.3.2 Lake Lisan chronology

Kaufman (1971) provided the first attempt to date the Lisan Formation through U/Th dating of sediments from the southwest side of the basin. Kaufmann divided the Lisan Formation into two members (Upper and Lower), similar to Begin *et al.* (1985), and reported ages of 22–39 cal ka BP for the Upper Member and 43–61 cal ka BP for the Lower Member. He found that although the ages were stratigraphically consistent, they showed poor agreement with climate data from lake records in California and deep-sea records. Kaufmann argued, therefore, that either the U/Th ages were incorrect, or the various climate histories might not be contemporaneous and might be more complex than originally thought.

Kaufman *et al.* (1992) revisited this study and dated aragonite samples from eight sections in an attempt to correlate the sections chronostratigraphically. The authors noted the error in assuming that lithostratigraphic changes were contemporaneous and that sections could easily be divided into Lower and Upper Members, as in Begin *et al.* (1985). Despite this, however, Kaufman *et al.* (1992) still relied on the lithostratigraphic concept to some degree by referring to dates in relation to changes in lithology from north to south across the basin. Nevertheless, results from their study indicated that Lake Lisan sedimentation began at 63 cal ka BP, but was not simultaneous in all parts of the basin, as would be expected. The youngest U/Th age obtained near the top of the Upper Member in the southern part of the Dead Sea Basin was 18 cal ka BP, giving an age estimate for the duration of Lake Lisan of 63–18 cal ka BP.

Schramm *et al.* (2000) provided further evidence for the suitability of Lisan carbonates for U/Th age determinations by comparing ^{14}C ages and U/Th ages between 20 and 52 cal ka BP in a 36-m section from Perazim Valley, Israel. They found that U/Th ages were stratigraphically consistent and agreed with calibrated ^{14}C ages for ages up to approximately 23 cal ka BP. In addition, older U/Th ages from Lisan aragonites gave insight into changes in atmospheric ^{14}C and allowed radiocarbon ages to be calibrated further back in time than had previously been possible (Schramm *et al.*, 2000). In their results, Schramm *et al.* dated the Lisan Formation at Perazim Valley from 62 cal ka BP, at 3 m above the base of the section, to 19 cal ka BP, 2 m below the top of the section, consistent with the dates of Kaufman *et al.* (1992). Hasse-Schramm *et al.* (2004) updated the ages of Schramm *et al.* (2000) and estimated that the duration of Lake Lisan was from 70 to14 cal ka BP. These authors also argued that the lithological Lower and Upper Members at Perazim Valley corresponded to Marine Isotope Stages (MIS) 4 and 2, respectively, while the Middle Member was related to MIS 3. From this it was concluded that the laminated Lower and Upper Members, associated with abundant freshwater influx, were deposited during cold climatic stages. In contrast, the Middle Member, commonly associated with less freshwater influx and lower lake levels, was thought to have been deposited in warmer interglacial times.

9.3.3 Lake levels

The reconstruction of Lake Lisan water levels has been used to gain insight into the lake's physical evolution and to infer palaeoclimatic trends in the area (Sneh, 1979, Begin *et al.*, 1985, Druckman *et al.*, 1987, Macumber and Head, 1991, Machlus *et al.*, 2000, Bartov *et al.*, 2002, Hazan *et al.*, 2005, Robinson *et al.*, 2006; Chapter 6, this volume). Many of these reconstructions are based on the sedimentology, stratigraphy and radiometric dating of palaeo-shoreline and offshore sediments, and the correlation of marginal and deeper water sections through absolute dating and physical tracing of specific marker beds or unconformities. The interpretation of sedimentological features observed in stratigraphic sections is often made through comparison with modern Dead Sea sedimentary environments and their associated sedimentological features (e.g. Machlus *et al.*, 2000). In this way, as an example, coarse-grained landward dipping clinoforms that characterise beach deposits today may be interpreted as palaeo-beach deposits in older sections (e.g. Bartov *et al.*, 2002). Marginal deposits are often the focus of these lake level reconstructions as these are potentially datable and direct indications of palaeo-shoreline levels.

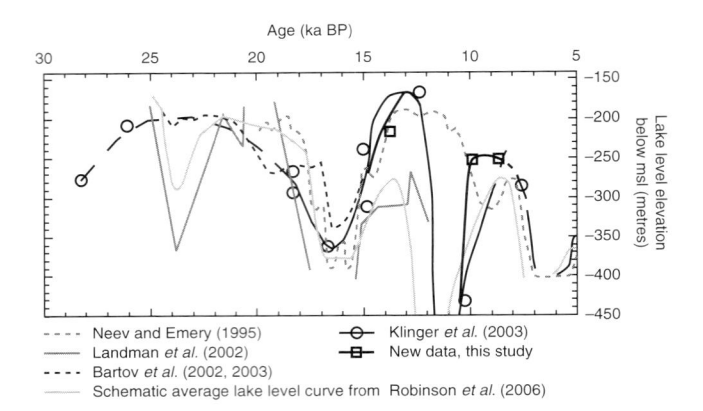

Figure 9.6. Compilation figure of previously published lake level data from Lisan sediments, predominately from the west side of the Jordan Valley together with information from this study coming from the east side.

Lake level reconstructions for Lake Lisan have yielded a number of different and sometimes opposing interpretations (e.g. Robinson *et al.*, 2006; Figure 9.6). The same is true for lake level and other independent palaeoclimatic reconstructions. For example, Abed and Yaghan (2000) suggested that deposition of the uppermost gypsum bed coincides with the Last Glacial Maximum (LGM), and thus inferred a dry climate in the area at that time. Some lake level curves, however, favour maximum levels at the LGM, and suggesting that to have been a wet period (Bartov *et al.*, 2003). The lake level picture is further complicated by the fact that, as suitable palaeo-shoreline indicators are patchy and spatially separated, some studies are forced to focus on discrete time intervals within the Lisan sequence. This produces a piecemeal appearance when different level curves are overlaid. Recently, Robinson *et al.* (2006) have attempted to combine previously published lake levels into a more coherent curve for the past 25 cal ka BP (Figure 9.6).

Machlus *et al.* (2000) reconstructed lake levels for the interval of 55–35 cal ka BP based on sequence stratigraphy and sedimentology of fan delta and lacustrine sediments in the Perazim Valley (Figure 9.1). These sediments correspond to the detrital-rich Middle Member in the tripartite division of the Lisan sequence commonly adopted for this area. Shallow water indicators from sediments dating to between 55 and 50 cal ka BP were interpreted as evidence for stable lake levels around 290 mbsl for this period. At approximately 47 cal ka BP, the formation of an unconformity and channel development mark a drop in the lake level of between 30 and 40 m. At 42 cal ka BP a short-lived lake level rise is inferred from the deposition of lacustrine aragonite and detritus laminations like those in the Lower and Upper Members. Lake level is thought to have fluctuated slightly between 40 and 38 cal ka BP, with an amplitude of a few metres. Based on the deposition of gypsum, halite and clastics in the Perazim Valley sections, lake level is thought to have dropped to an elevation of lower than 283 mbsl for the period between 37 and 35 cal ka BP. The occurrence of halite at this time is thought to reflect lagoonal deposition. Above and below the interval between 55 and 35 cal ka BP, the laminated aragonite and detritus of the Lower and Upper Members suggests that water levels were relatively high.

Bartov et al. (2002, 2003) also used sequence stratigraphy, sedimentology and absolute dating (U–Th, and ^{14}C) to reconstruct palaeo-lake levels (Figure 9.6). These authors looked at fan delta, near-shore and offshore sediments exposed within the dissected Ze'elim delta, north of the Perazim Valley. Correlation of these sections with the Perazim Valley sections was made through the use of absolute dates and traceable marker beds and unconformities. Along the Ze'elim delta transect, the first stratigraphic evidence for lake lowering occurred at 56 cal ka BP with the deposition of three massive gypsum beds. Before this time, from 70 to 55 cal ka BP, lake level may have been relatively stable and stood above 306 mbsl. According to Stein et al. (1997), deposition of gypsum layers (as at 56 cal ka BP) in the Lisan Formation is due to increased mixing between a lower Ca-chloridic brine and a sulfate-rich upper freshwater lens. Such mixing might occur with increased evaporation and associated lake level lowering. Consequently, a return to laminated aragonite and detritus above the gypsum beds signals rising water levels. From approximately 55 to 48 cal ka BP, shallow water indicators suggested that the lake levels had stabilised between 290 and 300 mbsl. Based on beach deposits in the Ze'elim sections, lake levels are thought to have dropped to 340 mbsl at 48 cal ka BP and remained relatively low (below 300 mbsl) until about 43 cal ka BP. The lake level drop at 48 cal ka BP is also coincident with the channel development in the Perazim Valley section (Machlus et al., 2000) and with an unconformity in the surrounding Masada plain. From 43 to 38 cal ka BP, lake level rose to within 280–290 mbsl. From about 38 to 36 cal ka BP lake levels lowered again to approximately 300 mbsl, as evidenced from near-shore deposits in the Ze'elim sections. Lake levels rose again to 280 mbsl starting at 36–33 cal ka BP, based on deposition of beach deposits in the upper (landward) parts of the Ze'elim delta complex and also in the Perazim Valley sections, and remained between 280 and 270 mbsl until 27 cal ka BP. Maximum lake levels were thought to have been attained at 27–23 cal ka BP, reaching elevations of greater than 165 mbsl. Lake level subsequently dropped to 220–190 mbsl from 23 to 17 cal ka BP. After 17 cal ka BP, lake level dropped to 270 mbsl and then further to 300 mbsl after 15 cal ka BP. Druckman et al. (1987) inferred highstand conditions at approximately 18–16 cal ka BP based on radiocarbon-dated ooid shoals found at an elevation of 180 mbsl in the southwest of the Lisan palaeomargin. Since ooids are shallow water deposits, their occurrence at 180 mbsl indicated water levels at least up to this elevation. Druckman et al. argued that at this time a humid phase dominated in the Lisan Basin, and that this humidity decreased after 16 cal ka BP. Their argument was based on the apparent decrease in freshwater diagenetic alteration of successively younger ooilitic shoals after 16 cal ka BP. This decrease in diagenetic alteration suggested that a freshwater aquifer had disappeared at the site, coincident with declining lake levels, and indicating an increase in aridity (lower humidity).

Begin et al. (1985) presented a lake level curve covering the last 40 cal ka BP based on published and unpublished radiocarbon dates from ooilitic shoals, organic-rich clays and stromatolites. According to their curve, three prominent highstands of 200 mbsl or more occurred at approximately 30 cal ka BP, 24 cal ka BP and 18 cal ka BP followed by lowstands to levels of approximately 370 mbsl (with the exception of the major post-Lisan/Younger Dryas lowstand; see below). Following the interpretation of Robinson et al. (2006), the lake level curve of Begin et al. (1985) shows that lake levels dropped drastically to 700 mbsl, associated with the Younger Dryas. Such conclusions are also consistent with interpretations by Yechieli et al. (1993) based on halite deposition between 13 and 11 cal ka BP, and Macumber and Head (1991) who relate the incision of wadi sequences to lowering base level as Lake Lisan shrank during Younger Dryas drying. Interestingly, Macumber and Head (1991) also suggest that Lake Lisan levels gradually rose without lowstand interruption in the Late Pleistocene from approximately 36 to 13 cal ka BP, in contrast to other interpretations that infer several lowstand intervals during this time (e.g. Landmann et al., 2002).

Robinson et al. (2006) suggested an integrated lake level curve for the interval from 25 to 5 cal ka BP, based on evaluation of previously published level curves (Figure 9.6). These authors suggest an overall decline in highstand lake levels from 25 cal ka BP to 14 cal ka BP, punctuated by three major lowstands at approximately 24 cal ka BP, 16 cal ka BP and 13 cal ka BP. At 25 cal ka BP lake levels were at their highest at an elevation of 160 mbsl, following interpretations by Begin et al. (1985), Bartov et al. (2002, 2003), Hazan et al. (2005) and Robinson et al. (2006). Lake level dropped to 290 mbsl at 24 cal ka BP before rising again to a highstand of approximately 200 mbsl at 22 cal ka BP. The second lowstand at 16 cal ka BP depicts a drop in lake level to 380 mbsl, before rising again to 280 mbsl at 14 cal ka BP. At approximately 13 cal ka BP, the most prominent lowstand occurs with lake levels plummeting to below 450 mbsl, consistent with low levels in the Younger Dryas inferred by previous investigators (Begin *et al.*, 1985, Macumber and Head, 1991, Yechieli *et al.*, 1993, Robinson *et al.*, 2006).

Hazan *et al.* (2005) reconstructed lake levels from Lake Kinneret, the furthest northern extension of Lake Lisan during highstand conditions. Lakes Kinneret and Lisan are thought to have remained separated for most of the lifespan of Lake Lisan, merging only when Lisan levels rose substantially (Hazan *et al.*, 2005). Water levels for Lake Kinneret were reconstructed through the recognition of shoreline or near-shore facies gathered from eight sections along the southern margin of the lake. From 43 to 40 cal ka BP the shoreline was approximately 220 mbsl. From 35 to 28 cal ka BP a depositional hiatus at the Ohalo II site suggested that lake levels were somewhere below 216 mbsl. Lake levels rose from 28 to 24 cal ka BP to their highest levels of 170 mbsl, coincident in timing with the Lake Lisan maximum highstand (Bartov *et al.*, 2002, Robinson *et al.*, 2006). During this highstand it is possible that the underlying brine also migrated northward as Lakes Lisan and Kinneret merged. A subsequent drop in lake level to about 215 mbsl was inferred from 24–23 cal ka BP, disconnecting Lake Lisan and Lake Kinneret once more. A lake level rise from approximately 23 to 22 cal ka BP was interpreted from beach ridge deposits, possibly indicating a lake level rise to above 205 mbsl.

Of importance to the Kinneret–Lisan level history is that several gaps appear in the Lake Kinneret record, which may represent periods of lake level decline and erosion or non-deposition. These depositional hiatuses occur at 40–36 cal ka BP and 32–28 cal ka BP and are coincident with Lake Lisan lowstands (Bartov *et al.*, 2003). This suggests a similar climatic control over the lake levels of Lisan and Kinneret.

Reconstructed lake levels for the Lisan period are incomplete and show a relatively wide range of variability in terms of timing and amplitude of change (Figure 9.6). This mainly results from the spatially patchy and incomplete exposure of shoreline indicators, as well as the difficulty associated with dating some near-shore deposits. The result is that the various Lake Lisan level reconstructions are not continuous, and are sometimes out of phase or even contradictory to each other. Despite these discrepancies, however, some general conclusions can be drawn. It appears that lake levels remained relatively stable from approximately 70 to 55 cal ka BP (Bartov *et al.*, 2002, 2003). From 55 to 12 cal ka BP the lake level was punctuated by lowstands at 48–47 cal ka BP, 38–35 cal ka BP, 24 cal ka BP (Machlus *et al.*, 2000, Bartov *et al.*, 2002, 2003), and during the Younger Dryas at approximately 13–12 cal ka BP (Yechieli *et al.*, 1993, Robinson *et al.*, 2006). Maximum highstand levels of up to 160 mbsl may have been attained at approximately 27–25 cal ka BP (Bartov *et al.*, 2002, 2003, Robinson *et al.*, 2006). In conclusion, although lake level reconstructions may be worthwhile for obtaining a broad picture of climatic and hydrological changes within an area, an approach which produces a more continuous record of environmental variability (e.g. stable isotopes, microfauna) may be more suitable and informative in this area.

9.3.4 Geochemical: lake evolution and palaeoclimate

Various geochemical proxies have been used to assess the chemical evolution of Lake Lisan as well as the palaeohydrologic and climatic conditions surrounding the basin. The proxies range from mineralogical analysis of Lisan sediment to elemental ratios (e.g. Sr/Ca, Na/Ca, Na/Cl) and stable isotopes in sediments and interstitial salts (Katz and Kolodny, 1989, Stein *et al.*, 1997, Stein, 2001, Landmann *et al.*, 2002, Begin *et al.*, 2004, Torfstein *et al.*, 2005, Kolodny *et al.*, 2007). These studies have provided insight into the limnological history of Lake Lisan including palaeosalinity (Begin *et al.*, 2004), precipitation gradients (Begin 2002, Landmann *et al.*, 2002, Begin *et al.*, 2004), sediment diagenesis (Katz and Kolodny, 1989), and sediment production (Stein *et al.*, 1997, 2000, Stein, 2001). Few of these studies, however, provide continuous palaeoenvironmental records that span the entire Lake Lisan interval (Kolodny *et al.*, 2007).

The geochemical records from Lake Lisan indicate at a crude scale that massive gypsum layers were deposited in times of decreased freshwater flux, while aragonite and detritus couplets reflect enhanced or stable freshwater influx (Stein *et al.*, 1997). In addition, north–south precipitation gradients, similar to today, existed in the area during the lake's evolution, although southern sources of water may have been more important in the past than today (Begin, 2002, Begin *et al.*, 2004). Finally, there is evidence for diagenetic alteration and chemical migration between aragonite and detritus laminations within the Lisan sequence (Katz and Kolodny, 1989).

Begin *et al.* (2004) investigated the palaeosalinity of Lake Lisan from 27 to 25 cal ka BP using both biological indicators (cyanobacteria, fish, ostracods) and geochemical indices (Sr/Ca, Na/Ca) from aragonite. The purpose of this work was to investigate whether there existed increased water input from southern sources in the past compared with today. These authors argued that the Sr/Ca ratio of aragonite in the Lisan sequence reflected the degree of mixing between the lower brine body and the upper freshwater lens. In areas where Sr/Ca ratios were high, increased mixing, due to higher salinities and unstable lake stratification, was inferred. The authors found that the area near the site of Masada had higher Sr/Ca ratios than more southern and northern sites and concluded that salinity was highest there. Their biological records were also consistent with this conclusion. The proposed implication of such a salinity gradient, therefore, was that the less saline areas to the south were due to an influx of freshwater from southern precipitation sources, possibly from a southern migration of storms during the last glacial period.

In the Perazim Valley, Stein *et al.* (1997) and Stein (2001) also used Sr/Ca ratios to infer palaeohydrologic conditions in Lake

Lisan, and to model aragonite and gypsum deposition. The Sr/Ca ratios were found to decline overall throughout the Perazim section and were related to freshwater input and lake level changes (Stein, 2001). In particular, the Lower Member of the sequence (dated at 67–55 cal ka BP in the Perazim Valley) showed moderate declines in Sr/Ca values, relative to the Upper Member (30–19 cal ka BP), while the Middle Member (55–30 cal ka BP) showed only small fluctuations. The Sr/Ca ratios in aragonite samples were related to the interaction of the lower brine with the upper freshwater lens. In periods of high freshwater influx, the precipitation of aragonite (as in the Upper and Lower Members) depleted some Ca and Sr from the lower brine. As a result, Sr from the lower brine started to decline, and through time this depletion was reflected in successive aragonite laminae. Declining values of Sr/Ca in the Lower and Upper Members, therefore, suggested rising lake levels (positive freshwater influx), while the small Sr/Ca fluctuations in the Middle Member were interpreted as reflecting low or negative freshwater influx, i.e. less Sr extraction from the lower brine. Sr/Ca and $^{87}Sr/^{86}Sr$ ratios from Lisan aragonites and interstitial salts were also used to gain insight into the process of aragonite and gypsum precipitation in Lake Lisan (Stein et al., 1997). During times of positive freshwater influx, a stratified water body developed with a freshwater lens overlying the lower brine. This freshwater supplied most of the SO_4^{2-}, and as aragonite precipitated from the freshwater lens (continually using incoming Ca and HCO_3^-), the sulfate concentrations increased. At times when freshwater input decreased, evaporation of the upper freshwater lens increased its density and caused the stratification of the lake to become unstable. The SO_4^{2-} in the fresh water then combined with the Ca of the underlying brine, and massive gypsum was precipitated. Therefore, according to Stein et al. (1997), aragonite deposition in the Lisan occurred at times of positive freshwater flux and lake stratification, while massive gypsum deposition occurred when freshwater flux decreased, or evaporation was enhanced.

The fate of sulfur and the origin of gypsum in the Lake Lisan system were recently investigated by Torfstein et al. (2005). These authors found that massive gypsum horizons had positive $\delta^{34}S$ values, while disseminated grains and thin gypsum laminae within aragonite layers had lower or negative $\delta^{34}S$ values. Torfstein et al. (2005) related these differences to different modes of deposition. Massive gypsum was deposited in the same manner as suggested by Stein et al. (1997), during those times when lower brine waters mixed substantially with the upper freshwater lens. The disseminated and thinly laminated gypsum, however, were thought to have been produced by bacterial sulfate reduction and post-Lisan sediment desiccation. Initially, sulfate reduction in the lower brine, during times of lake stratification and isolation of the brine from the atmosphere, produced ^{32}S-rich sulfide minerals. After the demise of Lake Lisan and subsequent desiccation of the Lisan sediments, these sulfides were thought to have oxidised, at which point the evolved sulfate combined with Ca to form disseminated and thinly laminated gypsum. Therefore, gypsum with positive isotopic values represented times of reduced freshwater input and brine–freshwater mixing, while gypsum laminations and disseminated grains with negative isotopic values were thought to reflect periods of increased freshwater flux leading to brine anoxia and bacterial sulfate reduction.

9.4 NATURE OF THE CURRENT INVESTIGATION

Despite the variety of attempts to unravel the palaeoclimatic history associated with the Lake Lisan system, some important gaps in our understanding and methodology are still present. The goal of this section is to highlight some of these gaps and provide some new data from our recent investigations to help fill these gaps. It is important to point out that the focus of this discussion will be on the localised climate reconstruction of the Lisan/Dead Sea system itself and not on comparing or extending that record out into regional or global context.

One major gap in our understanding of the Lake Lisan system deals with the chronology and stratigraphy of the basin as a whole. With the exception of the work by Begin et al. (1985), few studies have focused on constructing a detailed, basin-wide, chronostratigraphic framework of the Lisan Formation. Such a framework would be beneficial as it would allow for cross-correlation throughout the basin and allow for a more complete understanding of facies changes within the basin, as well as the timing of these changes. In addition, with a detailed chronostratigraphic framework, particular times in the history of the basin could be picked and studied in detail within contemporaneous facies, using different but complementary techniques. As an example, it might be possible to compare changes in diatom species within a northern sequence of the basin to an oxygen isotope stratigraphy in the south. Implicit in this previous point of developing a chronostratigraphic framework is a need for different exposures of the Lisan sediments to be dated at high resolution using U-series dating. At present, the best dated record is from the Perazim Valley (Schramm et al., 2000 and references therein), which has been used as the chronological basis for other palaeoclimatic studies in surrounding areas. Absolute dating of other exposures, however, would allow correlations to be built between more distant parts of the basin, which could then lead to an enhanced understanding of the timing and palaeoclimatic significance of depositional changes and hiatuses. For example, in the Perazim Valley, specific dates have been determined for the Lower, Middle and Upper Members of the Lisan Formation. Was the transition from laminated aragonite to gypsum time-transgressive or instantaneous across the basin? Such questions could be more easily addressed by increasing the number of well-dated sections.

Another point concerns the dearth of continuous palaeoclimate records through the entire Lisan sequence. Although somewhat continuous records of lake levels (e.g. Bartov et al., 2003), and isotopic variations related to palaeoclimate (e.g. Kolodny et al., 2007), have been obtained, more work at a greater number of sites is needed. A variety of sediment types are well exposed throughout the Lisan Basin, including finely laminated carbonate, diatomite and organic material, making the area suitable for constructing detailed stable isotopic records (O, C, H) that can be related to climatic indices. For example, the potential for constructing high-resolution and continuous records of oxygen and carbon isotopes from multiple well-dated sites provides the opportunity to gain insight into basin-wide precipitation/evaporation relationships, source water influx, catchment vegetation and lake productivity over time. Furthermore, adding hydrogen isotopes from organic material to the arsenal of palaeoclimatic proxies increases the potential for understanding the system. The end result may be a more coherent picture of palaeoclimatic evolution than is afforded by lake level reconstruction or single-site isotope records alone.

In addition to geochemical proxies of climate change, palaeoecological information needs to be better exploited in the Lisan sequence. For instance, high-resolution diatom palaeoecology, from biologically rich sediments present in some Lisan sections, may reveal changes in limnological or atmospheric parameters. Also, the potential to combine the quantitative palaeoecological reconstructions (e.g. the relative abundance patterns of diatom species through time) with oxygen isotopes from the diatoms within the same stratigraphic horizon, might provide an even more powerful tool for investigating palaeoenvironmental changes in the northern part of the Lisan Basin. Finally, there is a need to understand potential changes in input sources to the Lake Lisan system through time. It cannot simply be assumed that the relative proportions of Jordan River and spring input, for example, remained the same through time, as it is possible that the catchment hydrology of the Lisan and Dead Sea system has changed. Possible evidence for this comes from the observation of extensive tufa deposits around the margin of the Dead Sea, at various elevations and probably of different ages. The great expanse of these deposits in the area suggests that a large amount of water may have been contributed to the Lisan/Dead Sea system in the past from thermal spring sources. Higher contributions from springs, possibly hot and saline springs, would have consequences for the lake hydrology and its geochemical evolution. Clearly, these tufa deposits need to be dated and analysed further as a step towards gaining a more complete understanding of the hydrologic framework of Lake Lisan through its Late Pleistocene evolution.

9.5 METHODOLOGY

Given the nature of the previous work undertaken on the Lake Lisan sequences, we decided to undertake detailed U-series dating of sequences from the eastern shores of the Dead Sea to complement the extensive work undertaken on the western shores. Samples were collected from four locations on the eastern shore of the Dead Sea to represent the range of stratigraphy exposed in Jordan and the time range of materials available (Table 9.1, Figures 9.2, 9.3, 9.7 and 9.8). Samples were logged, sectioned and located with a portable GPS (Global Positioning System) module for latitude, longitude and elevation (Table 9.1); small sub-samples (100–500 mg) were taken from individual layers, as illustrated in Figures 9.7, 9.8 and 9.9. The U-series analysis was undertaken using a Perkin Elmer ELAN 6000 inductively coupled plasma mass spectrometer for ^{238}U, ^{232}Th, ^{230}Th and ^{226}Ra concentrations and (^{234}U/^{238}U) ratios together with a Harwell Instruments broad energy germanium detector (BeGe5030).

Table 9.1 *Uranium series ages for samples analysed as part of this study*

Sample number	Elevation (mbsl)	Age (ka BP)	Uncertainty
Site 2: 31° 23′ 42.82″ N 53° 33′ 16.90″ E			
LSRG 1	−170.0	24,880	827
LSRG 2	−170.3	13,060	526
LSRG 3	−195.5	13,350	531
LSRG 4	−200.3	20,250	860
LSRG 5	−201.0	22,540	701
LSRG 6	−220.5	13,860	555
LSRG 7	−221.3	13,871	683
LSRG 8	−224.0	24,150	641
LSRG 9	−233.5	18,340	874
Site 4: 31° 47′ 51.46″ N 35° 37′ 33.85″ E			
LSRG 10	−249.5	8,416	433
LSRG 11	−250.1	8,550	321
LSRG 12	−250.2	8,577	450
LSRG 13	−250.2	8,562	509
LSRG 14	−250.3	8,755	406
LSRG 15	−250.5	9,540	284
LSRG 16	−250.8	9,279	270
LSRG 17	−251.4	10,050	388
LSRG 18	−252.2	10,023	302
LSRG 19	−272.4	10,470	448
Site 3: 31° 14′ 12.25″ N 35° 31′ 03.78″ E			
LSRG 20	−273.4	10,921	608
LSRG 21	−304.3	14,780	604
Site 1: 30° 54′ 01.19″ N 35° 25′ 40.15″ E			
LSRG 22	−355.6	15,430	657
LSRG 23	−371.6	17,050	697
LSRG 24	−372.4	16,450	700

Figure 9.7. Site 4 used in this study. Inset shows stromatolite in cross-section, with intercalated gravels above and fine-grained sediments below.

Figure 9.8. Site 1. Inset shows laminated bands of aragonite (white) and grey (silicate-rich) units.

As the mass ratio of the ^{234}U/^{238}U is low (<1%) by mass spectrometry, the counts for the 234 mass peak were increased by running the mass spectrometer in isotope ratio mode using 10 replicate analyses, an increased dwell time (100 ms) and an average of 45 passes per replicate sample. This brought the uncertainty of the ratios to within a tolerable level (<1.8%). External reproducibility was checked using international standards (NIST SRM 3164) and by monitoring the (235/238) ratios in the samples to check that they were within the naturally abundant ratio (137.5). Concentrations of ^{230}Th and ^{226}Ra were determined after careful calibration using internal stock solutions at masses 230.0133 and 226.0254, respectively. The detection limits for ^{230}Th and ^{226}Ra were typically 1–2 and 4–5 pg L^{-1}, respectively and were determined using the instrument in

Figure 9.9. Site 3. Inset shows detailed layers of the sediments with annual bands.

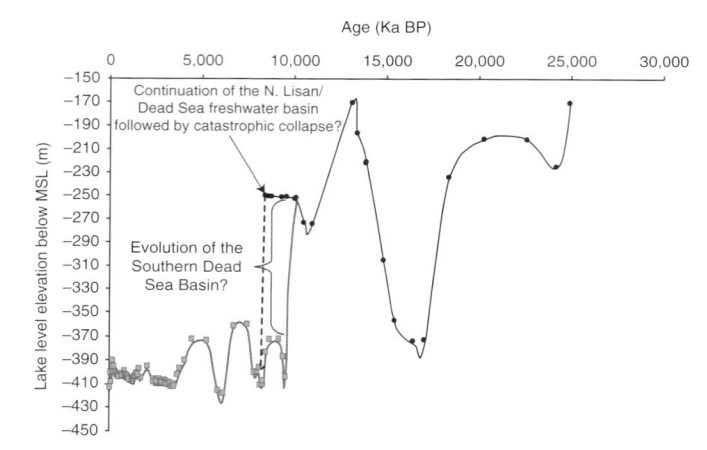

Figure 9.10. The data presented in this chapter (black circles) together with the elevation/age data from Enzel et al. (2003) (grey squares).

scanning mode. The accuracy of each of these peaks was checked using a lithium tetraborate glass disc made from USGS BCR-1 Basalt standard, subsequently digested in mineral acid. Analysis of this standard gave (^{230}Th/^{232}Th) ratios of 0.8733 +/− 0.032, within the reported standard range (0.8721–0.8866; $n = 4$), with Th concentrations of 6.04 +/−0.02 µg g^{-1} and U concentrations of 1.72 +/−0.03 µg g^{-1} which were also within the reported range of standard. (^{230}Th/^{232}Th) ratios were checked in the samples against the standards to verify the individual concentrations. Barium was also determined via mass spectrometry using the same instrument at masses 137 and 138.

9.6 RESULTS

The new U-series dates from the Lisan Formation can be seen in Table 9.1. Each of these dates is a product of isochrons whereby at least four sub-samples from each layer were taken and digested separately and analysed to compensate for the incorporation of detrital Th into the carbonate or gypsiferous sediments. The data are also plotted on Figure 9.6 together with elevation of the samples the contexts of which are discussed later. The U contents of the samples range from 0.9–9.0 µg g^{-1} with Th in the range 0.08–0.40 µg g^{-1}. The latter correlates strongly ($r^2 > 0.9$) in an exponential manner with (^{230}Th/^{232}Th) and detrital mineral (quartz and clay fraction) content. These ranges of data and observations fit with previously reported work on Lake Lisan (Landmann et al., 2002) and other detrital-rich carbonates (Candy et al., 2005, Candy and Black, 2009). (^{230}Th/^{232}Th) ratios also correlate strongly with detrital content of the samples ranging from those almost pure carbonate/gypsum with (^{230}Th/^{232}Th) ratios >30, Th < 0.05 µg g^{-1} and detrital components (from XRD) <1%, through to detrital-rich samples with (^{230}Th/^{232}Th) ratios < 1, Th > 0.35 μg g^{-1} and detrital components (from XRD) > 50%. Samples in the former category can be dated using U-series with very little trouble and require little correction, but samples in the latter category must be corrected using isochrons owing to the incorporation of ^{230}Th in the samples not from radioactive in-growth. Further details of the samples and methodologies for the Lake Lisan sequences can be found in Black, Robinson, Rambeau and Fitton (manuscript in preparation) and in Candy et al. (2005) and Candy and Black (2009).

9.7 DISCUSSION

The new sections of Lake Lisan material studied here were located in the Jordan Valley as shown on Figure 9.1 and in more detail on Figure 9.2 and were labelled sections 1–4. Site 1 shows a 3.5-m-thick sequence of finely laminated, alternate depositing carbonate-rich layers (up to 99% aragonite) inter-stratified with clastic-rich layers (up to 85% quartz, dolomite and kaolinite) consistent with the lower Laminated Member. Rip up clasts, folded beds and secondary Fe nodules are common in section together with abundant ripple structures.

There is a significant change in elevation of the Lake Lisan deposits through time as seen on Figure 9.6. Figure 9.6 also has plotted on it major lake level curves (summarised by Neev and Emery, 1967, Landmann et al., 2002, Bartov et al., 2002, 2003, Klinger et al., 2003, Robinson et al., 2006). The new data presented here fit well with most of the lake level curves previously published and add new resolution to the Lake Lisan reconstructions, particularly in the period 25–8 ka BP.

Figure 9.7 shows a section located at approximately −250 mbsl. This section has a transition of alternating aragonite and gypsum layers and clastic-rich layers at the base of the sequence, leading into laminated sediments with shallow-water algal stromatolites (Figure 9.7 inset). The sequence then coarsens upwards until gravel sequences are apparent at the top of the sequence. This coarsening upwards sequence is a sign of shallowing of the lake environment and also a change in conditions from low- to high-energy environments. The age of this shallowing event is significant as U-series dates from the base are 13.0±0.5 ka, whereas those at the top are 8.5±0.4 ka BP, indicating that the lake system was shallowing significantly during the early Holocene. This period coincides with key climatic events (e.g. 8.2 ka 'old' event), but the shallowing also occurs at a time of increased rainfall (Robinson et al., 2006), suggesting highly arid periods in the climate (i.e. increased seasonality). Figure 9.9 shows this sequence in detail. The presence of stromatolites and molluscs at the base of this section shows that this part of the lake was at least partly fresh at 13–10 ka BP. A freshwater north end of the Lake Lisan proto-Dead Sea would have been very influential in the region in terms of human habitation and food sources.

Figure 9.6 shows the data presented in this chapter together with the elevation/age data from Enzel et al. (2003). The large difference in elevation between Enzel's data and that presented in this chapter are consistent with large-scale changes in lake levels throughout time (Robinson et al., 2006). The data show that Lake Lisan has been fluctuating greatly over time with changes of several hundred metres in the space of a few thousand years. In Chapter 11 of this volume the fluctuations of the lake as a response to purely climatic condition changes are explored. Assuming a singular input (River Jordan), the rainfall implications for these lake levels are considerable (see Chapter 11). However, what is clear is that high lake levels at periods of the early Holocene are consistent with the review of Robinson et al. (2006), which suggests that the wettest period in the past 25 ka was the early Holocene period (9.5–7.0 ka BP). The early Holocene dates for the sequences measured as part of this study and the freshwater appearance are consistent with the Damya Formation (Damya Lake), post-Lisan, pre-Dead Sea formation in the early Holocene similar to that reported by Abed and Yaghan (2000). Palaeosol records in the Levant region also suggest wet periods in the early Holocene (both prior to 12.5 cal ka BP and 10–7.5 cal ka BP) with rainfall averages of 350–800 mm yr^{-1} (Gvirtzman and Wieder, 2001). Data from Wadi Faynan in southern Levant also suggest that the climate was wetter in the early Holocene (from meandering stream deposits; McLaren et al., 2004). Bar-Matthews et al. (2003) also concluded that the early Holocene was wetter and from the Soreq Cave deposits came up with a similar figure of 550–700 mm yr^{-1} of rainfall in the period 8.0–7.5 ka BP. Emeis et al. (2000) also suggest that during the early Holocene the eastern Mediterranean was becoming less saline and warmer as a result of increased freshwater input. All these data compare well to the modelled data for the rainfall records as described in Chapter 11 of this volume and the data presented here for the ages of the early part of the Lake Lisan system.

9.8 CONCLUSIONS

New data from Lake Lisan sequences around the present Dead Sea show ages in the range 25–8 ka BP. The elevations of these data fit very well with previously reported data from the Lake Lisan sequence and add to the existing database of Lake Lisan data. The new dates for the Lisan sequence suggest that there was at least a partially fresh sequence of lacustrine deposits at the northern end of the lake around the start of the Holocene.

The enormous changes in lake levels over the period 25–8 ka BP, demonstrated by this study being only the second set of U-series dates from the eastern shores of the Dead Sea, show a striking series of age–elevation relationships.

The Lake Lisan sequences remained stable and high from 30 to 22 ka BP as a result of the wet, cooler climates of the Late Pleistocene. During the LGM (20–18 ka BP) the lake levels fell to below −250 m, and then rose again during the warmer, wetter period of the Bølling–Allerød at around 14–12 ka BP. After this the lake fell to below −250 m again until the early Holocene (9.0–7.5 ka) when it recovered to a highstand of at least −190 m.

REFERENCES

Abed, A. M. and R. Yaghan (2000) On the paleoclimate of Jordan during the last glacial maximum. *Palaeogeography Palaeoclimatology Palaeoecology* **160**: 23–33.

Bar-Matthews, M., A. Ayalon, M. Gilmour, A. Matthews and C. J. Hawkesworth (2003) Sea–land oxygen isotopic relationships from planktonic foraminifera and speleothems in the Eastern Mediterranean region and their implication for paleorainfall during interglacial intervals. *Geochimica et Cosmochimica Acta* **67**: 3181–3199.

Bartov, Y., M. Stein, Y. Enzel, A. Agnon and Z. Reches (2002) Lake levels and sequence stratigraphy of Lake Lisan, the late Pleistocene precursor of the Dead Sea. *Quaternary Research* **57**: 9–21.

Bartov, Y., S. L. Goldstein, M. Stein and Y. Enzel (2003) Catastrophic arid episodes in the Eastern Mediterranean linked with the North Atlantic Heinrich events. *Geology* **31**: 439–442.

Begin, Z. B. (2002) Computer simulation of salinity in the Quaternary Lake Lisan. *Israel Journal of Earth Sciences* **51**: 225–232.

Begin, Z. B., A. Ehrlich and Y. Nathan (1974) Lake Lisan: the Pleistocene precursor of the Dead Sea. *Geological Survey of Israel Bulletin* **63**: 30.

Begin, Z. B., W. Broecker, B. Buchbinder et al. (1985) Dead Sea and Lake Lisan levels in the last 30,000 years: a preliminary report. Israel Geological Survey Report 29/85.

Begin, Z. B., M. Stein, A. Katz et al. (2004) Southward migration of rain tracks during the last glacial, revealed by salinity gradient in Lake Lisan (Dead Sea rift). *Quaternary Science* **23**: 1627–1636.

Bischoff, J. L. and J. A. Fitzpatrick (1991) U-series dating of impure carbonates – an isochron technique using total-sample dissolution. *Geochimica et Cosmochimica Acta* **55**: 543–554.

Bookman, R., Y. Enzel, A. Agnon and M. Stein (2004) Late Holocene lake levels of the Dead Sea. *Geological Society of America Bulletin* **116**: 555–571.

Bourdon, B., S. Turner, G. M. Henderson and C. C. Lundstrom (2003) Introduction to U series geochemistry. In *Uranium-Series Geochemistry, Reviews in Mineralogy and Geochemistry*, ed. B. Bourdon, S. Turner, G. M. Henderson and C. C. Lundstrom. Washington DC: Mineralogical Society of America.

Candy, I. and S. Black (2009) The timing of Quaternary calcrete development in semi-arid southeast Spain: investigating the role of climate on calcrete genesis. *Sedimentary Geology* **218**: 6–15.

Candy, I., S. Black and B. Sellwood (2005) U-series isochron dating of immature and mature calcretes as a basis for constructing Quaternary landform chronologies for the Sorbas basin, southeast Spain. *Quaternary Research* **64**: 100–111.

Druckman, Y., M. Magaritz and A. Sneh (1987) The shrinking of Lake Lisan, as reflected by the diagenesis of its marginal oolitic deposits. *Israel Journal of Earth Sciences* **36**: 101–106.

Edwards, R. L., C. D. Gallup and H. Cheng (2003) Uranium-series dating of marine and lacustrine carbonates. In *Uranium-Series Geochemistry, Reviews in Mineralogy and Geochemistry*, ed. B. Bourdon, S. Turner, G. M. Henderson and C. C. Lundstrom. Washington DC: Mineralogical Society of America.

Emeis, K. C., U. Struck, H. M. Schulz et al. (2000) Temperature and salinity variations of Mediterranean Sea surface waters over the last 16,000 years from records of planktonic stable oxygen isotopes and alkenone unsaturation ratios. *Palaeogeography Palaeoclimatology Palaeoecology* **158**: 259–280.

Enzel, Y., R. Bookman, D. Sharon et al. (2003) Late Holocene climates of the Near East deduced from Dead Sea level variations and modern regional winter rainfall. *Quaternary Research* **60**: 263–273.

Freund, R., Z. Garfunkel, I. Zak, M. Goldberg, T. Weissbrod and B. Derin (1970) The shear along the Dead Sea rift. *Philosophical Transactions of the Royal Society of London A*. **267**: 107–130.

Garfunkel, Z. (1981) Internal structure of the Dead-Sea leaky transform (rift) in relation to plate kinematics. *Tectonophysics* **80**: 81–108.

Garfunkel, Z. and Z. BenAvraham (1996) The structure of the Dead Sea basin. *Tectonophysics* **266**: 155–176.

Gvirtzman, G. and M. Wieder (2001) Climate of the last 53,000 years in the eastern Mediterranean, based on soil-sequence stratigraphy in the coastal plain of Israel. *Quaternary Science Reviews* **20**: 1827–1849.

Hasse-Schramm, A., S. L. Goldstein and M. Stein (2004) U-Th dating of Lake Lisan (late Pleistocene Dead Sea) aragonite and implication for glacial East Mediterranean climate change. *Geochimica et Cosmochimica Acta* **68**: 985–1005.

Hazan, N., M. Stein, A. Agnon et al. (2005) The late quaternary limnological history of Lake Kinneret (Sea of Galilee), Israel. *Quaternary Research* **63**: 60–77.

Katz, A. and N. Kolodny (1989) Hypersaline brine diagenesis and evolution in the Dead Sea–Lake Lisan system (Israel). *Geochimica et Cosmochimica Acta* **53**: 59–67.

Kaufman, A. (1971) U-series dating of Dead Sea Basin carbonates. *Geochimica et Cosmochimica Acta* **35**: 1269–1281.

Kaufman, A., Y. Yechieli and M. Gardosh (1992) Reevaluation of the lake-sediment chronology in the Dead Sea Basin, Israel, based on new ^{230}Th/U dates. *Quaternary Research* **38**: 292–304.

Klein-BenDavid, O., E. Sass and A. Katz (2004) The evolution of marine evaporitic brines in inland basins: the Jordan–Dead Sea Rift valley. *Geochimica et Cosmochimica Acta* **68**: 1763–1775.

Klinger, Y., J. P. Avouac, D. Bourles and N. Tisnerat (2003) Alluvial deposition and lake-level fluctuations forced by Late Quaternary climate change: the Dead Sea case example. *Sedimentary Geology* **162**: 119–139.

Kolodny, Y., M. Stein and M. Machlus (2007) Sea-rain-lake relation in the Last Glacial East Mediterranean revealed by δ^{18}O-δ^{13}C in Lake Lisan aragonites. *Geochimica et Cosmochimica Acta* **71**: 3926–3927.

Landmann, G., G. M. Abu Qudaira, K. Shawabkeh, V. Wrede and S. Kempe (2002) Geochemistry of the Lisan and Damya Formations in Jordan, and implications for palaeoclimate. *Quaternary International* **89**: 45–57.

Luo, S. and T. L. Ku (1991) U-series isochron dating: a generalized method employing total-sample dissolution. *Geochimica et Cosmochimica Acta* **55**: 555–564.

Machlus, M., Y. Enzel, S. L. Goldstein, S. Marco and M. Stein (2000) Reconstructing low levels of Lake Lisan by correlating fan-delta and lacustrine deposits. *Quaternary International* **73–4**: 137–144.

Macumber, P. G. and M. J. Head (1991) Implications of the Wadi al-Hammeh sequences for the terminal drying of Lake Lisan, Jordan. *Palaeogeography Palaeoclimatology Palaeoecology* **84**: 163–173.

Marco, S. and A. Agnon (2005) High-resolution stratigraphy reveals repeated earthquake faulting in the Masada Fault Zone, Dead Sea Transform. *Tectonophysics* **408**: 101–112.

McLaren, S. J., D. D. Gilbertson, J. P. Grattan et al. (2004) Quaternary palaeogeomorphological evolution of the Wadi Faynan area, southern Jordan. *Palaeogeography Palaeoclimatology Palaeoecology* **205**: 131–154.

Migowski, C., A. Agnon, R. Bookman, J. F. W. Negendank and M. Stein (2004) Recurrence pattern of Holocene earthquakes along the Dead Sea transform revealed by varve-counting and radiocarbon dating of lacustrine sediments. *Earth and Planetary Science Letters* **222**: 301–314.

Neev, D. and K. O. Emery (1967) The Dead Sea. *Geological Society of Israel Bulletin* **41**: 1–147.

Neev, D. and Emery, K. O. (1995) *The Destruction of Sodom, Gomorragh, and Jericho*. New York: Oxford University Press.

Prasad, S., H. Vos, J. F. W. Negendank et al. (2004) Evidence from Lake Lisan of solar influence on decadal- to centennial-scale climate variability during marine oxygen isotope stage 2. *Geology* **32**: 581–584.

Robinson, S. A., S. Black, B. Sellwood and P. J. Valdes (2006) A review of palaeoclimates and palaeoenvironments in the Levant and Eastern Mediterranean from 25,000 to 5000 years BP: setting the environmental background for the evolution of human civilisation. *Quaternary Science Reviews* **25**: 1517–1541.

Schramm, A., M. Stein and S. L. Goldstein (2000) Calibration of the C-14 time scale to > 40 ka by ^{234}U-^{230}Th dating of Lake Lisan sediments (last glacial Dead Sea). *Earth and Planetary Science Letters* **175**: 27–40.

Sneh, A. (1979) Late Pleistocene fan-deltas along the Dead Sea rift. *Journal of Sedimentary Petrology* **49**: 541–552.

Stein, M. (2001) The sedimentary and geochemical record of Neogene–Quaternary water bodies in the Dead Sea Basin – inferences for the regional paleoclimatic history. *Journal of Paleolimnology* **26**: 271–282.

Stein, M., A. Starinsky, A. Katz *et al.* (1997) Strontium isotopic, chemical, and sedimentological evidence for the evolution of Lake Lisan and the Dead Sea. *Geochimica et Cosmochimica Acta* **61**: 3975–3992.

Stein, M., A. Starinsky, A. Agnon *et al.* (2000) The impact of brine–rock interaction during marine evaporite formation on the isotopic Sr record in the oceans: evidence from Mt. Sedom, Israel. *Geochimica et Cosmochimica Acta* **64**: 2039–2053.

Stein, M., A. Torfstein, I. Gavrieli and Y. Yechieli (2010) Abrupt aridities and salt deposition in the post-glacial Dead Sea and their North Atlantic connection. *Quaternary Science Reviews* **29**: 567–575.

Torfstein, A., I. Gavrieli and M. Stein (2005) The sources and evolution of sulfur in the hypersaline Lake Lisan (paleo-Dead Sea). *Earth and Planetary Science Letters* **236**: 61–77.

Yechieli, Y., M. Magaritz, Y. Levy *et al.* (1993) Late Quaternary geological history of the Dead Sea area, Israel. *Quaternary Research* **39**: 59–67.

Part III
Hydrological studies of the Jordan Valley

10 The impacts of climate change on rainfall-runoff in the upper River Jordan: methodology and first projections

Andrew Wade, Emily Black, Nicola Flynn and Paul Whitehead

ABSTRACT

This chapter is concerned with the development and application of a modelling framework to investigate the sensitivity of the flows in the upper River Jordan to changes in daily and seasonal rainfall patterns, and the likely changes in the daily runoff in response to anthropogenic climate change. To this end, idealised climate scenarios as well as projections of future near-surface air temperature and precipitation from the HadRM3 climate model are used with a modelling framework incorporating the Pitman and INCA rainfall-runoff models to determine flow sensitivity to rainfall and the flow response to a scenario of projected climate change. The Pitman model was used to estimate the hydrologically effective rainfall (HER); the INCA model was used for simplified flood-routing. The INCA model was calibrated for the period 1989 to 1993, commensurate with the observed daily mean flows, and the goodness of fit coefficient R^2 was 0.7, indicating a good fit to the observed dynamics, although the simulated flows were an overestimate of those observed owing to unquantifiable river abstractions. On a seasonal timescale, comparison between sensitivity scenarios with long and short rainy seasons showed that rainfall in the winter months had the greatest impact on flow. In response to projected changes in the climate for 2071–2100, the modelled flood flows in the scenario period were reduced by approximately 31% for the largest simulated flood and 25% for the 10th percentile flow in comparison to those in the control period from 1961 to 1990. Future base flows were projected to remain unchanged because of maintained groundwater inputs. The discussion focuses on the utility of the modelling framework for assessing the rainfall-runoff in response to projected climate change.

10.1 INTRODUCTION

Between now and the end of the twenty-first century, the climate is forecast to change in the Middle East and these changes are characterised by increased near-surface air temperatures and reduced precipitation (Chapter 4, this volume). Anthropogenic emissions of greenhouse gases are believed to enhance the natural variations in the Earth's climate, leading in part to the increase in projected temperatures and reduced precipitation (IPCC Core Writing Team et al., 2007). The water resource of the upper Jordan and its tributaries is shared between Israel, Jordan, Lebanon, Syria and the West Bank, and these water resources are already under severe stress (US Geological Survey, 1998). Given the modelled increase in temperature and decrease in mean annual precipitation over the region, it is important to understand how this will affect the hydrology of the Jordan Valley and environs. To this end, other studies have looked at the water resources of the Jordan Valley and the likely changes, given climate projections (for example Kunstmann et al., 2005; Samuels et al., 2009).

On a country-wide scale, a simple but physically based model suggested that the water yield in Jordan would reduce by up to 60% if precipitation were to decrease by 10% and the region were to become 2 °C hotter (Oroud, 2008). These studies of the Middle East are consistent with a study of river runoff, using a globally implemented streamflow model, which suggests that, by the end of the twenty-first century, changes in precipitation and evaporation would lead to significant changes in runoff in much of the world – including the Middle East (Arnell, 2003).

A major initiative in the Jordan Valley is the Globaler Wandes des Wasserkreislaufs – Jordan River project (GLOWA JR) (Hoff et al., 2006). Work done to date within GLOWA JR has focused on constructing detailed hydrological models for the upper and lower Jordan river, with the upper Jordan defined as the catchment area draining into the Sea of Galilee and the lower Jordan

Water, Life and Civilisation: Climate, Environment and Society in the Jordan Valley, ed. Steven Mithen and Emily Black. Published by Cambridge University Press. © Steven Mithen and Emily Black 2011.

catchment defined as that draining to main channel downstream of the Sea of Galilee and to the Dead Sea itself. A model of the karst of the upper Jordan has been developed and used to forecast the likely flow response in the Dan, Hermon and Snir sub-catchments in addition to the flows in the Jordan River immediately downstream of the confluence between the Dan, Hermon and Snir tributaries. The results show that, while annual streamflow is proportional to total precipitation (provided annual precipitation exceeds 400 mm day^{-1}), increasing the frequency of wet spells lasting longer than 3 days leads to more frequent and more intense floods (Samuels et al., 2009).

The GLOWA results agree in some respects with other published work on the impact of climate change on water availability and river flow in the Middle East. In a study of the upper Jordan catchment that used a distributed hydrological model informed by regional model input, Suppan et al. (2008) predicted that, under a scenario in which there is a slowing of the present rates of increase in greenhouse gas emissions, the total runoff will decrease by 23% by the end of the twenty-first century – a conclusion consistent with the linear relationship between annual precipitation and streamflow proposed by Samuels et al. (2009). However, in contrast to the work of Samuels et al. (2009), which suggested little change in base flow, Suppan et al. (2008) suggested that groundwater recharge would decrease, resulting in a reduction in base flow.

The mean rainfall and temperature are the dominant controls on the volume of water that enters the hydrological system, and are thus likely to be the primary controls on river flow. It is nevertheless possible that the response of river flow to the climate is modulated by daily weather patterns (often referred to as the statistics of the weather). For example, a hydrological system might respond differently to a few intense rainfall events than it would to constant drizzle – even if the total amounts of seasonal rainfall were the same. The role of the statistics of the weather in controlling river flow is relevant to the studies presented in this volume for two reasons. Firstly, in climate change scenarios, variations in the mean climate may reflect different magnitudes of change in the individual statistics. For example, Chapter 4 of this volume suggests that the reduction in annual total rainfall projected for the Middle East occurs mainly at the peak of the rainy season and reflects a reduction in the number of rainy days rather than of rainfall intensity. Secondly, the climate model used for the studies described in subsequent chapters of this volume represents some aspects of the weather markedly better than others. The uncertainties in the projections of flow are therefore influenced by the sensitivity of the hydrological system to variability in the individual statistics of the weather.

Although previous work has mainly focused on the link between the mean climate and river flow, a few studies have looked at daily weather patterns. In addition to the work of Samuels et al. (2009) studies of regions other than the Middle East also demonstrate the importance of daily rainfall statistics.

For example, rainfall intensity was found to be a crucial factor for runoff over a humid equatorial catchment in Africa (Mileham et al., 2008, 2009), and a case study of two individual storms suggested that sudden intense events will have significant impact on the hydrological system (Lange et al., 2000). The importance of the daily rainfall time-series has also been emphasised in a methodological study, which found that the inability of a regional model to simulate the daily occurrence and amount of rainfall limits its ability to simulate variability in flow (Messager et al., 2006).

The study of Messager et al. (2006) raises an important question about the reliability of hydrological models driven with climate model data – a point also made by Dibike and Coulibaly (2005), who compared downscaling methods for the Saguenay watershed. Indeed, the question of how climate change is likely to affect river flow is challenging, not least because of the difficulty that climate models have with representing spatial and temporal variability in daily rainfall. These problems are exacerbated by a general lack of daily precipitation data for training hydrological models and evaluating climate models. Wilby et al. (2009) note that there is little consensus between general circulation models (GCM) in the Middle East and North Africa region, and that the process of taking GCM projections and using these to estimate the regional- or catchment-scale climate changes can introduce further error. The uncertainty in model chains that link climate projections to impact models is significant because of uncertainty in the emission scenarios and because of the structure and parameterisation of the GCM and the impact model (Oreskes et al., 1994; Beven and Freer, 2001; Wilby and Harris, 2006).

The aim of this work is to develop and test a new meteorological and hydrological model framework, and assess the output from this new model chain created to examine the likely impacts of climate variability and change on water availability in the upper Jordan. Specifically, the objectives are:

(1) to link a regional climate model, a weather generator, a rainfall-runoff model and a runoff routing model to create a framework with which to assess the impacts of climate variability and change;

(2) to collate a database to provide sufficient suitable data with which to develop and test a climate–hydrological model chain in a data-poor region;

(3) to assess the sensitivity of River Jordan flow to changes in annual total rainfall and to changes in individual statistics of the weather;

(4) to assess the impact of anthropogenic climate change on River Jordan flow and to compare the modelled outcomes with other studies;

(5) to consider the benefits and disadvantages of the approach within the context of other available methods.

The study has two parts. In the first, the effects of daily and seasonal rainfall patterns on the River Jordan are explored using idealised climate scenarios as inputs to the modelling framework. In the second part, we demonstrate a methodology for assessing the impact of anthropogenic climate change on River Jordan flow. Considered together, these two components give us insight into the mechanisms by which the projected changes in near-surface air temperature and precipitation affect River Jordan flow, and the degree to which climate model bias affects the predictions.

10.2 STUDY AREA AND DATA RESOURCE

The study area for this work is the upper Jordan from the headwaters to Obstacle Bridge gauging station on the upper Jordan (33.03° N, 35.62° E; Figure 10.1). This area was chosen because of its importance in terms of water provision and the dependence on this water resource of four countries: Syria, Lebanon, Israel and Jordan.

The upper Jordan to Obstacle Bridge covers an area of 1,752 km². The headwaters of the Jordan drain from Lebanon and from Mount Hermon in the Golan Heights which is the highest point in the catchment at 2,814 m. Given the altitude, precipitation can fall as snow on Mount Hermon during the winter. The key tributaries of the upper Jordan are the Dan, the Snir (of which the Hisbani is a tributary) and the Hermon (of which the Banias is a tributary). The geology of the upper Jordan is predominately limestone, which includes karst development. The springs, which form the Dan river, drain from the karst and are estimated to have a groundwater retention time of 2 to 3 years (Rimmer and Salingar, 2006). The Sea of Galilee is a well-known lake in the catchment. Downstream of the sea, the Yarmouk drains into the Jordan river. The King Abdullah canal is used to provide water for irrigation in northwest Jordan and water is diverted from the lower Yarmouk to supply the canal, lowering flows in the Yarmouk and the Jordan. Water is abstracted for irrigation in Israel using the National Water Carrier which draws water upstream of the inlet to the Sea of Galilee, and this also lowers flow to the Jordan river. Downstream of the confluence with the Yarmouk, the Jordan river flows southward to the Dead Sea.

Over the past 30 years, the water levels in the Dead Sea have dropped at the rate of 0.5 m per year as a result of a decline in the flow in the lower reaches of the River Jordan, downstream of the Sea of Galilee, due to over-abstraction to irrigate crops in Syria, Lebanon, Israel and Jordan, countries through which the Jordan and its tributaries flow. Flows in the River Jordan have declined from 10^9 m³ year^{-1} to less than 10% of this figure. As a result of this decline in freshwater input and the retreat of the water level in the Dead Sea, salt has been exposed along the shoreline, and

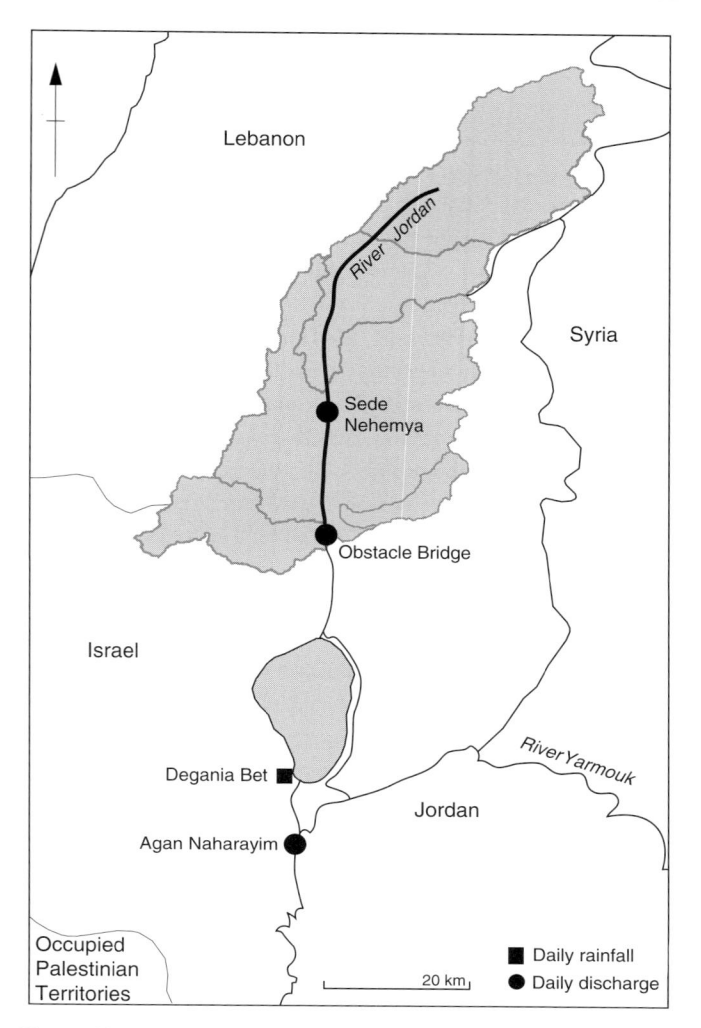

Figure 10.1. A schematic map of the upper River Jordan.

this has been dissolved by rainfall to create sink holes. The upper Jordan contains the Hula Valley which, prior to the 1950s, was an extensive wetland area. Like the Dead Sea, the water levels in the Hula Valley have declined owing to water diversion from the area and land drainage for agriculture. As a result there has been a loss of flora and fauna in the wetland. There are plans to restore this area but this depends on water availability and the competing demands for domestic, agricultural and industrial consumption. Whilst this damage to the landscape is unsightly in an area of tourism and is of ecological concern, the most pressing concerns in Jordan, with respect to the water resource, are the issues of water and food security.

Jordan is one of the most water-scarce countries in the world and the pressure on this resource is likely to rise in the future as the population is projected to increase from 5.10 million at the last census in 1 October 2004 to 8.55 million by 2030, owing to natural population increase and immigration from elsewhere in the region, in particular the West Bank and Iraq (United States Statistics Division, 2010). Furthermore, Jordan is looking for economic development. To this end, water is required for industrial growth and expansion of tourism. Surface waters are exploited to the full, and most of the wadis draining to the lower Jordan river in Jordan have dams built in them for water storage. The main water resource is groundwater. There are four main aquifers in Jordan, although the full extent and water capacity of these has yet to be mapped in detail (Puri, 2001; Puri et al., 2001; Puri and Aureil, 2005). The four aquifers are: the Syrian Steppe which spans Iraq, Jordan, Saudi Arabia and Syria, and is believed to extend over 1,600,000 km^2; the Hauran and Jabal Al-Arab, a basalt, Neogene and Quaternary aquifer which spans Jordan, Saudi Arabia and Syria and has an extent of 15,000 km^2; the Eastern Mediterranean aquifer, which spans Israel, Jordan, Lebanon, Palestinian Territory and Syria, and is a porous, fissured, karst aquifer with an extent of 48,000 km^2; and the Qa Disi aquifer which has an extent of 3,000 km^2 in Jordan and Saudi Arabia. Recharge varies from over 400 mm year^{-1} for the eastern Mediterranean aquifer and around the Sea of Galilee to less than 15 mm year^{-1} in the Qa Disi and Wadi Araba groundwater basins (US Geological Survey, 1998; Puri, 2001). The water in boreholes of the coastal aquifers of Israel is becoming increasingly saline owing to abstraction, irrigation, seawater intrusion and lateral transfers of groundwater from the south to the north of Israel (US Geological Survey et al., 1998). In Jordan, trends in groundwater salinity are less clear, and there are no data reported from observation boreholes in the upper Jordan. Elsewhere in Jordan there is evidence for both increasing and decreasing trends in salinity, reflecting local factors such as the natural salinity of the groundwater, over-pumping, dilution from high rainfall and leaching from irrigation waters.

Food security in Jordan is poor. In 2006, Jordan imported 93% of its annual wheat requirement and 95% of its annual barley requirement (United States Department of Agriculture, 2006). Wheat is used for bread production and barley is used as an animal feed. Vegetables are the main crop grown in Jordan in the northwest where the annual precipitation is highest. These crops are important to the national economy and an important food resource. Thus water is a vital resource, and the water and food security of Jordan is under threat from over-abstraction and potential climate change.

Precipitation patterns in Jordan show a strong gradient from west to east. The main storm track affecting northern Jordan is from west to east along the Mediterranean Sea. Precipitation in the south of Jordan can be affected by low pressure systems over the Red Sea, known as Red Sea lows (Alpert et al., 2004). The predominance of the Mediterranean storm tracks leads to mean annual precipitation over Israel of 50 to 900 mm and between 200 and 700 mm in the northwest of Jordan where the majority of the crops are cultivated. Wadi Araba, the valley along the Israel–Jordan border, is an extension of the Great Rift of Africa. This valley affects the precipitation annual total (Figure 2.4). To the east of the valley axis is a scarp slope. Along the ridge of the scarp slope and into northwest Jordan, which is at an elevation of between 400 m above sea level (masl) at Umm Qais in the north and 1,727 masl at Jebel Mubrak in the south, the rainfall tends to be higher than in the valley bottom which at the Dead Sea is approximately 400 m below sea level. The scarp slope causes orographic lifting of the moist air masses moving east over Israel and the valley, and this leads to greater precipitation over the ridge of the scarp. Further east, towards the desert centre, the rainfall is much lower at approximately 60 mm year^{-1} at Ma'an and along the 'pan-handle' of Jordan towards the border with Iraq.

The spatial coverage of readily available meteorological and hydrological data is sparse. For the purpose of this study, data were collated from 60 rainfall stations and seven discharge gauging stations across Israel, Syria, Lebanon and Jordan. These data are summarised in Table 10.1. These data were acquired from the Global Runoff Data Centre and the United States National Climatic Data Center. Daily rainfall data were purchased from the Israeli Meteorological Service (IMS) for the period 1984–2005 for nine sites and further daily rainfall data were available for seven stations in Jordan from 1937 to 1974 from the yearbooks of the Water Resources Division of the National Resources Authority. Monthly rainfall data were available from the United States National Climatic Data Centre for 11 sites in Israel (1846–1995), 11 sites in Jordan (1960–2000), 15 sites in Lebanon (1888–2000) and seven sites in Syria (1951–2000).

Daily flow data were, in general, difficult to find but these are needed to assess the flood extremes. Ideally 15-minute data should be used but no such data were available to this study. Daily data were available for flow gauges on the Jordan at Sede Nehemya (1984–1992), Obstacle Bridge (1973–1993) and Naharyim (1988–1993). No daily flow data could be found for the lower Jordan. Monthly flow data were available at six stations, and daily and monthly flows were reported for gauges in the 1963 Jordan Hydrological yearbook. This yearbook includes flows for the main channel of the Jordan and the side wadis in Jordan. It should be noted, however, that in many cases the flows for the side wadis were estimated using engineering calculations (flood hydrographs) rather than measurements. The flow in desert systems is notoriously difficult to measure, as rainfall can be infrequent and sporadic and large floods can damage measuring structures. Moreover, it is difficult to measure the flood flows because of high water velocities and the large sediment transport loads. As such, published flood flow values must be treated with caution.

The rainfall and runoff data were supplemented with data describing the local climate at 12 sites across Jordan. These data included monthly averages for the period 1983 to 2002 of measurements of near-surface air temperature, solar radiation, wind speed and sunshine hours.

Table 10.1 *A summary of rainfall and flow data collated for the application of the hydrological models to simulate the flows in the upper River Jordan*

Description	Country	Site	Period	Frequency	Notes
US National Climatic Data Center[a]					
Precipitation	Israel	Jerusalem	1846–1995	Monthly	Lat: 31.80 Long: 35.20 Elev: 809 m
		Beersheva	1951–1994	Monthly	Lat: 31.20 Long: 34.80 Elev: 275 m
		Eilat	1951–2000	Monthly	Lat: 29.60 Long: 35.00 Elev: 11 m
		Haifa Port	1881–1960	Monthly	Lat: 32.80 Long: 35.00 Elev: 5 m
		Jaffa	1880–1913	Monthly	Lat: 32.10 Long: 34.50 Elev: 20 m
		Jerusalem/Old City	1846–1960	Monthly	Lat: 31.80 Long: 35.30 Elev: 810 m
		Lod/Ben-Gurion AP.	1951–2000	Monthly	Lat: 32.00 Long: 34.90 Elev: 49 m
		Mt Kenaan	1951–1994	Monthly	Lat: 33.00 Long: 35.50 Elev: 936 m
		Tel Aviv/Sde Dov	1951–1994	Monthly	Lat: 32.10 Long: 34.80 Elev: 10 m
		Bet Dagan	1986–1995	Monthly	Lat: 32.00 Long: 34.80 Elev: 30 m
		Jerusalem Airport	1951–1967	Monthly	Lat: 31.90 Long: 35.20 Elev: 755 m
	Jordan	Amman Airport	1923–1998	Monthly	Lat: 32.00 Long: 35.90 Elev: 766 m
		Deir Alla	1952–1989	Monthly	Lat: 32.20 Long: 35.60 Elev: −224 m
		H-4/Ruwashed	1960–2000	Monthly	Lat: 32.50 Long: 38.20 Elev: 683 m
		Irbid	1955–1998	Monthly	Lat: 32.60 Long: 35.90 Elev: 619 m
		Ma'an	1960–2000	Monthly	Lat: 30.20 Long: 35.80 Elev: 1,069 m
		Upper Jordan	1962–1987	Monthly	Lat: 35.40 Long: 33.23 Elev: 918 m
		Aqaba Airport	1960–1989	Monthly	Lat: 29.60 Long: 35.00 Elev: 51 m
		Er Rabbah	1960–1989	Monthly	Lat: 31.30 Long: 35.80 Elev: 920 m
		Jordan University	1960–1989	Monthly	Lat: 32.00 Long: 35.90 Elev: 980 m
		Mafraq	1960–1989	Monthly	Lat: 32.40 Long: 36.30 Elev: 686 m
		Wadi Yabis	1960–1989	Monthly	Lat: 32.40 Long: 35.60 Elev: −200 m
	Lebanon	Aitaroun	1938–1971	Monthly	Lat: 33.80 Long: 35.90 Elev: 918 m
		Beirut Int. Airport	1888–1996	Monthly	Lat: 33.90 Long: 35.50 Elev: 24 m
		Hasbaya	1943–1971	Monthly	Lat: 33.80 Long: 35.90 Elev: 918.0 m
		Joub Jannine	1946–1971	Monthly	Lat: 33.80 Long: 35.90 Elev: 918.0 m
		Kfar Quoq	1961–1971	Monthly	Lat: 33.80 Long: 35.90 Elev: 918 m
		Ksara Obsy	1921–1990	Monthly	Lat: 33.80 Long: 35.90 Elev: 918 m
		Hasbani	1986–2003	Monthly	Lat: 33.50 Long: 35.70 Elev: 500 m
		Dier Al Achaya	1964–1971	Monthly	Lat: 33.80 Long: 35.90 Elev: 918 m
		Machghara	1938–1971	Monthly	Lat: 33.80 Long: 35.90 Elev: 918 m
		Marjeyoun	1944–1971	Monthly	Lat: 33.80 Long: 35.90 Elev: 918 m
		Rayack	1946–1971	Monthly	Lat: 33.80 Long: 35.90 Elev: 918 m
		Rayack	1951–1986	Monthly	Lat: 33.90 Long: 36.00 Elev: 921 m
		The Cedars	1960–1980	Monthly	Lat: 34.20 Long: 36.00 Elev: 1,915 m
		Tripoli	1951–1996	Monthly	Lat: 34.60 Long: 36.00 Elev: 10 m
		Yanyta	1964–1970	Monthly	Lat: 33.80 Long: 35.90 Elev: 918 m
	Syria	Aleppo	1951–2000	Monthly	Lat: 36.20 Long: 37.20 Elev: 393 m
		Damascus Int. Ap.	1951–2000	Monthly	Lat: 33.42 Long: 36.52 Elev: 605 m
		Deir Ezzor	1951–2000	Monthly	Lat: 35.30 Long: 40.20 Elev: 212 m
		Hama	1951–2000	Monthly	Lat: 35.10 Long: 36.80 Elev: 322 m
		Kamishli	1952–2000	Monthly	Lat: 37.10 Long: 41.20 Elev: 455 m
		Lattakia	1952–2000	Monthly	Lat: 35.60 Long: 35.80 Elev: 9 m
		Palmyra	1955–2000	Monthly	Lat: 34.60 Long: 38.30 Elev: 404 m
Global Runoff Data Centre[b]					
Daily or monthly discharge	Israel	River Jordan, Sede Nehemya	1984–1992	Daily	GRDC-No.: 6594070 Area (km^2): 1,659
		River Jordan, Sede Nehemya	1984–1992	Monthly	

Table 10.1 (*cont.*)

Description	Country	Site	Period	Frequency	Notes
		River Jordan, Obstacle Bridge	1973–1993	Daily	GRDC-No.: 6594050 Lat. (dec. degree): 33.03 Long. (dec. degree): 35.62 Area (km²): 1,376
		River Jordan, Obstacle Bridge	1976–1993	Monthly	
		River Jordan, Southern Jordan	1965–1972	Monthly	GRDC-No.: 6594060 Lat. (dec. degree): 33.03 Long. (dec. degree): 35.63
		River Jordan, Naharyim	1988–1993	Daily	GRDC-No.: 6594080 Area (km²): 11,900
		River Jordan, Naharyim	1988–1993	Monthly	
	Israel/ Golan	Banias	1962–1980	Monthly	
	Israel	Dan	1962–1980	Monthly	
	Israel/ Lebanon	Hasbani	1962–1980	Monthly	
Centre for Ecology and Hydrology archives					
Daily rainfall	Jordan	Karak	1938–1974	Daily	31° 11′, 35° 42′
		Tafileh	1937–1974	Daily	30° 50′, 35° 36′
		Buserah	1937–1974	Daily	30° 45′, 35° 36′
		Khanzeira	1945–1974	Daily	31° 03′, 35° 36′
		Mazar	1937–1974	Daily	31° 04′, 35° 42′
		Ain Bisas	1951–1974	Daily	31° 12′, 35° 40′
		Ghor Mazraa	1939–1974	Daily	31° 17′, 35° 32′
Bethlehem University					
Daily rainfall	Israel	Bethlehem	1993–1997	Daily	
Government of Transjordan Report on Water Resources					
Rainfall	Jordan	Amman RAF station	1929–1939	Monthly	Total (mm)
Discharge	Jordan	River Jordan at Deganiah	1921–1938	Monthly	Average (m³ s⁻¹)
		River Jordan at Jisr-el-Mujami	1926–1930	Monthly	Total (m³ s⁻¹)
		River Jordan at Allenby Bridge	1932–1936	Monthly	Totals and average (m³ s⁻¹)

[a] US National Climatic Data Center: http://www.ncdc.noaa.gov/oa/ncdc.html
[b] Global Runoff Data Centre: http://www.gewex.org/grdc.html

Land surface elevation data were obtained from the Shuttle Radar Topography Mission (SRTM) Digital Elevation Model (DEM). The data set has a resolution of 90 m and a vertical accuracy of 15 m. These data were stored and manipulated within the Geographical Information System, ESRI ArcGIS. Land cover was taken from the Global Land Cover Map 2000 (version 1.1) downloaded from the HYDE land cover database (Klein Goldewijk, 2001). The resolution of the land cover maps was 1°.

10.3 THE MODELLING FRAMEWORK

An overview of the modelling framework is shown in Figure 10.2. It can be seen that the framework consists of hydrological models driven by daily rainfall time-series derived using a statistical rainfall model (weather generator), and climatological potential evaporation (see Chapter 5 for a justification of this).

10.3.1 Climate component of the modelling framework

The regional climate model (RCM) used in this study was HadRM3. This is a regional-scale climate model with a spatial resolution of 0.44° (~50 km) for both latitude and longitude developed by the UK Hadley Centre. As such, the model has a finer spatial scale than GCMs such as HadCM3 which has a spatial resolution of 3.75° by 2.5° for longitude and latitude, respectively. The RCM is formally compared with observations in Chapters 4 and 8 of this volume and summarised here. A comparison of the HadRM3P simulated near-surface air temperatures and precipitation depths is shown in Figure 8.2. For the control period, 1961–1990, the modelled temperatures in the eastern Mediterranean are reasonably well represented. The modelled control-period precipitation data show that whilst the spatial pattern of precipitation over the 30-year control period is generally modelled well, the resolution of the HadRM3 model is too coarse to capture the subtle variations in rainfall across the

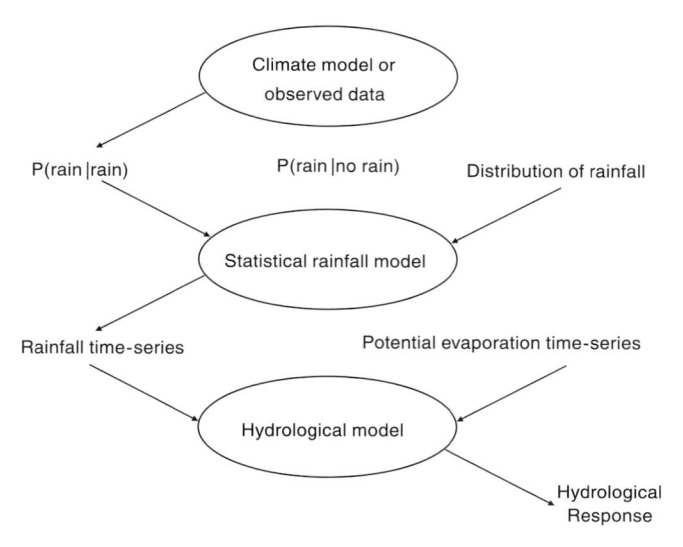

Figure 10.2. An overview of the modelling framework.

extension of the Rift Valley and northern Jordan. Moreover, the comparison with rain gauge measurements given in Figure 4.2 shows that the intensity of rainfall is underestimated, leading to the biases in the annual totals evident from Figure 8.2. Thus, these results highlight the need for the weather generator, both to interpolate the RCM simulations to a point and to correct the model bias (see Chapter 5 for further detail on this).

Climate projections were extracted from HadRM3 simulations of the 1961–1990 control and the 2071–2100 future periods. The A2 emission scenario was used for all the projections described in this chapter. This scenario represents a world with a slow technological response to mitigate climate change and where the economic differences between the industrial and developing worlds do not narrow (Nakicenovic et al., 2001). The greatest changes are expected in temperature and precipitation by the end of the century, so the scenario and the period considered represent a 'bad' case scenario. Current data suggest that global CO_2 emissions are following the 'worst' case A1F1 scenarios (le Quéré et al., 2009).

A weather generator was used for both the sensitivity studies and the future scenarios. The weather generator is described and evaluated in Chapter 5 of this volume, where it was shown that, while it was capable of reproducing the main features of the observed rainfall seasonal cycle, there were some biases – with more rain at the margins of the rainy seasons than is observed. The weather generator derives rainfall stochastically, according to some underlying patterns of daily rainfall. In the weather generator used here, the patterns of daily rainfall are described statistically through the mean rain per rainy day (rainfall intensity) and the probabilities of rain both given rain the day before (PRR) and given no rain the day before (PDR).

For the sensitivity studies, the rainfall probabilities were derived from the observed time-series and then adjusted for each case. For the future scenario integrations, the probabilities of rain given rain and of rain given no rain were determined from an RCM, for the control period, 1961–1990. The same probabilities were determined for the scenario period 2071–2100, and these projected future probabilities were bias-corrected using a factor derived by comparing the control model integration with the observed probabilities. The distribution of rainfall intensities (rain per rainy day) was based on the observed time-series, with an extra parameterisation for extreme rainfall events. The methods by which the rainfall probabilities and intensities were estimated from input rainfall time-series are described fully in Chapter 5.

10.3.2 Hydrological components of the modelling framework

The hydrological components of the model framework are the Integrated Catchments model of nitrogen (INCA-N version 1.11.10) and the Pitman rainfall-runoff model. INCA-N is a water quality model that incorporates a simple flow routing model that divides the main channel into a user-defined number of reaches. This model was then used to route water along the upper reaches of the River Jordan with a daily time resolution. To apply the INCA model it was necessary to determine the hydrologically effective rainfall (HER) in order to calculate the water volume contribution from the catchment each day. The HER is the rainfall that contributes to the river flow after evapotranspiration losses and replenishment of the soil moisture deficit are accounted for. In this study, HER was calculated using the Pitman rainfall-runoff model. Thus together the Pitman and INCA allowed the calculation of the runoff response to rainfall. The following sections describe Pitman and INCA in more detail.

The Pitman model is a conceptual, process-based model of the rainfall-runoff relationship (Pitman, 1973). The model was configured to simulate at a daily time-step the HER using daily rainfall calculated by the RCM and weather generator and the potential evaporation (PE) calculated using the Penman equation observed near-surface air temperatures for INCA calibration and RCM near-surface air temperatures as input for the control (1961–1990) and scenario (2071–2100) runs. For the sensitivity studies taken from the NCEP reanalysis (see Chapter 5). The Pitman model accounts for recharge of the soil moisture deficit, and the HER is estimated based on water contributing to the flow in the main channel and the water volume percolating to a deeper groundwater store and that lost through evaporation.

The INCA model is a catchment-scale, daily time-step hydrological and water quality model. In this application, only the hydrological model was used. The model is described in detail in Whitehead et al. (1998) and Wade et al. (2002). Briefly, the model is semi-distributed so that the catchment simulated is separated into a series of sub-catchments with HER input to each. Being semi-distributed, this HER is delivered instantaneously to the main channel rather than being routed using elevation differences, as would be done in a fully distributed model. Within each sub-catchment, different landscape units are specified according to soil, land use and geological types. The model has two reservoirs in each unit: one represents the flow of water through the unsaturated zone, incorporating the soil, and the other represents the groundwater. Different residence times can be associated, through user-defined model parameters, with each of the reservoirs and these residence times are further differentiated between landscape units. The unsaturated zone and groundwater boxes are differentiated using the base flow index which was developed in the UK to differentiate between catchments with permeable and impermeable geologies. For example, Cretaceous chalk and granite have indices of 0.95 and 0.2, respectively (Gustard et al., 1992).

Typically up to six different land use units are defined across all sub-catchments (Wade et al., 2002). The following land cover types, selected from the Global Land Cover Map 2000 (Global Land Cover Database, 2003), were included in the INCA application: broadleaved tree cover (open), shrub cover, cultivated, bare areas, inland water and urban. Shrub cover for the INCA application included shrub cover, both closed and open; deciduous; sparse herbaceous or sparse shrub cover; and regularly flooded shrub and/or herbaceous cover. In practice it was not possible to use these data to differentiate land cover types because of a lack of data to calculate the potential evaporation for each. Thus Pitman was set up for a compound single land cover, as was the INCA model.

The INCA water balance is computed on a 1 km × 1 km grid cell and this was then multiplied by the unit area in each sub-catchment to calculate the volume of water transferred from the unit to the main channel. Initially, the entire Jordan river basin was sub-divided into 19 reaches based on gauging stations and points just downstream of major confluences with tributaries and side wadi channels. The Dead Sea was included as a drainage basin, and of the defined reaches this had the largest drainage area of 49,000 km^2. During the study it became apparent that only limited daily time-step discharge data could be obtained. Some daily data were available for the lower Jordan River, namely on the main channel south of the Sea of Galilee at Naharayim, and monthly data were available at Allenby Bridge. Given that the purpose was to look at extremes in flow, this cannot be done with monthly flow data. The Naharayim gauging station is located downstream of the confluence of the Jordan and Yarmouk rivers, and whilst daily measurements are available for the period 1988–1993, the flows at this point are heavily modified by upstream abstractions. Thus, the INCA model was applied only to the upper four reaches defined in Figure 10.1 from the headwaters to the discharge gauging station at Obstacle

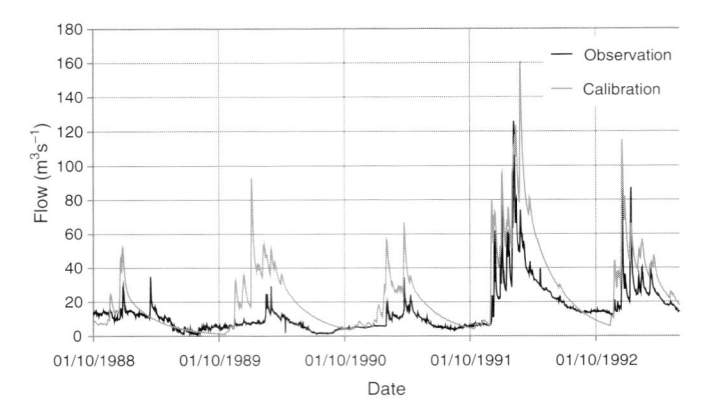

Figure 10.3. Modelled and observed daily mean flows in the Jordan river at Obstacle Bridge from 1 October 1988 to 30 September 1993.

Bridge in Israel, to use available daily data affected least by abstraction.

The INCA model was calibrated for the period 1 October 1988 to 30 September 1993. The purpose of the calibration was to determine the model parameters, and this period was chosen to provide the maximum overlap of available daily rainfall and flow data. The unsaturated and the groundwater zone residence times, the instream routing parameters that control the reach residence times and the baseflow indices were adjusted until the modelled output flow matched, as closely as possible, the observed flow time-series at Obstacle Bridge. The calibration was done using the Degania Bet rainfall time-series reduced by 55% to account for the actual evaporation (Kunstmann et al., 2005). There were insufficient data to perform a split-sample test.

The model performance was assessed using the R^2-value and the Nash–Sutcliffe criterion. The R^2-value for the calibration period was 0.7 and the Nash–Sutcliffe was negative. This result indicates that the pattern in the observed flows was simulated but the actual values were not replicated (Figure 10.3). This was due to an inability to quantify the volume of the abstractions in the upper Jordan. Despite this, the study is still useful as it provides an indication of how flow magnitude and frequencies will change relative to the present and the control period.

Once calibrated, the control-period precipitation data derived from HadRM3 and the weather generator were input to the Pitman model to estimate the HER for the control period. These HER data were input to INCA to derive a control-period flow series. This process was repeated but using the 2071–2100 scenario precipitation data, derived from the HadRM3 runs and the bias correction using the weather generator, to derive, through the Penman equation and the Pitman model, the HER for the INCA scenario run. Given the different calculation of the HER between the calibration and control-period run, there was a difference in the flows simulated between the calibration run and the control-period run.

10.4 THE SENSITIVITY OF THE RIVER JORDAN FLOW TO WEATHER STATISTICS

To investigate how the statistics of the weather affect flow, the hydrological modelling framework for the River Jordan described in the previous section was driven by observed daily time-series with modified statistics of the weather. The observed data were taken from Degania Bet, Israel, for the period 1984 to 2005, a site assumed to be representative of the upper Jordan. In all, eight climate scenarios were used:

- 1.5× probability of rain given no rain on the day before (herein referred to as PDR); 0.5× probability of PDR;
- 1.5× probability of rain given rain on the day before (herein referred to as PRR); 0.5× probability of PRR;
- 2× rain intensity; 0.5× rain intensity;
- longer duration rainy season; shorter duration rainy season.

This part of the study comprises two parts. In the first part of the study, the influence of the rainfall probabilities and their relationship with monthly rainfall totals is investigated. The second part describes how River Jordan flow was sensitive to modification of the underlying climate.

10.4.1 The influence of rainfall probabilities on River Jordan flow

There is considerable stochastic variability in rainfall, even given constant underlying statistics of the weather. To isolate the impact of changing individual rainfall statistics, instead of directly comparing the flow resulting from the climate scenarios described above, each month of each sensitivity study was considered individually. The rainfall probabilities, mean intensity and the resultant mean flow were calculated for each month. Using the sensitivity study data, rather than generating a long time-series with the observed statistics, ensured that a wide range of values of rainfall probabilities and intensities were created for analysis. Only the peak of the rainy season (November–February) was considered because in the margins of the season, there are too few rainy days to estimate meaningful rainfall probabilities for individual months.

The effect of changing the rainfall probabilities is shown in Figures 10.4 and 10.5. With respect to runoff, it is clear from these plots that the probabilities of rainfall occurrence are of secondary importance to the monthly rainfall totals. Nevertheless it can be seen that the PDR is important, with higher values associated with higher flows for a given rainfall total. In other words, a greater chance of rain given no rain leads to higher flow. This implies that the number of individual rainy events affects the monthly mean flow, so that a month with either only a few rainfall events or clumps of several consecutive rainy days is likely to have a lower mean flow (for a given rainfall total) than a month with many individual events.

This behaviour stems from the hot climate and the antecedent soil moisture conditions. Assuming a surface crust is not formed by the splash detachment of soil particles, rainfall will penetrate a dry soil more easily than one that is already wetted because there will be greater capillary suction and adsorption to dry soil surfaces. Darcy's law states that the rate at which water can infiltrate a permeable and unsaturated substance is a function of the hydraulic gradient and the vertical unsaturated hydraulic conductivity. On the first day of a rainy spell, assuming no antecedent soil moisture, the soil will be dry and the rainfall will infiltrate easily. On subsequent days, the soil will become progressively more difficult to infiltrate because the pressure force decreases as the wetting front descends, whilst gravity remains constant. The relationship between infiltration and soil moisture is nonlinear. The greatest rate of infiltration is observed at the initial transition from dry to a little wet. In practice, this means that the infiltration rate drops sharply between the first and second days of a rainy spell and then drops more slowly on subsequent days. The impact of the number of rainy spells on flow is thus greater than the length of rainy spells since the number of rainfall events has a greater impact on the period when the soil is wetted. The water which remains on the soil surface or in the channel and is not infiltrated will be evaporated and returned to the atmosphere. This is a particularly significant effect in regions where the climate is hot and the potential evaporation high, as in the upper Jordan.

If a surface crust forms, rain will not infiltrate the soil. Rather, infiltration-excess overland-flow will be generated and the water will flow downslope, possibly infiltrating in an unsaturated area. Typically this behaviour is observed on hill slopes in semi-arid and arid environments with colluviums found at the bottom of the hill slope. Infiltration and subsequent soil saturation resulting in saturation-excess overland-flow and infiltration-excess overland-flow are represented in the Pitman model, but in practice it is difficult to separate the two processes because of problems of parameter identification, and the complicating factor of transmission loss in the channel network.

These conclusions are based on hydrological modelling rather than observations, which raises the question of how applicable they are to the real world. This is particularly pertinent because, as has been described in previous sections, although the hydrological model has been carefully calibrated, it has some oversimplifications, particularly in the way it represents the soil. There are too few simultaneous observations of daily rainfall and flow to draw firm conclusions. However, comparison between flow and rainfall for high and low rainfall probabilities using the available data (four years) suggests that when the probability of rain given no rain is high, the mean flow for a given monthly total is higher than when the probability of rain

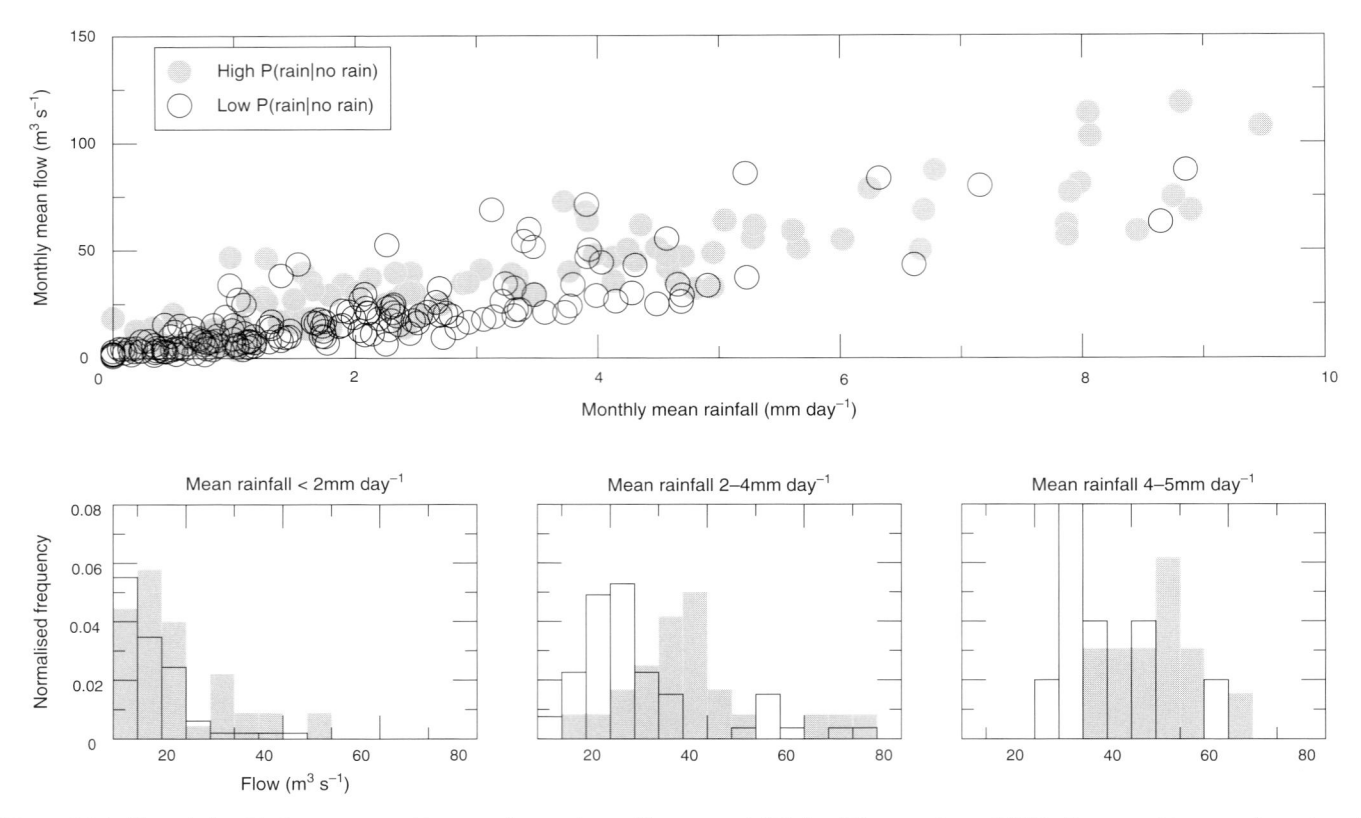

Figure 10.4. The relationship between monthly mean flow and monthly mean rainfall for different values of PDR. Top: monthly mean flow plotted against monthly mean rainfall. High PDR are filled circles and low PDR are unfilled circles. Bottom: histograms of flow for different ranges of monthly total rainfall for high PDR (filled bars) and low PDR (unfilled bars). The ranges are given on the figure.

given no rain is low. Specifically, when the PDR is high, the mean flow for rainfall between 2 and 4 mm day^{-1} is 15.0 m^3 s^{-1}; when PDR is low, the mean flow is 7.4 m^3 s^{-1}. The equivalent values for high and low PRR are 12.1 and 10.2 m^3 s^{-1}, respectively. Because of the lack of data, these results are not conclusive, but they are nevertheless consistent with the modelling.

10.4.2 The effect of changing the underlying climate on River Jordan flow

To assess the influence of changes in the underlying statistics of the weather on the flow of the River Jordan, Figure 10.6 presents the results of the hydrological modelling as flow duration curves, which show how the distribution of flows will change. A flow duration curve shows the percentage of time, for the observed or modelled period, that a given flow will be exceeded. As was stated above, although the underlying statistics for each case study are constant, stochastic variability causes considerable variation from month to month. This reflects the real world in which, for example, some months will be especially rainy, even in a future scenario where the mean climate is far drier than it is now.

The annual totals for each sensitivity study are different, and this explains some of the differences between the curves. For example, increasing PRR by a factor of 1.5 results in a mean total rainfall of 742 mm (compared with 352 mm observed); an equivalent increase in PDR increases the annual total to 578 mm. The difference in annual totals means that increasing PDR has less effect on the flow duration curve than increasing the PRR – despite the theoretically stronger influence of PDR.

In cases where the annual totals are similar, the impact of the individual daily statistics is evident. Figure 10.7 compares the flow duration curves in cases when the PRR and PDR are halved. It can be seen that, for low and moderate flows, reducing the PDR has a greater impact, with the flow lower for all time thresholds between 99% and 5%, than reducing the PRR. For example, the flow exceeded 95% of the time is 1.6 m^3 s^{-1} for the lower PDR case and 2.2 m^3 s^{-1} for the lower PRR case. For very high flows, the opposite is true with, for example, the flow exceeded 1% of the time being 48 m^3 s^{-1} for the lower PRR case and 52 m^3 s^{-1} for the PDR case. Although subtle, these differences, particularly for very high flows, are potentially relevant from an impacts point of view because they affect the frequency of floods. These results are consistent with Samuels *et al.* (2009), which showed that increasing the frequency of wet

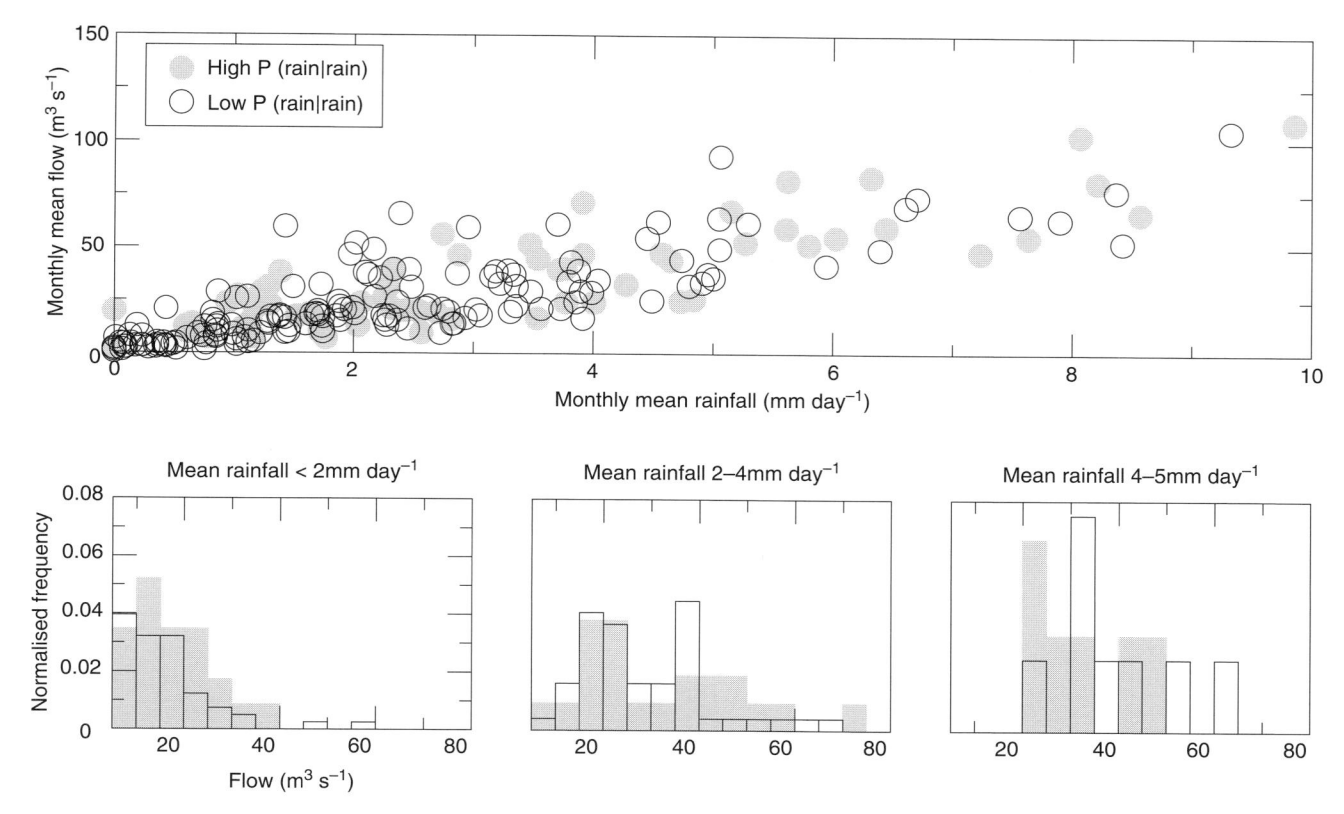

Figure 10.5. The relationship between monthly mean flow and monthly mean rainfall for different values of PRR. Top: monthly mean flow plotted against monthly mean rainfall. High PRR are shaded circles and low PRR are unfilled circles. Bottom: histograms of flow for different ranges of monthly total rainfall for high PRR (shaded bars) and low PRR (unfilled bars). The ranges are given on the figure.

spells (equivalent to increasing PRR) leads to more frequent and intense floods, even if the annual total rainfall is unchanged.

Figure 10.6 suggests some nonlinearities, which can be related to the climatological and hydrological features of the region. Increasing the intensity of rainfall has less effect on flow than decreasing it – an observation that cannot be related to differences in the changes in annual total. This is most likely to be because more intense rainfall leads to greater surface runoff. In reality, much of this additional runoff may re-infiltrate further downslope – an effect amplified by the drier climate and resulting lower antecedent soil moisture at lower elevations (Chapter 2 of this volume). This is neglected by the sensitivity studies, which relate to the impact of rainfall at a point on flow within a single reach. The hot climate means that a high proportion of water that is not infiltrated will be lost through evaporation.

The impact of changing the length of the seasonal cycle is also nonlinear, with a lengthening having far less effect on river flow than a shortening of the seasonal cycle. This is because the higher evaporation during the Middle East spring and early autumn (the margins of the rainy system) results in less rainfall entering the hydrological system than during the winter (the peak of the rainy season). In the sensitivity study, despite raising the annual total by 84 mm, lengthening the rainy season has little impact on flow; shortening the season, on the other hand, has a strong impact.

10.5 THE IMPACT OF ANTHROPOGENIC CLIMATE CHANGE ON RIVER JORDAN FLOW

The results of integrations using a regional model implemented for a large region encompassing southern Europe, the Mediterranean, North Africa and the Middle East are shown in Figure 4.4. It can be seen that there is a reduction in the total seasonal rainfall projected under the A2 scenario for the 2071–2100 period. The largest reductions (around 30% in the River Jordan region) are during December and January. At the margins of the rainy season, the signal is inconsistent and small with no significant changes in rainfall projected in some places and small increases in others. The overall picture is therefore of a longer rainy season with a less pronounced peak, with the total rainfall falling in the headwaters of the River Jordan projected to decrease. The reduction in rainfall is accompanied by an increase in temperature and hence evaporation.

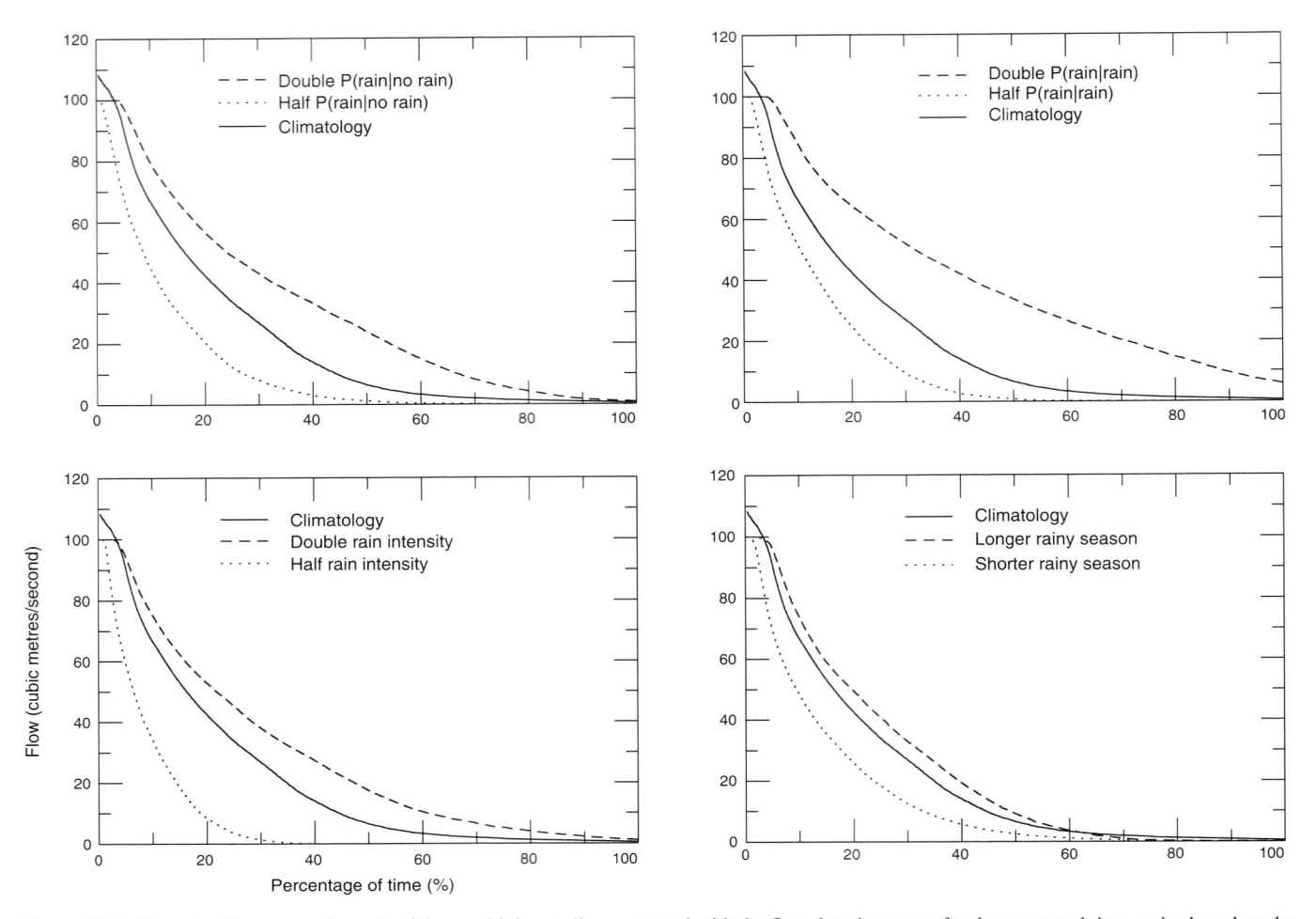

Figure 10.6. Flow duration curves for each of the sensitivity studies compared with the flow duration curve for the generated time-series based on the observed statistics of the weather (sensitivity studies labelled on the figure).

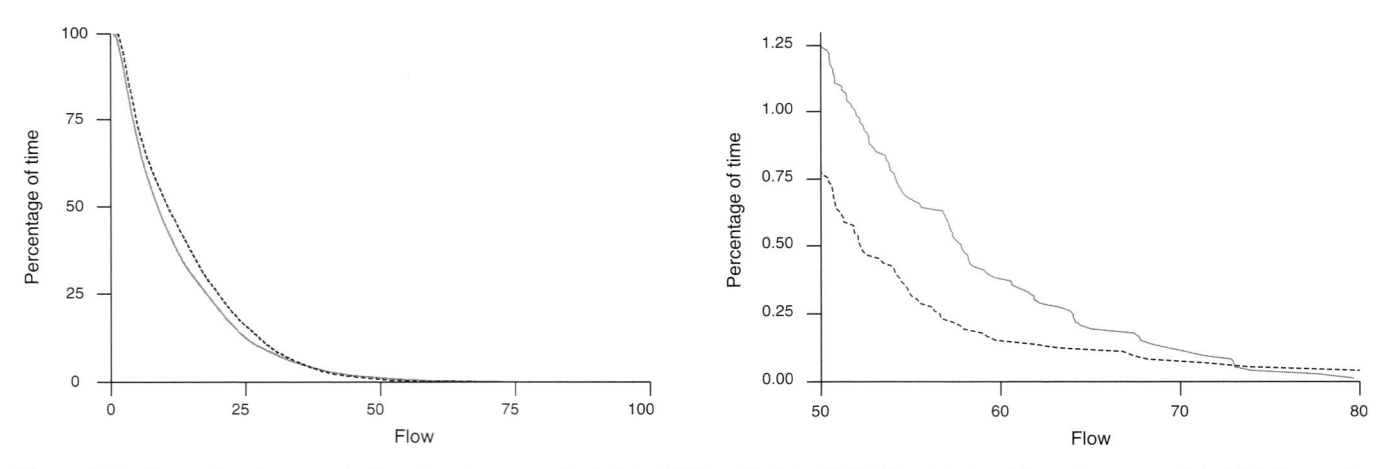

Figure 10.7. Comparison between the flow duration curves for halving PRR and halving PDR. The right-hand figure is a zoom of the high flow region of the left-hand figure, which shows all the data.

The sensitivity studies described in section 10.4 have ramifications for how the flow of the River Jordan is affected by climate variability and change; these can be applied to other catchments, such as Wadi Hasa and Wadi Faynan. It was shown that, although the annual rainfall total is the dominant factor influencing the River Jordan flow, the daily statistics of the weather are also important. The sensitivity of the River Jordan flow to the length of the seasonal cycle is particularly relevant

Table 10.2 *The mean monthly minimum and maximum near-surface air temperatures observed at Ramtha, Jordan, and those simulated by the HadRM3P regional climate model during a control run (1961–1990) and for the A2 scenario (2071–2100)*
The delta change values are calculated by taking the difference between the simulated temperatures for the scenario and control runs and adding this difference to the observation.

	Minimum temperature				Maximum temperature			
	Observed	Control	2050	ΔT_{min}	Observed	Control	2050	ΔT_{max}
Jan	4.1	3.3	6.3	7.1	13.0	14.6	18.6	17.0
Feb	4.3	4.1	8.0	8.2	14.2	16.6	20.7	18.3
Mar	6.0	7.6	10.7	9.1	17.5	21.4	24.5	20.6
Apr	9.4	11.0	14.4	12.8	23.4	26.2	29.6	26.8
May	12.6	14.2	18.8	17.2	28.3	29.9	35.2	33.6
Jun	15.4	16.4	21.3	20.3	30.9	32.4	38.0	36.5
Jul	17.6	18.9	24.2	22.9	32.2	35.0	41.1	38.3
Aug	18.0	19.3	25.2	23.9	32.6	35.4	41.5	38.7
Sep	16.3	16.7	21.8	21.4	30.8	31.7	36.2	35.3
Oct	13.6	14.2	18.3	17.7	26.9	28.3	32.1	30.7
Nov	9.5	10.1	14.9	14.3	20.7	22.6	26.9	25.0
Dec	6.0	5.6	9.0	9.5	15.1	16.8	20.5	18.8

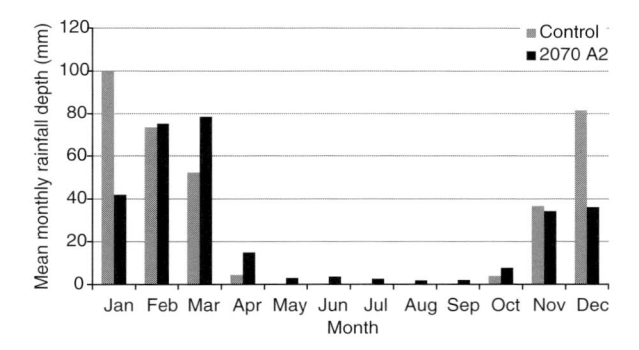

Figure 10.8. Projected changes in the monthly rainfall totals at Degania Bet, Israel, from the HadRM3 and weather generator models for 2070–2100 under the SRES A2 scenario.

because in the future scenarios presented in Chapter 4, and summarised by Figure 10.8, the rainy season is predicted to become longer, which partially offsets the marked decrease in rainfall projected at the peak of the rainy season. These results suggest that the flow will be even more reduced in the future than would be expected from the reduction in annual total rainfall, because the warmer temperatures (Table 10.2) at the margins of the rainy season lead to greater evaporation.

Although the climate change signal in winter rainfall can be related to changes in the large-scale circulation and is predicted by most climate models (see Chapter 4), the same cannot be said for the spring rainfall, which leads to large uncertainties in the predictions of rain in this season. However, the sensitivity studies carried out here imply that changes in spring rainfall have relatively little impact, and hence the uncertainties in our predictions of spring rainfall do not prejudice the reliability of the predictions of flow. The sensitivity of flow to the number of rainy days is also relevant to the climate change scenarios. At the peak of the rainy season, the number of rainy days is projected to decrease, reflecting reductions in both the PRR and PDR. The sensitivity studies suggest that this change from relatively frequent rainy events lasting a few days to less frequent, shorter events is likely to affect the flow strongly.

In comparison with the control period, the modelled outcome for the 2071–2100 A2 scenario is that the low (base) flows will remain similar to those occurring at present; there is little difference in the forecast flows at Q50 (the flow exceeded 50% of the time; Figure 10.9). The Q50 values in the control and scenario periods are 15 and 12 $m^3\ s^{-1}$, respectively. This lack of response is a result of the long residence time in the groundwater component of the INCA model, which suggests that groundwater acts to buffer changes in the rainfall amounts to maintain the low and intermediate flows. The flood response is different. There is a drop in the Q10 flow, namely the flow exceeded 10% of the time, from 76 to 57 $m^3\ s^{-1}$ between the control and scenario periods as a result of the reduced winter rainfall, and this indicates that flood magnitudes will be reduced. Increases in the flow extremes, in terms of flood magnitude and occurrence, are not evident – perhaps as a result of the assumptions that went into the parameterisations of extremes in the weather generator (which was used to drive the model).

The sensitivity studies provide some insight into the effect of climate model bias in individual statistics of the weather on the estimates of future River Jordan flow. Chapter 4 showed that although the rainfall probabilities are well simulated, the climate model did a poor job of simulating rainfall intensity; for this

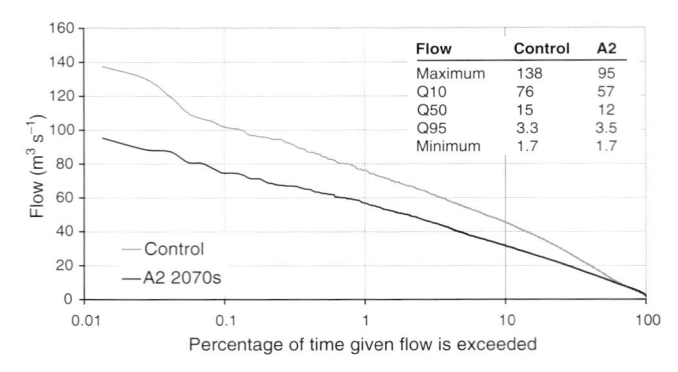

Figure 10.9. Modelled daily mean flows in the Jordan river at Obstacle Bridge for control (1961–1990) and scenario (2071–2100) periods.

Flow	Control	A2
Maximum	138	95
Q10	76	57
Q50	15	12
Q95	3.3	3.5
Minimum	1.7	1.7

reason, the projections of future intensity were inconclusive. Alpert et al. (2008) have suggested that in the future there will be a tendency towards more intense rainfall events. The sensitivity studies described in this chapter suggest that these intense rainfall events would not mitigate the reductions in flow because the additional rainfall would mostly be evaporated before entering the hydrological system. The nonlinearity in the response of the flow to rainfall intensity implies that, although increases in intensity have little impact, any reduction in intensity would significantly reduce the flow. This suggests that the projections of flow based on present-day rainfall intensity are an upper limit, with the possibility of the flow being markedly lower if the rainfall intensity drops. In contrast to Alpert et al. (2008), the future climate projections in Chapter 4 hint at a drop in intensity, although the noisy data and model error inhibit the conclusions that can be drawn.

10.6 DISCUSSION

The RCM projections suggest that in a world that does not work to find an integrated way to reduce greenhouse gas emissions then a temperature increase of 2 °C and a decrease in rainfall by 25% is projected in the eastern Mediterranean by the end of this century. In addition to a reduction in the mean annual rainfall, the seasonality of the rainfall will change also, as the start and end of the wet season are projected to become wetter but there will be less rainfall in December and January. This reduction in rainfall and increase in near-surface air temperatures suggest that irrigation requirements will increase, worsening the water shortage in the region. This suggestion is supported by preliminary applications of the CROPWAT model in the Water, Life and Civilisation study and applications of a soil–vegetation–atmosphere transfer (SVAT) model TRAIN, which indicate increases in evapotranspiration and water demand (Menzel et al., 2009). The preliminary predictions of the CROPWAT model suggest that at Ramtha in northwest Jordan the irrigation demand will increase from 62 to 132 mm of water when growing vegetables under the A2 scenario for 2071–2100 using HadRM3 and an assumed irrigation efficiency of 70%. The TRAIN model provides an overview of the Jordan Valley region, and the modelled outcomes suggest a 6% increase in the water demand for agriculture over the entire region and up to a 50% decrease in water availability in northwest Jordan, Israel and the West Bank (HadCM3, A1B scenario, 2021–2050 compared with 1961–1990 control period). Menzel et al. (2009) note that this region includes the Negev, which is already water-scarce, a factor that will in effect lessen the future projection of water demand since water is already extremely scarce there. These preliminary results highlight the local and regional differences that might be expected in irrigation demand and do not account for the possibility that the crop stomata may close in response to increased near-surface air temperatures, resulting in little difference in crop evapotranspiration, but lower yields due to the increased crop stress of an increased canopy temperature (Kimball and Bernacchi, 2006). An overall increase in local and regional irrigation demand has serious implications for Jordan since further stress will be put on the groundwater resource which, whilst the extent of the resource has not yet been quantified, could increase salinity. Israel has already invested in desalination of groundwater. Jordan may have to do likewise.

The modelling framework proposed has the same uncertainties as outlined by Wilby and Harris (2006). These uncertainties in model application are: the choice of SRES scenario; the subsequent regional climate projections; the probabilities and rainfall intensity distribution chosen for the weather generator; the structure and calibration of the hydrological models; and the sampling errors of the observed data used to define the structure and parameters of the model ensemble. In particular, there are limited daily flow data with which to calibrate and test the hydrological models, not only in the upper Jordan but also in the side wadis and other tributaries that comprise the Jordan drainage network. As such, this and other model chains cannot provide absolute changes in the rainfall-runoff response but rather give an indication of the distribution of flows. Further complications in the case of the Jordan river include an inability to quantify exactly the volume of water abstracted from different reaches, because of the numerous and diffuse nature of the abstractions. It is therefore difficult to separate this effect from that of changes in the groundwater contribution to the surface water flows.

This is one of the first studies to link RCM output with a weather generator to correct bias in the RCM projections, given that for many hydrological applications the catchment of interest is smaller than the cell size of the RCM. Further investigation is required to determine if the use of a weather generator approach introduces bias itself, as suggested here where more rain was predicted at the margins of the rainy season than observed. A potentially fruitful method for progress in the development of coupled climate–hydrological assessments would be the determination of what aspects of the climate or weather are most critical to the hydrological assessment. The climate and weather projections from a GCM, RCM, weather generator or a combination of these could then be tested, in terms of these aspects, against observation to assess reliability. In addition, further work is required to assess the ability of climate models to predict the frequency and duration of rainfall events, as well as the magnitude, in order to better determine antecedent moisture conditions and groundwater recharge.

Whilst coupling uncertainties together means that quantification of the runoff response is likely to be inaccurate, the modelling framework can be used to explore the hydrological system and assess the impact of different scenarios and management decisions. Thus, the methodology goes beyond an assessment of rainfall changes alone and allows us to begin to think about which stores and pathways in the hydrological system will change. At present, this study is limited in that only rainfall and runoff are considered (in the upper Jordan). Further work is required to understand the implications of the projected temperature and precipitation changes on other aspects of the water resource, such as groundwater recharge and soil moisture availability. This task has been started by Menzel et al. (2009) but further work is required to substantiate the initial projections of change.

It is recommended that an ensemble approach to climate and hydrological modelling be taken to account for structural uncertainty. In particular, it is useful that a number of studies do the same thing so that results can be compared (for example Samuels et al., 2009). Further applications of the framework proposed here and other coupled climate–hydrological approaches will allow a more extensive review of the potential outcomes to population and climate change. With such an approach, care must be taken to use the same SRES emission scenarios, time periods and spatial scale of comparison. This will require discussion and cooperation between hydrological modellers of the nature already achieved in the ENSEMBLES climate-modelling project (http://www.ensembles7eu.org).

The change in the low flows will depend on the volume of water stored in the karst system of the northern Jordan Valley and its spatial extent and recharge rate. The modelled outcomes from the conceptual Hydrological Model for the Karst Environment (HYMKE), described in Rimmer and Salingar (2006), corroborate the results from this study that the low flows will not change. It is unclear over what simulated time period the HYMKE model was run for the modelled scenarios. The INCA model was run for 30 years with a daily time-step, so this may be sufficient to examine long-term trends, but the relationship between groundwater recharge through percolation and soil moisture has not been modelled in detail. Thus, it is proposed that it will be necessary to run transient scenarios from present day to 2100 to see how the groundwater will change over the long term using both the HYMKE and INCA models, and that further consideration be given to the likely groundwater recharge mechanisms to determine if the current groundwater components of both models are a good representation of water storage and flow in the karst.

The results of this study suggest that the impact of rainfall decreases on flow may be, to a degree, mitigated by the contribution of groundwater. However, the combined effects of expected population increase and the changes in the projected climate are yet to be modelled. Whilst the groundwater levels appear to be maintained in response to climate change alone, it is likely that they will decrease if the population increases. Water security in Jordan, and Israel and the West Bank, will probably depend on how exploitable the groundwater reserves prove to be.

10.7 CONCLUSIONS

This study is one of the first to combine an RCM, a weather generator and hydrological models to project the likely rainfall-runoff response of the upper Jordan river, framing the results within other contemporary research. A substantial dataset has been collated to develop and test the modelling framework and in a data-poor region this represents a substantial undertaking. The modelled results provide a contribution to the debate about how the runoff response will change in the upper Jordan. Moreover, the methodology developed here of using sensitivity studies as an adjunct to future scenario projections could usefully be adapted for use in environmental and resource planning.

The results highlight how model bias, and omissions from the model framework, affect the projections of flow. Moreover, because of the uncertainties associated with the chosen greenhouse gas emissions scenario, the RCM, the weather generator and the hydrological models, the results can only be assumed indicative at this stage. Nevertheless, the modelled outcomes suggest that, although the mean annual flow of the River Jordan will reduce, the base flow of the upper Jordan will not change significantly in response to climate change – a result corroborated by other comparable studies.

REFERENCES

Alpert, P., S. Krichak, H. Shafir, D. Haim and I. Osentinsky (2008) Climatic trends to extremes employing regional modeling and statistical interpretation over the E. Mediterranean. *Global and Planetary Change* **63**: 163–170.

Alpert, P., I. Osetinsky, B. Ziv and H. Shafir (2004) Semi-objective classification for daily synoptic systems: application to the eastern Mediterranean climate change. *International Journal of Climatology* **24**: 1001–1011.

Arnell, N. W. (2003) Effects of IPCC SRES emissions scenarios on river runoff: a global perspective. *Hydrology and Earth System Sciences* **7**: 619–641.

Beven, K. and J. Freer (2001) Equifinality, data assimilation and uncertainty estimation in mechanistic modelling of complex environmental systems using the GLUE methodology. *Journal of Hydrology* 249: 11–29.

Dibike, Y. B. and P. Coulibaly (2005) Hydrologic impact of climate change in the Saguenay watershed: comparison of downscaling methods and hydrologic models. *Journal of Hydrology* 307: 145–163.

Global Land Cover 2000 database (2003) The Global Land Cover Map for the year 2000. GLC2000 database, European Commission Joint Research Centre. http://www-gem.jrc.it/glc2000.

Gustard, A., A. Bullock and J. M. Dixon (1992) Low flow estimation in the United Kingdom 1992. *Institute of Hydrology Report 108*. Wallingford: Institute of Hydrology.

Hoff, H., H. Küchmeister and K. Tielbörger (2006) The GLOWA Jordan River project – Integrated Research for Sustainable Water Management. *International Water Association Water Environment Management Series* 10: 73–80.

IPCC Core Writing Team, R. K. Pachauri and A. Reisinger (2007) Summary for Policymakers. In *Climate Change 2007: Synthesis Report. Contribution of Working Groups I, II and III to the Fourth Assessment Report of the Intergovernmental Panel on Climate Change*. Geneva: IPCC.

Kimball, B. A. and C. J. Bernacchi (2006) Evapotranspiration, canopy temperature and plant water relations. *Ecological Studies* 187: 311–324.

Klein Goldewijk, K. K. (2001) Estimating global land use change over the past 300 years: the HYDE database. *Global Biogeochemical Cycles* 15: 417–434.

Kunstmann, H., P. Suppan, A. Heckl and A. Rimmer (2005) Combined high resolution climate and distributed hydrological simulations for the eastern Mediterranean/Near East and the upper Jordan catchment. In *Conference on International Agricultural Research for Development (Tropentag)*, Stuttgart-Hohenheim. www.tropentag.de/2005/abstracts/full/461.pdf.

Lange, J., C. Liebundgut and A. P. Schick (2000) The importance of single events in arid zone rainfall-runoff modelling. *Physics and Chemistry of the Earth Part B – Hydrology Oceans and Atmosphere* 25: 673–677.

le Quéré, C., M. R. Raupach, J. G. Canadell *et al.* (2009) Trends in the sources and sinks of carbon dioxide. *Nature Geoscience* 2: 831–836.

Menzel, L., J. Kock, J. Onigkeit and R. Schaldach (2009) Modelling the effects of land-use and land-cover change on water availability in the Jordan River region. *Advances in Geosciences* 21: 73–80.

Messager, C., H. Gallee, O. Brasseur *et al.* (2006) Influence of observed and RCM-simulated precipitation on the water discharge over the Sirba basin, Burkina Faso/Niger. *Climate Dynamics* 27: 199–214.

Mileham, L., R. Taylor, J. Thompson, M. Todd and C. Tindimugaya (2008) Impact of rainfall distribution on the parameterisation of a soil-moisture balance model of groundwater recharge in equatorial Africa. *Journal of Hydrology* 359: 46–58.

Mileham, L., R. G. Taylor, M. Todd, C. Tindimugaya and J. Thompson (2009) The impact of climate change on groundwater recharge and runoff in a humid, equatorial catchment: sensitivity of projections to rainfall intensity. *Hydrological Sciences Journal – Journal des Sciences Hydrologiques* 54: 727–738.

Nakicenovic, N., J. Alcamo, G. Davis *et al.*, eds. (2001) *IPCC Special Report on Emissions Scenarios (SRES)*. Geneva: GRID-Arendal.

Oreskes, N., K. Shrade-Frechette and K. Belitz (1994) Verification, validation and confirmation of numerical models in the Earth Sciences. *Science* 263: 641–646.

Oroud, I. M. (2008) The impacts of climate change on water resources in Jordan. In *Climatic Changes and Water Resources in the Middle East and North Africa*, ed. F. Zereini and H. Hotzl. Berlin and Heidelberg: Springer pp. 109–123.

Pitman, W. V. (1973) A mathematical model for generating river flows from meteorological data in South Africa. Hydrological Research Unit Report 2/73. Johannesburg, South Africa: University of the Witwatersrand.

Puri, S. (2001) *Internationally Shared (Transboundary) Aquifer Resource Management: Their Significance and Sustainable Management*. IHP-VI Series in Groundwater No. 1. Paris: UNESCO.

Puri, S. and A. Aureli (2005) Transboundary aquifers: a global program to assess, evaluate, and develop policy. *Ground Water* 43: 661–668 (doi:10.1111/j.1745-6548.2005.00100.x).

Puri, S., B. Appelgren, G. Arnold *et al.* (2001) *Internationally Shared (Transboundary) Aquifer Resources Management, their Significance and Sustainable Management: A Framework Document*. IHP-VI, International Hydrological Programme, Non Serial Publications in Hydrology Sc-2001/WS/40. Paris: UNESCO.

Rimmer, A. and Y. Salingar (2006) Modelling precipitation-stream flow processes in Karst basin: the case of the Jordan River sources, Israel. *Journal of Hydrology* 331: 524–542.

Samuels, R., A. Rimmer and P. Alpert (2009) Effect of extreme rainfall events on the water resources of the Jordan River. *Journal of Hydrology* 375: 513–523.

Suppan, P., H. Kunstmann, A. Heckl and A. Rimmer (2008) Impact of climate change on water availability in the Near East. In *Climatic Changes and Water Resources in the Middle East and North Africa*, ed. F. Zereini and H. Hotzl. Berlin and Heidelberg: Springer pp. 45–57.

United States Department of Agriculture (2006) *Jordan – Grain and Feed Report*. Annual 2006 Global Agricultural Information Network (GAIN) Report JO6009.

United States Statistics Division (2010) http://unstats.un.org/unsd/default.htm, retrieved January, 2010.

US Geological Survey (1998) *Overview of Middle East Water Resources: Water Resources of Palestinian, Jordanian, and Israeli Interest*. Report compiled by the US Geological Survey for the Executive Action Team, Middle East Water Data Banks Project (EXACT). ISBN 0-607-91785-7.

Wade, A. J., P. Durand, V. Beaujouan *et al.* (2002) Towards a generic nitrogen model of European ecosystems: INCA, new model structure and equations. *Hydrology and Earth System Sciences* 6(3): 559–582.

Whitehead, P. G., E. J. Wilson and D. Butterfield (1998) A semi-distributed Integrated Nitrogen model for multiple source assessment in Catchments (INCA): Part 1 – model structure and process equations. *Science of the Total Environment* 210/211: 547–588.

Wilby, R. L. and I. Harris (2006) A framework for assessing uncertainties in climate change impacts: low-flow scenarios for the River Thames, UK. *Water Resources Research* 42: W02417.

Wilby, R. L., J. Troni, Y. Biot *et al.* (2009) A review of climate risk information for adaptation and development planning. *International Journal of Climatology* 29: 1193–1215.

11 Modelling Dead Sea levels and rainfall: past, present and future

Paul Whitehead, Dan Butterfield, Emily Black and David Plinston

ABSTRACT

The Dead Sea has played a crucial role in the past development of communities in the Jordan Valley, as evidenced by the wide range of archaeological sites close to the sea or potential old sea shorelines. There is also considerable debate concerning how levels have changed over the recent past and also how water resources in the Jordan Valley will be managed in the future. Over the past 50 years there has been a significant reduction in the level of the Dead Sea driven by abstractions from the Jordan River, the main source of fresh water to the sea. Falling levels have created problems for the tourism industry, and there are plans to restore levels using a water transfer from the Red Sea or the Mediterranean to the Dead Sea. A new model of the Dead Sea levels is described, based on historical rainfall and level data from 1860 to 1960. The model is used to simulate the impacts of abstractions on Dead Sea levels that have resulted in a 45-m reduction in levels since the 1960s. The model is also used to evaluate the impacts of future climate change: it is shown that the projected changes in rainfall have a far smaller impact on Dead Sea levels than do the abstractions. The model also shows that the only way to avoid this problem is to transfer water into the Dead Sea from either the Red Sea or the Mediterranean. The model suggests that this would be an effective way of restoring Dead Sea levels and could be used to balance the losses due to abstractions. Finally, the Dead Sea level model provides a method for estimating historical rainfall back to 25 ka BP, and the results indicate significantly higher rainfalls during the Early Bronze Age and particularly during the time before 9 ka BP when Lake Lisan was present. Rainfall is estimated to be approximately double the current rainfall during this period.

11.1 INTRODUCTION

The Dead Sea is a major feature of the landscape of the Near Middle East region (Figures 11.1 and 11.2) and has evolved from ancient water bodies as far back as the Neogene period, 23 million years ago. From 70,000 to 10,000 years ago the lake was 100 m to 250 m higher than the current level, and this ancient lake, called 'Lake Lisan', fluctuated greatly. It rose to its highest level around 25,000 years ago, indicating a wet climate in the Near Middle East. Soon after 10,000 years ago the lake level dropped steeply, probably to levels even lower than today (Chapter 9, this volume). During the past 25,000 years the Dead Sea and Lake Lisan have experienced some significant drops and rises, as shown in Figure 11.3 (Enzel et al., 2003; Chapter 9, this volume).

Over the past 50 years the change in Dead Sea level has been even more marked, with levels falling by 45 m. The level is now 423 m below sea level (Aqaba datum). This rapid decline has been caused by abstractions of water from the Jordan and Yarmouk rivers upstream as well as direct abstractions from Lake Kinneret (Greenbaum et al., 2006; Haddadin, 2006). There have also been abstractions from many of the groundwater systems that feed the upstream rivers or supply the sea directly. The rise and fall in the Dead Sea levels has always had an impact on the mineralogy and water quality of the sea, driven by the changing salinity of the Dead Sea (Gravrieli and Stein, 2006). The salinity increases as the water evaporates and decreases as the freshwater inflows rise. In recent years, tourism in the Dead Sea area has suffered, with hotels marooned some distance from the Dead Sea shore as the level has fallen. In addition, sink holes have formed as salt bridges have collapsed around the shores of the Dead Sea, making walking in the area extremely hazardous (Abelson et al., 2006). Industry located on the shores of the Dead Sea, such as the extensive potash and salt extraction industry in the southern basin, is also suffering as water levels drop and the costs of pumping the water to the industrial site increase (Haddadin, 2006).

Water, Life and Civilisation: Climate, Environment and Society in the Jordan Valley, ed. Steven Mithen and Emily Black. Published by Cambridge University Press.
© Steven Mithen and Emily Black 2011.

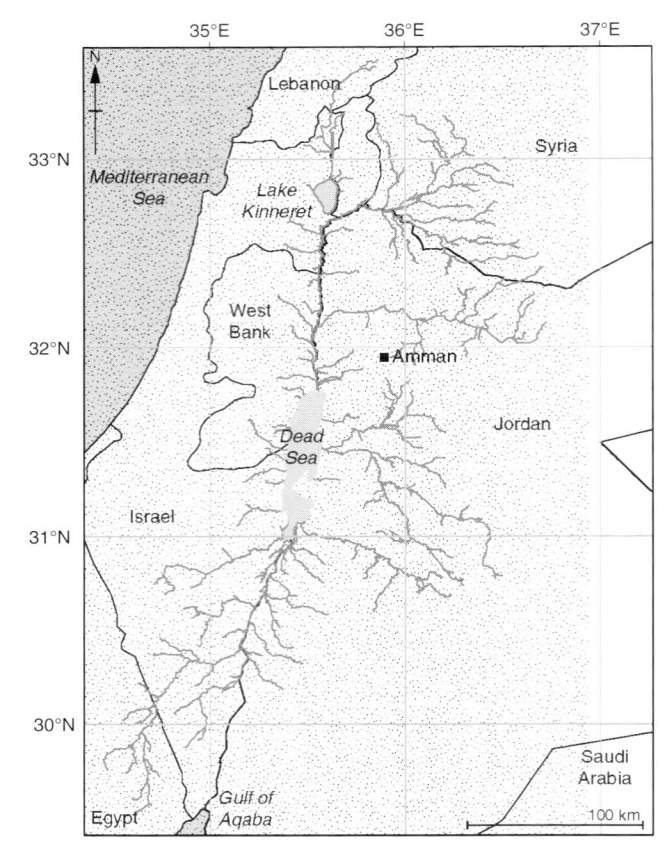

Figure 11.1. Map Showing the Dead Sea catchment area, with Lake Kinneret in the north.

Figure 11.2. A GIS representation of the digital terrain of the Jordan Valley and the Dead Sea.

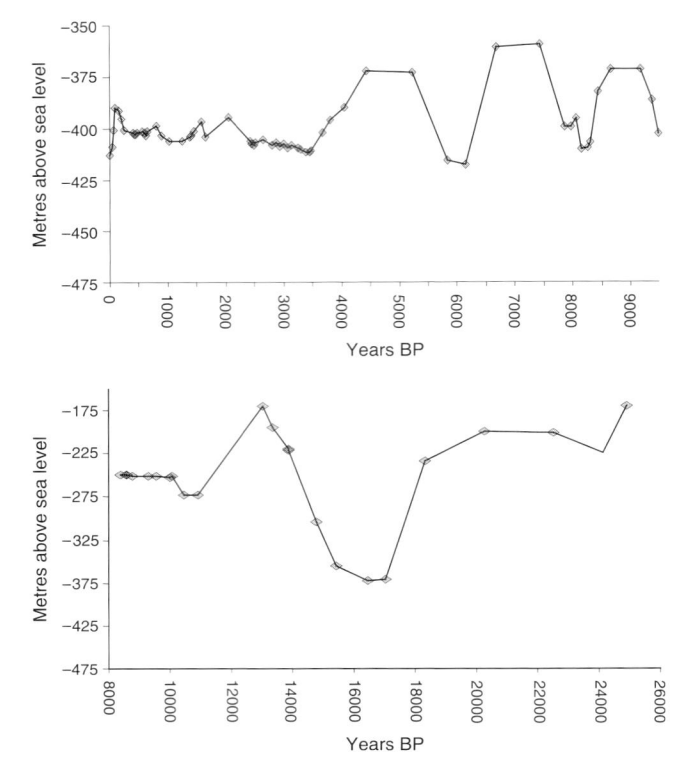

Figure 11.3. Estimates of changing Dead Sea levels over the past 25,000 years (Enzel *et al.* 2003 – upper graph; Black *et al.*, Chapter 9 of this volume – lower graph).

Future change in the Near Middle East region will be significant, with water rights becoming a major issue in future cultural and political discussion (Lipchin, 2006). However, water abstractions are likely to continue, and this is in addition to potential future climate change (Chapter 4, this volume). It is now accepted that human-induced climate change is occurring (IPCC, 2007), and there is now the question of how climate change will affect local rainfall patterns, river flows, abstractions and Dead Sea levels. Adapting to climate change will become a major issue in the Middle East, especially for water resource and water quality managers (Whitehead *et al.*, 2006). One strategy to cope with a declining Dead Sea and future climate change may be to consider water transfers from either the Mediterranean or the Red Sea (Al-Weshah, 2000; Lipchin, 2006).

In this chapter, we describe a model that simulates the Dead Sea levels driven by observed rainfall data. The model is used to simulate historical observed levels as well as the impacts of abstractions since the 1960s. The potential impacts of climate change on the water balance are explored, and we also investigate the potential impacts of water transfer schemes to restore the Dead Sea levels. Finally, we use the level model to recreate the historical rainfall in the Jordan Valley back to 25 ka BP. This is the first time that it has been possible to reconstruct the climate of this important Middle East region from field data.

11.2 THE DEAD SEA LEVELS, GIS AND SHORELINES

The Dead Sea acts as a terminal lake for the Jordan river and the other tributaries that arise in the Jordan Valley catchment, as shown in Figure 11.1. The total catchment area is 42,200 km (Greenbaum *et al.*, 2006) and the catchment extends from 100 km to the south of

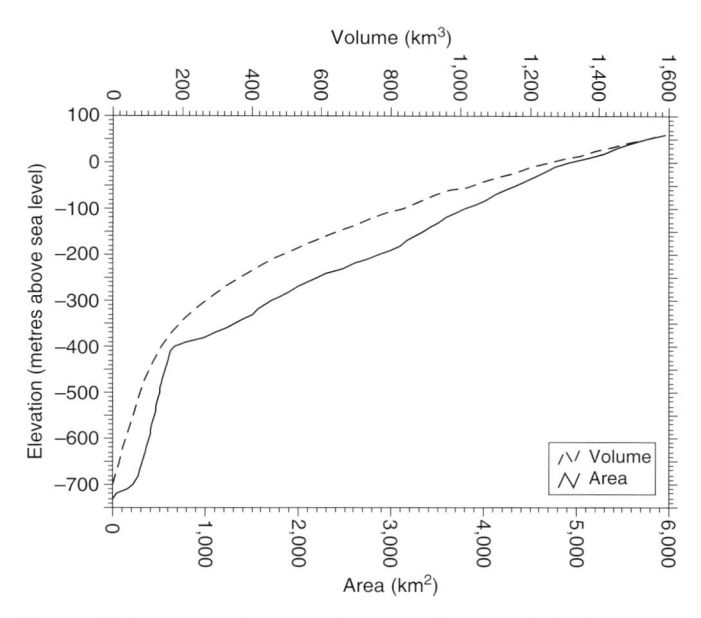

Figure 11.4. Dead Sea hypsometric curves showing relationships between sea elevation, surface area and volume.

the Dead Sea, near Eilat, up to Mount Herman in the north. As shown in the three-dimensional GIS (Geographic Information System) map of the region in Figure 11.2, the sea lies in a complex deep valley which is an extension of the Rift Valley with the steep sides adjacent to the Dead Sea. To the south is desert and from the north the Jordan and Yarmouk rivers feed the Dead Sea. In addition there are many wadi systems and groundwaters that supplement these river flows. As suggested in Figure 11.2 with the complex structure of the Jordan Valley, the bathymetry of the Dead Sea is also complex and gives rise to nonlinear relationships between levels, volumes and surface area. These are shown in Figure 11.4 (Hall, 1996) and indicate major changes in surface areas and sea volumes as the sea elevations change. Figure 11.3 shows estimated levels of the Dead Sea over the past 25,000 years and indicates a fluctuating sea level with much higher levels in the distant past (Enzel et al., 2003; Chapter 9, this volume), reflecting a wetter climate. These historical levels are important from an archaeological perspective as at each new level a different shoreline is exposed, which has provided locations for human settlement. For example Figure 11.5 shows the shoreline for the 170 m elevation. The 170 m level broadly equates to Lake Lisan at the Last Glacial Maximum at approximately 25 ka BP, and, as shown in Figure 11.5, the shoreline is extensive, covering the whole valley beyond the Dead Sea and incorporating Lake Kinneret to the north. The Dead Sea shorelines have changed significantly over time and the relationship of these to the archaeology of the Jordan Valley is explored in detail in a later chapter of this volume (Chapter 15).

As previously mentioned, the natural discharges from all sources are very difficult to estimate, given the large number of ungauged rivers, wadi systems and groundwater flows. However, the water balance has been estimated by several authors (Klein, 1998; Greenbaum et al., 2006; Haddadin, 2006) with the natural River Jordan flow at the inlet to the Dead Sea estimated to be of the order of 1,100 million m per year. In the past 50 years this inflow has been significantly reduced as water has been diverted for public water supply and irrigation (Klein, 1998; Haddadin, 2006). The volume abstracted upstream of the Dead Sea is difficult to estimate because of the large number of abstractions and irrigation schemes with few measured flows. However, Klein (1998) estimated the total abstractions by the end of the 1990s to be of the order of 890 million m^3 per year, which leaves only 210 million m^3 per year to discharge into the Dead Sea. With the surface area of the Dead Sea of approximately 820 km^2 in the 1990s and the high evaporation rates from the open water, the loss of water by evaporation from the sea surface has exceeded the reduced inflows, and hence sea levels have fallen significantly in recent years, as shown in Figure 11.6.

The elevation data from the 1860s to the 1960s, as shown in Figure 11.6 and Table 11.1, have been derived by Klein (1961) from a wide ranges of sources with 1860 levels at approximately 396 m below sea level (Aqaba datum) and a recent highstand of 388.5 m below sea level in the 1990s. The levels fell in the early part of the twentieth century because of below-average rainfall, but since the 1960s the levels have fallen rapidly because of the upstream abstraction of water. The abstractions have continued to the present time and the Dead Sea levels have continued to decline at a rate of 1 m per year to the current level of 423 m below sea level, as shown in Figure 11.6.

11.3 MODELLING THE DEAD SEA LEVELS

Modelling has become an established technique used by meteorologists, hydrologists, environmental chemists and biologists to analyse rivers, lakes and water resource systems. Modelling involves the representation of the system using equations coded as a computer program to simulate the behaviour of the system under study. There are a number of different approaches, and Wheater (2005) describes a range of alternative approaches to modelling hydrological systems in arid zones, including metric or data-based models, conceptual models which seek to represent the dominant processes operating in a catchment, and physics-based models, which try to describe the complex nature of the physical processes and stochastic approaches. Similarly for lake models, there are a wide range of approaches from empirical relationships through intermediate complexity models (Chapra, 1997; Whitehead et al., 1998) to complex physics-based models such as the DYRESM model (Gal et al., 2009). The major problem with the more physics-based models is that considerable

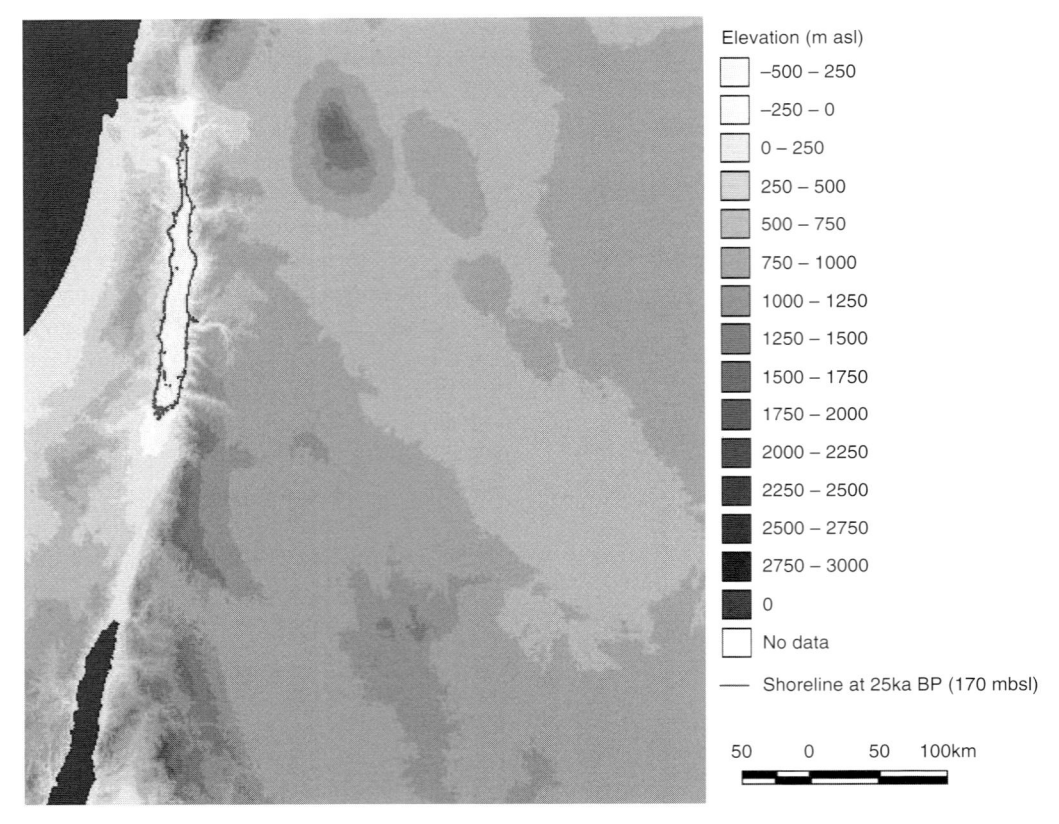

Figure 11.5. The shoreline of the Dead Sea reconstructed for four different depths. The 170 m depth equates to that at the Last Glacial Maximum, which occurred at approximately 20 cal. ka BP based on the palaeoenvironmental evidence summarised in Robinson *et al.* (2006).

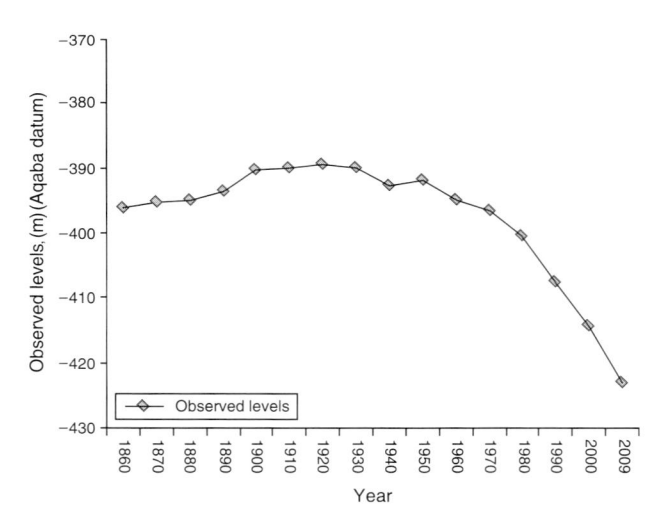

Figure 11.6. Dead Sea elevations from 1860 to 2009, showing rapid decline since the 1960s.

Table 11.1 *Mean rainfall and Dead Sea levels for the period 1860–1960*

	Jerusalem annual rainfall (mm year^{-1})	Dead Sea level (m) (Aqaba datum)	Average annual change in Dead Sea level (m)
1860		−396.28	
1860–1870	547.9	−395.28	0.1
1870–1880	598.5	−395.03	0.025
1880–1890	615	−393.63	0.14
1890–1900	658.3	−390.33	0.33
1900–1910	582.5	−390.18	0.015
1910–1920	562.5	−389.63	0.055
1920–1930	464.9	−390.08	−0.045
1930–1940	459.6	−392.75	−0.267
1940–1950	570.1	−392.03	0.072
1950–1960	448.4	−394.97	−0.294

information is required on inflows, climate, hydrodynamics and in-lake chemical and ecological processes. However, the DYRESM model has been applied recently to Lake Kinneret (Sea of Galilee) to simulate the density currents, water chemistry and ecology of the lake (Gal *et al.*, 2009). This is a very comprehensive modelling study providing a sound basis for evaluating environmental change in the lake systems. With all models there are also sources of uncertainty to consider. In the case of modelling the Dead Sea these uncertainties are multiple, including an incomplete knowledge of the inflows and abstractions, poor estimates of evaporation rates, complex hydraulic movements caused by large temperature and density gradients and even tectonic effects that are thought to have affected lake levels in the past (Chapter 9, this volume). Thus a physics-based model would be difficult to apply to the Dead Sea with any confidence, and inappropriate in the current application where we are concerned solely about water levels.

As previously discussed, modelling the Dead Sea levels is a matter of knowing the inflow rates, estimating evaporation rates and knowing the surface areas so that from a mass balance calculation it is possible to calculate the change in levels from year to year. Modelling the Dead Sea levels is at first sight straightforward as the sea acts as the terminal lake for all the rivers in the Jordan Valley, and thus all the rainfall falling into the catchment area will affect the sea levels. However, there are enormous transmission losses as rainfall is evaporated from the desert surfaces, as well as evapotranspiration losses from the agricultural and vegetated areas. There are also numerous groundwater systems that are recharged by the rainfall and are effectively lost to the River Jordan because of abstraction and irrigation schemes (Chapter 10, this volume). There is significant evaporation loss from the Dead Sea itself which has been estimated as 1.5 m per year (Greenbaum et al., 2006), and this evaporation rate will vary over time as the sea salinity concentrations change (Calder and Neal, 1984). Other factors that can affect the historical Dead Sea levels include tectonic movements, as the rift valley is one of the most geologically active areas in the world. Thus, estimating inflows into the Dead Sea and the consequent sea level changes is a difficult procedure, and again, a physics-based conceptual model is not practical in the current research project. Also, our interest is confined to the annual changes in level rather than the detailed seasonal patterns of behaviour. Thus a simple approach to modelling sea levels has been adopted based on observed data.

The water balance of the Dead Sea can be written as

$$\Delta L = Q/A - E \qquad (11.1)$$

where ΔL is the annual change in level (in m per year), Q is the volume of inflow water (m³ per year), A is the sea area (m²), and E is the annual evaporation rate (m per year). Rainfall on the sea is a relatively small quantity and can be considered to be incorporated into the evaporation term. If E and A are initially assumed to be constant, then ΔL is a linear function of the inflow Q. Further, if the balance is over a hydrological year such that the effects of changes in catchment storage can be neglected, it is reasonable to expect Q to be linearly related to an index of catchment rainfall. This leads to a relationship or model of the form

$$\Delta L = \omega (R - R_m) + \xi \qquad (11.2)$$

where ω is a weighting factor or parameter, R is the annual rainfall index (mm per year), R_m is the mean of the rainfall index (mm per year), and ξ is a factor to represent some of the uncertainty and errors associated with the relationship and the rainfall and level estimates. Thus, in this equation, ΔL should be linearly related to an index of catchment rainfall.

The longest-term index for catchment rainfall in the Jordan Valley is that of Jerusalem, for which there is an annual record from 1846 (Rosenan, 1955), as shown in Figure 11.7. Because of the completeness of the record the Jerusalem rainfall is probably the best index to use. Table 11.1 shows the average rainfall and the average change in sea level for each decade from 1860 to 1960 based on the Klein (1961) estimates of levels and the Jerusalem rainfall. The levels after 1960 are significantly affected by the abstractions and hence have been excluded from this initial analysis.

Table 11.1 gives the annual averages for rainfall and levels based on decadal time steps, and Figure 11.8 shows the regression relationship between these variables. As expected from Equation (11.2), there is a linear relationship with an R-squared fit of 0.81.

$$\Delta L = 0.0023 \, R - R_m + 0.0131 \qquad (11.3)$$

where the long-term mean (R_m) for the Jerusalem data over the period 1860–1960 is 550 mm per year. That such a simple equation can explain 81% of the variance in the data is interesting and suggests this is a fairly robust relationship, at least over the 100 years 1860–1960. Figure 11.9 compares the model simulation with the observed levels and indicates that the model gives a good representation of the Dead Sea levels. In order to validate this relationship, Equation (11.3) has been used to simulate the 50 years from 1960 to the present. In order to do this it is necessary to allow for the estimated abstractions over the period, and Figure 11.10 shows the simulation over the whole period 1860–2000. Again a good fit to the observed data is obtained and the importance of the abstractions is evident, as the levels fall significantly from 1960 to the present day, with the current rate of decline being approximately 1 m per year. This decline is due to the abstraction which is estimated as 100 million m³ per year for the 1970s, 300 million m³ per year for the 1980s, 800 million m³ per year for the 1990s and 800 million m³ per year for the past decade. This recent abstraction rate of 800 million m³ per year is slightly lower than the values estimated by Klein, but estimating discharge from the wide range of sources is difficult because of the many unmeasured sources of water. Estimating water loss rates for the whole range of irrigation schemes, public water supplies and water diversions is also extremely difficult. However, the 800 million m³ per year seems to fit the observed data for the level change and so must be more or less correct.

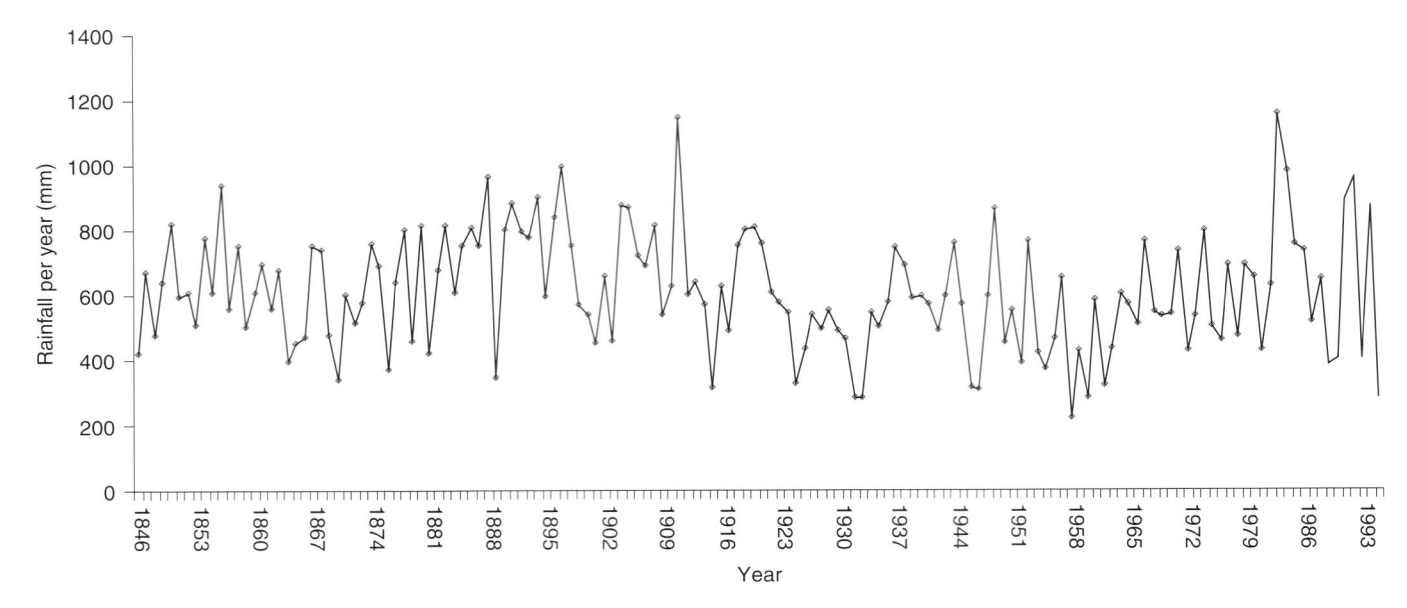

Figure 11.7. Jerusalem rainfall from 1846 to 1996.

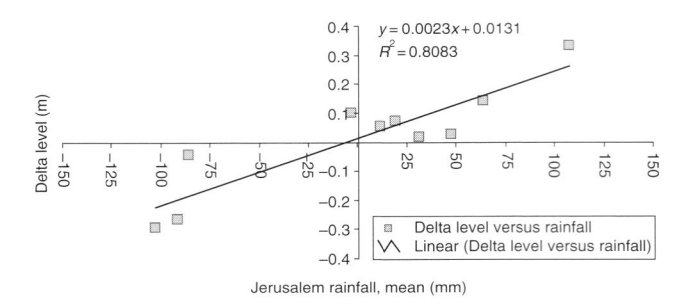

Figure 11.8. Regression of Dead Sea level change against Jerusalem rainfall for the period 1860–1960 using decadal averages.

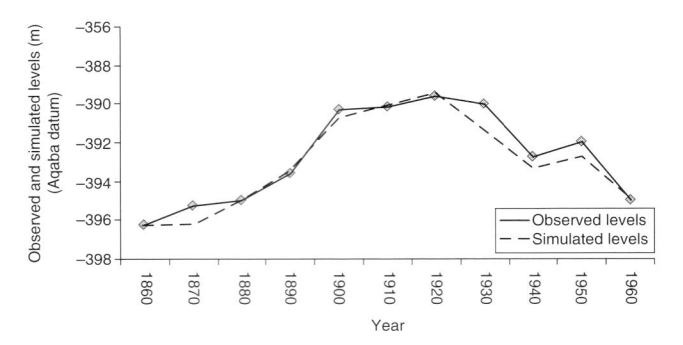

Figure 11.9. Observed and modelled Dead Sea levels 1860–1960.

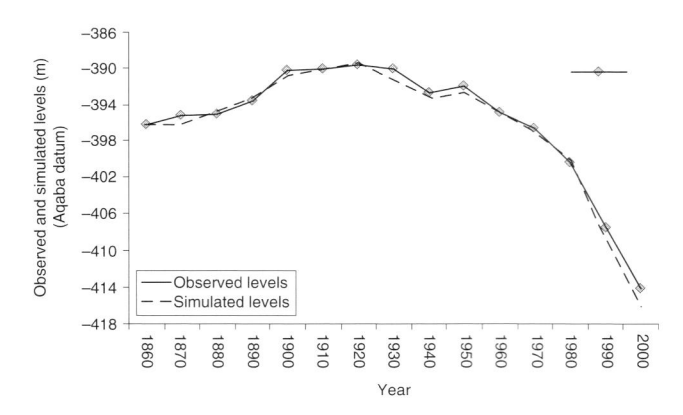

Figure 11.10. Extension of the observed and modelled data to include the recent period of abstraction and sea level decline.

11.4 MODELLING THE DEAD SEA LEVELS INTO THE FUTURE

11.4.1 Future abstractions

It is inevitable that abstractions will continue to reduce the Dead Sea levels. The situation looks quite grim in that, as shown in Figure 11.11, the levels will continue to decline into the future, reaching 100 m below the 1960s stable level by 2050. This will further reduce the viability of the Dead Sea as a tourist attraction and could create major health problems as toxic dust is blown across the Jordan Valley. The current potash and salt mining industries on the Dead Sea will be made uneconomic because of the high costs of pumping water up from the future low levels. Also, the chemistry, mineralogy and ecology of the Dead Sea will be changed forever, thereby destroying a unique environment (Gravrieli and Stein, 2006).

11.4.2 Future climate change

There has been significant climate change over the past 10,000 years, and given the current consensus (IPCC, 2007) that human-induced climate change will start to have significant impacts this century, it is timely to consider the potential future impacts. Chapter 6 of this volume analysed a wealth of palaeoclimate data to reconstruct past climate for the Levant and showed that the climate has changed significantly during

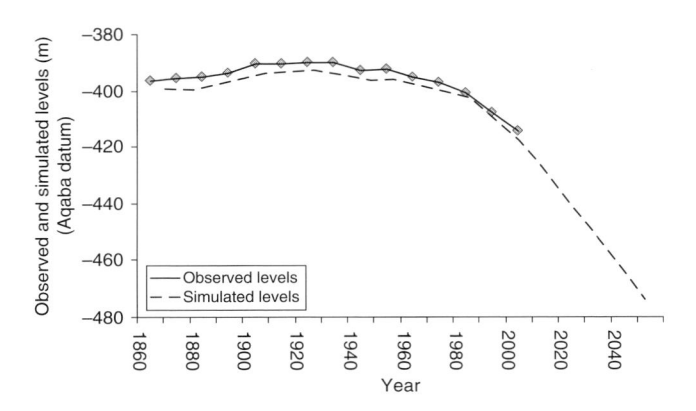

Figure 11.11. Predicted simulation of the levels to 2050 assuming continued abstraction from the River Jordan.

the Holocene with potentially wetter conditions prevailing in the Early Bronze Age at c. 5 ka BP (Sellwood and Valdes, 2006). Historical general circulation model (GCM) results have already been used on a study of hydrological change in the Jawa area of Jordan (Whitehead et al., 2008). These data were generated using the HadSM3 version of the coupled atmosphere–ocean GCM (Valdes et al., 1999). For the current analysis, we are more concerned with predicting future behaviours for the Jordan Valley region. The future climate time-series have been derived using a regional modelling/weather generator approach described in earlier chapters of this volume (Chapters 4 and 5). This approach uses a regional climate model (RCM) which is a high-resolution (0.44° horizontal grid spacing) implementation of the Hadley Centre atmosphere-only model HadAM3P, but applied over a reduced area. The model was applied over a large domain encompassing the Mediterranean, southern Europe, the Middle East and North Africa. The advantage of this approach is that high-resolution predictions can be made in a computationally efficient manner. For the future scenarios, time-slice integrations were performed for 1960–1990 and 2070–2100 under an A2 emissions scenario. The A2 scenario represents a business-as-usual scenario, in which greenhouse gas emissions continue on their current trajectory into the future (Nakicenovic et al., 2001). Detailed discussion of the set up and evaluation of these model integrations can be found in Chapters 4 and 8 of this volume.

11.4.3 Correcting climate model bias

The bias in climate model precipitation must be taken into account when climate model input is used to drive hydrological models. Comparison between observed and modelled precipitation shows that although the RCM captures the number of rainy days reasonably well, it grossly underestimates the rainfall intensity (amount of rain per rainy day), and this leads to underestimation of the annual rainfall. Comparison between the time slices reveals that most of the modelled decadal to centennial variability results from changes in the number of rainy days rather than from changes in rainfall intensity. There are two methods of correcting the bias in the model precipitation: applying a percentage change to the observed Jerusalem annual rainfall total; or using a weather generator to produce daily time-series based on corrected statistics of the weather. The weather generator approach is the only way to derive plausible daily time-series, and allows the impact of stochastic variability to be assessed. The percentage change approach can only be used to generate rainfall at seasonal or longer time-scales, but avoids errors arising from the assumptions implicit in the formulation of the weather generator. In this study both approaches were used.

The weather generator requires three inputs: the distribution of rain/rainy day, the probability of rain given no rain before (P(RNR)) and the probability of rain given rain the day before (P(RR)) (see Chapter 5 of this volume for further details about the weather generator). To correct the model bias, these inputs were obtained by combining modelled data with daily data from a rain gauge near Ramat David. The Ramat David station was chosen because it has a similar rainfall regime, and hence variability, to Jerusalem (on which the Dead Sea model is based), although the rainfall totals are smaller. It was not possible to obtain daily rainfall for Jerusalem itself. Because the changes in daily rainfall amounts were found to be small, and because the model does a poor job of simulating daily rainfall amounts, the present-day observed rainfall distribution was used for all the past and future time slices. To correct the small bias in rainfall probabilities, the percentage changes between the time slices and baseline scenarios were applied to the observed values. The daily data were then aggregated to yearly time intervals, and a percentage change correction was applied to account for the difference between Jerusalem and Ramat David. In order to account for stochastic variability, ten time-series based on the same statistics of the weather were generated.

Figure 11.12 shows the effects of ten realisations of the climate change model, with eight of these assuming no further abstractions into the future and with two assuming the current levels of abstraction. The future climate scenarios suggest that while there will be substantial reductions in precipitation at the peak of the rainy season, the impact on the Dead Sea will be offset by increases in the length of the rainy season and spring rainfall (Chapter 4 and Black, 2009). This means that there will be limited direct impact from precipitation on the residual flows into the Dead Sea and, hence, on sea levels. However, when abstraction is taken into account the levels will continue to fall and would reach a level of the order of 520 m below the Red Sea or Mediterranean Sea levels by 2050. This would be an environmental catastrophe for the Dead Sea.

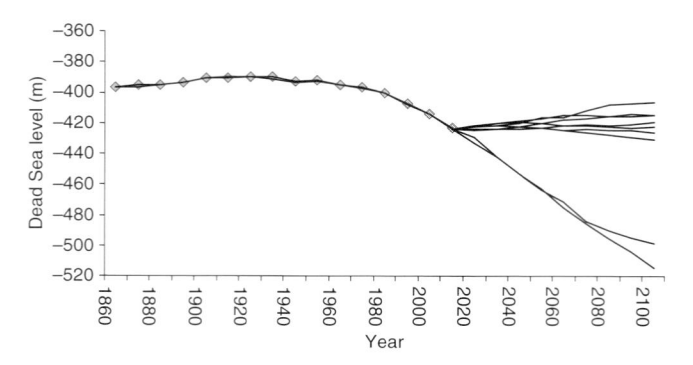

Figure 11.12. Predicted sea levels in the future assuming climate change for eight future realisations without the effects of abstraction, and two climate change scenarios that do include the abstraction.

11.4.4 Future water transfers

There have been proposals for water transfer schemes to top up the Dead Sea for many years and these have involved a transfer of water from the Mediterranean Sea or the Red Sea (Al-Weshah, 2000) into the Dead Sea, via a tunnel or channel. The idea is to use the water to restore the levels of the Dead Sea to the stable 1960s level and also take advantage of the difference in height to generate hydroelectric power. More recently, the World Bank has proposed a scheme that would also make use of the electricity generated to desalinate water so that a potable water supply could be made available. A current feasibility study is under way, but it is worth exploring the potential of such a scheme using the Dead Sea level model. Essentially what is required is sufficient flow to restore the Dead Sea to the 1960s levels and then to maintain this level by balancing the water losses from the abstractions upstream. Figure 11.13 shows the future recovery of the Dead Sea assuming three rates of water transfer (1,690, 1,900 and 2,150 million m³ per year) for the years 2020 to 2040 and the transfer rates then falling to match the abstraction rate of 800 million m³ per year. As shown in Figure 11.13, all three water transfer strategies would restore the Dead Sea to a higher level and, as might be expected, the first 20 years of flow will determine the rate of recovery and the stable level achieved. The flow rates are quite feasible and will generate significant quantities of power and desalinated water, although these quantities will reduce to a stable consistent and sustainable level after 2040 so that the abstractions are matched.

Although this looks attractive, there are environmental concerns with such water transfers, including the chemical reactions that might occur as a result of the mixing of the Red Sea or Mediterranean Sea waters or the addition of nutrients into the Dead Sea which might alter its ecology. However, all these issues can be thoroughly assessed and it is likely that the continuing decline in the Dead Sea will pose much more serious risks into the future.

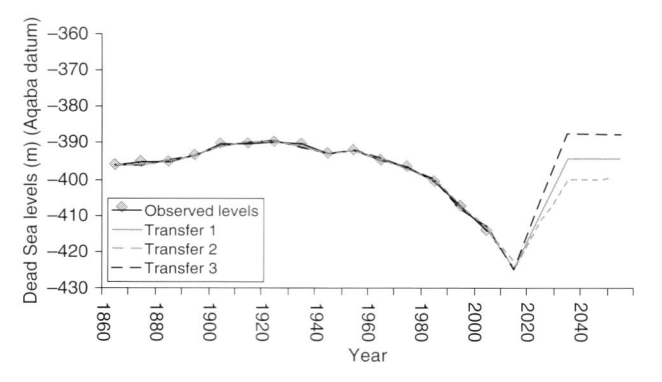

Figure 11.13. The effects on Dead Sea elevations assuming a major water transfer from either the Red Sea or the Mediterranean Sea into the Dead Sea. Shown are three water transfer rates of 1,690, 1,900 and 2,150 million m³ per year for the years 2020 to 2040, and then transfer rates falling to match the water abstraction rate of 800 million m³ per year.

11.5 ESTIMATING PAST RAINFALL FROM ELEVATION DATA

The past climate of the Jordan Valley is of considerable interest because of the effects of rainfall and evaporation on the Dead Sea elevations, the local vegetation and tree cover, the development of agriculture and the location and size of centres of population or settlement in the area. Chapter 2 of this volume reviews the climatology of the region, and Enzel *et al.* (2003) discuss the late Holocene climate and the changes that have occurred, using the Dead Sea level data to infer past climate conditions. In this study we have also used the Dead Sea level data from Enzel *et al.* (2003) and Black *et al.* (Chapter 9, this volume) together with the relationship between levels and rainfall (Equation (11.3)) to back-calculate the rainfalls over the past 25 kyr. To do so, it has been necessary to allow for three additional factors that affect sea levels over a long time period.

The first of these factors is the effect of changing sea surface area, as this will alter the volume of water evaporated from the sea surface: the larger the sea surface area, the larger the volume of evaporated water. We know from Equation (11.1) that the relationship between inflow and sea level rise implicitly assumes that the sea area and the evaporation rate (in depth terms) are constant. This is reasonable for relatively small fluctuations in sea level. However, should the sea area change significantly to, say, A_1, the new sea level change, ΔL_1, will differ from that predicted by the model for the same inflow or the same rainfall. Thus,

$$\Delta L_1 = Q/A_1 - E \qquad (11.4)$$

and, by substituting in for Q from the original expression (11.1),

$$\Delta L_1 = E(A/A_1 - 1) + \Delta L(A/A_1) \qquad (11.5)$$

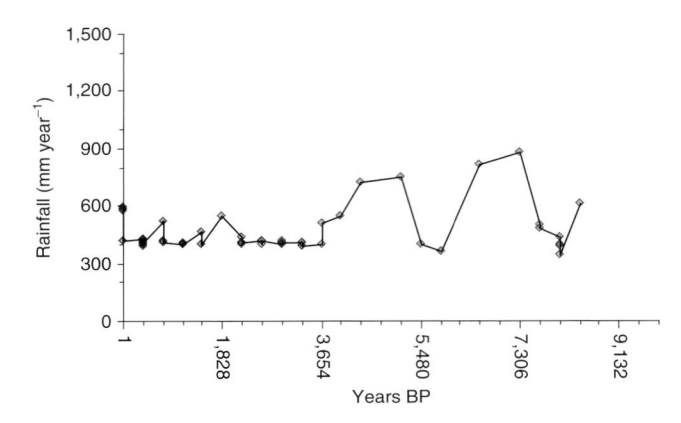

Figure 11.14. Estimated rainfall over the past 9,000 years based on the Enzel et al. (2003) Dead Sea elevations.

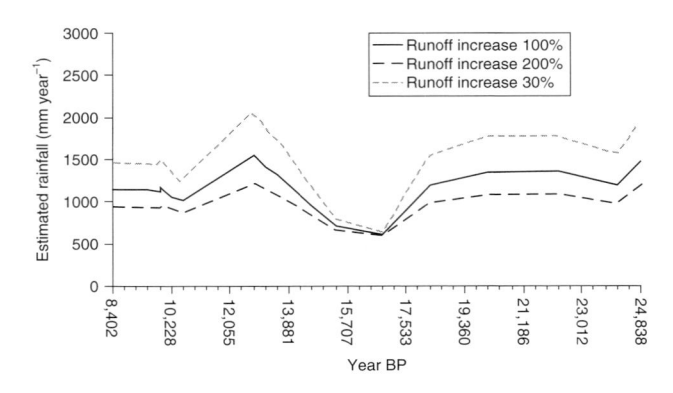

Figure 11.15. Estimated rainfall over the period 8–250 ka BP based on the Black et al. (Chapter 9, this volume) Dead Sea elevations.

Using this equation gives us a method of calculating the new level change and by substituting in the rainfall relationship with level from Equation (11.3), Equation (11.5) can be rearranged to give

$$R_e = \frac{1}{0.0023}\left(\frac{A_1}{A}\Delta L_1 - E\left(\frac{A}{A_1}-1\right) + R_m - 0.0131\right) \quad (11.6)$$

where R_e is the estimated rainfall in mm per year over the period in which the level change ΔL_1 is determined.

The second issue that affects sea level rise is the effect of changing density or salinity on the sea surface evaporation rate. Calder and Neal (1984) showed that the conventional evaporation equation can be modified to take account of changing salinity. The equation actually requires considerable information, but Calder and Neal provide estimates of evaporation for the Dead Sea under current saline conditions and under historical less saline conditions, with a 10% increase in evaporation under the less saline conditions. This effect can be incorporated into Equation (11.6) by varying E according to the expected salinity levels.

The third effect that needs to be considered for the Lake Lisan period is the hydrological conditions that would have prevailed under different land use and vegetation patterns. It is known that the area was thickly vegetated (Chapter 6, this volume), and with this type of soil cover and changed hydrological flowpaths, it would be expected that the runoff percentage would increase. The runoff percentage here is assumed to be the percentage of catchment rainfall converted into runoff flow that will reach the Dead Sea. The current runoff percentage is around 13%, according to Greenbaum et al. (2006). This is a typical low value for an arid desert or semi-desert landscape. However, given the vegetation cover in the Lake Lisan period (10–25 kaBP) the runoff could be a lot higher than 13%, and we need to take this into account when estimating rainfall.

To estimate past rainfall, the model Equation (11.6) has been applied to two sets of historical level data from Enzel et al. (2003) for the past 10 ka and from Black et al. (Chapter 9, this volume) for the period 9 ka to 25 ka BP. Figure 11.14 shows the estimated rainfalls over the Enzel et al. (2003) data period and Figure 11.15 shows the estimated rainfall over the Lake Lisan period, based on the Black et al. data period (Chapter 9, this volume). During the Early Bronze Age, rainfall is estimated as significantly higher, reflecting the wetter climate then, although there is considerable variability suggesting periods of drought and wetter conditions. This is reflected in the archaeology with the loss of population centres as the climate dries during the Bronze Age (Whitehead et al., 2008).

It is very interesting to note that the rainfalls over the Lake Lisan period show some significantly higher rainfalls than the current conditions, suggesting a much wetter climate. A range of rainfall estimates are given for this period, as shown in Figure 11.15, as it is difficult to determine the effects of the changed vegetation, soil cover and hydrological flowpaths on runoff and hence rainfall estimates. The three conditions shown represent a 30%, a 100% and a 200% increase in runoff. With the 100% change, a rainfall average of the order of 1,142 mm per year is estimated whereas the 200% extra runoff requires a rainfall of 940 mm per year over the period in order to match the estimated Lake Lisan levels. This rainfall estimate of 940 mm tends to agree with the increased rainfalls determined by the RCM modeling by Brayshaw et al. (Chapter 3, this volume).

11.6 CONCLUSION

This modelling study has proved to be an interesting and valuable exercise in evaluating the levels of the Dead Sea. A relatively simple model for predicting Dead Sea levels has been developed based on observed data and this model has been used to investigate past and future behaviour. The future response to climate change and the continuing abstraction is unequivocal. The dominant factor as far as the Dead Sea is concerned will be the abstractions. Thus the Dead Sea levels will continue to fall at about 1 m per year, with levels down to 433 m below sea level by the 2020s. By 2040 the level will be 443 m below sea level, with a sea surface area of approximately 600 km^2. This will be quite serious, as the sea salts will be further concentrated, thereby damaging the current ecology, the shores will be even further from the tourism facilities and the potash and salt industry will be incapacitated. Given that water is needed upstream for irrigation and public water supply, there is probably no alternative but to create some link with the Red Sea or the Mediterranean Sea to replenish and restore Dead Sea levels. There are numerous advantages to this, including the possibilities for power generation and desalination plants. There could also be detrimental effects such as changes in the Dead Sea chemistry. However, these changes are probably less severe than the Dead Sea drying out completely, which it is on track to do at the moment.

The research in this chapter has also explored the palaeohydrology and used the past estimated Dead Sea levels to calculate past rainfall. The past rainfalls reflect the lake levels, although into the distant past, other factors such as changing Dead Sea salinity, changing catchment land use and hence runoff conditions, and changing surface areas affect the rainfall estimation. Nevertheless, the model suggests that rainfall must have been significantly higher in the past, perhaps up to double the current rainfall rates in the Lake Lisan period.

REFERENCES

Abelson, M., Y. Yechieli, O. Crouvi *et al.* (2006) Evolution of the Dead Sea sinkholes. In *New Frontiers in Dead Sea Paleoenvironmental Research*, ed. Y. Enzel, A. Agnon and M. Stein. Boulder, CO: Geological Society of America.

Al-Weshah, R. A. (2000) The water balance of the Dead Sea: an integrated approach. *Hydrological Processes* **14**: 145–154.

Black, E. (2009) The impact of climate change on daily precipitation statistics for Jordan and Israel. *Atmospheric Science Letters* **10**: 192–200.

Calder, I. R. and C. Neal (1984) Evaporation from saline lakes: a combination equation approach. *Hydrological Sciences Journal* **29**: 89–97.

Chapra, S. (1997) *Surface Water Quality Modelling*. New York: McGraw Hill.

Enzel, Y., R. Bookman, D. Sharon *et al.* (2003) Late Holocene climates of the Near East deduced from Dead Sea level variations and modern regional winter rainfall. *Quaternary Research* **60**: 263–273.

Gal, G., M. R. Hipsey, A. Parparov *et al.* (2009) Implementation of ecological modeling as an effective management and investigation tool: Lake Kinneret as a case study. *Ecological Modelling* **220**: 1697–1718.

Gravrieli, I. and M. Stein (2006) On the origin and fate of the brines in the Dead Sea basin. In *New Frontiers in Dead Sea Paleoenvironmental Research*, ed. Y. Enzel, A. Agnon and M. Stein. Boulder, CO: Geological Society of America pp. 183–195.

Greenbaum, N., A. Ben-Zvi, I. Haviv and Y. Enzel (2006) The hydrology and paleohydrology of the Dead Sea tributaries. In *New Frontiers in Dead Sea Paleoenvironmental Research*, ed. Y. Enzel, A. Agnon and M. Stein. Boulder, CO: Geological Society of America pp. 63–95.

Haddadin, M. (2006) *Water Resources in Jordan. Evolving Policies for Development, the Environment, and Conflict Resolution*. Washington: RFF Press.

Hall, J. K. (1996) Digital topography and bathymetry of the area of the Dead Sea depression. *Tectonophysics* **266**: 177–185.

IPCC Core Writing Team, R. K. Pachauri and A. Reisinger (2007) Summary for Policymakers. In *Climate Change 2007: Synthesis Report. Contribution of Working Groups I, II and III to the Fourth Assessment Report of the Intergovernmental Panel on Climate Change*. Geneva: IPCC.

Klein, C. (1961) On the fluctuations of the Dead Sea since the beginning of the 19th century. Israel Hydrological Service Report number 7.

Klein, M. (1998) Water balance of the Upper Jordan River basin. *Water International* **23**: 244–248.

Lipchin, C. (2006) *A Future for the Dead Sea Basin: Water Culture among Israelis, Palestinians and Jordanians*. Fondazione Eni Enrico Mattei (FEEM) Working Paper pp. 22.

Nakicenovic, N., J. Alcamo, G. Davis *et al.*, eds. (2001) *IPCC Special Report on Emissions Scenarios (SRES)*. Geneva: GRID-Arendal.

Robinson, S. A., S. Black, B. Sellwood and P. J. Valdes (2006) A review of palaeoclimates in the Levant and Eastern Mediterranean from 25,000 to 5000 BP: setting the environmental background for the evolution of human civilisation. *Quaternary Science Reviews* **25**: 1517–1541.

Rosenan, N. (1955) One hundred years of rainfall in Jerusalem. *Israel Exploration Journal* **5**: 137–153.

Sellwood, B. and P. J. Valdes (2006) Mesozoic climates: general circulation models and the rock record. *Sedimentary Geology* **190**: 269–287.

Valdes, P., R. A. Spicer, B. Sellwood and D. C. Palmer (1999) *Understanding Past Climates: Modelling Ancient Weather*. Taylor & Francis Ltd (CD-ROM).

Wheater, H. S. (2005) Modelling hydrological processes in arid and semi-arid areas. In *Proceedings of the International G-WADI Modelling Workshops* http://www.gwadi.org.

Whitehead, P. G., E. J. Wilson and D. Butterfield (1998) A semi-distributed Integrated Nitrogen model for multiple source assessment in Catchments (INCA): Part I – model structure and process equations. *Science of the Total Environment* **210**: 547–558.

Whitehead, P. G., R. L. Wilby, D. Butterfield and A. J. Wade (2006) Impacts of climate change on in-stream nitrogen in a lowland chalk stream: an appraisal of adaptation strategies. *Science of the Total Environment* **365**: 260–273.

Whitehead, P. G., S. J. Smith, A. J. Wade *et al.* (2008) Modelling of hydrology and potential population levels at Bronze Age Jawa, Northern Jordan: a Monte Carlo approach to cope with uncertainty. *Journal of Archaeological Science* **35**: 517–529.

12 The hydrology of the Wadi Faynan

Andrew Wade, Paul Holmes, Mohammed El Bastawesy, Sam Smith, Emily Black and Steven Mithen

ABSTRACT

This chapter describes the hydrology of the Wadi Faynan and the potential changes to the hydrology that may occur under rainfall and vegetation change scenarios. The Wadi Faynan is a meso-scale (241 km^2), semi-arid catchment in southern Jordan, considered internationally important because of its rich archaeological heritage spanning the Pleistocene and Holocene. This is the first study to describe the hydrological functioning of the catchment, setting it within the framework of contemporary archaeological investigations. Historic climate records were collated and supplemented with new hydrological and water quality data. These data were analysed in relation to catchment geology and used to build a conceptual hydrological model. The analysis established the importance of the geology in determining the hydrological functioning of the catchment and suggested probable reasons for the initial settlement of the catchment. The potential impacts of rainfall and vegetation change were simulated using an application of the Pitman model. Results suggest that increased rainfall in the catchment where there is sparse vegetation can lead to increased flood magnitude and frequency. Increased rainfall alone does not necessarily imply better conditions for farming. The simulations also demonstrate that increased vegetation enhances infiltration, and a dense vegetation cover, when coupled with higher rainfall, leads to a more secure water resource characterised by a higher base flow and reduced flood frequency.

12.1 INTRODUCTION

Water is an essential requirement for life, and the relationship between climate and water availability has been fundamental to human activities in the past. Today, southern Jordan appears as a rather harsh, semi-arid environment, but there is evidence of a wetter early and mid-Holocene (Chapter 6, this volume). Moreover, the geology of southern Jordan includes limestone and sandstone, well known as potential aquifers, and these are exploited today for local water supply and transmission to Amman. To gain a complete understanding of water availability and security, it is necessary to appreciate the interplay between climate, geology and vegetation, since climate controls precipitation and evaporation, geology defines the groundwater resource and vegetation modifies flow pathways and can provide an additional water store (Bull and Kirby, 2002). The influence of climate, geology and vegetation on hydrology can be established at a regional scale, such as for the Jordan Valley, but to appreciate the subtlety of the interplay between these three factors and the resultant influence on past cultural developments, it is anticipated that a study at a more local scale would be instructive. As such, this work focuses on the hydrology of the Wadi Faynan, a principal site in studies that investigate the origins of farming (Finlayson and Mithen, 2007a).

Little is known of the hydrology of the Wadi Faynan. The work presented in this chapter will address this knowledge gap by analysing existing and newly collected hydrological and water quality data from the wadi and environs (latitude: 30° 38′ N, longitude: 35° 29′ E; Figure 12.1). These data are used to construct a conceptual model of the main water stores and pathways within the Faynan catchment. The conceptual model is then realised as a calibrated version of the Pitman hydrological model, a computer program devised for the hydrological simulation of semi-arid environments (Pitman, 1973). The Pitman model is subsequently used to explore the hydrological functioning of the wadi under different rainfall and vegetation regimes, and the results are interpreted in terms of water availability and security, and crop cultivation. In particular, the flood regime and the security of the groundwater source, which provides the perennial base flow, are considered. It has been proposed that the

Water, Life and Civilisation: Climate, Environment and Society in the Jordan Valley, ed. Steven Mithen and Emily Black. Published by Cambridge University Press.
© Steven Mithen and Emily Black 2011.

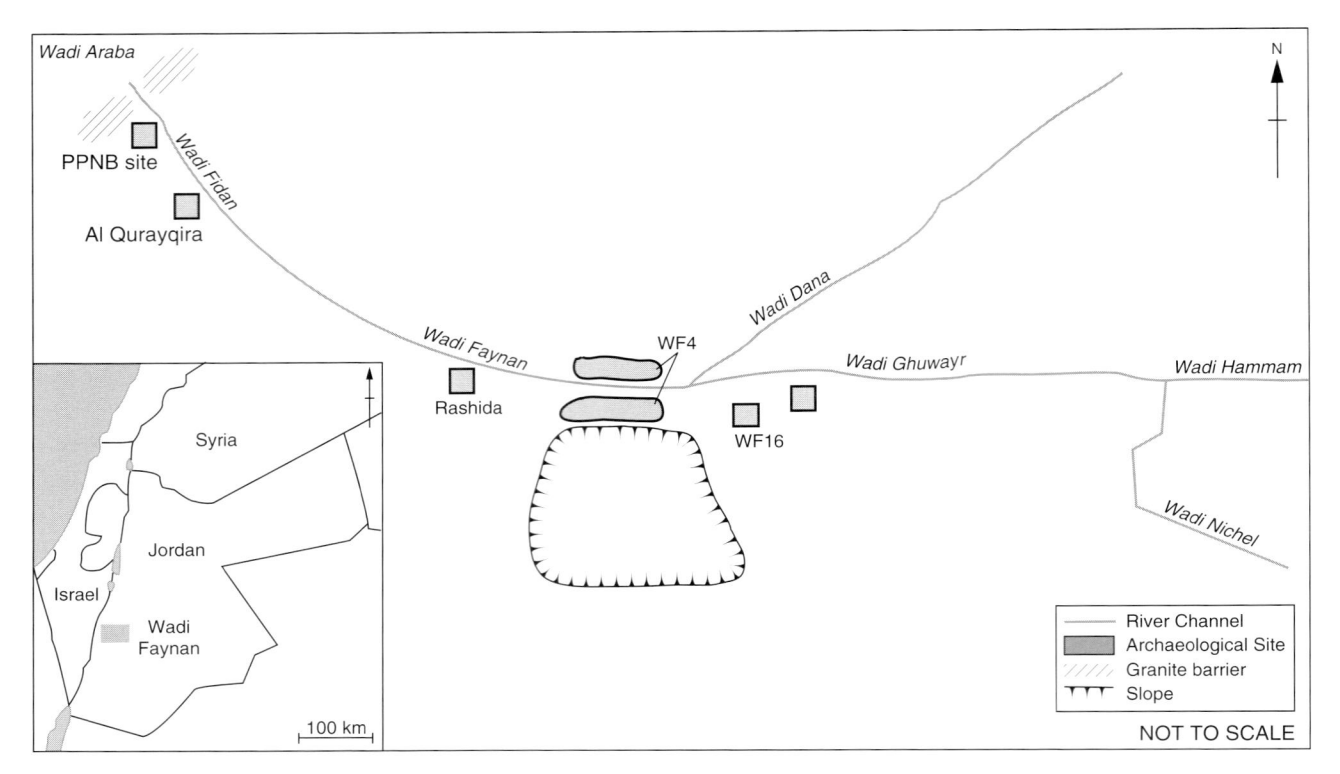

Figure 12.1. A schematic map of the Wadi Faynan, its major tributaries and settlements. PPNB, Pre-Pottery Neolithic B.

extent of woodland vegetation was much greater in the Neolithic than that evident today (Barker et al., 2007b; Mithen et al., 2007a). The effect of a more extensive vegetation cover on the catchment hydrology will be investigated.

Present and ancient water management technologies are not considered in detail, our focus being on the hydrology rather than the hydraulic constructs used to manipulate the water resource. Barker et al. (2007a, b) review archaeological investigations of the water management in Faynan and environs, Crook (2009) discusses the irrigation of the ancient field-system, and Darmame et al. (this volume, Chapter 27) review the contemporary situation. The interpretation of water availability and security in the context of cultural development in Wadi Faynan is considered further by Smith et al. in Chapter 15 of this volume.

12.2 STUDY AREA AND DATA RESOURCE

The Wadi Faynan drains the eastern scarp slope of Wadi Araba, south of the Dead Sea (Figure 12.1). The Wadi Faynan disgorges to Wadi Araba after passage through the Jebel Hamrat al Fidan, an aplite-granite mass located at the mouth of the Wadi Fidan; the Wadi Fidan is the name given to the extension of Wadi Faynan between Al Qurayqira and Jebel Hamrat al Fidan. Measured from the northwest side of Jebel Hamrat al Fidan, the catchment area of Faynan is 241 km². The climate of Wadi Faynan is currently classified as semi-arid since annual potential evaporation exceeds precipitation (Al-Qawabah et al., 2003).

The Wadi Faynan region has a rich archaeological heritage, covering much of the Pleistocene and Holocene periods, and this has been comprehensively described in several recent volumes (Barker et al., 2007a, b; Finlayson and Mithen, 2007b; Hauptmann, 2007). In addition, Chapter 15 of this volume provides a detailed consideration of the relationship between cultural developments and hydrology throughout the Holocene. In this context, the following paragraphs will provide only a very brief overview of the archaeology of the region.

Evidence of Lower and Middle Palaeolithic occupation occurs in the form of scatters of stone tools (including bifacial hand-axes and Levallois points) which have been found on many gravel terraces within the Faynan catchment (Rollefson, 1981; Mithen et al., 2007b). Evidence for later, Upper Palaeolithic and Epipalaeolithic occupation is much less common and no convincing evidence of Epipalaeolithic occupation has been found in the Faynan area (Mithen et al., 2007b), although as Epipalaeolithic occupation is common in the areas around Wadi Faynan, this absence of evidence may not be an accurate reflection of the settlement history of the area. Following this possible hiatus in settlement, the beginnings of the Holocene see clear evidence for occupation of the wadi by early (aceramic) Neolithic groups at sites such as WF16 (Finlayson and Mithen, 2007a) and

Ghuwayr 1 (Simmons and Najjar, 2000) located near the confluence of Wadis Dana, Ghuwayr and Shaqyr as well as in the upper reaches of the Wadi system at Wadi Badda (Fujii, 2007). There is also evidence that the region was occupied during later, Pottery Neolithic periods, with a major settlement at the site of Tell Wadi Faynan (Al-Najjar et al., 1990). Following the occupation of the region by these early farming groups, human activity in the wadi system begins to focus more intensively upon exploitation of the region's copper resources, with concurrent developments in water management systems.

Evidence for Chalcolithic occupation occurs at Tell Wadi Faynan (Al-Najjar et al., 1990) as well as in the form of lithic scatters (Mithen et al., 2007a) and evidence for Chalcolithic-period metallurgical activity in the area (Barker et al., 2007a). During the Early Bronze Age, there appears to have been a step change in the occupation of the Wadi Faynan, as shown by the large (c.12 ha) site at WF100 as well as by the presence of a plethora of smaller occupation, industrial, agricultural and ritual sites and lithic scatters (Barker et al., 2007a; Mithen et al., 2007b). During this period we also have the first evidence for water management, with the development of simple systems for managing floodwater for agricultural and domestic purposes (Barker et al., 2007a). Following a decline in human activity during the Middle and Late Bronze Ages (mirroring the general situation in the southern Levant), the wadi appears to have been intensively occupied once more during the Iron Age, when the large site of Khirbet an-Nahas was occupied, the field system (WF4) was developed and exploitation of copper resources greatly increased (Levy et al., 2004; Mattingly et al., 2007b). During the subsequent Nabataean and Roman/Byzantine periods the area was even more intensively exploited, with the growth of the large site of Khirbet Faynan, the industrial-scale exploitation of copper resources, and the further development and expansion of the WF4 field system (Mattingly et al., 2007a). Following the Classical period peak in human activity and occupation, evidence suggests that during the Islamic and Ottoman periods industrial activity rapidly declined and the region was mainly occupied by pastoral groups, a tradition which has continued up to the present day (Newson et al., 2007). Even from this brief review of the archaeology of the region, it is clear that the Wadi Faynan has been the scene of major periods of human activity for many millennia. Of particular interest, in the present context, is how human groups, drawn by the region's mineral resources, attempted to manage the slender water resources of this semi-arid location.

Today, the Bedouin located in the Wadi Faynan use the perennial water to irrigate their fields near the villages of Al Qurayqira and Rashida by conveying the water in plastic pipes under gravity from the Wadi Ghuwayr (Figure 12.1). There are no major urban centres in the catchment, with the population being sparse in comparison with that of northwest Jordan where the rainfall and topography are more favourable for irrigation. The Wadi Faynan fringes the Dana National Park, which is maintained by the Royal Society for the Conservation of Nature (RSCN), and for which a detailed characterisation of the quantity and quality of the water resource is planned (Al-Qawabah et al., 2003).

In this study, geological and hydrological information were derived from digital and paper maps. Hydrological measurements were collated from previous academic, government agency and engineering studies. These data were integrated with new field measurements of base flow, open-channel hydraulics (to estimate flood peaks) and water chemistry. There was recourse to satellite imagery to confirm the presence of specific geological structures and to derive the catchment boundaries of the study area. The following mapping sources were used:

- Topographic mapping at 1:50,000, Soviet military series sheets; source: Royal Jordanian Geographic Centre;
- Landsat imagery 28 m resolution;
- SPOT satellite imagery provided by the National Imagery and Mapping Agency (NIMA), containing 10 m resolution Digital Orthorectified Imagery (DOI-10M) derived from data obtained from the SPOT Image Corporation under an unrestricted licence;
- Geological Map of Jordan 1:250,000, prepared by Bundesanstalt für Geowissenschaften und Rohstoffe, Hannover 1968 (Sheet: Aqaba-Ma'an and Amman);
- Google Earth^{TM} (http://earth.google.com).

Academic studies of the hydrology of the Middle East and North Africa are typically limited by available data. Data gathered for engineering projects, if not proprietary, provide useful additional information. Rainfall and flow data were collated for wadis draining into the Dead Sea and the Jordan river just upstream of the Dead Sea. These data were produced from national monitoring programmes overseen by the Ministry of Irrigation, now subsumed by the Ministry of Water and Irrigation and the Jordan Valley Authority, and from engineering projects for planned irrigation schemes by Hazra Overseas Engineering Company (1978) and the development of the Arab Potash Company evaporation pans by Binnie and Partners (Overseas) Limited (1979). The historical data were supplemented by contemporary hydrological and water chemistry data gathered during field visits in 2006, 2007 and 2008.

12.3 TOPOGRAPHY, GEOLOGY AND HYDROGEOLOGY

The Rift Valley in Jordan, known as Wadi Araba, runs north-northeast/south-southwest between the African and Arabian tectonic plates and is an extension of the Great Rift. The movements of the two plates during the Cretaceous and Quaternary periods have resulted in both lateral and vertical displacements in the local geology (Tipping, 2007). The geology of the Wadi Faynan is constituted by fluvial deposits and aeolian sands from the Quaternary period; limestones from the Tertiary (Eocene/Palaeocene epochs) and Cretaceous periods; sandstones from the Cambrian period; and porphyrite and aplite-granite from the Precambrian. There is also an outcrop of basalt from the Quaternary on the northeast rim of the catchment which forms the Jebel al Afa'ita.

The Wadi Faynan is approximately 25 km long and flows in an east to west direction from the plateau down the scarp slope and past the town of Al Qurayqira. Beyond this, it flows into the Wadi Fidan and then on into the Wadi Araba which then drains towards the Dead Sea. Further south the Wadi Araba drains towards the Gulf of Aqaba. The Wadi Faynan has two major tributaries, the Wadi Ghuwayr and the Wadi Dana (Figure 12.1). These have contrasting geology, with the former sustaining a perennial flow fed by a series of springs in its middle and upper reaches. The Wadi Shagyr drains into the Wadi Faynan in a south to north direction and enters the Wadi Faynan channel immediately upstream of the confluence between the Wadi Ghuwayr and the Wadi Dana but downstream of WF16 (the Wadi Shagyr is not shown on Figure 12.1). The highest point in the Wadi Faynan catchment is Jebel al Afa'ita at 1,641 m above sea level. The elevation of the confluence with the Wadi Araba in the Rift Valley is 300 m below sea level. As noted by Tipping (2007), the geology of the Wadi Faynan catchment is dominated by two NE–SW trending faults, which break the line of the Rift escarpment and control the direction taken by the Wadi Dana and the Wadi Ghuwayr (Figure 12.2). The Wadi Ghuwayr and the Wadi Dana are steep. The range in altitude on the scarp slope varies from approximately 300 m asl at the Ghuwayr–Dana confluence on the alluvial plain to 1,300 m asl on the plateau. The Dana tributary is approximately 15 km long, giving a gradient of 0.067 m m^{-1}.

Springs in the Wadi Ghuwayr occur at the contact between the Cenomanian–Turonian (Cretaceous) limestones and the Cenomanian–Cretaceous sandstone, and at the fault-contact between the Cambrian sandstones and the underlying Precambrian volcanic rocks. It is also possible that springs occur at the contact between the Eocene–Palaeocene and Cenomanian–Turonian limestones. The contact between the Cenomanian–Turonian limestones and the Cenomanian Cretaceous sandstone runs along the eastern scarp slope, and springs on this interface are also found at Beidha and Petra, 30 and 35 km to the south, respectively. It is unclear if any of the springs in Faynan are artesian, as there is a lack of readily available borehole data.

To a first approximation, the extent of the Cretaceous geology probably demarcates the groundwater divide (Figures 12.2 and 12.3). The rocks of the scarp dip downwards towards the Rift Valley whilst, on the plateau side of the mountain crest, the drainage is either northeast or east-southeast towards the desert (Agrar und Hydrotechnik GmbH, 1977). Further investigation is required to determine whether there is a hydrological connection to the Disi aquifer and if so, what the water movements are. At this juncture, it is assumed that there is no flow of water from the Disi aquifer to the Wadi Faynan, because of the dip of geological strata away from escarpment (Figure 12.3). As such, the groundwater in Faynan does not appear to be derived from the large continental aquifers of eastern Jordan and therefore will not be controlled by regional-scale groundwater movements. Rather, the springs in the Wadi Faynan are fed by relatively small aquifers whose extent equates approximately to the area of the surface water catchment.

The aquifers in the Wadi Faynan outcrop in the middle and upper reaches of the Wadi Ghuwayr and the Wadi Dana, approximately 10 km upstream of the modern settlements of Residha and Al Qurayqira (Figure 12.3). No large aquifer extends beneath these settlements. Boreholes drilled at Al Qurayqira only reach the water table at approximately 200 m below the surface. The geological cross-section suggests that the limestone and sandstone, tapped by the borehole, is a small pocket surrounded by aplite-granite and therefore the borehole is unlikely to have a significant long-term yield (Figure 12.3). An outcrop of porphyrite also forms an adjacent hill southeast of the ancient field-system.

The volcanic ridge and associated faulting of the Jebel Hamrat al Fidan may have also influenced ancient settlement in the Wadi Faynan. The ridge of Hamrat al Fidan forms a barrier to the course of the Wadi Faynan channel, causing a constriction in the wadi along its pathway to the Rift floor in the Wadi Araba. This constriction has allowed a low-gradient alluvial plain to form between the volcanic ridge and the base of the escarpment. The alluvial plain may have improved the past fertility of the valley. Water is likely to be retained in the sand and gravel deposits for longer than would have occurred if the wadi emerged from the escarpment to the Rift floor unhindered.

The presence of the granitic barrier appears to be an isolated case along the Rift Valley south of the Dead Sea, although a few minor barriers are also present west of Petra. Other wadis generally meander their way across the Rift Valley floor unimpeded, forming widespread alluvial fans, on their route to the main carrier, the Wadi Araba. Satellite images of the Wadi Faynan show a number of Acacia (*Acacia* spp) trees in, and close to, the constriction, perhaps indicating a longer retention of moisture following a flood. Water flowing beneath the surface, originating from the springs in the Wadi Ghuwayr, may pond behind the barrier. Acacia is known to be capable of extracting water from depths of at least 50 m (de Vries *et al.*, 2000).

Inspection of a satellite image shows green patches of vegetation to the south of the barrier in northern Wadi Araba (bottom left, Figure 12.4) and the irrigated fields adjacent to Al Qurayqira in the Wadi Faynan (top centre). The dark area in the top left of the image shows the aplite-granite barrier of Hamrat al Fidan

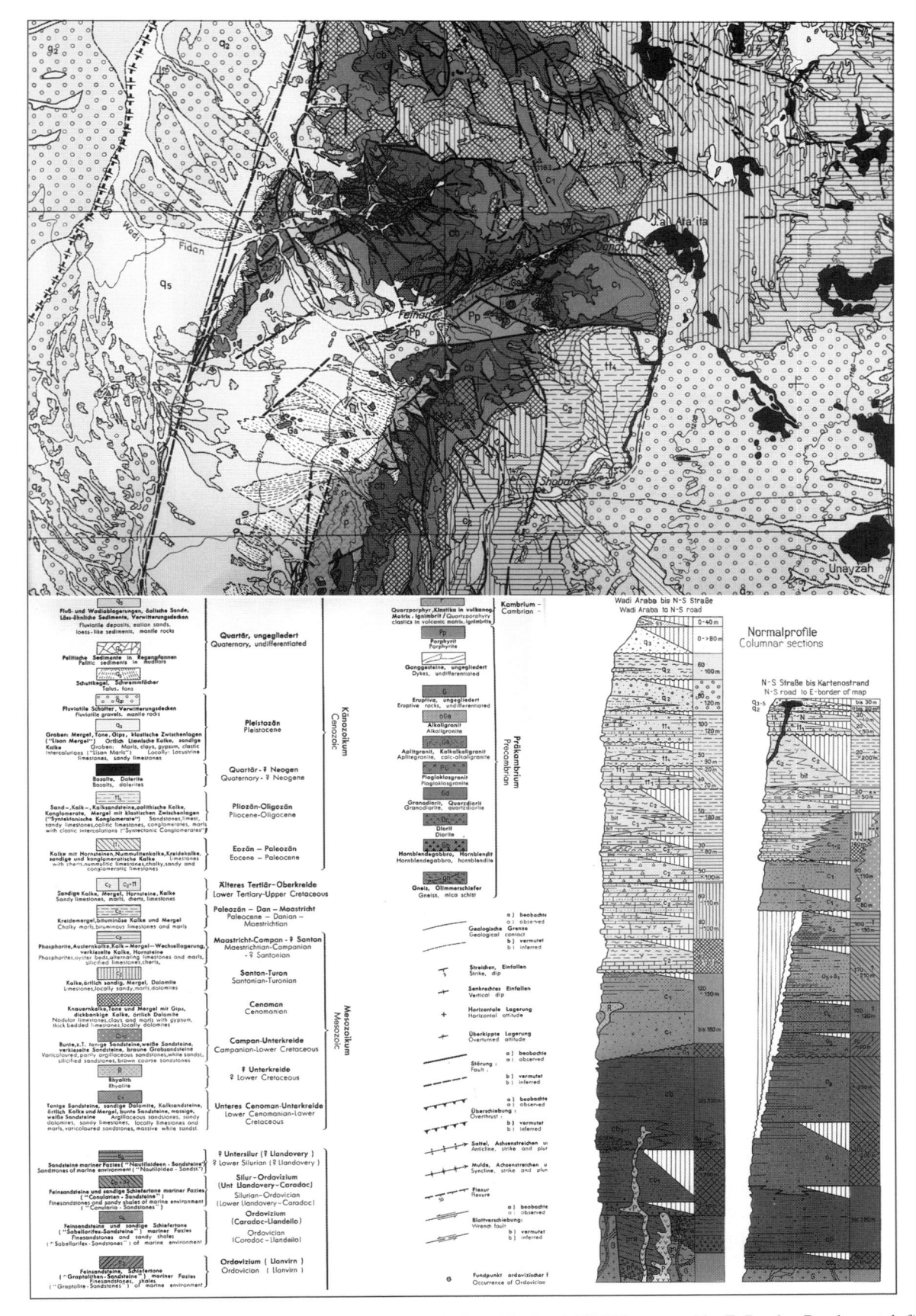

Figure 12.2. The geology of the Wadi Faynan area. Source Geological Map of Jordan 1:250,000, prepared by F. Bender, Bundesanstalt für Geowissenschaften und Rohstoffe, Hannover 1968 [Sheet: Aqaba-Ma'an and Amman]. Reproduced with permission. Not to scale. © Bundesanstalt für Geowissenschaften und Rohstoffe. See colour plate section.

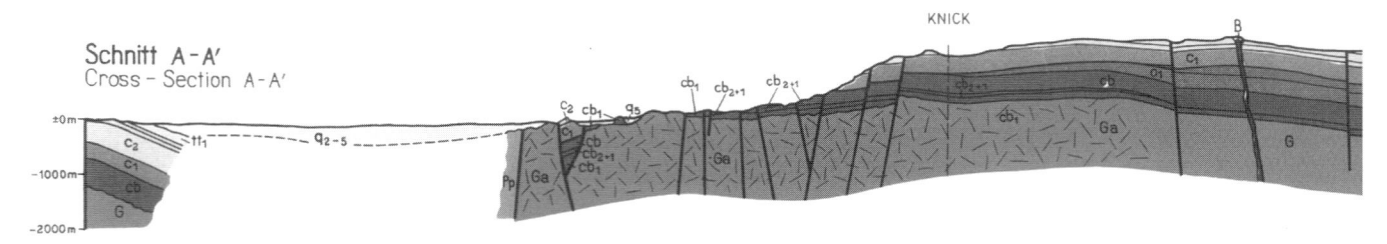

Figure 12.3. A geological cross-section made 5 km to the north of the Wadi Faynan. Source: Geological Map of Jordan 1:250,000, prepared by F. Bender, Bundesanstalt für Geowissenschaften und Rohstoffe, Hannover 1968 [Sheet: Aqaba-Ma'an and Amman]. Reproduced with permission. Not to scale. © Bundesanstalt für Geowissenschaften und Rohstoffe. See colour plate section.

Figure 12.4. A subset of the Landsat image of the Wadi Faynan area acquired on 08 March 2002.

which causes the formation of the large alluvial plain in the Wadi Faynan; we hypothesise that the barrier extends below the surface of Wadi Araba to the southwest of Hamrat al Fidan, acting as an impermeable layer. Where the alluvial overburden is shallow, vegetation is present in an otherwise arid environment as it is able to draw on residual moisture in the alluvium. Fine sediments, which remain on the surface after water has infiltrated into the wadi-channel gravels, have a high reflectance and thus the extent of the last flood in the Wadi Faynan before the image was taken can be seen in the image (top right). Field investigation determined that the green patches were areas of watermelon cultivation. A local farmer had dug a network of channels to a depth of approximately 2 m and a deep (4 m) sump from which to pump water (Figure 12.5). In April 2008 he was able to pump for 1 to 1.5 hours per day at a rate of 60 to 70 m^3 per hour before switching off the pump and allowing the sump and channels to refill over a 24-hour period. According to the farmer, this source of water is well known to the local people.

Figure 12.5. This picture was taken just to the south of the Jebel Hamrat al Fidan and shows how a farmer has tapped the groundwater held close to the surface by the granitic barrier by digging a network of trenches to expose the water. The water is pumped from the trench and used to irrigate fields of watermelon.

The course of the wadi floods is deeply incised into the bedrock in the upper reaches of both the Wadi Ghuwayr and the Wadi Dana. A flood would be less able to scour the resistant volcanic rocks, which form the lower reaches of the Wadi Ghuwayr and the Wadi Dana, than the unconsolidated material found in the alluvial plain in the Wadi Fidan. The energy of the flood will dissipate into the alluvial plain causing incision. This incision will be enhanced by the lowering of the water levels in the Dead Sea over the Holocene causing headward erosion (Chapters 6 and 7, this volume). The channels in the Wadi Ghuwayr and the Wadi Dana are narrow. At some points in the Wadi Ghuwayr, the gorge is little more than 2–3 metres wide. This contrasts with the alluvial plain of Fidan which spans approximately 1 km at its widest point. The rates of channel incision in the Wadi Faynan and the Wadi Fidan are difficult to estimate, but given the rapid drop in the water levels of the Dead Sea the incision may have been rapid. The uncertainty regarding the rate of wadi incision raises the question of whether the water used to irrigate the ancient field system was derived from the Wadi Faynan by flood irrigation, from the slopes behind the field system by rainfall harvesting or by abstraction from upstream in the Wadi Ghuwayr.

It is difficult to determine the key aquifer recharge mechanism in the wadi without recourse to a detailed hydrogeological study, in particular the collection of subsurface water level and soil moisture data from piezometers and tensiometers, respectively. The headwaters of both the Wadi Ghuwayr and the Wadi Dana are formed mainly from limestone with a partial mantle of colluvium on the plateau. These headwaters are likely to act as the main recharge area. During flood events, aquifer recharge will also take place through transmission loss whereby flood water percolates from the channel substrate, which is predominantly constituted by unconsolidated sands and gravels, to the aquifers below. A general review of hydrological processes in semi-arid regions by Wheater (2002) has noted the importance of transmission loss as a recharge mechanism. In a study in Saudi Arabia, 75% of the water entering the stream bed was shown to be transmitted to the underlying geology (Sorman and Abdulrazzak, 1993). In the upper reaches of the Wadi Ghuwayr and Dana such transmission losses may act to recharge the limestone and sandstone aquifers below. In the lower reaches, especially the alluvial plain in the Wadi Fidan, this water will be transferred deeper into the sedimentary detritus. These transmission losses are highly variable in space and time and can vary significantly between rainfall events. For example, in southern Africa, recharge was estimated to vary between 1 and 150 mm for a mean annual rainfall of approximately 400 mm year^{-1} (Selaolo, 1998).

12.4 SOILS AND VEGETATION

A soil pit was dug in Wadi Faynan as part of the project Jordan Soils and Land Management (2006). The soil pit was dug 9.5 km east of the Hamrat al Fidan (35° 27′ 37″ E; 30° 37′ 16″ N) in the alluvial terraces of the Wadi Fidan. The soil depth was recorded as 102 cm and the soil observed to be fine-loamy, calcareous, hyperthermic and from the family of Typic Calciorthids (United States Department of Agriculture (USDA), 1999). The parent material was sedimentary detritus (alluvium) consisting of limestone and sandstone. The surface was covered by stones and capped. Fine (1–2 mm) or very fine (<1 mm) roots were found in the top 54 cm and very fine (<0.5 mm) tubular pores were observed throughout the profile giving rise to a classification of moderately well drained with slow surface runoff.

A soil pit was also dug on the plateau to the south of the Wadi Faynan as part of the same survey. Whilst not in the catchment, the data are useful as they provide an indication of this soil type which extends into the southeast region of the catchment headwaters. Observations from the soil survey note that the soil had a depth of 96 cm at a location 1.5 km NE of Bir Khidad (35° 33′ 12″ E; 30° 27′ 10″ N). The soil was fine-silty, calcareous, thermic and of the family Calcixerollic Xerochrepts. Cereals were planted and stones were present on the soil surface. The soil was well drained and surface runoff classified as 'medium'; in the top 12 cm, fine and woody roots were present and from 12 to 96 cm fine roots were found. Fine and spherical pores (0.5–2 mm) were found from 0 to 38 cm. From 38 to 96 cm very fine (<0.5 mm) tubular pores and a few very fine (<1 mm) fibrous roots were found. Stones and gravels were found beneath the soil.

Both profiles suggest that the channel alluvium and the plateau soils are well drained, though the behaviour of all the catchment soils is likely to be very different under storm conditions in comparison with drainage following a storm. The infiltration capacity of soils in semi-arid regions can be reduced as the impact of raindrops will loosen the soil surface and rearrange the soil particles to form a crust. This crust promotes infiltration-excess overland flow, though the surface runoff generated may re-infiltrate as it flows to an area of higher permeability (Beven, 2002).

The 1:50,000 topographic mapping shows vegetation growth along the main channel in the Wadi Ghuwayr, and a widespread area with sparse vegetation cover in the upper catchment of both the Wadi Ghuwayr and the Wadi Dana. Hammam Adethni, a tributary of the Wadi Ghuwayr, was found to be densely vegetated close to the wadi channel in which the flow was sustained by spring water (Lancaster et al., 2007; Mithen et al., 2007a). Hammam Adethni lies on the more southerly of the two major faults in the Wadi Faynan. A major spring was found in Hammam Adethni during the 2007 field season though its location was difficult to identify exactly owing to a lack of available detailed mapping. An approximate location was determined on return from the field, using Google Earth™, as 30° 36′ 57″ N, 35° 32′ 49″ E. The channel of Hammam Adethni lies at the contact between the sandstone and porphyrite which is likely to form a spring line. The spring water temperature was 29.7 °C which is difficult to interpret. Such a temperature may be indicative of water derived from a deep source perhaps transmitted upward along the geological fault, or alternatively water derived from a shallower geological contact which has then ponded in a surface pool and been subsequently warmed by solar radiation.

The present-day woodland of Hammam Adethni and the archaeobotanical remains at WF16 are described by Mithen *et al.* (2007a) and the flora of the Dana Biosphere reserve by Al-Qawabah *et al.* (2003). The vegetation of Hammam Adethni was divided into four distinct zones by Mithen *et al.* (2007a): woodland, maquis, porphyry steppe and sandstone steppe. The densely vegetated valley bottom contains little diversity, and in the woodland the vegetation mainly consisted of *Salix acmophylla*, *Populus euphratica*, *Nerium oleander*, *Tamarix amplexicaulis* and *Ficus carica*. The area of maquis vegetation was dominated by *Phoenix dactylifera*, *Smilax aspera*, *Phragmites australis* and *Arundo donax*. The vegetation growing on the porphyry contains fewer species than on the sandstone steppe south of Hammam Adethni, and the vegetation on the steppes was sparser than in the valley bottom next to the running water. Mithen *et al.* (2007a) found evidence of a sufficient overlap between the species located at Hammam Adethni and those present in archaeobotanical assemblages from WF16 to imply that the riverine woodland provides an analogue for the predominant type of woodland in the Neolithic at the Wadi Faynan. Mithen *et al.* also go further, speculating that this type of woodland was more extensive in the Wadi Ghuwayr, especially in the lower areas and on the land immediately adjacent to WF16.

Today, in the area of the Wadi Faynan between the Khirbet Faynan, the ancient field-system and Hamrat al Fidan, the landscape is denuded and there is an absence of soil organic matter (Tipping, 2007). The true extent of the ancient woodland in the wadi is unknown. Archaeological evidence suggests that trees were present in the lower reaches, though it is debated whether these were removed predominantly during the Roman-Byzantine period (Barker *et al.*, 2007b), or in the early twentieth century by the Ottomans to build the Hijaz Railway (Palmer *et al.*, 2007). The former argument is supported by a reduction in tree pollen in the sediments (Hunt *et al.*, 2004).

Runoff coefficients are the ratio between surface flow and the generating precipitation. Studies of runoff coefficients measured at the hill-slope and catchment scales across a range of hill-slope lengths and catchments in semi-arid areas show a decrease as either the hill-slope length or catchment area increases (Beven, 2002). This occurs because the rainfall is often highly localised, and though runoff may be generated locally in response to the rainfall, as the water flows across the soil surface it re-infiltrates as it moves to an area of higher permeability. A vegetation canopy or groundcover helps to protect the soil surface from forming a crust during high-intensity rainfall events, and roots form conduits within the soil thereby aiding infiltration. Many studies have concluded that vegetation is the dominant control of runoff generation in semi-arid environments (Beven, 2002 and references therein). In addition, rain falling on a vegetation canopy may evaporate or may drip to the soil below either directly, as throughfall, or by trickling down the stems or trunks, as stem flow. Thus vegetation can reduce surface runoff by channelling water to the soil and by storing water which subsequently evaporates. Plants that grow in semi-arid or arid regions are adapted to harvesting throughfall and stem flow (Slayter, 1965).

12.5 PRECIPITATION

The predominant storm track for rainfall over Jordan is from the Mediterranean (Alpert *et al.*, 2004; Enzel *et al.*, 2008; this volume, Chapter 2). The rain gauge network in southern Jordan is too sparse to provide a detailed spatial characterisation of the rainfall patterns. A spacing of less than 1 km between gauges is recommended to provide sufficient detail to begin to relate rainfall to runoff in semi-arid and arid areas (Wheater, 2002). Despite the limitations of the existing data, a map of the spatial patterns in rainfall over Jordan has been drawn (Figure 12.6). This map provides an indication of the variations in rainfall amounts in the Wadi Faynan.

Rainfall patterns in the region of Faynan are dominated by the orographic effect of the Rift escarpment, and the area of highest annual rainfall follows a north–south line between Kerak, Tafilah and the Wadi Musa. Mean annual rainfall across the Wadi Faynan catchment decreases from 400 mm per year at El Atate on the plateau to 50 mm in the Rift Valley floor; the latter is in a rain shadow, being surrounded by highlands. The mean annual rainfall at Shawbak, which is located on the plateau on the southern boundary of the Faynan catchment, is 312 mm year^{-1} with a standard deviation of 136 mm year^{-1} (Tarawneh and Kadioglu, 2003). Rainfall generally occurs between October and May, as elsewhere in Jordan. During winter, the precipitation can fall as snow on the plateau (Al-Qawabah *et al.*, 2003).

12.6 TEMPERATURE

The temperature extremes in the Dana Reserve which are assumed representative of those in the Wadi Faynan are typically a mean of 9 °C and 27 °C during January and August, respectively (Al-Qawabah *et al.*, 2003). East of the Jordanian Highlands, first steppe and then desert result from the decrease in relative humidity and precipitation and the increase in temperature (Hunt *et al.*, 2004). The average relative humidity in the Jordan Valley ranges from 30% during the day to 70% during the night (Binnie and Partners (Overseas) Limited, 1979).

12.7 EVAPORATION

Potential evapotranspiration, estimated at Ghor Safi from December 1965 to 1979, varies from a minimum of 1.8 mm day^{-1} in December and January to 7.3 mm day^{-1} in June,

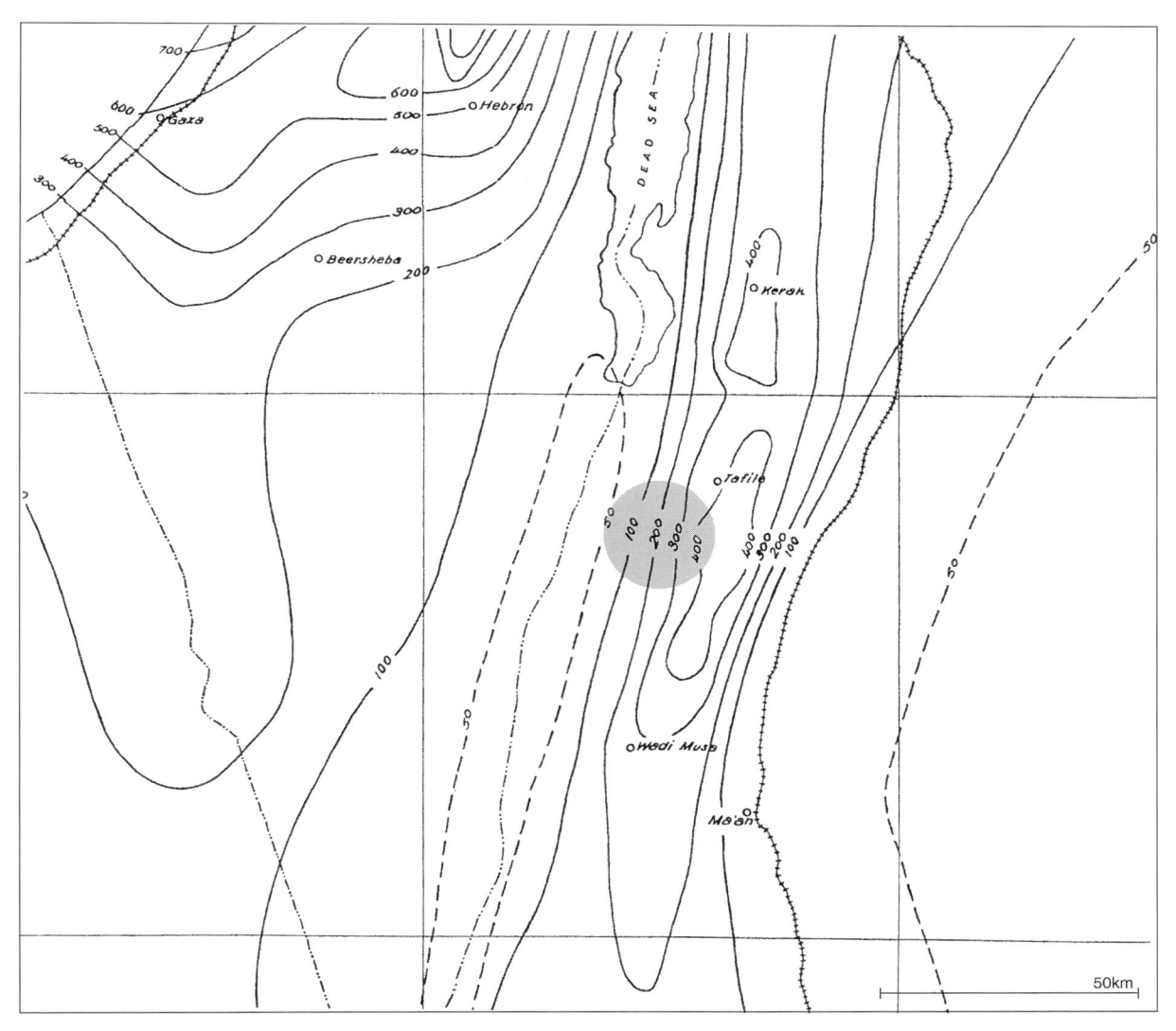

Figure 12.6. Rainfall patterns (isohyets) in the region of the Wadi Faynan (marked by the circle). Source: Department of Civil Aviation, Jerusalem, 1937–38.

July and August (Binnie and Partners (Overseas) Limited, 1979); the summer evapotranspiration rate is approximately four times that in winter. The mean annual potential evaporation measured at Tafilah during the period 1999 to 2003 was 1,978 mm year^{-1}. Compared with the mean annual precipitation of 186 mm year^{-1} over the same period at Tafilah, the annual potential evaporation is approximately 11 times greater (EMWATER, 2005; Hashemite Kingdom of Jordan Meteorological Department, 2006). Given the higher rainfall in winter and the lower evaporation rates, the optimum period for cropping is winter. A map of the Budyko dryness ratio, which calculates the number of times the mean annual net radiation could evaporate the local mean annual precipitation, shows that the Wadi Faynan has a dryness ratio between 3 and 10, which is classed as desert (Hare, 1997).

12.8 CONTEMPORARY HYDROLOGY AND HYDROCHEMISTRY

12.8.1 Methodologies

The base flows in the Wadi Ghuwayr and the Wadi Nichel were measured during the 2006, 2007 and 2008 field visits (Table 12.1) by the velocity–area method in which the flow velocities were measured at selected verticals of known depth across a measured section of the river. The velocity was measured using an OTT velocity meter at all sites apart from site WGDS3 where a Nixon Streamflo 422 micro flowmeter was used as the channel cross-section was wide but shallow (9 cm or less) and the OTT velocity meter was too large to be submerged fully. In total eight sites were surveyed. Changes in the stream bed cross-section occurred between years owing to the disturbance of the bed during flood events.

Table 12.1 *Baseflow measurements in the Wadi Ghuwayr made during the 2006, 2007 and 2008 field seasons*

Wadi	Site name	Latitude	Longitude	Sample date	Baseflow ($m^3 s^{-1}$)
Nichel	2008_4	30° 36' 34"	35° 31' 51"	07 May 2008	0.025
Nichel	2008_3	30° 36' 45"	35° 31' 45"	07 May 2008	0.016
Ghuwayr	2008_2	30° 36' 34"	35° 31' 51"	07 May 2008	0.032
Ghuwayr	2008_1	30° 37' 08"	35° 30' 47"	07 May 2008	0.020
Ghuwayr	WGDS3	30° 36' 45"	35° 31' 45"	29 Apr 2007	0.093
Ghuwayr	WGDS2	30° 37' 14.2"	35° 30' 33.8"	24 Apr 2006	0.018
Ghuwayr	WGDS1	30° 37' 10"	35° 30' 42.8"	24 Apr 2006	0.024
Fidan	2008_Fidan	30° 39' 11"	35° 23' 48"	09 May 2008	0.012

The peak flows were estimated using the Darcy-Weisbach technique as described by White (1995). This technique allows an estimation of the peak flood to be made based on the hydraulic properties of the channel which are the slope, the cross-section area and the streambed roughness.

Water chemistry data were collected during the 2007 and 2008 field visits. Samples were collected at each of the eight sites and filtered in the field using a 0.2 μm Whatman filter. The samples were kept as cool as possible and in the dark before being analysed upon return to Reading, UK. The samples were analysed for stable isotopes, the major and minor ions and metals. The major cations and trace elements were determined by inductively coupled plasma – optical emission spectroscopy (ICP-OES) using a Perkin-Elmer Optima 3000 Radial View to analyse the filtered samples prepped with nitric acid to 5% v/v. The anions (NO_3, NO_2, SO_4, Cl and F) were determined by ion chromatography using a DIONEX DX 500 system on the filtered samples. The stable isotopes were measured using a VG ISOGAS stable isotope mass spectrophotometer on the filtered samples. The data are limited in time and space but form a baseline survey and provide preliminary data on the flow pathways and water sources.

12.8.2 Results

Despite two flumes being present in the Wadi Faynan, one in the Wadi Ghuwayr and one in the Wadi Dana, no data from these are readily available. In light of the level of scour observed at the concrete gauging structure in the Wadi Dana it is uncertain that any rating equation would have remained valid during the course of a flood event. Given the lack of readily available discharge measurements in the wadi, the base flow and peak flow measurements were made to help characterise the flow range and to constrain the numerical hydrological model. Daily flow data were available for the Wadi Hasa at Ghor Safi from November 1962 to September 1963. The Wadi Hasa is located to the north of the Wadi Faynan. It is a much larger catchment (2,520 km^2) but drains a continuation of the same plateau and scarp slope as the Wadi Faynan. As such, it was hoped that the available flow measurements from the Wadi Hasa could be used to help understand the hydrological dynamics of the Wadi Faynan. This proved not to be the case since the Wadi Hasa and the Wadi Faynan are very different in terms of hydrology. Within the Wadi Hasa there is a large playa lake, Qa' el Jinz, which floods during the wet season and subsequently sustains the base flow in the Wadi Hasa. Given the differences in residence times between the lake and the aquifers of the scarp slope, it becomes uncertain how the hydrological response of the Wadi Hasa and the Wadi Faynan will compare without recourse to detailed hydrological measurements which are unavailable.

The observed base flow measured close to WF16 shows a range between 0.02 and 0.09 $m^3 s^{-1}$. Downstream of the springs in the Wadi Nichel, a transmission loss of approximately 0.014 $m^3 s^{-1} km^{-1}$ was estimated based on the measured base flow at Ghuwayr_2008_4 and Ghuwayr_2008_3 (Figure 12.7). Downstream of the confluence between the Wadi Nichel and Hammam Adethni, the flow increases when the two streams join. Downstream of this confluence, further transmission losses are evident from Ghuwayr_2008_2 to Ghuwayr_2008_1 as the water passes along the channel cut into the aplite-granite and porphyrite. As the igneous rocks are less permeable than the sandstone, the transmission loss drops to 0.004 $m^3 s^{-1} km^{-1}$. These estimates of the transmission losses relate to base flow. It is unclear how these transmission losses would change under flood conditions. Downstream of the aplite-granite and porphyrite on the alluvial plain, the flow disappears into the deep sands and gravels. The spring at the Wadi Fidan suggests that the water flows at depth in the alluvial plain only to be forced close to the surface at the granitic barrier of the Jebel Hamrat al Fidan.

The base flow measured in Wadi Ghuwayr on the 29 April 2007 was higher than the other measurements (Table 12.1). Analysis of rainfall data from the nearest available gauge at Ma'an available for the same period (30° 12' E 35° 48' N; National Climatic Data Centre, 2009) showed that the wet season of 2005–06 (October 2005 to April 2006) was wetter than equivalent periods for 2006–07 and 2007–08 with 64, 40 and

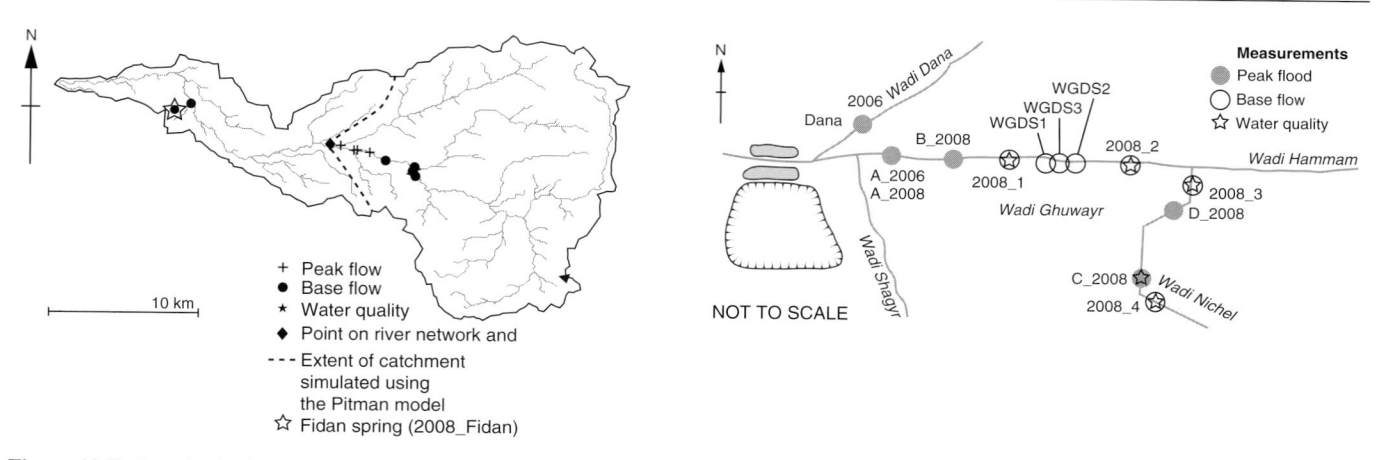

Figure 12.7. Sample site locations in the Wadi Faynan from the 2006, 2007 and 2008 field seasons.

27 mm of rainfall falling at Ma'an for each respective period. However, during the wet seasons of 2005–06 and 2007–08, the rain fell in January and February only. For the period October 2006 to April 2007, the rainfall distribution over these months was more even, as rain fell in each month from December 2006 to April 2007, and this recent rainfall gave rise to the higher base flow observed in April 2007. This observation suggests that, during the wet seasons, the baseflow component comprises spring water drawn from the deeper aquifers and a shallower source, most likely the channel alluvium, which acts as a short-term store of the recent rainfall and responds rapidly.

12.8.3 Water chemistry

The $\delta^{18}O$ ratios increase from −6.1 to −5.3‰ along the channel in the Wadi Nichel at the sandstone/carbonate contact as water flows from its spring source and evaporates as it progresses downstream to the confluence with Hammam Adethni. Immediately downstream of the confluence between Hammam Adethni and the Wadi Ghuwayr, the $\delta^{18}O$ values decrease to −7.5‰, owing to mixing with the waters from Hammam Adethni which have a lower proportion of ^{18}O due to the relatively short distance between the spring source and the confluence. The $\delta^{18}O$ ratio of the stream water then increases to approximately −5.8‰ as the water evaporates as it flows along the Wadi Ghuwayr before disappearing from the surface as it infiltrates into the alluvial plain downstream of WF16. The spring at Hamrat al Fidan also has a $\delta^{18}O$ ratio of −5.8‰ which provides evidence that the spring water is derived from water that infiltrates into the alluvial plain near WF16, and then moves at depth without further evaporation, to resurface as it follows the upthrusted granitic barrier. A deep source of the Fidan spring would be expected to cause a lower measured $\delta^{18}O$ ratio. The patterns in the $\delta^{18}O$ data from the Wadi Nichel to the Jebel Hamrat al Fidan spring are repeated in the $\delta^{2}H$ data.

The anions show a general increase in concentration between surface water samples taken in the Wadi Ghuwayr and the spring water at Jebel Hamrat al Fidan. The increase in anion concentrations is most apparent in the chloride measurements which increase from approximately 70 to 220 mg l^{-1} between the Wadi Ghuwayr and the Fidan spring, nitrate increases from 1.7 to 8.1 and sulphate from 75 to 226 mg l^{-1}. An increase is also evident in the cations: sodium increases from 51 to 103, potassium from 2.5 to 4.0, magnesium from 35 to 61 and calcium from 64 to 121 mg l^{-1}. A possible explanation for this increase is evaporation. However, if this is the mechanism for these increases it contradicts the interpretation of the stable isotope data, which suggests little or no evaporation along the sub-surface pathway from Wadi Ghuwayr to the Fidan spring. Another possible explanation, consistent with the observed water quality, is that evaporation is minimal but the anion and cation concentrations are modified by the Wadi Fidan alluvium or possibly altered by leachates from wastes at Al Qurayqira entering the groundwater during storm events.

A worrying aspect of the water chemistry at the Wadi Faynan is the high concentration of arsenic found in the Wadi Ghuwayr and the Jebel Hamrat al Fidan spring, at approximately 20 μg l^{-1}. This concentration exceeds the European health-based standard of 10 μg l^{-1}.

12.9 CONCEPTUAL MODEL

A conceptual model of the key water pathways and stores in the Wadi Faynan is shown in Figure 12.8. This model is built by integrating all the knowledge ascertained from the review of the existing and newly collected data. The following description and assumptions were used to build the conceptual model from the perceptual one following the methodology and definitions of Smith and Smith (2007). The model assumptions are as follows:

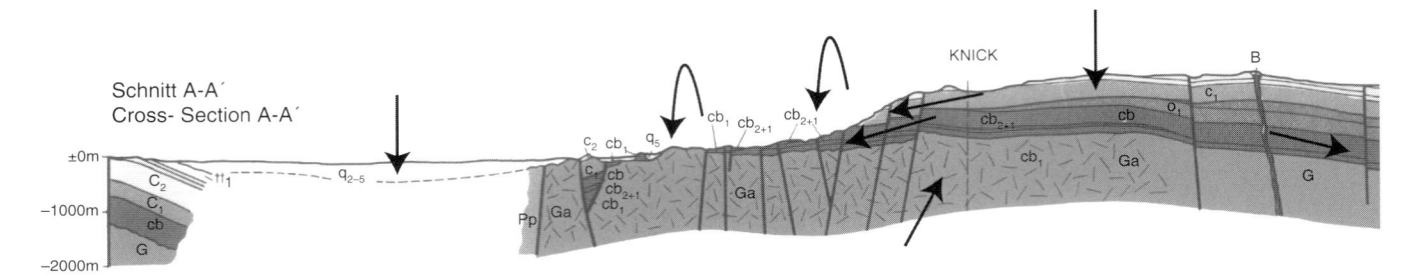

Figure 12.8. A conceptual model of the key water stores and pathways in the Wadi Faynan. Precipitation on the limestone and plateau soils in the upper reaches is likely to be the key aquifer recharge mechanism. The water will flow laterally through the limestone and sandstones until it emerges at a contact point between the two or at a contact between the sandstone and the aplite-granite. Re-infiltration (transmission loss) occurs as the water flows along the channel network. The Jebel Hamrat al Fidan forces the water to return close to the surface. Base map source: Geological Map of Jordan 1:250,000, prepared by Geological Survey of Germany, Hannover 1968 (Sheet: Aqaba-Ma'an). Reproduced with permission. Not to scale. See colour plate section.

- the major aquifers are defined by the catchment boundary and the major aquifers are the limestone and the sandstone;
- groundwater recharge of the limestone and sandstone aquifers occurs through the limestone and plateau soils in the upper reaches of the Wadi Faynan around Dana and Shawbak;
- this recharge is supplemented by transmission losses from the main wadi channels to the underlying aquifers and the shallow channel alluvium;
- springs occur at the contact between the limestone and sandstone and between the sandstone and Precambrian volcanic rocks;
- the Precambrian volcanic rocks act as an impermeable layer, keeping the water near the surface as it flows past WF16 before the sand and gravels in the channel deepen in the Wadi Fidan alluvial plain and the water flows beneath the surface, possibly along the contact with the underlying aplite-granite;
- the key pathways are lateral perennial flows through the limestone and sandstone with surface overland flow generated during rainfall events;
- snow does fall during winter in the headwaters of the catchment but it is assumed that this will infiltrate into the well-drained soils upon melting.

12.10 NUMERICAL MODEL

The conceptual model was realised as a numerical model through an implementation of the Pitman model (Pitman, 1973). The Pitman model was chosen as the basis to represent the hydrology of the Wadi Faynan because it is a model developed in South Africa for semi-arid hydrological conditions and forms a trade-off between model complexity, data requirements and useful model output appropriate for aims of this study. The Pitman model is designed to be applicable at the catchment scale and was applied at the daily time-step in this study to investigate flood characteristics. The purpose of this model application is not to quantify flood flows exactly. That cannot be achieved in this case because of a lack of observed time-series flow data with which to rigorously assess model performance. Rather, the purpose is to define a hydrological model as a best estimate of how the hydrology functions and then run scenarios to explore how changes in rainfall amounts and vegetation characteristics affect flood characteristics and the base flow. The latter provides a smaller but more secure water source.

12.10.1 Model set-up and calibration

The Pitman model was applied to the Faynan catchment, defined from a point on the channel network adjacent to the ancient field-system and immediately downstream of the confluence of the Wadi Ghuwayr and the Wadi Dana (Figure 12.7). At this point the upstream contributing area is 115 km^2. This was done since the lowest point at which the measurements of both the base flow and peak floods were made was the outflow of the Wadi Ghuwayr and the Wadi Dana, thus allowing estimates of the base flow and peak flow to be made at the confluence. It was extremely difficult to survey the channel downstream of the confluence where the alluvial plain flares to a width of approximately 1 km. Modelling the catchment flows to a point on the channel network adjacent to the ancient field-system also removes the complication of simulating the transmission losses and water residence (or transit) times in the alluvial plain of the Wadi Fidan and eliminates the need for a definitive understanding of the source of the Fidan spring within this model-based assessment.

This application of the Pitman model required daily estimates of rainfall and the potential evaporation. The use of the rainfall data from Tafilah was appropriate for the model application. Tafilah is located on the plateau 18 km north of Dana and receives similar rainfall to the upper reaches of the Wadi Faynan.

Moreover, a substantial daily record of rainfall was available from 1 October 1937 to 30 April 1974 at Tafilah, allowing the model to be run for a relatively long period which is important when making an assessment of flood magnitude and frequency.

An estimate of the annual variation in potential evaporation was derived from monthly measurements of wind speed, sunshine hours, relative humidity and air temperature available for a 2-year period using the Penman equation. The data were available for the meteorological station at Ma'an. This 2-year estimated time-series was repeated to form a time-series of 36-years, the same length as the observed rainfall time-series. This repetition of the 2-year time-series, whilst a clear over-simplification of the changes expected during the 36-year period, was done to allow progression with the model application.

It is practically impossible to separate the effects of transmission loss from those of infiltration and deep percolation to the underlying aquifer when estimating groundwater recharge, so no attempt was made to do so within this model-based assessment. Rather, the combined effects of transmission loss, infiltration to the soil and subsequent deep percolation were considered together as a single groundwater-recharge mechanism.

To apply the model, estimates of the groundwater volume and infiltration rate were also required. From the available geological mapping (Figures 12.2 and 12.3), the depth of the Cenomanian–Turonian limestones was estimated as 10 m and the horizontal extent as 60×10^6 m^2. Similarly, the depth of the sandstone was estimated as 40 m and the extent as 60×10^6 m^2. Assuming a porosity of 0.25, this gave an 'effective' groundwater depth of 6.5 m for the aquifers averaged over the 115 km^2 the catchment calculated using the equations of the Pitman model. The 'effective' depth is that in which the groundwater interacts with the surface channel network.

Observations from the soil pits dug in and close to the catchment suggest a soil depth of approximately 1 m in the alluvial terraces and on the plateau soils in the headwaters between Dana village and Shawbak. The soil surveyors also note that these soils were well drained. These observations contrast with those made during journeys along the Wadi Ghuwayr, Hammam Adethni and Wadi Nichel, where the landscape often consists of steep bare-rock faces or shallow soils on steep slopes covered with stones and gravel, and typically little vegetation cover. As such, the soil depth in the model application was assumed to be 50 cm. With an assumed soil porosity of 0.45, the maximum soil moisture capacity was 225 mm.

Infiltration is notoriously difficult to measure in the field. Double-ring infiltrometers were used during the 2006 and 2008 field visits, but it was felt that the results were misleading. The technique depends on filling the infiltrometer to a depth of approximately 20 cm and then recording the drop over a 30-second time interval. The infiltrometer was topped up after a 5 cm drop in water level to prevent differences in hydraulic head affecting the measurements. In reality, such a head of water would never exist during a rainfall event except, perhaps, in areas of depression storage. A recent study by Crook (2009) measured infiltration rates in the WF4 field-system, obtaining values of 29 and 90 mm hour^{-1}. The double-ring infiltrometer was not refilled during the experiment, potentially giving erroneous results but, despite this and a question about the likelihood of ponding, the rates still provide an indication of infiltration rate.

A review of the literature was also undertaken to help determine infiltration rates for use in the Pitman model application to the Wadi Faynan. In a study of a Typic Calciorthid soil (Aridisol) which is similar to the soils found in the lower Wadi Faynan, an experiment was done in which soil was packed into columns and then watered with simulated rainfall for 120 minutes (Mandal et al., 2008). It was observed that infiltration decreased as the cumulative rainfall increased owing to seal formation. After 20 mm of rainfall the infiltration rate was 37.5 mm hour^{-1}; after 40 mm of rainfall the rate dropped to 12 mm hour^{-1}; after 70 mm of cumulative rainfall, the infiltration rate was below 5 mm hour^{-1}. Other studies using single- and double-ring infiltrometers suggest a range for soils of 10 to 14 mm hour^{-1} for fine-crusted soil to 8 to 24 mm hour^{-1} on an interfluve slope (Abu-Awwad and Shatanawi, 1997; Al-Qinna and Abu-Awwad, 1998, 2001). Abu-Awwad and Shatanawi (1997) studied three field sites with sparse vegetation cover and observed a range of values from 9 to 35 mm hour^{-1}.

The model parameter representing the maximum infiltration rate, ZMAX, was set to 8 mm day^{-1}. This value and the minimum infiltration rate ZMIN, set to 0 mm day^{-1}, were determined through model calibration so as to obtain a runoff coefficient of 36%, thereby within the range of estimated runoff coefficients for studies in arid zones in Spain (Puigdefàbregas and Sánchez, 1996), in the southwestern USA (Dunne, 1978), and from a catchment in the semi-arid zone of Australia (Williams and Bonell, 1988), and within the range of runoff coefficients (6–80%) found in a general review by Wheater (2002). The total simulated runoff over 36 hydrological years (1 October 1937 to 30 September 1973) modelled was 3,640 mm, producing a long-term mean of 101 mm year^{-1}. This compared with a rainfall input over the same period of 10,019 mm, producing a long-term mean of 278 mm year^{-1}. In terms of the field and laboratory infiltration measurements, the calibrated value of 8 mm day^{-1} was lower than the values obtained from the literature review. However, it is difficult to interpret the laboratory and point estimates of infiltration in terms of a value representative of the entire catchment (Beven, 2009). Moreover, the concerns about the methodology for measuring infiltration are also valid for the studies of infiltration rate found in the literature. The Pitman model operates on a daily step, yet rainfall events may last only a few hours. If the modelled and measured infiltration rates are thought broadly representative of a rainfall event then the two rates are similar. Whilst the absolute value of infiltration rate representative of the catchment can be debated, the value of 8 mm day^{-1} is still useful because it provides a benchmark estimate for present-day conditions and to progress the study it can be assumed that sparser or denser vegetation will lower or increase the rate, respectively. The model was also calibrated to give a base flow of 0.03 m^3 s^{-1} immediately upstream of the ancient field-system, which was within the range of the recent observations (Table 12.1).

The threshold for the onset of simulated percolation from the bottom of the soil profile was set to 40 mm by calibration to achieve a situation whereby, over the simulated 36-year period, there was only a minimal change in the stored groundwater. It is unknown whether the groundwater store is truly stable or whether there is a net loss or gain of water with time. However, for the purposes of this study a stable groundwater store provides a basis with which to compare changes in the groundwater during scenario exploration.

Anecdotal evidence suggests that a good year in terms of water availability occurs when there are five rains over the headwaters of the catchment. In discussions regarding hydrological conditions in the Wadi Faynan and the lands around Shawbak and Dana, Lancaster and Lancaster (1999) note:

> People say five rains for the plateau and slopes, or five flowings for the low-lying land above the Wadi Araba, are needed for good crops and natural grazing; these occur on average one year in five. People expect a good year, two or three ordinary years, and one or two poor years.

(Lancaster & Lancaster, citing Bedouin informants, 1999 p. 118)

This anecdotal evidence was used to establish a definition of 'good hydrological conditions' in the Wadi Faynan for farming. Specifically a flow threshold, determined from the simulation results, was defined as 12 m^3 s^{-1}. This value was chosen to give a probability of 0.2 that five or more flows occurred in 1 year over a 5-year period.

To explore the catchment response to changes in the annual rainfall amounts and vegetation cover, simple scenarios were derived as follows:

1. *Rainfall*: The total rainfall was scaled by multiplying by the daily values in the time-series by the following percentages: 20, 40, 60, 80, 120, 140, 160 and 200%;
2. *Vegetation*: Vegetation is known to increase infiltration rates (Bevan, 2002) and to investigate this phenomenon, the model parameter ZMAX, which describes the maximum infiltration, was adjusted through the following series: 0, 5, 8, 35 and 100 mm day^{-1}.
3. *Combination*: To explore combinations of altered amounts of rainfall coupled with different vegetation cover the rainfall and maximum infiltration rates were varied in combination.

These scenarios are simplified and arbitrary. However, given the lack of a flow time-series with which to calibrate the model, it was deemed more appropriate to keep the scenarios simple. In this study the purpose was to gain an overview of the sensitivity of the hydrological response to a range of rainfall amounts and changes in vegetation density rather than precise flood estimates.

12.11 HYDROLOGICAL SIMULATION RESULTS

The output from the model shows that the simulated values fall within the base flow and peak flow constraints identified by field observations (Figure 12.9). The base flow in catchment was observed to be in the range 0.02 to 0.09 m^3 s^{-1} and the model replicates this (Table 12.2). The simulated annual flood ranges from 2 to 98 m^3 s^{-1} and floods that generally occur every year or every two years were simulated to have magnitudes in the range 2 to 17 m^3 s^{-1}, which is within the range estimated by survey. The extreme floods with return periods of 12 to 37 years also lie within the estimated flood flow range when the channel is flowing full.

The flow-duration curve also replicates the ephemeral characteristic of the catchment which is dominated for 95% of the time by base flow (Figure 12.9). For only 5% of the time-period considered do the flows increase in response to precipitation events, and this hydrological behaviour is typical of semi-arid catchments (Bull and Kirby, 2002). However, during floods the increase in flow is rapid with the simulated magnitudes reaching a maximum of 98 m^3 s^{-1} in response to the largest observed rainfall event at Tafilah of 90 mm day^{-1}.

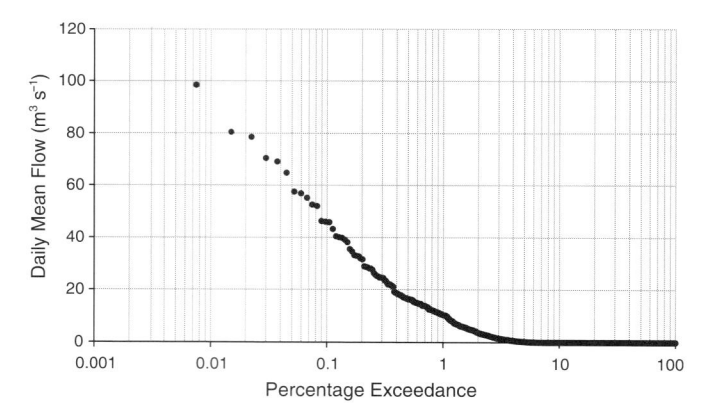

Figure 12.9. The flow duration curve based on the flows simulated in the Wadi Faynan over the period 1937 to 1973. The Pitman model was applied to simulate flows in the Wadi Faynan to a point on the channel network adjacent to the ancient field-system immediately downstream of the Ghuwayr–Dana confluence. The highest simulated flow of 98 m^3 s^{-1} occurred in response to a mean daily rainfall event of 90 mm.

Table 12.2 *Peak floods in the Wadi Ghuwayr and the Wadi Dana estimated using open-channel techniques*

Wadi	Site name	Northing	Easting	Sample date	Annual (m^3 s^{-1})	Bank full (m^3 s^{-1})	Overspill (m^3 s^{-1})
Ghuwayr	A_2006	30° 37' 31"	35° 29' 46"	26 Apr 2006	4	89	103
Ghuwayr	A_2008	30° 37' 30"	35° 29' 39"	05 May 2008	12	31	339
Ghuwayr	B_2008	30° 37' 26"	35° 30' 13"	05 May 2008	2	22	209
Ghuwayr	C_2008	30° 36' 39"	35° 31' 43"	05 May 2008	0.4	4.5	–
Ghuwayr	D_2008	30° 36' 49"	35° 31' 52"	05 May 2008	2	15	–
Dana	Dana_2006	30° 37' 41"	35° 29' 11"	26 Apr 2006	10	88	97

Table 12.3 *Surface and groundwater flow statistics for the Wadi Faynan generated by running the Pitman hydrological model with modified rainfall inputs and/or a modified infiltration rate parameter*

Rainfall (% present day)	Maximum infiltration rate (mm)	Flow statistics (m^3 s^{-1})				Groundwater		Peaks over threshold
		Max	Q10	Q50	Q95	Effective depth (m)	Flow (m^3 s^{-1})	Number of years with five or more flows >12 m^3 s^{-1}
Rainfall scenarios								
20	8	5	0.046	0.044	0.042	6.1	0.04	0
60	8	53	0.046	0.045	0.043	6.2	0.04	2
100	8	98	0.047	0.047	0.046	6.5	0.05	7
140	8	144	0.050	0.049	0.047	6.9	0.05	18
180	8	189	0.053	0.050	0.047	7.0	0.05	28
Infiltration scenarios								
100	0	109	0.047	0.044	0.042	6.0	0.04	26
100	5	101	0.047	0.045	0.044	6.2	0.04	13
100	35	84	0.057	0.053	0.048	7.5	0.06	2
100	100	40	0.061	0.054	0.048	7.9	0.06	1
Combination scenarios								
200	5	214	0.057	0.047	0.047	6.5	0.05	31
200	8	211	0.054	0.050	0.048	7.2	0.05	29
200	100	152	0.115	0.083	0.051	12.0	0.12	2

12.11.1 Rainfall scenarios

All the flow statistics show a prolonged period dominated by base flow, even those where the rainfall is 180 or 200% that of the observed time-series (Table 12.3). This occurs because the magnitude only was changed in the rainfall scenarios. Neither the frequency nor duration of rainfall occurrence was altered. Under the scenario of extreme low rainfall (56 mm year^{-1}), specifically 20% of the rainfall amount measured at Tafilah from 1 October 1937 to 30 September 1973, the groundwater depletes and there are few floods. Most of the rainfall that falls over the catchment infiltrates but this is insufficient to recharge the groundwater. Even at 60% of observed rainfall (169 mm year^{-1}), the flood occurrence is limited and the groundwater depletes, suggesting that this water resource is sensitive to changes in climate. In contrast, with 180% or 200% of the observed Tafilah rainfall (500 or 556 mm year^{-1}, respectively) which is more typical of present-day northwest Jordan, the hydrology is more responsive. The floods become larger and more frequent, and in at least 28 years of the 36, flood events exceed the 12 m^3 s^{-1} threshold (Table 12.3). Specifically, as the rainfall increases from 20 to 200% of the observed, then the number of years where floods are greater than 12 m^3 s^{-1} increases from 0 to 29. Between 100% and 180% of the observed mean annual rainfall, the relationship between mean annual rainfall and flood occurrence is linear. The number of floods then increases only marginally as rainfall is increased above a threshold of 500 mm year^{-1} (180% of the currently observed mean annual rainfall). This occurs because at this magnitude of rainfall, nearly every rainstorm produces channel runoff of 12 m^3 s^{-1} or greater. With increased rainfall the groundwater store also grows, as deep percolation, which recharges the aquifers, is enhanced (Table 12.3).

12.11.2 Infiltration scenarios

With no infiltration, the groundwater depletes from 6.5 m to 6 m owing to a lack of recharge, and surface runoff is the dominant hydrological pathway. Of the 36 years simulated, 26 produce events with a flow of 12 m^3 s^{-1} or more. As the infiltration rate is progressively increased from 0 mm day^{-1} representative of limited sparse vegetation to 35 and 100 mm day^{-1} indicative of greater vegetation abundance, then the groundwater recharge increases, and flood frequency and magnitude is reduced (Table 12.3). Specifically, with infiltration rates of 35 and 100 mm day^{-1}, the simulated effective groundwater depth increased to 7.5 and 7.9 m, respectively; the base flow increased from 0.046 m^3 s^{-1} at calibration to 0.057 and 0.061 m^3 s^{-1}; the maximum flood during the 36-year period decreased from 98 to 84 and 40 m^3 s^{-1}, and the number of years with events greater than 12 m^3 s^{-1} decreased from 7 to 2 and 1. As the infiltration rate increases, for a given annual rainfall amount, the flow-duration curve (not shown) becomes less steep as the floods become less extreme.

12.11.3 Combination scenarios

With low rainfall and low infiltration, the system is responsive with floods generated, albeit of low magnitude (Table 12.3). As the vegetation cover increases with a commensurate increase in infiltration rate, then groundwater recharge also increases, and flood magnitude and frequency decrease. If rainfall is high (200%) and the catchment vegetated (infiltration 100 mm day^{-1}), then the simulated maximum flood increases from 98 to 152 m^3 s^{-1} but the base flow and groundwater effective depth also increase to 0.12 m^3 s^{-1} and 12 m, respectively, and the flood frequency drops to 2 of 36 years having a flow of 12 m^3 s^{-1} or greater.

12.12 DISCUSSION

This examination of the hydrology and hydrogeology of the Wadi Faynan reveals why the location was potentially attractive for early settlement. The presence of springs, due to the geological faults and contacts between different lithologies, makes the locality attractive to sedentary farmers and transient herders because of the reliability of the flow in a semi-arid landscape. Faynan has other attractions based on the geology. These include a ready source of cherts (flints) in the limestone of the plateau which are washed down to the lower reaches, and veins of copper and manganese within the igneous rocks, the former exploited on an industrial scale by the Romans. Thus, following the initial attraction of perennial water, the subsequent finds of cherts and metals would have provided further reasons for settlers to have maintained their presence within the Wadi Faynan. The aplite-granite barrier of the Jebel Hamrat al Fidan is unique along the Rift Valley between Aqaba and the Dead Sea, and this feature is likely to aid moisture retention in the alluvial plain of the Wadi Fidan. The catchment also lies north of other settlements on the scarp slope of the Wadi Araba where spring sources are located, for example Beidha and Petra. Possibly together these would have provided a network of watering points for travellers or herders. The gorges of the Wadi Ghuwayr and the Wadi Dana also provide a route up the scarp slope to the plateau above, where the land is more productive than in the valley, having a soil depth of approximately 1 m and a higher rainfall.

The analysis of the collated data enabled a conceptual hydrological model to be created. This model is summarised in section 12.9, although it is not definitive. In particular, further water quality measurements are required to confirm the degree of evaporation in the alluvial plain of the Wadi Fidan and to provenance the Hamrat al Fidan spring water. The water temperature of the spring at Hamrat al Fidan of 25.6 °C is cooler than the spring in Hammam Adethni, but this single temperature measurement provides no clear evidence as to the depth of the spring-water source. Thus, as this time, it is not possible to confirm that water flows through the Wadi Fidan alluvial plain at depth before emerging at the spring adjacent to Jebel Hamrat al Fidan, being forced to rise to the surface or near surface by the thrusted aplite-granite. However, it is strongly suspected that this is the case.

There is a tendency in the region to dig and line irrigation ponds and pump in water from available surface and near-surface watersources. One such pond was present in the ancient field-system in May 2008. Given the high arsenic content in the Wadi Ghuwayr base flow, then the concentrations of potentially harmful metals and salts in these ponds should be monitored to determine the possible human health effects. This is important. Many ponds are filled and then left uncovered so evaporation will further concentrate the metals. Further reductions in the dilution capacity due to increased evaporation are likely, as temperature is projected to increase by 3 and 5 °C by 2040 and 2100, respectively (Cruz et al., 2007).

The hydrological simulations suggest that, with the same mean annual rainfall (278 mm year^{-1}) as observed from 1937 to 1973 at Tafilah, but a denser vegetation cover giving an increased infiltration rate, then the hydrology of the Wadi Faynan would have been much more conducive to sedentary farming and settlement. With a modelled maximum infiltration rate of 35 mm day^{-1} indicative of a denser vegetation cover than observed at present, water accumulates in the groundwater store producing a higher perennial base flow, and the flood magnitude and frequency are reduced. The simulated removal of the vegetation in the Wadi Faynan demonstrated a severe, negative impact on the water resource. Specifically, the removal had two key effects due to reduced infiltration. The first was a depletion of the groundwater resource through reduced recharge, and the second was greater flood magnitude and frequency. Increased flood frequency occurred in addition to increased flood magnitude because surface runoff that would have previously re-infiltrated into the soil remained on the surface and flowed to the channel network. Though not tested in the applications of the model, it is also possible that the removal of the vegetation canopy will reduce throughfall and stem flow, thus increasing flood magnitude and frequency. A vegetated Wadi Faynan would have been much easier to farm because the vegetation would have helped to provide a higher base flow and promoted the retention of soil moisture.

The results from the model-based assessment are suggestive only, simulating the key alterations in the water stores and pathways in the catchment. It is unclear whether the greatest effect on the present-day landscape of the Wadi Faynan has been a reduction in rainfall during the Holocene or the removal of vegetation. It is also unclear whether the reduction in rainfall caused the decline in vegetation, or if the vegetation was cleared by the Romans for pit props or smelting, or by the Ottomans for railway sleepers. These further details regarding the extent and timing of the changes in vegetation cover are required to provide a more detailed assessment of the hydrological functioning of the Wadi Faynan at key times during the Holocene. This is attempted in work described in Chapter 14 of this volume.

12.13 CONCLUSIONS

This study is amongst the first to both describe and model the hydrology of the Wadi Faynan. Prior to this, relatively little work had focused on the hydrological functioning of this archaeologically important, semi-arid catchment. This study is also one of the first to use a process-based hydrological model to investigate scenarios of changed rainfall and vegetation cover in order to understand how the hydrological functioning of a catchment may affect cultural development. A major component of this work was the collation of a substantial database with which to develop an understanding of the catchment hydrology and hydrogeology. Together, new and historic data were analysed to understand the catchment hydrological and geographic characteristics. This analysis has highlighted the importance of the climate and geology in the determination of the catchment hydrology, the vegetation cover and the establishment of sedentary communities in what is now a harsh landscape with low rainfall and sparse vegetation. Along the scarp slope of the Rift Valley between Aqaba and the Dead Sea, the Wadi Faynan is unique in terms of its geological setting. The tilting of the volcanic rocks and the associated faults gives rise to springs, such as in Hammam Adethni. Other springs occur at the contact between limestone and sandstone in the mid-reaches of the Wadi Ghuwayr. Thus, of the rain falling on the plateau, some will be used to water crops, some evaporated, some will run off to the channel, but most importantly some will recharge the aquifers below and ultimately reappear as perennial spring flow further down the Wadi Faynan.

Unbeknown to the earliest settlers, these were the causes of the availability of water and continued security of supply.

The construction of the perceptual hydrological model and the application of the Pitman model demonstrated the interrelationship between rainfall, vegetation and water availability in the Wadi Faynan. The results of the modelled scenarios show that increased rainfall can, in addition to providing more moisture for crop growth, result in larger, more frequent floods which may have an adverse impact on farming in a landscape denuded of vegetation. However, the presence of vegetation when coupled with higher rainfall is beneficial as more water is retained in the catchment soils and groundwater, and the spring flows are greater. A wetter, more densely vegetated Wadi Faynan would have been more favourable to sedentary farming than the barren landscape present today.

REFERENCES

Abu-Awwad, A. M. and M. R. Shatanawi (1997) Water harvesting and infiltration in arid areas affected by surface crust: examples from Jordan. *Journal of Arid Environments* **37**: 443–452.

Agrar und Hydrotechnik GmbH (1977) *National Water Master Plan of Jordan. Hashemite Kingdom of Jordan*. Amman: Natural Resources Authority.

Al-Najjar, M., A. Abu-Dayya, E. Suleiman, G. Weisgerber and A. Hauptmann (1990) Tell Wadi Faynan: the first Pottery Neolithic Tell in the South of Jordan. *Annual of the Department of Antiquities of Jordan* **34**: 27–56.

Al-Qawabah, M. S., C. Johnson, H. Ramez and M. Y. A. Al-Fattah (2003) Assessment and evaluation methodologies report: Dana Biosphere Reserve, Jordan. Sustainable management of marginal drylands (Sumamad). In *Second International Workshop on Combating Desertification: Sustainable Management of Marginal Drylands (SUMAMAD)*. Shiraz, Islamic Republic of Iran: UNESCO-MAB.

Al-Qinna, M. I. and A. M. Abu-Awwad (1998) Infiltration rate measurements in arid soils with surface crust. *Irrigation Science* **18**: 83–89.

Al-Qinna, M. I. and A. M. Abu-Awwad (2001) Wetting patterns under trickle source in arid soils with surface crust. *Journal of Agricultural Engineering Research* **80**: 301–305.

Alpert, P., I. Osetinsky, B. Ziv and H. Shafir (2004) Semi-objective classification for daily synoptic systems: application to the eastern Mediterranean climate change. *International Journal of Climatology* **24**: 1001–1011.

Barker, G., R. Adams, O. Creighton *et al*. (2007a) Chalcolithic (c.5000–3600 cal. BC) and Bronze Age (c. 3600–1200 cal. BC) settlement in Wadi Faynan: metallurgy and social complexity. In *Archaeology and Desertification: The Wadi Faynan Landscape Survey, Southern Jordan*, Wadi Faynan Series 2, Levant Supplementary Series 6. Oxford: Council for British Archaeology in the Levant/Oxbow Books, pp. 227–270.

Barker, G., D. Gilbertson and D. Mattingly, eds. (2007b) *Archaeology and Desertification: The Wadi Faynan Landscape Survey, Southern Jordan*, Wadi Faynan Series 2, Levant Supplementary Series 6. Oxford: Council for British Archaeology in the Levant/Oxbow Books.

Beven, K. (2002) Runoff generation in semi-arid areas. In *Dryland Rivers – Hydrology and Geomorphology of Semi-Arid Channels*, ed. L. J. Bull and M. J. Kirby. Chichester: Wiley pp. 57–105.

Beven, K. (2009) *An Introduction to Techniques for Uncertainty Estimation in Environmental Prediction*. London: Routledge.

Binnie and Partners (Overseas) Limited (1979) *Mujib and Southern Ghors Irrigation Project – Feasibility Report*. Amman: Ministry of Planning, Jordan Valley Authority.

Bull, L. J. and M. J. Kirby (2002) Dryland river characteristics and concepts. In *Dryland Rivers: Hydrology and Geomorphology of Semi-Arid Channels*, ed. L. J. Bull and M. J. Kirby. Chichester: Wiley pp. 3–15.

Crook, D. (2009) Hydrology of the combination irrigation system in the Wadi Faynan, Jordan. *Journal of Archaeological Science* **36**: 2427–2436.

Cruz, R. V., H. Harasawa, M. Lal et al. (2007) Asia. In *Climate Change 2007: Impacts, Adaptation and Vulnerability. Contribution of Working Group II to the Fourth Assessment Report of the Intergovernmental Panel on Climate Change*, ed. M. L. Parry, O. F. Canziani, J. P. Palutikof, P. J. van der Linden and C. E. Hanson. Cambridge, UK and New York, USA: Cambridge University Press pp. 469–506.

de Vries, J. J., E. T. Selaolo and H. E. Beekman (2000) Groundwater recharge in the Kalahari, with reference to paleo-hydrologic conditions. *Journal of Hydrology* **238**: 110–123.

Dunne, T. (1978) Field studies in hillslope flow processes. In *Hillslope hydrology*, ed. M. J. Kirby and R. J. Chorley. Chichester: Wiley pp. 227–293.

EMWATER (2005) *Prospects of Efficient Wastewater Management and Water Reuse in Jordan, Country Study*. Prepared within the Framework of the EMWATER-Project, 'Efficient Management of Wastewater, its Treatment and Reuse in the Mediterranean Countries'. Amman, Jordan: Al al-Bayt University, Mafraq, Jordan, InWEnt, Amman Office.

Enzel, Y., R. Arnit, U. Dayan et al. (2008) The climatic and physiographic controls of the eastern Mediterranean over the late Pleistocene climates in the southern Levant and its neighboring deserts. *Global and Planetary Change* **60**: 165–192.

Finlayson, B. L. and S. Mithen (2007a) Excavations at WF16. In *The Early Prehistory of Wadi Faynan, Southern Jordan: Archaeological Survey of Wadis Faynan, Ghuwayr and al-Bustan and Evaluation of the Pre-Pottery Neolithic A Site of WF16*, Wadi Faynan Series 1, Levant Supplementary Series **4**, ed. B. L. Finlayson and S. Mithen. Oxford: Council for British Archaeology in the Levant/Oxbow Books pp. 145–202.

Finlayson, B. L. and S. Mithen (2007b) *The Early Prehistory of Wadi Faynan, Southern Jordan: Archaeological Survey of Wadis Faynan, Ghuwayr and al-Bustan and Evaluation of the Pre-Pottery Neolithic A Site of WF16*. Wadi Faynan Series 1, Levant Supplementary Series **4**, ed. B. L. Finlayson and S. Mithen. Oxford: Council for British Archaeology in the Levant/Oxbow Books.

Fujii, S. (2007) Wadi Badda: A PPNB settlement below Fjaje escarpment near Shawbak. *Neolithics* **1/07**: 19–24.

Hare, K. F. (1997) Climate and desertification. In *Desertification: Its Causes and Consequences*. Nairobi, Kenya, Oxford: Pergamon Press pp. 65–167.

Hashemite Kingdom of Jordan Meteorological Department (2006). http://met.jometeo.gov.jo/portal/page?_pageid=113,1,113_56210:113_56242&_dad=portal&_schema=PORTAL (Retrieved August 2009.)

Hauptmann, A. (2007) *The Archaeometallurgy of Copper: Evidence from Faynan*. New York: Springer.

Hazra Overseas Engineering Company (1978) *Jordan Valley Irrigation Project, Stage II Feasibility Study*,Vol. **II**. Hashemite Kingdom of Jordan: Jordan Valley Authority.

Hunt, C. O., H. A. Elrishi, D. D. Gilbertson et al. (2004) Early-Holocene environments in the Wadi Faynan, Jordan. *The Holocene* **14**: 921–930.

Jordan Soils and Land Management (2006) http://alic.arid.arizona.edu/jordansoils/index.html (Retrieved July 2009.)

Lancaster, N., H. Emberson and S. Mithen (2007) The modern vegetation of Hammam Adethni and its palaeo-economic implications. In *The Early Prehistory of Wadi Faynan, Southern Jordan: Archaeological Survey of Wadis Faynan, Ghuwayr and al-Bustan and Evaluation of the Pre-Pottery Neolithic A Site of WF16*, Wadi Faynan Series 1, Levant Supplementary Series **4**, ed. B. L. Finlayson and S. Mithen. Oxford: Council for British Archaeology in the Levant/Oxbow Books. pp. 437–446.

Lancaster, W. and F. Lancaster (1999) *People, Land and Water in the Arab Middle East: Environments and Landscapes in the Bilad ash-Sham*. Amsterdam: Harwood Academic Publishers.

Levy, T. E., R. B. Adams, M. Najjar et al. (2004) Reassessing the chronology of Biblical Edom: new excavations and C-14 dates from Khirbat en-Nahas (Jordan). *Antiquity* **78**: 865–879.

Mandal, U. K., A. K. Bhardwaj, D. N. Warrington et al. (2008) Changes in soil hydraulic conductivity, runoff, and soil loss due to irrigation with different types of saline–sodic water. *Geoderma* **144**: 509–516.

Mattingly, D., P. Newson, O. Creighton et al. (2007a) A landscape of imperial power: Roman and Byzantine Phaino. In *Archaeology and Desertification: The Wadi Faynan Landscape Survey, Southern Jordan*, Wadi Faynan Series 2, Levant Supplementary Series **6**. Oxford: Council for British Archaeology in the Levant/Oxbow Books pp. 305–348.

Mattingly, D., P. Newson, J. Grattan et al. (2007b) The making of early states: the Iron Age and Nabatean periods. In *Archaeology and Desertification: The Wadi Faynan Landscape Survey, Southern Jordan*, Wadi Faynan Series 2, Levant Supplementary Series **6**. Oxford: Council for British Archaeology in the Levant/Oxbow Books pp. 271–303.

Mithen, S., P. Austen, A. Kennedy et al. (2007a) Early Neolithic woodland composition and exploitation in the southern Levant: a comparison between archaeobotanical remains from WF16 and present-day woodland at Hammam Adethni. *Environmental Archaeology* **12**: 49–70.

Mithen, S., B. L. Finlayson, A. Pirie, S. Smith and C. Whiting (2007b) Archaeological survey of Wadis Faynan, Ghuwayr, Dana and al-Bustan. In *The Early Prehistory of Wadi Faynan, Southern Jordan: Archaeological Survey of Wadis Faynan, Ghuwayr and al-Bustan and Evaluation of the Pre-Pottery Neolithic A Site of WF16*, Wadi Faynan Series 1, Levant Supplementary Series **4**, ed. B. L. Finlayson and S. Mithen. Oxford: Oxbow Books pp. 47–114.

National Climatic Data Centre (2009). *National Oceanic and Atmospheric Administration Satellite and Information Service, National Environmental Satellite, Data and Information Service (NESDIS)* Retrieved July, 2009. http://www.ncdc.noaa.gov/oa/climate/ghcn-monthly/index.php

Newson, P., D. Mattingly, P. Daly et al. (2007) The Islamic and Ottoman Periods. In *Archaeology and Desertification: The Wadi Faynan Landscape Survey, Southern Jordan*, Wadi Faynan Series 2, Levant Supplementary Series **6**. Oxford: Council for British Archaeology in the Levant/Oxbow Books pp. 349–368.

Palmer, C., D. Gilbertson, P. Newson et al. (2007) The Wadi Faynan today: landscape, environment, people. In *Archaeology and Desertification: The Wadi Faynan Landscape Survey, Southern Jordan*, Wadi Faynan Series 2, Levant Supplementary Series **6**. Oxford: Council for British Archaeology in the Levant/Oxbow Books pp. 25–57.

Pitman, W. V. (1973) A mathematical model for generating river flows from meteorological data in South Africa. Report 2/73: Hydrological Research Unit, University of the Witwatersrand, Johannesburg, South Africa.

Puigdefàbregas, J. and G. Sánchez (1996) Geomorphological implications of vegetation patchiness on semi-arid slopes. In *Advances in Hillslope Processes*, ed. M. G. Anderson and S. M. Brooks. Chichester: Wiley pp. 1027–1060.

Rollefson, G. (1981) The Late Acheulean site at Fjaje, Wadi el-Bustan. *Paleorient* **7**: 5–21.

Selaolo, E. T. (1998) Tracer studies and groundwater recharge assessment in the eastern fringe of the Kalahari. Unpublished PhD thesis: Vrije Universiteit.

Simmons, A. J. and M. Najjar (2000) Preliminary report of the 1999–2000 excavation season at the Pre-Pottery Neolithic settlement of Ghwair 1, Southern Jordan. *Neolithics* **1/00**: 6–8.

Slayter, R. O. (1965) Measurements of precipitation interception by an arid zone plant community (*Acacia ameura*). *UNESCO Arid Zone Research* **25**: 181–192.

Smith, J. and P. Smith (2007) *Environmental Modelling – An Introduction*. Oxford: Oxford University Press.

Sorman, A. U. and M. J. Abdulrazzak (1993) Infiltration – recharge through wadi beds in arid regions. *Hydrological Sciences Journal – Journal des Sciences Hydrologiques* **38**: 173–186.

Tarawneh, Q. and M. Kadioglu (2003) An analysis of precipitation climatology in Jordan. *Theoretical and Applied Climatology* **74**: 123–136.

Tipping, R. (2007) Long-term landscape evolution of the Wadis Dana Faynan and Ghuwayr. In *The Early Prehistory of Wadi Faynan, Southern Jordan: Archaeological Survey of Wadis Faynan, Ghuwayr and al-Bustan and Evaluation of the Pre-Pottery Neolithic A Site of WF16*, Wadi Faynan Series 1, Levant Supplementary Series **4**, ed. B. L. Finlayson and S. Mithen. Oxford: Council for British Archaeology in the Levant/Oxbow Books pp. 14–46.

United States Department of Agriculture (USDA) (1999) *Soil Taxonomy: A Basic System of Soil Classification for Making and Interpreting Soil Surveys*. Washington DC: United States Government.

Wheater, H. S. (2002) Hydrological processes in arid and semi-arid areas. In *Hydrology of Wadi Systems*, ed. H. S. Wheater and R. A. Al-Weshah. Paris: UNESCO pp. 5–22.

White, K. (1995) Field techniques for estimating downstream changes in discharge of gravel-bedded ephemeral streams: a case study of southern Tunisia. *Journal of Arid Environments* **30**: 283–294.

Williams, J. and M. Bonell (1988) The influence of scale of measurement on the spatial and temporal variability of the Philip infiltration parameters – an experimental study in an Australian savannah woodland. *Journal of Hydrology* **104**: 33–51.

13 Future projections of water availability in a semi-arid region of the eastern Mediterranean: a case study of Wadi Hasa, Jordan

Andrew Wade, Ron Manley, Emily Black, Joshua Guest, Sameeh Al Nuimat and Khalil Jamjoum

ABSTRACT

This chapter is concerned with a model-based assessment of the effects of projected climate change on water security in the rural west of Jordan. The study area is the Wadi Hasa, a large (2,520 km^2) catchment which drains from the Jordanian plateau to the Dead Sea at Ghor Safi. The Wadi Hasa is regionally important in terms of both water resources and archaeology. A substantial database was collated to describe the hydrological functioning of the catchment and a new monthly time-step hydrological model, HYSIMM, was developed and applied within a modelling framework, which also includes the HadRM3 regional climate model and a weather generator, to provide future projections of mean monthly flows. Under the A2 storyline, the climate in the region of Wadi Hasa in 2071–2100 was projected to become drier, with a mean annual precipitation 25% less than the present day, and warmer; winter and summer temperatures were projected to increase by approximately 4 and 6 degrees centigrade, respectively. Spatial differences in the projected precipitation depths and temperatures are apparent across the region. The modelled outcomes suggest that the mean monthly flows will decrease in winter because of the reduced precipitation, and the modelled flows were more sensitive to changes in precipitation than potential evapotranspiration. Overall, the monthly flood flows are predicted to decrease by 22% and the base flow by 7% by the end of the century under the A2 storyline. To maintain cultivation, farmers in the region will need to increase irrigation efficiency, further exploit groundwater resources or rely more on water conveyors to bring water from alternative sources, such as the Wadi Mujib or possibly the Red Sea.

13.1 INTRODUCTION

Water security in Jordan is poor and water availability is a major regional concern in the eastern Mediterranean. By the end of the twenty-first century, climate model projections suggest that mean annual rainfall over the eastern Mediterranean will decrease and near-surface air temperatures will increase by 2 °C (Chapter 4, this volume). These projected changes in climate must be considered within the context of Jordan's development. Between now and 2050, the Jordanian population is expected to increase from 5.94 million to between 8.86 and 13.11 million depending on the fertility scenario assumed (United Nations, 2009). Surface water resources are already exploited fully and the groundwater resources are being 'mined' with increasing vigour to provide water for irrigation, industry, tourism and domestic consumption, mainly within the urban centre of Amman (Chapter 25, this volume). Within the Water, Life and Civilisation project and other contemporary studies, the implications of climate change on the Jordanian water resource have been investigated in terms of the runoff response of the iconic River Jordan (Samuels et al., 2009; Chapter 10, this volume) and the palaeohydrology of the Wadi Faynan, Jawa and Humayma (Chapters 15, 18 and 19, this volume) and in terms of wider regional and global assessments of water resources in arid and semi-arid regions (Ragab and Prudhomme, 2002). This study aims to take these impact assessments further by considering how flow changes, caused by anthropogenic climate change, may affect rural west Jordan in the southern Ghors region, an area important in terms of farming and archaeology. Specifically, there are three objectives:

1. to collate a substantial database with which to assess the water resources of the southern Ghors, west Jordan;
2. to apply and test a new modelling framework for assessing monthly changes in rainfall-runoff using a new hydrological

Water, Life and Civilisation: Climate, Environment and Society in the Jordan Valley, ed. Steven Mithen and Emily Black. Published by Cambridge University Press.
© Steven Mithen and Emily Black 2011.

model, HYSIMM, which has a time step commensurate with typically available hydrological observations;
3. to use the modelling framework to begin to quantify the likely effect of climate change on the water resource of the southern Ghors, with a focus on the largest wadi in the region, the Wadi Hasa.

To achieve the aim and objectives, data were collated from government and engineering reports and stored within a database; this represented a major undertaking before the modelling work could begin. The data were used to apply a modelling framework, consisting of the HadRM3 Regional Climate Model (RCM), a weather generator and the HYSIMM rainfall-runoff model to assess the rainfall-runoff response of the Wadi Hasa under the A2 climate scenario for 2071–2100. Whilst it is recognised that a range of socio-economic storylines and modelled projections of climate futures are needed for a robust assessment, this study is exploratory in nature, linking for the first time an RCM, a weather generator and new monthly rainfall-runoff model.

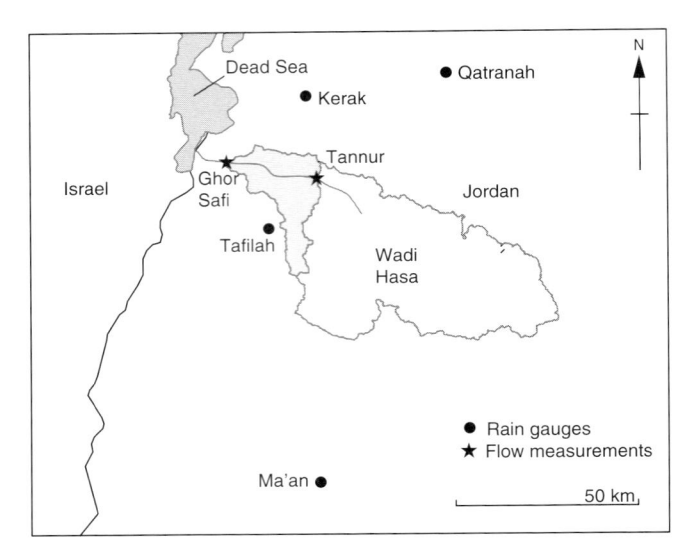

Figure 13.1. A schematic map of the southern Ghors which includes the Wadi Hasa.

13.2 STUDY AREA AND DATA RESOURCE

The southern Ghors is located in western Jordan (Figure 13.1). Situated on the southeast rim of the Dead Sea, the Ghors includes the communities of Safi, Fifa, Mazra'a and Haditheh. The environs of Ghor Safi are an area of intensive vegetable production. In addition to agriculture, the Arab Potash Company is a major employer in the region, its business being the extraction of minerals, mainly potash, from the waters of the Dead Sea. Within the southern Ghors, the Wadi Kerak (also known as Karak), Wadi Numeira, Wadi Hasa, Wadi Fifa and Wadi Khanzeira drain from eastern scarp slope of the Jordanian plateau to the Dead Sea Basin, all located to the south of the Wadi Mujib. This is the same scarp slope on which the Wadi Faynan is located, but the southern Ghors lie some 40 km to the north of the Wadi Faynan. Given their proximity to the Wadi Faynan, the climate and geology of the wadis draining from the plateau to the Dead Sea are similar to those of the Wadi Faynan. The Wadi Numeira (101 km^2), Wadi Fifa (155 km^2), Wadi Khanzeira (180 km^2) and Wadi Kerak (200 km^2) are similar in size to the upper part of the catchment of the Wadi Faynan (115 km^2) studied in other hydrological and archaeological components of the Water, Life and Civilisation project (Chapters 12 and 15, this volume). This study focuses on the Wadi Hasa, one of the largest wadis in Jordan, draining an area of 2,520 km^2. The Wadi Hasa extends from Safi, where it drains into the Dead Sea, to El Huseiniya in the north, Jurf ed Darawish in the south and Tuwayyil Ash Shihag in the southeast (not shown in Figure 13.1). Moreover, flow data are available for two flow gauging stations. The archaeology of the Wadi Hasa has been extensively studied (Marks, 1999; Delattre, 2000; Hill, 2004) but, in this study, the archaeology is not considered. Rather the focus is on an assessment of the present and future rainfall-runoff response.

The climate of the southern Ghors is characterised by hot, dry summers and cool winters during which the majority of the precipitation occurs between November and March. Owing to the increase in altitude of approximately 1,400 m between the valley floor (300 m below sea level) at Safi and the Eastern plateau (1,100 m above sea level), there is great variability in the patterns of precipitation over the southern Ghors (Vita-Finzi, 1966). The spatial variations in rainfall are dominated by the rain shadow in the Dead Sea valley and by the orographic effect of the Rift escarpment edge; the area of highest annual precipitation follows a north–south line between Kerak, Tafilah and Wadi Musa (Figure 12.6). The mean annual precipitation on the plateau is approximately 400 mm year^{-1} compared with 50 mm year^{-1} on the Dead Sea Valley floor and on the Jordanian plateau in the vicinity of Ma'an (Hashemite Kingdom of Jordan Meteorological Department, 2006).

The geology of the catchments draining to the southern Ghors consists mainly of fluvial deposits and aeolian sands from the Quaternary period in the Dead Sea Valley around Safi and to the southern extent of the Wadi Hasa on the Jordanian plateau; beneath the alluvium of the Safi plain are limestones from the Eocene/Oligocene; the scarp and dip slopes of the plateau comprise Cenomanian–Cretaceous (Burj) limestones and Cenomanian–Cretaceous sandstones; Cambrian sandstones; Upper Proterozoic Conglomerates; and undifferentiated, igneous rocks from the Precambrian (Bender, 1968). The igneous basement, the Aqaba granite complex, is found at depths of 1,000 to 3,000 metres below sea level (Bender, 1968). The area was faulted, mainly during the Cretaceous and Quaternary periods, resulting in both lateral and vertical displacements in the local geology, and the faults parallel to the main rift are wrench or shear faults (Food and Agriculture Organization (FAO), 1970; Tipping, 2007). Quaternary, possibly Neogene, basalt intrusions are evident and associated with vertical faults.

According to Tleel (1963), the sandstones in the Wadi Hasa are approximately 200 m thick and are 'poorly sorted, dark red, cross-bedded with canyons formed along jointing plains. Well rounded pebbles are found close to the base.' The 30-m band of marl and shale 'which contains no fossils and plant remains' (Tleel, 1963) appears in the geology maps of Bender (1968) in the cross-section which runs 6 km north of Karama to Sweilah and Zarqa. An outcrop of this marl and shale is evident on the geological map of Bender (1968) 5 km north of the point where the Wadi Mujib discharges to the Dead Sea, but this rock type does not appear in the cross-section of Bender (1968) which passes south of the Wadi Mujib and north of the Wadi Hasa through the Wadi Kerak. The main wadi channel of the Wadi Hasa is formed on the Wadi el Hasa fault, a shear fault with a vertical displacement of 600 m. There is geomorphic evidence that this and other vertical displacements have caused alternative aggradations and degradations of the major wadi channels (Binnie and Partners (Overseas) Limited, 1979). In the lower reaches, where the Wadi Hasa flows across the valley floor, alluvial sediment is extensive, and further channel erosion of the alluvium has formed terraces which can reach 5 m deep. The Wadi Afra enters the Wadi Hasa between Safi and Tannur, and springs are located in this tributary wadi providing base flow in the lower reaches of the Wadi Hasa.

The sandstone and limestone aquifers of the Wadi Hasa are recharged on the escarpment to the east and the Safi plain is recharged from periodic floods (Tleel, 1963). These aquifers are also recharged from the perennial flows which are diverted to irrigate the Safi plain, and the Safi plain aquifer is also recharged through faults in the alluvial fan. The alluvial fans between Safi and Fifa constitute the main aquifer which has been estimated to have a capacity of 75×10^6 m^3 assuming an alluvium volume of 500×10^6 m^3 and a porosity of 0.15 (FAO, 1970; Parker, 1970). In the alluvial fans, the water table is only 0 to 7 m below the surface. In 1962 there were five main springs: Ain Um El Thiker (300 m^3 hour^{-1}); Ain 'Abata (72 m^3 hour^{-1}); Ain El-Ytou (15 m^3 hour^{-1}); and Kanat Bab Ezzikanu Number 1 (353 m^3 hour^{-1}) and Number 2 (90 m^3 hour^{-1}) (Masri, 1962). The natural discharge in the Wadi Hasa from springs on the alluvial plain is approximately 700–900 m^3 hour^{-1} (Tleel, 1963). Other springs are found on the escarpment and the water from these springs infiltrates into the aquifers or feeds the perennial flow. The wadi becomes a perennial stream as opposed to the intermittent stream on the plateau, owing to springs in the lower reaches (Winer et al., 2006). Water was also drawn from hand-dug and pumped wells from depths between 10 and 70 m at the rate of 100–140 m^3 hour^{-1} (Masri, 1962).

As the Wadi Hasa flows from the plateau down the scarp slope, it incises into the rock and has created a gorge. East of the Hijaz railway, the Wadi Hasa forms a flat basin known as Qa' el Jinz, which is up to 6 km wide in places (Vita-Finzi, 1966). Qa' el Jinz is an ancient playa lake, formed in the Pleistocene. During storms this lake forms a major water store on the plateau. The Tannur Dam, located approximately 37 km downstream (at N30.97°, E35.71°) of Qa' el Jinz, was completed in 2001 as part of the Southern Ghors Integrated Development Project, and is 60 m high and 250 m long with a capacity of 17 million m^3 (Barakat et al., 2005). From the geological mapping available, it is unclear whether the sandstone and limestone aquifers of the Wadi Hasa are confined to the area defined by the surface catchment or whether they extend further, being part of a continental-scale aquifer (Bender, 1968; Sheet: Amman, Cross-section B). Of research papers published on the Wadi Hasa, the majority study the alluvial Hasa Formation (Vita-Finzi, 1966; Moumani et al., 2003), although Batayneh and Qassas (2006) studied the seasonal variability of the discharge from the Ghor Safi aquifer and the water quality of this discharge.

Fluviatile deposits and aeolian sands are the predominant land cover in the east and southeast of the catchment on the plateau in the vicinity of the Jebel Umm Rijam (1,073 m asl) and the Qa' el Jinz. West of the Hijaz railway and the Desert Highway, there is a sparse shrub cover. This shrub cover increases in density closer to the scarp slope where patches of grass, deciduous shrubs and cropland are found on the Santonian limestones. On the Cenomanian–Turonian limestones and Cambrian sandstones, where the Wadi Hasa begins to drop down the scarp slope from the plateau to the Dead Sea Valley below and the rainfall is higher, close to the main channel and larger tributaries, the land cover is a mosaic of grass, deciduous shrubs and cropland. A sparse shrub cover is found 5 to 20 km east of Tafilah. The precipitation of the Wadi Hasa valley at the point midway between Tafilah and Kerak, which is lower in altitude than either of the two towns, is 300 mm per year^{-1}; this is 100 mm per year^{-1} less than either Tafilah or Kerak.

The main crop grown in Ghor Safi is vegetables. Farming in the valley floor is difficult as the crop potential evapotranspiration (PET), estimated as 1,545 mm year^{-1}, is high and the mean annual rainfall at Ghor Safi is only approximately 70 mm year^{-1} (Hamdi et al., 2008). Ghor Safi is the hottest place in Jordan (Freiwan and Kadioglu, 2008).

According to Binnie and Partners (Overseas) Limited:

> The summer evapotranspiration rate in the southern Ghors is approximately four times that in winter. The optimum time for cultivation is the winter. There is a rapid change in evapotranspiration in spring and autumn. Planting and harvesting dates are thus critical to efficient water use.

Trial cropping patterns indicated that, with suitable planting dates, the maximum average winter crop requirement need not exceed 2 mm/day. However with water supplies limited to this amount, only a proportion of the land could be cropped in summer. It would thus be necessary to restrict the area of perennial crops.

(Binnie and Partners (Overseas) Limited, 1979, p. S.3)

The proposed conduit between the Dead Sea and the Red Sea to ameliorate the falling level of the Dead Sea is proposed to pass through Ghor Safi (The Royal Scientific Society, 2007). Irrigation of the farmland around Ghor Safi is from spring flow, seepages and hand-dug and drilled wells (Tleel, 1963). Following the inundation of the Wadi Mujib dam in winter 2003/4, water transported along the Mujib is collected at Sweimeh, together with water from other side wadis, and conveyed by a pumped pipeline to supply the Dead Sea hotels and Amman. In 2008, the government of Jordan envisaged that water would also be transferred to mining and processing industries in the southern Ghors and to irrigate the fields of Ghor Mazraa (Margane et al., 2008).

Of the two flow gauging stations in the Wadi Hasa, one is located on the plateau at Tannur near the road between Kerak and Tafilah (N30.97°, E35.72°); at this point the altitude is 300 m above sea level and the catchment area is 2,160 km². The other gauging station is at Ghor Safi (N31.01°, E35.50°) at 300 m below sea level and draining an area of 2,520 km². The flow records extend from 1945 to 1976 and 1937 to 1979 at Tannur and Ghor Safi, respectively, though the records at both sites are limited in quality. The flow data at Tannur are monthly flows measured by the Natural Resources Authority from 1968/69 to 1975/76. Flows were estimated for the hydrological years from 1945/46 to 1967/68 by Mott MacDonald & Partners from a relationship between rainfall and runoff (Binnie and Partners (Overseas) Limited, 1979).

The measured flows at Ghor Safi are predominantly base flows generated from spot gaugings from July 1938 through to May 1942. Stage readings were made at measuring weirs on two distribution canals from October 1943 through to September 1947. One current meter reading was made in each of October and November 1962, and 31 current meter readings were made from December 1962 to September 1963. The 1962/63 record may also be based partly on autographic water level recorder data, except for October and November 1962, which is based on spot flow measurements. The data from 1957/59 are based on spot flow measurements. Data from 1964–1977 are based on spot flow measurements only, since a satisfactory baseflow rating could not be derived to allow automated flow measurements (Binnie and Partners (Overseas) Limited, 1979). Despite these limitations, the measured and modelled flows at Tannur and Ghor Safi represent one of the few readily available long-term records of flow in west Jordan and give an indication of flow frequency and magnitude.

A flow profile, based on the 1942 Palestine Potash Co. survey and on other measurements made as part of an irrigation engineering project, suggests that, at the Tannur flow gauge, the base flow is very small or zero (Chart Section II, Fig. 3 of Binnie and Partners (Overseas) Limited, 1979). As part of the same engineering study, no significant trend was found in the baseflow data and no correlation was found between base flow and annual rainfall within the same, or preceding, years (Binnie and Partners (Overseas) Limited, 1979). This was thought to be due to the long lag between surface recharge and emergence at the springs feeding the base flow. No correlation was found between monthly rainfall and monthly baseflow variations. As part of the study it was noted that:

II.6.10 Flood flow data are available for the Wadi Hasa at Ghor Safi for the 6 years from 1962/63 to 1967/68. The maximum observed was estimated at 700 m³/s. The average of the observed annual maximum floods is 240 m³/s. Peak flows are available for Tannur for the period 1968/69 to 1975/76, but are of limited relevance to conditions in the Ghor.

(Binnie and Partners (Overseas) Limited, 1979, p. 13)

The reason that the peak flows measured at Tannur are of limited relevance to the conditions in the Ghor is mostly likely transmission loss. That is, the flood wave progressing along the wadi channel would re-infiltrate into the alluvial sediment and thereby the flood wave arriving at Ghor Safi would be diminished from that observed at Tannur.

Observations of flow are available for other wadis in the southern Ghors and for wadis draining into the northeast section of the Dead Sea. For Wadi Wala, limited flood flow data only are available. In the Wadi Mujib, base flows and flood flows were measured at a point immediately before the wadi enters the Dead Sea. The flows in the Wadi Mujib are heavily modified. Base flow is diverted for irrigation and two dams are located in the headwaters. Monthly spot measurements were made in the following: Wadi Isal, Wadi Numeira, Wadi Fifa, Wadi Khanzeira and Wadi Kerak. The measurements extend over the period from 1937/38 to 1976/77, but there are many months without measurements. Baseflow data, classified as 'Good' or 'Excellent' by the Central Water Authority (M. R. Masri, unpublished report on the Amman–Zarqa area, 1963), are available for three wadis which drain into the River Jordan, just north of the Dead Sea: Shu'eib, Kafrein and Hisban. 'Good' and 'Excellent' are defined as an accuracy of ±10 and ±5%, respectively. The reason for the high-quality data was the construction of permanent flow stations on the three wadis in 1956 (Shu'eib, Hisban) and 1961 (Kafrein). Prior to the installation of the weirs, synthesised base flows were generated by Baker and Hazra Engineering Company (1955); these data are thought to be of 'poor' quality (more than ±20% error) because of the limited flow data on which to base a rainfall-runoff relationship. Monthly flood flows were synthesised for the three wadis and are generally rated as 'poor', apart from Wadi Kafrein from 1944 to 1963 for which a rainfall-runoff correlation was established by Mott MacDonald & Partners (Binnie and Partners (Overseas) Limited, 1979). A dam was built on the Kafrein in 1968. Thus, given the length of the flow record available for two locations and the only recent dam construction with a long period of measurement preceding this, the Wadi Hasa was deemed the best study area for a model-based assessment of rainfall-runoff changes in west Jordan.

In addition to the flow data, other meteorological and hydrological data, available at a monthly time step, were collated for precipitation, near-surface air temperature, hours of sun, wind speed and humidity (Tables 13.1 and 13.2). Monthly precipitation data were available at four stations in or within 10 km of the Wadi Hasa. These were Tafilah, Kerak, Ma'an and Qatranah. Tafilah is located on the plateau 18 km north of Dana and receives similar rainfall to the upper reaches of the Wadi Faynan. Moreover, a substantial daily record of rainfall was available from 1 October 1937 to 30 April 1974 at Tafilah. Daily precipitation data were also available for Jerusalem from two different sites starting from 1846.

An estimate of the annual variation in potential evaporation was derived from monthly measurements of wind speed, hours of sun, relative humidity and air temperature available for a 20-year period (1983–2002) using the Penman equation. The data were available for meteorological stations at Tafilah, Ma'an and Qatranah. For near-surface air temperature, the earliest record starts in 1923 at Amman, Queen Alia International (QAI) airport.

13.3 THE MODELLING FRAMEWORK

The modelling framework consists of a hydrological model (Figure 13.2) driven by monthly precipitation time series derived using a statistical rainfall model (weather generator) and estimate of potential evaporation (see Wade et al., this volume, Chapter 10, for a justification of this).

13.3.1 Climate component of the modelling framework

The RCM used in this study was HadRM3 (see section 10.3.1 for details). As described in section 10.3.1, a weather generator was used to derive future scenarios of monthly precipitation depths. The weather generator derives rainfall stochastically, according to some underlying patterns of daily rainfall and these were integrated to give monthly projections. In the weather generator used here, the patterns of daily rainfall are described statistically through the mean rain per rainy day (rainfall intensity) and the probabilities of rain both given rain the day before (PRR) and given no rain the day before (PDR). The rainfall was

Table 13.1 *Meteorological data collated for the hydrological study of the Wadi Hasa*

Station	Start date	End date	Percent availability
Precipitation			
Al Hasan Tafilah	01/1983	12/2002	99.2
Al Hasan Tafilah	10/1937	04/1974	
Karak	01/1938	12/1976	99.1
Ma'an – 1	01/1983	12/2002	100.0
Beershava	01/1951	12/1994	41.5
Jerusalem – 1	01/1846	12/1995	99.7
Jerusalem - Old City	01/1846	12/1960	96.8
Amman	01/1923	12/1998	98.6
Deir Alla	01/1952	12/1989	98.7
Jerusalem Airport	01/1951	12/1967	81.9
Jordan University	01/1960	12/1989	99.7
Ma'an –2	01/1960	12/2000	96.1
Q.A.I. Airport	01/1983	12/2002	100.0
Qatranah	01/1983	12/2002	100.0
Wadi Musa	01/1983	12/2002	90.8
Near-surface air temperature			
Al Hasan Tafilah	01/1983	12/2002	98.8
Amman Q.A.I.	01/1923	12/1998	98.9
Aqaba	01/1971	12/1980	100.0
Eilat	01/1951	12/2000	97.2
Ma'an	01/1961	12/2002	99.4
Qatranah	01/1983	12/2002	95.0
Wadi Musa	01/1983	12/2002	83.3
Hours of sun			
Al Hasan	01/1983	12/2002	79.2
Ma'an	01/1983	12/2002	99.6
Qatranah	01/1983	12/2002	95.0
Wadi Musa	01/1983	12/2002	47.9
Wind speed			
Al Hasan Tafilah	01/1983	12/2002	93.3
Ma'an	01/1983	12/2002	99.6
Qatranah	01/1983	12/2002	85.0
Wadi Musa	01/1983	12/2002	10.0
Relative humidity			
Al Hasan Tafilah	01/1983	12/2002	94.2
Ma'an	01/1983	12/2002	100.0
Qatranah	01/1983	12/2002	85.4
Wadi Musa	01/1983	12/2002	58.8

characterised and projected for 2071–2100 at Tafilah, Kerak, Qatranah and Ma'an.

For the future scenario integrations, the probability of rain given rain and that of rain given no rain was first determined from an RCM for the control period, 1961–1990. The same

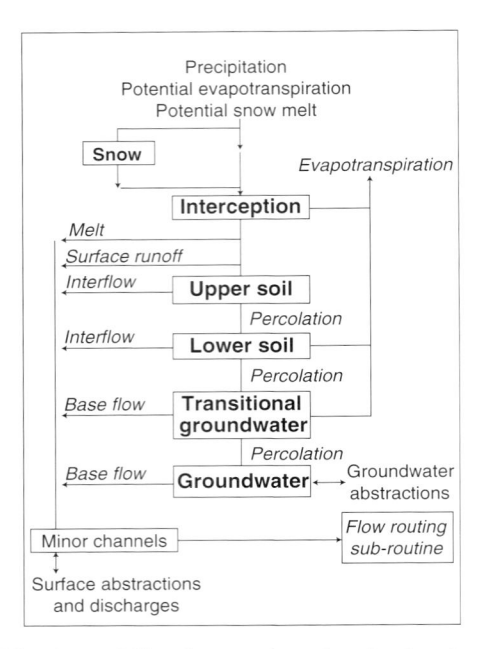

Figure 13.2. The modelling framework used to simulate the rainfall-runoff response in the Wadi Hasa catchment.

probabilities were determined for the scenario period 2071–2100, and these projected future probabilities were bias-corrected using a factor derived by comparing the control model integration with the observed probabilities. The distribution of rainfall intensities (rain per rainy day) was based on the observed time-series, with an extra parameterisation for extreme rainfall events. The method by which the rainfall probabilities and intensities were estimated from input rainfall time series is described fully in Chapter 5.

13.3.2 Hydrological components of the modelling framework

The rainfall-runoff model HYSIMM is a new hydrological model written for first use in the Water, Life and Civilisation project. It is based on the HYSIM model which is widely used in the United Kingdom and elsewhere for climate change impact assessment (Manley, 1975; Pilling and Jones, 2002; Murphy *et al.*, 2006; Manley *et al.*, 2008). HYSIMM uses data of, and runs at, a monthly, rather than daily time step. This is important. Monthly meteorological and hydrological records are more typically available in semi-arid and arid regions and, whilst care is needed when interpreting flood peaks, the model can be more readily applied than a detailed process-based, fully-distributed, rainfall-runoff model.

HYSIMM is a hydrologic simulation model, which uses mathematical relationships to simulate the hydrological cycle in a catchment on a continuous basis. The model uses rainfall and potential evaporation data as inputs and combines these with physical data defining the catchment surface, to determine the amount of direct runoff and contributions from soil and groundwater storage. Where available, potential evapotranspiration (PET), potential snowmelt, discharges to (positive) and abstractions from (negative) the river system, and abstraction from (positive) and augmentation of (negative) groundwater can be used as input. HYSIMM comprises five linear reservoirs to represent: vegetation canopy interception, the upper soil, the lower soil and shallow (transitional) and deep groundwater (Figure 13.2). The water-holding capacity of each store and the maximum rate of water transfer between them are defined by time-invariant parameters. The water volumes in each store and the rates of transfer vary with time. The relationships used in the model have been defined experimentally. Where complexity makes this impossible then a simplification has been used in preference to an empirical relationship.

One big difference between models which operate at a daily or shorter time step and monthly models is the degree of simplification. In daily models, factors such as channel hydraulics or the rate of transfer of moisture between different soil levels are important and have to be simulated. For a monthly model, except for a few very large or highly heterogeneous catchments, channel hydraulics are irrelevant and other short-term exchanges are difficult to simulate. Many of these short-term factors have an influence on monthly total runoff and the model has to take account of them. One clear example is storm runoff; a few short storms would produce different runoff from the same rainfall amount spread evenly over a month.

Another difficulty for monthly models is the fact that it is possible for evaporation to exceed precipitation over a month but for runoff still to occur. This can happen when all the precipitation occurs in a few short-duration events during which precipitation exceeds that rate of evaporation. This, and the accounting of the runoff response to short storms, is allowed for in HYSIMM by using rain days. If no local data are available, an estimate of the number of rain days can be downloaded from: http://www.cru.uea.ac.uk/cru/data/hrg/tmc/. HYSIMM divides the month into two periods, one with rain and one without, based on the ratio of the number of days with rain to the number of days in the month. It assumes that all the precipitation falls on days with rain (i.e. the very small amounts of rain recorded as 'trace' on other days are neglected) and that the rate of evaporation is constant throughout the month. For example, if the precipitation and evaporation were both 100 mm and there were 6 rain days in a 30-day month, then in the first period there would be 100 mm precipitation and 20 mm of evaporation. In the second period there would be no rain and 80 mm of evaporation.

To model the heterogeneity of a catchment it can be subdivided into two or three zones with different parameters and different input data applying to each. Similarly, the river channels can be divided into reaches with reasonably uniform hydraulic characteristics. Catchments can be 'cascaded' with the outflow from one catchment being input to a channel reach of another. The outflow from upstream may be totally simulated or it may be a combination of simulated and recorded discharge. The model also allows for a fixed proportion of precipitation to become runoff to represent the effect of urbanisation and other impermeable areas such as rocky outcrops.

HYSIMM is able to simulate snow accumulation and melt. A typical approach to snow is to assume that if the air temperature is above a critical value then precipitation is in the form of rain. If it is below that value precipitation is in the form of snow. Melt is assumed to take place at a rate that is proportional to the temperature above the critical value. Whilst this method is questionable for daily data, it is even more so for monthly data. To help resolve this problem, HYSIMM assumes that there is a critical temperature and a defined melt rate but one of the parameters is the temperature range. This allows both for variation of temperature with elevation and also for temperature variations during the month. If the monthly temperature is at the bottom of the temperature range, then all precipitation is snow and no melt takes place. If it is above the top end then all precipitation is rain. Between the lower and upper values, the proportion of precipitation which falls as snow and the proportion of the snow which can melt varies linearly with temperature. Any precipitation falling as snow is held in snow storage from where it is released into interception storage. The rate of release is equal to the potential melt rate.

Vegetation canopy storage of moisture on the leaves of trees and grasses is represented as a store. Moisture is added to this storage from rainfall or snowmelt. The first draw on this storage is for evaporation which, experiments have shown, can take place at more than the calculated potential rate particularly on the leaves of trees. This is allowed for in the model. Any moisture in excess of the storage limit is passed onto the next stage. A proportion of the moisture in excess of the interception storage limit is diverted to minor channel storage to allow for the impermeable proportion of the catchment.

In HYSIMM, the soil components are important. HYSIMM uses two soil stores to overcome the assumption that with a single soil reservoir, water runoff either is occurring or is not. Whilst such an assumption can be considered a close approximation to reality, the assumption that all the soils in a catchment are the same is not valid. Thus, to overcome this, HYSIMM assumes that 50% of the soils have a given moisture storage, 25% of them have half that storage and 25% have double that storage.

The upper soil horizon reservoir represents moisture held in the upper (A) soil horizon, i.e. top-soil. It has a finite capacity equal to the depth of this horizon multiplied by its porosity. A model parameter, known as the pore size distribution index, defines the relationship between moisture content and both effective permeability and capillary suction. In the case of effective permeability, which is the ratio between permeability for a given moisture content and the permeability at saturation, the value controls interflow and recharge to the groundwater. The value of the pore size distribution index varies from 0.09 for clays to 0.25 for sands and effective permeability is proportional to moisture content raised to the power of $(2 + 3\gamma)/\gamma$ where γ is a user-defined model parameter. The permeability is effectively zero until the moisture content is close to saturation.

A limit on the rate at which moisture can enter this horizon is applied, based on the potential infiltration rate. This rate is assumed to have a triangular areal distribution, as in the models of Crawford and Linsley (1966) and of Porter and McMahon (1971). The potential infiltration rate is based on Philip's equation:

$$x = \varphi t^{0.5} + \chi t^{1.0} + \omega t^{1.5} + \ldots \ldots \quad (13.1)$$

where x is the distance travelled downwards by the wetting front, t is time since $x = 0$, and φ, χ and ω are functions of soil type and condition. It has been shown by Manley that this relationship can be closely approximated to:

$$x = 2k\, Pt^{0.5} + kt \quad (13.2)$$

where P is the capillary suction (mm of water) and k the saturated permeability of the medium (mm hour^{-1}). This allows determination of the potential infiltration rate.

Corey and Brooks (1975) have shown that P can be expressed as:

$$P = \frac{P_b}{S_e}\frac{1}{\gamma} \quad (13.3)$$

where P_b is the bubbling pressure (mm of water), γ is a parameter (called the pore size distribution index) and S_e is the effective saturation defined as:

$$S_e = \frac{m - S_r}{1.0 - S_r} \quad (13.4)$$

where m is the saturation and S_r is the residual saturation, namely the minimum saturation that can be attained by dewatering the soil under increasing suction. By simulating the moisture content in the upper horizon, the forces causing movement of the water can therefore be simulated.

The first loss from the upper horizon is evapotranspiration which, if the capillary suction is less than 15 atmospheres, takes place at the potential rate (after allowing for any loss from interception storage). If capillary suction is greater than 15 atmospheres, evaporation takes place at a rate reduced in proportion to the remaining storage.

The next transfer of moisture that is considered is interflow (i.e. lateral flow). The rate at which this occurs is obviously a very complex function of the effective horizontal permeability, gradient of the layer and distance to a channel or land drain. Corey and Brooks (1975) have also shown that the effective permeability of porous media is given by:

$$k_e = k \cdot S_e^{\frac{2+3\gamma}{\gamma}} \quad (13.5)$$

where k_e is the effective permeability (mm hour^{-1}) and the other terms are as defined previously. Because of its complexity no attempt is made to separate the individual parameters for interflow and it is given as:

$$Interflow = R_{fac1} \cdot S_e^{\frac{2+3\gamma}{\gamma}} \quad (13.6)$$

where R_{fac1} is defined as the interflow runoff from the upper soil horizon at saturation. The final transfer from the upper horizon, percolation to the lower horizon, is given by:

$$Percolation = k_b \cdot S_e^{\frac{2+3\gamma}{\gamma}} \quad (13.7)$$

where k_b is the saturated permeability at the horizon boundary and S_e is the effective saturation in the upper horizon. By combining the above equations the rate of increase in storage is given by:

$$\frac{ds}{dt} = i - (R_{fac1} + k_b) S_e^{\frac{2+3\gamma}{\gamma}} \quad (13.8)$$

where i is the rate of inflow and S and t are moisture storage and time, respectively. Unfortunately this equation cannot readily be solved explicitly, so it has been assumed that the total change in storage in any time increment is small compared with the initial storage. In this case the equation can be simplified and an approximate solution obtained. As a check for extreme situations, the change in storage is constrained to lie within an upper and lower limit. The upper limit is defined by the level of storage at which the rate of outflow is equal to the rate of inflow. The lower limit results from setting i equal to zero in the above equation, in which case an explicit solution is possible.

The lower soil horizon reservoir represents moisture below the upper horizon but still in the zone of rooting (i.e. the B and C horizons). Any unsatisfied potential evapotranspiration is subtracted from the storage at the potential rate, subject to the same limitation as for the upper horizon (i.e. capillary suction less than 15 atmospheres). Similar equations to those in the upper horizon are used for interflow runoff and percolation to groundwater.

The transitional groundwater is an infinite linear reservoir and represents the first stage of groundwater storage. Particularly in karst limestone or chalk catchments, many of the fissures holding moisture may communicate with a stream rather than deeper groundwater, and the transitional groundwater represents this effect. Its operation is defined by two parameters: the discharge coefficient and the proportion of the moisture leaving storage that enters the channels. Being a linear reservoir, the relationship between storage and time can be calculated explicitly.

The deep groundwater is also an infinite linear reservoir, assumed to have a constant discharge coefficient. It is from this reservoir that groundwater abstractions are made. As in the above case the rate of runoff can be calculated explicitly. The model calculates the moisture excess. A fixed proportion of this goes to direct runoff and the balance to the first groundwater reservoir. From this reservoir a fixed proportion becomes runoff and a fixed proportion goes to the second groundwater reservoir. The two reservoirs are both linear reservoirs. A fixed proportion of their storage becomes runoff each month. The use of two reservoirs allows the simulation of nonlinear effects.

To use the model, parameter values need to be estimated. A first estimate is obtained by a study of the catchment type and this is refined, for a few of the most sensitive parameters, by adjustments based on a comparison of recorded and simulated flow. This process, called calibration, usually comprises both objective and subjective methods.

13.4 MODEL SET-UP AND CALIBRATION

Two model applications were prepared, one for the upper section of the Wadi Hasa draining the plateau to Tannur and a second for the entire catchment draining to Ghor Safi. This was done to investigate the flow response on the plateau and also to examine the influence of altered precipitation and temperature on the base flow. Input precipitation and potential evaporation datasets were prepared so that the model could be run from January 1923 to December 2002, which is the longest period for which both rainfall and temperature observations were available from at least one site. Potential evaporation was calculated from monthly temperature and hours of sun, wind and humidity at Tafilah and Ma'an which were assumed representative of the summit of the scarp slope and the eastern edge of the Wadi Hasa catchment, respectively. Since the hours of sun, wind and humidity data were available for 20 years, these datasets had to be replicated four times to provide an 80-year monthly time-series commensurate with the 80 years of air temperature generated for Tafilah and Ma'an; in each case the 20-year air temperature record at each of these stations was extended by quantifying a relationship with the 80-year record from QAI airport. Gaps in the precipitation records for the four stations were infilled by calculation of a matrix of rainfall monthly means for concurrent periods for all pairs of stations. The rainfall data from Jerusalem were included in this analysis. The matrix was then used to calculate the ratio of the means for all pairs of stations, for precipitation. These ratios were used to determine the standard errors of precipitation estimated using the relationships between all possible pairs of stations. Infilling was done using the monthly data from whichever station gave the lowest standard error and which was not itself infilled. The process was repeated for temperature, except that differences were used rather than ratios. Monthly flow data were used for Wadi Hasa at Tannur from 1945 to 1976 and Wadi Hasa at Safi from 1938 to 1979. Owing to the uncertainties in the observed flow, an objective function was not used to assess model fit, and split-sample testing was not used to assess model performance during calibration and test periods. The model application was designed to assess potential sensitivity to changes in precipitation and air temperature as a proof of concept in a semi-arid environment, which is known to be complex to model (Ragab and Prudhomme, 2002).

13.4.1 Catchment to Tannur

For the Wadi Hasa catchment to Tannur, an area-weighted estimate of monthly precipitation was based on the precipitation gauges at Tafilah (0.25) and Qatranah (0.25) and Ma'an (0.5). The weightings, given in the brackets, represent the area-weighted proportion of the total Wadi Hasa catchment, measured to Tannur, assigned to each rain gauge. The model was calibrated from 1923 to 2002. Potential evaporation PE was calculated using the Penman equation based on Ma'an monthly temperature, hours of sun, wind speed and humidity, infilled as necessary.

The modelled mean monthly flows at Tannur are shown in Figures 13.3a, 13.4a, 13.5a and 13.6a. The results show that the model is able to represent the seasonal changes in flow, though there is evidence that the model tends to over-estimate the observed flows. Given the known errors in the observed flows which from 1945/46 to 1967/68 are estimated by a runoff relationship, and the complexities in simulating the rainfall-runoff response in an environment with a high spatial variation in rainfall and geological type, then the model is unable to

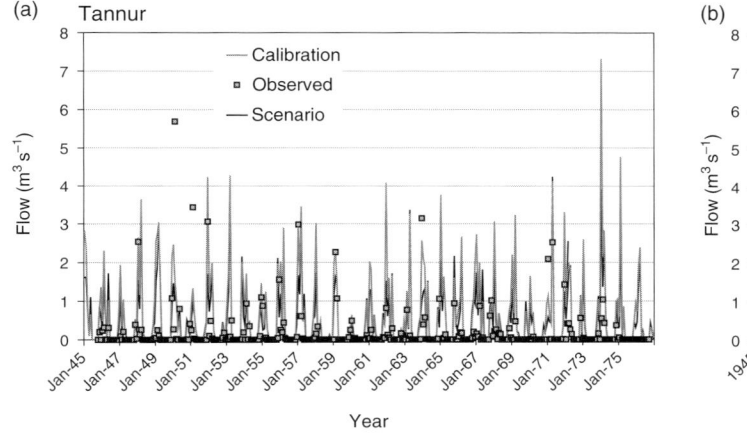

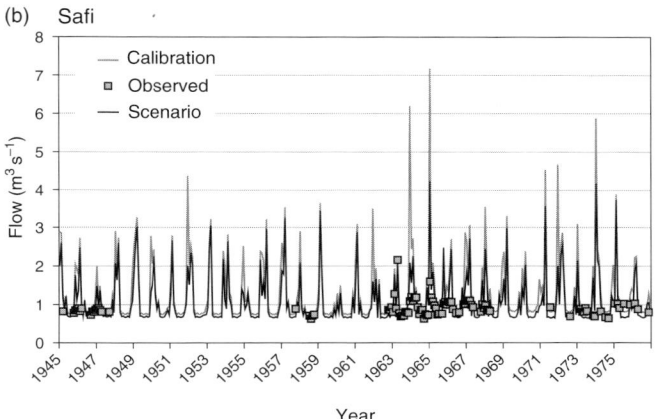

Figure 13.3. Observed and simulated mean monthly flows in the Wadi Hasa at (a) Tannur and (b) Safi. Simulated flows are shown for the calibration and scenario climate conditions.

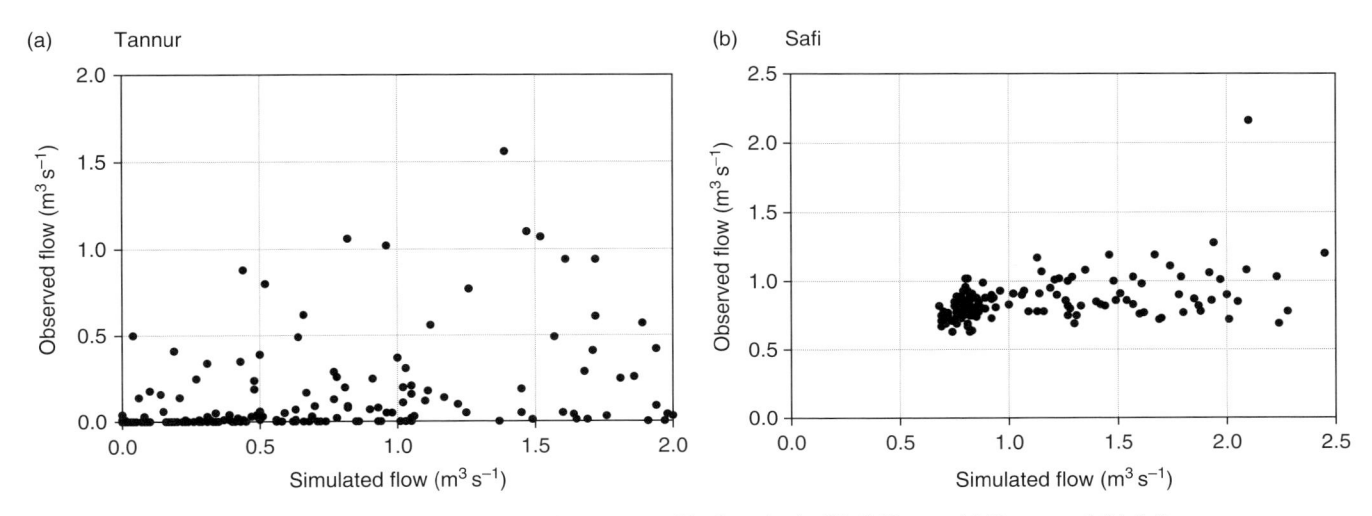

Figure 13.4. Observed versus simulated (calibration period) mean monthly flows in the Wadi Hasa at (a) Tannur and (b) Safi.

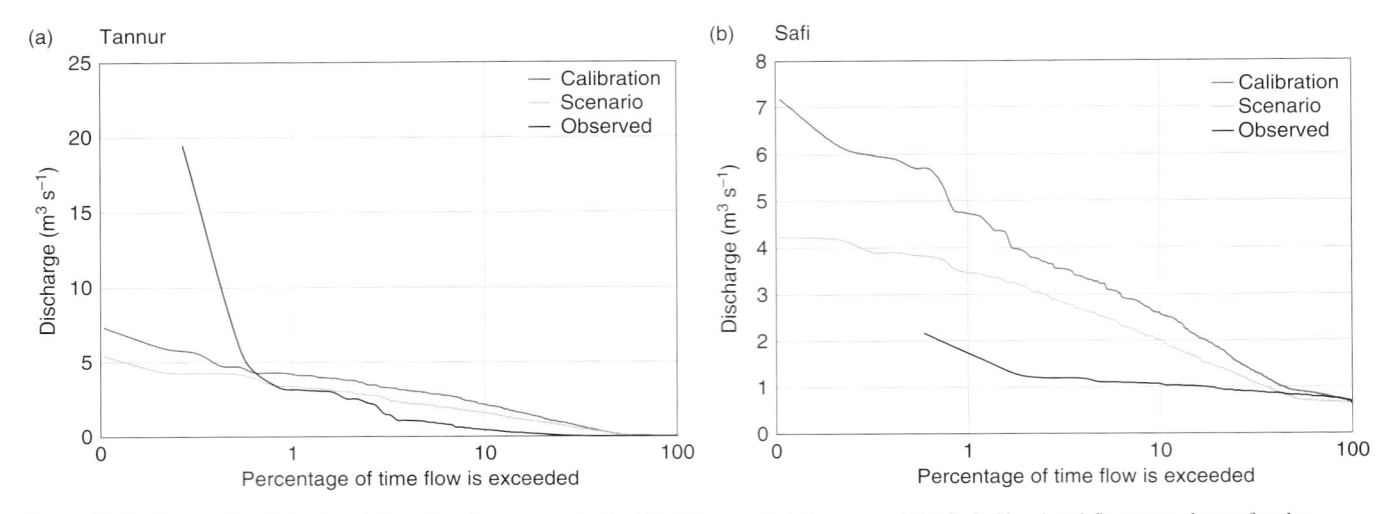

Figure 13.5. Observed and simulated flow-duration curves for the Wadi Hasa at (a) Tannur and (b) Safi. Simulated flows are shown for the calibration (1923–2002) and scenario climate conditions.

reproduce the flow magnitudes in the observed data series exactly. It should also be noted that a very large flood in January 1965 was excluded from Figure 13.3a. This flood, which was more than three times larger than any other observed flood, could not be simulated. Examination of the precipitation record showed that, although precipitation at Ma'an had been high in that month, it was by no means the highest in the record.

13.4.2 Catchment to Safi

For the Wadi Hasa catchment to Safi, an estimate of monthly precipitation was based on an average of the precipitation gauges at Tafilah, Kerak and Qatranah. The model was calibrated from 1923 to 2002. The PE was calculated using the Penman equation based on Tafilah monthly temperature, hours of sun, wind speed and relative humidity at the same site, infilled as necessary.

The modelled mean monthly flows at Safi are shown in Figures 13.3b, 13.4b, 13.5b and 13.6b. The results show that the model was again able to represent the seasonal changes in flow, though there is evidence that the model tends to overestimate the observed flows, as noted for Tannur. The observed flow record is far from complete; however, the simulated flows represent both the variation within a year and the general trend of groundwater movement (Figure 13.4a). The lower part of the catchment between Tannur and Safi is very different from the upper part. In the upper part, flows drop to zero very quickly after rainfall and at Tannur there is no discernable groundwater-supported base flow. The observed and simulated flows appear to show no long-term trend (Figure 13.4a). The modelled recession rate for groundwater is very slow and indicates that only 0.02% of the water in the aquifer flows out each month. The mean observed and simulated groundwater flow was $0.7 \text{ m}^3 \text{ s}^{-1}$.

The flow-duration curves (Figures 13.5a and b) show that at Tannur, the large flood observed in 1965 is not simulated and this reaffirms that the flood flows at Safi tend to be over-estimated. Because of the limitation of using spot samples to represent mean monthly flows and the general lack of observed flood flow data, then the observed flows most likely do not represent flood flow magnitude well. The anecdotal evidence of Binnie and Partners (Overseas) Limited (1979) (section 13.2) notes that flood flows had peaks of between 240 and 700 $\text{m}^3 \text{ s}^{-1}$. Despite this, the study is still useful as the sensitivity to projected changes in the climate can be investigated.

13.5 THE IMPACT OF ANTHROPOGENIC CLIMATE CHANGE ON THE WADI HASA FLOW

Delta change values were calculated between the control (1961–1990) and A2 scenario (2071–2100) periods in terms of ratios and differences for the monthly mean precipitation and temperature, respectively (Table 13.2). The control and scenario period mean daily precipitation were determined from the HadRM3 model simulations with bias correction applied using the weather generator for the Tafilah, Qatranah, Kerak and Ma'an precipitation measurement stations. The monthly delta change ratios were then applied to the infilled observed monthly precipitation timeseries used for the model calibration. As for the calibration period (1923–2002), the precipitation data were weighted, 0.25 Tafilah, 0.25 Kerak and 0.5 Ma'an for the upper Wadi Hasa catchment modelled at Tannur, and the average of Tafilah,

Table 13.2 *Delta change values calculated at four precipitation and two near-surface air temperature measurement stations to compare daily mean precipitation ratios and temperature differences between the HadRM3 modelled control period (1961–1990) and the A2 scenario period (2071–2100)*

	Precipitation ratio				Near-surface air temperature difference (K)	
	Ma'an	Tafilah	Qatranah	Kerak	Ma'an	Tafilah
Jan	0.38	0.63	0.69	0.66	3.5	3.5
Feb	0.45	0.83	0.51	1.07	3.9	3.9
Mar	0.78	0.98	1.14	0.85	3.2	3.3
Apr	0.59	1.31	0.72	0.53	3.2	3.3
May	1.91	1.11	1.81	1.17	5.0	5.0
Jun	1.44	2.28	1.25	1.32	5.6	5.5
Jul	2.43	0.73	1.15	0.55	6.2	6.0
Aug	1.22	2.07	1.26	1.25	6.2	6.1
Sep	1.45	0.96	1.11	1.35	5.0	4.9
Oct	1.28	1.25	1.82	1.94	4.0	4.0
Nov	0.96	1.15	0.99	0.75	4.7	4.7
Dec	0.16	0.47	0.35	0.31	3.5	3.6

Kerak and Qatranah was used for the model application to the entire Wadi Hasa catchment to Safi.

The difference in mean monthly air temperatures between the control and scenario periods at Ma'an (Tannur) and Tafilah (Safi) were used to re-scale Ma'an and Tafilah mean near-surface air temperatures for the calibration period data, respectively (Table 13.2). These adjusted temperature time-series were then used to recalculate the monthly PE time-series for the calibration period for Ma'an and Tafilah. For both stations, the same hours of sun, wind speed and humidity data as for the calibration period were used. The adjusted monthly PE time-series was then applied with the adjusted precipitation time-series to determine the changes to the flows in the upper and entire Wadi Hasa. In addition, the adjusted precipitation alone was applied to the entire Wadi Hasa model set-up to determine the sensitivity of the flow response to changes only in precipitation, rather than precipitation and PE.

The largest changes in precipitation between the control and scenario periods were projected for December and January. The delta change analysis highlighted the spatial variations in the projected rainfall response to climate change. The greatest decrease in December and January rainfall was at Ma'an with a decline to 16% of current mean annual rainfall; thus there is an apparent difference between rainfall changes to the east of the Wadi Hasa toward the desert centre and in the west on the ridge of the scarp slope. During the months May to October, increases in rainfall were predicted at all four rainfall stations. Given the low, often zero rainfall during the summer months of June, July and August, then the increases in these months are likely to be an artefact of modelling very small rainfall totals, and whilst the percentage increase in rainfall is large the change in the absolute mean monthly rainfall is not. The changes in the modelled precipitation also show an extension of the wet season with more rain falling in the months of March, April, October and November, though for March, April and November the pattern is not clear as, for each of these three months, only one station shows an increase. The mean annual rainfall decreases over the Wadi Hasa because of the large declines in the December and January mean monthly rainfall totals.

Large changes in the near-surface air temperatures were also estimated for the 2071–2100 period under the A2 scenario. For all months, at both Tafilah and Ma'an, the increase was greater than 3 °C. During July and August, the mean monthly increase in air temperature is 6 °C at both Tafilah and Ma'an. The effect of these temperature increases is to increase mean annual PE by an estimated 100 mm at Tafilah and Ma'an.

Under the current climate conditions, the observed base flow in the Wadi Hasa was stable in that the base flow neither increased nor decreased over time. The model reproduces this behaviour. Under the A2 scenario, the modelled base flow shows a gradual decrease as the Q95 flow, indicative of the base flow, declines from 0.70 to 0.65 m^3 s^{-1} (a 7% decrease). This reduction in base flow was a response to the simulated reduction in aquifer recharge, due to decreased rainfall and increased PE. The gradual reduction is a result of a long simulated residence time for the aquifer.

As a result of the decreases in precipitation, the flood flows also decrease at both Tannur and Safi. The Q10 flow, indicative of the flood flows, decreases at Tannur by 27% and at Safi by 22% (Figure 13.6a and b). The median flows (Q50) also decrease at Tannur and Safi by 15% and 19%, respectively and, whilst a base flow is maintained at Safi, the main channel at Tannur is likely to remain drier for longer as the number of months with flow decrease over the 80-year simulation period from 774 to 566 months. This will affect the replenishment of the Tannur Dam.

Figure 13.6a and b, which show the mean monthly flows at Tannur and Safi averaged over the simulation period, provide evidence that the modelling framework is able to reproduce the general seasonal flow response. The modelled flows during the calibration period are too high for the months of November, December, January, February and March but this is most likely due, in part, to the limitations of the observed data which do not represent well the flood flows in the Wadi Hasa as well as the structural and parameter uncertainty of the HYSIMM model. The modelled mean monthly flows are mainly sensitive to changes in rainfall. Evidence for this is shown in Figure 13.6b whereby the mean monthly flows derived from the scenario where precipitation alone is changed (ΔP) are within 2% of the flows derived from the scenario where both precipitation and PE are adjusted to take into account temperature change.

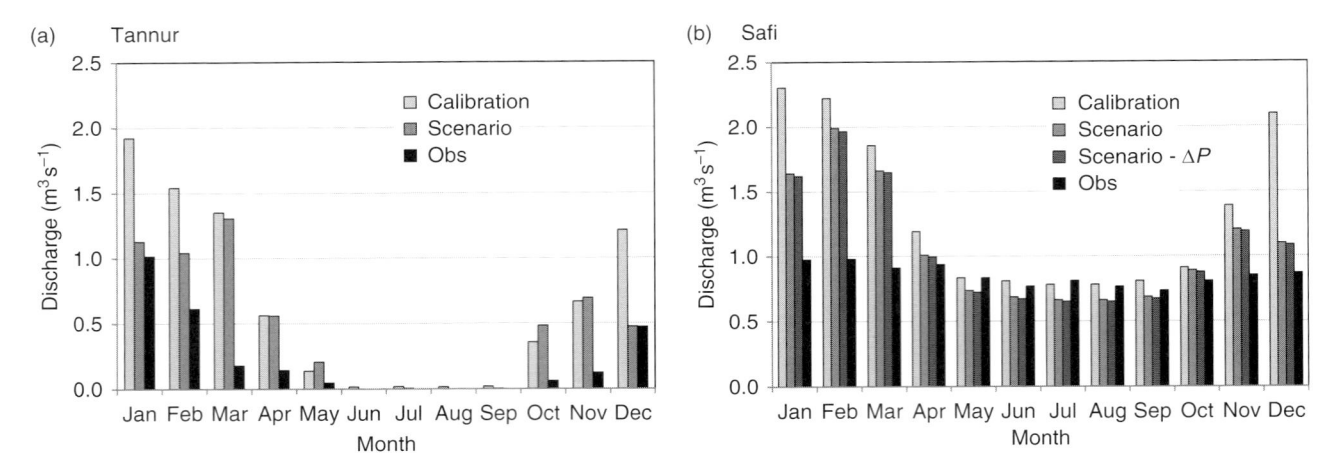

Figure 13.6. Observed and simulated mean monthly flows, averaged over the 80-year simulation period, in the Wadi Hasa at (a) Tannur and (b) Safi. Simulated flows are shown for the calibration (1923–2002) and scenario climate conditions where both precipitation and PET are adjusted and where precipitation alone is adjusted (Scenario $-\Delta P$) for the model simulations of flow at Safi.

13.6 DISCUSSION

As with applications of similar modelling frameworks to the upper River Jordan (Chapter 10, this volume) and the Wadi Faynan (Chapter 12), the framework has proved capable of providing an estimate of the likely rainfall-runoff response under projections of future climate change. The HYSIMM rainfall-runoff model was applicable in the Wadi Hasa to estimate flows given monthly precipitation, air temperature, hours of sun, wind speed and relative humidity data. Collation of these data required major effort and whilst precipitation and the other four meteorological measurements were obtained, there was difficulty in obtaining reliable estimates of flow. This is a major problem in hydrological assessments of arid and semi-arid environments. Flow data are difficult to collect because of the cost of installing and maintaining the infrastructure required, and moreover, given the velocity, magnitude and sediment load of flood waves in arid and semi-arid environments, such as the Wadi Hasa, then any flow measurement structure needs to be robust. There is evidence in the Wadi Dana that concrete flow structures in Jordan are easily eroded during flood events (Wade *et al.*, Chapter 12, this volume). Until sufficient investment is available to set up and maintain flow measurement infrastructures, or until existing data are made readily available to academics and engineers, then data and model-based assessments of flow changes in response to environmental change will continue to be limited to discussions of the likely relative changes in the flow magnitude and frequency; quantification will remain problematic. The lack of daily mean flow data meant that the study of how the catchment response to rainfall differed in the upper (Tannur) and lower (Safi) sections of the Wadi Hasa was limited and changes in flood and drought frequency could not be assessed. At present, model-based assessments of climate change remain useful tools for hydrological sensitivity assessments, but predictions of absolute change remain some way off.

The results of the study highlighted the spatial heterogeneity in terms of the projected changes in precipitation and groundwater. Reduced rainfall in the desert centre at Ma'an is indicative of less rainfall reaching the east of Jordan, because fewer rainstorms reach the eastern Mediterranean and as these storms contain less moisture they 'rain-out' over the scarp slope. The rate of groundwater decline is less than predicted at the Wadi Faynan (Wade *et al.*, Chapter 12, this volume). The Wadi Hasa is larger than the Wadi Faynan and therefore the aquifers contributing to the base flow observed at Safi are likely to be larger, extending further east into the desert centre. As with the hydrological assessment of the Wadi Faynan and the upper reaches of the River Jordan, quantifying the extent, water storage capacity, residence time and water quality (salinity) of the aquifers of Jordan is a key requirement to determine the water resource status. Given the spatial variability of rainfall, geology and topography, then at a local level the hydrological response is complicated given subtle variations in the runoff response caused by changes in the land cover, surface deposits, soils and geology. Despite this, the model-based assessment of the flow response highlights that the precipitation amount and aquifer size is the key control on the flood and baseflow response. Factors affecting potential evapotranspiration seem to be of secondary importance in determining the flow response because it is actual evaporation (and evapotranspiration) that is more important than PE. Actual evapotranspiration will be more important for the assessment of crop growth under a changed climate. A separate assessment is required to determine this.

The changes in temperature are also large, with temperature increases of 4 to 6 °C predicted under the A2 scenario. Current carbon dioxide emissions put the world on course to match this scenario. Reduced winter rainfall and shifts in the seasonal availability, higher temperatures and reduced base flow for use for irrigation are all detrimental to crop growth. The coupled effect of reduced rainfall delaying plant growth will be particularly damaging to yields, since growth during warmer months may increase transpiration or lead to an increase in canopy temperatures if crop transpiration is maintained. To combat these effects then it may be necessary to change to crop types requiring less water, improve irrigation efficiency or seek to exploit local groundwater resources more readily, possibly with a need to desalinate. At present, water is conveyed to Ghor Mazraa and Amman from the Wadi Mujib (Rumman et al., 2009). If climate change reduces water availability across the wadis of west Jordan then alternative groundwater sources will be required, especially if population growth, mainly in Amman, increases the water demand from the Wadi Mujib dam. If there is insufficient water to maintain the population of the southern Ghors then, to avoid widespread rural abandonment, the transfer of water from the Red Sea to the Dead Sea may be the only viable alternative, given that flows from the upper Jordan are unlikely to increase over the next century.

13.7 CONCLUSIONS

Runoff in the Wadi Hasa is projected to decrease by the end of the century in a world that does not attempt to provide global, cooperative solutions to help mitigate carbon dioxide emissions. The flow regime of the Wadi Hasa is projected to change to one of reduced flood and base flows, as rainfall in the future will not replenish the aquifers draining to the Safi plain, assuming the Wadi Hasa is not draining part of a large continental-scale aquifer. The modelled monthly flow regime is most sensitive to changes in the rainfall; potential evaporation changes cause only a secondary effect on the monthly mean flows. To improve model-based assessments of the rainfall-runoff response, then better flow data are required in arid and semi-arid environments, and a greater knowledge of the extent and storage capacity of aquifers is needed. Given the sensitivity of the monthly flow response to winter precipitation, further work is needed in climate modelling to better predict winter rainfall timing and amounts.

REFERENCES

Baker, M. and Hazra Engineering Company (1955) Yarmouk – Jordan Valley Project. *Master Plan Report*. Appendix V-A Hydrology and Groundwater: Hazra Engineering Company.

Barakat, S. A., A. I. Husein Malkawi and M. Omar (2005) Parametric study using FEM for the stability of the RCC Tannur dam. *Geotechnical and Geological Engineering* **23**: 61–78.

Batayneh, A. T. and H. A. Qassas (2006) Changes in quality of groundwater with seasonal fluctuations: an example from Ghor Safi area, southern Dead Sea coastal aquifers, Jordan. *Journal of Environmental Sciences – China* **18**: 263–269.

Bender, F. (1968) Geological Map of Jordan 1:250,000 (Sheets: Aqaba-Ma'an and Amman). Hanover: Bundesanstalt für Geowissenschaften und Rohstoffe.

Binnie and Partners (Overseas) Limited (1979) *Mujib and Southern Ghors Irrigation Project – Feasibility Report*. Amman: Ministry of Planning, Jordan Valley Authority.

Central Water Authority (1963) Review of stream flow data – prior to October 1963. Technical Paper No. 33. The Hashemite Kingdom of Jordan, Central Water Authority, Hydrology Division, Amman.

Corey, A. T. and R. H. Brooks (1975) Drainage characteristics of soils. *Soil Science Society of America Journal* **39**(2): 251–255.

Crawford, N. H. and R. K. Linsley (1966) *Digital Simulation in Hydrology: Stanford Watershed Model IV*. Tech. Rep. No. 39. Palo Alto, CA: Stanford University.

Delattre, A. (2000) Greek and Latin inscriptions from Syria, vol. 21: Inscriptions from Jordan, part A, Petra and southern Nabataea from the al-Hasa wadi to the Gulf of Aqaba. *Latomus* **59**: 950–951.

Food and Agriculture Organization (FAO) (1970) *The Hydrogeology of the Southern Desert of Jordan. Invesitgation of the Sandstone Aquifers of East Jordan*. Based on the work of J. W. Lloyd. United Nations Development Programme, Food and Agriculture Organization LA: SF/JOR 9, Technical Report 1. Rome: FAO.

Freiwan, M. and M. Kadioglu (2008) Spatial and temporal analysis of climatological data in Jordan. *International Journal of Climatology* **28**: 521–535.

Hamdi, M. R., A. N. Bdour and Z. S. Tarawneh (2008) Developing reference crop evapotranspiration time series simulation model using Class a Pan: a case study for the Jordan Valley/Jordan. *Jordan Journal of Earth and Environmental Studies* **1**: 33–44.

Hill, J. B. (2004) Land use and an archaeological perspective on socio-natural studies in the Wadi al-Hasa, west-central Jordan. *American Antiquity* **69**: 389–412.

le Quéré, C., M. R. Raupach, J. G. Canadell et al. (2009) Trends in the sources and sinks of carbon dioxide. *Nature Geoscience* **2**: 831–836.

Manley, R. (1975) A hydrologic model with physically realistic parameters. UNESCO Symposium, Bratislava.

Manley, R., L. Dimitrievski and S. Andovska (2008) Hydrology simulation of the Vardar River. Presented at the BALWOIS symposium.

Margane, A., A. Borgstedt, A. Subah et al. (2008) Delination of surface water protection zones for the Mujib Dam. Commissioned by Federal Ministry for Economic Cooperation and Development (Bundesministerium für wirtschaftliche Zusammenarbeit und Entwicklung, BMZ). Project: Groundwater Resources Management BMZ-No.: 2005.2110.4, BGR-Archive No.: 012600.

Marks, A. E. (1999) The archaeology of the Wadi Al-Hasa, west central Jordan, vol. 1: Surveys, settlement patterns and paleoenvironments. *Journal of Anthropological Research* **55**: 622–623.

Masri, M. (1962) Hydrogeological study of Ghor Es Safi. Amman: Arab Potash Company.

Moumani, K., J. Alexander and M. D. Bateman (2003) Sedimentology of the Late Quaternary Wadi Hasa Marl Formation of Central Jordan: a record of climate variability. *Palaeogeography Palaeoclimatology Palaeoecology* **191**: 221–242.

Murphy, C., R. Fealy, R. Charlton and J. Sweeney (2006) The reliability of an 'off-the-shelf' conceptual rainfall runoff model for use in climate impact assessment: uncertainty quantification using Latin hypercube sampling. *Area* **38**: 65–78.

Nakicenovic, N., J. Alcamo, G. Davis et al., eds. (2001) *IPCC Special Report on Emissions Scenarios (SRES)*. Geneva: GRID-Arendal.

Parker, D. H. (1970) Investigation of the sandstone aquifers of East Jordan. In *Jordan: The Hydrogeology of the Mesozoic-Cainozoic Aquifers of the Western Highlands and plateau of East Jordan*. LA: SF/JOR9. Technical Report number 2. Rome: United Nations Development Programme, Food and Agriculture Organization.

Pilling, C. G. and J. A. A. Jones (2002) The impact of future climate change on seasonal discharge, hydrological processes and extreme flows in the Upper Wye experimental catchment, mid-Wales. *Hydrological Processes* **16**: 1201–1213.

Porter, J. and McMahon, T. (1971) A model for the simulation of streamflow data from climatic records. *Journal of Hydrology* **13**: 297–324.

Ragab, R. and C. Prudhomme (2002) Climate change and water resources management in arid and semi-arid regions: prospective and challenges for the 21st century. *Biosystems Engineering* **81**: 3–34.

Rumman, M. A., M. Hiyassat, B. Alsmadi, A. Jamrah and M. Alqam (2009) A surface water management model for the Integrated Southern Ghor Project, Jordan. *Construction Innovation: Information, Process, Management* **9**(3): 298–22.

Samuels, R., A. Rimmer and P. Alpert (2009) Effect of extreme rainfall events on the water resources of the Jordan River. *Journal of Hydrology* **375**: 513–523.

The Royal Scientific Society (2007) An environmental and socioeconomic cost benefit analysis and pre-design evaluation of the proposed Red Sea/Dead Sea conduit. http://www.foeme.org/index_images/dinamicas/publications/publ73_1.pdf.

Tipping, R. (2007) Long-term landscape evolution of the Wadis Dana Faynan and Ghuwayr. In *The Early Prehistory of Wadi Faynan, Southern Jordan: Archaeological Survey of Wadis Faynan, Ghuwayr and al-Bustan and Evaluation of the Pre-Pottery Neolithic A Site of WF16*, Wadi Faynan Series 1, Levant Supplementary Series **4**, ed. B. L. Finlayson and S. Mithen. Oxford: Council for British Archaeology in the Levant/Oxbow Books, pp. 14–46.

Tleel, J. W. (1963) *Inventory and Groundwater Evaluation Jordan Valley Amman, Jordan*. The Hashemite Kingdom of Jordan: Central Water Authority.

United Nations (2009) World Population Prospects: The 2008 Revision. New York: UN Secretariat Department of Economic and Social Affairs, Population Division (advanced Excel tables). Last update in UNdata: 18 Jun 2009. http://data.un.org (Retrieved February 2010).

US Geological Survey *et al.* (2006) *Application of Methods for Analysis of Rainfall Intensity in Areas of Israeli, Jordanian and Palestinian Interest*. Reports of the Executive Action Team, Middle East Water Data Banks Project. http://www.exact-me.org/overview/p0405.htm (Retrieved August 2010).

Vita-Finzi, C. (1966) The Hasa Formation: an alluvial deposition in Jordan. *Man* **1**: 386–390.

Winer, E. R., J. A. Rech and N. R. Coinman (2006) Late Quaternary wetland deposits in Wadi Hasa, Jordan, and their implications for paleoenvironmental reconstruction. 2006 Philadelphia Annual Meeting (22–25 October 2006), Paper No. 126-4. Geological Society of America Abstracts with Programs. http://gsa.confex.com/gsa/2006AM/finalprogram/abstract_112203.htm (Retrieved August 2009).

Part IV
Human settlement, climate change, hydrology and water management

14 The archaeology of water management in the Jordan Valley from the Epipalaeolithic to the Nabataean, 21,000 BP (19,000 BC) to AD 106

Bill Finlayson, Jaimie Lovell, Sam Smith and Steven Mithen

ABSTRACT

This chapter reviews the archaeological evidence for water management in the Jordan Valley between the Last Glacial Maximum at 21,000 years ago and the annexation of the Nabataean Kingdom into the Roman Empire at AD 106 – the chronological bounds of the Water, Life and Civilisation project. It summarises the human need for water and available sources in the region before addressing the archaeological evidence for water management in the Epipalaeolithic, Neolithic, Bronze Age, Iron Age and Nabataean, with some consideration of the environmental, demographic social and economic influences that were either a cause or a consequence of changes in water management strategies. Greatest emphasis within the chapter is placed on the Neolithic period in light of relatively new archaeological discoveries that have not previously been drawn together in a review, and on the key role that water management may have played in the transition from hunting and gathering lifestyles. In contrast, the evidence for Nabataean water management has already received extensive consideration from other authors and is succinctly summarised towards the end of this chapter with a set of references leading to further information. As a whole, this chapter seeks to provide the archaeological background for the case studies that follow in Chapters 15–19 of this volume.

14.1 INTRODUCTION

14.1.1 The need for water

Water has been a key resource for human settlement in the Jordan Valley ever since the first traces of occupation more than one million years ago. Management of the water supply became one of the drivers of historical change as the demand for water increased because of the growth in population, the emergence of sedentism, and economic developments leading to water-intensive agriculture and industry. This chapter sets out a history of water management from the Epipalaeolithic to the Nabataean period to provide an archaeological framework for the case studies in the following chapters. Our focus is primarily on the Neolithic period as this marks the emergence of water management facilities for which the evidence has not previously been collated – much of it having been very recently discovered. By the time of the Bronze Age, water management had become widespread, and consequently this chapter seeks to identify key themes in its management rather than attempting to summarise the complete evidence. Water management in the Nabataean has been extensively reviewed by Oleson (e.g. Oleson et al., 1995; Oleson, 2001) and so we provide a summary of his work supplemented by the recent discoveries of the Nabataean pool complex of Petra (Bedal, 2003).

Almost all facets of human life require a reliable supply of water. All societies need water for drinking, a total demand that is governed by population size. Drinking water needs to be of good quality, although the amount required can be relatively small. While an average minimum demand for drinking water can be as little as 1–2 litres per day, high temperatures and activity rates can greatly increase this (West, 1985). Supplies of drinking water can be augmented by water-rich foods; indeed much of the daily water demands of the !Kung San are met by water contained in fruits and vegetables (Tanaka, 1976). The Bedouin tribes who inhabit the Black Desert region of Jordan supplement their water demands in a similar manner. In that water-scarce region, water is often found in brackish pools. Such water is unfit for human consumption, but is perfectly potable for goats. The Rwala Bedouin allow their herds to drink the brackish water and then drink the goats' milk. In this way the goats effectively act as mobile desalination devices (Lancaster and Lancaster, 1997).

Water, Life and Civilisation: Climate, Environment and Society in the Jordan Valley, ed. Steven Mithen and Emily Black. Published by Cambridge University Press.
© Steven Mithen and Emily Black 2011.

The quantity of water required by a community depends upon environmental, social, ideological and economic factors. The remarkable changes in these spheres which took place during the Holocene in the Jordan Valley created profound changes in the nature of the relationship between humans and water. As the primary economic base of the region shifted from hunting and gathering to agriculture, new demands were placed on the timing, the quality and, most obviously, the quantity of water required to grow crops, water herds and provide for a larger human population. This may have been further increased by the use of water in the preparation and cooking of dry, stored grain-based foods. The increasingly sedentary nature of rising populations would have made the task of obtaining suitable water for agriculture ever more challenging, and this would have become exacerbated with urbanisation from the Bronze Age onwards. The nature of living in large sedentary groups would also have demanded increasing amounts of water for basic hygiene needs, with the move from simple washing to sophisticated systems of sanitation and bathing. The amounts and quantities of water needed for any of these tasks is determined by both practical and social factors, including water use associated with ritual practices and ultimately the public display of water control in buildings such as baths and nymphaea (ornamental water fountains of the Classical period).

A further demand for water would have been made by the construction of the built environment itself. From the mud bricks and pisé of the earliest Neolithic settlements to the cements and plasters of Nabataean towns, vast quantities of water are needed for construction. Added to these demands are the requirements of industry, from ceramic manufacture to the intensive use of water in smelting. Wadi Faynan, for instance, became a major industrial centre for copper smelting which would have placed significant additional demands on the water supply within this semi-arid area. Water also becomes important in milling, both for the many grain mills that lie in the tributaries of the Jordan Valley, and, even more, for the large industrial sugar mills in the valley.

The challenge of achieving an adequate water supply in the region arises from its limited predictability and reliability as much as its overall quantity. Water management technology is often designed primarily to reduce risk and extend the availability of water, transforming unpredictable and fleeting wet events (such as floods) into more lasting static and reliable resources. To these ends humans have sought to find technological solutions in order to control or manage water more effectively. In general, technology has been used to provide improved access to water sources, enhanced water storage and the transportation of water from its source to location of need (Clark, 1944).

14.1.2 Water sources in the Jordan Valley

There were three main sources of water available to past populations in the Jordan Valley: direct precipitation, runoff and groundwater flows.

Rain and snow provide the primary source of water. In some areas, such as along the western edge of the Jordanian plateau, these provided sufficient water to support agriculture. One of the objectives of the climate research reported in earlier chapters of this volume has been to consider the varying quantities and seasonality of rainfall. It should be emphasised that in all (semi-)arid regions of Jordan, average annual rainfall values are of limited utility owing to the patchy nature and high interannual variability of rainfall (Lancaster and Lancaster, 1997; Betts, 1998). As a consequence of this variability, relying upon direct rainfall as the only source of water is inherently risky in most areas of Jordan and would probably have been so even during the wetter conditions postulated for some of the Holocene.

Runoff provides the second source of water, transporting and concentrating rainfall along natural drainage routes to specific points in the landscape. This can provide substantial quantities of water to low-lying areas that receive limited rainfall themselves. Wadi catchments can act to collect water from large areas, removing some of the uncertainty (spatial patchiness) associated with rainfall. Many areas of the Jordan Valley today experience large seasonal flash floods, channelled into wadi beds from highland catchments. The timing, intensity and duration of flooding during the early and middle Holocene is unclear, although insights for selected wadis are being gained by hydrological modelling (e.g. Wadi Faynan, Chapter 12 of this volume). It is likely that under the wetter climatic scenarios proposed for the early Holocene (e.g. Bar-Matthews et al., 2003), many wadi systems that are at present dominated by seasonal floods may have hosted more perennial streams or rivers.

Runoff water has several potential benefits for farming communities; it often involves vast quantities of water and also usually contains wet alluvial sediment, ideal for cultivation of crops. However, runoff water is often a fleeting resource, with most floods passing in days or even hours. The only exception to this is water stored in shallow gravels or natural pools. The timing and size of runoff events is also unpredictable, because they are directly related to seasonal rainfall patterns and suffer many of the same unpredictable characteristics. Large flood events can also be difficult to manage: even today with flood protection they can often be dangerous and destructive. The management of flood water thus involves a transformative process, manipulating the essentially transient and destructive nature of runoff into a static, more durable resource. In addition to the large quantities of water concentrated by major wadi systems, local runoff water that collects into relatively small tributary wadis or natural channels has also been an important resource throughout the Holocene.

The third source of water is groundwater, coming naturally from springs and, from the Neolithic onwards, via wells. Groundwater stores are often accumulated over long periods of time, providing a more perennial and predictable resource than rainfall and runoff. Set against this benefit are the frequent difficulty of accessing groundwater and its often poor quality because of salts and minerals which render it unsuitable for long-term use for irrigation or drinking.

Smaller, shallow groundwater reserves are also stored in silts and gravels in the bottom of seasonally flooded wadi channels, rather blurring the difference between groundwater and runoff. Such sources can be accessed through relatively simple technological means, such as digging shallow pits or *ghudran* (Lancaster and Lancaster, 1997). The exploitation of these shallow reserves is common in Jordan today. This form of groundwater exploitation most likely reaches back to the earliest Palaeolithic periods and provides an intermediary step between accessing natural springs and the sinking of formal wells.

14.1.3 The chronological framework

In this chapter, we provide an overview of the development of the relationship between society and water, exploring how the profound social and economic transformations of the Holocene period are intimately linked to the history of water management, and considering key themes which are expanded in the case studies that follow in Chapters 15–19. Whilst previous chapters in this volume have dealt with the history of climate and environmental change, here we begin to consider the impacts of climate change upon human communities. This is a complex relationship, primarily mediated by the hydrology of the region which in turn, is influenced by human activity itself including the manner in which the landscape is exploited and the character of water management systems that are developed.

The period described here is divided into the following general stages (with approximate date ranges); Figures 14.1, 14.2 and 14.3 provide the locations of sites mentioned in the text:

- Epipalaeolithic 21,000–12,000 BP. This is conventionally divided into a succession of phases, notably the Kebaran, Early Natufian and Late Natufian (see Goring-Morris, 1987, and Bar-Yosef, 2002, for explanation of cultural terms).
- Pre-Pottery Neolithic A (PPNA) 12,000–10,550 BP. See Bar-Yosef and Belfer-Cohen (1989) for an overview of the Pre-Pottery Neolithic A.
- Pre-Pottery Neolithic B (PPNB) 10,550–8300 BP. For the southern Levant, this is conventionally divided between the Middle Pre-Pottery Neolithic B (MPPNB) and the Late Pre-Pottery Neolithic B (LPPNB), while some archaeologists also identify a Pre-Pottery Neolithic C (PPNC) phase at the end of this cultural period. See Kujit and Goring-Morris (2002) for a full description of the phasing within the Pre-Pottery Neolithic.
- Pottery Neolithic 8,300–6,500 BP.
- Chalcolithic 6,500–5,600 BP; see Bourke (2008) for an overview.

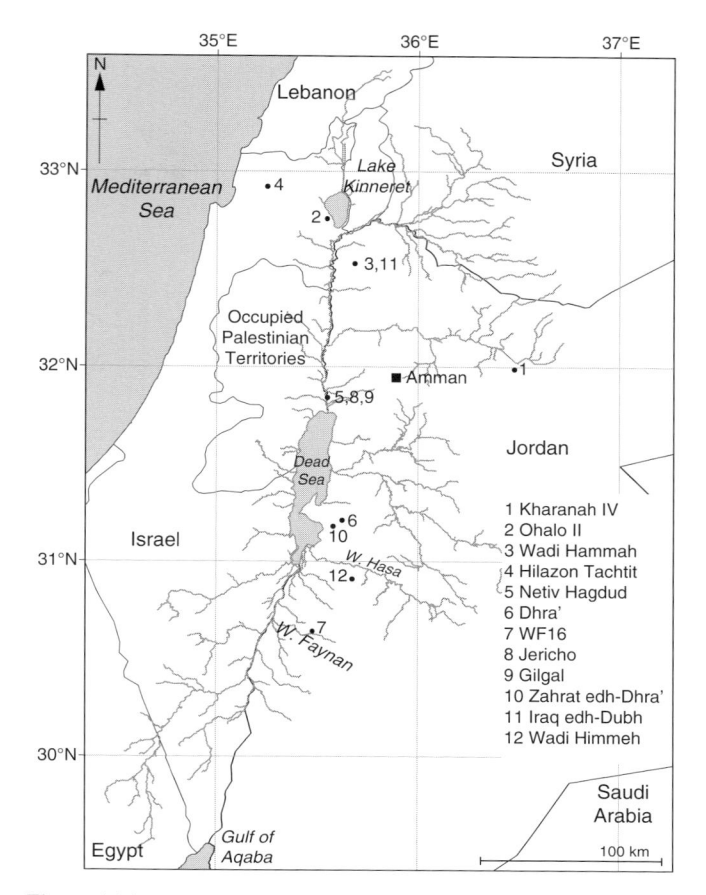

Figure 14.1. Map of the study region showing Epipalaeolithic and Pre-Pottery Neolithic A sites referred to in the text.

- Bronze Age 5,600–3,200 BP. For the southern Levant this is conventionally divided into the Early, Middle and Late Bronze Age, with the Early Bronze Age (EB) divided into a succession of phases, EBI to EBIV; see Philip (2008) for an overview.
- Iron Age 3,200–2,300 BP
- Nabataean 2,300–2,050 BP

14.2 WATER MANAGEMENT IN THE EPIPALAEOLITHIC?

The Epipalaeolithic is the final period when people lived exclusively by hunting and gathering, and is conventionally believed to end at the start of the Holocene. There are few Epipalaeolithic sites actually within the Jordan Valley but this is an important period for rising population levels, increasing sedentism, and plant exploitation (Goring-Morris and Belfer-Cohen, 2010). A recurring feature, mostly identified in the eastern desert of Jordan, is a tendency for people to return repeatedly to the same

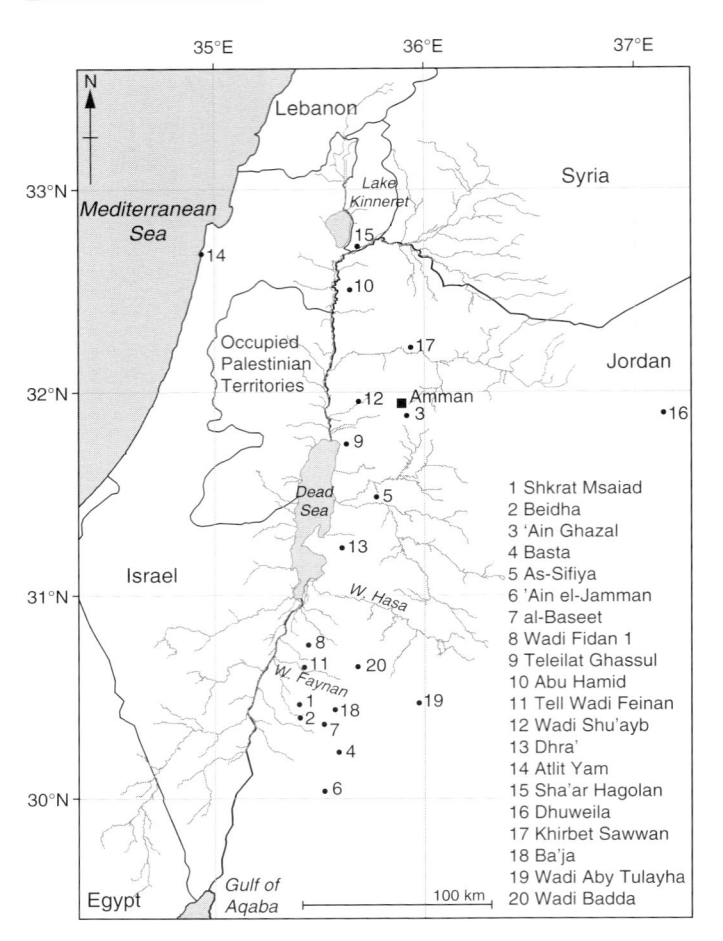

Figure 14.2. Map of the study region showing Pre-Pottery Neolithic B and Pottery Neolithic sites referred to in the text.

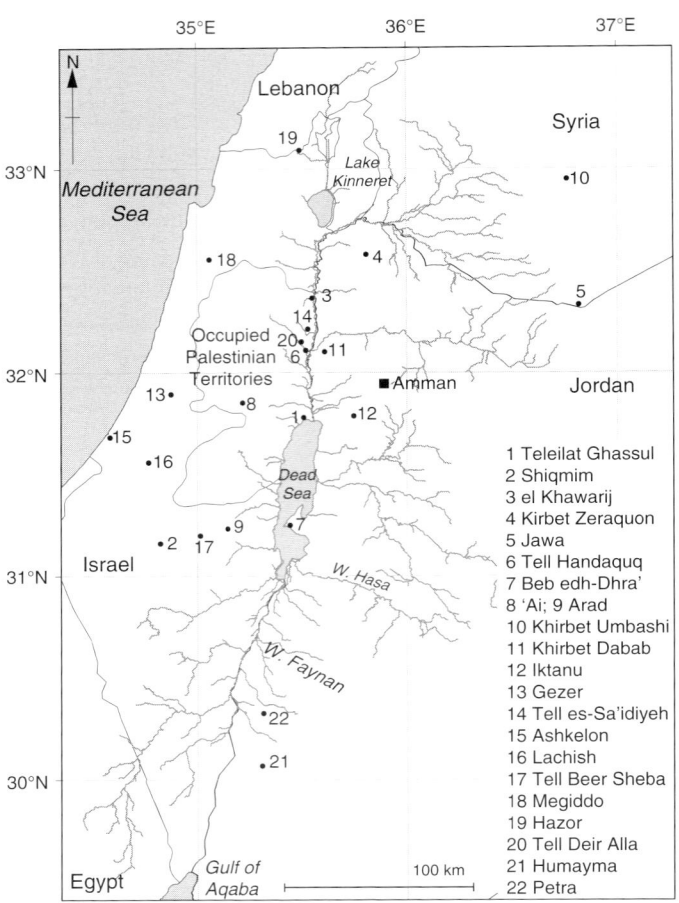

Figure 14.3. Map of the study region showing Bronze Age, Iron Age and Nabataean sites referred to in the text.

Figure 14.4. Massive concentration of chipped stone at Kharaneh IV, Wadi Jilat (© B. Finlayson), marked by the roughly oval, darkened area immediately in front of and next to the two figures.

location, resulting in large accumulations of debris, primarily surviving as chipped stone. The vast quantity of such debris at locations such as Kharaneh IV in Wadi Jilat (Figure 14.4) suggest that these may have been places where groups periodically gathered together (Maher, 2010) and would have been dependent on at least seasonally abundant water. To the north of the Jordan Valley, at Ohalo II, there is evidence of significant plant exploitation and the construction of shelters (Nadel and Hershkovitz, 1991; Nadel and Werker, 1999), both of which are likely to have been influenced by Ohalo II's lakeside location (Figure 14.5). By the late Epipalaeolithic, or Natufian period, there is evidence for substantial investment in settlement infrastructure, both in the form of architecture and possible storage features that some archaeologists interpret as sedentism, such as at Wadi Hammeh (Edwards et al., 1988). A key feature of such locations is their proximity to perennial springs (Goring-Morris and Belfer-Cohen, 2010) but none of these sites have any evidence for active water management. At the same time, however, people had begun to locate special-purpose sites away from water resources, such as the shaman burial in Hilazon Tachtit Cave (Grosman et al., 2008). This represents marking locations in the landscape and the use of hidden localities for ritual-based activities, a development that expands in the subsequent Neolithic. Although there is no direct evidence, an implication

Figure 14.5. Remnants of a brushwood hut at Ohalo II, showing location adjacent to Lake Tiberias (Kinneret) (© S. Mithen).

of such site locations is that people were transporting water, at least on the small scale required for immediate drinking purposes.

By the end of the Epipalaeolithic, the Late Natufian, the increasingly arid conditions of the Younger Dryas appear to have led to increased mobility (Bar-Yosef and Belfer-Cohen, 2002). Reconstructing such mobility patterns is challenging. In the Negev Desert, where a distinctive assemblage of artefacts and structures developed, known as the Harifian culture, a sophisticated movement pattern to exploit the vertical topography and seasonal rainfall to their maximum has been proposed (Goring-Morris, 1987).

14.3 WATER MANAGEMENT IN THE NEOLITHIC

14.3.1 Cultural developments

The Neolithic period can be seen as the crucial turning point in human history. This is conventionally divided into the Pre-Pottery Neolithic A (PPNA), the Pre-Pottery Neolithic B (PPNB) and the Pottery Neolithic, each with their own internal divisions and having their own cultural and economic characteristics. As a whole, the Neolithic is characterised by the development and growth of sedentary communities with complex social organisation, monumental architecture and religious institutions, all based upon the new economic practice of cultivating domestic plants and herding animals (e.g. Cauvin, 2000; Kujit and Goring-Morris, 2002; Mithen, 2003; Barker, 2006). Clearly the increasing size and sedentary nature of Neolithic populations, coupled with an increased reliance upon farming, would have had a significant impact upon the type and quantity of water needed by Neolithic communities.

Although the adoption of Neolithic lifestyles and economies was once referred to as a 'revolution' (e.g. Childe, 1936), the Neolithic period lasted for several thousand years and was one of gradual cultural and economic transition; several of the traits commonly associated with the Neolithic, such as pottery, did not appear until near the end of this period. There is no evidence for domesticated plants or animals until the PPNB period, while hunting continues to hold a prominent role until the Late PPNB (Becker, 1991; Rollefson, 2001). Pastoralism, another key economic practice in the region, probably appears at the very end of the PPN, most likely in its final PPNC phase (Köhler-Rollefson, 1992; Rollefson and Kohler-Rollefson, 1993). Large sedentary communities appear in the PPNB, leading to the so-called megasites of the Late PPNB Jordanian plateau edge (e.g. Bienert et al., 2004) which have been described as constituting modern 'village'-based society and even proto-urbanism. These should be viewed as a highly specific PPNB cultural development that lasted for only about 500 years before an apparent decline in sedentary settlement occupation in the PPNC (Rollefson, 2001); as such the PPNB communities did not lead smoothly to a recognisable agricultural landscape. That only begins to appear in the Pottery Neolithic, with the development of a scattered farming settlement pattern, as seen in the Wadi Ziqlab (Banning, 1996).

Underlying the social and economic changes of the Neolithic period are a series of profound climatic changes. These were outlined in Chapters 3 and 6 of this volume, and their hydrological implications are explored with regard to Wadi Faynan in Chapter 15 and Beidha in Chapter 16. Although considerable uncertainty still surrounds the precise nature of early Holocene climate, both proxy and modelling evidence indicates that this was relatively wet and warm, and had a more stable hydrological regime than currently exists within the region. Against this backdrop, burgeoning Neolithic communities would have been faced with the challenge of obtaining the reliable and predictable water supplies needed to support their sedentary and agricultural economies. In this light, it may be reasonable to expect that Neolithic groups may have been concerned with the management of their water resources, by storage, deflection and transport.

The archaeological evidence for such Neolithic water management is elusive. In part this may be because some water management features would have been constructed in volatile wadi beds and have long been washed away by wadi floods or buried under post-Neolithic alluvium. The traces that do survive of potential Neolithic water management are often very difficult to date. They are frequently isolated from the most likely associated settlement structures and contain limited evidence in the form of material culture for their date of construction and use. Moreover, by definition the locations of structures for water management are determined by landscape topography and consequently may have remained in use for a considerable time, being frequently remodelled and rebuilt, potentially masking or destroying their Neolithic origins.

14.3.2 Neolithic water demands

Neolithic populations not only may have required more water than their Epipalaeolithic predecessors, but may have needed the supply to be more predictable. The need for reliable water for agriculture is surely the most significant new water demand to emerge during this period. In the earliest phases of the Neolithic (PPNA), however, there is no evidence for agriculture with the likely cultivation of wild plants being made possible through a combination of increased rainfall and careful choice of site locations, most commonly alluvial fans adjacent to perennial springs (Sheratt, 1980; Bar-Yosef, 1989; Bar-Yosef and Belfer-Cohen, 1989). It is only later in the Neolithic, as agriculture developed and Neolithic communities sought to increase the extent of the landscape that could be farmed, that we might expect to see demand for irrigation. It is notable, however, that most PPNB settlement occurs both beside springs and within areas suitable for rain-fed farming.

The first evidence for an increasing Neolithic need for water is in the architecture of the settlements. PPNA settlements in the Jordan Valley, such as WF16 (Finlayson and Mithen, 2007), Netiv Hagdud (Bar-Yosef and Gopher, 1997; Finlayson et al., 2003) and Dhra' (Finlayson et al., 2003), typically use mud as a structural material, being used as pisé, as mud brick and in mud plasters (Figure 14.6). Experimental PPNB house building at Beidha (Figure 14.7) has shown that water is probably the most critical resource for this activity, in terms of its availability in adequate quantities and its transportation onto site. The mud component of the wall of the first reconstructed building alone required almost two tons (2,000 litres) of water (Dennis, 2003). Although there was some use of plaster in the preceding Natufian period, Natufian architecture used relatively little mud and was mostly based on pits and dry-stone construction.

Other new demands for water would include water for hygiene, which would seem to have been inevitable with rising sedentary populations within increasingly large and dense Neolithic settlements. Changing ideology and ritual practices may have also increased water demands, such as for making plastered skulls and statues. Some of the first buildings that appear to have been primarily designed for ritual activities have structures related to water, notably built-in channels as at 'Ain Ghazal (Rollefson, 2001), and in some cases large basins, as at Beidha (Byrd, 2005).

The Neolithic is therefore a critical period in the development of water use, with new and developing needs, most of which foreshadow in kind, if not scale, the requirements of later periods in the region. Several scholars (e.g. Gebel, 2004) have considered that this process should be seen as part of a wider Neolithic domestication agenda, which led to increasing human control of various aspects of the natural world, including plants, animals, landscapes and, crucially, water.

Figure 14.6. Extensive use of pisé and mud plaster at the Pre-Pottery Neolithic A settlement of WF16, Wadi Faynan, showing excavations of April 2009 (© S. Mithen).

Figure 14.7. Experimental PPNB buildings at Beidha (© B. Finlayson).

14.3.3 Neolithic settlement patterns

Within the Jordan Valley/Wadi Araba there is a substantial increase in settlement during the PPNA, from mostly West Bank sites in the north, notably Jericho (Kenyon, 1981), Netiv Hagdud (Bar-Yosef and Gopher, 1997) and Gilgal (Noy, 1989), to the PPNA sites in the south at Zahrat edh-Dhra', also known as Zahrat adh-Dhra (Edwards and Higham, 2001), Dhra' and WF16, as well as sites in the uplands at Iraq-ed-Dubb (Kujit, 1994) and Wadi Himmeh. During the PPNB there are sites within the Rift Valley, such as Middle PPNB Ghuwayr 1 (Simmons and Najjar, 1996), but more sites develop in the hills such as Shkarat Msaiad and Beidha (Kirkbride, 1966; Byrd, 2005). By the Late PPNB, so-called mega-sites emerge along the top edge of the Jordanian plateau, such as 'Ain Ghazal (Rollefson and Köhler-Rollefson, 1989), Basta (Gebel et al., 2006), As-Sifiya (Mahasneh, 1997), 'Ain el-Jamman (Waheeb and Fino, 1997) and

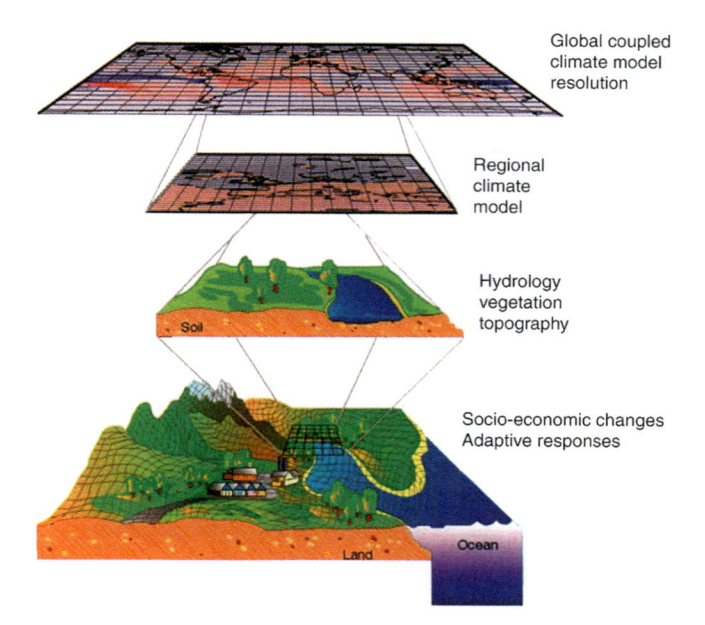

Figure 1.2. Hierarchical modelling from global circulation models to socio-economic impacts (courtesy of David Viner).

Figure 1.4. Water, Life and Civilisation team members during an orientation visit to Jordan in October 2004, here seen at the Iron Age tell of Deir 'Alla.

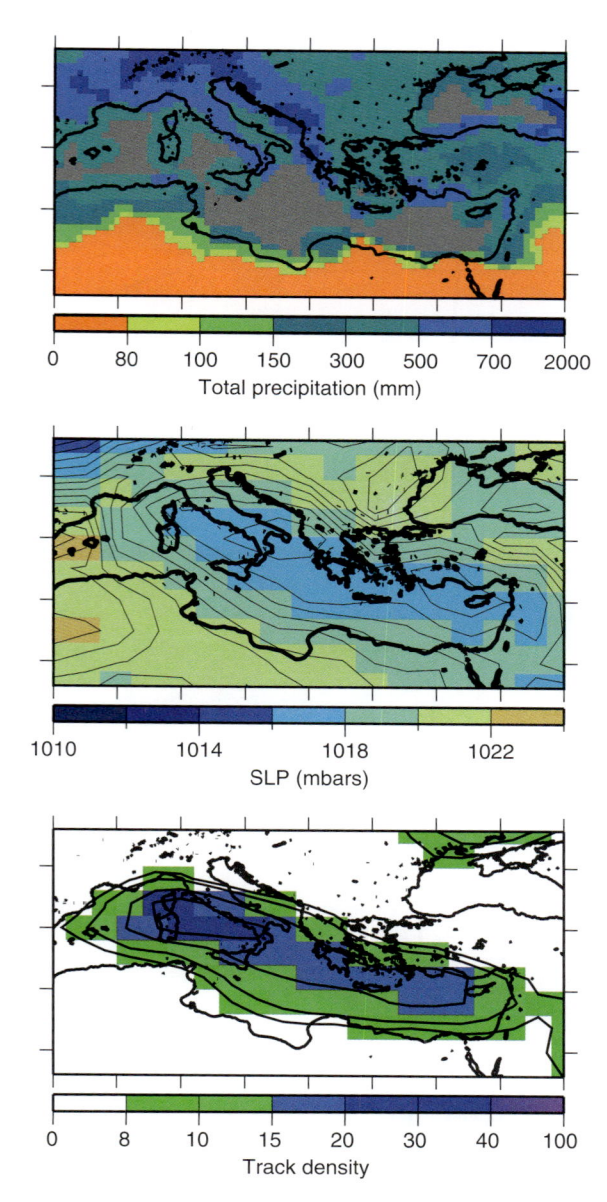

Figure 2.2. Mean climate over the Mediterranean. From top to bottom: December–February total precipitation; December–February SLP; December–February track density.

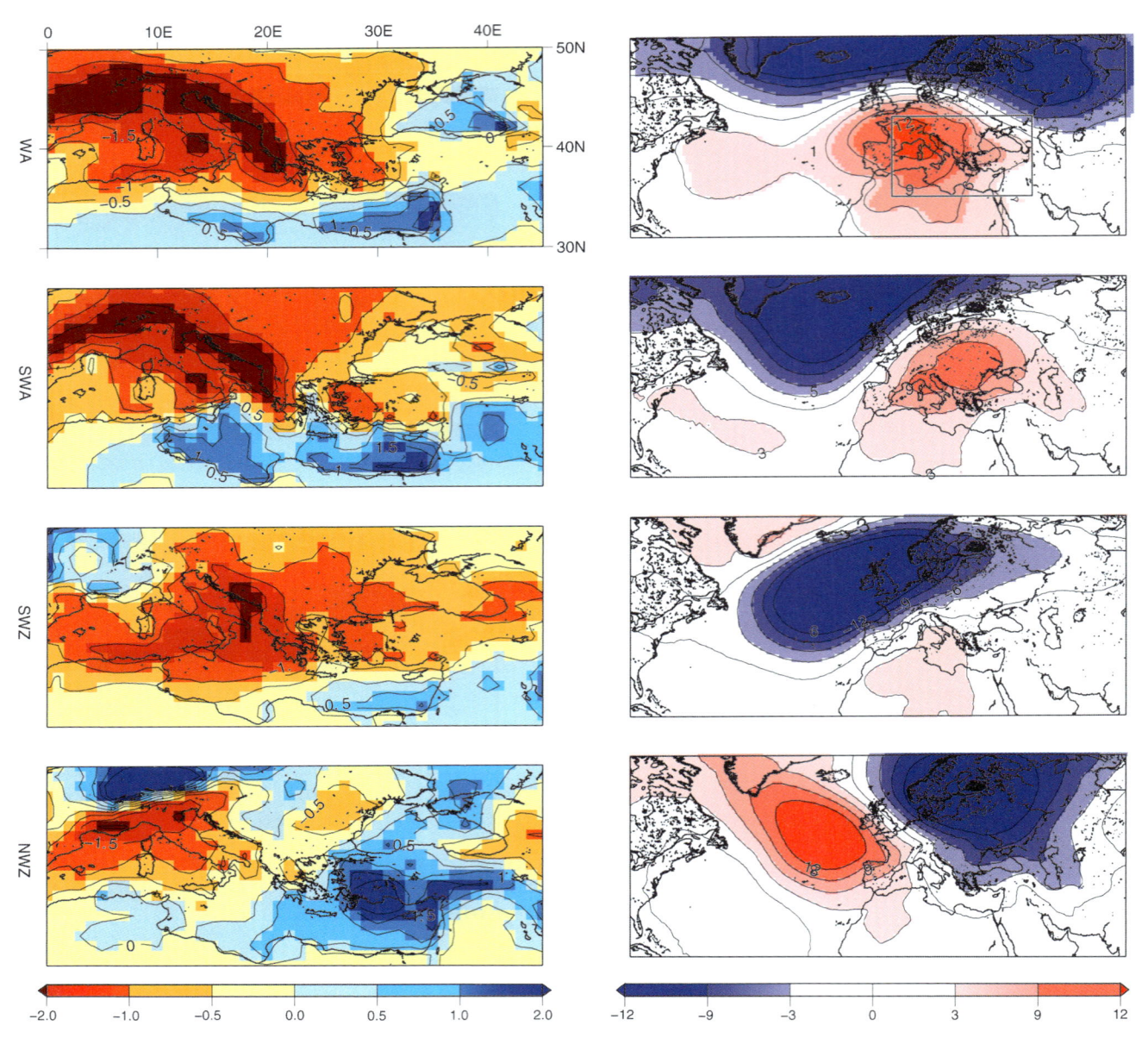

Figure 2.6. Composite daily anomalies during the four GWL regimes that favour rainfall most strongly (WA, SWA, SWZ and NWZ – abbreviations defined in Table 2.1). Left set: daily rainfall anomaly composites over the Mediterranean (box shown on the top right plot); right set: daily SLP anomaly composites over the Mediterranean and Atlantic.

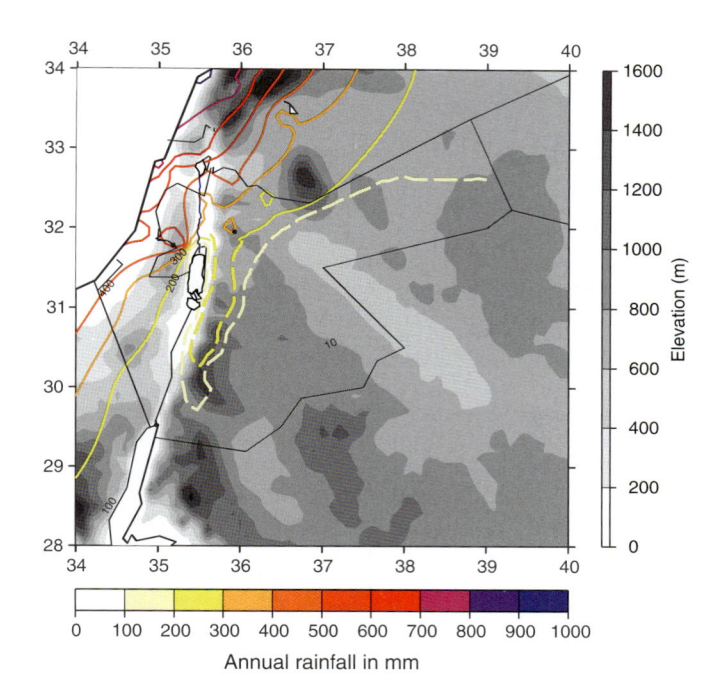

Figure 2.4. Annual total rainfall in Jordan and Israel superposed on the orography. The contours are based on the data from the gauges shown in Figure 2.1. The dashed contours are sketched from published sources (US Geological Survey, 2006) because we were unable to obtain suitable quality data in eastern Jordan.

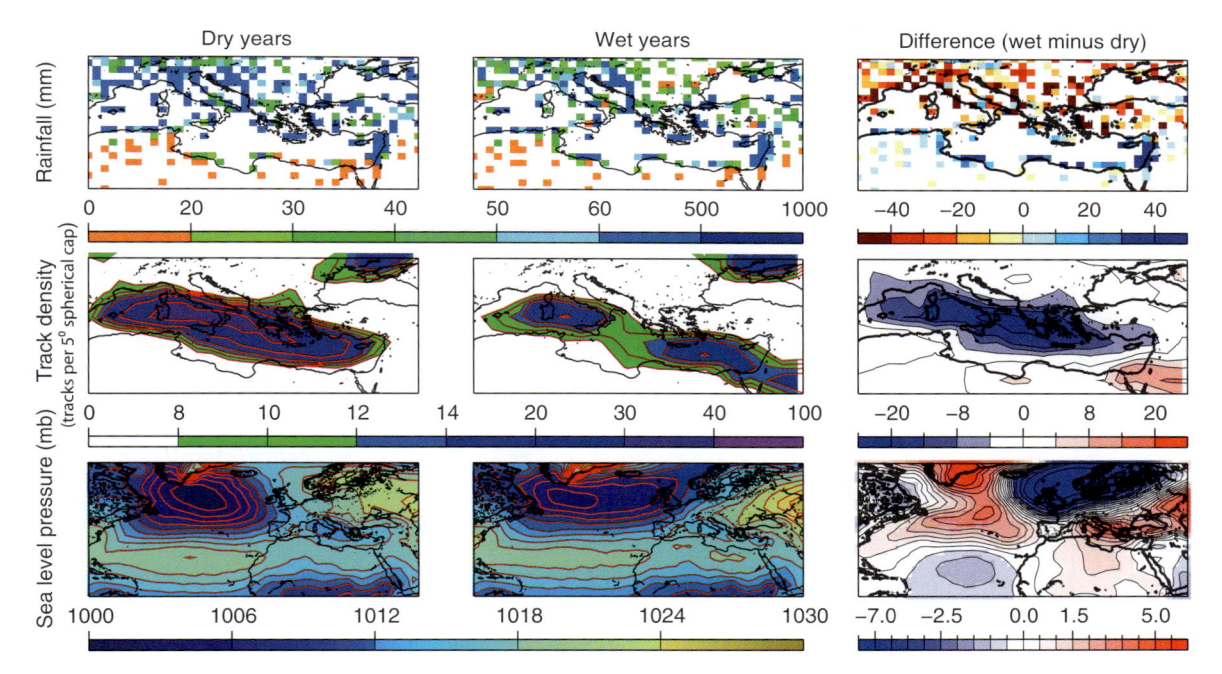

Figure 2.7. Composites of precipitation, track density and SLP during January based on the five wettest and driest Januaries in a box with minimum longitude 34°, maximum longitude 36°, minimum latitude 31°, maximum latitude 33°.

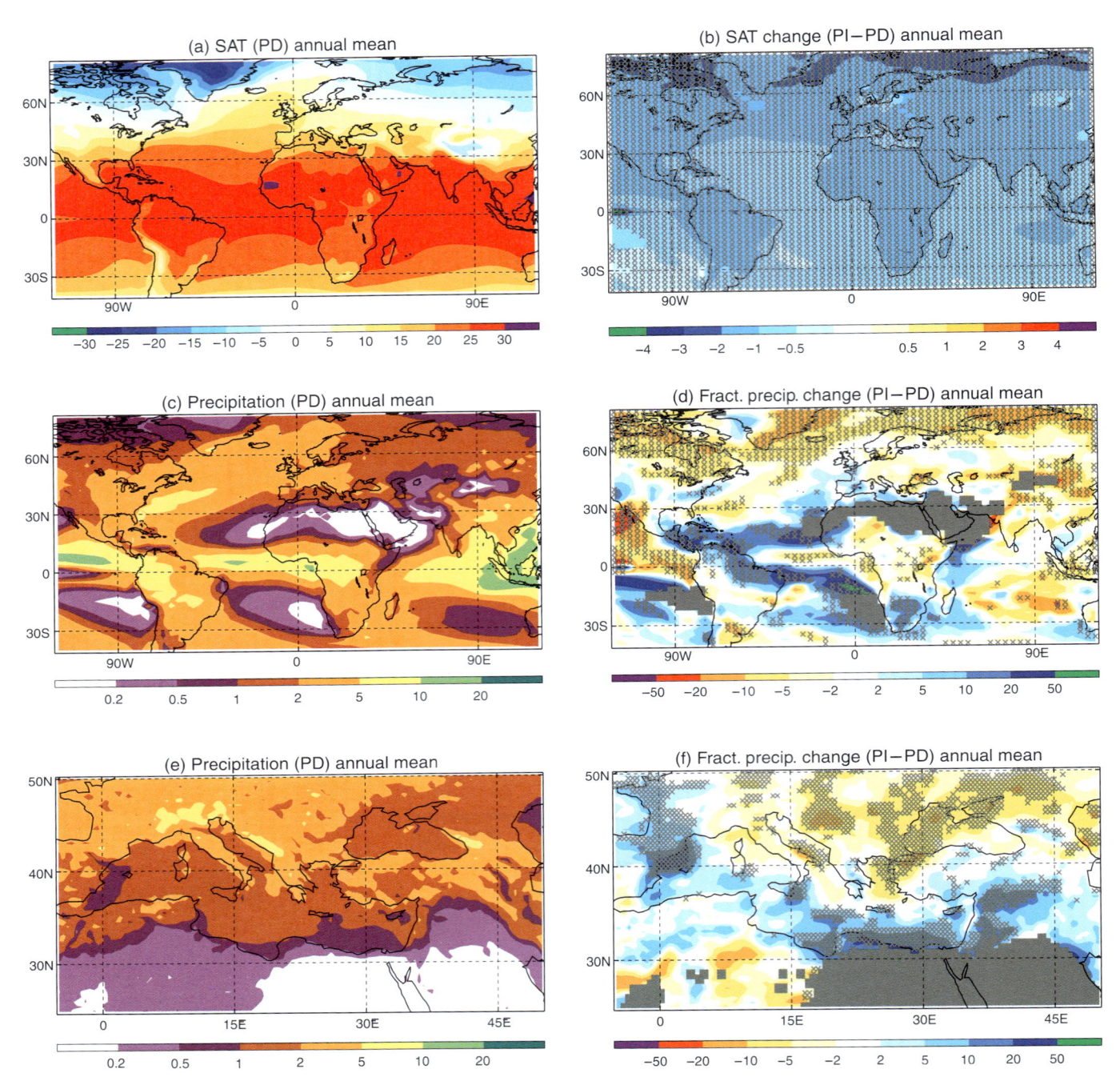

Figure 3.4. Annual mean SAT and precipitation during the pre-industrial period. The top row panels show (a) SAT (°C) in experiment PRESDAY, and (b) the difference (°C) found in experiment PREIND (i.e. PREIND − PRESDAY). The middle row panels show results from the global model where (c) is the precipitation in experiment PRESDAY (mm day^{-1}) and (d) is the fractional difference (%) found in experiment PREIND (i.e. [PREIND − PRESDAY] × 100/PRESDAY). The bottom row (e, f) is identical to the middle row but uses downscaled data from the regional model. For the difference plots (b, d, f), areas where the differences are statistically significant at (b) 99%, (d) 90% and (f) 70% confidence are indicated by grey crosses. Areas of extremely low precipitation (less than 0.2 mm day^{-1}) in experiment PRESDAY are greyed out in the difference plots.

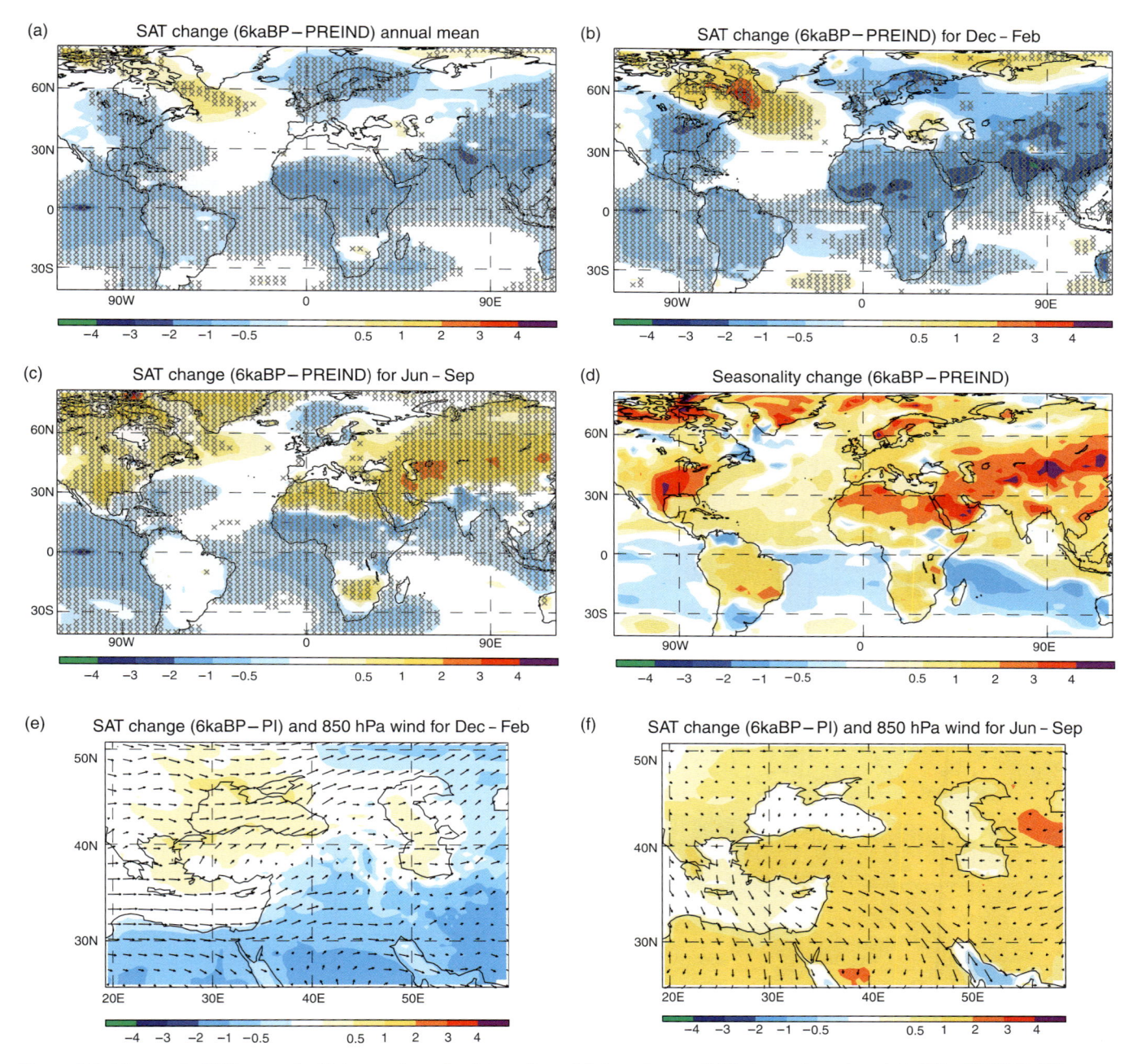

Figure 3.6. Changes in SAT across the Holocene time-slice integrations. (a) Annual mean SAT change (experiment 6kaBP – PREIND, °C). Panels (b) and (c) are as (a), but for boreal summer and winter seasons, respectively. (d) The change in the strength of the seasonal cycle of SAT between experiment 6kaBP and PREIND (the strength of the cycle is defined as the maximum monthly mean SAT minus the minimum monthly mean SAT, units °C). (e) Boreal winter SAT change (6kaBP – PREIND, colours, °C) in the regional model and lower tropospheric winds (850 hPa) in experiment PREIND (arrows, units m s^{-1}). (f) As (e) but for boreal summer. In panels (b) to (d), areas where the changes are statistically significant at the 90% level are marked with grey crosses.

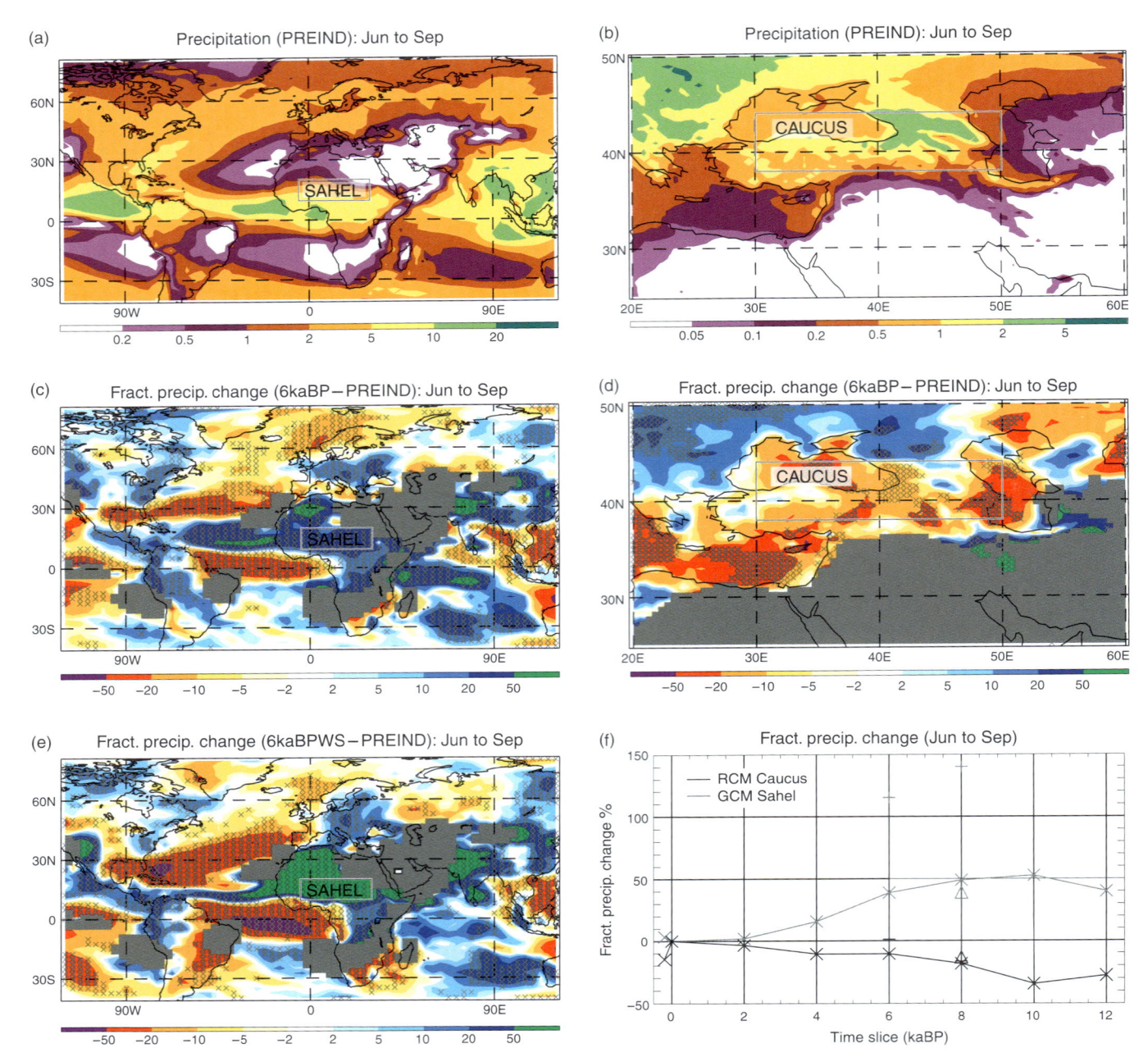

Figure 3.9. Differences in boreal summer precipitation across the Holocene time-slice integrations. The top row shows the precipitation in experiment PREIND (units mm day^{-1}) using data from (a) the global model and (b) the regional model. The middle row shows the fractional change in precipitation (units %) in experiment 6kaBP relative to experiment PREIND (i.e. [6kaBP−PREIND] × 100/PREIND), using data from (c) the global model and (d) the regional model. Panel (e) is similar to (c) but for experiment 8kaBP-WS. Panel (f) shows the fractional precipitation changes averaged over the SAHEL box (in the global model, as marked in panels (a) and (c)) and the CAUCUS box (in the regional model, as marked in panels (b) and (d)) in the time-slice experiments. Data points from the time-slice experiments are marked by × symbols whereas the + symbols mark data points from experiments 6kaBP-WS and 8kaBP-WS and triangles mark data points from experiment 8kaBPNOICE (experiment PRESDAY is shown at time = −0.2 kaBP). In panels (c) to (e), areas where the changes are statistically significant at the 90% level are marked with grey crosses. Areas of extremely low precipitation (less than 0.2 mm day^{-1} for the global model and 0.05 mm day^{-1} in the regional model) in experiment PREIND are greyed out in panels (b) to (e).

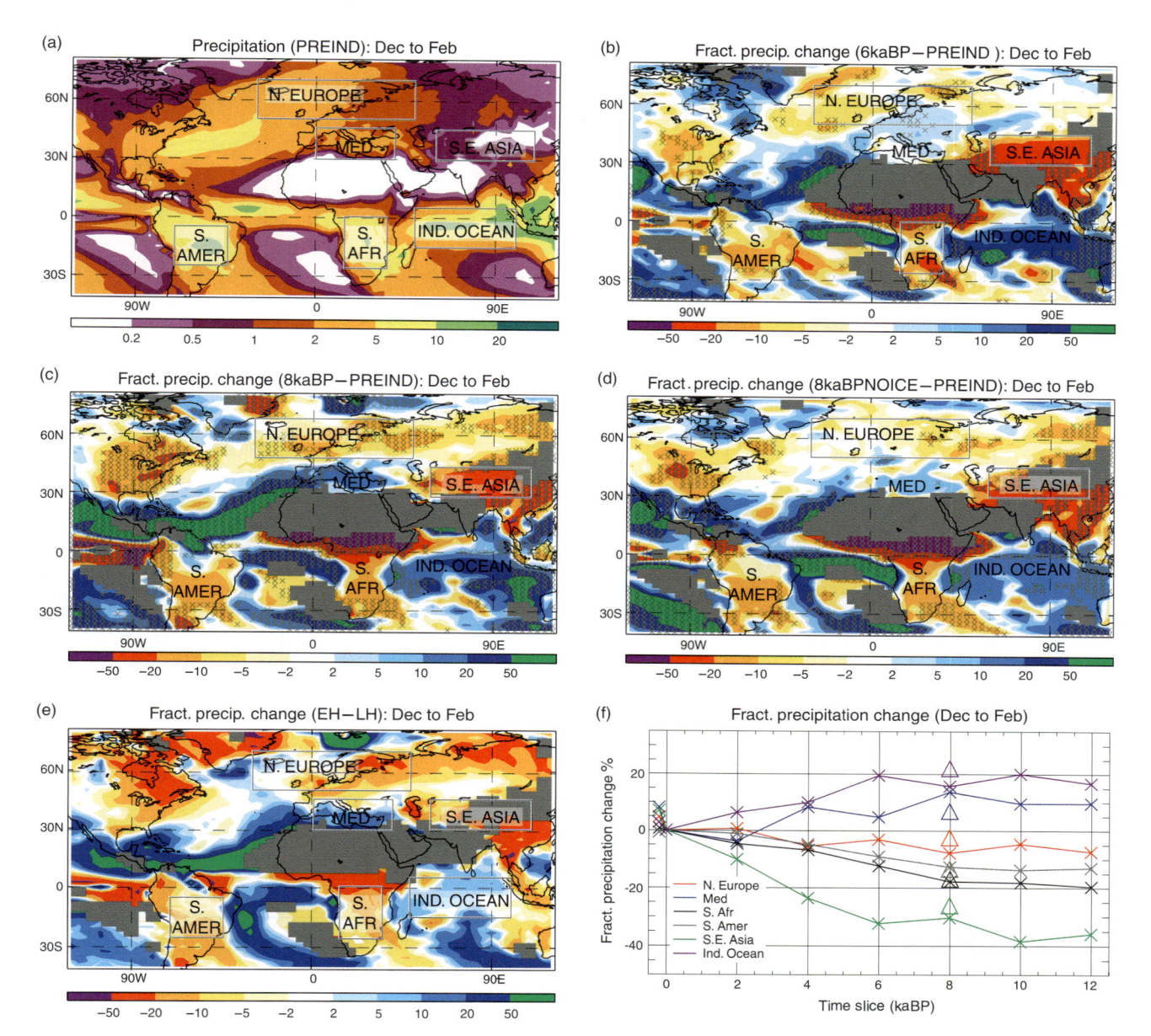

Figure 3.10. Differences in boreal winter precipitation across the Holocene time-slice integrations using the global model. (a) Experiment PREIND (units mm day^{-1}). (b) The fractional change in precipitation (units %) in experiment 6kaBP relative to experiment PREIND (i.e. [6kaBP − PREIND] × 100/PREIND). (c) As (b) but for experiment 8kaBP. (d) As (b) but for experiment 8kaBPNOICE. (e) As (b) but for Early Holocene experiments (8kaBP + 10kaBP + 12kaBP) minus the Late Holocene experiments (2kaBP + 4kaBP + 6kaBP). Panel (f) shows the fractional precipitation changes averaged over the boxes marked in panels (a) to (e). Data points from the time-slice experiments are marked by × symbols whereas triangles mark data points from experiment 8kaBPNOICE (experiment PRESDAY is shown at time = −0.2 kaBP). In panels (b) to (e), areas where the changes are statistically significant at the 90% level are marked with grey crosses. Areas of extremely low precipitation (less than 0.2 mm day^{-1}) in experiment PREIND are greyed out in panels (c) and (d).

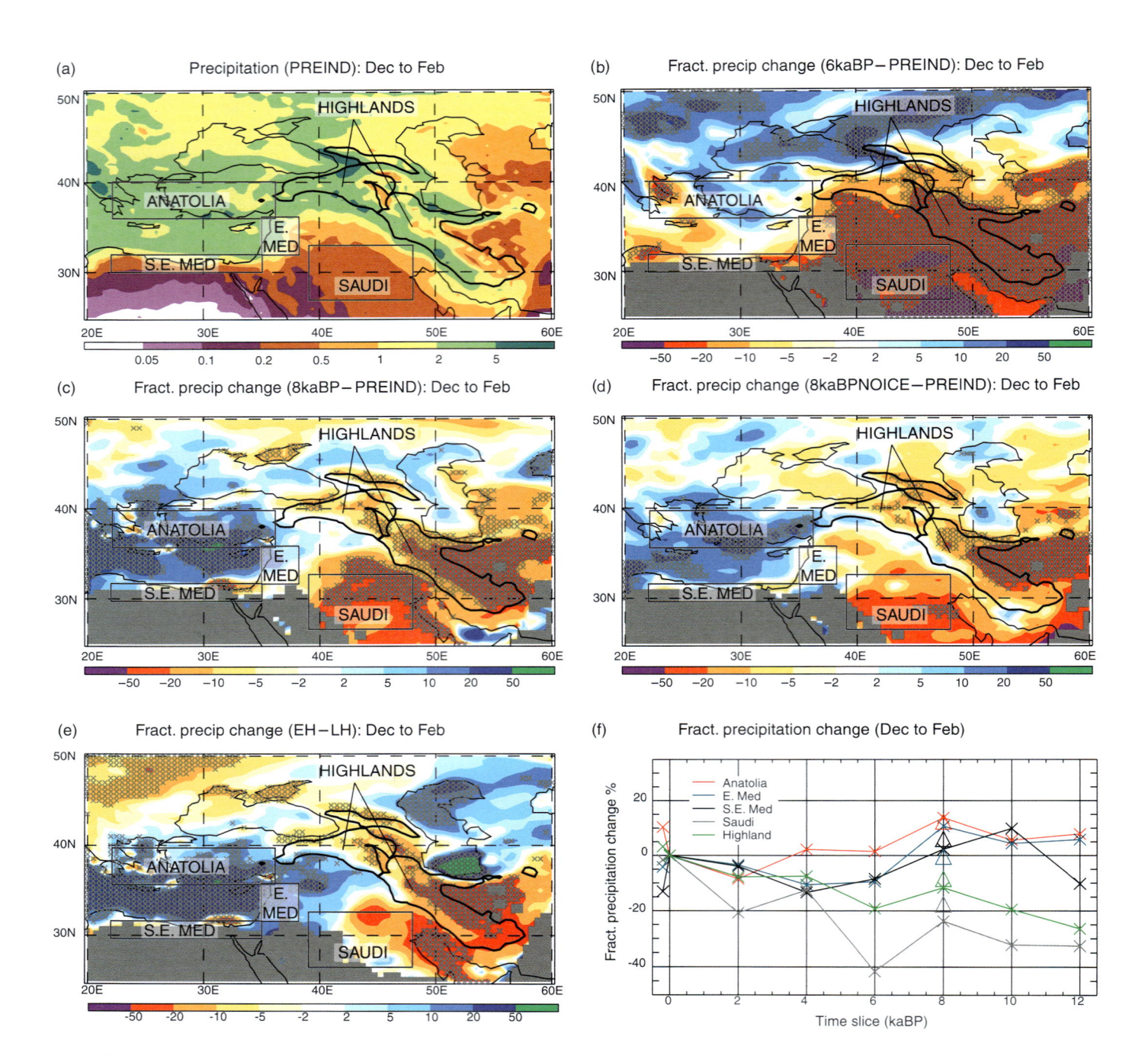

Figure 3.13. Differences in boreal winter precipitation across the Holocene time-slice integrations from the regional model. (a) Experiment PREIND (units mm day^{-1}). (b) The fractional change in precipitation (units %) in experiment 6kaBP relative to experiment PREIND (i.e. [6kaBP − PREIND] × 100/PREIND). (c) As (b) but for experiment 8kaBP. (d) As (b) but for experiment 8kaBPNOICE. (e) as (b) but for Early Holocene experiments (8kaBP + 10kaBP + 12kaBP) minus the Late Holocene experiments (2kaBP + 4kaBP + 6kaBP). (f) The fractional precipitation changes averaged over the boxes marked in panels (a) to (e). Data points from the time-slice experiments are marked by × symbols, whereas triangles mark data points from experiment 8kaBPNOICE (experiment PRESDAY is shown at time −0.2 kaBP). In panels (b) to (e), areas where the differences are statistically significant at the 90% level are marked with black crosses. Areas of extremely low precipitation (less than 0.2 mm day^{-1}) in experiment PREIND are blacked out in panels (c) and (d).

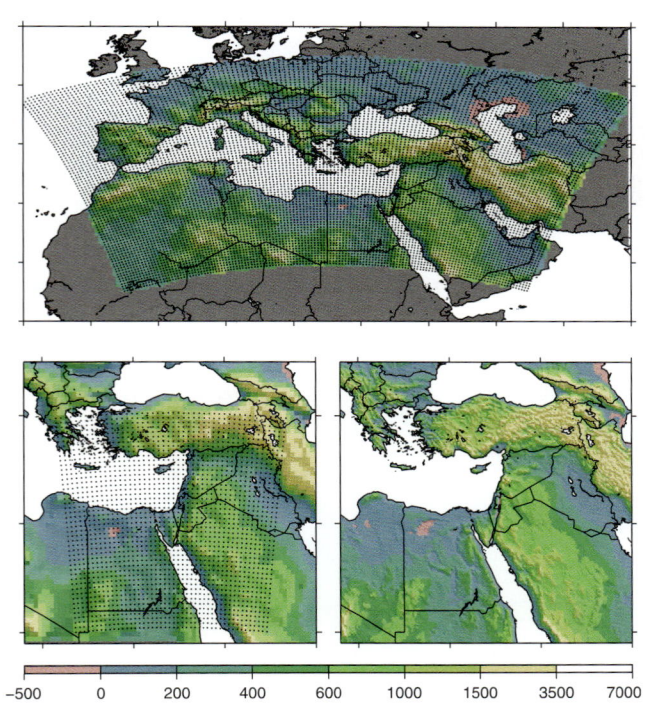

Figure 4.1. Model domains and modelled and observed topography. Top: model grid points (dots) on the model topography for the large domain. Bottom left: small domain (Middle East only) used for the ensembles. Bottom right: observed topography for the eastern Mediterranean and Middle East. Topography is given in metres above sea level for all the plots.

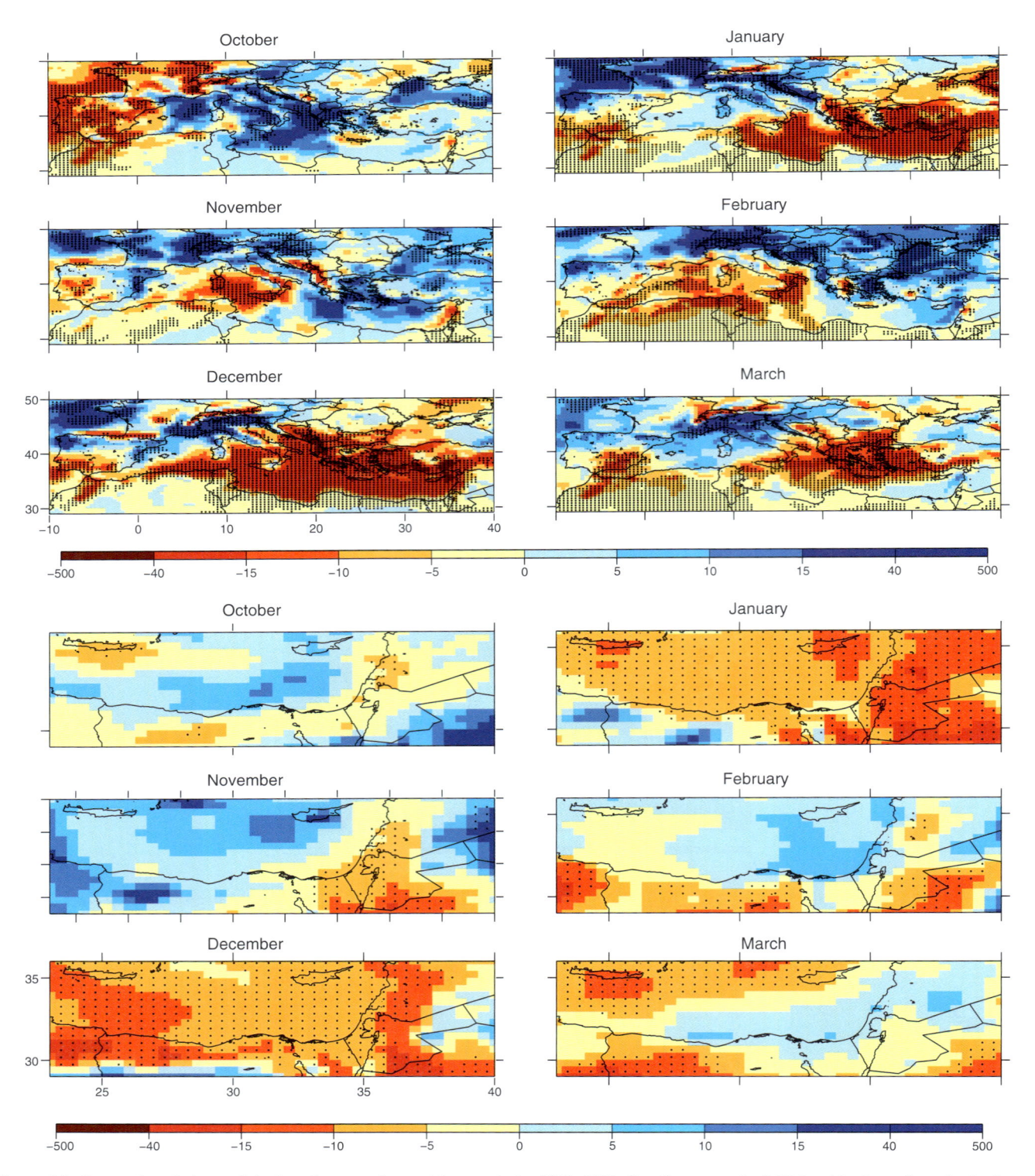

Figure 4.3. Seasonal cycle in precipitation change under an A2 scenario by 2070–2100. Significance at the 95% level is shown by a dot in the grid square. Top set: monthly mean absolute change in rainfall (mm) over the whole of the Mediterranean under an A2 scenario. Bottom set: monthly mean percentage change in rainfall (%) over the East Mediterranean only.

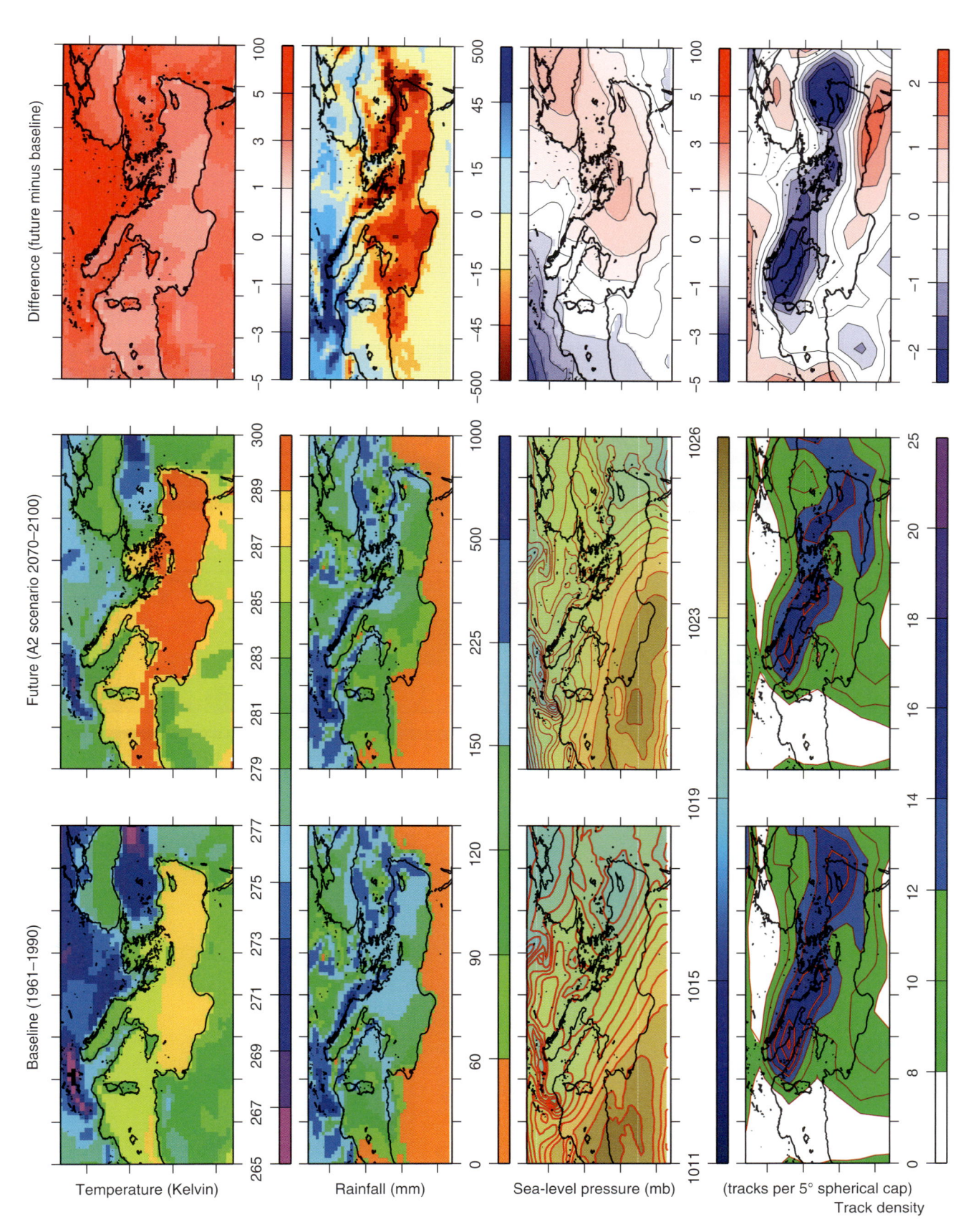

Figure 4.4. Change in the January climate (temperature, precipitation, sea-level pressure and 850 mb track density) over the Mediterranean under an A2 scenario by 2070–2100.

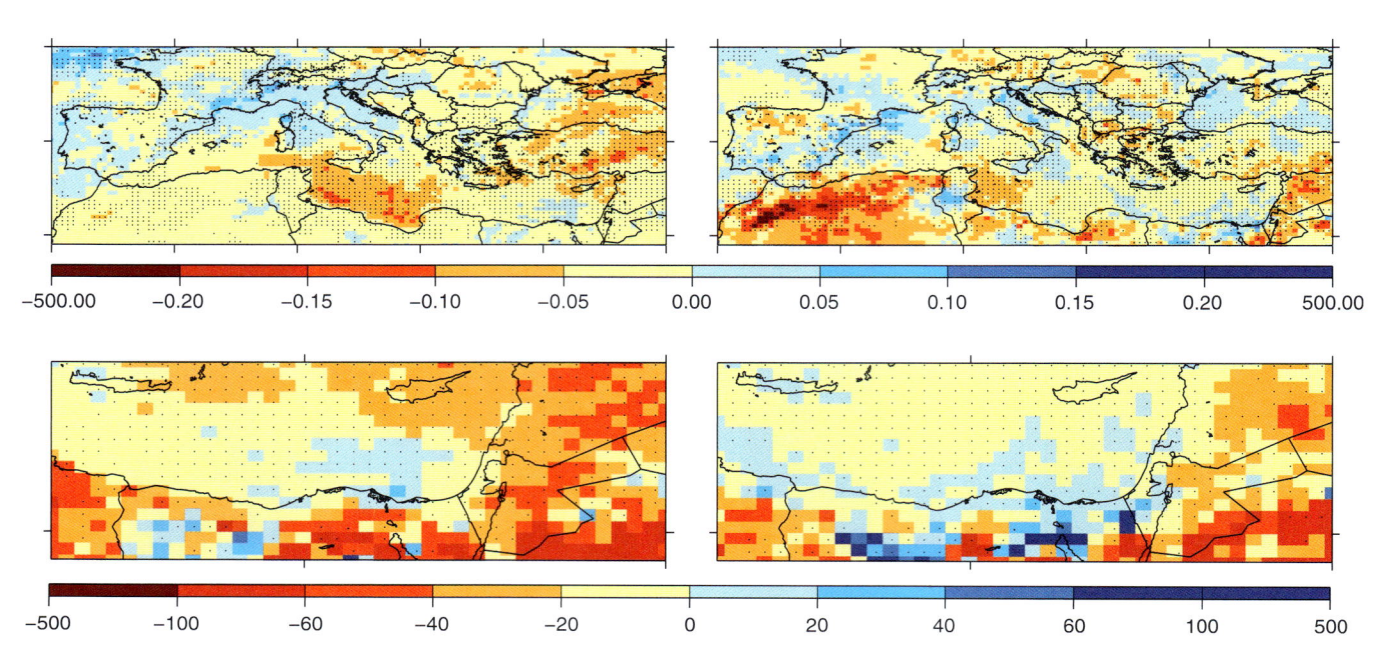

Figure 4.5. Change in daily rainfall probabilities. Significance at the 95% level is indicated by a dot within the grid square. Top row, left panel: absolute change in the probability of rain given no rain the day before; right: absolute change in the probability of rain given rain the day before. Bottom row: as above but for percentage changes for the southeast part of the region only.

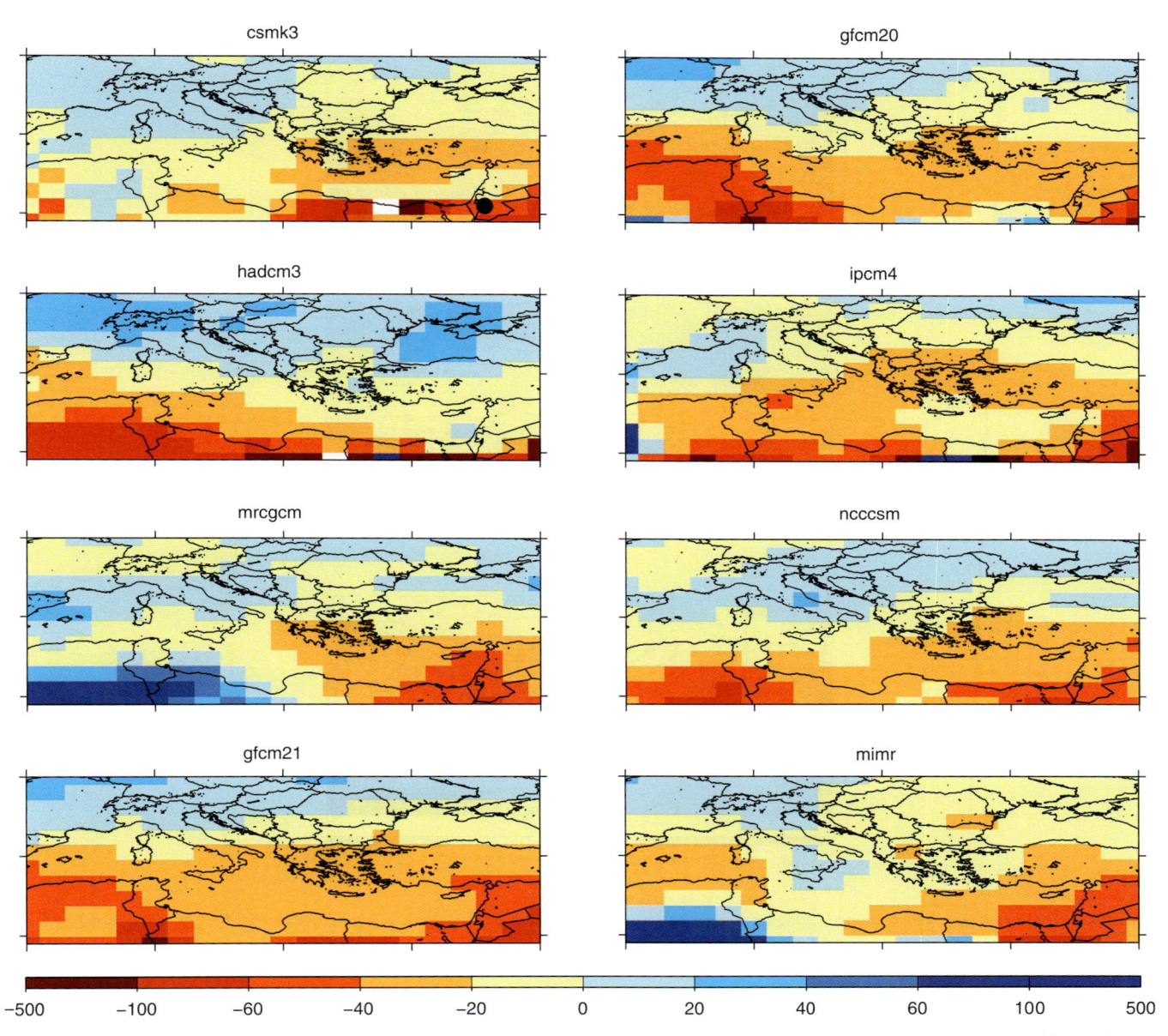

Figure 4.7. Percentage change in January precipitation under an A2 scenario by 2070–2100 for eight IPCC models. The model name abbreviations on the plots are: CSIRO Mark 3.0 (csmk3); GFDL CM 2.0 AOGCM (gfcm20); HadCM3 (hadcm3); IPSL CM4 (ipcm4); MRI-CGCM2.3.2 (mrcgcm); NCAR CCSM3 (nccsm); GFDL CM 2.1 AOGCM (gfcm21); MIMR MIROC3.2 (medium resolution).

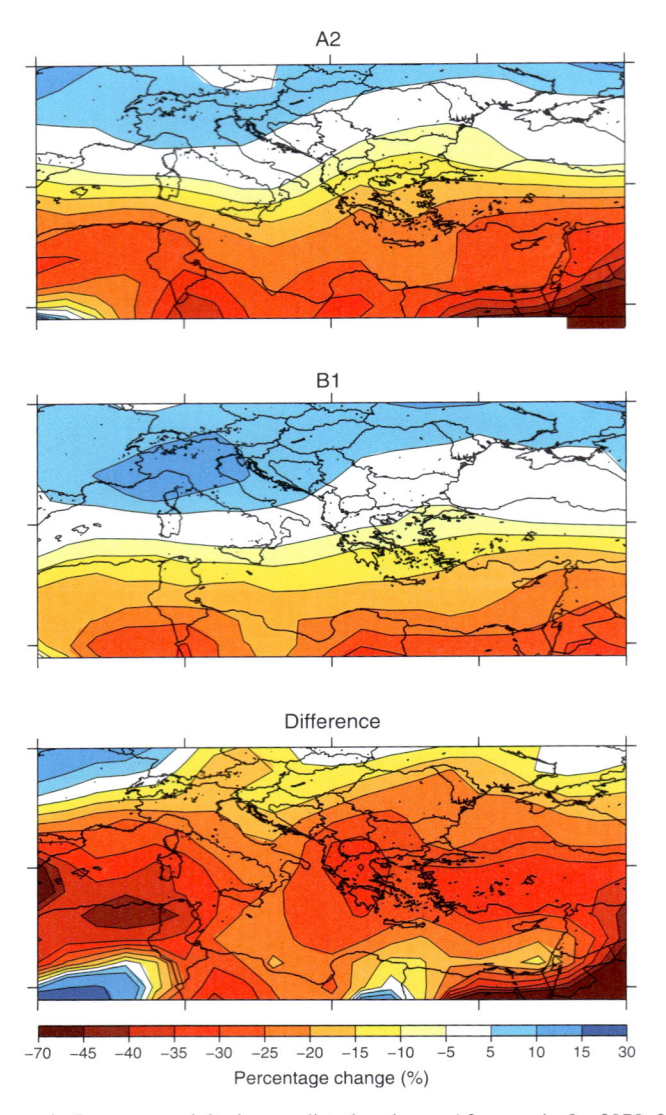

Figure 4.8. Top: mean percentage change in January precipitation predicted under an A2 scenario for 2070–2100 for the IPCC models shown in Figure 4.7; middle: mean percentage change in January precipitation predicted under a B1 scenario for 2070–2100 for the IPCC models shown in Figure 4.7. Bottom: difference in the mean percentage change between the A2 and B1 (B1 − A2).

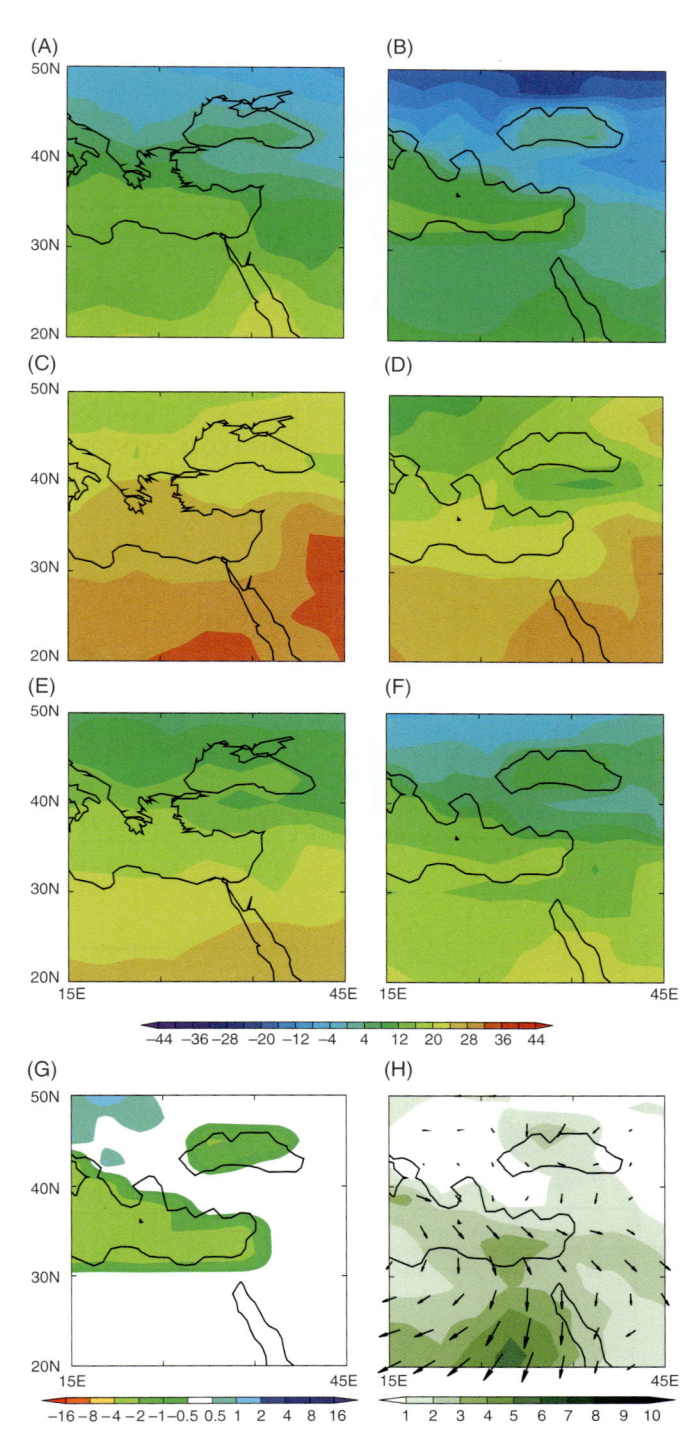

Figure 6.4. Climate model outputs for the LGM and present day. (A) Present-day winter (DJF) temperature (°C); (B) LGM winter (DJF) temperature; (C) present-day summer (JJA) temperature; (D) LGM summer (JJA) temperature; (E) present-day average annual temperature; (F) LGM average annual temperature; (G) LGM annual average precipitation minus evaporation in mm day^{-1}; (H) LGM annual average wind strength (in m s^{-1}) and vectors.

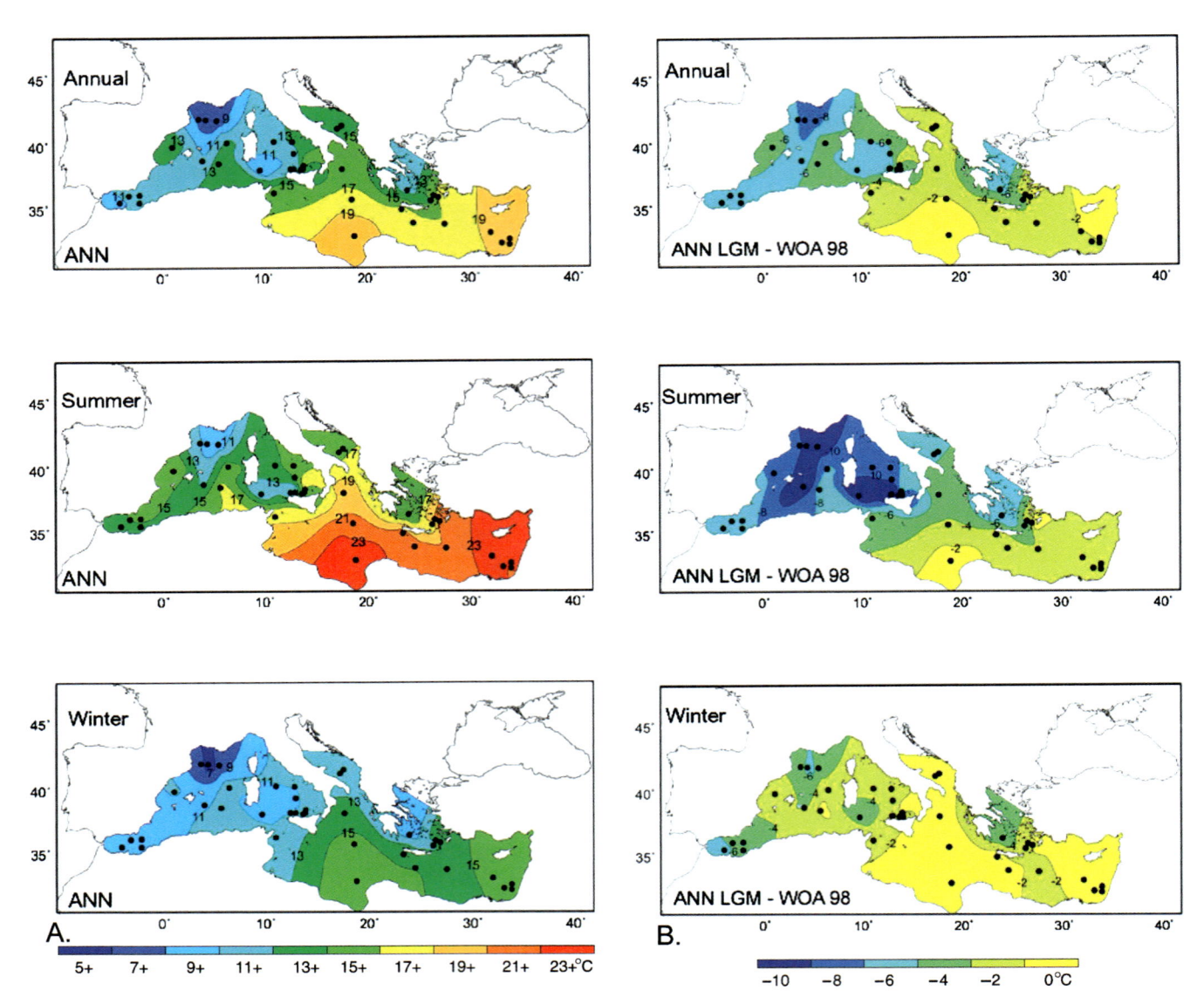

Figure 6.11. (A) Foraminiferal LGM annual, summer and winter SST reconstructions (Hayes *et al.*, 2005), calculated using an artificial neural network (ANN). (B) Temperature anomalies for annual, summer and winter SSTs during the Last Glacial Maximum, compared with modern-day values (Hayes *et al.*, 2005). Anomaly values were calculated by subtracting modern-day SSTs from the glacial values. The black dots represent the sites of the cores from which the LGM data were obtained. This figure is a reproduction of part of Figure 9 in Hayes *et al.* (2005).

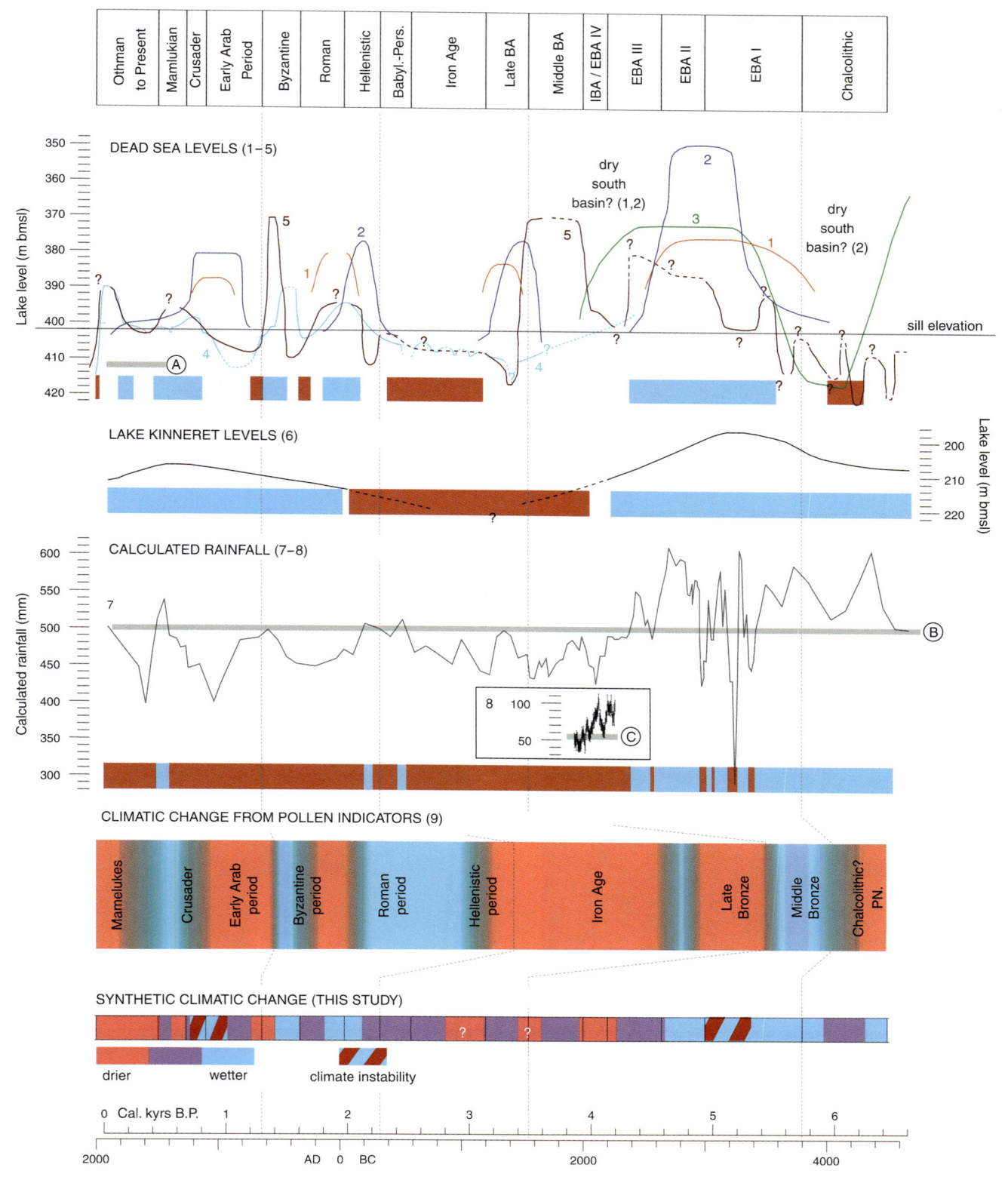

Figure 7.2. Compilation of several proxies for the middle to late Holocene in the southern Levant. Red and blue bars show interpreted climate fluctuations (wetter/drier conditions). Archaeological periods from Rosen (2007). Dead Sea levels: 1 – Frumkin and Elitzur (2002). 2 – Klinger *et al.* (2003). 3 – Enzel *et al.* (2003). 4 – Bookman *et al.* (2004). 5 – Migowski *et al.* (2006). (A) Dead Sea levels in 1997 (Migowski *et al.*, 2006). Lake Kinneret levels: 6 – Hazan *et al.* (2005). Calculated rainfall: 7 – From the Soreq cave record; Bar-Matthews and Ayalon (2004). (B) Present-day mean annual rainfall in Soreq area. 8 – From tamarisk wood, Mount Sedom cave; Frumkin *et al.* (2009). (C) Present-day mean annual rainfall at Mount Sedom. 9 – Climatic change from pollen indicators according to Neumann *et al.* (2007). Our synthesis is presented at the bottom of the figure.

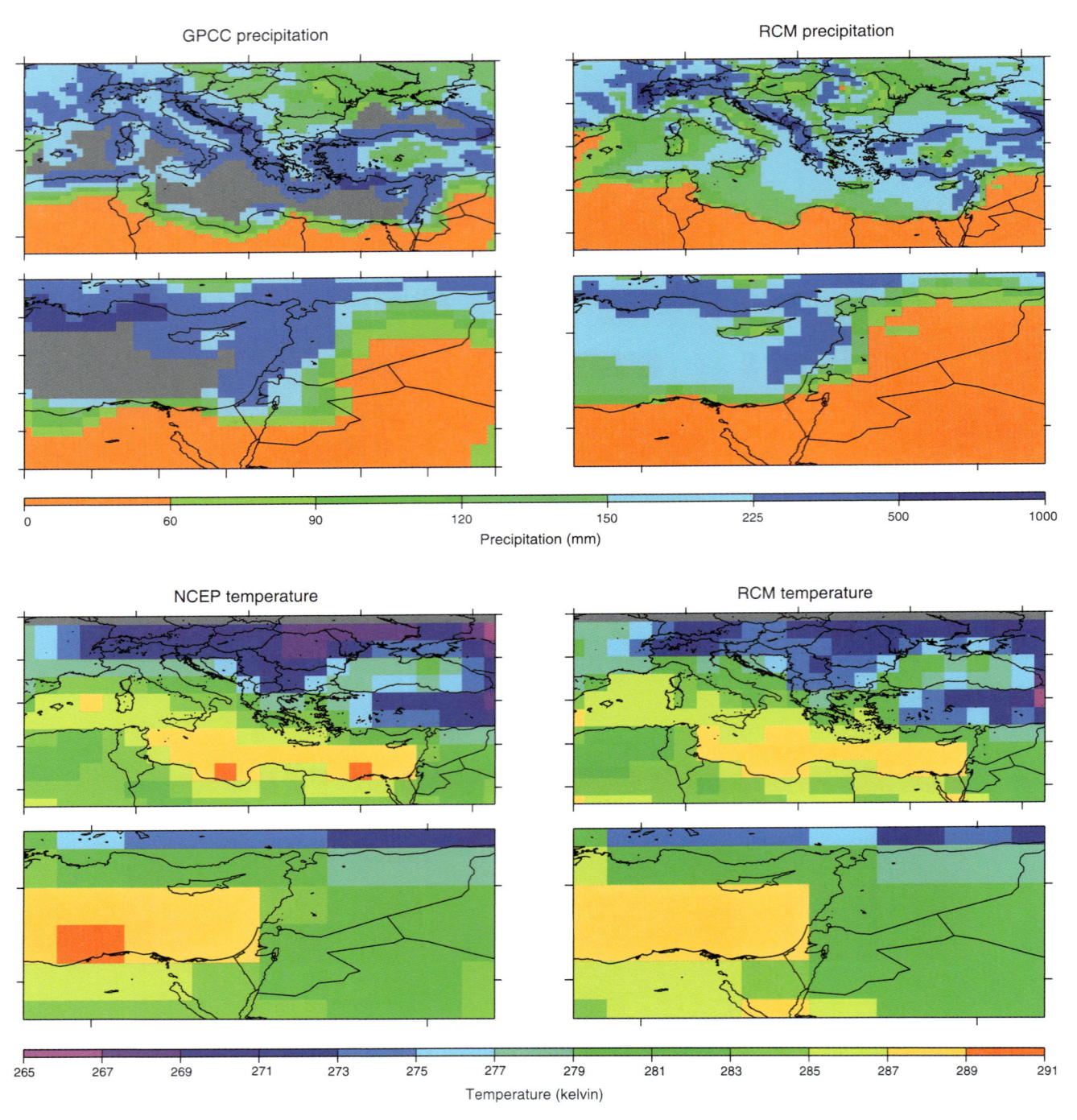

Figure 8.2. Comparison between the observed and modelled climate. Top set, left: GPCC precipitation for the whole Mediterranean (top) and for the Middle East only (bottom); right: RCM precipitation for the whole Mediterranean (top) and for the Middle East only (bottom). Bottom set, left: NCEP reanalysis temperature for the whole Mediterranean (top) and for the Middle East only (bottom); right: RCM temperature for the whole Mediterranean (top) and for the Middle East only (bottom).

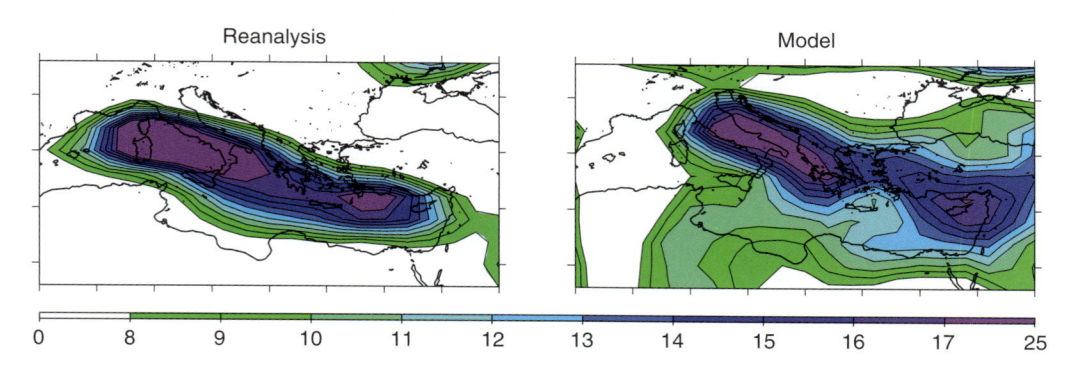

Figure 8.3. Comparison of modelled and observed track densities (in number of tracks per month per 5 degree spherical cap). Left: mean January track density in the reanalysis. Right: mean January track density in the RCM large-domain baseline scenario. Both figures are based on tracking of features in the 850 mb vorticity field.

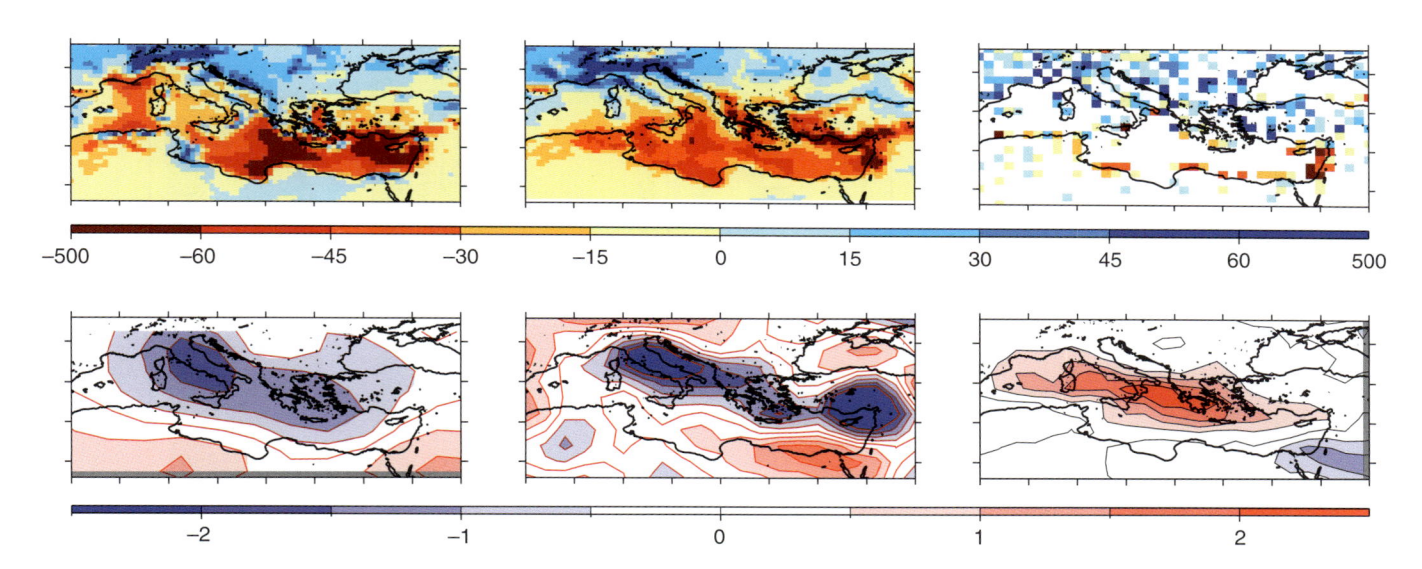

Figure 8.4. Modelled changes in October–March precipitation (top, in mm) and December–February track density (bottom, in number of tracks per month per 5 degree spherical cap for (from left to right) late Holocene minus early Holocene; future (2070–2100) – present (1961–1990); and driest years – wettest years from 1948–1999.

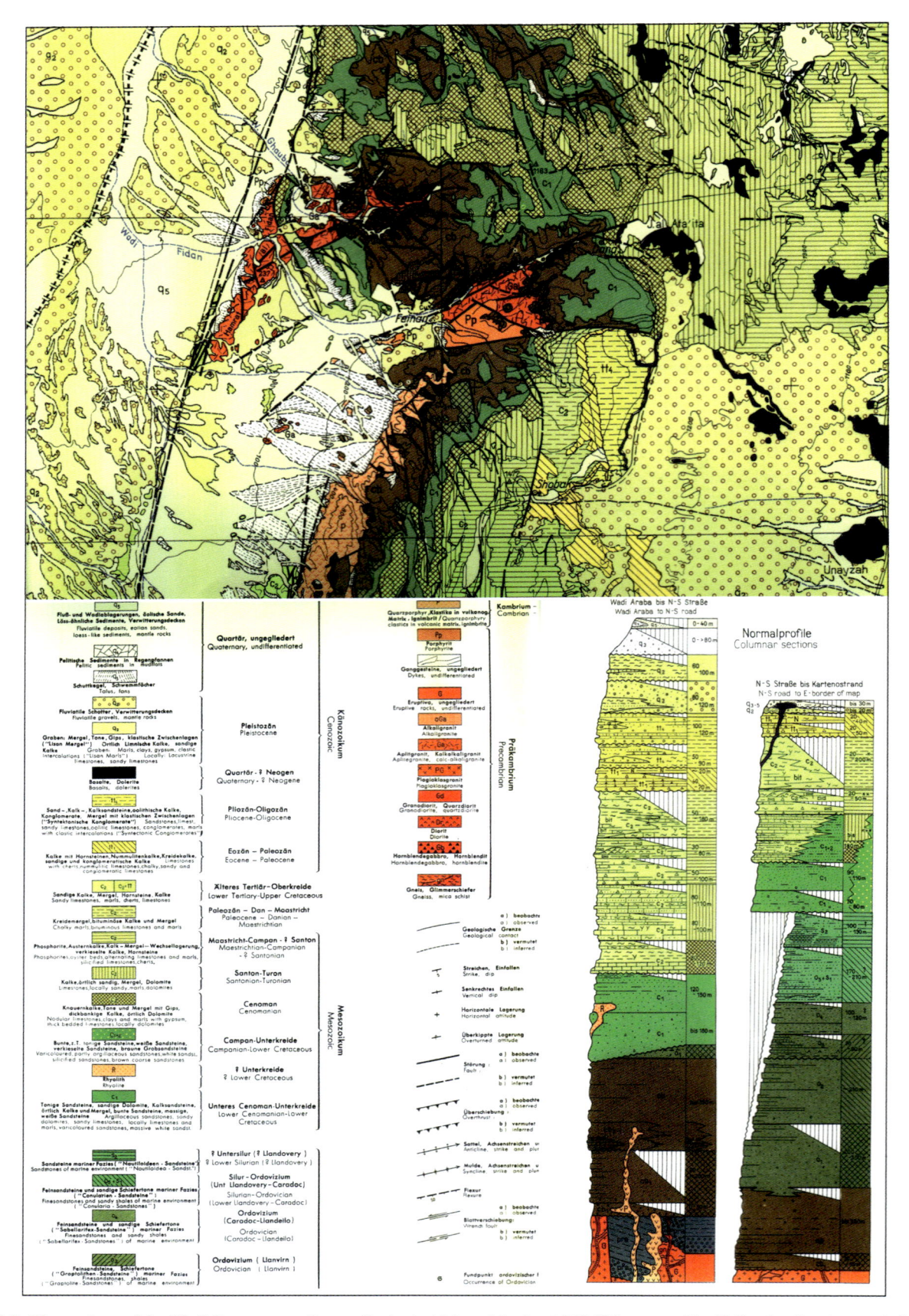

Figure 12.2. The geology of the Wadi Faynan area. Source Geological Map of Jordan 1:250,000, prepared by F. Bender, Bundesanstalt für Geowissenschaften und Rohstoffe, Hannover 1968 [Sheet: Aqaba-Ma'an and Amman]. Reproduced with permission. Not to scale. © Bundesanstalt für Geowissenschaften und Rohstoffe.

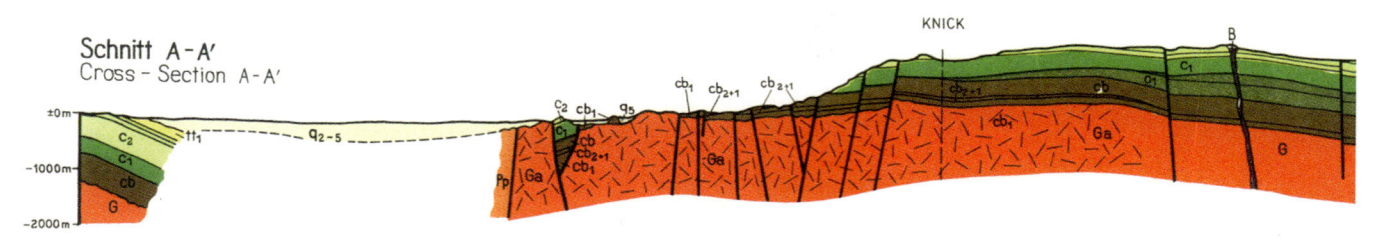

Figure 12.3. A geological cross-section made 5 km to the north of the Wadi Faynan. Source: Geological Map of Jordan 1:250,000, prepared by F. Bender, Bundesanstalt für Geowissenschaften und Rohstoffe, Hannover 1968 [Sheet: Aqaba-Ma'an and Amman]. Reproduced with permission. Not to scale. © Bundesanstalt für Geowissenschaften und Rohstoffe.

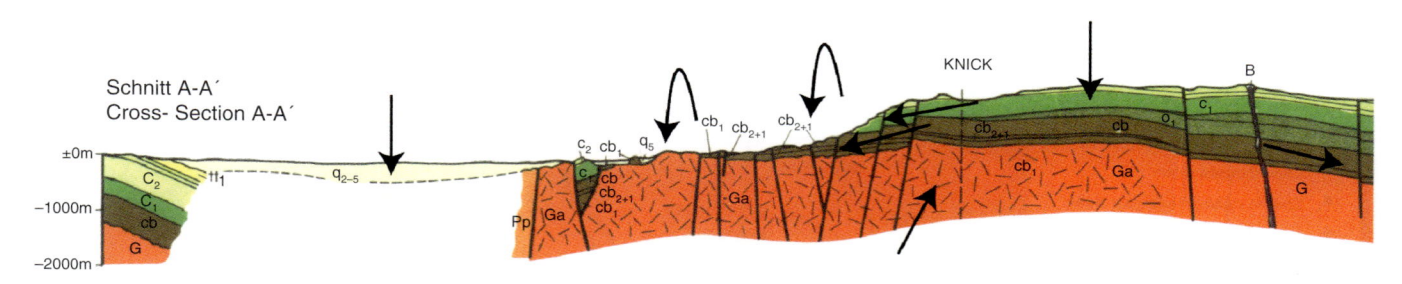

Figure 12.8. A conceptual model of the key water stores and pathways in the Wadi Faynan. Precipitation on the limestone and plateau soils in the upper reaches is likely to be the key aquifer recharge mechanism. The water will flow laterally through the limestone and sandstones until it emerges at a contact point between the two or at a contact between the sandstone and the aplite-granite. Re-infiltration (transmission loss) occurs as the water flows along the channel network. The Jebel Hamrat al Fidan forces the water to return close to the surface. Base map source: Geological Map of Jordan 1:250,000, prepared by Geological Survey of Germany, Hannover 1968 (Sheet: Aqaba-Ma'an). Reproduced with permission. Not to scale.

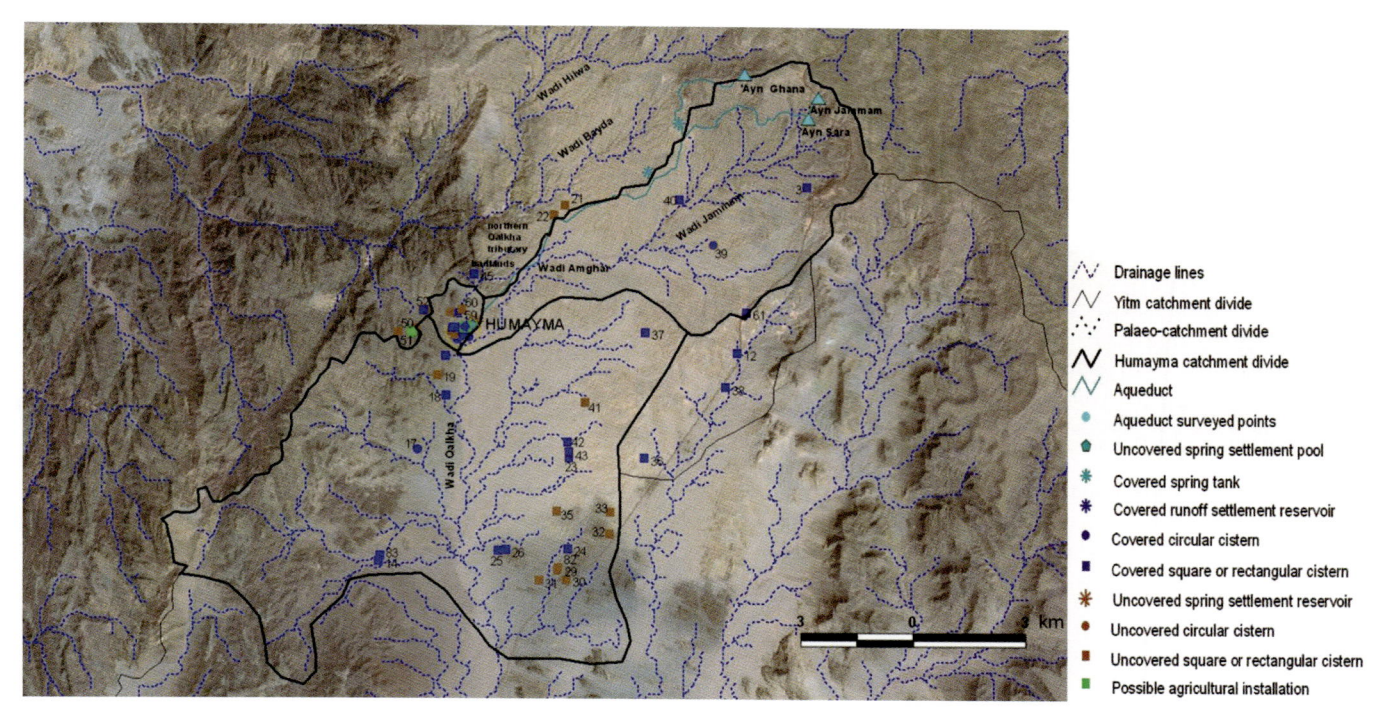

Figure 19.3. Humayma environs showing location of pertinent hydraulic structures surveyed in the Humayma Hydraulic Survey.

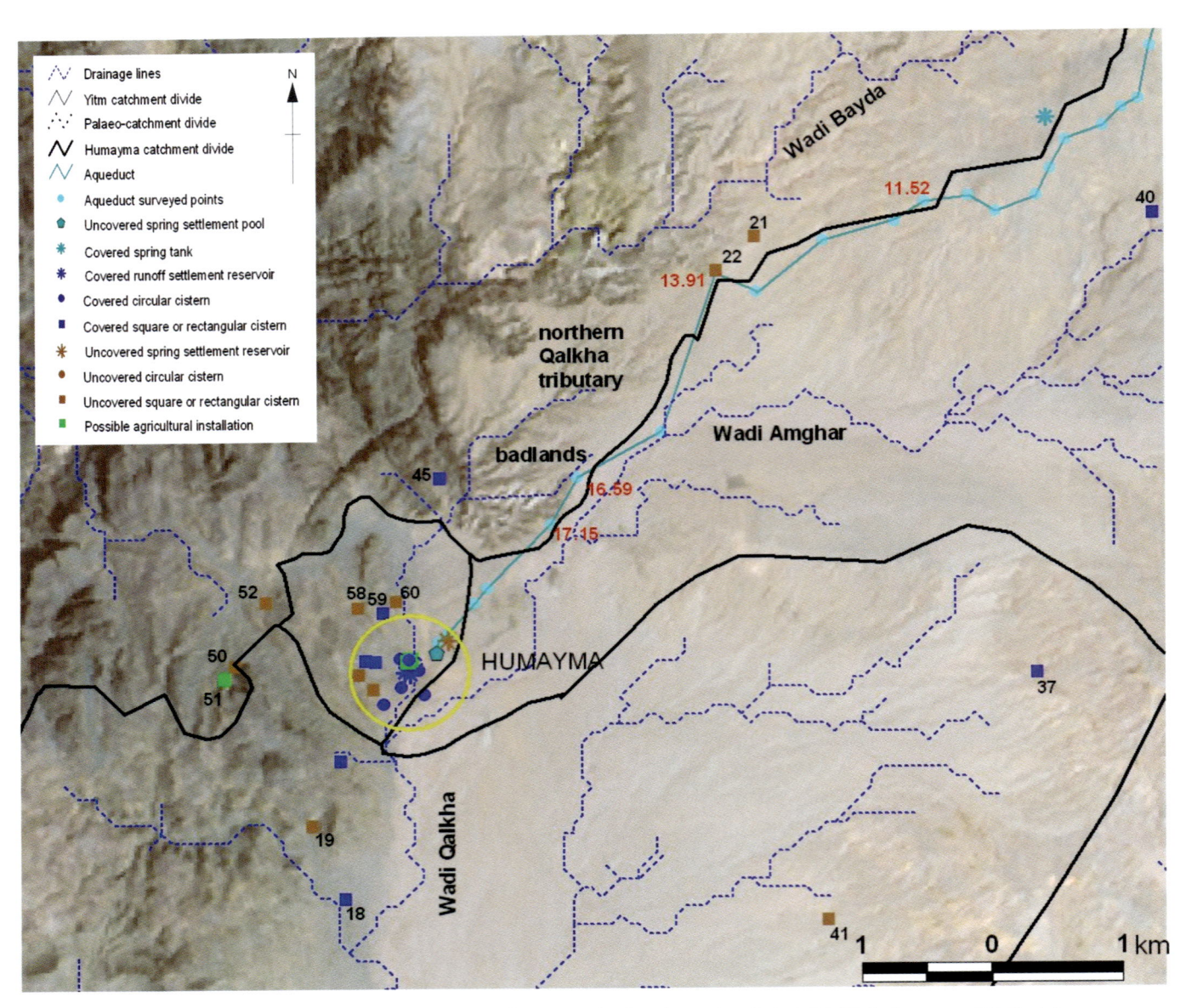

Figure 19.12. The interpolated path of the aqueduct between the surveyed points (turquoise circles) (M. El Bastawesy and R. Foote).

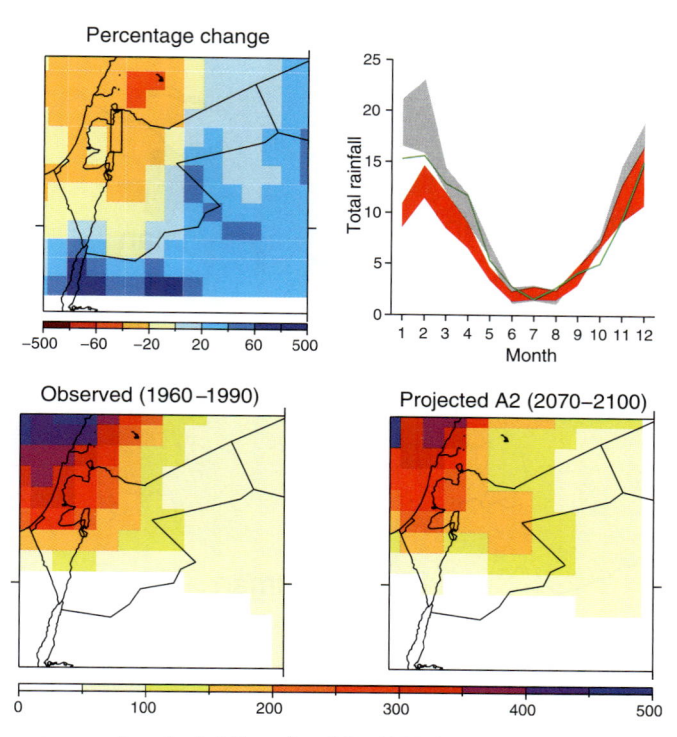

Figure 24.5 Top left: percentage change in annual total rainfall predicted for 2070–2100 under an A2 emissions scenario Top right: seasonal cycle in total rainfall in the box shown on the top left figure, for a group of model integrations for a present-day climate (light grey polygon); a 2070–2100 A2 scenario climate (red polygon) and a single B2 scenario model integration (line). Bottom left: observed annual total precipitation (based on the publicly available GPCC gridded dataset). Bottom right: percentage changes in annual total rainfall under an A2 scenario projected on to the observed present-day totals.

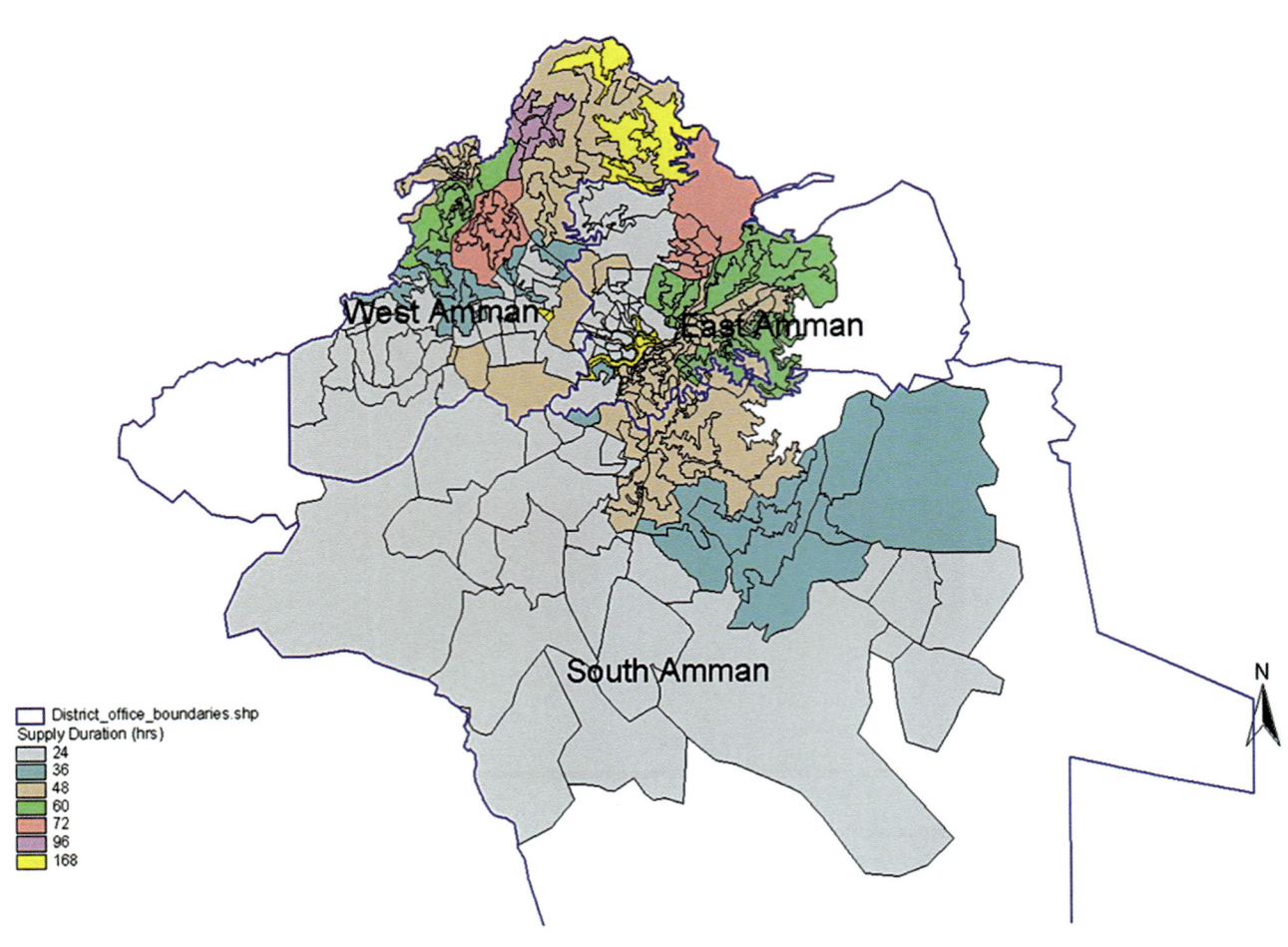

Figure 28.4. The total supply duration (hours per week) in Greater Amman, summer 2006. Source: LEMA Company, personal communication, Amman 2007.

Figure 14.8. Archaeological remains of the Pre-Pottery Neolithic B 'mega-site' of Basta (© S. Mithen).

al-Baseet (Fino, 1997) (Figure 14.8). Settlements also develop within the hilly regions such as Wadi Shu'ayb (Simmons et al., 1989) and in the Wadi Araba as at Wadi Fidan 1.

There is a significant change in distribution pattern and nature of settlement in the Pottery Neolithic period. Large sites such as Teleilat Ghassul (Bourke, 1997), Abu Hamid (Dolfus and Kafafai, 1988) and Tell Wadi Faynan (Al-Najjar et al., 1990) emerge in locations that have ready access to water for cultivation, although at other locations such as 'Ain Ghazal and Wadi Shu'ayb there is settlement continuity from the PPNB into the Pottery Neolithic (e.g. Rollefson, 1993). Pottery Neolithic settlement develops at Dhra', which had been occupied during the PPNA but then apparently abandoned during the PPNB/C (Finlayson et al., 2003). Where there is a Pottery Neolithic re-occupation of PPN sites, it is generally noticeably less densely packed in nature (Rollefson, 2001). The smaller size of many Pottery Neolithic settlements may have made them less easily identifiable than preceding PPNB settlements, but where research has targeted this period, as in the Wadi Ziqlab, numerous sites have been located (Banning, 1996).

Beyond the Rift Valley, there is no evidence for PPNA settlement in the more arid areas of Jordan, although there is some evidence for an early PPNB presence in Wadi Jilat, and then a rise in the number of sites, which may provide evidence for the development of a transhumant way of life such as in the Jafr Basin (Fujii, in prep–a). In the later PPNB/C and subsequent Pottery Neolithic there is a marked increase in site numbers, thought to be associated with the development of pastoral economies and the appearance of a symbiotic relationship between nomadic people and those living more sedentary, agricultural lives (cf. Betts, 1992). These sites, and indeed the spread of the Neolithic to the wider region, including Cyprus, are relevant here as they illustrate the development of a series of water management techniques during this period.

14.3.4 Groundwater exploitation

Perennial rivers are now rare in Jordan, and indeed throughout the Levant in general. With higher rainfall, different seasonality and more vegetation cover during the early and mid-Holocene, permanent stream flows may have been more available than they are today (Chapters 3, 6 and 7, this volume; McLaren et al., 2004; Barker et al., 2007; Hunt et al., 2007). Neolithic sites appear to have been located in areas where perennial water would have come from groundwater flows, which could have been accessed via springs or digging to reach the water table, although a formal study of such locations has not been undertaken.

During the PPNA there is little evidence that springs were manipulated or modified by human action, although there is a strong association between PPNA sites and spring locations (Sheratt, 1980). This is particularly clear in southern Jordan with examples at Dhra' (Finlayson et al., 2003), WF16 (Finlayson and Mithen, 2007; Chapter 14, this volume) and Zahrat edh-Dhra' (Edwards and Higham, 2001). The site of Jericho, located on the West Bank, provides one possible example of the manipulation of spring flows during the PPNA. Jericho is located next to the prodigious 'Ain el Sultan which produces a large flow of 26 m per second (Miller, 1980). Kenyon (1960) suggested that a series of mud plaster walls may have been used to manage and channel the spring flow, although Miller (1980) argued that these structures would have been ineffective and their identification as water management features is questionable.

Beidha provides a further example of a Neolithic settlement dependent upon local spring activity. Although the Natufian/PPNB site of Beidha is today 5 km from the nearest springs (e.g. Miller, 1980), recent research has shown that springs were flowing around the site during the entire PPNB, as well as the preceding Natufian period (Chapter 16, this volume). Gebel (2004) argues that shallow groundwater resources may have been exploited at Ba'ja during the LPPNB (Figure 14.9). In light of LPPNB exploitation of steppe and desert areas, where deep groundwater is often difficult to access, such as in the Black Desert, it seems unlikely that human groups did not regularly exploit shallow groundwater resources. The archaeology evidence remains, however, elusive, other than for a few striking instances that we will describe.

As Clark (1944) described, human exploitation of groundwater ranges from accessing natural springs and digging pits to sinking formal wells, and creating channels and dams to store water. Such activity is, as seen in the case of Jericho, likely to be difficult to identify and interpret archaeologically. However, such small-scale modifications are common in the contemporary world (e.g. see Figure 15.2, this volume) and it seems relatively safe to assume that these were normal parts of the Neolithic water management repertoire. In terms of more formal

Mylouthkia, the most thoroughly described of the Cypriot Neolithic wells, both had diameters of *c.* 2 m and were sunk directly into the natural sediments and bedrock to reach (now dry) underground water courses which occurred in the form of small pipe-like channels flowing towards the sea (Peltenberg *et al.*, 2000; Croft, 2003b; Peltenberg, 2003). Well 116 was at least 8 m deep, while well 133 has survived to a depth of 7 m below the present ground surface. The tops of both wells have been truncated by erosion and/or modern quarrying so these are minimum depths. These wells were simple constructions, the shafts lacking any built structures and being no more than deep holes dug into the ground. The shafts of both wells are marked with several small niches, interpreted as hand/footholds to allow access during construction and for maintenance and cleaning.

An intriguing feature of these wells is that there would have been no evidence of the location of underground water channels from the ground surface. Consequently, the methods used to select their location remains unclear. One suggestion put forward by the excavators is that locations were chosen 'by using water divining above small, underground streams' (Peltenberg *et al.*, 2000, p. 848).

Accelerator mass spectrometry (AMS) radiocarbon dates have been recovered from the fills of both wells. Well 133 has two dates recovered from the upper section of the well infill deposit, which are both in good agreement and span the period 9,295–8,645 BP. Three further dates have been obtained from the fills of Well 116, one of which derives from an upper fill and covers the period 10,573–10,246 BP whilst two dates derived from the top of a lower fill cover the period 10,684–10,178 BP (Peltenberg *et al.*, 2000; Peltenberg, 2004). As these dates are recovered from infill contexts they provide a minimum date for the manufacture and functioning of the wells. These dates suggest that Well 116 should be attributed to the Cypro Early PPNB, whereas Well 133 was constructed about a millennium later and belongs to the Cypro Late PPNB period (Peltenberg, 2003).

A range of evidence suggests that the construction, use and eventual infilling of wells at Mylouthkia had additional significance for the Neolithic community beyond their use as water harvesting devices. The deposits infilling the wells at Mythloukia contain an unusually large number of limestone vessel fragments and possible ritual depositions. For example, the fill of Well 133 included the remains of five human skeletons in addition to 22 complete, un-butchered caprine skeletons (Peltenberg *et al.*, 2000). The internal structure of the wells shows little sign of wear or erosion, which has been interpreted as evidence that the wells were infilled shortly after construction (Peltenberg *et al.*, 2000; Peltenberg, 2003). Finally, the site of Mylouthkia is located in an area where springs are common, at least in the present day, which lends further support to the argument that the construction, use and infilling of wells was not a purely functional process (Peltenberg, 2003).

Figure 14.9. Archaeological remains of the Pre-Pottery Neolithic B site of Ba'ja, located within a steep-sided siq (© S. Mithen).

exploitation of groundwater resources, there is now incontrovertible evidence that Neolithic groups had the sophisticated hydrological and technical knowledge to construct large wells for tapping underground water resources. Whilst no Neolithic wells have as yet been discovered in Jordan, evidence from the wider region shows that the sinking of wells was a widespread aspect of the early Neolithic technical repertoire. Although identifying the function of wells is relatively easy, and the dates of their manufacture and use can often be ascertained, it is unusual for these features to be discovered during excavation. This may simply be because they were often located away from the main areas of occupation (cf. Garfinkel *et al.*, 2006) and hence it is possible that many Neolithic settlements had access to wells which have, thus far, eluded discovery (Gebel, 2004).

To date, the earliest evidence for Neolithic wells comes from the Aceramic Neolithic on Cyprus where several wells have now been identified at the sites of Mylouthkia and Shillourokambos (Peltenberg *et al.*, 2000; Guilane and Briois, 2001; Croft, 2003a; Croft, 2003b; Peltenberg, 2003). Wells 116 and 133 at

As the Cypriot PPNB is culturally similar to the PPNB on mainland Levant and Anatolia, the origins of well technology may have roots in mainland traditions, although the Cypriot wells are earlier than any wells yet discovered on the mainland (Peltenberg, 2003). However, some support for this idea has recently come from excavations at the site of Tell Seker al-Aheimar, where a PPNB period well has recently been discovered (Nishiaki and Kadowaki, 2009). This 4-m-deep well is in an inland setting and, as with the Cypriot examples, the fill contained apparently ritually deposited stone tools, reinforcing the idea that wells were regarded with special significance in this period.

Further evidence for Neolithic well use on the mainland dates to the end of Pre-Pottery Neolithic period. The site of Atlit-Yam, which covers an area of $c.$ 40,000 m^2, is located around 300 m off the present-day Carmel coast, submerged under 8–12 m of sea (Galili et al., 1993). On the basis of radiocarbon dates and material culture, which includes Byblos and Amuq points, the site has been dated to the PPNC, being occupied between 9,500 and 8,000 BP (Kislev et al., 2004). The site probably lay in a coastal setting on a drying swamp, possibly on the edge of a lagoon formed by a sandstone ridge (Galili et al., 1993). The excavators argue that the site was abandoned as a result of rising sea levels and was covered by sand dunes immediately following its abandonment before being submerged beneath the rising tides. In terms of freshwater resources, the occupants of the site could have made use of seasonal flood water from the nearby Oren River, and may also have exploited springs which occur in the vicinity of the site (Galili and Nir, 1993). It is possible that one of the large walls at the site (structure 15) acted as a defence against the seasonal flooding of the Oren River (Galili et al., 1993). The remains of at least three wells have been discovered. Structure 11, located outside the main area of the Atlit-Yam settlement, has been described in greatest detail (Galili et al., 1993; Kislev et al., 2004; Garfinkel et al., 2006).

Structure 11 is a circular feature with a diameter of 1.5 m which extends into the clay and bedrock ground to a depth of 5.7 m, where it would have reached the level of the Neolithic water table. Unlike the Mylouthkia wells, its sides are lined with undressed stone blocks to a depth of 4.7 m, at which depth it cuts through bedrock. Four courses of stonework survive above the Neolithic ground surface. The well was filled with artefact-rich sediment, which has yielded four radiocarbon dates spanning the period 8,540–7,933 BP, confirming a late PPNC date (Galili et al., 1993). The excavators suggest that the well was abandoned as a result of rising sea levels, which would have contaminated the fresh water with salt, and that the well was then used as a rubbish pit. At this time, the residents of Atlit-Yam may have begun to use alternative wells including the two other possible wells identified (Kislev et al., 2004; Garfinkel et al., 2006),

Figure 14.10. Excavation at the Pottery Neolithic settlement Sha'ar Hagolan (© Y. Garfinkel).

although changes in groundwater hydrology caused by rising sea levels may have created new springs in the area, removing the need for wells (Kislev et al., 2004). Several other submerged Neolithic settlements have been identified along the Carmel coast. Although details are scarce, various circular features at these sites may be the remains of further Neolithic wells, suggesting that groundwater exploitation had become relatively commonplace in the region (Galili and Nir, 1993).

In the Pottery Neolithic, a well has been discovered at the site of Sha'ar Hagolan in the Jordan Valley (Garfinkel et al., 2006). This large ($c.$ 20 ha) site dates to the Yarmoukian, the first phase of the Pottery Neolithic (Garfinkel et al., 2006) (Figure 14.10). As at Atlit-Yam, the well at Sha'ar Hagolan was a chance discovery located outside the main excavated area of the settlement. Unlike at Atlit-Yam, where excavation was hampered by its underwater location, the well at Sha'ar Hagolan was exposed in vertical cross-section, allowing its construction to be clearly understood (Figure 14.11). The first phase involved digging a stepped pit through the natural sediments to reach the water table at a depth of 4.26 m. The well reached a gravel deposit which would have been the source of fresh water (Garfinkel et al., 2006). The second phase of construction involved lining the upper 2.5 m with stone blocks; as at Atlit-Yam the basal portion of the well was not lined with stones. As the step trench widened towards the ground surface, the space between the pit edge and the stone lining was filled with artefact-rich sediment. There is a 'mushroom'-shaped widening of the shaft at a depth of 3.3–3.8 m which Garfinkel et al. (2006) attribute to small waves on the water in the well eroding the sides of the shaft. This suggests that the well was open for a reasonable length of time, unlike the Cypriot examples (Peltenberg, 2004), although radiocarbon date ranges from construction deposits (8,383–8,200 BP) and post-abandonment fill (8,340–8,179 BP) cover virtually the same range, suggesting that the period of use cannot have been very long.

Figure 14.11. Cross-section of the Pottery Neolithic well at Sha'ar Hagolan (© Y. Garfinkel).

As with the other Neolithic well sites discussed above, the site of Sha'ar Hagolan is located near an alternative source of fresh water (the Yarmouk river), suggesting possible non-functional motivations for the construction of this feature (Garfinkel et al., 2006). Indeed, it may be significant that alternative sources of fresh water appear to have been available at most, if not all, of the Neolithic sites with wells. This raises several issues regarding the motivation for well building. Although the Neolithic people may have valued water from wells because of practical factors such as reliability, cleanliness and ease of access, it is possible that the decision to construct wells also involved a range of social and ideological motivating factors. The presence of 'ritual' depositions within the fills of several wells suggests that wells were, at least some of the time, imbued with significance beyond their function as water harvesting devices.

Taken together, the wells on Cyprus, the Carmel coast and Sha'ar Hagolan show that well building was an established facet of the Neolithic technical repertoire from at least PPNB times. Equally importantly, the level of technical sophistication and social organisation required for the construction and maintenance of these large wells suggests that simpler (and less visible or durable) methods of managing groundwater resources were probably rather widespread.

14.3.5 Flood water/runoff exploitation

Exploitation of runoff water can take many forms. The most basic is the exploitation of water 'traps' such as alluvial fans, natural pools or *qa* (mud playas). We should expect that human groups have exploited these resources ever since the first Palaeolithic occupation, and their opportunistic exploitation continues amongst the Bedouin today (Lancaster and Lancaster, 1991). Indeed, there is good evidence that these resources were widely exploited during the Neolithic period, particularly in eastern Jordan. For example, in the Jabal Tharwa, at a location known as Wisad pools, substantial evidence of Pottery Neolithic settlement is clustered around the pool edges (Wasse and Rollefson, 2005). Similarly, the Pottery Neolithic occupation at Dhuweila, located in the Black Desert, is located adjacent to a large natural pool (Betts, 1998). Also in eastern Jordan there is a series of Neolithic (and indeed earlier) settlements clustered around natural pools in the bed of the Wadi Jilat (Garrard, 1994). The LPPNB site of 'Ain Abu Nukayla, which is located in a similarly arid setting, in southern Jordan has strong evidence for agriculture, which is thought to have been feasible only by exploiting runoff water trapped in a natural *qa* (Albert and Henry, 2004).

In terms of more direct exploitation and manipulation of flood/runoff flows, evidence from the earliest Neolithic is scant. Indeed, the only direct evidence for PPNA management of flood waters may come from Jericho where a series of walls and ditches around the site, originally interpreted as defensive in nature, have been reinterpreted as features to divert seasonal flood water and sediments away from the settlement, and onto adjacent land suitable for agriculture (Bar-Yosef, 1986).

It is useful to note there that beyond the southern Levant, Pottery Neolithic groups in Mesopotamia had begun to use runoff water for irrigation. At the site of Choga Mami (eighth millennium BP), Oates discovered several water channels whose morphology and appearance suggest they were anthropogenically constructed (Oates and Oates, 1976). These above-ground channels are often cited as the earliest evidence of irrigation agriculture (Oates and Oates, 1976; Wilkinson, 2003). Recently, the evidence from Choga Mami has been supported by the discovery of a possibly artificial channel at the Iranian site of Tepe-Pardis (Gillmore et al., in press). This small, 'v'-shaped channel is securely dated to between 7170 and 6950 BP and associated with Late Neolithic horizons. If we accept the evidence for irrigation agriculture in Late Neolithic Mesopotamia, then it is likely that the origins of runoff management may lie earlier in the Neolithic.

In the southern Levant, one possible example of PPNB management of flood water is the artificially enhanced rock pools at PPNB Khirbet Sawwan, located in the Syrian Hawran (Braemer *et al.*, 2009), but details of these are yet to be published. A further possible example comes from the site of Ba'ja in Jordan (Figure 14.9). Ba'ja is a large LPPNB settlement which lies in a unique type of location 'hidden' between steep cliffs at the top of a narrow siq in the sandstone hills near Petra (Gebel, 2004). No springs have been identified in the immediate vicinity of the site and the nearest modern springs are at least an hour's walk away. Although there is no direct evidence for water management at the site, Gebel argues that the occupants would have used the hydrological and topographic setting of the site for the management of flood waters. The siq al Ba'ja has many sharp turns (greater than 90 degree) and would have acted to slow flood water, allowing the formation of shallow pools of relatively static water. Gebel suggests that LPPN inhabitants would probably have constructed small dams to enhance the water-retaining nature of these locations (Gebel, 2004). He has no direct evidence, but it is difficult to imagine how else the inhabitants of Ba'ja could have obtained sufficient fresh water for their needs.

The best, albeit still controversial, evidence for PPNB runoff water management in Jordan, and indeed the wider Near East, comes from the Jafr Basin Prehistoric Project (Fujii, 2007a; 2007b; 2007c; 2008; Fujii, in prep–a; Fujii, in prep–b). The Jafr basin is a large depression located on the Transjordan plateau. From the present perspective the most significant discoveries have been made in and around the Wadi Abu Tulayha area where a PPN settlement was discovered in 2001 and excavated between 2005 and 2008 (Fujii, in prep–a) (Figure 14.12, 14.13).

The present environment of the Wadi Abu Tulayha region is hyper-arid, averaging less than 50 mm rainfall per year; no springs are present in the vicinity of the site and vegetation is sparse, except following occasional winter rains (Fujii, 2008). The present-day lack of rainfall in the region is a result of the interaction between westerly rain-bearing storm systems and the topography of Jordan, and it is thus unlikely that the Jafr region would have been wet enough to support dry farming during the Holocene. However, given the regional evidence for increased rainfall during the early Holocene (e.g. Bar-Matthews *et al.*, 2003), it seems safe to assume that the region would have been wetter during the LPPNB and would probably have supported a richer and more varied range of plant and animal life than at present. In this sense the present-day landscapes, which are also potentially overgrazed by modern goat herds, may not be entirely reflective of the environments faced by Early Neolithic groups. From an archaeological standpoint, the arid climate of the Jafr region, coupled with consequent low sedimentation rates and limited human activity, will have improved both the preservation and the visibility of water management features (Fujii, 2007a). However, as in the Black Desert, low rates of sedimentation also pose problems, as archaeological deposits often take the form of palimpsests, with limited opportunity for stratigraphic control and dating (Fujii, 2007a).

The site of Wadi Abu Tulayha is located on a small tributary wadi flowing into the main Wadi Abu Tulayha, in the vicinity of a natural *qa*. Occupation at the site appears to have been seasonal, occurring during spring and summer, taking advantage of water and vegetation from winter rains, and the dominance of young gazelle bones amongst the fauna supports this interpretation (Fujii 2007a; 2007b). The site covers an area of *c*. 1.5 ha and is largely composed of 60 semi-subterranean structures, in the form of large subcircular or subrectangular pits lined with stone walls which extend several courses above the Neolithic ground surface (Fujii, 2008 Fujii, in prep–b). Many of the structures have no clear evidence for plaster flooring, appearing to make use of the natural limestone bedrock instead. Hearths and storage bins are commonly associated with the structures, whilst clearly defined entrances are rare, perhaps suggesting that many of the structures served as storage features rather than dwellings (Fujii, 2008). Interestingly, where entrances are present, they are often blocked off with stones, which, coupled with the frequent discovery of upturned grinding stones, supports the idea that the site was seasonally occupied (Fujii, 2007a; 2008). Significantly, there is no clear evidence for animal pens at the site, suggesting that the management of herds played a minimal role in the economy of the site (Fujii, in prep–b).

In terms of chronology, Fujii suggests the settlement belongs primarily to the Late PPNB period. This is supported by three radiocarbon dates, argued to be from the latest phase of occupation at the site, which cover the period 9,536–9,318 BP. Material culture from the site also supports a LPPNB date for the site, with an abundance of Byblos and Amuq points, although as Jericho points are also present, the foundation of the site may have occurred in the preceding Middle PPNB period. Significantly, other evidence of human activity in the area is scarce, being limited to a few Early Bronze Age burial cairns and limited evidence for transitory Nabataean and later period activity (Fujii, 2007a).

Aspects of material culture include flint sickles, grinding stones, several (25) gaming boards, petroglyphs, several bilaterally notched weights and an anthropogenic figurine (Fujii, in prep–b). The relatively rich nature of the material culture suggests that occupation at the site was fairly intense and of a reasonable duration. Faunal evidence includes domestic sheep and goat, although in very low numbers (less than 1% of the assemblage) and is dominated by the remains of wild sheep and goat (Fujii, in prep–b). Plant remains including wheat, barley and lentils have also been recovered (Fujii, 2007b; Fujii, in prep–b). Based upon this, Fujii suggests that the economy of the site was based upon a mixture of hunting, herding domestic animals and small-scale plant cultivation. The evidence that cultivation was

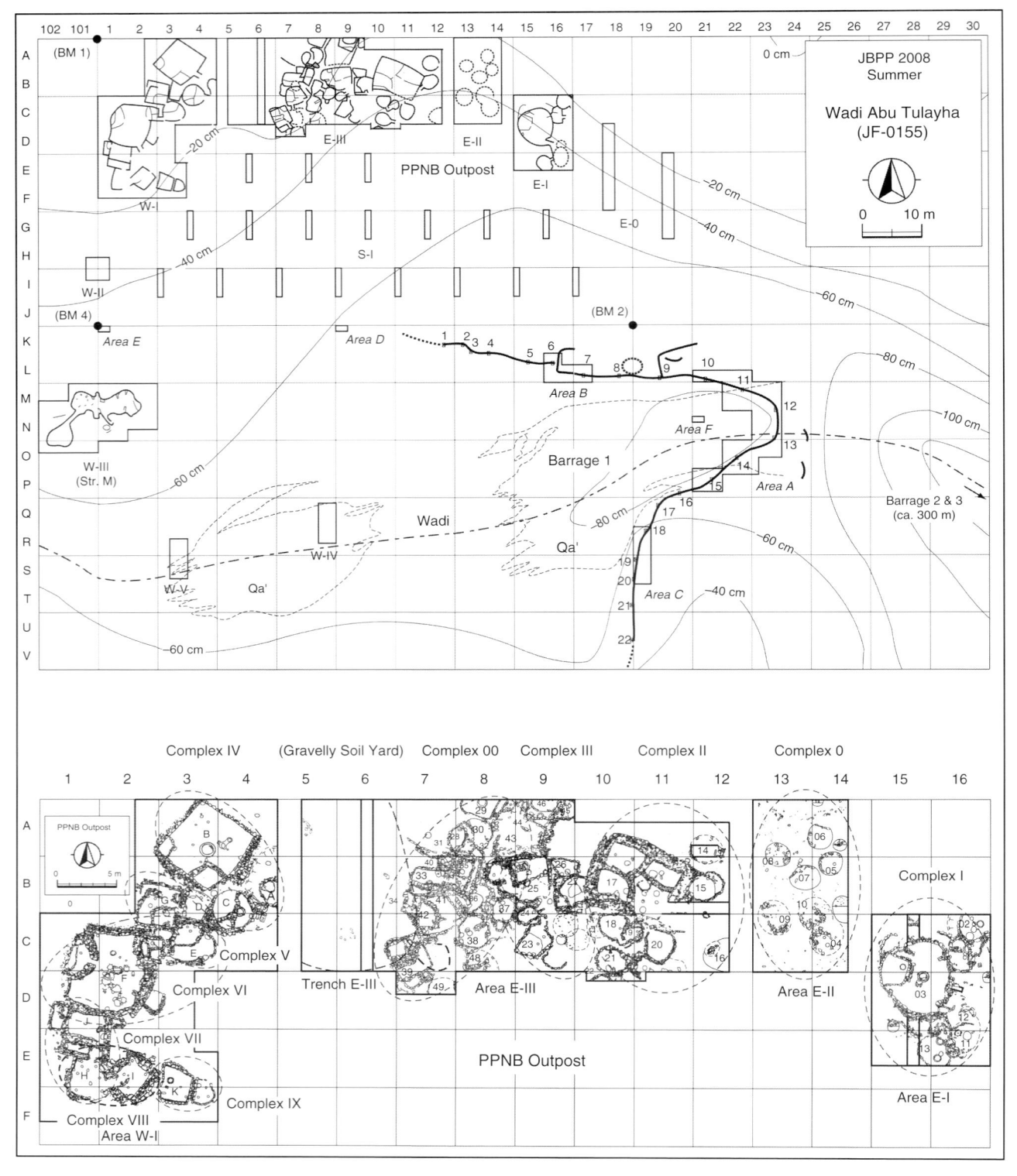

Figure 14.12. Plan of the Pottery Neolithic B settlement at Wadi Abu Tulayha, showing the relationship between the settlement structures, the barrage and the proposed cistern (Str M, W-III) (© S. Fujii).

practised at the site is strong, with remains of plants and of tools, such as sickles and quern stones, for the harvesting and processing of crops. Given the arid setting of this site, the evidence for plant cultivation provides indirect evidence for Neolithic water management at the site. Indeed, while hunting and herding may have been possible at Wadi Abu Tulayha during the wetter climatic regimes suggested for the Neolithic period (Chapters 3, 6 this volume; Bar-Matthews *et al.*, 2003), it seems unlikely that

Figure 14.13. The Pre-Pottery B 'outpost' settlement at Wadi Abu Tulayha (© S. Fujii).

Figure 14.14. The proposed Pre-Pottery Neolithic B barrage in Wadi Abu Tulayha (© S. Fujii).

dry farmed crop cultivation could have taken place at Wadi Abu Tulayha without some form of human manipulation of water. In this light, the discovery of several possible water management features, in the form of barrages and cisterns, is of considerable interest.

Several barrages have been identified in the Wadi Abu Tulayha area, the most thoroughly investigated of which is barrage 1 (Figure 14.14). This is located roughly 50 m to the south of the main PPN settlement and is a 'v'-shaped structure (total length 120 m) constructed of stone blocks. The two wings of the 'v' converge in the centre of the tributary wadi, where the barrage is reinforced by a semi-circular wall. Four courses of stonework are preserved in the centre of the barrage while the 'tails' or 'wings' of the 'v' are much simpler and are often formed from a single row of stones (Fujii, in prep–b). Given the form and location of this structure, its interpretation as a water management feature appears secure. Because the stonework of the structure is not waterproof and the barrage is built in a flat location on a permeable soil matrix it is unlikely that this structure served as a dam to create a reservoir for drinking water. In this light, following Fujii, it seems most likely that this barrage was designed to allow shallow flooding over a large area, creating an exaggerated or artificial *qa*. In addition to water, the barrage would also trap sediment; in this way it would have functioned rather like a terrace wall (Fujii, in prep–a). Fujii convincingly argues that the water and soil trapped behind this barrage during winter floods would have allowed cultivation of the crop species identified in the adjacent PPN settlement. In this way the barrage is understood as a wadi barrier for small-scale basin irrigation agriculture (Fujii, in prep–a) to support seasonal occupation of this arid area. Fujii argues that this basic irrigation method would have allowed a few hectares of land to be cultivated, or at the very least (in a dry year) would have improved grazing for herds (Fujii, in prep–a). Significantly, the efficacy of the barrage is shown by the sudden appearance of diatoms in ponded sediments behind the barrage, not present in pre-barrage levels (Fujii, 2007b).

The function of the barrage is thus reasonably clear. More controversial, however, is its affiliation to the PPN period, which relies on several lines of rather circumstantial evidence (Fujii, 2007a; Fujii, in prep–b). Firstly, Fujii argues that the barrage is stratigraphically correlated with the PPN settlement. However, in reality this probably means little given the (near-)surface location of both settlement and barrage; certainly the barrage is not sealed by PPN stratigraphy. A second argument is that the barrage shares similar architectural features with the structures in the outpost, namely a 'protruded reinforcement wall' (Fujii, in prep–a). In a similar vein, the barrage and much of the outpost have foundations built with stone arranged in 'stretcher bond' whilst 'header bond' is used for above-ground structures (Fujii, 2007a). A further line of circumstantial evidence comes from the lack of pre- or post-PPN settlements in the Wadi Abu Tulayha region. Perhaps a more convincing line of evidence is that the inhabitants of the adjacent PPN settlement clearly engaged in

Figure 14.15. The proposed Pre-Pottery Neolithic B barrage in Wadi Ruweishid (© S. Fujii).

agricultural activity and it seems unlikely that this could have been achieved without such a barrage. The best evidence for a PPN date for this feature comes from the discovery of a large bilaterally notched and grooved stone weight, found incorporated into the barrage structure and similar to items found in the PPN settlement. This weight (which measures 45 × 35 cm and weighs 25 kg) was discovered built into the right-hand corner of the reinforcing wall (Fujii, 2007a). It is possible, of course, that the weight was manufactured in the PPN and then incorporated in the barrage structure during a later period. But the discovery of a similar weight in another potential PPN barrage in the Jafr region, that in Wadi Ruweishid (Fujii, 2007a) (Figure 14.15), bolsters the argument that this find provides a definite link between the construction of the barrage and the PPN settlement at Wadi Abu Tulayha.

A further water management feature at Wadi Abu Tulayha is structure M, which is interpreted as a cistern by Fujii (Fujii, 2008; Fujii, in prep–a; Fujii, in prep–b) (Figure 14.16). Unlike the barrage described above, the dating of this structure is uncontroversial; it clearly belongs to the PPN. However, its interpretation as a water management feature is less certain. Structure M takes the form of three subcircular, semi-subterranean 'rooms', creating an elongated, irregular complex of structural elements with a total width of 18 m (Fujii, in prep–b). The structure is located between the settlement and barrage 1, near to the natural *qa*, in an area potentially liable to occasional flooding. The excavation of these structures revealed a limited range of material culture objects and internal structures. Unlike many of the structures at the site, there are no hearths, or ashy deposits associated with the floor layers, which are of natural limestone bedrock. The lower fill is unique amongst structures at the site, being comprised of hard silty sand, only found elsewhere in the fills built up behind the barrage. The upper fills do, however, contain some hearths and other evidence for human occupation. It is in part the location of this structure, near the *qa*, and the unusual, clean, basal fills which suggest that this complex of interconnected structures may have a different function from the other, superficially similar, structures at the site.

Figure 14.16. Interior of the proposed Pre-Pottery Neolithic B cistern in Wadi Abu Tulayha (© S. Fujii).

The floors of structure M are located nearly 2 m below the Neolithic ground surface, more than twice the depth of most other structures at the site. Construction of this structure would thus have involved considerable effort, digging not only through natural silts but also through at least two layers of solid limestone. A clay coating/mortar with inset limestone slabs was identified covering parts of the upper, fragmentary, limestone layer in the western part of the complex. Fujii suggests that this may have been undertaken in order to increase the waterproofing of this layer. There are several other features which Fujii believes support the interpretation of this structure as a cistern. These include an irregular, stepped entrance to the central room, which is lined with clay mortar and interpreted as a water intake channel. Additionally a pit, dug through limestone bedrock, below the entrance may have served as a sludge tank. A final feature of note is a possible channel, again lined with clay mortar and slabs, which may have allowed water to run between the western and central rooms. If this complex did really function as a cistern, the total storage capacity (up to the top of the impermeable limestone layers at the base of the masonry walls) would have been approximately 50 m^3.

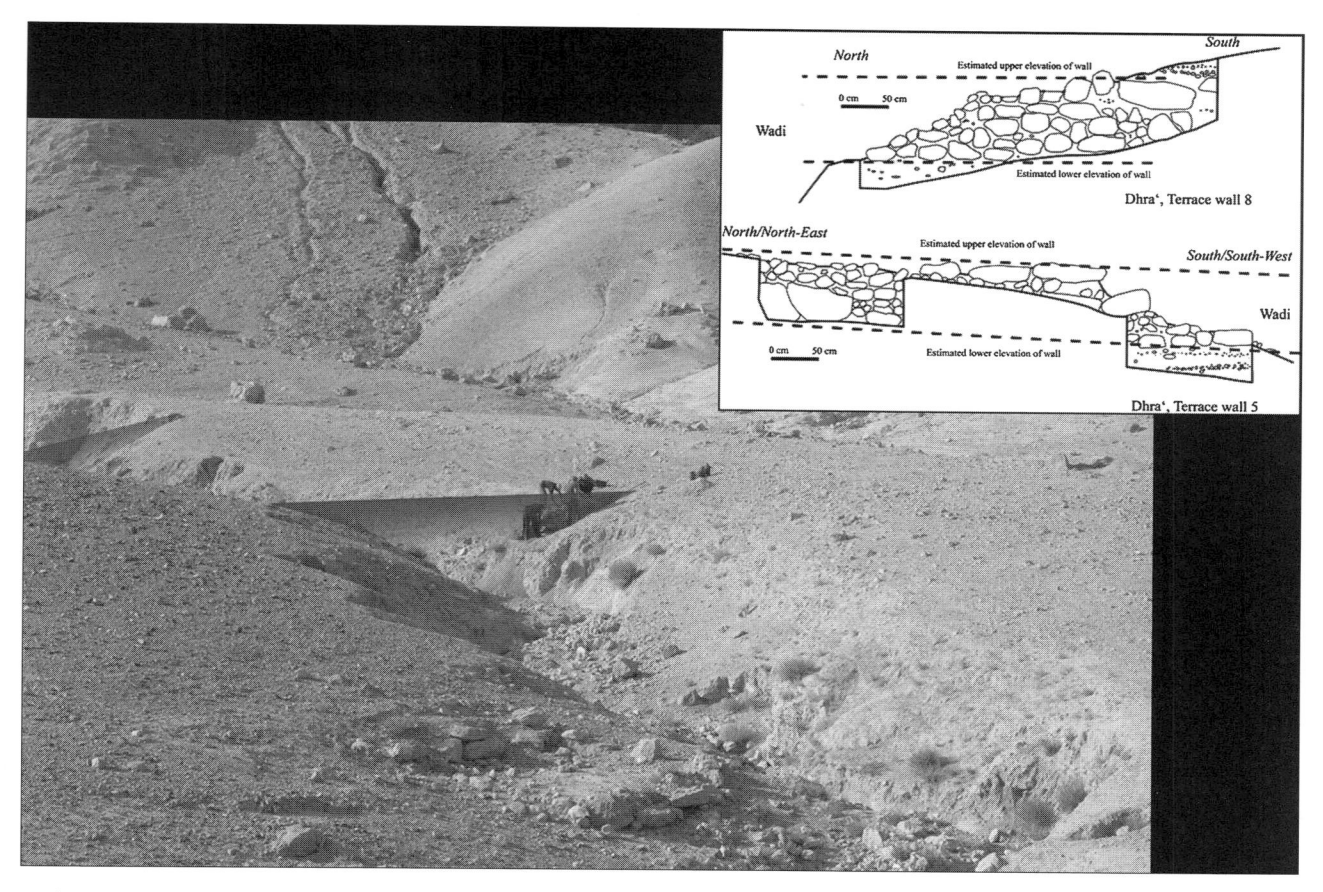

Figure 14.17. Excavation of a remnant of a terrace wall in the vicinity of the Pottery Neolithic settlement of 'Dhra and projected extent of wall (© B. Finlayson).

The evidence from Wadi Abu Tulayha thus provides tantalising glimpses of potential management of runoff/flood water during the PPNB, albeit plagued by uncertainties surrounding the chronology of the barrage and the function of the so-called cistern. Indeed, the evidence from Wadi Abu Tulayha suggests that water management was an essential aspect of PPNB expansion into the arid regions of Jordan.

A further possible LPPNB water management feature, again discovered by Fujii and colleagues, is located in the Wadi Badda, at the top of the Wadi Faynan, near the modern day town of Shawbak. During reconnaissance visits (no excavation has yet taken place at the site) to Wadi Badda, Fujii (2007c) discovered archaeological deposits containing masonry walls and LPPNB artefacts, including Amuq points and naviform debitage. Of present interest is the discovery of a barrage estimated to have been 15–20 m in length running across the wadi below the site, the centre of which has been destroyed by floods. The surviving ends of the barrage, buried by more than a metre of sediment, illustrate that it was at least 0.8 m high and was manufactured of locally available basalt cobbles. As at Wadi Abu Tulayha, providing firm evidence that the barrage is contemporary with the LPPNB settlement is problematic. At present the evidence for this comes from the lack of other period settlement and, more convincingly, the fact that the present wadi channel is located several metres below the base of the barrage, indicating significant down-cutting since its construction.

A further line of evidence for Neolithic management of runoff water comes from the Pottery Neolithic period at the site of 'Dhra (Kujit et al., 2007). At this site a series of large buildings were exposed and dated to the Jericho IX phase of the Pottery Neolithic, with a radiocarbon date spanning the period 7941–7680 BP. Of particular interest is the discovery of a series of terrace/barrage walls running across a small wadi (Figure 14.17). These walls appear similar to those at Wadi Badda, in that only the ends of the barrages survive. One wall of the barrage, Wall 8, was constructed by excavation of a foundation trench which was filled with large foundation boulders on top of which a wall (surviving up to four courses high) of smaller wadi cobbles was constructed. The walls at 'Dhra may have stood up to 0.75 m high and been around 20 m in length (Kujit et al., 2007). Dating of these walls is largely circumstantial, but the fact that the bases of the walls are located several metres above the present wadi level suggests that they were constructed before wadi down-cutting commenced, probably in the late Pottery Neolithic. This interpretation is also

Figure 14.18. Reconstruction of Pottery Neolithic cultivation plots supported by terrace walls at 'Dhra (© B. Finlayson).

supported by the dearth of finds dated to any period later than the Pottery Neolithic in the vicinity of the walls. The excavators of the site argue that the walls would have been used to trap runoff water and associated sediments, thus improving and reducing the annual variability of crop yields (Figure 14.18) (Kujit *et al.*, 2007). Interestingly, the excavators also argue that the wide spread of Pottery Neolithic artefacts around the terrace walls (particularly marked when compared to the tight boundaries of the PPNA site) may indicate that during the Pottery Neolithic people were spreading waste onto the field to improve soil fertility.

The evidence described above demonstrates the difficulty of identifying runoff water management features dating to the Neolithic. The fact that such features have to be placed in recently volatile wadi bed locations does not usually allow for their preservation. A second major problem is that dating these features with any absolute confidence is often impossible. Despite these issues, the weight of evidence discussed above strongly suggests that Neolithic groups had begun to explore means of harnessing and manipulating runoff water, and that these techniques were key to the expansion of Neolithic settlement into the arid regions of Jordan. The importance of this should not be understated, as agricultural water demands can quickly become huge: today they dominate our demand for water (Allan, 2001). In many ways the ability to use, manage or domesticate runoff water not only allows the expansion of agricultural economies into previously inaccessible regions, but also provides a basis for reducing interannual variability of water supply and hence crop yields: they must be regarded as a critical element of the Neolithisation process.

14.3.6 Summary

The remarkable societal and economic transformations which took place during the Neolithic period created a range of new water supply demands. The development of large sedentary settlements would have needed significant quantities of water for construction, hygiene and ritual as well as the more obvious demands created by rapid population growth and the adoption of agriculture. Indeed, it seems certain that Neolithic communities would not just have required more water, but would have required more predictable and reliable water to meet the needs of their increasingly sedentary and agricultural lifestyles. Moreover, from the PPNB onwards the expansion of Neolithic settlement into the arid regions of Jordan and beyond may have demanded not only social and economic development but also new ways of manipulating water resources. In part, these demands may have been met by the wetter climates and more stable hydrological regimes of the early Holocene, and in part, through the development of pastoralism. However, there is now clear evidence that Neolithic communities were actively engaged in attempts to manipulate and control water resources, in a way which foreshadows many later developments in the region. In this way, we suggest that the 'domestication' of water should be regarded as a critical feature of the Neolithic world. Indeed, it seems apparent that many of the more widely acknowledged Neolithic achievements, such as the adoption of sedentary agriculture, would not have been possible without achieving control of water resources.

14.4 WATER MANAGEMENT IN THE CHALCOLITHIC, BRONZE AND IRON AGES

The Chalcolithic, 6,450–5,500 BP, has been termed the 'end of prehistory' (Joffe, 2003). However, while 'history' conventionally begins with the advent of writing, in the southern Levant evidence for writing is minimal until the Middle Bronze Age (MBA) (3,950–3,450 BP). Basic markings on pot sherds represent the 'writing' of the Early Bronze Age (EBA) (5,500–3,950 BP), which characterises the period as belonging to proto-history rather than history. Nevertheless, the EBA has been described as representing the dawn of 'urban systems', complete with fortifications and palace economies (de Miroschedji, 1999). It is clear that the EBA inhabitants of the southern Levant were engaged in sophisticated economic activity involving long-range trade and serious investment in agriculture, horticulture and viticulture. The antecedents of these developments are evident in the Chalcolithic, with its 'rich symbolic landscape' (Bourke, 2008) and increasing reliance on horticultural products (Lovell, 2008). All of these developments must have required significant increases in water requirements and should be seen in the context of the climatic fluctuations of the mid-Holocene.

Palaeoclimatic data for the eastern Mediterranean during the Holocene all point to recurring climatic fluctuations, although there is an overall trend towards aridity from 6.0 cal ka BP or earlier (Brooks, 2006; Chapters 3 and 6, 7, 8 and 9, this volume). The most significant of these fluctuations are thought to be a centennial-scale wet event *c*. 5.0 cal ka BP (around or at the end of the EBI) and a major arid event *c*. 4.2 cal ka BP. The latter is certainly associated with the end of the EBIII (Rosen, 2003).

14.4.1 Chalcolithic

While skins can provide adequate storage for water at 11 °C, ceramics, developed during the Pottery Neolithic, are uniquely suitable for providing cool water storage via the slow evaporation through vessel walls (Matson, 1965, pp. 204–205; Miller, 1980, p. 334). Ceramic jars can also provide greater capacity than skins and waterproofed baskets, although at the expense of greater weight. Chalcolithic pits lined with lime plaster (Perrot, 1957, p. 9; Bourke *et al.*, 1998) may have been used for water storage.

The production of ceramics themselves requires a significant water investment, with the three main materials required being: clay, non-plastic inclusions and water. Water for ceramic manufacture does not have to be of high quality; indeed, adding salt to the mix, in the form of saline water, can result in a higher fired, better quality vessel in some circumstances (Arnold, 1985, pp. 26–28). This was probably the case at Teleilat Ghassul, situated in a saline environment, where fluxing is noted in the ceramic repertoire despite the fact that the firing temperatures achievable during the Chalcolithic are thought to be lower than required for this effect (Lovell, 2001).

Settlement patterns in the Jordan Valley reflect a strong correlation of Chalcolithic sites to wadi systems, with sites concentrated in the upper reaches, but also very significantly in the lower reaches which are suitable for floodwater irrigation (Chapter 17, this volume). Bourke has noted that the dearth of settlement between the Wadis Zerqa and Nimrin is due to the lack of major lateral wadis between these two systems. He further suggests that because more than 90% of sites are less than 2 ha, and the larger sites are associated with floodplains, 'growth is likely to be correlated with irrigated agricultural potential' (Bourke, 2008, p. 115). Rosen has argued for irrigation at Shiqmim on the basis of phytolith studies (Jenkins, 2009) although the evidence from these may be more problematic than first thought (Rosen, 1987; Chapter 20, this volume).

In upland sites, Chalcolithic water management strategies are clearly connected with natural springs and water catchment within rock-cut and natural rock installations. At el Khawarij, in the Wadi Rayyan, Chalcolithic wall structures demarcated and governed access to natural cavities in the rock which tapped the water table (Lovell *et al.*, 2007), providing evidence for small-scale localised management of water facilities. The walls, as preserved, appear to be low and they were perhaps aimed at keeping the water source unpolluted by animals, rather than barring human access. At the same site, rock-cut and natural

Figure 14.19. Khirbet Zeraqoun, in the northern highlands of Jordan (© S. Mithen).

installations collected water from surface runoff and rainwater that could be used for flocks, agriculture and human consumption (Lovell *et al.*, 2005). These kinds of installations are no doubt related to the Chalcolithic phenomena of 'cupmarks' (van den Brink *et al.*, 2001) and small rock-cut installations in general.

14.4.2 Early Bronze Age

Water storage structures have been documented for the Early Bronze Age, particularly in the context of walled town centres (Ben-Tor, 1992, p. 104). Corporate building activities within villages and towns would have required greater water supplies for the new scale of construction. The earliest known defensive walls, dating to the EBI, 5,550–4,850 BP, are built of mud brick (Shaub, 2007) and would have required a very considerable amount of water for their construction. Such corporate building projects would have required a significant level of management of the water resource itself.

At Khirbet Zeraqoun (Figure 14.19) in the northern highlands of Jordan, rock-cut tunnels below the EBA walled settlement are accepted as EBA in date (Philip, 2008). These consist of three shafts cut into the bedrock, in the northeastern part of the settlement, which are presumed to link up with the ancient tunnel system which descends around 60 m below the surface (Bienert, 2004). The average width of the tunnel (Bienert, 2004, fig. 2), as explored, is 1.3 m but the configuration of the various shafts and tunnels is unconfirmed because rockfall prevents full investigation. Bienert (2004) has proposed that the shafts functioned as construction shafts for cutting the tunnel, which itself functioned as a perennial supply of water. Cut marks are visible on the tunnel walls, which would appear to date to a later period – one with iron tools – and some traces of wall plaster were found, leading some to postulate an Iron Age date despite the lack of corroborative material culture on the settlement itself. However, other tunnels are also present beyond the immediate boundaries of the site, closer to Iron Age settlements (Bienert, 2004). These tunnels appear to be of similar construction, and therefore presumably date to the same period, and probably connect to that on the tell itself.

Tunnels cut in the EBA may well have been re-cut in later periods (as at Ta'annek; Albright, 1944, see also Lapp, 1969) and there are other cited examples of water systems in the EBA (see below). As Miller points out, 'there is no reason to restrict the ability to construct such underground water systems to later periods, given the knowledge of hard rock mining techniques as early as the Chalcolithic in Palestine at Timna' (Miller, 1980, p. 339).

Philip argues that it is 'probably no coincidence that Early Bronze Age settlements in the Jordan Valley are concentrated along the major side wadis [Ibrahim et al., 1988: p. 171], the places most suitable for water capture techniques' (Philip, 2008, p. 171), a proposition evaluated by Lovell and Bradley (Chapter 17, this volume). The existence of water catchment techniques in the EBA is a frequent assumption (see e.g. Prag, 2007 and Bourke, 2008, discussed above) but there is limited archaeological evidence for these systems. One such example may be the capture of seasonal flood waters in the Wadi Sarar by the EBA population living in and around Tell Handaquq (Figure 14.20; Mabry et al., 1996).

As Philip notes in relation to industrial activities, the archaeological predilection for intra-mural excavation (Philip, 2008, p. 172) has no doubt inhibited the discovery of other such systems. In addition, recent geomorphological processes (Mabry, 1989; Banning, 1996; Donahue, 2003) will have also played their part in reducing both the survival and visibility of these systems. Indirect evidence for floodwater systems comes from the discovery of 'large seeded' flax from Bab edh-Dhra which is thought to have grown 'almost certainly' under irrigation (McCreery, 2003, p. 457).

Intra-mural water storage systems are known from EBIII 'Ai (et-Tell) in the occupied Palestinian territories (oPt) excavated by the American Schools of Oriental Research in the late 1960s (Callaway, 1970, figs. 15–16). The method of construction, primarily the care taken to prevent seepage loss and water damage to the fortifications, suggested to the excavators that the builders were imported for the job (Callaway, 1978, p. 51). This probably reflects the lack of known parallels in the 1970s rather than any real lack of expertise on the part of the EBIII inhabitants of 'Ai. The system consisted of a 3-m-deep, 25-m-wide, above-ground pool with carefully constructed dikes around three sides to contain the water (as much as 1,815 m^3) (Callaway, 1978, p. 51). Other water systems are known from Arad, where runoff was channelled via streets to a central reservoir, located in a depression with a capacity of 2,500–

Figure 14.20. Tell Handaquq and the Wadi Sarar (© S. Mithen).

3,000 m^3. (Amiran, 1978, pp. 13–14) and on the Madaba plain (Harrison, 1997, p. 17).

At Bab edh-Dhra a plaster-lined cistern provides a smaller-scale example of intra-mural water storage (Rast and Schaub, 1974). This type of storage facility was no doubt widespread for larger family units (or small communities), especially in the uplands. Cisterns dating to the EBA are found throughout the region (e.g. Horvat Tittora West, Rast and Schaub, 2003). This type of arrangement is entirely consistent with community storage of other types of commodities (grain, fruits, etc.) found at EBA sites like Bab-adh-Dhra' (Kogan-Zehavi, 2005).

Beyond the fringes of the Jordan Valley, sophisticated water management in the desert is documented at Khirbet Umbashi and Khirbet Dabab (Braemer et al., 2009) and probably also at Jawa (Chapter 18, this volume). Braemer et al. (2009, p. 51) argue that it was only through the installation of extensive hydraulic works that the Harra was occupied, and that these installations made the 'conquest of new territory' possible for the following millennia. Certainly, as Philip (2008) notes, the evidence for large-scale livestock herding in this region is growing, and more intensive exploration of the region on the Jordanian side (e.g. Wasse and Rollefson, 2005) will be valuable in contextualising the larger sites and placing the exploitation of the region in a deep time, environmental framework.

Further north, at Byblos in Lebanon, a marshy area around the spring was dug out to provide a small reservoir at the intersection of two major streets (Dunand, 1960; Dunand, 1973). As the town developed, the pool walls were reinforced and a flight of stairs was added for access. Another reservoir was added to the north c. 2500 BC. Thus the technological developments of the Early Bronze Age southern Levant were part of a broader eastern Mediterranean/Near Eastern trend.

14.4.3 EBIV or Intermediate Bronze Age

As stated above, the beginning of the EBIV corresponds to a major arid event at $c.$ 4.2 cal ka BP (Rosen, 2003). The period is widely discussed as a 'rural interlude' (Palumbo, 2008, p. 227) characterised by a lack of clear settlements but with a very large number of cemeteries. There are, however, important exceptions to this rule, particularly in Jordan in light of its EBIV settlement record (Palumbo, 2008, p. 233).

Sites in the Jordan Valley are generally situated on low ground and, with a few exceptions (e.g. Iktanu), with proximity to water. Archaeological evidence for EBIV water management strategies is elusive, although Prag (2007) has attempted to reconstruct the resources and methods available to the inhabitants of southern Jordan around Iktanu. The area is subject to flash flooding, and as such all systems were relatively fragile and required regular maintenance in any period. Prag notes that in this area there is no evidence for cisterns pre-dating the Hellenistic/Roman period.

14.4.4 Middle Bronze Age

The Middle Bronze Age has been called the 'Age of Internationalism' (Ilan, 1998), and a number of new technological developments can be highlighted. As Mediterranean trade intensified in this period (Sheratt, 2003), so too would the demands on the materials required to fuel this new mercantile world. This would certainly have included water, which was a major component in the ceramics which formed the shipping containers for many goods and also played a part in other manufacturing processes, such as tanning, dyeing and metalworking. Add to this a growing population, and competent management of this precious resource, especially regarding access during the dry summer months, would have been vital. The difficulty of accessing water depends upon distance from the water table. On the coastal plain, wells can provide access to water, such as that found at Tel Gerisa (Tsuk and Herzog, 1992). But as Weinberger et al. (2008) note, ancient cities built on high defensive positions suffer from the disadvantage of distance to natural water sources. Ilan (1998) notes that while runoff catchment and open reservoirs were a development of the EBA, the MBA also saw the development of closed cisterns, complete with clay pipes that drew rainwater down from roofs, and sealed stone-built and carved channels for drainage. At MBA Hazor, for instance, there are a number of underground cisterns, waterproofed with plaster and each with capacity of $c.$ 23 m^3. These collected water from roofs of adjacent buildings (Yadin, 1972, pp. 38–40).

Finkelstein notes that plastered water cisterns were 'mastered' by the MBA inhabitants of the oPt, if not earlier in the EBA, and were the result of the penetration of settlement into arid areas (Finkelstein, 1989, pp. 136–144). Gezer had such cisterns in the MBA (Dever, 2003) as part of the urban settlement. These were superseded at the end of the MBA or early in the LBA by a large reservoir which was dug down to the water level in a system which included a tunnel and a rock-cut cistern of 150 m^3 capacity. The dating of the tunnel has been much discussed, with Reich and Shukron (2003) concluding that it is MBA. Their argument is based primarily upon the opinions of the first excavators, Macalister and Vincent (who are the only ones to have explored the entire system), and the fact that a re-dating of 'Warren's shaft' at Jerusalem to the MBA (Reich and Shukron, 1999) now provides a parallel for the Gezer system. It should be noted that Jerusalem's Siloam Tunnel has been dated to the Iron Age (Frumkin and Shimron, 2006; see below).

The development of long-range exchange systems and international relations during the MBA was a further factor in the development of palace-style economies. Mudbrick public buildings were common, such as the temple at Tell el-Hayyat, as were extensive fortification systems such as at Tel Da (Biran, 1981) which required heavy labour investment and, of course, large amounts of water (Bunimovitz, 1992).

14.4.5 Late Bronze Age

Around 3,500 BP the climate was apparently drier, evidenced by a significant drop in olive production and a lowering of the water table (Lipschitz, 1986; Chapter 7, section 7.6, this volume). The change in settlement pattern that can be observed in Jordan is probably a result of such aridity, which would have affected marginal areas more severely, and no doubt contributed to the economic decline during the Late Bronze Age (LBA) (Strange, 2008, p. 284).

It therefore comes as no surprise that security of water sources was a prime concern at LBA sites (Wright, 1985, 166ff.), one factor leading to the construction of massive fortifications at a number of sites throughout the region. As noted above, a water system is known from Gezer dating to the LBA (although originally dated to the MBA; for a discussion see Dever, 1969). While it appears that the inhabitants of Gezer may have made use of a natural tunnel formed by a spring, there is evidence to suggest modifications, if not a total excavation (Dever, 1969).

A water system at Tell es-Sa'idiyeh (Pritchard, 1985; Miller in Tubb, 1988; Tubb et al., 1996, pp. 35–36) consists of a staircase from the top of the tell to a spring-fed pool (the level of the aquifer, −276 m asl). The supply of water appears to have been facilitated by a 12-cm diameter conduit (Miller in Tubb, 1988, pp. 84–85). While defence is often cited as the reason for intra-mural water systems, Miller stresses that there would have been considerable peacetime benefits to having a close, secure water supply, including control over pollution (Miller in Tubb, 1988, pp. 86–87). Ensuring that dirt carried on people's feet did not pollute the pool could have been achieved via a separate access point. Such an access point may have been provided by the gallery excavated on the south side in 1995 (Tubb et al., 1996, p. 36).

14.4.6 Iron Age

The Iron Age is the first period where archaeologists openly discuss different ethnic groups within the archaeological record, e.g. the Philistines (primarily coastal dwellers), the Israelites and the Canaanites. Most of these terms are biblical and real territorial boundaries are more blurred. The 'Philistines' brought many new technologies with them from their Mediterranean antecedents, such that much of early Iron Age material culture shows innovations, such as the mudbrick and plaster buildings at Netzer (1992) and at Ashkelon where the large mudbrick fortifications included the construction of towers (Dever, 1998).

In upland areas to the west of the Jordan Valley, smaller unfortified settlements are sometimes associated with the early Israelites (Finkelstein, 1988). Upland settlement required different agricultural technologies and water strategies, including terracing for agriculture. Dever (2003) maintains that it is only in the Iron Age that lime plastered cisterns were first implemented in an intensive manner. However, many of these technologies had long been in gestation (see Finkelstein, 1998, p. 364). In the later Iron Age larger civic centres did exist, although there is no evidence to suggest that in the regions known historically as 'Israel' and 'Judah' there was anything more elaborate than a network of 'peasant farmers' who were operating from 'decentralised settlements sprawled across the broken highlands and dissected slopes' (Holladay, 1998). Wells continued to be dug in the Iron Age, although not always successfully. A 44-m-deep well at Lachish, dug at the lowest point on the mound's surface (Tufnell, 1953; Ussishkin, 1982), apparently sought the gravel aquifer but bottomed in Oligocene limestone yielding insufficient quantities of water (Weinberger et al., 2008, p. 3036). Similarly, at Tel Beer Sheba a 69-m-deep well apparently sought the groundwater table in the gravel bed which unfortunately did not stretch laterally underneath the mound (Herzog, 2002).

In the Iron Age II, monumental water access facilities are known from sites in royal centres such as Megiddo and Hazor, and at other slightly lower-lying sites with access to good farmland, such as Gibeon (Dever, 1998, plate 4). These can be shafts or water tunnels leading to a source inside or outside the city or tunnels and feed channels supplying reservoirs (Shiloh and Herzog, 1992). It is clear that these cities probably grew from earlier settlements which were most likely located on natural water tunnels created by natural springs (as with Gezer, above). The most extensively studied water system is at Hazor, constructed between 2,900 and 2,800 BP. The original excavator noted the particular direction of the sloping tunnel built to access the groundwater within the mound, which he felt indicated superior knowledge of hydro-engineering (Yadin, 1969). However, Weinberger et al. (2008) have demonstrated via a geo-hydrological survey that the tapping of the groundwater was more likely a lucky accident. They surmise that the Iron Age inhabitants intended to reach the springs via a sloping tunnel, but at about 195 m above mean sea level the diggers hit the Dead Sea Fault while quarrying roughly parallel to the city walls. Weinberger et al. suggest that

> This accidental success, as well as the failure to draw significant quantities of water in deep water wells dug in the contemporary cities of Lachish and Beer Sheba, imply limited hydrogeological understanding at the beginning of the first millennium B.C.E.
>
> (Weinberger et al. 2008, p. 3041)

Of the historical geographic divisions east of the Jordan Valley, only in the area known as 'Ammon' and further north in 'Gilead' was it possible to conduct dry farming. The more arid regions to the south, known as 'Moab' and 'Edom', are more arid. The settlement system in these areas appears to have been highly flexible and therefore able to cope with more uncertain conditions. The system of tribal pastoralism required quite different water resources from those required by farmers, and it is presumed that the ancient inhabitants of these more marginal regions managed a complex interplay of both strategies (LaBianca and Younker, 1998). An underground water system known at the citadel of Amman (Conder, 1889; Dornemann, 1983; Zayadine et al., 1989) presumably fed the needs of more sedentary, land-tied groups. In this area we find walled settlements, while in Moab there are very few and none in the heartland south of the Wadi Mujib. Meanwhile, further south in Edom, there is a real scarcity of settlements, and again, of these very few are walled. Rock-cut cisterns and associated channels in the sandstone formation of es-Sela and Ba'ja III are loosely attributed to the Iron II period by Lindner (Lindner, 1992).

The study by Barker and Gilbertson (2000) of Wadi Faynan found extensive Iron Age occupation suggesting floodwater farming. Further north in the Jordan Valley itself, the site of Tell Deir Allah has produced a large amount of Iron Age material (Figure 14.21). We know that a dam and water canals were built to bring water from the Wadi Zerqa, also known as Wadi Zarqa (c. 3 km to the east), for irrigation (LaBianca and Younker, 1998, p. 415). These irrigation systems are now being extensively studied by a multidisciplinary team in relation to climate, hydrology and archaeology (see Van der Kooij, 2007 for an outline) and the project promises new insights into Iron Age water management.

14.4.7 Summary

It can be argued that it was during the Chalcolithic that the inhabitants of the southern Levant developed their characteristic mixed farming economy, supplemented by horticulture which required a long-term commitment to place (see Chapter 17, this

Figure 14.21. Iron Age Tell Deir Allah, looking across the Jordan Valley where water canals were built to supply the settlement (© S. Mithen).

volume). The development of land tenure systems that was the inevitable co-requisite for this would have resulted in community cooperation to manage vital resources. Water, of course, was of prime importance as it had always been. During the urbanising Bronze Age and the following Iron Age the methods for water management are only as sophisticated as the concurrent sociopolitical structures which made them necessary.

14.5 WATER MANAGEMENT IN THE NABATAEAN

Water management in the Nabataean requires less extensive treatment within this chapter than that provided for the Neolithic, Bronze and Iron Ages because of the existing reviews by John Oleson (1995, 2007) and the detailed study concerning the water supply of Petra by Ortloff (2005). Consequently, this section will provide a succinct summary of the key developments concerning water management during the Nabataean period, primarily drawing on Oleson (2007), this acting as a prelude to the hydrological modelling study of Nabataean Humayma in Chapter 19 of this volume.

The Nabataean culture emerged within the region around Petra in the third century BC from an integration of indigenous Iron Age communities with nomadic tribal groups that had spread north from Arabia and were familiar with caravan-based trade activities. During the second and first centuries BC the Nabataeans established an extensive trading empire with Petra as its commercial centre, a Nabataean state emerging in 64 BC. Petra was the node for caravan trade from Arabia, Africa and the Eastern Mediterranean; its power, wealth and population grew rapidly, eventually reaching 30,000 people, leading to a concomitant increase in the needs of a water supply to fund the resident and visiting population. Numerous Nabataean settlements developed in the region, notably Humayma, while the Nabataean influence extended northwards as far as Damascus. Throughout this period there was a turbulent relationship with Rome, which ultimately led to Petra falling under Roman control and the Nabataean Kingdom being formally annexed into the Roman Empire in AD 106 under Trajan.

With regard to water management it must first be stressed that the Nabataean Kingdom flourished in a region of hyper-aridity, one that received an average annual precipitation of 25–75 mm. That meagre rainfall was highly localised, as discussed elsewhere in this volume (Chapters 3, 15). Strong sunshine, winds and low humidity led to high rates of evaporation. There were no rivers in the core Nabataean area; springs were infrequent and had low rates of flow, while the topography was rugged.

14.5.1 Methods of water management

Oleson (2007) suggests that Nabataean water management can be divided into two phases. The first involved utilisation of the systems that had developed within the region during the Neolithic, Bronze and Iron Ages, as has been documented within this chapter. While these may have been used with greater intensity, there were no striking innovations. As such, dating is especially difficult, it often being unclear whether structures were first constructed in the Nabataean period or reused from earlier times. The second phase of Nabataean water management occurs during the first century BC and involves several new developments, partly reflecting the needs of a rapidly increasing population and partly from new techniques of hydraulic engineering acquired from Hellenistic influences.

During the first of these two phases, the most prominent means of water collection that survive today are rock-cut cisterns, often made in a bottle shape with a diameter of 3–4 m, the narrow entrance hole acting to minimise evaporation and the risk of pollution, such as from animal faeces entering the water. Such cisterns were cut into bedrock close to natural catchments with feeder channels to collect runoff water. These were often constructed in large numbers around Nabataean settlements. Some had plaster linings to reduce percolation and erosion of the cistern by the water itself. Water was also collected from natural springs and seepages, sometimes with modifications being made to facilitate access. Oleson (2007) describes how a natural basin below the Muqawwar Cascades of water at the foot of the al-Shera escarpment southeast of Ras en Naqb was enlarged to increase its storage capacity. Conduits were sometimes constructed in the vicinity of springs to distribute water, these being made from stone blocks with a deep longitudinal channel. But such conduits rarely ran for more than a few metres, aqueducts being a later development. The topography and geology of the

Nabataean region was not conducive to the sinking of wells, but, when possible, these were constructed by the Nabataeans, notably in the western and southern parts of the Negev. The Nabataeans continued to construct terrace walls on hill-slopes to capture runoff and prevent soil erosion, as can now be dated back to the Pottery Neolithic at 'Dhra (as described in section 14.3.5 above) and to build barriers across wadi floors themselves, as now dated back to the PPNB in the Jafr Basin (Fujii, 2007a).

Oleson (2007) describes numerous developments in Nabataean water management that occurred in the first century BC, although dating hydraulic structures remains problematic. The developments were a response to increased population levels, requiring more extensive irrigation and drinking water, and a reflection of Hellenistic influences also seen in various other aspects of Nabataean culture, such as the adoption of coinage. They included increasing the surface runoff from slopes by using geometric arrangements of stones in heaps, enabling some fields to benefit from a catchment of 30 times their own areas. Wadi barriers were now used to divert excess water flow into channels to irrigate fields, while new containment dams were constructed to supplement the many smaller dams built during the early phase of Nabataean culture. One such dam at Humayma was 10.7 m long and 4.4 m thick, and was used to contain 1,400 m^3 of water within a reservoir. A much larger dam was built at Mampsis in the Negev which contained 10,000 m^3.

Another key development was the construction of aqueducts. Unlike Roman aqueducts, Nabataean examples were all primarily built at ground level and were dependent upon natural slopes for the flow of water. This is not to deny a considerable engineering challenge because the rugged topography often required the use of bridges across gullies, channels cut across bedrock and, at Petra, an arched bridge across a gorge. The Nabataean aqueducts were built from blocks of stone, limestone, sandstone or marl, between 60 and 90 cm long which had longitudinal channels 12 cm wide and 15 cm deep. These were placed end to end and held in position by framing walls. The longest such aqueduct was at Humayma, running for 26.5 km and carrying spring water from the crest of the ash-Shara escarpment to a reservoir at the centre of the settlement (Figure 14.22, see further discussion in Chapter 19, this volume). The main aqueduct in Petra, that which carried water from 'Ain Musa to the city was 8 km. Oleson (2007) stresses that the scale of these aqueducts was modest compared with those in the Hellenistic and later Roman world: the Humayma aqueduct could have discharged no more than 148.6 m^3 cubic metres per day while the daily discharge of the Roman aqueduct at the Pont de Gard, Nimes, amounts to no less than 20,000 m^3. Aqueducts fed into basins that often had several outlets to take water for different purposes – to reserves, cisterns, baths or agricultural plots. At Humayma, for instance, the aqueduct filled a large shallow reservoir that had a capacity of

Figure 14.22. Aqueduct at Humayma (© R. Foote).

633 m^3. Overflow from this was carried away in conduits and terracotta pipes to serve a bath building, private cisterns and a shrine, and to irrigate trees and vegetables.

The construction of aqueducts may have arisen from knowledge of such hydraulic engineering methods gained by direct travel or contacts with traders from the Hellenistic world. Oleson (2007) suggests that this was also the cause of the adoption of transverse arches to cover cisterns as a means to support a roof. Covers for cisterns were critical to avoid excessive evaporation and pollution of the water from waste, notably animal faeces. Bottle-shaped cisterns could only be located in suitable bedrock, while the size of cisterns that could be covered with rock slabs and timber (if any was available) was highly limited. Oleson suggests that Nabataean merchants may have seen Hellenistic Greek cistern design when visiting Delos and other arid Aegean islands. This design involved the use of transverse arches which could then support roofing slabs, enabling cisterns of substantial size to be entirely covered. This was adopted and gained extensive use, allowing greater freedom as to where cisterns could be constructed. Oleson describes how two public cisterns were built

Figure 14.23. Cistern at Humayma with an arched roof (scale provided by Dr Claire Rambeau) (© R. Foote).

Figure 14.24. Nabataean water channel in the siq at Petra (© S. Mithen).

at Humayma, with roofs supported by 16 arches that supported stone slabs enabling people and animals to move around and even over the cisterns (Figure 14.23).

The planning, construction and management of long aqueducts, arched-roofed cisterns and other hydraulic engineering works would have been significant undertakings. Some evidence for the administrative systems has survived in the form of inscriptions. One found at the cult centre of Khirbet at-Tannur refers to a person with the title of 'Master of the Springs of La 'ban', a nearby site (Oleson, 2007). It is clear from the titles bestowed on the Nabataean kings that their patronage, and most likely direct control, was essential for the major water management projects undertaken at Petra and Humayma (Oleson, 2007).

14.5.2 Water supply, distribution and display in Petra

Ortloff (2005) has provided a detailed description and interpretation of the water supply and distribution system of Petra. This had originally been based on an open channel aqueduct from the spring at 'Ain Mousa that followed a path down the siq to the centre of the settlement (Figure 14.24). This involved the construction of a dam and a tunnel to divert the intermittent floods of the Wadi Mousa river, that would have otherwise periodically flooded the siq. From this basis a remarkable complex system of water management developed within Petra, making use of all of the methods referred to above: surface cisterns, deep underground cisterns, rock-cut channels, storage reservoirs, dams and tunnels which exploited multiple springs within and outside the city. These acted to capture every conceivable amount of the meagre seasonal rainfalls, distribute water throughout the city and provide protection from seasonal and potentially destructive floods. Terracotta pipes were employed extensively as part of the distribution system, these appearing to have been designed to promote the stable, regular supply of water. Ortloff (2005) notes that the unique achievement of the hydraulic planners and engineers of Petra was to create a system in which water conservation was practised on a much larger scale than in any other contemporary city, with the intermittent seasonal rainfall being managed to sustain a continuous supply of water throughout the year. He stresses how there was redundancy within the system so that if one supply line failed, such as from seasonal runoff, there would be a back up, such as from a spring. He notes that when Petra came under Roman control with the potential for use of new forms of hydraulic engineering, the Nabataean system remained effectively unchanged, indicating that the exploitation of water supply had been optimised.

One of the most significant features of the water management system within Petra was the 'Petra garden and pool complex' (Bedal, 2003; Bedal and Schryver, 2007). This was located in the centre of the city, adjacent to the Great Temple complex and overlooking the Colonnaded Street, in an area that was originally thought to have been a market place. Excavations since 1998 have revealed a suite of architectural and hydraulic features that suggest that water had been used for ostentatious display, perhaps the most dramatic sign possible of the power of any ruler of an intensely arid region.

The centre of the complex was a pool, 43 m × 23 m and 2.5 m deep, which would have held over two million litres of water. It had a 16-m-high sandstone cliff as a backdrop while the water was contained by a massive decorated wall built of limestone blocks along the northern edge of the sandstone shelf. This wall contained channels and pipes that carried water from a tank perched on an escarpment to the east, from which water also dramatically cascaded into the pool. It is assumed that the pool was used for swimming, with a flight of steps appearing to provide easy access into the water. An artificial island had been built in the centre of the pool that provided support for what is termed a pavilion. This appears to have been painted in bright colours and would have provided impressive views across extensive gardens that ran for 50 m to the edge of the street in the centre of the city.

Bedal and Schryyer (2007) argue that the development of the garden and pool complex at Petra provides another example of Hellenistic influence. Such displays had become important political statements of wealth and power in the late second and first centuries BC in the Hellenised Mediterranean world. They suggest that the plan of the Petra example had been directly modelled on the garden pool complex of Herod's summer palace-fortress at Herodium, built 23–20 BC. Oleson (2007) suggests that the large shallow reservoir at Humayma may have been another, less ostentatious but nevertheless symbolic display of water.

14.6 SUMMARY

Although the extraordinary water management systems evident from Petra and Humayma partly derived from Hellenistic influence, they were also dependent upon innovations in water management that occurred within the southern Levant itself from the Neolithic onwards. Although the distribution and character of human activity within this region was influenced by the distribution of water from the time of its first occupation more than one million years ago, it is only in the Neolithic that the first evidence appears for the management of the water supply. There should be no doubt that Epipalaeolithic and indeed Palaeolithic communities did manipulate water flows for their own need, such as by enlarging natural pools and diverting water flows by the use of barriers made from pebbles or sand. Such structures leave no archaeological trace and would only have been able to serve short-term needs for small groups. It should not be surprising that the first evidence for substantial levels of water management is only found when the people are living in permanent villages containing several hundred, and in some cases thousands of inhabitants. But the scarcity of wells, wadi barriers and terrace walls remains striking, with the earliest farming communities seeming to remain largely dependent on springs and seasonal river flows. Indeed, the absence of a significant level of water management may have been a key reason for the depopulation of many PPNB towns. Other than in the Jafr Basin and on Cyprus, the first evidence for substantial water management comes in the form of wells and terrace walls in the Pottery Neolithic, which is probably the first time that an 'agricultural landscape' develops in the southern Levant. The subsequent development of economic and social complexity in the Chacolithic and then proto-urbanisation in the Bronze and Iron Ages were dependent upon increasing investment in water management to enable the capture, storage and transport of water. This provided the hydraulic foundations for the development and cultural achievement of the Nabataean Kingdom.

REFERENCES

Al-Najjar, M., A. Abu-Dayya, E. Suleiman, G. Weisgerber and A. Hauptmann (1990) Tell Wadi Faynan: the first pottery Neolithic Tell in the south of Jordan. *Annual of the Department of Antiquities of Jordan* **34**: 27–56.

Albert, R. M. and D. O. Henry (2004) Herding and agricultural activities at the early Neolithic site of Ayn Abu Nukhayla (Wadi Rum, Jordan). The results of phytolith and spherulite analyses. *Paleorient* **30**: 81–92.

Albright, W. F. (1944) The Prince of Taanach in the 15th century BC. *Bulletin of the American School of Oriental Research* **94**: 12–27.

Allan, J. A. (2001) *The Middle East Water Question: Hydro-Politics and the Global Economy*. London: I. B. Tauris.

Amiran, R. (1978) *Early Arad Jerusalem*. Jerusalem: Israel Exploration Society.

Arnold, D. E. (1985) *Ceramic Theory and Cultural Process*. Cambridge: Cambridge University Press.

Banning, E. B. (1996) Highlands and lowlands: problems and survey frameworks for rural archaeology in the Near East. *Bulletin of the American Schools of Oriental Research* **301**: 25–46.

Bar-Matthews, M., A. Ayalon, M. Gilmour, A. Matthews and C. J. Hawkesworth (2003) Sea-land oxygen isotopic relationships from planktonic foraminifera and speleothems in the Eastern Mediterranean region and their implication for paleorainfall during interglacial intervals. *Geochimica et Cosmochimica Acta* **67**: 3181–3199.

Bar-Yosef, O. (1986) The walls of Jericho: an alternative interpretation. *Current Anthropology* **27**: 157–162.

Bar-Yosef, O. (1989) The PPNA in the Levant, an overview. *Paleorient* **15/1**: 57–63.

Bar-Yosef, O. (2002) Natufian. A complex society of foragers. In *Beyond Foraging and Collecting. Evolutionary Change in Hunter-Gatherer Settlement Systems*, ed. B. Fitzhugh and J. Habu. New York, NY: Kluwer Academic/Plenum Publishers pp. 91–149.

Bar-Yosef, O. and A. Belfer-Cohen (1989) The origins of sedentism and farming communities in the Levant. *Journal of World Prehistory* **3**: 447–498.

Bar-Yosef, O. and A. Belfer-Cohen (2002) Facing environmental crisis – societal and cultural changes at the transition from the Younger Dryas to the Holocene in the Levant. In *The Dawn of Farming in the Near East*, ed. R. T. J. Cappers and S. Bottema. Berlin: ex Oriente.

Bar-Yosef, O. and A. Gopher, eds. (1997) *An Early Neolithic Village in the Jordan Valley. Part I: The Archaeology of Netiv Hagdud*. Cambridge, MA: American School of Prehistoric Research, Peabody Museum.

Barker, G. (2006) *The Agricultural Revolution in Prehistory: Why Did Foragers become Farmers?* Oxford: Oxford University Press.

Barker, G. and D. Gilbertson, eds. (2000) *The Archaeology of Drylands: Living at the Margin*. London and New York: Routledge.

Barker, G., D. Gilbertson and D. Mattingly, eds. (2007) *Archaeology and Desertification: The Wadi Faynan Landscape Survey, Southern Jordan*. Wadi Faynan Series Volume 2, Levant Supplementary Series. Oxford, UK and Amman, Jordan: Council for British Research in the Levant in Association with Oxbow Books.

Becker, C. (1991) The analysis of mammalian bones from Basta, a pre-pottery Neolithic site in Jordan: problems and potential. *Paleorient* **17/1**: 59–75.

Bedal, L.-A. (2003) *The Petra Pool Complex: A Hellenistic Paradeisos in the Nabatean Capital*. Piscataway, NJ: Georgias Press.

Bedal, L.-A. and J. G. Schryyer (2007) Nabatean landscape and power: Evidence from the Petra garden and pool complex. In *Crossing Jordan: North American Contributions to the Archaeology of Jordan*, ed. T. E. Levy, P. M. Daviau, R. W. Younker and M. Shaer. London: Equinox Publishing pp. 376–383.

Ben-Tor, A. (1992) The Early Bronze Age. In *The Archaeology of Ancient Israel*, ed. A. Ben-Tor. New Haven, CT: Yale University Press.

Betts, A. (1992) Eastern Jordan: Economic choices and site location in the Neolithic periods. In *Studies in the History and Archaeology of Jordan IV*, ed. M. Zaghoul. Amman: Department of Antiquities.

Betts, A., ed. (1998) *The Harra and the Hamad: Excavations and Surveys in Eastern Jordan*. Sheffield: Sheffield University Press.

Bienert, H.-D. (2004) The underground tunnel system in wadi ash-Shellalah, Northern Jordan. In *Men of Dikes and Canals: The Archaeology of Water in the Middle East*, ed. H.-D. Bienert, H. G. K. Gebel and R. Neef. Rahden, Westphalia: Verlag Orient-Archaologie Band 13 pp. 43–60.

Bienert, H.-D., H. G. K. Gebel and R. Neef, eds. (2004) *Central Settlements in Neolithic Jordan: Studies in Early Neolithic Production, Subsistence and Environment*. Berlin: ex Oriente.

Biran, D. (1981) The discovery of the Middle Bronze Age gate at Dan. *The Biblical Archaeologist* **44/3**: 139–144.

Bourke, S. (1997) The 'Pre-Ghassulian' sequence at Tleilat Ghassul. In *The Prehistory of Jordan II, Perspectives from 1997*, ed. H. G. K. Gebel, Z. Kafafai and G. Rollefson. Berlin: ex Oriente pp. 395–417.

Bourke, S. (2008) *The Chalcolithic in Jordan: An Archaeological Reader*, ed. R. Adams. London: Equinox.

Bourke, S., R. T. Sparks, K. N. Sowada, P. B. Mclaren and L. D. Mairs (1998) Preliminary report on the University of Sydney's sixteenth and seventeenth seasons of excavations at Pella (Tabaqat Fahl) in 1994/5. *Annual of the Department of Antiquities of Jordan* **42**: 179–211.

Braemer, F., D. Genequand, C. D. Maridat et al. (2009) Long-term management of water in the Central Levant: the Hawran case (Syria). *World Archaeology* **41**: 36–57.

Brooks, N. (2006) Cultural responses to aridity in the middle Holocene and increased social complexity. *Quaternary International* **151**: 29–49.

Bunimovitz, S. (1992) The Middle Bronze Age fortifications as a social phenomenon. *Tel Aviv* **19**: 221–234.

Byrd, B. F. (2005) *Early Village Life at Beidha, Jordan: Neolithic Spatial Organization and Vernacular Architecture*. Oxford: Oxford University Press.

Callaway, J. A. (1970) The 1968–1969 'Ai (et-Tell) excavations. *Bulletin of the American School of Oriental Research* **198**: 7–31.

Callaway, J. A. (1978) New perspectives on Early Bronze III in Canaan. In *Archaeology in the Levant*, ed P. R. S. Moorey and P. J. Parr. Warminster: Aris and Phillips pp. 46–58.

Cauvin, J. (2000) *The Birth of the Gods and the Origins of Agriculture*. Cambridge: Cambridge University Press.

Childe, V. G. (1936) *Man Makes Himself*. London: Watts.

Clark, G. (1944) Water in antiquity. *Antiquity* **18**: 1–15.

Conder, C. R. (1889) *Survey of Eastern Palestine: Memoirs of the Topography, Orography, Hydrography and Archaeology: The Adwan Country*. London: Palestine Exploration Fund.

Croft, P. (2003a) Current activity: Mylouthkia again. In *The Colonisation and Settlement of Cyprus. Investigations at Kissonerga-Mylouthkia 1976–1996*, ed. E. Peltenberg. Savedalen: Astroms.

Croft, P. (2003b) The wells and other vestiges. In *The Colonisation and Settlement of Cyprus. Investigations at Kissonerga-Mylouthkia 1976–1996*, ed. E. Peltenberg. Savedalen: Astroms pp. 3–11.

de Miroschedji, P. (1999) Yarmouth: the dawn of city-states in Southern Canaan. *Near Eastern Archaeology* **62**: 2–19.

Dennis, S. (2003) The experimental reconstruction of a Pre-Pottery Neolithic B structure at Beidha in Jordan. *Levant* **35**: 39–48.

Dever, W. (1969) The water systems at Hazor and Gezer. *The Biblical Archaeologist* **32**: 71–78.

Dever, W. (1998) Social structure in Palestine in the Iron II Period on the eve of destruction. In *The Archaeology of Society in the Holy Land*, ed. T. E. Levy. Leicester: Leicester University Press pp. 416–430.

Dever, W. (2003) *Who Were the Early Israelites, and Where Did They Come From?* Grand Rapids, MI: William B. Eerdmans Publishing Company.

Dolfus, G. and Z. Kafafai (1988) *Abu Hamid, village du IVe millénaire de la Vallée du Jourdain*. Amman: Economic Press.

Donahue, J. (2003) Geology and geomorphology. In *Bab edh-Dhra: Excavations at the Town Site (1975–81)*, ed. W. E. Rast and R. T. Shaub. Winona Lake, IN: Eisenbrauns pp. 18–55.

Dornemann, R. H. (1983) *The Archaeology of the Transjordan in the Bronze and Iron Ages*. Milwaukee: Milwaukee Public Museum.

Dunand, M. (1960) Histoire d'une source. *Mélanges de l'Université St. Joseph* **37**: 37–53.

Dunand, M. (1973) *Fouilles de Byblos. Tome V. L'architecture, les Tombes, le Matériel Domestique, des Origines Néolithiques à l'Avènement Urbain*. Paris: P. Guethner.

Edwards, P. C., S. Bourke, S. Colledge, J. Head and P. G. Macumber (1988) Late Pleistocene prehistory in the Wadi al-Hammeh, Jordan Valley. In *The Prehistory of Jordan*, ed. A. Garrard and H. G. K. Gebel. Oxford: British Archaeological Reports.

Edwards, P. D. and T. Higham (2001) Zahrat adh-Dhra' 2 and the Dead Sea Plane at the dawn of the Holocene. In *Australians Uncovering Ancient Jordan, 50 Years of Middle Eastern Archaeology*, ed. A. Walmsley. Sydney: University of Sydney pp. 139–152.

Finkelstein, I. (1988) *The Archaeology of the Israelite Settlement*. Jerusalem: IES.

Finkelstein, I. (1989) The Land of Ephraim Survey 1980–1987: preliminary report. *Tel Aviv* **15/16**: 117–183.

Finkelstein, I. (1998) The Great Transformation: the conquest of the highland frontiers and the rise of territorial states. In *The Archæology of Society in the Holy Land*, ed. T. E. Levy. London: Leicester University Press pp. 349–365.

Finlayson, B. L., T. Kuji, M. Arpin et al. (2003) Dhra' Excavation Project 2002, interim report. *Levant* **35**: 1–38.

Finlayson, B. L. and S. Mithen (2007) *The Early Prehistory of Wadi Faynan, Southern Jordan: Archaeological Survey of Wadis Faynan, Ghuwayr and Al Bustan and Evaluation of the Pre-Pottery Neolithic A Site of WF16*, ed. B. L. Finlayson and S. Mithen. Oxford and Amman, Jordan: Oxbow Books and the Council for British Research in the Levant.

Fino, N. (1997) Al-Baseet, a new LPPNB site found in Wadi Musa, southern Jordan. *Neolithics* **3/97**: 13–14.

Frumkin, A. and A. Shimron (2006) Tunnel engineering in the Iron Age: geoarchaeology of the Siloam Tunnel. *Jerusalem Journal of Archaeological Science* **33**: 227–237.

Fujii, S. (2007a) Barrage systems at Wadi Abu Tulayha and Wadi Ar-Ruwayshid Ash-Sharqi: a preliminary report of the 2006 spring field season of the Jafr Basin Prehistoric Project, Phase 2. *Annual of the Department of Antiquities of Jordan* **51**: 403–427.

Fujii, S. (2007b) Wadi Abu Tulayha: a preliminary report of the 2006 summer field season of the Jafr Basin Prehistoric Project, Phase 2. *Annual of the Department of Antiquities of Jordan* **51**: 373–401.

Fujii, S. (2007c) Wadi Badda: A PPNB settlement below Fjaje escarpment near Shawbak. *Neolithics* **1/07**: 19–24.

Fujii, S. (2008) Wadi Abu Tulayha: a preliminary report of the summer field season of the Jafr Basin Prehistoric Project, Phase 2. *Annual of the Department of Antiquities of Jordan* **52**: 445–478.

Fujii, S. (in prep–a) Domestication of runoff water: current evidence and new perspectives from the Jafr Pastoral Neolithic. *Neolithics*, in prep.

Fujii, S. (in prep–b) Wadi Abu Tulayha: a preliminary report of the 2008 summer final field season of the Jafr Basin Prehistoric Project, Phase 2. *Annual of the Department of Antiquities of Jordan*, in press.

Galili, E. and Y. Nir (1993) The submerged Pre-Pottery Neolithic water well of Atlit-Yam, Northern Israel and its palaeoenvironmental implications. *The Holocene* **3**: 265–270.

Galili, E., M. Weinstein-Eyron, M. Hershkovitz et al. (1993) Atlit-Yam: a prehistoric site on the sea floor off the Israeli coast. *Journal of Field Archaeology* **80**: 133–157.

Garfinkel, Y., A. Vered and O. Bar-Yosef (2006) The domestication of water: the Neolithic well at Sha'ar Hagolan, Jordan Valley, Israel. *Antiquity* **80**: 686–696.

Garrard, A. (1994) Prehistoric environment and settlement in the Azraq basin: an interim report on the 1987 and 1988 excavation seasons. *Levant* **26**: 73–109.

Gebel, H. G. K. (2004) The domestication of water. Evidence from Early Neolithic Ba'ja. In *Men of Dikes and Canals. The Archaeology of Water in the Middle East*, ed. H.-D. Bienert and J. Häser Rahden: Marie Leidorf pp. 25–36.

Gebel, H. G. K., H. J. Nissen and Z. Zaid (2006) *Basta II: The Architecture and Stratigraphy*. Berlin: ex Oriente.

Gillmore, G. K., R. A. E. Conningham, H. Fazeli et al. (in press) Irrigation on the Tehran Plain, Iran: Tepe Pardis – The site of a possible Neolithic irrigation feature? *CATENA* **78**: 285–300.

Goring-Morris, A. N. (1987) *At the Edge: Terminal Pleistocene Hunter-Gatherers in the Negev and Sinai*. Oxford: British Archaeological Reports.

Goring-Morris, A. N. and A. Belfer-Cohen (2010) Different ways of being, different ways of seeing… Changing worldviews in the Near East. In *Hunters in Transition*, ed. B. L. Finlayson and G. Warren. Oxford, UK and Amman, Jordan: Oxbow and the Council for British Research in the Levant.

Grosman, L., N. D. Munro and A. Belfer-Cohen (2008) A 12,000-year-old Shaman burial from the southern Levant (Israel). *Proceedings of the National Academy of Sciences of the United States of America* **105**: 17665–17669.

Guilane, J. and F. Briois (2001) Parekklisha Shillourokambos: an early Neolithic site in Cyprus. In *The Earliest Prehistory of Cyprus. From Colonization to Exploitation*, ed. S. Swiny. Boston, MA: Boston American Schools of Oriental Research pp. 37–54.

Harrison, T. (1997) Shifting patterns of settlement in the highlands of central Jordan during the Early Bronze Age. *Bulletin of the American School of Oriental Research* **306**: 1–38

Herzog, Z. (2002) Water supply at Tel Beer Sheba. In *Proceedings of the Cura Aquarum of Israel in Memoriam, 11th International Water Conference on the History of Water Management and Hydraulic Engineering in the Mediterranean Region*. Leuvan: Peeters pp. 15–22.

Holladay, J. S. (1998) Israel and Judah: political and economic centralization in the Iron Age II A–B (ca. 1000–750 BCE). In *The Archaeology of Society in the Holy Land*, ed. T. E. Levy. Leicester: Leceister University Press pp. 368–398.

Hunt, C. O., D. D. Gilbertson and H. A. El-Rishi (2007) An 8000-year history of landscape, climate, and copper exploitation in the Middle East: the Wadi Faynan and the Wadi Dana National Reserve in southern Jordan. *Journal of Archaeological Science* **34**: 1306–1338.

Ibrahim, M., K. Yassine and J. A. Sauer (1988) The East Jordan Valley survey 1975 (parts 1 and 2). In *The Archaeology of Jordan: Essays and Reports*, ed. K. Yassine. Amman: Department of Archaeology, University of Jordan pp. 159–207.

Ilan, D. (1998) The dawn of internationalism – the Middle Bronze Age. In *The Archaeology of Society in the Holy Land*, ed. T. E. Levy. Leicester: Leceister University Press pp. 297–319.

Jenkins, E. (2009) Phytolith taphonomy: a comparison of dry ashing and acid extraction on the breakdown of conjoined phytoliths formed in Triticum durum. *Journal of Archaeological Science* **36**: 2402–2407.

Joffe, A. H. (2003) Slouching toward Beersheva: Chalcolithic mortuary practices in local and regional context. In *The Near East in The Southwest, Essays in Honor of William G. Dever*, ed. B. Alpert-Nakhai. Oxford: Oxbow pp. 45–67.

Kenyon, K. M. (1960) *Archaeology in the Holy Land*. London: Benn.

Kenyon, K. M. (1981) *Excavations at Jericho, Vol. III: The Architecture and Stratigraphy of the Tel*. London: British School of Archaeology in Jerusalem.

Kirkbride, D. (1966) Five seasons at the Pre-Pottery Neolithic village of Beidha in Jordan. *Palestine Exploration Quarterly* **98**: 8–72.

Kislev, M. E., A. Hartmann and E. Galili (2004) Archaeobotanical and archaeoentomological evidence from a well at Atlit-Yam indicates colder, more humid climate on the Israeli coast during the PPNC period. *Journal of Archaeological Science* **31**: 1301–1310.

Kogan-Zehavi (2005) Horvat Tittora West sheet 117 (Hadashot Arkheologioyot – Excavations and Surveys in Israel). Jerusalem: Israel Antiquities Authority.

Köhler-Rollefson, I. (1992) A model for the development of nomadic pastoralism on the Transjordanian plateau. In *Pastoralism in the Levant*, ed. O. Bar-Yosef and A. Khazanov. Madison: Prehistory Press pp. 11–18.

Kujit, I. (1994) Pre-Pottery Neolithic A period settlement systems of the southern Levant: new data, archaeological visibility, and regional site hierarchies. *Journal of Mediterranean Archaeology* **7**: 165–192.

Kujit, I. and B. L. Finlayson (2009) New Evidence for Food Storage and Pre-Domestication Granaries 11,000 years ago in the Jordan Valley *Proceedings of the National Academy of Science* **106**: 10966–10970.

Kujit, I. and A. N. Goring-Morris (2002) Foraging, farming and social complexity in the Pre-Pottery Neolithic of the South-Central Levant: a review and synthesis. *Journal of World Prehistory* **16**: 361–439.

Kujit, I., B. L. Finlayson and J. Mackay (2007) Pottery Neolithic landscape modification at Dhra'. *Antiquity* **81**: 106–118.

LaBianca, Ø. S. and R. W. Younker (1998) Kingdoms of Ammon, Moab and Edom: the archaeology of society in Late Bronze Age/Iron Age Transjordan. In *The Archaeology of Society in the Holy Land*, ed. T. E. Levy. Leicester: Leceister University Press pp. 399–415.

Lancaster, W. and F. Lancaster (1991) Limitations on sheep and goat herding in the Eastern Badia of Jordan: an ethno-archaeological enquiry. *Levant* **28**: 125–138.

Lancaster, W. and F. Lancaster (1997) Indigenous water management systems in the badia of the Bilad ash-Sham. *Journal of Arid Environments* **35**: 367–378.

Lapp, P. W. (1969) The 1968 excavations at Tell Ta'aneck. *Bulletin of the American Schools of Oriental Research* **195**: 2–49.

Lindner, M. (1992) Edom outside the famous excavations: evidence from surveys in the Greater Petra area. In *Early Edom and Moab: The Beginning of the Iron Age in Southern Jordan*. Sheffield: Sheffield Archaeological Monographs pp. 143–166.

Lipschitz, N. (1986) Overview of the dendrochronological and dendroarchaeological research in Israel. *Dendrochronologia* **4**: 37–58.

Lovell, A. I., J. L. Richter, P. B. Mclaren and A. I. Abu Shmeis (2005) The first preliminary report of the Wadi Rayyan Archaeological Project: the survey of el Khawarij. *Annual of the Department of Antiquities of Jordan* **49**: 189–200.

Lovell, J. L. (2001) *The Late Neolithic and Chalcolithic Periods in the southern Levant: New data from Teleilat Ghassul, Jordan*. Oxford: Archaeopress.

Lovell, J. L. (2008) Horticulture, status and long-range trade in Chalcolithic southern Levant: early connections with Egypt. Proceedings of the international conference 'Origin of the State. Predynastic and Early Dynastic Egypt', Toulouse, 5–8 September 2005, ed. B. Midant-Reynes and M. Y. Tristant. Toulouse: Peeters Publishers pp. 739–760.

Lovell, J. L., D. C. Thomas, H. L. Miller et al. (2007) The third preliminary report of the Wadi Rayyan Archaeological Project: the second season of excavations at el-Khawarij. *Annual of the Department of Antiquities of Jordan* **51**.

Mabry, J. M. (1989) Investigations at Tell el-Handaquq, Jordan (1987–88). *Annual of the Department of Antiquities of Jordan* **32**: 59–95.

Mabry, J. M., L. Donaldson, K. Gruspier et al. (1996) Early town development and water management in the Jordan Valley: investigations at Tell el-Handaquq North. *Annual of the Department of Antiquities of Jordan* **53**: 115–154.

Mahasneh, H. (1997) Es-Sifiya: a Pre-Pottery Neolithic B site in the Wadi el-Mujib, Jordan. In *The Prehistory of Jordan II: Perspectives from 1996*, ed. H. G. K. Gebel, Z. Kafafai and G. Rollefson. Berlin: ex Oriente pp. 203–214.

Maher, L. (2010) People and their places at the end of the Pleistocene: evaluating perspectives on physical and cultural landscape change. In *Hunters in Transition*. ed. B. L. Finlayson and G. Warren. Oxford, UK and Amman, Jordan: Oxbow and the Council for British Research in the Levant.

Matson, F. R. (1965) Ceramic ecology: an approach to the study of the early culture of the Near East. In *Ceramics and Man*, ed. F. R. Matson. Chicago, IL: Aldine Publishing Company pp. 202–217.

McCreery, D. W. (2003) The paleoethnobotany of Bab edh-Dhra. In *Bab edh-Dhra: Excavations at the Town Site (1975–81)*, ed. W. E. Rast and R. T. Schaub. Winona Lake, IN: Eisenbrauns pp. 449–463.

McLaren, S. J., D. D. Gilbertson, J. P. Grattan et al. (2004) Quaternary palaeogeomorphologic evolution of the Wadi Faynan area, southern Jordan. *Palaeogeography Palaeoclimatology Palaeoecology* **205**: 131–154.

Miller, R. (1980) Water use in Syria and Palestine from the Neolithic to the Bronze Age. *World Archaeology* **11/3**: 331–341.

Mithen, S. J. (2003) *After the Ice: a global human history, 20,000–5000 BC*. London: Weidenfeld and Nicolson.

Nadel, D. and I. Hershkovitz (1991) New subsistence data and human remains from the earliest Levantine Epi-palaeolithic. *Current Anthropology* **32**: 631–635.

Nadel, D. and E. Werker (1999) The oldest ever brush hut plant remains from Ohalo II, Jordan Valley, Israel (19,000 BP). *Antiquity* **73**: 755–764.

Nesbitt, M. (2002) When and where did domesticated cereals first occur in southwest Asia? In *The Dawn of Farming in the Near East*, ed. R. T. J. Cappers and S. Bottema. Berlin: ex Oriente pp. 113–132.

Netzer, E. (1992) Domestic architecture in the Iron Age. In *The Architecture of Ancient Israel from the Prehistoric to the Persian Periods*, ed. Aharon, R. Reich and R. Kempinki. Jerusalem: Biblical Archaeology Society pp. 193–201.

Nishiaki, Y. and S. Kadowaki (2009) The PPNB water well at Tell Seker al-Aheimar, Upper Khabur, Northeast Syria. *Conference on Interpreting the Late Neolithic of Mesopotamia*, Leiden, the Netherlands, 26–28 March 2009.

Noy, T. (1989) Gilgal I, a Pre-Pottery NeolithicA site, Israel. The 1985–1987 seasons. *Paleorient* **15**: 11–18.

Oates, D. and J. Oates (1976) Early irrigation agriculture in Mesopotamia. In *Problems in Economic and Social Archaeology*, ed. G. Sieveking, I. H. Longworth and K. E. Wilson. London: Duckworth pp. 109–135.

Oleson, J. P. (2001) Water supply in Jordan through the ages. In *The Archaeology of Jordan*, ed. B. MacDonald, R. Adams and P. Bienkowski. Sheffield: Sheffield University Press pp. 603–624.

Oleson, J. P. (2007) Nabatean water supply, irrigation and agriculture: an overview. In *The World of the Nabataeans*, ed. K. D. Poltis. Stuttgart: Franz Steiner Verlag pp. 217–249.

Oleson, J. P., K. Amr, R. Schick and R. Foote (1995) Preliminary report of the Humeima Excavation Project, 1993. *Annual of the American Schools of Oriental Research* **39**: 317–354.

Ortloff, C. R. (2005) The water supply and distribution system of the Nabatean city of Petra (Jordan), 300 BC–AD 300. *Cambridge Archaeological Journal* **15**: 93–109.

Palumbo, G. (2008) The Early Bronze IV. In *Jordan: An Archaeological Reader*, ed. R. Adams. London: Equinox pp. 227–262.

Peltenberg, E. (2003) Conclusions: Mylouthkia 1 and the early colonists of Cyprus. In *The Colonisation and Settlement of Cyprus. Investigations at Kissonerga-Mylouthkia 1976–1996*, ed. E. Peltenberg. Savedalen: Astroms pp. 83–99.

Peltenberg, E. (2004) Introduction: a revised Cypriot prehistory and some implications for the study of the Neolithic. In *Neolithic Revolution. New Perspectives on Southwest Asia in Light of Recent Discoveries on Cyprus*, ed. E. Peltenberg and A. Wasse. Oxford: Oxbow Books.

Peltenberg, E., S. Colledge, P. Croft *et al.* (2000) Agro-pastoralist colonization of Cyprus in the 10th millennium BP: initial assessments. *Antiquity* **74**: 844–853.

Perrot, J. (1957) Les fouilles d'Abou Matar pres de Beersheba. *Syria* **34**: 1–38.

Philip, G. (2008) The Early Bronze Age I–III. In *Jordan: An Archaeological Reader*, ed. R. Adams. London: Equinox pp. 161–226.

Prag, K. (2007) Water strategies in the Iktanu region of Jordan. In *Studies in the History and Archaeology of Jordan IX*, ed. F. al-Khraysheh. Amman, Jordan: Department of Antiquities pp. 405–412.

Pritchard, J. B. (1985) *Tell es-Sa'aidiyeh, Excavations on the Tell, 1964–1966*. Philadelphia: University Museum.

Rast, W. E. and R. T. Schaub (1974) Survey of the southeastern plain of the Dead Sea. *Annual of the Department of Antiquities of Jordan* **19**: 5–53.

Rast, W. E. and R. T. Schaub (2003) *Bab edh-Dhra': Excavations at the Town Site (1975–1981)*. Winona Lake, IN: Eisenbrauns.

Reich, R. and E. Shukron (1999) Light at the end of the tunnel. *Biblical Archaeology Review* **25/1**: 22–33.

Reich, R. and E. Shukron (2003) Notes on the Gezer water system. *Palestine Exploration Quarterly* **135/1**: 22–29.

Rollefson, G. (1993) Neolithic chipped stone technology at 'Ain Ghazal, Jordan: the status of the PPNC. *Paleorient* **16**: 119–124.

Rollefson, G. (2001) The Neolithic period. In *The Archaeology of Jordan*, ed. B. MacDonald, R. Adams and P. Bienkowski. Sheffield: Sheffield University Press pp. 67–105.

Rollefson, G. and I. Kohler-Rollefson (1993) PPNC adaptations in the first half of the 6th millennium B.C. *Paleorient* **19**: 33–42.

Rollefson, G. and I. Köhler-Rollefson (1989) The collapse of early Neolithic settlements in the southern Levant. In *People and Culture in Change: Proceedings of the Second Symposium on Upper Palaeolithic, Mesolithic and Neolithic Populations of Europe and the Mediterranean Basin*, ed. I. Hershkovitz. Oxford: Oxbow pp. 73–89.

Rosen, A. (1987) Phytolith studies at Shiqmim. In *Shiqmim I: Studies Concerning Chalcolithic Societies in the Northern Negev Desert, Israel (1982–1984)*. ed. T. E. Levy. Oxford: Biblical Archaeology Reviews International Series pp. 243–249.

Rosen, A. (2003) Paleoenvironments of the Levant. In *Near Eastern Archaeology: A Reader*, ed. S. Richard. Winona Lake, IN: Eisenbrauns pp. 10–16.

Shaub, T. (2007) Mud-brick town walls in the EBI-II southern Levant and their significance for understanding the formation of new social institutions. In *Studies in the History and Archaeology of Jordan IX*, ed. F. al-Khraysheh. Amman: Department of Antiquities pp. 247–252.

Sheratt, A. (1980) Water, soil and seasonality in early cereal cultivation. *World Archaeology* **11**: 313–330.

Sheratt, A. (2003) The Mediterranean economy: 'globalization' at the end of the second millennium B.C.E. In *Symbiosis, Symbolism, and the Power of the Past. Canaan, Ancient Israel, and their Neighbors from the Late Bronze Age through Roman Palaestina*, ed. W. Dever and S. Gitin. Winona Lake, IN: Eisenbrauns pp. 37–62.

Shiloh, Y. and Z. Herzog (1992) Underground water systems in the land of Israel in the Iron Age. In *The Architecture of Ancient Israel from the Prehistoric to the Persian Periods*, ed. A. Kempinki and R. Reich. Jerusalem: Biblical Archaeology Society pp. 275–293.

Simmons, A. J. and M. Najjar (1996) Current investigations at Ghwair I, a Neolithic settlement in southern Jordan. *Neolithics* **2**: 6–7.

Simmons, A. J., G. Rollefson, Z. Kafafai and K. Moyer (1989) Test excavations at Wadi Shu'eib, a major Neolithic settlement in Central Jordan. *Annual of the Department of Antiquities of Jordan* **33**: 27–42.

Strange, J. (2008) *The Late Bronze Age in Jordan: An Archaeological Reader*. London: Equinox pp. 281–310.

Tanaka, J. (1976) Subsistence ecology of the central Kalahari San. In *Kalahari Hunter-Gatherers. Studies of the !Kung San and Their Neighbours*, ed. R. Lee and I. DeVore. Cambridge, MA: Harvard University Press pp. 98–119.

Tsuk, T. and Z. Herzog (1992) The water system of Tel Gerisa (Israel) and its contribution to the dating of underground water systems. In *Geschichte der Wasserwirtschaft und des Wasserbaus im Mediterranen*. Raum: Braunschweig pp. 333–356.

Tubb, J. N. (1988) Tell es-Sa'idiyeh: preliminary report on the first three seasons of renewed excavations. *Levant* **20**: 23–88.

Tubb, J. N., P. G. Dorell and F. Cobbing (1996) Interim report on the eighth (1995) season of excavations at Tell es-Sa'idiyeh. *Palestine Exploration Quarterly* **128**: 16–40.

Tufnell, O., ed. (1953) *Lachish III (Tell ed-Duweir): The Iron Age*. London, New York, Toronto: Oxford University Press.

Ussishkin, D. (1982) *The Conquest of Lachish by Sennacherib*. Tel Aviv: Tel Aviv University.

van den Brink, E. C. M., N. Lipschitz, D. Laza and G. Dorani (2001) Chalcolithic dwelling remains, cup marks and olive (*Olea europaea*) stones at Nevallat. *Israel Exploration Journal* **51/1**: 36–43.

Van der Kooij (2007) Irrigation systems at Dayr Alla. In *Studies in the History and Archaeology of Jordan IX*, ed. F. al-Khraysheh. Amman, Jordan: Department of Antiquities pp. 133–144.

Waheeb, M. and N. Fino (1997) Ayn elJammam: a Neolithic site near Ras e-Naqb, southern Jordan. In *The Prehistory of Jordan II, Perspectives from 1997*, ed. H. G. K. Gebel, Z. Kafafai and G. Rollefson. Berlin: ex Oriente pp. 215–220.

Wasse, A. and G. Rollefson (2005) The Wadi Sirhan Project: report on the 2002 archaeological reconnaisance of Wadi Hudruj and Jabal Tharwa, Jordan. *Levant* **37**: 1–20.

Weinberger, R., A. Sneh and E. Shalev (2008) Hydrogeological insights in antiquity as indicated by Canaanite and Israelite water systems. *Journal of Archaeological Science* **35**: 3035–3042.

West, J. B., ed. (1985) *Best and Taylor's Physiological Basis of Medical Practice*, 11th Edition. Baltimore: Williams and Wilkins.

Wilkinson, T. J. (2003) *Archaeological Landscapes of the Near East*. Tucson: University of Arizona Press.

Wright, G. R. H. (1985) *Ancient Building in South Syria and Palestine*. Netherlands: Brill Academic Publishers.

Yadin, Y. (1969) The fifth season of excavations at Hazor 1968–1969. *Biblical Archaeology* **32**: 50–71.

Yadin, Y. (1972) *Hazor*. London: Oxford University Press.

Zayadine, F., J. B. Humbert and M. Najjar (1989) The 1988 excavations on the Citadel of Amman, Lower Terrace, Area A. *Annual of the Department of Antiquities of Jordan* **33**: 357–363.

15 From global climate change to local impact in Wadi Faynan, southern Jordan: ten millennia of human settlement in its hydrological context

Sam Smith, Andrew Wade, Emily Black, David Brayshaw, Claire Rambeau and Steven Mithen

ABSTRACT

Wadi Faynan, southern Jordan, provides an archaeological record of human settlement from the Lower Palaeolithic to the Islamic period, and indeed into the present day. As for any long-term record of settlement, an understanding of the changes in economy and society requires knowledge about the impacts of climate and environment change on human communities, especially when dealing with settlement in arid landscapes. This chapter attempts to place the 10,000 years of Holocene settlement in Wadi Faynan between c. 12,000 and 2,000 years ago into its hydrological context. A rainfall-runoff model is used to examine the potential impacts of both Holocene climatic change and human behaviour on the hydrological behaviour of the wadi and then on human settlement. Wade et al. (this volume, Chapter 12) have shown that rainfall-runoff models can successfully simulate the behaviour of the present-day wadi system, demonstrating how such behaviour is sensitive to variability in rainfall and infiltration rates. Here we use the results of regional climate modelling to determine statistical properties of palaeo-rainfall for the Wadi Faynan and then use a stochastic weather generator (this volume, Chapter 5) to create a rainfall series which is used to drive the hydrological model. Results are used to explore the potential impacts of climatic variability on human communities from 12,000 to 2,000 years ago, demonstrating that palaeohydrology may provide a bridge between regional-scale climate data and local-scale cultural developments.

15.1 QUESTIONS OF SCALE AND UNCERTAINTY IN HUMAN–CLIMATE INTERACTIONS

The relationships between climate and society are too often considered in rather broad terms, using regional (or even global) records of climatic variability in an attempt to understand cultural developments at a local scale (e.g. Byrd, 2005). In the Levant, where climatic gradients are steep and major changes occur over very short distances, such regional approaches are unable to provide the detail needed for local-scale analyses of the archaeological record. Often this leads to generalised conclusions where postulated increases in rainfall are equated to 'good' climatic conditions, which allow communities to flourish, whilst reductions in rainfall are seen to provide more challenging or simply bad conditions for society.

In contrast to such coarse-scale approaches, our study considers how global changes in climate may have influenced local weather patterns and how these in turn would have impacted upon the hydrology and human communities of our chosen study area, Wadi Faynan. To this end we use a series of hydrological and climatological models to develop a series of new palaeoenvironmental scenarios for this wadi, building on the previous research concerning the past environments of Wadi Faynan. Our new scenarios include not only estimates of local palaeo-rainfall but also estimates of palaeohydrological conditions aiming to provide a more nuanced understanding of the impacts of climatic variability upon the Holocene communities of the region.

The methodology used in this study employs a hydrological model developed for the Wadi Faynan (this volume, Chapter 12), driven with estimates of the amount and timing of palaeo-rainfall derived from a weather generator (this volume, Chapter 5) which in turn is driven with precipitation statistics from the regional circulation model (RCM) (this volume, Chapter 3).

Water, Life and Civilisation: Climate, Environment and Society in the Jordan Valley, ed. Steven Mithen and Emily Black. Published by Cambridge University Press. © Steven Mithen and Emily Black 2011.

In order to simulate past infiltration rates, which have been shown to have a significant impact upon hydrology, we use regional and local reconstructions of vegetation (e.g. Hunt et al., 2007). Using this methodology we develop a series of time-slice scenarios (12, 10, 8, 6, 4, 2 ka BP and the Present Day) for the Holocene evolution of the Faynan hydrological system. As we explain in section 15.3.1 below, our 12 ka BP time slice is thought to be more applicable to the earliest Holocene conditions at 11.6 ka BP rather than the condition of the Younger Dryas. We are not attempting to quantify past wadi flows exactly. Rather the purpose is to provide a best estimate of how the hydrology functions, to develop a series of sensitivity scenarios exploring how changes in rainfall and ground surface properties affect water availability and to examine the potential impact of such changes upon the Holocene communities of Wadi Faynan.

Whilst the use of GCM data is becoming a key tool is assessing palaeoclimates (e.g. COHMAP members, 1988; Braconnot et al., 2007) and the application of hydrological modelling is increasingly recognised as essential to understand past water availability (e.g. Eadie and Oleson, 1986; Whitehead et al., 2008; Crook, 2009) it is rare for both simulation techniques to be integrated within a single archaeological study. From an archaeological perspective, this chaining of RCM output, with a weather generator, to a rainfall-runoff model represents a methodological advance in the reconstruction of the hydrological context of past communities. As such, this chapter has the joint aims of presenting a new methodology which has general applicability for archaeological studies and to make a substantive contribution to the understanding of long-term change in the human settlement of Wadi Faynan.

15.2 WADI FAYNAN

Chapter 12 of this volume has described the location, geology and present-day hydrology of the Wadi Faynan system, whilst Palmer et al. (2007) have described the wadi's contemporary settlement (see also Chapter 27, this volume). Here we need to supplement those descriptions by reviewing the present-day availability of water in the wadi system, considering how the different sources of water provide both challenges and opportunities to its present-day inhabitants. In addition, we also need to review the present-day vegetation of the Faynan region, as this plays a significant role in the hydrological system.

15.2.1 Present-day water availability

Water is available in Wadi Faynan from a variety of sources, including rainfall, groundwater flow and wadi floods. The present-day inhabitants also use imported water which is trucked in to the area and boreholes which tap deep groundwater resources. These sources have their own characteristics with regard to reliability, manageability, quality and predictability. Each source provides different types of opportunities and challenges for its exploitation. Although Wadi Faynan itself currently receives an average annual rainfall of between 100 and 150 mm year^{-1} (Raikes, 1967; Hunt et al., 2007), its large catchment (c. 250 km^2) acts to considerably amplify this amount, augmenting rainfall through wadi floods and groundwater flow. Figure 15.1 shows the extent of this catchment together with present-day precipitation isohyets, indicating that much of the catchment receives considerably more rainfall than the Wadi Faynan itself. In this way the catchment acts to integrate, or amplify, the amount of water available in the low-rainfall lower reaches of Wadi Faynan.

The scant rainfall in the Wadi Faynan itself, most of which falls in winter storms, is considerably below that needed for dry farming (c. 250 mm yr^{-1}) and creates semi-arid conditions in much of the wadi system. As is the case in many semi-arid regions of Jordan (e.g. Lancaster and Lancaster, 1991) the low annual averages for rainfall are exacerbated by high levels of interannual variability and a considerable spatial patchiness, which combine to make rainfall a highly unpredictable resource in Wadi Faynan.

In light of such rainfall, the groundwater flow is an essential resource for those living in the wadi today. As described in Chapter 12 of this volume, groundwater flow emanates from springs in the upper reaches of the Wadis Ghuwayr and Dana where it issues from underground aquifers. Although the amount of water provided by this source is relatively small, measurements suggest flow rates between 0.01 and 0.09 m^3 s^{-1} (see this volume Chapter 12, Table 12.1). The supply is perennial and the storage of water in aquifers acts to buffer this resource against the vagaries of rainfall. The quality of groundwater is high and this source is currently used for a range of purposes including drinking and irrigation. An advantage of the groundwater flow is its ease of management because of its low and predictable flow rates. As such, it can be diverted and stored by using simple interventions of a type that are unlikely to leave any archaeological trace. For example, Figure 15.2 shows a Bedouin dam constructed in 2007 from wadi cobbles and sandbags. This was used to divert groundwater into plastic pipes for transport to the lower wadi where it was stored in reservoirs and then mainly used to irrigate fields (Figure 15.3).

Under present conditions, the greatest volume of water arrives in Wadi Faynan in the form of large wadi floods, which occur exclusively during winter months. Such floods are the direct and rapid result of rainfall in the upper catchment and occur several times each year, with the average year receiving around five floods in excess of 12 m^3 s^{-1} (see this volume, Chapter 12). Despite the vast quantities of water supplied by flooding, the utility of this resource is limited because of the unpredictability in timing of flood events, their short duration and the sheer physical force of the flow. Today, no attempt is made to manage or directly use flood flows. But the legacy of flooding, in terms of water trapped in wadi gravels and silts, is an essential source of

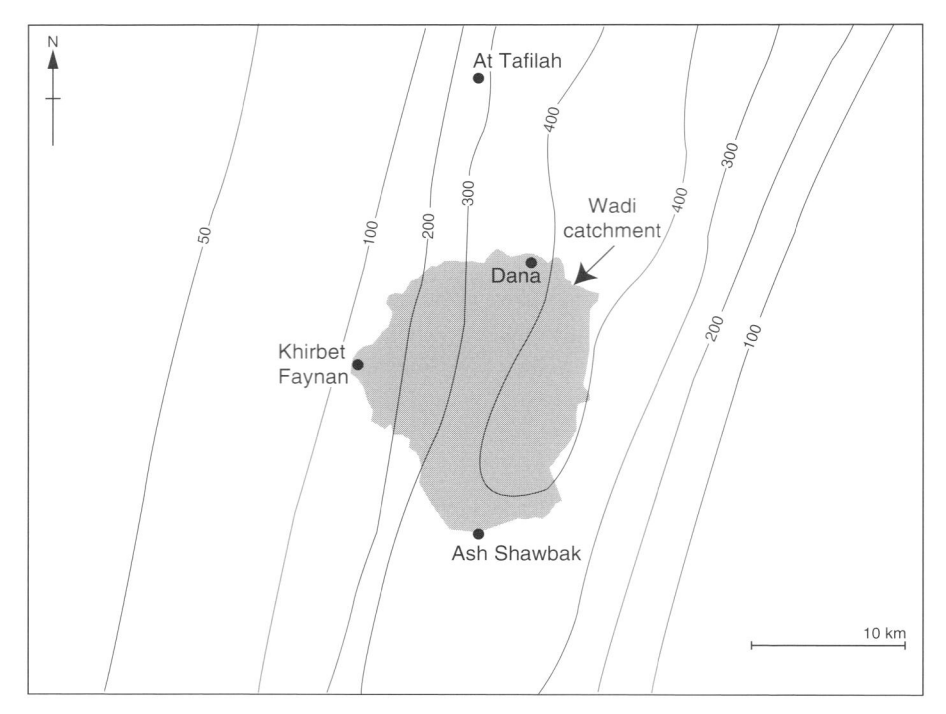

Figure 15.1. Location of Wadi Faynan catchment in relation to Khirbet Faynan and present-day precipitation isohyets (mm yr^{-1}).

Figure 15.2. Contemporary Bedouin dam in Wadi Ghuwayr, used to help channel water into plastic pipes for transport to fields to west.

Figure 15.3. Contemporary Bedouin reservoir in Wadi Faynan. The water here is drawn from groundwater flow in the Wadi Ghuwayr using plastic pipes (as shown in Figure 15.2).

moisture. This encourages plant growth which provides an important source of grazing for goats, and also allows opportunistic cultivation of flooded soils (Lancaster and Lancaster, 1999). With regard to their physical force, the floods pose a threat to any activity which might be taking place in the wadi channel. For example, small-scale dams like that illustrated in Figure 15.2 are destroyed every year by wadi flooding. In addition to the large floods of the main Wadi Faynan channel, smaller flows in tributary wadis and runoff from local slopes provide a further source of water. As these flows are derived from smaller catchments located in lower rainfall areas, they have smaller, more manageable flows than those of the main channel. However, the fact that these floods are dependent upon the timing and intensity of the sparse and unreliable local rainfall, as opposed to the more predictable rainfall of the upper catchment, the tributary wadi floods are, like the rainfall, an unpredictable water resource.

To summarise, although the relative lack of rainfall in the lower reaches of the Wadi Faynan is offset by the size and amplifier effect of the large Faynan catchment, wadi floods make a small contribution to the direct water supply for local inhabitants. The relatively small but easily managed groundwater flow provides the key water resource for modern-day inhabitants.

15.2.2 Present-day vegetation

The extent and nature of vegetation cover, along with other ground surface properties, has been shown to be an important component of semi-arid hydrological systems, exerting a strong impact on infiltration and runoff rates (e.g. Yair and Kossovsky, 2002). The present-day vegetation of the Wadi Faynan catchment, which in general becomes more dense with increasing altitude and rainfall (Figure 15.4), is a consequence of complex interactions between many factors including climate, topography, geology and human agency.

Present-day vegetation in the lower reaches of the Wadi system, below the Khirbet Faynan, is sparse, consisting mainly of desertic shrubs punctuated with occasional *acacia* and *tamarix*. This is primarily a result of low annual rainfall; denser vegetation exists along wadi channels (Palmer *et al.*, 2007). In contrast, the highest regions of the Faynan catchment, above *c.* 1,100 m altitude, receive significantly higher rainfall (see Figure 15.1) and vegetation is consequently more diverse and can be characterised as of degraded Mediterranean type. Significant tree cover is present, including *Quercus*, *Phoenix* and *Cupressus*. In between these two extremes, vegetation is characterised as degraded, wooded steppe including occasional trees but also featuring frequent areas of bare rock and rocky soils (Palmer *et al.*, 2007). The sparse and degraded vegetation which covers much of the catchment is in large part responsible for the low infiltration rates (~ 8 mm day^{-1}) which characterise the present-day hydrological system. The Holocene evolution of the vegetation cover is discussed below (section 15.3.3).

15.2.3 Overview of the Holocene archaeology of Wadi Faynan

Wadi Faynan has a rich archaeological heritage, covering much of the Pleistocene and Holocene periods (Barker *et al.*, 2007b; Finlayson and Mithen, 2007b; Hauptmann, 2007). The following section will provide a brief review of the Holocene archaeology (Figure 15.5), highlighting the significance of understanding water availability for its interpretation.

The early Holocene (11.6 ka BP to 8 ka BP), which broadly equates with the Pre-Pottery Neolithic (PPN) period, was a time of tremendous social and economic change throughout the Levant. This included the origins of agricultural economies and the development and expansion of villages and towns, along with concurrent changes in technology, ideology and social organisation (Bar-Yosef and Belfer-Cohen, 1989; Cauvin, 1994; Kuijt and Goring-Morris, 2002). Most explanations of the nature and timing of the 'neolithisation' process include a climatic component (e.g. Mithen, 2003; Byrd, 2005) and much recent thinking has explicitly emphasised the significance of water management during the Neolithic period. Such studies stress the importance of the domestication of water resources, alongside the domestication of other facets of the natural world such as plants and animals (Gebel, 2004; Garfinkel *et al.*, 2006; Fujii, in prep; this volume, Chapter 14).

In Wadi Faynan, this process of 'neolithisation' is best represented by the PPNA (*c.* 11,600–10,500 BP) site of WF16 (Finlayson and Mithen, 2007a), the PPNB (10,500–8,700 BP) village of Ghuwayr 1 (Simmons and Najjar, 1998) and the Pottery Neolithic (8,300–7,500 BP) occupation at Tell Wadi Feinan (Al-Najjar *et al.*, 1990) (Figure 15.6). Early Neolithic settlement is also present in the upper reaches of the wadi system at Wadi Badda (Fujii, 2007). Although the precise nature of the PPNA economy has yet to be established, and there is no reliable evidence of domesticated plants or animals at this time, the increase in plant processing tools, such as grinding stones and sickles, together with possible food storage facilities or silos suggest that PPNA populations were developing an ever more intensive relationship with plants and that intensive cultivation of crop species was under way (Kuijt and Finlayson, 2009). By the PPNB, this process has led to a fully fledged agricultural economy based upon food production, underpinned by the cultivation of domesticated crops and management of domestic herds (Rollefson, 2008). Many researchers have argued that population levels grew rapidly during this period (e.g. Kuijt, 2000),

Figure 15.4. Schematic representation of present-day vegetation cover in Wadi Faynan catchment (after Palmer *et al.*, 2007, figure 2.11).

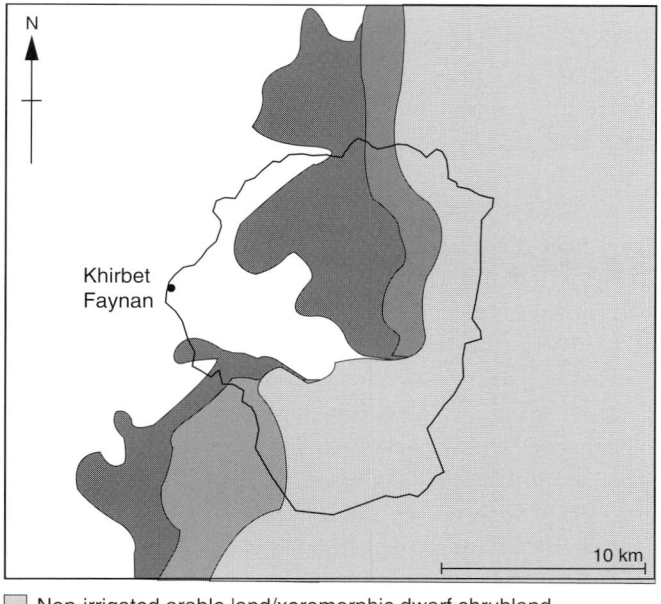

Legend:
- Non-irrigated arable land/xeromorphic dwarf shrubland
- Deciduous broad leaved woodland
- Evergreen broad/needle leaved woodland
- Desert and xeromorphic dwarf shrubland

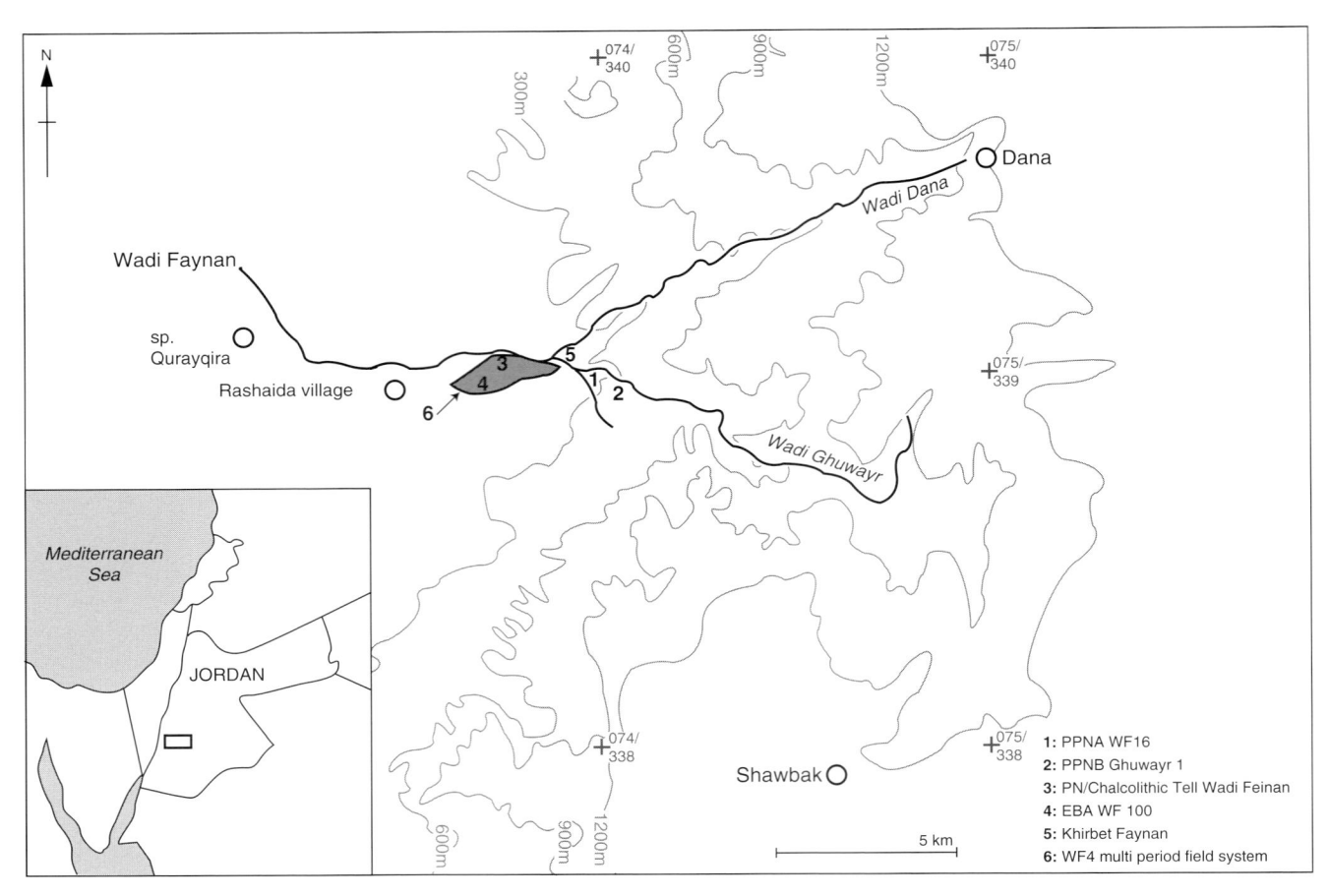

Figure 15.5. Map of main archaeological sites discussed in this chapter in relation to modern settlements and principal wadi systems.

supported by the intensive exploitation of plant and animal resources. Underlying this process, most researchers argue that the relatively wet conditions of the early Holocene allowed sedentary communities to obtain increasingly reliable and predictable yields from their crops (e.g. Mithen, 2003; Byrd, 2005).

Following the occupation of Wadi Faynan by Neolithic communities, human activity began to focus more intensively upon exploitation of the local copper resources, with concurrent developments in water management systems. The Chalcolithic and Bronze Age (c. 7,000–3,200 BP) are periods of major societal transformations in the Levant. In Wadi Faynan these periods mark the beginning of the intensive exploitation of the region's copper resources (Hauptmann, 2007). Evidence for Chalcolithic occupation is relatively sparse in the Wadi Faynan, although there is evidence for Chalcolithic activity at Tell Wadi Feinan (Al-Najjar et al., 1990). In addition, many Chalcolithic lithic scatters have been identified during landscape surveys, while there is also evidence for Chalcolithic period metallurgical activity in the area (Barker et al., 2007a; Mithen et al., 2007a). During the Early Bronze Age (from ~ 5,600 BP) there is evidence for an 'explosion in population' (Barker et al., 2007a, p. 236) in Wadi Faynan. A substantial settlement appears, known as WF100, covering an area of c.11 ha and which may have had a population in excess of 2,500 people (Barker et al., 2007a). In addition, there is evidence for a range of smaller settlement, smelting and ritual sites scattered around the Wadi system, as well as increasing use of the hinterland by pastoral groups (Barker et al., 2007a).

In terms of water management, there are significant developments during the EBA at both the local and regional scales (see this volume, Chapters 14, 17 and 18). In Wadi Faynan the archaeological record of the EBA shows the first definitive evidence of water management in the forms of 'cisterns' for water storage, as well as check dams, deflection walls, field systems and terraces used for the purpose of floodwater farming (Barker et al., 2007a). There are at least two water harvesting systems (e.g. WF148 and WF24) which channel runoff into large storage cisterns (Barker et al., 2007a). There are also many examples of cross-wadi walls (e.g. WF1628 and WF409), all located on small tributary wadis, which would have acted to deflect runoff along with eroded soils onto field systems (Figure 15.7) (Barker et al., 2007a).

The intensive occupation of the wadi during the EBA appears to have been short-lived and there are limited traces of occupation between 4 ka BP and 3 ka BP (Barker et al., 2007a). The apparent absence of Middle and Later Bronze Age occupation is not unique to Faynan, this mirroring a regional pattern of declining settlement (Rosen, 2007). Whilst the reasons for the collapse of EBA social

Figure 15.6. PPN sites in Wadi Faynan. (a) PPNA WF16; (b) PPNB Ghuwayr 1.

Figure 15.8. WF4 field-system looking northwest.

and economic systems are undoubtedly complex and involve a range of social and natural factors, there is good evidence that climatic and hydrological variability may have played a key role (Arz *et al.*, 2006; Barker *et al.*, 2007a; Rosen, 2007).

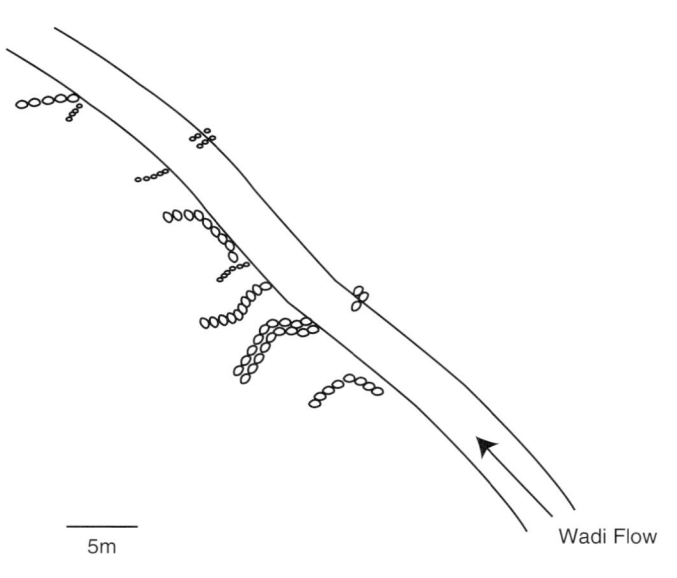

Figure 15.7. Schematic representation of EBA water harvesting system in Wadi Faynan (WF 1628) showing several cross-wadi walls and check dams built to deflect surface runoff onto surrounding landscape (after Barker *et al.*, 2007b, figure 8.24).

In the subsequent Iron Age and Nabataean periods, human activity again begins to intensify and there is evidence of substantial communities exploiting the local copper resources (Hauptmann, 2007; Mattingly *et al.*, 2007b). A range of water management facilities were developed at this time, which reached their culmination during Byzantine and Roman periods (Mattingly *et al.*, 2007a). At this period, roughly equivalent to our 2 ka BP scenario, the landscape of Faynan became dominated by mines and smelting sites, under the central control of Khirbet Faynan (WF1), believed to be the site of Phaino within classical texts (Hauptmann, 2007; Mattingly *et al.*, 2007b). Following this peak in human activity and occupation, archaeological evidence suggests a decline in industrial activity during the Islamic and Ottoman periods during which the region became mainly occupied by pastoral groups, a tradition which has continued up to the present day (Newson *et al.*, 2007b).

Given the large population sizes (~1,000 people) and intensive smelting activity in the Wadi Faynan during the Classical period, supplying food and water for domestic and industrial purposes must have posed a major challenge (Mattingly *et al.*, 2007a). It is not surprising, therefore, that we see the advanced systems of water management in the Faynan during this period. These include the construction of aqueducts and channels to capture groundwater flows from the Wadi Ghuwayr, the establishment of a large reservoir, a barrage adjacent to Khirbet Faynan, a water mill and a series of elaborately designed irrigated field-systems including the large WF4 system (Figure 15.8) (Hauptmann, 2007; Mattingly *et al.*, 2007a; Mattingly *et al.*, 2007b; Newson *et al.*, 2007a; Newson *et al.*, 2007b). Several of these systems are likely to have originated in the Iron Age, Nabataean or even earlier periods, and became incorporated within a complex and sophisticated system of water management during the Classical period (Mattingly *et al.*, 2007a). One aim of the present study is to provide a hydrological context within which to examine the impressive water management features which arose during the Classical period. Although these have already been intensively studied, there remains considerable uncertainty about how they had functioned (see section 15.5, and Barker *et al.*, 2007c; Crook, 2009).

Even from this very brief review of the archaeology of the region, it is clear that there has been a persistent presence of human activity within Wadi Faynan throughout much of the Holocene. Of particular interest, in the present context, is how human communities of various sizes and levels of social and economic development attempted to manage the slender water resources of this semi-arid location. In this context, section 15.5 re-examines the archaeological evidence from Wadi Faynan in light of our hydrological simulations, providing a range of hydrological scenarios which explore how water availability may have acted as the backdrop to human achievement in this locality.

15.3 DEVELOPING INTEGRATIVE METHODOLOGIES

15.3.1 The hydrological model

The set-up and calibration of the hydrological model used in this chapter is described in detail in Chapter 12 of this volume. In brief, the Pitman model was used because of its applicability to semi-arid situations and the balance it provides between model complexity, data requirements and useful model output. The Pitman model was applied to the Faynan catchment, defined from a point on the channel network just below the Khirbet Faynan (Figure 15.1). The main input parameters of relevance were rainfall and infiltration, and varying these allowed the behaviour of the Wadi Faynan system to be modelled under a range of palaeoenvironmental scenarios. As demonstrated in Chapter 12, variability in infiltration rates can have a major impact upon the hydrology of the wadi system, with increased infiltration acting to reduce runoff. Increased infiltration also increases the potential for percolation into aquifers in the upper reaches of the catchment, leading to increased groundwater flow. It should be noted that in the present study we do not explore the effect of temperature (which is held steady in all simulations) nor do we simulate evapotranspiration. Model outputs, important for this study, include the size and frequency of flood flows (at a daily time-step) together with simulations of groundwater flow and groundwater storage in aquifer(s). Together, these outputs allow examination of the sensitivity of the Faynan system to changes in infiltration and rainfall which are explored in a series of time-slice experiments.

The time slices chosen for analysis are 12, 10, 8, 6, 4, 2 ka BP and the Present Day. The Present Day simulation provides a control against which to compare other time slices, allowing us to explore the Holocene evolution of the Wadi Faynan system. It should be noted that the 12 ka BP simulation is problematic as our GCM modelling did not attempt to simulate Younger Dryas conditions, which are widely acknowledged to have been relatively cold and dry in the region (Bar-Matthews *et al.*, 2003; Robinson *et al.*, 2006; but see also Stein *et al.*, 2010). The Younger Dryas is believed to have been caused by an influx of meltwater into the North Atlantic which reduced Atlantic meridional overturning circulation (Carlson *et al.*, 2007). Because this process is not incorporated in the climate simulations underlying the present study, the 12 ka BP slice should not be regarded as representative of the climate during the Younger Dryas. In this way the precise attribution of simulations to specific time periods must be treated with some caution. In the case of the 12 ka BP simulation, this may in fact provide a more accurate reconstruction of early Holocene climate after the termination of the Younger Dryas at $c.$ 11.5 ka BP.

15.3.2 Simulated palaeo-rainfall

We drive the hydrological model with estimates of palaeo-rainfall derived from the RCM experiments described in Chapter 3 of this volume. However, driving the model using raw daily precipitation output from the RCM is not acceptable because of the large errors associated with these data (see this volume, Chapter 4). To account for these biases the weather generator – a statistical rainfall model – described in Chapter 5 of this volume was used. This takes as input (i) the probabilities of rain given rain the day before and rain given no rain the day before (the rainfall probabilities), and (ii) the distribution of rain per rainy day (the rainfall distribution). These inputs are then used to generate 100-year daily rainfall time-series using the statistical rainfall model.

We use rainfall at the modern town of Tafilah (Figure 15.1) as representative of the catchment area, both RCM and observed data being available for this location. In this case, rainfall probabilities were derived from the appropriate 50×50 km^2 RCM grid square for each time-slice experiment, whilst we maintain the present-day Tafilah rainfall distribution for all model runs. This simple methodology allows us to overcome the biggest bias in the RCM – the actual amount of rainfall – whilst utilising the data on the frequency of rainfall events, about which we have more confidence. It should be noted, however, that because rainfall distributions for past time slices may be different from that observed today, the rainfall data derived within our model may underestimate the degree of change from the present-day situation.

Table 15.1 *Summary of synthetic Tafilah rainfall series for time slices (results of 100-year simulation)*

	12 ka	10 ka	8 ka	6 ka	4 ka	2 ka	0 ka control
Annual mean (mm yr^{-1})	342	361	318	308	342	365	288
Standard deviation (mm yr^{-1})	86	117	94	100	98	100	79
Median (mm yr^{-1})	332	352	318	300	324	350	284
Max. (mm yr^{-1})	538	614	579	578	680	647	532
Min. (mm yr^{-1})	167	101	77	106	82	186	137
Percent of control	118	125	110	107	118	126	100
Annual mean at Khirbet Faynan (mm yr^{-1})	142	150	132	128	142	151	120

This downscaling method is applied to each of the time-slice climate model simulations to produce a synthetic 100-year daily time series of precipitation (Table 15.1 and Figure 15.9). As discussed in Chapter 3 of this volume, changes in precipitation between adjacent time slices are generally small when compared with the interannual variability within an individual time slice. As such, the time-evolution of Holocene precipitation at Tafilah should be considered as an indicator of likely general trends rather than a deterministic reconstruction (further discussion is provided in Chapter 3). A final point to note is that subtle problems with the climate model configuration, as discussed in Chapter 3, this volume, may over-emphasise the simulated changes between the 6 ka BP and 8 ka BP time slices, but it is not believed that this will significantly affect the general patterns discussed in this chapter.

According to our methodology, and consistent with the discussion in Chapter 3, the mean annual rainfall at Tafilah has been higher throughout the Holocene than at the present day. The wettest periods are the early Holocene (near 10 ka BP) and the late Holocene (near 2 ka BP) which have approximately 125% of the present-day annual mean rainfall. The lowest rainfall occurs between these two periods (6 ka BP) which has just a seven per cent increase on present-day values (as noted above, the peculiarity of any particular time slice should be interpreted with caution given the modelling framework used). It is interesting to note, however, that the evolution of the precipitation at this particular site is slightly different from the 'East Mediterranean' box (Figure 3.13 of this volume). This indicates that this site (at the southeasternmost part of that box) may be experiencing slightly different meteorological conditions, perhaps connected with Red Sea Trough activity (Ashbel, 1938) rather than the behaviour dominated by Mediterranean storm tracks as seen over the eastern Mediterranean coastline and discussed extensively in Chapter 3.

Today, Wadi Faynan (around the Khirbet Faynan) receives 40% of the rainfall of Tafilah. Because this relationship is a consequence of the interaction between storm tracks and the topography of the region, it seems valid to assume that this relationship would have persisted throughout the Holocene.

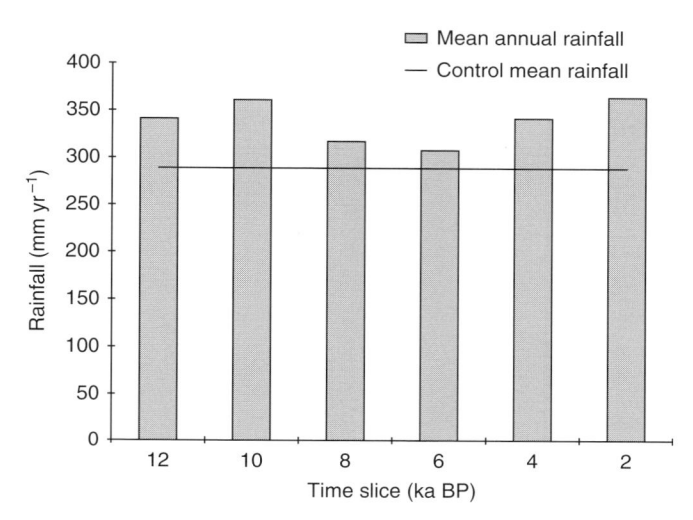

Figure 15.9. Mean annual rainfall at Tafilah for each time slice against 0 ka (control) mean.

Table 15.1 also shows the mean annual rainfall at Khirbet Faynan (based on 40% of that at Tafilah) for the time slices, indicating that maximum Holocene rainfall at Khirbet Faynan (at 10 ka BP and 2 ka BP) would have been in the region of 150 mm yr^{-1}.

In addition to producing estimates of rainfall, our methodology allows us to consider the timing and seasonal cycle of rainfall events. Figure 15.10 and Table 15.2 provide a summary of the monthly rainfall averages for the six time slices together with the control period. This shows that the present-day pattern of wet winters and dry summers is retained in all experiments. Despite the retention of this broad pattern, the data suggest that there may have been some variability in the timing of winter rains during the Holocene. For example, comparison of the 10 and 8 ka BP simulations shows that in the 10 ka BP simulation most rain fell in the early winter (November–January) whilst in the 8 ka BP simulation the majority of rain fell during the later months (February–March) of the winter. However, further climate simulations and a more detailed analysis of the data would be required to assess the robustness and significance of these results.

Table 15.2 *Synthetic monthly mean rainfall (mm per month) at Tafileh for each time slice (mean of 100-year total)*

Month	12 ka	10 ka	8 ka	6 ka	4 ka	2 ka	0 ka control
January	60.46	68.48	63.93	48.04	51.53	75.58	61.18
February	72.03	47.02	74.61	53.21	59.15	57.55	54.31
March	67.89	60.42	63.87	65.70	53.27	58.52	49.47
April	35.19	40.43	36.12	22.75	35.94	31.03	20.54
May	5.04	5.72	7.00	4.46	7.40	7.81	3.44
June	2.15	2.63	4.38	2.84	2.20	2.97	1.67
July	4.39	3.51	1.26	2.46	4.23	2.42	3.70
August	3.30	4.04	3.38	3.67	3.71	2.92	2.09
September	2.32	2.99	2.25	3.39	3.49	2.47	3.58
October	3.99	2.49	2.93	4.95	5.16	4.63	4.89
November	27.14	27.06	21.49	34.00	51.06	51.30	28.80
December	57.50	96.49	36.30	62.61	65.00	67.31	55.28
Total	**341.38**	**361.28**	**317.52**	**308.08**	**342.13**	**364.51**	**288.95**

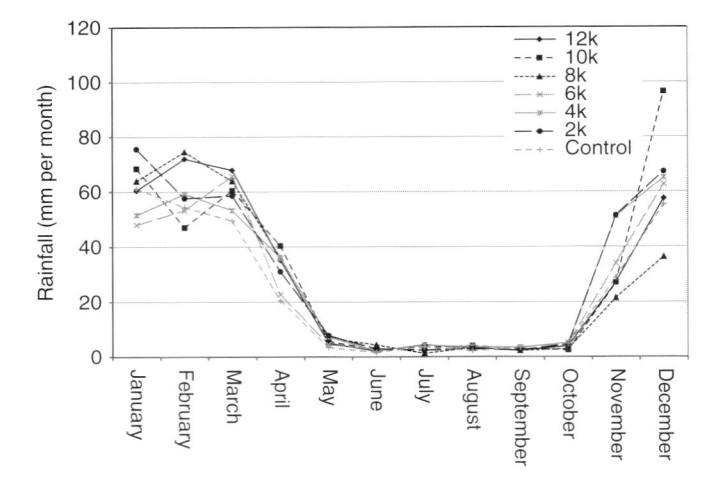

Figure 15.10. Mean monthly rainfall amounts for each time slice and control experiments.

15.3.3 Evaluating palaeo-rainfall estimates

Although using model data has several advantages over using proxy evidence, it is imperative that the potential accuracy of the data is assessed by comparison with proxy evidence. Chapters 3 and 8 of this volume have evaluated the RCM data in this way whilst Chapter 5 has provided a detailed discussion of the weather generator methodology. In this section we will evaluate the estimates of palaeo-rainfall derived from our simulations against the backdrop of the general changes in Holocene rainfall proposed in Chapter 3 and also against published attempts to derive palaeo-rainfall estimates from proxy records in the southern Levant, especially for the Wadi Faynan region.

Reconstructing palaeo-rainfall is a challenging task as most proxy records do not reflect rainfall amount alone, being a consequence of the interaction between a range of climatic variables. In the southern Levant the most thorough attempt to provide precise estimates of palaeo-rainfall derive from an interpretation of the high-resolution $\delta^{18}O$ isotopic sequence from speleothems at Soreq Cave, located near Jerusalem (Bar-Matthews *et al.*, 1997, 2003), in which variation in $\delta^{18}O$ was interpreted mainly in terms of variation in rainfall amount. It should be noted, however, that several authors have challenged the interpretation of the Soreq sequence, arguing that variability in speleothem $\delta^{18}O$ is likely to also be a reflection of variability in $\delta^{18}O$ of Mediterranean source water (Frumkin *et al.*, 1999, 2000; Enzel *et al.*, 2008).

Despite this issue, the Soreq sequence remains one of the best estimates of palaeo-rainfall in the southern Levant and it is informative to examine the results of our present simulations in this light. Figure 15.11 provides a comparison of our data with estimates of palaeo-rainfall derived from Soreq Cave (Bar-Matthews *et al.*, 1997, 2003) whilst Table 15.3 also includes palaeo-rainfall estimates derived from pollen sequences recovered in Wadi Faynan (Hunt *et al.*, 2007). These sources show similar overall patterns of a wetter early Holocene declining unevenly until the present day, with possible rapid short-lived fluctuations. These estimates suggest that during the early Holocene (10–7 ka BP) rainfall may have reached almost double present levels at Soreq Cave (Bar-Matthews *et al.*, 1997). Minimal rainfall estimates suggest that at Soreq Cave, at around 5 ka BP, rainfall may have been only *c*. 40% of present-day amounts (Bar-Matthews *et al.*, 2003). At Wadi Faynan, rainfall before 8 ka BP is argued to have been in the region of 160% of present-day amounts, again declining unevenly from 8 ka BP to the present day (Hunt *et al.*, 2007).

Examination of Figure 15.11 illustrates some of the difficulties inherent in comparing model data with that from proxy records. One key factor is that the RCM simulations only provide a series of static time slices as a means to illustrate the general trend of Holocene climatic change. Because the simulations

Table 15.3 *Comparison of palaeo-rainfall estimates*

Estimates are from: this study; Soreq Cave 1 (Bar-Matthews *et al.*, 2003); Soreq Cave 2 (Bar-Matthews *et al.*, 1997) and Wadi Faynan (Hunt *et al.*, 2007). All dates in ka BP

	Bar-Matthews *et al.* (2003)		Bar-Matthews *et al.* (1997)		Hunt *et al.* (2007)		This chapter	
	Rainfall (mm yr^{-1})	% of change from present-day rainfall	Rainfall (mm yr^{-1})	% of change from present-day rainfall	Rainfall (mm yr^{-1})	% of change from present-day rainfall	Rainfall (mm yr^{-1})	% of change from present-day rainfall
Pre-12 ka	no data		between 15 and 12 ka: 550–750	110–150	no data	–	–	–
12 ka	no data		between 12 and 10 ka: 680–850	136–170	no data	–	341	118
10 ka	no data		between 10 and 7 ka: 675–950	135–190	pre-8 ka: 200	160	361	125
8 ka	550	110	between 10 and 7 ka: 675–950	135–190	*c.* 8 ka: 150 Similar until *c.* 7 ka	120	318	110
6 ka	480	96	between 7 and 1 ka: 450–580	90–116	no data	–	308	107
4 ka	480	96	between 7 and 1 ka: 450–580	90–116	no data	–	342	118
2 ka	400	80	between 7 and 1 ka: 450–580	90–116	at *c.* 2.5 ka: 100, possibly dropping to 70 between 2 and 0.3 ka. Minima at 0.3 ka of *c.* 30 mm yr^{-1}	25–80	365	126
0 ka	500	–	500	–	100–125	–	289	–

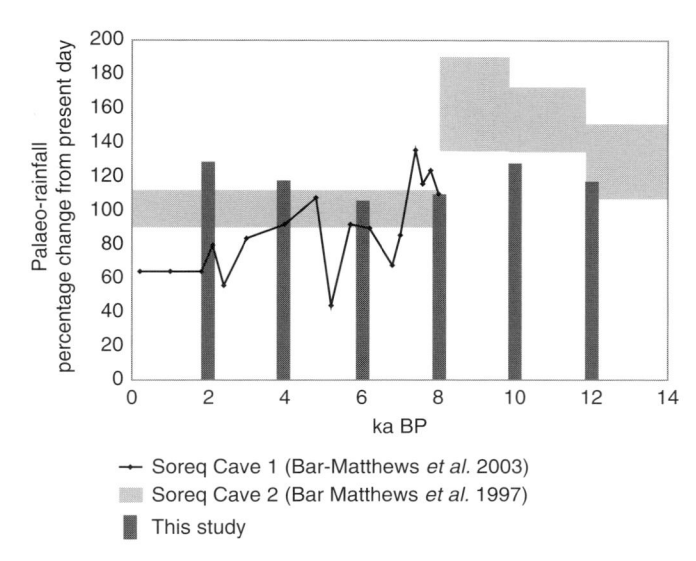

Figure 15.11. Comparison of palaeo-rainfall estimates used in this study with those derived from Soreq Cave sequence.

occur at 2000-year time steps they leave plenty of room for lengthy, even bi-millennial scale, variability. Given that most proxies (e.g. Bar-Matthews *et al.*, 2003) suggest that Holocene climatic factors, notably rainfall, were variable over relatively short timescales it is indeed likely that much Holocene variability is being missed by our time-slice approach. For example, the Soreq record suggests a pronounced dry spell at *c.* 5.2 ka BP (Figure 15.11) which is clearly not captured by our time-slice experiments.

Despite this difficulty, it is possible to compare the general trend in Holocene rainfall derived from our simulations and the proxy evidence. As shown in Figure 15.11, the general picture of a relatively wet early Holocene is a feature of both RCM and proxy data. However, Figure 15.11 also shows that the estimates of rainfall produced in the simulation are at the lower end of the estimates provided by Bar-Matthews *et al.* (1997). Moving to the mid-Holocene (8–4 ka BP) we can compare our results with the higher-resolution estimates provided by Bar-Matthews *et al.* (2003). This shows that at 8 ka BP our estimates of a 20% increase (compared with present-day values) is in line with estimates from Soreq and that the general pattern of drying by 6 ka BP followed by slight increase in rainfall by 4 ka BP are common to both studies.

The modelled evolution of precipitation in the late Holocene is rather more difficult to reconcile with Bar-Matthews *et al.* (2003): the simulation suggests this period is consistently wetter than the present day, whereas the Bar-Matthews *et al.* record suggests it would have been relatively dry. This discrepancy between model and Soreq data is further illustrated by comparison of estimates at 2 ka BP, for which we estimate an increase in rainfall of *c.* 25% (compared with present day), whilst Bar-Matthews *et al.* (1997) suggest a *c.* 20% decrease in rainfall. It is difficult to resolve this issue, which may suggest that we have overestimated rainfall at 2 ka BP. This conclusion is supported by comparison with data from Wadi Faynan (Hunt *et al.*, 2007), which suggests that 2 ka may have been very dry and by the fact that this figure seems anomalous compared to other grid squares in our RCM simulation. However, it should be noted that there is a wet spike in the Soreq data at this time (albeit from very dry baseline) and that other studies have suggested that 2 ka BP may have been a very wet period in the southern Levant (Bookman *et al.*, 2004), perhaps receiving twice present-day rainfall amounts (Yakir *et al.*, 1994; Isaar and Yakir, 1997).

Seasonality of rainfall is often even more difficult to interpret from proxy evidence, limiting the potential to provide a comparison for our model. However, a few authors have suggested that there are indications of summer rainfall during the early Holocene, possibly as a result of the influence of monsoonal systems (Simmons and Najjar, 1997; Rossignol-Strick, 1999). Whilst this conclusion is not supported by our simulation and seems unlikely from climate models in general (Rodwell and Hoskins, 2001), summer rainfall cannot be completely ruled out by modelling studies of this type (see Chapter 3 of this volume for a discussion of this issue).

This brief comparison has served to illustrate the complexity of comparing simulated and proxy data, highlighting some of the differences between the two types of evidence along with their respective shortcomings. Despite these issues, the above comparisons suggest that the broad trends indicated by our data are not contradicted by the proxy evidence, although it seems likely that our data may underestimate early Holocene rainfall whilst overestimating absolute rainfall amounts for the later Holocene.

15.3.4 Simulation of palaeo-infiltration

The second variable with which we drive the hydrological model is infiltration. In order to simulate present-day conditions in Wadi Faynan, infiltration rate was set at 8 mm day^{-1} through calibration of the Pitman model (Chapter 12). This is an estimate of the mean infiltration rate over the entire catchment of the Wadi Faynan system; in reality the catchment is a mosaic of different ground surface types and vegetation coverages with varying infiltration rates. Moreover, due to the mismatch in scale between the duration of a storm-event, typically of the order of 3–4 hours, and the daily model time-step, it is justifiable to consider the simulated infiltration to relate to the duration of the storm rather than the day.

Several factors have a strong influence on infiltration rates. These include: vegetation cover (including type and density), soil properties (especially the presence of surface crusts), antecedent soil moisture together with general surface features such as roughness, slope and stoniness (Cerda, 1997; Leonard and Andrieux, 1998). As such, a knowledge of changes in Holocene ground surface cover is essential for simulating the behaviour of the Wadi system. Several lines of evidence suggest that there has been substantial change in the distribution of vegetation and soil cover during the Holocene in the southern Levant and in Wadi Faynan in particular (e.g. Roberts and Wright Jr, 1993; Tchernov, 1994; Rossignol-Strick, 1999; Roberts, 2002; Cordova, 2007; el-Rishi *et al.*, 2007; Hunt *et al.*, 2007; Palmer *et al.*, 2007). Regional evidence, largely based upon palynology, indicates that early Holocene woodland was more widespread than at present, while a reduction in forest cover began during the mid-Holocene as precipitation levels decreased and human impacts, such as grazing and tree felling, increased (e.g. Rossignol-Strick, 1999; Roberts, 2002; Rosen, 2007).

Evidence of past vegetation from Wadi Faynan generally mirrors the regional picture of abundant early Holocene woodland, which declines from the middle Holocene onwards. The most complete studies of changes in Holocene vegetation patterns in Wadi Faynan are provided by Hunt *et al.* (2007) and el-Rishi *et al.* (2007), although other studies (such as Mithen *et al.*, 2007a and Palmer *et al.*, 2007) provide additional information. Hunt *et al.* (2004, 2007), using pollen recovered from sediments within the wadi, split the Holocene record into a series of biozones, providing a summary of plant species present in each zone together with an interpretation of prevailing patterns of vegetation cover and estimations of palaeo-precipitation (Table 15.4).

Hunt *et al.* (2004) define two biozones dating to around, or earlier than, 8 ka BP (Table 15.4). Both these biozones suggest significantly more forest cover than is present in the Wadi Faynan today with high proportions of tree pollen including *Juniperus*, *Pinus* and *Ulmus* and also *Olea* and *Quercus*, the latter possibly growing adjacent to watercourses. Hunt *et al.* suggest that this environment is similar to that which currently occurs at altitudes of 1,000–1,100 m asl, implying that early Holocene Wadi Faynan was a densely vegetated steppe lying just outside the margin of the Mediterranean forest zone

Table 15.4 *Summary of results from Hunt* et al. *(2007) (table 12) showing changing patterns of Holocene vegetation in Wadi Faynan*

Biozone	POP	PPA	PAP	PCPJ	CLP	CPE	C	CL
Dates (uncal BP)	Pre-7,200	7,200–6,400	6,400–5,700	5,700–3,000	3,000–2,000	2,000–350	350–100	100–0
Dates cal BP (approx.)	Pre-8,000	~8,000–7,500	~7,500–6,500	~6,500–3,200	~3200–2000	~2,000–400	~400–150	~150–0
Analogue	Around villages of Dana and Tafilah	Around Dana village	Around Dana village	?	Present-day Wadi Faynan	Present-day Wadi Faynan	Around town of Ghor Safi	Present-day Wadi Faynan
Dominant environmental conditions	Steppic	Steppic with disturbance	Steppic with less disturbance	Steppic	Probably degraded	Very degraded	Extremely degraded steppe/desert	Degraded steppe
Estimated annual precipitation (mm yr^{-1})	~200	~150	~150	No data	~100	~70–100	~30–70	~100–125

(Hunt *et al.*, 2004, p. 926). During the middle Holocene the PAP biozone (between ~5.7 and 3 ka BP) provides evidence for a marked reduction in the amount and diversity of tree pollen alongside a reduction in desertic species such as *Chenopodiaceae* (Hunt *et al.*, 2004). Moving into the later Holocene, the CLP and CPE biozones (dated to between ~3 and 0.35 ka BP) provide further evidence of severe environmental degradation and soil erosion with high percentages of *Chenopodiaceae* and other desertic species such as *Ephedra* (CLP, *Chenopodiaceae–Lactuceae–Poaceae* assemblage biozone; CPE, *Chenopodiaceae–Pinus–Ephedra* assemblage biozone). Hunt *et al.* (2007) argue that the dominant conditions in Wadi Faynan at this time would be a degraded steppe. The final biozones proposed by Hunt *et al.* (2007) relate to the very recent past and indicate further removal of soil cover and an increased dominance of desert species including *Chenopodiaceae*.

To summarise, the work of Hunt and colleagues (Hunt *et al.*, 2004; el-Rishi *et al.*, 2007; Hunt *et al.*, 2007; Palmer *et al.*, 2007) provides good evidence for the development of Holocene vegetation patterns in Wadi Faynan. The early Holocene (until *c*. 6 ka BP) appears to have been markedly more wooded than at present, with conditions around Khirbet Faynan similar to those present today around Dana village. This suggests that the present-day distribution of biogeographic zones has seen an eastward (and upward) retreat of vegetation cover into the higher altitude, wetter areas of the Faynan catchment. Following *c*. 6 ka BP there appears to have been a relatively rapid decrease in both the amount and diversity of tree cover in the area, with a corresponding increase in indicators of desertic or degraded steppic conditions together with signs of soil erosion. Conditions appear to have shifted towards those of the present degraded steppe from around 3 ka BP with apparent peaks in aridity and vegetation degradation during medieval times. It is worth noting that the majority of species identified throughout the Holocene are still present in the region today, suggesting that the main changes in vegetation cover have been in the extent to which species were able to colonise the drier, lower altitude reaches of the catchment (el-Rishi *et al.*, 2007). Significantly, the denser, tree-rich vegetation postulated for the early Holocene is in agreement with analysis of charcoal remains from the early Holocene PPNA site of WF16 (Mithen *et al.*, 2007a).

Whilst the reduction in tree cover from the early Holocene is in line with regional trends (e.g. Rossignol-Strick, 1999; Roberts, 2002; Robinson *et al.*, 2006; Chapter 6, this volume), and is in part caused by increasing aridity, there is clear evidence that the reduction in the number of trees in Wadi Faynan was also partly a consequence of human action. There are at least two examples of massive, industrial scale removal of trees by humans. The first of these occurred at around 2 ka BP, between the Iron Age and the end of the Byzantine period, when millions of tonnes of wood would have been needed to support smelting activity in Wadi Faynan strongly suggesting that the inhabitants of Wadi Faynan must have ranged far afield, presumably to the forested upper reaches of the catchment, to collect wood (Mattingly *et al.*, 2007). A second major cause of deforestation, which would have particularly affected the upper regions of the catchment, was the cutting of upland forests for timber for the Hijaz railway, which had a Shawbak branch between 1908–1917. Timber was needed for both railway construction and fuel (Raikes, 1967; Palmer *et al.*, 2007).

A further line of evidence suggesting that recent major changes in soil cover and vegetation have occurred in the Holocene Wadi Faynan is provided by archaeological survey in the Wadi al Bustan (upper Faynan catchment) area by Mithen et al. (2007b). This survey found Lower and Middle Palaeolithic artefacts on extremely steep rocky slopes. These artefacts were very fresh in appearance, with no evidence for rolling or batter, suggesting the artefacts were largely in situ. Given the steep nature of the slopes where these were found, it seems unlikely that they had lain in situ for very long. Consequently, it is most plausible that these artefacts had recently been buried within sediments, providing further evidence of major soil erosion in the upper catchment in the relatively recent past.

As the pollen records examined by Hunt et al. (2004, 2007) do not encompass the earliest Holocene periods we need to draw on evidence from the wider region to provide estimates of vegetation cover for our 12 and 10 ka BP simulations. As discussed above (section 15.3.1), we interpret the 12 ka BP simulation as reflecting post-Younger Dryas conditions. In this context the evidence that vegetation responded relatively slowly to the dramatic climate change at the end of the Younger Dryas is significant (e.g. Roberts and Wright Jr, 1993; Roberts, 2002; Steffensen et al., 2008). This suggests that vegetation cover would have been relatively sparse in the immediate aftermath of the Younger Dryas period (represented by our 12 ka BP simulation) in Wadi Faynan. By 10 ka BP, however, more dense and wooded vegetation, rather like that attaining in Wadi Faynan at 8 ka BP, is likely to have been present in the region.

To summarise, it is evident that the ground surface cover of the Wadi Faynan system has undergone massive changes during the Holocene, with the early Holocene (pre 8 ka BP) being characterised by much more dense, diverse and tree-rich vegetation types which have retreated upslope as vegetation loss and erosion have taken place during the middle and late Holocene periods. From the present hydrological perspective, the postulated changes in ground surface cover are extremely significant and suggest the potential for major reductions in infiltration rates during the Holocene. This also suggests that the low infiltration rates which characterise the present-day hydrological system are a consequence of many millennia of declining rainfall and human induced removal of vegetation and soils. While we can state with confidence that early Holocene infiltration rates would have been considerably higher than present and that rapid declines in infiltration rates probably began in the middle Holocene, providing a precise estimate of infiltration rates is difficult. In this context we propose to take an extreme, and rather arbitrary, infiltration rate of 100 mm hr^{-1} as reflecting an upper limit on Holocene infiltration rates for the Wadi system. We will then explore a range of infiltration scenarios for each time slice to examine the potential impacts of changes in infiltration on Holocene water availability.

15.4 RESULTS: SCENARIOS FOR THE HOLOCENE DEVELOPMENT OF THE WADI FAYNAN SYSTEM

15.4.1 The impact of changes in Holocene rainfall

In this section we drive the hydrological model with 100-year simulated series of palaeo-rainfall described above (section 15.3.2). The results are provided in Table 15.5, summarising the wadi flows (which include both groundwater and flood flows) and the groundwater flow alone for each of our time slices. The results indicate that the impact of the proposed changes in rainfall is small, with the wadi system being characterised by small perennial flows (minimum flows), with occasional large floods (maximum flows) in all simulations. Indeed, the median flow levels are constant at 0.05 m^3 s^{-1}. Given the uncertainty regarding extreme flood events, the differences in maximum flow sizes should be treated with caution. Nevertheless, very large floods, with flows between 50 and 70 m^3 s^{-1}, occur at least once in each simulation. In terms of smaller, more typical flood flows, the data again suggest limited differences between the simulations, with the frequency of floods being strongly related to rainfall amount. Floods in excess of 12 m^3 s^{-1}, the equivalent of a typical winter flood in the present day, occur most frequently at 2 and 10 ka BP – the periods with the highest rainfall.

In terms of groundwater flow there is again limited difference between the time slices. Minimum groundwater flow levels for all palaeo-simulations show a slight increase from Present-Day levels, whilst median groundwater flow levels (0.05 m^3 s^{-1}) are the same in all simulations. However, the depth of stored groundwater (which is initially set at 6.5 m) increases (over the 100-year simulation period) in all simulations other than that of the Present Day. This suggests that groundwater is building up within the aquifer(s) and will lead flows to gradually increase over the course of the simulation. These data indicate that none of the rainfall scenarios make a significant difference to groundwater flow levels. They suggest that small perennial flows, of similar size to those of the present day, would have been present in the Wadi Faynan under all our Holocene rainfall scenarios.

Figure 15.12 provides a graphical summary of monthly flow rates (mean daily flow rates for each month), showing how monthly flows vary throughout the year. These data again show that all simulations are similar to the Present Day, and are characterised by large winter floods augmented by small perennial groundwater flows, which provide the only water during summer months. As in the present day, the wadi responds rapidly to rainfall events and therefore flood flows mirror rainfall timing and intensity (compare Figure 15.12 with Figure 15.10).

In summary, these data suggest that under the range of palaeo-rainfall scenarios proposed in section 15.3.2 the behaviour of the Wadi Faynan is similar to that of the present day. The rather

Table 15.5 *Summaries of simulated wadi flows under changing precipitation scenarios*
Flow rates, m^3 s^{-1}; rainfall, mm yr^{-1}. GW = groundwater

	12 ka	10 ka	8 ka	6 ka	4 ka	2 ka	0 ka control
Max. flow	60.13	50.57	54.62	67.25	55.24	66.33	66.19
Min. flow	0.05	0.05	0.05	0.05	0.05	0.05	0.04
Median flow	0.05	0.05	0.05	0.05	0.05	0.05	0.05
n floods over 12 m^3 s^{-1}	251	290	256	221	237	270	204
n floods over 6 m^3 s^{-1}	447	492	431	368	436	467	340
n floods over 1 m^3 s^{-1}	1,708	1,772	1,521	1,482	1,655	1,776	1,366
Max. GW flow	0.05	0.05	0.05	0.05	0.05	0.06	0.05
Min. GW flow	0.05	0.05	0.05	0.05	0.05	0.05	0.04
Median GW flow	0.05	0.05	0.05	0.05	0.05	0.05	0.05
Final day GW store	6.99	7.12	6.62	6.52	6.76	7.22	6.24
Percentage of present rainfall	119	125	111	107	119	127	100

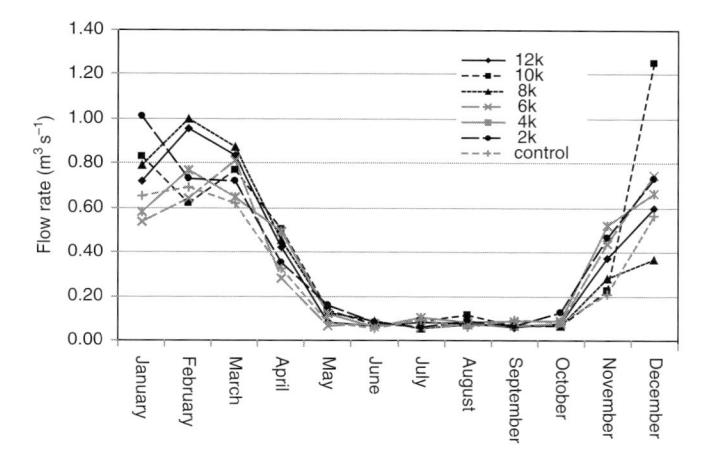

Figure 15.12. Summary of mean daily flow rates for each month of time-slice and control experiments. Infiltration rate is 8 mm per day for all time slices.

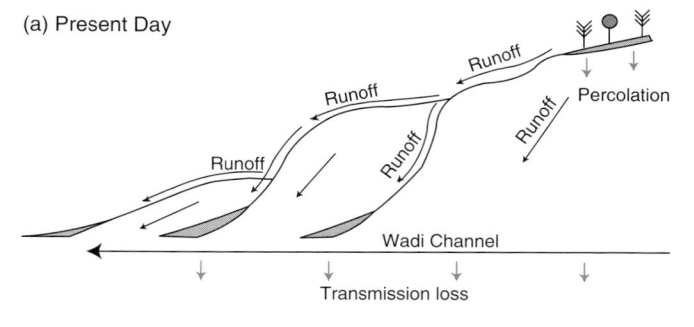

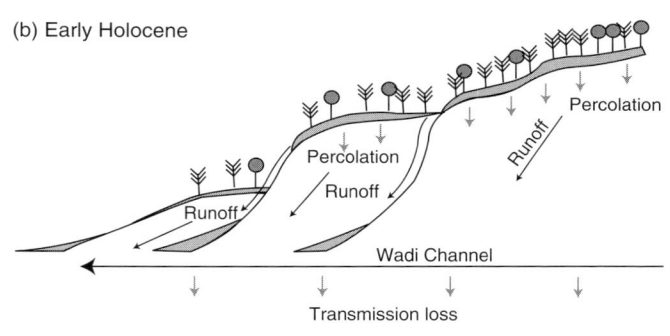

Figure 15.13. Schematic diagram of Wadi Faynan showing the potential impact of changes in vegetation on hydrological processes. (a) Present day. Infiltration is limited to areas of colluvium, small areas of vegetation and soil cover in upper catchment (coloured grey), and the (saturated) wadi channel. The majority of the wadi comprises bare, rocky slopes and gravel terraces with very low infiltration rates, which induce high runoff. (b) Early Holocene. Dense vegetation cover and increased soil cover in the wadi system increases areas of infiltration (coloured grey). This increases potential percolation in upper catchment and reduces surface runoff generation.

small changes in the annual amount of rainfall make limited difference to the behaviour of the system, which is in all cases characterised by large winter floods supplemented by a small yet stable and perennial groundwater flow.

15.4.2 The impact of changes in Holocene ground cover

In this section we build upon the above results and explore the impact of varying the infiltration rates over the catchment, in an attempt to simulate past changes in vegetation and soil cover as described in section 15.3.3. Figure 15.13 provides a schematic overview of the potential effects of major increases in vegetation and soil cover on the hydrological system in Wadi Faynan. In the first instance we explore the impact of increasing infiltration rates to 100 mm day^{-1} for all time slice simulations. This is estimated to be the maximum infiltration rate for this type of catchment (see this volume, Chapter 12). Following this, we provide a series of alternative scenarios in which we explore the impact of a range of infiltration rates. Specifically, we explore the effect of infiltration rates of 80 mm day^{-1} on the 6 ka BP scenarios, and infiltration rates of 35 mm day^{-1} on late

Table 15.6 *Summaries of simulated wadi flows under changing precipitation and infiltration scenarios*
Flow rates, m^3 s^{-1}; rainfall, mm yr^{-1}. GW = groundwater

Time slice	12 ka	10 ka	8 ka	6 ka	4 ka	2 ka	6 ka	4 ka	2 ka
Infiltration (mm day^{-1})	100	100	100	100	100	100	80	35	35
Max. flow	6.64	6.26	5.00	12.31	6.91	11.3	18.69	31.23	43.82
Min. flow	0.05	0.05	0.05	0.05	0.05	0.05	0.05	0.05	0.05
Median flow	0.07	0.08	0.07	0.07	0.07	0.07	0.07	0.07	0.07
n floods over 12 m^3 s^{-1}	0	0	0	2	0	0	2	22	38
n floods over 6 m^3 s^{-1}	4	2	0	4	2	4	4	72	70
n floods over 1 m^3 s^{-1}	45	84	59	46	54	60	90	262	293
Max. GW flow	0.09	0.09	0.08	0.07	0.08	0.09	0.07	0.08	0.09
Min. GW flow	0.05	0.05	0.05	0.05	0.05	0.05	0.05	0.05	0.05
Median GW flow	0.07	0.08	0.07	0.07	0.07	0.07	0.06	0.07	0.07
Final day GW store	9.71	10.19	9.18	8.80	9.31	10.22	8.76	8.91	9.71
Percentage of 0 ka (control) rainfall	119	125	111	107	119	127	107	119	127

Holocene (4 and 2 ka BP) simulations. The chosen infiltration rates are rather arbitrary, but in the absence of direct evidence for palaeoinfiltration we believe that our chosen rates reflect the maximum plausible range of Holocene variability, allowing us to assess the overall sensitivity of the Wadi Faynan system.

Table 15.6 summarises the results of the infiltration experiments, clearly showing that variation in the infiltration rate has the potential to cause a profound alteration in the behaviour of the system. This is most clearly seen in the 100 mm day^{-1} scenarios, although the effects are substantial even for the much lower infiltration rates of 35 mm day^{-1}.

Under the 100 mm day^{-1} infiltration scenarios, the size of maximum wadi flows is severely reduced in all time-slice simulations. As shown in Table 15.6 the maximum flow achieved under these conditions (at 6 ka BP) reaches only 12 m^3 s^{-1}. This is the size of a typical winter flood in the present day and is much smaller than the largest floods (67.25 m^3 s^{-1}) achieved at 6 ka BP under present-day (8 mm day^{-1}) infiltration conditions. Moreover, even relatively small floods (≥ 6 m^3 s^{-1}) are rare, and very small flood flows (≥ 1 m^3 s^{-1}) occur far less frequently than under present-day (low infiltration) scenarios. These data indicate that as infiltration is increased to this level, runoff generation within the catchment is severely reduced, leading to reductions in the size and number of flood flows. Higher infiltration acts to increase the amount of rainfall needed to initiate a flood flow and also reduces the size of flows as runoff levels are reduced. In this way the direct and rapid relationship between catchment rainfall and wadi floods, which characterises the present-day hydrology of the Faynan system, is weakened.

In contrast to maximum flows, minimum and median flows are generally either stable or increased under high infiltration scenarios. This is largely a consequence of increasing groundwater flow levels in these simulations and suggests that under high infiltration scenarios percolation into aquifers is increased,

Table 15.7 *Comparison of simulated median groundwater flows between first and last decade of simulations, for 12 ka scenarios (8 mm hr^{-1}, 100 mm hr^{-1}) and 0 ka (control) period*

		Years 1–10	Years 91–100
Time slice	Infiltration rate (mm day^{-1})	Median groundwater flow	Median groundwater flow
12 ka	8	0.047	0.052
12 ka	100	0.048	0.084
Control	8	0.047	0.044

leading to increased levels of groundwater flow. Table 15.6 illustrates this phenomenon and it is clear that the final level of stored groundwater increases significantly (from an initial level of 6.5 m) in all time slices under the 100 mm day^{-1} scenarios.

As groundwater recharge is a continuous process over the 100 years of the simulation, groundwater flow levels may vary throughout the simulation period. In the present examples groundwater flow increases throughout the simulation period, with the minimum groundwater flow rates (0.05 m^3 s^{-1}) occurring in the first few years of the simulation. Table 15.7 provides an example of this process, showing how groundwater flows develop between the first decade and last decade of the 100-year simulation period. This shows that while groundwater levels are fairly stable at both Present Day and 12 ka BP when infiltration is set to 8 mm day^{-1}, when infiltration rates are increased to 100 mm day^{-1} groundwater levels increase substantially during the course of the simulation as percolation into aquifers is increased. It should be noted that because we do not know the maximum size of the aquifer(s) which provide groundwater, it is not possible to determine the maximum possible groundwater flow for the system (see this volume, Chapter 12) and consequently the precise data should be treated with

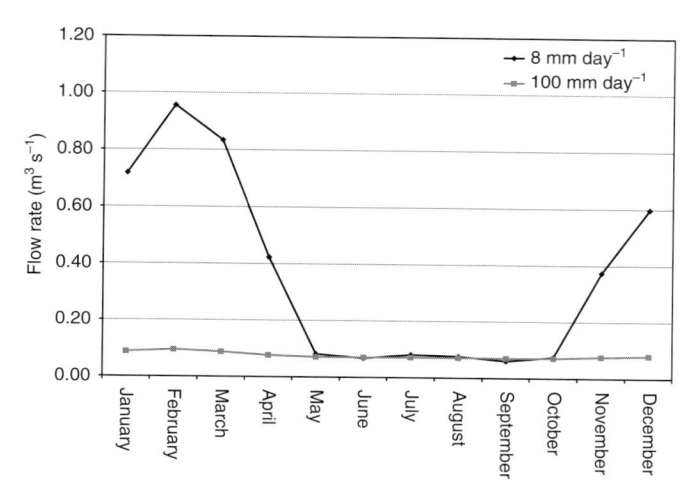

Figure 15.14. Comparison of monthly flows for 12 ka BP time slice under different infiltration scenarios.

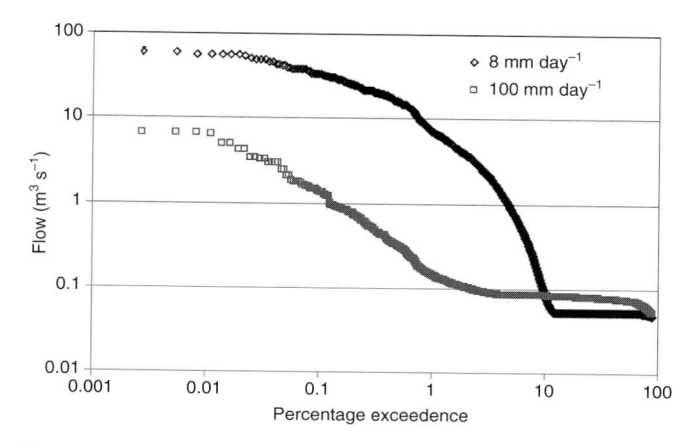

Figure 15.15. Comparison of flow duration curves for 12 ka BP simulation under high and low infiltration scenarios.

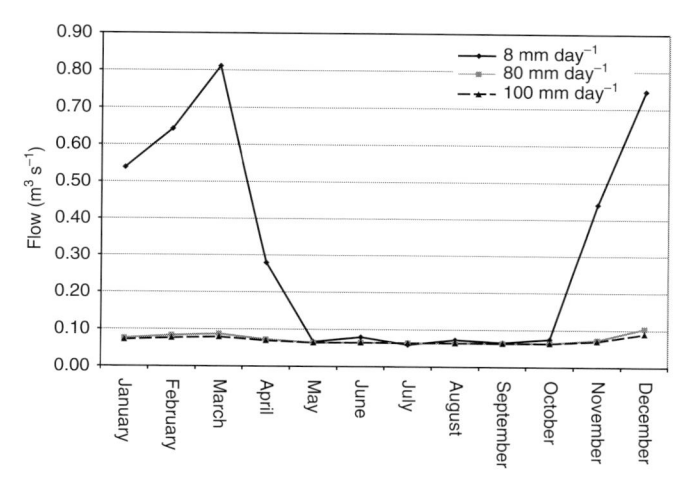

Figure 15.16. Comparison of monthly flows for 6 ka BP time slice under different infiltration scenarios.

caution. However, the broad picture of increased percolation under increased infiltration is a robust feature of our simulations.

It is easiest to examine the detail of the hydrological changes caused by increasing infiltration rates from 8 to 100 mm day^{-1} by considering a single time-slice simulation. Focusing upon the 12 ka BP simulations, (see Tables 15.5 and 15.6) we can see that maximum flow rates (winter floods) are reduced from 60 to 7 m s^{-1}. In terms of groundwater, storage increases from 6.99 m to 9.71 m with a consequent increase in maximum groundwater flow rates from 0.05 to 0.09 m s^{-1} (achieved at the end of the 100-year simulation). Such an increase in groundwater flow levels represents an increase of 144,000 litres per hour under 100 mm day^{-1} infiltration conditions. In terms of the annual distribution of water, Figure 15.14 shows that under the 100 mm day^{-1} infiltration scenario winter flooding is severely reduced, and the hydrological system is dominated by relatively stable perennial flow. Figure 15.15 provides a comparison of flow duration curves for the two scenarios, clearly illustrating that while flood magnitudes are reduced under the high infiltration scenario, this also creates an increase in groundwater flow during most of the year.

If we shift attention to the 6 ka BP simulations (Figure 15.16) we can see that the differences discussed above between the 100 mm day^{-1} and 8 mm day^{-1} scenarios are maintained. Moreover, this also shows that when infiltration rates are reduced to 80 mm day^{-1} there is little difference from the 100 mm day^{-1} scenario. Examining the 2 ka BP simulations (Figure 15.17) we can see that even when infiltration rates are reduced to 35 mm day^{-1} the wadi system is substantially less prone to flash floods than under 8 mm day^{-1} infiltration rates. In contrast to the limited changes caused by our Holocene rainfall scenarios, coupling variability in rainfall with variability in infiltration causes radical changes to the hydrology of the Faynan system. Under high infiltration scenarios (100 mm day^{-1}), the frequency and size of wadi floods are severely reduced whilst groundwater flow levels gradually increase over the course of the simulations. As there is uncertainty over the precise level of infiltration rates for each time slice, it is significant that our simulations of 80 mm day^{-1} and even 35 mm day^{-1} infiltration also cause profound changes to the hydrology of the system. This suggests that the present-day behaviour of the wadi system is largely a consequence of the very low contemporary infiltration rates and that even relatively small increases in infiltration will significantly modify wadi behaviour.

15.4.3 Scenarios for the Holocene palaeohydrology of Wadi Faynan

We now draw upon the simulations presented above and the vegetation history presented in section 15.3.3 to define a series of simulations which provide the most plausible infiltration and rainfall scenarios for the Holocene development of the Faynan

Table 15.8 *Infiltration rates and rainfall values used in proposed simulation of Holocene evolution of Wadi Faynan*

	12 ka	10 ka	8 ka	6 ka	4 ka	2 ka	Control
Infiltration rate (mm day^{-1})	8	100	100	80	35	8	8
Percentage of present rainfall	119	125	111	107	119	127	100

Table 15.9 *Summary of proposed wadi flows for Holocene Wadi Faynan. Flow rates, $m^3\ s^{-1}$; rainfall, $mm\ yr^{-1}$. GW = groundwater*

Time slice	12 ka	10 ka	8 ka	6 ka	4 ka	2 ka	Control
Infiltration rate	8	100	100	80	35	8	8
Max. flow	60.13	6.26	5.00	18.69	31.23	66.33	66.19
Min. flow	0.05	0.05	0.05	0.05	0.05	0.05	0.04
Median flow	0.05	0.08	0.07	0.07	0.07	0.05	0.05
n floods over 12	251	0	0	2	22	270	204
n floods over 6	447	2	0	4	72	467	340
n floods over 1	1708	84	59	90	262	1776	1366
Max. GW flow	0.05	0.09	0.08	0.07	0.08	0.06	0.05
Min. GW flow	0.05	0.05	0.05	0.05	0.05	0.05	0.04
Median GW flow	0.05	0.08	0.07	0.06	0.07	0.05	0.05
Final day GW store	6.99	10.19	9.18	8.76	8.91	7.22	6.24
Percentage of present rainfall	119	125	111	107	119	127	100

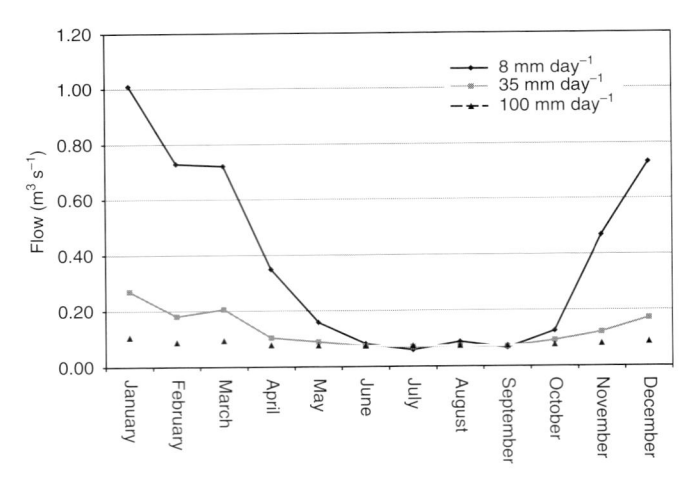

Figure 15.17. Comparison of monthly flows for 2 ka BP time slice under different infiltration scenarios.

system. Table 15.8 provides a summary of estimates for infiltration rates, together with rainfall scenarios, which are applied to each time slice. These estimates are based on the vegetation and erosional history of the Wadi Faynan outlined in section 15.3.2. Although considerable uncertainty surrounds the validity of the precise infiltration rates used for each time slice, as the present-day system is a consequence of very low infiltration, we believe that the broad pattern of Holocene hydrological development outlined below is robust.

A summary of the results for each time slice is presented in Table 15.9 and Figure 15.18. These suggest that there have been significant changes in the hydrology of the Wadi Faynan during the Holocene, with regard to the timing, size and number of floods as well as the amount of groundwater flow. Significantly, the radical changes in wadi hydrology are primarily a consequence of variability in infiltration rates rather than changes in rainfall. Figure 15.19 provides a summary of monthly flows alongside monthly rainfall amounts for each time slice, graphically illustrating the proposed changes in the hydrology of the wadi system. This shows that between 10 ka BP and 4 ka BP the wadi hydrology is radically different to that of today, with a steady base flow maintained throughout the year and minimal flood surging in winter. The situations at 2 and 12 ka BP are much more similar to that of the present day, with sparse soil and vegetation cover providing limited infiltration which causes the wadi to respond quickly to winter peaks in rainfall. The following section will briefly review the prevailing hydrological conditions for the early, middle and late Holocene periods.

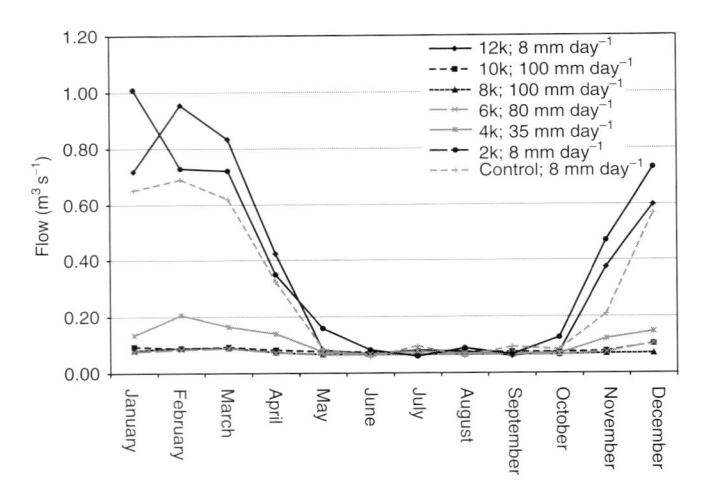

Figure 15.18. Comparison of monthly flows for each time slice and 0 ka BP (control) simulations under infiltration scenarios proposed for the Holocene. The key indicates time; infiltration rate.

EARLY HOLOCENE (12–8 KA BP)

Figure 15.20 shows the monthly flows for the early Holocene periods. It is clear that under the proposed scenarios there is a radical change of the wadi hydrology between the 12 and 10 ka BP simulations. At 12 ka BP, the wadi responds to rainfall in

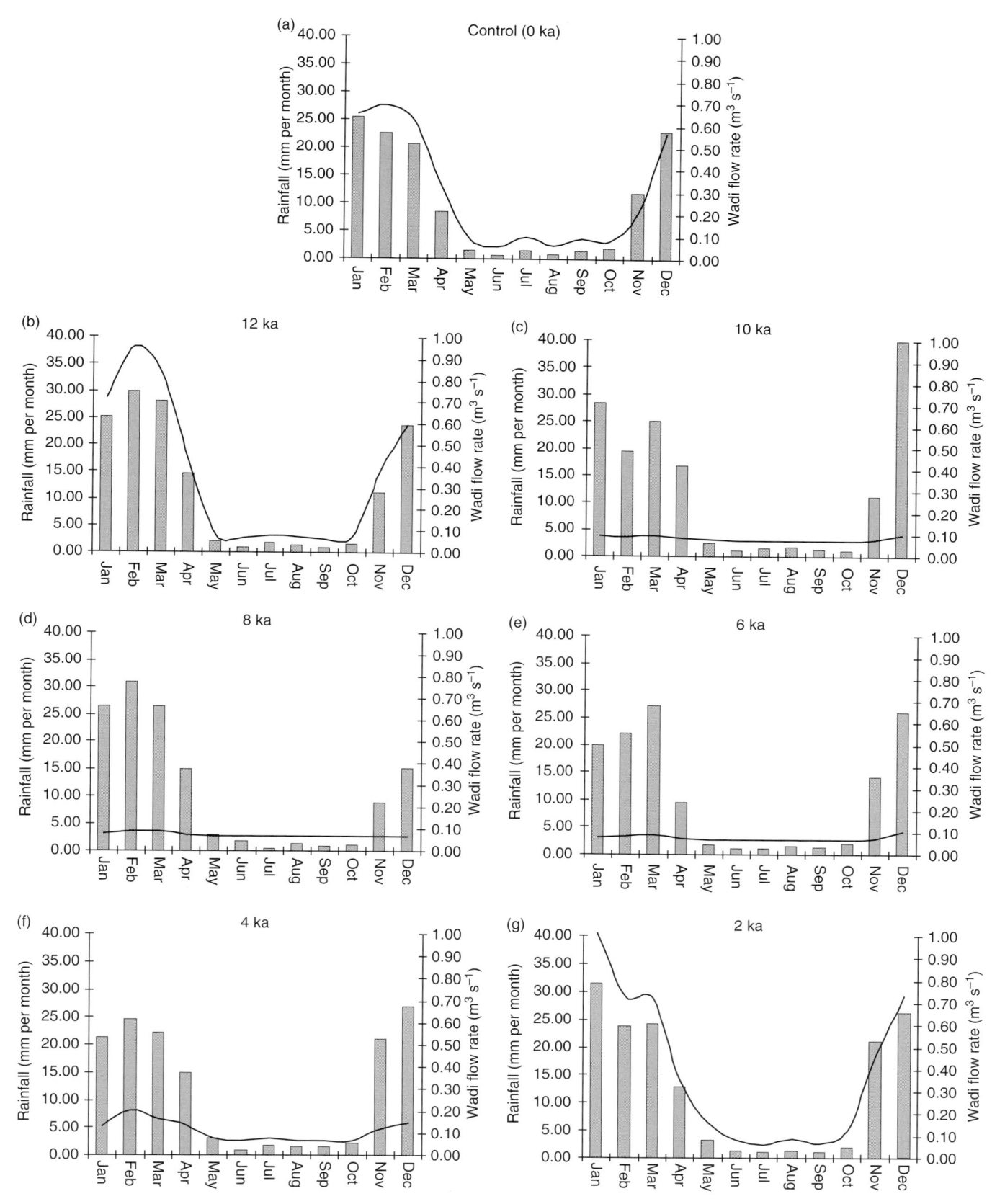

Figure 15.19. Palaeo-rainfall (bars) and palaeo-flows (line) for Holocene scenarios and 0 ka (control) simulations for Wadi Faynan.

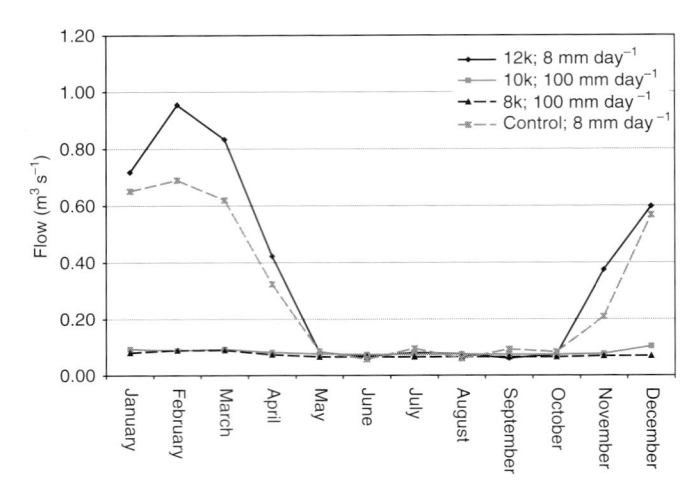

Figure 15.20. Summary of monthly flows for proposed early Holocene (12–8 ka BP) and 0 ka BP (control) simulations. The key indicates time; infiltration rate.

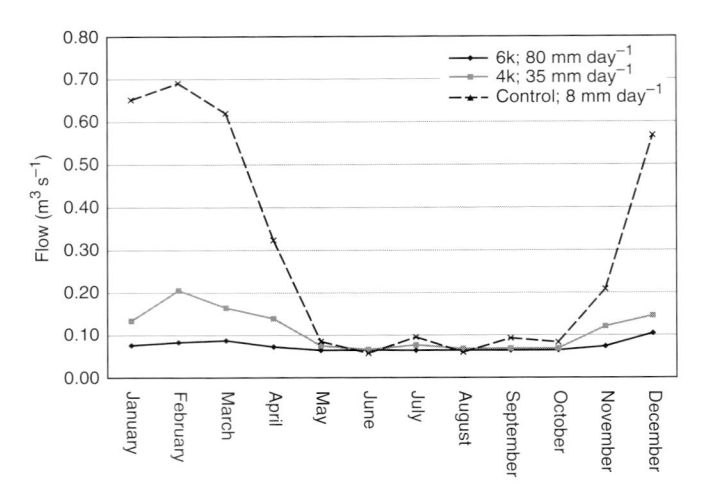

Figure 15.21. Summary of monthly flows for proposed mid-Holocene (6–4 ka BP) and 0 ka BP (control) simulations. The key indicates time; infiltration rate.

similar way to the present day. Interestingly, the maximum floods at this time are smaller than those of the Present Day control run, despite increased rainfall at this time. However, the number of floods over 12, 6 and 1 $m^3 s^{-1}$ is increased. Groundwater flow at this time is relatively stable at around 0.05 $m^3 s^{-1}$, and groundwater store increases only marginally throughout the simulation. Peak rainfall and peak flows occur in February with rainfall at Khirbet Faynan averaging 142 mm yr^{-1}.

In the subsequent 10 ka BP simulation the situation is very different. Even though rainfall is only slightly increased from the 12 ka BP simulation, the Khirbet Faynan now receiving c. 150 mm yr^{-1}, the wadi's response to rainfall is radically changed. Owing to the increase in vegetation and soil cover, the system now responds slowly to rainfall events, and is dominated by perennial groundwater flow. In contrast to the 12 ka BP simulation, groundwater increases rapidly during the simulation and reaches a maximum value of 0.09 $m^3 s^{-1}$ during the last years of the simulation. It is worth reiterating that although the differences between maximum groundwater flows between these simulations may sound trivial, 0.05 $m^3 s^{-1}$ at 12 ka BP and 0.09 $m^3 s^{-1}$ at 10 ka BP, this equates to an increase of more than 140,000 litres of water per hour. In terms of floods, the 10 ka BP simulation has a maximum flow of only 6 $m^3 s^{-1}$, and even very small floods of 1 $m^3 s^{-1}$ occur less than once a year on average. The 8 and 10 ka BP simulations are broadly similar in terms of wadi flows, although the reduction in rainfall at the Khirbet Faynan to 131 mm yr^{-1} at 8 ka BP means that both maximum and median baseflow levels are slightly reduced at this time.

MIDDLE HOLOCENE (6 KA BP)

Figure 15.21 provides a summary of monthly flows for this period, showing that under proposed infiltration rates the wadi becomes a slightly more 'flashy' system than in preceding periods. Although the reduction in infiltration at 6 ka BP has a limited effect on median flows and on groundwater flow, the system does now occasionally receive rather large floods, reaching 19 $m^3 s^{-1}$, and the frequency of smaller floods is also increased compared to 8 ka BP. It is also notable that the 6 ka BP simulation has the lowest rainfall of all the time-slice simulations with Khirbet Faynan receiving an average of only 128 mm yr^{-1}.

LATE HOLOCENE (4–2 KA BP)

Figure 15.22 provides a summary of monthly flows for this period, illustrating the transformation of the wadi from its relatively stable early and mid-Holocene state to the system of the present day, which is characterised by minimal base flow and severe winter flooding. At 4 ka BP, when infiltration rates are set to 35 mm day^{-1}, maximum floods are increased to 31 $m^3 s^{-1}$. In addition to the increase in maximum flows, we also see an increase in smaller flood events, with floods of 12 $m^3 s^{-1}$ (or larger) occurring 72 times during the 100-year simulation. Very small flows of 1 $m^3 s^{-1}$ also occur in increasing frequency, happening several times each year. In terms of groundwater flow, conditions are now actually rather better than at 6 ka BP and the maximum flow values of 0.08 $m^3 s^{-1}$ are similar to those prevailing at 8 ka BP with stored groundwater reaching a depth of 9 m by the end of the simulation. In essence, the reduction in infiltration rates (from 6 ka BP levels) is offset by the increase in rainfall and the 4 ka BP system appears transitional between the early and late Holocene states.

The wettest of the time-slice simulations occurs at 2 ka BP, when the Khirbet Faynan receives an annual mean rainfall of 151 mm yr^{-1}. This is also the time slice where we suggest that

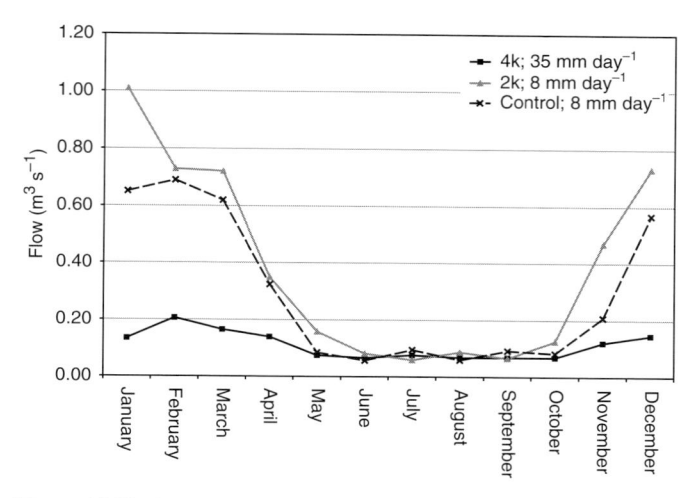

Figure 15.22. Summary of monthly flows for proposed late Holocene (4, 2 and 0 ka BP (control)) simulations. The key indicates time; infiltration rate.

infiltration rates would have declined to present-day levels, creating a wadi system very like that of the present day with a direct and rapid response to rainfall. In terms of wadi floods, maximum flows now reach 66 $m^3 s^{-1}$ whilst relatively large floods of 12 $m^3 s^{-1}$ occur several times per year. However, median flows are now reduced to present-day levels (0.05 $m^3 s^{-1}$) and maximum groundwater flow is only 0.06 $m^3 s^{-1}$. Stored groundwater reaches a level of 7 m, more than under control conditions as a result of increased rainfall, but substantially less than at any point since 12 ka BP.

SUMMARY

The simulation results suggest that the hydrology of the Wadi Faynan has altered radically during the Holocene, primarily as a result of changes in infiltration rates caused by alteration of ground surface properties. At the beginning of the Holocene (12 ka BP) we propose that the system was rather similar, albeit slightly wetter, to that of the present day, being dominated by rapid flood responses to rainfall events in the catchment. As vegetation and soils were re-established following the end of the Younger Dryas interval, our simulations suggest that the hydrology became significantly less 'flashy' and increasingly characterised by perennial groundwater flow and our data show that from *c.* 10 ka BP onwards groundwater levels were rapidly rising. This suggestion is supported by the geomorphological evidence for relatively large braided streams carrying perennial flow in the Wadi Faynan (McLaren *et al.*, 2004).

During the middle Holocene (6 ka BP) we postulate a slight reduction in infiltration rates, coupled with, and in part caused by, a slight decrease in rainfall. This appears to have a limited effect on the hydrology of the system, which retains much of its early Holocene character. In the later Holocene we suggest that infiltration rates would have decreased rapidly, primarily as a consequence of human action, action which has profound hydrological consequences. At 4 ka BP, the hydrological system appears to be in a transitional state, with an increase in the size and frequency of winter flooding against a backdrop of relatively high groundwater flows. By 2 ka BP we suggest that infiltration levels would have reached those of the present day and that at this time the Wadi would have taken on most of its present characteristics; being dominated by large winter floods which occur as a rapid response to rainfall in the catchment. The reduction in infiltration at this time also causes groundwater flows to decline towards present-day levels.

15.5 FROM LOCAL TO GLOBAL: POTENTIAL IMPACTS OF CLIMATIC VARIABILITY ON HOLOCENE COMMUNITIES IN WADI FAYNAN

The 'waterscape' of people living in Wadi Faynan was the result of a combination of factors including rainfall in the upland catchment, local rainfall, groundwater flows and floods. These all presented different opportunities and challenges to those inhabiting the wadi. They required different responses and had different impacts, all of which were mediated by the socio-economic circumstance of the particular community in question. Under early Holocene (10 ka BP, 8 ka BP) conditions, there would have been a reliable, perennial supply of groundwater, supplemented by a *c.* 25% increase in rainfall compared to the present day. During this period there is limited flooding and the total amount of water which reached the lower reaches of the catchment is reduced compared to today. By the late Holocene (2 ka BP) the system supplies far more water than for most of the Holocene, but most of this arrives in floods, which are susceptible to the vagaries of rainfall and pose significant water management and storage challenges. In some ways, any such problems may have been offset by the fact that, in our time slices at least, groundwater flow persists throughout all periods. Indeed, the lowest level of groundwater flow occurs in the present day, for which we estimate this source still to be providing more than 100,000 litres of fresh water per hour, perhaps providing the key support to life and activity in Wadi Faynan.

15.5.1 Neolithic communities

Our simulation scenarios suggest that in the immediate aftermath of the Younger Dryas (12 ka BP), the earliest Holocene inhabitants of the Wadi Faynan, the initial Pre-Pottery Neolithic

A community at WF16, would have had to contend with a very 'flashy' wadi system, a consequence of high rainfall and low infiltration. Over the ensuing millennia we anticipate an increase in infiltration rates, caused by increasing vegetation cover in the catchment, itself a consequence of climatic amelioration. By 10 ka BP, around the time of the beginning of occupation at the Pre-Pottery Neolithic B village of Ghuwayr 1, we suggest that the system would have undergone profound changes, with winter flooding reduced and increased perennial groundwater flows. This radical rearrangement of water resources is significant, particularly given the relatively small (~ 25% more than in the present day) increases in rainfall which we propose for this period.

The early Holocene scenarios presented here suggest that the Early Neolithic communities in Wadi Faynan may have been located in a hydrological setting which was ideally suited to meet the needs of the expanding, increasingly sedentary and agricultural Neolithic populations. The increase in easily managed, predictable groundwater would have provided sufficient water for the immediate needs of the relatively small Pre-Pottery Neolithic populations of WF16 and Ghuwayr 1. Moreover, during the winter months, groundwater flows would have been significantly augmented by gentle floods, with maximum annual flows of around 1 m^3 s^{-1}. In many ways, the perennial groundwater supply flowing past the sites of WF16 and Ghuwayr 1 may be regarded as 'pre-domesticated' as this source is predictable, buffered and perennial in its natural state. Indeed, it is tempting to suggest that the increased stability and predictability of early Holocene groundwater coupled with reliable, easily managed overbank flooding may have been a key factor in the development of sedentism and adoption of agriculture in this locality. Our scenarios for the early Holocene differ radically from some previous work (e.g. Kuijt, 1994; Kujit and Goring-Morris, 2002) which suggested that Pre-Pottery Neolithic Wadi Faynan was likely to have been an arid and marginal area, with its communities existing on the margins of the Neolithic world. Indeed, the benevolent hydrology outlined in our scenarios may help explain the location of major Neolithic sites in presently hostile Wadi Faynan.

With regard to plant cultivation, the predictable, perennial and buffered groundwater supply, boosted in winter months by rainfall and small flood flows, may have enabled opportunistic cultivation of plants on seasonally flooded soils. In this way the Wadi Faynan may have provided an excellent niche for Pre-Pottery Neolithic communities whose agricultural experiments appear to have depended upon the availability of such conditions (Sheratt, 1980). It is also possible that the irrigation of riverside soils would have been facilitated through the construction of simple check dams and barrages. Whilst there is no direct evidence of such practices in Neolithic Wadi Faynan, evidence from the wider region (see this volume, Chapter 14) suggest that Early Neolithic populations had an excellent grasp of hydrology and were fully capable of such achievements.

We are confident that even if our scenarios are only correct in broad outline, the predictable early Holocene waterscapes of Wadi Faynan would have provided many of the essential hydrological features necessary for the establishment of early Neolithic agricultural economies. Nevertheless, given the relatively small increases in rainfall which we propose, water availability in the Faynan area would have been restricted to areas immediately surrounding the groundwater flow channel. As such, we support the palaeoenvironmental reconstructions of Mithen *et al.* (2007a) who suggest that during the early Holocene, vegetation in lower Wadi Faynan would have been strongly riverine in nature, mostly limited to the banks of the stream channel. In this light, we propose that while Wadi Faynan was far from being a marginal setting in the early Holocene, its human communities are likely to have been strongly dependent upon a localised water source, being tethered to the groundwater supply.

15.5.2 Bronze Age developments

Our hydrological scenarios suggest that at 6 ka BP the hydrology of Wadi Faynan was similar to that which existed during the earlier Holocene. The slight reduction in infiltration rates (from 100 to 80 mm day^{-1}) makes little difference to the hydrology of the system. Our reconstructions of palaeo-rainfall suggest that this may have been the driest period of the Holocene, with Tafilah receiving only marginally higher rainfall than at present. As discussed in section 15.3.2, however, most records suggest that this period is marked by highly variable rainfall, and the general trend to aridity is punctuated by several pronounced periods of drought, as well as some rather wetter interludes (Bar-Matthews *et al.*, 2003; Arz *et al.*, 2006; Robinson *et al.*, 2006). Whilst our simulations of hydrology do not include these short-term events, it is important to bear in mind their possible impact.

The most thoroughly documented middle Holocene activity in Wadi Faynan derives from the Early Bronze Age (EBA) (from *c*. 5600 BP). During this period there appears to have been a short-lived 'explosion' in the population of the region centred on the settlement of WF100 (Barker *et al.*, 2007a) and the first clear evidence of water management systems are found in the Wadi (see section 15.2.3). It appears that the majority of these systems were based upon collection, diversion and storage of runoff water, derived from relatively local rainfall. Whilst EBA groups would almost certainly also have utilised groundwater flows, it appears that this predictable and buffered resource may have been insufficient, or inappropriate for their needs. Similarly, there is no evidence in Wadi Faynan that wadi floods in the main channel were intensively utilised or managed in any way and, as today, they may have been exploited primarily through the remnant moisture of floods providing grazing for animal herds.

It should be noted, however, that evidence from the wider region suggests that floodwater farming was a key feature of the EBA economy (Rosen, 2007). As such, it would seem unlikely that the gentle overbank floods which we suggest would have characterised the hydrology of the middle Holocene in Wadi Faynan would not have contributed in some way to the EBA economy of this locality. In this light, the suggestion of Rosen (2007) that the end of the EBA was characterised by the downcutting of wadis may be relevant. If downcutting did indeed occur in Wadi Faynan, then the gentle floods for this period may not have been sufficient to overtop the channel banks. As such, the main channel floods would have been lost as a source of irrigation water, perhaps forcing EBA groups to seek alternative water supplies or reduce population size, as is suggested at regional level by Rosen (2007). Whatever the reason for the decision, archaeological evidence suggests that the EBA populations in Wadi Faynan did indeed become increasingly dependent upon local rainfall to supplement the larger and more predictable flows emanating from the upper catchment area.

From a hydrological viewpoint this seems an exceptionally high risk strategy for several reasons. First, there appears to have been a substantial growth in population (from a few hundred to several thousand) which may have stretched the carrying capacity of the Wadi Faynan system. Producing sufficient food for such a large population would have posed a severe challenge. Second, the location of most EBA settlement is in the open plain of the Wadi Faynan, rather than in the gorge occupied for most of the Neolithic (see Figure 15.5). Whilst this was presumably a necessary move, in order to accommodate the increased population and agricultural activities of the EBA populations, the effect would have been to move away from the groundwater flow channel, which is unlikely to have flowed far beyond the Khirbet at this time because of high transmission losses in the wadi gravels. Under the relatively wet conditions suggested for the beginning of this period (*c*. 6 ka BP) this may not have posed a significant problem. But under the variable and declining rainfall regimes suggested by proxy evidence for this period (e.g. Bar-Matthews *et al.*, 2003) a system based upon unpredictable local rainfall and runoff, rather than one designed to exploit the amplifier effect of the large, well-watered upper catchment, looks an ever more risky strategy.

In this context, a consideration of infiltration further emphasises the risks inherent in the EBA system and suggests that the inter-relationships between human agency, climatic deterioration and infiltration may be highly complex at this time. We have argued that infiltration rates at 6 ka BP would have been slightly lower than those of the early Holocene, a consequence of declining vegetation caused by reduced rainfall and increased human activity. Critically, the proposed reduction in infiltration rates, particularly in the lower wadi (around the site of WF100), may have been essential in forming the EBA waterscape and economy, as this may have increased the generation of local runoff and opened the possibility for floodwater farming using this resource. There is, however, evidence for wind-blown sand deposits in the lower wadi at this time (Barker *et al.*, 2007a). Wind-blown sand can accumulate very rapidly and has a high infiltration rate (Yair and Kossovsky, 2002). As a consequence, any such deposition in areas needed to generate runoff would have severely reduced the efficiency of EBA runoff farming systems. As such, the security of the food supply to EBA communities in Wadi Faynan may have been vulnerable to the impacts of both decreasing rainfall and locally increasing infiltration in the lower Wadi Faynan.

In summary, the population growth – some argue 'explosion' – associated with the EBA may have demanded more water than could be supplied by groundwater flow alone. This led to the development of simple systems of floodwater farming, using local rainfall. This system was probably unsustainable in the long term in the face of a general trend towards increasing aridity, probably punctuated by periods of drought. Moreover, this system also relies upon low infiltration rates in water harvesting areas; the deposition of wind-blown sand, itself a further consequence of increasing aridity and de-vegetation, may have acted to further reduce the sustainability of the EBA economic system. It thus seems plausible, that declining rainfall associated with later Bronze Age periods, together with possible downcutting of wadi systems (Rosen, 2007), and changes in infiltration rates may have created a water crisis in the Wadi Faynan, forcing abandonment of the area.

15.5.3 Later Holocene archaeology

As discussed above (section 15.3.3) the late Holocene was a period of intensive and rapid stripping of vegetation, both in the lower wadi and the upper catchment. In Roman/Byzantine times, between 80 and 258 tonnes of charcoal per year would have been required for smelting activities, much of which appears to have been imported from the upper catchment of the wadi system (Mattingly *et al.*, 2007a). This would have led to a rapid and marked decrease in infiltration rates. Consequently, this is the most likely time when the wadi took on its present characteristics and became dominated by seasonal floods with relatively small groundwater flows.

The key feature of the landscape in the late Holocene with regard to hydrology was the low infiltration rate caused by removal of tree cover in the wadi catchment. This radically changed the nature of the wadi hydrology, increasing flooding and gradually reducing groundwater. As with the EBA populations discussed above, much late Holocene activity took place downstream of the Khirbet Faynan, below the point at which groundwater flows would have been available. The Classical period occupants of the wadi addressed this problem by transporting water from the Wadi Ghuwayr along a channel/ aqueduct system to a large reservoir located on the bank of the wadi opposite the Khirbet. Interestingly, Mattingly *et al.* (2007a) suggest that the water stored in this reservoir would have been polluted by metallurgical activity. Rather than providing a source of drinking water, they suggest that the reservoir was primarily used to drive a water mill for processing ores, with the wastewater directed away from the field-system into the wadi channel. While this is a persuasive interpretation, it nevertheless seems implausible that the groundwater flow would not also have been an essential supply of drinking water.

As in EBA times, the main challenge facing the Classical population in Wadi Faynan would have been production of a reliable food supply. It is unsurprising, therefore, that the WF4 field-system became extensive during this period, covering an area of more than 200 ha and comprising at least 800 individual fields (Figure 15.8; Mattingly *et al.*, 2007a). In terms of food supply, Mattingly *et al.* (2007a) argue that the WF4 field-system could have supplied food for only 300 people, suggesting a substantial amount of food imports to the wadi. Given the size of the field-system, this relatively low carrying capacity appears surprising, requiring nearly three fields to support each individual, and reflects a view that the farming system was highly inefficient. Indeed, from the present perspective it is difficult to see how the WF4 system would have functioned for crop growing at all if, as suggested by several researchers (Mattingly *et al.*, 2007a; Newson *et al.*, 2007b; Crook, 2009) the water supply to the WF4 system was based purely upon trapping local runoff water and supplies from tributary wadis. Whilst the suggestion that the Classical engineers, who otherwise demonstrate considerable hydrological knowledge, would have designed a system that ignores completely the amplifier effect of the large, well-watered Faynan catchment is surprising, this nevertheless appears to be the most plausible interpretation of the archaeological evidence (Mattingly *et al.*, 2007a; Newson *et al.*, 2007a). If so, the Classical system would have been under similar stresses to those of the EBA.

When Crook's 2009 hydrological analysis of the WF4 systems is combined with the hydrological models we have developed, the ineffectiveness of that field-system becomes particularly apparent. The high infiltration rates proposed by Crook (2009) for the runoff-generating slopes around the WF4 system mean that very large rainstorms (c. 100 mm hr^{-1}) are required to generate very low volumes of runoff. This may suggest that either the runoff system itself, or current interpretations of the way this system functioned, are flawed. Moreover, when the easily managed groundwater flow is producing year-round flows of around 100,000 litres an hour, and millions of litres are available annually from main channel floods, it seems almost implausible that irrigation would be based upon such an inefficient, unpredictable and vulnerable local runoff system. A further problem with this scenario is that the system, as described by Crook, would have required

Figure 15.23. Present-day cultivation in the WF4 field-system, irrigated using groundwater flow captured by plastic pipes in Wadi Ghuwayr (see Figures 15.2 and 15.3).

predictable and regular intensive rain events with rainfall events of c. 100 mm hr^{-1} (Crook, 2009, p. 2430). According to our palaeoclimatic reconstructions, rainstorms providing c. 100 mm hr^{-1} are extremely unlikely to have occurred in Wadi Faynan at this time, or indeed at any point in the Holocene. Even allowing for the fact that our reconstructions may underplay extreme events (see this volume, Chapter 4) it seems unlikely that this system would have ever received sufficient rain to provide even the most negligible benefit to the region's crops.

It is thus difficult to reconcile the high rainfall amounts needed to make the WF4 system function effectively, with our palaeoenvironmental reconstructions. Moreover, in a situation where rainstorms of 100 mm hr^{-1} occurred regularly, it is not likely that any form of rainfall harvesting would have been necessary for crop production. Given these difficulties, allied to the suggestions that the system would in any case have made negligible difference to the amount of water available in the fields, it is possible that the proposed design of the Classical water catchment system requires some revision, or that the hydrological modelling of Crook (2009) is flawed, such as by over-estimating infiltration rates. Alternatively, Classical farmers may have found ways to artificially reduce infiltration rates, such as by moving stone cover, which has been shown to reduce infiltration rates by up to 250% in the Negev (e.g. Lavee *et al.*, 1997). Given the massive effect of infiltration upon runoff generation that we have shown, it is possible that by greatly reducing infiltration rates the WF4 system could have been made much more effective. The most plausible solution to this problem, however, is that local runoff water was supplemented by alternative sources of water, such as the vast main channel floods or groundwater flow, as Crook (2009) has also suggested. Indeed, in the present day the local Bedouin community make use of groundwater flow to irrigate substantial portions of the WF4 system (Figure 15.23).

A further possibility, following Mattingly *et al.* (2007b), is that the WF4 system never really functioned effectively, requiring the Wadi Faynan population to have been sustained by drawing upon the infrastructure of empire and importing food from other areas. Following this line of thought, and bearing in mind the massive demand for wood at this time, it may also be possible that the field-system was never intended to produce water-hungry crops but could have been used primarily to grow less water-demanding tree species, perhaps for the purposes of charcoal manufacture.

In sum, evidence suggests that the Classical inhabitants of Wadi Faynan developed a complex range of water management techniques, which used both groundwater flow and local runoff. However, present interpretations of the irrigation system are highly problematic. They suggest that by ignoring the ample water sources derived from the upper catchment and concentrating solely upon locally generated runoff, the Classical water harvesting systems would have been of very limited efficiency.

15.6 SUMMARY AND DISCUSSION

The principal aim of this chapter has been to develop a methodology for relating patterns of regional climate change, derived from an RCM and a weather generator, to patterns of social and economic change in Wadi Faynan. In order to achieve this, hydrological modelling was employed to act as a bridge between climatic and archaeological data in a series of time-slice simulations. In the case of the Wadi Faynan, this is particularly significant as the majority of the water reaching the inhabitants of the area is not generated by local rainfall, but by rainfall in the upland catchment which is then redistributed by hydrological processes in the form of flood and groundwater flows. The way in which the sources have responded to climatic variability, impacts upon their utility for human populations.

To summarise our key results:

- In terms of palaeo-rainfall, results suggest that the Wadi Faynan received more rainfall than at present during all time-slice simulations. However, results also suggest that a *c.* 25% increase in rainfall, at 10 and 2 ka BP, represents the maximum degree of change from the present. Comparison with proxy data (e.g. Bar-Matthews *et al.*, 2003), suggest that these results may underestimate the degree of Holocene rainfall variability.
- In terms of hydrology, the simulation modelling suggests that our postulated rainfall changes would have a minimal impact upon water availability in Wadi Faynan. Under all the time-slice rainfall scenarios, the Wadi Faynan system behaves much as in the present day, being characterised by large winter floods and a small, stable groundwater flow. However, changes in the ground surface cover of Wadi Faynan during the Holocene may have had a more significant impact.
- A review of regional and local vegetation histories suggests that major landscape changes have occurred during the Holocene. There appears to be consensus that the early Holocene in the Levant was far more wooded than the present day (Rossignol-Strick, 1999; Roberts, 2002; Hunt *et al.*, 2007), as a consequence of small human populations and prevailing warm, wet climates. Since mid-Holocene times, the combined trends of increasing aridity and human impact upon the landscape have combined to cause deforestation and erosion of soils. In Wadi Faynan, there is clear evidence that Classical period industrial activity would have played a significant role in this process. We propose that these changes would have greatly reduced infiltration rates in Wadi Faynan since the middle Holocene.
- When our palaeo-rainfall and palaeoinfiltration scenarios are used in tandem, we see major variability in the hydrology of the Wadi Faynan system. Even relatively small changes in rainfall can lead to radical changes in hydrology if accompanied by changes in ground surface cover and infiltration.
- Under high infiltration conditions, reaching 100 mm day^{-1}, the hydrology of the Wadi Faynan is radically different from that of the present day.

Key features of early Holocene hydrology are:

- High infiltration rates reduce runoff generation in the Wadi Faynan catchment. Consequently the frequency and, particularly, size of winter flooding are severely diminished.
- Percolation into aquifers is increased, raising perennial groundwater flow rates.
- This hydrological regime would have provided an ideal niche for the development of early agriculture, providing a predictable, reliable and perennial groundwater supply, augmented by gentle winter (overbank?) flooding.
- Early Holocene communities may have been 'tethered' to the wadi channel, as surrounding areas may have been arid.

Middle Holocene (6 ka BP) hydrology appears to have been similar to that of the Early Holocene.

- Large EBA populations began, for whatever reason, to supplement water derived from the catchment with local rainfall and runoff for agricultural activity – a high risk strategy under middle Holocene climatic conditions.
- Wind-blown sand in the lower wadi and periods of drought would have had detrimental impacts on the EBA farming systems.

During the late Holocene (4, 2 ka BP), the hydrology of the Wadi Faynan was similar to that of today, a consequence of reduced infiltration caused by industrial-scale deforestation to support metallurgical activity. The highest Holocene rainfall amount produced by our model occurred in the 2 ka BP simulation which suggests a c. 25% increase on Present Day rainfall amounts at this time.

- Large Classical period populations placed a significant strain on water resources and a series of elaborate and sophisticated water management features were developed.
- Present interpretations suggest that, as in the EBA, Classical period water management for drinking and irrigation was based around management of local rainfall and runoff.
- Such systems would have been incapable of supporting large populations, except under exceptionally high rainfall conditions.
- We believe that re-evaluation of the functioning of Classical period water management systems may suggest that more use was made of groundwater and main channel floods – more predictable and reliable sources of water which are derived from the large upland catchment.

In summary, we hope that we have demonstrated the potential of our methodology to provide new datasets against which to evaluate the long-term relationships between climate, water and society in Wadi Faynan. The palaeohydrological scenarios presented in this chapter provide nuanced data which allow us to assess the impact of climatic variability upon water availability at a specific time in a specific place. As such this method provides a novel means of assessing the impact of climatic variability upon society.

Perhaps the most significant aspect of the present study is the potential impact of changes in ground surface cover upon water availability. Although ground surface cover has been recognised as a major variable in semi-arid hydrological systems (e.g. Yair and Kossovsky, 2002), the potential implications of the interaction between climate variability, vegetation cover and human action have not been widely discussed in archaeology (but see Lavee et al., 1997). It is, however, difficult to untangle the complex interrelationships between human action and natural climate change. Whilst it seems reasonable to suggest that during the early Holocene the small size of human communities may have limited the impact of their activities upon the landscape, it also seems likely that the large populations and increasing demand for timber which characterised the middle and later Holocene periods had a severe and rapid impact upon vegetation cover and infiltration rates. At the very beginning of the Holocene, however, changes in climate may have played a more significant role. For example, as the region emerged from the arid period of the Younger Dryas, moderate increases in precipitation may have led to the gradual development of more abundant vegetation and a gradual increase in the infiltration rate. In this manner, changes in vegetation may act as a feedback or amplifier of the local impacts of climate change. Indeed, changes in infiltration (either anthropogenic or climatic in origin) may establish a positive feedback mechanism, where lower infiltration causes more flooding, which creates more erosion, leading to still lower infiltration rates (Yair and Kossovsky, 2002). The suggestion that the impact of relatively small changes in climate on hydrological systems may be significantly amplified by changes in vegetation cover is significant for our understanding of the impact of climate change on society. Examining the relationships between human activity, vegetation growth and climatic amelioration at local and regional scales should be a priority for future research.

The novel and multi-disciplinary nature of the work in this chapter which draws together meteorologists, hydrologists and archaeologists means that there are significant areas where this study must be regarded as preliminary. Specifically, the hydrological modelling is rather basic, and more sophisticated representation of both infiltration and percolation would provide greater confidence in our results. In the same vein, incorporating temperature changes and variability in evapotranspiration into our modelling would improve our palaeohydrological scenarios. However, whilst the absolute numbers provided by present simulations should be treated with extreme caution given the uncertainties surrounding several of our parameters, we have reasonable confidence in the overall picture of the evolution of the wadi system outlined above. Moreover, the methodology presented in this chapter has a general significance for providing a means to downscale regional climate trends in order to assess their impact upon water availability and communities in semi-arid regions.

REFERENCES

Al-Najjar, M., A. Abu-Dayya, E. Suleiman, G. Weisgerber and A. Hauptmann (1990) Tell Wadi Faynan: the first Pottery Neolithic Tell in the South of Jordan. *Annual of the Department of Antiquities of Jordan* **34**: 27–56.

Arz, H. W., F. Lamy and J. Patzold (2006) A pronounced dry event recorded around 4.2 ka in brine sediments from the northern Red Sea. *Quaternary Research* **66**: 432–441.

Ashbel, D. (1938) Great floods in Sinai Peninsula, Palestine Syria and the Syrian Desert, and the influence of the Red Sea on their formation. *Quarterly Journal of the Royal Meteorological Society* **64**: 635–639.

Bar-Matthews, M., A. Ayalon, M. Gilmour, A. Matthews and C. J. Hawkesworth (2003) Sea–land oxygen isotopic relationships from planktonic foraminifera and speleothems in the Eastern Mediterranean region and their implication for paleorainfall during interglacial intervals. *Geochimica et Cosmochimica Acta* **67**: 3181–3199.

Bar-Matthews, M., A. Ayalon and A. Kaufman (1997) Late quaternary paleoclimate in the eastern Mediterranean region from stable isotope analysis of speleothems at Soreq Cave, Israel. *Quaternary Research* **47**: 155–168.

Bar-Yosef, O. and A. Belfer-Cohen (1989) The origins of sedentism and farming communities in the Levant. *Journal of World Prehistory* **3**: 447–498.

Barker, G., R. Adams, O. Creighton et al. (2007a) Chalcolithic (c. 5000–3600 cal. BC) and Bronze Age (c. 3600–1200 cal. BC) settlement in Wadi Faynan: metallurgy and social complexity. In *Archaeology and Desertification: The Wadi Faynan Landscape Survey, Southern Jordan*, ed. G. Barker, D. Gilbertson and D. Mattingly. Oxford, UK and Amman, Jordan: Council for British Research in the Levant in Association with Oxbow Books pp. 227–270.

Barker, G., D. Gilbertson and D. Mattingly, eds. (2007b) *Archaeology and Desertification: The Wadi Faynan Landscape Survey, Southern Jordan*. Wadi Faynan Series Volume 2, Levant Supplementary Series. Oxford, UK and Amman, Jordan: Council for British Research in the Levant in Association with Oxbow Books.

Barker, G., D. Gilbertson and D. Mattingly (2007c) The Wadi Faynan landscape survey: research themes and project development. In *Archaeology and Desertification: The Wadi Faynan Landscape Survey, Southern Jordan*. Wadi Faynan Series Volume 2, Levant Supplementary Series, ed. G. Barker, D. Gilbertson and D. Mattingly. Oxford, UK and Amman, Jordan: Council for British Research in the Levant in Association with Oxbow Books pp. 3–23.

Bookman, R., Y. Enzel, A. Agnon and M. Stein (2004) Late Holocene lake levels of the Dead Sea. *Geological Society of America Bulletin* **116**: 555–571.

Braconnot, P., B. Otto-Bliesner, S. Harrison et al. (2007) Results of PMIP2 coupled simulations of the Mid-Holocene and Last Glacial Maximum – Part 1: experiments and large-scale features. *Climate of the Past* **3**: 261–277.

Byrd, B. F. (2005) Reassessing the emergence of village life in the Near East. *Journal of Archaeological Research* **13**: 231–290.

Carlson, A. E., P. U. Clark, B. A. Haley et al. (2007) Geochemical proxies of North American freshwater routing during the Younger Dryas cold event. *Proceedings of the National Academy of Sciences of the United States of America* **104**: 6556–6561.

Cauvin, J. (1994) *Naissance des divinites, Naissance de l'agriculture*. Paris: CNRS.

Cerda, A. (1997) Seasonal changes of the infiltration rates in a Mediterranean scrubland on limestone. *Journal of Hydrology* **198**: 209–225.

COHMAP members (1988) Climatic changes of the last 18,000 years: observations and model simulations. *Science* **241**: 1043–1052.

Cordova, C. (2007) *Millennial Landscape Change in Jordan: Geoarchaeology and Cultural Ecology*. Tucson: University of Arizona Press.

Crook, D. (2009) Hydrology of the combination irrigation system in the Wadi Faynan, Jordan. *Journal of Archaeological Science* **36**: 2427–2436.

Eadie, J. W. and J. P. Oleson (1986) The water-supply systems of Nabataean and Roman Humayma. *Bulletin of the American Schools of Oriental Research* **262**: 49–76.

el-Rishi, H., C. Hunt, D. Gilbertson et al. (2007) The past and present landscapes of the Wadi Faynan: geoarchaeological approaches and frameworks. In *Archaeology and Desertification: The Wadi Faynan Landscape Survey, Southern Jordan*. Wadi Faynan Series Volume 2, Levant Supplementary Series, ed. G. Barker, D. Gilbertson and D. Mattingly. Oxford, UK and Amman, Jordan: Council for British Research in the Levant in Association with Oxbow Books pp. 59–96.

Enzel, Y., R. Arnit, U. Dayan et al. (2008) The climatic and physiographic controls of the eastern Mediterranean over the late Pleistocene climates in the southern Levant and its neighboring deserts. *Global and Planetary Change* **60**: 165–192.

Finlayson, B. L. and S. Mithen (2007a) Excavations at WF16. In *The Early Prehistory of Wadi Faynan, Southern Jordan: Archaeological Survey of Wadis Faynan, Ghuwayr and Al Bustan and Evaluation of the Pre-Pottery Neolithic A Site of WF16*. Wadi Faynan Series 1, Levant Supplementary Series **4**, ed. B. Finlayson and S. Mithen. Oxford and Amman, Jordan: Oxbow Books and the Council for British Research in the Levant pp. 145–202.

Finlayson, B. L. and S. Mithen (2007b) *The Early Prehistory of Wadi Faynan, Southern Jordan: Archaeological Survey of Wadis Faynan, Ghuwayr and Al Bustan and Evaluation of the Pre-Pottery Neolithic A Site of WF16*. Wadi Faynan Series 1, Levant Supplementary Series **4**, ed. B. Finlayson and S. Mithen. Oxford and Amman, Jordan: Oxbow Books and the Council for British Research in the Levant.

Frumkin, A., D. C. Ford and H. P. Schwarcz (1999) Continental oxygen isotopic record of the last 170,000 years in Jerusalem. *Quaternary Research* **51**: 317–327.

Frumkin, A., D. C. Ford and H. P. Schwarcz (2000) Paleoclimate and vegetation of the last glacial cycles in Jerusalem from a speleothem record. *Global Biogeochemical Cycles* **14**: 863–870.

Fujii, S. (2007) Wadi Badda: A PPNB settlement below Fjaje escarpment near Shawbak. *Neolithics* **1/07**: 19–24.

Fujii, S. (in prep) Domestication of runoff water: current evidence and new perspectives from the Jafr Pastoral Neolithic.

Garfinkel, Y., A. Vered and O. Bar-Yosef (2006) The domestication of water: the Neolithic well at Sha'ar Hagolan, Jordan Valley, Israel. *Antiquity* **80**: 686–696.

Gebel, H. G. K. (2004) The domestication of water. Evidence from early Neolithic Ba'ja. In *Men of Dikes and Canals. The Archaeology of Water in the Middle East*, ed. H.-D. Bienert and J. Häser. Rahden: Marie Leidorf pp. 25–36.

Hauptmann, A. (2007) *The Archaeometallurgy of Copper: Evidence from Faynan*. New York: Springer.

Hunt, C. O., H. A. Elrishi, D. D. Gilbertson et al. (2004) Early-Holocene environments in the Wadi Faynan, Jordan. *The Holocene* **14**: 921–930.

Hunt, C. O., D. D. Gilbertson and H. A. El-Rishi (2007) An 8000-year history of landscape, climate, and copper exploitation in the Middle East: the Wadi Faynan and the Wadi Dana National Reserve in southern Jordan. *Journal of Archaeological Science* **34**: 1306–1338.

Isaar, A. and D. Yakir (1997) Isotopes from wood buried in the roman siege ramp of Masada: the Roman period colder climate. *Biblical Archaeologists* **60**: 101–106.

Kuijt, I. (1994). Pre-Pottery Neolithic A settlement variability: evidence for socio-political developments in the Neolithic Levant. *Journal of Mediterranean Archaeology* **7/2**: 165–192.

Kuijt, I. (2000) People and space in early agricultural villages: exploring daily lives, community size, and architecture in the late pre-pottery neolithic. *Journal of Anthropological Archaeology* **19**: 75–102.

Kuijt, I. and B. L. Finlayson (2009) New evidence for food storage and pre-domestication granaries 11,000 years ago in the Jordan Valley. *Proceedings of the National Academy of Science* **106**: 10966–10970.

Kuijt, I. and A. N. Goring-Morris (2002) Foraging, farming and social complexity in the Pre-Pottery Neolithic of the South-Central Levant: a review and synthesis. *Journal of World Prehistory* **16**: 361–439.

Lancaster, W. and F. Lancaster (1991) Limitations on sheep and goat herding in the Eastern Badia of Jordan: an ethno-archaeological enquiry. *Levant* **28**: 125–138.

Lancaster, W. and F. Lancaster (1999) *People, Land and Water in the Arab Middle East. Environments and Landscapes in the Bilad ash-Sham*. Amsterdam: Harwood Academic Publishers.

Lavee, H., J. Poesen and A. Yair (1997) Evidence of high efficiency water-harvesting by ancient farmers in the Negev desert, Israel. *Journal of Arid Environments* **35**: 341–348.

Leonard, J. and P. Andrieux (1998) Infiltration characteristics of soils in Mediterranean vineyards in southern France. *Catena* **32**: 209–223.

Mattingly, D., P. Newson, O. Creighton et al. (2007a) A landscape of Imperial Power: Roman and Byzantine Phaino. In *Archaeology and Desertification: The Wadi Faynan Landscape Survey, Southern Jordan*. Wadi Faynan Series Volume 2, Levant Supplementary Series, ed. G. Barker, D. Gilbertson and D. Mattingly. Oxford, UK and Amman, Jordan: Council for British Research in the Levant in Association with Oxbow Books pp. 305–348.

Mattingly, D., P. Newson, J. Grattan et al. (2007b) The making of early states: the Iron Age and Nabatean periods. In *Archaeology and Desertification: The Wadi Faynan Landscape Survey, Southern Jordan*. Wadi Faynan Series Volume 2, Levant Supplementary Series, ed. G. Barker, D. Gilbertson and D. Mattingly. Oxford, UK and Amman, Jordan: Council for British Research in the Levant in Association with Oxbow Books pp. 271–303.

McLaren, S. J., D. D. Gilbertson, J. P. Grattan et al. (2004) Quaternary palaeogeomorphologic evolution of the Wadi Faynan area, southern Jordan. *Palaeogeography Palaeoclimatology Palaeoecology* **205**: 131–154.

Mithen, S., P. Austen, A. Kennedy et al. (2007a) Early Neolithic woodland composition and exploitation in the southern Levant: a comparison between archaeobotanical remains from WF16 and present-day woodland at Hammam Adethni. *Environmental Archaeology* **12**: 49–70.

Mithen, S., B. Finlayson, A. Pirie *et al.* (2007b) Archaeological survey of Wadis Faynan, Ghuwayr, Dana and al-Bustan. In *The Early Prehistory of Wadi Faynan, Southern Jordan: Archaeological Survey of Wadis Faynan, Ghuwayr and al-Bustan and Evaluation of the Pre-Pottery Neolithic A Site of WF16*, ed. B. L. Finlayson and S. Mithen. Oxford: Oxbow Books pp. 115–133.

Mithen, S. J. (2003) *After the Ice: A Global Human History, 20,000–5000 BC*. London: Weidenfeld and Nicolson.

Newson, P., G. Barker, P. Daly, D. Mattingly and D. Gilbertson (2007a) The Wadi Faynan field system. In *Archaeology and Desertification: The Wadi Faynan Landscape Survey, Southern Jordan. Wadi Faynan Series* Volume 2, Levant Supplementary Series, ed. G. Barker, D. Gilbertson and D. Mattingly. Oxford, UK and Amman, Jordan: Council for British Research in the Levant in Association with Oxbow Books pp. 141–176.

Newson, P., D. Mattingly, P. Daly *et al.* (2007b) The Islamic and Ottoman Periods. In *Archaeology and Desertification: The Wadi Faynan Landscape Survey, Southern Jordan. Wadi Faynan Series* Volume 2, Levant Supplementary Series, ed. G. Barker, D. Gilbertson and D. Mattingly. Oxford, UK and Amman, Jordan: Council for British Research in the Levant in Association with Oxbow Books pp. 349–368.

Palmer, C., D. Gilbertson, H. el-Rishi *et al.* (2007) The Wadi Faynan today: landscape, environment, people. In *Archaeology and Desertification: The Wadi Faynan Landscape Survey, Southern Jordan. Wadi Faynan Series* Volume 2, Levant Supplementary Series, ed. G. Barker, D. Gilbertson and D. Mattingly. Oxford, UK and Amman, Jordan: Council for British Research in the Levant in Association with Oxbow Books pp. 22–57.

Raikes, R. (1967) *Water, Weather and Prehistory*. London: John Baker.

Roberts, N. (2002) Did prehistoric landscape management retard the postglacial spread of woodland in Southwest Asia? *Antiquity* 76: 1002–1010.

Roberts, N. and H. E. Wright Jr (1993) Vegetation, lake-level and climatic history of the Near East and Southwest Asia. In *Global Climates since the Last Glacial Maximum*, ed. H. E. Wright Jr, J. E. Kutzbach, T. Webb III *et al.* University of Minnesota Press pp. 194–220.

Robinson, S. A., S. Black, B. Sellwood and P. J. Valdes (2006) A review of palaeoclimates and palaeoenvironments in the Levant and Eastern Mediterranean from 25,000 to 5000 years BP: setting the environmental background for the evolution of human civilisation. *Quaternary Science Reviews* 25: 1517–1541.

Rodwell, M. J. and B. J. Hoskins (2001) Subtropical anticyclones and summer monsoons. *Journal of Climate* 14: 3192–3211.

Rollefson, G. (2008) *The Neolithic Period in Jordan: An Archaeological Reader*, ed. R. Adams. London: Equinox pp. 71–108.

Rosen, A. (2007) *Civilizing Climate: Social Responses to Climate Change in the Ancient Near East*. Plymouth: AltaMira Press.

Rossignol-Strick, M. (1999) The Holocene climatic optimum and pollen records of sapropel 1 in the eastern Mediterranean, 9000–6000 BP. *Quaternary Science Reviews* 18: 515–530.

Sheratt, A. (1980) Water, soil and seasonality in early cereal cultivation. *World Archaeology* 11: 313–330.

Simmons, A. J. and M. Najjar (1997) Ecological changes in Jordan during the Late Neolithic in Jordan: a case study. In *The Prehistory of Jordan II: Perspectives from 1997*, ed. H. G. K. Gebel, Z. Kafafai and G. Rollefson. Berlin: ex Oriente pp. 309–318.

Simmons, A. J. and M. Najjar (1998) Al-Ghuwayr I. A pre-pottery Neolithic village in Wadi Faynan, southern Jordan: a preliminary report on the 1996 and 1997/98 seasons. *Annual Report of the Department of Antiquities of Jordan* 42: 91–101.

Steffensen, J. P., K. K. Andersen, M. Bigler *et al.* (2008) High-resolution Greenland ice core data show abrupt climate change happens in few years. *Science* 321: 680–684.

Stein, M., A. Torfstein, I. Gavrieli and Y. Yechieli (2010) Abrupt aridities and salt deposition in the post-glacial Dead Sea and their North Atlantic connection. *Quaternary Science Reviews* 29: 567–575.

Tchernov, E. (1994) *An Early Neolithic Village in the Jordan Valley, II: The Fauna of Netiv Hagdud*. Cambridge, MA: Peabody Museum of Archaeology and Ethnography, Harvard University.

Whitehead, P. G., S. J. Smith, A. J. Wade *et al.* (2008) Modelling of hydrology and potential population levels at Bronze Age Jawa, Northern Jordan: a Monte Carlo approach to cope with uncertainty. *Journal of Archaeological Science* 35: 517–529.

Yair, A. and A. Kossovsky (2002) Climate and surface properties: hydrological response of small and semi-arid watersheds. *Geomorphology* 42: 43–57.

Yakir, D., A. Issar, J. Gat *et al.* (1994) ^{13}C and ^{18}O of wood from the Roman siege rampart in Masada, Israel (AD 70–73): evidence for a less arid climate of the region. *Geochimica et Cosmochimica Acta* 58: 3535–3539.

16 Palaeoenvironmental reconstruction at Beidha, southern Jordan (c. 18,000–8,500 BP): Implications for human occupation during the Natufian and Pre-Pottery Neolithic

Claire Rambeau, Bill Finlayson, Sam Smith, Stuart Black, Robyn Inglis and Stuart Robinson

ABSTRACT

The Beidha archaeological site in Southern Jordan was occupied during the Natufian (two discrete occupation phases, c. 15,200–14,200 cal. BP and c. 13,600–13,200 cal. years BP) and Pre-Pottery B Neolithic periods (c. 10,300–8,600 cal. years BP). This chapter reconstructs the palaeoenvironments at Beidha during these periods, using sedimentological observations and the stable isotopic composition (oxygen and carbon) of carbonate deposits. Age control is provided by uranium-series and radiocarbon dating. Detailed analysis of a carbonate stratigraphic section related to a fossil spring close to the site, and a sequence of carbonate nodules from a section on the western edge of the archaeological site, permits a reconstruction of climatic variations between c. 18,000 and c. 8,500 years BP. The results of the palaeoenvironmental study are compared with the archaeological evidence, to explore the relationship between human occupation and climatic variability at Beidha. The results indicate a marked correspondence between more favourable (wetter) environmental conditions and phases of occupation at Beidha, and provide clues to the likely sources of water that sustained the settlement during the Late Pleistocene and early Holocene.

16.1 INTRODUCTION

Climate change during the Late Pleistocene–early Holocene is often seen as a key factor in the transition to sedentism and stable, agricultural societies in the Middle East, given the background of major events such as the start of the Younger Dryas and the Holocene (e.g. Moore and Hillman, 1992; Mithen, 2003; Cordova, 2007, see also Feynman and Ruzmaikin, 2007). Even more abrupt climatic events, such as the '8.2 ka BP cooling event', recognised in various parts of the globe, may also have had an influence on Levantine societies (e.g. Staubwasser and Weiss, 2006; Weninger et al., 2006; Berger and Guilaine, 2009).

The regional and local impacts of past climatic events on societies remain largely unknown, and detailed studies are required to compare palaeoenvironmental data and archaeological information. The southern Levant is a key area for this type of investigation because of its transitional character between arid zones to the south and east and more favourable, wetter climates to the north and west (e.g. EXACT (Executive Action Team Middle East Water Data Banks Project), 1998; Cordova, 2007). The various regions of the southern Levant receive contrasting amounts of mean annual rainfall (Figure 16.1). Inter-annual variations in rainfall also increase with decreasing annual precipitation (e.g. Sanlaville, 1996). The eastern and southern parts of the Levant experience low annual precipitation rates that are highly variable from year to year. This is produced by a dominant source of moisture from the Mediterranean (e.g. Enzel et al., 2008), combined with extreme topographic gradients, and results in the juxtaposition of different bioclimatic zones within a limited geographical area (Zohary, 1962; Al-Eisawi, 1985, 1996; EXACT, 1998).

It is likely that past populations living on the fringe of arid areas were especially susceptible to the impact of climate change, since such locations are more vulnerable to environmental variations (e.g. Neumann et al., 2007). As such, the southern Levant provides a particularly good case study for exploring the relationship between the transition to settled farming societies and climate change, especially in marginal areas. In this chapter we present a study centred on the Beidha

Water, Life and Civilisation: Climate, Environment and Society in the Jordan Valley, ed. Steven Mithen and Emily Black. Published by Cambridge University Press.
© Steven Mithen and Emily Black 2011.

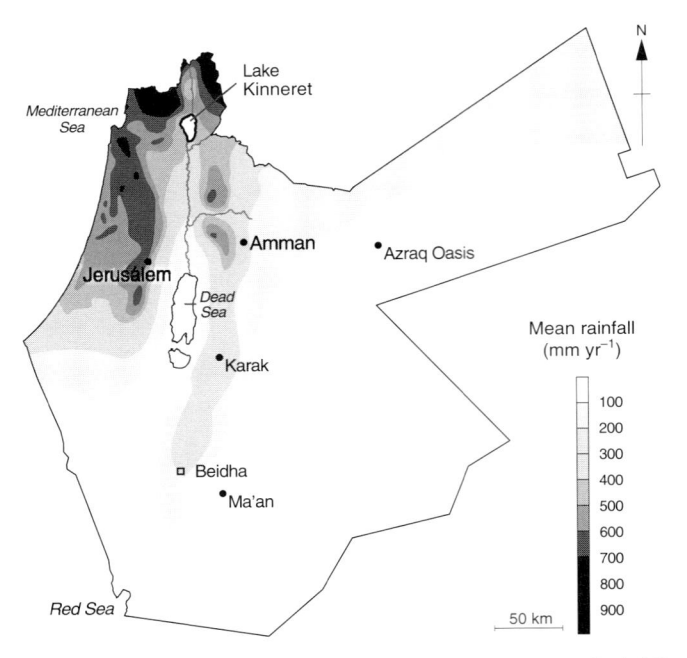

Figure 16.1. Map of the southern Levant showing average annual rainfall (modified from EXACT, 1998) and the location of the study area.

archaeological site, located in the arid zone of southeast Jordan (Figure 16.1), which compares localised changes in the environment across the Late Pleistocene/early Holocene transition with the known history of occupation.

Beidha is an important prehistoric archaeological site situated c. 4.5 km north of the well-known Nabataean city of Petra, with occupation phases during the Natufian (Levantine term corresponding to the later Epipalaeolithic) and the Pre-Pottery Neolithic B (PPNB) (e.g. Byrd, 1989, 2005). At the time of its discovery by Diana Kirkbride in 1956, the site appeared as a low tell, created by Neolithic deposits on a remnant alluvial terrace, and was covered by Nabataean(/Roman) agricultural terraces (Kirkbride, 1989). When Kirkbride began excavations at Beidha in 1958, very little was known about the early Neolithic of the Near East. One of Kirkbride's major objectives was to 'spread sideways rather than to dig deeply' (Kirkbride, 1960, p. 137). This modern approach of open area excavation is one of the reasons her excavation remains so important because a relatively large area of the settlement was uncovered. Her other chief objective was to reconstruct the Neolithic economy, looking at the 'practice of agriculture and the domestication of animals' (Kirkbride, 1960, p. 137). This of course remains an important research issue. The project ran under Kirkbride's direction with field seasons (and publication of interim reports) almost every year between 1958 and 1967, with a final season in 1983. She was joined in that year by Brian Byrd, who subsequently took on the role of bringing the project to publication (Byrd, 1989, 2005). Beidha is now one of many PPNB sites excavated in the southern Levant, but Kirkbride's fieldwork has made the site one of the best-known early villages in the world (Byrd, 1989). The site remains a key location for our understanding of Neolithic origins and development.

Previous studies at Beidha have tentatively reconstructed Late Pleistocene–Holocene palaeoenvironmental changes and correlated these with periods of human occupation (Raikes, 1966; Field, 1989; Comer, 2003; see also Helbaek, 1966 and Fish, 1989, and section 16.4 below). Such studies were, however, reliant on limited sources and quantities of information about the palaeoenvironmental evolution and had to contend with limited knowledge about the chronology of settlement.

Our new study uses a combination of geomorphological and sedimentological observations and stable isotopic (C, O) analysis of spring-carbonates (tufas) and pedogenic nodules collected around the site, to examine past environments at Beidha during the Late Pleistocene to early Holocene (c. 18,000–8,000 BP). Age control is provided by uranium-series dating of carbonate deposits and radiocarbon dating of organic remains. The new data are compared to, and integrated with, previous studies from Beidha and elsewhere in the southern Levant in order to provide a multi-proxy reconstruction of past environments that can be compared with changes in the archaeological record over the same period.

16.2 PRESENT-DAY ENVIRONMENTAL SETTING

Beidha is situated at c. 1,020 m above mean sea level (amsl) and c. 4.5 km north of the Nabataean city of Petra, within the alluvial valley created by the Wadi el-Ghurab ('Valley of the Ravens'; also spelled Wadi Ghuraib, Wadi Gharab, Wadi Ghrab; e.g. Byrd, 1989, 2005; Helbaek, 1966; Kirkbride, 1966; Kirkbride, 1968). The valley is bordered by steep cliffs of Cambrian/Ordovician sandstone (Figure 16.2) and is dissected by a modern wadi bed currently dry except during major rain (flash-flood) events (Comer, 2003). The valley drains the Jebel Shara (also spelled Jebel Shara' or Gebl Sharah; Mount Seir of the Bible), an upland area of Cretaceous limestone, situated a few kilometres to the east of Beidha, with altitudes up to 1,700 m amsl (Byrd, 1989; Comer, 2003). The Wadi el-Ghurab, generally flowing northeast to southwest, reaches the Wadi Araba (Kirkbride, 1985; Byrd, 1989, 2005) after a drop of over 400 m less than 2 km downstream of the site (Kirkbride, 1985).

Generally, the sandstone area in which Beidha and Petra lie forms a shelf interrupting the east-to-west abrupt altitudinal descent from the high Jordanian plateau (including Jebel Shara) to the low, desertic lands of the Wadi Araba. The N–S orientated sandstone shelf is about 4 km wide at Petra and 6 km wide at Beidha (Kirkbride, 1985; Byrd, 1989). Beidha is therefore situated in the midst of an area of abrupt variations in elevation and geology that determine a variety of natural plant (and animal)

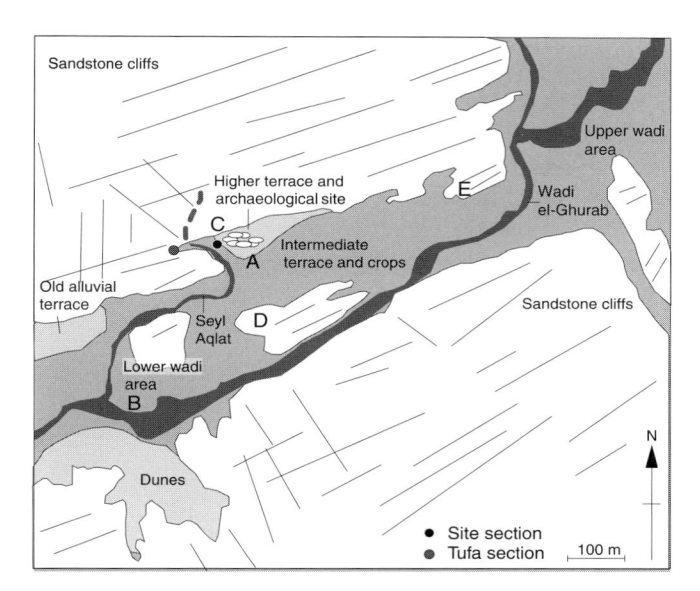

Figure 16.2. Schematic map showing the surroundings of the archaeological site of Beidha. The letters A to E refer to the locations from which the panoramic pictures presented in Figure 16.3 were taken.

communities: from the forested, Mediterranean highlands of Jebel Shara (Zohary, 1962; Helbaek, 1966), to the steppic zone of the sandstone and alluvial valleys where Beidha itself is situated, and ultimately to the desert settings of the Wadi Araba (Byrd, 1989). This particular situation is believed to have played a major role in the initial human occupation of the site, as it would have provided for a variety of edible plants and herd animals, distributed along an altitudinal gradient within a reduced catchment area (Byrd, 1989).

The alluvial valley harbouring Beidha is aligned on a roughly east–west axis (Figure 16.2). The site itself is situated on the northern side, on an elevated area interpreted to be an ancient alluvial terrace (the 'upper terrace' of Raikes, 1966; Field, 1989). Above the surface of this terrace the Neolithic village has created a low tell (Kirkbride, 1989; Byrd, 1989). An intermediate valley base is located south of the site, with elevations dropping gently from the upper to this lower alluvial terrace, where the cultivable soils are located (Raikes, 1966; Byrd, 2005). At present the Wadi el-Ghurab dissects these lower alluvial deposits and runs alongside the sandstone cliffs on the southern side of the valley (Figure 16.2).

A seasonal channel, the Seyl Aqlat, additionally drains the sandstone cliffs just northwest of the site (e.g. Kirkbride, 1960, 1989) before joining the Wadi el-Ghurab. The Seyl Aqlat cuts over 18 m through the high bank upon which the archaeological site is located and appears to have dissected or removed the western part of the site (Figure 16.3), probably after its abandonment in $c.$ 8,500 BP (Kirkbride, 1968; Kirkbride, 1985; Byrd, 1989; Kirkbride, 1989; Field, 1989). Two large sand dunes have developed against sandstone cliffs in the southwest of the valley (Figures 16.2 and 16.3).

Springs are nowadays scarce in the area near Beidha (Kirkbride, 1966) and occur notably at the geological transition between the limestone and sandstone formations. About four kilometres east of Petra is the strong spring of 'Ain Musa (the biblical spring of Moses; Byrd, 1989). With the exception of the 'Ain Musa spring, none of the springs described in the Beidha area are particularly powerful nowadays (Kirkbride, 1966). The nearest water source to the site today is the spring of Dibadiba (or Dibadibah), situated $c.$ 3 km east of the Beidha settlement on the slope towards the Cretaceous limestone uplands, at an elevation of 1,320 m amsl and near the contact between the limestone and the Cambrian/Ordovician sandstone (Kirkbride, 1966, 1968; Byrd, 1989). Kirkbride (1968) (see also Raikes, 1966) had assumed this to be the nearest perennial source of water during the Neolithic occupation, although she mentioned the possibility of rock pools and catchment areas near the village to provide for additional, intermittent water sources.

Mean annual rainfall in the area varies greatly between the Jebel Shara ($c.$ 300 mm year^{-1}) and the Wadi Araba (less than 50 mm year^{-1}). It is $c.$ 170 mm year^{-1} at Wadi Musa and has been estimated to average 170–200 mm year^{-1} at Beidha (Raikes, 1966; Byrd, 1989; see also Banning and Kohler-Rollefson, 1992). In the region, rainfall occurs principally in winter, with a dry summer (e.g. Henry, 1997; EXACT, 1998). Raikes (1966) further indicates the extreme range of variation in mean annual rainfalls over a 20-year period at Wadi Musa (1940–1960 period, −60% to +82% around the mean value of 170 mm year^{-1}). Predominant winds are from the west (Banning and Kohler-Rollefson, 1992).

Present-day vegetation at Beidha belongs to the steppe category (Helbaek, 1966; Gebel and Starck, 1985), but may have been strongly affected by human-induced environmental degradation during the past 9,000 years (Byrd, 1989). On the sandstone shelf, the alluvial valleys themselves contain no trees, although better-watered niches within the sandstone massifs harbour tree species such as oaks and junipers. The soil of the valley is composed of a mix of calcareous alluvium from Jebel Shara and wind-blown particles from the sandstone cliffs and Wadi Araba, and is considered as highly permeable and unable to retain winter rain for long (Helback, 1966). Dry farming in settings similar to that of Beidha is usually associated with a minimum of 200 mm average annual rainfall (Raikes, 1966), and Helback (1966) notes that nowadays 'only hollow patches with a comparatively high groundwater table can bear a crop and then only in years of a favourably distributed winter rain and snow'. Comer (2003) indicates that water is now brought to the Beidha area and used to replenish Nabataean cisterns. However, this water is not used for irrigation. Instead, it provides drinking water for goats, and cooking and washing water for the Bedouin families. Agriculture at Beidha is therefore still dependent, as it was $c.$ 40 years ago, on sufficient and well-distributed rainfall. On a good year, herds of goats owned by the Bedouin tribes will be allowed to eat stubble after the harvesting; the entire plants will be consumed by the animals if the absence of rainfall at the appropriate moment induces crop failure.

Figure 16.3. (A) Panoramic view from the site, looking southwest. (B) View of ancient gravel terraces, looking south-southwest. Robyn Inglis for scale. (C) View from the site on the tufa section and the dunes, looking southwest. (D) View from a sandstone outcrop on the site and the Seyl Aqlat, looking north-northwest. (E) View from a sandstone outcrop on the alluvial plain and the present-day bed of the Wadi el-Ghurab. Looking southeast. Picture credits: C. Rambeau and R. Inglis, May 2006.

16.3 HUMAN OCCUPATION AT BEIDHA: CHRONOLOGICAL FRAMEWORK

The site of Beidha appears to have been occupied by an Early Natufian camp, with some later Natufian occupation after a short break (Byrd, 1989). There was then a long period of abandonment before its reoccupation in the Pre-Pottery Neolithic (Byrd, 2005).

Complementary to archaeological observations, such as the nature of the lithic assemblage and architectural types, the absolute chronology of occupation at Beidha (Figure 16.4; Table 16.1) has been previously determined by radiocarbon dating techniques (Byrd, 1989; Byrd, 2005). All dates are presented here as calendar years before present (cal. BP) re-calibrated from the original radiocarbon dates using the Oxcal 4.0 programme online (IntCal 04, 95.4%).

Table 16.1 *All radiocarbon dates from Beidha*

In bold: this study (three dates). All other dates from Byrd (1989, 2005a), recalibrated. Calibration was carried out using the Oxcal 4.0 programme online (IntCal 04, 95.4%).

Period	Deposit	Lab. reference	Date uncal. BP	±	Date cal. BP	±
Natufian (1)		AA-1463	12,910	250	15,155	861
Natufian (1)		AA-1465	12,450	170	14,549	540
Natufian (1)		AA-1464	12,130	190	14,225	605
Natufian (2)	**site section**	**beta 235214**	**11,820**	**70**	**13,645**	**176**
Natufian (2)	**site section**	**beta 235215**	**11,260**	**60**	**13,156**	**120**
Natufian (2)		AA-1462	10,910	520	12,584	1315
Neolithic	**site section**	**beta 235216**	**9,110**	**50**	**10,302**	**108**
Neolithic	L4:11	P-1380	9,128	103	10,262	321
Neolithic	F4:300	GrN-5062	9,030	50	10,104	166
Neolithic	L4:13	K-1086	8,940	160	9,984	428
Neolithic	F4:300	P-1382	8,892	115	9,925	315
Neolithic	L4:11	GrN-5136	8,810	50	9,913	242
Neolithic	E2/P:24	BM-111	8,790	200	9,911	469
Neolithic	J4:16	AA-13036	8,830	70	9,907	272
Neolithic	K3:16	K-1410	8,850	150	9,892	344
Neolithic	K3:17	K-1411	8,770	150	9,866	336
Neolithic	L4:13	P-1381	8,765	102	9,851	304
Neolithic	J13:41	AA-13038	8,765	80	9,850	300
Neolithic	E1:30	K-1082	8,710	130	9,842	323
Neolithic	E2/P:24	K-1084	8,730	160	9,842	356
Neolithic	E1:30	P-1378	8,715	100	9,841	310
Neolithic	K3:17	K-1412	8,720	150	9,838	344
Neolithic	L4:11	K-1083	8,640	160	9,752	432
Neolithic	L4:51	AA-14109	8,646	69	9,694	194
Neolithic	F4:300	K-1085	8,550	160	9,643	507
Neolithic	E1:30	GrN-5063	8,640	50	9,630	103
Neolithic	E1:30	P-1379	8,546	100	9,593	293
Neolithic		AA-1461	8,390	390	9,398	978
Neolithic	J13:30	AA-13037	7,720	130	8,649	333
Neolithic	J4:16	AA-13035	6,535	70	7,443	127

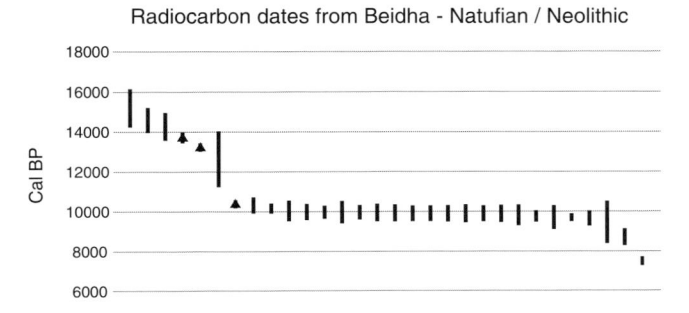

Figure 16.4. Chronological framework at Beidha. Dates marked by triangles: this study. Black lines: data from Byrd (1989, 2005), recalibrated. Ages calculated using Oxcal 4.0 (Int Cal 04, 95.4%).

16.3.1 The Natufian period

Natufian (e.g. Byrd, 1989; Bar-Yosef, 1998; Byrd, 2005; Boyd, 2006) populations have been described as semi-sedentary complex hunter-gatherer groups, exploiting natural resources such as wild cereals and the products of hunting. While there are a small number of apparently long-term Natufian settlements that indicate an increase in sedentary behaviour, many sites still appear to fall into the classic hunter-gatherer range of base camps with smaller, shorter-lived special purpose camps (Olszewski, 1991). It appears that the increase in sedentism and complexity mostly occurred during the Early Natufian, whereas during the Late Natufian there was an increase in mobility and a reduction in group size. This is often thought to reflect the harsher environmental conditions of the Younger Dryas (e.g. Mithen, 2003).

At Beidha, soundings below the Neolithic village and trenches revealed the presence of Natufian deposits in six areas of the site. During the earlier seasons (1958–1967; Byrd, 1989) a continuous sequence of about 0.4–0.6 m thickness was recorded, and subdivided into three depositional phases. In 1983, a trench was

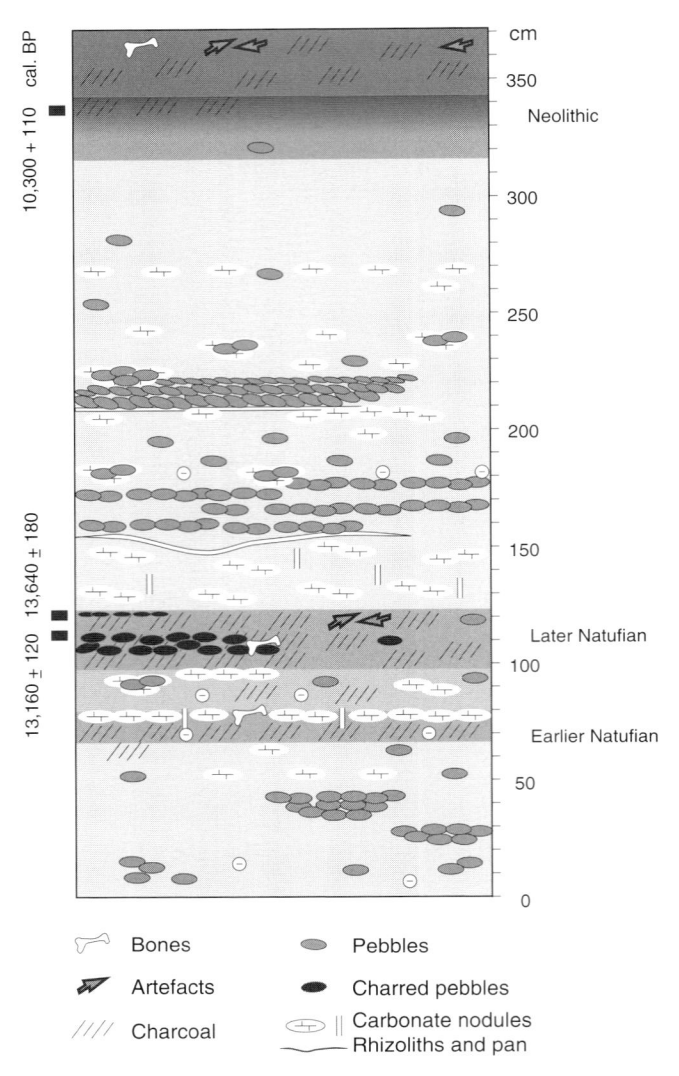

Figure 16.5. Schematic succession of sedimentary layers at the site section, as illustrated in Figures 16.2 and 16.3, and location of the levels from which new radiocarbon dates have been obtained.

opened on the northwestern (eroded) slope of the site (Figure 16.3A and D). There, two discrete horizons of Natufian occupation were discovered (c. 0.15 m and 0.4 m thick) separated by a sterile layer (0.4 m of paler sand with diffused charcoal and carbonate nodules but no artefacts or bones; Figure 16.5). The lower, thinner occupation layer contains a lithic assemblage comparable to the Natufian deposits discovered in the other excavation areas, whereas the upper layer had an extremely low density of artefacts with a different character. Five dates were obtained from this trench (Byrd, 1989). The oldest occupation horizon was dated between 15,155 ± 861 cal. BP and 14,225 ± 605 cal. BP (three dates; Figure 16.4; Table 16.1), placing this layer within the Early Natufian. The younger Natufian horizon yielded two dates, one of which (12,584 ± 1,315 cal. BP) definitely corresponds to the later Natufian settlement: the other (9,398 ± 978 cal. BP) has been considered as intrusive from the Neolithic (sample AA-1461; Figure 16.4; Table 16.1). Byrd (1989) therefore placed the later Natufian occupation at Beidha at the juncture between the Early and Late Natufian. The date used to justify this is, however, associated with a significant error bar of ±1,315 years (Figure 16.4; Table 16.1).

The Early Natufian at Beidha has been described as composed of at least three phases of semi-permanent, complex hunter-gatherer settlement (Byrd, 1989). Evidence for hunting is abundant, whereas evidence for plant collecting and consumption is limited; no architectural features such as walls, storage facilities or burials were discovered. Byrd (1989) suggests that the Early Natufian occupation Beidha represents a short-term/seasonal campsite, with recurrent occupation during an extended time period. Byrd (1989) proposes that occupation occurred either during winter or mid-summer, at times when plant resources were less abundant and hunting may have represented a favoured procurement strategy.

The later Natufian period has only been recorded on the northern edge of the site and seems to correspond to more ephemeral occupation – maybe representing a short period of transient occupation. No definite human activity is then recorded for c. 3,500 years, but the presence of a small structure made of sandstone slabs – 'such as children at play might make', c. 20 cm below the start of PPNB occupation layers, hints at an occasional occupation of the site (Kirkbride, 1968).

16.3.2 Pre-Pottery Neolithic A (PPNA) – Early Pre-Pottery Neolithic B (EPPNB)

There has been considerable debate regarding the early Neolithic in the southern Levant, especially concerning the character of the transition from the PPNA to the PPNB. No structural phase clearly pre-dating the Middle PPNB (MPPNB) was reported from Beidha, but there is some evidence for earlier occupation. Byrd notes that some material from the site initially thought to be Natufian is Neolithic in character, and 'may well predate the Middle PPNB' (Byrd, 2005, p. 17). This was associated with a mudbrick structure, unlike the stone architecture associated with the MPPNB. There are further structural remains that pre-date the first well-known architectural phase (i.e., wall-less, outdoor courtyards deposits, extending to c. 20 cm below the first houses; Kirkbride, 1968). In addition El Khiam points (diagnostic of the PPNA) and Helwan points (diagnostic of EPPNB) were discovered during Kirkbride's excavation in the earliest layers (Mortensen, 1970). Byrd suggests that examination of the earliest Neolithic layers still requires further field investigation and sampling. Even the well-known first architectural phase (A, characterised by curvilinear architecture) conventionally associated with the MPPNB poses some questions as the style of architecture is typical of that associated with the EPPNB in the northern

Levant, for example at Jerf el-Ahmar (Stordeur et al., 1997). Furthermore, the earliest radiocarbon date for the Neolithic occupation from the excavations (Phase A) appears to lie on the boundary between the EPPNB and MPPNB (Figure 16.4; Table 16.1).

16.3.3 Pre-Pottery Neolithic B (PPNB)

In the Levant, the PPNB is a period of settlement expansion and development of animal herding strategies, following the beginning of plant cultivation that occurred during the PPNA. Large permanent settlements became established during the MPPNB in the Levantine region, Beidha being a classic example although others, such as 'Ain Ghazal, became substantially larger, especially in the Late PPNB (Rollefson and Köhler-Rollefson, 1989). It has been argued that around 8,500 BP, large PPNB settlements in the southern Levant abruptly collapsed, and were replaced by the smaller settlements of the PPNC, which marked a return towards nomadic pastoralism (Rollefson and Köhler-Rollefson, 1989).

There is no evidence to indicate abandonment and reoccupation at any point in the Neolithic occupation of the site (Byrd, 2005). Byrd argues that occupation was largely restricted to the MPPNB (ibid.) although the radiocarbon dates suggest a longer sequence from c. 10,300 and 8,600 cal. BP, with a recent review suggesting that the MPPNB ran from 10,100 to 9,250 cal. BP, with the Late PPNB dating to 9,250–8,700 cal. BP (Kujit and Goring-Morris, 2002). Kirkbride recorded a complicated stratigraphic sequence, which Byrd (2005) simplified into three major phases, reflecting both stratigraphic relationships and architectural styles. The earlier Phase A is characterised by semi-subterranean single-room round stone buildings built around wooden pole frames. In Phase B the walls become straighter with rounded corners and the buildings more subrectangular, but they are still single-room and generally semi-subterranean. In Phase C the buildings become rectangular, with ground (or possibly basement) floors divided into a series of cells by massive piers that probably supported upper floors. Beidha has always been seen as an archetypal model of the transition from round single-roomed to rectangular multi-roomed buildings that occurred during the Neolithic. This probably reflects a greater density of settlement occupation, increasing population, and more storage on a household level.

Barley was the main cereal recovered from the farming community of PPNB Beidha, but domestic emmer wheat has also been identified in the plant assemblage (Helback, 1966; Perkins Jr, 1966); wild plant species were also collected. There are strong indications for the domestication of goats, while hunting remained an important subsistence strategy (Perkins Jr, 1966).

Most of the published dates (22 out of 23) related to this period fall between $10,262 \pm 321$ cal. BP and $8,649 \pm 333$ cal. BP (Figure 16.4; Table 16.1). This implies an abandonment of the site shortly after the youngest date. One date has been, however, obtained at Beidha at $7,443 \pm 127$ cal. BP (Figure 16.4; Table 16.1). This is an outlier with no other supporting dates or cultural remains distinctive to this date, and hence its validity is questionable. It must be noted here, additionally, that Rollefson (1998, 2001) has argued that Beidha Phase C is PPNC (c. 8,800–8,200 cal. BP; Staubwasser and Weiss, 2006), mostly on the base of architectural arguments.

It is difficult to determine whether the abandonment of the village was a gradual or sudden event. Refuse dumping patterns within the uppermost layers of the Neolithic occupation seem to indicate a gradual abandonment of buildings and, more generally, of the site (Byrd, 2005), yet the destruction of part of the younger Neolithic deposits by Nabataean terracing renders the pattern of abandonment difficult to interpret. After the site's abandonment, the Seyl Aqlat cut through the western part of the site, probably removing a considerable area of both the Natufian and Neolithic occupation (e.g. Kirkbride, 1966; Kirkbride, 1985; Byrd, 1989; Field, 1989; Byrd, 2005). Although the extent of the erosion is impossible to determine, it has been estimated that up to half of the Neolithic village may have been destroyed (Field, 1989).

16.3.4 Nabataean and Roman time periods

There is no evidence for subsequent use of the valley until the Nabataean/Roman period. During the Nabataean, the Wadi el-Ghurab was terraced with a series of stone walls, marking fields that are still used today by local Bedouin communities for rain-fed cultivation. The Nabataeans also terraced the tell area more than 6,000 years after its abandonment, destroying part of the upper cultural layers in the process (Kirkbride, 1968; Byrd, 2005). There are badly weathered remains of rock-cut aqueducts, also presumed to date from this period, running around the sandstone outcrops of Seyl Aqlat. Numerous Nabataean cisterns, linked to an elaborate system of runoff water collection and storage, are present in the Beidha area (Helback, 1966; Comer, 2003) and are still used today as water storage facilities by the local Bedouin communities.

16.4 THE PALAEOENVIRONMENTAL CONTEXT

16.4.1 Environmental conditions in the southern Levant (c. 20,000–7,000 cal. BP)

The general background for this study is provided by a regional assessment of climate and palaeoenvironmental changes in the eastern Mediterranean, derived from various critical reviews of the available evidence (Chapter 6, this volume; see also Sanlaville, 1996; Henry, 1997; Sanlaville, 1997). It seems that the Last Glacial Maximum (LGM, c. 23,000 to 19,000 cal. BP) period in the Levant was characterised by cooler temperatures than present; there is still some debate about whether it was also a drier period. After a period of transition possibly marked by a short-term climatic event at c. 16,000 cal. BP (Heinrich Event 1), the Bølling–Allerød warm interval (c. 15,000 to 13,000 cal. BP) was a period of increased rainfall and higher temperatures in the eastern Mediterranean region, as evidenced by palynology, speleothems and marine sediments from the Red Sea. The Bølling–Allerød was directly followed by the Younger Dryas (c. 12,700 to 11,500 cal. BP), a cold and arid period of major importance in the North Atlantic region. It has been argued that this event had a major impact on Levantine climates and populations (e.g. Bar-Matthews et al., 1997, 1999, 2003; Mithen, 2003; Robinson et al., 2006; Cordova, 2007; Chapter 6 of this volume). The dominant view of an arid and cold Younger Dryas in the eastern Mediterranean, and its influence on human communities, are, however, disputed by several authors (Tchernov, 1997; Stein et al., 2010; Stein et al., in press). The warm and wet characteristics of the following early Holocene period (c. 9,500 to 7,000 cal. BP) are clearly shown in pollen records, isotopic records, fluvial deposits and soil sequences (Chapter 6, this volume). Around 8,200 cal. BP, cold and arid conditions seem to settle in the eastern Mediterranean (e.g. Staubwasser and Weiss, 2006; Weninger et al., 2006; Berger and Guilaine, 2009) although, considering the scarcity of data in the region, it is still unclear if this relates to an abrupt climatic event or is part of a more gradual trend towards climate deterioration starting globally at c. 8,600 cal. BP (Rohling and Palike, 2005).

Of all the proxies available for the reconstruction of palaeoclimates in the Levant during the periods of occupation at Beidha, speleothem isotopic records are possibly the most informative, as they offer a continuous and well-dated sequence. In particular, speleothems from the Soreq Cave (Bar-Matthews et al., 1996, 1997, 1999, 2000, 2003) have allowed for high-resolution records dated by ^{230}Th–^{234}U (TIMS) methods. The oxygen isotopes variations at Soreq clearly show variations that probably correspond to major climatic events such as the LGM, the Bølling–Allerød and the Younger Dryas (see also Chapter 6, this volume). Bar-Matthews et al. (2003) indicate a close variation between planktonic foraminiferal and speleothem δ^{18}O values during the past 250,000 years, which suggest a link between marine and terrestrial records. They therefore used sea surface temperature (SST) reconstructions as an approximation for land surface temperature variations. This information was employed to refine Bar-Matthews and colleagues' previous interpretations of the oxygen-isotopic record and calculate both the isotopic composition (δ^{18}O$_{rain}$), and, for certain periods, the amount, of 'palaeo annual rainfall'. The calculated δ^{18}O$_{rain}$ is more positive during the Younger Dryas, suggesting low rainfall and enhanced aridity. Prior to this period, and during the early Holocene (11,000–7,500 cal. years BP), wetter and warmer conditions are suggested by a more negative δ^{18}O$_{rain}$ record. Recently, Affek et al. (2008) have used the 'clumped isotope' thermometry technique (which is based on the number of ^{13}C–^{18}O bonds within the carbonate lattice and can be utilised to calculate temperatures at the time of the carbonate growth) on Soreq speleothems. The temperatures calculated have been proved similar to those estimated for the eastern Mediterranean SST. The calculated temperatures are 6–7 °C colder than for the present-day during the LGM (20,000–19,000 cal. BP), whereas they are slightly higher than, or similar to, those of today for the early Holocene (10,000–7,000 cal. BP).

In climatic settings closer to those at Beidha, i.e. in the semi-arid and arid regions of the southern Levant, a series of proxies, including soil and fluvial deposits and pollen information from archaeological sites, indicate a similar history. In the Wadi Faynan area (western Jordan, c. 30 km north of Beidha), drier conditions are attested by aeolian deposits in the Wadi Dana around the LGM (McLaren et al., 2004). During the period c. 18,300–12,900 cal. BP, flora and fauna assemblages from archaeological sites in Eastern and Southern Jordan (Azraq, Judayid) and within the Jordan Valley (Wadi Fazaël) seem to indicate overall favourable climatic conditions (Sanlaville, 1996 and references therein; dates recalibrated), interrupted by a short, dry incursion (dated at c. 13,900 cal. BP at Judayid; Sanlaville, 1996). A short, moderately dry episode at c. 15,400–13,900 cal. BP following better climatic conditions is also recorded in pollen studies from Wadi Hisma (southern Jordan; Emery-Barbier, 1995; Henry, 1997; Sanlaville, 1996; dates recalibrated).

In the Negev, dune-dammed lakes that appeared during this phase of overall increased humidity (coeval with a phase of pedogenesis) were transformed into sebkhas at c. 12,900–11,500 cal. BP (Sanlaville, 1996 and references therein). The size and isotopic compositions of terrestrial gastropods in the Negev also hint at an arid episode during the interval c. 12,900–12,500 cal. BP (Magaritz and Heller, 1980; Sanlaville, 1996). The accumulation of drift sand on top of Early Natufian layers, at both Wadi Judayid and Beidha, has been interpreted as suggesting a dry episode (Henry, 1997).

During the early Holocene, the expansion of PPNB sites (in particular, Middle and Late PPNB sites) in settings that are now hyper-arid (Azraq Basin, the Black Desert in Eastern Jordan, Wadi Hisma) seem to correspond to a major pluvial episode (Henry, 1997 and references therein). In the Azraq basin, domestic-type einkorn and barley were exploited since the early PPNB (Jilat 7; Garrard et al., 1996; Henry, 1997) in locales receiving far less rainfall (i.e., about half) than necessary for rain-fed agriculture, and without spring-water supply. C3 grasses are present at the site of Ain Abu Nukhayla in Wadi Rum,

c. 90 km south of Beidha, during the PPNB, whereas such plants are uncommon in the modern, steppic and desertic vegetation of the area (Portillo et al., 2009). A southward shift in the distribution of pure C3-plant communities is recorded by the $\delta^{13}C$ composition of fossil land snails in the northern Negev during the early Holocene (c. 11,000–6,800 cal. BP), indicating wetter conditions then than during the middle Holocene and today (Goodfriend, 1999). In Wadi Faynan, palynological, plant macrofossils and molluscs studies indicate a much more humid environment during the early Holocene than at present (Hunt et al., 2004). A rainfall value of about 200 mm (compared with the modern 120 mm; see Chapter 15, this volume) is proposed for the time period before 8,000 BP. Decreased rainfall and desiccation, coeval with variations in global temperature, is evidenced in Wadi Faynan at c. 8,000 BP, followed by a return to increased rainfall after 7,400 BP (the then absence of coeval forest regeneration may have been due to human inhibition). Meandering, perennial rivers are recorded at Wadi Faynan prior to 6,000 cal. BP (Hunt et al., 2004). The early Holocene optimum seems to be marked in the Jordan Valley by warm and humid conditions compatible with a geomorphological record of marshlands, wide flood plains and travertine deposition. The following phase begins with a marked climate instability with torrential flows and erosion, then stability under a cold and humid climate, before renewed erosion and the establishment of a semi-arid, Mediterranean climate in the Jordan Valley in the mid-Holocene (Hourani and Courty, 1997).

16.4.2 Previous environmental studies at Beidha

Most of the detailed studies about the environmental changes at Beidha were produced during the excavation seasons. In particular, Raikes (1966) and Field (1989) produced preliminary reports on the geomorphology and palaeoenvironments at Beidha. Fish (1989) presented a preliminary study of the pollen record. Observations made during the excavations by Kirkbride and her colleagues provide further information about the palaeoenvironment. Later studies around the Beidha area (Comer, 2003) also provide useful remarks and observations.

Geomorphologically, the archaeological site of Beidha is located on an ancient alluvial fill terrace bordered by sandstone cliffs on the northern edge of the valley. To the south, elevations slope gently to an intermediate terrace level (Figure 16.2). To the west, the steep descent (c. 18 m) towards the modern bed of the Seyl Aqlat results from the considerable erosion that this seasonal watercourse inflicted to the western part of the site (Byrd, 2005; Figure 16.3).

Overall, the sedimentation at the site can be summarised as follows: a sandy sediment is deposited prior to the first level of the Natufian occupation (on average 0.4–0.6 m throughout the site, c. 0.15 m on the western edge where the second Natufian level is recorded; Byrd, 1989), then c. 0.4 m of sterile sand overlies this cultural layer in the northwestern part of the site, then c. 0.4 m related to the second Natufian occupation. This is surmounted by over 1.5 m of sterile sand, on which the Neolithic village was constructed (Figure 16.5). The pollen assemblage, albeit patchy, with probable localised biases and difficult to interpret, indicates an open, steppe-like vegetation in both sterile and cultural horizons, with arboreal pollen always less than 20% (Fish, 1989).

16.4.3 Environmental conditions during the Earlier Natufian occupation

Raikes (1966) identifies two terrace systems around Beidha. The upper terrace bears the archaeological site, whereas the lower terrace (intermediate valley base) contains the ancient agricultural soils (Figure 16.2). The upper terrace, following Raikes' interpretation, would be the remnant of a valley fill created during an interpluvial phase, at a 'very remote period of long duration' (Raikes, 1966). Raikes identifies an erosional phase, postponing the creation of the upper terrace and responsible for the creation of the lower terrace. He attributes this phase to downward faulting activity in the rift, which would have resulted in lowered drainage datum and increased erosion. The lower terrace was then itself eroded during another episode of downfaulting. The upper and lower terraces, following Raikes' scenario, may have already been isolated during the Natufian.

Travertine deposits have been discovered at Bir Abu Roga, less than 1 km downstream of Beidha (Kirkbride, 1966; Raikes, 1966). Raikes postulates that the spring at Bir Abu Roga – situated at the level of the lower terrace – and/or other springs in the region may have represented the drinking water supply for the Natufian people of Beidha. He further suggests that this period may be characterised by higher rainfall, perhaps as much as 400 mm year^{-1} on average.

Field (1989) proposes a slightly different story. He studied a composite sedimentary section from two areas of the site and interprets the Natufian occupation to have occurred during a time of aggradation of the Wadi el-Ghurab. Sediments would have been accumulated by stream deposition. Interestingly, a lens of gravels, pebbles and sand related to wadi deposition, uncovered 3 m below the surface on the southeastern gentle slope of the tell, contains artefacts of probable Natufian origin; it has been suggested that they may originate from the erosion of Natufian layers further to the north or the west (Byrd, 1989). This deposit confirms wadi deposition at the same elevation as, and thus probably contemporaneous with, Early Natufian deposits on the western slope (c. 2.3 m under the Neolithic).

Field relates the creation of a higher valley floor, prior to the erosion event that isolated the terrace on which the site is located, to a period of greater sediment supply to the wadi system, such as may occur during a time of less intense (fewer storm events, and more regular but not necessarily greater) rainfall. Field therefore postulates that different rainfall patterns during the Natufian may have created a more favourable environment, with enhanced vegetation cover. It has been suggested that during the Early Natufian, vegetation cover was more developed in the alluvial valleys of the sandstone shelf, and that the Mediterranean forest of Jebel Shara extended further down its slopes (Byrd, 1989). Low contents of phosphorus, nitrogen and organic matter indicate limited human impact on the environment during Natufian times (compatible with low intensity occupation; Field, 1989).

Comer (2003), in his 'cultural site analysis' of Beidha and its surrounding, notes a close correspondence between the Natufian occupation and the Bølling–Allerød amelioration with increased temperature and humidity, and suggests (as well as for the Neolithic) increased runoff as a means to provide enhanced water supply to the valley.

16.4.4 The Later Natufian and the Natufian/Neolithic sterile interval

Raikes (1966) interprets the sterile layer between the Natufian and PPNB occupation layers, and probably the sterile layer covering the Neolithic as well, as being aeolian in origin, within which current-bedded forms were created by storm-related outwash from the sandstone cliff. In contrast, Field (1989) postulates that the sterile interval between the Natufian and the Neolithic was deposited by ephemeral streams, and not by aeolian deposition. He bases his argument on the fact that several fining-upward sequences, made of gravelly sands and capped up by siltier deposits and carbonate build-up, are observable in the sedimentological section: such sequences form in stream deposition settings, with a short period of non-deposition at the end of the sequence. There is a non-negligible gravel component, and cross-bedded sands are visible near the base of the sterile interval; all these indications point towards ephemeral stream deposition.

Fish (1989) studied pollen assemblages within the same areas investigated by Field (1989). She noticed higher frequencies of chenopod and amaranth pollen in three samples, one from the intermediate layer between the two Natufian occupations, and two from the base of the sterile horizon overlying the upper Natufian layer (highest frequencies). This may represent the local expression of a more widespread cool, dry interval during the Natufian in the southern Levant (Fish, 1989). Interestingly, Kirkbride (1968) mentions an erosion gully cutting through Early Natufian levels, which was first interpreted as being a human-made structure (Kirkbride, 1968). Byrd (1989) further mentions that this erosional gully may cut from the top of the Early Natufian deposit, or from a slightly higher level. At the base of the sterile interval, pollen also indicates permanently damp habitats in the vicinity.

At the top of the sterile interval, below the Neolithic occupation horizons, the pollen assemblage is quite different from the base – no increase in chenopod or amaranth is recorded, whereas relatively high grass values suggest a rather mesic steppe assemblage (Fish, 1989). Raikes (1966) notes equally that the top layer of the sterile sand between the Natufian and the PPNB bears marks of organic material, hinting at relatively important vegetation coverage at the site before the Neolithic settlement. Kirkbride (1968) similarly records shadows in the sand just underneath the Neolithic layers, suggestive of areas where a cover of shrubs has been removed prior to the village construction.

16.4.5 The Neolithic

According to Raikes (1966), at the time of the Neolithic occupation, the settlement was located on a high terrace remnant, adjacent to a wide valley floor composed of loamy sand, with a more intensive vegetation cover protecting it from erosion.

Field (1989) identifies an erosional phase, after the Natufian but prior to or coeval with Neolithic occupation, that isolated, among others along the Wadi el-Ghurab, the alluvial terrace on which the Neolithic village was constructed. This erosion phase may have been related to increased rainfall intensity and a flash-flood regime. Field suggests that such climatic conditions may have reigned during the PPNB occupation; the settlement, situated on a higher terrace, may have been protected from flash-flood events (Field, 1989). In this context, the presence of a village wall constructed during the earliest phase of the village (Phase A), with steps connecting the wall to an upper layer where the houses themselves were situated (Kirkbride, 1968; Byrd, 2005), seems to indicate that the terrace was already isolated at the start of the Neolithic occupation. Bar-Yosef (1986) suggested that this wall may have been a protection from floods, although Byrd (2005) notes that flooding may not have represented a considerable issue if the Wadi el Ghurab was following a course on the southern side of the valley during the PPNB as it does today.

According to Field (1989), no aggradation occurred during the Neolithic occupation itself, but thin cross-bedded sands, probably caused by stream deposition at minor tributaries coming from the sandstone cliffs bordering the site, are observable in a few places. The higher concentrations of phosphorus, nitrogen and organic matter during the PPNB are coherent with reduced sedimentation and increased human occupation.

No water harvesting or storage techniques were apparently in use during the Neolithic. In light of this, Helbaek (1966) proposed a palaeo-rainfall of 300–350 mm year^{-1} during the occupation, based on estimated requirements for the growth of wild wheat (wild species are considered to require substantially higher annual rainfall than the 200 mm limit usually associated with dry farming; Raikes, 1966). Raikes (1966) alternatively proposed that changes in the characteristics of soil cover, or a similar/slightly higher rainfall (but less intense and with a better distribution throughout the year, and accompanied by higher soil retention), would have made agriculture at PPNB Beidha possible without a major change in the climate conditions. Kirkbride (1968) reiterated this conclusion and declared that since the Neolithic village was first constructed there has probably been no major climate change; enhanced human activity (clearance, grazing and cultivation) and intensified erosion (maybe worsened by down-faulting of the drainage datum) are made responsible for gradual environmental degradation. Raikes (1966) and Kirkbride (1966, 1968) postulate that the tributary drainage of the Seyl Aqlat, responsible for the destruction of the western side of the site after its abandonment, was not in existence during the time of occupation or was just starting, and was therefore of minor influence to the settlement.

Raikes (1966) views the wadi system during the Neolithic as sub-perennial at best (mid-winter to early summer), even if the local rainfall had increased to 300 mm yearly average. He suggests that most of the drinking water during the Neolithic must come from the spring at Dibadiba. Permanent sources of water (springs or pools) during the Neolithic are, however, attested by the pollen record (*Typha*/*Sparganium*; Fish, 1989). Moreover, trench excavations near the sandstone cliff edge, *c*. 80 m east of the village, have uncovered possible travertine deposits just over a metre under the surface (Kirkbride, 1966, 1968; Byrd, 1989, 2005). The presence of a spring in the immediate vicinity of the settlement is interpreted as a factor that may have influenced settlement location (Byrd, 1989).

Comer (2003) notes a good correspondence between the time of Neolithic occupation at Beidha and the wetter and warmer conditions of the early Holocene Climatic Optimum, suggesting that water may have been present near the site owing to increased runoff. His study demonstrates that, considering the geographical position of Beidha, any runoff produced by sufficient rainfall would be channelled towards the site. Comer (2003) argues that the depletion of certain natural resources, as they were increasingly and more effectively exploited, triggered the abandonment of Beidha. He bases this proposal on the claim that Beidha was deserted well before a period of climate deterioration in terms of significant decreases in temperature and rainfall, which he places at roughly 7,000 BP.

16.5 MATERIAL AND METHODS

The area around Beidha offers numerous sedimentary outcrops and carbonate precipitations that can be used for palaeoenvironmental reconstruction. This study focused on two major sedimentary sections:

1. A sedimentary section containing a series of pedogenic nodules, located directly on the western side of the site (Figure 16.5);
2. A sequence of carbonate deposits, related to an ancient spring, outcropping near the site in a small canyon (Figure 16.6).

We also use observations resulting from a geomorphological survey conducted around Beidha in 2006, which aimed to identify major episodes of wadi aggradation/downcutting along with changes in the sedimentation patterns.

16.5.1 Sedimentary sections and sample preparation

A section of 372 cm, directly situated on the western side of the Beidha archaeological site, where it has been exposed by stream-cutting erosion, was cleaned and sampled for carbonate concretions, bulk sediment and charcoal. It is mainly composed of sandy deposits and comprises black/grey layers with charcoal, burnt pebbles and bones that were believed to correspond to Natufian and EPPNB occupation layers (Figure 16.5). A complementary section of 251 cm, situated inside the archaeological site and encompassing the rest of the PPNB occupation layers, was sampled for carbonate concretions and bulk sediment.

A 2-m-thick sequence of spring carbonates outcrops less than 100 m away from the site on the western side of the Seyl Aqlat (Figures 16.3 and 16.6). It consists of thin layers of sand-rich carbonates, with occasional reed imprints. Collected samples were mechanically cleaned, including the removal of all altered surfaces, then cut into *c*. 1-cm-thick slices.

Carbonate samples from both the site section and the spring carbonate sequence were ground to a fine powder using an agate planetary ball mill.

16.5.2 Dating methods

Three fragments of charcoal from the site section have been dated by radiocarbon methods (AMS analysis, Beta Analytic Radiocarbon Dating Laboratory, Miami, Florida). Two of the samples originate from a level attributed to the Later Natufian horizon described by Byrd (1989), and the third belongs to the start of the Neolithic occupation layers (Figure 16.5). The two samples related to the Natufian horizon (beta 235214 and beta 235215; Table 16.1) failed to yield separable charcoal fractions and were analysed as organic sediment fractions following pre-treatment by acid washes. The sample from the start of the Neolithic occupation layers was treated as charred material with a pre-treatment by acid and alkali washes, following Beta Analytic pre-treatment standards.

Selected samples from the spring sequence have been dated using uranium series (U-series). Using the ^{230}Th–U disequilibrium

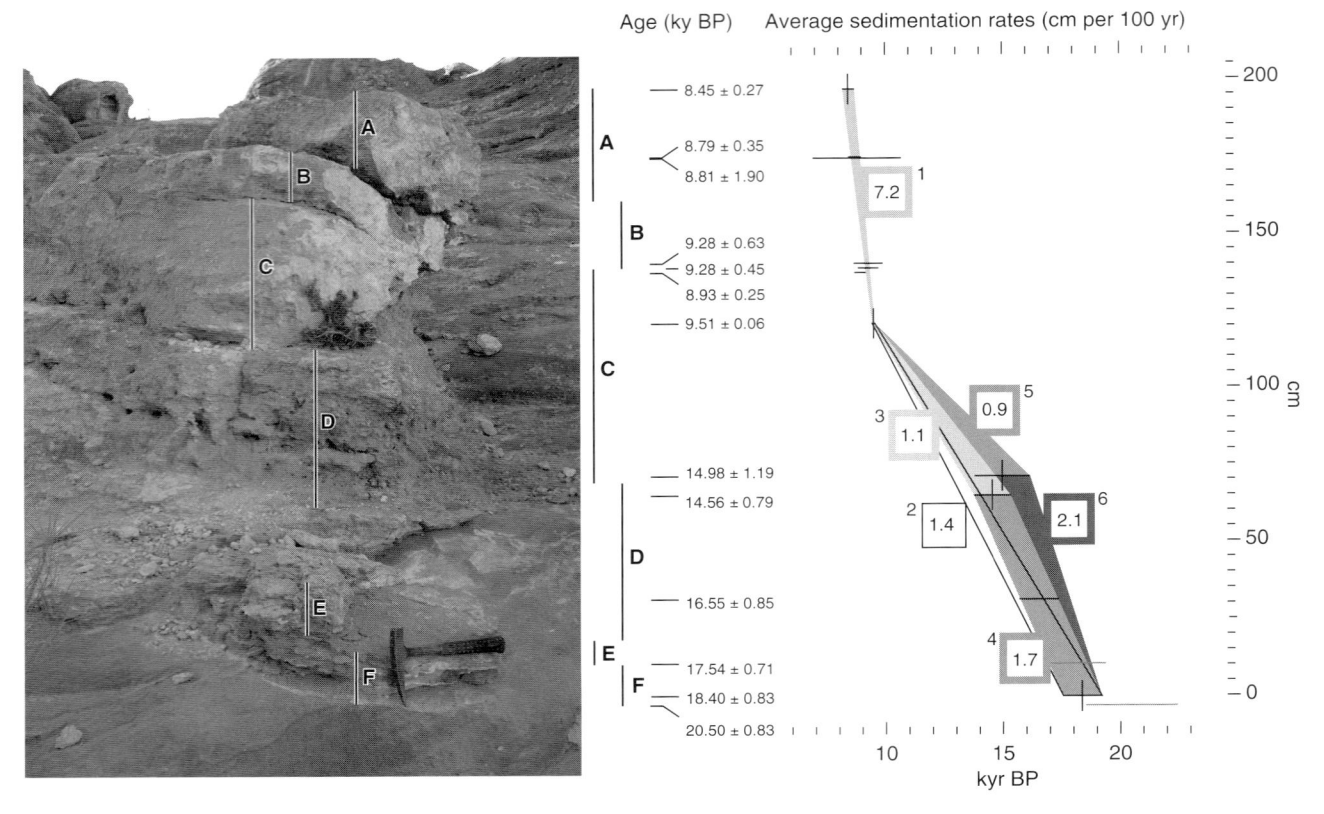

Figure 16.6. Spring carbonate (tufa) sequence, U-series dates and average sedimentation rates. Italic numbers refer to dates obtained on one sub-sample only; other dates were obtained with the isochron technique on multi sub-samples. Numbers 1–6 relate to sedimentation rates as calculated in Table 16.3.

method to date carbonates implies that at the time of precipitation the sample was free from ^{230}Th, which, along with ^{232}Th, predominantly adsorbs onto clay minerals. The spring carbonate ^{230}Th/^{232}Th ratios indicate extensive detrital contamination (Table 16.2). We therefore used an isochron method to correct for the detrital component. Isochron methods for impure carbonates described in Candy et al. (2005) and Candy and Black (2009) (see also Ludwig and Titterington, 1994; Ludwig, 2001; Hercman and Goslar, 2002 and references therein; Garnett et al., 2004 and references therein. We used several sub-samples from the same layer, which supposedly contain different ratios of a single carbonate and a single detrital phase. A 2-D isochron is constructed with the results for each sub-sample (Table 16.2), using ^{238}U/^{232}Th and ^{230}Th/^{232}Th activity ratios (S. Black; see also Chapter 9 of this volume for further details about instrumentation).

16.5.3 Granulometric analyses

Selected sediment samples from the site section were analysed for their granulometry (Coulter LS 230 laser granulometer) to determine potential changes in the sedimentation type within the sedimentological sequence, notably to discriminate between aeolian and fluvial deposits. Selected bulk sediments were also sampled from different settings (dunes, ancient terraces, wadi sedimentary sequences) for granulometry analysis and comparison with sediments from the site section.

16.5.4 Stable isotope (C, O) analyses

Bulk samples from both the site and the spring carbonate section were analysed for stable isotopes (carbon and oxygen) using a SIRA II stable isotope ratio mass spectrometer (SIRMS) after careful calibration with international standards. A maximum error of ±0.1‰ has been determined for the measurements of calcite δ^{18}O, using a series of duplicated samples and systematic measures of an internal standard (CAV-1, standardised against NBS-19).

16.5.5 Mineralogy and facies analyses

Mineralogical compositions of spring carbonate samples were determined by X-ray diffraction using a Siemens D5000 X-ray diffraction spectrometer, together with major and trace elements after dissolution of powders with mineral acids and analysis via a Perkin Elmer Elan 6000, inductively coupled plasma mass spectrometer (ICP-MS). Six thin sections were also obtained from different levels of the spring carbonate sequence.

Table 16.2 *Uranium-series dates from the tufa series*

Uranium/thorium isochron ages (in bold; e.g. **SB1**) are calculated using information from a series of sub-samples (e.g. SB1a-d). Dates based on one sample only (uncorrected ages; e.g. Bei-Ee) are shown not in bold. Uncertainties on U and Th concentrations (average SDs calculated on all the data) are ± 0.45% and ± 0.67%, respectively.

Sample	Distance from the base (mm)	U (µg kg^{-1})	Th (µg kg^{-1})	^{234}U/^{238}U	^{230}Th/^{238}U	^{230}Th/^{232}Th	U/Th isochron age
SB1a	200	908	1,366	1.015±0.012	0.031±0.002	0.064±0.019	
SB1b	200	820	830	2.288±0.014	0.043±0.005	0.131±0.024	
SB1c	200	637	423	3.261±0.014	0.056±0.007	0.257±0.026	
SB1d	200	1,145	551	3.753±0.025	0.060±0.005	0.382±0.029	
SB1							**8,435±270**
SB2a	178	95	174	1.382±0.037	0.035±0.023	0.059±0.011	
SB2b	178	118	344	0.766±0.021	0.008±0.030	0.010±0.010	
SB2c	178	128	282	1.126±0.020	0.027±0.021	0.038±0.013	
SB2d	178	90	161	1.403±0.018	0.037±0.029	0.066±0.015	
SB2							**8,790±350**
Bei-A3a	178.3	949	1,506	2.332±0.029	0.073±0.000	0.141±0.003	
Bei-A3a-1	178.3	472	999	1.843±0.039	0.071±0.001	0.102±0.002	
Bei-A3a-2	178.3	773	1,687	2.646±0.050	0.072±0.062	0.101±0.088	
Bei-A3a-4	178.3	423	906	1.774±0.040	0.068±0.051	0.098±0.073	
Bei-A3a-5	178.3	659	1,616	2.931±0.018	0.072±0.035	0.090±0.045	
Bei-A3a							8,811±1,899
Bei-B3	143.8	539	965	1.822±0.008	0.081±0.001	0.140±0.002	
Bei-B3–1	143.8	2,498	7,136	2.651±0.066	0.082±0.002	0.088±0.002	
Bei-B3–2	143.8	1,385	2,608	3.004±0.008	0.083±0.001	0.135±0.003	
Bei-B3							**9,282±628**
Bei-BC	142.3	1,216	994	1.725±0.009	0.083±0.000	0.313±0.001	
Bei-BC-1	142.3	972	2,110	1.299±0.029	0.083±0.001	0.118±0.001	
Bei-BC-2	142.3	798	1,328	1.626±0.030	0.087±0.001	0.160±0.003	
Bei-BC							**9,282±450**
Bei-C1	140.8	675	965	1.860±0.013	0.086±0.001	0.184±0.003	
Bei-C1–1	140.8	930	2,017	1.751±0.024	0.089±0.001	0.125±0.002	
Bei-C1–2	140.8	865	1,683	1.496±0.030	0.088±0.001	0.139±0.002	
Bei-C1							**8,928±248**
Bei-C2–1	124.3	805	1,001	1.600±0.045	0.087±0.003	0.215±0.005	
Bei-C2–2	124.3	1,249	623	2.169±0.048	0.084±0.001	0.518±0.009	
Bei-C2–3	124.3	763	1,684	2.275±0.005	0.085±0.006	0.118±0.010	
Bei-C2–4	124.3	884	1,093	1.442±0.024	0.085±0.004	0.210±0.011	
Bei-C2							**9,507±59**
Bei-C5a	74.3	748	1,106	2.360±0.027	0.123±0.001	0.255±0.000	
Bei-C5a-1	74.3	877	1,426	1.516±0.018	0.135±0.001	0.255±0.004	
Bei-C5a-2	74.3	827	1,617	1.943±0.017	0.128±0.001	0.200±0.002	
Bei-C5a-3	74.3	641	1,454	1.472±0.016	0.127±0.001	0.171±0.003	
Bei-C5a							**14,984±1,190**
SB3a	68	639	524	2.120±0.002	0.038±0.009	0.141±0.024	
SB3b	68	814	1,010	1.634±0.020	0.051±0.010	0.125±0.009	
SB3c	68	1,181	230	2.902±0.021	0.019±0.018	0.293±0.007	
SB3d	68	744	381	2.504±0.022	0.029±0.018	0.171±0.009	
SB3e	69	637	1,068	1.153±0.017	0.066±0.012	0.121±0.014	
SB3f	70	451	431	2.044±0.021	0.042±0.019	0.134±0.019	
SB3g	71	492	430	2.055±0.029	0.040±0.023	0.140±0.020	
SB3							**14,555±790**
SB4a	34	773	214	3.304±0.029	0.010±0.008	0.109±0.028	
SB4b	35	834	328	3.007±0.024	0.010±0.010	0.078±0.021	

Table 16.2 (cont.)

Sample	Distance from the base (mm)	U (µg kg^{-1})	Th (µg kg^{-1})	^{234}U/^{238}U	^{230}Th/^{238}U	^{230}Th/^{232}Th	U/Th isochron age
SB4c	36	821	119	3.677±0.029	0.012±0.008	0.259±0.018	
SB4d	37	717	613	1.833±0.015	0.003±0.008	0.010±0.012	
SB4e	38	564	442	1.955±0.020	0.003±0.002	0.013±0.020	
SB4f	39	206	131	2.365±0.020	0.006±0.003	0.029±0.008	
SB4g	40	487	340	2.207±0.019	0.005±0.002	0.020±0.007	
SB4h	41	423	231	2.625±0.021	0.007±0.001	0.038±0.006	
SB4							**16,550±850**
Bei-Ee	13.5	914	497	1.3156±0.0061	0.1493±0.0004	0.8419±0.0081	17,543±709
Bei-F6a	3	1,098	1,697	1.6696±0.0435	0.1573±0.0027	0.3121±0.0077	
Bei-F6a-1	3	1,993	5,908	1.1380±0.0320	0.1621±0.0025	0.1676±0.0037	
Bei-F6a-2	3	580	2,021	2.2713±0.0795	0.1564±0.0044	0.1377±0.0033	
Bei-F6a							**18,400±826**
Bei-F7	0	733	699	1.9167±0.0105	0.1722±0.0005	0.5532±0.0015	20,504±829

16.6 RESULTS AND INTERPRETATION

16.6.1 Site section: chronology of occupation

Three new ^{14}C dates have been obtained from bulk charcoal collected within the slope section on the western edge of the Beidha settlement (Figures 16.3, 16.5). A black layer corresponding to the later Natufian phase previously identified (Byrd, 1989) yielded two dates of 13,645 ± 176 and 13,156 ± 120 cal. BP (Figures 16.3 and 16.6). The sample from the base of the subsequent PPNB occupation layer gave an age of 10,302 ± 108 cal. BP (Figures 16.3 and 16.6). These new dates refine our knowledge of the timescale of human settlement at Beidha, particularly the later Natufian occupation, which falls within the Early Natufian time period (Bar-Yosef and Belfer-Cohen, 1999), contrary to Byrd's (1989) preliminary conclusions about a second occupation at the transition between the Early and the Late Natufian. The new ^{14}C dates also support arguments for an earlier start for Neolithic occupation than previously thought. The new Neolithic date obtained at Beidha falls within the same range of previously acquired dates following their recalibration (Figure 16.3), but is better constrained. It actually suggests an EPPNB, or very early MPPNB occupation at Beidha. This is compatible with the architectural style of the earliest Beidha Neolithic, and with the dates obtained for similar architecture in the northern Levant.

16.6.2 Site section: sedimentological analysis, and geomorphological observations around the site

Granulometry analysis and sedimentological observations have been performed on a composite sedimentological section on the western side of the Beidha terrace and within the Beidha tell. Results of the granulometry analysis are presented in Figure 16.7. Overall, these results are rather inconclusive, as samples from the dunes and certain levels of the wadi sequence are indistinguishable. They are mostly composed of a fine fraction usually attributed to wind-blown deposits, and are potentially the result of local weathering of the sandstone. This renders the differentiation between aeolian and fluvial deposits very difficult on the basis of the granulometry alone (Figure 16.7). However, a shift in sedimentation type is noticeable between the Natufian occupations (sediments aligned on the same composition trend than sediments from the main wadi system and dunes; Figure 16.7) and later deposits (sediments richer in clay; Figure 16.7). This apparent shift in sedimentation seems at odds with Field's (1989) conclusions, which suggest a similar type of deposition before, during and after the Natufian, with continuous aggradation. It is, however, compatible with an erosional phase before the Neolithic levels (also suggested by Field, 1989), which would have induced incision in the wadi bed and isolation of the terrace from the wadi system. The sedimentation mechanism of the sterile interval remains unclear – possible interpretations include deposition from ephemeral streams washing material from the cliffs and/or an increase in aeolian input (Raikes, 1966; Field, 1989).

The presence of pebbles within the site terrace sedimentary sequence aligned on a major flow direction towards the west to west-northwest indicate the episodic incursion of meandering channels under Natufian deposits. Before the Early Natufian, the sedimentary system thus appears to function mostly in an aggradation mode, as suggested by Field (1989). Root concretions just under the Natufian levels at the site section reflect vegetation colonisation and possibly relatively mesic conditions, maybe indicative of a riverine setting.

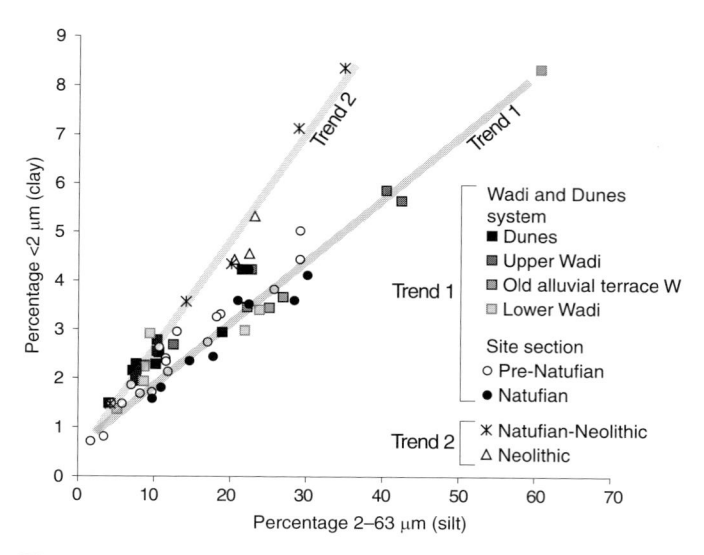

Figure 16.7. Granulometry analysis from the site section and other sediments from the valley.

Geomorphological features west of the site, in particular rounded sandstone terraces at a lower elevation than the Natufian layers (Figure 16.3), suggest previous burial under fluvial sediments. These features are compatible with a valley filled to Natufian occupation levels, then subjected to erosional processes at a later time.

16.6.3 Spring carbonate section: chronology, sedimentology and sedimentation rates

The oxygen isotopic composition of the spring carbonate sequence suggests that these carbonates precipitated from ambient-temperature waters (see, e.g., Andrews, 2006). This is further supported by the low sedimentation rates and abundance of plant encrustations, such as reed imprints, observable in the spring carbonate sequence of Beidha (Ford and Pedley, 1996; Pentecost, 1995). The term tufa, following the definition of Pedley (1990) (see also Ford and Pedley, 1996, and Pedley et al., 2003), can be used to describe the Beidha carbonate sequence. The mineralogical composition of the carbonate phase of the tufa deposits is essentially calcitic, with traces of dolomite for certain samples. An important detrital phase is always present, dominated by quartz grains. Thin sections derived from the Beidha spring carbonate series show a predominantly micritic fabric, including a significant amount of wind-blown sand particles. The depositional setting at the Beidha tufa sequence can be described as a spring-fed shallow-water marshy environment, probably in close proximity to the spring outlet, but corresponding to a calm depositional environment with low-velocity water flows.

The use of U-series to date impure carbonates such as tufa has been widely discussed in the literature (e.g. Geyh, 2001; Mallick and Frank, 2002; Garnett et al., 2004; Candy et al., 2005; Geyh, 2008 and references therein). Tufa deposits have been recently dated using U-series (e.g. Eikenberg et al., 2001; Soligo et al., 2002; O'Brien et al., 2006). It has been suggested that the best kind of tufa deposits for dating purposes are those of low porosity (typically, micritic facies), thus avoiding problems linked with secondary sparitic cementation within big pores (e.g. Garnett et al., 2004). At Beidha, although the micritic facies is an advantage for U-series dating techniques, the presence of a significant detrital phase forced the use of the isochron dating method.

The independent dating of vertically adjacent samples from the spring carbonate section (separated by no more than a few centimetres; Figure 16.6) shows very good consistency. However, large error bars obtained with the isochron method for certain layers (Figure 16.6) hint at complicating processes within the sedimentary sequence. One possible explanation is the mixing of successive layers of precipitation during the sample preparation. If the sedimentation rate is particularly low, grouping these layers will result in an average, less accurate date. Another possibility is the presence of secondary carbonate, after the main phase of precipitation, filling voids. Considering the dominantly micritic facies at the spring carbonate sequence, this hypothesis is less likely. Other explanations include phases of meteoric alteration/reprecipitation, especially for the upper layers directly exposed to the atmosphere and for now-internal layers that may have been exposed for an unknown amount of time following a precipitation hiatus and mixing in groundwater sources. However, no correlation was found between Mg, Sr, Mg/Ca and either the carbon or oxygen isotopic ratios. This suggests an open and fluid system, with no significant changes in the groundwater source through time.

Dates obtained from the Beidha tufa series using U-series techniques indicate that sedimentation probably started just after the LGM (Figure 16.6), although the older date obtained by U-series at Beidha is based on one sub-sample only and is therefore less reliable than dates obtained from multiple sub-samples and the isochron technique. Carbonate precipitation then continued with variable sedimentation rates until c. 8,450 BP. The spring carbonate record at Beidha therefore spans more than 10,000 years, which is unusual for tufa deposits (Andrews, 2006), although a Czech Republic tufa sequence has been shown to span c. 7,000 years (Zak et al., 2002). Tufas in more northern latitudes are usually rapidly accumulating systems and can, therefore, record short-time variations such as seasonality. At Beidha the sequence is much more condensed and such a resolution is probably impossible to attain.

Sedimentation rates at the Beidha tufa sequence can be calculated using the U-series dates obtained (Figure 16.6; Table 16.3), although large error bars associated to some of the dates render them less accurate. Additionally, a potential error of ±5 cm has been attributed to sampling depths. Sedimentation rates calculated using mean dates and sampling depths are shown in

Table 16.3 *Sedimentation rates calculated for various intervals of the tufa section*

Intervals are numbered 1–6 as presented in Figure 16.6. Ranges of sedimentation rates include potential errors on both dates and sampling depth.

	Sample age (years BP)	Distance from the base (cm)	Sedimentation rate (cm per 100 yr)	
			Average	Range
1	8,450 ± 270 9,507 ± 59	200.3 ± 5 124.3 ± 5	7.2	4.8–11.8
2	9,507 ± 59 18,400 ± 826	124.3 ± 5 3.0 ± 5	1.4	1.1–1.6
3	9,507 ± 59 14,555 ± 790	124.3 ± 5 68.0 ± 5	1.1	0.8–1.6
4	14,555 ± 790 18,400 ± 826	68.0 ± 5 3.0 ± 5	1.7	1.0–1.9
5	9,507 ± 59 14,984 ± 1190	124.3 ± 5 74.3 ± 5	0.9	0.6–1.4
6	14,984 ± 1190 18,400 ± 826	74.3 ± 5 3.0 ± 5	2.1	1.1–5.8

Figure 16.6. Values calculated using both extremities of the ranges given by uncertainties on dates and sampling depths are also presented in Table 16.3.

Sedimentation rates are significantly different between the lower and upper parts of the sections (Figure 16.6; Table 16.3). The higher sedimentation rates at the upper part of the section (*c.* 9,500–8,450 cal. BP) reflect higher spring activity, probably in link with higher meteoric precipitation and aquifer recharge during the early Holocene wet period. The lower sedimentation rates related to the lower part of the section (*c.*18,500–9,500 cal. BP) may reflect aquifer recharge after the end of the LGM (during which the spring did not appear to have been flowing) but drier conditions than during the following early Holocene. The period between *c.* 15,000/14,500 and 9,500 cal. BP seems to show a relative decrease in sedimentation rates compared to the period *c.* 18,500–15,000 cal. BP (Figure 16.6; Table 16.3). This may reflect the influence of a dry period, possibly related to the Younger Dryas, although potential errors related to dating imprecision render this difficult to assess with certainty.

16.6.4 Spring carbonate and site section: stable isotopic composition of carbonates

Oxygen and carbon isotopes have been measured both at the site and the spring carbonate sections. The oxygen curves of both sections exhibit significant variations, which may be related to fluctuations of the local environmental conditions (Figure 16.8). The major excursions seem to be synchronous in both records,

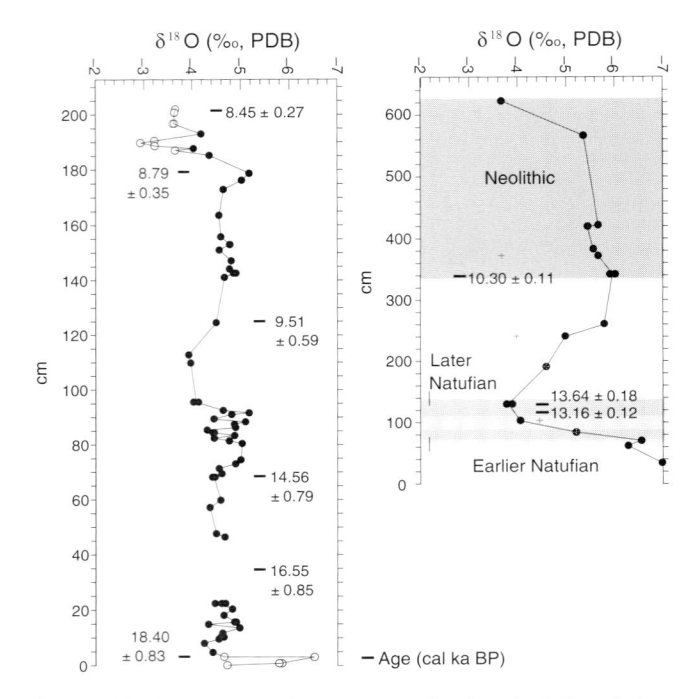

Figure 16.8. Oxygen stable isotope curves for the tufa (left) and site (right) sections. Grey highlights refer to times of occupation at the archaeological site (Natufian and Neolithic). White dots in the tufa section correspond to isotopic compositions potentially influenced by disequilibrium effects (e.g. evaporation).

considering the U-series and radiocarbon dates available for the spring-carbonate and site section, respectively. The more noticeable events are a marked increase in both oxygen isotopic records at *c.* 13,000 BP, followed by more negative values from *c.* 10,500–9,500 BP. This positive excursion seems to correspond in time with the Younger Dryas period (*c.* 12,700–11,500 BP), whereas the more negative values before and after this period can be associated with the Bølling–Allerød interval (*c.* 15,000–13,000 BP) and the onset of the early Holocene. These three major divisions of the record seem to generally correspond to specific times of occupation (Early Natufian during the Bølling–Allerød, PPNB during the early Holocene) and abandonment of the site. A shift towards more favourable climatic conditions (more negative $\delta^{18}O$) seems to be recorded at the site section even before the first known date of occupation (*c.*10,300 BP; Figure 16.8). A shift towards more positive ratios around 8,800–8,500 BP is coeval with the end of PPNB occupation at the site (Figure 16.8). Overall, the good correlation between occupation phases and the evolution of the oxygen isotopic composition of carbonates suggest a climatic control over periods of settlement at Beidha. In this context, more negative $\delta^{18}O$ values would correspond to more favourable climatic conditions at Beidha, whereas more positive values would mark periods of climatic deterioration, inducing the abandonment of the site (see discussion).

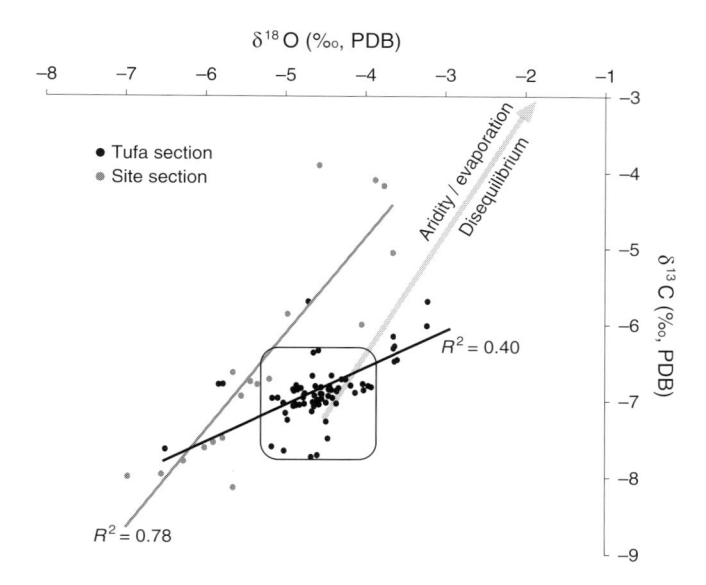

Figure 16.9. Oxygen and carbon isotopic composition and covariation (trend lines) for the site and tufa sections at Beidha. Evaporation/disequilibrium trend from Andrews (2006). The black box contains the samples that may be considered independent from disequilibrium effects.

A maximum in $\delta^{18}O$ values, following a sharp increase of the ratio, at the site section seems coeval with the last level of Natufian occupation (Figure 16.8). This may suggest that the Natufian population, facing increasingly degraded climatic conditions, nevertheless came back to Beidha and tried to settle for a short period (the more ephemeral occupation of the Later Natufian noticed by Byrd, 1989). Alternatively, it is possible that the soil carbonate nodules, from which the isotopic record is derived, developed slightly after the site abandonment (see discussion).

At about that time, increased climate variability is recorded in the tufa series by alternating lower and higher $\delta^{18}O$ values in the carbonates after an initial shift towards more negative ratios (Figure 16.8). This may indicate quickly changing conditions that would have allowed a brief return to Beidha for the Natufian population, during a short period of climatic improvement within a general trend towards a less favourable environment.

A marked correlation between C and O isotopic composition of carbonate precipitates is usually interpreted as reflecting the influence of evaporative processes (e.g. Smith *et al.*, 2004; Andrews, 2006; O'Brien *et al.*, 2006). At Beidha, a strong correlation ($R^2 = 0.78$; Figure 16.9) between the carbon and oxygen isotopic records is noticeable for soil carbonates deposited within the site section, whereas the spring carbonate series shows a far less marked covariation (Figure 16.9). Most of the samples from the spring carbonate sequence seem actually independent of an evaporative trend (black box, Figure 16.9). However, the uppermost samples seem to record an increasing influence of evaporative processes, probably linked with the cessation of the spring flow shortly afterwards.

16.7 DISCUSSION

16.7.1 Beidha spring: related aquifer and groundwater residence time

Springs in Jordan today are related to either water-table or confined aquifers. Springs related to confined aquifers tend to have steadier, larger flows than water-table springs and show little sensitivity to climate change. Water-table springs generally present small and highly variable flows and show great sensitivity to climatic conditions; they may stop flowing during times of low rainfall (EXACT, 1998). The former spring studied at Beidha only flowed during a limited period of time (*c.* 20,000–8,500 cal. BP) and presents slow and variable sedimentation rates throughout this period. Considering the average sedimentation rates of *c.* 1.4 cm per 100 years and *c.* 7.2 cm per 100 years for, respectively, the lower part and the upper part of the section (Figure 16.6; Table 16.3), and the average thickness of the samples analysed (1 cm), each sample should represent between *c.* 14 and 71 years of deposition, with a maximum of 167 years given by the lowest sedimentation rate calculated at the section (Table 16.3). The fact that clear variations are observable in the isotopic record therefore suggests that the residence time of the aquifer related to the relict spring is lower than this upper figure.

This points to a spring related to a water-table aquifer, highly sensitive to climatic variations. This would also explain the good correlation between the spring isotopic record and the isotopic record derived from the carbonate nodules at the site section, which do not relate to spring waters. We therefore take the view that, even if the exact time of residence of the aquifer waters is unknown, the spring system reacted promptly to climate changes and the isotopic record derived from the spring carbonates reflects these variations.

16.7.2 The isotopic record at Beidha: potential interpretations and palaeoenvironmental implications

ISOTOPIC VARIATIONS IN CARBONATES
Carbonate precipitates can reflect a range of the environmental conditions present during the time of their formation. These include the temperature and isotopic composition of the parent water from which the carbonate formed, as well as other factors that can affect the carbonate precipitation equilibrium, such as evaporation.

The relation between water temperature, the isotopic composition of the parent water and the isotopic composition of calcite precipitates can be summarised, in equilibrium conditions, by equations derived from the Craig palaeotemperature relationship (Craig, 1965; Andrews, 2006) such as (amongst others) presented in O'Neil *et al.* (1969) or Hays and Grossman (1991). The isotopic composition of the parent water, however, is itself dependent on a series of parameters including mean air temperature, the source and amount of precipitations, evaporation, and in the case of precipitates related to groundwater systems, on residence time and groundwater mixing. A good correlation, for example, has been proved between surface air temperatures and the oxygen isotopic composition of rainfall (Rozanski et al., 1993).

Soil carbonate nodules have been interpreted to reflect changes in the bulk soil water composition, and, for shallow soils, evaporation (e.g. in the Rio Grande area, New Mexico; Deutz et al., 2001). Palaeoclimatic changes, however, may only be recorded in carbonate nodules related to rapidly buried soils (Deutz et al., 2001 and references therein). Several studies have suggested that pedogenic carbonate $\delta^{18}O$ compositions can be indicative of the $\delta^{18}O$ values of local rainfall (e.g. Cerling, 1984; Cerling and Quade, 1993; Amit et al., 2007), although other parameters influencing the $\delta^{18}O$ composition of soil water must not be neglected (e.g. Deutz et al., 2001 and references therein). Air temperatures, for example, can have an important impact: Zanchetta et al. (2000) interpreted an abrupt shift of about $-1‰$ in $\delta^{18}O$ of pedogenic carbonates of the Somma-Vesuvius in Italy as related to a cooling of about 2 °C after the Avellino eruption (3.8 ka BP). Quade et al. (2007) listed three major factors influencing the $\delta^{18}O$ of soil carbonates in the Atacama desert: the $\delta^{18}O$ value of local rainfall (itself dependent on temperature, and potentially other parameters such as the amount and source of rainfall, altitude and continentality; Rozanski et al., 1993; Rowe and Maher, 2000 and references therein; see also Zanchetta et al., 2000); soil temperature; and soil water evaporation before the formation of soil carbonates. In arid environments, evaporation seems to be the main control on soil carbonate $\delta^{18}O$ composition (Quade et al., 2007). Zanchetta et al. (2000) also mention dewatering soil processes prior to carbonate precipitation, owing to evaporation, as a way of increasing pedogenic $\delta^{18}O$ values in arid and semi-arid environments, potentially to a point masking the rainfall signature (Zanchetta et al., 2000 and references therein). On lake margins in an evaporative system in Tanzania, higher $\delta^{18}O$ values in rhizoliths have been recorded during dry periods while lower values have been interpreted as reflecting meteoric water composition in times of increased precipitation (Liutkus et al., 2005). The data of Liutkus et al. (2005) also shows a strong covariation between $\delta^{13}C$ and $\delta^{18}O$ values.

In tufa deposits, variations of the oxygen isotopic records can be interpreted in terms of temperature variations and/or variations of the isotopic composition of the parent water, including precipitation sources and air temperature variations (e.g. Andrews et al., 2000; Andrews, 2006; O'Brien et al., 2006 and references therein; Smith et al., 2004; see also the discussion about interpretations of the isotopic record in Chafetz and Lawrence, 1994).

The consensus view seems to be that short-term variations (e.g. seasonal) in the isotopic record of tufas are dominantly controlled by stream temperatures (i.e. lower isotopic values corresponding to higher temperatures), when the isotopic composition of waters can be considered essentially invariant (Andrews et al., 2000; Matsuoka et al., 2001; Andrews, 2006; O'Brien et al., 2006; Brasier et al., 2010). By contrast, longer-term variations (on millennial scales, such as, for example, between the mid-Holocene and the present day) are thought to be mainly controlled by changes in the meteoric water composition in the aquifer catchment area (e.g. Andrews et al., 1997; Smith et al., 2004). Andrews et al. (1997) have demonstrated that modern tufa $\delta^{18}O$ values are closely linked to the isotopic composition of precipitations, while moderated by other factors such as water temperature, evaporation and residence time. Rainwater composition is in turn controlled by, amongst others, latitude, changes in the air-mass sources, air-mass temperature, altitude and rainout/amount effects (Andrews et al., 2000). Higher values of $\delta^{18}O_{calcite}$ may then be caused by higher $\delta^{18}O_{recharge}$ (higher air temperatures or variations in the source of airmasses; Andrews et al., 2000).

Overall, it has been suggested that mean air temperature variations exert the strongest influence on the long-term $\delta^{18}O$ composition of tufas (e.g. Andrews et al., 1994, 2000; Andrews, 2006; O'Brien et al., 2006). The relationship between temperature and the $\delta^{18}O$ composition of tufas is then complicated by the fact that air temperature and calcite precipitation temperature dependence exert opposite influences on the isotopic composition ($\delta^{18}O_{water}$ increases with rising air temperature; $c.\ +0.58‰$ per °C in Europe; Andrews, 2006; whereas $\delta^{18}O_{calcite}$ decreases with rising water temperature; $c.\ -0.24‰$ per °C; Andrews, 2006). The final signal is dominated by the air temperature effect but 'damped' by the water temperature influences (Andrews et al., 1994, 2000; Andrews, 2006). In Europe, it has been claimed that $c.$ 40% of the air temperature variation is finally recorded in the tufa $\delta^{18}O$ signal, with rising air temperatures inducing an increase in $\delta^{18}O$ values (Andrews, 2006). Following the same line of interpretation, O'Brien et al. (2006) have inferred the mean isotopic change in their tufas from the Grand Canyon area between the mid-Holocene and the modern day ($-0.5‰$) to correspond to a shift of $c.\ -1$ °C, while Andrews (2006) cites variations of $-1‰$ in the $\delta^{18}O_{tufa}$ corresponding to $c.\ -3.5‰$ between 9.5 and 8.2 ka BP in the United Kingdom and Czech republic.

Short-term variations in the $\delta^{18}O$ of precipitations can also result from events such as intense rainfall (the 'amount effect'), seasonal variability (more negative precipitations during the cooler period of the year), and temporal changes in the air-mass sources (Andrews, 2006). In strongly evaporative settings, evaporation processes can also influence the oxygen isotopic composition of the tufas (e.g. O'Brien et al., 2006; Andrews, 2006; Chafetz and Lawrence, 1994 and references therein). It is believed that seasonal evaporation in arid climates can shift

$\delta^{18}O$ compositions of tufas up by at least 1‰ (Andrews and Brasier, 2005). Andrews et al. (2000) also indicate that O and C isotopes covariation can be related to climatic factors such as aridity.

THE ISOTOPIC RECORD AT BEIDHA

Overall, both the relict spring carbonate system near the archaeological site of Beidha, and the carbonate nodules sampled directly at the site, seem to coevally record changes in climatic conditions. However, it may be difficult to determine precisely which parameters are mainly responsible for the fluctuations of the Beidha record. The interpretation of the spring carbonate sequence at Beidha is further complicated by the fact that the spring itself is no longer in existence (e.g. Smith et al., 2004; Andrews, 2006 and references therein). This means that a comparison between the composition of modern carbonate precipitates, spring water values and climate parameters such as temperature and rainfall (as notably realised at Soreq Cave; Bar-Matthews et al., 1996), is not directly possible at Beidha.

Pedogenic nodules at Beidha form part of a relatively fast-accumulating sequence (Figures 16.5 and 16.8) and therefore may be recording environmental conditions at the time of their formation. In this study, we take the view that isotopic variations in the pedogenic nodules and carbonates are mainly controlled by evaporation, as evidenced by the significant correlation between C and O isotopes (Zanchetta et al., 2000; Liutkus et al., 2005; Quade et al., 2007). As such, positive shifts of the isotopic record will correspond to periods of more intense evaporation linked to increased aridity, while lighter compositions will correspond to wetter periods (such as suggested by Liutkus et al., 2005, for lake-margin rhizoliths in Tanzania). This interpretation is coherent with the increased presence of aridity indicators in the pollen record during and after the Later Natufian occupation (Fish, 1989), the presence of a later erosion gully cutting through Early Natufian levels (Byrd, 1989; Kirkbride, 1989), and the change in sedimentation suggesting a shift in the climate regime following a time of wadi aggradation at the site (Figure 16.7).

Both shifts towards heavier isotopic values at the top of the tufa and site sequence are consistent with increased evaporation ($\delta^{13}C$ and $\delta^{18}O$ covariation; Figures 16.7, 16.8) just before the site's abandonment, coherent with the cessation of spring activity shortly afterwards. A major phase of erosion took place after the site abandonment, also suggestive of a shift towards a drier climate. Considering the covariation of tufa and pedogenic carbonate oxygen isotopes, and the indication of increased aridity and evaporation at the top of the tufa sequence, a cessation of spring activity linked with a sudden tectonic event that would have changed the groundwater trajectory is unlikely.

In the rest of the tufa record, it is difficult to know exactly how the other isotopic shifts, corresponding in time to the Bølling–Allerød, the Younger Dryas and the start of the early Holocene, relate to climatic variations. At Beidha, the sampling resolution (samples of c. 1 cm thickness), coupled with low sedimentation rates, means that decades/centuries of climatic variations are averaged out; therefore an interpretation in terms of water temperature changes (mainly valid on a seasonal timescale) is probably invalidated. An overall control of mean air temperature variations is possible in the long term, i.e. between the start and the end of sedimentation at the Beidha spring. It is possible (although difficult to prove) that the general trend towards less negative $\delta^{18}O$ (c. 1‰ over the whole section), on which the shorter time variations such as related to the Bølling–Allerød, the Younger Dryas and the early Holocene are superimposed, is linked with generally increasing mean air temperatures through time (c. 3.5 °C, following the estimates of Andrews, 2006). Since the spring at Beidha seems to have started flowing after the end of the LGM, such an interpretation would be coherent with the general increase in temperatures (6–7 °C; Affek et al., 2008) recorded at Soreq between the LGM (20,000–19,000 cal. BP) and the early Holocene (10,000–7,000 cal. BP). However, it is difficult to interpret the medium-term variations (during the Bølling–Allerød and Younger Dryas, and at the start of the early Holocene) as purely dependent on air temperature variations, since they would imply higher temperatures during the Younger Dryas and colder temperatures during both the Bølling–Allerød and the early Holocene, at odds with all other palaeoclimatic interpretations of the region.

We therefore suggest, as a working hypothesis, that medium-term variations in the Beidha tufa record are related to changes in the $\delta^{18}O$ composition of spring water, themselves related to shorter-term effects than mean air temperature changes, such as the amount of rainfall, seasonality or changes in the rainfall sources. The marked shift towards lighter compositions during the early Holocene clearly corresponds to more extended sedimentation rates (more recharge, hence wetter conditions). Earlier shifts towards, successively, lighter and heavier compositions may therefore correspond to alternate wetter and drier periods: respectively, the Bølling–Allerød and the Younger Dryas. In this context, the correlation between times of lighter isotopic composition and times of increased sedimentation points in favour of a general control by rainfall amounts or distribution. The very negative values recorded at the start of the Beidha sequence (Figure 16.8) may be related to isotopically light snow-melt waters feeding the groundwater system after the termination of the LGM.

On this topic, certain similarities between the Beidha and Soreq $\delta^{18}O$ isotopic records are worth considering (Figure 16.10). In particular – and even if the magnitude of variations is quite different – both records show a shift towards heavier $\delta^{18}O$ during the time period corresponding to the Younger Dryas, bracketed by two periods of lighter isotopic values. The Soreq isotopic record has been interpreted by Bar-Matthews and

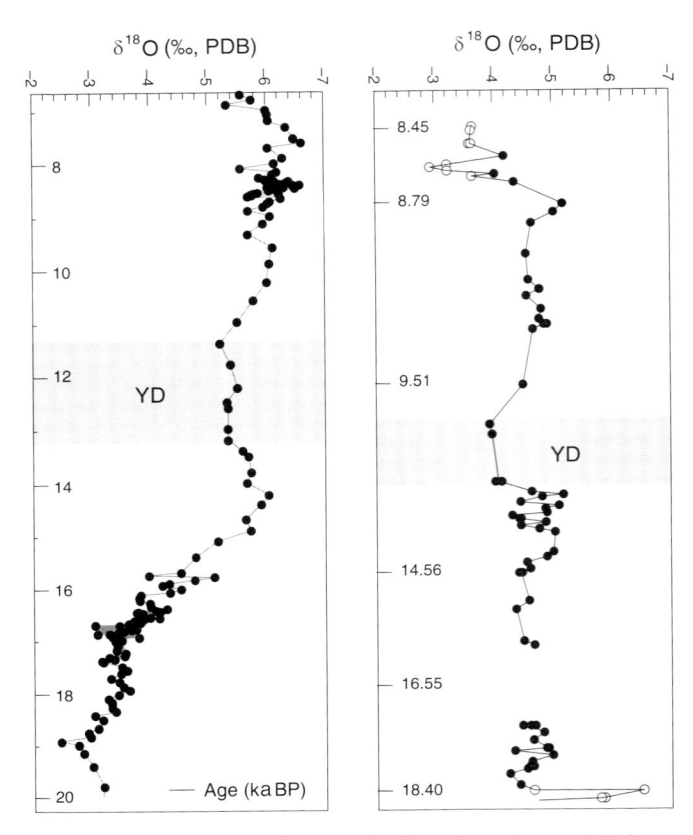

Figure 16.10. Comparison between the Soreq isotopic record (left; modified from Bar-Matthews et al., 1999) and the Beidha spring-carbonate record (right; this study). Open dots correspond to samples from the Beidha record that show potential disequilibrium effects (e.g. evaporation). Light grey highlights indicate the probable time of the Younger Dryas (YD).

colleagues as primarily reflecting changes in the isotopic composition of the cave waters, related to rainfall amounts (Bar-Matthews et al., 1996, 1997, 1999, 2003). The fact that speleothem records (Soreq), pedogenic carbonates and tufa deposits (Beidha) seem to record similar trends in oxygen isotopes may be an important piece of information for the interpretation of such records, and hints at a common controlling climatic parameter. Further studies are needed to better understand this control.

16.7.3 Source of water to the settlements

It appears unlikely that the former spring studied in this work was ever sufficient to support a large Neolithic settlement by itself. However, the presence of other (albeit undated) former springs in the area has been attested by previous studies; they may have represented collectively a significant source of water to both the Natufian and Neolithic settlements. Geomorphological and sedimentological studies seem to suggest that the wadi system may have been wider and more perennial during occupation periods. The sedimentation rate of the tufa, and possibly the oxygen isotope variations, suggest wetter conditions during both times of occupation; this may have resulted into a more stable riverine environment, with more vegetation cover and less flash-floods (see also the study at Wadi Faynan by Smith et al., this volume, Chapter 15). During the earlier Natufian level at least, wadi-type sedimentation, with meandering channels, is attested in the immediate vicinity of the archaeological site. This suggests a wider, less-incised valley floor, possibly supporting a more widespread river system. The position of the wadi itself is unknown during Neolithic times, but there are suggestions that the major erosive processes occurred after the site abandonment at c. 8,500 BP, coeval with climate change, and are marked, for example, by the deep incision of the Seyl Aqlat. In this case, it is possible that the valley floor, albeit slightly lower than during Natufian time, was still relatively large during the Neolithic, and supported a river system not confined to a single, flash-flooding, deeply incised wadi bed as it is at present. During the entire occupation history of the site, and even (at least partially) during the interval between Natufian and Neolithic occupations, the spring near Beidha was flowing, indicating overall wetter conditions at the site, or better recharge in the highlands, than those of today. Permanent sources of water near the settlement are suggested by the pollen record (Fish, 1989) and other tufa deposits may have deposited in the immediate vicinity of the site (Kirkbride, 1966, 1968; Byrd, 1989, 2005). In the light of this new study and observations, we therefore postulate that it is unlikely that the people at Beidha were travelling the c. 5 km to the spring at Dibadiba to get their water; they probably had water resources, either from the wadi itself or from the numerous springs in the area, closer at hand. It is also possible that a wider, less incised and better-watered floodplain was present, which was attractive to early agricultural practices.

16.7.4 Neolithic settlement at Beidha

A more favourable climate seems to have settled at Beidha sometime before the first known time of occupation (c. 10,300 BP; Figure 16.4; Table 16.1), as indicated by more negative $\delta^{18}O$ in carbonate nodules from the site section (Figure 16.8). This change in environmental conditions may have encouraged people to settle at Beidha earlier than previously thought. This hints at the possibility of an earlier than MPPNB occupation at Beidha, coeval with an improvement in the environmental conditions. This may explain the earliest, undated structures found at Beidha (Kirkbride, 1968; Byrd, 2005) as well as some features of the lithic assemblage from the lower levels (Mortensen, 1970). The new date obtained from the section, at c. 10,300 cal. BP (EPPNB/very early MPPNB), also revives the debate about the architectural style of Phase A, which may be related to EPPNB rather than MPPNB.

16.8 SUMMARY AND CONCLUSIONS

This new integrative study, based on oxygen isotopic data from two sections (pedogenic carbonate and tufa deposits) as well as sedimentological observations, and rigorously time-constrained (U-series at the tufa section, new radiocarbon dates and archaeological levels at the site section), allows for a better understanding of the palaeoenvironmental conditions at Beidha during the Natufian (Early Natufian) and Neolithic (PPNB) times of occupation. We have interpreted the isotopic record as showing primarily changes in the water isotopic composition and evaporative processes, with more negative values corresponding to wetter periods and more positive values related to increased aridity. This is in accordance with observations from the sedimentation rates at the tufa section and previous observations at the site, notably the pollen record (Fish, 1989). Temperature information remains inconclusive, although we suggest that the spring section may record an overall increase of temperature of $c.$ 3.5 °C between the start and the end of the depository sequence, a gradual change generally coherent with previous studies (Affek et al., 2008). Overall, the coincidence between times of occupation and environmental changes recorded in the isotopic composition of carbonates, as well as in other proxies, suggests that the population at Beidha responded strongly to environmental variations during both the Natufian and Neolithic.

The new radiocarbon dates suggest that both Natufian occupations at Beidha occurred during the Early Natufian. The first period of Natufian occupation occurred during a time of wadi aggradation, as suggested by the granulometric analyses (Figure 16.7) and previous observations at the site (Field, 1989; contrary to the views of Raikes, 1966). Environmental conditions seem to have been more favourable, with higher rainfall (less positive $\delta^{18}O$ values at both the site and the tufa sections; Figure 16.8). These conclusions are in accordance with previous studies at the site and elsewhere in the Levant that suggested wetter conditions, with a more extended vegetation cover at Beidha (Byrd, 1989), corresponding to the Bølling–Allerød wet and warm interval (e.g. Bar-Matthews et al., 2003; Robinson et al., 2006; Chapter 6, this volume). The presence of a wider, more perennial wadi system is proposed for that time period.

A shift towards more arid conditions is recorded around the time of the second Natufian occupation, with more positive $\delta^{18}O$ values in the pedogenic carbonates hinting at increased evaporation (site section, Figure 16.8). This is coherent with the increase of aridity indicators in the pollen record before and after the second Natufian occupation layer (Fish, 1989). In the tufa section, increased climate variability is recorded by alternating lower and higher $\delta^{18}O$ values in the carbonates, after an initial shift towards more negative ratios (Figure 16.8). This climate variability during the Early Natufian at Beidha may correspond to the short, dry spell indicated by flora and fauna records in the southern Levant that interrupts overall favourable conditions (e.g. Sanlaville, 1996). It is possible that, in a climate of generally degrading conditions, the brief, later Natufian occupation represents a last try for the population to settle at Beidha – maybe during a short period of climatic improvement – before the environment became too hostile.

The sterile interval between the Natufian and Neolithic occupation seem to correspond to an arid event that settled in by $c.$ 13,000 BP (more positive $\delta^{18}O$ values at both the tufa and site section; Figures 16.6 and 16.8). This is coherent with the change of sedimentation at the site section (granulometry analyses, Figure 16.7), the pollen record (Fish, 1989) showing higher frequencies of aridity indicators at the base of the interval, the isolation of the terrace on which the Neolithic site is built, and the marks of erosion in the sterile sediments between the two levels of occupation (Byrd, 1989; Kirkbride, 1989), but at odds with Field's (1989) interpretation of continuous wadi aggradation. Our results are in accordance with previous studies indicating drier (and probably cooler) conditions during the Younger Dryas in the Levant (e.g. Bar-Matthews et al., 2003; Robinson et al., 2006; Chapter 6, this volume).

The site section records a shift towards less positive isotopic composition even before the first known date of occupation (i.e. $c.$ 10,300 cal. BP, this study; Byrd, 2005), coherent with other observations of renewed, denser and more mesic vegetation cover (Raikes, 1966; Kirkbride, 1968; Fish, 1989). The hypothesis of an earlier (PPNA/EPPNB) settlement at Beidha would fit with some of the archaeological evidence, such as the lithic material uncovered from the earlier levels, and possibly early architectural structures (Kirkbride, 1968; Mortensen, 1970; Byrd, 2005).

Generally, during the early Holocene and the PPNB occupation, high sedimentation rates in the tufa sequence, as well as more negative $\delta^{18}O$ values at both the site and the tufa sections (Figures 16.6, 16.8) indicate wetter conditions at Beidha, such as suggested for the Levant in general during the Climatic Optimum (e.g. Chapter 6, this volume; Bar-Matthews et al., 2003; Henry, 1997). We propose, contrary to Field's (1989) interpretation of a flash-flood regime during the PPNB, that the wadi flow at that time may have been more perennial and could have supplemented water resources available from the spring we investigated, and others in the area (see Kirkbride, 1966; Raikes, 1966; Kirkbride, 1968; Byrd, 1989, 2005). A wide floodplain, with better water availability, may have been decisive for the start of agricultural practices at Beidha.

Towards the end of the tufa and site sections, the isotopic records indicate a shift towards more arid conditions. Shortly after that, the spring stopped flowing, coeval with the Neolithic site final abandonment, at around 8,500 cal. BP. We interpret this shift to be responsible for the progressive abandonment of Beidha (Byrd, 2005). Subsequently, a major phase of erosion destroyed part of the site, probably in relation with the new, drier climatic conditions. A similar phase of aridity, generalised erosion and flash-flood regime has been recorded elsewhere in the region (e.g. Hourani and Courty, 1997; Hunt *et al.*, 2004).

Overall, the conditions at Beidha seem to have been wetter than today during both Natufian and Neolithic times of occupation, and probably during part of the interval in between, as attested by the spring activity and the pollen indicators of perennial water sources around the site (Fish, 1989). It is unlikely, therefore, that the populations at Beidha used the spring at Dibadiba as a main supply for their water, as suggested by previous studies (Raikes, 1966; Kirkbride, 1968), or that greater water supply was provided solely by increased runoff (Comer, 2003).

The stable isotopic composition of carbonates has proved to be a powerful tool for the reconstruction of palaeoenvironments in a semi-arid setting. However, the exact parameters controlling their variations in such an environment remain hidden. In this context, the similarity of trends, in the southern Levant, between oxygen isotopic variations from speleothems (Soreq, Central Israel; Bar-Matthews *et al.*, 1997, 1999, 2003), tufas and pedogenic carbonates (Beidha, this study) is an important observation. This should trigger additional studies about the specific climatic variations (in particular, changes in rainfall amounts, rainfall sources, or seasonality) that can direct such covariations.

REFERENCES

Affek, H. P., M. Bar-Matthews, A. Ayalon, A. Matthews and J. M. Eiler (2008) Glacial/interglacial temperature variations in Soreq cave speleothems as recorded by 'clumped isotope' thermometry. *Geochimica et Cosmochimica Acta* **72**: 5351–5360.

Al-Eisawi, D. (1985) Vegetation in Jordan. In *Studies in the History and Archaeology of Jordan II*, ed. A. Hadidi. Amman: Department of Antiquities pp. 45–58.

Al-Eisawi, D. (1996) *Vegetation of Jordan*. Cairo: UNESCO Regional Office for Science and Technology for the Arab States.

Amit, R., J. Lekach, A. Ayalon, N. Porat and T. Grodek (2007) New insight into pedogenic processes in extremely arid environments and their paleoclimatic implications – the Negev Desert, Israel. *Quaternary International* **162**: 61–75.

Andrews, J. E. (2006) Palaeoclimatic records from stable isotopes in riverine tufas: synthesis and review. *Earth-Science Reviews* **75**: 85–104.

Andrews, J. E. and A. T. Brasier (2005) Seasonal records of climatic change in annually laminated tufas: short review and future prospects. *Journal of Quaternary Science* **20**: 411–421.

Andrews, J. E., M. Pedley and P. F. Dennis (1994) Stable isotope record of palaeoclimate change in a British Holocene tufa. *Holocene* **4**: 349–355.

Andrews, J. E., R. Riding and P. F. Dennis (1997) The stable isotope record of environmental and climatic signals in modern terrestrial microbial carbonates from Europe. *Palaeogeography Palaeoclimatology Palaeoecology* **129**: 171–189.

Andrews, J. E., M. Pedley and P. F. Dennis (2000) Palaeoenvironmental records in Holocene Spanish tufas: a stable isotope approach in search of reliable climatic archives. *Sedimentology* **47**: 961–978.

Banning, E. B. and I. Kohler-Rollefson (1992) Ethnographic lessons for the pastoral past: camp locations and material remains near Beidha, Southern Jordan. In *Pastoralism in the Levant: Archaeological Materials in Anthropological Perspectives*, ed. O. Bar-Yosef and A. Khazanov. Madison WI: Prehistory Press pp. 181–204.

Bar-Matthews, M., A. Ayalon, M. Gilmour, A. Matthews and C. J. Hawkesworth (2003) Sea-land oxygen isotopic relationships from planktonic foraminifera and speleothems in the Eastern Mediterranean region and their implication for paleorainfall during interglacial intervals. *Geochimica et Cosmochimica Acta* **67**: 3181–3199.

Bar-Matthews, M., A. Ayalon and A. Kaufman (1997) Late quaternary paleoclimate in the eastern Mediterranean region from stable isotope analysis of speleothems at Soreq Cave, Israel. *Quaternary Research* **47**: 155–168.

Bar-Matthews, M., A. Ayalon and A. Kaufman (2000) Timing and hydrological conditions of Sapropel events in the Eastern Mediterranean, as evident from speleothems, Soreq cave, Israel. *Chemical Geology* **169**: 145–156.

Bar-Matthews, M., A. Ayalon, A. Kaufman and G. J. Wasserburg (1999) The Eastern Mediterranean paleoclimate as a reflection of regional events: Soreq cave, Israel. *Earth and Planetary Science Letters* **166**: 85–95.

Bar-Matthews, M., A. Ayalon, A. Matthews, E. Sass and L. Halicz (1996) Carbon and oxygen isotope study of the active water-carbonate system in a karstic Mediterranean cave: implications for paleoclimate research in semiarid regions. *Geochimica et Cosmochimica Acta* **60**: 337–347.

Bar-Yosef, O. (1986) The walls of Jericho: an alternative interpretation. *Current Anthropology* **27**: 157–162.

Bar-Yosef, O. (1998) The Natufian culture in the Levant, threshold to the origins of agriculture. *Evolutionary Anthropology* **6**: 159–177.

Bar-Yosef, O. and A. Belfer-Cohen (1999) Facing environmental crisis – societal and cultural changes at the transition from the Younger Dryas to the Holocene in the Levant. In *The Dawn of Farming in the Near East*, ed. R. T. J. Cappers and S. Bottema. Berlin: ex Oriente pp. 55–66.

Berger, J. F. and J. Guilaine (2009) The 8200 cal BP abrupt environmental change and the Neolithic transition: a Mediterranean perspective. *Quaternary International* **200**: 31–49.

Boyd, B. (2006) On 'sedentism' in the later Epipalaeolithic (Natufian) Levant. *World Archaeology* **38**: 164–178.

Brasier, A. T., J. Andrews, A. D. Marca-Bella and P. F. Dennis (2010) Depositional continuity of seasonally laminated tufas: implications for $\delta^{18}O$ based palaeotemperatures. *Global and Planetary Change* **71**: 160–167.

Byrd, B. F. (1989) *The Natufian Encampment at Beidha: Late Pleistocene Adaptation in the Southern Levant Moesgard*. Arhus: Jutland Archaeological Society.

Byrd, B. F. (2005a) *Early Village Life at Beidha, Jordan: Neolithic Spatial Organization and Vernacular Architecture*. Oxford: Oxford University Press.

Byrd, B. F. (2005b) Reassessing the emergence of village life in the Near East. *Journal of Archaeological Research* **13**: 231–290.

Candy, I. and S. Black (2009) The timing of Quaternary calcrete development in semi-arid southeast Spain: investigating the role of climate on calcrete genesis. *Sedimentary Geology* **218**: 6–15.

Candy, I., S. Black and B. Sellwood (2005) U-series isochron dating of immature and mature calcretes as a basis for constructing Quaternary landform chronologies for the Sorbas basin, southeast Spain. *Quaternary Research* **64**: 100–111.

Cerling, T. E. (1984) The stable isotopic composition of modern soil carbonate and its relation to climate. *Earth and Planetary Science Letters* **71**: 229–240.

Cerling, T. E. and J. Quade (1993) Stable carbon and oxygen isotopes in soil carbonates. In *Proceedings of the Chapman Conference, Jackson Hole, Wyoming*, ed. J. Swart, J. A. McKenzie and K. C. Lohman. Oxford: Elsevier Science Ltd pp. 217–231.

Chafetz, H. S. and J. R. Lawrence (1994) Stable isotopic variability within modern travertines. *Geographie Physique et Quaternaire* **48**: 257–273.

Comer, D. C. (2003) Environmental history at an early prehistoric village: an application of cultural site analysis at Beidha, in southern Jordan. *Journal of GIS in Archaeology* **1**: 103–116.

Cordova, C. (2007) *Millennial Landscape Change in Jordan: Geoarchaeology and Cultural Ecology*. Tucson: University of Arizona Press.

Craig, H. (1965) The measurement of oxygen isotope palaeotemperatures. In *Stable Isotopes in Oceanographic Studies and Palaeotemperatures*, ed. E. Tongiorgi. Pisa: Consiglio Nazionale Della Richerche, Laboratorio de Geologia Nucleare pp. 161–182.

Deutz, P., I. P. Montanez, H. C. Monger and J. Morrison (2001) Morphology and isotope heterogeneity of Late Quaternary pedogenic carbonates:

implications for paleosol carbonates as paleoenvironmental proxies. *Palaeogeography Palaeoclimatology Palaeoecology* **166**: 293–317.

Eikenberg, J., G. Vezzu, I. Zumsteg *et al.* (2001) Precise two chronometer dating of Pleistocene travertine: the Th-230/U-234 and Ra-226(ex)/Ra-226 (0) approach. *Quaternary Science Reviews* **20**: 1935–1953.

Emery-Barbier, A. (1995) Pollen analysis: environmental and climatic implications. In *Prehistoric Cultural Ecology and Evolution: Insights from Southern Jordan*, ed. D. O. Henry. New York: Plenum Press pp. 375–384.

Enzel, Y., R. Arnit, U. Dayan *et al.* (2008) The climatic and physiographic controls of the eastern Mediterranean over the late Pleistocene climates in the southern Levant and its neighboring deserts. *Global and Planetary Change* **60**: 165–192.

EXACT (Executive Action Team Middle East Water Data Banks Project) (1998). *Overview of Middle East Water Resources – Water Resources of Palestinian, Jordanian, and Israeli Interest.* http://water.usgs.gov/exact/overview/index.htm

Feynman, J. and A. Ruzmaikin (2007) Climate stability and the development of agricultural societies. *Climatic Change* **84**: 295–311.

Field, J. (1989) Appendix A: geological setting at Beidha. In *The Natufian Encampment at Beidha, Late Pleistocene Adaptation in the Southern Levant*, ed. B. F. Byrd. Aarhus: Jutland Archaeological Society pp. 86–90.

Fish, S. K. (1989) Appendix B: The Beidha pollen record. In *The Natufian Encampment at Beidha, Late Pleistocene Adaptation in the Southern Levant*, ed. B. F. Byrd. Aarhus: Jutland Archaeological Society pp. 91–96.

Ford, T. D. and Pedley, H. M. (1996) A review of tufa and travertine deposits of the world. *Earth Science Reviews* **41**: 117–175.

Garnett, E. R., M. A. Gilmour, P. J. Rowe, J. E. Andrews and R. C. Preece (2004) Th-230/U-234 dating of Holocene tufas: possibilities and problems. *Quaternary Science Reviews* **23**: 947–958.

Garrard, A., S. Colledge and L. Martin (1996) The emergence of crop cultivation and caprine herding in the 'marginal zone' of the southern Levant. In *The Origins and Spread of Agriculture and Pastoralism in Eurasia*, ed. D. R. Harris. London: University College pp. 204–226.

Gebel, H. G. K. and J. M. Starck (1985) Investigations into the Stone Age of the Petra Area (Early Holocene Research): a preliminary report on the 1984 campaigns. *Annual of the Department of Antiquities of Jordan* **29**: 89–114.

Geyh, M. A. (2001) Reflections on the ^{230}Th/U dating of dirty material. *Geochronometria* **20**: 9–14.

Geyh, M. A. (2008) Selection of suitable data sets improves ^{230}Th/U dates of dirty material. *Geochronometria* **30**: 69–77.

Goodfriend, G. A. (1999) Terrestrial stable isotope records of Late Quaternary paleoclimates in the eastern Mediterranean region. *Quaternary Science Reviews* **18**: 501–513.

Hays, P. D. and E. L. Grossman (1991) Oxygen isotopes in meteoric calcite cements as indicators of continental palaeoclimate. *Geology* **19**: 441–444.

Helbaek, H. (1966) Appendix A – Pre-Pottery Neolithic farming at Beidha: a preliminary report. *Palestine Exploration Quarterly* **98**: 61–66.

Henry, D. O. (1997) Prehistoric human ecology in the southern Levant east of the Rift from 20 000–6 000 BP. *Paleorient* **23**: 107–119.

Hercman, H. and T. Goslar (2002) Uranium-series and radiocarbon dating of speleothems – methods and limitations. *Acta Geologica Polonica* **52**: 35–41.

Hourani, F. and M.-A. Courty (1997) L'évolution climatique de 10 500 à 5500 B.P. dans la vallée du Jourdain. *Paleorient* **23**: 95–105.

Hunt, C. O., H. A. Elrishi, D. D. Gilbertson *et al.* (2004) Early-Holocene environments in the Wadi Faynan, Jordan. *The Holocene* **14**: 921–930.

Kirkbride, D. (1960) The excavation of a Neolithic village at Seyl Aqlat, Beidha, near Petra. *Palestine Exploration Quarterly* **92**: 136–145.

Kirkbride, D. (1966) Five seasons at the Pre-Pottery Neolithic village of Beidha in Jordan. *Palestine Exploration Quarterly* **98**: 8–72.

Kirkbride, D. (1968) Beidha: Early Neolithic village life south of the Dead Sea. *Antiquity* **42**: 263–274.

Kirkbride, D. (1985) The environment of the Petra region during the Pre-Pottery Neolithic. In *Studies in the History and Archaeology of Jordan II*, ed. A. Hadidi. Amman: Department of Antiquities pp. 117–124.

Kirkbride, D. (1989) Preface. In *The Natufian Encampment at Beidha, Late Pleistocene Adaptation in the Southern Levant*, ed. B. F. Byrd. Aarhus: Jutland Archaeological Society pp. 7–8.

Kujit, I. and A. N. Goring-Morris (2002) Foraging, farming and social complexity in the Pre-Pottery Neolithic of the South-Central Levant: a review and synthesis. *Journal of World Prehistory* **16**: 361–439.

Liutkus, C. M., J. D. Wright, G. M. Ashley and N. E. Sikes (2005) Paleoenvironmental interpretation of lake-margin deposits using δC-13 and δO-18 results from early Pleistocene carbonate rhizoliths, Olduvai Gorge, Tanzania. *Geology* **33**: 377–380.

Ludwig, K. R. (2001) *ISOPLOT/Ex Rev. 2.49.* Boulder, CO: United States Geological Survey.

Ludwig, K. R. and D. M. Titterington (1994) Calculation of ^{230}Th/U isochrons, ages, and errors. *Geochimica et Cosmochimica Acta* **58**: 5031–5042.

Magaritz, M. and J. Heller (1980) A desert migration indicator – oxygen isotopic composition of land snail shells. *Palaeogeography Palaeoclimatology Palaeoecology* **32**: 153–162.

Mallick, R. and N. Frank (2002) A new technique for precise uranium-series dating of travertine micro-samples. *Geochimica et Cosmochimica Acta* **66**: 4261–4272.

Matsuoka, J., A. Kano, T. Oba *et al.* (2001) Seasonal variation of stable isotopic compositions recorded in a laminated tufa, SW Japan. *Earth and Planetary Science Letters* **192**: 31–44.

McLaren, S. J., Gilbertson, D. D., Grattan, J. P. *et al.* (2004) Quaternary palaeogeomorphologic evolution of the Wadi Faynan area, southern Jordan. *Paleogeography Paleoclimatology Paleoecology* **205**: 131–154.

Mithen, S. J. (2003) *After the Ice: A Global Human History, 20,000–5000 BC.* London: Weidenfeld and Nicolson.

Moore, A. M. T. and G. C. Hillman (1992) The Pleistocene to Holocene transition and human economy in Southwest Asia: the impact of the Younger Dryas. *American Antiquity* **57**: 482–494.

Mortensen, P. (1970) A preliminary study of the chipped stone industry from Beidha. *Acta Archaeologica* **41**: 1–54.

Neumann, F. H., E. J. Kagan, M. J. Schwab and M. Stein (2007) Palynology, sedimentology and palaeoecology of the late Holocene Dead Sea. *Quaternary Science Reviews* **26**: 1476–1498.

O'Brien, G. R., D. S. Kaufman, W. D. Sharp *et al.* (2006) Oxygen isotope composition of annually banded modern and Mid-Holocene travertine and evidence of paleomonsoon floods, Grand Canyon, Arizona, USA. *Quaternary Research* **65**: 366–379.

O'Neil, J. R., R. N. Clayton and T. K. Mayeda (1969) Oxygen isotope fractionation of divalent metal carbonates. *Journal of Chemical Physics* **30**: 5547–5558.

Olszewski, D. I. (1991) Social complexity in the Natufian? Assessing the relationship of ideas and data. In *Perspectives on the Past: Theoretical Biases in Mediterranean Hunter-Gatherer Research*, ed. G. Clark. Philadelphia: University of Pennsylvania Press pp. 322–340.

Pedley, H. M. (1990) Classification and environmental models of cool freshwater tufas. *Sedimentary Geology* **68**: 143–154.

Pedley, M., J. A. G. Martin, S. O. Delgado and M. G. Del Cura (2003) Sedimentology of Quaternary perched springline and paludal tufas: criteria for recognition, with examples from Guadalajara Province, Spain. *Sedimentology* **50**: 23–44.

Pentecost, A. (1995) The Quaternary travertine deposits of Europe and Asia Minor. *Quaternary Science Reviews* **14**: 1005–1028.

Perkins Jr, D. (1966) Appendix B – Fauna from Madamagh and Beidha: a preliminary report. *Palestine Exploration Quarterly* **98**: 66–67.

Portillo, M., R. M. Albert and D. O. Henry (2009) Domestic activities and spatial distribution in Ain Abu Nukhayla (Wadi Rum, Southern Jordan): the use of phytoliths and spherulites studies. *Quaternary International* **193**: 174–183.

Quade, J., J. A. Rech, C. Latorre *et al.* (2007) Soils at the hyperarid margin: the isotopic composition of soil carbonate from the Atacama Desert, Northern Chile. *Geochimica et Cosmochimica Acta* **71**: 3772–3795.

Raikes, R. (1966) Appendix C – Beidha: prehistoric climate and water supply. *Palestine Exploration Quarterly* **98**: 68–72.

Robinson, S. A., S. Black, B. Sellwood and P. J. Valdes (2006) A review of palaeoclimates and palaeoenvironments in the Levant and Eastern Mediterranean from 25,000 to 5000 years BP: setting the environmental background for the evolution of human civilisation. *Quaternary Science Reviews* **25**: 1517–1541.

Rohling, E. J. and H. Palike (2005) Centennial-scale climate cooling with a sudden cold event around 8,200 years ago. *Nature* **434**: 975–979.

Rollefson, G. (1998) The Aceramic Neolithic of Jordan. In *The Prehistoric Archaeology of Jordan*, ed. D. O. Henry. Oxford: British Archaeology Reports pp. 102–126.

Rollefson, G. (2001) The Neolithic period. In *The Archaeology of Jordan*, ed. B. MacDonald, R. Adams and P. Bienkowski. Sheffield: Sheffield University Press pp. 67–105.

Rollefson, G. and I. Köhler-Rollefson (1989) The collapse of early Neolithic settlements in the southern Levant. In *People and Culture in Change: Proceedings of the Second Symposium on Upper Palaeolithic, Mesolithic and Neolithic Populations of Europe and the Mediterranean Basin*, ed. I. Hershkovitz. Oxford: Oxbow pp. 73–89.

Rowe, P. J. and B. A. Maher (2000) 'Cold' stage formation of calcrete nodules in the Chinese Loess Plateau: evidence from U-series dating and stable isotope analysis. *Palaeogeography Palaeoclimatology Palaeoecology* **157**: 109–125.

Rozanski, K., L. Araguas-Araguas and R. Gonfiantini (1993) Isotopic patterns in modern global precipitation. In *Climate Change in Continental Isotopic Records*, ed. P. K. Swart, K. C. Lohman, J. A. McKenzie and S. Savin. Washington, DC: American Geophysical Union pp. 1–36.

Sanlaville, P. (1996) Changements climatiques dans la région levantine à la fin du Pléistocène supérieur et au début de l'Holocène. Leurs relations avec l'évolution des sociétés humaines. *Paleorient* **22**: 7–30.

Sanlaville, P. (1997) Les changements dans l'environnement au Moyen-Orient de 20 000 BP à 6 000 BP. *Paleorient* **23**: 249–262.

Smith, J. R., R. Giegengack and H. P. Schwarcz (2004) Constraints on Pleistocene pluvial climates through stable-isotope analysis of fossil-spring tufas and associated gastropods, Kharga Oasis, Egypt. *Palaeogeography Palaeoclimatology Palaeoecology* **206**: 157–175.

Soligo, M., P. Tuccimei, R. Barberi *et al.* (2002) U/Th dating of freshwater travertine from Middle Velino Valley (Central Italy): paleoclimatic and geological implications. *Palaeogeography Palaeoclimatology Palaeoecology* **184**: 147–161.

Staubwasser, M. and H. Weiss (2006) Holocene climate and cultural evolution in late prehistoric-early historic West Asia – Introduction. *Quaternary Research* **66**: 372–387.

Stein, M., A. Torfstein, I. Gavrieli and Y. Yechieli (2010) Abrupt aridities and salt deposition in the post-glacial Dead Sea and their North Atlantic connection. *Quaternary Science Reviews* **29**: 567–575.

Stordeur, D., D. Helmer and G. Willcox (1997) Jerf el ahmar, un nouveau site de l'horizon PPNA sur le moyen Euphrate syrien. *Bulletin de la Société préhistorique française* **94**: 282–285.

Tchernov, E. (1997) Are late Pleistocene environmental factors, faunal changes and cultural transformations causally connected? The case of the southern Levant. *Paleorient* **23**: 209–228.

Weninger, B., E. Alram-Stern, E. Bauer *et al.* (2006) Climate forcing due to the 8200 cal yr BP event observed at Early Neolithic sites in the eastern Mediterranean. *Quaternary Research* **66**: 401–420.

Zak, K., V. Lozek, J. Kadlec, J. Hladikova and V. Cilek (2002) Climate-induced changes in Holocene calcareous tufa formations, Bohemian Karst, Czech Republic. *Quaternary International* **91**: 137–152.

Zanchetta, G., M. Di Vito, A. E. Fallick and R. Sulpizio (2000) Stable isotopes of pedogenic carbonates from the Somma-Vesuvius area, southern Italy, over the past 18 kyr: palaeoclimatic implications. *Journal of Quaternary Science* **15**: 813–824.

Zohary, M. (1962) *Plant Life of Palestine: Israel and Jordan*. New York: Roland Press.

17 The influence of water on Chalcolithic and Early Bronze Age settlement patterns in the southern Levant

Jaimie Lovell and Andrew Bradley

ABSTRACT

While the environment cannot be considered a '*deus ex machina*' for any event in human history, it is becoming increasingly clear to prehistorians that the extraordinary developments in human social complexity documented in the archaeological record since the beginnings of sedentism in the Late Pleistocene occurred in concert with profound climatic and environmental changes. This chapter investigates settlement patterns in Jordan, Palestine and Israel during a key archaeological transition in the southern Levant, the Chalcolithic to the Early Bronze Age (EBA). We summarise regional settlement location in relation to potential forcing factors which themselves may be indicative of societal and climatic change, such as springs, wadis, routes and permanent sites. We make a geospatial analysis and calculate 'cost distance' values between these forcing factors and the settlement data. We then analyse how these cost distance values change over time in different altitudinal belts. We find that the cost distance patterns varied at different altitudes during the transition period. In some altitudinal belts in the EBI and EBA, springs appear to determine settlement location, showing the importance of climate, while in other altitudinal belts formalisation of settlements around the longer-standing sites and routes suggest that socio-political changes may have been more influential in the EBA. The value of this method and the implications of our results are then discussed in the light of existing and emerging research on the transition from the 'prehistoric' to the 'historic' period.

17.1 INTRODUCTION

The Chalcolithic, 6,450–5,550 BP, has been termed the 'end of prehistory' (Joffe, 2003). The changes that took place in the transition from the Chalcolithic to the Early Bronze Age (EBA), 5,550–4,350 BP, and the dawn of 'urban systems' have, therefore, broad implications for the way in which we understand adaptations to environment and ecosystems in the southern Levant. As a result, previous investigations have focused upon settlement systems in the key transitional sub-period, the EBI, 5,550–5,050 BP (e.g. Finkelstein and Magen, 1993). Such work has been stymied by protracted discussions regarding EBI material culture and arguments over early and late EBI ceramics. Braun (1996) suggests that part of the problem is that the diagnostic criteria for ceramic identification are often regionally specific or simply inadequate.

Consequently, since the review by Hanbury-Tenison (1986) of the nature of the Chalcolithic/EBA transition in the southern Levant, in-depth analyses have been few and primarily concerned with the later phases of the transition. An example is that of Esse (1991) who put together a synthesis of urbanisation in the Jezreel Valley in the EBA. More recent research has served to highlight the view that EBA material culture is a more generic and, in some respects, a de-skilled version of the previous Chalcolithic artefact set (Braun, 2010). In spite of these difficulties of definition and identification, the EBI is recognised as representing a pivotal point in the archaeology of the southern Levant. It follows a period of Chalcolithic artistic creativity and precedes the nucleated, walled settlements that led into the 'internationalised' Bronze Age.

The classic reading has been that the transition involved a settlement collapse (Gophna and Portugali, 1988) accompanied by a significant reduction in quality and individuality of artistic craftsmanship. While some recent scholarship acknowledges a more complex past reality, the concept of a 'collapse in settlement' has been reiterated in recent general syntheses (e.g. Levy, 1995), and all would agree that '[the] rich symbolic landscape of the Chalcolithic vanishes forever' (Bourke, 2008 p. 147). Various explanations have been offered for this 'collapse' (see Joffe, 1993), including a failure of the 'chiefdom'-style socio-political

Water, Life and Civilisation: Climate, Environment and Society in the Jordan Valley, ed. Steven Mithen and Emily Black. Published by Cambridge University Press. © Steven Mithen and Emily Black 2011.

system (Gilead, 1995; Levy, 1995; cf. Kerner, 1997), the impact of colonialist expansion from Egypt (de Miroschedji, 2002; cf. Lovell, 2008) and a consequence of a climatic shift (Lovell, 2001; cf. Burton and Levy, in press; see also Robinson et al., 2006).

New data blur such simplistic narratives by indicating that higher-altitude sites were more widespread in the Chalcolithic than previously believed (Gibson et al., 1991; van den Brink et al., 2001; Lovell, 2008), and were the precursor to a more intensive occupation of the hill country in the following EBA (Lovell, 2002; Gophna and Tsuk, 2005). There are, therefore, a series of socio-political and environmental factors that need to be considered as potential causes of the collapse, while recognising that there may be more continuity than previously acknowledged. Researching these changes on a regional scale presents some challenges. At smaller scales one may be able to identify evidence for socio-political change, either via artefactual evidence or via settlement clustering. Scaling up localised evidence is more difficult given the current issues in EBI ceramic interpretation. However, if environmental factors, as opposed to socio-political factors, played a part in the transition, then this ought to be reflected in the settlement data. That is, the locations of settlements should be an indication of how they were positioned to take advantage of natural resources such as water during periods of environmental change. Of course, it is likely that socio-political factors were also at play, but broad patterns of settlement change should reflect environmental factors operating at the larger scale.

Our aim is to investigate the available archaeological survey data for Jordan, the occupied Palestinian territories (oPt) and Israel to test whether an altitudinal shift in settlement location occurred during the transition. Our secondary aim is to build an understanding of the environmental and socio-political factors operating during the transition from the Chalcolithic to the EBA. An analysis of altitudinal shifts in conjunction with potential factors influencing settlement location will afford insight into technological development and predominant land-use practices in the region. Despite discussions regarding the importance of regional environments in the Chalcolithic, there has been limited discussion of the different requirements for each zone in terms of water, land use and communications, although ongoing debates stress the relative importance of horticulture and pastoralism in the EBA economy (Philip and Baird, 1993; Lovell, 2002, 2008; Bourke, 2008). In the EBA there is clearer evidence for technological advances in water management (see below) and thus more explicit academic interest in water (Philip, 2008). We aim to understand patterns of settlement in relation to the exploitation of water, using spatial analysis of settlement locales with a Geographical Information System (GIS).

Spatial analysis using a GIS populated with the location and duration of settlements within a continuous field of surface elevation can provide quantitative and qualitative evidence of altitudinal shifts through time. A few GIS studies have been carried out in the Levant, for example by Fletcher (2008), and Fletcher and Winter (2008), who made an in-depth analysis of settlement patterns in relation to suitable locations for farming in the Negev. On a longer timescale, Al-Shorman (2002) considered the number of archaeological sites in Jordan with respect to wet and dry periods between the Palaeolithic and Ottoman. We expand on Fletcher and Winter (2008) by considering a transitional time period, and we expand on Al-Shorman (2002) by considering altitudinal shifts and localised controlling factors on site location. In this chapter we have endeavoured to overcome a number of challenges with respect to the quality of the survey data and evenness in the spatial coverage on the regional scale, to describe and explain the patterns of settlement change.

17.2 METHOD

To identify variation in the altitudinal location of settlements across the Chalcolithic/EBA transition requires: (i) the use of archaeological survey data consisting of geo-referenced archaeological sites (subsection 17.2.1); (ii) collation of the archaeological data into appropriate time periods (17.2.2); (iii) the use of GIS to divide the landscape, to determine which parts of the landscape contain the best archaeological 'blanket coverage', and then calculate cost distance values for all sites with respect to potential 'forcing factors' (17.2.3); and (iv) analysis of the characteristics of settlement patterns with respect to time, altitude and proximity to potential forcing factors (section 17.3).

17.2.1 The survey data

The data for this study were taken from the continually updated Jordan Archaeological Data Information System (JADIS) (Palumbo, 1994) and the most recently published maps produced by Archaeological Survey of Israel conducted by the Israel Antiquities Authority (IAA) (Cohen, 1981; Cohen, 1985; Hirschfeld, 1985; Haiman, 1986; Lender, 1990; Ne-eman, 1990; Gal, 1991; Haiman, 1991; Avni, 1992; Dagan, 1992; Haiman, 1993; Patrich, 1994; Rosen, 1994; Gazit, 1996; Frankel and Getzov, 1997; Gophna and Beit-Arieh, 1997; Gal, 1998; Gophna and Ayalon, 1998; Haiman, 1999; Raban, 1999; Ne-eman et al., 2000; Beit-Arieh, 2003; Olami and Gal, 2003; Olami et al., 2003; Baumgartel, 2004; Berman et al., 2004; Olami et al., 2004; Weiss et al., 2004; Berman and Barda, 2005; Ne-eman et al., 2005a; Ne-eman et al., 2005b; Stark et al., 2005). In Jordan, each survey project has been required to supply grid references and basic data to the Department of Antiquities in order to update the JADIS system; the coverage is therefore uneven and dictated by individual survey interests. We should note that the Department of Antiquities has very recently moved to a new system, MEGA (Middle Eastern Geo-Database for Antiquities), which is designed to overcome many of JADIS's shortcomings.

In the case of the Israeli survey data, numbered grid squares, termed 'maps', are published sporadically and therefore provide an incomplete patchwork of data. To compensate for these gaps in the IAA survey data we have included ancillary data from surveys in the vicinity of the Jezreel Valley (Esse, 1991); the coastal plain (Gophna and Portugali, 1988); the 'hill country' (Finkelstein and Gophna, 1993; Finkelstein and Magen, 1993; Finkelstein et al., 1997); and Galilee (Frankel et al., 2001). In theory the IAA data should provide blanket coverage (i.e. every square metre of land is covered on foot by a surveyor); in reality the spatial coverage is variable. Coverage in the oPt is poor because, except in the case of Jerusalem, the IAA surveys generally confine themselves to 1948 borders. In order to mitigate the effects of inconsistent spatial coverage with as few gaps on the ground as possible we looked for areas with the best 'blanket' survey coverage. We considered 'blanket' survey coverage to be contiguous surveys, sometimes referred to as 'full combing' or with a reasonable spatial sampling strategy over wide areas. Mapping the extent of these surveys allowed assessment of coverage in individual drainage basins. We focused our study on those basins with the best 'blanket' survey coverage (see 17.2.3 below).

17.2.2 Dating

There are no standard period divisions into which survey data are classified in Jordan, Israel or the oPt, and not all surveys distinguish between sub-periods, e.g. in some cases sites are listed as EBA rather than EBI or EBII. For the purposes of this study we have taken the surveyor to be correct in their classification of a site to a period or sub-period. We are aware that in many cases the diagnostic criteria for sub-periods are far from adequate, and therefore certain regions are more reliable than others for comparison across sub-periods. Survey material is notoriously difficult to date, made up as it is of rolled and degraded surface material, such that surveyors sometimes list sites simply as 'Chalc/EBA'. Furthermore, it is generally not possible to check attributions because surveyors commonly publish a small selection of their material as a reference, if at all. We assumed that when most surveyors allocate a site to the Chalcolithic they mean classic or Late Chalcolithic, c. 6,450–5,950 BP, and occasionally as recent as 5,750 BP, giving a 500- to 700- year span. Sites recorded as 'Early Chalcolithic' were not included in our analysis. We did not confine our examination to the EBIa or 'Early EBI' because few surveyors define sub-periods within EBI; instead all EBI sites are grouped together. The EBI runs from 5,550 to 5,050 BP, a 500- year span. In most cases when surveyors designate a site as simply 'EBA', they generally mean EBII–III. For this study, sites listed as EBA are those where no sub-period was given, or where a surveyor allocated the site to the EBII–III (5,050–4,350 BP, dates following Chesson and Philip, 2003), which indicates a 700- year window. Therefore these data probably include a small number of EBI sites that were listed as EBA by the surveyor. To address the period of the transition we used sites representing the Chalcolithic, EBI and EBA from the IAA dataset and used corresponding time windows from the JADIS database. This involved merging the 'Late Chalcolithic' sites with 'Unspecified Chalcolithic' sites in the JADIS data and IAA datasets and comparing these with the 'Early Bronze Age I' data and 'Unspecified Early Bronze Age' data from the JADIS database and IAA datasets. The ancillary data (see above) were classified into these three periods as appropriate.

17.2.3 GIS analysis

The GIS was used as a tool to perform three main tasks: (i) to merge the two sources of archaeological surveys, filter data into specific time periods and plot the results on a map; (ii) find the best archaeological coverage to analyse and restrict the analysis to drainage basins with a wide range of altitudinal belts; and (iii) to identify the influence of potential factors i.e. wadis, springs, permanent sites and routes that contribute to settlement location using cost surface analysis.

Each of these three tasks will now be elaborated.

(i) Each database has detailed information on archaeological surveys, in particular information on dating and location. The two databases were merged and the grid coordinates in each database were standardised to decimal degrees (Savage, 2002) and plotted with ESRI Geographical Information Software, Arc-Map v9.0 (ArcGIS). All sites were then divided into the archaeological periods of Chalcolithic, EBI and EBA (see 17.2.2). We assume that a continuity of settlement reflects a degree of resilience to changes in conditions or circumstances over time since Al-Shorman (2002) had indicated that a period of unfavourable climate can cause the abandonment and then the re-occupation of sites. To reflect the continuity of settlement, sites occupied during all three time periods were allocated to a category 'permanent sites'. Where the site was occupied in two consecutive time periods, e.g. Chalcolithic and EBI or EBI and EBA, or where the site was abandoned during the EBI, they were categorised into 'multi-period', and where the site was occupied only once in any one of the three time periods it was defined as 'single-period' occupation.

(ii) A major problem in the use of archaeological data in a spatial database is that it is only possible to plot data that have been recorded at a particular location (often without knowing the extent to which the surrounding area has been surveyed). This could potentially influence any analysis as there is no way of knowing the extent to which surrounding areas have been

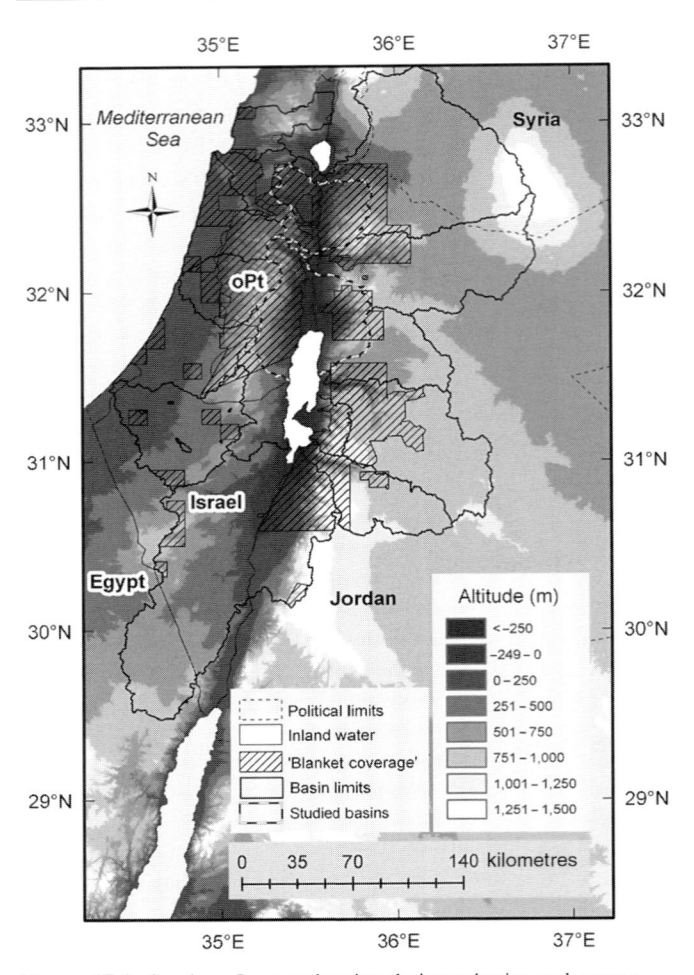

Figure 17.1. Southern Levant showing drainage basins and survey coverage, with the North Rift Basin and North Dead Sea Basin indicated. oPt refers to occupied Palestine territory.

surveyed without checking the original research publications. To overcome this we revisited the original survey publications, digitised their maps and the IAA survey squares and then plotted them in the ArcGIS to represent theoretical blanket coverage (Figure 17.1). In many cases survey boundaries are notional rather than actual, and sectors are sampled rather than fully covered (e.g. Palumbo et al., 1990). To select which basins might be considered as comprehensively sampled we had to make a judgement call on the basis of the published information about the survey itself and on the strength of the dataset. The final study area (see also Figure 17.1) was chosen on the basis of the best 'blanket' coverage from those areas where all ranges of altitude were covered.

To limit the boundaries of the best survey/altitude coverage we divided the area according to the surface hydrology of the region and chose the drainage basins with the best survey/altitude coverage. The drainage basins were calculated from the Shuttle Radar Topography Mission Digital Elevation Model (DEM) (Jarvis et al., 2008) using spatial analyst functions in ESRI ArcGIS.

(iii) The potential forcing factors, permanent sites, major routes, springs, and wadis were mapped and created from several sources.

Permanent sites. The permanent sites, as defined in section 17.2.3 (i), were extracted from the merged archaeological database.

Major routes. The routes were calculated as the most repeated paths across the study region from each permanent site to every other settlement in each time period using a cost surface calculated from the DEM (see below). This produced a complex network of routes (or 'roadmap'). To avoid the inclusion of anomalous routes the roadmap required some cleaning, i.e. those routes duplicated fewer than eight times were discarded.

Wadis. ESRI Arcinfo was used to calculate wadi drainage networks from the DEM. To produce a representative wadi network it was found that a threshold of greater than third order (Strahler) streams worked well. Because wadis are ephemeral, this network represents potential water sources, not permanent surface drainage.

Springs. To identify spring locations we used a 1:250,000 topographic map of Israel (Survey of Israel 1988). These data were plotted in decimal degrees in the ArcGIS. For the purposes of this exercise we assume that these spring locations existed during the transition period.

Cost surfaces were then calculated with respect to each forcing factor, and at each settlement the value of the cost surface was recorded. Cost surfaces indicate how costly (in terms of effort) it would be to move from a particular source location (across the surface of a region) to a target variable. In simple terms, travel within the immediate area of the target would be cheapest; travel would become more costly as the distance in any direction increased. In addition, ArcGIS can accommodate obstructions on the surface and calculate the modified cost surface. Previous studies for the Negev in Israel have employed amalgamated cost surfaces (i.e. cost distances to water, alluvial flats for farming, grasses, Mediterranean maquis and loess soils, merged together as one value) in order to understand site location in the Chalcolithic (Fletcher, 2008; Fletcher and Winter, 2008). In this chapter we have deliberately kept each variable separate in order to investigate the changing roles of socio-political and environmental factors. We also extend our analysis into the temporal domain and take the opportunity of examining how these factors relate to site location through a transitional time period rather than simply considering a snapshot in time.

These cost surface calculations were based on topographic characteristics such as slope and aspect, and the assumption that water bodies (such as the Dead Sea) would be circumnavigated.

The cost surfaces were calculated from the source destinations of potential wadi networks, spring locations, permanent sites and potential major routes (Figure 17.2a–c). At each individual site (i.e. the target point) the cost surface values were then extracted. High cost surface values indicate that it is difficult to access the source, low values that it is easy to reach. We can use these values to infer environmental and socio-political conditions in each time period. The cost surface value to wadis may indicate how the occupants of settlements decided between harvesting of surface water from watercourses and locating near to springs to access groundwater sources. For the socio-political organisation of settlements, cost surface values may indicate some kind of settlement organisation and show whether settlements have located near to permanent sites and/or close to permanent route ways.

To understand the changes in settlement patterns within each basin we divided the chosen study areas into altitudinal belts, according to the topographic characteristics of the basins (see section 17.3, 'Results'). To provide a statistical summary of settlement characteristics in each of the altitude belts over time, the mean and variance were calculated for cost to wadi, spring, site and routes. To identify and compare any altitude- or time-related trends, each group of cost surface values for all time periods was then plotted against altitude, and a linear regression was applied to each individual time period within each altitudinal belt.

To identify any trends and biases in particular driving factors within the altitudinal belts over time, the cost surface values to wadi, spring, site and routes for all three time periods were analysed in pairs using scatter plots with a best fit regression and with the 1:1 line of equal value. If one of the forcing factors is stronger the points will fall to one side of the 1:1 line, and if one forcing factor has a strong determinate influence on settlement location the data points should cluster. We should also be able to detect a temporal change in forcing factors by assessing if there was a major change in the gradient of the linear regression between time periods or if there were any changes in position of data clusters between each time period. Changes in the relationships between each driving factor were then analysed for evidence of changes in socio-political and environmental factors and contextualised using the maps (Figure 17.2 a–c). We then compared these results to existing scholarship on the Chalcolithic to EBA transition.

17.3 RESULTS

The area with the best 'blanket' survey coverage occurred in the Jordan Valley, north of the Dead Sea (Figure 17.1). According to the results of our DEM basin calculation, this area was mostly covered by our Northern Dead Sea Basin (NDSB) and North Rift Basin (NRB). Sites within these two basins were divided according to the three main topographic characteristics common to both basins: Belt 1, less than −200 m (the valley bottom); Belt 2, between −200 m and 300 m (the rift escarpment/foothills); and Belt 3, greater than 300 m (the highlands to the edge of the watershed basin).

17.3.1 Character of the two basins and background data

The NDSB has been covered by numerous surveys. These include surveys around major sites like Jerusalem, Iraq el Amir, Teleilat Ghassul and Hesban (Ibach, 1987); surveys undertaken by the Jordanian Department of Antiquities on the eastern shore of the Dead Sea and surveys by the IAA around the areas of Jerusalem, extending into occupied East Jerusalem (Kloner, 2001); surveys under the aegis of the Military Governor in the oPt (Finkelstein and Magen, 1993; Finkelstein et al., 1997); and sporadic surveys and explorations undertaken by earlier researchers working in Palestine (de Vaux, 1971).

The major permanent sites that fall within the surveyed areas are Jerusalem/Al Quds and Hesban. Other sites such as Muraygat show evidence of permanent use, and continual occupation of small subregions is clear from the presence of various small sites dating from all periods relevant here. There is a gap in blanket survey coverage in the centre of this basin, resulting in gaps in data: for instance, the site of Jericho (Tell es Sultan) does not fall within the surveyed areas for this analysis. Evidence for the changes in Dead Sea levels (Migowski et al., 2006) indicates that some of this area may also have been submerged during the EBI, which may account for the lack of data in this region. Data on the fluctuating levels of the Dead Sea are various (Chapters 6, 9 and 11, this volume), with Frumkin et al. (1991) suggesting a highstand during the EBA of up to −280 m, and then revising this in subsequent papers (1997, 2001). Bruins, using the same data, postulates a highstand of −300 m during the EBII (Bruins, 1994). Our data here show levels of −330 m for the EBI and a contrasting lowstand in the EBA, essentially representing the rapid drop postulated for the end of the period and start of the EBIV (Migowski et al., 2006). However, the EBA level could be higher at various points throughout the period, and a change in the Dead Sea level will affect spring sources, which may be covered by a higher stand.

There are known sites in this region picked up in ad hoc surveys, e.g. Mellaart's Jordan Valley sites (Leonard, 1992), although some of these have since been found to have been misdated (Anfinset et al., in press). The basin takes in the Wadis Shu'eib (Wright et al., 1989) and Hesban (Ibach, 1987), cutting into the plateau and greater Amman (Abu Dayyah et al., 1991), in Jordan, and the hinterland of Tell Balatah (Shechem), the Wadi el Farah, Wadi Auja, Wadi Makkuk/Nueima, Wadi Qilt

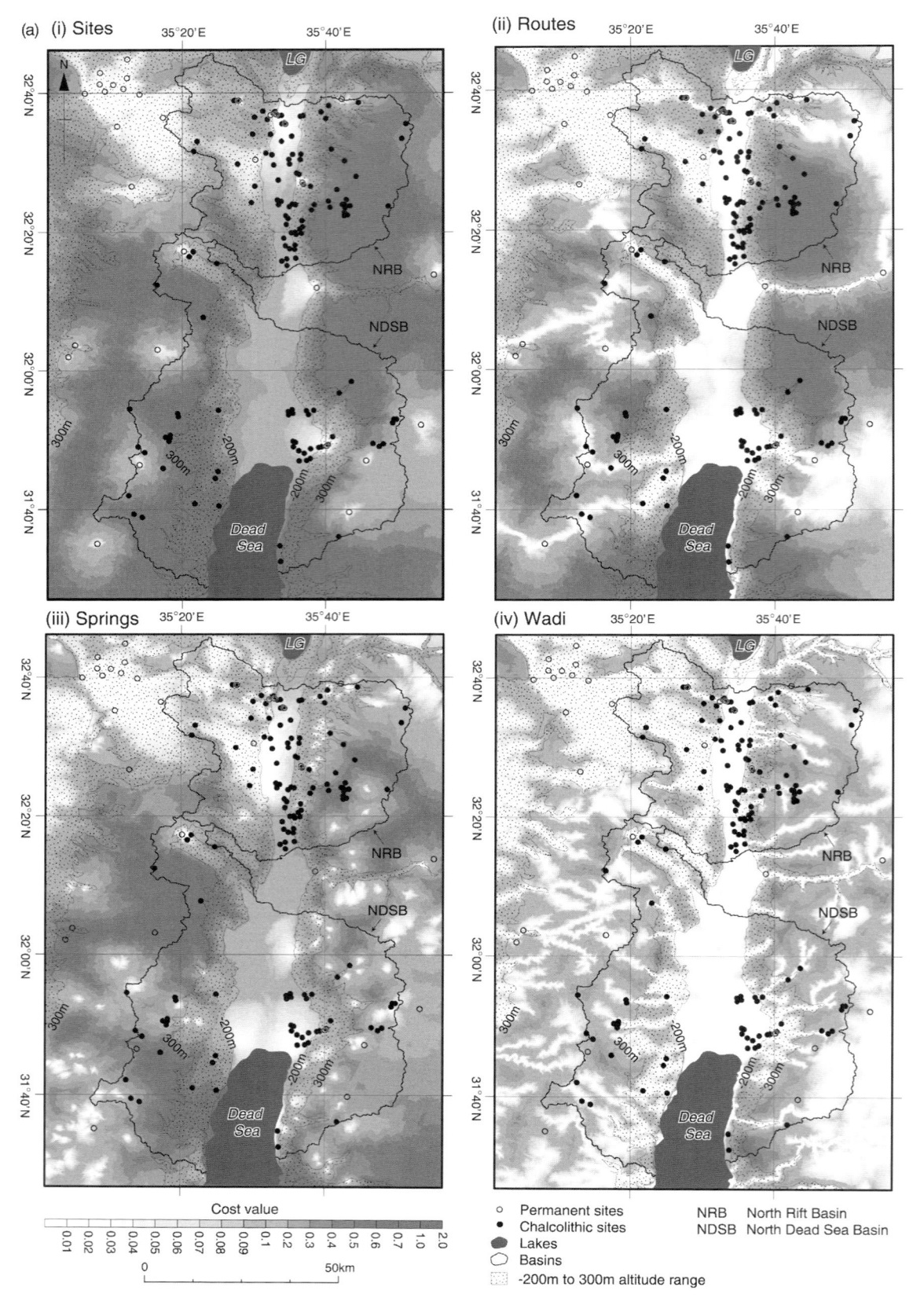

Figure 17.2. Cost distance maps and site locations for the three time periods: (a) Chalcolithic, (b) EBI, (c) EBA. Top left, to permanent sites; top right to routes; bottom left, to springs; bottom right, to wadi.

INFLUENCE OF WATER ON CHALCOLITHIC AND EBA SETTLEMENT PATTERNS

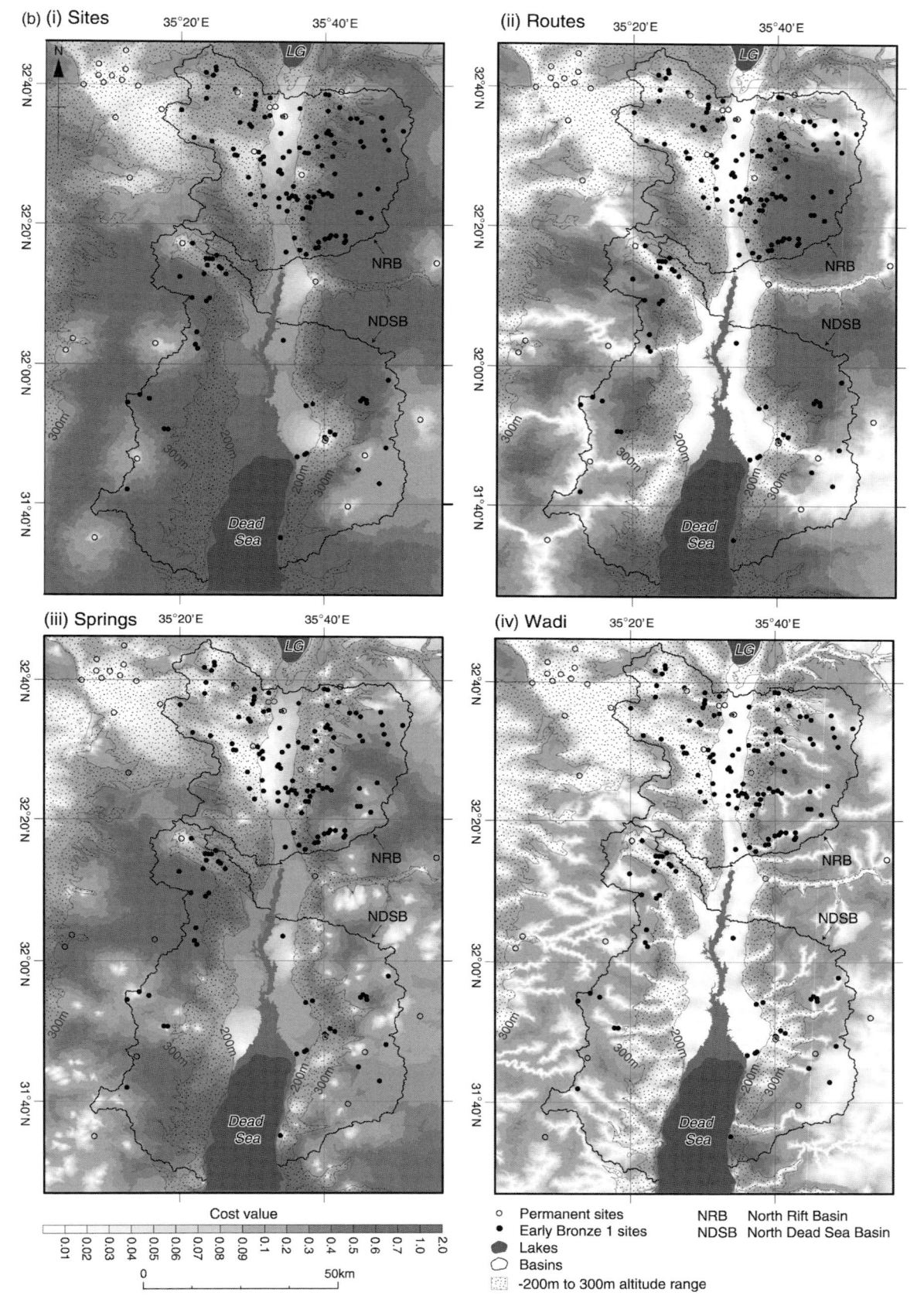

Figure 17.2. (*cont.*)

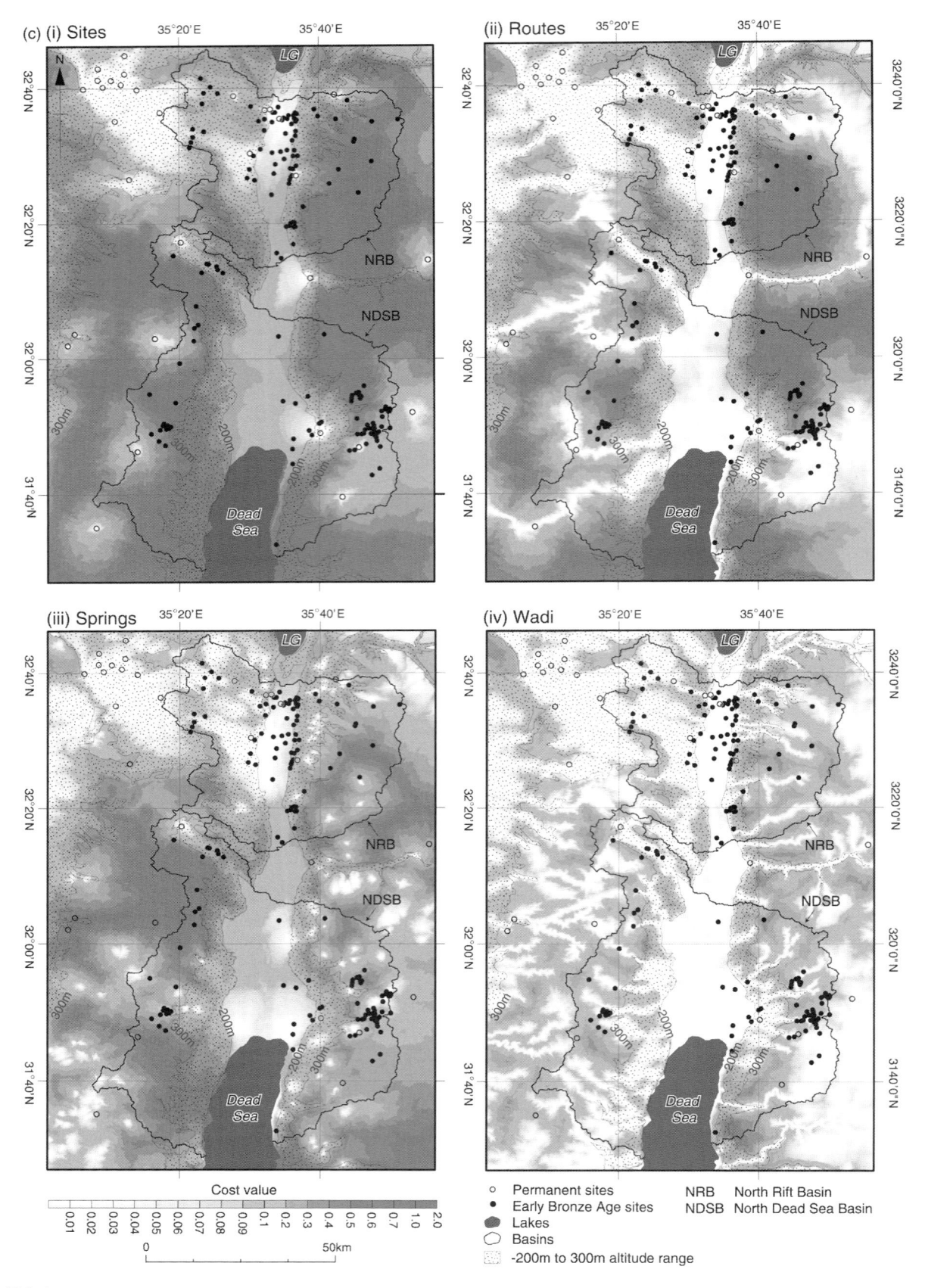

Figure 17.2. (*cont.*)

Table 17.1 *Number of sites and average altitude for the North Dead Sea Basin for all sectors and time periods (includes all survey data; sites may be listed in more than one column)*

Belt	Altitude range	Chalco		EBI		EBA	
		n	Mean	n	Mean	n	Mean
1	<− 200 m	18	−268 m	6	−282 m	9	−266 m
2	−200 m to 300 m	12	−4 m	16	8 m	10	−26 m
3	>300 m	25	615 m	25	638 m	68	632 m

Table 17.2 *Number of sites and average altitude for the North Rift Basin for all sectors and time periods (includes all survey data; sites may be listed in more than one column)*

Belt	Altitude range	Chalco		EBI		EBA	
		n	Mean	n	Mean	n	Mean
1	<− 200 m	28	−249 m	32	−240 m	32	−243 m
2	−200 m to 300 m	41	−39 m	85	34 m	84	−21 m
3	>300 m	19	567 m	68	517 m	44	516 m

and smaller wadis to the west of the Dead Sea in the oPt. In the NDSB the table of site numbers (Table 17.1) shows that sites occur in all three time periods. Below −200 m settlement activity seems to decrease in the EBI and EBA, but in the >300 m belt is steady in the EBI and then increases by the EBA. During the transition there appears to have been a mean increase in occupation at higher altitudes and, potentially, a reduction in activity or occupation at the lower altitudes.

Data from this basin in particular highlight that site positioning is inevitably affected by topography and landscape. At the lower altitudes settlement is confined to wadis which cut the steep-sided valley escarpment (−200 to 300 m asl belt) and also control the location of routes (these are the easiest pathways between the high and low altitudes). This topographic effect would almost certainly force settlement into the wadis and encourage clustering of sites. There is a particular cluster of sites in the Wadi el Farah, which suggests that this was an important corridor of movement. Occupation appears to push out to the east and west to higher altitudes in the EBI (Figure 17.2b; cf. 17.2a), probably as the Dead Sea expanded. Clustering is evident at the higher altitudes during the EBA (Figure 17.2c). This shift may be reflected by the data in the NRB.

The NRB has been covered by various surveys including the Jordan Valley survey (Ibrahim *et al.*, 1976), the Wadi Yabis and Jerash regional surveys (Hanbury-Tenison, 1987; Mabry and Palumbo, 1988; Palumbo *et al.*, 1990), Mittmann's (1970) survey of the northern highlands and smaller surveys of other lateral wadis (e.g. Gordon and Villiers, 1983). The western sector has been covered by the IAA with ancillary data provided by Esse (1991). The major permanent sites include Beth Shean (Mazar and Mullins, 2007) and Pella (Bourke *et al.*, 2003). In addition to the Jordan Valley, below sea level, the drainage basin takes in the northern highlands and the Irbid plateau of Jordan, the lower Jezreel Valley in Israel and parts of the Samarian hills in the oPt. Very high altitudes (up to 1,000 m) exist only in the east of the basin, but a large portion of the basin consists of land >300 m. In the lower reaches, there are open flat areas that break the −200 to 300 m belt into habitable steps. Both Figure 17.2 and Table 17.2 show that the NRB had consistent occupation below the <−200 m belt in all time periods, but in the −200 to 300 m and the >300 m belts there is increased activity from the EBI onwards, i.e. a general increase in settlement at altitude whilst maintaining occupation at low altitudes. Intense Chalcolithic exploitation of the Wadi Rayyan contrasts with wadis to the north and south. Although this may be a function of different surveyors, it would appear that the data come from a variety of sources, including Mittmann (1970). By contrast, further south, Gordon and Villiers (1983) do not distinguish between sub-periods in their survey of the Wadi Zerqa and this probably inflates the numbers of EBA sites in this region.

17.3.2 Results of the cost distance analysis

Because sites are often occupied for two or more periods (Table 17.1) it is possible to analyse the results according to period and according to duration of occupation. Single-period statistics prove useful because sites occupying a single time period show sensitivity to change when favourable conditions lasted for shorter time periods whereas sites that cover two time periods are assumed to have had some resilience to change. In each altitude belt (see Tables 17.3 and 17.4) the average cost distance and variance around the mean generally increase with altitude for wadi, site, route and spring in each time period, with occasional exceptions, e.g. NDSB altitude vs. spring in the EBI (a decrease from 0.364 to 0.304). In the NRB the general trend for average least cost was to wadi, then generally spring or route, followed by site. In the NDSB the least average cost was also to wadi then usually route, followed by spring then site.

Further information can be drawn from variation within the altitudinal belts: Figure 17.3 shows the trends in cost distance for each basin, altitude sector and time period. In the NRB, as with the averages, the trend indicated a general increase in cost distance values for all factors within altitude belts. In the NDSB, however, there is a reversal in the cost distance trend, as cost distances actually begin to fall at the higher altitudes.

Table 17.3 *Statistics of cost distance values for the North Rift Basin for all sectors and time periods (single time period data)*

North Rift		<−200 m		−200 m to 300 m		>300 m	
		Mean	Variance	Mean	Variance	Mean	Variance
Chalco	SITES	0.116	0.003	0.187	0.028	0.674	0.034
EB1		0.095	0.001	0.207	0.014	0.622	0.047
EBA		0.046	<0.001	0.125	0.008	0.516	0.044
Chalco	WADI	0.012	<0.001	0.072	0.006	0.121	0.005
EB1		0.012	<0.001	0.048	0.003	0.098	0.005
EBA		0.009	<0.001	0.049	0.002	0.142	0.006
Chalco	ROUTE	0.026	0.001	0.120	0.012	0.618	0.044
EB1		0.011	<0.001	0.130	0.015	0.437	0.091
EBA		0.010	<0.001	0.062	0.005	0.246	0.082
Chalco	SPRING	0.012	0.005	0.123	0.005	0.385	0.015
EB1		0.035	<0.001	0.111	0.010	0.188	0.007
EBA		0.058	0.001	0.116	0.003	0.227	0.018

Table 17.4 *Statistics of cost distance values for the North Dead Sea Basin for all sectors and time periods (single time period data)*

North Dead Sea Basin		<−200 m		−200 m to 300 m		>300 m	
		Mean	Variance	Mean	Variance	Mean	Variance
Chalco	SITES	0.121	0.013	0.389	0.048	0.444	0.059
EB1		0.440	0.297	0.309	0.068	0.478	0.025
EBA		0.116	0.005	0.217	0.023	0.361	0.031
Chalco	WADI	0.015	0.001	0.026	0.000	0.115	0.006
EB1		0.117	0.038	0.038	0.001	0.102	0.007
EBA		0.027	0.002	0.006	<0.001	0.092	0.004
Chalco	ROUTE	0.028	<0.001	0.153	0.040	0.200	0.060
EB1		0.307	0.266	0.131	0.027	0.303	0.054
EBA		0.033	0.001	0.097	0.017	0.280	0.037
Chalco	SPRING	0.064	0.001	0.283	0.032	0.307	0.033
EB1		0.121	0.003	0.364	0.052	0.304	0.067
EBA		0.063	0.001	0.163	0.003	0.226	0.024

The relative position of the best fit lines indicates whether there are changes in site and cost distance relationships to the four variables in each time period. Generally gradients between time periods are similar within each belt, indicating that over time trends in forcing factors do not change much between the Chalcolithic, EBI and EBA at different altitudes and there are consistent increases or decreases in cost value within each belt. However, there are variations which may indicate different trends or distinct changes in forcing factors for specific belts at specific times. (a) There are reasonably low or 'flat' gradients, e. g. in the NRB (Figure 17.3a, iv) e.g. altitude vs spring, belt 2, indicating that the springs were consistently close to settlements regardless of altitude and time; (b) a downward fanning of trend lines, e.g. NRB (Figure 17.3a, iii) altitude vs route, belt 3, indicating that in the EBA at lower altitudes closeness to a route became a more significant forcing factor as opposed to conditions in the Chalcolithic; (c) an upward fanning in gradients, NDSB (Figure 17.3b iv) altitude vs spring, belt 2, indicating that during the EBI springs were a less important forcing factor at the higher altitudes in this belt as opposed to conditions in the EBA; and (d) a reversed trend, NDSB (Figure 17.3b iv) altitude vs spring, belt 3. For the EBI and EBA at the higher altitudes a reliance on springs may be reflected by low cost distance values, and at low altitudes springs are less important as cost distances are much higher. In the Chalcolithic it appears the opposite may be the case, since the trend line gradient is reversed, implying the opposite conditions to the other two time periods.

In both basins, for each altitudinal belt, most of the trend lines in each time period only have subtle variations in gradient, but in some cases there are definite gradient changes. In the

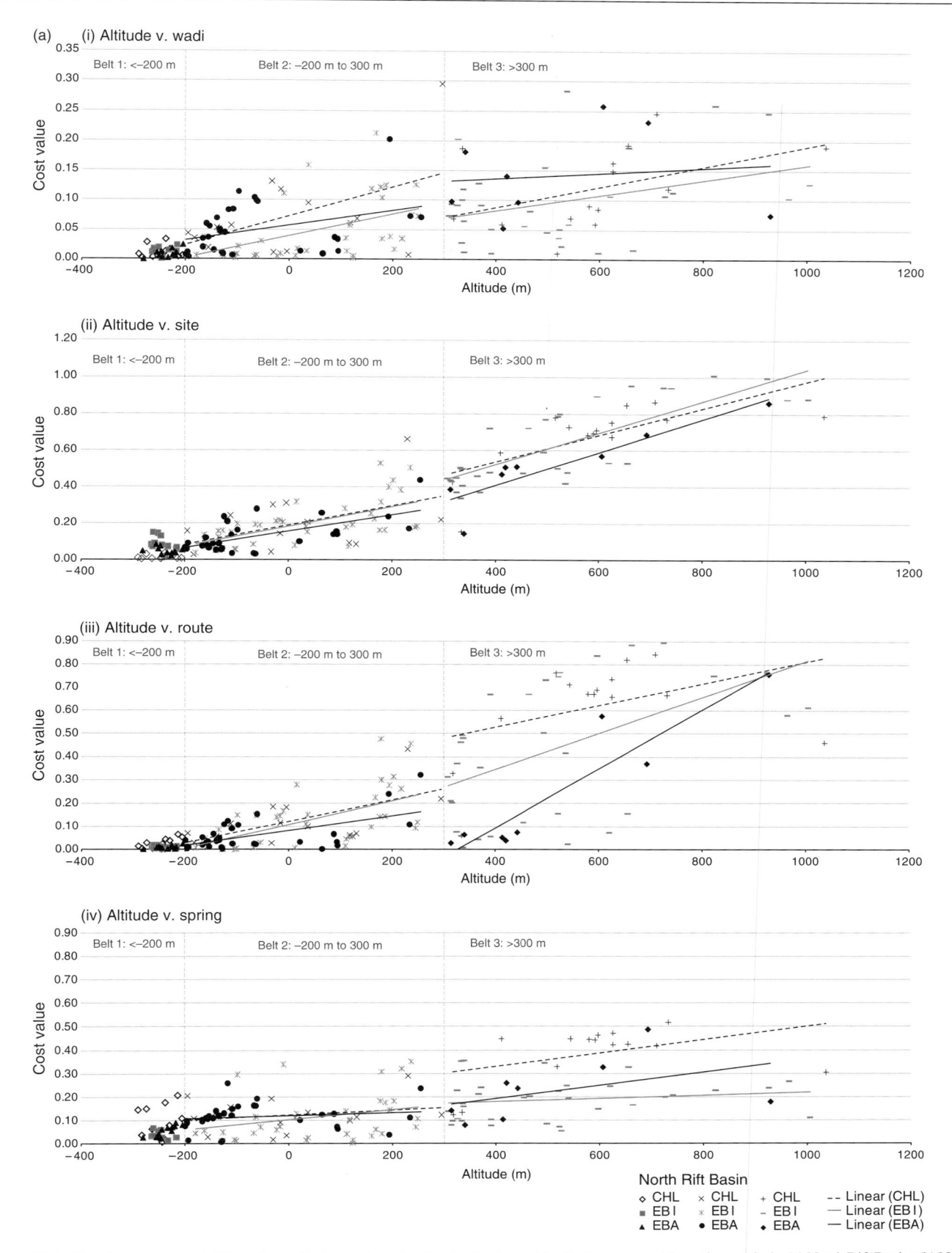

Figure 17.3. Cost distance against altitude for wadi, site, route and spring for the three altitudinal sectors and three time periods. (a) North Rift Basin; (b) North Dead Sea Basin. Best fit lines represent the general trends in each sector. Altitudinal sectors, <-200 m, 200 m to 300 m and >300 m. CHL, Chalcolithic.

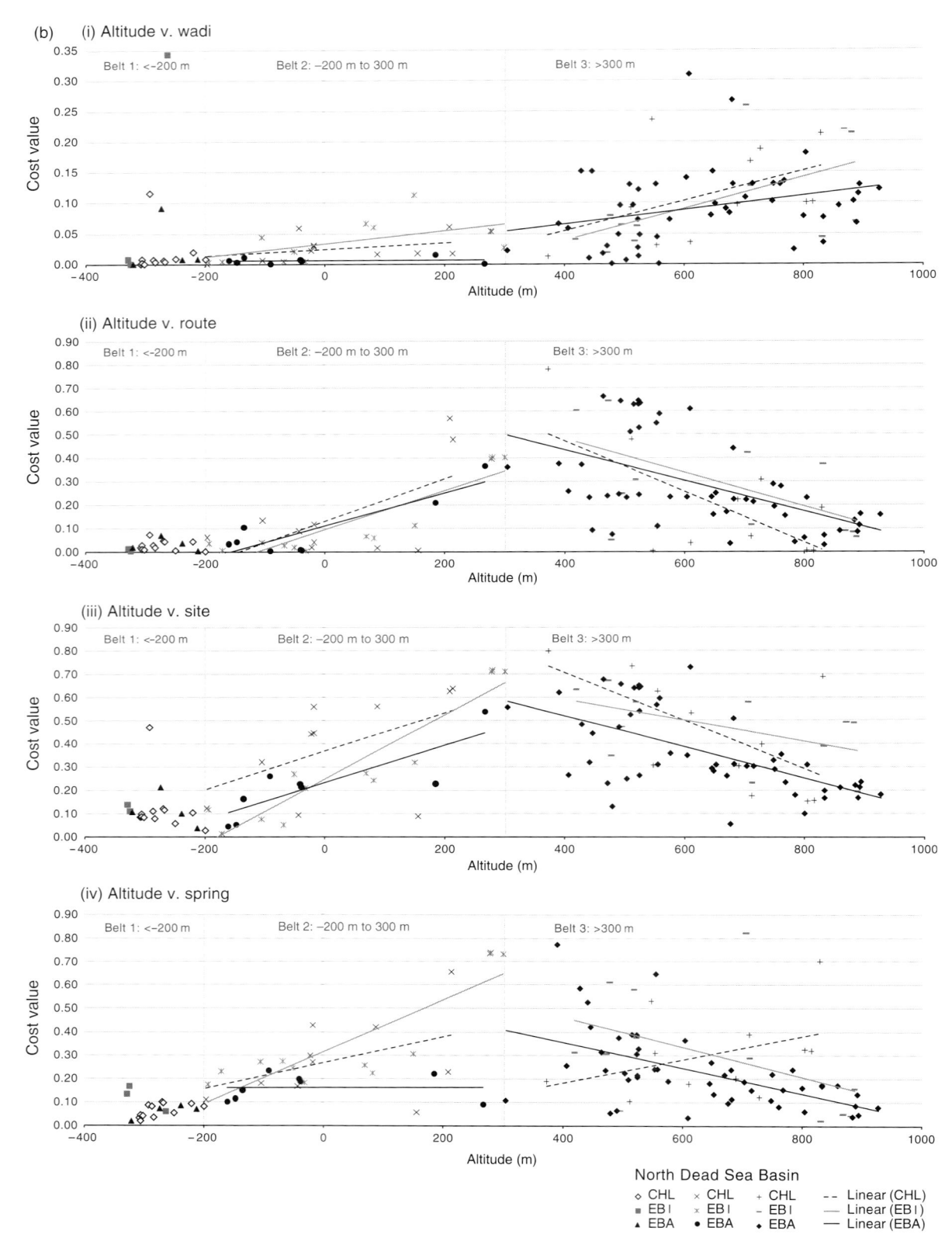

Figure 17.3. (*cont.*)

NRB (spring, >300 m sector, EBI), the best fit line is flatter, showing a more equal reliance on springs at all altitudes (300 m to 1,000 m; Figure 17.3 a, iv).

The results of the paired cost distances show the changes in relationships between two factors and how each factor can change relative to another through time (Figure 17.4a and b). If cost distances are similar they will be plotted on the line of equal value and both factors are considered equally important. If one cost distance is lower the point will be plotted to the opposite side of the 1:1 line to the axis and that factor is considered more important than the other. If one factor is consistently more dominant then the points will form a cluster. If clusters are located in different sections of the plot in time periods then the relationship between these factors is considered dynamic through time. In the NRB (Figure 17.4a), site and route have a roughly similar importance in the 300 m belt (plot iii), but this tends towards routes in belt 2 (plot ii) and it is much clearer that routes are favoured over sites in belt 1 (plot i). Clustering tends to occur when data pairs involve springs (plots iv–ix), and springs tended to be more influential than sites or routes, with the exception of certain altitude bands (plot vii) and time periods (plot viii). However, the influence of springs has varied. In the >300 m belt of the NRB, when comparing cost distance to springs and site and cost distance to springs and route (vi and ix), the cost distances were greatest to springs in the Chalcolithic but decreased in the EBI. This lower cost distance to springs in the EBI will be discussed below (17.3.3 and 17.4.1). In the −200 m to 300 m belt (plots v, viii and xi), spring, wadi, site and route relationships are more complex and forcing factors vary because cluster positions change through time. In the <−200 m belt it is difficult to ascribe a controlling factor since cost distances are extremely low. There is, however, a notable exception demonstrated by one of two clusters that contain lower site cost values relative to spring cost values in the EBA (iv). Examining cluster patterns for belt 2, sites and routes may have had a greater role than springs in the EBA (plots v and viii). Plots x to xii demonstrate that wadis are more favoured than springs for all time periods but in the EBI and EBA springs were more important because the clusters tend to show lower cost distance values than the Chalcolithic (plots xi and xii) (see below, sections 17.3.3 and 17.4.2).

In the NDSB (Figure 17.4b) the pairs show some different patterns. At >300 m we see a similar linear site vs. route relationship to the NRB, plot iii, but in the −200 to 300 m sector routes have lower cost distances and in general are more favourable than sites, plot ii. In this basin, in the >300 m belt, it is difficult to ascribe a single factor between site–spring and route–spring relations, plots vi and ix, because in each time period data are spread across the plot and are divided into more than one cluster. For example, some sites in the >300 m belt in the EBA have lower cost distances to springs, and other sites lower cost distances to sites or routes, but a majority have similar site–spring and route–spring cost distance values. In the −200 m to 300 m belt, plot v, sites and springs have similar cost distances with the exception of lower cost distances to spring in the EBA, while in plots viii and xi, routes and wadis tend to have lower cost distance values than springs. The <−200 m belt is difficult to interpret because of the low cost distance values (plots i, iv, vii, x) except for noticeable outliers for the three time periods, particularly in the EBI where springs appear to be supporting individual isolated sites (see below, 17.4.1). In the NDSB the domination of wadi over spring is well marked at all altitudes, although the wadi domination is reduced with altitude (plots x–xii).

17.3.3 Spatial patterns and controlling factors

It is in the NRB where the survey coverage is strongest and the data are more widespread and robust (Figure 17.2). Unsurprisingly, in the belt below −200 m, route consistently produces a lower cost value than permanent sites and springs, but it does appear that sites cluster over time (see also the gradual reduction in cost values for altitude vs. site in NRB, Figure 17.3a, ii). In addition, in the EBI particularly, springs are favoured over permanent site (see Figure 17.4a, iv–vi). In the −200 to 300 m asl belt in the EBI there is evidence of a reliance on springs as opposed to wadis (see Figure 17.2b; Figure 17.4a, x–xii). In the >300 m asl belt, concentrated in the east of the basin, the cost distances to springs are much less than those for permanent sites and routes in the EBI, although this is mitigated slightly in the EBA (Figure 17.4a, iv–vi). And again, when compared with wadi cost distances, springs are more important in the EBI (although there is considerable variation across belts; see Figure 17.3a). However, looking at the <300 m belt spatially in the EBI, it seems that routes and wadis are perhaps more influential in the north, whereas springs appear very influential in the south of the altitude belt.

There are some changes in settlement pattern in the NDSB, but consideration has to be given to the quality of coverage which is poorest in the <−200 m belt. In this same belt Chalcolithic settlements are clustered to the south around springs and routes, during the EBI and the EBA settlement becomes more widespread. It is difficult to determine the main forcing factors as cost distances are generally quite low, springs and wadis may be most influential. In belt 2, −200 to 300 m asl site numbers are low. In the south of the NDSB there are more sites in the Chalcolithic and occupation diminishes during the EBI and EBA. The cost distance values (Table 17.4) suggest importance of routes and sites increased whilst wadis remained similar and spring was inconsistent (Table 17.4). Examining the cost distance pairs, wadi and route were most important through all time periods (Figure 17.4b). In the highest belt, >300 m asl, following an upward shift of sites in the EBI driven by wadi (plot xii) and perhaps spring (plot ix), there is an increase in settlement sites in central areas in the EBA, driven by wadi (plot xii) and site (plot vi).

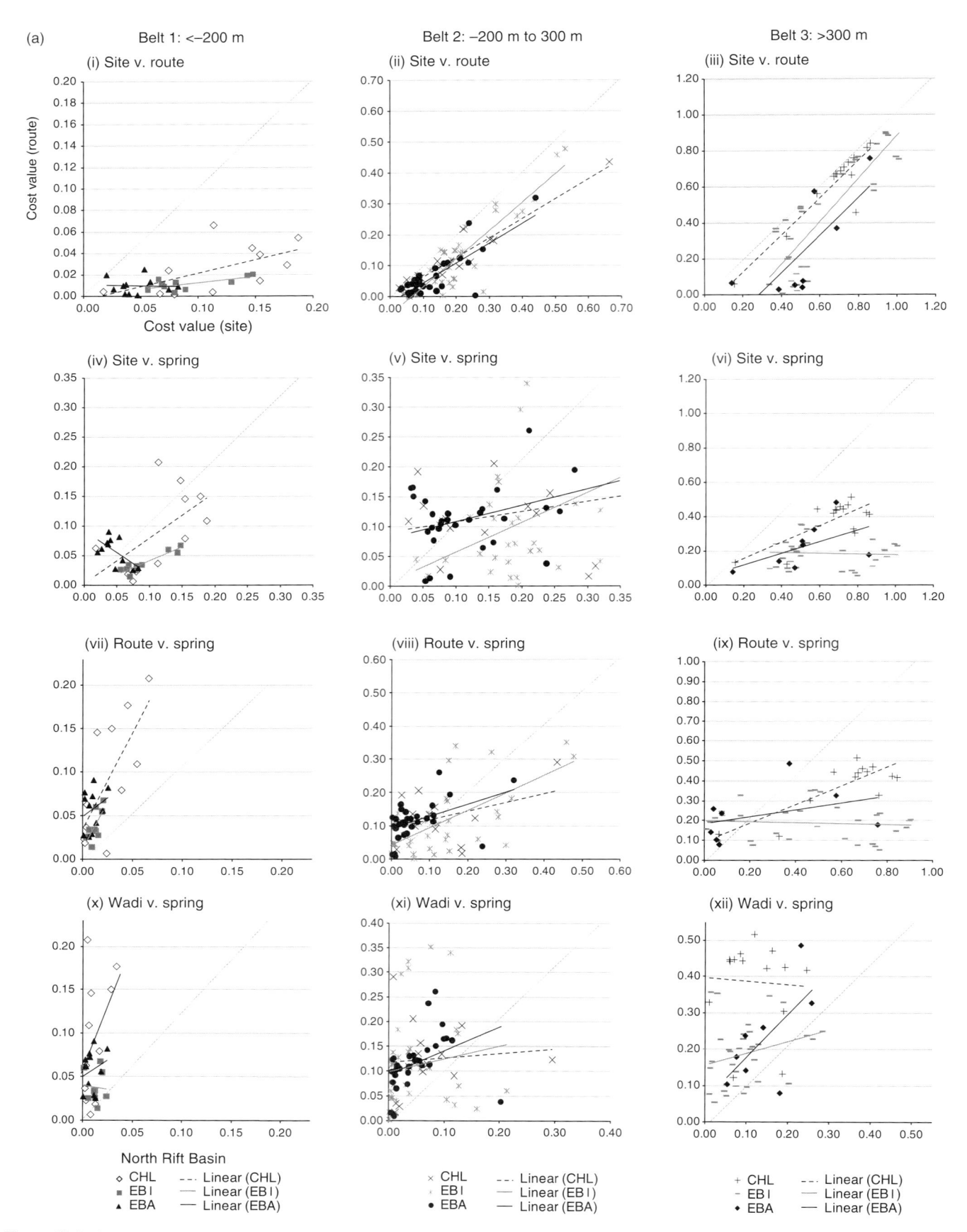

Figure 17.4. Cost distance pairs, site v. route, site v. spring, route v. spring and wadi v. spring regression plots for the three altitudinal sectors and three time periods. The first variable is on the horizontal axis, second on the vertical, dotted line shows line of equal value. (a) North Rift Basin; (b) North Dead Sea Basin.

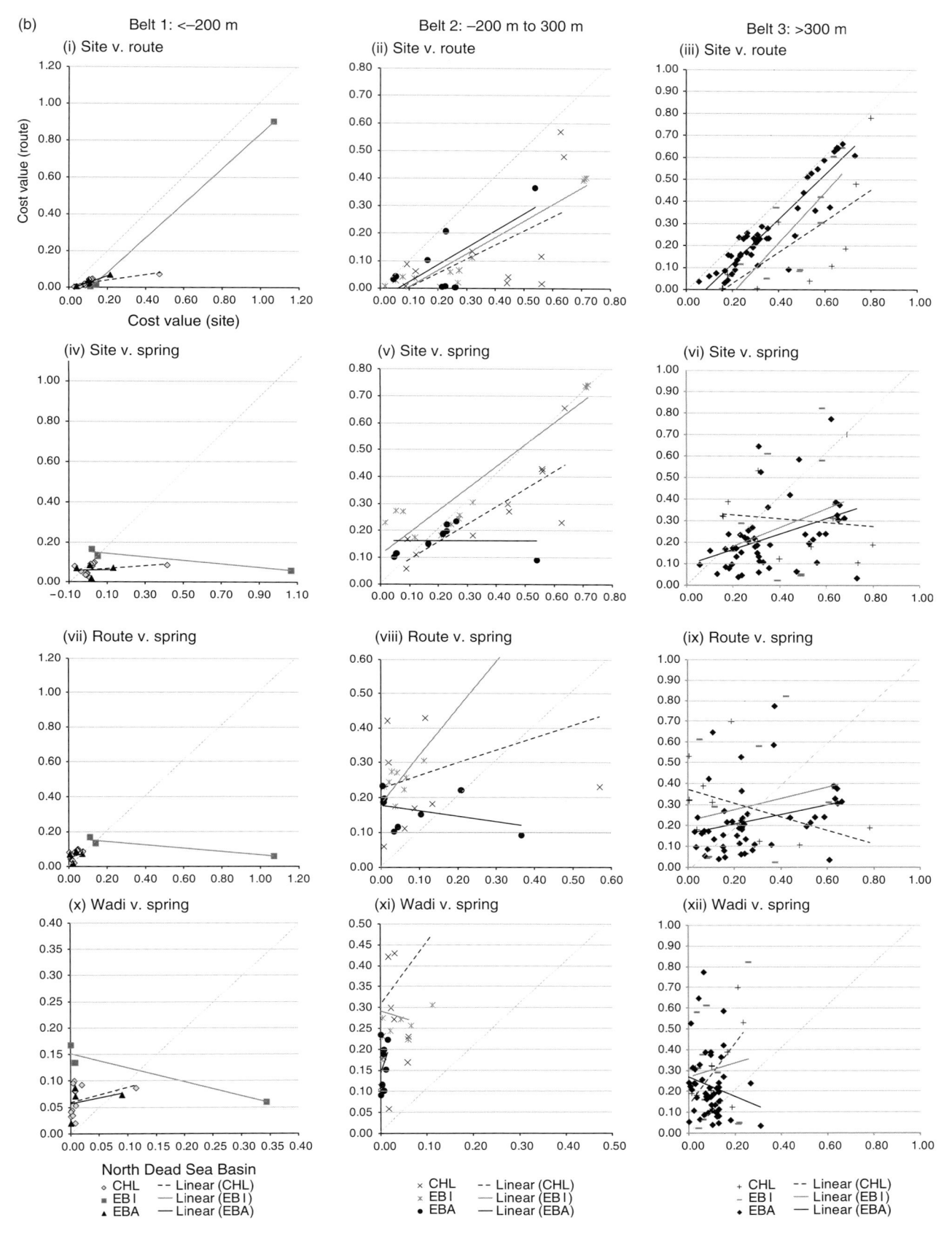

INFLUENCE OF WATER ON CHALCOLITHIC AND EBA SETTLEMENT PATTERNS

Figure 17.4. (*cont.*)

17.4 DISCUSSION

We have identified and considered altitudinal changes and trends for two basins with relatively sound archaeological evidence. We were able to identify variation in site numbers within altitudinal belts, which can be interpreted as changes in human activity. We also considered four potential forcing factors through the use of cost distance surfaces. As with site numbers, cost distance values vary with altitude belts, but there is also variability within the altitudinal belts. The variability can be considered in two ways: (i) as a change in trend through the whole belt, and (ii) as clusters within the altitudinal belts. Overall we found that there was no one distinct trend for all belts at the same time, which indicates that causes of settlement pattern have been more subtle than the 'classic' explanations imply.

The most marked variation in cost distances was the overall increase in cost distance with altitude. Some of this may well be a function of the characteristics of the landscape. In the <-200 m belt in the valley bottom, sites are more confined and closer to the four target factors, whereas at the higher altitudes the basin is less confined, and it is possible for sites to be positioned further away from the four target factors. However, of all the target factors, wadi was most important, followed by (depending on the basin or altitude belt) springs or permanent sites. We acknowledge that the cost distance values to wadi may have been biased by survey technique, which often favour wadis over less accessible surrounding territory; however, we selected our basins based on the best available coverage, and this ought to mitigate against serious sample biases.

The more interesting results are found in the spring data (see Figure 17.3). Cost distances to wadis remain low and do not increase much with altitude. In the >300 m belt in the EBI and EBA the cost distances to springs in the NRB do not increase much with altitude, and in the NDSB there is a marked decrease at the highest altitudes. In these two basins there is a strong suggestion that in the EBI and EBA in the >300 m belt, it was more important for sites to be located closer to springs. This is particularly marked for sites above 400 m. Rainfall and surface runoff collection may have been a less effective strategy in these later periods because sources were less reliable than they were in the Chalcolithic period. Conversely, springs may have become more reliable because of changes in the water table as a result of climatic fluctuations (see below). It is also likely that sites in the 300–800 m zone may have been better connected to alternative water sources and/or subsistence strategies up and down the system that relied upon harvesting rainfall.

In the NDSB the decline in cost distances above 800 m (Figure 17.3b, iv) might be a result of sites being more isolated at the top of the settlement system (and therefore located right on water sources). Higher altitude settlements are likely to be reliant on a mixture of rain-fed agriculture/horticulture and pastoralism, but latitudinal gradients affect rainfall in this region (see Chapter 2 of this volume) and the northern sector usually has higher rainfall per annum.

Limestone is one of the few geological bases where rock-cut installations can be effective water collection systems. In the better-watered limestone region of the NRB water catchment it would probably have been easier to cut and maintain cisterns than in the steeper, more marginal south, where sandstone and other geological formations predominate. This corresponds with the accepted wisdom that rock-cut cisterns (Kenyon, 1960; Miller, 1980; Mabry and Palumbo, 1988), groundwater installations (Bienert, 2004), and water reservoirs and other water catchment systems (Mabry et al., 1996; and for examples outside the Jordan Valley see Helms, 1982; Rollefson and Wasse, in press) were a development of the EBA (although aspects of these were no doubt developed earlier). Better water catchment technology may have been a requirement of new, less favourable environmental conditions.

At the lower altitudes the concentration of settlements along the Jordan Valley in the EBA might be related to greater opportunities for water capturing (Philip, 2001); equally, Mabry notes that intramural reservoirs have been located inside contemporaneous fortified sites, such as at Ai (et Tell) and Arad (Mabry et al., 1996). These two observations point to a re-ordering of priorities by the EBA population in an environment of reduced water availability. Comparisons with the region's modern political situation, where ongoing conflict is intimately linked with access to water, are hard to ignore. In fact it is tempting to interpret the development of fortified towns in the EBI/II as a response to a mobilisation of the population around reduced but nonetheless vital resources. Our data show that in the EBA the site arrangement is orientated in a more linear fashion, north to south, suggesting a more formalised arrangement. EBA sites in upland territory were more sparsely spread and were often located at a small distance from wadis, mirroring the observations made by Philip (2001, pp. 110–11). Meanwhile, settlements in the lower belts had distinct clusters with low cost distance to springs (Figure 17.4a, plot iv).

Interestingly, in the NDSB there were few sites in the -200 to 300 m asl belt (except during the Chalcolithic). Wadi erosion (Donahue, 1981; Mabry, 1989; Banning, 1997) in different climatic regimes may have played a role in discouraging settlement at the foothills (-200 to 300 m asl belt), which perhaps explains why spring and routes were more influential than sites (Figures 17.4b ii; see 17.3.2 above). If this is the case, then the failure of the so-called 'chiefdom' organisation may well have its origins in climatic instability (consistent with Hourani and County, 1997). However, what the data presented here show is the 'reorganisation' of the population around new and existing eco-anchors.

Palaeoclimatic data for the eastern Mediterranean during the Holocene all point to recurring climatic fluctuations, although there is an overall trend towards aridity from 6.0 cal ka BP or earlier (Brooks, 2006; Robinson et al., 2006). The most significant of these fluctuations are thought to be a centennial-scale wet event around 5.0 cal ka BP (around or at the end of the EBI) and a major arid event around 4.2 cal ka BP. The latter is certainly associated with the end of the EBIII (Rosen, 1989).

Considering our data spatially, there is a clear 'formalisation' of the data over time, but also a noticeable depopulation in the NDSB. In the Chalcolithic the sites are more scattered in both basins, albeit closely aligned to wadis, while in the EBI there is an expansion of settlement into the >300 m asl belt, possibly at the expense of the more marginal south. This may have been because the Dead Sea expanded to a maximum at about the EBI (Figure 17.2b). Note that a very high stand in the Dead Sea may have meant that there was higher standing and fresher groundwater adjacent to the sea (Donahue, 2003) because a rapid rise in the Dead Sea may well have resulted in the development of a less saline upper level (Neev and Emery, 1995). Conversely, a reduction in the Dead Sea will have a direct effect upon salinity (Neev and Emery, 1995; Donahue, 2003).

Rather than a settlement collapse in the EBI there is actually an expansion of settlement at higher altitudes. Given that we also see a slight increase in cost values to permanent sites and routes in the higher altitude belt (Figure 17.4a, b), it is tempting to see some of these sites as 'pioneering' settlements. In the following period, the EBA, the network of smaller settlements in foothills of the NRB (−199 to 299 m asl belt) appears much diminished and the settlement pattern appears formalised into two clear belts: 0 to −200 m and >300 m asl (Figure 17.2c). The cost distances to routes and major sites are reduced in the EBA in the NDSB, suggesting a more formalised settlement structure. Of course, permanent sites have not yet *become* permanent in the Chalcolithic – the first period under consideration in this study. However, there is a clear difference in cost distance values to the sites which later became key sites.

As stated above (17.4.1), there is supporting evidence for environmental instability at the end of the Chalcolithic but our data here show that traces of continuity are also in evidence. The approach we have adopted here has highlighted the reorganisation of the landscape and suggests that elements of continuity will be more readily identified in upland sites, as is being demonstrated by new excavations in Israel (van den Brink et al., 2001).

This new, formalised EBA landscape would have been reliant on more widespread medium-distance animal transport. The widespread introduction of the donkey would clearly have had an effect in light of the role of donkey transport in the EBA (Milevski, 2005). The introduction of the donkey is conventionally dated to the EBA (*see* Ovadia, 1992; cf. Lovell, 2001), although recent discoveries of larger numbers of donkey bones from Chalcolithic (Grigson, 1995; Bourke et al., 2000; Mairs in Bourke et al., 2000) and EBIa contexts (Whitcher Kansa, 2004) are gradually lowering the horizon. Of course, distances of sites to springs will be mediated by effective water transport, but better terrestrial transport may have over-ridden the need for settlements in the −200 to 300 m belt, which may previously have assisted the distribution system of goods. Additionally, the socio-political process widely held to have transformed EBA culture into what Philip has termed a 'corporate village organisation' (Philip, 2003) is reflected in spatial patterning of the data. In the valley bottom itself, larger-scale water management was possible and practised by what Mabry describes as 'irrigation communities' (Mabry et al., 1996), equivalent to Philip's 'corporate villages' (Philip, 2008). These captured surface runoff and formed a counterweight to new and emerging economies based in the uplands.

Following Lovell (2002), upland sites were a significant portion of the settlement system in the Chalcolithic, although it is said to have become more prevalent in the EBI (see Finkelstein and Gophna, 1993; cf. Gophna and Tsuk, 2005). Higher-altitude sites were more widespread in the Chalcolithic than previously believed (Gibson et al., 1991; van den Brink et al., 2001; Lovell, 2008), and are perhaps the precursors to a more intensive occupation of the hill country in the following EBA (Lovell, 2002; Gophna and Tsuk, 2005). These altitudinal shifts may be related to use of the horticultural zone. Without the aid of irrigation systems, the cultivation of olive (*Olea europaea*) is normally constrained to altitudes of 300 to 800 m asl, and within the precipitation range of 300 to 500 mm; irrigation may have been in use in the EBA but clear evidence is lacking (see Philip, 2003; this volume, Chapter 14). Previous arguments regarding olive cultivation and shifts in resources at the transition to the EBA (Lovell, 2002, 2008; Philip, 2003) have made reference to the fact that the cultivation of olives has important implications for developing socio-economic systems. The key element is that olive trees are a long-term investment: it normally takes 10–12 years before a tree will fruit and they can produce a good harvest only every 2 years. As a consequence, olive cultivation ties villagers to certain parts of the landscape.

While both basins show a reasonably similar spread of territory across the altitude belts (for both NDSB and the NRB, *c*. 20% of the total falls in the belt below −200 m; 30–40% in the −200 to 300 m belt; and 30–45% in the +300 m asl belt), these data show an interesting contrast in the cost distance values basin to basin. Cost values to springs in the NDSB are generally higher than their NRB counterparts. It is tempting to suggest that pastoralism may have played a greater role in the south (where animal herding can be carried out simultaneously with water collection), while horticultural practice may have been carried out more intensively in the NRB. This would certainly fit with current rainfall gradients. Our data point to an increase in the number of sites in the 300–800 m asl sector over time in both basins.

We do not assume that every site located in the >300 m belt was associated with horticulture – there are other agricultural, industrial or pastoral practices that could be associated with habitation/exploitation of these areas (herding, gathering, hunting). However, we note that in basins where land in higher altitudes is available, settlements nevertheless show strong clusters in the 300–800 m band (exactly that required for olive production). We therefore argue that overall increases are a potential indicator of more intensified use of a horticultural lifestyle. This is particularly credible in the light of the heavy presence of olive within the Chalcolithic botanical record (Meadows, 2005).

Although we can only provide broad accounts in the altitudinal belts, we identified several occurrences of data clusters, with respect to altitude and in the pairings (e.g. Chalcolithic occupation in the Wadi Rayyan). This indicates that there are groups of sites sharing similar factors of control within the altitudinal belts which provide small 'niches' where factors encouraging settlement location are advantageous – in these cases, an in-depth analysis like that of Fletcher and Winter (2008) may be appropriate for further investigating the reasons for such settlement clusters. Further, this paper also highlights areas where further survey would fill gaps. For example, the Zerqa region and the Wadi Shu'eib (Wright *et al.*, 1989) region are under-surveyed and would produce useful comparative data for these two basins.

17.5 CONCLUSIONS

The research presented here provides supporting data for the impact of climate change on settlement patterns during a key transition in the archaeology of the protohistoric periods. While settlement studies cannot provide primary data on actual palaeoclimatic events, they can propose that observed patterns are the results of human responses to new environmental conditions. Using data from Israel, oPt and Jordan has allowed us to scale up our analysis to fit larger questions of regional climate and temporal socio-political change.

The value of this type of analysis is that it can throw broad trends into high relief. It is also more powerful than traditional studies of rank size distribution (Falconer, 1987) which mask the particular ways of traversing a heavily dissected landscape by treating regions as essentially monolithic. Use of a GIS-based model allows more flexible analysis as higher-resolution data become available.

We have attempted to comment on both environmental factors and socio-political influences. In examining cost distance according to separate and paired factors we have been able to examine the relationship of the data to different variables. We deliberately avoided blending these together into a combined cost distance figure, and this allowed us to test more qualitative statements made by researchers in the past and to quantify shifts and trends. We determined the limits of our altitude belts on the basis of what we felt were natural breaks in the data, but conditions do change within the belts and shifting the limits might produce different results. These belts are also, no doubt, dynamic through time.

Our study has identified two key shifts – one is a shift in the importance of springs during the periods of the EBI and EBA, perhaps a result of a rise in the water table brought about by the wet event identified by palaeoclimatologists at 5.0 cal ka BP. The other is a more formalised settlement pattern, more aligned with routes and permanent settlements in the EBA. In addition we have been able to tie this to the prevalence of settlement in the 300–800 m asl zone, which we have associated with Chalcolithic/EBA horticultural practice.

These three aspects highlight the importance of the upland region in the new settlement landscape of the 'urbanising' EB villages, which were clearly reorganised in response to a shift in water resources and were developing new techniques to exploit new environments. It would appear that the climatic and economic environment of the Chalcolithic–EBA transition contained two key push and pull factors. One was an environmental shift which affected water availability (a push factor) and the other was a powerful commitment to place, in the form of horticulture (the pull factor). Both of these factors were operating in the context of shifting water availability. Technological innovations for consolidating these late prehistoric lifestyles, including means of water catchment, and donkey transport, were no doubt accelerated by this tension.

REFERENCES

Abu Dayyah, A. S., J. A. Greene, I. H. Hassan and E. Suleiman (1991) Archaeological Survey of Greater Amman, Phase 1: Final Report. *Annual of the Department of Antiquities of Jordan* **35**: 361–414.

Al-Shorman, A. A. (2002) Archaeological site distribution in Jordan since the Palaeolithic and the role of climate change. *Adumantu* **5**: 7–26.

Anfinset, N., H. Taha, M. al-Zawahra and J. Yasine (in press) Societies in transition: contextualising Tell el-Mafjar, Jericho. In *Culture, Chronology and the Chalcolithic: Theory and Transition*, ed. J. L. Lovell and Y. M. Rowan. Oxford: Oxbow.

Avni, G. (1992) Map of Har Saggi Northeast (225). Jerusalem: Archaeological Survey of Israel.

Banning, E. B. (1997) Recovery from landslides in Wadi Ziqlab. *American Journal of Archaeology* **101**: 504.

Baumgartel, Y. (2004) Map of Shivta (166). Jerusalem: Archaeological Survey of Israel.

Beit-Arieh, I. (2003) Map of Tel Malhata (144) Jerusalem: Archaeological Survey of Israel.

Berman, A. and L. Barda (2005) Map of Nizzanim (West and East) (87 and 88). Jerusalem: Archaeological Survey of Israel.

Berman, A., H. Stark and L. Barda (2004) Map of Ziqim (91) Jerusalem: Archaeological Survey of Israel.

Bienert, H.-D. (2004) The underground tunnel system in wadi ash-Shellalah, Northern Jordan. In *Men of Dikes and Canals: The Archaeology of Water in the Middle East*, ed. H.-D. Bienert, H. G. K. Gebel and R. Neef. Rahden, Westphalia: Verlag Marie Leidorf.

Bourke, S. (2008) The Chalcolithic. In *Jordan: An Archaeological Reader*, ed. R. Adams. London: Equinox.

Bourke, S., J. L. Lovell, R. T. Sparks *et al.* (2000) A second and third season of renewed excavation by the University of Sydney at Tulaylat al-Ghassul (1995–1997). *Annual of the Department of Antiquities of Jordan* **44**: 37–89.

Bourke, S., R. T. Sparks, P. B. Mclaren *et al.* (2003) Preliminary report on the University of Sydney's eighteenth and nineteenth seasons of excavations at Pella (Tabaqat Fahl) in 1996/7. *Annual of the Department of Antiquities of Jordan* **47**: 335–388.

Braun, E. (1996) Cultural diversity and change in the Early Bronze I of Israel and Jordan. Unpublished PhD thesis: Tel Aviv University.

Braun, E. (2010) The transition from Chalcolithic to EB I in the southern Levant, a 'lost horizon' slowly revealed. In *Culture, Chronology and the Chalcolithic: Theory and Transition*, ed. J. L. Lovell and Y. M. Rowan. Oxford: Oxbow pp. 160–177.

Brooks, N. (2006) Cultural responses to aridity in the middle Holocene and increased social complexity. *Quaternary International* **151**: 29–49.

Bruins, H. J. (1994) Comparative chronology of climate and human history in the southern Levant from the late Chalcolithic to the Early Arab Period. In *Late Quaternary Chronology and Palaeoclimates of the Eastern Mediterranean*, ed. O. Bar-Yosef and R. S. Kra. Tucson: Radiocarbon pp. 301–314.

Burton, M. and T. E. Levy (in press) The end of the Chalcolithic period (4500–3600 BCE) in the northern Negev desert, Israel. In *Culture, Chronology and the Chalcolithic: Theory and Transition*, ed. J. L. Lovell and Y. M. Rowan. Oxford: Oxbow.

Chesson, M. and G. Philip (2003) Tales of the city? 'Urbanism' in the early Bronze Age Levant from Mediterranean and Levantine perspectives. *Journal of Mediterranean Archaeology* **16/1**: 3–16.

Cohen, R. (1981) Map of Sede Boqer-East (168) 13–03. Jerusalem: Archaeological Survey of Israel.

Cohen, R. (1985) Map of Sede Boqer-west (167) 12–03. Jerusalem: Archaeological Survey of Israel.

Dagan, Y. (1992) Map of Lakhish (98). Jerusalem: Archaeological Survey of Israel.

de Miroschedji, P. (2002) The socio-political dynamics of the Egyptian–Canaanite interaction in the Early Bronze Age. In *Egypt and the Levant: Interrelations from the 4th through the Early 3rd Millennium B.C.E.*, ed. E. C. M. van den Brink and T. E. Levy. London: Leicester University Press.

de Vaux, R. (1971) Palestine in the Early Bronze Age. In *Early History of the Middle East 3rd Edition*, ed. I. E. S. Edwards, J. Gadd and N. G. L. Hammond. Cambridge: Cambridge University Press pp. 208–237.

Donahue, J. (1981) Geologic investigations at Early Bronze sites. In *The Southeastern Dead Sea Plain Expedition: An Interim Report of the 1977 Season*, ed. W. E. Rast and R. T. Schaub. Boston, MA: American Schools of Oriental Research pp. 137–154.

Donahue, J. (2003) Geology and geomorphology. In *Bab edh-Dhra: Excavations at the Town Site (1975–81)*, ed. W. E. Rast and R. T. Shaub. Winona Lake, IN: Eisenbrauns pp. 18–55.

Esse, D. L. (1991) *Subsistence, Trade, and Social Change in Early Bronze Age Palestine*. Chicago, IL: Oriental Institute of the University of Chicago.

Falconer, S. (1987) Heartland of villages: reconsidering early urbanism in the southern Levant. Unpublished PhD thesis: University of Arizona.

Finkelstein, I. and R. Gophna (1993) Settlement, demographic and economic patterns in the highlands of Palestine in the Chalcolithic and Early Bronze Periods and the beginning of urbanism. *Bulletin of the American Schools of Oriental Research* **289**: 1–22.

Finkelstein, I., Z. Lederman, S. Bunimovitz and R. Barkai (1997) *Highlands of Many Cultures*. Tel Aviv: Institute of Archaeology of Tel Aviv University Publications.

Finkelstein, I. and Y. Magen, eds. (1993) *Archaeological Survey of the Hill Country of Benjamin*. Jerusalem: Archaeological Survey of Israel.

Fletcher, R. (2008) Some spatial analyses of Chalcolithic settlement in Southern Israel. *Journal of Archaeological Science* **35**: 2048–2058.

Fletcher, R. and R. Winter (2008) Prospects and problems in applying GIS to the study of the Chalcolithic archaeology in southern Israel. *Bulletin of the American Schools of Oriental Research* **352**: 1–27.

Frankel, R., N. Gertov, M. Aviam and A. Degani (2001) *Settlement Dynamics and Regional Diversity in Ancient Upper Galilee*. Jerusalem: Israel Antiquities Authority.

Frankel, R. and N. Getzov (1997) Map of Akhziv (1); Map of Hanita (2). Jerusalem: Archaeological Survey of Israel.

Frumkin, A., M. Magaritz, I. Carmi and I. Zak (1991) The Holocene climatic record of the salt caves of Mount Sedom. *The Holocene* **1**: 191–200.

Frumkin, A., M. Magaritz, I. Carmi and I. Zak (1997) The Holocene climatic record of the Salt Caves of Mount Sedom. *The Holocene* **1**: 191–200.

Frumkin, A., G. Kadan, Y. Enzel and Y. Eyal (2001) Radiocarbon chronology of the Holocene Dead Sea: attempting a regional correlation. *Radiocarbon* **43**: 1179–1189.

Gal, Z. (1991) Map of Gazit (46). Jerusalem: Archaeological Survey of Israel.

Gal, Z. (1998) Map of Har Tavor (41); Map of 'En Dor (45). Jerusalem: Archaeological Survey of Israel.

Gazit, D. (1996) Map of Urim (125). Jerusalem: Archaeological Survey of Israel.

Gibson, S., B. Ibbs, and A. Kloner, (1991) The Sataf Project of Landscape Archaeology in the Judean Hills: a preliminary report on four seasons of excavation and survey. *Levant* **23**: 29–54.

Gilead, I. (1995) *Grar: A Chalcolithic Site in the Northern Neg*. Tel Aviv: Ben Gurion University of the Negev Press.

Gophna, R. and A. Ayalon (1998) Map of Herziliyya (69). Jerusalem: Archaeological Survey of Israel.

Gophna, R. and I. Beit-Arieh (1997) Map of Lod (80). Jerusalem: Archaeological Survey of Israel.

Gophna, R. and J. Portugali (1988) Settlement and demographic processes in Israel's coastal plain from the Chalcolithic to the Middle Bronze Age. *Bulletin of the American Schools of Oriental Research* **269**: 11–28.

Gophna, R. and T. Tsuk (2005) Chalcolithic settlements in the Western Samaria foothills. *Tel Aviv* **32/1**: 3–19.

Gordon, R. L. and L. E. Villiers (1983) Tulul edh-Dhahab and its environs. Survey of 1980 and 1982, a preliminary report. *Annual of the Department of Antiquities of Jordan* **27**: 285–289.

Grigson, C. (1995) Plough and pasture in the early economy of the southern Levant. In *Archaeology of Society in the Holy Land*, ed. T. E. Levy. London: Leicester University Press pp. 245–268.

Haiman, M. (1986) Map of Har Hamran – Southwest (198). Jerusalem: Archaeological Survey of Israel.

Haiman, M. (1991) Map of Mizpé Ramon – Southwest (200). Jerusalem: Archaeological Survey of Israel.

Haiman, M. (1993) Map of Har Hamran – Southeast (199). Jerusalem: Archaeological Survey of Israel.

Haiman, M. (1999) Map of Har Ramon (203). Jerusalem: Archaeological Survey of Israel.

Hanbury-Tenison, J. (1986) *The Late Chalcolithic to Early Bronze I Transition in Palestine and Transjordan*. Oxford: British Archaeological Reports International Series 311.

Hanbury-Tenison, J. (1987) Jerash region survey, 1984. *Annual of the Department of Antiquities of Jordan* **31**: 129–157.

Helms, S. (1982) Paleo-bedouin and Transmigrant Urbanism. In *Studies in the History and Archaeology of Jordan I*, ed. A. Hadidi. Amman: Department of Antiquities pp. 97–113.

Hirschfeld, Y. (1985) Map of Herodium (108/2). Jerusalem: Archaeological Survey of Israel.

Hourani, F. and M. -A. County (1997) L'évolution climatique de 10 500 à 5500 B.P. dans la vallée du Jourdain. *Paleorient* **23**: 95–105.

Ibach, R. (1987) *Hesban 5: Archaeological Survey of the Hesban Region: Catalog of Sites and Characterization of Periods*. Berrien Springs: Andrews University Press.

Ibrahim, M., J. A. Sauer and K. Yassine (1976) The East Jordan Valley Survey. *Bulletin of the American Schools of Oriental Research* **222**: 41–66.

Jarvis, A., H. I. Reuter, A. Nelson and E. Guevara (2008). http://srtm.csi.cgiar.org.

Joffe, A. H. (2003) Slouching toward Beersheva: Chalcolithic mortuary practices in local and regional context. In *The Near East in The Southwest, Essays in Honor of William G. Dever*, ed. B. Alpert-Nakhai. Oxford: Oxbow pp. 45–67.

Joffe, A. H. (1993) *Settlement and Society in the Early Bronze Age I and II, Southern Levant: Complementarity and Contradiction in a Small-Scale Complex Society*. Sheffield: Sheffield University Press.

Kenyon, K. M. (1960) Excavations at Jericho, 1957–58. *Palestine Exploration Quarterly* **92**: 88–108.

Kerner, S. (1997) Status, perspectives and future goals in Jordanian Chalcolithic research. In *The Prehistory of Jordan II, Perspectives from 1997*, ed. H. G. K. Gebel, Z. Kafafai and G. Rollefson. Berlin: ex Oriente pp. 465–474.

Kloner, A. (2001) *Survey of Jerusalem (The Northeastern Sector)*. Jerusalem: Archaeological Survey of Israel.

Lender, Y. (1990) Map of Har Nafha (196). Jerusalem: Archaeological Survey of Israel.

Leonard, A. (1992) *The Jordan Valley Survey, 1953: Some Unpublished Soundings Conducted by James Mellaart*. Boston, MA: Annual of the American Schools of Oriental Research.

Levy, T. E. (1995) Cult, metallurgy and rank societies – Chalcolithic period. In *Archaeology of Society in the Holy Land*, ed. T. E. Levy. Leicester: Leicester University Press pp. 226–244.

Lovell, J. L. (2001) *The Late Neolithic and Chalcolithic Periods in the Southern Levant: New Data from Teleilat Ghassul, Jordan*. Oxford: Archaeopress.

Lovell, J. L. (2002) Shifting subsistence patterns: some ideas about the end of the Chalcolithic in the southern Levant. *Paleorient* **28**/1: 89–102.

Lovell, J. L. (2008) Horticulture, status and long-range trade in Chalcolithic southern Levant: early connections with Egypt. *Proceedings of the International Conference on Origin of the State. Predynastic and Early Dynastic Egypt*, ed. B. Midant-Reynes and M. Y. Tristant. Toulouse: Peeters Publishers pp. 739–760.

Mabry, J. M. (1989) Investigations at Tell el-Handaquq, Jordan (1987–88). *Annual of the Department of Antiquities of Jordan* **32**: 59–95.

Mabry, J. M., L. Donaldson, K. Gruspier et al. (1996) Early town development and water management in the Jordan Valley: investigations at Tell el-Handaquq North. *Annual of the Department of Antiquities of Jordan* **53**: 115–154.

Mabry, J. M. and G. Palumbo (1988) The 1987 Wadi el Yabis survey. *Annual of the Department of Antiquities of Jordan* **32**: 275–305.

Mazar, A. and R. A. Mullins (2007) *Excavations at Tel Beth-Shean 1989–1996, Vol. 2. The Middle and Late Bronze Age Strata in Area R Jerusalem*. Jerusalem: Israel Exploration Society.

Meadows, J. (2005) Early farmers and their environment: archaeobotanical studies of Neolithic and Chalcolithic sites in Jordan. Unpublished PhD thesis: La Trobe University.

Migowski, C., M. Stein, S. Prasad, J. F. W. Negendank and A. Agnon (2006) Holocene climate variability and cultural evolution in the Near East from the Dead Sea sedimentary record. *Quaternary Research* **66**: 421–431.

Milevski, I. I. (2005) Local Exchange in Early Bronze Age Canaan. Unpublished PhD thesis: Tel Aviv University.

Miller, R. (1980) Water use in Syria and Palestine from the Neolithic to the Bronze Age. *World Archaeology* **11**/3: 331–341.

Mittmann, S. (1970) *Beiträge zur Siedlungs- und Territorialgeschichte des nördlichen Ostjordanlandes*. Wiesbaden: Harrassowitz.

Ne-eman, Y. (1990) Map of Ma'anit (54). Jerusalem: Archaeological Survey of Israel.

Ne-eman, Y., S. Sender and E. Oren (2000) Map of Mikhmoret (52); Map of Hadera (53). Jerusalem: Archaeological Survey of Israel.

Ne-eman, Y., S. Sender and E. Oren (2005a) Map of Binyamina (48). Jerusalem: Archaeological Survey of Israel.

Ne-eman, Y., S. Sender and E. Oren (2005b) Map of Dor (30). Jerusalem: Archaeological Survey of Israel.

Neev, D. and K. O. Emery (1995) *The Destruction of Sodom, Gomorrah, and Jericho*. New York: Oxford University Press.

Olami, Y. and Z. Gal (2003) Map of Shefar'am (24). Jerusalem: Archaeological Survey of Israel.

Olami, Y., A. Ronen and A. Romano (2003) Map of Haifa – West (22). Jerusalem: Archaeological Survey of Israel.

Olami, Y., S. Sender and E. Oren (2004) Map of Yagur (27). Jerusalem: Archaeological Survey of Israel.

Ovadia, E. (1992) The domestication of the ass and pack transport by animals: a case of technological change. In *Pastoralism in the Levant: Archaeological Materials in Anthropological Perspectives*, ed. O. Bar-Yosef and A. Khazanov. Madison, WI: Prehistory Press pp. 181–204.

Palumbo, G. (1994) *The Jordan Antiquities Database and Information System, A Summary of the Data*. Amman: Department of Antiquities.

Palumbo, G., J. M. Mabry and I. Kujit (1990) The Wadi el-Yabis Survey: report on the 1989 field season. *Annual of the Department of Antiquities of Jordan* **34**: 95–118.

Patrich, J. (1994) Map of Deir Mar Saba (109/7). Jerusalem: Archaeological Survey of Israel.

Philip, G. (2001) The Early Bronze I–III Ages. In *The Archaeology of Jordan*, ed. B. MacDonald, R. Adams and P. Bienkowski. Sheffield: Sheffield University Press pp. 163–232.

Philip, G. (2003) The Early Bronze Age of the southern Levant: a landscape approach. *Journal of Mediterranean Archaeology* **16**: 103–132.

Philip, G. (2008) The Early Bronze Age I–III. In *Jordan: An Archaeological Reader*, ed. R. Adams. London: Equinox pp. 161–226.

Philip, G. and D. Baird (1993) Preliminary report on the second (1992) season of excavations at Tell esh-Shuna North. *Levant* **15**: 13–36.

Raban, A. (1999) Map of Mishmar Ha-'Emeq (32). Jerusalem: Archaeological Survey of Israel.

Robinson, S. A., S. Black, B. Sellwood and P. J. Valdes (2006) A review of palaeoclimates and palaeoenvironments in the Levant and Eastern Mediterranean from 25,000 to 5000 years BP: setting the environmental background for the evolution of human civilisation. *Quaternary Science Reviews* **25**: 1517–1541.

Rollefson, G. and A. Wasse (in press) Wissad, a Late Prehistoric necropolis in Jordan's Eastern Badia. In *Archaeology in Jordan*, ed. B. Porter. American Journal of Archaeology.

Rosen, A. (1989) Environmental change at the end of the Early Bronze Age Palestine. In *L'urbanisation de la Palestine à l'âge du Bronze Ancien*, ed. P. de Miroschedji. Oxford: British Archaeological Reports International Series pp. 247–256.

Rosen, S. (1994) Map of Makhtesh Ramon (204). Jerusalem: Archaeological Survey of Israel.

Savage, S. H. (2002) http://archaeology.asu.edu/jordan/Reproject.html.

Stark, H., L. Barda and A. Berman (2005) Map of Ashdod (84). Jerusalem: Archaeological Survey of Israel.

van den Brink, E. C. M., N. Lipschitz, D. Laza and G. Dorani (2001) Chalcolithic dwelling remains, cup marks and olive (*Olea europaea*) stones at Nevallat. *Israel Exploration Journal* **51**/1: 36–43.

Weiss, D., B. Zissu and G. Solimany (2004) Map of Nes Harim (104). Jerusalem: Archaeological Survey of Israel.

Whitcher Kansa, S. (2004) Animal exploitation at Early Bronze Age Ashqelon, Afridar: what the bones tell us—initial analysis of the animal bones from areas E, F and G. *Atiqot* **45**: 279–329.

Wright, K., R. Schick and R. Brown (1989) Report on a preliminary survey of the Wadi Shu'eib. *Annual of the Department of Antiquities of Jordan* **33**: 343–350.

18 Modelling water resources and climate change at the Bronze Age site of Jawa in northern Jordan: a new approach utilising stochastic simulation techniques

Paul Whitehead, Sam Smith and Andrew Wade

ABSTRACT

Water in arid zones is at a premium, and this applied to past populations living on scarce water resources as it does today. An annual water balance model for Wadi Rajil, in northern Jordan, is used to simulate the ancient water supply system for the Early Bronze Age site of Jawa. The model includes water delivery from the catchment, local pond storage, and water demand for people, animals and irrigation. A Monte Carlo approach is used to incorporate the uncertainty associated with a range of factors including rainfall, evaporation, water losses and use. The stochastic simulation provides estimates of population levels sustainable by the water supply system. Past rainfall estimates from a global circulation model (GCM), with uncertainty bounds, are used to reconstruct the climate at Jawa in the Early Bronze Age (EBA). Model results indicate that the population levels in the predicted wetter conditions in the EBA could have risen to ~6,000 and may have been higher in wet years. Pond storage sustained the population during drought years. The GCM results suggest that prolonged droughts occurred in the later Bronze Age during which the water management system was unable to provide adequate supply for a population of 6,000. The utility of Monte Carlo based hydrological modelling as a tool within archaeological science is discussed.

18.1 INTRODUCTION

Populations living in the near Middle East are particularly vulnerable to drought: water is often the controlling factor in the stability of a settlement (Weiss and Bradley, 2001). There are many historical and contemporary examples of how drought or flood influences population behaviour, and how people have planned or engineered their water resources to sustain supply (Biswas, 1970). Hydrologists have developed a range of modern techniques for simulating water resource systems and assessing the recurrence of droughts and floods. Few studies have sought to use the new techniques to simulate the behaviour of ancient water systems, although Shaw and Sutcliffe (2001) investigated irrigation development in India using modern hydrological techniques. These new techniques are readily applied because of improved computing power in recent years. Given the important influence of climatic change on people and development over the past 12,000 years, it is now timely to consider the application of modern hydrological techniques to assess water resource issues at key archaeological sites. Such an approach should be able to assist archaeologists with the interpretation of sites and the behaviour of the people occupying them, and to assess the water-driven factors that might have influenced the development of such locations.

Our aim in this chapter is to show how an understanding of previous hydrological regimes may prove particularly useful in deriving estimates of potential population sizes for specific sites. Whilst archaeologists frequently speculate on the population sizes of prehistoric settlements and employ several techniques to achieve this end, these are generally based on some form of ethnographic analogy, may be overly deterministic and often incorporate a high degree of uncertainty (see Hassan, 1978; Zorn, 1994 for reviews). The present chapter demonstrates a methodology specifically designed to deal with the uncertainty surrounding prehistoric use of water resources. This approach incorporates Monte Carlo statistics allied to the results of a computer-based general circulation model (GCM) to simulate the water balance in a catchment over a range of climatic conditions. We apply these techniques to the site of Jawa, Jordan (Figure 18.1). Through an investigation of the influence of climate-driven changes in rainfall and, hence, water availability we examine the potential population levels sustainable at the site during the Early Bronze Age (EBA). The estimates of potential

Water, Life and Civilisation: Climate, Environment and Society in the Jordan Valley, ed. Steven Mithen and Emily Black. Published by Cambridge University Press.
© Steven Mithen and Emily Black 2011.

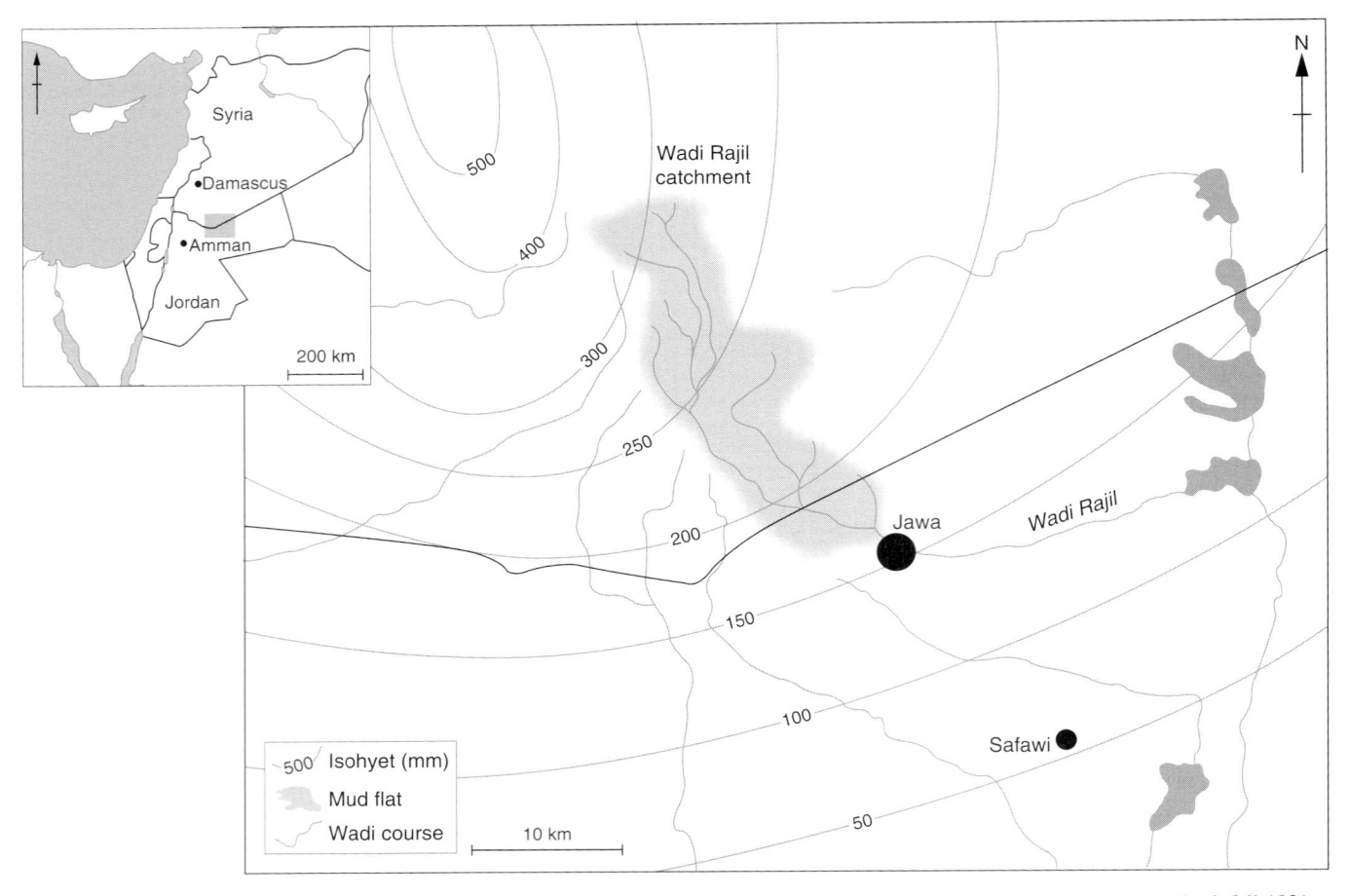

Figure 18.1. Location map of Jawa showing catchment of Wadi Rajil with rainfall isohyets (rainfall data represent average annual rainfall 1931–1960). Source: NRA Jordan. Rajil catchment after Helms (1981).

population sizes provided by this method may be regarded as being at the upper limit, because they estimate the carrying capacity of the water supply and storage systems, which may be able to support a significantly higher population than actually existed at the time.

The procedures proposed in this chapter take account of the inevitable uncertainty surrounding the mechanics of an ancient water system and uncertainties about rainfall patterns and climatic change. The stochastic modelling approach proposed is ideally suited to many archaeological investigations where there may be alternative explanations and there is a need to investigate a range of possible outcomes. A hydrological model of Jawa was developed and used to estimate population levels under a range of climates, reflecting the spread of climate in the EBA as predicted by a GCM. From this analysis it was possible to tentatively quantify population levels and investigate the occupation history.

18.2 JAWA

The large (12 ha) walled site of Jawa is located on the Wadi Rajil (catchment area 270 km^2) at (32.336° N, 37.002° E) on the basalt *harra*, close to the present Syria–Jordan border (Figure 18.1). The original excavations at Jawa were first published in a popular science format (Helms, 1981) and in a series of short reports (Roberts, 1977; Helms, 1987, 1989; Betts *et al*., 1991). Following this, a detailed monograph was produced (Betts, 1991b), although a planned second volume providing details of the palaeoeconomy and water management systems is still awaited. In the present context, the lack of radiometric dates from the site and the limited attention paid to palaeoclimatic and palaeoenvironmental indicators are particularly problematic (Philip, 1995). Despite these issues we base our current modelling upon the work and conclusions of Svend Helms, namely that Jawa was a short-lived, walled settlement and that all visible water management features were in use concurrently at some point during the EBI (Helms, 1981, 1991b). Figure 18.2 shows the layout of EBI Jawa according to Roberts (1977) and Helms (1981, 1991a); the walled town is located adjacent to the wadi, with storage ponds and potential fields surrounding the town. The water management system functioned by diverting water from the Rajil into ponds for subsequent supply to people, to their animals and perhaps for irrigation. In addition, runoff of

MODELLING WATER RESOURCES AND CLIMATE CHANGE AT BRONZE AGE JAWA

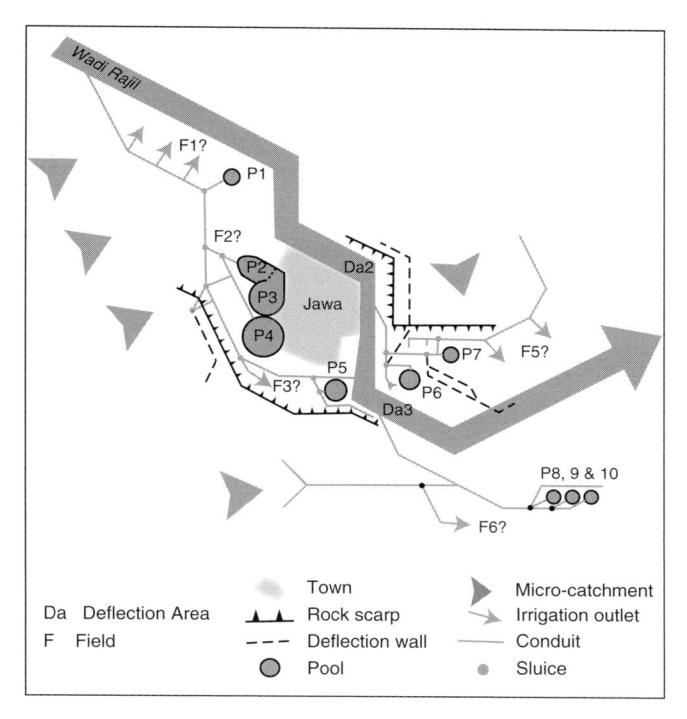

Figure 18.2. Detail of the Wadi Rajil, storage ponds and Jawa. (After Helms, 1981 p. 157.)

local rainfall from microcatchments may have provided some additional top up of water resources in wet years.

The steppic location of Jawa provides both advantages and problems for the study of the archaeology and hydrology. The intense deflation of deposits, allied to thin soil cover, prevent the build up of deep stratigraphy. This allowed Helms and colleagues to survey the spatial extent of the town and its associated water management systems rapidly, with limited excavation. However, the lack of stratigraphy has limited understanding of the phasing of the site and has led to debate about the contemporaneity of the archaeology. The lack of available groundwater resources makes reconstructing prehistoric hydrological regimes simpler than in other areas of Jordan where water supply is often augmented by spring flow. Under present climatic regimes, Jawa receives less than 100 mm per year; for example, the nearby station of Safawi, shown in Figure 18.1, received on average 72.6 mm per year between 1943 and 2003 (http://met.jometeo.gov.jo). There are no springs around Jawa, and any groundwater resources probably could not have been exploited by prehistoric populations as the aquifers are located beneath the basalt. A key feature of the modern climate is that rainfall is patchy and unpredictable, and although the region as a whole is classified as arid/semi-arid, certain areas may receive sufficient rainfall, when stored in natural pools, to support year-round occupation (Lancaster and Lancaster, 1991).

The *harra* has a long, if sporadic, history of human settlement and has formed an integral part of seasonal pastoralist systems from the late Neolithic to the modern day (Betts, 1991a; Philip, 2001). In addition, many 'desert kites', used for the hunting of gazelle, litter the basalt landscape (Helms and Betts, 1987). Present-day occupation in the region is mainly limited to Bedouin groups who practise cereal cultivation in tandem with pastoral grazing of sheep and goats (Lancaster and Lancaster, 1991). In the long-term context of an arid landscape, occupied seasonally by nomadic groups, the large, walled EBI settlement of Jawa has long been regarded as anomalous (Helms, 1981). Where other EBI occupations are present, such as the rockshelter site of Tell el Hibr, they are more obviously small-scale and temporary, with quite different material culture from that found at Jawa (Betts, 1992). This has been taken to suggest that the settlement at Jawa represents an intrusive population and that Jawa functioned as part of a wider economic system with links to both Syria and the Jordan Valley (Philip, 2001). This position is supported by finds of Jawa-type pottery along the Wadi Zarqa and into the Jordan Valley at the sites of Tell Um Hammad (Helms, 1987, 1991b, 1992) and Jebel Abu Thawwab amongst others (Douglas and Kafafai, 2000), and also by the presence of imported Palestinian flint tool types and knapping techniques in the Jawa assemblage (Betts, 1991c; Betts *et al.*, 1991). In addition, survey around Jawa has revealed a rich range of Early and Middle Bronze Age settlement, often characterised by elaborate water management systems and featuring similar pottery types to Jawa (McClellan and Porter, 1995; Betts *et al.*, 1996; Braemer *et al.*, 2004).

In terms of chronology, Helms (1981, 1987, 1991b) claims, upon the basis of correlation of the Jawa pottery sequence with that recovered from the stratified deposits at Tell Um Hammad, that the principal settlement and water management systems should be attributed to a short phase (perhaps as short as 3–50 years) at the start of the EBI ~5,600 BP – this and all other dates within this chapter are in calibrated calendar years BP. According to his model, the site was reoccupied, again rather briefly, during the Middle Bronze Age 2 (~4,000 BP), when a single 'citadel' building was constructed upon the top of the EBI settlement (Helms, 1989). Owing to the limited extent of the Middle Bronze Age settlement, the present modelling will focus upon the EBI occupation.

Helms' assertion that the majority of Jawa's architectural and water management features should be dated to the EBI initially proved controversial. However, most researchers now appear to accept an EBI date for at least some of the site (Hanbury-Tenison, 1989; Philip, 1995; Braemer and Echallier, 2000). Despite this apparent consensus, there is still considerable debate regarding the nature and extent of the Jawan occupation(s), and Philip's assertion that the evidence at Jawa could 'support alternative positions' to those of Helms (Philip, 1995 p. 161) is well made. For example, Braemer and Echallier (2000) state that some of the Jawan pottery has equivalent forms in the Syrian

EBII (i.e. fifth millennium BP), a position supported by other researchers (e.g. McClellan and Porter, 1995). In this light, researchers have argued that the settlement of Jawa may not represent a single phase of occupation but may represent successive, spatially overlapping settlement phases and that many features (architectural and hydrological) which Helms interpreted as contemporary may relate to different phases of EB occupation (McClellan and Porter, 1995; Philip, 1995; Braemer and Echallier, 2000).

Publication of palaeoeconomic data from Jawa remains incomplete and detailed reconstruction of the economy of the settlement is not possible. However, the significance of plant foods in the diet is attested by the large number of grain processing tools (Helms, 1981) and by the presence of glossed sickle blades which appear to have been used for harvesting cultivated cereals (Unger-Hamilton, 1991). This picture is supported by analysis of plant remains, which include six-row hulled barley, and are suggestive of a climate at least as wet as that at present and possibly of irrigation (Wilcox, 1981). In addition, preliminary reports suggest that the faunal assemblage (~2,500 analysed bone fragments) is dominated by sheep and goat bones (86.7%) and that within this group, sheep remains may have outnumbered goat remains by a ratio of 4:1 (Köhler, 1981). This may suggest that the population of Jawa was not under severe water stress; although sheep provide superior meat, they also require more water than goats (Köhler, 1981). This hypothesis may draw additional support from the fact that domestic cattle, which require far more water than either sheep or goat, are the third most abundant species, accounting for 8.5% of sampled remains (Köhler, 1981). Taken together this evidence suggests that, under a climate similar to that of the present day, the growing of crops and the provisioning of animals may have placed significant demands upon the settlement's water reserves. Indeed, Helms tentatively identifies several of the surrounding pools as animal watering stations and suggests that there may have been irrigated fields (19 ha) associated with the EBI water system. Significantly, it is unclear whether irrigation would have used stored water only or surplus water (from wadi flooding and local runoff) or a combination of both.

In terms of population levels in the EBI, Helms (1981) estimated the population of Jawa based upon the assumption that the entire settlement was occupied simultaneously and that the structure and layout revealed in small excavated areas was representative of the remainder of the settlement. Based upon these assumptions, Helms estimated the population of the site by first calculating the number of dwellings, then assuming (based upon ethnographic data from the Negev) that each dwelling would have contained a 'family' of between four and six people. According to this system the population would have been between 3,078 and 5,066 people. Using these figures, Helms then demonstrated that if the pools were filled each year there would be sufficient water to support these population levels. However, Helms reconstructed the water balance at the site on the assumption that the EBA climate was the same as that of the modern day, an assumption which does not concur with more recent understanding of late Holocene climatic shifts (e.g. Goodfriend, 1988; Rosen, 1995; Frumkin et al., 2001; Bar-Matthews et al., 2003). As such, it is appropriate to revisit the work of Helms in order to test his propositions using a modelled estimate of climate change.

Despite the considerable ambiguities associated with the interpretation of Jawa, we base our hydrological modelling largely upon the Helms version of events: we assume that the visible architectural and hydrological features were broadly contemporary and that all date to the EBI. The difference between ourselves and Helms is that, following the suggestion of Braemer and Echallier (2000) and McClellan and Porter (1995), we base our model upon the premise that Jawa was occupied, perhaps intermittently, throughout much of the EBI, which we take to span the period 5,600–5,000 BP (Chesson and Philip, 2003). The key advantage of the hydrological model presented is that we are able to test a range of different scenarios easily and can also incorporate GCM output data to consider the EBI occupation in the context of a dynamic climate. Thus, we examine the relationship between population level and climate and assess the potential role of climatic change on settlement possibilities in this seemingly inhospitable region.

18.3. METHODOLOGY: HYDROLOGICAL MODELLING AND MONTE CARLO ANALYSIS

Modelling has become an established technique used by hydrologists to analyse water resource systems; it is the representation of the system, either natural (a river) or constructed, using equations often coded as a computer program. Whilst there is a wealth of information about hydrological modelling techniques in general, there is relatively little concerning modelling of arid zones, especially in the Middle East and Jordan. A recent review paper on arid zone hydrology by Wheater (2005) describes a range of alternative approaches, including metric- or data-based models (Jakeman et al., 1990; Young, 2005), conceptual models which seek to represent the dominant processes operating in a catchment, and physics-based models, which try to describe the complex nature of the physical processes and stochastic approaches. A good example of a physics-based model is KINEROS (Woolhiser et al., 1990); this has been applied to model wadi flows (Semmens et al., 2005). The major problem with such physics-based models is that considerable information on rainfall, runoff and the processes is required to calibrate and validate the model adequately. In Wadi Rajil none of these data are available, so there is considerable uncertainty about the processes operating: the transmission losses (i.e. losses down the channel due to re-infiltration); the soil infiltration rates; and the evaporation rates in the catchment. In the Wadi Rajil, runoff is often dominated by overland flow, as the basalt geology of the Jawa region creates an impermeable layer which prevents significant infiltration. Thus a physics-based model would be difficult to apply with any confidence and inappropriate for Wadi Rajil where there are many other uncertainties in addition to the catchment characteristics. These uncertainties include the rainfall, the size of the ponds at Jawa, the timing and synchronicity of their construction, the efficiency of the dam deflection, the contributions of the micro-catchments and the water use by the people, the animals and the irrigation system. For example, water use by different animals can vary significantly, as can the water requirements for different crops in the irrigated areas. Another form of uncertainty is associated with climatic change. GCMs produce a scenario of what might have happened and this is based on the physics of global weather patterns. As such, assumptions are made regarding solar output, carbon dioxide in the atmosphere, volcanic activity, Earth orbit and the palaeogeography, which includes bathymetry, land ice extent and land elevation. There is significant uncertainty associated with these GCM simulations and there is a need to reflect this in the modelling study.

18.3.1 Modelling strategy for Jawa

To model the water supply at Jawa, an annual water-balance model has been established based on flows from Wadi Rajil and storage in the ponds, within a Monte Carlo framework. Monte Carlo is a well-established technique for dealing with uncertainty in hydrological models (Whitehead and Young, 1979; Clarke, 2004). The general approach is to set up a basic hydrological model for a catchment or water resource system, defining the inputs in probabilistic terms rather than as a specific value or time series. Thus rainfall can be represented, for example, as a normal distribution with a given mean and standard deviation to define the range of behaviour. In a Monte Carlo simulation all the model inputs and parameters are described in such a statistical manner and the model is run hundreds or thousands of times to generate multiple outputs, such as runoff flows for a catchment or pond volumes in the case of Jawa. These multiple outputs are then analysed statistically to produce output distributions, which reflect the ensemble behaviour of the hydrological system. Thus the runoff flows will have a mean and a distribution around the mean, with the range of the distribution reflecting the uncertainties in the inputs (e.g. rainfall) and the model parameters (e.g. water use by people, animals etc.). Given limited data availability, such an approach, based on an annual water balance within a Monte Carlo framework and linked to output from a GCM, is a pragmatic methodology for determining the relationship between climate and the sustainable population.

GCM precipitation and temperature data hindcast for a range of climatic conditions from 5,600 BP to 4,000 BP were used to simulate the effects of changing climate on water availability. GCMs have been successfully applied to past climate simulations (e.g. Valdes et al., 1999; Sellwood and Valdes, 2006). In this work, the UK Meteorological Office HadSM3 GCM downscaled using HadRM3 is used to investigate the changing climate at Jawa (see section 18.4). This research version of the Hadley Centre model was used to generate historical climate for the specific northern Jordan region between 36 and 38 degrees longitude and 32 and 33 degrees latitude.

The potential amount of stored water is estimated from the measured dimensions of the ponds. By calculating the pond storage, it is possible to assess the water availability in these ponds and to calculate how many people, animals and hectares of irrigated land could have been supported. This gives an independent method of determining possible population levels at Jawa which is a completely different approach from that used by Helms.

18.3.2 Model structure

The annual water-balance model utilised for the Jawa study uses the following equations and conditions.

The hydrological effective rainfall, HER (mm), is calculated as

$$\text{HER} = K_1 \cdot P - \text{ET}/100 \qquad (18.1)$$

where P and ET are the precipitation (mm) and evapotranspiration (mm), respectively, per year; K_1 is the percentage rainfall delivered to the wadi after allowing for infiltration.

The runoff, Q (m^3 yr^{-1}), is estimated from the HER and catchment area, A_c (m^2)

$$Q = \text{HER} \cdot A_c / 1000 \qquad (18.2)$$

The flow diverted to the ponds, D (m^3 yr^{-1}), is estimated as a fraction K_2 of the total runoff, Q:

$$D = K_2 \cdot Q / 100 \qquad (18.3)$$

The total water demand over the year, W_u (m^3 year^{-1}) is based on water use for irrigation ($A_i W_i$) and use by humans ($N_p W_p$) and animals ($N_p W_a$):

$$W_u = N_p W_p + N_p W_a + A_i W_i \qquad (18.4)$$

where N_p is the number of people living at Jawa, W_p and W_a are the water uses in m^3 yr^{-1} for the people and the animals associated with each person, and A_i is the irrigated area (m^2) and W_i the water irrigation rate (m yr^{-1}).

The volume of water stored in the pond depends on the amount of water diverted from the wadi, D, the evaporation from the pond during the year, E_p (m^3 yr^{-1}) and is limited by the maximum volume of the pond, $P_{s,max}$ (m^3):

$$\text{If } D < P_{s,max} \; P_s = D - E_p \quad (18.5)$$

Else

$$P_s = P_{s,max} - E_p \quad (18.6)$$

Assuming that all the water in the pond is used for irrigation or by humans or animals by the maximum population ($N_{p,max}$) then the water used equates to the water available:

$$P_s = N_{p,max}W_p + N_{p,max}W_a + A_iW_i \quad (18.7)$$

By re-arranging Equation (18.7), the maximum population that can be sustained by the water stored is given by

$$N_{p,max} = (P_s - A_iW_i)/(W_p + W_a) \quad (18.8)$$

The values of K_1, K_2, P, ET, A_c, N_p, W_p, W_a, A_i, W_i, E_p and $P_{s,max}$ can all be defined as constants or distributions for the Monte Carlo simulations.

The model is steady-state, meaning that for each model run, the inputs are time invariant. As such, the model does not have a dynamic component and is unable to account for the seasonality of wadi flows and demand for irrigation water due to crop growth, nor can the effect on the population of successive drought years be modelled. This would require a more sophisticated rainfall-runoff model operating, at least, at monthly time-step. Given that this is a preliminary analysis of water availability at Jawa at key times in the past and one of the first studies to link predictions of the palaeoclimate from a GCM and palaeoenvironmental evidence with an assessment of water availability, it was decided to use a simple hydrological model. Understanding the limitations of the GCM predictions and the dating of the palaeoenvironmental and archaeological evidence on the model linkages would be confused by considering a more sophisticated hydrological model in the first instance. The model presented is still useful; it considers the water available, stored and used within a single year. When the model is incorporated within a Monte Carlo based framework of sampling from the input ranges, then it can be used to determine whether a surplus or deficit of water exists for a wide range of climate conditions and population sizes, and thereby how a population is pressured by water availability, how it can respond by building ponds and where thresholds exist at which the climate conditions become such that the population can no longer alleviate water-induced stress through technological change. When run in the Monte Carlo framework, the model can produce a probability for the occurrence of drought conditions, and this probability can be interpreted in terms of likely recurrence; for example, drought conditions may exist 1 year in every 10.

18.3.3 Model assumptions of supply and demand

The main factors involved in modelling the water balance at Jawa are supply and demand. Supply refers to the amount of water from the catchment and the pond system that was available to the population, while the demand is the amount of water needed for domestic supply (drinking, washing, cooking) by each individual, for the watering of animals and for irrigation.

In terms of the amount of drinking water required by each adult individual, modern medical science suggests that in order to maintain health, adults require between 1 and 2 litres per day, but this can range (depending upon individual circumstance and environmental factors) from less than 1 litre to more than 20 litres (West, 1985). This is a large range and can perhaps be extended further. For example, Kung San people of the Kalahari in Namibia survive for most of the year with no standing water, drawing the required water from plants (Tanaka, 1976). Other ethnographic studies suggest that water use can range from between 3 to 363 litres per capita per day, depending upon location, environment and social factors and tending to increase greatly for industrialised societies (Wilkinson, 1974). Perhaps the most appropriate source of analogy is provided by Evenari *et al.* (1971) who studied water use by contemporary Bedouin groups in the Negev and suggested that the average adult requires around 1.5 m per year (approximately 4.2 litres per day). There is considerable uncertainty regarding the precise amount of water required by any individual, and cultural and social practices will influence the amount of water regarded as sufficient to sustain an acceptable quality of life. Although some water will also have been required for industrial processes, such as making ceramics, we have assumed that total industrial usage will have been negligible compared with other factors.

The amount of water required by animals is similarly uncertain. For example, different species require different quantities and qualities of water and respond to periods of no water in differing ways. With regard to the species known to be present at EBI Jawa, Evenari *et al.* (1971) state that donkeys require 1 m^3 per year, whilst sheep and goats require around 0.5 m^3 per year. Tivy (1990 p. 151) suggests that cattle require between 1.8 and 8.4 m^3 per year depending upon factors such as temperature, lactation and movement. In terms of the timing of watering, there are differences between species. For example, whilst camels can go without water for 12 days and sheep/goats for 3–5 days, cattle need to be watered twice every day (Tivy, 1990).

In addition to the uncertainty regarding the amount of water required for each animal there are several other complicating factors when modelling demand on the water resources at Jawa. Firstly, one has to assume how many animals are associated with each human, an assumption which implies some knowledge of the economic system. For example, an economy based upon herding sheep would have a higher animal to person ratio than an economy based upon grain production. In the present example we again turn to Evenari *et al.* (1971) who suggest that a typical desert family in the Negev consists of six people, two camels, one donkey, ten sheep/goats and two dogs. The validity of using these figures at EBI Jawa remains unclear, particularly as we know there were no camels at Jawa and if the population is assumed to be of intrusive farmers, rather than desert-adapted pastoral nomads. A final complication is that one must assume how much these animals depend upon the stored water. For example, flocks may spend much of the year grazing away from the site and only make demands upon the stored water at certain times of year. In the present example we assume that all animal water requirements were met by the storage pools; again, a questionable assumption given the long-term history of pastoralism in the area but one necessary to progress the application of the hydrological model.

The final demand on water use, in the current model, is water required for irrigation. There are 19 ha of fields in the vicinity of Jawa. The crop species present at the site suggest that either these fields were irrigated, or the climate was wetter during the EBI than it is today (Wilcox, 1981). At present it is unclear whether these fields were irrigated using water from the storage pools and resolving this issue is an aim of this work.

18.3.4 Model assumptions of water transfer and storage

Helms (1981) describes in detail the hydrological system at Jawa and proposes a range of engineered structures for regulating the transfer and storage of water. These include low dams on the Wadi Rajil which deflect water into a series of channels which, in turn, feed a series of ponds or pools constructed close to the city and adjacent to the potentially irrigated field systems, as shown in Figure 18.2. In addition, the Jawan water system also utilised runoff rainwater from local micro-catchments. Together, these two seasonal and unpredictable water sources provide the only water supply. As shown in Figure 18.2, there are three discrete water supply systems, each incorporating a deflection dam, local micro-catchments, storage pools and perhaps irrigated fields. Understanding the phases of construction of the water management systems is complicated by the difficulties inherent in dating such unstratified features and this is compounded by a lack of radiometric dates. However, on the basis of the similarity of construction and water management techniques in each system, Helms (1981) proposes that all three systems should be attributed to the EBI and that, at some point, all three systems would have been in concurrent use. Helms suggested a developmental sequence of phasing which outlines the order in which pools were added over the course of the EBI occupation. While we accept Helms' assertion that at some point all three systems were in simultaneous use, we prefer to treat each separate system as a developmental phase. As such, the first system includes pools 1–5, the second system added pools 6 and 7, whilst in the third system pools 8–10 were added. The total storage capacity of all three systems is in the region of 52,000 m^3. Helms (1981) categorises each pool as either for human consumption or for watering animals, but the basis for this is unclear and we do not differentiate between pools for humans and animals. The success of this storage pond system depended on sufficient flow in Wadi Rajil to refill the ponds each year.

18.3.5 Rainfall simulation

Helms (1981) performed water balance calculations based on rainfall data from Jordan archives for the period 1931–60, assuming that this represented rainfall levels in the EB. However, recent advances in our understating of patterns of Holocene climates (e.g. Bar-Matthews *et al.*, 2003) suggest that rainfall levels have fluctuated during the Holocene and that major fluctuations may have occurred at the time of Jawa's occupation. Here we use proxy records and GCM data to model the water system at Jawa against a backdrop of dynamic climatic variability.

In this study, the annual rainfall has been characterised by analysing the statistical properties of the Jerusalem rainfall, for which a 148-year record is available, beginning in 1846. A statistical package, Crystal Ball (Decisioneering, 2006), has been used to analyse these data. Fitting distributions to this annual time-series generates almost equally statistically significant fits for both normal and Weibal distributions. Both distributions have similar statistics, where the normal distribution has a mean of 607 mm and a standard deviation of 181 mm (or 33.3% of the mean). Jerusalem is approximately 180 km from Jawa, so the rainfall has to be scaled to the dryer Jawa region. Figure 18.1 shows the rainfall contours for the Wadi Rajil catchment, and an area-weighted rainfall average for Wadi Rajil is approximately 233 mm $year^{-1}$, or 27% of the annual Jerusalem rainfall. In the Jawa model application, we have assumed that the rainfall is normally distributed with the standard deviation set to 33.3% of the mean.

Table 18.1 shows a list of all parameters, inputs, distribution types and scale factors. In each Monte Carlo simulation, 5,000 model runs are made; within each a random sample is drawn from the rainfall distribution and the other model parameters which describe the water supply and demand. At the end of each run, the sustainable population is calculated. Running 5,000 simulations allows a high confidence level for the posterior distributions, and the Kalmorgarov statistic (Kendall *et al.*, 1964) indicates that the posterior distributions should be within 2% of the true distribution with 95% confidence level (Whitehead and

Table 18.1 *Distribution types and characteristics for key parameters used in the Monte Carlo analysis for the high rainfall condition*

Parameter	Units	Distribution type	Range or distribution characteristics
Catchment delivery rate, K_1	%	Rectangular	2–4
Dam deflection rate, K_2	%	Rectangular	3–5
Actual evaporation rate	mm yr^{-1}	Rectangular	10–50
Rainfall	mm yr^{-1}	Normal	Mean 329 St. Dev. 98.7
People water use	mm^3 yr^{-1}	Normal	Mean 3.0 St. Dev. 2.0
Animal water use	mm^3 yr^{-1}	Normal	Mean 1.5 St. Dev. 0.15
Irrigation water use	mm^3 yr^{-1} ha^{-1}	Normal	Mean 2,500 St. Dev. 300
Irrigation area	ha	Rectangular	5–8

Young, 1979). In other words, statistically significant results are generated with 5,000 simulation runs. The hydrological model can be run inside the Monte Carlo wrapper 5,000 times in approximately 15 seconds on a modern computer with 512 Mbytes of RAM and a 1.8 GHz processor.

18.4 MODEL RESULTS, CLIMATE CHANGE AND DISCUSSION

Before we can consider this analysis for Jawa, it is necessary to consider potential climatic changes in the region during the EB period. Chapters 6 and 7 of this volume analysed a wealth of palaeoclimate data to reconstruct past climate for the Levant and have shown that the climate has changed significantly during the Holocene with potentially wetter conditions prevailing in the EBA at c. 5,000 BP. The GCM results presented here represent rainfall and temperature changes generated using the HadSM3 version of the coupled atmosphere–ocean GCM. The model was developed at the Hadley Centre for Climate Prediction and Research, which is part of the UK Meteorological Office. The model outputs such as rainfall and temperature are available from the BRIDGE website (www.bridge.bristol.ac.uk). Aspects of the workings of the model are described in Pope et al. (2000) and its use in palaeoclimate studies is illustrated in Haywood et al. (2002).

The GCM has been run at key points over the past 20,000 years to provide a time-series of climatic change. Figure 18.3 shows the rainfall and temperature variations for the past 14,000 years; the temperature in the region increases after the last ice age and reaches a plateau with relatively little change thereafter. Rainfall, however, shows significant variation with higher rainfall in the early Holocene dropping irregularly towards modern levels from around 8,000 BP. This general pattern of declining rainfall is interrupted by a brief wet phase at around 5,000 BP with wet conditions pertaining at the time that Jawa was thought by Helms (1981) to be established. A rapid decline in rainfall is evident after 5,000 BP leading to a longer period of dry conditions. It is likely that the GCM predicted increase in precipitation at 5,000 BP is related to changes in levels of solar insolation, which suggests an altered route of rain-bearing storm tracks in the Atlantic Ocean (this volume, Chapter 3).

The general picture of irregularly declining rainfall after 8,000 BP, with significant fluctuation during the EBA, is supported by a range of palaeoclimatic evidence (Chapters 6 and 7, this volume). For example, Frumkin et al. (2001) provide evidence of high Dead Sea levels (peaking at −370 m asl) between 6,000 and 4,900 BP, followed by a rapid lowering of sea levels as shown in Figure 18.3. The high-resolution palaeo-rainfall record developed from analyses of speleothems at Soreq cave in Israel (Bar-Matthews et al., 2003) also agrees with this general picture (Figure 18.3). The Soreq record is, however, chronologically out of sequence with the GCM records, positing a short-lived but extreme dry phase centred on ~5,200 BP with a rapid return to wetter conditions by ~5,000 BP. These differences may be a consequence of the chronological accuracy and spatial resolution of the different records, especially the GCM which cannot match the precision of the Soreq record. Although resolving the different predictions for the timing of climatic events in the mid-Holocene is of importance for our understanding of the EBA, for the purposes of the present study the salient facts are that fairly extreme fluctuations in rainfall levels during the EBI are predicted by several datasets, and such fluctuations may have had a significant impact upon the inhabitants of Jawa.

As discussed above, there is considerable uncertainty over the accuracy of GCM simulations. For example the ENSEMBLES project (www.ensembles-eu.org) has a large team of modellers evaluating a number of GCMs investigating model uncertainties due to forcing inputs, boundary conditions, process parameters, process understanding and numerical solution techniques. All of these give rise to uncertainty and will generate a range of

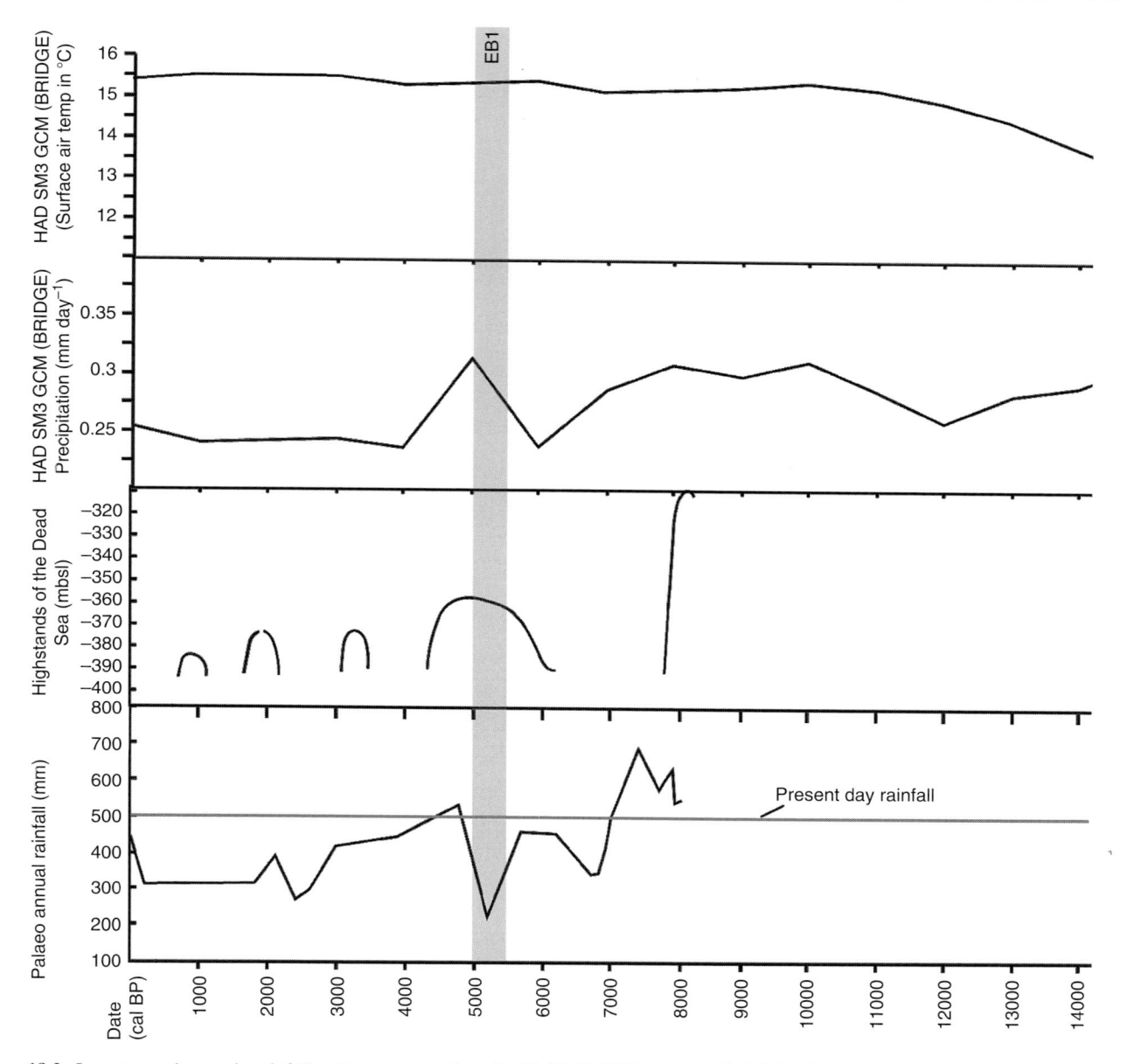

Figure 18.3. Long-term changes in rainfall and temperature from the HadSM3 GCM, compared with Dead Sea levels (after Frumkin *et al.*, 2001) and palaeo-rainfall estimates derived from analyses of Soreq isotope sequence (after Bar-Matthews *et al.*, 2003).

behaviours within the models. The BRIDGE team are also assessing uncertainty and, for example, provide statistics such as the standard deviation of interannual variability generated by the GCM. This gives a measure of the uncertainty in the model reconstructions of climate. In the case of precipitation, the standard deviation is estimated as 0.1 mm day^{-1} or 36.5 mm year^{-1}, which is close to the observed interannual variability of the 146-year Jerusalem rainfall record discussed previously. Thus variability in the GCM matches the observed variability and gives further confidence in the GCM predictions. However, it should be emphasised that this does not take account of all the uncertainties in the GCMs, and we must await the results of the major research projects currently under way to better understand the uncertainty in GCM predictions. Despite this, the hydrological method presented here is still useful as it describes how to deal with uncertainty when linking climate and archaeology.

In this study we use three estimates of rainfall over the Rajil catchment and combine these with different water storage volumes at Jawa. The rainfall scenarios used here include high (329 mm yr^{-1}), medium (234 mm yr^{-1}) and low (123 mm yr^{-1}) scenarios developed from the GCM data. Storage scenarios include high (52,000 m^3), medium (42,000 m^3) and low (28,000 m^3) volumes and are based upon the development of water storage systems outlined by Helms (1981). Figure 18.4

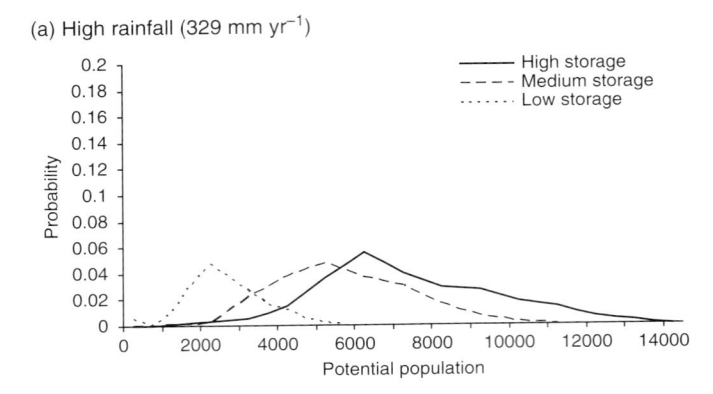

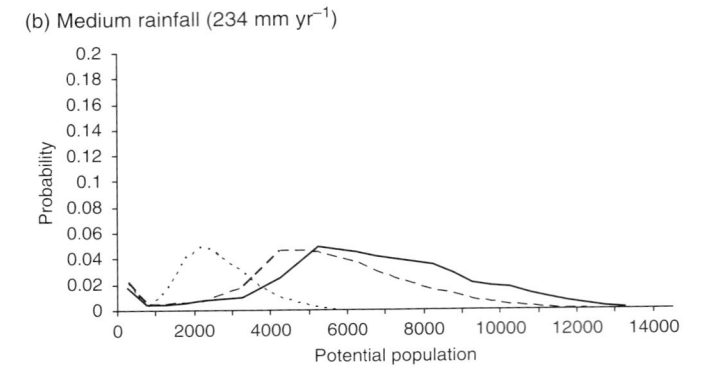

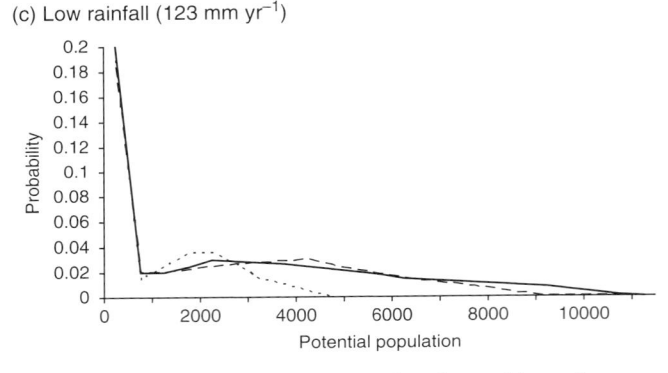

Figure 18.4. Sustainable population levels at Jawa with varying water storage volumes for (a) high, (b) medium and (c) low rainfall scenarios.

provides the distribution of sustainable population sizes modelled under different combinations of these scenarios.

Figure 18.4b provides the distribution of population sizes sustainable when rainfall is set to a medium (close to modern day) value. When a medium storage of water is assumed (combined storage of stages 1 and 2 ponds), the number of people that can theoretically live at Jawa ranges from zero to 10,000, with a mode of 4,200. The distribution represents the range of uncertainties specified in the model run. The model shows that, as might be expected, the drier the conditions, the lower the sustainable population and the wetter the conditions, the higher the population. Thus the lowest populations simulated occur in the driest years, and the highest are in the wettest. For approximately 2% of the time, the sustainable population level is zero and drought conditions would have prevented anyone living in Jawa.

Under wetter conditions (Figure 18.4a), with a pond volume set to the maximum of 52,000 m^3 (all ponds) the mean population levels rise to about 6,000 people, which matches the estimates of Helms (1981). At the other extreme, the low rainfall/ low pond storage (system 1 only) condition indicates much lower levels of potential population, with a significant (20%) chance of a sustainable population of zero (Figure 18.4c). Under this rainfall scenario, whilst increasing storage levels may allow a larger population in some years, it does not alleviate the significant risk of total failure of the system caused by drought.

18.4.1 Impacts of climate change

To model the changing climate conditions in a systematic manner, the hydrological model was run assuming a range of rainfall conditions taken from the GCM outputs over the EBA period, with four rainfall conditions being simulated ranging from low to high rainfall: 123 to 329 mm yr^{-1}, respectively. This range corresponds to very dry conditions through to the wetter conditions indicated by the GCM over the past 6,000 years. The rainfall levels for Wadi Rajil have been obtained by scaling the GCM mean daily rainfall given in Figure 18.3, to the Wadi Rajil catchment. This scaling is difficult because of the lack of actual rainfall measurements in the Jawa region. Also, the GCM grid square covers a very large area and does not take into account topographic features such as Jebel Druze which dominates the hydrology of the upper reaches of the wadi. Thus only an approximate scaling can be given. For example, if the rainfall isohyets averages for the wadi are used for scaling purposes, then the rainfall at Wadi Rajil is approximately 2.5 times the GCM levels. In other words the mountain Jebel Druze is considerably affecting the rainfall patterns. However, if we use the rainfall levels for Safawi which is located close to Jawa, then the ratio is much less at 0.78 the GCM rainfall estimates. As a consequence, the conditions in the town of Jawa would have been much drier than the upper wadi catchment conditions, making local rainfed irrigation and crop production much more difficult.

Given the uncertainty in estimating the rainfall in the Wadi Rajil and at Jawa a range of rainfall conditions have been investigated from 123 to 329 mm year^{-1}. Figure 18.5 shows the model results for the simulated mean populations for the four different rainfall levels from wet to dry conditions and assuming that we have three different pond storage systems, corresponding to large, medium and small storages of 52,000, 42,000 and 28,000 m^3. The mean populations rise as the pond storage increases, especially under wet conditions, as shown in the top line in Figure 18.5. This rapid increase in population becomes moderated as the rainfall levels fall until, under low rainfall, the

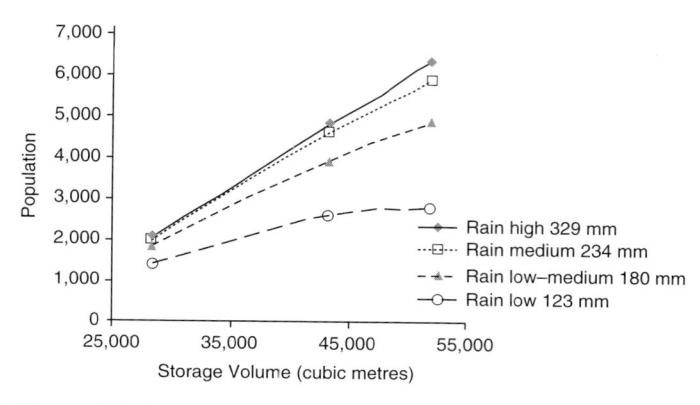

Figure 18.5. Sustainable population levels as a function of pond storage under a range of rainfall conditions.

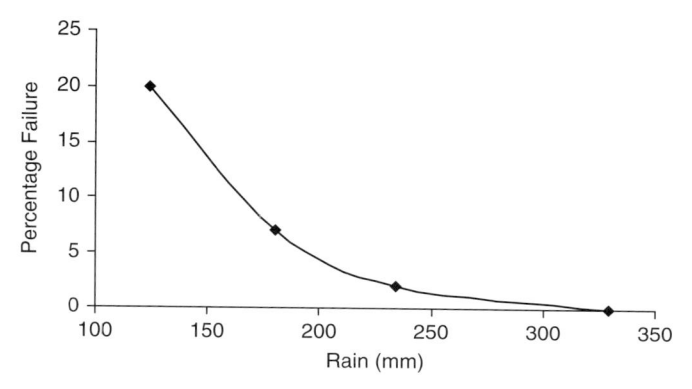

Figure 18.6. Percentage population failure at Jawa as a function of rainfall.

rise is minimal despite the increased pond storage. Thus, even extra pond storage under the very dry conditions is insufficient to sustain large population numbers. These results give a feeling for the vulnerability of Jawa to drought and the high dependence on pond storage. This dependence can be further evaluated by calculating the percentage failure rates against the rainfall. Figure 18.6 shows the failure rate, namely the percentage of time when no people can survive at Jawa, plotted against the rainfall, assuming large pond storage. The failure rate increases significantly as the rainfall is reduced. This shows the vulnerability to climate change.

18.5 CONCLUSION

This modelling study has proved to be an interesting and valuable exercise in evaluating the water resource at Jawa. In particular, two aspects of the relationship between water and population were highlighted.

Firstly, the sensitivity of the population to rainfall behaviour was demonstrated, with rapid climatic fluctuations around 5,000 BP leading to severe water supply problems at Jawa. Increasing pond storage would have helped sustain the population, but eventually, under low rainfall scenarios, more frequent droughts would have forced the abandonment of the site. The model suggests that population levels of 6,000 during the wetter climate at 5,000 BP would have been feasible, and this does agree with the Helms (1981) estimate, but is at the upper end of the potential scale and, furthermore, assumes all the ponds were in use and no stored water was diverted for crop irrigation. The GCM rainfall estimates and the hydrological model results also suggest a fairly rapid depletion of population at Jawa after 5,000 BP. Using the higher-resolution pattern of palaeo-rainfall from Soreq cave (Bar-Matthews *et al.*, 2003) would indicate that larger populations would have been sustainable during the very early EBI, the late Chalcolithic period and also during the early EBII, but that it would have been very difficult to sustain a population of any reasonable size during much of the EBI. This perhaps suggests that the site may be the result of several relatively short periods of occupation separated by periods of abandonment. It is equally possible that the combination of maximum rainfall and maximum water storage did not coincide and that as rainfall declined more ponds were added to the system, which would suggest a lower mean population. The possible staged nature of pond construction may support this model, which would also have made the system more sustainable as less food was required.

Secondly, the sensitivity of the water supply system to irrigation was apparent. It was shown that the population would have been much reduced if water from the ponds was being used in a major way to irrigate fields. Moreover, to supply sufficient food for a population of ~6,000, a much larger area of cultivated fields would have been required than the 19 ha of fields adjacent to the site today. This suggests that food must have been imported from seasonally inundated mud flats some distance from Jawa (perhaps through trade), that rainfall was sufficient to support dry farming closer to Jawa than is currently possible, or that water for irrigation may have utilised excess wadi flood water which was not needed to fill the storage pools. While the symbiosis of nomadic pastoralists with settled farmers is well known, the establishment of a large, settled, walled community unable to support itself in food would require some considerable explanation within the larger EBI economy.

Finally, the fluctuations in precipitation suggested by the GCM and proxy records may have far-reaching consequences for our understanding of human settlement at Jawa and adjacent areas during the EBI. During drier periods of the intensity predicted by the Soreq record (Bar Matthews *et al.*, 2003), sustaining any permanent population would have been impossible. During wetter periods not only would wadi flows be substantially increased but local rainfalls, runoff, and water stored in natural pools would have been more predictable and reliable than at present. It is important to stress that in such a marginal area as the *harra*, relatively small changes in precipitation can have a significant effect on the possibilities for human settlement and economy. Taken together, these lines of evidence indicate that climate-driven changes in water supply have played a key role in the occupation history of Jawa.

It is proposed that this methodology be applied to other archaeological sites where climate is a potential factor in controlling water supplies and hence population levels. This approach is particularly useful where considerable uncertainty exists and these uncertainties need to be incorporated into the analysis. This is generally true, by its very nature, in most archaeological studies.

REFERENCES

Bar-Matthews, M., A. Ayalon, M. Gilmour, A. Matthews and C. J. Hawkesworth (2003) Sea-land oxygen isotopic relationships from planktonic foraminifera and speleothems in the Eastern Mediterranean region and their implication for paleorainfall during interglacial intervals. *Geochimica et Cosmochimica Acta* **67**: 3181–3199.

Betts, A. (1991a) Appendix A: The Jawa area in prehistory. In *Excavations at Jawa 1972–1986. Stratigraphy, Pottery and Other Finds*, ed. A. Betts. Edinburgh: Edinburgh University Press pp. 181–190.

Betts, A., ed. (1991b) *Excavations at Jawa 1972–1986. Stratigraphy, Pottery and Other Finds*. Edinburgh: Edinburgh University Press.

Betts, A. (1991c) The chipped stone assemblage. In *Excavations at Jawa 1972–1986. Stratigraphy, Pottery and Other Finds*, ed. A. Betts. Edinburgh: Edinburgh University Press pp. 140–144.

Betts, A. (1992) Tell el-Hibr: a rock shelter occupation of the fourth millennium B.C.E. *Jordanian Badiya Bulletin of the American Schools of Oriental Research* **289**: 5–23.

Betts, A., S. Eames, S. Hulka et al. (1996) Studies of Bronze Age occupation in the Wadi al-'Ajib, Southern Hauran. *Levant* **28**: 27–39.

Betts, A., S. Helms, W. Lancaster and F. Lancaster (1991) The Burqu/Ruweishid project: preliminary report on the 1989 field season. *Levant* **23**: 7–28.

Biswas, A. K. (1970) *History of Hydrology*. Amsterdam, New York: Elsevier.

Braemer, F. and J. C. Echallier (2000) A summary statement on the EBA ceramics from southern Syria and the relationship of this material with that of neighbouring regions. In *Ceramics and Change in the Early Bronze Age of the Southern Levant*, ed. G. Philip and D. Baird. Sheffield: Sheffield Academic Press pp. 403–410.

Braemer, F., J. C. Echallier and A. Taraqji (2004) *Khirbet al Umbashi. Villages et Campments de Pasteurs dans le 'Desert Noir' (Syrie) à l'âge du Bronze*. Beyrouth: Institut Français du Proche-Orient.

Chesson, M. and G. Philip (2003) Tales of the city? 'Urbanism' in the early Bronze Age Levant from Mediterranean and Levantine perspectives. *Journal of Mediterranean Archaeology* **16/1**: 3–16.

Clarke, R. T. (2004) *Statistical Modelling in Hydrology*. Chichester: Wiley.

Decisioneering (2006) *The Crystal Ball Simulation Package*. New York: Decisioneering.

Douglas, K. and Z. Kafafai (2000) The main aspects of the Early Bronze I Pottery from Jebel Abu Thawwab, North Jordan. In *Ceramics and Change in the Early Bronze Age of the Southern Levant*, ed. G. Philip and D. Baird. Sheffield: Sheffield Academic Press pp. 101–112.

Evenari, M., L. Shanan and N. Tadmor (1971) *The Negev, Challenge of a Desert*. Cambridge, MA: Harvard University Press.

Frumkin, A., G. Kadan, Y. Enzel and Y. Eyal (2001) Radiocarbon chronology of the Holocene Dead Sea: attempting a regional correlation. *Radiocarbon* **43**: 1179–1189.

Goodfriend, G. A. (1988) Mid-Holocene rainfall in the Negev Desert from 13C of land snail shell organic matter. *Nature* **333**: 757–760.

Hanbury-Tenison, J. (1989) Desert urbanism in the fourth millennium. *Palestine Exploration Quarterly*: 55–63.

Hassan, F. A. (1978) Demographic archaeology. In *Advances in Archaeological Method and Theory*, ed. M. B. Schiffer. New York: Academic Press pp. 49–103.

Haywood, A. M., P. J. Valdes, B. Sellwood and J. O. Kaplan (2002) Antarctic climate during the middle Pliocene: model sensitivity to ice sheet variation. *Palaeogeography Palaeoclimatology Palaeoecology* **182**: 93–115.

Helms, S. (1981) *Jawa: Lost City of the Black Desert*. New York: Cornell University Press.

Helms, S. (1987) Jawa, Tell Um Hammad and the EB 1/Late Chalcolithic landscape. *Levant* **29**: 49–81.

Helms, S. (1989) Jawa at the beginning of the Middle Bronze Age. *Levant* **21**: 141–168.

Helms, S. (1991a) Introduction. In *Excavations at Jawa 1972–1986. Stratigraphy, Pottery and Other Finds*, ed. A. Betts. Edinburgh: Edinburgh University Press pp. 6–19.

Helms, S. (1991b) The pottery. In *Excavations at Jawa 1972–1986. Stratigraphy, Pottery and Other Finds*, ed. A. Betts. Edinburgh: Edinburgh University Press pp. 55–105.

Helms, S. (1992) *The 'Zarqa Triangle': A Preliminary Appraisal of Protohistorical Settlement Patterns and Demographic Episodes*. Amman: SHAJ IV.

Helms, S. and A. Betts (1987) Desert kites of the Badiyat esh-Sham and North Arabia. *Paleorient* **13**: 41–67.

Jakeman, A. J., I. G. Littlewood and P. G. Whitehead (1990) Computation of the instantaneous unit hydrograph and identifiable component flows with application to two upland catchments. *Journal of Hydrology* **117**: 275–300.

Kendall, M., A. Stuart and J. K. Ord (1964) *The Advanced Theory of Statistics*. High Wycombe: Griffin.

Köhler, I. (1981) Appendix E. Animal remains. In *Jawa: Lost City of the Black Desert*, ed. S. Helms. New York: Cornell University Press.

Lancaster, W. and F. Lancaster (1991) Limitations on sheep and goat herding in the Eastern Badia of Jordan: an ethno-archaeological enquiry. *Levant* **28**: 125–138.

McClellan, T. L. and A. Porter (1995) *Jawa and North Syria*. Amman: SHAJ V.

Philip, G. (1995) Jawa and Tell Um Hammad: two early Bronze Age sites in Jordan. Review article. *Palestine Exploration Quarterly* **127**: 161–170.

Philip, G. (2001) The Early Bronze I–III ages. In *The Archaeology of Jordan*, ed. B. MacDonald, R. Adams and P. Bienkowski. Sheffield: Sheffield University Press pp. 163–232.

Pope, V. D., M. L. Gallani, P. R. Rowntree and R. A. Stratton (2000) The impact of new physical parametrizations in the Hadley Centre climate model: HadAM3. *Climate Dynamics* **16**: 123–146.

Roberts, N. (1977) Water conservation in ancient Arabia. *Seminar for Arabian Studies* **7**: 134–146.

Rosen, A. M. (1995) The social response to environmental change in Early Bronze Age Canaan. *Journal of Anthropological Archaeology* **14**: 26–44.

Sellwood, B. and P. J. Valdes (2006) Mesozoic climates: general circulation models and the rock record. *Sedimentary Geology* **190**: 269–287.

Semmens, D. J., D. C. Goodrich, C. L. Unkrich et al. (2005) KINEROS 2 and the AGWA modelling framework. In *Hydrological Modelling in Arid and Semi-Arid Areas*, eds. H. S. Wheater, S. Sorooshian and K. D. Sharma. Cambridge, Cambridge University Press, pp. 41–48.

Shaw, J. and J. Sutcliffe (2001) Ancient irrigation works in the Sanchi area: an archaeological and hydrological investigation. *South Asian Studies* **17**: 55–75.

Tanaka, J. (1976) Subsistence ecology of the central Kalahari San In *Kalahari Hunter-Gatherers. Studies of the !Kung San and Their Neighbours*, ed. R. Lee and I. DeVore. Cambridge, MA: Harvard University Press pp. 98–119.

Tivy, J. (1990) *Agricultural Ecology*. UK: Longman.

Unger-Hamilton, R. (1991) The microwear analysis of scrapers and 'sickle blades'. In *Excavations at Jawa 1972–1986. Stratigraphy, Pottery and Other Finds*, ed. A. Betts. Edinburgh: Edinburgh University Press pp. 149–154.

Valdes, P., R. A. Spicer, B. Sellwood and D. C. Palmer (1999) *Understanding Past Climates: Modelling Ancient Weather* (CD-ROM).

Weiss, H. and R. S. Bradley (2001) What drives societal collapse? *Science* **291**: 609–610.

West, J. B., ed. (1985) *Best and Taylor's Physiological Basis of Medical Practice*, 11th Edition. Baltimore: Williams and Wilkins.

Wheater, H. S. (2005) Modelling hydrological processes in arid and semi-arid areas. In *Proceedings of the International G-WADI Modelling Workshops* http://www.gwadi.org

Whitehead, P. and P. Young (1979) Water quality in river systems: Monte Carlo analysis. *Water Resources Research* **15**: 451–459.

Wilcox, G. H. (1981) Appendix D. Plant remains. In *Jawa: Lost City of the Black Desert*, ed. S. Helms. New York: Cornell University Press pp. 247–248.

Wilkinson, J. (1974) Ancient Jerusalem: its water supply and population. *Palestine Exploration Quarterly* **106**: 33–51.

Woolhiser, D. A., R. E. Smith and D. C. Goodrich (1990) *KINEROS. A Kinematic Runoff and Erosion Model: Documentation and User Manual*. ARS-77. Washington, DC: US Department of Agriculture, Agricultural Research Service.

Young, P. (2005) Real time flow forecasting. In *Proceedings of the International G-WADI Modelling Workshop* http://www.gwadi.org

Zorn, J. R. (1994) Estimating the population size of ancient settlements: methods, problems, solutions and a case study. *Bulletin of the American Schools of Oriental Research* **295**: 31–48.

19 A millennium of rainfall, settlement and water management at Humayma, southern Jordan, c. 2,050–1,150 BP (100 BC to AD 800)

Rebecca Foote, Andrew Wade, Mohammed El Bastawesy, John Peter Oleson and Steven Mithen

ABSTRACT

Humayma is an extensive archaeological site located in the hyper-arid region of southern Jordan. It was principally occupied between the Nabataean and early Islamic periods, c.100 BC to 800 AD, and provides evidence for the most complex water management system known outside of Petra. Since 1986 John Oleson has led extensive survey and excavation at Humayma, documenting the hydraulic structures and examining the various settlement remains that include a Nabataean cultic centre, Roman fort and early Islamic *qasr*. By making estimates of annual precipitation, surface runoff, water storage capacity and the extent of cultivated land, he has recently proposed that a total of 448 persons, 3,008 ovicaprids and 300 camels or donkeys could have been sustained within the Humayma catchment. This chapter draws on Oleson's research to develop a water balance model for Humayma to further explore the relationship between rainfall, runoff, storage capacity, agriculture and population levels. To do so, it reviews the textual and archaeological evidence for the settlement, undertakes an evaluation of the site's catchment using remote-sensing and digital elevation modelling, and then undertakes multiple simulations of the hydrological and hydraulic systems under a range of rainfall scenarios. Its results are in broad agreement with Oleson's estimates for past population levels, although various additional constraints on population are highlighted. Overall, a more thorough understanding of the hydrological dynamics of Humayma is developed.

19.1 INTRODUCTION

19.1.1 Research aims

Humayma is a settlement located in the northern uplands of the Hisma desert of southern Jordan (Oleson, 1997a). This is within the hyper-arid region, one that receives an average of not more than 80 mm of rainfall per annum (Figure 19.1). Although there are traces of an early prehistoric presence in the vicinity of Humayma, substantial occupation only began during the Nabataean period at c. 100 BC. Humayma then became a major settlement of the Roman, Byzantine and Islamic periods up until c. AD 800 (Figure 19.2). During this millennium of settlement, stock raising and farming were undertaken, while the settlement functioned as a regional market, a long-distance way station and a military post. All such activity was dependent upon a complex water management system that provided drinking water for humans and animals, sanitation, water for agriculture and a means for social display – a luxury pool and bath were constructed in this intensely arid location (Oleson, 1992). The archaeology of Humayma, with its numerous hydraulic structures, has been studied by John Oleson and his colleagues for more than two decades (Oleson, 1986, 1988, 1990, 1991, 1992, 1995, 1997a, 1997b, 2001a, 2001b, 2004, 2007a, 2007b, 2008, 2010; Oleson et al., 1993, 1995, 1999, 2002, 2003, 2008). This chapter seeks to further our understanding of Humayma by the use of hydrological modelling of the local catchment and water management system.

While the settlement of Humayma poses a large number of archaeological questions, one of the most basic and yet most challenging to address is the size of its past population. Oleson made estimates of annual precipitation, surface runoff, water storage capacity and the extent of cultivated land to suggest that a total of 447 persons, 3,008 ovicaprids and 300 camels or donkeys could have been sustained within the Humayma

Water, Life and Civilisation: Climate, Environment and Society in the Jordan Valley, ed. Steven Mithen and Emily Black. Published by Cambridge University Press.
© Steven Mithen and Emily Black 2011.

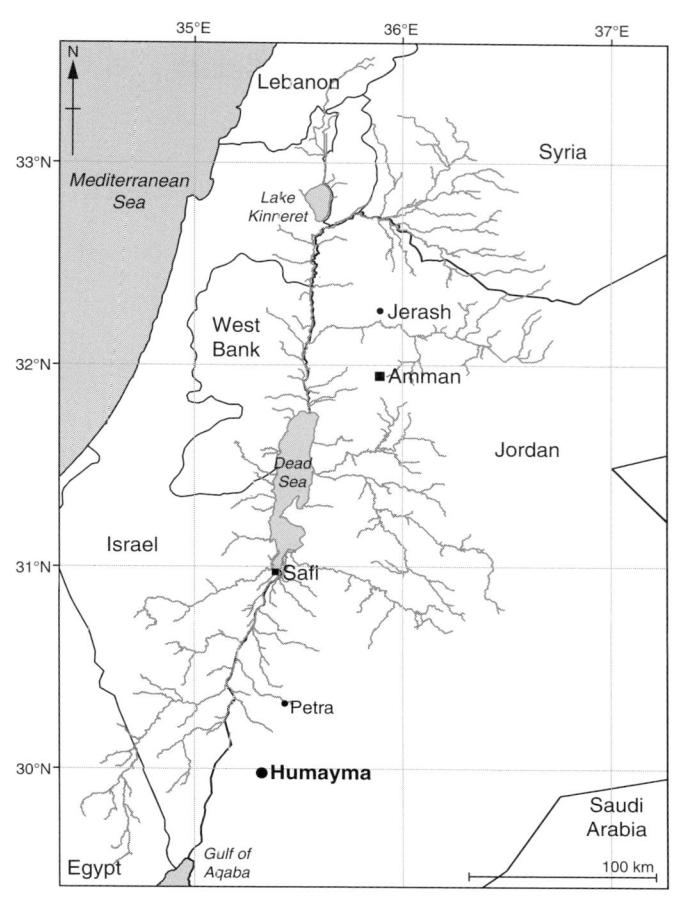

Figure 19.1. Location of Humayma in the hyper-arid zone of southern Jordan.

catchment. Our primary aim in this chapter is to evaluate these figures and to further understand the interaction between rainfall, runoff and storage capacity by developing a water balance model, similar to that used to examine the Bronze Age settlement of Jawa elsewhere in this volume (Chapter 18).

To develop this model, we combine the data and methods of archaeology, hydrogeology and hydrology, further developing the methodology used for the Water, Life and Civilisation study of Jawa (Whitehead *et al.*, 2008; this volume Chapter 19). Following a description of the environment setting of Humayma, section 19.2 summarises previous studies of the settlement concluding with a summary of the structural elements and phasing of its hydraulic system. In section 19.3 remote sensing and digital elevation modelling (DEM) propose a more detailed and accurate understanding of the Humayma catchment than has hitherto been available. More specifically, we compare a 1926 aerial photograph and a 2008 satellite image of the Humayma catchment to identify the rate and nature of erosion of the landscape. We then use this information to model the palaeo-landscape at the time of the settlement occupation at approximately 2,100 years ago. In section 19.4 a hydrological water balance model is incorporated within a Monte Carlo framework, the latter being used to capture the uncertainty in the model inputs and parameters. This model estimates the likely sustainable population at Humayma under different scenarios of rainfall and water storage.

Humayma (Nabataean Hawar/a, Greek Auara, Latin Hauarra) is located in the south of modern Jordan (E: 35° 20′ 43.44″, N: 29° 57′ 4.59″, *c.* 950 m asl), in the northern uplands of the Hisma Desert (Figure 19.1). The Hisma begins just below (south of) the Shara Mountains, which rise to nearly 1,700 m asl, and extends southward and eastward across the Jordanian border into northwestern Saudi Arabia. The Great Rift escarpment, rising to *c.* 1,200 m asl, frames the desert along the west, southward to the Gulf of Aqaba. Positioned in a unique and dramatic water catchment, with the exception of the smaller settlement of Wadi Ramm, Humayma is the only historic inland settlement of a significant size between Ma'an on the plateau north of and leading to the Shara Mountains (to the north) and the oasis of Tabuk (to the south, in Arabia).

Today, the Humayma environs receive about 80 mm of annual rainfall. Temperatures range from approximately 44 °C in summer to 10 °C in winter. The area is considered hyper-arid owing to the low rainfall and high evaporation caused by strong, dry wind and sunshine (Jordan Meteorological Department, 1971, Table II.01; Cordova, 2007).

The light, sandy and loessal alluvium in the basin south of the Shara Mountains and east of the Rift, which surrounds the Humayma settlement to the north, east and south, is surprisingly fertile for a desert. Oleson (1997b) determined that the key chemical properties of the local soils, such as alkalinity, salinity and soda content, are all within the acceptable levels to allow the farming of cereals and pulses, and possibly even grape, fig, almond and olive. The physical properties of the soil samples that affect moisture retention, such as the ratios of gravel, sand, silt and clay, ranged more widely, suggesting that water availability would have had a significant impact on agricultural potential within the catchment basin.

19.2 THE SETTLEMENT OF HUMAYMA

19.2.1 Textual information

Written historical references to Humayma are few, and those which shed light on the rainfall, water harvesting and uses of the water by the inhabitants are especially sparse. The Nabataeans of the third century BC Nabataea (before Humayma was founded) reportedly established subterranean reservoirs lined with plaster filled by precipitation runoff but did not construct houses or cultivate grains and fruit-bearing trees (*Diodorus* XIX.94.2–10 [Geer, 1954]). In contrast, reports of the late first century BC describe Nabataeans living in unfortified settlements with masonry homes, tending herds of sheep and cattle, and probably cultivating grape vines (described by Strabo – see Jones and Sterrett [translators], 1946–1963). The shift implies that

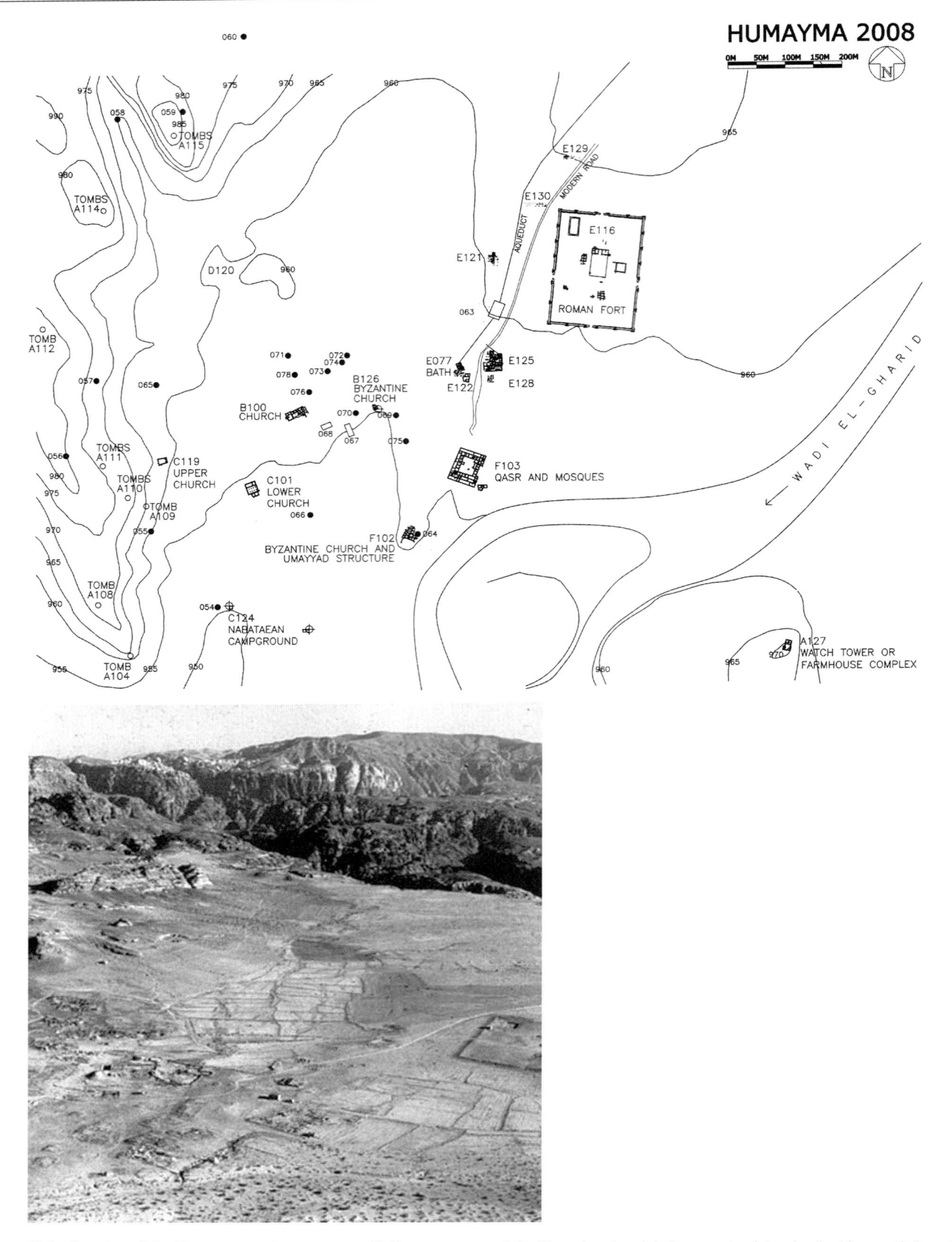

Figure 19.2. Site plan of the Humayma settlement centre (S. Fraser, courtesy J. P. Oleson) and aerial photograph of the site, looking north (courtesy D. Kennedy).

there was either higher rainfall to make agriculture possible in their mainly arid territory, or a more active harvesting and management of scarce water resources.

The establishment of Humayma is likely to have been a royal initiative by King Aretas III in the 80s BC (Ouranios Arabica, fragment A.1., FGrHIIIC, 340). In the mid-sixth century AD, Humayma was levied the second highest tax of all Transjordanian settlements under the *dux Palaestina*, which essentially covered southern Jordan and the Negev (Beersheba Edict = Di Segni, 2004). Because taxation during the Roman and Byzantine periods was primarily based on agricultural output, this implies that Humayma was second only to Udhruh in agricultural productivity. As such, it ranked above many locations on the plateau to the north that enjoyed higher rainfall and had subterranean water to tap, such as Ma'an and the equally arid but highly productive Negev Desert to the west.

During the early Islamic period, in the late seventh or early eighth century, the Abbasid family bought the Humayma village. 'Ali, the elder and head of the Abbasid family at the time, reportedly prayed twice in front of each of 500 olive trees there (Anonymous, 1971 [al-Duri and al-Muttalaabi; ed.] pp. 144–145). It is not clear whether he planted the grove or it pre-existed. The account is aimed at conveying 'Ali's religiosity and the number of trees mentioned is correlated to the number of prostrations he performed. Although some hyperbole may be involved in this account, there is no reason to think that the number of 500 with respect to the trees is an exaggeration or that this necessarily represents the total number at the site.

A large volume of travellers requiring food and water are reported as arriving at Humayma during 'Ali's tenure at the site, which probably would have included traders, soldiers, pilgrims and postmen. They were travelling north and south, between the Arabian Peninsula, where the religious hubs of Mecca and Medina were (and still are) located, and the Levant, where the political capital was located at Damascus. 'Ali received them hospitably, but required a tremendous quantity of supplies (Anonymous 1971 [al-Duri and al-Muttalaabi, ed.] p. 142). A time when many travellers would have passed through southern Jordan during the Islamic and earlier periods was between February and May, particularly traders transporting goods from the south, carried from the Indian subcontinent on the monsoon winds (Groom, 1981, pp. 146–148). This was during the rainy season, and hence water resources could have been drawn down from the storage structures at Humayma (and sites along other routes) at a time when they would have been refilled by runoff.

Humayma was noted by only a few of the European and North American travellers to Transjordan during the late Ottoman period, British Mandate and early years after independence (e.g. Laborde, 1830; Maughan, 1874; Musil, 1926; Frank, 1934; Alt, 1936). Historically, north–south travel between Petra and Aqaba via Humayma was not the easiest route. Even if choosing to journey up the Wadi Yitm and across the Shara Mountains, travellers often went through the Hisma east of Humayma, crossing between Guwayra and Naqb Ishtar, which was more direct and offered spring water (e.g. Morris, 1842; Glueck, 1935). The information offered by such travellers and by visitors to the site itself during the nineteenth and early twentieth centuries is quite useful with respect to understanding land use. Importantly, only some who passed through the area in springtime observed agricultural and pastoral activity in the Humayma environs. Variations in rainfall and inter-tribal territorial disputes appear to have affected the extent of settlement. Tribal affairs also affected the ability to travel through the Hisma (for extracts of travellers' reports and further discussion, see Oleson, 2010, Chapter I).

19.2.2 Archaeological research

It was not until the mid-1970s that Humayma received archaeological attention. John P. Oleson initially joined a research project led by John Eadie and David Graf because of his expertise in Greco-Roman hydraulic systems. Ultimately Oleson directed fieldwork at the site himself, leading a small team for the extensive Humayma Hydraulic Survey between 1986–89 (Oleson, 1986, 1988, 1990, 1991, 1992, 1995, 2008, 2010, Oleson et al., 1995), designating each hydraulic feature with a unique identifying number (Figure 19.3). He began systematic excavation of the habitation centre in 1991 in order to further understand the water-supply system and the historical development of the ancient settlement. Extensive excavations have identified and examined many aspects of the ancient settlement, including built drainage systems in the Nabataean cultic centre, and the Roman civilian settlement (E125), the Roman fort (E116), the early Islamic *qasr* (F103) and many phases of the bath complex (E077). For preliminary reports, topical studies and synthetic articles, see work by Oleson et al. (Oleson et al., 1993, 1995, 1999, 2003, 2008; Oleson, 1997b, 2001a, 2004, 2007b).

Large covered reservoirs (Figure 19.4, Figure 19.5) appear to have been placed at the centre of the settlement from the start. Indeed, a hydraulic system must have been essential to the original planning of the settlement. Oleson argues that the effort and expense to collect and store drinking water indicate that the reservoirs were a project of the Nabataean monarchy for public access to drinking water. The construction of the aqueduct (Figure 19.6) and pool at its terminus further suggest ostentatious display of governmental/royal control over the precious resource in such an arid environment.

Within its first few hundred years, Humayma grew to cover 10 hectares containing military, religious/cultic, residential, recreational and industrial buildings (Figure 19.2). Most of the hydraulic system documented by Oleson appears to have been established from the earliest Nabataean occupation and then continued in use throughout the Roman, Byzantine and early Islamic periods.

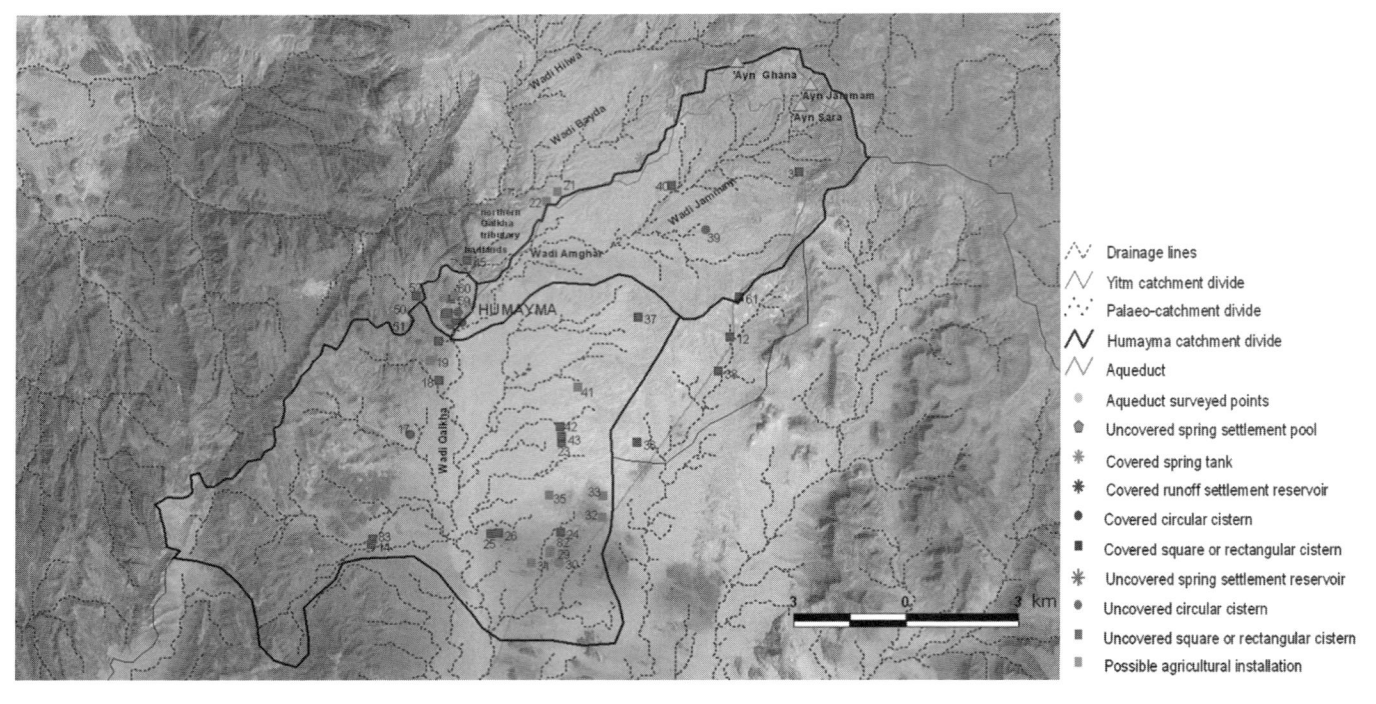

Figure 19.3. Humayma environs showing locations of pertinent hydraulic structures surveyed in the Humayma Hydraulic Survey. See colour plate section.

Figure 19.4. Reservoir at Humayma (after modern restoration, originally covered). (Courtesy E. De Bruijn).

Figure 19.5. Covered reservoir at Humayma (R. Foote).

The scale of occupation appears to have changed significantly during the early Islamic period. Although many existing buildings were modified, no new structures dating between the eighth century and the Ottoman peroid (sixteenth–eighteenth centuries) have been identified. Within the memory of the Bedouin descendants, the inhabitants lived mostly in tents until 1979, after which the local Bedouin community relocated to Humayma al-Jadida/ New Humayma, 8 km to the southeast astride the Desert Highway.

The initial investment and then growth of Humayma related to its key location regarding regional and longer-distance travel, including trade and military activity. The site was located on the so-called King's Highway, which later became the Via Nova Traiana, a main historic north–south route from Syria through Jordan to Aqaba, and on to Arabia. Traces of Roman-period paving near the site were still extant until the end of the last century, and several associated milestones survive (Graf, 1983). The Wadi Humayma also provides a natural direct route westward from the site to the Wadi Araba, through the Great Rift highlands.

19.2.3 Agricultural viability and archaeobotanical evidence

Another inference of the Humayma Hydraulic Survey is that agriculture was viable and pursued in the Jammam-Amghar

Figure 19.6. Section of the aqueduct at Humayma (J. P. Oleson).

and the Qalkha environs but not in the wadi system east of the Disi sandstone. Although three cisterns were located east of the Disi, the terrain is not suitable for cultivation (Figure 19.3, nos. 12, 36, 38).

A Landsat™ satellite image that covers the Wadi Yitm (in which the Humayma sub-catchments lie) enabled us to trace alluvial deposits beyond what is actually cultivated today in order to consider potential historic areas for cultivation. This identified three areas suitable for agriculture:

1. Jammam-Amghar – a 62.4 km^2 area of which actual cultivated land is 470,000 m^2 and potential alluvial land is 3,850,000 m^2;
2. North Qalkha – 7.1 km^2 palaeo- and 3 km^2 present-day areas of which actual cultivated land is 43,900 m^2 and potential alluvial land is/was 834,000 m^2; and
3. South Qalkha – a 132.4 km^2 area of which actual cultivated land is 109,000 m^2 and potential alluvial land is 27,260,000 m^2.

Archaeobotanical evidence from excavations by Oleson *et al.* suggests that the crops grown intensively in the catchment during the Late Roman, Byzantine and early Islamic periods were six-rowed barley (*Hordeum vulgare* L.), bread wheat (*Triticum aestivum* L.) and lentil (*Lens culinaris* Medik) (Ramsey, 2004). The presence of hydrophilic (water-loving) weed species in the post Nabataean-period botanical samples suggests either that more water was harvested and concentrated to saturate field areas during the later periods or that rainfall was higher than it is today.

Millets (foxtail (*Setaria* sp. [L.] Beauv.) and broomcorn (*Panicum* sp. L.)), as well as rye (*Secale* sp. L.) were cultivated during the Byzantine and early Islamic period, possibly being introduced from Africa (Ramsey, 2004, pp. 12–13). Fig, olive, date and grape are among the fruits found in the archaeobotanical remains. While all could have been imports, the interpretation of a structure (No. 51) as a wine press suggests that at least grapes may have been locally grown.

The Water, Life and Civilisation project collected and analysed phytolith samples from the catchment field north of the Humayma settlement (see this volume, Chapter 23). The phytoliths are mostly those that form in the leaves of dicotyledons (trees and shrubs), suggesting cultivation of a tree crop, possibly olives or figs. Neither have been grown extensively at Humayma in the memory of the local Bedouin and the phytolith samples had come from a sufficient depth to be of a historic nature. Two samples yielded sponge spicules that suggest the fields had been either moist from heavy rainfall or inundated through water harvesting.

Soil sampling by the Humayma Hydraulic Survey within the three agricultural areas revealed similar chemical properties, indicating the potential for arable farming. There were, however, different physical properties: the North Qalkha had the highest clay content, and therefore the best ability to retain moisture; Jammam-Amghar had above average moisture retention, while that of the South Qalkha was quite low (Oleson, 1997a). Yet, with the significant surface water naturally passing across the slopes of the vast catchment, the entire Humayma basin appears as a prime site for agricultural activity. Water naturally collects in certain lower locales, which substantially increases the real volume of water available for crops. Infiltration rates of water into the soil can be enhanced by the construction of artificial terraces on gentle slopes, the building of barriers in tributary wadis, and collection of stones in piles (see Evenari *et al.*, 1982, pp. 95–119, for experiments in the Negev that used such methods to maximise water availability).

These types of structures were used to create fields in each of the three areas – in the Jammam-Amghar, North Qalkha and South Qalkha – as documented within Oleson's Humayma Hydraulic Survey (e.g. Figure 19.7). Such structures are, however, more ephemeral than other elements of the hydraulic system and there were more than those documented in the survey. There are, for instance, at least two sets of wadi barriers in the North Qalkha clearly visible in a 1926 aerial photo in the David Kennedy archive (Figure 19.8), but which were not

Figure 19.7. Terraced hillside at Humayma (J. P. Oleson).

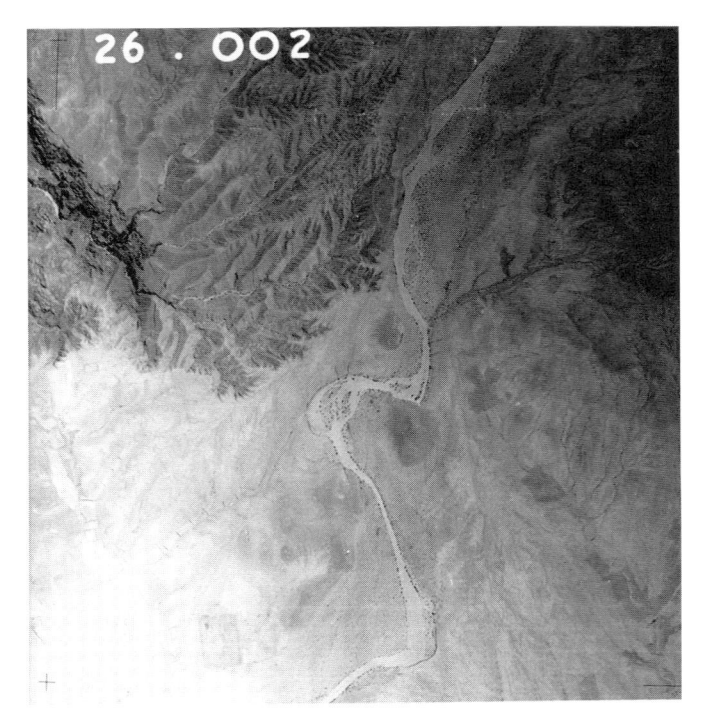

Figure 19.8. Aerial photograph of the area north and east of the Humayma Settlement Centre (courtesy D. Kennedy archive, 26.002).

detected in the 1980s Humayma Hydraulic Survey. The fact that traces of these barriers had disappeared by the time that Oleson was conducting his survey underscores their ephemeral nature and points to the possibility of a more extensive water harvesting regime than now recorded in the archaeological record.

19.2.4 The Humayma water catchment and hydraulic system

The Nabataeans and later inhabitants prospered at Humayma primarily because it is located in an exceptionally large natural water drainage area within the arid zone of the Hisma (Figure 19.3; see section 19.3 for full discussion of the sub-catchments). Its elevation at $c.$ 950 m is $c.$ 750 m lower than the highest peak of the Shara Mountains just to the north and $c.$ 250 m lower than the Great Rift escarpment along the west.

The presence of the highlands on two sides and the vast depression below, measuring $c.$ 200–240 km^2, create an extraordinary rainfall watershed and collection basin.

The Humayma Hydraulic Survey provided evidence for the harvesting of seasonal runoff water across the slopes and in the drainage networks of four wadi sub-catchments. The settlement centre itself was established at the conflux of the two major wadi systems, just above where the Jammam-Amghar flows into the Qalkha, which is just east of the Rift escarpment. This prime location enabled the North Qalkha and western slopes to supply the significant amount of water required to fill the large reservoirs within and just north of the settlement.

The area exploited by the inhabitants of Humayma offers more than surface water. Subterranean sources to tap were located along the Shara scarp, $c.$ 14 km north and northeast of the settlement centre. An aqueduct network nearly 27 km long was constructed to deliver water from three of the four springs from the scarp to the settlement (Figure 19.6). One major limitation of the settlement location, however, is that the water table is so deep south of the scarp that wells could not offer additional supply.

The Humayma Hydraulic Survey concluded that most of the hydraulic system is Nabataean in origin but continued in use at least into the eighth century AD. A few individual hydraulic structures were modified over time, and a few new structures were added after the Nabataean foundation. At its greatest extent this system included the following: (see Figure 19.3 and Table 19.1, for details on features related to drinking water storage):

- Three springs ('Ayn Qana, 'Ayn Jammam and 'Ayn Sara) supplying:
  - 26.508 km of aqueduct armature. This had two main branches, Qana and Jammam, with 'Ayn Sara augmenting the Jammam branch;
  - two off-take tank cisterns drawing water from the Qana conduit line approximately 10 km before it entered the settlement centre. One is unnumbered at the 6.512 km point from 'Ayn Qana, and the other (No. 20) is at the 9.597 km point from 'Ayn Qana. Neither is roofed;
  - one unroofed ornamental pool in the settlement centre (No. 63);
  - one bathing complex in the settlement (Field E077);
  - one reservoir in the Roman fort, built in the second century AD at the northeast fringe of the settlement (No. 62). This was unroofed and supplied by a conduit spur off the aqueduct. Its water was used for multiple purposes in the fort.
- Precipitation runoff filling:
  - one roofed reservoir outside the settlement centre (No. 53);
  - forty-one (if counting each in the set of seven at No. 33) cisterns outside the settlement centre (23 were

Table 19.1 *Humayma water storage features, divided by sub-catchment*

Structure number & runoff sub-catchment	Covered/ uncovered (c/u)	Type	Length or diameter (in m)	Width (in m)	Depth (in m)	Surface area (m^2)	Capacity (m^3)	Capacity used (m^3)	Micro catchment area (ha)
JAMMAM/AMGHAR northeast and east of the settlement									
21	u	cistern	1.28	1.25	0.4	1.6	0.6	0.6	17.01
22	u	cistern	2.98	1.28	0.94	3.8	3.6	3.6	8.1
3	c	cistern	c. 4	c. 3	>3.27		>39.2	39.2	17.01
39	c	cistern	may be circular and have a diameter of at least 3		4		28.3	28.3	39.69
40	c	cistern	3.2	3.2	2.5		25.6	25.6	102.87
61	c	cistern	5.9	5.4	4.3		137	137	1.62
64	c	cistern	c. 3.8		>4.50, >3.80 below overflow		>43	43	25.92
							5.4144 (total for sub-catchment)		
QALKHA north of the settlement PALAEO									
45	c	cistern	7.9	4	3.9		123.2	123.2	
QALKHA north/north west of the settlement and Qalkha in the settlement, filled from the north and west runoff									
52	c	cistern	8.42	5.64	4.28	47.5	203.3	203.3	4.05
53	c	reservoir	17.53	6.54	4.5		515.9, stairs	515.9	1.62
58	c	cistern	26.8	23	?		Unknown		1.62
59	c	cistern	5.2	5	3.75		97.5	97.5	26.73
60	u	cistern	N side: 4.75	W side: 2.34	1.65	11.1	18.3	18.3	3.24
67	c	reservoir	19.74	7.04	depth 3.83; original floor possibly at the level of 3.75 below the top of the wall, and extended downward either in antiquity or during the modern reno.		If so, the original volume would have been 521.14, while the present volume is 730.84	486.4	144.18
68	c	reservoir	20.05	6.95–7.05	below top of wall 3.83		537.5	488.4	0.81
57	c	cistern	4	3.95	>3.50		55.3	55.3	1.62
65	c	cistern	4.12	3.87	>1.3		>20.7	20.7	1.62
66	c	cistern	circular: assuming 7.4		assuming a depth of at least 5		the capacity of the cistern would be at least 215	215	176.58
69	c	cistern	circular: c. 7.0		>5.92		>227.8	227.8	144.18
70	c	cistern	circular: >5.0		>4.02		>78.9	78.9	144.18

Table 19.1 (cont.)

Structure number & runoff sub-catchment	Covered/ uncovered (c/u)	Type	Length or diameter (in m)	Width (in m)	Depth (in m)	Surface area (m²)	Capacity (m³)	Capacity used (m³)	Micro catchment area (ha)
71	c	cistern	appears to be circular: >5.0		>4.25		>83.5	83.5	31.59
72		cistern or grain storage	circular: 5.5		>3.90		>92.7	92.7	0.81
73		cistern or grain storage	circular: 4.9		assumed >3.90		>73.5	73.5	31.59
74	c	cistern	circular: 5.1		>4.54		>92.7	92.7	31.59
54	c	cistern	circular: 5.80–5.83		6.25		166	166	4.05
55	c	cistern	irregular rectangular chamber: c. 8 to 10 m	5.2	6.65		276.6–345.8, catalogued as 300	300	0.81
56	u	cistern	6.5	5.95	1.6	38.7	61.9	61.9	0.81
							97.3 (total for sub-catchment)		
QALKHA catchment, south of the settlement									
14	c	cistern	8.65	3.88	>3.32		>111.4	111.4	3.24
17	c	cistern or grain storage	oval opening: 1.93 × 1.15; provides access to a roughly spherical storage chamber with a D of c. 3		depth?	7.1	c. 10.6	10.6	1.62
18	c	cistern	10.95	5.7	>2.75		>171.6	171.6	8.91
19		cistern	Two of them are approximately 3. The last pothole larger and more regular: 5	Two @ 2 4	Two have a depth of 1 and may preserve traces of trimming or enlargement; Largest is 2		c. 5; c. 17.9; c. 35 in all? Surface area = 65.188 047 56 (but not used)	35	26.73
23	c	cistern	4.62	4.63	>3.83		>81.9	81.9	18.63
25	c	cistern	13.7 (N wall), 11.15 (S wall) AVR: 12.42	3.65 (E wall), 5.4 (W wall) AVR: 4.53	>4.24		>237.9, stairs	237.9	1.62
26	c	cistern	at least 7.5	?	assuming a height of 10 ft., 3.05		approx. 153	153	0.81
35	u	cistern	>4	2	undetermined	8	>16	16	1.62
37	c	cistern	4.95	3.85	2.6		49.6	49.6	1.62
41	u	cistern	5.7	3.23	3.9	18.4	71.8	71.8	9.72

Table 19.1 (cont.)

Structure number & runoff sub-catchment	Covered/ uncovered (c/u)	Type	Length or diameter (in m)	Width (in m)	Depth (in m)	Surface area (m²)	Capacity (m³)	Capacity used (m³)	Micro catchment area (ha)
42	c	cistern	6	3	4		72	72	0.81
43	c	cistern	3.76	2.8	2.44		25.7	25.7	3.24
48	c	cistern	c. 8	c. 5	>4		>160	160	18.63
83	c	cistern	3.94	3.5	>3.6		49.6	49.6	4.05
24	c	cistern	11.2	6.2	>4		>277.8	277.8	1.62
29	u	cistern	11.55	2.02	2.7	23.3	63	63	28.35
30	u	cistern	10	1.55	2.1	15.5	32.6	32.6	0.81
31	u	cistern	19	0.85–2.5	4.4	31.8	140	140	0.81
82	u	cistern	very small: 1.30	0.83	>0.30	1.1	>0.3	0.3	10.53
32	u	cistern	5.65	3.35	4	4.4	75.7	75.5	2.43
33	u	cistern	9.95 7.30 10.35 8.90 (E)/ 10.18 (W) 7.59 12.17 6.61	3.90 3.10 2.90 2.53 (S)/ 3.70 (N) 3.67 1.97 6.70	4 >1.5 4.40 3.50 3.64 >1.0 4.44	215.3	155 >34 123.1 c. 104.5 101.4 >24 196.6 total of the group: 738.6	738.6	2.43
34	c	cistern	circular: c. 3		4		28.3	28.3	0
50	u	cistern	2.98	2.45	1.05	7.3	7.7, stairs	7.7	0.81
51		cistern or wine press	tank 1 (L 2.36; W 2.30; depth 0.50; cap. 2.7 m³); tank 2 (L 0.88; W 0.75; depth 0.75; cap. 0.50 m³)					3.2	9.72
44		dammed pool				380		300	
27		walled pool	20	20	2.5			1000	3.5
28		walled pool	26		2.3			1069	4
						717 (total for sub-catchment)			
East of the Disi sandstone ridges/outside the Humayma catchment proper									
12	c	cistern	6.7	7.15	even 5 below the present doorsill		would give a capacity of 240, stairs	240	1.62
36	c	cistern × 2	7.4 7.3	52.55	>3.2 3.5		>118.4 61.7	180.1	10.53
38	c	cistern	25	c. 1.7	6	c. 255; traces of stairs		255	1.62

roofed [Nos. 3, 12, 14, 18, 23, 24, 25, 26, 34, 36.1, 36.2, 37, 38, 39, 40, 42, 43, 45, 48, 52, 59, 61, 83]; 18 were possibly unroofed [Nos. 21, 22, 29, 30, 31, 32, 33.1, 33.2, 33.3, 33.4, 33.5, 33.6, 33.7, 35, 41, 50, 60, 82]) (Nos. 51.1 and 51.2 are probably a grape press and related activity, and too little is known to include No. 58);

- two roofed reservoirs in the settlement (Nos. 67 and 68);
- eleven cisterns in the settlement (10 roofed [Nos. 54, 55, 57, 64, 65, 66, 69, 70, 71, 74]; one unroofed [No. 56]). (Two further structures, Nos. 72 and 73, are unlikely to have been used for water storage);
- one impoundment dam (No. 44) and two barrier walls situated in rock clefts creating dammed pools (Nos. 27, 28); enhanced natural pools including a modified and roofed pothole (No. 17), seven unroofed potholes (No. 19), and several unroofed potholes (near No. 44 and No. 39) outside the settlement area, most likely providing water for animals;
- three sets of wadi barrier areas of terracing, outside the settlement (Nos. 5, 16, 79);
- eight sets of artificial terraces or cleared fields in which stones were gathered in piles, outside the settlement (Nos. 8, 9, 15, 46, 47, 49, 80, 81).

19.2.5 Exclusions from stored drinking water capacity

To develop the water balance model described below we have for the most part included the hydraulic structures as recorded by Oleson and described in the above list. In some cases, however, we re-evaluated the archaeological evidence to assess the likelihood that the structures listed above had indeed been used for water storage.

Three structures were excluded from the runoff drinking water storage capacity: Nos. 51.1, 51.2 and 19. The first of these is most likely a grape press and the second constructed for a related activity. Structure No. 19 is a set of seven shallow potholes, unroofed, with a fairly small total volume $c.$ 35 m^3. Six of the seven are 1.0 m deep or less. We are concerned that the combined exposed surface area could skew the figure used as input to the model for evaporation loss from total uncovered surface area.

We also excluded from the spring flow storage capacity the two tanks drawing water from the aqueduct: the unnumbered one at the 6.512 km point from 'Ayn Qana (44.5 m^3) and No. 20, at the 9.597 km point from 'Ayn Qana (40.7 m^3). Their locations in surroundings not easily inhabited and far from the settlement centre, 12.3 km and 9.2 km, respectively, as well as the fact that they are shallow, 0.72 m and 0.80 m deep, respectively, filled by means of sluice gates off the aqueduct, drained by means of a drain, and not roofed, suggest only periodic and temporary filling for campers or passers-by, who may have chosen this route for travel across the Shara scarp. It would be difficult, therefore, to assess how often these tanks were filled on an annual basis. Nonetheless, because filling could be frequent when seasonal travel or nearby pasturing was heavy, and possibly used by inhabitants of the settlement, we may include the tanks in a future model run to see the effect. This was not done in this study, to limit the number of scenarios for first inspection.

Finally we should note that cistern No. 34 is not plotted on our maps, because its specific location was not recorded. It is nonetheless included in calculations for water storage.

19.2.6 Volume stored in drinking water hydraulic structures

The hydrological model that we develop for Humayma inputs stored water as fixed numbers, differentiated by covered and uncovered with an evaporation rate applied to the uncovered. We also consider a scenario in which the cisterns are partially refilled within a single year, assuming that water is being drawn from them for drinking during the rainy season. For this we chose the figure of 25% of the total capacity as a plausible seasonal top up increase. In other words, a cistern may have supplied a total of 125% of its total capacity during any one year. This scenario contrasts with Oleson's view that cisterns were only filled to their maximum capacity every two years because of given interannual rainfall variability.

19.2.7 Hydrological sub-phases

Combining information from the Humayma Hydraulic Survey and the Humayma Excavation Project, three hydrological phases can be identified:

In the initial phase, Hydrological Phase I, the three springs on the Shara escarpment $c.$ 14 km to the north ('Ayn Qana, 'Ayn Jammam and 'Ayn Sara) supplied fresh water via an aqueduct to a large shallow pool (No. 63) in the settlement centre, which probably and importantly was not a reservoir for drinking water. The pool overflowed and discharged freely through a pipe that probably supplied a building thought to be a Nabataean bath or shrine (beneath the later Roman bath E077). The two off tank cisterns (No. 20 and unnumbered) along the aqueduct north of settlement were most likely only periodically and temporarily filled with the perennial flow. The reservoirs in the settlement (Nos. 67 and 68) and outside (No. 53), and all cisterns both in the settlement and outside, were filled by precipitation runoff. The many presumably private and mostly rock-cut cisterns may have been added to over time, or gone in and out of use, but most are not possible to date. For simplicity all are considered to be of this initial phase. The same is assumed for the impoundment dam, barrier walls and enhanced natural pools. It is unclear how much local cultivation was carried out. Most likely barley, wheat and lentil were farmed to a limited extent. Other activities at the site during Phase I include Nabataean communities living in a housing and shrine complex (E125) south of the pool (No. 63) and northeast of the reservoirs, and in a campsite to the south-southwest (C124).

With regard to estimating the capacity for drinking water, for both human and animal need, Phase I provided:

- Three large reservoirs filled by runoff (two in the settlement centre Nos. 67 and 68, one outside No. 53).
- Fifty-two cisterns filled by runoff (11 in the settlement, 41 outside).
- One impoundment dam and two barrier walls creating open pools of water, all collecting water, most likely for livestock.
- Several enhanced natural pools (at least Nos. 17 and 19, the later not included in calculations).
- Two cistern off tanks, No. 20 and unnumbered, likely filled only periodically by the spring water flowing to the settlement through the aqueduct and located well north of the settlement (neither figuring in the calculations).

With regard to agriculture, Phase I involved field areas created by barriers in tributary wadis, field areas created by terracing slopes and cleared fields in which stones were gathered in piles in the Jammam-Amghar, North Qalkha and South Qalkha sub-catchments.

Water also appears to have been used for social purposes in terms of the pool (No. 63) that received the aqueduct flow. Discharge from the pool may have been used for a bath or shrine (under E077) and/or a surrounding garden. Water may also be involved at the early cultic centre/shrine/temple (the south end of E125) in the later part of this phase.

During Hydrological Phase II the Nabataean cultic centre developed into a Roman civilian settlement (*vicus*) (E125). When the Roman fort (E116) was constructed *c*. 100–150 years after the aqueduct was built (i.e. early second century AD), a spur conduit was constructed off the aqueduct to draw some of the spring water from the pool into a reservoir built within the fort (No. 62). It was designed from the beginning as an unroofed storage tank that may have had some flow-through capacity. As such, the Phase II capacity for drinking water was the same as Phase I plus this reservoir (No. 62) in the Roman fort (E116). Probably in this same phase, the pool (No. 63) was plugged by a pressurised pipe and became the tank to supply the new Roman bath constructed on the site of perhaps an earlier bath or a shrine (E077). All other components in the system, the reservoirs, cisterns, dams and various techniques to concentrate runoff in field areas, continued to function.

The presence of hydrophilic (water-loving) weed species in the post Nabataean-period botanical samples suggests that more water was harvested and concentrated to saturate field areas during the later periods of Phase II. Terracing areas and techniques probably expanded by the end of this phase. But to reiterate a point previously made, these features are constantly in flux and we do not have a full sequence documented.

During Hydrological Phase III arrangements were made to intercept the spring flow into the pool *cum* bath tank with a pipeline, which carried some of the water to a destination farther along the ridge. The bath complex (E077) itself was expanded through the late Byzantine/early Islamic periods. All other components in the system – reservoirs, cisterns, dams and various terracing techniques – continued to function as in Phase II. Cultivation expanded and more water infiltrated the fields. Rye and millet were introduced, possibly from Africa.

Although there was limited change in the hydraulic system, there were significant changes at the settlement. The Roman fort (E116) was abandoned. Christianity spread to Humayma, expressed in the building of four churches scattered within the settlement centre and up the lower slope of the western escarpment (B100, B126, C101, C119). A large residence with a small mosque complex was built at the southeastern fringe of the settlement (F103). A new church was constructed nearby (F102) and there was a great deal of refashioning of existing/earlier buildings throughout the site. It is likely that additional nodes for water use to enable Christian baptism and ablutions before Muslim prayer were added to the system.

19.3 HYDROGEOLOGICAL REMOTE SENSING AND DIGITAL ELEVATION MODELLING

19.3.1 Introduction to the Humayma water catchment

The geology and geomorphology of the northern Yitm basin, including the Humayma environs (i.e. the area between the Shara Mountains to the northeast and the village of Guwayra to the south of the site) have been previously studied (e.g. Osborn and Dunford, 1981). In short, the Precambrian igneous rocks, aplite-granite, granodiorite and quartz diorite, form the mountain ridges west of the Guwayra settlement and appear in a few places around the site of Humayma. The Palaeozoic (Cambrian to Silurian) sandstone hills and inselbergs around Humayma overlay these basement rocks and are capped by Upper Cretaceous (Cenomanian) limestones at the Shara escarpment. The wadis and flood plains are covered by alluvial and aeolian deposits of sandy loam and loess, which were cultivated in antiquity and are so today. The surrounding sandstone delivering great runoff to the areas of soil is quite impermeable. While the softer Qa Disi

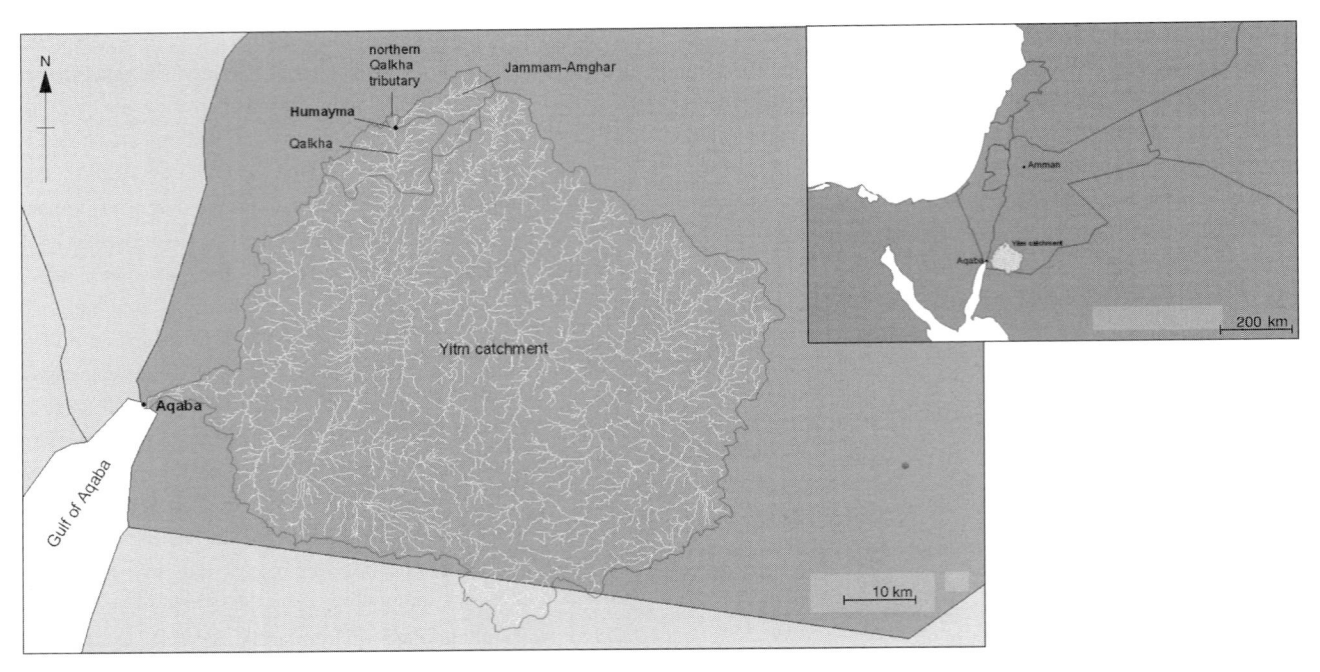

Figure 19.9. The Wadi Yitm catchment in southern Jordan (M. El Bastawesy).

formation to the east still carries runoff, it is often in channels naturally cut in the rock.

The water catchment for the settlement of Humayma consists of two principal wadi systems: the Jammam-Amghar, to the northeast and east of the settlement, measuring 62.4 km^2, and the Qalkha, mostly to the south and southeast, measuring 132.4 km^2. A third but less significant system is located to the north. Both of the former wadi systems are drainage sub-catchments of the vast Wadi Yitm catchment, which covers an area of 4,545 km^2 east of the Great Rift escarpment and flows southward into the Gulf of Aqaba (Figure 19.9). The Jammam-Amghar sub-catchment is located at the northern tip of the Wadi Yitm catchment and is bounded on the northeast by the Shara escarpment, from which it gains the seasonal rain runoff from some of Jordan's highest peaks. The Wadis Bayda and Hilwa bound the Humayma sub-catchments at the northwest.

Quaternary fluvial and lakebed deposits have sustained agriculture activity in the Humayma catchment historically and at present. These deposits also form the badlands, a barren landscape with deep gullies and ravines, to the north of the settlement (Figure 19.10). The badlands are separated from the Jammam-Amghar wadi system to the east by a low ridge of soft fluvial sediments, which is only a few tens of metres across in some places. Importantly, the Quaternary lakebed deposits cap some of the inselbergs to the west of the badlands, clearly indicating that they have been separated from the badlands and the rest of the Humayma catchment to the east. These inselbergs form part of the Wadi Araba drainage system that eventually empties into the Gulf of Aqaba. Clearly the Quaternary lakebed deposits originally covered a large area before the separation, and when the break occurred a considerable area of these deposits was captured into the Wadi Araba/Great Rift system. This separation of Quaternary deposits began in the Quaternary, when a deep wadi/canyon formed in the western section of the lakebed and an area vulnerable to erosion and further cutting eastward – what became the badlands – formed at the eastern terminus of the new wadi/canyon. The separation may have been the result of tectonic activity, where the wadis of the badlands and inselbergs areas are structurally controlled and follow the fault trends in the Rift system (NE–SW, NW–SE and E–W).

The original northernmost headwaters of the northern tributary of the Wadi Qalkha, which is located north of the Humayma settlement, have been beheaded from the Qalkha system by the badlands. The Wadi Qalkha now drains into the deep wadi/canyon flowing to the west, which ultimately joins the Wadi Araba catchment. The original headwaters, current badlands and deep east–west flowing wadi/canyon form an independent sub-catchment of the Wadi Araba, created between the Qalkha (to the south), Bayda (to the north) and Jammam-Amghar (to the east). Prior to the formation of the badlands, greater seasonal rain runoff from the north travelled through the Humayma settlement. Today, only a small part of this northern tributary remains north of the settlement, connected to the Qalkha network south of the site and hence to the Wadi Yitm catchment. The original northernmost headwater area has become disconnected by the intervening badlands.

As stated above, the principal wadi bed of the Jammam-Amghar network is separated from the badlands by a low ridge

Figure 19.10. The badlands at Humayma (R. Foote).

of soft alluvium deposits, in places only 32 m across at present. This ridge is subjected to intermittent fluvial erosion from both sides. When it is fully eroded, most of the Jammam-Amghar runoff will drain into the Wadi Araba (Eadie and Oleson, 1986). This will radically change the patterns of flow in the Humayma catchment, also significantly affecting the upper Yitm flow.

The objective of the Water, Life and Civilisation hydrogeological study was to evaluate the losses to the low ridge of soft alluvium deposits from the badlands to the west and potential future losses from there and along the Wadi Amghar to the east by:

1. estimating a rate of erosion of the alluvium deposits in the badlands in order to calculate how much may have eroded since the Nabataeans established the water management system in the Humayma catchment during the first century BC; and
2. assessing erosion in the Wadi Amghar.

Archaeological information documented in the Humayma Hydraulic Survey 1986–89 was combined with hydrogeological information both documented in previous studies and observed by comparing a 1926 aerial photo (Figure 19.8) with a WorldView 1 Panchromatic satellite image taken on 15 May 2008 (Figure 19.11). This enabled us to determine when the badlands may have cut the northern Qalkha tributary headwaters from the Humayma catchment network, and to evaluate changes to the main channel of the Wadi Amghar.

19.3.2 Erosion of the local Quaternary lakebed

Figure 19.11 is a satellite image of the Humayma catchment taken on 15 May 2008 (WorldView 1 Panchromatic satellite). This clearly shows that the tributary network north of the badlands, the badlands and the remaining trunk of the northern Qalkha tributary south of the badlands/north of the settlement were part of a continuous landmass prior to the deep canyon and badlands development.

Several stretches of the Nabataean-period aqueduct have been lost in landslips, rockslides and gully erosions. The Nabataean engineers apparently anticipated trouble from erosion all along its course. Whenever it was necessary for the aqueduct to cross a gully they constructed solid bridging structures (Oleson, 2010, Chapter III). Many of the bridges have completely collapsed since antiquity, their stone blocks having been scattered down the slopes by runoff and associated erosion. The aqueduct was still intact in the early Islamic period, but it is uncertain how much longer thereafter it functioned. The well-preserved parts of the aqueduct are mostly in the northern area of the Jammam-Amghar sub-catchment, where it lies directly on sandstone or marl bedrock providing significant support against slides and gullying. In the southern area, starting 13 km distance from 'Ayn Qana, the aqueduct was directed whenever possible away from the most unstable and vulnerable area of the northern Qalkha tributary. This was an already eroding region, ultimately becoming the present-day badlands.

Merging Oleson's surveyed points of the aqueduct (Oleson, 2010, Chapter III) with our delineated Jammam-Amghar catchment outline (from the SRTM DEM, as described below) it is clear that the aqueduct was originally located almost wholly within the Jammam-Amghar catchment (Figure 19.12). A few stretches of the aqueduct have slipped into the badlands indicating that the Quaternary lake deposits onto which the aqueduct was constructed have suffered severe erosion since the Nabataean period, and some considerable parts of the Jammam-Amghar catchment area have been lost by erosion and gulying into the Wadi Bayda and the badlands sub-catchments.

19.3.3 Data and methods

Our study created GIS layers for each of the hydrological structures documented in Oleson's survey and then integrated the layers with remote sensing images in order to simulate, model and interpret the palaeohydrology and palaeogeography of Humayma.

The Shuttle Radar Topography Mission (SRTM) DEM was used to derive the current catchment area and drainage networks following the widely used D-8 method embedded in Environmental System Research Institute (ESRI) Geographic Information Systems (GIS) software (Jenson and Domingue, 1988) (Figure 19.13, left). This method is based on the following: first, the DEM is filled to eliminate the spurious sinks and to obtain a continuity of flow direction between the cells. The flow direction is calculated for each cell in the DEM based on the highest drop of elevation into one of the neighbouring eight cells (hence the

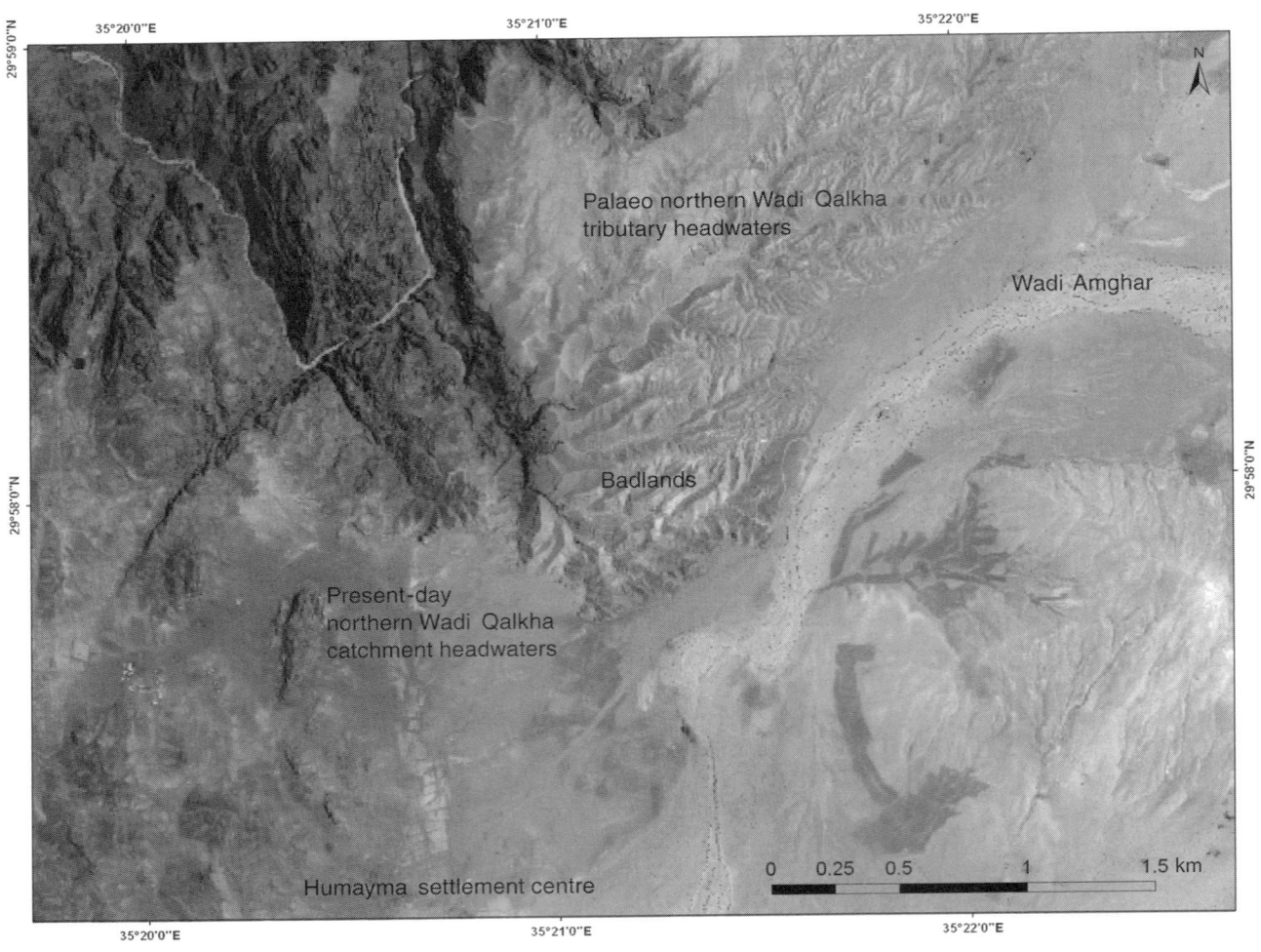

Figure 19.11. WorldView 1 Panchromatic satellite image from Eurimage, Jordania WV01 taken in 2008, showing the Wadi Amghar, the palaeo northern Wadi Qalkha tributary head waters, the badlands, the present-day northern Wadi Qalkha catchment headwaters and northern part of the Humayma settlement centre.

method is called the D-8). Thereafter, the flow is accumulated along the estimated downstream flow paths to reach the catchments outlets (flow accumulation). The drainage network is portrayed by setting a threshold for a minimum number of cells (i.e. flow accumulation) required to initiate a stream segment. The larger the flow accumulation threshold the coarser the drainage network density will be.

The aerial photograph taken in 1926 (Figure 19.8) was compared with the WorldView 1 Panchromatic satellite taken on 15 May 2008 (Figure 19.11). The two images were rectified and given the WGS84 UTM coordinate system using Erdas Imagine software. Both of the images are of high and comparable resolution (despite the graininess of the 1926 image). They were used to analyse the texture and extent of erosion in the Quaternary lakebed area, in an attempt to model how the loss of sediments affected the hydrological evolution of the catchment. A possible rate of gullying during the past 82 years in the Humayma badlands was determined by comparing channel widths in both images. This estimated rate was extrapolated to 2,100 years ago to define how much area (i.e. how many pixels on the DEM) eroded from the periphery of the Quaternary fluvial and lakebed substrate and between the ridges of the badlands. In addition, changes to the lower Jammam-Amghar wadi bed and banks were observed.

The elevations of the estimated eroded pixels were restored in the SRTM to their neighbouring pixels in the badland ridges to create the Palaeo-DEM 2,100 years ago. This allowed us to recreate the historic flow pathways of the wadi network in what is now the badlands area. The D-8 method was applied again to estimate the palaeo catchment and palaeo drainage networks, particularly in the badlands area (Figure 19.13, right). Unfortunately, there are no field measurements for runoff and associated sediment load that could verify our estimate of the erosion rate and hence of the proposed Palaeo-DEM.

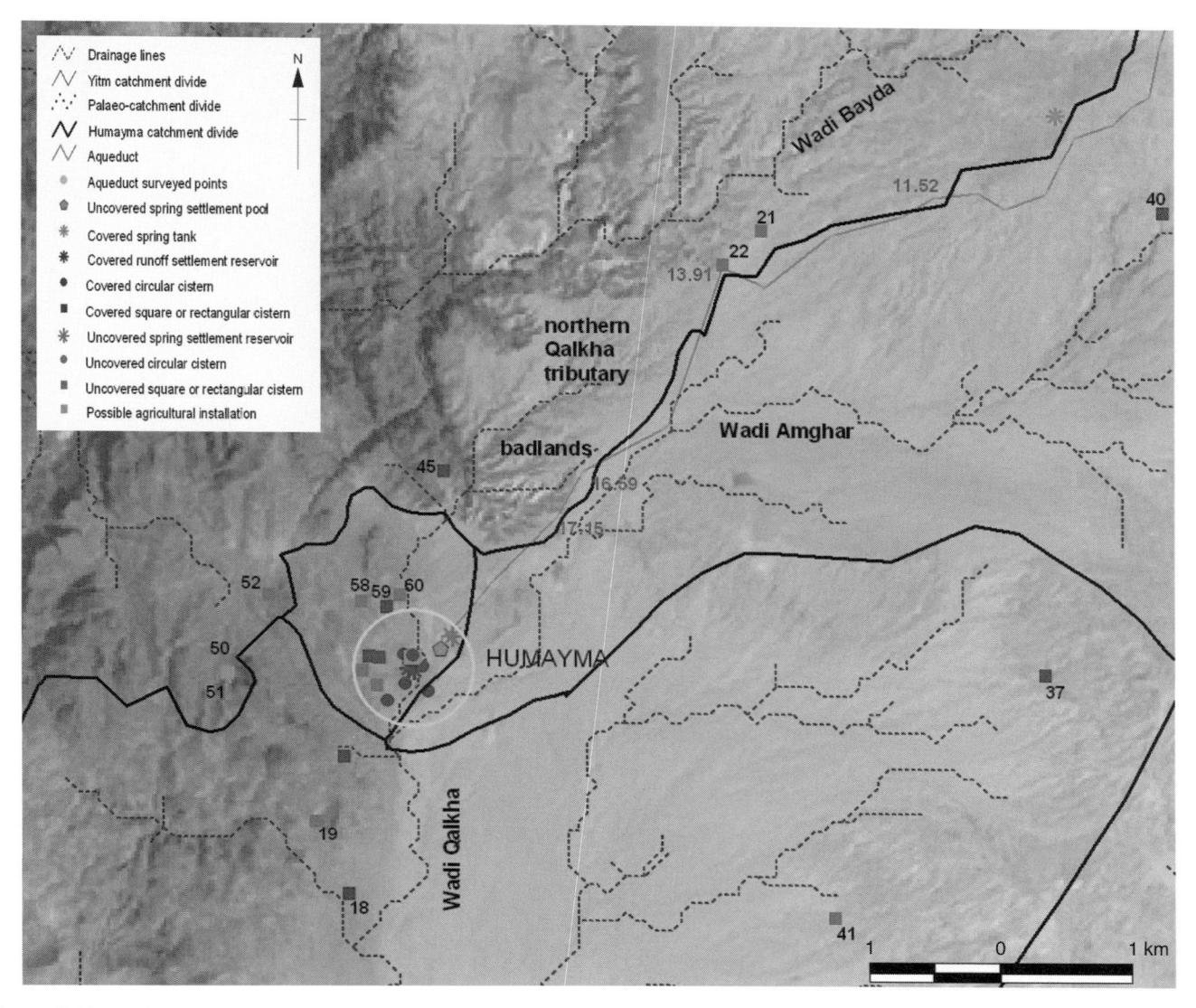

Figure 19.12. The interpolated path of the aqueduct between the surveyed points (turquoise circles) (M. El Bastawesy and R. Foote). See colour plate section.

19.3.4 Results

In visually comparing the 2008 and the 1926 images to analyse the rills and drainage network patterns in the badlands area it is clear that: (1) there are some new well-developed gullies in the 2008 image that did not exist in the 1926 image; (2) some gullies extend further up slope and some areas have suffered an overall erosion; and (3) the up-slope areas of the gullies appear coarser in surface texture in 2008 than in 1926 owing to sediment erosion and development of small rills on these high surfaces (Figure 19.14).

Certain gullies in the badlands had clearly widened and extended upstream. Measuring the difference using Erdas Imagine software gave a rate of nearly 4 m in the past 82 years. This gave an average of stream gullying and erosion of nearly 5 m per 100 years. This rate allowed us to create a scenario for a Palaeo-DEM 2,100 years ago, when the Nabataeans established the hydraulic system. This scenario assumed that the erosion rate was more or less constant across the Quaternary lakebed, with the badlands advancing 105 m (21 centuries × 5) since the first century BC, when the aqueduct was laid. It should be noted that the erosion rate is not uniform in the alluvium of the Humayma badlands. Some gullies do not appear to have changed. This may be related to local conditions, such as how much runoff fills each gully following rainfall, as well as the slope, gradient and hydraulic cross-section parameters of each. This must be accounted for when assessing the palaeohydrology. Yet particularly at the eastern extent of the badlands, we propose that the upslope retreat rate

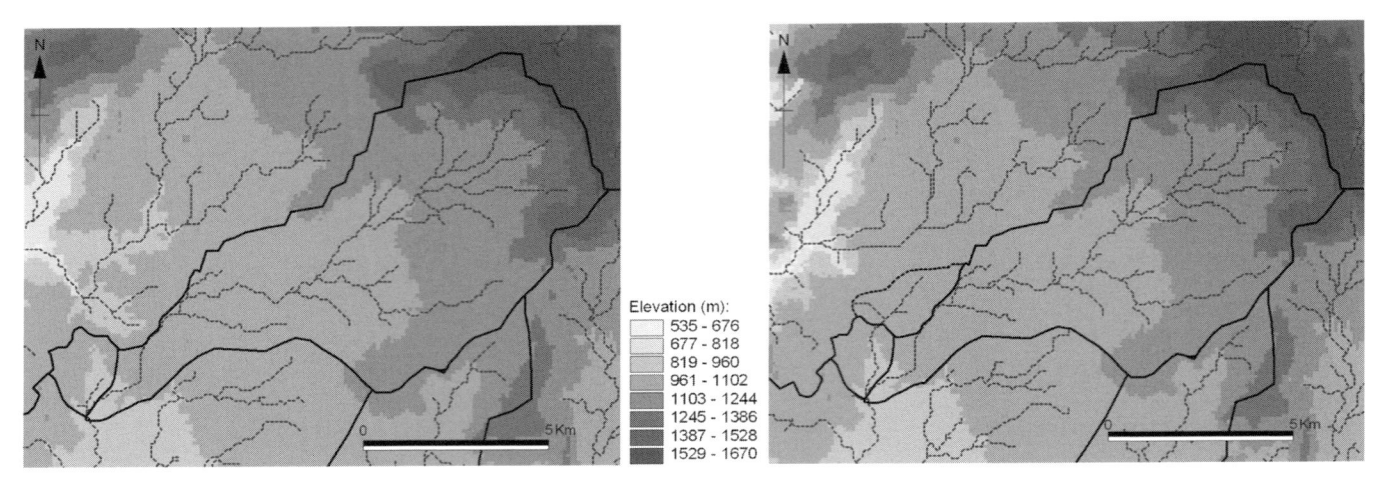

Figure 19.13. Digital elevation models, comparing palaeo- and present-day northern Qalkha sub-catchment elevations and wadi pathways. Present-day, left; palaeo, right panel (M. El Bastawesy).

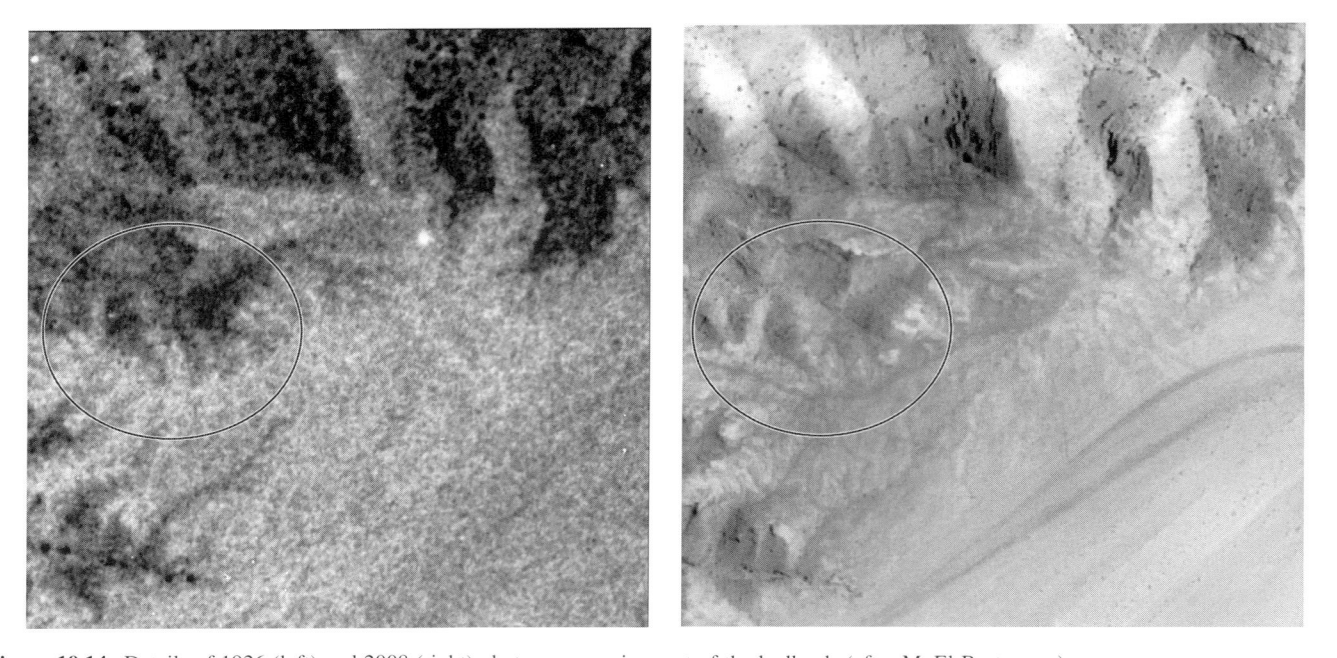

Figure 19.14. Details of 1926 (left) and 2008 (right) photos, comparing part of the badlands (after M. El Bastawesy).

of the gullies is 5 m per 100 years. As a consequence, the badlands area has lost a significant amount of sediments during the past 2,100 years.

Noticeably in the SRTM DEM, certain ridges of the badlands stand at similar or higher elevations and in a similar pattern to elevations in the wadi network to the south and southwest. This means that before the gully lines in the badlands carried the water flow westward into the deep canyon, the waters collected from north-northeast flowed to south-southwest, contributing significantly more seasonal waters into the Humayma settlement centre within the Wadi Qalkha network of the Humayma catchment. This was also confirmed by the application of conventional automatic drainage network delineation from the created palaeo-DEM of Humayma. The present catchment area of the northern Qalkha tributary feeding the settlement is estimated to equal 3 km^2. The estimated area of the northern Qalkha tributary feeding the settlement when the delineated palaeo-DEM network has been added is 7.1 km^2.

The cross-section profile of the Amghar's main wadi bed is much wider in 2008 than it was in 1926. In comparing the extent of channel bed tone and shrubs and small trees visible on both images, the eastern (left) bank of the channel appears to have significantly eroded while the western (right) bank of the channel is more or less fixed and has not changed its position since 1926

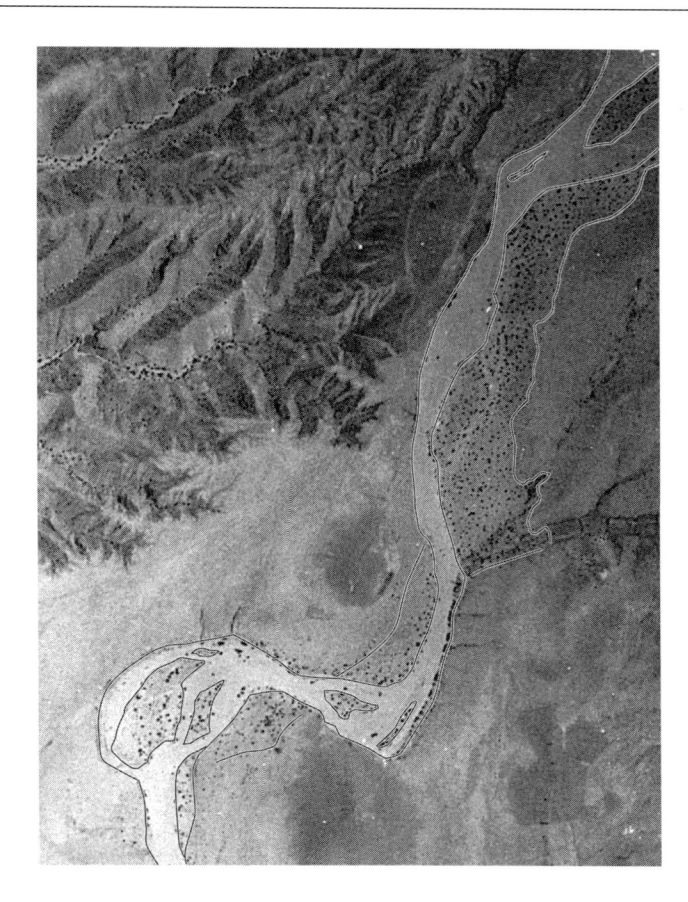

Figure 19.15. Details of 1926 and 2008 photos, comparing a section of the Wadi Amghar. The fine lines sketch as it was in 1926 (after M. El Bastawesy).

(Figure 19.15). The straight segments of the wadi bed are less altered, but the meandering sections have changed significantly, and it seems that the channel bends are shifting away from the badlands. These observations imply the occurrence of at least one severe flash flood event sometime between 1926 and 2008, when the peak flow caused extensive inundation and removal of considerable sediments from the eastern bank and the wadi bed gained more width. The effect of extreme flash flood events is usually more significant to the swing and extent of channels over alluvial plains than the slow and gradual changes that accompany moderate flows through a catchment.

This hydrogeological study of Humayma has concluded that its catchment morphometric parameters such as area and main flow directions cannot be considered constant when modelling its palaeohydrology. As such, this study agrees with others that have reported Quarternary cycles of stream aggradations and incisions for several wadi systems of the Levant (e.g. Donahue, 1985; Goldberg, 1994; Hill, 2004; Hunt *et al.*, 2004). At Humayma we suggest that an area of nearly 4.1 km^2 of Quaternary lake deposits was cut from the northern head of the Wadi Qalkha and captured into the Rift system over the past 2,100 years.

19.3.5 Summary

Our geomorphological evaluation has concluded that the hydrological catchment servicing the Humayma settlement and environs where hydraulic structures were constructed was comprised of four wadi sub-catchments (Figures 19.3 and Table 19.2):

1. the entire Jammam-Amghar (northeast and east of the settlement centre);
2. the North Qalkha (north of the settlement centre);
3. the South Qalkha (south of the settlement centre);
4. the northern tributaries of the wadi just east of the Disi.

In addition, two small parts of the Bayda sub-catchment were utilised near two of the four sub-catchment perimeters:

1. north of the settlement, where two cisterns (Nos. 21 and 22) were constructed just west of and contiguous to the Jammam-Amghar but were not technically in it;
2. west of the settlement, where two cisterns, Nos. 52 and 50, the probable press (No. 51) and terraced fields (No. 46) were constructed just west of and contiguous to the North and South Qalkha but were not technically in them.

Table 19.2 *Catchment characteristics of the four sub-catchments used in the model-based water balance assessment rendered in present-day and proposed historic areas*

Catchment	Catchment area (km^2)	Cultivation potential (km^2)		Porosity (m^3 m^{-3})	Micro-catchment area for rainfall harvesting (km^2)
		Min.	Max.		
Jammam-Amghar	62	0.5	3.8	0.363	21.2
South Qalkha	132	0.1	0.8	0.293	1.5
East of Disi	22	0.0	0.0		0.1
Palaeo North Qalkha	7.1	0.04	0.35	0.403	7.1^c
Sum^a	**223**	**0.64**	**4.95**		**29.9**
Current North Qalkha	3.1	0.04	0.35	0.403	3.1^c
Sum^b	**219**	**0.64**	**4.95**		**25.9**

^{a} Sum is determined using the estimated area of the North Qalkha palaeo-catchment.
^{b} Sum is determined using the estimated area of the present North Qalkha catchment after the upper reaches of the tributaries were captured and now flow westward to Wadi Araba.
^{c} Estimated as area of land for rainfall harvesting for potable water plus micro-catchment area draining to cisterns Nos. 50 and 52. This area is maintained for the present-day catchment area since the rainfall harvesting area for collection in the cisterns within Humayma plus Nos. 50 and 52 remain largely unaffected by the tributary capture. No. 51 may be a cistern or a wine press – so not included because of this uncertainty.

For simplicity we have included those technically in Bayda in the adjoining sub-catchments, i.e. Nos. 21 and 22 in Jammam-Amghar, No. 52 in North Qalkha, and No. 50 in South Qalkha.

The most important difference from Oleson's evaluation is that we attribute the North Qalkha with a palaeo-area measuring 7.1 km^2 rather than its present-day 3 km^2. The additional 4.1 km^2 area north-northeast of the present North Qalkha includes the badlands and the original headwaters that it cut.

19.4 HYDROLOGICAL MODELLING

The purpose of the hydrological modelling is two-fold. Firstly, the model is used to estimate the population that could be sustained at Humayma based on crop irrigation and the provision of drinking water for people and animals. Secondly, the model provides a tool with which to explore how water provision may have changed in response to developments in the hydraulic structures of Humayma set against a backdrop of a range of mean annual rainfall volumes that characterise dry and wet climates. Specifically, an annual water balance model is constructed similar in form to that applied to explore the water resource availability at Jawa, north Jordan, during the Early Bronze Age (this volume, Chapter 18; Whitehead *et al.*, 2008). As in the case of the Jawa model, the water balance model is set within a Monte Carlo framework so that uncertainty in the model parameters can be explicitly accounted for.

Three scenarios are considered in this model-based study, each relating to a successive phase of hydraulic development at Humayma as described in section 19.2.7 above. During all three phases, cisterns are filled for drinking and farming is carried out to the northeast and east in the Jammam-Amghar, the north in North Qalkha and south and southeast in South Qalkha sub-catchments. To explore the impact of climate change on water provision, seven rainfall scenarios are considered for phase I with the mean annual rainfalls of: 30 ($\pm$10), 80 ($\pm$27), 130 ($\pm$43), 180 ($\pm$60), 250 ($\pm$83), 350 ($\pm$117) and 450 ($\pm$150) mm.

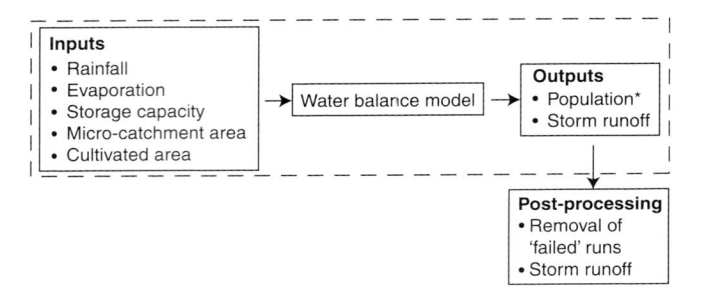

Figure 19.16. Schematic diagram of the Humayma water balance model (A. Wade).

19.4.1 Methodology: model equations

An annual water balance model was created for the Nabataean settlement at Humayma (Figure 19.16). As with the hydrological model of Jawa, the following aspects of the hydrology are considered: water delivery from the catchment; storage in reservoirs, cisterns and behind an impoundment dam and wall barriers; water demand for people, animals and cultivation. A Monte

Table 19.3 *Summary of the parameter distributions used in the model simulations*

Parameter	Unit	Distribution	Distribution statistics			
			Mean	Std. dev.	Min.	Max.
Rainfall	mm year^{-1}	Normal	80	27		
Water use per person	m^3 person^{-1} year^{-1}		3.0	2.0	1.5	
Crop water requirement	mm year^{-1}		250	83	100	500
Animal water use per person	m^3 person^{-1} year^{-1}		1.5	0.15		
		Rectangular	Min.		Max.	
Actual evaporation from soil	mm year^{-1}		0.0		30	
Initial population*			10		1,000	
Runoff coefficient, K_1	%		10		30	
Initial volume of water in stores as % of the total storage capacity	%		0		100	
Percentage of wadi flow diverted to cultivated area, K_2	%		70		100	
Percentage of wadi flow diverted to reservoirs and cisterns, K_3	%		70		100	

* In all sub-catchments except north Qalkha for which the initial population range was 10 to 3,000.

Carlo approach is used to capture the uncertainty associated with the climate, water storage and demand. The model for Humayma represents an advance on the Jawa model as it considers crop production in addition to water availability and differences in evaporation from covered and uncovered cisterns and reservoirs.

The initial population range is specified as a rectangular distribution so that no bias is introduced as to the final outcome. Rectangular (or uniform) distributions are those where each value has an equal probability of selection. This contrasts with the Normal distribution as used for the Jawa model, often described as having a 'bell-shape', which has a mean value (of the input data or parameter in this case) located at the centre of the distribution, and for which the mean has highest probability of being chosen. The distributions used to define the inputs and parameter values are given in Table 19.3. Normal distributions were chosen to define the mean annual rainfall, water use per person and crop water requirements because a most probable value could be defined in each case based on observations for rainfall, literature values for water use per person and the assumption that 250 mm yr^{-1} is the water volume generally accepted as the defining quantity required for crop agriculture. Rectangular distributions were chosen for those parameters where there was no clear evidence for the most likely value (Table 19.3).

The Humayma model computes storm runoff, the volume of water available for human and livestock consumption and for farming. From these it calculates the population that can be sustained. The equations for the hydrological and population model of Nabataean Humayma follow.

The annual storm runoff, Q (mm yr^{-1}) is calculated as

$$Q = \frac{K_1.P}{100} \quad (19.1)$$

where P is the rainfall per year (mm yr^{-1}) and K_1 is the runoff coefficient, which is the percentage rainfall flowing to the wadi outlet. The coefficient accounts for transmission losses along the wadi channel network, actual evapotranspiration and the prevailing soil moisture deficit (Yair and Kossovsky, 2002).

The wadi channel runoff, Q_w (m^3 yr^{-1}), is estimated from the annual storm runoff and the catchment area, A_c (m^2)

$$Q_w = \frac{Q.A_c}{1000} \quad (19.2)$$

The annual contribution to soil moisture, M (mm yr^{-1}), from direct precipitation is

$$M = P - Q - L \quad (19.3)$$

where L is the loss to percolation when the soil reaches field capacity (mm yr^{-1}).

The water available to the crops from the soil moisture, W_M (m^3 yr^{-1}) is given by

$$W_M = A_i \left(\frac{M}{1000}\right) \quad (19.4)$$

where A_i is the cultivated area (m^2). This is a simple expression that does not account for the wilting point of plants. However, given the limited data available to understand the system and the need for a parsimonious model, it was assumed that all soil moisture was available. As such, the model represents a 'best case' scenario for water available for crop uptake from the catchment soils.

The water available for cultivation in natural depressions in the landscape and from water harvesting through terracing and barriers in tributary wadis, W_F (m^3), is given by

$$W_F = A_C \cdot \frac{Q}{1000} \frac{K_3}{100} \quad (19.5)$$

where K_3 is a coefficient used to represent the fraction of surface runoff captured in natural depressions and through water harvesting within the cultivated areas.

The agricultural water balance, AWB (m^3), is given by

$$\text{AWB} = W_M + W_F - A_i \frac{\text{CWR}}{1000} \quad (19.6)$$

where CWR is the crop water requirement (mm yr^{-1}).

The water captured in reservoirs and cisterns for human and animal consumption, W_{RC} (m^3) is given by:

If

$$\frac{Q}{1000} > d.p$$

$$W_{RC} = A_m (\frac{Q}{1000} - d.p) \cdot \frac{K_2}{100} + W_I + B \quad (19.7)$$

Else

$$W_{RC} = W_I + B \quad (19.8)$$

where d is the effective soil depth (m) and p is the porosity (m^3 m^{-3}), W_I (m^3) is the amount of water stored in the reservoirs and cisterns at the start of the simulated year, B is the base flow from the springs (m^3) (if and when applicable) and K_2 is the percentage of overland flow either diverted or channelled by terraces to fill cisterns and reservoirs for potable supply. Data are available to describe the clay, sand and gravel fractions and the porosity (Table 1 of Oleson, 1997b). It is difficult to know the soil depth without recourse to a soil survey across the catchment area and thereby determine the water-holding capacity of the soil. Given this, it was deemed more appropriate to estimate the soil moisture capacity available each year for cropping by estimating the effective soil depth in which water is held and the porosity. Base flow is not included in the first phase of Nabataean habitation of Humayma. The storage of water is limited by the capacity available, this is modelled as:

$$\text{if } W_{RC} \geq \text{Store}_{\max} \text{ then } W_{RC} = \text{Store}_{\max} - E \quad (19.9)$$

$$\text{else } W_{RC} < \text{Store}_{\max} \text{ then } W_{RC} = W_{RC} - E \quad (19.10)$$

where Store$_{\max}$ is the maximum capacity of the reservoirs and cisterns in a catchment (m^3) and E is the evaporation from the uncovered cisterns (m^3). The maximum capacity and surface area of the uncovered cisterns in each catchment were estimated from the hydraulic surveys by Oleson (2010, Chapters II–IV) and this chapter, Table 19.1.

The Humayma model also differs from that at Jawa in that two water balances are calculated for Humayma, an agricultural water balance and a drinking water balance. Another notable advance over the Jawa model is that the Humayma model can be applied to simulate multiple sub-catchments within a system, with an agricultural and drinking water balance being applied to each and the sustainable population estimates summed to provide a total value for the whole system. This is advantageous because it allows exploration of the relative importance of urban and agricultural areas in Humayma and environs, and the importance of the possible extension of the catchment draining to Humayma from the North Qalkha before erosion may have diverted the wadi flow westward to the Wadi Araba (detailed in section 19.3 above).

The total drinking water demand over the year, D_D (m^3 year^{-1}) is based on the consumption by humans ($N_p W_p$) and animals ($N_p W_a$) and is given by

$$D_D = N_p W_p + N_p W_a \quad (19.11)$$

where N_p is the population and W_p is the annual water demand per person (m^3) and W_a the annual water demand per animal per person (m^3). The Drinking Water Balance (DWB, m^3) is given by

$$\text{DWB} = W_{RC} - D_D \quad (19.12)$$

If AWB ≥ 0 and DWB ≥ 0 then the model run is accepted as behavioural.

An additional condition can be set in order to consider the model run as behavioural based on the amount of cultivated land per person. Specifically the amount of cultivated land per person (m^2 per person) can be calculated as

$$A_{ip} = \frac{A_i}{N_p} \quad (19.13)$$

The value of A_{ip} is checked against a user-defined threshold to determine if food production would be sufficient to sustain the population.

If all conditions are met and the model run is considered behavioural then population used in the calculations is stored as a possible outcome.

An estimate of the maximum potential population, N_{sus} (persons) is also estimated as

$$N_{sus} = \frac{W_{RC}}{W_p + W_a} \text{ if } AWB \geq 0 \quad (19.14)$$

As such, three estimates of the population are calculated:

- The population which equals the value of the median population parameter used in the run when both the drinking- and cultivation-water balances are in surplus (i.e ≥ 0). The population is calculated from the sum of the populations in the North Qalkha, South Qalkha, Jammam-Amghar and East of Disi sub-catchments. It should be noted that the median population for the Humayma catchment does not equal the sum of the median populations for North Qalkha, South Qalkha, Jammam-Amghar and East of Disi taken over all runs because of different distributions. There is also a condition that each person must have a minimum amount of land (0.4 ha) to grow food, based on the total population in the four sub-catchments divided by the total amount of cultivated land.

- The *potential population* equals the value of N_p used in Equation (19.11) when both the drinking- and agricultural-water balances are in surplus (i.e. ≥ 0);
- The *drinking water only population* equals the value of N_p used in Equation (19.11), and the cultivation water balance is not taken into account. This population estimate therefore does not account for food availability.

19.4.2 Methodology: model assumptions and initial parameter values

To run the model it is necessary to specify the values of fixed parameters or the ranges for those parameters described by distributions. This was done based on the best information available, which for some parameters, such as storage volume available and spring discharge, was from observation, and for others, such as land area required per person to grow food, from literature values. The latter is a pragmatic response to data scarcity and, whilst not ideal, the approach allows the methodology to be developed. Moreover, the Monte Carlo basis for the model is designed to incorporate such uncertainty into the model and deal with it explicitly.

The four sub-catchments considered are Jammam-Amghar, North Qalkha, South Qalkha and East of Disi, with a present-day and proposed historic sub-catchment area figure set for the Palaeo North Qalkha used in an alternate scenario designed to determine the importance of the additional drainage area (Table 19.2). The micro-catchment areas contributing runoff to individual cisterns and reservoirs were derived from a Digital Elevation Model using the ESRI ArcGIS Geographical Information System. The minimum and maximum micro-catchment areas were used to specify the range of the rectangular distribution used as input to the model. The Humayma settlement, which lies in the North Qalkha, contains a large number of cisterns which have overlapping contributing micro-catchments. When summed these give an area greater than the total catchment area in the present day. As such, the contributing area to the cisterns in North Qalkha was estimated as the total catchment area plus the micro-catchment area of cisterns No. 50 and No. 52, which lie outside, but close by, the North Qalkha catchment (distances from cistern No. 67 are 1.68 and 1.49 km, respectively) and would have been used as drinking water supply for the inhabitants living on this outskirt of the urban centre of Humayma.

Table 19.4 *Estimates of the storage capacity available in the cisterns, reservoirs and created pools throughout the Humayma catchment*

Storage	Storage capacity (m³)			
	J-A	NQ	SQ	EoD
Covered storage volume in cisterns/reservoirs fed by runoff	273	3,031	1,429	675
Uncovered storage volume in cisterns/reservoirs fed by runoff	4	80	3,214	0
Spring-fed reservoir volume	0	1,615[a]	0	0
Uncovered dam-pool volume	0	0	1,400	0
Total covered storage	273	3,031	1,429	675
Total uncovered storage	4	80	4,614	0
Total storage	277	3,111	6,043	675

Derived from an analysis of the dimensions of the hydraulic structures surveyed by Oleson (2010, Chapters III and IV).
[a] Included in Phase II only.

In present-day Africa, approximately 0.2 to 0.4 ha is cultivated per person with a greater area of land required per person in drier climes (Ker, 1995). Where the soil fertility is poor then the cultivated area required per person rises to approximately 0.7 ha to gain sufficient yield. For this study, a figure of 0.4 ha per person, which equates to an area of 63×63 m², was used to provide an estimate of the land area required to grow sufficient food for each of the inhabitants. This compares with a land area required per person of 0.6 ha derived from the cultivated area (288 ha) needed to feed a hypothetical minimum population of 448 with 250 kg wheat per person per year proposed by Oleson (2010, Chapter VII, referring to Broshi, 1980; Bruins, 1986, pp. 175–181).

The initial Humayma population was specified as rectangular distributions with a minimum of 10 and a maximum of 1,000 for Jammam-Amghar, South Qalkha and East of Disi, and a minimum of 10 and a maximum of 3,000 for North Qalkha and Palaeo North Qalkha in which the settlement centre of Humayma is located. The population ranges were chosen because they were thought to cover the likely ancient population size at Humayma based on the populations estimates of Oleson (1997b; 2010). Following preliminary model runs, it was determined that most populations where the water balances were in equilibrium or in surplus occurred when the initial population was in the range 10 to 1,000. Thus, the populations ranges specified were deemed appropriate for the simulations.

The estimates of the storage volumes were taken from the detailed site surveys by Oleson (2010, Chapters III and IV); and relevant data presented in this chapter in Table 19.1. The aggregated capacities for the covered and uncovered cisterns, reservoirs and created pools are given in Table 19.4 and the

Table 19.5 *Estimates of the surface area of the uncovered cisterns, reservoirs and created pools throughout the Humayma catchment*

Surface area of uncovered cisterns	Surface area (m²)			
	J-A	NQ	SQ	EoD
Uncovered surface area cisterns and reservoirs filled by runoff	5	97	337	0
Uncovered surface area reservoir volume filled by spring water	0	0	380	0
Uncovered dam-pool surface area	0	0	717	0

surface areas in Table 19.5. The latter are used in the calculation of annual evaporation from open water stores.

19.4.3 Methodology: other assumptions

The mean annual rainfall was assumed to be 80 ($\pm$ 27) mm which is the mean value used by Oleson (2010) in his assessment of the water resource of Humayma; the standard deviation was estimated as a third of the mean value. The mean annual rainfall was assumed to have a Normal distribution, as is the case for the 160-year-long rainfall record available for Jerusalem (Whitehead et al., 2008). Estimates of water use per person and per animal were assumed to be the same as those used in the study of the water resource at Jawa. This drew on the study of Evenari et al. (1982), which suggested that the average adult requires 1.5 m³ per year, which is approximately 4 litres per day. This figure is below the modern estimate of 15 litres per person per day for short-term emergency situations and 100 litres per person per day as a minimum for a reasonable standard of life (The Sphere Project, 2004). The distribution describing the water usage by animals was also assumed to be the same as that for Jawa given similarities in the species present in historic Humayma: sheep, goats, donkeys and camels (Helms, 1981; Whitehead et al., 2008). It is assumed that industrial consumption is negligible, although there is some evidence for pottery, leather and woollen production but not on a major industrial scale.

The flow in Wadi Jammam-Amghar is ephemeral in nature, being entirely reliant on rainfall events. The rainfall in semi-arid regions, such as southern Jordan, is characterised by low-frequency, high-magnitude events which are highly localised in spatial extent (Wheater, 2002). This gives rises to a very 'flashy' flow regime whereby a large fraction of the rainfall is transformed to channel flow through the mechanism of infiltration of excess overland flow with subsequent transmission loss into the channel bed and colluvium where present. Such floods are typically highly competent for conveying sediment. The infrequency of the flow events and the high sediment loads make it difficult to measure the flows in such channels. Similar channels in southern Jordan, such as Wadi Dana, have proved difficult to gauge either because no one is present when the flood occurs or structures built in the river bed are damaged by the flood, thereby rendering their geometry different from that for which the structures were calibrated. The flows in the main channel of the Wadi Jammam-Amghar were estimated from open-channel measurement techniques using the Darcy–Weisbach equations and a hydraulic survey on 1 May 2008 of the Jammam channel at Humayma at a point adjacent to the main settlement (35° 20' 58" E; 29° 56' 57" N) with an upstream contributing area of 61 km². The resultant flows were 3 m³s^{-1} for the flood prior to May 2008 and 266 m³s^{-1} with the channel flowing-full. These data allow comparison with the modelled runoff estimates (Equation 19.2).

The actual evapotranspiration from the soil is uncertain for Humayma. The annual potential evaporation at Humayma was estimated at 3,300 mm from available monthly Class A evaporation pan measurements made between 1983 and 2002 at Ma'an, which is approximately 45 km northeast of Humayma. Whilst this evaporation pan is located in a different environment, up on the plateau, this evaporation rate was used in the model to estimate the loss of water from uncovered reservoirs and cisterns over a 1-year period; it is similar to that used by Oleson (Natural Resources Authority – Amman, 1977, Map SW-5). An assumption was made that the actual evapotranspiration ranged from 0 to 0.33 m yr^{-1}, the upper bound defining the maximum rate of evapotranspiration. Other ground cover will affect actual evapotranspiration but given the difficult task of quantifying this loss of water from the soil this effect was assumed to be included in the estimated range of 0 to 0.33 m yr^{-1}.

The figure of 250 mm yr^{-1} is commonly used as a threshold for viable farming since it is the crop water requirement of bread wheat (*Triticum aestivum* L.) (Allen et al., 1998). Barley grows at about 200 mm. A Normal distribution with a mean of 250 mm yr^{-1} was used to describe the crop water requirement within the model. In the modern period, with little effort in water harvesting, the Bedouin grow barley and some wheat in the fields east and northeast of Humayma because the runoff naturally collects in some places. The historic barriers constructed to concentrate water in fields could easily increase the amount of water infiltrating to the root zone. This is accommodated in the model through the estimation of water entering the soil.

The estimated range of runoff coefficients used to define the distribution were taken from studies in arid zones in Spain (Puigdefàbregas and Sánchez, 1996), the southwestern USA (Dunne, 1978), a catchment in the semi-arid zone of Australia (Williams and Bonell, 1988) and the work reported in Oleson (2010, Chapter II) for Wadi Rumman catchment and the Negev. The studies of semi-arid-zone hydrology in Spain, south-western USA and Australia suggest a range of 10 to 30% for the runoff coefficient, K_1, which was set with minimum and maximum values of 10 and 30%. The 2% runoff presented in the National Water Master Plan of Jordan (Natural Resources Authority – Amman, 1977) for the Wadi Ramman catchment was considered conservative by Oleson (2010, Chapter II), who proposed that 15% is a more appropriate figure based on studies from the Negev (Kedar, 1957; Hillel, 1982, p. 97). As such, Oleson's estimate is consistent with the range proposed for use in the model based on the studies undertaken in Spain, southwestern USA and Australia, and supports the use of a 10–30% range used within our model. Given an absence of data with which to describe the percentage of catchment runoff diverted to cultivation areas or to reservoirs and cisterns, an assumption of 70 to 100% was made in each case. Even today, without any maintenance, many of the surviving reservoirs and cisterns fill during rain storms. This suggests high efficiency in the capture of runoff. Moreover, the very reason for the construction of terracing walls and wadi barriers is to ensure infiltration into defined agricultural fields.

Rectangular distributions were defined for the cultivated areas in each of the four sub-catchments. The minimum value of the cultivated area was defined according to the area farmed visible from remote sensing. The maximum area of potential cultivation in each sub-catchment was estimated using the GIS as the area of alluvium identified from remotely sensed images of the wadi network. The present-day fields are located on alluvium and thereby provide an indication that the alluvial deposits were farmed in the past. Oleson (2010, Chapters II and VII) estimates that 288 ha was cultivated (which is 9% of the present-day extent of the alluvium). This estimate is based on the required yield of wheat per hectare, after accounting for seed, spoiling and loss, needed to feed Oleson's estimation of 448 persons living in the Humayma catchment.

To allow the possibility that more than 288 ha were cultivated, the model assumes that eight times the present-day cultivated land may have been farmed. This gives a total maximum cultivated area of 500 ha, which is approximately 15% of the total potentially cultivable land. In a given year, owing to the need to rotate land to maintain soil fertility and provision areas for grazing, it is typical that not all potentially cultivable land is farmed. Thus, whilst the actual total maximum cultivated land in the past is unknown, the use of 500 ha as a maximum allows exploration of greater investment in farming than Oleson considered whilst recognising that not all potentially available land is farmed. Therefore, to progress the methodology it was assumed that the area where infiltration was enhanced equals that cultivated.

19.4.4 Mathematical realisation

The model was coded as a Microsoft EXCEL © spreadsheet and Oracle Crystal Ball 7 Professional (Academic License) was used to provide the Monte Carlo, sensitivity analysis capability. For each scenario the model was run 25,000 times taking a sample of parameters from the specified parameter distributions on each model run. Sampling of parameter space was undertaken by using Latin Hypercube to maximise the even exploration of the specified range. On a machine with a 2.8 GHz dual core processor and 2 GBytes RAM the 25,000 model runs were completed in 3 seconds. Post-processing of the population and storm runoff data was undertaken by using S-plus from Insightful software ©.

The Crystal Ball software was also used to provide an indication of the most sensitive model parameter for each scenario. For each scenario, the respective influence of each parameter on the response of the output variables was ranked according to correlation coefficients. The larger the absolute value of the correlation coefficient relating the output variable of interest to a parameter, the stronger is the relationship between the output value and the parameter. This sensitivity analysis only works if the relationships between outputs and parameters is a monotonic function, namely as the parameter value increases, the output value either continually increases or continually decreases.

19.4.5 Scenarios for Phases I, II and III

For Phase I, the model was run for a range of precipitation conditions with the mean of the rainfall distribution adjusted from 30 ($\pm$10), 80 ($\pm$27), 130 ($\pm$43), 250 ($\pm$83), 350 ($\pm$117) and 450 ($\pm$150) mm and the available water storage set to the capacity of the stores listed in section 19.4.2 above. Further model runs were undertaken with the mean annual rainfall set to 30 ($\pm$10), 80 ($\pm$27) and 250 ($\pm$83) mm but with water storage capacity set to 125% of that estimated from the hydraulic survey of Oleson. The purpose of these three additional model runs was to investigate the possibility that the water stores were refilled during the course of the year. An additional model run investigates the effect of increasing the area of the North Qalkha catchment, in which Humayma is situated, from 3.1 to 7.1 km^2. The former area is that estimated for the area of the present-day catchment and the latter is the palaeo-catchment at the time of Phase I (Table 19.2).

For phase II, the model was run for six cases. Three mean annual rainfall scenarios were used: 30 ($\pm$ 10), 80 ($\pm$ 27) and 250 ($\pm$ 83) mm. For each the water storage was initially determined by the dimensions of the cisterns and reservoirs thought to be in use and then the water storage was increased to 125% of the volume estimate to account for refilling within the year. This is the only set of simulations in which spring water is included in stored drinking water capacity, because of the construction of a new reservoir within the fort drawing water from the aqueduct.

For phase III the model was also run for six cases. Three mean annual rainfall scenarios were used, 30 ($\pm$ 10), 80 ($\pm$ 27) and 250 ($\pm$ 83) mm, and again the scenarios differed in that the model was run for 100 and 125% of the estimated water storage capacity. The spring flow was again removed because the reservoir of Phase II went out of use.

19.5 RESULTS

19.5.1 Population estimates under different rainfall scenarios

This section reports the results of applying the annual water balance model to explore three archaeological scenarios related to water storage capacity and management for six rainfall scenarios. Whilst there is little evidence to suggest that the rainfall at Humayma was different in the past from that inferred from nearby gauges at Ras en Naqb, Guwayra and Wadi Ramm between 1963 and 1975, it is instructive to determine how resilient the hydraulic and rainfall harvesting system was to different mean annual rainfall. Rainfall in desert environments is known to be highly variable from year to year. For example, Oleson (2010, Chapter II) notes the large interannual variability in annual rainfall totals at the Ras al-Naqb, Guwayra and Wadi Ramm precipitation measurement stations. At these three stations, the mean annual rainfall and ranges measured between hydrological years 1963/4 and 1974/5 by the Natural Resources Authority – Amman (1977) varied as follows: Ras al-Naqb (mean 143, range 58–304 mm); Guwayra (87, 17–335 mm); Wadi Ramm (82, 14–176 mm). Thus, the highest annual rainfall totals can exceed three times the long-term mean.

On 1 May 2008, a section of the Wadi Qalkha channel at Humayma was surveyed to estimate flood volumes during the last storm and when the channel flows full. Using a GIS, the catchment area draining to the survey point was determined, using a DEM, as 61 km^2. Assuming that the flow following rainfall in the catchment lasts for one day then the estimated surface channel runoff for the last storm prior to May 2008 was 0.3 million cubic metres (MCM). With a mean annual rainfall of 80 ($\pm$ 27) mm, the annual runoff, derived from the water balance model and assumed runoff coefficients in the range 10 to 30%, was 1 MCM year^{-1} (Table 19.5). Thus the annual runoff estimate simulated by the model was three times larger than that estimated by survey. Given the uncertainty in the model assumptions and errors in applying the Open-Channel method to reconstruct flood flows, the three-fold difference was considered acceptable in terms of a reliable model fit.

The three median population measures were calculated from the output distributions generated from the 25,000 simulations. In each of the three cases, the median population was presented rather than the mean given the negative skew in the output data, rendering the distributions of the population measures non-Normal. When the model was run with a mean annual rainfall of 80 ($\pm$27) mm, considering the annual water balance alone, the estimated median *drinking water only population* was 484 (Table 19.6) with an accompanying 1,452 ovicaprids (sheep and goats) also provisioned, and 334 persons with 1,002 ovicaprids if including consumption by agriculture. As discussed above, population figures embed an animal population: the drinking needs of roughly three goats or sheep are included with that for each person based on estimated water consumption of 1.5 m^3 yr^{-1} per person and 0.5 m^3 yr^{-1} per sheep or goat (Evenari *et al.*, 1982). We acknowledge that the animal population would have included camels and donkeys, which require more water than ovicaprids, but for simplicity a number of animals is only presented in sheep and goat.

The archaeological evidence is unclear with regard to all cisterns being covered or not. Some of the cisterns surveyed were so ruinous that potential original roofing arrangements could not be identified. Temporary covers, such as palm fronds or thatching, for some of the smaller or less regularly shaped structures may have been lost from the archaeological record. Thus to explore such a scenario it was assumed that all cisterns were covered apart from the pool behind the impoundment dam (No. 44) and two pools that formed behind the barrier walls in rock clefts (Nos. 27 and 28) in another model run. With reduced evaporation, the population increased slightly from a median of 147 to 180, while the drinking water-only population increased from 484 to 759 indicating that such a change chiefly affects the provision of drinking water because the water harvested and stored is mostly for drinking. In other words, this does not help cultivation because the cisterns and reservoirs are not used in large-scale irrigation. A discussion of the populations potentially supported for each development phase follows.

19.5.2 Population estimates for Phase I

As the mean annual rainfall increases the population and potential population that can be sustained also increases. Considering only the drinking water balance (Equation 19.11), the drinking water-only population sustained increased, as a non-linear function of rainfall, to a maximum of 1,422 people when the mean annual rainfall was 450 ($\pm$150) mm (Table 19.6). This drinking water population increases to approximately 1,713 people for rainfall of 250 ($\pm$83) mm if an allowance is made for the cisterns and reservoirs being replenished through the wet season, assuming more than one storm event, giving an effective capacity of 125% of the actual storage capacity. When the storage capacity was increased to 125% it was assumed that the surface area of the cisterns and reservoirs remained constant.

If we follow Oleson (2010, Chapter VII) and consider that the cisterns and reservoirs only filled every two years, therefore implementing the 50% figure, the population is very different: a

Table 19.6 *The modelled mean and median populations at Humayma and modelled runoff for a succession of rainfall and water management scenarios*

Scenario	Rainfall (mm)	Storage (%)	Population median	Potential population median	Drinking water-only median	Annual runoff (mm yr^{-1}) Mean	Sd	Maximum
Phase I								
Zero Initial	30	100	0	0	0	6	3	19
	30	100	108	360	443	6	3	19
Some uncovered	80	50	57	100	137	16	7	51
Some uncovered	80	100	147	334	484	16	7	51
Some uncovered	80	125	188	426	733	16	7	51
All covered except 27, 28, 44	80	50	109	201	261	16	7	51
All covered except 27, 28, 44	80	100	180	416	759	16	7	51
All covered except 27, 28, 44	80	125	193	507	1,007	16	7	51
	130	100	172	474	814	26	12	82
	180	100	211	561	1,083	36	16	114
	180	125	284	735	1,469	36	16	114
	250	100	279	605	1,271	50	23	158
	250	125	325	800	1,713	50	23	158
	350	100	319	619	1,377	70	32	221
	450	100	360	624	1,422	90	41	284
Palaeo	80	100	171	344	484	16	7	51
Phase II								
	30	100	110	378	861	6	3	19
	30	125	121	500	1,171	6	3	19
	80	100	180	394	882	16	7	51
	80	125	213	498	1,200	16	7	51
	250	100	311	616	1,401	50	23	158
	250	125	347	812	1,886	50	23	158
Phase III								
	80	100	147	334	484	16	7	51

median 109 and the drinking water population increasing to 261. Given that there are 57 cisterns and three large reservoirs throughout the catchment storing a total 10,107 m^3 (Table 19.1) with 3,111 m^3 in the North Qalkha catchment in which Humayma is centred, it is questionable as to why the population supported could be this small. Inspection of the model results shows that, with a mean annual rainfall of 80 (±27) mm, there is sufficient water flowing across the landscape to fill the cisterns but this only happens for 1% of the 25,000 model runs. In the other 99% of runs, there is insufficient water to fill the cisterns because of the infiltration of water into the cultivated soils. In the North Qalkha catchment, the cisterns and reservoirs were filled to a mean volume of 1,241 (± 939) m^3. This compares with the maximum storage capacity available of 3,111 m^3, or 2,790 m^3 when annual evaporation from the uncovered stores is factored in, and therefore the mean volume filled was 40% of capacity. As the annual rainfall amount increased then the mean volume filled also increased, as did the reliability of filling the cisterns to capacity. With a mean annual rainfall of 130 (±83) mm yr^{-1}, which approximates to the rainfall over the escarpment to the north of Humayma, the mean volume filled was 1,414 (±980) m^3 and the maximum capacity was achieved for 12% of the model runs. For 250 (±83) mm yr^{-1} then the mean volume filled was 2,195 (±939) m^3 and the maximum capacity was achieved for 62% of the model runs. Capacity is achieved for 90% of model runs when the mean annual rainfall is 450 (±150) mm yr^{-1}.

Agricultural potential is limited by the water holding capacity of the soil and the crop water requirement. This results in lower population and potential population estimates than for the comparative drinking water-only populations. Effective rainfall is increased by terracing and flood inundation irrigation. If the model parameter relating to this diversion (i_2) is set to zero, then the crop demand for water is never met with a mean annual rainfall of 80 (±27) mm and therefore the use of terracing and deflection dams is essential to meet a mean crop water requirement of 250 mm per year, which is typical of wheat. The model is not able to reflect the very small percentage of the landscape that may naturally receive enough water, given topographic features that direct and contain water and thereby offer some concentrations to pool and enable inundation and crop growth.

When consideration is given to the need to grow food *in situ* then the population that can be sustained is less than that where provision of drinking water is the only concern. When sufficient water is delivered to an area of 0.4 ha per person to satisfy a mean crop requirement of 250 mm yr^{-1} then the median population that can be sustained with an annual rainfall of 80 ($\pm$27) mm is 147 with a range, estimated from the 25,000 simulations, from 0 to 591. The potential population was higher with a median of 334 and a range of 0 to 2,064. This result indicates that, with measures to enhance rainfall harvesting, such as deflection, barrier and terrace walls and collection of surface runoff in land-surface depressions, all of which can increase the volume of water available for crop growth, a mean annual rainfall of 80 ($\pm$27) mm is sufficient to sustain crops requiring 250 mm of water per year and possibly more. The population would be 416 persons and 1,248 ovicaprids if all but Nos. 44, 27 and 28 were covered and there were losses to agriculture in the catchment field due north of the settlement, over which most of the runoff passes before filling the many cisterns and large reservoirs in the settlement centre.

When the mean annual rainfall is low, namely 30 ($\pm$3) mm then the water stored from the previous year becomes the most important factor affecting the *drinking water-only population*. At this volume of rainfall, agriculture is unsustainable and the population is zero. Thus if the mean annual rainfall at Humayma were highly variable between years, as is apparent today, then the cisterns and reservoirs would form an important buffer against interannual rainfall variability. In conditions of such a low mean rainfall life would be unsustainable at Humayma if there was no water remaining from the previous years.

The increase in the size of the North Qalkha catchment from 3.1 (present day) to 7.1 km^2 (proposed palaeo) increased the *population* (from 147 to 171) and the *potential population* (from 334 to 344) but not the *drinking water-only population* (which remained at 484) with a mean annual rainfall of 80 ($\pm$27) mm (Table 19.6). This occurs because the runoff from the catchment is increased from 0.04 to 0.10 MCM year^{-1} and thereby allows an increase in the potential cultivable area but there remains insufficient water reaching the drinking storage facilities in the settlement. The extent of the potentially cultivable area is unknown but to serve as an example of how the population might increase, it was assumed that the maximum extent of cultivated land in the Palaeo North-Qalkha catchment increased four-fold, from 35 to 140 ha. This resulted in a small increase in population.

19.5.3 Population estimates for Phases II and III

When the reservoir in the Roman fort (No. 62) is added to the hydraulic system, we assume that this additional volume of 417 m^3 is filled by spring water carried via a spur conduit off the main aqueduct; the water is stored and uncovered but periodically replenished from the aqueduct flow. That flow is also supplying the pool (No. 63) and bath (E077) and civilian settlement (E125). Thus within the model it is assumed that a flow of 0.005 m^3 s^{-1} is available from the aqueduct for capture and storage in the reservoir (Oleson, 2010, Chapter VII). With the inclusion of this water store, the drinking water population potentially increases from 484 to 882, suggesting 400 soldiers and 45 horses could be provisioned. This assumes a daily water requirement of 36 litres for horses in place of the 0.5 m^3 yr^{-1} for three ovicaprids per person, this water being provided from the reservoir, which is constantly replenished from the aqueduct. The number of soldiers provisioned with water may have been lower since the calculation does not account for other activities drawing from the reservoir, such as latrines or light industry, or for more horses. This reservoir provides an important buffer against years of low (<80 mm) rainfall. It also provides water to sustain the fort during times of conflict. However, this water source could be easily fouled through the addition of a contaminant to the open aqueduct. Consequently the reservoir would have to be isolated from the aqueduct if the fort lay under siege.

When the Roman fort is abandoned in Phase III and the reservoir (No. 62) is no longer in use, the population estimates revert to those of Phase I. The archaeobotany indicates wetter growing environments for crops, which implies more aggressive and successful water harvesting. Given the uncertainty in the cropped area and the efficiencies of rainfall harvesting, the Phase III scenario was not explored further.

19.5.4 Comparison with Oleson's model and results

Oleson calculated that the water resources available to the inhabitants throughout the Humayma catchment could probably sustain a total of 448 persons, 3,008 ovicaprids and 300 camels or donkeys. He operated on the suppositions that the total annual precipitation across the Humayma catchment could be as high as 16 MCM on average, the surface runoff would approximate to between 0.32 and 2.4 MCM, and that the lower amount could fill all the documented water storage features. In particular, he proposed that the Pleistocene lakebed north of the settlement centre, approximately 1 km square, would have delivered somewhere between 12,000 and 1,600 m^3 yr^{-1} directly to the public and private reservoirs (and cisterns) in the centre of the settlement.

Oleson considered the runoff to the reservoirs and cisterns in detail. Although he recognised that agriculture was the activity consuming the largest amount of water in the catchment, he did not explicitly calculate the water available for cultivation. Instead, he calculated the area of land necessary to produce enough wheat for the 478 persons he estimated for the population based on drinking water capacity at 288 ha (Oleson, 2010, Chapter VII). He assumed that the rainfall that did not enter storage structures was either lost from the catchment through wadi flow or captured in natural depressions that became field areas, or was directed to areas for cultivation. This water was thought to be sufficient to grow the food necessary for the population while its livestock could be sustained from the water stored in the cisterns, reservoirs and behind the dam and barrier walls. An assumed catchment mean annual rainfall of 80 ($\pm$ 27) mm is indeed sufficient to fill the cisterns if it is assumed, as Oleson did, that 15% (equivalent to an effective depth of 12 mm) of this rainfall flows across the surface of the present-day North Qalkha catchment to recharge the cisterns and reservoirs, losing only whatever infiltration rate already is embedded in the runoff coefficients.

Among the crucial differences between Oleson's approach and the annual water balance model developed here when determining the sustainable population at Humayma is the attempt in the latter to nuance the calculation of the water volume available to cisterns and reservoirs by considering the amount of water also available to cultivate crops in each sub-catchment. In the water balance model, it is assumed that some rain falling on the catchment will replenish the soil moisture (Equation 19.3) or will evaporate or percolate to depth (if contained) and the remainder will flow across the land surface either to the cisterns and reservoirs, beyond these to the wadi channel network, or partially infiltrate when moving over unsaturated ground away from the area where the rainfall is occurring. For example, rain may fall over the escarpment and the overland flow generated will travel downslope over the dry or drier lower slopes of the catchment (Yair and Kossovsky, 2002). This infiltration of the overland flow is known as a transmission loss. Rainfall in desert environments has been observed to be highly localised, perhaps occurring with a spatial extent less than 1 km^2 (Wheater, 2002). Therefore, it is feasible that the rain may fall on the higher ground of the Humayma catchment and flow downslope, infiltrating as it goes.

In the annual water balance model, it is assumed that land was cultivated in a fallow/farmed regime. The presence of nine wadi barriers in the 1926 aerial photograph, which were gone by the time of the Humayma Hydraulic Survey in the 1980s, exemplifies the ephemeral nature of the terrace walls and wadi barriers to create fields in which surface water collects and infiltrates. We disagree with Oleson's view that the rarity of such walls indicates that the labour needed to prepare and maintain them only infrequently justified the effort involved; we suspect that many have simply been lost through erosion.

The simulated runoff value of 1 MCM resultant from a mean annual rainfall of 80 ($\pm$27) mm falls squarely in the same range as Oleson and is similar to the flood estimate determined by measurement of the channel hydraulics on 1 May 2008. It is unsurprising that the annual water balance model produces a similar surface runoff to that estimated by Oleson. His estimate of surface runoff, equivalent to 12 mm depth over the catchment, was based on a runoff coefficient of 15% and that of 16 mm for the water balance model was based on a rectangular distribution of runoff coefficients with a range 10 to 30%.

By considering only the fill capacities of cisterns and reservoirs and not the flow over the landscape, Oleson effectively assumes zero transmission loss of the 12 mm of surface runoff. The cisterns and reservoirs in the North Qalkha sub-catchment (where the settlement centre lies) can be filled to capacity with an 80 mm rainfall event under this assumption. With the annual water balance model this is not the case, since some surface runoff is assumed to infiltrate as a transmission loss of overland flow into the fields (where dry) or as further infiltration (where forced by dams or barriers). The North Qalkha sub-catchment runoff, which fills most of the cisterns and all of the reservoirs in the settlement centre, travels mainly across soil which has a porosity of 0.4 m^3 m^{-3}. We are unsure how much of the 1 km^2 runoff field due north of the settlement centre was cultivated, but it was the most fertile soil and the easiest field area to reach by inhabitants of the centre; therefore, it is proposed that, to supply sufficient food to the population in Humayma, it was significantly farmed. This leads to a situation whereby the cisterns are only completely filled during rainfall events with a probability of 0.01 for a mean annual rainfall distribution of 80 ($\pm$ 27) mm, though it is possible for the cisterns to be filled to the brim on occasion with this rainfall distribution.

If the rainfall increases to a mean annual value of 130 mm yr^{-1} (more typical it is believed of that found at the foothills of the escarpment), then the probability of filling the cisterns increases to 0.12. With a mean annual rainfall 450 ($\pm$150) mm yr^{-1}, which given the episodic nature of desert rainfall is not possible on an annual basis over the long term but may happen on occasion, then the probability of filling the cisterns to capacity becomes 0.9. However, given the gap between large rainfall events, then there would be an apparent dependency on stale water between rainfall events. Thus, the results of the annual water balance model demonstrate that, owing to transmission loss, the cisterns in the Humayma centre may not always fill and therefore *population* increases with rainfall, since increased rainfall gives an increased probability of sufficient rainfall to fill the cisterns. This is an important difference from the studies of Oleson.

In this study, the annual water balance model employs distributions to handle vagaries, as above with respect to losses in cultivable field areas, while Oleson chose values for consumption against an assumption that all cisterns and reservoirs supplying drinking water for the population and their animals fill only every two years, as a safety margin. At first glance our estimate of 484 persons and 1,452 ovicaprids as the catchment total, if agriculture is not pursued within the micro-catchments draining to the cisterns and reservoirs, is, especially in the area north of the settlement, in broad agreement with that calculated by Wheater (2002) and Oleson (2010, Chapter VII) of 448 persons, 300 camels or donkeys and 3,008 ovicaprids with a mean annual average rainfall of 80 mm. Yet with the 100% safety margin and, further, with no evaporation loss applied to the water stored in uncovered cisterns or the pools behind the dam and barrier walls, the differences in the population estimates between the two approaches increase.

We also ran the annual water balance model at this safety margin of 100%, reflected as 50% in Table 19.6, both with and without evaporation applied to the cisterns for which roofing could not be identified in the Humayma Hydraulic Survey (except the three pools, which surely were never covered). The drinking water population falls by more than half to 137 with the safety margin figure and evaporation and to 261 if no evaporation is applied. The numbers from our many runs that are comparable to Oleson are a *potential population* of 334 people and 1,002 ovicaprids (if we assume the cisterns without evidence for roofing indeed were not roofed) and 416 people and 1,248 ovicaprids (if those cisterns were indeed covered), which also assumes agriculture (in sub-catchments, based on surface water drainage patterns, that restrict the areas filling storage structures rather than applying broad catchment-wide calculations) and applies a distribution of 70–100% to account for inefficiencies in filling (such as failure to capture all the surface runoff due to variations in the topography directing water away from the cisterns).

A major difference remains in the number of animals. We have a similar number of people but a lower number of animals and their variety because the Evenari *et al.* (1982) calculation for sheep and goat water consumption is one-third that of humans (so the ratio in the model outputs will always be 3:1). But a lower sustainable total number of people and animals primarily results from less water available to the storage units in the settlement centre as stated above.

Taken together, the results of Oleson's calculations and those of the Humayma water balance model suggest that the latter applies greater restrictions. As described above, we exclude the spring-fed, off-take tanks (No. 20 and one without a number), because they were probably filled only periodically; we also exclude No. 19, because they are shallow pot holes and could have skewed the evaporation rate even further; and we have considered No. 51 a wine press with related activity. Therefore, we have used an input of 149 m^3 (their combined total) less into the runs and have applied losses to evaporation, which can account for some of the difference. Yet Oleson's annual filling rate is lower, only 100% every two years, while we used an annual distribution of between 70–100%.

An accurate description of water movement and storage in the landscape of Humayma probably falls somewhere between the models of Oleson and the annual water balance model presented here. Oleson does not account for transmission loss whereas the water balance model may over-estimate transmission loss when the rainfall is directly over the settlement centre and fields to the north of the populated centre. Rainfall directly over the field and settlement centre would easily saturate the soils, leading to high runoff, whereas surface runoff generated by rainfall some distance away would then flow over drier soils leading to higher transmission loss.

In the next phase of work, improvements to the modelling of water movement and storage at Humayma should be possible through further use of a GIS model to determine the area contributing to ancient field systems. With a rendering of the areas contributing to the cultivated fields it would be possible to determine the areas contributing to the fields and cisterns or both. A sound basis would also be available on which to make assumptions about whether surface runoff in each micro-catchment of each storage unit would flow over bare rock or soil. Once done, a fully distributed, physically based model could be implemented for a range of design storms with different frequency, intensity and duration characteristics both directly above the settlement centre and fields and at different locations over the catchment to understand how the runoff responded. Ideally such modelling should be confirmed by field observation though this is notoriously difficult because of the problems of measuring flood events owing to water velocity and the large sediment loads that move in the main channels. Both the velocity and sediment damage flow measurement equipment and structures.

19.6 CONCLUSIONS

Even with the application of assumed distributions for the model inputs and parameters, uncertainty with regard to the population numbers at Humayma remains and therefore the model, at this stage, is unable to provide a full solution to the difficult problem of determining population numbers based on water balance or otherwise. Despite this, this modelling exercise has been useful as a further development of the methodology to estimate population levels from a water balance approach. Specific further developments required are greater consideration of the frequency–duration–intensity of rainfall events and the runoff coefficients that need to be applied at different scales: catchment, hill-slope and field. These problems are central to the science of hydrology and, to be addressed, further hydrological measurements are required to develop a better understanding of how semi-arid and arid landscapes respond to rainfall events, particularly in terms of runoff generation and subsequent transmission loss. To this end, it is suggested that the next phase of the hydrological study of Humayma should include the detailed mapping of ancient field systems to determine the extent of the micro-catchment areas draining to the fields. This would determine the amount of water available to each and help determine how much water was available for agriculture and how much to fill the cisterns and reservoirs.

Even if the hydrological issues were solved, if we assumed a higher ratio of animals to people, which may indeed have been the case historically, the potential population figure could go down within the model runs even further, and possibly to a number that does not make sense in a system with so many (nearly 60) cisterns, reservoirs and pools to store drinking water. Culturally, one way to explain this would be to assume a more pastoral than settled community of inhabitants. Despite the permanence of the masonry used in the many well-built structures within the 10 ha Humayma settlement, the residents may only have occupied the site periodically, probably seasonally, from when the fields were sown through pasturage. Or perhaps the resident population expanded during this period. Because the site is located along a major historic artery of travel through Transjordan, another possibility is that foodstuffs were also imported. Foods may even have been bartered by travellers passing through for access to the local water stores. Therefore, because Humayma no doubt received transients, and in effect shared its water resources, and because the population that belonged may have come and gone throughout the year, such variations would be difficult to handle within a model designed as an annual water balance, which assumes a closed system in which a fixed number of year-round residents and their animals utilise the available water resources to be self-sustaining. Nonetheless, with further reflection, research and experimentation in model inputs and runs, these greater complexities could possibly be handled by distributions.

With respect to the hydrogeology and the proposed effect of long-term erosion changing the North Qalkha sub-catchment size, the impact can be considered either minimal or highly significant. When the sub-catchment area increases in the proposed palaeo-scenario, the area for surface runoff generation also increases and yet the drinking water-only population in the model run remains the same. This occurred because the cultivated area increased in proportion to the increase in total catchment area and therefore the additional runoff generated was infiltrated into the field with no increase in water reaching the cisterns and reservoirs in the settlement centre. If it is instead assumed that a reasonably small area was cultivated and was not proportionally greater given the greater size of the palaeo-catchment, for example only the 1 km^2 area north of the settlement centre, the greater water flow could not only have been contained in this field area, but also more water could have reached the cisterns and reservoirs in the settlement centre, having the same effect as discussed regarding higher rainfall scenarios in the model runs.

The phytolith evidence suggests that a tree crop, possibly olive trees, was cultivated in these environs. The soil character could retain the necessary water requirement, upwards of the 400 mm required depending upon how densely the trees would have been planted, and it therefore appears that this higher volume of water passing down through the sub-catchment was aggressively harvested and concentrated satisfying this greater water demand. There may even have been more water reaching the cisterns and reservoirs in the settlement centre and therein sustaining a greater population, either resident or transient or both, and possibly even a greater population than the Oleson rendering or present model runs have calculated. This greater amount of water available from the proposed palaeo-catchment size perhaps also helps understand why the perennial source of cool clear spring water was not connected to any of the reservoirs or cisterns in the original Nabataean system. One cistern, No. 45, is situated outside the present-day parameters of the North Qalkha tributary, in the palaeo-area, hence such placement further supports the argument for the palaeo-catchment.

In conclusion, we recognise there remain unresolved problems in the hydrological estimations utilised in the water balance model. Questions also remain about the nature of the population at Humayma, notably the ratio of transient visitors to permanent inhabitants, and the economy, such as the extent of pastoralism and importation of food. Nevertheless, the approach we have developed has furthered our understanding about the intriguing settlement of Humayma and provided the basis for further archaeological and hydrogeological studies.

REFERENCES

Allen, R. G., L. S. Pereira, D. Raes and M. Smith (1998) *Crop Evapotranspiration – Guidelines for Computing Crop Water Requirements*. FAO Irrigation and drainage papers 56. Rome: Food and Agriculture Organisation of the United Nations.

Alt, A. (1936) Der südliche Endabschnitte der römischen Strasse von Bostra nach Aila Zeitschrift des Deutschen. *Palästina-Vereins*: 92–111.

Anonymous (1971) *Akhbār al-dawlah al-'abbāsīyah*, ed. A. a.-A. al-Dūrī and A. a.-J. al-Muṭṭalabī. Beirut.

Broshi, M. (1980) The population of western Palestine in the Roman–Byzantine Period. *Bulletin of the American Schools of Oriental Research* **236**: 1–10.

Bruins, H. J. (1986) Desert environment and agriculture in the Central Negev and Kadesh–Barnea during historical times. Unpublished PhD thesis, Agricultural University of Wageningen, MIDBAR Foundation.

Cordova, C. (2007) *Millennial Landscape Change in Jordan: Geoarchaeology and Cultural Ecology*. Tucson: University of Arizona Press.

Di Segni, L. (2004) The Beersheba Tax Edict reconsidered in the light of a newly discovered fragment. Scripta Classica Israelica **23**: 131–158.

Donahue, J. (1985) Hydrologic and topographic change during and after Early Bronze occupation at Bab edh-Dhra and Numeira. In *Studies in the History and Archaeology of Jordan II*, ed. A. Hadidi. Amman: Department of Antiquities pp. 131–140.

Dunne, T. (1978) Field studies in hillslope flow processes. In *Hillslope Hydrology*, ed. M. J. Kirby and R. J. Chorley. Chichester: Wiley pp. 227–293.

Eadie, J. W. and J. P. Oleson (1986) The water-supply systems of Nabataean and Roman Humayma. Bulletin of the American Schools of Oriental Research **262**: 49–76.

Evenari, M., L. Shanan and N. Tadmor (1982) *The Negev, Challenge of a Desert*. Cambridge, MA: Harvard University Press.

Frank, F. (1934) Aus der Araba I: Reiseberichte. Zeitschrift des Deutschen. Palästina-Vereins **57**: 191–280.

Geer, R. M. (1954) *Diodorus of Sicily*. Cambridge, MA: Harvard University Press.

Glueck, N. (1935) Explorations in Eastern Palestine. Annual of the American Schools of Oriental Research **2**: 1934–1935.

Goldberg, P. (1994) Interpreting Late Quaternary continental sequences in Israel. In *Late Quaternary Chronology and Palaeoclimates of the Eastern Mediterranean*, ed. O. Bar-Yosef and R. S. Kra. Tucson, AZ: Radiocarbon pp. 89–102.

Graf, D. (1983) The Nabateans and the Hisma: in the footsteps of Glueck and beyond. In *The Word of the Lord Shall Go Forth: Essays in Honor of David Noel Freedman*, ed. C. L. Meyers and M. O'Connor. Winona Lake, IN: Eisenbrauns pp. 647–664.

Groom, N. (1981) *Frankincense and Myrrh: A Study of the Arabian Incense Trade*. London: Longman.

Helms, S. (1981) *Jawa*. London: Methuen.

Hill, J. B. (2004) Land use and an archaeological perspective on socio-natural studies in the Wadi al-Hasa, west-central Jordan. American Antiquity **69**: 389–412.

Hillel, D. (1982) *Negev: Land, Water and Life in a Desert Environment*. New York: Praeger.

Hunt, C. O., H. A. Elrishi, D. D. Gilbertson et al. (2004) Early-Holocene environments in the Wadi Faynan, Jordan. The Holocene **14**: 921–930.

Jenson, S. K. and J. O. Domingue (1988) Extracting topographic structure from digital elevation data for geographical information system analysis. Photogrammetric Engineering and Remote Sensing **54**: 1593–1600.

Jones, H. L. and J. R. S. Sterret (1946–1963) *The Geography of Strabo*. Michigan: Heinemann.

Jordan Meteorological Department (1971) *Climatic Atlas of Jordan*. Amman: Hashemite Kingdom of Jordan, Ministry of Transport, Meteorological Department.

Kedar, Y. (1957) Ancient agriculture at Shivtah in the Negev. Israel Exploration Journal **7**: 178–189.

Ker, A. (1995) *Farming Systems of the African Savanna – A Continent in Crisis*. Ottawa, Canada: International Development Research Centre.

Laborde, L. d. (1830) *Voyage de l'Arabie Pétrée*. Paris: Girard.

Maughan, W. C. (1874) *The Alps of Arabia: Travels in Egypt, Sinai, Arabia and the Holy Land*, 2nd Edition. London: H. S. King.

Morris, E. J. (1842) *Notes of a Tour through Turkey, Greece, Egypt, Arabia Petraea, to the Holy Land*. Philadelphia: Carey and Hart.

Musil, A. (1926) *The Northern Hedjaz*. New York: American Geographical Society Oriental Explorations and Studies.

Natural Resources Authority – Amman (1977) *National Water Master Plan of Jordan*. Amman: Natural Resources Authority, in cooperation with the German Agency for Technical Cooperation, Federal Republic of Germany.

Oleson, J. P. (1986) The Humayma Hydraulic Survey: preliminary report of the 1986 season. Annual of the Department of Antiquities of Jordan **30**: 253–260.

Oleson, J. P. (1988) The Humayma Hydraulic Survey: preliminary report of the 1987 season. Annual of the Department of Antiquities of Jordan **32**: 155–169.

Oleson, J. P. (1990) The Humayma Hydraulic Survey: preliminary report of the 1989 season. Annual of the Department of Antiquities of Jordan **34**: 285–311.

Oleson, J. P. (1991) Aqueducts, cisterns, and the strategy of water supply at Nabataean and Roman Auara (Jordan). In *Future Currents in Aqueduct Studies*, ed. A. T. Hodge. Leeds: Francis Cairns.

Oleson, J. P. (1992) The water-supply system of Ancient Auara: preliminary results of the Humeima Hydraulic Survey. In *Studies in the History and Archaeology of Jordan IV*, ed. M. Zaghoul Amman: Department of Antiquities pp. 269–276.

Oleson, J. P. (1995) The origins and design of Nabataean water-supply systems. In *Studies in the History and Archaeology of Jordan V*, ed. G. Bisheh. Amman: Department of Antiquities pp. 707–719.

Oleson, J. P. (1997a) Humeima. In *Oxford Encyclopedia of Archaeology in the Near East*, ed. E. Meyers. Oxford: Oxford University Press pp. 121–122.

Oleson, J. P. (1997b) Landscape and cityscape in the Hisma: the resources of Ancient Al-Humayma. In *Studies in the History and Archaeology of Jordan VI*, ed. G. Bisheh. Amman: Department of Antiquities pp. 175–188.

Oleson, J. P. (2001a) King, emperor, priest, and caliph: cultural change at Hawara (Ancient al-Humayma) in the first millennium AD. In *Studies in the History and Archaeology of Jordan VII*, ed. G. Bisheh. Amman: Department of Antiquities pp. 569–580.

Oleson, J. P. (2001b) Water Supply in Jordan through the ages. In *The Archaeology of Jordan*, ed. B. MacDonald, R. Adams and P. Bienkowski. Sheffield: Sheffield University Press pp. 603–624.

Oleson, J. P. (2004) 'Romanization' at Hawara (al-Humayma)? The character of 'Roman' culture at a desert fortress. In *Studies in the History and Archaeology of Jordan VIII*. Amman: Department of Antiquities pp. 353–360.

Oleson, J. P. (2007a) From Nabataean king to Abbasid caliph: the enduring attraction of Hawara/al-Humayma, a multi-cultural site in Arabia Petraea. In *Crossing Jordan: North American Contributions to the Archaeology of Jordan*, ed. T. E. Levy, P. M. Daviau, R. W. Younker and M. Sha'er. London: Equinox Publishing pp. 447–456.

Oleson, J. P. (2007b) Nabataean water supply, irrigation and agriculture: an overview. In *The World of the Nabataeans*, ed. K. D. Poltis. Stuttgart: Franz Steiner Verlag pp. 217–249.

Oleson, J. P. (2008) Social and technological strategies for the design of Nabataean water-supply systems in hyper-arid environments. In *L'eau Comme Patrimoine, de la Méditerranée à l'Amerique du Nord*, ed. E. Hermon. Québec: Les Presses de L'Université Laval.

Oleson, J. P. (2010) *Humayma Excavation Project*, final report. I: *The Site and the water-supply system. Annual of the American Schools of Oriental Research*, 15. Boston: ASOR.

Oleson, J. P., K. Amr, R. Foote et al. (1999) Preliminary report of the Al-Humayma Excavation Project, 1995, 1996, 1998. American Schools of Oriental Research Excavation Reports **43**: 411–450.

Oleson, J. P., K. Amr, R. Schick and R. Foote (1995) Preliminary report of the Humeima Excavation Project, 1993. American Schools of Oriental Research Excavation Reports **39**: 317–354.

Oleson, J. P., K. Amr, R. Schick, R. Foote and J. Somogyi-Csizmazia (1993) The Humeima Excavation Project, Jordan: preliminary report of the 1991–1992 seasons. American Schools of Oriental Research Excavation Reports **37**: 461–502.

Oleson, J. P., G. Baker, E. de Bruijn et al. (2003) Preliminary report of the Al-Humayma Excavation Project, 2000, 2002. American Schools of Oriental Research Excavation Reports **47**: 411–450.

Oleson, J. P., M. B. Reeves, G. Baker et al. (2008) Preliminary report on excavations at al-Humayma, Ancient Hawara, 2004 and 2005. American Schools of Oriental Research Excavation Reports **52**: 309–342.

Oleson, J. P., M. B. Reeves and B. J. Fisher (2002) New dedicatory inscriptions from Humayma (Ancient Hawara), Jordan. Zeitschrift für Papyrologie und Epigraphik **140**: 103–121.

Osborn, G. and M. Dunford (1981) Geomorphological processes in the inselberg region of S.W. Jordan. Palestine Exploration Quarterly **113**: 1–17.

Puigdefàbregas, J. and G. Sánchez (1996) Geomorphological implications of vegetation patchiness on semi-arid slopes. In *Advances in Hillslope Processes*, ed. M. G. Anderson and S. M. Brooks. Chichester: Wiley pp. 1027–1060.

Ramsey, J. (2004) *An analysis of the archaeobotanical remains from Humaima, Jordan*. Unpublished report.

The Sphere Project (2004) *Humanitarian Charter and Minimum Standards in Disaster Response*. Geneva: Published online. (http://www.sphereproject.org/)

Wheater, H. S. (2002) Hydrological processes in arid and semi-arid areas. In *Hydrology of Wadi Systems*, ed. H. S. Wheater and R. A. Al-Weshah. Paris: UNESCO pp. 5–22.

Whitehead, P. G., S. J. Smith, A. J. Wade *et al.* (2008) Modelling of hydrology and potential population levels at Bronze Age Jawa, Northern Jordan: a Monte Carlo approach to cope with uncertainty. *Journal of Archaeological Science* **35**: 517–529.

Williams, J. and M. Bonell (1988) The influence of scale of measurement on the spatial and temporal variability of the Philip infiltration parameters – an experimental study in an Australian savannah woodland. *Journal of Hydrology* **104**: 33–51.

Yair, A. and A. Kossovsky (2002) Climate and surface properties: hydrological response of small arid and semi-arid water sheds. *Geomorphology* **42**: 43–57.

Part V
Palaeoeconomies and developing archaeological methodologies

20 The reconstruction of diet and environment in ancient Jordan by carbon and nitrogen stable isotope analysis of human and animal remains

Michela Sandias

ABSTRACT

Diet reconstruction by carbon and nitrogen stable isotope analysis of human and faunal skeletal remains can shed light on exploitation of resources in ancient Jordan and may contribute to the understanding of past environments. This contribution reports on the results of a diachronic investigation of dietary and environmental change in the north of Jordan by stable isotope analysis. Dates of samples range between the Middle Bronze Age and the Early Islamic period, and the archaeological sites included in the discussion are Gerasa, Ya'amūn, Yajūz and Sa'ad, in the Western Highlands, and Pella, in the Jordan Valley. Results indicate the predominance of foods derived from C_3 plants in the human diet over all periods, as part of a mixed diet of plant and animal protein. Raised carbon stable isotope ratios for some domestic herbivores show that C_4 plants were consumed. These were sometimes combined with high $\delta^{15}N$ values, which suggest animal husbandry strategies that made use of arid environments.

20.1 THE ISOTOPIC APPROACH IN THE STUDY OF DIET AND ECOLOGY OF ANCIENT JORDAN

Stable isotope analysis of bone collagen is an established methodology that allows diet reconstruction at an individual level (Ambrose, 1993; Katzenberg, 2000; Sealy, 2001). It can, therefore, complement information on food resources obtained through the study of faunal and botanical remains, as well as information from textual sources. Moreover, as bone collagen isotopic composition is related to certain habitat characteristics, such as aridity (Heaton, 1987; Ambrose and DeNiro, 1989), this technique may also improve the understanding of past environments. The climatic variability that characterises Jordan is reflected in a diversity of ecosystems. Cultural factors and ecological features have shaped the way people have interacted with their environment. Food choices and foddering strategies, in particular, are examples of human behaviours that are influenced by environmental conditions as well as by cultural factors. By reconstructing patterns of human and animal diet, carbon and nitrogen stable isotope analysis of skeletal remains can contribute to the understanding of the relationship between human populations, their animals and the surrounding habitat (see for example Richards et al., 2003; Thompson et al., 2008). For this study, bone samples were taken from five archaeological sites (see map in Figure 20.1), of which Pella is located in the Jordan Valley, while Gerasa, Khirbet Yajūz, Ya'amūn and Sa'ad are in the Northern Western Highlands. Dates of bone samples span the period from the Middle Bronze Age to the Early Islamic period.

The ratios of the stable isotopes of carbon and nitrogen ($^{13}C/^{12}C$ and $^{15}N/^{14}N$, respectively) are amongst the most frequently measured in ancient skeletal remains for diet and environmental reconstruction. The stable isotope ratios measured in human and animal bone collagen reflect the isotopic composition of the foods, and especially of dietary protein (Ambrose and Norr, 1993; Tieszen and Fagre, 1993), consumed over several decades of life of an individual (Hedges et al., 2007). Different foods, in turn, show a range of stable isotope ratios related to the isotopic composition of the environment they are produced in.

When referred to a standard, the stable isotope ratios of the carbon and nitrogen are indicated as $\delta^{13}C$ and $\delta^{15}N$, respectively. Such delta values are expressed in ‰ and calculated as

$$\delta = [(Rx - R_{std})/R_{std}] \times 1000$$

where R is the ratio of the abundance of the heavy to the light isotope, x is the sample and std stands for 'standard' (Sharp, 2007, pp. 17–20). Routinely, carbon isotope ratios are reported relative to VPDB, a marine carbonate, and nitrogen isotope ratios are

Water, Life and Civilisation: Climate, Environment and Society in the Jordan Valley, ed. Steven Mithen and Emily Black. Published by Cambridge University Press.
© Steven Mithen and Emily Black 2011.

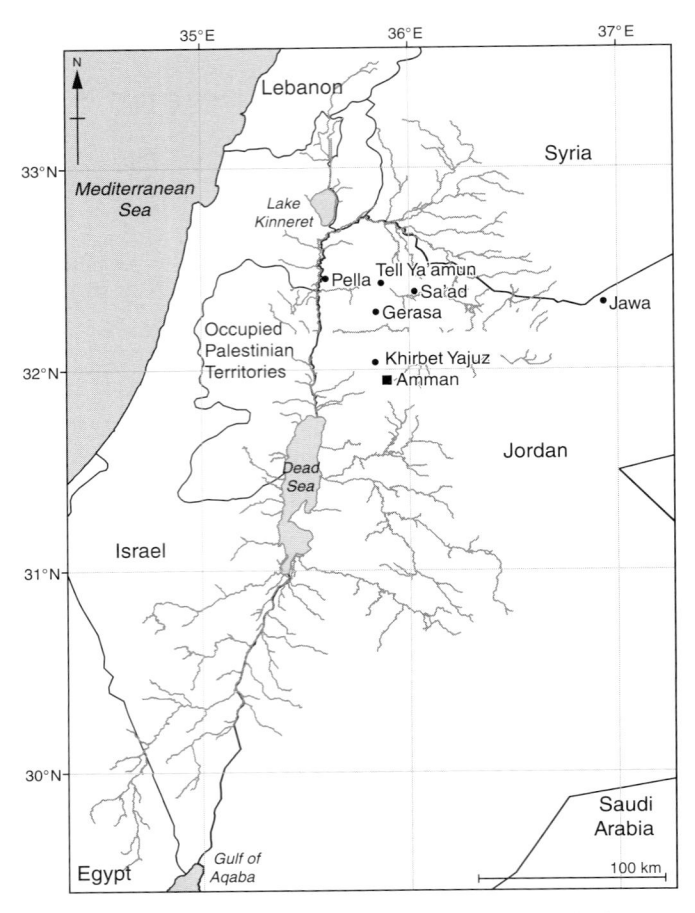

Figure 20.1. Map of Jordan indicating the sites mentioned in the text.

expressed relative to atmospheric nitrogen or AIR (ambient inhalable reservoir).

Physical, chemical and biological processes are responsible for the unequal distribution of the isotopes of carbon and nitrogen between different compartments (e.g. compounds in soils and waters, as well as tissues or fractions of tissues in plants and animals). Processes such as phase transitions and biologically mediated reactions may cause 'isotope fractionation' (Hoefs, 1997, p. 5) and hence may be responsible for the isotopic characterisation of different food chains (Hoefs, 1997, pp. 5–11; Pollard and Wilson, 2001).

20.1.1 Carbon

Plants are at the base of the food web, and the transformation of atmospheric CO_2 into organic compounds during photosynthesis in plants is the main source of carbon isotopic variation in terrestrial ecosystems. Plants can be distinguished into C_3 and C_4 plants, according to their photosynthetic pathway. C_3 plants discriminate more strongly against the heavy isotope ^{13}C in the atmosphere and therefore display more negative $\delta^{13}C$ values than C_4 plants, i.e. C_3 plants are more ^{13}C-depleted than C_4 plants. Mean $\delta^{13}C$ ratios are –26‰ and –12.5‰ for C_3 and C_4 plants, respectively (Smith and Epstein, 1971). C_3 plants are typical of temperate environments. Most plants used for human consumption are C_3 plants and examples are wheat, most fruits, legumes and nuts. C_4 plants are usually adapted to high light intensity, high temperatures and frequent water shortages. They include many tropical grasses and the cultural crops millet, maize, sorghum and sugar cane. In the Middle East, a survey of modern grass species of Sinai, Negev and Judea showed that, overall, C_3 plants predominate. However, C_4 species occur with higher frequency in areas with lower precipitation and represent, for instance in the Arava Valley, more than 70% of grasses (Vogel et al., 1986). Amongst non-grass species, C_4 photosynthesis is a relatively common adaptation in species of the Chenopodiaceae and Polygonaceae families that grow in modern Israel and Jordan (Winter, 1981). C_4 chenopods, for instance, are a frequent occurrence in the steppe and desert habitats of Israel (Shomer-Ilan et al., 1981). Several such plant types are components of the Jordanian present-day vegetation (Al-Eisawi, 1996). The differences in the carbon isotopic values of C_3 and C_4 plants are passed on to the body tissues of their consumers. Therefore, the $\delta^{13}C$ ratios of herbivore bone collagen may provide an indication of the relative contributions of C_3 and C_4 plants in their diet. For humans, who in most cases consume an omnivorous diet, they may reflect either direct consumption of plants or the isotopic composition of meat or dairy products derived from C_3- or C_4-fed animals (Ambrose, 1993). Measurements of $\delta^{13}C$ ratios in bone collagen may therefore show whether any C_4 plants were included in the diet of ancient Jordanian herbivores and indirectly contributed to human diet. An additional factor that may influence carbon isotope values in animals is trophic level. The $\delta^{13}C$ ratios of consumers can become ^{13}C-enriched relative to food (DeNiro and Epstein, 1978). In particular, for modern carnivores that feed on herbivores this enrichment has been estimated to range between 0 and 2‰ (Bocherens and Drucker, 2003). Because of this 'trophic level effect' of carbon, which results in a small increase in consumer $\delta^{13}C$ in exclusively C_3-plant-based diets, it can be difficult to trace the consumption of small amounts of C_4-based protein in consumer isotope signals with confidence. Within the environmental context of Jordan, variation between C_3 and C_4 plants at the base of the food chain is likely to explain most carbon isotope variation in animal and consumer tissues. The $\delta^{13}C$ ratios are also used to identify consumption of marine resources (Schoeninger and DeNiro, 1984; Richards and Hedges, 1999).

20.1.2 Nitrogen

The $\delta^{15}N$ ratios of bone collagen are related to the trophic position of the organism in the food web. This relationship consists in the enrichment in ^{15}N moving up along a food chain. Field and laboratory observations have shown that ample variability exists in the size of the enrichment between food and consumer. Results of these studies suggested the use of an average incremental enrichment of 3‰ at each stage of the food chain to estimate the $\delta^{15}N$ ratio of the diet (DeNiro and Epstein, 1981; Minagawa and Wada, 1984; Schoeninger and DeNiro, 1984). Yet, following observations and isotopic analyses of the members of a modern terrestrial food web, the range 3–5‰ was proposed as more appropriate than a fixed value, because it accounts for the complexity and species-specificity of the trophic level shift in $\delta^{15}N$ (Bocherens and Drucker, 2003). Thanks to the trophic level effect, when $\delta^{15}N$ ratios of humans and animals from comparable contexts are available, relative contributions of plant and animal protein to human diet can be suggested (Hedges and Reynard, 2007). However, $\delta^{15}N$ ratios cannot distinguish between different foods of animal origin such as meat and dairy products (O'Connell and Hedges, 1999; Petzke et al., 2005).

Many factors other than trophic level have been shown to affect the nitrogen stable isotope composition of body tissues. A marked negative correlation has been demonstrated between bone collagen $\delta^{15}N$ ratios and amount of rainfall, so that herbivores living in arid environments have elevated $\delta^{15}N$ ratios. High $\delta^{15}N$ ratios in low rainfall habitats have been found to be related to physiological mechanisms of water conservation, low protein diets and ^{15}N-enriched soil resulting from phenomena such as volatilisation of ^{15}N-depleted ammonia (Heaton et al., 1986; Sealy et al., 1987; Ambrose, 1991; Gröcke et al., 1997; Schwarcz et al., 1999). As aquatic food chains are longer than terrestrial ones, resulting in greater enrichment in ^{15}N of aquatic organisms, higher $\delta^{15}N$ ratios can also indicate consumption of marine and freshwater resources (Schoeninger et al., 1983).

20.2 PREVIOUS CARBON AND NITROGEN ISOTOPE STUDIES IN THE MENA REGION

At present, examples of applications of stable isotope analysis for diet reconstruction to skeletal remains from southwest Asia and northern Africa are not numerous. Work in Jordan has been carried out by Al-Shorman (2004) who analysed tooth enamel carbonate from the Middle and Late Bronze Age cemetery of Ya'amūn. The isotope data suggested a diet mostly based on C_3 plants. No differences were found between the two Bronze Age phases even though changes in climate had occurred (Al-Shorman, 2004). Carbonate analysis does not allow for the measurement of nitrogen stable isotope ratios, so no inferences could be drawn about trophic level or aridity. Within the wider context of the Middle East, Richards and colleagues (2003) examined remains from Neolithic Çatalhöyük, Turkey. This study suggested a diverse animal diet that included C_4 plants. The human dietary signals also suggested a contribution from C_4 plant-based protein, although this was interpreted as being due to the consumption of herbivore products rather than direct consumption of C_4 plants.

Stable isotope data also demonstrated that cattle did not make a significant contribution to human diet (Richards et al., 2003). The development across time in the herding strategy of sheep and goats was the focus of an isotope study on skeletal remains from Neolithic Çatalhöyük and Aşıklı Höyük in south-central Anatolia. Carbon and nitrogen stable isotope ratios in sheep and goats showed the wider range of plants (including C_4 plants) consumed by herbivores at Çatalhöyük when compared with the homogeneous diet at earlier Aşıklı Höyük. Isotope data of sheep and goats from the former site reflect exploitation of wider territories characterised by heterogeneous vegetation. Such data are in contrast with isotope ratios in Aşıklı Höyük sheep and goats. Such animals, if representative of a wild population, may have been artificially restricted to certain feeding areas devoid of C_4 plants, which were consumed by contemporary wild cattle. Alternatively, Aşıklı Höyük sheep/goats, if representative of protodomestic ungulates, may have been managed as a single flock and possibly kept close to the settlement. Additionally, at both Anatolian sites, sheep were indistinguishable from goats in their carbon and nitrogen stable isotope signatures, suggesting consumption of a similar diet (Pearson et al., 2007). Many isotope studies from the region, up until now, have been carried out on material from the Nile Valley. The patterns of human diet that emerged from the studies by Iacumin and colleagues (1996) and Thompson and colleagues (2005) both indicate that resources in the Egyptian Nile Valley were dominated by C_3 plant-derived foods. The observed $\delta^{13}C$ and $\delta^{15}N$ ratios were also compatible with the archaeologically attested importance of freshwater fish consumption, although environmental conditions, and aridity in particular, offer an alternative explanation for the high nitrogen isotopic values (Iacumin et al., 1996; Thompson et al., 2008). Further south, in Nubia, the proportion of C_4 plants that directly or indirectly contributed to human diet was greater than in Egypt (Iacumin et al., 1998; Thompson et al., 2008). A study of carbon and nitrogen isotopic values in humans from Wadi Halfa, Lower Nubia, was carried out by White and co-workers (2004). They established that C_4 plants formed a greater proportion of human diet at that site between AD 350 and 550 than between AD 500 and 1400. These results could be correlated with data on changes in water levels of the Nile river, which were relatively low in the earlier period, explaining the need to grow more arid-adapted C_4 crops (White et al., 2004).

20.3 THE ARCHAEOLOGICAL SITES

20.3.1 Pella

Ancient Pella lies on the eastern side of the Jordan Valley, 5 km from the Jordan River and about 75 km from the Dead Sea. The site consists of two hills separated by the small valley of Wadi Jirm, characterised by a perennial spring and stream (Walmsley *et al.*, 1993; Bourke *et al.*, 2006). Since the Late Lower Palaeolithic, the region of Pella/Tabaqat Fahl has been settled with few interruptions until the present day. Carbon and nitrogen isotopic data from humans and animals dated to the Middle and Late Bronze Age and to the Late Roman, Byzantine and Umayyad periods are presented in this chapter (abbreviations in text and figures are MB, LB, LR, Byz and Um, respectively).

During the Middle Bronze Age, Pella was a flourishing city, possibly inhabited by 2,000 people, featuring large-scale constructions, city walls and a temple, and terracing, these structures having characteristics in common with constructions at, for instance, Megiddo (Smith, 1987; Bourke, 1997; Bourke *et al.*, 2006, Falconer, 2008). Imported or 'foreign inspired' items are present in great number during this period and attest to lively trade interactions. In particular, Egyptian and Syrian sources suggest that Pella was involved in trade of products such as wine and olive oil, wool and textiles, as well as cattle and sheep (Knapp, 1993, p. 38; Bourke *et al.*, 2006, p. 48). During the Late Bronze Age, Pella is described in several texts and inscriptions as a thriving city-state having strong relationships with Egypt, which at that time ruled over Transjordan (Bourke, 2006). Intensive building activities, imports and luxury goods, and elaboration of tombs characterise the earlier phase of the LB (Bourke *et al.*, 1994, p. 9; Bourke, 2006).

Since Roman times Pella is listed as a member of the Decapolis. This term indicates a group of cities of Hellenistic origin which, besides identifying a geographical area (Bourke, 2006, p. 13), starting from the first century AD, may have also formed a political union (Schmid, 2008). Late Roman Pella featured the public buildings that characterised all Graeco-Roman cities of that period, intensive construction activities demonstrating economic vitality (Smith and Day, 1989; Freeman, 2008). Additionally, during the second century AD, the Roman administration constructed or improved many roads in this region (Kennedy, 2007, pp. 88–95), so that, included within such a network of secure routes, Pella's economic vitality increased. Smith (1987) suggests that the city derived part of its wealth from being an intermediate station for camel caravans of traders. During the Byzantine period, Pella continued to derive much of its wealth from being one of the centres through which camel caravans passed on their way to and from the Mediterranean coast and the west in general. Overall, the city thrived in the context created by a central administration, a state religion guaranteeing social cohesion, and possibly a slightly cooler/wetter climate than has been experienced previously (Watson, 2008). After the Muslim Conquest (AD 636/640) Pella was inhabited by a population composed of Muslims and Christians. Excavation of Early Islamic domestic buildings bore evidence of private wealth, testified by material culture and abundance of remains of domestic animals (McNicoll *et al.*, 1982, pp. 130–139). From Pella were sampled 20 MB/LB and 16 AD 3rd–4th humans; 6 MB/LB, 12 LR/Byz, and 9 Um sheep/goats; 6 LR/Byz and 10 Um pigs.

20.3.2 Gerasa/Jerash

Gerasa lies 58 km north of Amman in the Highlands of Ajloun, northwest Jordan. The city is located along the perennial course of the Wadi Jerash. This region is characterised by rich soil, abundant rainfall and, since Hellenistic time, high density of settlements (Kennedy, 2007). The Hellenistic city of Gerasa was founded around the middle of the second century BC. The size of the original settlement was probably small and limited to the elevated area to the west of the river. Initially the settlement seems to have consisted of houses and agricultural installations surrounding the acropolis on the hilltop, while a small temple dedicated to Zeus was located on a nearby hill (Seigne, 1992; Kennedy, 2007; Schmid, 2008).

The Late Hellenistic and Early Roman periods (first century BC to first century AD) represent a time of expansion of Gerasa, one of the Decapolis cities. At this time, large areas around the city were colonised and Gerasa grew exponentially (Seigne, 1992; Freeman, 2008). Samples of human bone from Gerasa were obtained from the mass burials at the Hippodrome, which suggest that a catastrophic event, most likely a plague, afflicted the city in the mid-seventh century AD. At this time, Gerasa's population was a mixture of Muslims and Christians, the latter probably representing the majority of inhabitants (Gawlikowski, 2001; Walmsley and Damgaard, 2005). In this period, aspects of the economy, such as the production of domestic pottery, were still flourishing in continuity with the Byzantine period. Furthermore, the presence of a mint confirms the administrative importance of the city, which is mentioned in some texts as the capital of an Early Islamic district (Walmsley, 1997; Walmsley and Damgaard, 2005). Twenty-two individuals from the Hippodrome mass burial were sampled.

20.3.3 Tell Ya'amūn

Ya'amūn is located north of Gerasa and 25 km southeast of Irbid in the hills bordering the fertile southern Hawrān. Elevation of the site is 828 metres above sea level (m asl). Excavations at Ya'amūn which started in 1999 and are still ongoing found evidence of occupation spanning from the Early Bronze Age to the Islamic periods (Burke and Rose, 2001; El-Najjar *et al.*, 2001; Al-Shorman, 2004; Rose *et al.*, 2007). However, so far, excavations and publications have focused on exploring and describing Late Roman and Byzantine remains. During Late Antiquity, Tell Ya'amūn was a densely inhabited village which relied on intensive agriculture. This is testified by more than 50 water cisterns, two olive presses and four wine presses. One of the wine presses, in particular, was a sophisticated facility of considerable size which excavators associate with commercial production of wine (El-Najjar and Rose, 2003; Rose *et al.*, 2007). Additionally, the architectural remains around the Byzantine basilica church suggest that an important religious complex might have existed at Ya'amūn. Mosaics decorating the building are of high quality and comparable to those found in contemporaneous churches at the near Decapolis cities (El-Najjar *et al.*, 2001; Rose *et al.*, 2007). The tombs at Ya'amūn show great diversity in size and decoration and suggest significant differences in the amount of resources that inhabitants of this community could afford to invest in funerary artefacts. The remains of 45 individuals were sampled from Ya'amūn tombs. Of these, 22 date to the Middle and Late Bronze Age, while 23 were from tombs dating to the Late Roman/Byzantine period. In addition, isotope data from 11 samples of sheep or goat remains dating from the Middle and Late Bronze Age and from Late Roman/Byzantine time will be presented.

20.3.4 Khirbet Yajūz

Khirbet Yajūz is located about 11 km northwest of the centre of Amman. The first scholars who visited the site in the late nineteenth century described Khirbet Yajūz as a large Romano-Byzantine town strategically placed along the way connecting Amman/Philadelphia and Gerasa (Suleiman, 1996). Amongst the main structures excavated so far are a basilica church and a small chapel. The basilica had mosaic floors and plastered and painted walls, all features found in similar contemporary churches of other towns of the region. The chapel overlies an underground cemetery with loculi and built-up graves. Further structures consist in industrial facilities, in particular a mill and two wine presses (Khalil, 1998). During the sixth and seventh centuries, when basilica, chapel and cemetery, manifest signs of prosperity, were founded, and when the whole area around Amman/Philadelphia was densely inhabited (Watson, 2008), Yajūz may have been a large public centre for wine making (Khalil and al-Nammari, 2000). Artefacts and objects recovered from the cemetery were unevenly distributed amongst the burials, so that some social heterogeneity seems to have existed at Yajūz (Khalil, 1998). Human bone samples were taken from 14 individuals recovered from the cemetery.

20.3.5 Sa'ad

Sa'ad is located in the eastern extension of the Ajloun mountains at around 900 m asl. Soil in this area shows a red-brown colour and is rich in minerals so that it supports, at present, numerous crops. Furthermore, a permanent spring is present in the vicinity (Williams *et al.*, 2004; Rose *et al.*, 2007, p. 179). The archaeological site is 27 km west of the city of Mafraq and 17 km northeast of Gerasa. Since ancient times this region was included in the network of trade routes linking the cities and towns of Syria, Palestine, Jordan and Egypt (Figure 1 of Rose and Burke, 2004). Remains of Late Roman buildings represent the earliest evidence for occupation, so Sa'ad could be one example of the many new agricultural settlements that were established during the Late Antiquity in the southern Levant (Parker, 1999, p. 157; Rose and Burke, 2004; Rosen, 2007; Watson, 2008). During the Byzantine period the town had a church and a wine press. The dimensions of this wine press would have been suitable for large-scale grape and wine production intended for local consumption as well as for export. The objects recovered from the tombs include pieces of jewellery made in iron or copper alloy whose designs imitate more expensive items (Rose *et al.*, 2004, 2007). The skeletal remains of 21 individuals dating to Late Roman/Byzantine period were sampled for isotopic analysis.

20.4 ANIMAL AND HUMAN ISOTOPE DATA FROM NORTH JORDAN

20.4.1 Herding strategies and human diet at Pella and Ya'amūn

Successful extraction of good-quality bone collagen from faunal skeletal remains was very variable within and between the different sites. Well preserved animal bones were found at Pella and at Ya'amūn. Figures 20.2 and 20.3 show mean δ^{13}C and δ^{15}N values for animals and humans from Pella and Ya'amūn, respectively, distinguished on the basis of species and time period. Error bars represent one standard deviation, a measure of dispersion of values around the mean. In the legend, figures in parentheses indicate number of individuals. At Pella (Figure 20.2), mean δ^{13}C values for sheep/goats range across time from −18.9‰ for earlier animals to −19.4‰ for Umayyad specimens. Mean δ^{15}N values range between 8.1‰ for Middle/Late Bronze (MB/LB) sheep/goats and 6.7‰ for Late Roman–Byzantine (LR/Byz) animals. Mean δ^{13}C and δ^{15}N ratios in LR Byz and Umayyad pigs are very close to each other, plotting around −20.2‰ for carbon, and assuming the values 6.0 and 6.5‰ for nitrogen. All humans from Pella have mean δ^{13}C value of −19.4‰. As for nitrogen, Late Antiquity people have mean δ^{15}N value of 8.9‰, whereas the single individual dating to the MB/LB is slightly higher in its δ^{15}N value.

At Ya'amūn (Figure 20.3), mean δ^{13}C values for sheep/goats are more positive in MB and LB animals (−18.5‰) than in specimens dating to the Late Antiquity period (−19.3‰). The two groups of sheep/goats differ markedly in their mean δ^{15}N ratio. MB and LB herbivores have mean δ^{15}N of 8.5‰, while LR/Byz sheep/goats plot around a δ^{15}N of 5.3‰.

The most ^{13}C-depleted values were found in pigs, while some of the sheep/goats had the most ^{13}C-enriched values. In pigs (Figure 20.2), δ^{13}C values indicate a diet consisting almost exclusively of C_3 foodstuff. This diet composition persisted at

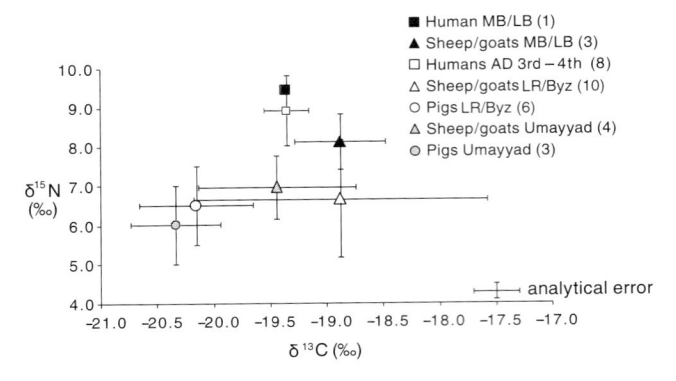

Figure 20.2. Mean isotopic values (±1 standard deviation) for humans and domestic animals from Pella. Figures in parentheses indicate number of individuals. Abbreviations: MB/LB = Middle Bronze/Late Bronze; LR/Byz = Late Roman/Byzantine.

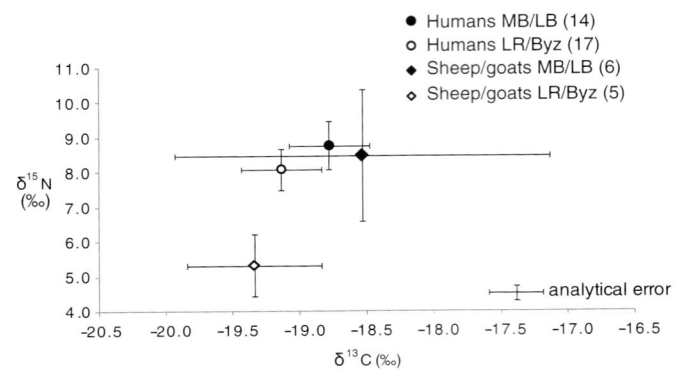

Figure 20.3. Mean isotopic values (±1 standard deviation) for humans and domestic animals from Ya'amūn. Figures in parentheses indicate number of individuals. Abbreviations: MB/LB = Middle Bronze and Late Bronze; LR/Byz = Late Roman/Byzantine.

Pella at the transition between Byzantine and Umayyad time. In contrast, particularly during Late Roman/Byzantine time, sheep/goats at Pella show a much wider spread of carbon values than pigs, indicating varying contributions of C_4 plants to their diet. Additionally, the spread of $\delta^{15}N$ ratios is larger in sheep/goats than in pigs. Overall, the $\delta^{13}C$ and $\delta^{15}N$ values of domestic animals from Pella reflect the difference in the feeding orientation and ecological requirements between sheep/goats and pigs. Pigs need to drink more often than sheep/goats, and do not tolerate the high light intensity of the open landscapes (Grigson, 2007). Here, their $\delta^{13}C$ values indicate that they were fed a monotonous diet based on C_3 plants. Their $\delta^{15}N$ values, similar to those of the sheep/goats, indicate a plant-based diet rather than an omnivorous diet that included also some kind of animal protein. As the $\delta^{15}N$ values of these pigs are substantially lower than values in contemporary humans, it is unlikely that they were fed scraps from the human table. Therefore, it is possible that the pigs slaughtered at Pella were kept close to the settlement, in shaded enclosures; however, these enclosures may have been at some distance from human habitations so that the practice of emptying the household kitchen bin into the pigs' trough may have not been common.

In contrast, the wider spread of the $\delta^{13}C$ and $\delta^{15}N$ values in sheep/goats shows that the more flexible ecology of these small herbivores was exploited and that animals were herded in the wider landscape. The $\delta^{13}C$ values indicate that during all periods sheep/goats consumed various amounts of the low rainfall-adapted C_4 plants. Spread of $\delta^{13}C$ values in the LR/Byz sheep/goats indicates that some of these animals consumed substantial amounts of C_4 plants. In contrast, lower amounts of this type of plants seem to have composed the diet of sheep/goats during the Umayyad period. Additionally, plants having particularly high $\delta^{15}N$ values, possibly as a consequence of soil salinity and aridity (Heaton, 1987; Schwarcz et al., 1999), contributed to the diet of LR/Byz as well as MB/LB sheep/goats. At present, C_4 plants and soil and climate conditions that may promote high $\delta^{15}N$ values in plants' tissues exist, for instance, in the Jordan Valley as well as in the steppe areas west and east of the Highlands (Winter, 1981; Al-Eisawi, 1996, Figure 2.11; Cordova, 2007). Both Pella and Ya'amūn are within short distances of semi-arid habitats such as the Jordan Valley and zones characterised by steppe vegetation.

At Ya'amūn, the homogeneity in $\delta^{13}C$ and $\delta^{15}N$ values among the sheep/goats from Late Roman/Byzantine time is in sharp contrast with the variability shown by isotope values of MB/LB sheep/goats. Mean $\delta^{13}C$ ratio shows that amongst the MB/LB sheep/goats there were individuals that consumed C_4 plants in significant amounts. As for nitrogen, earlier herbivores, in addition to a greater intra-group variability, have higher mean $\delta^{15}N$ ratios. The wide range of $\delta^{13}C$ and $\delta^{15}N$ values shown by these sheep/goats may indicate that at Ya'amūn during the Middle-to-Late Bronze Age more varied ways of managing sheep/goats were in use than during Roman and Byzantine times. During the later period, the animals may have been kept in proximity to the site and may have grazed the plants that grew around the irrigated fields (Rose et al., 2007) and that did not have the high $\delta^{15}N$ values reflecting soil aridity and salinity. Alternatively, the MB/LB animals with a more marked C_4 signature may have been brought to the site from more arid ecological settings with which trade connections existed (Rose, 2001). A further possible scenario is linked to the modifications in climate over time that have been documented for the Near East. In particular, the Middle and Late Bronze Age period was characterised, in the Levant, by numerous fluctuations in the amount of precipitation, while more regular rainfall seems to have characterised Late Antiquity (Rosen, 2007, p. 165 onwards). Therefore, $\delta^{13}C$ indicating C_4 foodstuff in the diet, coupled with higher $\delta^{15}N$ values possibly due to aridity, could reflect a period of lower rainfall and an advancement of the vegetation adapted to aridity towards the areas where Middle and Late Bronze Age people herded their animals. In contrast, during Late Antiquity, the more stable patterns of precipitation may have led to strategies of animal raising that kept the animals safe from water stress. Unfortunately, data from MB/LB Pella (Figure 20.2) are for the moment too sparse to support the scenario of changed environmental conditions over time in the north of Jordan. However, the mean δ^{15}N value of the MB/LB sheep/goats from Pella is consistent with the interpretation of herbivores grazing in semi-arid environments.

The ways of managing sheep/goats at Pella and Yaʿamūn during Late Antiquity appear rather different. At the former site, it seems that the herbivores were herded in diverse environments, including areas characterised by a certain degree of aridity and hence by presence of C_4 plants and of plants having higher δ^{15}N values. In contrast, at Yaʿamūn, sheep/goats seem to have enjoyed much less mobility while possibly taking advantage of the vegetation growing around the cultivated plots.

The availability of carbon and nitrogen isotopic data from both domestic fauna and humans from the same site and period allows an attempt to outline the human diet. At Pella (Figure 20.2), all sampled humans have δ^{13}C values suggesting a diet mostly based on C_3 plants, such as cultivated cereals and vegetables, which constituted, for the majority of the people, the greatest part of the food intake (Wilkins and Hill, 2006). For their supply of animal protein, the MB/LB human specimen may have consumed products from the contemporary sheep/goats. However, the moderate C_4 signal of the sheep/goats is not discernible in this human specimen as, very probably, most of its intake consisted of C_3 plant food (e.g. cereals) and possibly products from C_3-fed animals. The Late Roman–Byzantine inhabitants of Pella may have derived part of their C_3-plant signature from eating pork, while meat or secondary products from sheep/goats, which have in some instances a C_4-plant isotopic signature, played a lesser role.

All humans sampled from Yaʿamūn consumed a predominantly C_3 diet (Figure 20.3). During the Middle and Late Bronze Age, human diet included some animal protein. The comparison between the isotope ratios in humans and contemporary herbivores suggests that foods of animal origin mostly derived from the specimens which, among the sheep/goats that compose the sampled heterogeneous group, have the lowest δ^{13}C and δ^{15}N ratios. The C_4-fed animals, too, probably contributed to the human diet; however, the larger intake of C_3-derived products hides the C_4 signal. The later inhabitants of Yaʿamūn equally consumed a C_3-based diet, the sampled sheep and goats being one of the likely sources of animal proteins.

20.4.2 Human diet in the north of Jordan between the third and the seventh centuries AD

Good-quality collagen for isotope analysis was obtained from human skeletal remains from Pella, Gerasa, Yaʿamūn, Saʿad and

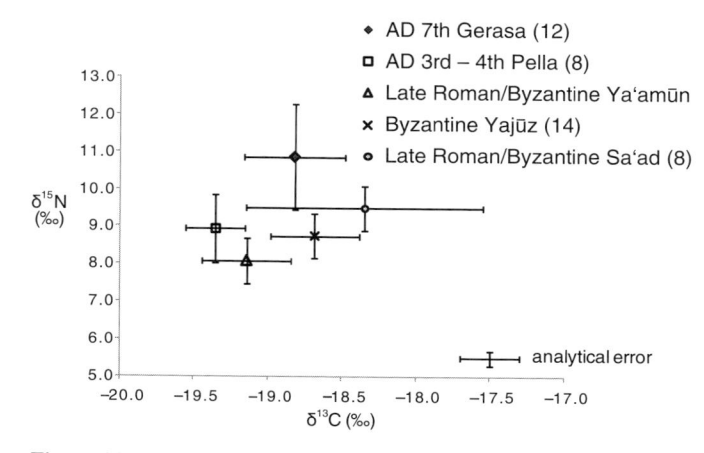

Figure 20.4. Mean isotopic values (± 1 standard deviation) for humans from Gerasa (seventh century AD), Pella (third–fourth centuries AD), Yaʿamūn (Late Roman/Byzantine); Yajūz (Byzantine) and Saʿad (Late Roman/Byzantine) in the north of Jordan.

Yajūz. Such samples range in date from the Late Roman to the Early Islamic periods. Mean δ^{13}C and δ^{15}N values for different groups are plotted in Figure 20.4. Mean values of the δ^{13}C ratios for these sites range between −19.3‰ and −18.3‰, while mean values of the δ^{15}N ratios fall in the interval 8.1–10.8‰. For most groups in Figure 20.3, intra-group variability for the δ^{13}C values is low, and diet was based on resources derived from C_3 plants. Saʿad is an exception, however, as at that site some C_4-plant-derived food was consumed in substantial amounts. At all sites intra-group differences in the δ^{15}N ratios are only slightly higher than differences found in the δ^{13}C ratios. The exception to this pattern is Gerasa where, among the individuals buried in the mass grave, δ^{15}N ratios are much more variable owing to the presence in the group of specimens having particularly high δ^{15}N ratios.

At both types of site, rural towns and cities, C_3-derived resources such as cereals represent the most important component of the diet. Error bars (Figure 20.4) show the spread of human carbon and nitrogen isotope data. Within site, heterogeneity of δ^{13}C and δ^{15}N ratios is low at the city of Pella as well as at the towns of Yaʿamūn and Yajūz. There were no differences in the isotopic quality of the foods consumed by the components of these communities. Differences in wealth between the inhabitants of the same site that are, for instance, reflected in the unequal distribution of grave goods at Yaʿamūn and Yajūz, are not visible as diversity in isotopically reconstructed diet.

In contrast to the homogeneity of Pella, Yaʿamūn and Yajūz, there is a wider spread in values seen at Late Roman/Byzantine Saʾad and Umayyad Gerasa (Figure 20.4). Food resources available at these two sites were more varied. At Saʿad δ^{13}C values are on average more positive, indicating that more C_4-derived foods were consumed. In particular, the difference in mean δ^{13}C values between Saʿad and Pella is relatively large and could be due to the general greater consumption at the former site of products from animals that fed in a semi-arid and low-rainfall environment, such as the region located east of Sa'ad. Carbon and nitrogen stable isotope ratios from faunal remains from archaeological sites located in that area will help in clarifying whether the causes of the more positive $\delta^{13}C$ ratios recorded at Sa'ad are due to environmental features.

At Gerasa, the $\delta^{15}N$ values are particularly variable (Figure 20.4). The highest $\delta^{15}N$ values could be due to consumption of foods particularly enriched in ^{15}N, such as plant or animal products from arid environments. Inclusion in the diet of protein from animals from higher trophic levels (e.g. pigs or poultry) could also be an explanation. For instance, protein with an isotopic signature close to that of sheep/goats from Umayyad Pella (Figure 20.2) could explain some of the highest among the $\delta^{15}N$ values measured in Gerasa's people. On the other hand, the lower average nitrogen isotopic signature of sheep/goats from Late Roman/Byzantine Ya'amūn (Figure 20.3) and of pigs from Umayyad Pella could account for the lowest human $\delta^{15}N$ values observed at Gerasa.

The heterogeneity in the $\delta^{13}C$ and $\delta^{15}N$ values at Sa'ad and Gerasa is in agreement with the picture of a society which was diverse in terms of the ethnicity, religion and language of its members (Cameron, 1993; Parker, 1999; Gawlikowski, 2001; Walmsley and Damgaard, 2005), the cultural identity of each group potentially influencing food preferences. In particular, the skeletal assemblage from Gerasa's Hippodrome was the result of a deadly pandemic, and it is likely that few distinctions by social status or creed were made when burying the dead. Therefore, and more than other burials in the dataset, the samples from the Hippodrome may be reflecting a true cross-section of Gerasa's heterogeneous population in the seventh century AD. Carbon and nitrogen isotope data of domestic herbivores from Umayyad Gerasa would certainly aid in discriminating between the environmental and cultural factors that may be causing the observed intra-site variability. For the moment, the preliminary comparison with the faunal $\delta^{13}C$ and $\delta^{15}N$ values from Umayyad Pella and Late Roman/Byzantine Ya'amūn offer the possibility to interpret the human isotopic data from Gerasa as reflecting multiple sources of food supply. Such provisions may have come from areas of the north of Jordan that differed in the herding strategies in use, the management of water resources or the environmental conditions.

20.5 CONCLUSIONS

Although the results presented here may be biased because of the small number of samples that yielded good-quality collagen, some provisional comments on herding strategies, environmental characteristics and patterns of human diet in ancient Jordan can be made. Owing to the differences in feeding behaviour and mobility, sheep and goats, more than pigs, appear to reflect the environmental variability of the surroundings of the settlements. Although the general indication emerging from these data is of a diet based on C_3 resources, some of the animals appear to have had access to substantial amounts of aridity-adapted C_4 plants and grasses. The high $\delta^{15}N$ ratios displayed by some of the sheep/goats are equally evidence of feeding in low-rainfall environments. Overall evidence from the sites analysed here suggests that the contribution of semi-arid habitats to the diet of domestic animals was greatest during the Middle and Late Bronze Age, and decreased during the following periods. More data will be of help in discriminating between the different scenarios that could explain the diversity between periods. For instance, additional $\delta^{13}C$ and $\delta^{15}N$ values from present-day wild and domestic herbivorous species that fed in the various regions and habitats from Jordan might aid the testing of the existence of a diachronic trend linked to changed environmental conditions. With regard to the diversity in the $\delta^{13}C$ and $\delta^{15}N$ values amongst the MB/LB sheep/goats from Ya'amūn in particular, the alternative scenarios of variable herding strategies and of a foreign origin for some of the animals could be tested by oxygen and strontium isotope analysis.

In the humans from the north of Jordan, the C_3 signal predominates. The animal data in part explain it; however, this signal is also the consequence of a diet in which C_3 plants and, in particular, the staple foods prepared from cereals prevail. At some sites, diet is very homogeneous and, therefore, it seems to reflect an inward-oriented system for procurement of food. This is in contrast with the picture given by archaeological evidence for agricultural production which, for instance at Ya'amūn and Yajūz, had reached the level of industrial output and must therefore have represented contacts with other communities. In contrast, the diversity in isotopic values seen among the humans from the mass burial of Gerasa's Hippodrome could be reflecting the diverse cultural backgrounds of its inhabitants and their resultant varied food choices. Alternatively, the isotopic characterisation of the people may be showing the environmental diversity between the regions that supplied such a big urban area.

Despite the limitation of sample size, this combined study of human and faunal carbon and nitrogen stable isotope values from the north of Jordan has demonstrated patterns of human diet and foddering strategies that reflect the diverse environmental and cultural contexts where human and animal populations lived. This study represents an exploratory step. Many more data need to be collected to improve our understanding of the interactions between civilisations and environment in ancient Jordan.

REFERENCES

Al-Eisawi, D. M. (1996) *Vegetation of Jordan*. Cairo: UNESCO Regional Office for Science and Technology for the Arab States.

Al-Shorman, A. (2004) Stable carbon isotope analysis of human tooth enamel from the Bronze Age cemetery of Ya'amoun in Northern Jordan. *Journal of Archaeological Science* **31**: 1693–1698.

Ambrose, S. H. (1991) Effects of diet, climate and physiology on nitrogen isotopes abundances in terrestrial foodwebs. *Journal of Archaeological Science* **18**: 293–317.

Ambrose, S. H. (1993) Isotopic analysis of palaeodiets: methodological and interpretative considerations. In *Investigation of Ancient Human Tissue: Chemical Analysis in Anthropology*, ed. M. K. Sandford. Langthorne: Gordon and Breach pp. 59–130.

Ambrose, S. H. and M. J. DeNiro (1989) Climate and habitat reconstruction using stable carbon and nitrogen isotope ratios of collagen in prehistoric herbivore teeth from Kenya. *Quaternary Research* **31**: 407–422.

Ambrose, S. H. and L. Norr (1993) Experimental evidence for the relationship of the carbon isotope ratios of whole diet and dietary protein to those of bone collagen and carbonate. In *Prehistoric Human Bone. Archaeology at the Molecular Level*, ed. J. B. Lambert and G. Grupe. Berlin: Springer-Verlag pp. 1–37.

Bocherens, H. and D. Drucker (2003) Trophic level isotopic enrichment of carbon and nitrogen in bone collagen: case studies from recent and ancient terrestrial ecosystems. *International Journal of Osteoarchaeology* **13**: 46–53.

Bourke, S., R. Sparks and M. Schroder (2006) Pella in the Middle Bronze Age. In *The Chronology of the Jordan Valley During the Middle and Late Bronze Ages: Pella, Tell Abu Al-Kharaz, and Tell Deir 'Alla*, ed. P. M. Fischer. Wien: Verlag der Österreichishen Akademie der Wissenschaften pp. 9–58.

Bourke, S., R. Sparks, K. N. Sowada and L. Mairs (1994) Preliminary report of the University of Sydney's fourteenth season of excavations at Pella (Tabaqat Fahl) in 1992. *Annual of the Department of Antiquities of Jordan* **38**: 81–126.

Bourke, S. J. (1997) Pre-Classical Pella in Jordan: a conspectus of ten years' work (1985–1995). *Palestine Exploration Quarterly* **129**: 94–115.

Bourke, S. J. (2006) Pella and the Jordanian Middle and Late Bronze Age. In *The Chronology of the Jordan Valley During the Middle and Late Bronze Ages: Pella, Tell Abu Al-Kharaz, and Tell Deir 'Alla*, ed. P. M. Fischer. Wien: Verlag der Österreichishen Akademie der Wissenschaften pp. 243–255.

Browning, I. (1982) *Jerash and the Decapolis*. London: Chatto & Windus Ltd.

Cameron, A. (1993) *The Mediterranean World in Late Antiquity*. London: Routledge.

Cordova, C. E. (2007) *Millennial Landscape Change in Jordan*. Tucson: The University of Arizona Press.

DeNiro, M. J. and S. Epstein (1978) Influence of diet on the distribution of carbon isotopes in animals. *Geochimica et Cosmochimica Acta* **42**: 495–506.

DeNiro, M. J. and S. Epstein (1981) Influence of diet on the distribution of nitrogen isotopes in animals. *Geochimica et Cosmochimica Acta* **45**: 341–351.

El-Najjar, M. and J. C. Rose (2003) Preliminary report of the 2003 field season at Ya'mun by the joint Yarmouk University/University of Arkansas Project. *Annual of the Department of Antiquities of Jordan* **47**: 491–492.

El-Najjar, M., J. C. Rose, N. Atallah et al. (2001) First season of excavation at Ya'mun (1999). *Annual of the Department of Antiquities of Jordan* **45**: 413–417.

Falconer, S. (2008) The Middle Bronze Age. In *Jordan. An Archaeological Reader*, ed. R. B. Adams. London: Equinox pp. 263–280.

Freeman, P. (2008) The Roman Period. In *Jordan. An Archaeological Reader*, ed. R. B. Adams. London: Equinox pp. 413–441.

Gawlikowski, M. (2001) From Decapolis to al-Urdunn: the cities of Jordan in the Early Islamic times. *Studies in the History and Archaeology of Jordan* **7**: 655–658.

Grigson, C. (2007) Culture, ecology, and pigs from the 5th to the 3rd millennium BC around the Fertile Crescent. In *Pigs and Humans. 10,000 Years of Interaction*, ed. U. Albarella, K. Dobney, A. Ervynck and P. Rowley-Conwy. Oxford: Oxford University Press pp. 83–108.

Gröcke, D. R., H. Bocherens and A. Mariotti (1997) Annual rainfall and nitrogen-isotope correlation in macropod collagen: application as a palaeoprecipitation indicator. *Earth and Planetary Science Letters* **153**: 279–285.

Heaton, T. H. E. (1987) The $^{15}N/^{14}N$ ratios of plants in South Africa and Namibia: relationship to climate and coastal/saline environments. *Oecologia* **74**: 236–246.

Heaton, T. H. E., J. C. Vogel, G. v. la Chevallerie and G. Collet (1986) Climatic influence on the isotopic composition of bone nitrogen. *Nature* **322**: 822–823.

Hedges, R. E. M. and L. M. Reynard (2007) Nitrogen isotopes and the trophic level of humans in archaeology. *Journal of Archaeological Science* **34**: 1240–1251.

Hedges, R. E. M., J. G. Clement, C. D. L. Thomas and T. C. O'Connell (2007) Collagen turnover in the adult femoral mid-shaft: modeled from anthropogenic radiocarbon tracer measurements. *American Journal of Physical Anthropology* **133**: 808–816.

Hoefs, J. (1997) *Stable Isotope Geochemistry*. Berlin: Springer.

Iacumin, P., H. Bocherens, L. Chaix and A. Marioth (1998) Stable carbon and nitrogen isotopes as dietary indicators of ancient Nubian populations (Northern Sudan). *Journal of Archaeological Science* **25**: 293–301.

Iacumin, P., H. Bocherens, A. Mariotti and A. Longinelli (1996) An isotopic palaeoenvironmental study of human skeletal remains from the Nile Valley. *Palaeogeography Palaeoclimatology Palaeoecology* **126**: 15–30.

Katzenberg, M. A. (2000) Stable isotope analysis: a tool for studying past diet, demography and life history. In *Biological Anthropology of the Human Skeleton*, ed. M. A. Katzenberg and S. R. Saunders. New York: Wiley-Liss pp. 305–327.

Kennedy, D. (2007) *Gerasa and the Decapolis. A 'Virtual Island' in Northwest Jordan*. London: Duckworth.

Khalil, L. (1998) University of Jordan excavations at Khirbat Yajuz. *Annual of the Department of Antiquities of Jordan* **XLII**: 457–472.

Khalil, L. A. and F. M. al-Nammari (2000) Two large wine presses at Khirbet Yajuz, Jordan. *BASOR* **318**: 41–57.

Knapp, B. A. (1993) *Society and Polity at Bronze Age Pella: an Annales Perspective*. Sheffield: Sheffield Academic Press.

McNicoll, A., R. H. Smith and B. Hennessy (1982) *Pella in Jordan 1, An interim report on the joint University of Sydney and The College of Wooster excavations at Pella 1979–1981*. Canberra: Australian National Gallery.

Minagawa, M. and E. Wada (1984) Stepwise enrichment of ^{15}N along food chains: further evidence and the relation between ^{15}N and animal age. *Geochimica et Cosmochimica Acta* **48**: 1135–1140.

O'Connell, T. C. and R. E. M. Hedges (1999) Investigation into the effect of diet on modern human hair isotopic values. *American Journal of Physical Anthropology* **108**: 409–425.

Parker, T. S. (1999) An Empire's new Holy Land: the Byzantine Period. *Near Eastern Archaeology* **62**: 134–180.

Pearson, J. A., H. Buitenhuis, R. E. M. Hedges et al. (2007) New light on early caprine herding strategies from isotope analysis: a case study from Neolithic Anatolia. *Journal of Archaeological Science* **34**: 2170–2179.

Petzke, K. J., H. Boeing and C. C. Metges (2005) Choice of dietary protein of vegetarians and omnivores is reflected in their hair protein ^{13}C and ^{15}N abundance. *Rapid Communications in Mass Spectrometry* **19**: 1392–1400.

Pollard, A. M. and L. Wilson (2001) Global biogeochemical cycles and isotope systematics – how the world works. In *Handbook of Archaeological Sciences*, ed. D. R. Brothwell and A. M. Pollard. Chichester: John Wiley and Sons pp. 191–201.

Richards, M. P. and R. E. M. Hedges (1999) Stable isotope evidence for similarities in the types of marine foods used by Late Mesolithic humans at sites along the Atlantic coast of Europe. *Journal of Archaeological Science* **26**: 717–722.

Richards, M. P., J. A. Pearson, T. I. Molleson, N. Russell and L. Martin (2003) Stable isotope evidence of diet at Neolithic Çatalhöyük, Turkey. *Journal of Archaeological Science* **30**: 67–76.

Rose, J. C. (2001) *Tell Ya'amun 2001 Excavations*. Last accessed on 29 June 2007, http://www.uark.edu/~jcrose/yaamun1/.

Rose, J. C. and D. L. Burke, eds. (2004) *Sa'ad: A Late Roman/Byzantine Site in North Jordan*. Yarmouk: Yarmouk University Publications.

Rose, J. C., D. L. Burke and K. L. Johnson, (2004) Tombs. In *Sa'ad: A Late Roman/Byzantine Site in North Jordan*, ed. J. C. Rose and D. L. Burke. Yarmouk: Yarmouk University Publications pp. 99–108.

Rose, J. C., M. El-Najjar and D. L. Burke (2007) Trade and the acquisition of wealth in rural Late Antique North Jordan. In *Studies in the History and Archaeology of Jordan IX*, ed. R. Harahsheh, Q. Fakhoury, H. Taher and S. Khouri. Amman: Department of Antiquities of Jordan pp. 61–70.

Rosen, A. M. (2007) *Civilizing Climate. Social Responses to Climate Change in the Ancient Near East*. Lanham, MD: Altamira Press.

Schmid, S. G. (2008) The Hellenistic Period and the Nabataeans. In *Jordan. An Archaeological Reader*, ed. R. B. Adams. London: Equinox pp. 353–411.

Schoeninger, M. J. and M. J. DeNiro (1984) Nitrogen and carbon isotopic composition of bone collagen from marine and terrestrial animals. *Geochimica et Cosmochimica Acta* **48**: 625–639.

Schoeninger, M. J., M. J. DeNiro and H. Tauber (1983) Stable nitrogen isotope ratios of bone collagen reflect marine and terrestrial components of prehistoric human diet. *Science* **220**: 1381–1383.

Schwarcz, H. P., T. L. Dupras and S. I. Fairgrieve (1999) ^{15}N enrichment in the Sahara: in search of a global relationship. *Journal of Archaeological Science* **26**: 629–636.

Sealy, J. (2001) Body tissue chemistry and palaeodiet. In *Handbook of Archaeological Sciences*, ed. D. R. Brothwell and A. M. Pollard. Chichester: John Wiley and Sons pp. 269–279.

Sealy, J. C., N. van der Merwe, J. A. Lee Thorp and J. L. Lanham (1987) Nitrogen isotopic ecology in southern Africa: implications for environmental and dietary tracing. *Geochimica et Cosmochimica Acta* **51**: 2707–2717.

Seigne, J. (1992) Jérash romaine et byzantine: développement urbain d'une ville provinciale orientale. *Studies in the History and Archaeology of Jordan* **4**: 331–341.

Sharp, Z. (2007) *Stable Isotope Geochemistry*. Upper Saddle River, NJ: Pearson Prentice Hall.

Shomer-Ilan, A., A. Nissenbaum and Y. Waisel (1981) Photosynthetic pathways and the ecological distribution of the Chenopodiaceae in Israel. *Oecologia* **48**: 244–248.

Smith, B. N. and S. Epstein (1971) Two categories of ^{13}C/^{12}C ratios for higher plants. *Plant Physiology* **47**: 380–384.

Smith, R. H. (1987) Trade in the life of Pella of the Decapolis. In *Studies in the History and Archaeology of Jordan*, ed. A. Hadidi. Amman: Department of Antiquities.

Smith, R. H. and L. P. Day (1989) *Pella of the Decapolis*, Vol. **2**. *Final Reports on The College of Wooster Excavations in Area IX, The Civic Complex, 1979–1985*. Wooster, OH: The College of Wooster.

Suleiman, E. (1996) A short note on the excavations of Yajuz 1994–1995. *Annual of the Department of Antiquities of Jordan* **XL**: 457–463.

Thompson, A. H., L. Chaix and M. P. Richards (2008) Stable isotopes and diet in Ancient Kerma, Upper Nubia (Sudan). *Journal of Archaeological Science* **35**: 376–387.

Thompson, A. H., M. P. Richards, A. Shortland and S. R. Zakrzewski (2005) Isotopic palaeodiet studies of Ancient Egyptian fauna and humans. *Journal of Archaeological Science* **32**: 451–463.

Tieszen, L. L. and T. Fagre (1993) Effect of diet quality and composition on the isotopic composition of respiratory CO_2, bone collagen, bioapatite and soft tissues. In *Prehistoric Human Bone, Archaeology at the Molecular Level*, ed. J. B. Lambert and G. Grupe. Berlin: Springer pp. 121–155.

Vogel, J. C., A. Fuls and A. Danin (1986) Geographical and environmental distribution of C_3 and C_4 grasses in the Sinai, Negev, and Judean deserts. *Oecologia* **70**: 258–265.

Walmsley, A. G. (1997) Land, resources and industry in Early Islamic Jordan (seventh–eleventh century). Current research and future directions. *Studies in the History and Archaeology of Jordan* **6**: 345–351.

Walmsley, A. G., P. G. Macumber, P. C. Edwards, S. J. Bourke and P. M. Watson (1993) The eleventh and twelfth season of excavations at Pella (Tabaqat Fahl) 1989–1990. *Annual of the Department of Antiquities of Jordan* **37**: 165–231.

Walmsley, A. G. and K. Damgaard (2005) The Umayyad congregational mosque of Jarash in Jordan and its relationship with early mosques. *Antiquity* **79**: 362–378.

Watson, P. M. (2008) The Byzantine Period. In *Jordan. An Archaeological Reader*, ed. R. B. Adams. London: Equinox pp. 443–482.

White, C., F. J. Longstaffe and K. R. Law (2004) Exploring the effect of environment, physiology and diet on oxygen isotope ratios in ancient Nubian bones and teeth. *Journal of Archaeological Science* **31**: 233–250.

Wilkins, J. M. and S. Hill (2006) *Food in the Ancient World*. Oxford: Blackwell.

Williams, K. D., M. Y. El-Najjar, J. C. Rose *et al.* (2004) Skeletal biology. In *Sa'ad: A Late Roman/Byzantine Site in North Jordan*, ed. J. C. Rose and D. L. Burke. Yarmouk: Yarmouk University Publications pp. 149–180.

Winter, K. (1981) C_4 plants of high biomass in arid regions of Asia – occurrence of C_4 photosynthesis in Chenopodiaceae and Polygonaceae from the Middle East and USSR. *Oecologia* **48**: 100–106.

21 Irrigation and phytolith formation: an experimental study

Emma Jenkins, Khalil Jamjoum and Sameeh Al Nuimat

ABSTRACT

It has been proposed that phytoliths from archaeological sites can be indicators of water availability and hence inform about past agricultural practices (Rosen and Weiner, 1994; Madella et al., 2009). Rosen and Weiner (1994) found that the number of conjoined phytoliths from cereal husks increased with irrigation while Madella et al. (2009) demonstrated that the ratio of long-celled phytoliths to short-celled phytoliths increased with irrigation. In order to further explore these hypotheses, wheat and barley were experimentally grown from 2005 to 2008 in three different crop growing stations in Jordan. Four different irrigation regimes were initially employed: 0% (rainfall only), 80%, 100% and 120% of the optimum crop water requirements, with a 40% plot being added in the second and third growing seasons. Each plot measured 5 m × 5 m and a drip irrigation system was used. Environmental variables were measured on a daily basis, and soil and water samples were taken and analysed at the University of Reading. Phytoliths from the husks of these experimentally grown plants were extracted using the dry ashing method. Results demonstrate that although the number of conjoined cells increases with irrigation, there were considerable inter-site and inter-year differences suggesting that environmental variables other than water availability affect phytolith uptake and deposition. Furthermore, analytical experiments demonstrated that conjoined phytoliths are subject to change or breakage by external factors, making this methodology problematic to apply to archaeological phytolith assemblages that have an unknown taphonomic history. The ratio of long cells to short cells also responded to increased irrigation, and these forms are not subject to break up as are conjoined forms. Our results from the modern samples of durum wheat and six-row barley show that if an assemblage of single-celled phytoliths consists of over 60% dendritic long cells then this strongly suggests that the crop received optimum levels of water. Further research is needed to determine if this finding is consistent in phytolith samples from the leaves and stems, as suggested by Madella et al. (2009), and in other species of cereals. If this is the case then phytoliths are a valuble tool for assessing the level of past water availability and, potentially, past irrigation.

21.1 INTRODUCTION

21.1.1 Archaeology, irrigation and phytoliths

The development of water management systems in southwest Asia has long been recognised as important for understanding socio-economic change. Although Wittfogel's 'hydraulic hypothesis' (Wittfogel, 1957) of irrigation management as the prime mover for the emergence of early states may no longer be tenable, the management requirements of irrigation systems and the potential increase in surplus that can arise from their use remain key issues for understanding the emergence of social complexity (Scarborough, 2003). Direct archaeological evidence for water management in Jordan takes numerous forms, including wells, cisterns, field systems and irrigation ditches (Chapter 14, this volume; Oleson, 2001). Such evidence is often substantial for proto-historic and historic periods, such as the sophisticated Nabataean modifications to the siq at Petra (Bellwald and al-Huneidi, 2003) or the Roman/Byzantine reservoir, aqueduct and field system in Wadi Faynan (Barker, 2000). Structural evidence for water management is both more elusive and more difficult to interpret for the prehistoric periods, when it is likely that water management, including the irrigation of cereals, began.

The earliest known structural evidence in the Water, Life and Civilisation study region has been summarised in Chapter 14 of

Water, Life and Civilisation: Climate, Environment and Society in the Jordan Valley, ed. Steven Mithen and Emily Black. Published by Cambridge University Press.
© Steven Mithen and Emily Black 2011.

this volume: the PPNB wadi barriers and cistern in the Jafr Basin; the Pottery Neolithic well at Sha'ar Hagolan, Israel (Garfinkel *et al.*, 2006); and the set of small terrace walls at the Neolithic site of Dhra', Jordan, which have been interpreted as functioning to minimise soil erosion, controlling water runoff during wet periods of the year, and as field systems for growing wild and domesticated plants.

Whether or not the Dhra' walls functioned to provide additional water to crops remains unclear. Indeed, such uncertainty often exists even when a complex water management system is evident. At the Early Bronze Age site of Jawa, for instance, there appears to have been a system of channels, dams and pools to collect and store winter flood water from the adjacent Wadi Rajil (Helms, 1981; Helms, 1989). Hydrological models have been developed to estimate the size of the animal and human populations that could be sustained by such water storage (Whitehead *et al.*, 2008). But the estimates are highly dependent upon whether the stored water had also been used to irrigate surrounding fields for the growth of cereals, for which there is no direct evidence.

At other archaeological sites there may be circumstantial evidence for agricultural intensification requiring irrigation, but a complete absence of any structural evidence for water management. At the Chalcolithic site of Ghassul, for instance, Bourke (2001, p. 119) proposes that there had been 'elite-regulated exploitation of flood-water irrigation systems', but has no evidence for ditches, walls or dams. Such structures may not have been necessary or simply insufficiently substantial to have survived in the archaeological record. Indeed, in some cases effective water management requires no more than minor and ephemeral adjustments to water courses, as we have observed among the Bedouin in Wadi Faynan (Figure 15.2, this volume). They use small walls of pebbles and mud to divert seasonal streams and create substantial pools of water; these are constructed in a few minutes, frequently modified and then simply washed away leaving no archaeological trace. Far more substantial evidence may have been destroyed: Philip (2001) notes that the down-cutting of wadis and the deposition of colluvium in the Jordan Valley may have removed or buried structural evidence for water management of the Early Bronze Age, for which there is only direct evidence from Tall Handaquq (Mabry, 1989) and Jawa (Helms, 1981, 1989).

As the structural evidence for water management is difficult to interpret and may simply not exist, it would be of considerable value to have a methodology for the inference of crop irrigation directly from archaeobotanical remains. Helback (1960) proposed that the size of charred flax seeds could be used to this end, while Mabry *et al.* (1996) suggested that the size of wheat grains from Tall Handaquq implied irrigation agriculture. Such arguments are problematic because of the impact that charring itself may have on the size of seed grains, and the numerous other factors that may also influence grain size. Studies by Jones *et al.* (1995) and Charles *et al.* (2003) demonstrated that when modern-day fields and crops are available for study, their weed floras can be indicators of past water availability. Unfortunately, sufficiently well-preserved assemblages of charred plant remains from prehistoric sites for such studies are rarely recovered from the archaeological record.

Carbon isotope analysis is another method which has been proposed for identifying irrigation in arid regions (Araus *et al.*, 1997, 1999). Studies of carbon isotopes from cereal grains and rachises have demonstrated that ratios change as a result of increased water availability. However, there are problems in applying this method to archaeological remains. Firstly, as with the method proposed by Jones *et al.* (1995) and Charles *et al.* (2003) it is often impossible to recover a large enough sample from an archaeological site to use this method. Secondly, there are unknown environmental variables that could have affected the carbon isotope signature such as climate and soil chemistry. Thirdly, the impacts of charring and diagenesis on carbon isotope ratios are poorly understood (Tieszen, 1991; Heaton, 1999; Codron *et al.*, 2005). Finally, a highly specialised laboratory is needed in order to conduct analysis. However, despite these problems this method has potential for identifying past water availability, and research into the effect of irrigation and taphonomy on carbon isotope ratio in cereal grains is being conducted as part of the Water, Life and Civilisation project (Chapter 22, this volume).

Another proposed method for identifying irrigation is phytolith analysis. Rosen and Weiner (1994) found that the number of conjoined phytoliths from cereal husks increased with irrigation while Madella *et al.* (2009) demonstrated that the ratio of long-celled phytoliths to short-celled phytoliths increased with irrigation. If phytoliths could be used to identify ancient irrigation, this would be a very valuable tool for archaeologists; phytoliths are often abundant on archaeological sites, samples are easy to take and processing is straightforward and relatively inexpensive compared with the hours of flotation time needed to recover macroscopic remains or the cost involved in setting up a stable isotope laboratory.

In order to further explore the hypothesis that phytoliths can be indictators of water availability, wheat and barley were experimentally grown from 2005 to 2008 in three different crop growing stations in Jordan under different irrigation regimes. The phytoliths from the husks of these plants were then extracted for analysis. This chapter outlines the methodology used in the experiment and discusses the results in light of their implications for the use of phytoliths as indicators of past water availability.

21.1.2 Phytoliths and their formation

Phytoliths are composed of opaline silica which is taken up as monosilicic acid by plants through their roots into the vascular system during transpiration and deposited in a solid state as silicon dioxide in inter- and intra-cellular spaces. The reasons that plants produce phytoliths are only partially understood. One reason appears to be protection from disease (Williams and Vlamis, 1957; Yoshida et al., 1959), herbivory, pathogenic fungi and insect attack (Takijima et al., 1949; Djamin and Pathak, 1967; Heath, 1979). Phytoliths also protect plants from the harmful impacts of trace elements such as aluminium and manganese (Jones and Handreck, 1967; Horiguchi, 1988; Hodson and Evans, 1995; Hodson and Sangster, 1999). It has also been demonstrated that phytoliths slow down the rate of transpiration in cuticular and epidermal cells (Mitsui and Takatoh, 1959; Yoshida et al., 1959; Okuda and Takahashi, 1965; Raeside, 1970), inhibit the plant's uptake and translocation of sodium in saline conditions (Ahmad et al., 1992; Liang et al., 1996) and increase the oxidising power of roots (Okuda and Takahashi, 1965). Phytoliths also increase the strength and yield capabilities of plants and even the fertility of pollen in certain species (Mitsui and Takatoh, 1959; Yoshida et al., 1959; Vlamis and Williams, 1967; Raeside, 1970; Miyake and Takahashi, 1983, 1986; Ahmad et al., 1992). In Japan, silica from slag has been used as a form of fertiliser for rice plants since the 1950s because it increases dry matter production and grain yield (Agarie et al., 1996).

Two modes – passive and active – of silicon uptake in plants have been proposed, while some species have also been identified as silicon rejectors (Richmond and Sussman, 2003). Jones and Handreck (1965) were able to predict the weight percent of phytoliths to original plant matter processed (phytolith weight % = weight of phytoliths/weight of plant matter processed × 100) if they knew the concentration of silicon in the soil solution and the amount of water transpired, suggesting that the uptake of silicon was passive (Jones and Handreck, 1965). Okuda and Takahashi (1965) demonstrated that silicon uptake can occur against a concentration gradient in rice, suggesting that uptake can also be active. Barber and Shone (1966) argued that while passive uptake explained the entry of silicon into the roots of barley it did not adequately explain its uptake into the transpiration stream (Barber and Shone, 1966). It has also been suggested that some plants such as tomato and faba bean are silicon excluders or rejecters and actively prohibit the uptake of silicon (Liang et al., 2006). Generally the evidence for silicon uptake in grasses and sedges now suggests that both passive and active components coexist in many species (Jarvis, 1987; Walker and Lance, 1991; Mayland et al., 1993; Ernst et al., 1995; Liang et al., 2006) but, unlike diatoms, a suite of genes responsible for the transportation of silicon has not yet been identified in plants (Richmond and Sussman, 2003).

Phytolith research is a relatively new topic within archaeology, the potential of which is still being explored (see Chapter 23, this volume). One role that they may play is as a proxy for past water availability and possible irrigation (Rosen and Weiner, 1994; Rosen, 1999). This is because silicon uptake and hence deposition is influenced by the rate of transpiration, which in turn is dependent on water availability. Richardson found a correlation between water transpired by barley grown in controlled greenhouse conditions and the silicon content of the plant (Richardson's results reported in Hutton and Norrish, 1974, p. 204). Jones and co-workers (Jones and Milne, 1963; Jones et al., 1963; Jones and Handreck, 1965) demonstrated that water transpiration affected silicon uptake in oats, while Hutton and Norrish (1974) showed that the amount of silica found in the husks of wheat was proportional to water transpired.

Additional variables also affect silicon uptake and deposition. One important factor is the amount of silicon available to the plant in the growing medium (Parry and Smithson, 1958, 1964, 1966; Yoshida et al., 1959; Blackman, 1968a, 1968b; Blackman and Parry, 1969). This can come from two different sources, the soil and the water. Rain water contains little silicon, with silicon accretion and input from rain water being ≤ 1 kg ha^{-1} yr^{-1} (Alexandre et al., 1997). The level of silicon in the water used for irrigation can vary according to the source of water used; if a form of rainwater harvesting is employed, then the water will contain little silicon, whereas if the water comes from a wadi or river, the silicon level is likely to be higher (Imaizumi and Yoshida, 1958; Meybeck, 1987; Bluth and Kump, 1994; White and Blum, 1995). The amount of silicon in the soil varies according to geology and land use (Alexandre et al., 1997). If land is under cultivation, silicon levels in soils can be reduced by plant uptake. If plants remain in situ, silicon will eventually be released back into the soil through phytolith dissolution. However, if plants are harvested, and thus removed from the growing site, then silicon levels will be reduced. The level to which this occurs is partly dependent on plant species: monocots produce between 14 and 20 times as much weight percent of phytolith as dicots, thus leaving less silicon available in the growing medium (Albert et al., 2003).

Another variable which affects the uptake and deposition of silicon is the transpiration rate. This is climatically dependent, being much faster in arid and semi-arid regions than in temperate and humid climes (Jones and Handreck, 1965; Barber and Shone, 1966; Raeside, 1970; Hutton and Norrish, 1974; Rosen and Weiner, 1994; Webb and Longstaffe, 2002). It has been suggested, however, that one of the roles of silicon in plants is to slow down transpiration rates: in both rice and (to a lesser extent) barley, the transpiration rate is faster in silicon-deprived plants than in the non-silicon-deprived control samples (Yoshida et al., 1959; Okuda and Takahashi, 1965). Lewin and Reimann (1969) suggest this increased transpiration rate in silicon-deficient plants could be due to a lack of silica gel associated with the cellulose in the cell walls of epidermal cells which helps reduce water loss (Lewin and Reimann, 1969).

Soil texture and chemistry also affect phytolith production. For example, a clay-rich soil will retain more water than other soil types, while aluminium oxides, iron oxides and alkaline soils are adsorbers of silicon, making less available to the plant

(Okamoto *et al.*, 1957; Parry and Smithson, 1958, 1964, 1966; McKeague and Cline, 1963). When nitrogen is added to soil, the percentage of silicon in wheat decreases (Hutton and Norrish, 1974), while sodium fluoride inhibits silicon uptake in rice (Mitsui and Takatoh, 1959; Okuda and Takahashi, 1965). Other variables that affect silicon uptake are species (Parry and Smithson, 1958, 1964, 1966; for a review of silicon uptake in different species see Lewin and Reimann, 1969, p. 294) and plant age (Sangster, 1970; Bartoli and Souchier, 1978; Perry *et al.*, 1984).

21.1.3 Phytoliths as a proxy for water availability

Rosen and Weiner (1994) explored the possibility of using phytoliths as a proxy for past irrigation by analysing dry-farmed and irrigated samples of emmer wheat (*Triticum turgidum* subsp. *dicoccum*) and bread wheat (*T. aestivum*). They hypothesised that the increased level of transpiration in arid and semi-arid regions would affect silicon uptake and deposition to such an extent that it would be discernible in the archaeological record. They set up a field experiment at the Gilat Agricultural Research Station, Israel, where emmer wheat was planted in two plots, one irrigated, the other non-irrigated, each measuring approximately 3 m × 1 m. The topography was flat and the growing medium was a light loessial soil. Rainfall for the growing season was 224 mm and the irrigated plot received an additional 200 mm of water. In addition, samples of bread wheat (*T. aestivum*) were collected from irrigated and dry-farmed fields in Gilat and northern and central Israel as well as from dry-farmed fields in Germany and eastern Washington State, USA. A limited number of wild (*Hordeum vulgare* subsp. *spontaneum*) and domestic (*H. vulgare* subsp. *vulgare*) barley samples were also collected. Phytoliths were isolated from the samples using acid extraction following the methodology of Piperno (1988) (Rosen and Weiner, 1994, p. 127).

The results showed that the samples grown under irrigation not only had a greater yield of phytoliths but also a greater number of conjoined cells. The percentage of phytoliths with 10 or more conjoined cells was only 2.1% for the non-irrigated plants but 13% in the irrigated plants (Rosen and Weiner, 1994). A similar pattern was evident from bread wheat, with the yield from irrigated wheat collected from Germany and the USA consisting almost entirely of single-celled phytoliths, whereas the irrigated Israeli-grown wheat contained a greater number of conjoined forms. Owing to the limited number of barley samples counted, it was not possible to obtain a statistically viable result, but preliminary findings suggested that barley may respond to irrigation in a similar way, with irrigated barley having a greater number of conjoined cells than the dry-farmed barley (Rosen and Weiner, 1994). Rosen and Weiner (1994) propose that when dealing with archaeological samples from arid and semi-arid regions, the presence of at least 10% of phytoliths with 10 or more conjoined cells, or any phytoliths with 100 or more conjoined cells, provides an indication of past irrigation. This was used to infer that irrigation had been used for growing emmer wheat at two Chalcolithic sites in the northern Negev: Gilat and Shiqmim (Rosen and Weiner, 1994).

Although the work of Rosen and Weiner was pioneering, experimentation was on a small scale in order to establish whether the methodology had potential. As such, many variables such as soil chemistry and climate were not accounted for. In addition, only two irrigation regimes were employed, irrigated and non-irrigated, and as such they were not able to determine whether the size and number of conjoined phytoliths increases linearly with irrigation or whether there is an exponential relationship between conjoined phytoliths and water availability.

Webb and Longstaffe (2002) reported that the weight percent of phytoliths in Prairie grass (*Calamovilfa longifolia*) was higher in plants grown in arid conditions than those grown in regions with a high relative humidity (Webb and Longstaffe, 2002; Madella *et al.*, 2009). A recent study by Madella *et al.* (2009) also explored the possibility of using phytoliths as indictors of past water availability. Their study involved five different cereals: bread wheat (*Triticum aestivum*), emmer wheat (*T. dicoccum*), spelt wheat (*T. spelta*), two-row barley (*Hordeum vulgare*) and six-row barley (*H. distichon*). These were grown under two different climatic regimes: Middle East climatic conditions, which were simulated using a growing chamber, and a North European climatic condition, i.e. open fields in Cambridge, UK. The plants in the Middle Eastern climatic conditions were grown under two different irrigation regimes: wet and dry. The wet regime involved keeping the pots at water-holding capacity, with water being administered on a daily basis, and the dry regime was irrigated to 50% of the water-holding capacity.

Madella *et al.* (2009) classified phytoliths according to their method of silicification as either fixed forms or sensitive forms. Fixed forms were defined as cells whose silicification is under genetic control (presumably equivalent to the passive silicon uptake described above) and would therefore be less influenced by water availability; these comprise all short cells e.g. dumbbell/bilobate, rondel, trapezoid, crenate trapezoids, cross, keeled, conical etc. (Madella *et al.*, 2009, p. 35). The sensitive forms are phytoliths formed in cells whose silicification is assumed to be under environmental control (or active silicon uptake) and which would, therefore, be indicative of past water and other climatic variables; these consist of all grass long cells. Phytoliths from the leaves of all plants were analysed, while phytoliths from the stems were analysed for emmer and spelt wheat. Madella *et al.* also used X-ray micro-chemical analysis on bread wheat to measure the elemental concentration of silicon and oxygen in the silica to help gain an understanding of water availability versus evapo-transpiration by detecting differences in the ratios of oxygen and silicon in the plant (Madella et al., 2009).

Madella et al. (2009) compared the ratios of fixed to sensitive forms from both the dry and wet regimes grown under Middle Eastern climatic conditions. In the leaves, they found an increase in the number of sensitive forms relative to fixed forms under the wet regime for bread wheat, emmer wheat and two-row barley, while an overlap was seen in the values between fixed and sensitive forms in spelt wheat and six-row barley. The analysis of phytoliths from plant stems showed that the mean for sensitive forms from the dry regime was higher than for the wet regime, although the error bars indicate that there was considerable overlap in the values between the two regimes. Results from the X-ray micro-analysis of phytoliths from bread wheat found that the level of oxygen was higher in the wet grown samples than in the dry regime samples (Madella et al., 2009).

While these results are valuable for the potential use of phytoliths as indicators of past water availability, they are not without their limitations.

The first concern is that the plants grown under the Middle Eastern climatic regime were cultivated in pots in a greenhouse rather than in open fields and thus natural growing conditions would not have been emulated. These plants would have received little competition for water and nutrients from other plants because of the restricted growing area and would presumably have been weed-free, all of which could affect silicon uptake. In addition, greenhouses increase humidity levels and, unless dehumidifiers were used (which is not stated in the paper), it is probable that the humidity would have been higher than natural for a Middle Eastern arid environment, affecting transpiration rates. A second issue is that the amount of water given to the plants, and how this relates to their known crop water requirements, is not stated and so it is unclear how these irrigation systems relate to plant water requirements. For example, what percentage of the crop water requirements is represented by the 50% of the pot holding capacity, and would results have differed if a regime supplying 25% of the pot holding capacity been included in the experiment? A third concern is that the soil silicon levels were not measured and so we do not know if they were higher in the soils used for the pot experiments than in the open fields.

21.2   AIMS OF THE EXPERIMENTAL CROP GROWING STUDY

To explore in more detail the hypothesis that phytoliths can be indicators of past water availability, crop growing experiments were established in Jordan as part of the Water, Life and Civilisation project in collaboration with the National Centre for Agricultural Research and Extension (NCARE), Jordan. The aims of the experiment were threefold: (1) to determine whether the differences in irrigated and non-irrigated phytoliths observed by Rosen and Weiner (1994) are apparent in other species of wheat; (2) to determine if these differences are also observable in other cereals (barley); and (3) to assess whether variables such as climate and soil and water chemistry affect silica deposition.

21.3   MATERIALS AND METHODS

21.3.1   Experimental conditions

Two crops were grown for phytolith analysis, both of which were native land races: durum wheat (*T. durum*) (ASCAD 65) and six-row hulled barley (*H. vulgare*) (ASCAD 176). These were grown at three different NCARE crop growing stations: (1) Khirbet as Samra, which is on the Jordanian Plateau to the northeast of Amman, (2) Ramtha, which is in the north of Jordan, 5 km from the Syrian border and (3) Deir 'Alla, which is in the Jordan Valley (see Figure 21.1 and Table 21.1). The experiment involved three growing seasons. Each experimental plot measured 5 m by 5 m and was surrounded by a soil bund with a 1.5 m

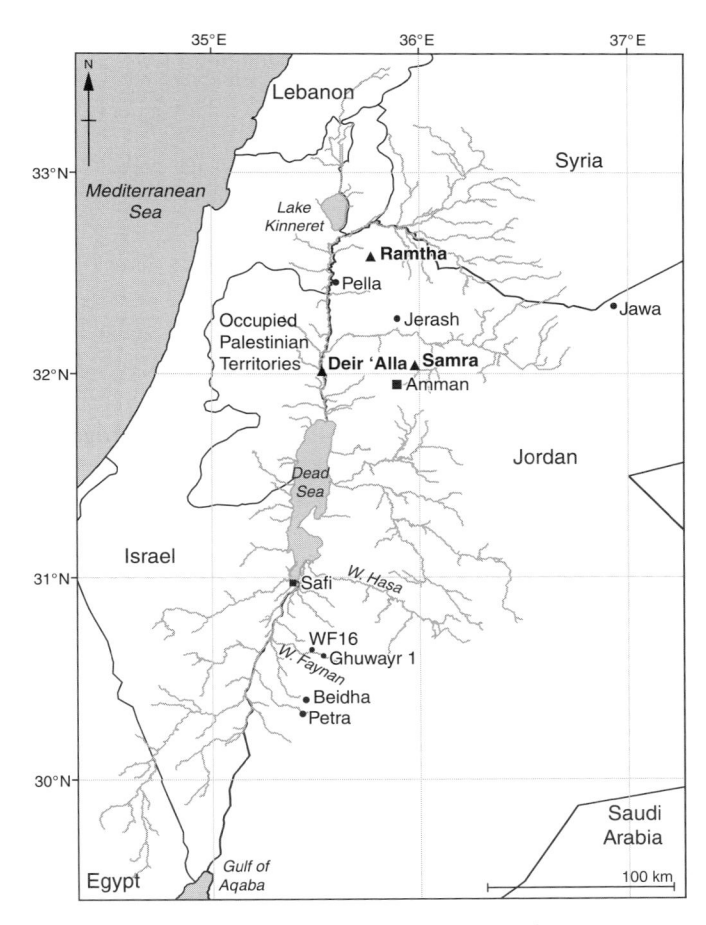

Figure 21.1. Map showing location of crop growing sites.

Table 21.1 *Description of crop growing sites*

	Khirbet as Samra	Ramtha	Deir 'Alla
Location (latitude and longitude)	N 32° 08.890'	N 32° 34'	N 32° 11.483'
	E 36°08.710'	E 36° 1'	E 035° 37.167'
Altitude (metres)	567 m above sea level	500–600 m above sea level	−192 m below sea level
Slope (at soil sampling localities)	<3°	<3°	<3°
Precipitation average (mm year^{-1})	150	300–350	250

After Carr (2009)

separation from the adjacent plot. In the first season, four different irrigation regimes were employed: (1) no irrigation (0% of crop water requirements); (2) under-irrigated (80% of crop water requirements); (3) irrigated (100% of crop water requirements); and (4) over-irrigated (120% of crop water requirements). In the second and third seasons an additional under-irrigated plot was added which was given 40% of crop water requirements. The calculations for irrigation levels were based on knowledge of crop water requirements estimated by using Class A – Pan evaporation readings (Allen *et al.*, 1998). Daily rainfall and evaporation were taken into account and allowed for when irrigation was calculated, with less irrigation water being applied during periods of higher rainfall and low evaporation. The total amount of irrigation water applied to each of the experimental plots is shown in Table 21.2, and total rainfall and evaporation for all three sites by plant growth over the three growing seasons is shown in Figure 21.2 (for more information on crop growth stages see Allen *et al.*, 1998). Water was provided by a drip irrigation system with a 60 cm spacing between water pipes and a 40 cm spacing between the drippers on each pipe. Each irrigation plot had eight lines, with reclaimed wastewater being used for irrigation. The water used for irrigation was treated wastewater at Khirbet as Samra and Ramtha and a mixture of treated wastewater and fresh water at Deir 'Alla. The water used was within the Jordanian standards for the irrigation of restricted crops. Samples of irrigation water from all three crop growing stations were collected and analysed at the Department of Soil Science, University of Reading. Table 21.3 provides a list of the analyses undertaken (see chapter 3 of Carr, 2009, for methodologies).

Crops were planted in November of each year and harvested in May. Figure 21.3 shows the harvesting of barley from the third growing season at Khirbet as Samra. All plots, including the non-irrigated ones, were given 25 mm of water after sowing to encourage germination. No pesticides or fertilisers were employed and the plots were not weeded. Bird attack was an ongoing problem at Deir 'Alla and Ramtha, with some plots having to be entirely covered with mesh for protection (Table 21.4), although this was not applied until the plants were reaching maturity.

Table 21.2 *Amount of applied irrigation (mm per year) by crop, year and growing site*

Irrigation regime	Deir 'Alla	Ramtha	Khirbet as Samra
Year 1 (2005–6)	barley		
0%	25.0	25.0	25.0
80%	74.1	101.5	124.2
100%	92.6	126.8	155.2
120%	111.1	152.2	186.3
Year 1 (2005–6)	wheat		
0%	25.0	25.0	25.0
80%	91.2	150.7	176.0
100%	114.0	188.4	220.1
120%	136.8	226.1	264.1
Year 2 (2006–7)	wheat and barley		
0%	25.0	25.0	25.0
40%	28.4	41.0	41.3
80%	56.8	82.0	82.5
100%	71.0	102.5	103.2
120%	85.2	123.0	123.8
Year 3 (2007–8)	wheat and barley		
0%	25.0	25.0	25.0
40%	88.9	60.2	79.2
80%	177.8	120.4	158.4
100%	222.3	150.5	198.1
120%	266.7	180.6	237.7

A grid system was used to collect the plants. This involved running a tape measure diagonally across the plot and collecting the plants along the diagonal transect from six 50 cm intervals: 0–50 cm, 50–100 cm, 100–150 cm, 150–200 cm, 200–250 cm and 250–300 cm. This was done to avoid edge effect. Plants were placed inside paper bags after collection. In addition to the plants taken for phytolith analysis a 1 m × 1 m square was sampled for

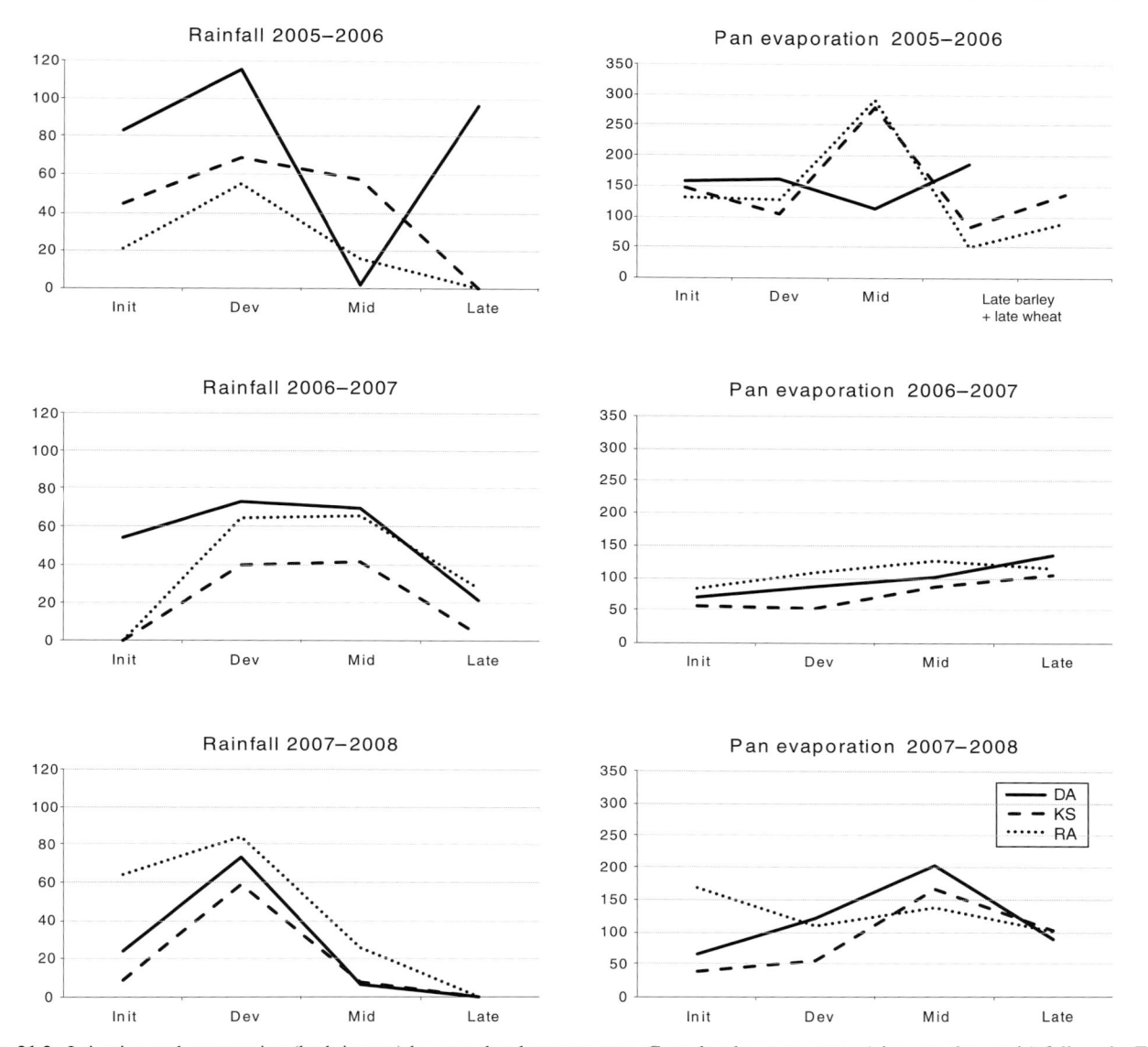

Figure 21.2. Irrigation and evaporation (both in mm) by crop development stage. Crop development stages (given on the *x* axis) follow the Food and Agricultural Organisation convention of Initial (Init.), Crop development (Dev.), Mid-Season (Mid.) and Late. Late barley and late wheat are shown separately, reflecting differences in their development. DA, Deir 'Alla; KS, Khirbet as Samra; RA, Ramtha.

yield for each of the irrigation plots which was analysed by the scientists at the NCARE crop growing stations. The area sampled for yield was selected at random by throwing a 1 m × 1 m square into the 5 m × 5 m plot. The only occasions when this procedure was not observed were when parts of the plot had been eaten by birds. In these instances the square was placed into an uneaten area to avoid biasing results.

21.3.2 Soil analysis

Soil samples were taken from each plot at three different depths: 0–5 cm, 5–25 cm and >25 cm, and characterised and analysed at the Department of Soil Science, University of Reading (Carr, 2009). Table 21.5 provides a list of the analyses undertaken (see chapters 4 and 5 of Carr, 2009, for methodologies). Soil samples were also taken after the first and last years of experimentation to test for plant-available silicon. The extraction was done using 0.025M citric acid (details of this methodology are provided in Table 21.6). Analysis was conducted using a PE Optima ICP-OES in the Department of Soil Science, University of Reading.

21.3.3 Phytolith processing and counting

Before the processing of the modern plants began, an experiment was conducted to compare the impact that analytical methods have on conjoined phytoliths (Jenkins, 2009). This was undertaken using the husks from the 100% irrigated wheat from the first season of crop growing at Khirbet as Samra. Two different

Table 21.3 Selected water quality parameters for reclaimed water from the crop growing stations based on samples collected in the field and available data

Location	Potentially plant-beneficial ions in the irrigation water (mg L^{-1})					Potentially plant-toxic ions in the irrigation water (mg L^{-1})			Additional parameters			
	Sulphate (SO$_4$)	Calcium (Ca)	Potassium (K)	Magnesium (Mg)	Phosphate (PO$_4$)	Chloride (Cl)	Sodium (Na)	Boron (B)	EC (dS m^{-1})^a	pH	Total organic carbon (TOC) (ppm)	Sodium adsorption ratio (SAR)
Khirbet as Samra measured data (May 2006 and May 2007)	65.09	44.55	31.32	27.22	14.41	374.81	261.14	0.94	1.82	8.14	12.11	7.67
Khirbet as Samra average (based on measured and published data) (1995–2007)	65.09	44.55	31.32	27.22	35.63	364.61	261.14	0.91	2.14	7.86		7.67
Ramtha measured data (May 2007)	104.66	44.16	27.25	31.12	6.80	398.76	195.53	0.73	1.50	8.49	10.73	5.69
Ramtha average (based on measured and published data) (2005–2007)	104.66	49.71	32.97	34.19	6.80	398.76	232.46	0.73	1.71	8.21		6.30
King Talal Reservoir measured at Deir 'Alla (May 2007)	15.47	19.21	4.11	3.68	8.07	47.55	35.13	0.16	1.49	7.67	7.18	1.92
King Talal Reservoir average for Deir 'Alla (based on measured and published data) (1995–2007)	90.78	50.09	15.74	31.39	21.74	276.66	125.67	0.54	1.91	7.85		

After Carr (2009); other data GTZ (2005); Al-Zu'bi (2007); Ammary (2007); Bashabsheh (2007).
^{a} EC, electrical conductivity

Table 21.4 *Record of plots covered with mesh*

		Year 1	Year 2	Year 3
Deir 'Alla	barley	not covered	covered	not covered
	wheat	covered	covered	covered
Ramtha	barley	covered	covered	covered
	wheat	covered	covered	not covered
Khirbet as	barley	not covered	not covered	not covered
Samra	wheat	not covered	not covered	not covered

Figure 21.3. Harvesting barley at Khirbet as Samra after the third growing season.

processing methods were employed: dry ashing and acid extraction. The former method involves burning the plant samples in a muffle furnace to remove organic matter while the latter uses nitric acid to remove organic matter.

The results demonstrated that dry ashing produces a higher weight percent of phytoliths to original plant matter and a far greater number of conjoined cells than the acid extraction method. This is in agreement with earlier studies, such as those by Jones and Milne (1963) and Raeside (1970), both of which reported a greater number of conjoined cells with dry ashing than acid extraction. Two explanations are proposed for this. The first is that the oxidation of the organic matter during acid extraction forces the phytoliths apart and causes a mechanical breakdown of conjoined forms that does not occur with dry ashing or that acid extraction destroys the silica gel holding the phytoliths together (Hayward and Parry, 1980). The second is that dry ashing causes the silica to dehydrate, as proposed by Jones and Milne (1963), causing fusion between forms resulting in a stronger structure (Jenkins, 2009).

The finding that the analytical procedure employed can change the structure of the phytoliths has implications for analysis of archaeological assemblages. Frequently phytoliths are recovered from ashy deposits or hearths and are the product of plants that have been burnt in the past. It is presumed that such phytoliths would resemble modern plants that have been dry ashed. Samples are often taken, however, that do not appear to have been burnt. These samples could either resemble those that have been wet ashed based on the premise that dry ashing causes fusion, or resemble those that have been dry ashed based on the premise that acid extraction forces phytoliths apart. For the purpose of this experiment it was decided to process the modern plants using the dry ashing method. This is because phytoliths are frequently recovered from ashy deposits and these can be sampled in isolation for the application of the proposed methodology; given that it is still unclear why the morphology of the phytoliths changes with processing, it is most reliable to compare archaeological samples that are presumed to have been heated with modern samples that have also been heated. Until the exact cause of the differences resulting from processing can be pinpointed, there is limited value in proceeding with wet ashing and employing this methodology to unburnt deposits. The methodology employed for extracting phytoliths from modern plants is provided in Table 21.7. Only husks were analysed because they have been found to have a higher silica content than other parts of the plant. Hutton and Norrish (1974) have suggested that the percentage of silica in the husks is closely related to the amount of water transpired in wheat and hence more accurately reflects water availability during growth.

Slides were counted using a Leica DME at $\times 400$. Phytoliths were counted according to the number of dendritic long cells in each conjoined form and the following broad counting categories were used: single cell, 2 to 5, 6 to 10, 11 to 15, 16 to 20, 21 to 30, 31 to 50, 51 to 70, 71 to 100, 101 to 150, 151 to 200, 201 to 250, 251 to 300, 301 or over. Frequently, forms were found that were not properly silicified. In these forms the cork-silica cells and papillae were silicified but the dendritic long cells were either poorly silicified or unsilicified. Examples of well silicified and poorly silicified forms are shown in Figure 21.4. In these cases the same counting categories were used as above, but they were classified as unclear/poorly silicified forms and the number of dendritics was estimated by counting their outlines between the silicified cork-silica cells and papillae. In addition, single cork-silica cells were counted.

Ten slides were counted from the samples from the first growing season and five from the second and third seasons. Occasionally it was not possible to count the total target number of slides, for example when the crop had failed to grow successfully or if it had been eaten by birds. Many of the barley samples contained a large number of fused phytolith forms, even when the ashing temperature was reduced to 400 °C, making it impossible to count these slides. It is assumed that the phytolith forms

Table 21.5 *Soil physical and chemical properties at the crop growing stations*

	Khirbet As Samra – non-irrigated soil			Ramtha – non-irrigated soil			Deir 'Alla – irrigated soil (100% of the crop water demand for 1 year)		
	Surface (0–5 cm)	Middle (5–25 cm)	Bottom (>25 cm)	Surface (0–5 cm)	Middle (5–25 cm)	Bottom (>25 cm)	Surface (0–5 cm)	Middle (5–25 cm)	Bottom (>25 cm)
Soil classification (World Reference base)	Calcisol			Calcisol			Cambisol		
Soil texture	Silty clay loam			Silty clay loam			Silty clay loam		
Soil colour	Hue 7.5YR 6/6 – reddish yellow			Hue 2.5YR 4/8 – red			Hue 2.5YR 6/4 – light yellowish brown		
Parent material	Limestone			Limestone			Quaternary sediments		
Sand (%)	16.1	16.1	18.5	19.9	19.9	13.8	17.1	17.1	20.9
Silt (%)	60.2	60.2	64.9	66.4	66.4	74.0	64.6	64.6	64.4
Clay (%)	23.8	23.8	16.6	13.7	13.7	12.2	18.3	18.3	14.7
CEC to clay ratio (CCR)	0.86			4.34			1.28		
Clay mineralogy	Smectite, kaolinite, illite (Khresat and Taimeh, 1998)			Smectite, kaolinite, illite (Khresat, 2001)			Smectite, kaolinite (Neamen *et al.*, 1999)		
pHe	8.30	8.16	7.96	8.27	8.21	8.20	8.41	8.63	8.63
ECe (dS m^{-1})	0.77	1.49	6.48	0.77	0.59	0.73	1.89	1.28	1.21
Organic carbon (%)	1.18	0.53	0.14	0.77	0.65	n/a	0.68	0.57	0.49
Organic matter (%) (assuming OM contains 0.58 g C per g organic matter)	2.03	0.91	0.24	1.33	1.12	n/a	1.17	0.97	0.84
CEC (cmol$_c$ kg^{-1})	20.50			59.50			23.50		
Ex Ca (cmol$_c$ kg^{-1})	9.76			32.82			9.89		
Ex Mg (cmol$_c$ kg^{-1})	2.99			6.94			6.49		
Ex Na (cmol$_c$ kg^{-1})	0.67			0.66			1.66		
Ex K (cmol$_c$ kg^{-1})	1.23			1.79			2.76		
ESP (%)	3.26			1.10			7.08		

After Carr (2009)
CEC = Cation exchange ratio; ECe = electrical conductivity of a saturation paste extract; pHe = pH of extract; Ex Ca = Exchangeable calcium; Ex Na = Exchangeable sodium; Ex Mg = Exchangeable magnesium; Ex K = Exchangeable potassium.

in these samples were only weakly silicified and were not robust enough to withstand the ashing process. Over 250 forms were counted per slide and in total, 90,855 barley phytolith forms were counted from 245 slides and 104,360 wheat forms from 264 slides.

21.4 RESULTS

21.4.1 Soil and water

All three growing sites had silty clay loam soils and a pH in excess of 8.1 with Deir 'Alla resulting in the highest reading with 8.63, indicating that the soil was alkaline (Table 21.5; see Carr, 2009, for the analyses). The organic carbon and organic matter levels are low at all sites. Salinity levels and exchangeable sodium percentage are highest at Deir 'Alla, probably because the soil had been irrigated with a mixture of treated wastewater and fresh water for a year.

Results for extractable silicon are presented in Figure 21.5. The greatest difference in levels of extractable silicon was between sites, with Ramtha having the greatest amount of extractable silicon and Khirbet as Samra the lowest. The levels of silicon increased throughout experimentation at Ramtha and Khirbet as Samra but remained the same at Deir 'Alla. Results from the water analysis demonstrate that the water from Khirbet as Samra and Ramtha has a higher concentration of both plant-beneficial ions and potentially plant-toxic ions than the water used for irrigation at Deir 'Alla. However, solutes of toxic metals such as arsenic, lead, zinc, nickel, cadmium and copper were below detection limits in the water at all sites (Carr, 2009).

Table 21.6 *Methodology used for analysis of extractable silicon*

Stage	Procedures followed for available plant Si analysis
1	Air dry soils
2	Grind and pass through 2 mm mesh
3	Weigh 3 g into a 50 ml polypropylene centrifuge tube
4	Add 30 ml, 0.025M citric acid by pipette
5	Shake samples end over end at 14 rpm for 6 hours at 30 °C
6	Centrifuge samples at 3,000 rpm for 15 minutes (a Mistral 3000i machine was used)
7	Filter samples through filter paper (Whatman no 1 papers were used)
8	Store extracts at 4 °C until measurement
9	Dilute extracts 1:20 with water before measurement
10	Measurements of Si concentrations were obtained using a PE Optima ICP-OES
	All storage, measuring and dispensing of solutions was carried out with plastic ware

Table 21.7 *Dry ashing methodology employed for extracting phytoliths from modern plants*

	Procedure followed for dry ashing
1	Weigh empty crucibles
2	Put dried plant samples in crucibles and weigh them
3	Ash samples in muffle furnace for 3 hr at 500 °C
4	Transfer ashed samples into centrifuge tube
5	Add HCl 10% (up to 6 ml) and shake tube
6	Wait ~5 min
7	Level samples with distilled water (up to 10 ml), tighten lid and shake tubes
8	Centrifuge 5 min at 2,000 rpm
9	Discard supernatant
10	Repeat three times
11	Transfer into weighed beakers
12	Put in drying cupboard at less than 50 °C until dry
13	Remove samples, allow to cool and weigh beakers and sample
14	Zero a balance with a labelled slide
15	Weigh 1 mg ± 0.1 mg of sample on the slide
16	Add mounting agent (we used *Entellan*) and mix thoroughly before covering with cover slip

21.4.2 Crop yield

Crop yield could not be recorded for wheat from Ramtha for the third growing season because the crop had been decimated by birds. Also, it should be noted that although more water was given to wheat than barley in the first growing season, to meet the higher water requirements of wheat, the levels of irrigation for both crops was the same for the second and third growing seasons (see Table 21.2). However, a corresponding decrease in

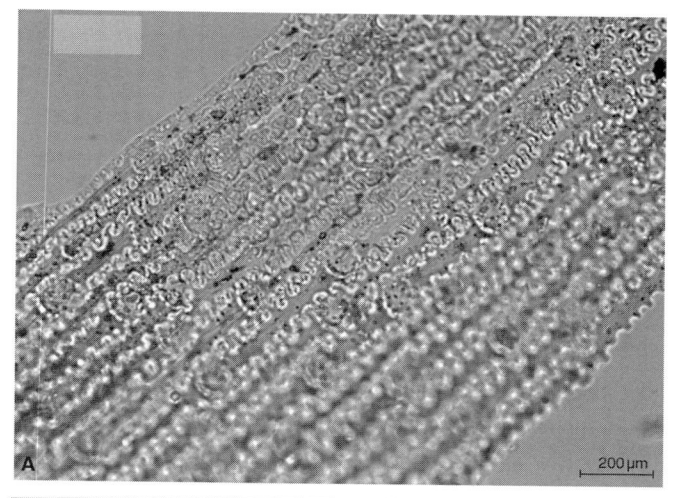

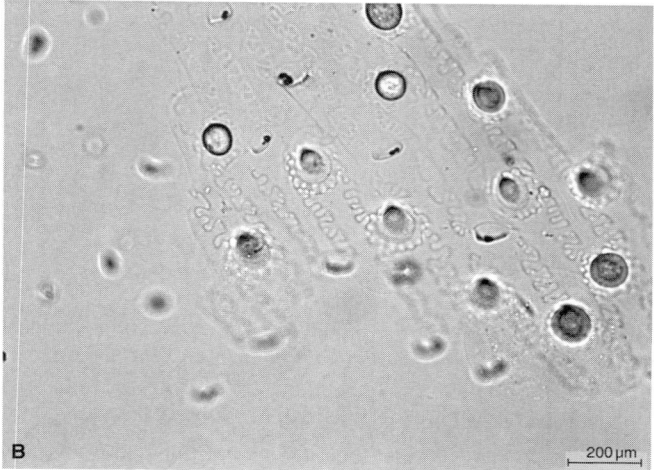

Figure 21.4. (A) Well silicified conjoined phytolith. (B) Poorly silicified conjoined phytolith.

yield from the first to the second season is not found as a result of this, as shown in Figure 21.6. The most substantial difference in the wheat yields was between the non-irrigated and 40% irrigated plots, which had low yields, and the other irrigation regimes which had higher yields. The lowest yields were from the non-irrigated plots at Khirbet as Samra which failed to produce any grains. This is probably attributable to the low rainfall at this site which, in total, was less than 100 mm in all three years.

A more significant increase in yield was observable between the non-irrigated barley and the 40% irrigated barley than was seen in wheat, as illustrated in Figure 21.7. A similar result to wheat was found with the non-irrigated barley from Khirbet as Samra, which produced neither grains nor inflorescences, preventing any phytolith analysis. It is interesting that non-irrigated wheat produced inflorescences when non-irrigated barley did not, because barley has a lower water requirement than wheat. The non-irrigated plot from the third growing season at Deir 'Alla also had a low grain yield with only 0.2 tonnes per hectare.

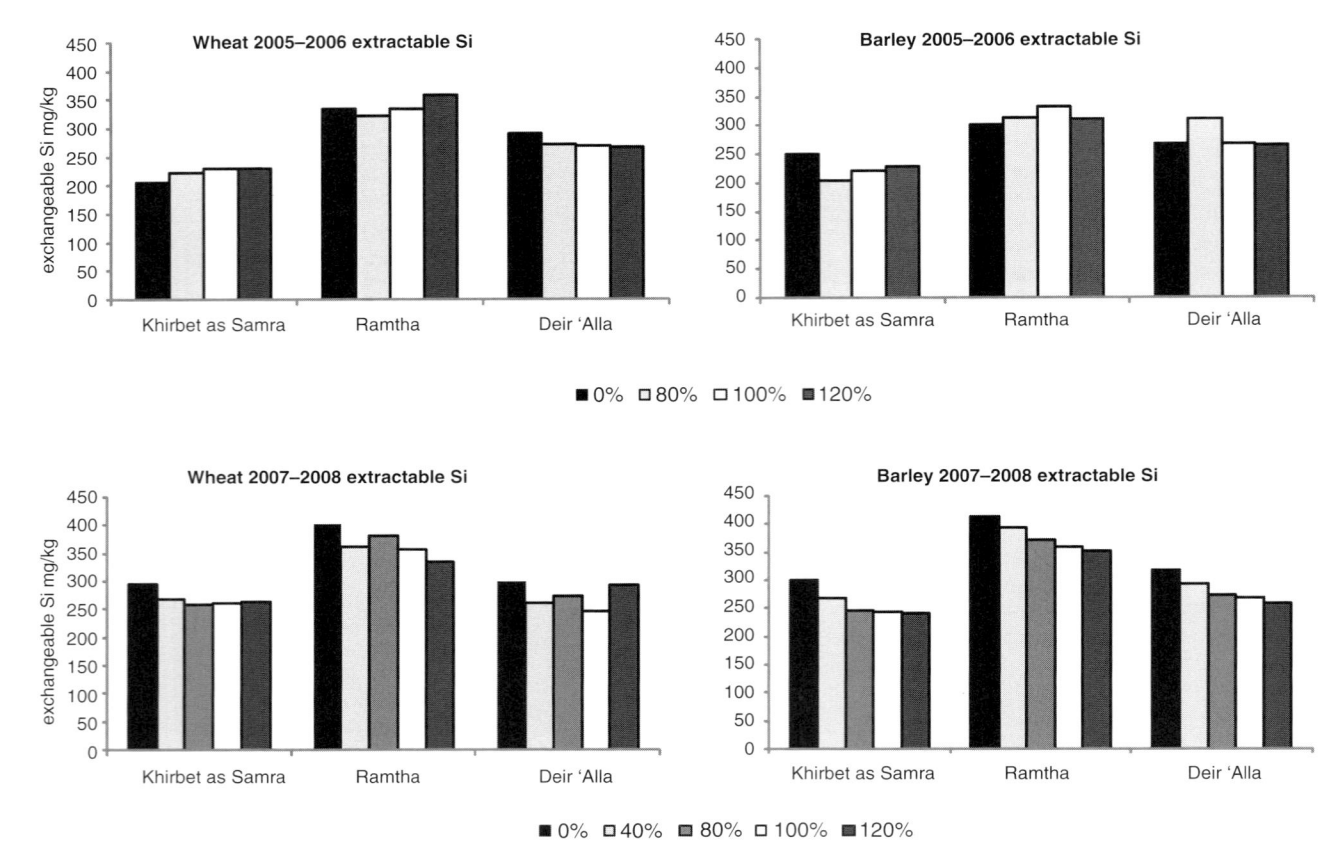

Figure 21.5. Extractable silicon from soil samples taken before and after experimentation.

The total rainfall at Deir 'Alla was less in the third growing season than in the previous two seasons and it is notable that a similarly low yield was not observed for wheat as for barley. The barley yields from Ramtha in the irrigated plots were less than for the other two sites but this may be a product of bird attack which, as stated above, severely damaged the wheat. Field notes record that the non-irrigated barley was least affected by bird attack, probably because it ripened after the irrigated plots; this would also explain why the yield from Ramtha was higher in comparison to the other two sites for the non-irrigated plot than for the irrigated ones.

21.4.3 Phytolith analysis

Weight percent was calculated by expressing the weight of phytoliths to original plant matter processed (phytolith weight % = weight of phytoliths/weight of plant matter processed × 100). This is useful for determining the level of silicon uptake and resulting phytoliths in the plant. A comparison of the results from plants grown in irrigated and non-irrigated conditions can establish if the uptake is increased with irrigation. Figure 21.8 shows the mean weight percent of phytoliths for the wheat samples and illustrates that the non-irrigated plot has the lowest weight percent and that the highest values are from the samples from the 80% irrigated plot. Deir 'Alla has a greater mean weight percent of phytoliths than the other two sites for the first and third growing season but not for the second growing season. The exception is the 40% plot. Generally Deir 'Alla has greater weight percents than the other two sites.

The most striking observation that can be made for the results of the mean weight percent of phytoliths from barley is that the values are much lower than for wheat (see Figure 21.9), Deir 'Alla has the greatest weight percent for the non-irrigated barley samples in all three years, followed by Ramtha and lastly Khirbet as Samra. Values also rise in all of the non-irrigated plots with each growing season. When this is plotted against rainfall it is apparent that the increase in weight percent for non-irrigated barley correlates with increased rainfall, with the exception of the third year at Deir 'Alla (see Figure 21.10). The results of the samples from the irrigated plots show that the mean of those from Ramtha is the highest, and that there is an increase in weight percent of phytoliths in each growing season. This correlates with the results of the extractable silicon analysis which demonstrated that at Ramtha and Khirbet as Samra the level of silicon in the soil increased from the beginning to the end of the experiment. Figure 21.11 shows the mean value of extractable silicon from all three sites plotted against phytolith weight percent which demonstrates that barley has a positive correlation, while wheat has a negative one. This increase in silicon levels is

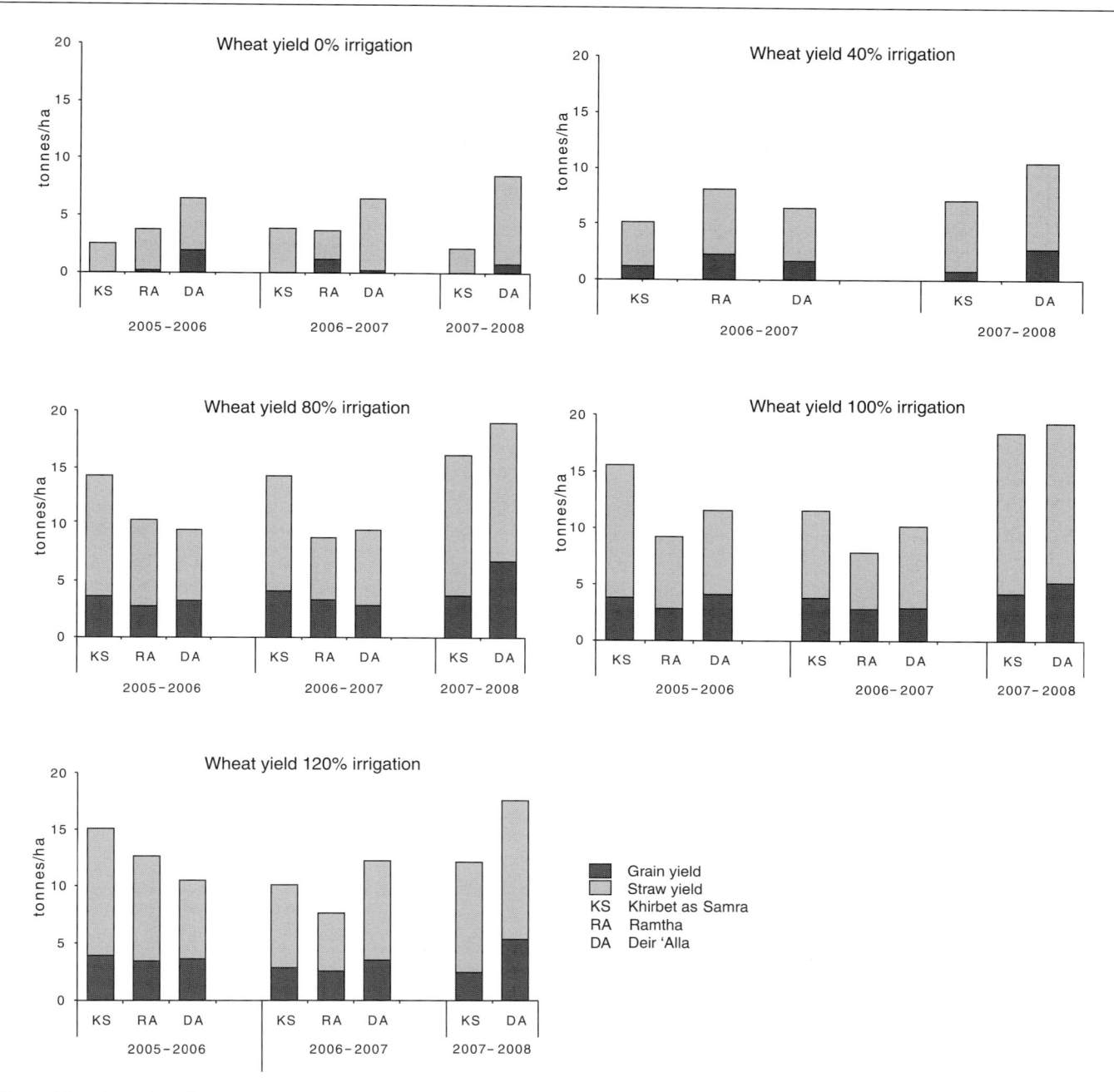

Figure 21.6. Crop yield for wheat.

unexpected because the plants are taking up silicon from the soil which is not being returned in phytolith form because the crops are harvested and removed from the sites. Silicon is not entering the soil through the irrigation water because the results show that the greatest rise in silicon levels was in the non-irrigated plots. Further tests are needed to check if other changes occurred in the soil through time. For example, it is possible that the pH or cation exchange capacity of the soil changed through time in a manner that resulted in more available silicon. For example, the soil may have become more alkaline which would have caused greater dissolution of silicates in the soil. It is also possible that the clay mineral fraction was being washed down through the soil profile by the irrigation water, removing the clay silicates from the rooting zone of the plants. This would explain why the non-irrigated samples have more available silicon than the irrigated ones.

A comparison was made between long dendritic cells, those termed by Madella *et al.* (2009) as *sensitive forms*, and cork-silica cells or *fixed forms* (Madella *et al.*, 2009). Silica cells and cork cells form in pairs and for the purposes of this study are grouped together as one category (Kaufman *et al.*, 1970). These cells are known by a variety of names; for example Blackman and Parry (1968) refer to them as silico-suberous couples. Images

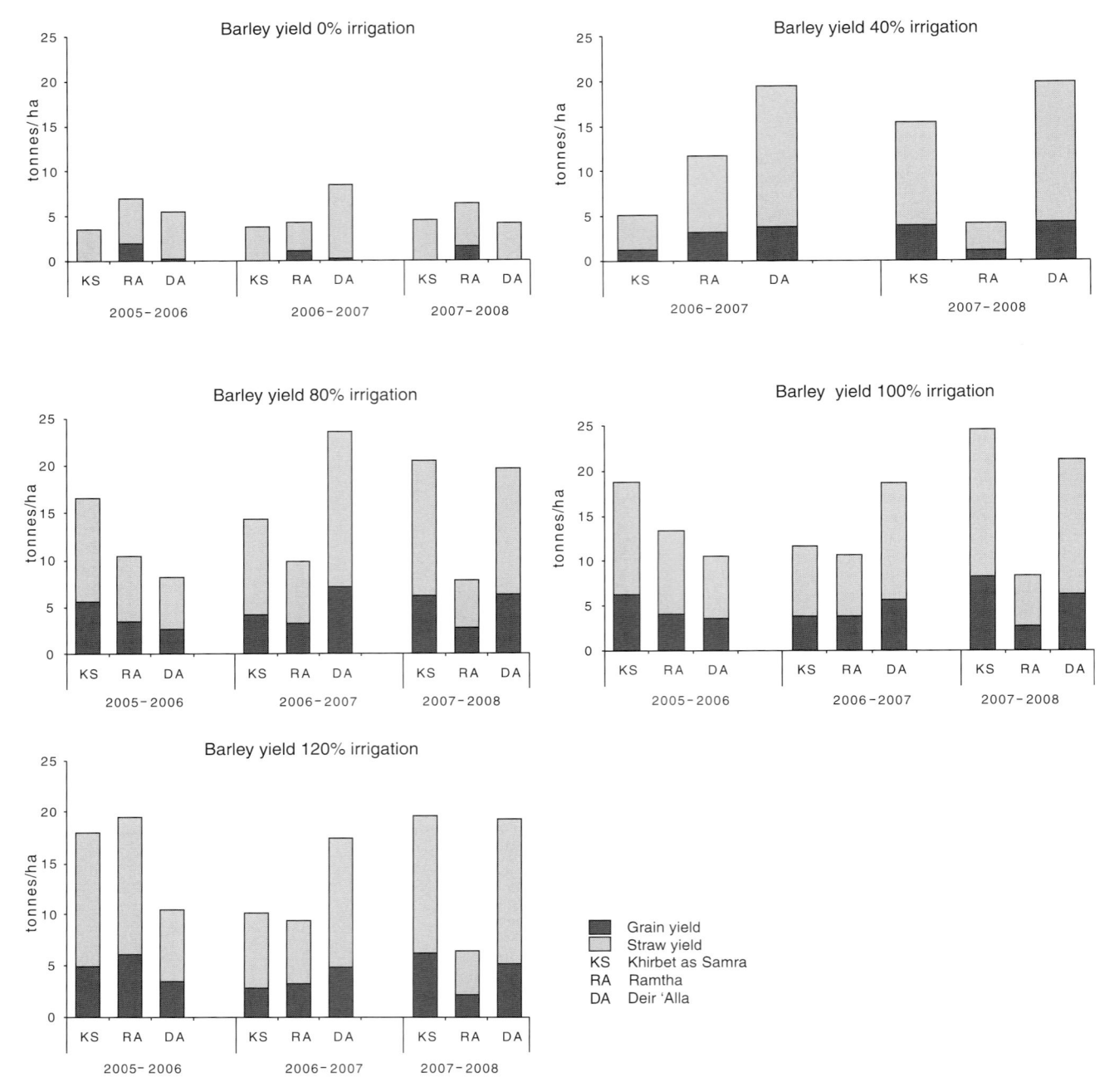

Figure 21.7. Crop yield for barley.

of both these phytolith types are shown in Figure 21.12. Results of the comparison of long dendritic cells to cork-silica cells from the wheat samples are shown in Figure 21.13. Only samples with a combined total of over 200 cork-silica cells and dendritics were used in this analysis. The results demonstrate that, with the exception of the first growing season at Ramtha, the mean percent of dendritics in the irrigated samples is always higher than the percent for the non-irrigated ones. Despite this trend, it is also apparent that there is often a significant overlap between the values for the irrigated and non-irrigated samples, with the exception of Khirbet as Samra where the percent of dendritics in the non-irrigated samples is always lower than in the irrigated ones. This correlates with rainfall which is significantly lower at Khirbet as Samra than the other two sites. This negates the claim of Rosen and Weiner (1994) that changes in phytolith assemblages reflect irrigation and not rainfall. It is also clear from Figure 21.13 that there is a decrease in the mean percent of dendritics from both the irrigated and non-irrigated samples through time: the first growing season has an average of 48%, the second season 28%, and the third season only 15%. This

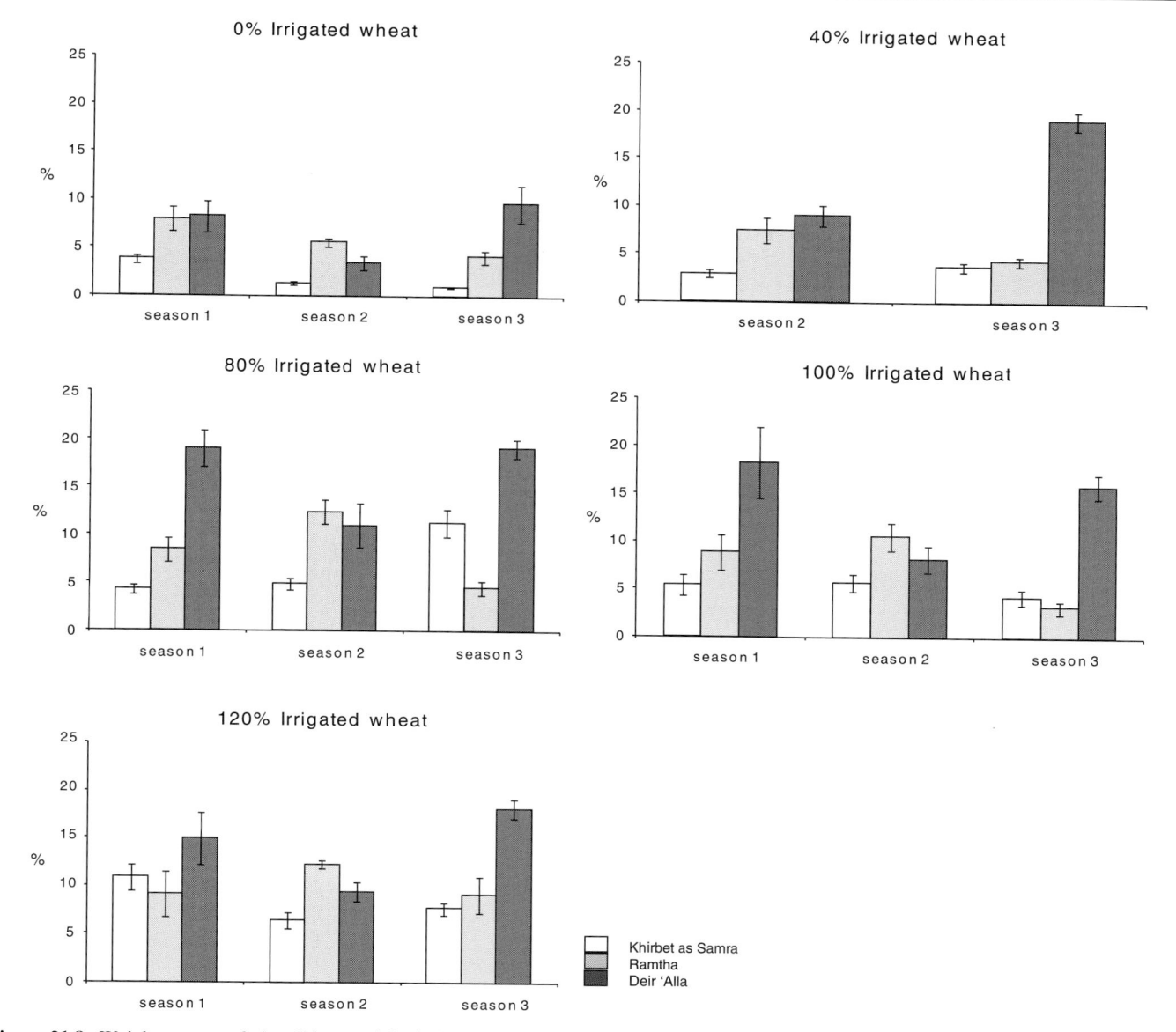

Figure 21.8. Weight percent of phytoliths to original plant matter processed for wheat.

correlates with weight percent of phytoliths which also decreases, but not with the results of the extractable silicon analysis which increases throughout experimentation.

The results from the barley analysis are provided in Figure 21.14. These show that the mean percent of dendritics in the irrigated samples is always greater than the percent for the non-irrigated ones. There is also a decrease through time in the percent of dendritics to cork-silica cells for barley, but the main decrease happens between the second and third growing seasons. In the samples from the first growing season, dendritics make up 50% of the total single cells. This decreases to 49% in the second season and to 25% in the third growing season. This result, however, could partly be influenced by sample size which was reduced in the third growing season because many of the samples fused during ashing. As with the wheat, the non-irrigated samples from Khirbet as Samra have low percentages of dendritics, although this is based on results from only one slide in the third growing season because the phytoliths from the other four samples were all fused and could not be counted. However, the levels of dendritics in the irrigated barley are also lower at Khirbet as Samra than in the samples from the other two sites.

A comparison of the number of well silicified to poorly silicified conjoined wheat phytoliths can be found in Figure 21.15 and the absolute counts with standard deviations are shown in Table 21.8. A difference can be seen between the sites in the first growing season; the non-irrigated samples from Khirbet as Samra produced far fewer well-silicified conjoined phytoliths than the non-irrigated samples from the other two crop growing

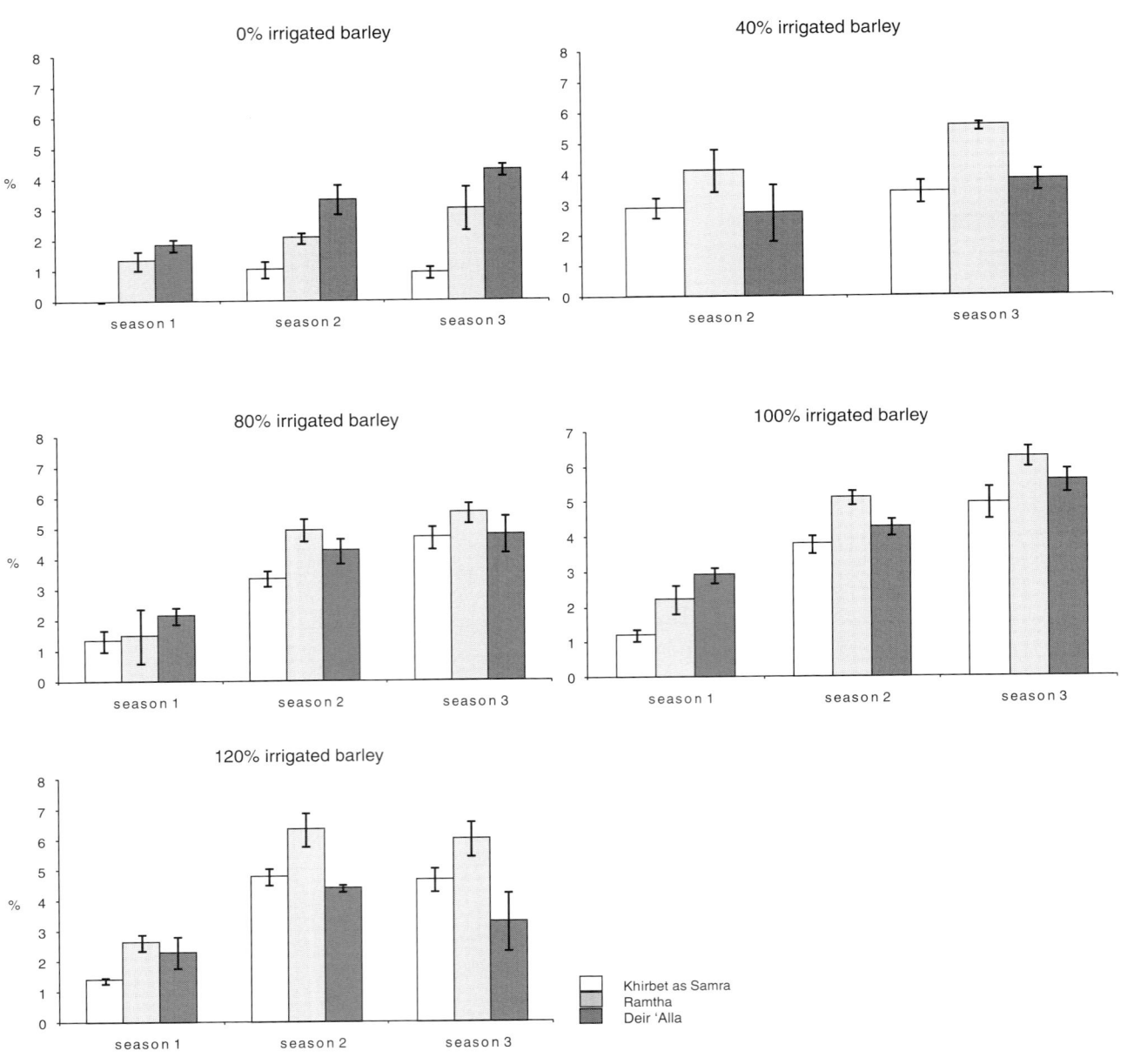

Figure 21.9. Weight percent of phytoliths to original plant matter processed for barley.

stations, while Deir 'Alla had a greater percent of well silicified forms in the non-irrigated samples (63%) than in the mean of the irrigated samples (60%). However, in all other seasons at the three sites the irrigated samples had a greater mean of well silicified conjoined forms than the non-irrigated ones. There is an increase in the number of conjoined forms in the Khirbet as Samra non-irrigated samples in the second year which reaches 32% but falls in the third year to 16%. A decrease in well silicified forms is apparent in the third year in the irrigated samples from both Khirbet as Samra and Ramtha. Figure 21.16 shows the comparison of well silicified and poorly silicified forms for the barley samples with Table 21.9 showing the absolute counts and standard deviations. From these it is clear that there is a gradual decline in the number of well silicified forms over time. It is also apparent that, unlike wheat, the non-irrigated samples from Deir 'Alla have far fewer well silicified forms than the irrigated ones and that there is a decrease in the number of well silicified forms from the second to the third growing season in the non-irrigated samples from Khirbet as Samra.

Figure 21.17 shows the percent of conjoined dendritics for the wheat samples by four different categories: 2–15 cells, 15–50 cells, 51–100 cells and 100 cells and over. It is clear that in all of the plots the 2–15 cell category has the highest percent of phytoliths. However, it is also apparent that this varies between both sites and years. The most consistent site over the three years was Khirbet as Samra. Here the phytoliths from the

IRRIGATION AND PHYTOLITH FORMATION

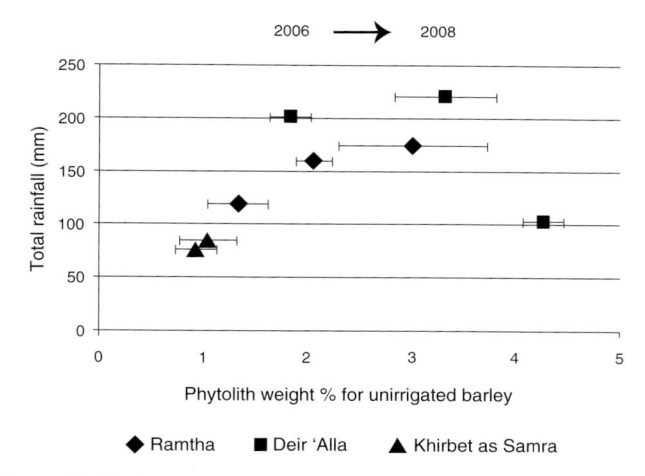

Figure 21.10. Correlation between non-irrigated barley and rainfall.

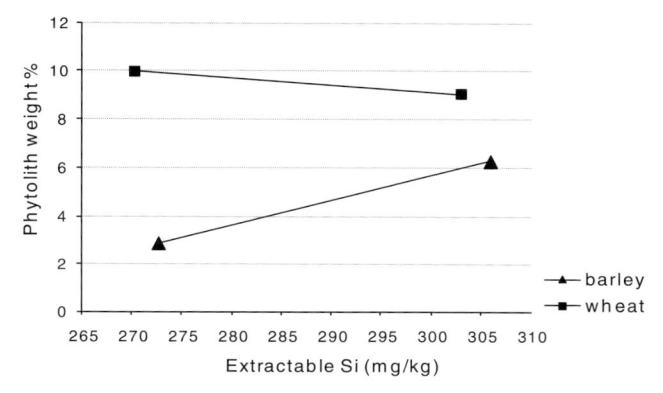

Figure 21.11. Correlation between mean weight percent of phytoliths and levels of extractable silicon.

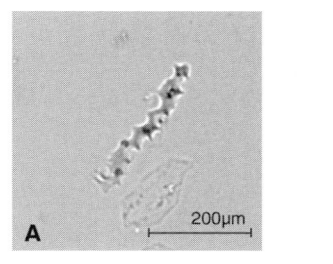

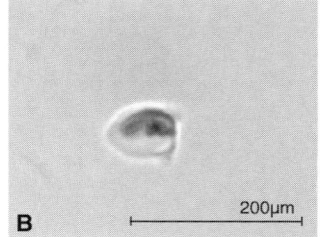

Figure 21.12. (A) Single dendritic long cell. (B) Single cork-silica cell.

non-irrigated plots are largely dominated by forms that consist of between 2 and 15 cells, with the first season having 80%, the second season 79% and the third season 78%. Khirbet as Samra also has the lowest percent of forms from the non-irrigated plots in the over 100 cells category. The greatest variation between years is found in the samples from Ramtha which has a more even distribution of numbers of phytoliths over all four counting categories than in the previous two years. Both Ramtha and Deir 'Alla have an increase in the percent of forms in the 2–15 cell category in the second growing season which then decreases in the third season, and in the third growing season both Ramtha

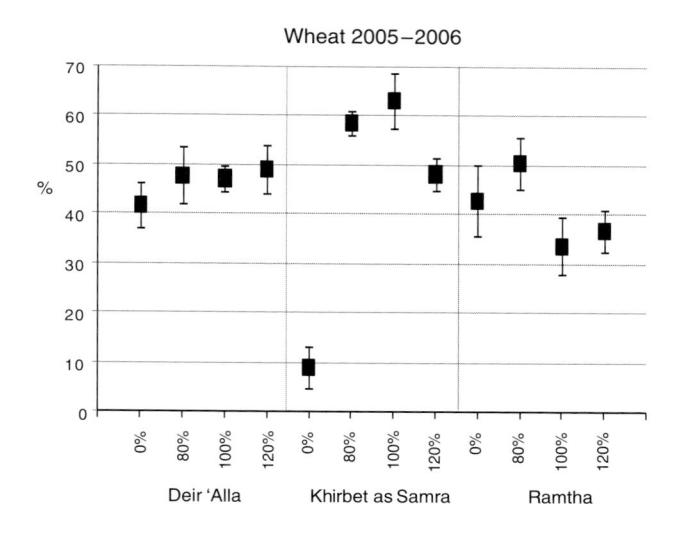

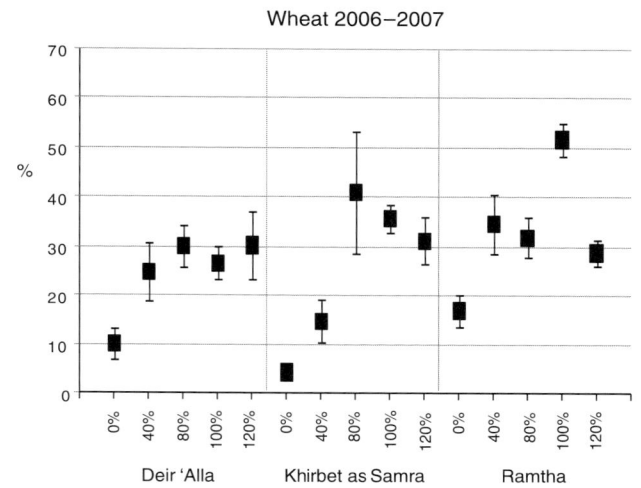

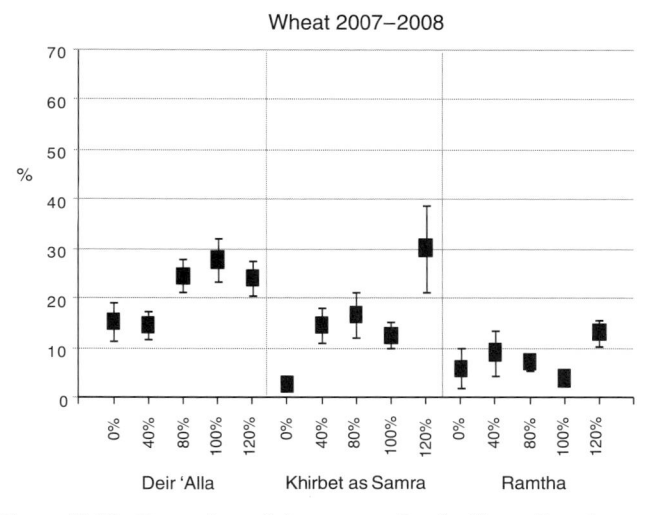

Figure 21.13. Comparison of the percent of cork-silica cells and dendritic long cells for the wheat samples.

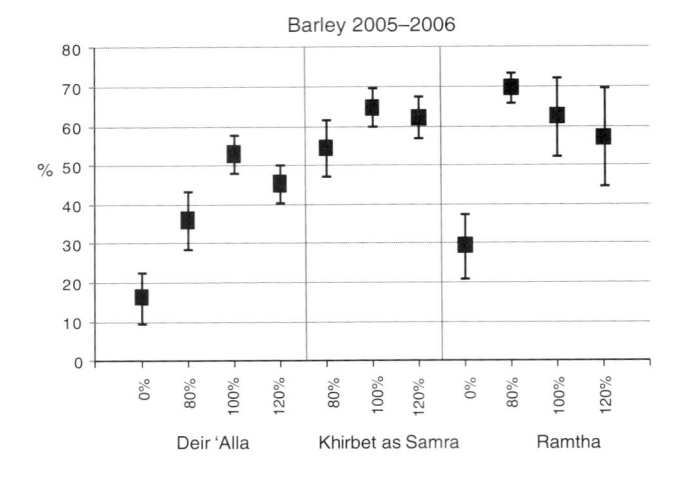

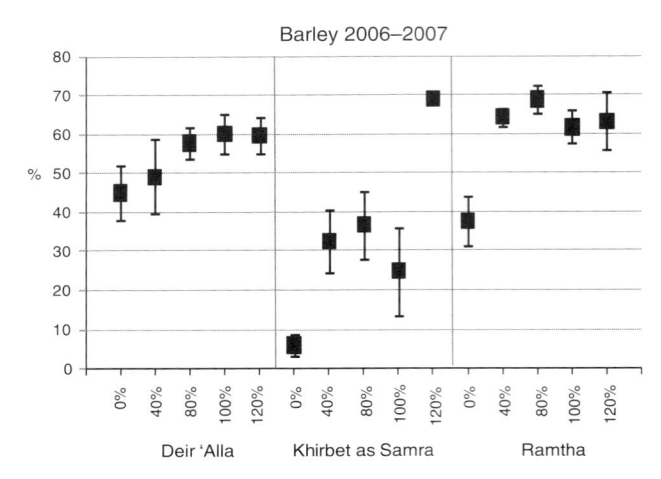

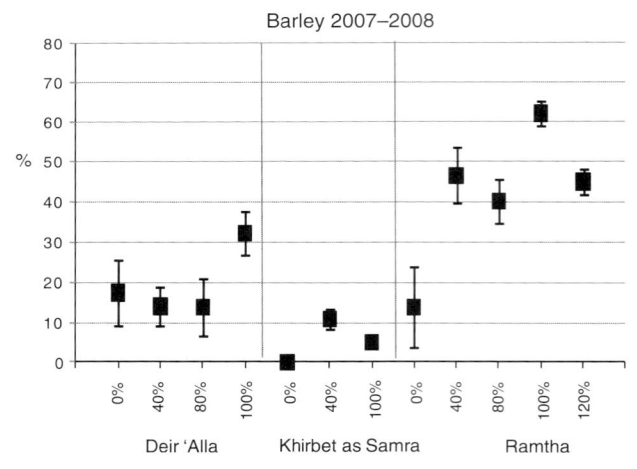

Figure 21.14. Comparison of cork/silica cells and dendritic long cells for the barley samples.

and Deir 'Alla have a greater percent of non-irrigated samples in the over 15 conjoined cell categories than irrigated ones. However, overall the number of conjoined cells increases with irrigation; the average number of forms with over 15 conjoined cells

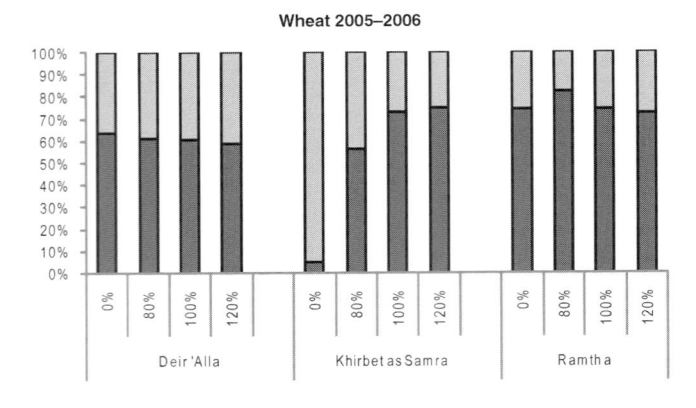

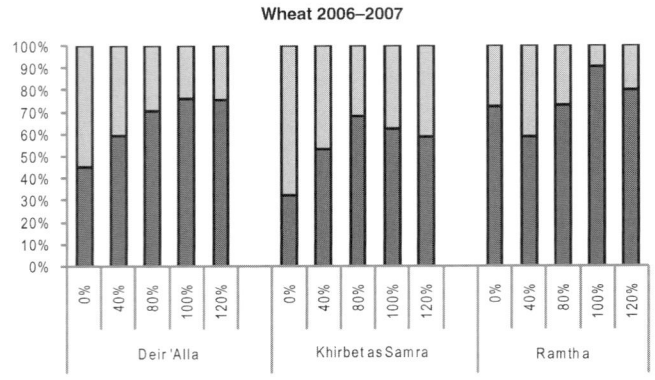

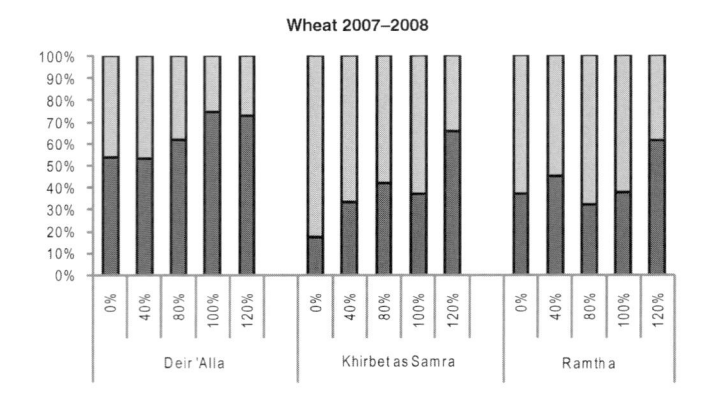

Figure 21.15. Comparison of well silicified to poorly silicified conjoined forms from the wheat samples.

from the irrigated samples is 47% but is only 41% for the non-irrigated samples.

Figure 21.18 shows the same comparison for barley. As with wheat, there is an increase in the number of forms falling in the 2–15 cell category at Ramtha and Deir 'Alla in 2006 to 2007 which decreases again in the third season. Unfortunately, there was no sample for barley from the first year of crop growing at Khirbet as Samra, and in the third season only one replicate was analysed because, as stated above, the phytoliths in the other four

Table 21.8 *Absolute numbers of well silicified and poorly silicified conjoined phytoliths counted with standard deviations from the wheat samples (all standard deviations were calculated using the STDEV function in Excel 2007). DA, KS, RA = Deir 'Alla, Khirbet as Samra and Ramtha, respectively*

Site/regime	Phytolith type	n phytos counted	Mean	n slides	Standard deviation	Phytolith type	n phytos counted	Mean	n slides	Standard deviation
Wheat 2005–2006										
DA 0%	poorly silicified	562	56	10	35.24	well silicified	1,498	150	10	47.66
DA 80%	poorly silicified	880	88	10	25.24	well silicified	1,373	137	10	27.46
DA 100%	poorly silicified	942	94	10	36.34	well silicified	1,436	144	10	33.13
DA 120%	poorly silicified	953	95	10	43.29	well silicified	1,356	136	10	32.57
KS 0%	poorly silicified	1,189	119	10	72.13	well silicified	59	6	10	6.56
KS 80%	poorly silicified	746	75	10	33.15	well silicified	957	96	10	24.91
KS 100%	poorly silicified	562	56	10	35.24	well silicified	1,498	150	10	47.66
KS 120%	poorly silicified	571	57	10	25.37	well silicified	1,668	167	10	42.15
RA 0%	poorly silicified	558	56	10	21.21	well silicified	1,574	157	10	20.08
RA 80%	poorly silicified	378	38	10	14.19	well silicified	1,760	176	10	16.44
RA 100%	poorly silicified	541	54	10	18.51	well silicified	1,532	153	10	45.66
RA 120%	poorly silicified	634	63	10	27.28	well silicified	1,666	167	10	23.02
Wheat 2006–2007										
DA 0%	poorly silicified	173	35	5	22.53	well silicified	142	28	5	17.52
DA 40%	poorly silicified	237	47	5	17.56	well silicified	349	70	5	18.32
DA 80%	poorly silicified	197	39	5	6.50	well silicified	470	94	5	18.25
DA 100%	poorly silicified	146	29	5	7.22	well silicified	463	93	5	23.36
DA 120%	poorly silicified	171	34	5	14.96	well silicified	528	106	5	15.90
KS 0%	poorly silicified	106	21	5	9.44	well silicified	50	10	5	7.00
KS 40%	poorly silicified	347	69	5	27.48	well silicified	394	79	5	16.32
KS 80%	poorly silicified	186	37	5	9.26	well silicified	392	78	5	21.22
KS 100%	poorly silicified	216	43	5	11.03	well silicified	361	72	5	20.39
KS 120%	poorly silicified	272	54	5	14.52	well silicified	390	78	5	13.34
RA 0%	poorly silicified	186	37	5	5.89	well silicified	481	96	5	19.61
RA 40%	poorly silicified	217	43	5	13.87	well silicified	304	61	5	14.69
RA 80%	poorly silicified	222	44	5	19.73	well silicified	593	119	5	25.51
RA 100%	poorly silicified	78	16	5	12.10	well silicified	701	140	5	39.32
RA 120%	poorly silicified	166	33	5	3.56	well silicified	637	127	5	25.75
Wheat 2007–2008										
DA 0%	poorly silicified	252	50	5	13.90	well silicified	293	59	5	18.68
DA 40%	poorly silicified	299	60	5	20.73	well silicified	336	67	5	12.83
DA 80%	poorly silicified	274	55	5	12.56	well silicified	441	88	5	24.99
DA 100%	poorly silicified	194	39	5	8.47	well silicified	566	113	5	18.31
DA 120%	poorly silicified	218	44	5	16.88	well silicified	591	118	5	16.84
KS 0%	poorly silicified	44	N/A	1	N/A	well silicified	9	N/A	1	N/A
KS 40%	poorly silicified	366	73	5	18.29	well silicified	185	37	5	23.31
KS 80%	poorly silicified	292	58	5	21.79	well silicified	212	42	5	37.11
KS 100%	poorly silicified	304	61	5	26.11	well silicified	178	36	5	18.61
KS 120%	poorly silicified	319	64	5	15.51	well silicified	495	99	5	23.99
RA 0%	poorly silicified	219	44	5	10.99	well silicified	129	26	5	8.93
RA 40%	poorly silicified	312	62	5	18.60	well silicified	256	51	5	31.75
RA 80%	poorly silicified	392	78	5	18.98	well silicified	187	37	5	10.57
RA 100%	poorly silicified	185	37	5	23.66	well silicified	113	23	5	18.39
RA 120%	poorly silicified	140	47	3	10.79	well silicified	218	73	3	22.68

phytolith forms with over 15 conjoined cells increasing from 33% for non-irrigated samples to 39% for the mean of the irrigated ones.

21.5 DISCUSSION

21.5.1 Water availability and phytolith formation

Results from the crop yield analysis demonstrate that the yield capability of the crops grown in this experiment were, as would be expected, positively affected by irrigation but were not affected by the increasing levels of salinity that were found to have built up in the soil over the three years of experimentation (Carr, 2009).

The results from the analysis of the available silicon and weight percent of phytoliths produced interesting results. For barley it is clear that, with the exception of the non-irrigated plots, the results correlate with those from the extractable silicon analysis, with Ramtha having both the greatest silicon levels and highest phytolith yields. Wheat, however, does not respond in the same way. In the second growing season the wheat from Ramtha produced the highest weight percent of phytoliths in all plots except the 40% plot, but in the first and third seasons, Deir 'Alla has the highest weight percent. It is also notable that, with the exception of the third growing season at Deir 'Alla, the phytolith weight percent and soil silicon levels increase linearly with barley, but the weight percent of phytoliths from wheat decrease as silicon levels increase. These results, coupled with the significantly lower weight percent for barley, suggest that the processes that govern silica uptake and deposition differ between the two crops. It appears that while wheat is more efficient in its uptake of silicon than barley, there are factors other than available silicon in the growing medium that affect phytolith production and deposition in wheat.

Deir 'Alla is located in the Jordan Valley, is below sea level and, as such, records higher temperatures than either of the other two crop growing stations. However, records of evaporation show that Ramtha has the highest rates of evaporation, because Deir 'Alla is more humid owing to its proximity to the Dead Sea. This greater phytolith weight percent from Ramtha for wheat is not due to the use of mesh to cover the crops during maturation because the wheat crops at both Deir 'Alla and Ramtha were covered in the first two growing seasons. Furthermore, the use of mesh would increase humidity resulting in slower transpiration rates and presumably decreased silicon uptake. Similarly there is no correlation with this result and rainfall because while Deir 'Alla had the highest rainfall in the first and second seasons Ramtha recorded the greatest amount in the third growing season. Another factor which could affect silicon uptake is high concentrations of nitrogen in the soil. Unfortunately nitrogen was

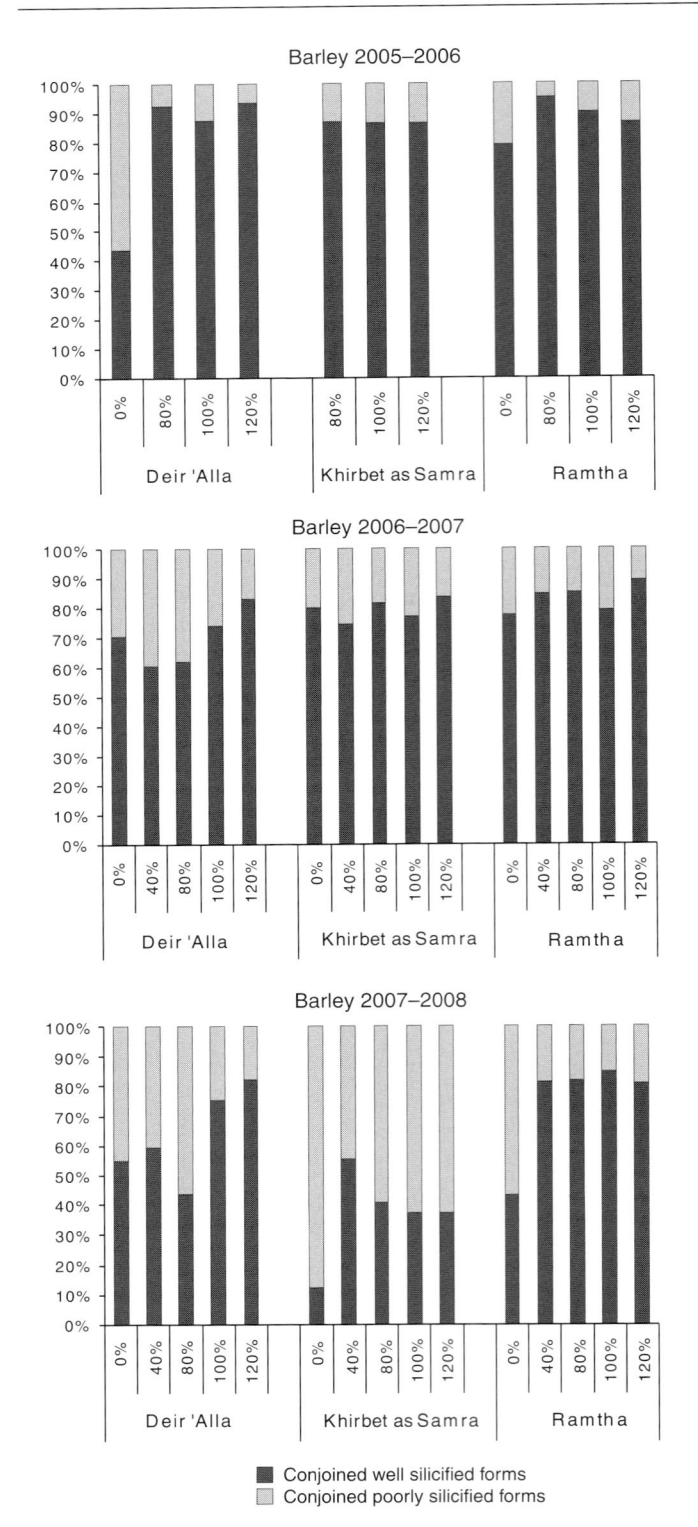

Figure 21.16. Comparison of well silicified to poorly silicified conjoined forms from the barley samples.

samples all fused during processing. Khirbet as Samra has the highest percent of forms with over 100 cells while Deir 'Alla has the most variation in results over time. Overall, the number of conjoined cells does increase with irrigation, with the number of

Table 21.9 *Absolute numbers of well silicified and poorly silicified conjoined phytoliths counted with standard deviations from the barley samples*

Site/regime	Phytolith type: poorly silicified				Phytolith type: well silicified			
	n phytos counted	Mean	n slides	Standard deviation	n phytos counted	Mean	n slides	Standard deviation
Barley 2005–2006								
DA 0%	1,372	137	10	56.70	1,069	107	10	22.46
DA 80%	239	24	10	10.57	1,638	164	10	36.76
DA 100%	158	16	10	13.37	1,105	111	10	9.86
DA 120%	89	9	10	10.56	1,323	132	10	34.35
KS 80%	168	17	10	11.48	1,123	112	10	47.57
KS 100%	105	11	10	9.66	661	66	10	31.09
KS 120%	133	13	10	8.33	840	84	10	29.54
RA 0%	393	44	9	30.55	1,488	165	9	38.32
RA 80%	48	5	10	2.82	913	91	10	26.17
RA 100%	135	14	10	18.40	1,232	123	10	43.30
RA 120%	162	18	9	32.49	1,021	113	9	40.69
Barley 2006–2007								
DA 0%	64	13	5	6.69	153	31	5	8.91
DA 40%	50	17	3	10.02	84	28	3	13.86
DA 80%	74	15	5	9.15	121	24	5	5.93
DA 100%	31	6	5	4.82	89	18	5	8.70
DA 120%	41	8	5	5.50	199	40	5	5.85
KS 0%	31	6	5	3.11	124	25	5	13.22
KS 40%	66	13	5	6.69	195	39	5	8.94
KS 80%	77	15	5	13.90	344	69	5	14.31
KS 100%	125	25	5	15.03	413	83	5	31.61
KS 120%	35	7	5	2.92	179	36	5	17.14
RA 0%	72	14	5	5.77	246	49	5	10.33
RA 40%	33	7	5	3.21	180	36	5	7.07
RA 80%	28	6	5	3.36	157	31	5	5.27
RA 100%	42	8	5	7.02	158	32	5	5.94
RA 120%	24	5	5	0.84	194	39	5	10.08
Barley 2007–2008								
DA 0%	156	31	5	12.52	190	38	5	21.95
DA 40%	115	29	4	10.47	169	42	4	13.82
DA 80%	335	67	5	25.80	261	52	5	16.50
DA 100%	102	26	4	9.29	306	77	4	16.34
DA 120%	16	N/A	1	N/A	75	N/A	1	N/A
KS 0%	14	N/A	1	N/A	2	N/A	1	N/A
KS 40%	327	65	5	21.56	405	81	5	15.15
KS 80%	366	92	4	28.72	251	63	4	35.35
KS 100%	285	95	3	11.31	170	57	3	36.77
KS 120%	442	88	5	37.04	262	52	5	32.67
RA 0%	242	48	5	19.37	184	37	5	21.99
RA 40%	96	19	5	5.54	411	82	5	12.40
RA 80%	111	22	5	5.76	493	99	5	15.98
RA 100%	76	15	5	5.17	416	83	5	15.74
RA 120%	73	15	5	7.23	308	62	5	9.71

not tested for during soil analysis. However, after the first growing season plant samples of both wheat and barley were taken, oven dried and crushed and then tested for nitrogen, potassium and phosphorus by NCARE. The results for both wheat and barley found that the level of nitrogen was lower for the plants from Ramtha than plants grown at the other two stations,

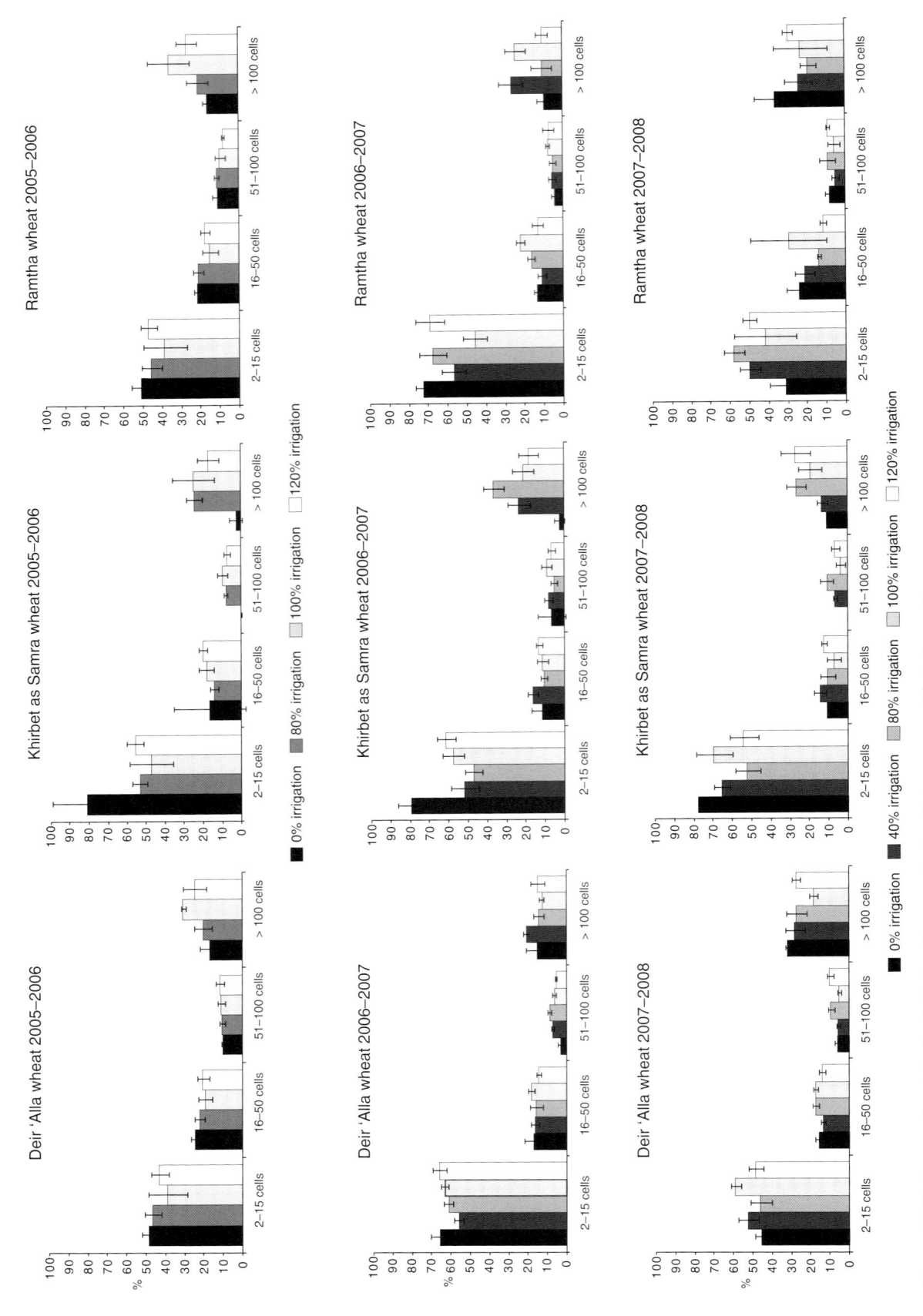

Figure 21.17. Percent of conjoined cells in the wheat samples for all sites, years and irrigation regimes.

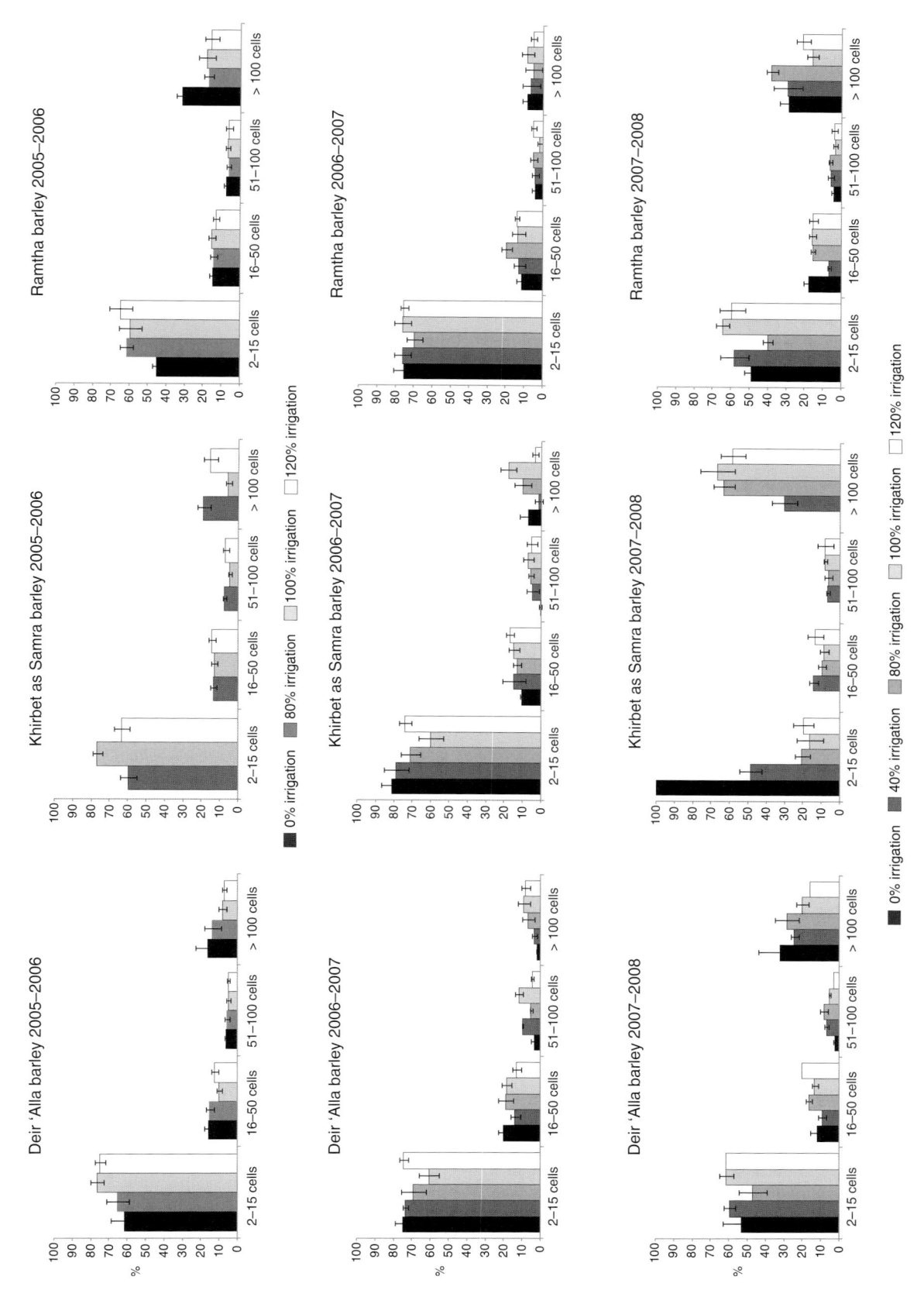

Figure 21.18. Percent of conjoined cells in the barley samples for all sites, years and irrigation regimes.

suggesting that the plants from Ramtha did not take higher levels of nitrogen from the growing medium than plants from Deir 'Alla and Khirbet as Samra.

Results from the comparison of the cork-silica cells to dendritic long cells, and from the comparison of well silicified conjoined forms to the poorly silicified conjoined forms, show a decrease in indicators of water availability (dendritic long cells and well silicified conjoined forms) over time. This is interesting because an increase in sensitive forms due to irrigation was not found in the six-row barley analysed by Madella et al. (2009). With wheat, the lowest percent of both these forms is found in the non-irrigated plots from Khirbet as Samra. As this is the site with the least rainfall, it confirms that the production of dendritic long cells is governed by water availability.

While this study found that the ratio of dendritic long cells to cork-silica cells is an indicator of water availability for both wheat and barley, there were significant inter-site and, more notably, inter-year differences. This demonstrates that local climatic and environmental conditions affect results. However, the decrease in dendritics over time is the biggest difference seen in the results and affects all three sites. It is possible that this is due to a build up of salinity in the soil caused by the use of treated wastewater for irrigation, as demonstrated by the work of Carr (2009). Ahmad et al. (1992) showed that silicon increases the tolerance of bread wheat to salinity while Liang et al. (1996) report similar findings for common barley, though they found that barley was more salt-tolerant than bread wheat. Ahmad et al. (1992) suggest that this tolerance occurs because there is an interaction between freely available sodium and silicon ions which reduces their uptake into the plants, as observed in the roots of the wheat they studied. The fact that common barley exhibits a greater tolerance for salt stress than wheat could explain why in our experiments the percent of dendritics decreased after one growing season for wheat but only after two seasons for barley. It may also explain why phytolith weight percent increases over time in barley samples but decreases in wheat.

21.5.2 Implications for archaeological study

The results from this research demonstrate that phytolith assemblages are altered by increased water availability. The most effective method for identifying water availability is the ratio of long cells to short cells. This is because these forms are single celled and not subject to break up, a problem with the conjoined phytolith method proposed by Rosen and Weiner (1994). Analytical experiments conducted by Jenkins (2009) demonstrated that conjoined phytolith forms do not remain stable from the time of formation to the time of analysis, and using this method to indicate water availability in assemblages with unknown taphonomic pathways is problematic.

Although the proportion of dendritic long cells to cork-silica cells is a more reliable indicator of water availability than changes in the number of conjoined cells, this method is not without its pitfalls. It is clear that the inter-year differences found in the percent of dendritics are sometimes greater than that found between the irrigated and non-irrigated samples. Further work is planned to establish if the decrease in dendritics from the second to the third growing season is the result of smaller sample size or a real change in phytolith ratios. Work is also needed to determine if this method is applicable to the leaves and stems as well as the husks as suggested in the work of Madella et al. (2009). This is important because while the types of long cells found in the leaves and stems have smooth edges and so are distinct from the wavy edged dendritics found in husks, the corresponding short cells formed in leaves (rondels) are morphologically very similar to cork-silica cells. This means that in an archaeological assemblage derived from a mixture of plant parts, it would be difficult to isolate short cells formed in husks from those formed in leaves.

Results from this study also show that the source of water, i.e. rainfall or irrigation, is unimportant in its effects on changes in phytolith formation and deposition. It is possible that the greater amount of available silica coming from wadi water may increase the level of silicon uptake, but our results found that the important factor for affecting changes to phytolith formation was water availability. This inability to detect the difference between rainfall and irrigation is of course true for many methodologies which claim to be able to identify irrigation, such as the FIBS method (Charles et al., 2003). The changes they identify occur as a result of increased water availability and not necessarily as a result of rainfall. However, if other proxies, such as carbonate deposits or stable isotopes, indicate that the site under excavation was occupied during an arid period, then phytoliths can potentially be used to infer past irrigation. Results from this experiment suggest that an archaeological phytolith assemblage consisting of 60% dendritics would indicate that water was abundant. Chapter 22 shows an example of the application of this methodology to archaeological and modern phytolith assemblages with encouraging results.

21.6 CONCLUSION

This study has confirmed that the uptake and deposition of phytoliths is affected by water availability, as suggested by Rosen and Weiner (1994) and Madella et al. (2009). But it also demonstrates that these changes reflect increased water availability which could be from precipitation or irrigation, a claim previously refuted by Rosen and Weiner (1994). The change in phytolith composition is discernible in an increase in the ratio of dendritic long cells to cork-silica cells in the husks of durum wheat and six-row barley, supplementing the work of Madella et al. (2009). However, results for six-row barley contrast with those of Madella et al. (2009) who found that six-row barley responded negatively to increased irrigation, with the percent of long cells in leaves decreasing, not increasing. The method proposed by Rosen and Weiner (1994) which suggested that the number of conjoined phytoliths increased with irrigation was found to be less reliable for identifying past water availability. Our results suggest that an assemblage consisting of over 60% dendritic long cells indicates a level of water availability sufficient to meet the crop requirements of cereals. If this method is found to be consistent in all plant parts, as suggested by the work of Madella et al. (2009), phytoliths could be a valuable tool for estimating past water availability and, potentially, irrigation.

REFERENCES

Agarie, S., W. Agata, H. Uchida, F. Kubota and P. B. Kaufman (1996) Function of silica bodies in the epidermal system of rice (*Oryza sativa* L.): testing the window hypothesis. *Journal of Experimental Botany* **47**: 655–660.

Ahmad, R., S. H. Zaheer and S. Ismail (1992) Role of silicon in salt tolerance of wheat (*Triticum aestivum* L.). *Plant Science* **85**: 43–50.

Albert, R. M., O. Bar-Yosef, L. Meignen and S. Weiner (2003) Quantitative phytolith study of hearths from the Natufian and middle palaeolithic levels of Hayonim Cave (Galilee, Israel). *Journal of Archaeological Science* **30**: 461–480.

Alexandre, A., J. D. Meunier, F. Colin and J. M. Koud (1997) Plant impact on the biogeochemical cycle of silicon and related weathering processes. *Geochimica et Cosmochimica Acta* **61**: 677–682.

Al-Zu'bi, Y. (2007) Effect of irrigation water on agricultural soil in Jordan Valley: an example from arid area conditions. *Journal of Arid Environments* **70**: 63–79.

Allen, R. G., L. S. Pereira, D. Raes and M. Smith (1998) *Crop Evapotranspiration – Guidelines for Computing Crop Water Requirements*. FAO Irrigation and drainage papers 56. Rome: Food and Agriculture Organisation of the United Nations.

Ammary, B. Y. (2007) Wastewater reuse in Jordan: present status and future plans. *Desalination* **211**: 164–176.

Araus, J. L., A. Febrero, R. Buxo et al. (1997) Identification of ancient irrigation practices based on the carbon isotope discrimination of plant seeds: a case study from the south-east Iberian Peninsula. *Journal of Archaeological Science* **24**: 729–740.

Araus, J. L., A. Febrero, M. Catala et al. (1999) Crop water availability in early agriculture: evidence from carbon isotope discrimination of seeds from a tenth millennium BP site on the Euphrates. *Global Change Biology* **5**: 201–212.

Barber, D. A. and M. G. T. Shone (1966) The absorption of silica from aqueous solutions by plants. *Journal of Experimental Botany* **52**: 569–578.

Barker, G. (2000) Farmers, herders and miners in the Wadi Faynan, southern Jordan: a 10,000-year landscape archaeology. In *The Archaeology of Drylands*, ed. G. Barker and D. Gilbertson. London: Routledge pp. 63–85.

Bartoli, F. and B. Souchier (1978) Cycle et Rôle du Silicium d'Origine Végétable dans les Ecosystèmes Forestiers Tempérés. *Annales Des Sciences Forestieres* **35**: 187–202.

Bashabsheh, I. M. (2007) Impact of treated wastewater on soil properties and cut flower production under drip irrigation systems. Unpublished MSc thesis, Jordan University of Science and Technology.

Bellwald, U. and M. al-Huneidi (2003) *The Petra Siq: Nabataean Hydrology Uncovered*. Amman: Petra National Trust.

Blackman, E. (1968a) Opaline silica deposition in rye (*Secale cereale* L.). *Annals of Botany* **32**: 199–206.

Blackman, E. (1968b) The pattern and sequence of opaline silica deposition in rye (*Secale cereale* L.). *Annals of Botany* **32**: 207–218.

Blackman, E. and D. W. Parry (1968) Opaline silica deposition in rye. (*Secale cereale* L.). *Annals of Botany* **32**: 199–206.

Blackman, E. and D. W. Parry (1969) Observations on the development of the silica cells of the leaf sheath of wheat (*Triticum aestivum*). *Canadian Journal of Botany* **47**: 827–838.

Bluth, G. J. S. and L. R. Kump (1994) Lithologic and climatologic controls of river chemistry. *Geochimica et Cosmochimica Acta* **58**: 2341–2359.

Bourke, S. (2001) The Chalcolithic Period. In *The Archaeology of Jordan*, ed. B. MacDonald, R. Adams and P. Bienkowski. Sheffield: Sheffield University Press pp. 109–162.

Carr, G. (2009) Water reuse for irrigated agriculture in Jordan. Unpublished PhD thesis: University of Reading.

Charles, M., C. Hoppe, G. Jones, A. Bogaard and J. G. Hodgson (2003) Using weed functional attributes for the identification of irrigation regimes in Jordan. *Journal of Archaeological Science* **30**: 1429–1441.

Codron, J., D. Codron, J. A. Lee-Thorp et al. (2005) Taxonomic, anatomical, and spatio-temporal variations in the stable carbon and nitrogen isotopic compositions of plants from an African savanna. *Journal of Archaeological Science* **32**: 1757–1772.

Djamin, A. and M. D. Pathak (1967) Role of silica in resistance to Asiatic Rice Borer, *Chilo suppressalis* (Walker), in rice varieties. *Journal of Economic Entomology* **60**: 347–351.

Ernst, W. H. O., R. D. Vis and F. Piccoli (1995) Silicon in developing nuts of the sedge *Schoenus nigricans*. *Journal of Plant Physiology* **146**: 481–488.

Garfinkel, Y., A. Vered and O. Bar-Yosef (2006) The domestication of water: the Neolithic well at Sha'ar Hagolan, Jordan Valley, Israel. *Antiquity* **80**: 686–696.

GTZ (2005) *Long-Term Groundwater and Soil Monitoring Concept – Final Report 2005*. Amman, Jordan: Reclaimed Water Project of the Deutsche Gesellschaft für Technische Zusammenarbeit.

Hayward, D. M. and D. W. Parry (1980) Scanning electron microscopy of silica deposits in the culms, floral bracts and awns of barley (*Hordeum sativum* Jess.). *Annals of Botany* **46**: 541.

Heath, M. C. (1979) Partial characterization of the electron-opaque deposits formed in the non-host plant, French bean, after cow-pea rust infection. *Physiological Plant Pathology* **15**: 141–148.

Heaton, T. H. E. (1999) Spatial, species, and temporal variations in the C-13/C-12 ratios of C-3 plants: Implications for palaeodiet studies. *Journal of Archaeological Science* **26**: 637–649.

Helback, H. (1960) Ecological effects of irrigation in ancient Mesopotamia. *Iraq* **22**: 186–196.

Helms, S. (1981) *Jawa: Lost City of the Black Desert*. New York: Cornell University Press.

Helms, S. (1989) Jawa at the beginning of the Middle Bronze Age. *Levant* **21**: 141–168.

Hodson, M. J. and D. E. Evans (1995) Aluminium/silicon interactions in higher plants. *Journal of Experimental Botany* **46**: 161–171.

Hodson, M. J. and A. G. Sangster (1999) Aluminium/silicon interactions in conifers. *Journal of Inorganic Biochemistry* **76**: 89–98.

Horiguchi, T. (1988) Mechanism of manganese toxicity and tolerance of plants. IV. Effects of silicon on alleviation of manganese toxicity of rice plants. *Soil Science and Plant Nutrition* **34**: 65–73.

Hutton, J. T. and K. Norrish (1974) Silicon content of wheat husks in relation to water transpired. *Australian Journal of Agricultural Research* **25**: 203–212.

Imaizumi, K. and S. Yoshida (1958) Edaphological studies on silicon-supplying power of paddy fields. *Bulletin of the National Institute of Agricultural Science (Japan)* **B8**: 261–304.

Jarvis, S. C. (1987) The uptake and transport of silicon by perennial ryegrass and wheat. *Plant and Soil* **97**: 429–437.

Jenkins, E. (2009) Phytolith taphonomy: a comparison of dry ashing and acid extraction on the breakdown of conjoined phytoliths formed in *Triticum durum*. *Journal of Archaeological Science* **36**: 2402–2407.

Jones, G., M. Charles, S. Colledge and P. Halstead (1995) Towards the archaeobotanical recognition of winter-cereal irrigation: an investigation of modern weed ecology in northern Spain. In *Res Archaeobotanicae— 9th Symposium IWGP*, ed. H. Kroll and R. Pasternak. Kiel: Institut für Ur- und Frühgeschichte der Christian-Albrecht-Universität.

Jones, L. H. P. and K. A. Handreck (1965) Studies of silica in the oat plant. III. Uptake of silica from soils by the plant. *Plant and Soil* **23**: 79–96.

Jones, L. H. P. and K. A. Handreck (1967) Silica in soils, plants and animals. *Advances in Agronomy* **19**: 107–149.

Jones, L. H. P. and A. A. Milne (1963) Studies of silica in the oat plant I. Chemical and physical properties of the silica. *Plant and Soil* **18**: 207–220.

Jones, L. H. P., A. A. Milne and S. M. Wadham (1963) Studies of silica in the oat plant II. Distribution of silica in the plant. *Plant and Soil* **18**: 358–371.

Kaufman, P. B., L. B. Petering and J. G. Smith (1970) Ultrastructural development of cork-silica cell pairs in Avena Internodal Epidermis. *Botanical Gazette* **131**: 173.

Khresat, S. A. (2001) Calcic horizon distribution and soil classification in selected soils of north-western Jordan. *Journal of Arid Environments* **47** (2): 145–152.

Khresat, S. A. and A. Y. Taimeh (1998) Properties and characterization of vertisols developed on limestone in a semi-arid environment. *Journal of Arid Environments* **40**: 235–244.

Lewin, J. C. and B. E. F. Reimann (1969) Silicon and plant growth. *Annual Review of Plant Physiology* **20**: 289–304.

Liang, Y. C., H. Hua, Y.-G. Zhu et al. (2006) Importance of plant species and external silicon concentration to active silicon uptake and transport. *The New Phytologist* **172**: 63–72.

Liang, Y. C., Q. R. Shen, Z. G. Shen and T. S. Ma (1996) Effects of silicon on salinity tolerance of two barley cultivars. *Journal of Plant Nutrition* **19**: 173–183.

Mabry, J. M. (1989) Investigations at Tell el-Handaquq, Jordan (1987–88). *Annual of the Department of Antiquities of Jordan* **32**: 59–95.

Mabry, J. M., L. Donaldson, K. Gruspier et al. (1996) Early town development and water management in the Jordan Valley: investigations at Tell el-Handaquq North. *Annual of the Department of Antiquities of Jordan* **53**: 115–154.

Madella, M., M. K. Jones, P. Echlin, A. Powers-Jones and M. Moore (2009) Plant water availability and analytical microscopy of phytoliths: implications for ancient irrigation in arid zones. *Quaternary International* **193**: 32–40.

Mayland, H. F., D. A. Johnson, K. H. Asay and J. J. Read (1993) Ash, carbon isotope discrimination, and silicon as estimators of transpiration efficiency in crested wheat-grass. *Australian Journal of Plant Physiology* **20**: 361–369.

McKeague, J. A. and M. G. Cline (1963) silica in soils. *Advances in Agronomy* **15**: 339–396.

Meybeck, M. (1987) Global chemical weathering of surficial rocks estimated from river dissolved loads. *American Journal of Science* **287**: 401–428.

Mitsui, S. and H. Takatoh (1959) Role of silicon in rice nutrition. *Soil Science and Plant Nutrition* **5**: 127–133.

Miyake, Y. and E. Takahashi (1983) Effect of silicon on the growth of solution-cultured cucumber plant. *Soil Science and Plant Nutrition* **29**: 71–83.

Miyake, Y. and E. Takahashi (1986) Effect of silicon on the growth and fruit production of strawberry plants in a solution culture. *Soil Science and Plant Nutrition* **32**: 321–326.

Neaman, A., A. Singer and K. Stahr (1999) Clay mineralogy as affecting disaggregation in some palygorskite containing soils of the Jordan and Bet-She'an Valleys. *Australian Journal of Soil Research* **37**: 913–928.

Okamoto, G., T. Okura and K. Goto (1957) Properties of silica in water. *Geochimica et Cosmochimica Acta* **12**: 123–132.

Okuda, A. and E. Takahashi (1965) The role of silicon. In *Proceedings of the Symposium of the International Rice Research Institute 1964*. Baltimore, MD: Johns Hopkins University Press pp. 123–146.

Oleson, J. P. (2001) Water supply in Jordan through the ages. In *The Archaeology of Jordan*, ed. B. MacDonald, R. Adams and P. Bienkowski. Sheffield: Sheffield University Press pp. 603–624.

Parry, D. W. and F. Smithson (1958) Techniques for studying opaline silica in grass leaves. *Annals of Botany* **22**: 543–549.

Parry, D. W. and F. Smithson (1964) Types of opaline silica depositions in the leaves of British grasses. *Annals of Botany* **28**: 169–185.

Parry, D. W. and F. Smithson (1966) Opaline silica in the inflorescence of some British grasses and cereals. *Annals of Botany* **30**: 525–538.

Perry, C. C., S. Mann, R. J. P. Williams et al. (1984) A scanning proton microprobe study of macrohairs from the lemma of the grass *Phalaris canariensis* L. *Proceedings of the Royal Society of London Series B-Biological Sciences* **222**: 439–445.

Philip, G. (2001) The Early Bronze Age of the southern Levant: a landscape approach. *Journal of Mediterranean Archaeology* **16/1**: 103–132.

Piperno, D. R. (1988) *Phytolith Analysis: An Archaeological and Geological Perspective*. San Diego, CA: Academic Press.

Raeside, J. D. (1970) Some New Zealand plant opals. *New Zealand Journal of Science* **13**: 122–132.

Richmond, K. E. and M. Sussman (2003) Got silicon? The non-essential beneficial plant nutrient. *Current Opinion in Plant Biology* **6**: 268–272.

Rosen, A. (1999) Phytoliths as indicators of prehistoric farming. In *Prehistory of Agriculture: New Experimental and Ethnographic Approaches*, ed. P. Anderson. Los Angeles: UCLA, Institute of Archaeology pp. 193–198.

Rosen, A. M. and S. Weiner (1994) Identifying ancient irrigation: a new method using opaline phytoliths from emmer wheat. *Journal of Archaeological Science* **21**: 125–132.

Sangster, A. G. (1970) Intracellular silica deposition in mature and senescent leaves of *Sielingia decumbens* L. 'Bernh'. *Annals of Botany* **34**: 557–570.

Scarborough, V. L. (2003) *The Flow of Power: Ancient Water Systems and Landscapes*. Santa Fe, NM: SAR Press.

Takijima, Y., M. Shiojima and K. Kanno (1949) Studies on soil of peaty paddy fields. Effect of silica on the growth of the rice plant and its nutrient absorption. *Journal of Soil Science* **30**: 544–550.

Tieszen, L. L. (1991) Natural variation in carbon isotope values of plants: implications for archaeology, ecology and paleoecology. *Journal of Archaeological Science* **18**: 227–248.

Vlamis, J. and D. E. Williams (1967) Manganese and silicon interaction in Graminae. *Plant and Soil* **27**: 131–140.

Walker, C. D. and R. C. M. Lance (1991) Silicon accumulation and 13C composition as indices of water-use efficiency in barley cultivars. *Australian Journal of Plant Physiology* **18**: 427–434.

Webb, E. A. and F. J. Longstaffe (2002) Climatic influences on the oxygen isotopic composition of biogenic silica in prairie grass. *Geochimica et Cosmochimica Acta* **66**: 1891–1904.

White, A. F. and A. E. Blum (1995) Effects of climate on chemical weathering in watersheds. *Geochimica et Cosmochimica Acta* **66**: 1891–1904.

Whitehead, P. G., S. J. Smith, A. J. Wade et al. (2008) Modelling of hydrology and potential population levels at Bronze Age Jawa, Northern Jordan: a Monte Carlo approach to cope with uncertainty. *Journal of Archaeological Science* **35**: 517–529.

Williams, D. E. and J. Vlamis (1957) Manganese and boron toxicities in standard culture solutions. *Soil Science Society of America Journal* **21**: 205–209.

Wittfogel, K. A. (1957) *Oriental Despotism: A Comparative Study of Total Power*. New Haven, CT: Yale University Press.

Yoshida, S., Y. Onishi and K. Kitagishi (1959) The chemical nature of silicon in rice plants. *Soil and Plant Food (Tokyo)* **5**: 127–133.

22 An investigation into the archaeological application of carbon stable isotope analysis used to establish crop water availability: solutions and ways forward

Helen Stokes, Gundula Müldner and Emma Jenkins

ABSTRACT

Carbon stable isotope analysis of charred cereal remains is a relatively new method employed by archaeological scientists to investigate ancient climate and irrigation regimes. The aim of this study was to assess the effect of environmental variables on carbon isotope discrimination (Δ) in multiple environments to develop the technique and its archaeological application, using crops grown at three experimental stations in Jordan. There are two key results: (1) as expected, there was a strong positive relationship between water availability and Δ; (2) site, not water input, was the most important factor in determining Δ. Future work should concentrate on establishing ways of correcting Δ for the influence of site specific environmental variables and on assessing how well carbon isotope discrimination values are preserved within the archaeological record.

22.1 INTRODUCTION

Carbon stable isotope analysis of archaeological cereals is a relatively new approach for the investigation of ancient irrigation systems (Araus and Buxo, 1993; Araus et al., 1997b; Ferrio et al., 2005; Araus et al., 2007; Riehl et al., 2008). The approach relies on the basic principle that plants under water stress discriminate less against the 'heavier' carbon-13 (^{13}C) isotope than plants whose water requirements are fully met (Farquhar and Richards, 1984; Farquhar et al., 1989). As a consequence, the tissues of water-stressed plants contain more ^{13}C than those which have received sufficient water through rainfall or irrigation. Although this general principle regarding water availability and carbon isotope discrimination (Δ) is fully established within modern plant sciences, using carbon stable isotope analysis of plant tissues to investigate archaeological questions has its own specific challenges. A central question concerns the extent to which environmental variables other than water stress affect ^{13}C discrimination in plants and whether these effects can be sufficiently accounted for in past environments. Addressing this issue was one of the key aims of the Water, Life and Civilisation project's experimental crop growing programme in Jordan (Mithen et al., 2008). This chapter reports on the first results of this investigation, by first discussing how carbon stable isotope analysis of cereal remains is currently used within archaeology and then presenting the outcomes of a pilot study into the effect of water stress on carbon stable isotope signatures under different environmental conditions using wheat from three crop growing stations in Jordan.

22.2 PRINCIPLES OF THE METHOD

Isotopes are atoms of the same element which have different atomic masses. Carbon has three naturally occurring isotopes: carbon-12 (^{12}C), which makes up ~99% of carbon atoms, carbon-13 (^{13}C) and carbon-14 (^{14}C). Unlike ^{14}C (radiocarbon), ^{12}C and ^{13}C isotopes are stable and not subject to radioactive decay. The ratio of ^{12}C to ^{13}C (expressed as $^{13}C/^{12}C$ or $\delta^{13}C$) is widely used within environmental sciences to understand natural pathways and food webs within ecosystems (see Rundel et al., 1987; Lajtha and Michener, 1994; Dawson and Siegwolf, 2007).

During photosynthesis, plants discriminate against the 'heavier' ^{13}C isotope in favour of the 'lighter' ^{12}C isotope. In plants using the C_3 or Calvin–Benson photosynthetic pathway (most land plants and major cultigens) this results in plant tissues containing a far lower proportion of ^{13}C compared with ^{12}C isotopes than the atmosphere (Lajtha and Michener, 1994). The size of the difference between atmospheric $\delta^{13}C$ and the isotopic composition of plant tissues (carbon isotope discrimination or Δ) depends on a number of different factors.

Water, Life and Civilisation: Climate, Environment and Society in the Jordan Valley, ed. Steven Mithen and Emily Black. Published by Cambridge University Press.
© Steven Mithen and Emily Black 2011.

Farquhar and colleagues (Farquhar and Richards, 1984; Farquhar et al., 1989; Ehleringer et al., 1993) were, however, the first to establish a relationship between water status and carbon isotope discrimination in plants. The basic principle they observed was that, as a result of stomatal closure and associated changes in intercellular partial pressure, water-stressed plants discriminate less against ^{13}C than plants which have their water needs met (see Ehleringer, 1989; Farquhar et al., 1989; Ferrio et al., 2003). Put very simply, as water stress increases, the difference (Δ) between the $\delta^{13}C$ of atmospheric CO_2 and plant tissue decreases. The value of Δ can therefore act as a record of water availability during plant growth. Small Δ values should reflect poor water availability and high Δ values greater water availability. This relationship between water stress and Δ is being used by archaeological scientists to track changes in climate and instances of agricultural irrigation in prehistory.

22.3 APPLICATIONS IN ARCHAEOLOGY

22.3.1 Carbon stable isotope analysis for reconstructing climate change

Climate change has been identified as an important factor driving social change in early societies during the Holocene (e.g. Dalfes et al., 1997; Rosen, 2007). As carbon stable isotope data from C_3 plants reflect moisture conditions, they can be expected to track changes in climate over time. In the first applications of this methodology to charred cereal remains from archaeological sites, Araus and colleagues (Araus and Buxo, 1993; Araus et al., 1997a) compared Δ values for wheat and barley kernels from southeastern and northeastern Spain dating from the Neolithic to the Iron Age (7,000–2,200 BP). They concluded that the southeast had been consistently drier than the northeast and that, although climate was generally wetter than today, there was a trend towards more arid conditions throughout the periods studied (see also Araus et al., 2003). More recently, Riehl et al. (2008, 2009) demonstrated that stable carbon isotopes in plants record broad-spectrum environmental changes. The main focus of Riehl et al. (2009) was to assess the reliability of pedogenic carbonates from southeastern Turkey as palaeoenvironmental indicators. To achieve this they compared the carbonate data with three local palaeoenvironmental records including archaeological seed assemblages, $\Delta^{13}C$ evidence from plant remains and the predictions of macrophysical climatic models (MCM). Seven early Bronze Age to Classical (5,400–2,000 cal BP) sites from northern Mesopotamia and the northern Levant were considered along with previously published data, giving a total sample size of 200 barley grains. Riehl et al. (2009) found that there was a good general correlation between $\Delta^{13}C$ values and other palaeoclimate indicators, particularly in the identification of a temporary increase in moisture levels during the mid-Holocene followed by a decrease thereafter. However, the authors are quick to point out that considerable further research is still needed to improve the chronology of these proxy data in order to produce a more accurate picture of climate change within the northern Fertile Crescent.

22.3.2 Carbon stable isotope analysis for reconstructing irrigation practices

Water management systems are at the centre of the debate surrounding the emergence of complex societies in arid regions of the world (e.g. Scarborough, 2003). Unfortunately, they are often difficult to detect archaeologically (see Mithen et al., 2008). A reliable method for establishing whether artificial crop irrigation existed at an ancient settlement would therefore be invaluable. Initial investigations, especially by Araus and his group (Araus and Buxo, 1993; Araus et al., 1997a, 1997b, 1999) suggest that plant stable carbon isotope analysis has great potential in this respect, as it should, theoretically, be possible to infer from the carbon stable isotope composition of crops whether they received more water than was 'naturally' available, and therefore whether a system of artificial irrigation was in place. In order for this approach to be viable, it is first essential to distinguish between the effects on carbon isotope discrimination caused by irrigation and those of climate fluctuations. Secondly, it is necessary to establish how much water was available through natural means (precipitation) or rather, what Δ values would be expected in crops that were grown at the same site, i.e. under the same environmental and climatic conditions, yet without any artificial irrigation. As a possible solution to these issues, researchers have started contrasting carbon isotope discrimination data for a number of different plant species from the same archaeological site. The aim is to identify differences in Δ between these crops which could then be explained in terms of differences in irrigation and crop management techniques (Araus et al., 1997b; Ferrio et al., 2005; Riehl et al., 2008). In relation to this, Ferrio et al. (2005) proposed comparing isotope data from major food crops with the remains of non-cultivated plants such as trees or with cultigens which are relatively well adapted to arid conditions and are therefore unlikely to have received any irrigation (see also Ferrio et al., 2005, 2006). Following this suggestion, Riehl et al. (2008) used Δ values for barley as baseline data for non-irrigated crops. They combined these data with the results of climate modelling to infer that irrigation was practised at some of the investigated sites. While the strategies employed to detect irrigation through carbon stable isotope signals of plants are therefore relatively sophisticated, all researchers concede that the trends observed in their datasets may well be explained by differences in the growing cycles or water use efficiency of individual species rather than by any water management strategies employed by humans (Ferrio et al., 2005; Riehl et al., 2008).

22.3.3 The importance of modern reference data

Modern reference data from plants grown under known climatic and hydrological conditions are essential for the correct interpretation of carbon stable isotope values obtained from archaeological specimens. Apart from establishing the importance of environmental parameters other than water availability on the carbon isotope discrimination of plants (see discussion below), these modern analogues are used to provide typical Δ values for rain-fed and irrigated plants in a given geographical region. These can then be compared with archaeological samples from the same area. The majority of work establishing modern reference data has been undertaken by Araus and colleagues (Araus and Buxo, 1993; Araus et al., 1997b; Ferrio et al., 2005) and has centred on the Mediterranean, especially Spain. They employed the reference values from modern crops mainly in two ways: as baseline data for relative statements about past climate conditions ('wetter/drier than today') (Araus et al., 1997a; Ferrio et al., 2003, 2005), or for establishing relationships between the Δ of different types of crops grown under the same environmental conditions. As we have seen above, such data are essential for identifying irrigated crops by comparison with other, probably non-irrigated, plant species from the same site (Araus et al., 1997b, 1999; Ferrio et al., 2005). Araus and colleagues (Araus et al., 1997b, 1999; Ferrio et al., 2003) also used modern Δ values from crops grown under a variety of known watering regimes to formulate non-linear regression equations for calculating total water input from Δ values.

22.4 OTHER ENVIRONMENTAL PARAMETERS AFFECTING Δ

The importance of water availability, and especially total water input during grain filling, for Δ is not in doubt. It is a well-established principle in modern plant science and emerged, for example, as the most important environmental parameter when Araus et al. (2003) tested samples of 25 genotypes of durum wheat from 12 experimental fields in different regions of Spain.

Nevertheless, there are a number of other environmental factors which contribute to the overall plant carbon isotope signature. Modern studies of carbon isotope discrimination in crops are primarily aimed at refining the use of Δ as a selection criterion for crop improvement (e.g. Monneveux et al., 2006). They have the benefit of large sample sizes and are, in consequence, affected by sources of small-scale variation in Δ. In contrast, when studying archaeological materials sample sizes are restricted by what is available and therefore typically very small. At the same time,

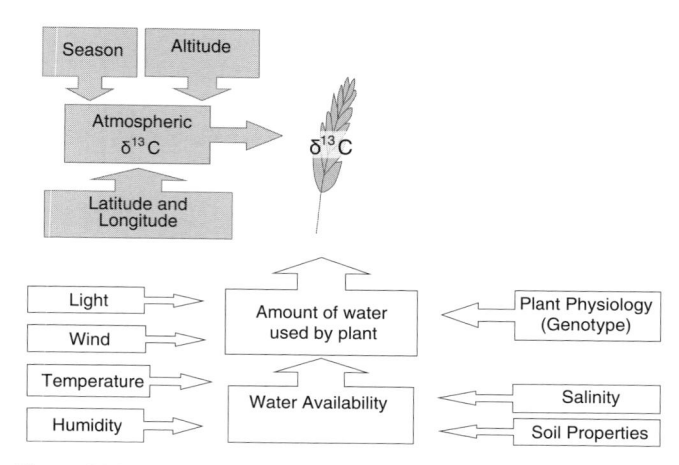

Figure 22.1. Flowchart illustrating the different environmental parameters determining the carbon stable isotope composition ($\delta^{13}C$) of C_3 plants (for references, see text).

palaeo-data describing environmental parameters such as rainfall, temperature or moisture conditions are usually sporadic. In these cases, it is pertinent to ask whether the effect on Δ of environmental variables other than water availability is large enough to influence the interpretation of carbon isotope datasets. The main variables in question are discussed below.

22.4.1 Evapotranspiration and local growing conditions

While water input, whether from precipitation or artificial irrigation, is usually the most important factor influencing Δ, concurrent water loss through evaporation (from the soil) and transpiration (through the plant) is also a key variable for determining the water status of plants. It therefore also has a known effect on Δ. The rate of transpiration is the result of complex interplay between plant physiology and transpiration efficiency as well as light intensity, air movement, temperature and humidity (see Figure 22.1 and Allen et al., 1998; see also Farquhar et al., 1989; Condon et al., 1992; Ferrio et al., 2003; Monneveux et al., 2006). The presence of dense tree cover can have a particularly strong effect on the isotopic composition of plant tissues, probably as the combined result of reduced light levels and CO_2 'recycling' under the canopy (van der Merwe and Medina, 1991; Heaton, 1999; Ferrio et al., 2003). The difference between $\delta^{13}C$ values of a plant in the canopy and one in the understorey can be as much as 3–4‰ (Heaton, 1999). The relative position of the plant within the overall environment is therefore very important when interpreting carbon stable isotope data obtained from plants.

The nature of soil is also crucial to understanding differences in plant Δ values. Apart from evaporation, other properties such as soil depth, water holding capacity, salinity, soil resistance and nitrogen content can either limit the water available to crops or affect photosynthesis rates in other ways (see Masle and Farquhar, 1988; Farquhar et al., 1989; Condon et al., 1992; Shangguan et al., 2000;

Shaheen and Hood-Nowotny, 2005; Cabrera-Bosquet *et al.*, 2007). The variable nature of soil properties means that they can change considerably within a relatively small area, affecting noticeable intra- and interspecies differences in plant stable isotope composition within the same habitat (Tieszen, 1991; Heaton, 1999).

22.4.2 Variations in atmospheric $\delta^{13}C$ and CO_2 concentrations through time

Throughout the Holocene there have been fluctuations in the concentration and the ^{13}C content of atmospheric CO_2. Both result in variations in the $\delta^{13}C$ of plant tissues even where other environmental factors are constant (Heaton, 1999; Ferrio *et al.*, 2003). It is therefore necessary to correct for any changes in atmospheric composition before samples from different time periods can be compared. This is normally done through the use of ice core data which provide information on the isotopic composition and concentration of atmospheric CO_2 for the period in question. These data are then used as a basis to calculate Δ for archaeological samples (see Ferrio *et al.*, 2005; Araus *et al.*, 2007; Riehl *et al.*, 2008).

22.4.3 Geographical variables

Two key geographical variables are altitude and latitude, which affect plant Δ probably as a result of changes in atmospheric partial pressure and variations in temperature (Korner *et al.*, 1991). Korner *et al.* (1988) studied the effect of altitude on carbon isotope discrimination of C_3 plants across major mountainous regions and found that Δ was reduced in magnitude by 2–3‰ per 1,000 m altitude on a community level and by 1.2‰ within species. After surveying a number of studies, Heaton (1999) reported an average decrease in Δ of between 0.5 to 1.5% for every 1,000 m of elevation. In a subsequent study, Korner *et al.* (1991) also established that latitude influences Δ values. Plants at higher-latitude sites (Arctic and sub-Arctic lowland) typically exhibit lower Δ values and therefore discriminate less against ^{13}C than vegetation at tropical sites at similar altitudes. These results demonstrate that care must be taken when conducting inter-site comparisons or even pooling samples from different locations, especially in regions with marked differences in altitudes. For larger settlements which may have been sustained from a sizeable hinterland, the question must also be asked whether all crops were grown locally.

22.5 DEVELOPING A MODERN REFERENCE DATASET FOR ARCHAEOLOGICAL STUDIES IN JORDAN

The above discussions highlight the importance of establishing good regional reference values for any given area before conducting carbon stable isotope analysis of charred plant remains. They also raise the question of whether we really can sufficiently control variation at a local level to allow meaningful comparison between sites from different time-periods. Few modern crop growing experiments are designed with archaeological samples in mind, and, for want of better alternatives, modern samples are often assembled from a number of sites which differ not only in water availability but also in other environmental parameters such as altitude (see, for example, Araus and Buxo, 1993; Araus *et al.*, 1997b, 1999).

The primary aim of this chapter is to explore the sensitivity of carbon stable isotopes in cereal kernels to water stress in different environments but where water input should be the only considerable variable at a site level. A land race of durum wheat (*Triticum durum*) was chosen for analysis as it is one of the earliest known cultivates within the Near East. In undertaking this study we are seeking to assess the validity of comparing datasets from different sites, even from within a small geographical area, with the aim of establishing differences in water management strategies. At the same time, modern reference values for carbon isotope discrimination of wheat grown in this region will also be established. These are essential before further isotope studies can explore the antiquity of active water management in crop production in one of the key areas for early agriculture.

22.5.1 Experimental crop growing sites

The Water, Life and Civilisation project's crop growing experiments in Jordan are described in detail by Jenkins *et al.* (this volume, Chapters 21 and 23) and Mithen *et al.* (2008). In brief, native land races of durum wheat were grown over three consecutive years (2005–2008) at three crop growing stations on the Jordanian plateau (Khirbet as Samra and Ramtha) and in the Jordan Valley (Deir 'Alla) (Figure 22.2). At each site the cultivation area was divided into five 5 m × 5 m irrigation plots, varying from 0% irrigation (rainwater only) to 40% (second and third year only), 80%, 100% and 120% of the plants' optimum water requirements. These were estimated on a weekly basis using readings from Class A Evaporation Pans. The environmental conditions at each site were monitored closely, including minimum and maximum daily temperature, rainfall, evaporation and relative humidity.

22.5.2 Sampling strategy

In order to explore the relationship between Δ values, water availability and the environment, wheat kernels from each of the five irrigation regimes and all three years of cultivation were sampled at each of the three sites. This strategy was also meant to make it possible to assess the quantitative effect of varying water input on plant Δ values.

22.5.3 Processing and methods.

Each sample consisted of 10 wheat kernels from one water regime and year of cultivation. All grains were washed in deionised water, frozen for at least 24 hours, freeze-dried for 48 hours and then homogenised using a pestle and mortar. After exploring measurement reproducibility with multiple replicates of the same sample, it was found that analysing samples in duplicates provided sufficient precision. All analyses were conducted by continuous-flow isotope ratio mass spectrometry (analytical error $\pm 0.1‰$ (1σ)). Values for $\delta^{13}C$ (relative to VPDB) were converted into positive Δ values using the formula:

$$\Delta = \frac{\delta_{air} - \delta_{plant}}{1 + \delta_{plant} \div 1000}$$

22.5.4 Statistical analysis

In order to establish the relationships between the major variables (water regime and site) and Δ values statistically, regression analysis was conducted using the GenStat® software package. The statistical analysis was conducted on $\delta^{13}C$ values before they were converted into Δ values.

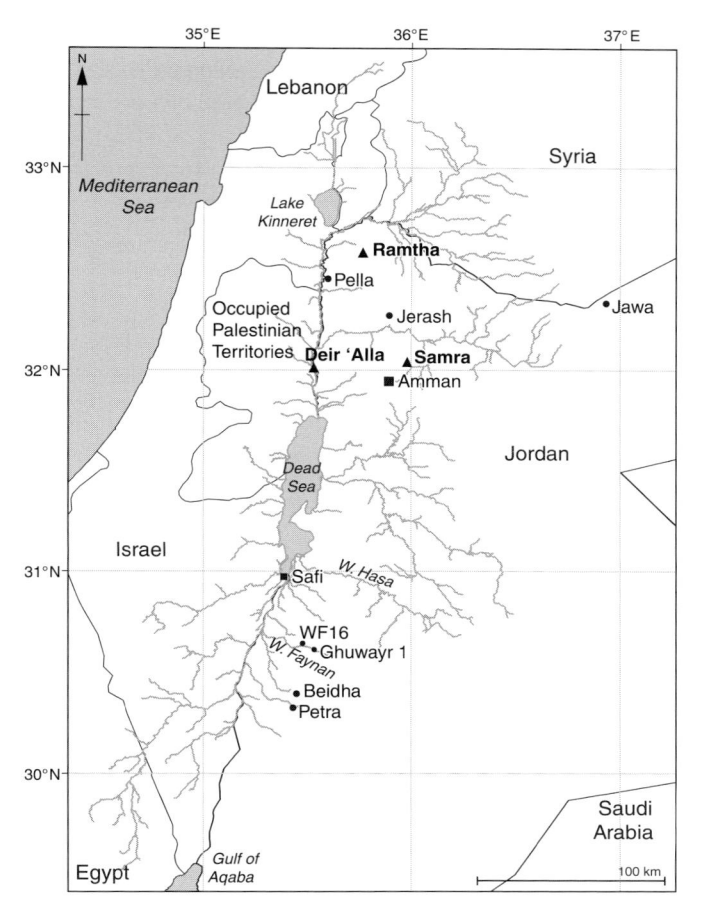

Figure 22.2. Location of crop growing stations in Jordan used in the WLC experiments (map reproduced from Fig. 1 of Mithen *et al.* 2008). Khirbet as Samra N 32° 08′, E 36° 09′, 564 m asl, Ramtha N 32° 33′, E 32° 36′, 510 m asl and Deir 'Alla N 32° 13′, E 35° 37′, 234 m bsl. Altitudes were obtained from the Google Earth database and were checked against OS maps (Ordnance Survey 1949a and b and 1950).

22.6 RESULTS

The results are displayed in Figure 22.3 and Table 22.1. For the purpose of this chapter, we have averaged data from different years of cultivation. As expected, there were strong positive relationships between Δ and water availability at each of the three sites (Figure 22.3). More unusually, however, the curves follow different trajectories. While the one for Deir 'Alla levels out at 40% water availability, Δ values at Ramtha and Khirbet as Samra are still increasing at 120%. Importantly, there also appears to be an offset in Δ between the sites which is relatively consistent through the different watering regimes. Although there is some variation, the average difference between sites is about 2.1‰, with Deir 'Alla exhibiting the highest values (overall mean 18.9±1.7‰, 1σ), followed by Ramtha (16.8±1.6‰) and finally Khirbet as Samra, where Δ are smallest (14.7±1.0‰).

We applied a General Linear Model using the GenStat® program in order to test the importance of the two major variables, site and irrigation regime, for Δ. The analysis confirmed the visual impression from Figure 22.3 that the relationships of Δ with both of these variables were statistically significant (accumulated ANOVA, $p < 0.001$ in both cases).

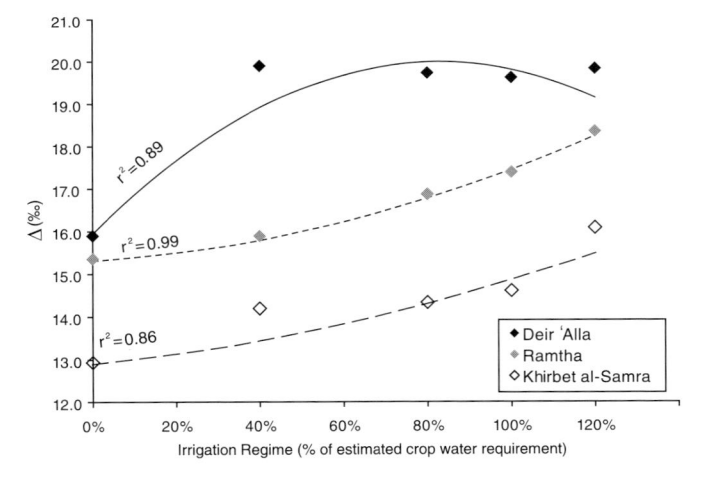

Figure 22.3. Graph showing mean Δ (‰) values for each irrigation regime at the three sites with trendlines fitted by polynomial regression.

Table 22.1 *Table showing mean Δ values in ‰.*

Irrigation regime	Deir 'Alla (234 m bsl)			Ramtha (523 m asl)			Khirbet as Samra (564 m asl)		
	Δ (mean)	1σ	n (years)	Δ (mean)	1σ	n (years)	Δ (mean)	1σ	n (years)
0%	15.9	0.5	3	15.4	1.2	3	12.9	–	1
40%	19.9	0.5	2	15.9	0.9	2	14.2	0.6	2
80%	19.7	0.3	3	16.9	2.1	3	14.3	0.3	3
100%	19.6	0.2	3	17.4	0.9	3	14.6	0.2	3
120%	19.8	0.4	3	18.4	0.7	3	16.1	0.1	3
Overall mean	18.9	1.7	14	16.8	1.6	14	14.7	1.0	12

Values shown are average of up to three years of cultivation and overall average of all available samples, and standard deviation (1σ) for each site across all irrigation regimes. Some years did not produce sufficient cereal crop for analysis because of the dry conditions and attack by birds (see Mithen *et al.*, 2008).

22.7 DISCUSSION

22.7.1 Water availability and Δ

Wheat kernels from all three sites show the expected positive relationship with water availability, with an increase in Δ of ~3‰ (Ramtha, Khirbet as Samra) to 4‰ (Deir 'Alla) from the lowest to the highest value. The trends exhibited are not uniform. The curve for Deir 'Alla levels out at 40% of the calculated crop water requirement, indicating that water is no longer the limiting factor to photosynthesis. Conversely, Δ values at Ramtha and Khirbet as Samra are still increasing at 120%, suggesting that the plants' water requirements have not yet been fully met.

The easiest explanation for these different trends is that estimates of optimal water requirements were not fully accurate, possibly because these were calculated on a weekly basis but not taking into account precipitation and humidity patterns of previous weeks and months which might still have affected water availability during grain filling. Deir 'Alla is situated in the Jordan Valley and, from Figure 22.3 which shows Δ values levelling out at 40% water input, more water was available here than either at Khirbet as Samra or Ramtha. Detailed analysis of all environmental parameters is still ongoing. However, for now it is suggested that the differences between sites can be explained by differences in soil water holding capacity, surface runoff or local climate which could have produced much wetter conditions at Deir 'Alla. It is possible that similar local parameters explain the continuing water stress evident at Ramtha and especially Khirbet as Samra, the most arid of the three sites. It is worth mentioning here that preliminary results suggest significant inter-annual variation in crop response to irrigation regime at Ramtha, and that continuing water stress may have affected the plants only in the second year of cultivation, but skewed the overall mean values (as expressed in the high standard deviation in comparison with other sites, see Table 22.1). However, further analyses are necessary to verify this.

22.7.2 Site-specific variables

As Figure 22.3 illustrates, although the relationship of Δ with water availability is highly significant, location also has a very considerable effect on the magnitude of carbon isotope discrimination. Only some of these differences can be easily explained by the relatively large differences in altitude between Deir 'Alla (234 m below sea level), Ramtha (523 m asl) and Khirbet as Samra (564 m asl). As stated earlier, a difference in altitude has been found to decrease Δ by 0.5–1.5‰ with each 1,000 m of elevation (Heaton, 1999). Even the difference of almost 800 m between Deir 'Alla and Khirbet as Samra could therefore only account for about a quarter of the ~4‰ offset observed between average Δ values from the two sites. It does not explain the ~2‰ difference in mean Δ values between Khirbet as Samra and Ramtha which are almost at the same altitude. Other site-specific factors which are not as easily controlled in prehistoric samples therefore need to be considered.

In order to guard the crops from bird attack it was necessary to cover the fields at Deir 'Alla and Ramtha with protective netting for part of the year. This may have resulted in increased humidity and may also have introduced a 'canopy effect' (see above), both possibly leading to higher calculated Δ values for the analysed plants, although the magnitude of this effect is difficult to quantify with the environmental data obtained. Other likely causes of variation are differences in rainfall patterns, soil properties or transpiration rates. Which, if any, of these factors was the most important will have to be determined by detailed analysis of the environmental data collected.

22.8 CONCLUSIONS

The results presented here highlight the importance of site-specific factors for the carbon stable isotope composition of cereal kernels. This observation is neither surprising nor new, given the multitude of parameters that impact on carbon isotope discrimination in plants. Nevertheless, it is important to emphasise that the variation observed in our study, between plants grown at different sites but with nominally the same water availability and in a relatively small geographical area, is as great as, and sometimes significantly larger than, differences which have been considered meaningful in inter-site comparisons of archaeological plant data (see Araus et al., 1997b; Ferrio et al., 2005; Riehl et al., 2008). While some of the causes for variation, such as differences in altitude, can be relatively easily accounted for, even for prehistoric sites (although it is not always made explicit in publications whether their effects are actually corrected for), others, such as soil conditions and rainfall patterns, are much more elusive for the past. Even carefully modelled values can only ever be approximations (see, for example, Riehl et al., 2008), and care should therefore be taken to avoid over-interpreting relatively small differences in calculated Δ values, even where sites are from the same geographical area.

Stable isotope analysis of plant remains has the makings of a powerful tool for the investigation of past crop husbandry practices. However, future research must centre on exploring new ways of establishing site-specific baselines, of assessing the magnitude of normal background variation ('noise'), and of investigating in greater detail than before if and when the isotopic composition of plant tissues is changed through taphonomic processes and burial in the ground. Only if these issues are resolved can the potential of the method be fully realised. The ongoing larger project, of which this pilot study has been only the first step, aims to make a significant contribution to addressing these open questions. Further work will systematically seek to assess and characterise the nature of the relationship between carbon stable isotope values and environmental parameters, not only in wheat but also in barley and sorghum. In addition to carbon stable isotopes, nitrogen isotopes and their response to different environments will also be analysed and complemented by the outcomes of charring and burial experiments which will address the issue of alteration of the stable isotope signal through taphonomic processes. Combined, these results will enable us to interpret data from archaeological plant remains with greater confidence and to maximise the information that can be gleaned from them, in order to significantly improve our understanding of water management strategies in the past.

REFERENCES

Allen, R. G., L. S. Pereira, D. Raes and M. Smith (1998) *Crop Evapotranspiration – Guidelines for Computing Crop Water Requirements.* FAO Irrigation and drainage papers 56. Rome: Food and Agriculture Organisation of the United Nations.

Araus, J. L. and R. Buxo (1993) Changes in carbon isotope discrimination in grain cereals from the North-West Mediterranean basin during the past seven millennia. *Australian Journal of Plant Physiology* **20**: 117–128.

Araus, J. L., A. Febrero, R. Buxo et al. (1997a) Changes in carbon isotope discrimination in grain cereals from different regions of the western Mediterranean Basin during the past seven millennia. Palaeoenvironmental evidence of a differential change in aridity during the late Holocene. *Global Change Biology* **3**: 107–118.

Araus, J. L., A. Febrero, R. Buxo et al. (1997b) Identification of ancient irrigation practices based on the carbon isotope discrimination of plant seeds: a case study from the south-east Iberian Peninsula. *Journal of Archaeological Science* **24**: 729–740.

Araus, J. L., A. Febrero, M. Catala et al. (1999) Crop water availability in early agriculture: evidence from carbon isotope discrimination of seeds from a tenth millennium BP site on the Euphrates. *Global Change Biology* **5**: 201–212.

Araus, J. L., J. P. Ferrio, R. Buxo and J. Voltas (2007) The historical perspective of dryland agriculture: lessons learned from 10,000 years of wheat cultivation. *Journal of Experimental Botany* **58**: 131–145.

Araus, J. L., D. Villegas, N. Aparicio et al. (2003) Environmental factors determining carbon isotope discrimination and yield in durum wheat under Mediterranean conditions. *Crop Science* **43**: 170–180.

Cabrera-Bosquet, L., G. Molero, J. Bort, S. Nogues and J. L. Araus (2007) The combined effect of constant water deficit and nitrogen supply on WUE, NUE and Delta C-13 in durum wheat potted plants. *Annals of Applied Biology* **151**: 277–289.

Condon, A. G., R. A. Richards and G. D. Farquhar (1992) The effect of variation in soil water availability, vapour pressure deficit and nitrogen nutrition on carbon isotope discrimination in wheat. *Australian Journal of Agricultural Research* **43**: 935–947.

Dalfes, H. N., G. Kukla and H. Weiss, eds. (1997) *Third Millennium BC Climate Change and Old World Collapse.* Berlin and Heidelberg: Springer.

Dawson, T. E. and R. T. W. Siegwolf, eds. (2007) *Stable Isotopes as Indicators of Ecological Change.* London, Amsterdam, Oxford, Burlington & San Diego: Academic Press.

Ehleringer, J. R. (1989) Carbon isotope ratios and physiological processes in arid land plants. In *Stable Isotopes in Ecological Research*, ed. P. W. Rundel, J. R. Ehleringer and K. A. Nagy. Heidelberg: Springer.

Ehleringer, J. R., J. K. Hall and F. D. Farquhar, eds. (1993) *Stable Isotopes and Plant Carbon–Water Relations.* San Diego and London: Academic Press.

Farquhar, G. D., J. R. Ehleringer and K. T. Hubick (1989) Carbon isotope discrimination and photosynthesis. *Annual Review of Plant Physiology and Plant Molecular Biology* **40**: 503–537.

Farquhar, G. D. and R. A. Richards (1984) Isotopic composition of plant carbon correlates with water-use efficiency of wheat genotypes. *Australian Journal of Plant Physiology* **11**: 539–552.

Ferrio, J. P., N. Alonso, J. B. Lopez, J. L. Araus and J. Voltas (2006) Carbon isotope composition of fossil charcoal reveals aridity changes in the NW Mediterranean Basin. *Global Change Biology* **12**: 1253–1266.

Ferrio, J. P., J. L. Araus, R. Buxo, J. Voltas and J. Bort (2005) Water management practices and climate in ancient agriculture: inferences from the stable isotope composition of archaeobotanical remains. *Vegetation History and Archaeobotany* **14**: 510–517.

Ferrio, J. P., J. Voltas and J. L. Araus (2003) Use of carbon isotope composition in monitoring environmental changes. *Management of Environmental Quality* **14**: 82–98.

Heaton, T. H. E. (1999) Spatial, species, and temporal variations in the C-13/C-12 ratios of C-3 plants: implications for palaeodiet studies. *Journal of Archaeological Science* **26**: 637–649.

Korner, C., G. D. Farquhar and Z. Roksandic (1988) A global survey of carbon isotope discrimination in plant from high altitude. *Oecologia* **74**: 623–632.

Korner, C., G. D. Farquhar and S. C. Wong (1991) Carbon isotope discrimination by plants follows latitudinal and altitudinal trends. *Oecologia* **88**: 30–40.

Lajtha, K. and H. Michener, eds. (1994) *Stable Isotopes in Ecology and Environmental Science.* Oxford: Blackwell Scientific Publications.

Masle, J. and G. D. Farquhar (1988) Effects of soil strength on the relation of water-use efficiency and growth on carbon isotope discrimination in wheat seedlings. *Plant Physiology* **86**: 32–38.

Mithen, S., E. Jenkins, K. Jamjoum et al. (2008) Experimental crop growing in Jordan to develop methodology for the identification of ancient crop irrigation. *World Archaeology* **40**: 7–25.

Monneveux, P., D. Rekika, E. Acevedo and O. Merah (2006) Effect of drought on leaf gas exchange, carbon isotope discrimination, transpiration efficiency and productivity in field grown durum wheat genotypes. *Plant Science* **170**: 867–872.

Riehl, S., R. Bryson and K. Pustovoytov (2008) Changing growing conditions for crops during the Near Eastern Bronze Age (3000–1200 BC): the stable carbon isotope evidence. *Journal of Archaeological Science* **35**: 1011–1022.

Riehl, S., K. E. Pustovoytov, S. Hotchkiss and R. A. Bryson (2009) Local Holocene environmental indicators in Upper Mesopotamia: pedogenic carbonate record vs. archaeobotanical data and archaeoclimatological models. *Quaternary International* **209**: 154–162.

Rosen, A. (2007) *Civilizing Climate: Social Responses to Climate Change in the Ancient Near East*. Plymouth: AltaMira Press.

Rundel, P. W., J. R. Ehleringer and K. A. Nagy, eds. (1987) *Stable Isotopes in Ecological Research*. Heidelberg: Springer.

Scarborough, V. L. (2003) *The Flow of Power: Ancient Water Systems and Landscapes*. Santa Fe, NM: SAR Press.

Shaheen, R. and R. C. Hood-Nowotny (2005) Effect of drought and salinity on carbon isotope discrimination in wheat cultivars. *Plant Science* **168**: 901–909.

Shangguan, Z. P., M. A. Shao and J. Dyckmans (2000) Nitrogen nutrition and water stress effects on leaf photosynthetic gas exchange and water use efficiency in winter wheat. *Environmental and Experimental Botany* **44**: 141–149.

Tieszen, L. L. (1991) Natural variation in carbon isotope values of plants: implications for archaeology, ecology and paleoecology. *Journal of Archaeological Science* **18**: 227–248.

van der Merwe, N. J. and E. Medina (1991) The canopy effect, carbon isotope ratios and foodwebs in Amazonia. *Journal of Archaeological Science* **18**: 249–259.

23 Past plant use in Jordan as revealed by archaeological and ethnoarchaeological phytolith signatures

Emma Jenkins, Ambroise Baker and Sarah Elliott

ABSTRACT

Ninety-six phytolith samples were analysed from seven archaeological sites ranging from the Pre-Pottery Neolithic to the Classical period and from two ethnoarchaeological sites in Jordan. The aims were to test the possibility of detecting past irrigation with the methodology outlined by Madella *et al.* (2009) and Jenkins *et al.* (Chapter 21, this volume) and to study the contextual and temporal variation of plant use in Jordan. We utilised a water availability index using the proportion of phytolith types and ordination statistical methods to explore the similarities between the phytolith assemblages. The result of applying the irrigation methodology was promising, with contexts from water channels showing the greatest indication of water availability. Changes in plant use through time were also apparent with regard to phytolith densities and taxonomy. Date palm was identified in the Pottery Neolithic, providing one of the earliest records for this taxon in Jordan. This study shows the potential of both the water availability index and the value of inter-site comparison of phytolith assemblages.

23.1 INTRODUCTION

In this chapter phytolith analysis will be used to gain an understanding of past plant exploitation on a range of archaeological sites in Jordan dating from the Pre-Pottery Neolithic A (PPNA) to the Classical period. The broad aim of our analysis is to provide insights into past plant use in Jordan and to determine whether general patterns are observable through the phytolith record. One issue we wish to address is whether chronological changes in plant economy can be discerned. For example, in light of our current understanding of the evolution of crop cultivation and domestication in the Levant,

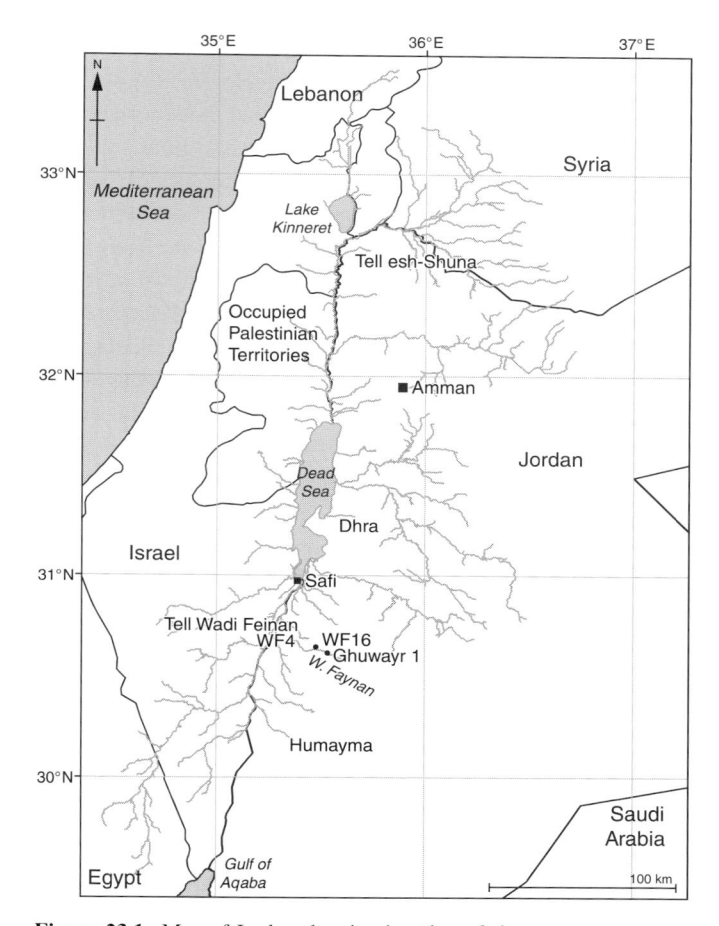

Figure 23.1. Map of Jordan showing location of sites.

we wanted to compare the phytolith assemblages from the three Neolithic sites in Wadi Faynan: WF16 (Pre-Pottery Neolithic A), Ghuwayr 1 (Pre-Pottery Neolithic B (PPNB)), and Tell Wadi Feinan Pottery Neolithic) (see Figure 23.1 for site locations). Previous phytolith and macrobotanical analysis from WF16 was focused on samples from the small trenches excavated during evaluation. Our aim was to confirm if the low density of cereals found in these assemblages was representative of the site as a whole or was the

Water, Life and Civilisation: Climate, Environment and Society in the Jordan Valley, ed. Steven Mithen and Emily Black. Published by Cambridge University Press.
© Steven Mithen and Emily Black 2011.

product of limited sampling. We also wished to compare the assemblages from WF16 with those from Ghuwayr 1 and Tell Wadi Feinan to establish if we could observe a change in palaeoeconomy throughout the Neolithic in a single region. Similarly, we were interested in comparing the samples from the Chalcolithic deposits at Tell esh-Shuna with those from the Early Bronze Age. We also included assemblages from two Classical period field systems in our analysis, Humayma and WF4. Our aim was to determine if phytoliths can be used to identify the crops under cultivation. In addition, two assemblages from ethnoarchaeological sites, one occupied and one abandoned, were included in this analysis. The purpose of this was two-fold. The first aim was to see how well certain context categories such as floors and hearths in the ethnoarchaeological study compared with archaeological samples presumed to be from the same context type, and the second was to assess how much the phytolith record is altered by time.

A more specific aim in analysing the archaeological and ethnographic samples was to apply the methodology for identifying past water availability developed in Chapter 21 and to see how effectively it can be applied to assemblages with unknown or mixed taphonomic pathways. Jordan is the ideal region to test this methodology because, although there were environmental fluctuations over time, throughout much of its history southern Jordan would have been semi-arid and irrigation would have been needed to sustain agriculture. The level of complexity of irrigation practices may have intensified over time, ranging from simple dams and barrages in prehistoric periods, to the complex water management systems found in the Nabataean period. Nonetheless, to ensure the success of cultivated crops some form of water management would probably have been necessary in southern Jordan even during the wetter times such as the Neolithic, Early Bronze Age and Byzantine periods. Having a method to test whether plants were irrigated allows us to infer cultivation practices and also to assess how effectively people managed agriculture in the past.

In this chapter we will firstly give an outline of the types of botanical proxies that are often used in southwest Asian archaeology and discuss the type of evidence that can be derived from phytolith assemblages. This will be followed by a review of archaeological and ethnoarchaeological phytolith studies. We will then provide a general overview of past plant use in Jordan ranging from the PPNA to the Classical period followed by a brief discussion of the sites from which our phytolith assemblages derive. We will then give details of the methodology employed, followed by the results and their discussion.

23.2 PLANT PROXIES IN SOUTHWEST ASIAN ARCHAEOLOGY

Macrobotanical remains are traditionally used to reconstruct the palaeoeconomy of a site – the intentional exploitation of plants for food, fuel, fodder, building materials, textiles, and medicinal and ritual purposes. Because they can be accurately identified, macrobotanical remains are especially useful for addressing site-specific questions or for the compilation of databases covering long time periods and broad geographical areas. For instance, Hillman et al. (2001) inferred the condition and mechanism for the initiation of wild cereal pre-domestication cultivation at the site of Abu Hureyra, Syria. Similarly, using a database covering a long time period, Colledge et al. (2004) retraced the origin of farming in the eastern Mediterranean and contributed a significant insight into the question of 'population movement versus cultural diffusion'. In addition, as discussed by Stokes et al. in Chapter 22 of this volume, the biochemistry of macrofossils can be used for reconstructing the conditions in which plants grew (Araus et al., 1997).

Pollen is frequently used as a proxy for palaeoenvironmental reconstruction but is less suited as a palaeoeconomic indicator. This is because pollen grains are relatively mobile and can be introduced to a site through a number of taphonomic pathways. Therefore, the source of the pollen recovered in archaeological samples is extremely difficult to establish. The problems are exacerbated by the fact that pollen is produced in very low quantities in arid areas and offers poor taxonomic distinction between plants of economic importance like the Poaceae (cereal family) (Faegri and Iversen, 1989; Moore et al., 1991). Although pollen grains from cereals can be identified in certain situations, they do not represent an absolute identification, and can be the source of discord even amongst the most renowned specialists (e.g. the recent argument between Tinner et al., 2007, 2008 and Behre, 2008, regarding the possible identification of pre-Neolithic agriculture in Switzerland on the basis of palynology alone).

23.3 PHYTOLITH ANALYSIS IN ARCHAEOLOGY

Phytoliths are a useful proxy for past plant use because they are extremely durable and survive well in both aerobic and anaerobic conditions. As such, they are frequently recovered from archaeological deposits in arid regions. The only conditions that phytoliths cannot tolerate are alkaline deposits in which they have been reported to dissolve (Krauskopf, 1956; Jones and Handreck, 1967; Bartoli and Wilding, 1980). Phytoliths are more abundant and distinctive in monocotyledons than in dicotyledons, which allows archaeologists to track plant groups of great economic importance such as the cereal family (Poaceae) and the palm family (Arecaceae). The terms monocotyledons and dicotyledons will be used as informal taxa in this chapter, following the classification of Cronquist (1981). Phytoliths and plant macroremains provide complementary information about past plant use on archaeological sites, and thus the most complete reconstruction of the palaeoeconomy can be gleaned by using these lines of evidence in conjunction (Warnock, 1998). They can also be used for palaeoenvironmental reconstruction if taken from non-habitation contexts such as lake cores (Piperno, 2006).

Phytolith research is a rapidly expanding field in archaeology, and phytoliths have been used to provide a range of information that increases our understanding of archaeological contexts and material. For example, they have been used to differentiate between the parts of plants present in assemblages, providing information on crop processing (Rosen, 1999; Harvey and Fuller, 2005; Portillo et al., 2009). Moreover, they have been used to inform about past land use. Pearsall and Trimble (1984), for instance, used the ratios of phytoliths from monocotyledons and dicotyledons as evidence for deforestation and the creation of terraced fields in Hawaii. Another application has been the differentiation between wild and domesticated strains of plants, allowing researchers to retrace the use of cultivated plants in archaeological contexts (e.g. rice: Piperno, 1988; Pearsall et al., 1995; Zhao et al., 1998; Piperno et al., 2000, 2002; Pearsall et al., 2003; Piperno and Stothert, 2003; and maize: Piperno, 1988; Piperno and Pearsall, 1993; Pearsall et al., 2003). In some instances, phytolith preservation is so complete that clear phytolith impressions of artefacts such as reed matting and baskets are visible *in situ*, providing information about past craft processes and the use of space in archaeological sites (Rosen, 2005).

In addition to domestic plant use, phytoliths can be indicators of past water availability through two botanically distinct means. Firstly, as identified by Jenkins et al. (Chapter 21, this volume) and by Madella et al. (2009), differential silicification occurs in cereals depending on water availability. As a consequence, it may be possible to use a phytolith index based on different morphologies of Poaceae phytolith to infer the presence of past irrigation in the absence of structural remains. Secondly, the morphology of the different phytolith types allows researchers to infer their taxonomical origin. Thus, the presence of plants whose ecology indicates water availability can be pinpointed. For instance, members of the Sedge family (Cyperaceae) are restricted to wet habitat like swamps and river banks worldwide (Heywood et al., 2007). The date palm is a reliable indicator of hot and dry climate but only grows where groundwater supply is abundant, e.g. in oasis or irrigated situations (Barrow, 1998). Therefore, the recovery of phytoliths that formed in these plants would imply sufficient water availability for them to grow.

23.4 ETHNOARCHAEOLOGICAL PHYTOLITH ANALYSIS

In contrast, there have been few ethnoarchaeological phytolith studies conducted. Shahack-Gross et al. (2003) analysed phytoliths from penning deposits from one occupied and four abandoned Masai settlements, in the Kajiado District, southern Kenya. The duration of abandonment varied from 1 to 40 years so that the taphonomic history varied. They compared these with regional sediments sampled from outside the settlements. They found that the density of phytoliths was generally higher in the penning deposits with samples typically containing 20 million phytoliths per gram of sediment while those from the regional sediment samples contained below 10 million phytoliths per gram. This analysis was supplemented by micromorphological and mineralogical analyses which, when used in combination, demonstrated that penning areas were distinct from regional sediments and from features inside the settlements such as hearths.

Tsartsidou et al. (2008) conducted an ethnoarchaeological study of phytolith assemblages from the small village of Sarakini in the mountains of northern Greece. Their aim was to explore the potential of using phytolith assemblages to reconstruct aspects of spatial organisation in an agro-pastoral community. Sediment samples were collected from four houses, three barns, a water mill and a smith's house. All of these structures were abandoned and different feature types were sampled, for example wall fill, roof coating, hearths, floors and yards. Open areas from the village were also sampled, including paths, fields, threshing floors, fodder piles and fresh dung (goat, cow and mule/donkey). They also gathered control samples from open grasslands and forested areas near the village. Tsartsidou et al. (2008) then devised a method they call the Phytolith Difference Index (PDI) which involved dividing the quantity of each phytolith morphotype in the sample, measured per thousand phytoliths, by the average control or natural value for that morphotype. The authors argue that the Phytolith Difference Index, together with phytolith concentrations, allows functional spaces to be differentiated and subsistence practices to be identified (Tsartsidou et al., 2008).

Tsartsidou et al. (2009) subsequently compared their ethnoarchaeological results to archaeological samples from the Pottery Neolithic village site of Makri, Greece. They found that the overall level of phytolith diversity in the Neolithic phytolith assemblage was much lower than in the modern assemblage, which was largely attributable to the lack of phytoliths from dicotyledon leaves. They also used the phytolith evidence to suggest that the subsistence economy at Neolithic Makri was a mixed agricultural and pastoral economy and that the site was occupied on a permanent basis (Tsartsidou et al., 2009).

23.5 PAST PLANT USE AND AGRICULTURAL PRACTICES IN JORDAN

The Levant is an important region for archaeobotanists; it is here that we see the earliest examples of domesticated plants and can chart the development of many important agricultural practices such as irrigation. Zohary (1992) identifies eight founder crops of Neolithic agriculture. These include three cereals: emmer wheat (*Triticum turgidum* subsp. *dicoccum*), barley (*Hordum vulgare*) and einkorn wheat (*T. monococcum*); and five other taxa: lentil (*Lens culinaris*), pea (*Pisum sativum*), chickpea (*Cicer arietinum*), bitter vetch (*Vicia ervilia*) and flax (*Linum usitatissimum*). Domesticated cereals are morphologically distinct from their wild ancestors, but it is more problematic to recognise domestication in pulses. It was originally believed that domestication was a relatively rapid process, which would have occurred over a couple of hundred years at the most (Hillman and Davies, 1992). More recent archaeobotanical research, however, has demonstrated that there was a long history of pre-domestication cultivation (Colledge, 1998; Harris, 1998; Willcox, 1999; Hillman, 2000; Colledge, 2001; Hillman *et al.*, 2001; Colledge, 2002; Nesbitt, 2002; Willcox, 2002). Harris (1989) suggests four stages in the history of plant exploitation that eventually led to domestication: (1) wild plant food procurement, i.e. hunter-gathering; (2) wild plant food production, i.e. the initial stages of cultivation; (3) systematic cultivation of morphologically wild plants; and (4) agriculture based on morphologically domesticated plants. Similarly, our geographical understanding of how domestication occurred has also evolved; previously it was believed that there was one centre of domestication within the Levant from where the domesticates spread (Zohary, 1996). This hypothesis has now largely fallen out of favour, with researchers currently proposing a polycentric evolution with multiple independent centres of domestication within the Levant (Gebel, 2004; Willcox, 2005).

Jordan, situated as it is in the southern Levant, has been a key region for charting these developments. The Pre-Pottery Neolithic site of Dhra, next to the Dead Sea, has evidence for grain storage as early as 11,300–11,175 cal. BP. This is based on the discovery of four structures at Dhra that appear to be granaries or silos. These are circular structures constructed out of pisé or mud, which would have had suspended floors (Kujit and Finlayson, 2009). Phytolith analysis of floor deposits from one of these granaries, Structure 4, found high densities of phytoliths formed in barley husks. This density was unrivalled in any of the other contexts analysed from the site (Jenkins and Rosen, forthcoming). In addition, the micromorphological analysis detected the presence of voids thought to be from straw and glumes of a comparable size to barley (Arpin, 2005). Preliminary analysis of the macrobotanical remains from Dhra indicates that those of barley are morphologically more similar to wild than to domesticated strains (Purugganan and Fuller, 2009). Similarly, barley was abundant at Zahrat adh-Dhra 2 (ZAD 2) (Zahrat adh-Dhra is also known as Zahrat edh-Dhra') and the size of the grains coupled with the presence of arable weeds suggests that barley was at least cultivated. Fig and *Pistacia* remains were found at this site demonstrating that foraging was an important subsistence strategy (Meadows, 2005). At Iraq-ed-Dubb both wild and domesticated-type wheat and barley were recovered, as well as large-seeded legumes, pistachio and walnuts, figs and several herbs often associated with cultivation. The results from Iraq-ed-Dubb should, however, be treated with caution. The excavators noted that there was much mixing of PPNA and Natufian deposits (Colledge, 2001) and it is therefore possible that there was also mixing of PPNA and later Neolithic deposits (Nesbitt, 2002).

In addition to the problem of chronologically contaminated samples, macroremains of domesticated-type grains and chaff can occur in low frequencies in wild assemblages (Willcox, 1999). As a result, a large sample size is needed to ascertain whether an assemblage is predominantly wild or domesticated. Willcox (1999) estimates that over 10% of domesticated-type chaff is needed in an assemblage to reliably identify it as domesticated. In addition, identification based on grain morphology alone is problematic because different genera within the grass family often resemble one another; for example, wild einkorn is very similar to wild rye, and modern-day wild barley closely resembles domesticated barley. This problem is exacerbated by the fact that grains can distort with charring (Nesbitt, 2002). Taking all of the evidence and these factors into account, it appears that the PPNA people in Jordan had a mixed plant diet which included wild cereals, pulses, nuts and fruits.

In the PPNB, more secure evidence for cereal domestication is found in Jordan, particularly from the middle PPNB and the final PPNB/PPNC (Nesbitt, 2002; Willcox, 2002; Colledge *et al.*, 2004). Domesticated cereals were found at: Wadi el-Jilat 7 and 13; Azraq 31 (Colledge, 2001); Ain Ghazal (Donaldson, 1985); Beidha (Helback, 1966; Colledge, 2001); possibly Basta (Neef, 1987); Ghuwayr 1 (Simmons and Najjar, 1998); and Wadi Fidan A and C (Colledge, 2001). The macrobotanical remains from these sites also contained wild cereals, nuts, pulses and fruits, demonstrating that wild food sources remained an important constituent of the diet during this time.

In the Late or Pottery Neolithic period, there is little macrobotanical evidence from Jordan. The site of Tel Rakan I, in northwest Jordan, has PPNB and Late Neolithic deposits, both of which have provided macrobotanical assemblages. The PPNB samples contained an abundance of barley, pulses and gathered plant foods such as fig and *Pistacia*. In contrast, samples from the Late Neolithic were rich in *Lolium*, a genus frequently found in irrigated fields (Meadows, 2005). The Late Neolithic site of ash-Shalaf was rich in small-seeded legumes, probably *Astragalus*, which could have been used as a source of fuel. In addition, glume wheat, barley, lentil and possibly *Pistacia* were found. A small macrobotanical assemblage was found at Dhuweila which included small-seeded legumes and wild grasses but no evidence of domesticates (Colledge, 2001). Most sites with Late Neolithic deposits such as Jebel Abu Thawwab, Wadi Rabah and 'Ain Rahub had macrobotanical assemblages that partly comprised domesticated cereals, though non-domesticated (probably cultivated) cereals were also present in Late Neolithic sites, as were domesticated flax, lentils, peas, small-seeded legumes and arable weeds (Meadows, 2005). It is also in the Late Neolithic that we see the first evidence for the exploitation of olives in Jordan (Bourke, 2002; Meadows, 2005).

At some sites, such as Pella XXXII, little difference in palaeoeconomy can be seen between the Late Neolithic and the Chalcolithic, whereas at other sites such as Teleilat Ghassul there is compelling evidence for changes in subsistence (Meadows, 2005). The evidence from Teleilat Ghassul indicates that by the Chalcolithic, olive exploitation had evolved into full arboriculture with evidence for cultivation (possibly domestication) and oil extraction, while at Abu Hamid olive remains were present in 87% of samples (Neef, 1990; Bourke, 2002; Meadows, 2005). Furthermore, dates and grape pips are found in the Chalcolithic deposits in Jordan, although Meadows (2005) argues that some of these finds could be intrusive.

Meadows (2005) suggests that the archaeobotanical assemblage from Teleilat Ghassul reflects a change in agricultural practices from the Early to the Late Chalcolithic. Early Chalcolithic farming at Teleilat Ghassul was based on the cultivation of emmer wheat, two-row barley and lentils, with other pulses, free-threshing wheat and six-row barley as secondary crops. By the Late Chalcolithic, farming still relied on emmer wheat, but six-row barley now rivalled the two-rowed variety, and other pulses were as important as lentils. Significant changes were also observable in the wild/weed species in the assemblages which, according to Meadows (2005), may reflect a shift in agricultural practices as a response to population growth. He argues that if the weeds present in the assemblages were those of cultivation, then the Early Chalcolithic assemblage reflects cultivation of gardens in relatively damp areas of the wadi bed, whereas the Late Chalcolithic assemblage is indicative of agricultural expansion into more marginal areas (Meadows, 2005). Agriculture in such lowland areas in the Jordan Valley could have been supported by floodwater irrigation systems similar to those found in the Negev (Levy, 1995), expanding the types of crops that could have been grown in this area (Meadows, 2005; Bourke, 2008). Plough cultivation was another important development that occurred during the Chalcolithic (Levy, 1995; Bourke, 2008).

The EBA I–III saw an intensification of many activities and practices already found in the Chalcolithic, such as irrigation agriculture (Grigson, 1995; Rosen, 1995). There is structural evidence for the use of irrigation systems at Jawa located in the Black Desert in north Jordan and from Tell el-Handaquq (North) in the Jordan Valley (Helms, 1981; Mabry et al., 1996). The macrobotanical evidence supports the structural evidence for irrigation; grains of bread wheat of a size large enough to suggest irrigation were recovered from Tell el-Handaquq (North), while six-row barley, which would have been difficult to grow in the Black Desert without irrigation, was found in the macrobotanical assemblage from Jawa (Willcox, 1981; Rosen, 1995). Rosen (1995) suggests that in Palestine, EBA sites were ideally positioned to take advantage of seasonal runoff, and it is probable that the same irrigation technique was used in the Jordan Valley. This is supported by the macrobotanical evidence from Bab-adh-Dhra (also known as Bab edh-Dhra) which found flax seeds of a size that suggests irrigation, coupled with the fact that flax, along with other elements of the assemblage such as six-row barley and emmer wheat, could not have been successfully cultivated without some form of irrigation (Rast and Schaub, 1978; McCreery, 2003). In addition, wheat and barley, as well as domesticated fig, olive and grape, were all found at Tell Abu an-Ni'aj and Tell el-Hayyat, both of which are located in the Jordan Valley and would have required additional irrigation (Falconer et al., 2007). The cultivation of tree crops such as olives and grapes increases during this time, probably partly owing to trade (Cartwright, 2002; Falconer et al., 2007; Philip, 2008). It should be borne in mind, however, that inferences of irrigation using the cultural requirement of specific crop plants depend on our knowledge of past precipitation, which cannot be fully quantified, either in amount or in frequency.

By the Middle Bronze Age, the three prevailing arboreal crops in areas where rain-fed agriculture was undertaken were olives, figs and grapes. At sites in more arid areas these were not always present. At ZAD 1 (a Middle Bronze Age village situated 200 m from the PPNA site ZAD 2 referred to earlier) the remains of hulled and naked barley, emmer and bread wheat, fig and grape were all found but olive was absent (Fall et al., 2007). Fall et al. (2007) state that this is typical for Bronze Age sites in the more arid regions of the southern Levant, despite the fact that olive is a relatively drought-resistant crop compared with cereals. Pollen analysis from a core taken in the Sea of Galilee shows that there was an increase in olive pollen coupled with a decrease in oak pollen during the Middle Bronze Age and again during the Iron Age, suggesting that land was being cleared for cultivation, presumably for trade with the Roman Empire (Falconer, 2008). Grape pollen also appears in the core during the Iron Age, demonstrating that the cultivation of this tree crop was greatly increasing, and by the Classical period, grapes were an important crop.

During the Nabataean period some areas, such as those around Amman and near Wadi Musa, would have received enough precipitation to allow dry farming, while other areas, such as Petra and Humayma, were more arid and required direct irrigation or the harvesting of runoff water to sustain agriculture (Oleson, 2007b). At Nabataean Petra, a diverse macrobotanical assemblage was found which included wheat, barley, bitter and

Table 23.1 *Chronology of the sites studied*

Period	Approximate dates (cal.)	Sites
PPNA	9750–8550 BC	WF16
PPNB	8550–6750 BC	Ghuwayr 1
Pottery Neolithic	6300–5850 BC	Tell Wadi Feinan
Chalcolithic	4600–4000 BC	Tell esh-Shuna North
Early Bronze Age I–III	3600–2350 BC	Tell esh-Shuna North
Classical (Hellenistic/Nabataean/Roman/Byzantine)	322 BC–AD 640	WF4/Humayma

After Kuijt and Goring-Morris (2002); Philip (2008); Watson (2008)

common vetch, lentil, pea, olive, grape, fig, date and walnut as well as a variety of weed species (Bedal and Schryver, 2007). Furthermore, according to evidence from the Petra papyri, cultivation of grains and tree crops continued into the Byzantine period (Caldwell and Gagos, 2007).

A survey of the Pella region found hillsides full of small wine presses, and wine from Um Qays (also known as Umm Qais) is mentioned in sixth-century AD poetry (Watson, 2008). Other crops found in the Classical period are cereals such as wheat and barley, chickpeas, a variety of root and leaf vegetables, and fruits such as olive, fig, date, plum and apple. Obviously, the types of crops grown depended largely on the region. At the Byzantine site of Qasr al-Bulayda located on the northern bank of Wadi al-Karak, agriculture was sustained with a complex aqueduct system. As a result, it was possible to cultivate bread and emmer wheat, hulled barley, lentil and pea in fields close to the site (Fall *et al.*, 2007).

In summary, the major crops which comprised the mainstay of the Jordanian palaeoeconomy in later prehistory and the historic periods – wheat, barley, lentil, pea, chickpea, bitter vetch, flax, fig, date, olive and grape – were in evidence by the Early Bronze Age and possibly even by the Chalcolithic period. In light of the climatic conditions and arid landscapes during these periods (see Chapters 3 and 5, this volume), irrigation would have been necessary for Chalcolithic and Early Bronze Age agriculture.

23.6 THE STUDY SITES

23.6.1 Study sites in the Wadi Faynan region

Phytolith assemblages from six archaeological and two ethnoarchaeological sites were included in this study. Four of the archaeological sites (WF16, Ghuwayr 1, Tell Wadi Feinan and WF4), and both of the ethnoarchaeological sites are located in the Wadi Faynan region in southern Jordan. Therefore, this section will begin with a description of the Wadi Faynan region and a discussion of the sites studied from this area. This will be followed by a description of the remaining two archaeological sites.

Wadi Faynan has a dry arid climate and is ecotonal between the desert of Wadi Araba and the residual steppe-land of the Edom mountains (Hunt *et al.*, 2004). The Wadi Faynan is formed at the confluence of three tributary wadis, the Dana, Ghuwayr and Shayqar, which flow down from the western edge of the Jordanian Plateau and south from the Edom Mountains. The Wadi Faynan then continues westwards towards Wadi Fidan and eventually out into the Wadi Araba (Barker *et al.*, 2007).

The archaeology in Wadi Faynan is remarkable in both its preservation and its magnitude. The region is most famous for being an early copper smelting area and for its Roman and Byzantine settlements, but it also has important Palaeolithic and Neolithic sites (Barker *et al.*, 1997; Mithen *et al.*, 2007). The earliest site included in this study is the PPNA site of WF16 (see Table 23.1 for information about site chronologies). This is located on a small knoll adjacent to Wadi Ghuwayr. It was discovered as part of the Dana–Faynan–Ghuwayr Early Prehistory project and was subject to four seasons of evaluation from 1997 to 2000 (Finlayson and Mithen, 2007). Work at this site recommenced in 2007 with an open area excavation, and the samples included in this study are from this phase of research (for information about the phytolith results from the evaluation, see Jenkins and Rosen, 2007). The site largely comprises small circular pisé structures, with a larger structure being partially uncovered in the northern side of the excavation area. Analysis has been conducted only on the macrobotanical and phytolith assemblages from the evaluation, though the preservation of the plant macroscopic remains was poor (Kennedy, 2007). Fig seeds, cereal grain fragments (one of which was identified as barley), small-seeded legumes, fragments of pulses (some of which were classified as lentil/pea/vetch), *Pistacia* and nutshell fragments were all found (Austin, 2007; Kennedy, 2007). It was not possible to identify the barley or the pulses as wild or domesticated, and Kennedy (2007) concluded that WF16 was not an agricultural village. This hypothesis was supported by the phytolith assemblage from the evaluation. This was dominated by phytoliths formed in dicotyledons and contained very few monocotyledons. Those that could be identified were largely from reeds (*Phragmites*) rather than cereals (Jenkins and Rosen, 2007).

Phytolith samples were also taken from the adjacent PPNB site of Ghuwayr 1. This was first excavated in 1993 by Najjar (1994) revealing a site of approximately three acres. Excavation continued for three more seasons in 1996, 1997 and 1998 under the joint direction of Simmons and Najjar (Simmons and Najjar, 1996, 1998, 2003, 2006, 2007). Since the end of these seasons, Najjar has conducted small-scale excavation in an area of the site that was damaged by flood water (Simmons and Najjar, 2006). Excavation uncovered a modest-sized PPNB village comprising rectilinear buildings made of stone which were well preserved. It has deeply stratified architectural remains which, in some areas, are in excess of 5 m, representing multiple occupation phases. It was originally believed that Ghuwayr 1 was a Late PPNB site but radiocarbon dates now place it in the Middle PPNB. The macrobotanical report has yet to be published, but the assemblage is said to comprise: barley, emmer wheat, einkorn wheat, pea, fig, pistachio, caper and date palm, the last of which is the earliest record of this species in Jordan (Simmons and Najjar, 2003, 2007).

Tell Wadi Feinan, 5 km west of Ghuwayr 1, was also sampled for phytoliths. It is situated directly next to Wadi Faynan and there are ongoing concerns about its gradual erosion. It was originally excavated in the late 1980s by a joint Jordanian and German team (al-Najjar *et al.*, 1990). In addition, further test pits were dug by Simmons and Najjar in 2000. This site has Pottery Neolithic, Chalcolithic and Byzantine deposits. Excavation of the Pottery Neolithic deposits revealed a large village consisting of rectangular drystone houses (Barker and Gilbertson, 2000). Domesticated cereals (glume wheat and six-row barley) were identified from the Pottery Neolithic deposits, as well as pulses, fig, possibly *Pistacia* and grass seeds. A broad range of wood taxa was identified including juniper, oak, caper, olive, fig and pistachio (unpublished data analysed by Austin and Kennedy, cited in Meadows, 2005, p. 80).

The Wadi Faynan field system consists of several field systems that date back to at least the Early Bronze Age (Newson *et al.*, 2007). It did not reach its zenith, however, until the Late Roman/ Byzantine period when the separate systems became a single bounded field group of approximately 800 fields (Mattingly *et al.*, 2007). One of these field systems, WF4, which covers an area of about 209 ha, was the subject of a survey conducted as part of the Wadi Faynan landscape survey (Newson *et al.*, 2007), and small-scale excavations were also undertaken as part of the Water, Life and Civilisation project in 2005 (Smith *et al.*, this volume, Chapter 15). These fields are still visible in the landscape today and each of them would have been surrounded by low stone walls. As rainfall in this region is low and sporadic, irrigation would have been needed to sustain agriculture. Newson *et al.* (2007) suggest irrigation would have been from runoff from the immediate vicinity and from tributary streams of the main wadi, an idea which has since been disputed by Crook (2009) and Smith *et al.* (this volume, Chapter 15). Mattingly *et al.* (2007) propose that in some areas of the field system, irrigation water was directed to the field systems via deliberately constructed stone-lined channels whereas in other areas a simpler runoff system was used. This latter option would have been used to irrigate the southern part of the WF4 field system which Mattingly *et al.* (2007) propose would have supported tree crops such as vines and olives. In contrast, the more northerly part of the system would have received channel-fed runoff irrigation and, as a result, Mattingly *et al.* (2007) suggest this area would have supported cereals. The samples discussed here are from the small-scale excavations conducted by Smith *et al.* (this volume, Chapter 15) which covered both the northern and the southern parts of the field system and are believed to date from the Nabataean period onwards.

Two modern Bedouin tent sites were included in this study in order to explore the phytolith signature of known contexts. Both of these sites were those of Juma Ali, from the Azazma tribe, and his family (Figure 23.2). One of these sites was the winter tent site which was occupied at the point of sampling; the other site was the summer tent site, which had been unoccupied for approximately six months when the samples were taken. The two sites are adjacent to one another, with the winter site being on a raised terrace in a small tributary wadi, while the summer site is located to the north of the winter tent above the wadi. Bedouin tents are known in Arabic as *beit-as-sha'r* which translates as house of hair (Palmer *et al.*, 2007). This is because the tents are made of goat hair and would have traditionally been handwoven. Juma's winter tent is north facing and comprises two main areas: the public area, where visiting males are entertained, known as the *shigg*, and the private or domestic area, the *mahram*, where the women and children sleep, the cooking and domestic activities take place and food and belongings are stored (*Ibid.*). Juma's tent is home to nine people: Juma, his wife (Um Ibrahim) and seven children. In addition, Juma has three donkeys and 150 goats which are not kept inside the tent but in penning areas close by. The winter tent arrangement was slightly different, with a tent being used for the domestic area and a temporary structure, constructed using tent material but with a roof thatched with reed, being used for the public area.

The public or men's area has a central hearth, which is used for making tea and coffee for guests and a sleeping area for the men. This hearth is square and is delineated with a series of stones; in the summer tent it is external, presumably to prevent the tent from becoming too hot. The domestic area also has a hearth, but

Figure 23.2. (A) Juma in front of his winter tent. (B) Inside the public part of the tent. (C) Inside the domestic area of the tent. (D) Um Ibrahim (Juma's wife) cooking bread on the domestic hearth.

this is round (Figure 23.2). It is used to cook food, most frequently bread, potatoes, tomatoes, chickpeas, lentils, rice and meat (mostly goat and chicken). A combination of dung and wood are used as fuel, and the fodder is brought to the region from Aqaba.

23.6.2 Study sites from other regions of Jordan

Tell esh-Shuna North is located in the north Jordan Valley, southeast of Lake Kinneret and south of the Wadi Yarmouk (Figure 23.1). The first excavations at Shuna were undertaken by Henri de Contenson and Hasan Abu Awad in 1953 and revealed Chalcolithic occupation (de Contenson, 1960). Mellaart also conducted excavations at the western side of the site and uncovered evidence for Early Bronze Age occupation (Mellaart, 1962), while Gustavson-Gaube found Chalcolithic occupation in his excavations in 1984 and 1985 (Gustavson-Gaube, 1985, 1986). Our samples were collected from the most recent excavations, which were undertaken by Philip and Baird between 1991 and 1994 (Philip and Baird, 1993). These uncovered both Chalcolithic and Early Bronze Age I occupation. The macrobotanical remains from both periods included emmer wheat, possible einkorn wheat, six-row barley, olive, broad bean and lentil. Free-threshing wheat, bitter vetch, flax, fig and grass pea were present in the Chalcolithic period while grape was found only in the Early Bronze Age samples. The presence of grape in the Early Bronze Age assemblage is interesting because some form of irrigation would almost certainly have been needed to grow this species in the Jordan Valley. However, a shift in weeds was seen between the Chalcolithic and the EBA at Shuna, with small-seeded legumes replacing *Lolium* (often found in irrigated fields) as the dominant taxa (Holden, forthcoming). Therefore, it seems possible that some form of irrigation was used for agriculture in both periods.

The final site of our study is Humayma or ancient Hawara, situated in the hyper-arid desert region of southern Jordan, approximately 80 km south of Petra and 80 km north of Aqaba (Figure 23.1). It was founded by the Nabataeans as a desert trading post and caravan way-station but was also occupied through the Roman, Byzantine and Early Islamic periods. Oleson (2007a) proposes that it was established in an attempt to sedentarise the nomadic Nabataean pastoralists who occupied the region. The settlement prospered and, through careful water management, agriculture was practised (Oleson, 2007a). The macrobotanical assemblage was dominated by cereals, particularly barley, though wheat, rye and millet were also found. Legumes were common, particularly lentil, and remains of figs, olives, grapes and dates suggest that arboriculture flourished (Foote, 2007). It was originally excavated by Oleson between 1986 and 1989 and again from 1991 to 2005. During this time, excavation included a range of areas and structures in an attempt to encompass all periods of occupation (Oleson, 2007a). Further excavation was conducted by Foote as part of the Water, Life and Civilisation project. This was focused on a collapsed, domed structure (Structure 72) which was adjacent to a field used for cultivation (for full details see Foote, Chapter 19, this volume). The samples included in this study are from this recent phase of excavation and are derived from Structure 72 and the adjacent field.

23.7 METHODS

23.7.1 General approach

In this work, the results of several phytolith analyses of archaeological and ethnoarchaeological samples from Jordan are combined.

The initial interpretation of the phytolith assemblages was based on a cross analysis of, firstly, the contextual and chronological information of the samples and, secondly, the anatomical and taxonomic significance of the phytolith types. Moreover, indices of wood content (Albert et al., 2003) and plant water availability (this volume, Chapter 21; and Madella et al., 2009) were calculated.

In addition, the level of similarity between the phytolith assemblages was numerically assessed by ordination analysis. This analysis takes into account the different phytolith assemblages as a whole, calculates the similarity between the samples and thus allows us to put their botanical content into a wider context. In fact, the level of similarity between the samples provides an opportunity to interpret the phytolith assemblages into more quantitative information. This approach is particularly justified because the current knowledge of phytolith formation, deposition and taphonomy does not allow researchers to assess insightfully the significance of either a proportion or a concentration of a phytolith type within a sample. Our approach circumvents this problem by formulating the simple assumption that similar phytolith assemblages are likely to originate from similar plant material, and thus in turn to have deposited in a similar context. Therefore, comparing phytolith assemblages for which the deposition processes are well understood (ethnoarchaeological samples) with phytolith assemblages from unknown contexts (archaeological samples) can enlighten us about past plant use. This approach is known as Middle Range Theory in archaeology (Binford, 1977) or as analogue matching in palaeoecology (Birks and Gordon, 1985; Jackson and Williams, 2004).

This comparative approach using ordination methods has certain limitations. As discussed by Jackson and Williams (2004), the lack of a modern analogue for certain past assemblages can bias the result of such analysis. In addition, the phenomenon named redundancy or equifinality (i.e. when two different situations produce the same microfossil assemblage) can also hamper the interpretation. Finally, the comparison of phytolith assemblages represents a numerical challenge partly resolved by ordination analysis (Lepš and Šmilauer, 2003), but also biased by the human choices that remain prevalent behind the numerical analysis. For instance, decisions regarding the importance attributed to the different phytolith types can have a marked effect on the results and must thus be made with extreme care.

Far from addressing all of the limitations induced by this approach, this chapter innovatively attempts to apply analogue matching to archaeological phytolith analysis, using the experience gathered by the field of Quaternary science.

23.7.2 Sampling strategy

At WF16 the samples were taken during excavation, while at Ghuwayr 1 and Tell Wadi Feinan the samples were taken from exposed and cleaned sections from Late Neolithic deposits under the guidance of Mohammed Najjar who had directed the excavations. Samples were taken using a clean trowel and were double bagged. Samples were allowed to dry out in the CBRL laboratories adjacent to the Wadi Faynan Ecolodge before being sealed for export. The samples from Shuna, WF4 and Humayma were collected for us by the excavators.

23.7.3 Laboratory work

Samples were processed using the following methodology:

1. Samples of dried raw sediment (2 to 0.8 g) were weighed out.
2. Each sample was screened through a 0.5 mm mesh to remove any coarse-sized particles.
3. Calcium carbonates were dissolved using a dilution of 10% hydrochloric acid and then washed in distilled water three times.
4. Clay was removed using a settling procedure and sodium hexametaphosphate (Calgon) as a dispersant. Distilled water was added and the samples left for 75 minutes before pouring off the suspension. This was repeated at hourly intervals until the samples (i.e. the portion left at the bottom after the clay is poured off in the suspension) were clear. Samples were then transferred into crucibles and left to dry at a temperature of less than 50 °C.

5. After drying, samples were placed in a muffle furnace for two hours at 500 °C to remove any organic matter present.
6. Phytoliths were then separated from the remaining material using a heavy liquid calibrated at 2.3 specific gravity. Phytoliths were transferred to centrifuge tubes and washed three times in distilled water. They were then placed in small Pyrex beakers and left to dry.
7. Two milligrams of phytoliths per sample were mounted onto microscope slides, using the mounting agent Entellan.

23.7.4 Counting, identification and nomenclature

Counting was done using a Nikon Optiphot 2 microscope at ×400 magnification. Single phytoliths and conjoined phytoliths were counted up to a sum of 300 and 100 units, respectively. The identification of phytolith types was carried out using Twiss *et al.* (1969), Brown (1984) and Piperno (2006). More specific identification was carried out for the following taxa using the references stated:

1. Reeds (Metcalfe, 1960; Ollendorf *et al.*, 1988)
2. Sedges (Ollendorf, 1992)
3. Cereals (Rosen, 1992; Tubb *et al.*, 1993)
4. Palms (Rosen, 1992)
5. Dicotyledons (Albert *et al.*, 1999)

All identifications were compared with the phytolith reference collection of the Archaeology Department, University of Reading. Figures 23.3 and 23.4 illustrate the main phytolith types identified. The phytolith nomenclature follows Madella *et al.* (2005).

23.7.5 Numerical analysis

The weight percent of phytoliths was calculated using the weight of phytoliths extracted at the end of the laboratory procedure divided by the weight of dry sample material. This method has the disadvantage of not taking into account the possibility of post-depositional formation of calcrete and the leaking of organic matter (see the acid insoluble fraction of Albert *et al.*, 1999). It was estimated, however, that these samples did not contain a high proportion of these types of components from post-depositional origin.

The phytolith countings were transformed into percents of each phytolith type per sample. Percent data is also named relative data, because the values obtained for each phytolith type are not absolute in that they depend on the amount of other phytolith types. Despite this, the use of phytolith concentration (number per gram) as an alternative was deemed inappropriate in a situation where the sedimentation rate of the samples is unknown (see Birks and Gordon, 1985). The calculation of percent per sample was calculated separately for single and conjoined phytoliths. Special treatment was reserved for the siliceous aggregates, because according to our experience, the size and the number of siliceous aggregate are highly sensitive to mechanical breakdown and thus vary depending on the taphonomic history of the sample. Siliceous aggregates were not taken into account in the percent calculation of single phytoliths. However, the siliceous aggregate percent was calculated as follows: $100 \times$ siliceous aggregate/(sum of single + siliceous aggregate). This approach follows the custom in palynology to exclude some pollen types like aquatic plants or to accommodate samples in which the overwhelming presence of one type obscures the variation of others when expressed as relative data. The water availability index was calculated by expressing the percentage of long-cell phytoliths relative to the sum of short and long cells. This was done in preference to dividing total long cells to total short cells which, owing to the small sample size of some context categories, resulted in a disproportionately high value for the water channels.

Ordination analysis was carried out on the single phytolith percent data only. Samples containing fewer than 200 single phytoliths were excluded from this analysis because it was estimated that they would carry too much statistical uncertainty (Stromberg, 2009), although the exact calculation of the error margin goes beyond the scope of this study. The data was centred and standardised (division by the standard deviation) and square root transformed in order to minimise the impact of very common phytoliths versus rarer ones. Underlying this choice is the idea that in the absence of any real knowledge of differential production, deposition and preservation of the various types, the same level of significance should be given to all phytolith types.

An initial DCA (Detrended Correspondence Analysis) (Hill and Gauch, 1980) performed in Canoco 4.5 (ter Braak and Šmilauer, 2003) showed that the first gradient length was smaller than 2.5. This suggests that most of the phytolith types have a linear spread across the samples, and therefore a PCA (Principal Component Analysis) was preferred for the analysis. This was carried out with the archaeological samples, while the Bedouin ethnoarchaeological samples were added passively. The passive, or silent, addition of samples during the course of an ordination analysis permits the distinction between the samples that are to be explained (archaeological samples) and the explanatory samples (ethnoarchaeological samples with known context). In addition, this means of analysis addresses some of the limitations of the analogue matching technique in that it reduces the chances of misinterpreting the archaeological samples because of a lack of appropriate ethnoarchaeological samples.

Samples from the PPNA and Pottery Neolithic site of Dhra were included in the ordination analysis but not in any of the other analyses. This was done in order to establish whether the similarities found in the results from the Neolithic sites,

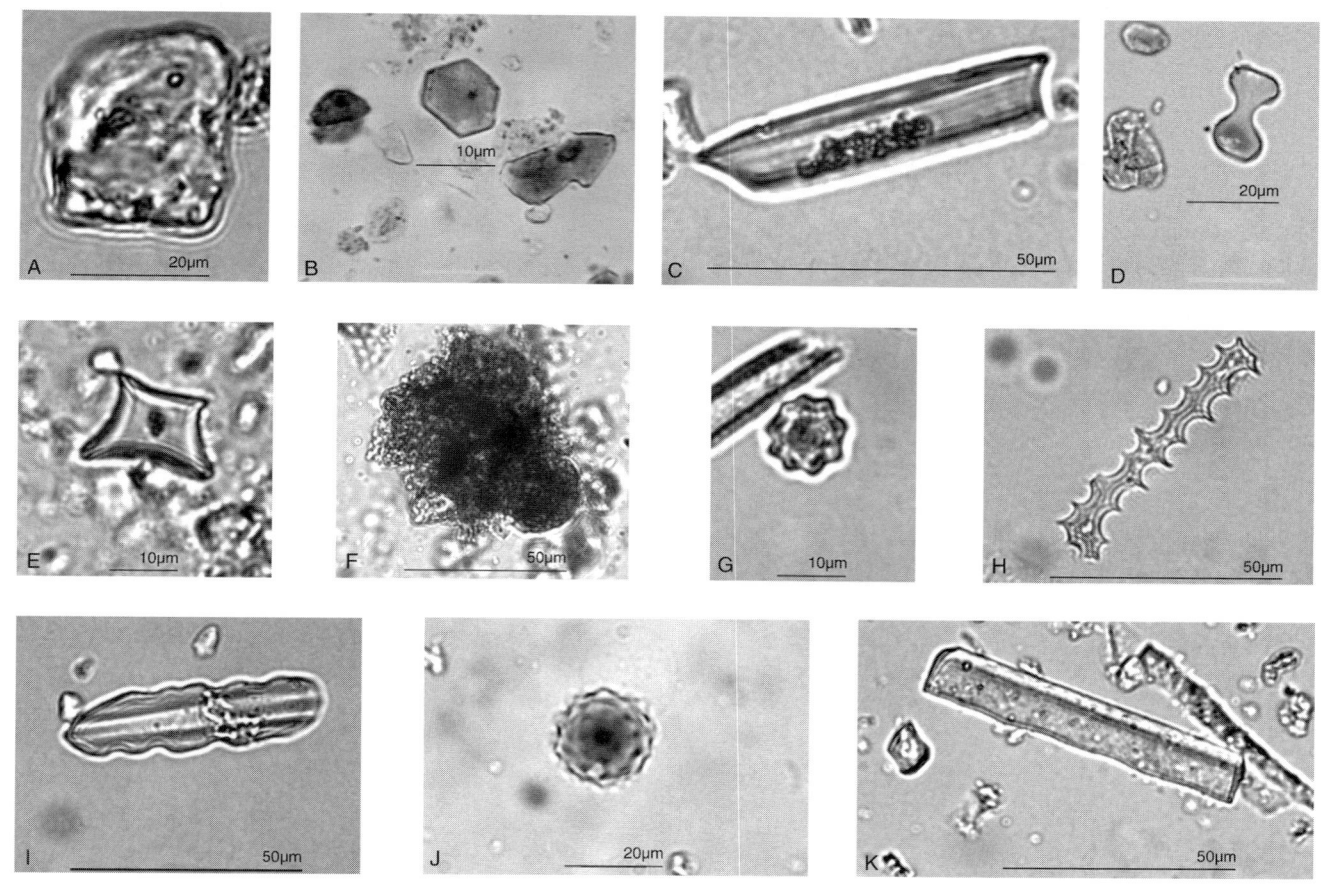

Figure 23.3. Images of the main single-celled phytolith morphotypes identified: (A) irregular shaped dicotyledon (platey) from Tell esh-Shuna; (B) polyhedral shaped psilate from Tell Wadi Feinan; (C) trapeziform psilate from Tell esh-Shuna; (D) bilobate from WF16; (E) rondel from Tell esh-Shuna; (F) siliceous aggregate (usually bright orange) from Tell esh-Shuna; (G) globular echinate (date palm) from Tell esh-Shuna; (H) elongate dendriform from Tell esh-Shuna; (I) trapeziform crenate from Tell esh-Shuna; (J) globular echinate (date palm) from Tell Wadi Feinan; (K) elongate psilate from WF16.

which were all from the Wadi Faynan region, were attributable to geography rather than chronology. As the site of Dhra is located to the north of Wadi Faynan near the Dead Sea, samples from here were used as control samples. The results from Dhra, however, are not discussed in detail in this chapter as they form part of an earlier study conducted by Jenkins and Rosen (forthcoming).

23.8 RESULTS

Table 23.2 and Figure 23.5 illustrate the presence and absence of any taxa identified in the 96 samples of this study, while the weight percent of the samples can be seen in Figure 23.5. They reflect the concentration of phytoliths within the samples; however, in the absence of any known sedimentation period of time they cannot be related to an amount of plant material per units of time. Thus, a large number of phytoliths may represent a low rate of plant accumulation during a long period of time or,

contrastingly, a high rate in a short period of time. The indices of water availability and woody plant material are presented in Figures 23.6 and 23.7. The two first axes of the PCA performed on the archaeological samples explains 37.6% of the variation between the samples, and some successful aspects of this numerical analysis are highlighted in Figures 23.8, 23.9, 23.10 and 23.11.

23.9 DISCUSSION

23.9.1 Methodological issues

The 96 samples used in this study provided a good opportunity to apply the methodology of Madella *et al.* (2009) and Jenkins *et al.* (Chapter 21, this volume), both of whom found that in some cereal species the production of long celled phytoliths increased with irrigation. This method is used as an alternative to the

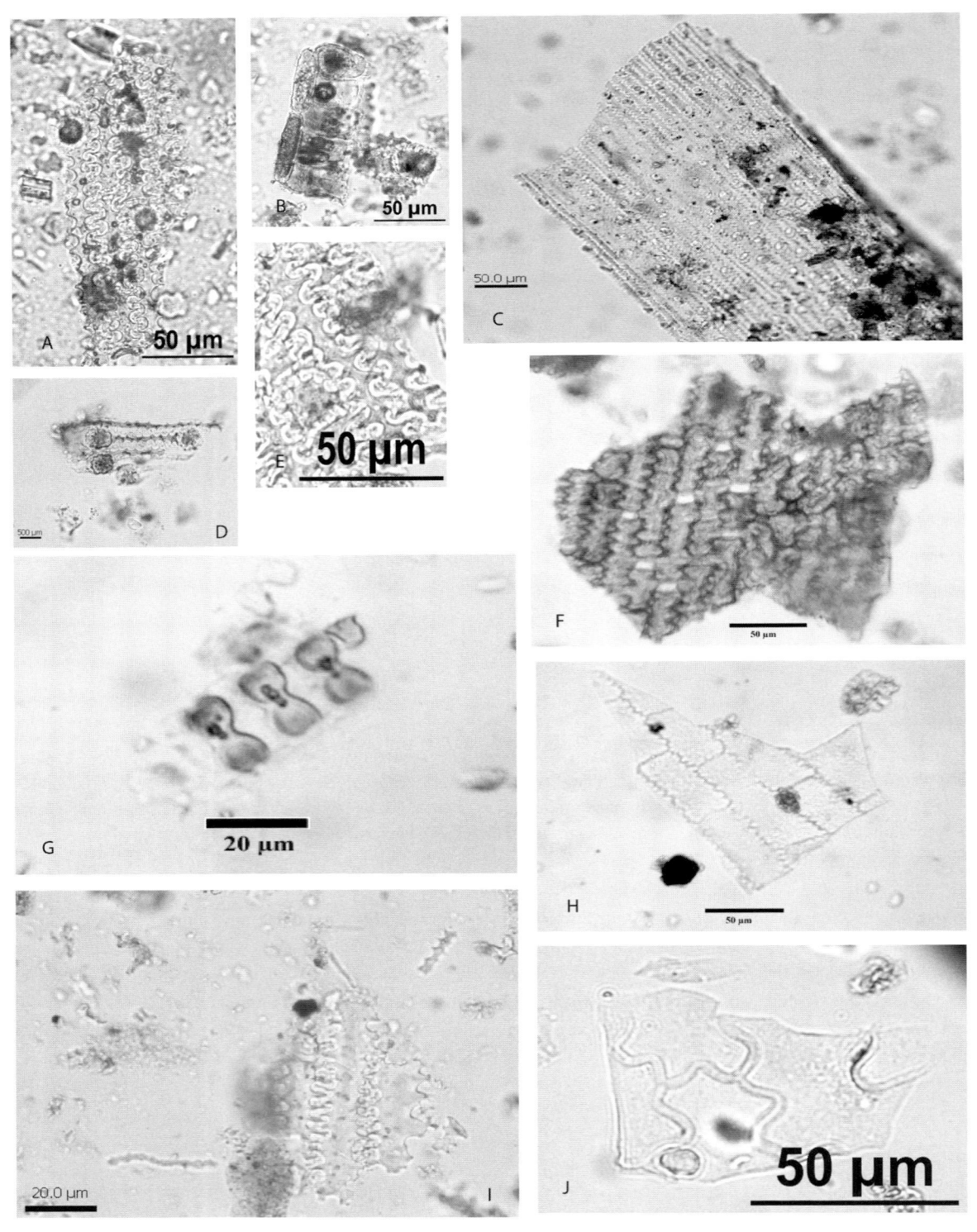

Figure 23.4. Images showing conjoined phytoliths. (A) Elongate dendriform and short cells formed in a wheat husk from Tell esh-Shuna; (B) conjoined parallepipedal bulliforms from Tell-esh Shuna; (C) conjoined elongate sinouate from a reed stem from Juma's summer tent; (D) elongate dendriform and short cells formed in a barley husk from Humayma; (E) elongate dendriform and short cells formed in a barley husk from Tell esh-Shuna; (F) conjoined elongate sinouate from a reed stem from Ghuwayr 1; (G): bilobates from Tell Wadi Feinan; (H) sinuous long cells from Ghuwayr 1; (I) elongate dendriform and short cells formed in a wheat husk from Juma's summer tent; (J) jigsaw puzzle from Tell esh-Shuna.

Table 23.2 *Taxa present by chronological period*

Plant name	Taxa	Taxonomic level	PPNA	PPNB	Pottery Neolithic	Chalcolithic	Early Bronze Age I	Classical	Modern
Barley	*Hordeum* sp.	Genus		+				+	+
Wheat	*Triticum* sp.	Genus					+		
Reed	*Phragmites* (I)	Genus		+	+				+
	Chloridoideae	Subfamily		+		+	+		+
	Panicoideae	Subfamily	+	+	+	+	+	+	+
	Pooideae	Subfamily	+	+	+	+	+	+	+
Palms	Arecaceae	Family			+	+		+	+
Sedges	Cyperaceae	Family					+	+	+
Grasses	Poaceae	Family	+	+	+	+	+	+	+
Dicots	Dicotyledons	Dicot/monocot	+	+	+	+	+	+	+

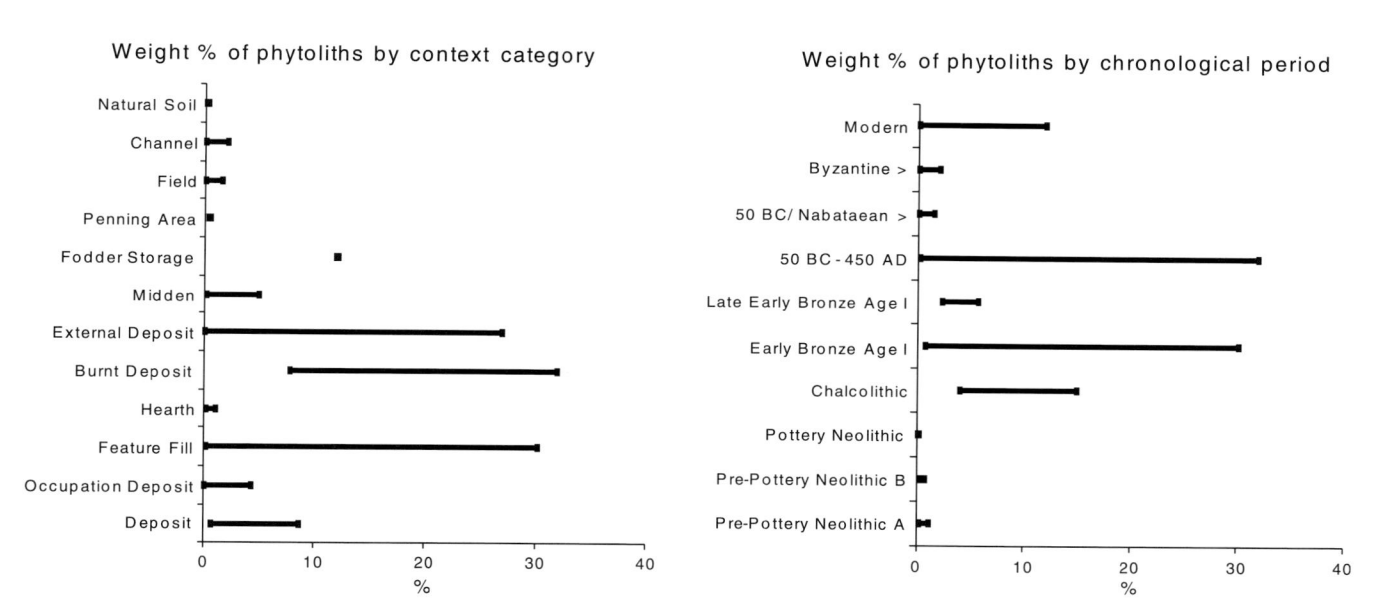

Figure 23.5. Weight percent of phytoliths by context category and chronological period.

methodology proposed by Rosen and Weiner (1994) and was applied by the creation of a plant water availability index. The highest water availability was obtained for the channel category, a context type that is in fact the most likely to have had high water contents. This example, amongst others, convinces the authors of the reliability of a plant water availability index based on the work described in Chapter 21 of this volume. The applicability of this index in other climatic regions of the world could be the subject of further studies.

The most useful way to present the water availability index proved to be the percentages of long-cell phytoliths relative to the sum of short and long cells. It must, however, be stressed that the error associated to this calculation is sensitive to the number of phytoliths counted. Future research should thus estimate the minimum number of short and long cells to count in order to obtain a statistically reliable water availability index. It should be noted that this estimation will be specific to the index and not relevant to other types of analysis (Stromberg, 2009).

The application of ordination analysis to phytolith assemblages provided equally interesting results. As for the water availability index, however, the error margins associated with this numerical approach remain to be assessed (Lytle and Wahl, 2005). The fact that phytolith types from similar taxonomic origin remain grouped in the ordination analysis (Figure 23.8) gives confidence in the ability of phytolith analysis to capture signals from plants. In fact, there is a long series of bias from plant growth to the analysis that could conceal or blur this signal. This in turn informs us that the distinction between monocotyledon and dicotyledon phytoliths is a basic but reliable way to classify the assemblages. Because dicotyledons comprise all

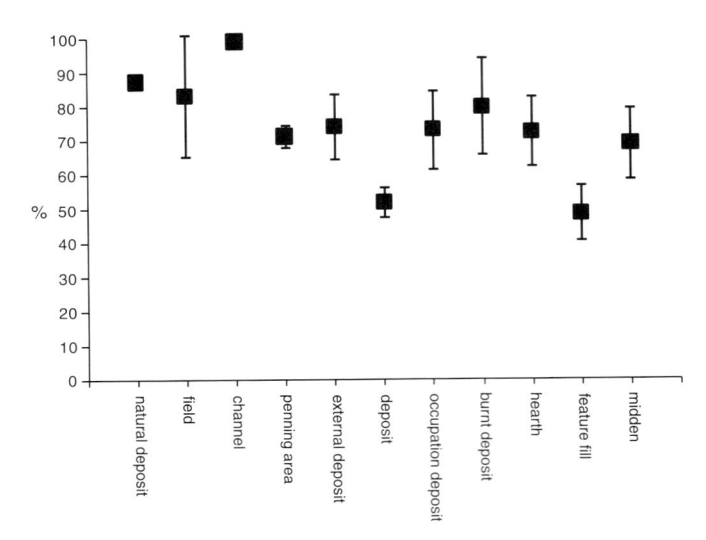

Figure 23.6. Water availability index showing percentage of long-cell phytoliths relative to the sum of short and long cells by context category.

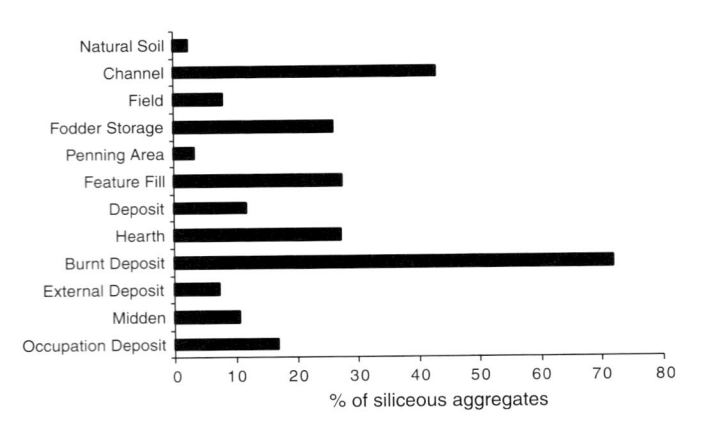

Figure 23.7. Woody material index showing percent of siliceous aggregates by context category.

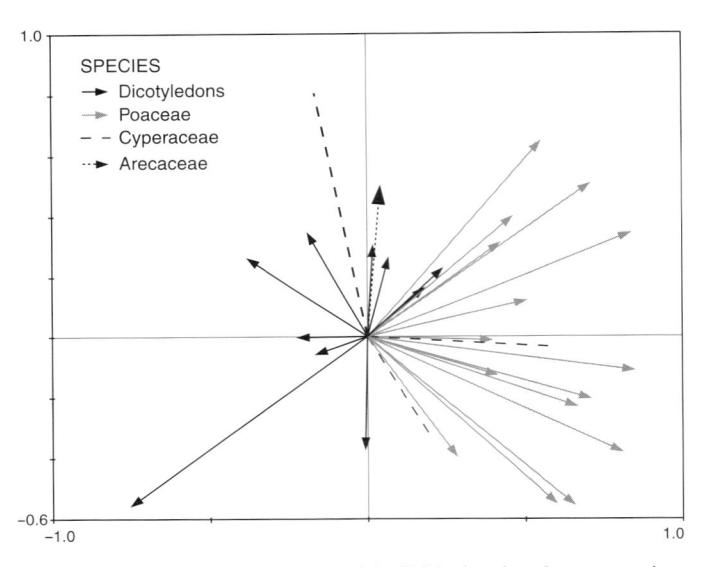

Figure 23.8. Plot of two first axes of the PCA showing the taxonomic origin of the phytolith types. The individual vectors (phytolith types) are unlabelled for clarity reasons.

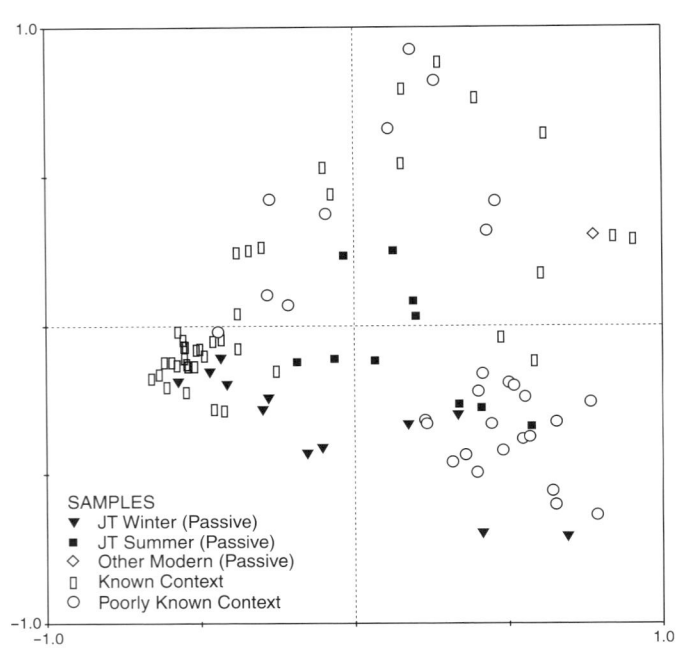

Figure 23.9. Plot of two first axes of the PCA showing the archaeological samples (used for the analysis) and the ethnoarchaeological samples (passively plotted) (JT, Juma's tent).

trees (but the palms), shrubs and forbs, their abundance can suggest wood material or agriculture weeds, for instance. Within the monocotyledons, the cereal family (Poaceae) shows the most consistent signal. This family has domesticates but also numerous wild species. The sedge family (Cyperaceae) does not show a very consistent pattern with the ordination analysis. This may indicate that certain part of the sedges had been brought preferentially to the sampling localities, but our knowledge in this field is very limited (Ollendorf, 1992) and could be the subject of further research. There are, for instance, archaeological and ethnographical evidences for the use of sedge *Cyperus esculentus* roots as food, and the leaves of most members of the sedge family can be used for basket making.

The results from the comparison of the Bedouin tent sites suggest that the two tents systematically have different assemblages. However, these differences appear very subtle, and it may be the case that the methodologies employed accumulate a larger error margin than the variation observed. At the time of sampling, Juma was living in the winter tent and the summer tent had been unoccupied for approximately six months. If the differences highlighted in the PCA prove reliable, this would demonstrate that the phytolith record can be altered even in a relatively short period of time.

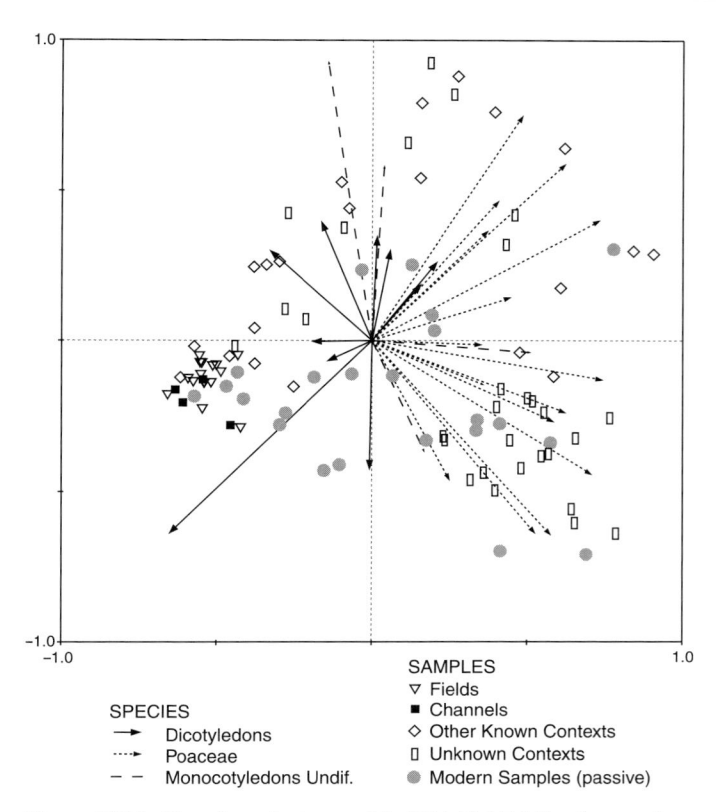

Figure 23.10. Plot of two first axes of the PCA highlighting the samples from fields and channel.

23.9.2 Phytoliths by context category

In Figure 23.6, it is apparent that the channels have the consistently highest percent of elongate dendriforms, indicating a higher availability of water. This category comprises samples from the channels at WF4 and also modern samples taken from Wadi Dana in the Wadi Faynan region. It is likely that the samples taken from the modern wadi bed would have benefited not only from the increased water availability but also from higher levels of silica as a result of phytolith dissolution from plant material carried in the wadi flow. This result lends confidence to the method developed in Chapter 20 of this volume and demonstrates that the percent of elongate dendriforms is increased in assemblages taken from areas of high water availability.

PCA found that the context category that clustered most closely together, and hence showed the most similarities in samples, was the field category (see Figure 23.10). This comprises samples from WF4 and Humayma which largely consisted of phytoliths from dicotyledons (shrubs and trees) and contained very few morphotypes formed in monocotyledons (grasses and sedges). This supports the hypothesis that olives were being grown at Humayma but contradicts the theory of Mattingly et al. (2007) which proposes that some of the fields at WF4 were being used to grow cereals. Instead, it suggests that some tree crops were being grown at both Humayma and WF4.

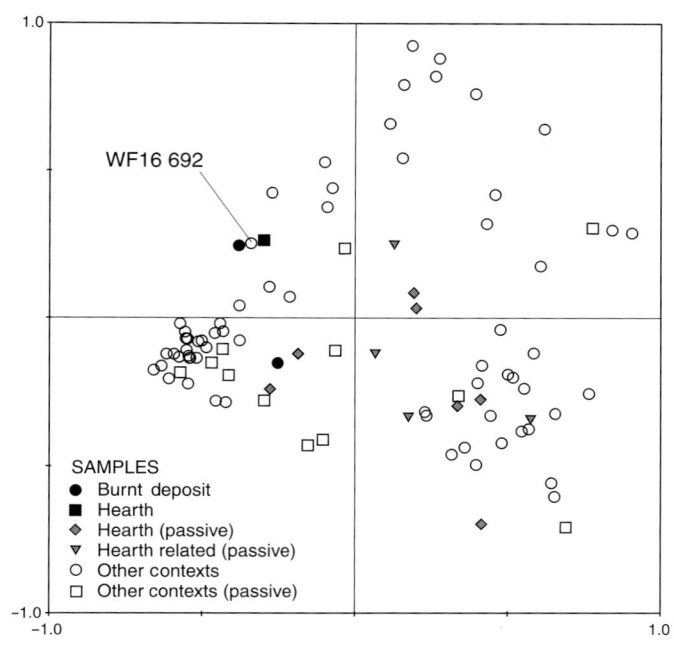

Figure 23.11. Plot of two first axes of the PCA highlighting the affinites of hearth and other fire-related contexts.

These dicotyledon phytoliths found in the field samples consisted largely of phytoliths such as polyhedral shaped psilates and irregular shapes from dicotyledons (plateys) but contained few siliceous aggregates (see Figure 23.3 for images of these forms). It is probable that these two former types are formed in the leaves of the dicotyledons rather than in the wood or bark (Albert et al., 2003). Siliceous aggregates, on the other hand, are believed to form in the wood (Albert et al., 2003). Figure 23.7 shows the percent of siliceous aggregates by context category. From this it is apparent that the burnt deposits, which are comprised of burning deposits from the domed structure at Humayma, have the highest value. This suggests that some of the building material used for the domed structure must have been wood. These deposits also have the greatest weight percent of phytoliths as demonstrated in Figure 23.5. Channels have the second highest level of siliceous aggregates which could be the result of woody material washed down with the wadi flow.

Hearths do not have particularly high value for siliceous aggregates. Nine hearth deposits were sampled for phytoliths, seven of which came from the modern Bedouin tents where dung is used for fuel. Unfortunately, there was no fodder available for us to examine, but a phytolith sample taken from the fodder storage areas produced an assemblage which consisted of dicotyledons (see Table 23.3). Analysis of the penning areas, however, produced samples that also contained grass and sedge phytoliths. This demonstrates that the animals were grazing on grasses while freely roaming during the day time. Only two archaeological samples were taken from presumed hearth contexts, both of which came from WF16. These two samples also consisted of a

Table 23.3 *Single-cell phytoliths by family and subfamily for the various context categories*

	Poaceae	Cyperaceae	Arecaceae	Dicotyledons	Panicoid.	Pooid.	Chloridoid./Arundinoid.
Occupation deposit	39.5	0.0	0.0	59.6	0.5	8.7	0.4
Midden	60.9	0.0	0.0	35.7	0.2	14.0	0.4
External deposit	45.7	0.0	1.5	50.1	0.2	11.4	0.2
Burnt deposit	18.6	0.0	0.0	81.4	0.0	5.2	0.0
Hearth	38.6	0.2	0.0	61.1	0.6	6.8	1.8
Deposit	65.4	0.1	0.0	24.3	0.6	27.1	1.1
Feature fill	68.0	0.1	0.0	25.4	0.6	28.9	1.8
Penning area	90.6	0.4	0.0	9.0	2.4	21.0	1.8
Fodder storage	0.0	0.0	0.0	100.0	0.0	0.0	0.0
Field	0.5	0.0	0.0	99.5	0.0	0.0	0.0
Channel	23.3	0.0	0.0	76.6	0.1	0.0	0.0
Natural soil	84.9	0.0	0.0	15.1	1.2	2.8	4.4

variety of dicotyledon phytoliths containing forms from both leaves and wood/bark. This suggests that: (1) animal dung was also used for fuel at WF16; or (2) smaller branches and twigs with the leaves still attached were used, possibly as kindling; or (3) phytoliths from leaves became incorporated into hearths. PCA of the hearths and burnt deposit demonstrated that one of the samples taken from WF16 (692) which was from a midden context has affinities with one of the hearths and one of the burnt deposits, suggesting that this was a hearth rake-out deposit that was dumped in the midden.

23.9.3 Chronological changes

The comparison of weight percent of phytoliths by chronological period demonstrates that the samples from the Neolithic levels have a lower weight percent than the later periods (Figure 23.5). This suggests that there are preservational issues due to the long time span, a hypothesis which is supported by the high number of single-celled phytoliths that were pitted and degraded (data not shown). The intensity and nature of plant use may also be a factor in explaining the low phytolith densities, and it is possible that Neolithic people may have been exploiting plants less intensively than people in later periods. The lower weight percent of phytoliths in the Neolithic samples could also be simply attributable to a lower degree of activity on site. It is possible that the Neolithic settlements in Wadi Faynan were only seasonally occupied by relatively small groups of people as compared with those of later periods. The lower weight percent seen in the samples from the Classical period is probably attributable to the fact that these come mainly from field systems which had low densities of phytoliths compared with other context categories (see Figure 23.5).

The chronological changes in taxa found between the samples contribute to our understanding of plant use through time in Jordan. It is notable that in the PPNA no phytoliths from cereals were found, while in the samples from the PPNB barley was present. This is supported by the macrobotanical and phytolith evidence from the evaluation of WF16 which also found little evidence for cereal exploitation (Jenkins and Rosen, 2007; Kennedy, 2007). In contrast, as stated above, barley, emmer wheat and einkorn wheat were all found at Ghuwayr 1 (Austin and Kennedy, cited in Meadows, 2005). However, the absence of findings for the earlier samples cannot be interpreted as an absence of cereal as it may also represent a sampling bias.

PCA demonstrated that in addition to the clustering observed for the Classical period sites discussed above, the Early Bronze Age and Chalcolithic samples show little variability (see Figure 23.12). The Neolithic samples are particularly interesting because a progression can be observed from the PPNA to the PPNB, with the latter plotting more closely with the Pottery Neolithic samples, showing the gradual change in plant use through time. In contrast, many of the PPNA samples show similarities with the Classical period sites, presumably because of the high levels of dicotyledon phytoliths found in both time periods. This may also be due to the fact that the earliest samples have a higher component of natural vegetation, as open fields would have. For instance, Rosen (2005, p. 211) refers to this phenomenon as 'overall phytolith noise' and Tsartsidou et al. (2008) deal with this issue by subtracting natural vegetation assemblages from their archaeological samples in order to highlight the anthropogenic component of the assemblages.

The presence of palm phytoliths interpreted as date palm (*Phoenix dactylifera*), in the Pottery Neolithic, as seen in Table 23.2, is represented by 27 phytoliths from Tell Wadi Feinan. This is one of the earliest records of this species in Jordan with the only earlier record coming from Ghuwayr 1 (Simmons and Najjar, 2003). Other than this evidence from Wadi Faynan, the earliest date palms recorded are from the Chalcolithic period,

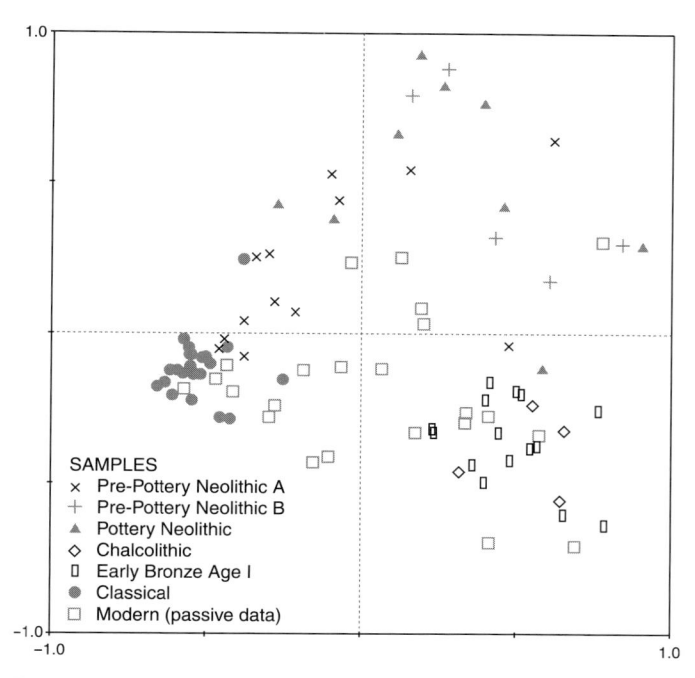

Figure 23.12. Plot of two first axes of the PCA showing the chronological patterns.

although this is based on only two date stones from Teleilat Ghassul (Meadows, 2005). As discussed above, Tell Wadi Feinan has deposits dating to the Chalcolithic and the Byzantine periods and so the intrusion of these phytoliths from later deposits is a possibility. Twenty-two of the date palm phytoliths came from eight samples from Tell Wadi Feinan, taken from a cleaned section. Nineteen of these phytoliths came from the bottom three samples making post-depositional downwards movement of the phytoliths from Chalcolithic deposits unlikely and contamination from Byzantine deposits most improbable. A further 15 samples came from the bottom of a small sondage that was dug into what were presumed occupation deposits within a Pottery Neolithic structure which had been left uncovered after the excavation. If these are secure occupation deposits then a Byzantine or Chalcolithic origin can be discounted.

The original natural distribution of *P. dactylifera* is unknown, with cultivation extending the distribution of this species far beyond its presumed original range (Barrow, 1998). However, the recovery of date palm phytoliths in Wadi Faynan in both the Middle PPNB and the Pottery Neolithic suggests that they were naturally occurring in the region. This would have been possible, because Wadi Faynan would have been wetter in the Neolithic than it is at present (see Chapter 15, this volume). An Arab proverb describes the date palm's favoured growing conditions: 'its feet shall be in a stream of water and its head in the furnace of heaven'. This is because it thrives in hot arid conditions providing that there is some groundwater available, such as in wadis,

crevices and rocky ravines (Barrow, 1998). There is some evidence to suggest that it could have originated in southwest Asia. Solecki and Leroi-Gourhan (1961) found *Phoenix* pollen which compared favourably with *P. dactylifera* in Mousterian deposits at Shanidar cave (Northern Iraq). Obviously the fruit of the date palm is of great nutritional value but the leaves can also be useful, for example as a building material or for making baskets, rope and thatch. In addition, all parts of the plants, but particularly the leaf bases, are used as fuel.

23.10 CONCLUSION

One of the aims of this study was to assess how effectively the water availability index outlined in Chapter 21 can be applied to the archaeological and ethnoarchaeological assemblages. Through our analysis of the phytolith assemblages by context category, it has been demonstrated that this method is relatively effective for this purpose. Elongate dendriforms (the type that formed in high numbers in plants that received optimum water) were most abundant in water channels, lending confidence to the use of this methodology on archaeological assemblages. This study also proved successful in reconstructing the palaeoeconomy with distinct changes being seen in taxa through time and species such as date palm being recovered from sites which had no prior evidence for the use of this plant. Phytolith assemblages from the fields from Humayma and WF4 suggest that tree crops were being grown at both locations, supporting the hypothesis that olives were being grown at Humayma (see Foote, Chapter 19, this volume) while questioning that of Mattingly *et al.* (2007) who suggested that cereals were being grown in some of the fields at WF4. The inclusion of ethnoarchaeological phytolith assemblages proved informative about how the phytolith record can be altered by taphonomic factors over only a short period of site abandonment. It is our future aim to continue working on ethnoarchaeological phytolith assemblages, not only to address taphonomic issues, but also to explore the phytolith signatures of known contexts to help identify these in the archaeological record.

REFERENCES

Albert, R. M., O. Bar-Yosef, L. Meignen and S. Weiner (2003) Quantitative phytolith study of hearths from the Natufian and middle palaeolithic levels of Hayonim Cave (Galilee, Israel). *Journal of Archaeological Science* **30**: 461–480.

Albert, R. M., O. Lavi, L. Estroff *et al.* (1999) Mode of occupation of Tabun Cave, Mt Carmel, Israel during the Mousterian Period: a study of the sediments and phytoliths. *Journal of Archaeological Science* **26**: 1249–1260.

Al-Najjar, M., A. Abu Dayya, E. Suleiman, G. Weisgerber and A. Hauptmann (1990) Tell Wadi Faynan: the first Pottery Neolithic tell in the south of Jordan. *Annual of the Department of Antiquities of Jordan* **34**: 27–56.

Araus, J. L., A. Febrero, R. Buxo *et al.* (1997) Identification of ancient irrigation practices based on the carbon isotope discrimination of plant seeds: a case study from the south-east Iberian Peninsula. *Journal of Archaeological Science* **24**: 729–740.

Arpin, T. (2005) Micromorphological analysis of four early Neolithic sites. Unpublished PhD thesis: University of Boston.

Austin, P. (2007) The wood charcoal remains. In *The Early Prehistory of Wadi Faynan, Southern Jordan: Archaeological Survey of Wadis Faynan, Ghuwayr and al-Bustan and Evaluation of the Pre-Pottery Neolithic A Site of WF16*, ed. B. L. Finlayson and S. Mithen. Oxford: Oxbow Books pp. 408–419.

Barker, G., O. Creighton, D. Gilbertson *et al.* (1997) The Wadi Faynan Project, southern Jordan: a preliminary report on geomorphology and landscape archaeology. *Levant* **29**: 19–40.

Barker, G. and D. Gilbertson, eds. (2000) *The Archaeology of Drylands: Living at the Margin*. London, UK; New York: Routledge.

Barker, G., D. Gilbertson and D. Mattingly (2007) The Wadi Faynan landscape survey: research themes and project development. In *Archaeology and Desertification: The Wadi Faynan Landscape Survey, Southern Jordan*, ed. G. Barker, D. Gilbertson and D. Mattingly. Oxford, UK and Amman, Jordan: Council for British Research in the Levant in Association with Oxbow Books pp. 3–23.

Barrow, S. D. (1998) A monograph of *Phoenix* L. (Palmae: Coryphoideae). *Kew Bulletin* **53**: 513–575.

Bartoli, F. and L. P. Wilding (1980) Dissolution of biogenic opal as a function of its physical and chemical properties. *Soil Science Society of America Journal* **44**: 873–878.

Bedal, L.-A. and J. G. Schryyer (2007) Nabatean landscape and power: evidence from the Petra garden and pool complex. In *Crossing Jordan: North American Contributions to the Archaeology of Jordan*, ed. T. E. Levy, P. M. Daviau, R. W. Younker and M. Sha'er. London: Equinox Publishing pp. 373–383.

Behre, K. E. (2008) Comment on: 'Mesolithic agriculture in Switzerland? A critical review of the evidence' by W. Tinner, E. H. Nielsen and A. F. Lotter. *Quaternary Science Reviews* **27**: 1467–1468.

Binford, L. R. (1977) General Introduction. In *For Theory Building in Archaeology*, ed. L. R. Binford. New York: Academic Press pp. 1–13.

Birks, H. J. B. and A. D. Gordon (1985) *Numerical Methods in Quaternary Pollen Analysis*. London: Academic Press.

Bourke, S. (2002) The origins of social complexity in the southern Levant: new evidence from Teleilat Ghassul, Jordan. *Palestine Exploration Quarterly* **134**: 2–27.

Bourke, S. (2008) The Chalcolithic. In *Jordan: An Archaeological Reader*, ed. R. Adams. London: Equinox.

Brown, D. A. (1984) Prospects and limits of a phytoliths key for grasses in the central United States. *Journal of Archaeological Science* **11**: 345–368.

Caldwell, R. C. and T. Gagos (2007) Beyond the rock: Petra in the sixth century CE in the light of the papyri. In *Crossing Jordan: North American Contributions to the Archaeology of Jordan*, ed. T. E. Levy, P. M. Daviau, R. W. Younker and M. Sha'er. London: Equinox Publishing pp. 425–433.

Cartwright, C. (2002) Grape and grain: dietary evidence from an Early Bronze Age store at Tell es-Sa'idiyeh, Jordan. *Palestine Exploration Quarterly* **134**: 98–117.

Colledge, S. (1998) Identifying pre-domestication cultivation using multivariate analysis. In *The Origins of Agriculture and Crop Domestication*, ed. A. B. Damania, J. Valkoun, G. Willcox and C. O. Qualset. Aleppo, Syria: ICARDA pp. 121–131.

Colledge, S. (2001) *Plant Exploitation on Epipalaeolithic and early Neolithic site in the Levant*. Oxford: British Archaeological Reports.

Colledge, S. (2002) Identifying pre-domestication cultivation in the archaeobotanical record using multivariate analysis: presenting the case for quantification. In *The Dawn of Farming in the Near East*, ed. R. T. J. Cappers and S. Bottema. Berlin: ex Oriente pp. 141–151.

Colledge, S., J. Conolly and S. Shennan (2004) Archaeobotanical evidence for the spread of farming in the eastern Mediterranean. *Current Anthropology* **45**: S35–S58.

Cronquist, A. (1981) *An Integrated System of Classification of Flowering Plants*. New York: Columbia University Press.

Crook, D. (2009) Hydrology of the combination irrigation system in the Wadi Faynan, Jordan. *Journal of Archaeological Science* **36**: 2427–2436.

de Contenson, H. (1960) Three soundings in the Jordan Valley. *Annual of the Department of Archaeology* **4–5**: 12–98.

Donaldson, M. L. (1985) The plant remains. In *Excavation at the Pre-Pottery Neolithic B Village of 'Ain Ghazal (Jordan) 1983*, ed. G. Rollefson, A. J. Simmons, M. L. Donaldson *et al.* Berlin: Mitteilungen der Deutsche Orient-Gesellschaft.

Faegri, K. and J. Iversen (1989) *Textbook of Pollen Analysis*. Chichester: Wiley.

Falconer, S. (2008) The Middle Bronze Age. In *Jordan: An Archaeological Reader*, ed. R. Adams. London: Equinox pp. 262–280.

Falconer, S., P. L. Fall and J. E. Jones (2007) Life at the foundation of Bronze Age civilisation: agrarian villages in the Jordan Valley. In *Crossing Jordan: North American Contributions to the Archaeology of Jordan*, ed. T. E. Levy, P. M. Daviau, R. W. Younker and M. Sha'er. London: Equinox Publishing pp. 261–268.

Fall, P. L., S. Falconer and P. C. Edwards (2007) Living on the edge: settlement and abandonment in the Dead Sea Plain. In *Crossing Jordan: North American Contributions to the Archaeology of Jordan*, ed. T. E. Levy, P. M. Daviau, R. W. Younker and M. Sha'er. London: Equinox Publishing pp. 225–232.

Finlayson, B. L. and S. Mithen (2007) The Early Prehistory of Wadi Faynan, Southern Jordan: Archaeological Survey of Wadis Faynan, Ghuwayr and Al Bustan and Evaluation of the Pre-Pottery Neolithic A Site of WF16. In *Wadi Faynan Series 1, Levant Supplementary Series 4*, ed. B. Finlayson and S. Mithen. Oxford and Amman, Jordan: Oxbow Books and the Council for British Research in the Levant.

Foote, R. (2007) From residence to revolutionary headquarters: the early Islamic Qasr and Mosque complex at al-Humyma and its 8th century context. In *Crossing Jordan: North American Contributions to the Archaeology of Jordan*, ed. T. E. Levy, P. M. Daviau, R. W. Younker and M. Sha'er. London: Equinox Publishing pp. 457–465.

Gebel, H. G. K. (2004) There was no centre: the polycentric evolution of the Near Eastern Neolithic. *Neolithics* **1/04**: 28–32.

Grigson, C. (1995) Plough and pasture in the early economy of the southern Levant. In *Archaeology of Society in the Holy Land*, ed. T. E. Levy. London: Leicester University Press pp. 245–268.

Gustavson-Gaube, C. (1985) Tell Esh-Shuna North 1984: a preliminary report. *Annual of the Department of Archaeology* **29**: 43–87.

Gustavson-Gaube, C. (1986) Tell Esh-Shuna North 1985: a preliminary report. *Annual of the Department of Archaeology* **30**: 65–114.

Harris, D. R. (1989) An evolutionary continuum of people-plant interaction. In *Foraging and Farming: The Evolution of Plant Exploitation*, ed. D. R. Harris and G. C. Hillman. London: Routledge.

Harris, D. R. (1998) The origins of agriculture in southwest Asia. *The Review of Archaeology* **19**: 5–11.

Harvey, E. L. and D. Q. Fuller (2005) Investigating crop processing using phytolith analysis: the example of rice and millets. *Journal of Archaeological Science* **32**: 739–752.

Helback, H. (1966) Appendix A – Pre-Pottery Neolithic farming at Beidha: a preliminary report. *Palestine Exploration Quarterly* **98**: 61–66.

Helms, S. (1981) *Jawa: Lost City of the Black Desert*. New York: Cornell University Press.

Heywood, V. H., R. K. Brummitt, A. Culham and O. Sceberg (2007) *Flowering Plant Families of the World*. London: Royal Botanic Gardens, Kew.

Hill, M. O. and H. J. Gauch (1980) Detrended Correspondence Analysis: an improved ordination technique. *Vegetation* **42**: 47–58.

Hillman, G. C. (2000) Abu Hureyra 1: The Epipalaeolithic. In *Village on the Euphrates: From Foraging to Farming at Abu Hureyr*, ed. A. M. T. Moore, G. C. Hillman and A. J. Legge. New York: Oxford University Press pp. 327–398.

Hillman, G. C. and T. D. Davies (1992) Measured domestication rates in wild wheats and barley under primitive cultivation, and their archaeological implications. *Journal of World Prehistory* **4**: 157–222.

Hillman, G. C., R. Hedges, A. M. T. Moore, S. Colledge and P. Pettitt (2001) New evidence of Late Glacial cereal cultivation at Abu Hureyra on the Euphrates. *The Holocene* **11**: 383–393.

Holden, T. (in press) The plant remains from Tell esh-Shuna North, Jordan Valley. In *The Tell esh-Shuna Monograph* Chapter 14.

Hunt, C. O., H. A. Elrishi, D. D. Gilbertson *et al.* (2004) Early-Holocene environments in the Wadi Faynan, Jordan. *The Holocene* **14**: 921–930.

Jackson, S. T. and J. W. Williams (2004) Modern analogue in Quaternary palaeoecology: here today, gone yesterday, gone tomorrow? *Annual Review of Earth Planetary Science* **32**: 494–537.

Jenkins, E. and A. Rosen (2007) The phytoliths. In *The Early Prehistory of Wadi Faynan, Southern Jordan: Archaeological Survey of Wadis Faynan, Ghuwayr and al-Bustan and Evaluation of the Pre-Pottery Neolithic A Site of WF16*, ed. B. L. Finlayson and S. Mithen. Oxford: Oxbow Books pp. 429–436.

Jones, L. H. P. and K. A. Handreck (1967) Silica in soils, plants and animals. *Advances in Agronomy* **19**: 107–149.

Kennedy, A. (2007) The plant macrofossils. In *The Early Prehistory of Wadi Faynan, Southern Jordan: Archaeological Survey of Wadis Faynan, Ghuwayr and al-Bustan and Evaluation of the Pre-Pottery Neolithic A Site of WF16*, ed. B. L. Finlayson and S. Mithen. Oxford: Oxbow Books pp. 420–428.

Krauskopf, K. B. (1956) Dissolution and precipitation of silica at low temperatures. *Geochimica et Cosmochimica Acta* **10**: 1–26.

Kuijt, I. and B. L. Finlayson (2009) New evidence for food storage and pre-domestication granaries 11,000 years ago in the Jordan Valley. *Proceedings of the National Academy of Science* **106**: 10966–10970.

Kuijt, I. and A. N. Goring-Morris (2002) Foraging, farming and social complexity in the Pre-Pottery Neolithic of the South-Central Levant: a review and synthesis. *Journal of World Prehistory* **16**: 361–439.

Lepš, J. and P. Šmilauer (2003) *Multivariate Analysis of Ecological Data using CANOCO*. Cambridge: Cambrige University Press.

Levy, T. E. (1995) Cult, metallurgy and rank societies – Chalcolithic period. In *Archaeology of Society in the Holy Land*, ed. T. E. Levy. Leicester: Leicester University Press pp. 226–244.

Lytle, D. E. and E. R. Wahl (2005) Palaeoenvironmental reconstruction using the modern analogue technique: the effects of sample size and decision rules. *The Holocene* **15**: 554–566.

Mabry, J. M., L. Donaldson, K. Gruspier *et al.* (1996) Early town development and water management in the Jordan Valley: investigations at Tell el-Handaquq North. *Annual of the Department of Antiquities of Jordan* **53**: 115–154.

Madella, M., A. Alexandre, T. Ball and ICPN Working Group (2005) International Code for Phytolith Nomenclature 1.0. *Annals of Botany* **96**: 253–260.

Madella, M., M. K. Jones, P. Echlin, A. Powers-Jones and M. Moore (2009) Plant water availability and analytical microscopy of phytoliths: implications for ancient irrigation in arid zones. *Quaternary International* **193**: 32–40.

Mattingly, D., P. Newson, O. Creighton *et al.* (2007) A landscape of Imperial Power: Roman and Byzantine Phaino. In *Archaeology and Desertification. The Wadi Faynan Landscape Survey, Southern Jordan*, ed. G. Barker, D. Gilbertson and D. Mattingly. Oxford: Oxbow Books pp. 305–348.

McCreery, D. W. (2003) The paleoethnobotany of Bab edh-Dhra. In *Bab edh-Dhra: Excavations at the Town Site (1975–81)*, ed. W. E. Rast and R. T. Schaub. Winona Lake, IN: Eisenbrauns pp. 449–463.

Meadows, J. (2005) Early farmers and their environment. Archaeobotanical research at Neolithic and Chalcolithic sites in Jordan. Unpublished PhD thesis: La Trobe University, Australia.

Mellaart, J. (1962) Preliminary report of the archaeological survey in the Yarmouk and Jordan Valley. *Annual of the Department of Archaeology* **6–7**: 126–157.

Metcalfe, C. R. (1960) *Anatomy of the Monocotyledons. I. Gramineae*. Oxford: Clarendon Press.

Mithen, S., B. L. Finlayson, A. Pirie, S. Smith and C. Whiting (2007) Archaeological Survey of Wadis Faynan, Ghuwayr, Dana and al-Bustan. In *The Early Prehistory of Wadi Faynan, Southern Jordan: Archaeological Survey of Wadis Faynan, Ghuwayr and al-Bustan and Evaluation of the Pre-Pottery Neolithic A Site of WF16*, ed. B. L. Finlayson and S. Mithen. Oxford: Oxbow Books pp. 47–114.

Moore, P. D., J. A. Webb and M. E. Collinson (1991) *Pollen Analysis*, 2nd edition. Oxford: Blackwell Scientific.

Najjar, M. (1994) *Ghwair I, A Neolithic Site in Wadi Feinan*, ed. S. Kerner. Amman: Al Kutba pp. 75–85.

Neef, R. (1987) Palaeoethnobotany. In Report on the First Two Seasons of excavations at Basta (1986–1987), ed. H. J. Nissen, H. J. Muheisen and H. G. K. Gebel. *Annual of the Department of Antiquities of Jordan* **31**: 79–119.

Neef, R. (1990) *Introduction, Development and Environmental Implications of Olive Culture: The Evidence from Jordan. Man's Role in the Shaping of the Eastern Mediterranean Landscape*, ed. S. Bottema, G. Entjes-Nieborg and W. van Zeist. Rotterdam: Balkema pp. 295–306.

Nesbitt, M. (2002) When and where did domesticated cereals first occur in southwest Asia? In *The Dawn of Farming in the Near East*, ed. R. T. J. Cappers and S. Bottema. Berlin: ex Oriente pp. 113–132.

Newson, P., G. Barker, P. Daly, D. Mattingly and D. Gilbertson (2007) The Wadi Faynan field system. In *Archaeology and Desertification: The Wadi Faynan Landscape Survey, Southern Jordan*, ed. G. Barker, D. Gilbertson and D. Mattingly. Oxford, UK and Amman, Jordan: Council for British Research in the Levant in Association with Oxbow Books pp. 141–176.

Oleson, J. P. (2007a) From Nabataean king to Abbasid caliph: the enduring attraction of Hawara/al-Humayma, a multi-cultural site in Arabia Petraea. In *Crossing Jordan: North American Contributions to the Archaeology of Jordan* ed. T. E. Levy, P. M. Daviau, R. W. Younker and M. Sha'er. London: Equinox Publishing.

Oleson, J. P. (2007b) Nabataean water supply, irrigation and agriculture: an overview. In *The World of the Nabataeans*, ed. K. D. Poltis. Stuttgart: Franz Steiner Verlag **2** pp. 217–249.

Ollendorf, A. (1992) Towards a classification scheme of sedge (Cyperaceae) phytoliths. In *Phytolith Systematics: Emerging Issues*, ed. G. Rapp and S. Mulholland. Berlin: Springer pp. 91–106.

Ollendorf, A. L., S. C. Mulholland and G. Rapp Jr (1988) Phytolith analysis as a means of plant identification: *Arundo donax* and *Phragmites communis*. *Annals of Botany* **61**: 209–214.

Palmer, C., H. Smith and P. Daly (2007) Ethnoarchaeology. In *Archaeology and Desertification. The Wadi Faynan Landscape Survey, Southern Jordan*, ed. G. Bakes. Oxford: Oxford Books pp. 369–395.

Pearsall, D. M., K. Chandler-Ezell and A. Chandler-Ezell (2003) Identifying maize in neotropical sediments and soils using cob phytoliths. *Journal of Archaeological Science* **30**: 611–627.

Pearsall, D. M., D. R. Piperno, E. H. Dinan *et al.* (1995) Distinguishing rice (*Oryza sativa* Poaceae) from wild Oryza species through phytolith analysis: results of preliminary research. *Economic Botany* **49**: 183–196.

Pearsall, D. M. and M. K. Trimble (1984) Identifying past farm activity through soil phytolith analysis: a case study from the Hawaiian Islands. *Journal of Archaeological Science* **11**: 119–133.

Philip, G. and D. Baird (1993) Preliminary report on the second (1992) season of excavations at Tell esh-Shuna North. *Levant* **15**: 13–36.

Philip, G. (2008) The Early Bronze Age I–III. In *Jordan: An Archaeological Reader*, ed. R. B. Adams. London/Oakville: Equinox pp. 161–226.

Piperno, D. R. (1988) *Phytolith Analysis: An Archaeological and Geological Perspective*. San Diego, CA: Academic Press.

Piperno, D. R. (2006) *Phytoliths: A Comprehensive Guide for Archaeologists and Palaeoecologists*. New York: Altamira Press.

Piperno, D. R., T. C. Andres and K. E. Stothert (2000) Phytoliths in Cucurbita and other neotropical curcurbitaceae and their occurrence in early archaeological sties from the Lowland American tropics. *Journal of Archaeological Science* **27**: 193–208.

Piperno, D. R., I. Holst, L. Wessel-Beaver and T. C. Andres (2002) Evidence for the control of phytolith formation in Cucurbita fruits by the hard rind (Hr) genetic locus: archaeological and ecological implications. *Proceedings of the National Academy of Sciences of the United States of America* **99**: 10923–10928.

Piperno, D. R. and D. M. Pearsall (1993) Phytoliths in the reproductive structures of maize and teosinte: implications for the study of maize evolution. *Journal of Archaeological Science* **20**: 337–362.

Piperno, D. R. and K. E. Stothert (2003) Phytolith evidence for Early Holocene Cucurbita domestication in Southwest Ecuador. *Science* **299**: 1054–1057.

Portillo, M., R. M. Albert and D. O. Henry (2009) Domestic activities and spatial distribution in Ain Abu Nukhayla (Wadi Rum, Southern Jordan): the use of phytoliths and spherulites studies. *Quaternary International* **193**: 174–183.

Purugganan, M. D. and D. Q. Fuller (2009) The nature of selection during plant domestication. *Nature* **457**: 843–848.

Rast, W. E. and R. T. Schaub (1978) A preliminary report from of excavations at Bab adh-Dhra, 1975. *Annual of the American Schools of Oriental Research* **43**: 1–32.

Rosen, A. (1999) Phytolith analysis in near eastern archaeology. In *Phytoliths: Applications in Earth Sciences and Human History*, ed. J. D. Meunier, F. Colin and A. A. Lisse. Netherlands: Balkema Publishers pp. 183–198.

Rosen, A. (2005) Phytolith indicators of plant and land use at Çatalhöyük. In *Inhabiting Çatalhöyük: Reports from the 1995–1999 Seasons*, ed.

I. Hodder. Cambridge and London: McDonald Institute for Archaeological Research and British Institute at Ankara Monographs.

Rosen, A. M. (1992) Preliminary identification of silica skeletons from Near Eastern archaeological sites: an anatomical approach. In *Phytolith Systematics*, ed. G. Rapp and S. C. Mulholland. New York: Plenum Press pp. 129–148.

Rosen, A. M. (1995) The social response to environmental change in Early Bronze Age Canaan. *Journal of Anthropological Archaeology* **14**: 26–44.

Rosen, A. M. and S. Weiner (1994) Identifying ancient irrigation: a new method using opaline phytoliths from emmer wheat. *Journal of Archaeological Science* **21**: 125–132.

Shahack-Gross, R., F. Marshall and S. Weiner (2003) Geo-ethnoarchaeology of pastoral sites: the identification of livestock enclosures in abandoned Maasai settlements. *Journal of Archaeological Science* **30**: 439–459.

Simmons, A. H. and M. Najjar (1996) Current investigations at Ghwair I, a Neolithic settlement in southern Jordan. *Neolithics* **2**: 6–7.

Simmons, A. H. and M. Najjar (1998) Al-Ghuwar I, a pre-Pottery Neolithic village in wadi Faynan, southern Jordan: a preliminary report of the 1996 and 1997/98 seasons. *Annual of the Department of Antiquities of Jordan* **42**: 91–101.

Simmons, A. J. and M. Najjar (2003) Ghuwayr I, a pre-pottery Neolithic B settlement in southern Jordan: report of the 1996–2000 campaigns. *Annual of the Department of Antiquities of Jordan* **47**: 407–430.

Simmons, A. H. and M. Najjar (2006) Ghwair I: a small, complex Neolithic community in southern Jordan. *Journal of Field Archaeology* **31**: 77–95.

Simmons, A. H. and M. Najjar (2007) Is big really better: life in the resort corridor-Ghwair I, a small but elaborate Neolithic community in southern Jordan. In *Crossing Jordan: North American Contributions to the Archaeology of Jordan*, ed. T. E. Levy, P. M. Daviau, R. W. Younker and M. Sha'er. London: Equinox Publishing pp. 233–244.

Solecki, R. and A. Leroi-Gourhan (1961) Paleoclimatology and archaeology in the Near East *Annals of the New York Academy of Sciences.* **95**: 729–739.

Stromberg, C. A. E. (2009) Methodological concerns for analysis of phytolith assemblages: does count size matter? *Quaternary International* **193**: 124–140.

ter Braak, C. J. F. and P. Šmilauer (2003) *CANOCO for Windows Version 4.51*. Wageningen: Plant Research International.

Tinner, W., E. H. Nielsen and A. F. Lotter (2007) Mesolithic agriculture in Switzerland? A critical review of the evidence. *Quaternary Science Reviews* **26**: 1416–1431.

Tinner, W., E. H. Nielsen and A. F. Lotter (2008) Evidence for Late-Mesolithic agriculture? A reply to Karl-Ernst Behre. *Quaternary Science Reviews* **27**: 1468–1470.

Tsartsidou, G., S. Lev-Yadun, N. Efstratiou and S. Weiner (2008) Ethnoarchaeological study of phytolith assemblages from an agro-pastoral village in Northern Greece (Sarakini): development and application of a phytolith difference index. *Journal of Archaeological Science* **35**: 600–613.

Tsartsidou, G., S. Lev-Yadun, N. Efstratiou and S. Weiner (2009) Use of space in a Neolithic village in Greece (Makri): phytolith analysis and comparison of phytolith assemblages from an ethnographic setting in the same area. *Journal of Archaeological Science* **36**: 2342–2352.

Tubb, H. J., M. J. Hodson and G. C. Hodson (1993) The inflorescence papillae of the Triticeae – a new tool for taxonomic and archaeological research. *Annals of Botany* **72**: 537–545.

Twiss, K. C., E. Seuss and R. M. Smith (1969) Morphological classification of grass phytoliths. *Soil Science of America Proceedings* **33**: 109–115.

Warnock, P. (1998) From plant domestication to phytolith interpretation: the history of Paleoethnobotany in the Near East. *Near Eastern Archaeology* **61**: 238–252.

Watson, P. (2008) The Byzantine period. In *Jordan: An Archaeological Reader*, ed. R. Adams. London: Equinox pp. 443–482.

Willcox, G. (1999) Agrarian change and the beginnings of cultivation in the Near East: evidence from wild progenitors, experimental cultivation and archaeobotanical data. In *The Prehistory of Food: Appetites for Change*, ed. C. Gosden and J. Hather. London: Routledge pp. 478–500.

Willcox, G. (2002) Charred plant remains from a 10th millennium BP kitchen at Jerf el Ahmar (Syria). *Vegetation History and Archaeobotany* **11**: 55–60.

Willcox, G. (2005) The distribution, natural habitats and availability of wild cereals in relation to their domestication in the Near East: multiple events, multiple centres. *Vegetation History and Archaeobotany* **14**: 534–541.

Willcox, G. H. (1981) Appendix D. Plant remains. In *Jawa: Lost City of the Black Desert*, ed. S. Helms. New York: Cornell University Press pp. 247–248.

Zhao, Z. J., D. M. Pearsall, R. A. Benfer and D. R. Piperno (1998) Distinguishing rice (*Oryza sativa* poaceae) from wild Oryza species through phytolith analysis, II: Finalized method. *Economic Botany* **52**: 134–145.

Zohary, D. (1992) Domestication of the Neolithic Near eastern crop assemblage. In *Préhistoire de l'agriculture: nouvelles approches expérimentales et ethnographiques*, ed. P. Anderson-Gerfaud. Paris: CNRS pp. 81–86.

Zohary, D. (1996) The mode of domestication of the founder crops of southwest Asian agriculture. In *The Origins and Spread of Agriculture and Pastoralism in Eurasia*, ed. D. R. Harris. London: UCL Press Ltd pp. 142–158.

Part VI
Society, economy and water today

24 Current water demands and future strategies under changing climatic conditions

Stephen Nortcliff, Emily Black and Robert Potter

ABSTRACT

Jordan is currently one of the world's 10 water-poorest nations, a situation which has been exacerbated by a rapidly growing population. This chapter reviews the current status of the water resources in the Hashemite Kingdom of Jordan in the context of the global pattern of increasing demand for water linked to population growth which is increasingly urban-based, increasing economic progress and the potential impact of climate change. Jordan's climate ranges from Mediterranean to arid with approximately 80% of the country receiving less than 100 mm of rainfall annually. Evaporation ranges from 2,000 mm in the north to 5,000 mm in the south. Current demand for water exceeds available renewable water resources, with the shortfall met by exploiting non-renewable reserves and water rationing. An additional water supply available for agricultural use is treated wastewater, but there are concerns and limitations to the use of this resource. The demand for water is predicted to continue to grow, and current predictions of future climate in the region indicate no change or a reduction in the quantity of rainfall and changes in the distribution through the year. These predictions of growth in water demand and shifts in rainfall patterns highlight the need to make more efficient use of the available water and to use other sustainable sources such as treated wastewater in a more effective manner.

24.1 INTRODUCTION

In 1987, The World Commission on Environment and Development (also known as the Brundtland Commission) warned in the final report, *Our Common Future*, that water was being polluted and water supplies were being overused in many parts of the world (World Commission on Environment and Development, 1987). Since this time there has been a steep increase in water use for food and energy production to meet the demands of a rapidly growing population and to enhance human wellbeing (WWAP (World Water Assessment Programme), 2006). These demands for water continue to rise, and there is increasing evidence of water scarcity in many parts of the world. This scarcity could seriously affect economic development, environmental quality, sustainable human activities and a wide range of economic, environmental and societal objectives (UNEP, 2007). In addition, in many countries, particularly in the less developed world, there is a rapid growth in the proportion of the population based in urban communities. This growth in urban populations places substantial pressures on water resources and the infrastructure to deliver fresh water to the population and to remove wastewater. It is projected that 75% of the population growth to 2015 will take place in cities, predominantly in the developing world (UNDP, 1998), and consequently the need to develop and maintain the infrastructure to deliver water efficiently and of the required quality to both established and new urban dwellers will place tremendous strain on existing facilities and require major investments in new facilities. In addition to the requirements for water provision there is also a rapidly growing requirement to develop and maintain systems to remove wastewater and treat it before returning it to the hydrological cycle. The nature and magnitude of these problems are highlighted in the United Nations Development Programme's *Human Development Report* (UNDP, 2006).

Available water resources to meet these demands have been declining for a number of years and continue to do so as a result of excessive withdrawal of surface water and groundwater, as well as decreased water runoff caused by reduced precipitation and increased evaporation attributed to global warming (Parry *et al.*, 2007; UNEP, 2007). In many parts of the world (for example, West Asia and the Indo-Gangetic Plain in South Asia) human consumption of water exceeds annual replenishment. It is anticipated that by 2025 almost one-fifth of the global population

Water, Life and Civilisation: Climate, Environment and Society in the Jordan Valley, ed. Steven Mithen and Emily Black. Published by Cambridge University Press.
© Steven Mithen and Emily Black 2011.

will be living in countries or regions with absolute water scarcity and two-thirds of the population will be living under conditions of water stress (UNWater, 2007). Jordan in the Middle East and North Africa (MENA) region provides an excellent example of a water-poor country which has seen rapidly changing demands on water resources in recent years. The case of Jordan highlights the need to manage water resources carefully, setting clear programmes for identifying priorities for their use. Zhang et al. (2007) also note that in many areas of the world, including the eastern Mediterranean, human actions have had a substantial influence on changing the precipitation patterns in the past century. These human-induced changes, coupled with the predictions of significantly drier conditions in this region under the IPCC A2 scenario, reinforce the need for an urgent strategy to ensure efficient and sustainable use of water resources.

24.2 THE HASHEMITE KINGDOM OF JORDAN

Jordan, or more properly the Hashemite Kingdom of Jordan, lies in the eastern part of the Mediterranean region to the east of Israel, and has a land area of approximately 90,000 km^2 (Figure 24.1). Topographically Jordan is diverse and the major topographic and geomorphologic features in Jordan control the drainage pattern. The overall drainage system in Jordan consists of two main flow patterns. The first drains rainfall towards the Jordan Rift Valley, through deeply incised wadis and rivers dissecting the Jordan Valley/Dead Sea escarpments, to discharge ultimately into the Dead Sea. The second drains rainfall through shallow streams and washes, which generally run eastwards from the western highlands towards the internal desert depressions and mudflats.

The climate of Jordan ranges from Mediterranean to arid. The Rift Valley and the highlands belong to the semi-arid to arid climate zone, which is largely affected by moist westerly air masses in winter. In summer, dry easterly and northeasterly desert winds affect the Kingdom. Winds are generally westerly to southwesterly. A Mediterranean climate dominates most of the highlands on both sides of the Jordan River and in the mountain chains east of the Dead Sea and Wadi Araba, extending as far south as Ras en Naqb. Dry summers with an average maximum annual temperature of 39 °C occur between April and September. In the winter months, from October or November until March, the average minimum annual temperature is 0–1 °C. In winter, the average mean daily temperatures recorded at Amman Airport and Deir 'Alla were 10 °C and 17 °C, respectively, for the period 1981–98.

The average temperature in the wet season is generally higher in the Jordan Valley than along the Mediterranean coast and falls

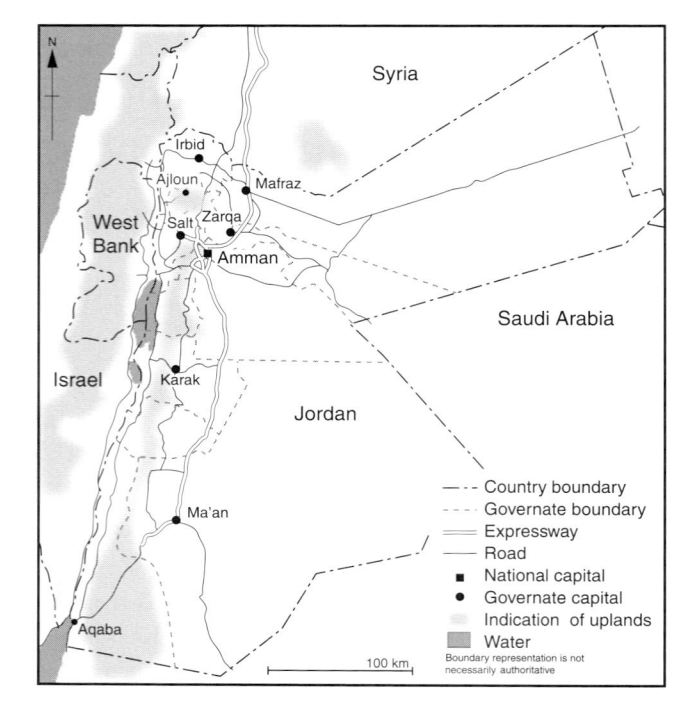

Figure 24.1. The general geography of Jordan.

again over the highlands and within the eastern plateau. The average annual evaporation rate ranges from 2,042 mm in Zarqa to 5,038 mm in Ma'an and from 2,594 mm in the Jordan Valley to 3,516 mm in the eastern hills.

Seasonal, uneven and fluctuating rainfall affects the country between October and May. Eighty per cent of the annual rainfall occurs between December and March. Average annual rainfall in Jordan (Figure 24.2) ranges from <50 mm in the eastern desert to approximately 600 mm over the Ajloun heights. Approximately 80% of the country receives less than 100 mm per year and less than 5% of the country receives more than 300 mm, which is considered to be the threshold below which it is not possible to grow wheat in the region. Table 24.1 presents the categories into which Jordan may be classified based on the rainfall distribution. For more details about the climate within Jordan, see Chapter 2 of this volume.

Rainfall is the only source of water supply in Jordan to recharge the groundwater aquifers – as described in Chapters 10, 12 and 13 for specific case studies. Rainfall is scarce and unevenly distributed over the country. The mountainous highlands along the Jordan Valley/Dead Sea/Wadi Araba depression receive the majority of total rainfall volume. Estimates of long-term records (1937/38–2000/01) of rainfall distribution over Jordan indicate that the average annual rainfall volume over the country is equivalent to around 8,360 million cubic metres (MCM), although probably in excess of 90% may be lost immediately to evaporation and makes little contribution to the overall water budget.

Table 24.1 *Climatic classification according to rainfall distribution*

Zone	Annual rainfall (mm year^{-1})	Area (km^2)	Area as a percentage of the total area of Jordan
Semi-humid	500–600	620	0.7%
Semi-arid	300–500	2,950	3.3%
Marginal	200–300	2,030	2.2%
Arid	100–200	20,050	22.3%
Desert	< 100	64,350	71.5%
Total		**90,000**	**100%**

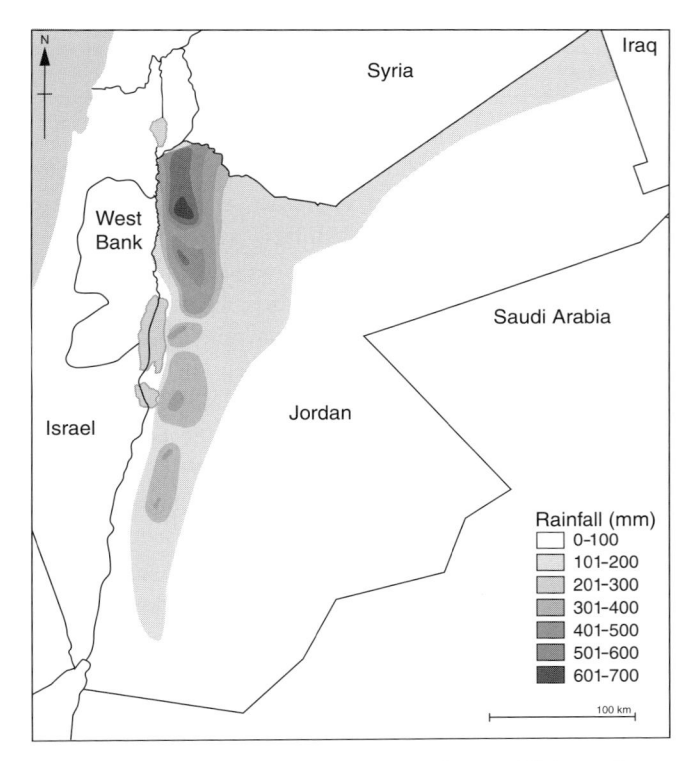

Figure 24.2. Rainfall distribution in Jordan (Source: Ministry of Environment, 2006).

The population of Jordan has seen a steep rise over the past 60 years, growing from an estimated figure of around 470,000 in the early 1950s to 900,000 in 1961 (1961 census), 2,150,000 in 1979 (1979 census) and 4.14 million in 1994 (1994 census). The Department of Statistics estimated the total population in 2002 to be 5.3 million, but recent influxes of refugee migrants from Iraq have significantly increased this number. The present indigenous rate of growth is estimated to be in the region of 2.8%. It is estimated that the total population will be around 10 million by *c.* 2020 if growth continues at the present rate. Whilst the indigenous growth rate is expected to slow gradually in the coming years, the influx of refugees is markedly increasing the rate of growth (Potter and Darmame, 2010).

24.3 WATER AND ITS MULTIPLE USES

Water is a key, indeed an essential element in biological, social and economic systems. Human beings, plants and animals require a minimum level of water supply to be able to survive, and their ability to function is often directly related to the extent to which this minimum requirement is exceeded. Within households the livelihoods, both with respect to domestic consumption and in many cases productive capability, depend upon the provision of adequate water. In the broader context water plays a key role in environmental systems, maintaining their productivity and providing the context for many other activities. Water cycles through the atmosphere, soil, plants and animals, and its absence introduces a major break in the food chain. But often water is scarce, and the intense competition for these resources means it will not be able to satisfy all these alternative uses simultaneously. As a consequence it will be necessary to prioritise the use, or alternatively, if there are water resources of different qualities, consider allocating water of different qualities to different uses. In any such allocation process a key consideration is whether access to water is based on need or on which demand sector is able to pay the highest price. Additionally, the geographical location of the water often dictates how the water can be used.

The hydraulic disconnectivity of the water resources of Jordan with the water demands has been discussed by Molle and Berkoff (2006). They argue that apparent bias in the allocation of water resources towards agriculture results from the source and quality of this water (reclaimed wastewater or periodic flood flows) which makes it unsuitable for supplying to the municipality network. The geographical location of the water in relation to the urban sites of demand also plays a key role in sectoral allocation. Fresh surface water and groundwater that is located far from urban areas is often not exploited for drinking water supply because of the high related pumping and transfer costs. This water is therefore available to be used for irrigation at its source. Whilst water for irrigation is often essential for the production of food, there is often a tendency to transfer water from agricultural use to municipal and, in some cases, industrial uses. These decisions are made on the basis that irrigation is of 'low value', and municipal and industrial uses are 'high value' (Meizen-Dick, 1997). Economists often promote the strategy that water should be allocated to the sector where the highest economic returns can be gained from the water inputs. This often results in a tendency to advocate the transfer of water from agriculture, where economic returns per cubic metre are low, to industry, tourism and municipality uses where economic returns are seen to be higher.

24.4 CURRENT WATER RESOURCES AND USES

In a global context, Jordan is characterised as a 'water-scarce' country (Winpenny, 2000). A 'water-scarce country' is one with less than 1,000 cubic metres of fresh water per person per year. Jordan is recognised as one of the 10 most water-deprived countries of the world; currently the *per capita* share of water is estimated to be of the order of 140 cubic metres *per capita* per year, well below the threshold of 1,000 cubic metres, and on current trends this will fall to less than 90 cubic metres per year by 2025 (not taking into account any change or variability in the climate). In contrast, the average citizen of the USA has more than 9,000 cubic metres of fresh water available per year (Winpenny, 2000).

Current annual water consumption in Jordan is estimated to be 955 MCM whilst the renewable freshwater resources (surface and groundwater) are estimated to be only in the region of 780 to 850 MCM per year, with approximately 65% derived from surface waters and 35% from groundwaters (see Table 24.2). There is consequently a gap between demand and supply which evidence suggests has been widening rapidly in recent years (Table 24.3). Some of these scarce resources are shared with neighbouring countries and as a consequence their management is not solely under national control. Surface water resources are spread across 15 major basins. Of these the Yarmouk River which forms part of the border with Syria to the north accounts for approximately 40% of the country's surface water resources (Figure 24.3). In the 1970s the total flow of the Yarmouk River was estimated to be in the region of 400 MCM, but because of developments in the upstream Syrian part of the catchment, the flow arriving in Jordan in the past 20 years has been substantially reduced and is now estimated at 150 MCM, and in addition, through international treaty agreements, Israel is allowed to access 100 MCM. The Yarmouk River is the main source of water for the King Abdullah canal in the Jordan Valley. The other principal basins include Zarqa, wadis adjacent to the Jordan River, Mujib, Hasi and Wadi Araba.

Whilst surface water resources are important and must be managed carefully, where there is no surface water, groundwater is the only source. Whilst there are 12 groundwater basins, most comprising several inter-related aquifer systems, identified within the country, approximately 80% of the known reserves are concentrated in the Yarmouk, Amman-Zarqa and Dead Sea basins (Figure 24.4). Current exploitation of these groundwater resources is at maximum capacity and in some cases is well above what is recognised as a safe yield (nationally it is estimated that extraction from groundwater resources is currently 50% above the safe extraction rate). In some cases this over-exploitation has resulted in significant decline in the quality of the groundwater resources. There has also been some contamination of near-surface aquifers as a result of the over-application of pesticides and fertilisers as part of intensive agricultural practices and through the seepage of septic tanks. Table 24.2 shows the allocation of water resources by source for the three major sectors of agricultural, municipal and industrial use.

Table 24.2 *Sources of water used by sectors in Jordan in 2000 (million cubic metres, MCM)*

Source	Municipal sector	Industrial sector	Agricultural sector	Total
Surface water				
Jordan Valley	38.5	2.5	121.2	162.2
Springs	14.9	0	38.0	52.9
Base flow and floods	0	0	56.5	56.5
Groundwater				
Renewable	176.4	29.6	206.0	412.0
Non-renewable	9.4	4.6	47.7	61.7
Treated wastewater	0	0	72.0	72.0
Total	239.2	36.7	541.4	
% of Total	29.3	4.4	66.3	

Recent trends have seen a rapid increase in the water consumption for municipal use, from around 20% in 1992 to in excess of 29% in 2000 (MWI (Ministry of Water and Irrigation), 2004). This has resulted from both the higher number of people in the urban areas and the greater water demand resulting from raised standards of living. Increases are expected to continue as urban populations rise and people demand more water to meet their needs. In 2000 approximately 94% of the water used for industry was from groundwater, with approximately 12.5% from non-renewable sources. With the planned economic expansion of the industrial sector this demand is likely to increase in the coming years, so there is a need to seek access to reliable sustainable water resources to service this industrial growth.

The demand for water has been growing rapidly in recent years and the current estimated demand is in excess of the total used (Table 24.3). There are few non-utilised conventional freshwater resources capable of development, although following the successful completion of the Unity Dam on the Yarmouk River in the north there are a number of proposals to construct dams in the eastern desert to promote groundwater recharge (MWI, 1997, 2004). In part this is a component of a national programme to encourage groundwater recharge, but there is anecdotal evidence that the relatively high price paid for groundwater is encouraging some landowners to recharge areas as an income

Table 24.3 *Jordan water use (MCM) by sector according to the Ministry of Water and Irrigation*

Year	Domestic	Industrial	Irrigation	Total	Domestic and industrial
1985	116	22	501	639	138
1986	135	23	461	619	158
1987	150	25	570	744	174
1988	165	39	613	817	204
1989	170	36	625	830	206
1990	176	37	657	870	212
1991	173	42	618	833	215
1992	207	35	709	951	251
1993	214	33	737	984	257
1994	216	25	669	909	250
1995	250	33	606	878	272
1996	236	36	610	882	272
1997	236	37	603	876	273
1998	236	38	561	835	274
1999	232	38	532	801	207
2000	239	37	541	817	276
2001	256	33	495	774	279
2005[a]	281	76	750	1107	357
2010[a]	380	93	746	1219	473
2015[a]	463	112	704	1279	575
2020[a]	517	130	665	1312	647

[a] Estimated by MWI (Source: MWI)

generation process. There is also a proposed scheme to transfer treated wastewater from Irbid to a proposed dam in one of the wadis leading to the Jordan Valley, with subsequent transfer of this water to supplement irrigation water needs for agriculture in the Valley (MWI, 1997, 2004). The pressures on water are marked in Amman, the capital, where the vast majority of households receive water only on one or two days per week (Potter et al., 2007a, 2007b).

Whilst the total supply of water is a critical problem with predicted demand well in excess of sustainable resources, the allocation of fresh water resources is equally problematic. To date approximately two-thirds of the country's water has been used for agricultural crop production, occasionally in priority over urban consumers, industrial needs and the growing demand from tourism. Whilst this may seem peculiar given the high value given to domestic and industrial uses, frequently the problem is geographical or perhaps more appropriately as a result of hydraulic disconnectivity; the water is not necessarily found where it can be made available to these high value uses. This is further exacerbated by the apparent 'loss' of approximately half of Amman's water in the distribution network, both through leakages and unaccounted usage. One immediate response to these shortfalls has been increased extraction from groundwater resources, with many groundwater aquifers extracted at more than double their sustainable yield (MWI, 2004). This over-extraction further exacerbates the problems faced in planning for future demands. Jordan's rapid population growth and the increasing importance of the industrial sector, coupled with the very limited available water resources, has resulted in water scarcity being identified as one of the major constraints to sustained economic growth and development. Addressing the demands for water and producing an appropriate framework for the sustainable provision of water has become a national priority of the highest order.

24.5 WATER IN JORDAN AND THE FUTURE

The MWI is the government body with responsibility for monitoring the water sector and managing the supply of fresh water and the wastewater system. The Ministry has recently established a Water Demand Management Unit to provide support for the optimisation of water use and the recommendation of enforcement and regulatory measures to prevent water misuse and wastage. The MWI embraces the two most important bodies dealing with water in Jordan: the Water Authority of Jordan with responsibilities for water supply and sewerage management, and the Jordan Valley Authority, with responsibilities for the

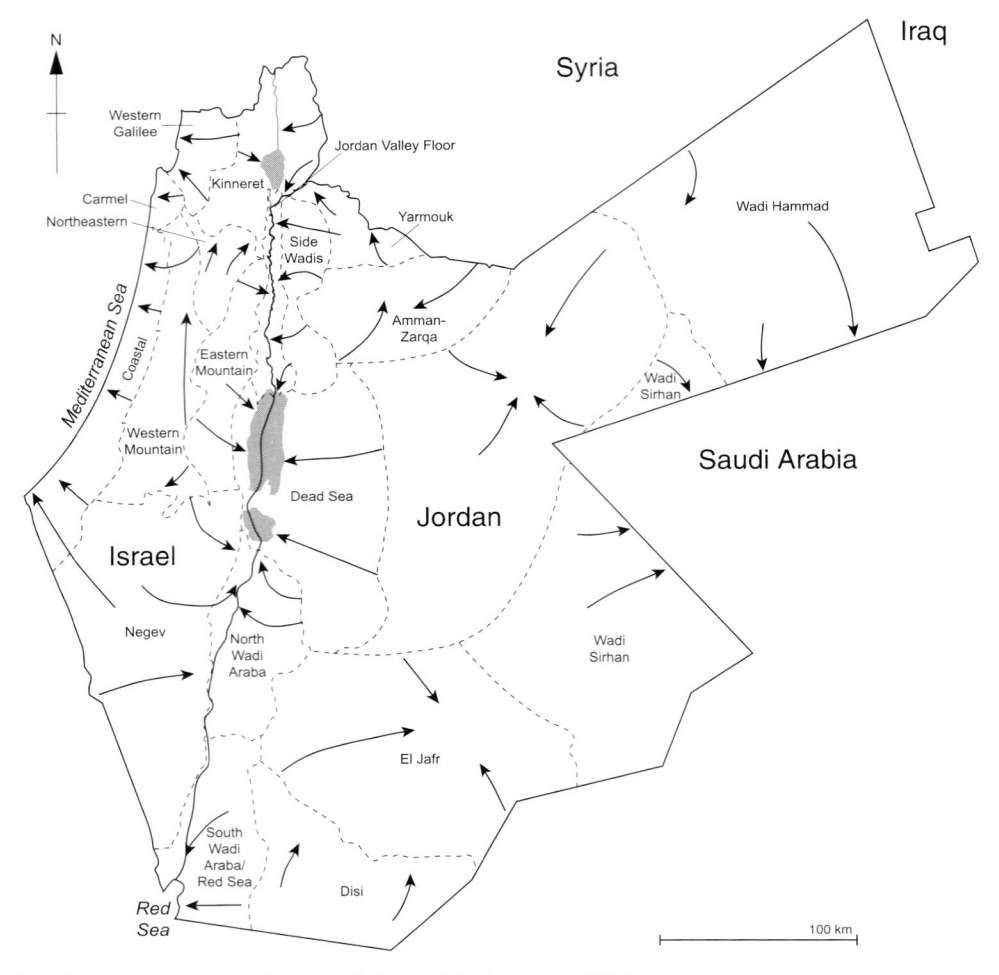

Figure 24.3. Jordan's surface water resources (Source: Ministry of Environment, 2006).

socio-economic development of the Jordan Rift Valley, and a particular responsibility for water development and irrigation which is key to continued economic wellbeing in this part of the country.

In 1997 the Government of Jordan, recognising the critical importance of water to the future of the kingdom and under the guidance of the World Bank, committed the MWI to the development of an integrated policy of water resource management. The Water Strategy for Jordan was given the undertaking to consider sustainability of water resource management in the context of the economic and social development of the country, but also in the broader environmental context (MWI, 1997).

The key features of the strategy are:

- to protect surface and groundwater resources by careful management of their use;
- to improve the efficiency in the management of both urban water and irrigation water;
- to develop an institutional capacity capable of managing water resources, supported by a legislative framework;
- to involve the private sector in the development of utilities with efficiency of water use and good financial management structures;
- to introduce a socially acceptable tariff system which might vary depending upon both the nature of the water use (e.g. domestic, industrial, irrigation, etc.) or type of water (e.g. surface water, groundwater, reclaimed wastewater).

Whilst these proposals seek to address the contemporary issues, there is, however, no consideration of the potential changes in water resources resulting from predicted shifts in climate. The principal outcome of this strategy statement was to move Jordan towards a situation where privatisation of the water sector was a real possibility, as witnessed by the last two points above. The provision in Amman was privatised in 1999. A key element of this strategy, developed subsequently, is the National Water Master Plan of Jordan (NWMP) which was developed in 2004 by the MWI with institutional support from the Deutsche Gesellschaft für Technische Zusammenarbeit (MWI, 2004) following from a national review of water resources in 1997

Figure 24.4. Jordan's groundwater resources (Source: Ministry of Environment, 2006).

(MWI, 1997). The NWMP is a dynamic document which exists in electronic form, providing the opportunity for regular updating both to provide a context for changing conditions and for forecasting outcomes in scenario analysis. The NWMP has no specific legal framework within which it operates, but the key role is to ensure 'Sustainable use of scarce water resources in line with a continuous improvement in living conditions for the country's population' (MWI, 2004). The NWMP provides forecasts and scenarios for 2005 through to 2020, but because of its dynamic nature it is capable of being iteratively upgraded, and current practice is to operate a 5-year rolling planning programme. A notable exception from the planning document, however, is any attempt to consider or take account of potential changes in rainfall as a result of changes in regional climatic conditions. Following broad guidance from the United Nations on plans to address the optimal use of water resources (UN, 1989), the principal objective of the Plan is to provide a framework for ensuring organised and integrated planning and implementation of a programme for water resource allocation and use which is consistent with the national objectives of Jordan.

The NWMP recognises that within Jordan water resources are administered by a number of governmental authorities and private organisations, although the overall control lies with MWI. The Water Authority of Jordan is responsible for public water supply and sewerage; the Jordan Valley Authority is concerned with the overall development of the Jordan Valley and within that development programme a key component is the management and development of water resources for domestic, irrigation and industrial uses. Within Amman, from 1999 to 2007, water supply and management was the responsibility of the private sector via LEMA (Lyonniase de Eaux–Montgomery–Arabtech), a venture established specifically for this purpose. Within Aqaba water provision and sewerage management is under a public venture, run in the style of the private sector, known as the Aqaba Water Company. Since 2007 a similar approach has been used in Amman.

The NWMP seeks to address the provision of water under three broad strategies: those concerned with water resources; those concerned with water allocation; and those concerned with the control of water quality. Each of these is reviewed below.

24.5.1 Water resource strategies

Within the broad remit of Water Resource Strategies the NWMP identifies three 'sub-strategies':

- *Use natural water resources sustainably.* In particular there is a need to protect and sustainably manage groundwater resources which constitute more than half of the current water supply. For a number of years, extraction rates from the aquifers have often exceeded the sustainable yield which has resulted in decreasing groundwater tables and increasing salinity in several areas. In addition there are plans to extract 'fossil' water from non-renewable resources.
- *Defend and protect the rightful share of international waters* through bilateral and multilateral contracts, negotiations and agreements.
- *Recycle wastewater* and allow its use in unrestricted agriculture, groundwater recharge and other non-domestic purposes in accordance with WHO and FAO guidelines (Pescod, 1992; JISM (Jordanian Institute of Standards and Metrology), 2002)

24.5.2 Water allocation strategy

The priority under this strategy is the provision of water for domestic use and for this reason, the strategy indicates that this will be the focus for resource investment. Agricultural water provision is clearly identified as of lower priority and the NWMP allocates 'unwanted' water (e.g. treated wastewater) to this sector. If groundwater which is not required for domestic use may be extracted sustainably for agricultural use, this may be used in the agricultural sector. An underlying principle in the strategy for allocating water resources, and ensuring that available water resources meet the demands in the key sectors identified, is the need to reduce demand and wastage coupled with the reallocation of water resources from low priority sectors (principally agriculture) to higher priority sectors (municipal, industry and tourism).

24.5.3 Water quality control

Any water strategy not only needs to provide an adequate volume of water, but also to ensure that this water is of sufficient quality for the projected uses. For example, Ayers and Westcott (1994) provide general guidance on water quality for agriculture. The strategy emphasises the need to enforce existing national health standards for municipal water quality (JISM, 2002) and stresses the importance of an efficient and effective system of wastewater collection and treatment. Chapter 25 in this volume considers water reuse in agriculture, reviewing the perceptions of water quality amongst direct and indirect users. Direct users of treated wastewater have developed strategies for ensuring careful use of the water and soil management strategies to avoid accumulation of potential toxic substances in the soil. In urban areas, surveys recently undertaken amongst consumers have shown that poor water quality (taste, smell and pathogens) is a major issue amongst consumers regardless of their income levels and place of residence (Potter and Darmame, 2010). Whilst ensuring the quality of water being used is an essential element of the overall strategy for water resources, the management of these resources must also ensure that there is no reduction in the quality of the water resources themselves. For example, there is a need to prevent the decline in the quality of groundwater resources through recharge with poor quality water and, particularly in shallow groundwater aquifers, to prevent contamination of the water resources through inappropriate activities at the surface (e.g. over-application of fertilisers and pesticides) with consequent leaching in to the aquifer.

24.6 THE DEMAND FOR WATER

There is consistent evidence of an increasing gap between the demand for water use and the supply of water. This situation has resulted in considerable competition between competing sectors. Within the NWMP the four basic water demands have been emphasised and prioritised:

1. Domestic/municipal
2. Tourism
3. Industry
4. Agriculture

Domestic/municipal water is given the highest priority. The trend in water provision has been downwards with a daily per capita provision of 103 litres in 1996 falling to 86 in 2001. The short-term target is to return to 100 litres per person per day by 2005, with an eventual aim of reaching an internationally accepted minimum standard of 150 litres by 2015. A significant tool in achieving these targets is the reduction in water loss from poorly maintained infrastructure (physical loss) and unauthorised abstraction, known locally as 'unaccounted for water'. Levels of physical water loss were estimated at 30% of the total water provision in 2004 and it is anticipated that current water loss reduction programmes will reduce this to 15 to 20% of the total. It is recognised that future developments in urban water provision must involve improved water distribution networks and less interrupted supplies to ensure that water is used effectively and sustainably (Potter et al., 2007b).

Tourism is a rapidly growing sector within the national economy, with significant developments in Amman and Aqaba and small concentrated developments on the shores of the Dead Sea at Sweimeh and Zara. The water demands associated with tourism are not simply the demands of humans but also include demands from the need to maintain attractive landscapes. Tourism is principally attracting visitors from outside the national boundaries, both from within the MENA region and beyond. Many tourists will have no perception of the status of Jordan's water supply and will frequently use water in volumes well in excess of local rates. The water demand for tourism is currently estimated at 5 MCM per year, and short-term projects suggest this will quadruple by 2020. With careful planning and the development of water allocation strategies, water of different qualities and different sources should be used to meet these demands. For example, good quality treated wastewater is an appropriate source for many landscaping needs, and is being used as such in Aqaba and Dead Sea resorts.

Traditionally industry in Jordan has focused on extractive industries, such as mining and quarrying (phosphate and potash) and the industrial production of cements, fertilisers and refined petroleum. Of the 32 MCM consumed by the industrial sector in 2001, some 86% was associated with these large traditional industrial activities. In recent years the industrial sector has begun to increase in importance within the national economy, with developments planned at Aqaba, Irbid and Amman. Whilst many of these new industries are not characterised as large users of water, the development plans suggest that by 2020, industrial needs will be approximately 120 MCM, of which 54 MCM will be required by the traditional heavy industrial users, some 13 MCM by other established industries and 53 MCM by the new industrial users.

Historically agriculture has been a major activity among the population, either as a principal activity or, often, in conjunction with other forms of employment. In 2000 agriculture used approximately 66% of water resources (Table 24.2) although by 2002 the agricultural sector's contribution to GDP was below 4%. Whilst this suggests that the reduction in agricultural production will not have significant impacts on the national economy, it is clear that the agricultural sector has an important social and economic role. Agriculture provides employment both directly within the agricultural sector and in a vast number of agricultural support services such as fertiliser manufacture, transportation, irrigation supply and maintenance. Agriculture also contributes to food security and nutrition within the country and plays a role in generating tax revenue for central and local governments. The possible social impacts of abandoning agriculture are likely to be particularly pronounced in rural areas where a large proportion of the population are active in the agricultural sector. Yet despite the benefits of agricultural production to the economy, the agricultural water demand is considered to be of low priority nationally, and the access to surface and groundwater is most likely to be increasingly restricted. To maintain profitable agriculture there is a need to look at the strategy for the provision of replacement resources such as reclaimed wastewater, brackish water, water harvesting and the desalination of groundwater. The provision of resources must be coupled with a strategy to maximise the efficiency of water use. This requires careful water scheduling by farmers and correspondingly appropriate scheduling and provision by state water providers.

Whilst there have been substantial improvements in the provision of drinking water to the population, with more than 98% of the population having access, there are still major problems concerning the reliability and adequacy of the water supply. Many households receive water only two days per week.

As was stated above, not all Jordan's water resources are directly under national control. The Peace Treaty of 1994 guarantees Jordan rights to 215 MCM of water annually through new dams, diversion structures and pipelines and also a desalination/purification plant. Currently less than 30% of this water is received. The water supply from the Yarmouk River, which has its headwaters in Syria, is also currently subject to complex bilateral discussions. The recent completion of the Unity Dam on the Yarmouk will allow greater control of the flow of this river.

24.7 OTHER SOURCES OF WATER

As stated previously, economically developable renewable water resources in Jordan are finite and, even taking the upper estimate of 850 MCM, fall well short of the projected increases in demand (Table 24.3). In part, the increased demand has to date been met by abstraction from groundwater sources at levels, nationally, in excess of 1.5 times what are considered renewable rates; and locally some rates of extraction are considerably greater than 1.5 times the renewable rates. These extraction rates are clearly not sustainable and provide only a short-term solution to the problem.

Recently, agreement has been reached to extract water from the Disi aquifer that lies on Jordan's border with Saudi Arabia (Figure 24.3), connecting the aquifer to Amman with a 325 km pipeline. Additionally, non-renewable groundwater resources are being exploited from the Ran aquifer (approximately 43 MCM). Even excluding these non-renewable resources the current rates still result in a major shortfall of renewable water resources and other sources must be considered; these include:

1. *Treated wastewater* – in 2000 approximately 72 MCM of treated wastewater were used in the agricultural sector, although much of this water was not of high quality. With recent investments in water treatment facilities (for example, at Al Zarqa and Ramtha) the availability of treated wastewater will increase markedly, and with the shift from predominantly primary and secondary treatment to tertiary and better treatment technologies there will be a marked improvement in water quality. Current estimates from the MWI are that the amounts of wastewater used for irrigation should reach 232 MCM by 2020 (Mohsen, 2007).
2. *Brackish groundwater* – Jordan has a number of brackish groundwater resources. Whilst there is some limited extraction for agricultural use, the salinity of these waters places some limitations on crop production. Currently proposals have been made to extract brackish groundwater in the Jordan Valley (50 MCM per year) and treat this at a desalination plant to provide water for municipal and industrial use. The GTZ (Deutsche Gesellschaft für Technische Zusammenarbeit GmbH) have worked extensively on brackish water use in the Jordan Valley. Several farms in the Jordan Valley are irrigated with brackish water and some large farms also operate their own desalination plants privately.
3. *Seawater desalination* – technologies to desalinate seawater are being considered to supplement Jordan's water resources, initially at Aqaba, Jordan's outlet to the Red Sea. In addition there are tentative proposals to build a pipeline to transfer water from the Read Sea to the Dead Sea (Beyth, 2007). Whilst part of the aim of this pipeline would be to replenish the decline in saline water in the Dead Sea, coupled with plans for desalination facilities this water could supplement other freshwater sources.

Whilst not an alternative source of water, it is estimated that addressing the losses of water in the water supply system, the so called 'unaccounted for water', could result in a substantial saving of resources. In 2000 it was estimated that approximately 50% of the municipal water supply across the country was 'unaccounted for', and within the Jordan Valley some 35% of the irrigation water is unaccounted for. While unaccounted does not mean unused, it is thought that significant losses from poorly maintained infrastructure could be avoided through pipe and canal improvement that would reduce leakage.

24.8 WATER SUPPLY IN A CHANGING CLIMATE

A survey of IPCC models suggests that under an A2 scenario, the Middle East may become significantly drier by the end of the twenty-first century. Chapter 4 of this volume explored this projection in more detail using a regional climate model implemented for the study area. Unlike standard resolution global climate models, the regional climate model is capable of making predictions for the Middle East at the country level. Figure 24.5 summarises the results from the regional model study. It can be seen that, under an A2 emissions scenario, Israel and Jordan are likely to become drier by the end of the twenty-first century, with most drying occurring at the peak of the rainy season in Israel and western Jordan. The plot of seasonal cycles shows that, under a B2 scenario, the predicted changes are far smaller.

The model was found to represent the underlying controls on rainfall well, lending credibility to the predictions. However, significant uncertainties remain – most notably arising from the emissions scenario. It was found that significant changes in rainfall during the twenty-first century are contingent on a business-as-usual type scenario, in which current trends in emissions continue, leading to a doubling in CO_2 levels by the end of the twenty-first century. If emissions increase more slowly or stabilise, the changes in rainfall within the eastern Mediterranean are predicted to be small and statistically insignificant. This

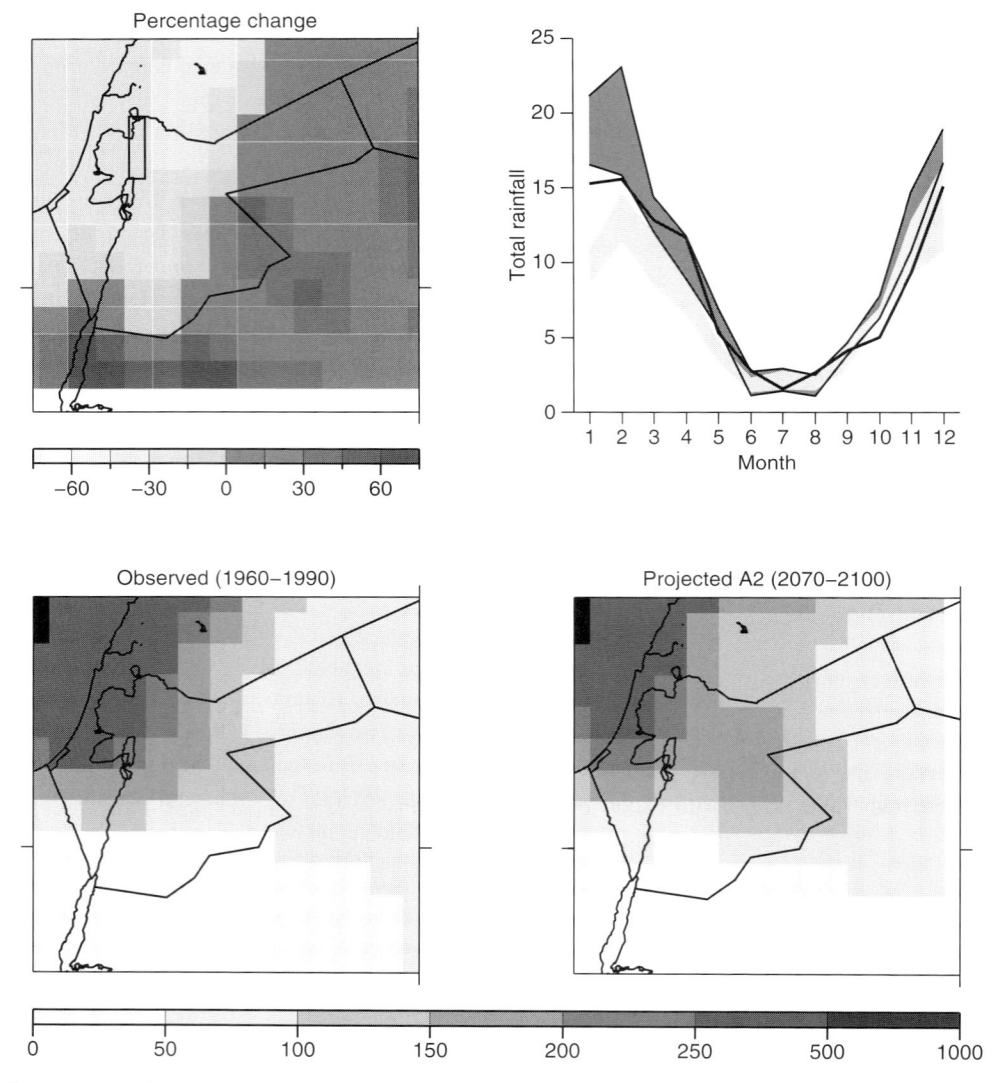

Figure 24.5. Top left: percentage change in annual total rainfall predicted for 2070–2100 under an A2 emissions scenario. Top right: seasonal cycle in total rainfall in the box shown on the top left figure, for a group of model integrations for a present-day climate (dark grey polygon); a 2070-2100 A2 scenario climate (pale grey polygon) and a single B2 scenario model integration (line). Bottom left: observed annual total precipitation (based on the publicly available GPCC gridded dataset). Bottom right: percentage changes in annual total rainfall under an A2 scenario projected on to the observed present-day totals. See colour plate section.

implies that the devastating rainfall decreases that are projected if we continue current trends in emissions may be prevented if emissions are reduced.

It should be noted that the regional model projections described here and in Chapter 4 of this volume are for the end of the twenty-first century, while the time frame for the planning processes described in previous sections of this chapter is 20 to (at most) 40 years hence. Global simulations for the middle of the twenty-first century predict smaller but significant temperature rises (on average 1.4 °C in comparison to 4 °C by the end of century) and some decrease in precipitation – although this is projected not to be statistically significant until the end of the century (Evans, 2009).

24.9 CONCLUSIONS

Ensuring that Jordan has a supply of water that is both sufficient to meet the nation's needs and is of appropriate quality is one of the major issues facing the country in the twenty-first century. The history of water use in Jordan has been one of poor control and often wasteful use, with, in recent years, consumption consistently in excess of sustainable supply, the shortfall being met by over-extracting from renewable groundwater supplies or extracting water from non-renewable groundwaters. Following national and international investment there is now a satisfactory monitoring system which provides the database on the nature and extent of the country's water resources, and how the water is used between different regions and different sectors. This monitoring and the data provided are essential components of any future strategy for water resource management.

With water being such a scarce resource, coupled with the very rapid expansion of population, it is essential that water use is planned and prioritised. In the past, agriculture has been a major consumer of Jordan's water resources. The plans for the future indicate that the limited water resources available for agriculture need to be allocated appropriately. Consideration must be given to suitable high value crops, with export potential, appropriate irrigation methods, scheduling and cultivation locations. Within the municipal context a priority is to address the infrastructure to ensure that the losses of water, either through poorly maintained delivery systems or through unauthorised accessing of water supplies, are reduced. Given the paramount importance of relatively low-cost, high-quality water for human use and some industrial uses, it is imperative that Jordan ensures that the high-quality resource is not used wastefully. For example, where there is a need to irrigate gardens and clean road surfaces there is a strong case for using alternative resources such as treated wastewater through grey water reuse systems, and roof-top water harvesting. Water demand management is imperative for every water use sector within the kingdom.

In addition to the predicted growth demand for water arising from economic development and population growth, there is the additional problem that the most widely accepted future predictions of rainfall in the region, for example the IPCC A2 scenario of 'business-as-usual', suggest a decrease in total amounts of rainfall and possible small shifts in the distribution of rainfall during the winter months. If these predicted outcomes are correct the reduction in fresh water supplies in Jordan are likely to be substantial, and hence the need to manage and prioritise Jordan's water resources from all sources becomes even more essential. Prioritising the use of fresh water for the rapidly growing population and the use of treated wastewater to support agricultural production and non-essential domestic and industrial uses become imperative, as is the need to manage water resources effectively to ensure that deep and shallow groundwaters and available surface waters are not polluted.

REFERENCES

Ayers, R. S. and D. W. Westcott (1994) *Water Quality for Agriculture Irrigation and Drainage Paper*. Rome: Food and Agriculture Organisation of the United Nations.

Beyth, M. (2007) The Red Sea and the Mediterranean–Dead Sea canal project. *Desalination* **214**: 365–371.

Evans, J. P. (2009) 21st century climate change in the Middle East. *Climatic Change* **92**: 417–432.

JISM (Jordanian Institute of Standards and Metrology) (2002) *Technical Regulation – Water – Drinking Water*: JS893. Amman, Jordan.

Meizen-Dick, R. (1997) Valuing the multiple uses of irrigation water. In *Water: Economics, Management and Demand*, ed. M. Kay, T. Franks and L. Smith. London: E&FN Spon pp. 50–58.

Ministry of Environment (2006) National Strategy and Action Plan to Combat Desertification. Amman, Jordan: Ministry of Environment.

Mohsen, M. S. (2007) Water strategies and potential of desalination in Jordan. *Desalination* **203**: 27–46.

Molle, F. and J. Berkoff (2006) *Cities Versus Agriculture: Revisiting Intersectoral Water Transfers, Political Gains and Conflicts*. Comprehensive Assessment Research Report 10. Colombo, Sri Lanka: IWMI.

MWI (Ministry of Water and Irrigation) (1997) *Water Strategy in Jordan*. Jordan: Ministry of Water and Irrigation.

MWI (Ministry of Water and Irrigation) (2004) *National Water Master Plan of Jordan*. Amman: GTZ and Ministry of Water and Irrigation.

Parry, M. L., O. F. Canziani, J. P. Palutikof, P. J. van der Linden and C. E. Hanson, eds. (2007) *Contribution of Working Group II to the Fourth Assessment Report of the Intergovernmental Panel on Climate Change, 2007*. Cambridge and New York: Cambridge University Press.

Pescod, M. B. (1992) *Wastewater Treatment and Use in Agriculture*. FAO Irrigation and Drainage Paper. Rome: FAO.

Potter, R., K. Darmane, N. Barham and S. Nortcliff (2007a) *An Introduction to the Urban Geography of Amman, Jordan*. Reading Geographical Papers **182**: University of Reading.

Potter, R., K. Darmane and S. Nortcliff (2007b) The provision of urban water under conditions of 'water stress', privatization and deprivatization in Amman. *Bulletin of the Council for British Research in the Levant* **2**: 52–54.

Potter, R. B. and K. Darmame (2010) Contemporary social variations in household water use, management strategies and awareness under conditions of 'water stress': the case of Greater Amman, Jordan. *Habitat International* **34**: 115–124.

UN (1989) *Guidelines for the Preparation of National Water Master Plans.* Water Resources Series Geneva: United Nations.

UNDP (1998) *Global Human Development Report 1998.* Oxford: Oxford University Press.

UNDP (2006) *Human Development Report 2006. Beyond Scarcity: power, poverty and the global water crisis.* New York: United Nations Development Programme.

UNEP (2007) *GE04 Global Environment Outlook (Environment for Development).* New York: United Nations Environment Programme.

UNWater (2007). *Coping with Water Scarcity: Challenges of the Twenty First Century.* Prepared for World Water Day, 2007. Retrieved December, 2008. http://www.unwater.org/wwd07/download/documents/escarcity.pdf

Winpenny, J. T. (2000) *Managing Water Scarcity for Water Security.* Retrieved January, 2007. http://www.wca-infonet.org/servlet/BinaryDownloaderServlet?filename=documents/1354.Managing_water_scarcity_for_water_security. 2001–11–13.pdf&refID=1354

World Commission on Environment and Development (1987) *Our Common Future.* Oxford: Oxford University Press.

WWAP (World Water Assessment Programme) (2006) *Water: A Shared Responsibility.* The United Nations World Water Development Report. Paris: UNESCO.

Zhang, X. B., F. W. Zwiers, G. C. Hegerl *et al.* (2007) Detection of human influence on twentieth-century precipitation trends. *Nature* **448**: 461–465.

25 Water reuse for irrigated agriculture in Jordan: soil sustainability, perceptions and management

Gemma Carr

ABSTRACT

Treated domestic wastewater offers a valuable contribution to Jordan's water balance. Water reuse is particularly well-suited to irrigated agriculture, for which water is in constant demand. The Jordanian reclaimed water generally has a high concentration of both plant-beneficial ions (nutrients) and potentially plant-toxic ions. Suitable soil management strategies such as periodic leaching and nutrient management are therefore required to maintain productivity. In this research the effect of water reuse on soil sustainability has been investigated by means of soil analysis combined with interviews with farmers and organisations working with reclaimed water. The soil analysis results suggest that irrigation does lead to the accumulation of plant-toxic solutes, but soil analysis from farms which have been irrigated with reclaimed water for several decades reveals that solute accumulations have been avoided through water management strategies on the farm. Interviews with farmers have shown that on-farm water management is influenced by a range of factors, which include the decisions taken by organisations with regards to water allocation and management. Interviews with organisations have shown that the priorities of decision-makers are often different from those of farmers and that they have a limited awareness of the challenges faced by farmers. Maintaining soil sustainability is imperative and can be achievable through water management both on and off the farm.

25.1 INTRODUCTION

25.1.1 Background to water reuse in Jordan

Water scarcity is a growing problem across the world as greater pressure is placed on the available water resources. The rising global population requires more fresh water for drinking, domestic usage and particularly for food production (Postel, 1992). Irrigated agriculture is essential to sustain the increasing populations, and in arid and semi-arid lands it commonly absorbs the majority of freshwater resources (Allan, 2001; Molle and Berkoff, 2006). The countries of the Middle East are facing some of the greatest water shortage problems in the world, with their available resources being significantly lower than the potential demands arising from industry, agriculture, tourism and domestic needs (MWI (Ministry of Water and Irrigation), 2004). Jordan, at the centre of the Middle East, is one of the most water-scarce countries of the world. With only about 160 cubic metres per person per year (FAO, 2008) Jordan is classified as experiencing absolute scarcity according to the Falkenmark indicator of water stress (Falkenmark et al., 1989).

There are three principal ways by which water scarcity can be alleviated. The first is through developing new resources via the construction of dams, desalination of sea water or exploiting groundwater resources. Such schemes typically involve hard engineering and technological developments and tend to be expensive. Secondly, scarcity can be confronted through water demand management which aims to lower the demand for water by encouraging water-saving devices and reducing water use in the home, in industry and in agriculture. The third approach is by reusing urban domestic wastewater after it has been treated.

Treated wastewater (reclaimed water) is particularly well suited for agriculture as it often contains significant quantities of plant essential nutrients such as nitrogen, phosphorus and potassium and can have high amounts of suspended and dissolved organic matter. The release of such nutrient-rich water to natural water bodies can have detrimental ecological effects, while its reuse for irrigation offers a means of waste disposal that potentially avoids such pollution of the natural environment (Feign et al., 1991; Shelef and Azov, 1996). Crop growing experiments in Saudi Arabia found that the use of reclaimed water reduced the need for 50% of the inorganic nitrogen usually required for maximum crop productivity (Hussain and Al-Saati,

Water, Life and Civilisation: Climate, Environment and Society in the Jordan Valley, ed. Steven Mithen and Emily Black. Published by Cambridge University Press.
© Steven Mithen and Emily Black 2011.

1999). A survey of farmers in Jordan by Abu-Madi (2004) found that farmers could reduce their fertiliser expenditure by 65% when irrigating fruit trees with reclaimed water compared with groundwater.

Despite the benefits, there are disadvantages associated with the use of reclaimed water. Public health concerns have arisen because the water has the potential to carry human pathogens such as *Escherichia coli* and intestinal worm eggs, although treatment processes will remove these (WHO, 2006). These present a degree of risk to the farmer in contact with the water as well as the consumer who may ingest contaminants present on the surface of the irrigated produce. However, the risk is very low provided that the water is adequately treated and careful management is undertaken on the farm to prevent contact between the produce and the water (Oron *et al.*, 1999; Raschid-Sally *et al.*, 2005; Jimenez, 2008).

Another area for concern is the risk that the water presents to the sustainability of the soil. While the water contains nutrients it also contains solutes of sodium chloride and boron which are potentially toxic to plants if they accumulate in the soil to levels beyond the tolerance threshold of the crop (Shainberg and Oster, 1978). Sodium also has the potential to damage the soil structure under certain conditions. These solutes enter domestic wastewater from household cleaning products such as dishwasher and laundry detergents. Sodium chloride is added to these as a manufacturing agent and water softener (Patterson, 2001) and sodium perborate is added as a bleaching agent (Barth, 1998).

In arid regions, irrigated soils are particularly sensitive to the potentially detrimental effects of slightly saline, reclaimed water (Feign *et al.*, 1991; Carr *et al.*, 2008). The low input of fresh water due to limited precipitation means that accumulated ions are rarely removed naturally from the soil profile by flushing or leaching. Significant long-term problems for soil productivity can occur if irrigation with slightly saline water continues without additional water being applied to leach the solutes. The problems of solute accumulation include salinity (high salt content), sodicity (high sodium content), anion toxicity (high concentrations of chloride, carbonate or bicarbonate) or specific toxicity by minor elements (e.g. boron).

Consideration must also be given to the nutrient additions from the irrigation water. While these are generally beneficial, the application of some nutrients in quantities beyond the requirement of the crop at certain times in the growing season can be detrimental to productivity. A high concentration of solutes in the soil solution, including beneficial solutes, can reduce water availability to the plant if the osmotic pressure in the soil solution exceeds that in the root. High availability of nitrogen during certain plant growth stages can result in reduced fruiting, while excesses of some nutrients can lead to deficiency in others. For example, magnesium can replace calcium, leading to calcium deficiency in the plant. Careful management of mineral fertiliser inputs is therefore needed, and this can only be obtained through awareness of the nutrient content of the irrigation water and adjusting the fertiliser schedule accordingly to prevent crop damage. By reducing chemical fertilisers and applying only the required quantity of irrigation water to meet the crop requirements when the crop is sensitive to excess nutrients, an attempt can be made by the farmer to manage these risks.

Despite the potential hazards from irrigation with reclaimed water, the benefits for arid region soils are substantial. Irrigation raises productivity, the soil organic matter should increase with commensurate improvements in the soil structure and aggregate stability (Vogeler, 2009) and the addition of beneficial ions helps to meet the nutrient demands of the crop and reduce the quantities of non-renewable mineral fertilisers that are needed (Lazarova *et al.*, 2005).

In Jordan, water reuse is already taking place on a substantial scale. Sixty-five per cent of households are connected to a central sewage system (Water Authority of Jordan, 2007) which means that about 80 million cubic metres (MCM) of wastewater are produced annually (Vallentin, 2006). This constitutes just under 10% of Jordan's 867 MCM of renewable water annually (MWI, 2009). The majority of the reclaimed water is produced at the Khirbet as Samra wastewater treatment plant located 30 km east of Amman (the capital of Jordan).

Figure 25.1 is a schematic diagram showing the movement of domestic and reclaimed water in northwest Jordan. Domestic water which enters Amman from a variety of sources and the sewage water is then transferred to Khirbet as Samra where it is treated before being released into the Zarqa River. Some water is abstracted directly from the Zarqa River for irrigation by riparian farmers but the majority of the water flows, along with natural runoff, down to the King Talal Reservoir. From the reservoir the water is released as required to be used for irrigation in the Jordan Valley (a major agricultural area). The use of reclaimed water is therefore split between that consumed directly around the treatment plant and that used indirectly (after significant transportation and the seasonal addition of natural surface runoff) in the Jordan Valley.

Two different types of agriculture take place at the direct-reuse and the indirect-reuse sites. Directly around the treatment plants the land is used for the supplementary irrigation of fodder crops such as alfalfa and barley. Flood irrigation is used and the farmers have low outlay costs in cultivating these cash crops. In the Jordan Valley, the blended mixture of reclaimed water and fresh surface water is used for the irrigation of high value fruits and vegetables. The farmers have high outlay costs and must invest in drip irrigation to maximise productivity with the available water resources. Plastic mulch (plastic soil covering) is commonly used to reduce evaporative losses; high cost fertilisers

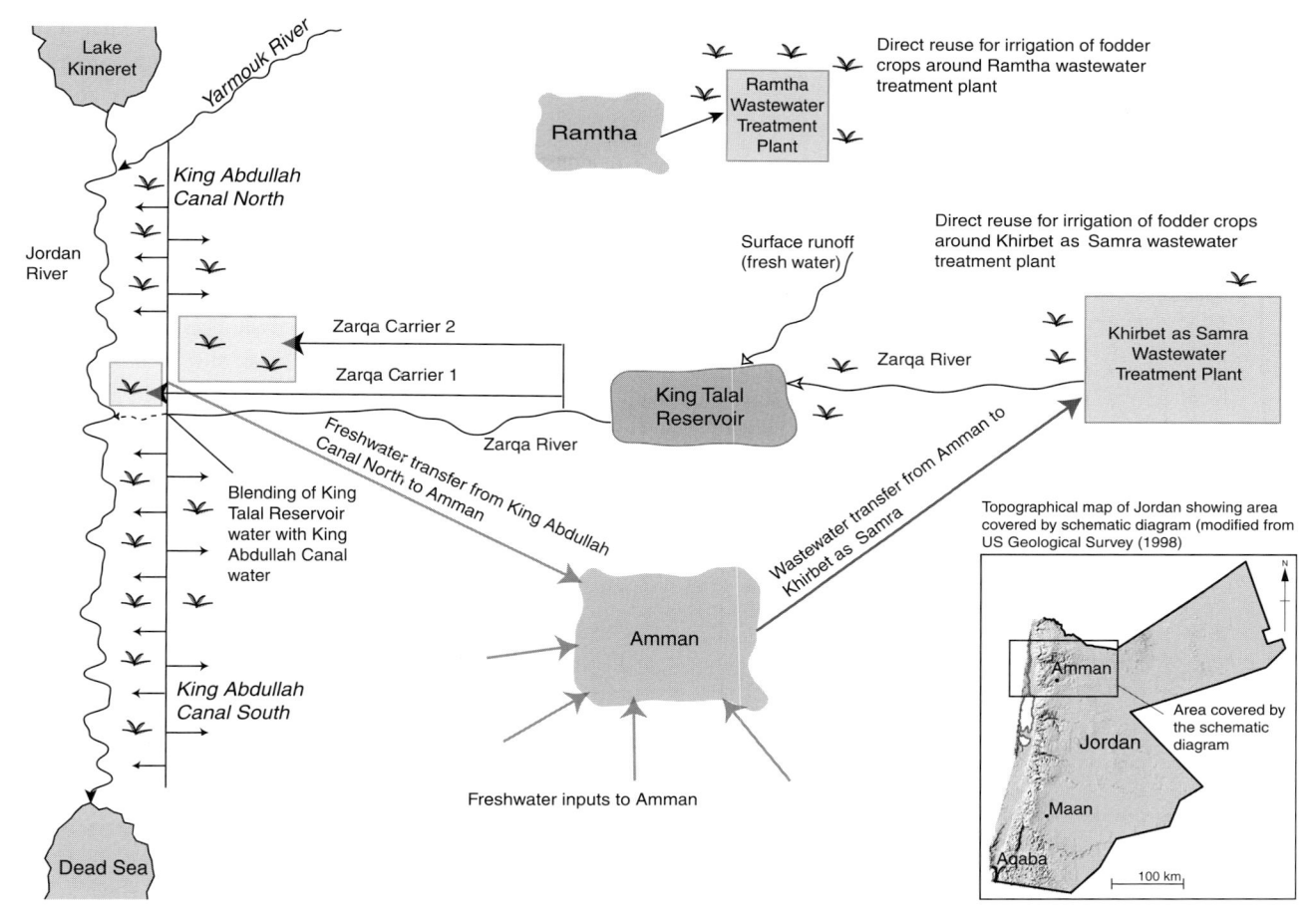

Figure 25.1. Schematic diagram and corresponding map of the pathway of water reuse in northwest Jordan.

and pesticides are typically needed because of the intensity of cropping.

25.1.2 Aims of the research

Water has been reused successfully in Jordan for several decades, and the aim of this research was to investigate the sustainability of the practice in terms of both the effects on soil and the ability of people to manage and mitigate these effects.

Water reuse is a complex and fragmented process. It affects the environment and includes a variety of stakeholders such as farmers and various decision-making institutions. As such, the aims of this work were three-fold, with the overall objective of investigating how and why water reuse can be managed sustainably: firstly, to explore how soil sustainability is affected by irrigation with reclaimed water; secondly, to identify how farmers are managing reclaimed water on their farms to enhance productivity and soil sustainability; and thirdly, to describe how organisations are affecting soil sustainability under irrigation with reclaimed water through their actions, decisions and strategies. The synthesis of research from these three areas leads to a holistic view of water reuse which should be of value in advancing our understanding of sustainable water resource management.

25.2 RESEARCH METHODS

To reflect the interdisciplinary nature of the work, several different research methods were used: soil sampling and laboratory analysis to explore how reclaimed water alters the soil chemistry; mathematical modelling to further understand the processes occurring in the soil; and semi-structured interviews with farmers and organisations involved with water reuse to investigate the management strategies, priorities and concerns of these stakeholder groups.

25.2.1 Soil and water sampling and analysis

Soil and water samples were taken from three research sites, Khirbet as Samra, Ramtha (also known as Ar Ramtha or Al-Ramtha) and Deir Alla, the locations of which are shown on Figure 25.2. At each site experimental barley (*Hordeum*

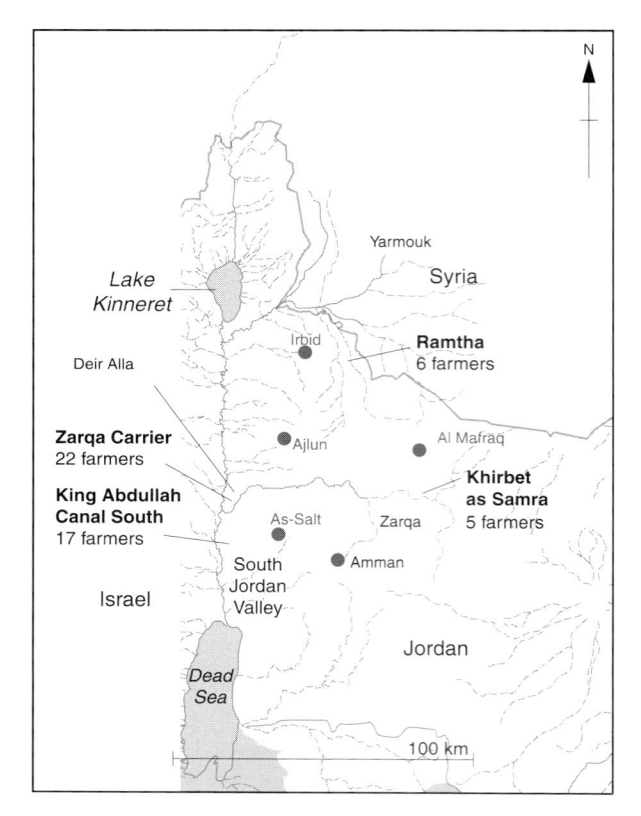

Figure 25.2. Map of northwest Jordan showing the locations of soil sampling and farmer interviews.

vulgaris) was being grown by the WLC Project with reclaimed water under carefully regulated water supply conditions (see this volume, Chapter 21). Data were therefore available on the exact quantities of irrigation water applied to the crop which had been calculated according to the crop water demand (Allen *et al.*, 1998). At each locality soils were available which had been irrigated to meet 100% and 120% of the crop water demand. This meant that the soils irrigated to 120% had received 20% more water than required by the plant, resulting in percolation of excess water through the soil profile and potential leaching of soil solutes. This made it possible to examine how water reuse affected soils which had been leached compared with those which had not been leached. Soil samples were taken with an auger to extract material from a number of set depths in the profile and were collected at the end of the cropping season (May) over two consecutive years (2006 and 2007). Composite samples were formed by mixing soil taken from four places in the irrigated area in order to reduce the effect of localised variation in the soil.

As well as sampling from the research sites it was also possible to take soil samples from farms which had been irrigated with reclaimed water for extensive periods of time. At one site in the Jordan Valley land had been irrigated for 28 years with water from the King Talal Reservoir. The farmer of this land was able to give information on the management methods used, which had included covering the soil with plastic mulch during the growing season, and applying excess reclaimed water periodically to leach the soil and maintain low salinity. At Khirbet as Samra it was possible to sample soil from an olive plantation which had been irrigated with effluent from the wastewater treatment plant for 18 years.

Soil analysis was conducted to determine the soil salinity through the measurement of the electrical conductivity of the soil saturation extract (ECe) measured in deci-siemen per metre (dS m^{-1}). This was performed using the method of Rowell (1994) whereby a sample of soil was moistened with ultra-pure water until it reached saturation. After allowing the mixture to equilibrate for 16 hours the solution was removed from the soil using a vacuum pump. The solution was then analysed to determine the pH of the extract (pHe), the ECe and the concentration of soluble cations (sodium, calcium, magnesium, potassium and boron) and anions (chloride, sulphate and phosphate) using inductively coupled plasma optical emission spectroscopy (ICP-OES) (Optima 3000) and ion chromatography (Dionex), respectively. The water samples taken in the field were analysed in a similar manner. The water analysis data were then integrated with historic data collected from published and grey literature from Jordan (GTZ, 2005; Al-Zu'bi, 2007; Ammary, 2007; Bashabsheh, 2007) to obtain a more complete picture of the water quality.

25.2.2 Mathematical modelling

A pre-developed mathematical model of solute movement through a variably saturated medium (HYDRUS-1D, Šimůnek *et al.*, 2005) was used to further understand solute movement through the soil and to attempt to develop scenarios of appropriate management methods to maintain soil sustainability under irrigation with reclaimed water. For simplicity, only chloride (Cl) was modelled using HYDRUS. This ion was chosen because it is negatively charged which means it is not retained by the negatively charged soil particles and so it travels easily through the soil in solution with the water. Chloride is also closely correlated to total salinity (Hajrasuliha *et al.*, 1991) and therefore knowledge of the chloride concentration is a valuable indicator of soil salinity.

The HYDRUS model was calibrated so that the modelled soil solution chloride concentration at the end of the growing season in 2006 and 2007 would fit as closely as possible to the measured soil solution Cl concentration which had been determined in the soil samples. The calibration was done through careful adjustment of the soil and solute numerical parameters required by the model. This allowed a reasonably good fit to be achieved between modelled and measured data from Deir Alla. Using the determined soil parameters and the available data on irrigation, rainfall, evaporation and transpiration from this location, the model was run for a variety of scenarios whereby the amounts and timings of leaching water were changed while all other parameters remained the same. This allowed the identification of how changes in the irrigation schedule would affect the concentration of Cl at different depths in the soil profile and at different times in the season.

25.2.3 Interviews with farmers irrigating with reclaimed water

Extensive semi-structured interviews were conducted with 39 farmers irrigating with reclaimed water. The aims of these interviews were to gauge the farmers' opinions of the water, to identify perceived and real problems and benefits and to document local knowledge as to how reclaimed water is managed to minimise adverse effects on crops and soil. Through interviewing farmers who were using reclaimed water directly around the wastewater treatment plants at Ramtha and Khirbet as Samra and those using reclaimed water in the Jordan Valley, it was possible to compare the perceptions and management strategies adopted by direct reusers to those of indirect reusers. The locations and numbers of farmers interviewed are shown in Figure 25.2.

It was anticipated that large- and small-scale farmers might have had differing views and employed different management methods. To ensure that the dataset was unbiased, an attempt was made to interview an equal number of small-, medium- and large-scale farmers. The interviews were conducted in the field via a translator with advanced knowledge of Jordanian agricultural systems. Notes were taken and often a tour of the farm was provided to clarify and explain management methods and issues. Interview notes were typed up and expanded according to memory immediately after the interviews. Each data bit (piece of information) within the interview transcript was then coded and grouped according to its code. For example, all comments related to fertiliser application were grouped together. This method meant that similarities and discrepancies between the farmers' views became clear, and factors that were frequently mentioned or ignored were highlighted (Kitchin and Tate, 2000).

25.2.4 Interviews with organisations working with reclaimed water

Semi-structured interviews were held with individuals working actively in the field of water reuse in the private sector, international agencies, governmental, non-governmental and research institutions in Jordan. These interviews focused on the perception of the organisation towards water reuse, but also covered wider management and implementation challenges, and priorities for the future, and attempted to explore how the organisation regarded the farmers and accommodated their involvement in the management of water reuse. These interviews were conducted in English and either were recorded or notes were taken during the interview. They were transcribed immediately after the interview with additional observations and comments. For analysis they were coded according to various themes and the coded data were analysed to identify similarities and differences between the individuals and the organisational groups. Quotes were identified for each theme and compiled together in order that narratives could be developed on water reuse from the perspective of the organisations.

25.3 RESEARCH FINDINGS AND DISCUSSION

25.3.1 Water quality

Table 25.1 shows the solute content of irrigation water at Khirbet as Samra, Ramtha and Deir Alla (Jordan Valley) as determined by the water samples collected and the long-term averages identified in the available literature. The data in Table 25.1 show that both potentially plant-beneficial and potentially plant-toxic ions are present in the irrigation water.

Potentially plant-toxic solutes in the irrigation water can be coarsely estimated by the EC of the water. The reclaimed water at all sites has an EC of around 2 dS m^{-1} which classifies it as having a slight to moderate restriction to use according to FAO guidelines (Ayers and Westcott, 1985). Of greater concern is the Cl concentration which can reach almost 400 mg L^{-1} and the boron concentration which can reach almost 1 mg L^{-1}. These concentrations can cause a specific toxicity hazard for some crops (Ayers and Westcott, 1985). There is some difference between the water qualities at each of the research sites. The water of Deir Alla has a lower solute concentration than the water of Khirbet as Samra and Ramtha, owing to the addition of fresh, surface water to the Zarqa River, the source of the irrigation water at Deir Alla.

Careful management of the soil is therefore required to ensure that potentially toxic ions do not accumulate to levels that serve to reduce plant productivity and that the addition of mineral fertilisers does not lead to excessive nutrient availability which has the potential to reduce productivity.

25.3.2 Crop nutrients and farmer awareness of 'free' fertilisers

The nutrient content of the water has been extrapolated to show the quantities of nutrients added to the soil through the irrigation water when sufficient water is applied to meet 100% of the water requirement of a typical barley crop (Figure 25.3). The data suggest that the irrigation water meets the nutrient requirements of the crop with regards to potassium, phosphorus, sulphur and

Table 25.1 *Water quality parameters at Khirbet as Samra, Ramtha and Deir Alla based on data determined through water sampling and published data*

Location	Potentially plant-beneficial ions in the irrigation water (mg L^{-1})				Potentially plant-toxic ions in the irrigation water (mg L^{-1})			Additional parameters		
Average determined from all available data	K	P	S	Mg	Cl	Na	B	EC (dS m^{-1})	pH	SAR
Khirbet as Samra	31.32	35.63	65.01	27.22	364.61	261.14	0.91	2.14	7.86	7.67
Ramtha	32.97	6.80	104.66	34.19	398.76	232.46	0.73	1.71	8.21	6.30
Deir Alla (water from the King Talal Reservoir)	15.74	21.74	90.78	31.39	276.66	125.67	0.54	1.91	7.85	1.92

Source: (GTZ, 2005; Al-Zu'bi, 2007; Ammary, 2007; Bashabsheh, 2007).
EC, electrical conductivity. SAR, Sodium Adsorption Ratio.

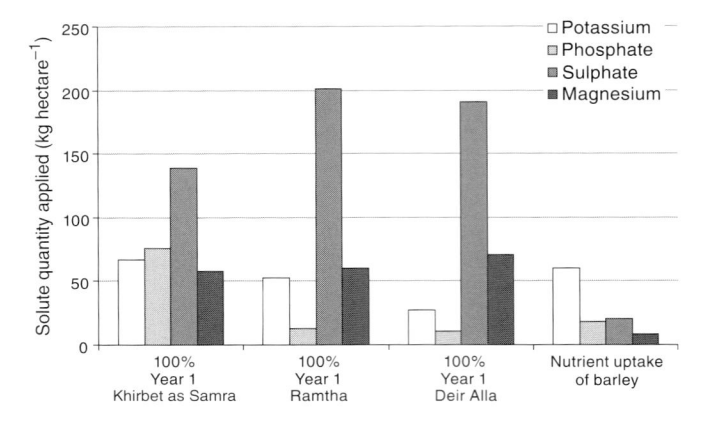

Figure 25.3. Nutrient inputs and nutrient requirements of barley at the research sites (nutrient uptake of barley is based on data from Cooke, 1982).

magnesium at Khirbet as Samra. At Ramtha all the nutrient requirements are met by the water except for phosphorus, and at Deir Alla the requirements of phosphorus and potassium are not met by the water. The exact crop nutrient requirements will vary depending on the type of crop and its yield but the available data give an indication of the high nutrient capacity of the water, particularly at Khirbet as Samra.

The interviews with farmers explored their awareness of the nutrients provided by the reclaimed water. The interviews revealed that 74% of the indirect-reuse farmers in the Jordan Valley were aware that the water contained nutrients and subsequently reduced their use of chemical fertilisers. However, one farmer explained:

> I know that I over fertilise. When the price of cucumbers is high I'll add more fertilisers. I'll try and use less fertiliser. But when the price of cucumbers is high it's a psychological thing to add fertiliser as I wish for the best productivity.

Another farmer explained:

> I went to a training course about how the water has fertilisers in it just a few days ago. But then the fertiliser company sales representatives came here and persuaded me to buy fertilisers!

To indirect-reuse farmers the nutrient content of the water appears to be considered, but many factors influence whether mineral fertilisers are also applied. Information to farmers on the exact nutrient content of the water would be of value in helping farmers to plan their fertilisation schedules, but the psychological factors and persuasiveness of sales agents also need to be considered when providing information to farmers in order that awareness of these issues is acknowledged.

Several of the farmers commented that while they can save money on nutrients they felt that these savings are offset by greater spending on buying, installing and operating water filters to reduce the suspended load of the irrigation water. The organic and mineral constituents of the irrigation water clog the irrigation pipes and drip emitters through mineral precipitation and algal growth (Duran-Ros *et al.*, 2009). To overcome this challenge the farmers use water filters and replace components of the drip irrigation systems every few years. One-third of all the indirect-reuse farmers interviewed spoke of having additional irrigation costs due to filters and irrigation pipe/emitter replacement as a direct result of the quality of the irrigation water.

The farmers directly reusing water around the wastewater treatment plants also had an enhanced awareness of the nutrients in the irrigation water. In contrast to the indirect-reuse farmers, they tended to recognise the financial benefits. This was particularly apparent at Ramtha where the treatment plant had been upgraded in 2004, resulting in a greater quantity of the nutrients being removed from the irrigation water. Here the farmers commented how this had resulted in them needing to buy mineral fertilisers:

> Since the upgrade of the wastewater treatment plant from primary to secondary the natural fertiliser in the water has

decreased. This means there is now an additional expense of adding fertiliser.

The wastewater treatment plant was upgraded two years ago to include anaerobic digestion and chlorination. This results in less nitrogen being available in the water and now I need to apply mineral fertiliser.

Before the upgrade of the plant we didn't add manure to the soil. Now we have to add manure as there is less fertiliser in the water.

The water is good. But now we need to use much more fertiliser since the water treatment plant was upgraded two years ago.

Before, when I was irrigating alfalfa the plants survived for five years. Now I have to plant a new crop every two years because the water is better treated now and so the plant productivity is lower.

These findings emphasise the importance of the water chain concept developed by Huibers and Van Lier (2005), which emphasises consideration for the end use and user of reclaimed water when planning wastewater collection and treatment. For direct reuse around Ramtha it could be argued that advanced treatment with nitrogen and phosphorus removal is unnecessary. Around Khirbet as Samra, where the water is released to the environment, the case for advanced treatment is much stronger in order to protect the natural surface water.

The direct-reuse farmers appeared to have a higher regard for the beneficial nutrients than the indirect-reuse farmers. This is likely to be a result of the type of agriculture and the investment requirements at each site. The clogging of drip irrigation systems by reclaimed water leads to additional expense which is outside the control of the farmer. It would seem that in areas of intensive, high value agriculture, the farmers' inability to control the negative effects of the water outweigh the recognised benefits of free nutrients.

25.3.3 Solute accumulation from irrigation with reclaimed water

It was hypothesised that irrigation with reclaimed water to meet 100% of the crop water requirement would lead to an increase in the solute concentration in the soil solution as the number of years of irrigation with reclaimed water increased. Figure 25.4 shows the soil salinity through the soil profile at Ramtha following one and two years of irrigation compared with a non-irrigated control. The data show that the soil salinity at the surface is much higher following two years of irrigation than after one year of irrigation which supports the hypothesis that soil salinity increases with the length of the irrigation period, though there is no difference between the salinity after one and two years of irrigation below 5 cm depth.

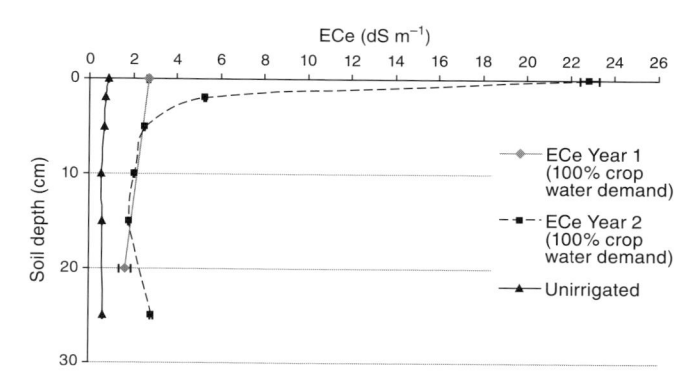

Figure 25.4. Salinity of the saturation extract (ECe) from Ramtha after one and two years of irrigation with reclaimed water (error bars give standard error from the mean).

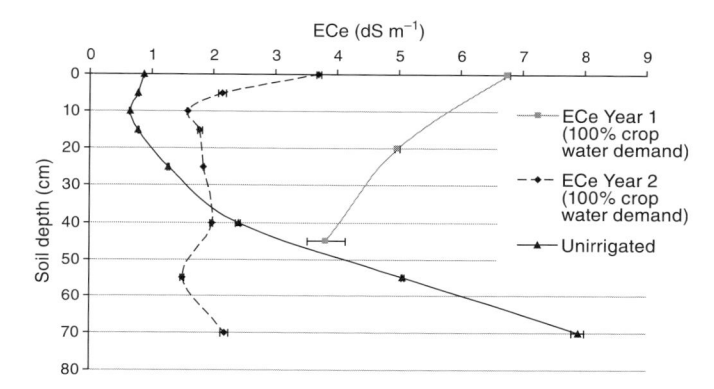

Figure 25.5. Salinity of the saturation extract (ECe) from Khirbet as Samra after one and two years of irrigation with reclaimed water (error bars give standard error from the mean).

The data from Khirbet as Samra show that the relationship between irrigation period and soil salinity is less straightforward (Figure 25.5). The data show that soil irrigated for two years has lower salinity than soil irrigated for one year. This is likely to be due to the mobilisation and upward transfer of solute during the first year of irrigation from deeper in the soil profile driven by evaporation from the soil surface. The non-irrigated soil shows a high solute concentration at depth which is likely to be present because of the movement of saline groundwater. During the second year of irrigation it appears that the solute is transferred back down through the soil profile. The reasons for this are unclear, possibly being connected to the timing and magnitude of rainfall events in conjunction with irrigation.

Soil samples were also taken from two sites which had been irrigated with reclaimed water for 18 years (where flood irrigation had been used for olive cultivation) and 28 years (where drip irrigation had been used for vegetable production). The site irrigated for 28 years had received water from the King Talal Dam (blended fresh water and reclaimed water) with a lower solute concentration than that applied to the site irrigated for

Table 25.2 *Comparison between the average ECe in the root zone (10–40 cm depth) of soil irrigated to 100% and 120% of the crop water demand for two years*

Location	Unleached soil (irrigation to 100% of the crop water demand) Average ECe of the root zone (10–40 cm) (dS m^{-1})	Leached soil (irrigation to 120% of the crop water demand) Average ECe of the root zone (10–40 cm) (dS m^{-1})	ECe reduction in the root zone as result of leaching as a percentage
Khirbet as Samra	1.74 (0.06)	1.78 (0.10)	−2.69
Ramtha	2.24 (0.20)	1.76 (0.05)	21.47
Deir Alla	1.28 (0.04)	1.24 (0.05)	2.63

Standard error from the mean is given in brackets.

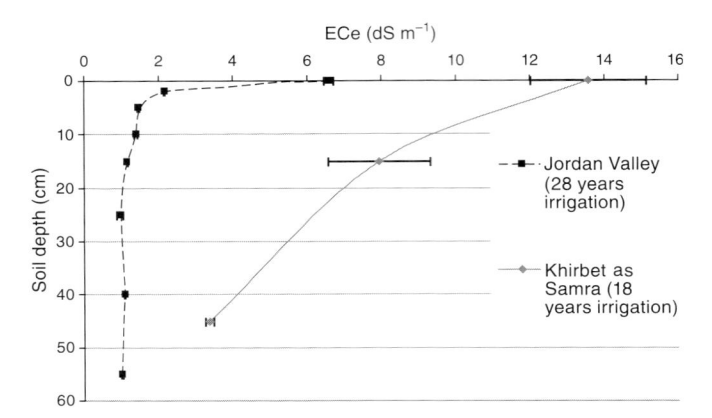

Figure 25.6. Salinity of the saturation extract (ECe) from sites irrigated with reclaimed water for extensive periods of time (error bars give standard error from the mean).

18 years with water directly from the Khirbet as Samra wastewater treatment plant (Table 25.1). Both sites were equally well drained but the irrigation quantities applied and the soil management differed to reflect the different crop types grown. The 28-year site had received no rainfall inputs but it had been leached annually with reclaimed water, manure had been applied regularly and plastic mulch had been used to prevent evaporative losses. It is estimated that approximately 900 mm of irrigation water were applied annually to meet the water requirements of tomatoes (Harmanto et al., 2005). The 18-year site had received supplementary irrigation to meet the requirements of olives (estimated at 600 mm per year (Testi et al., 2006)), but owing to the deeper rooting depth of the olives, leaching of the surface soil had not intentionally taken place. The two sites therefore offered an opportunity to compare a site that had been managed through leaching and one which had not.

Figure 25.6 gives the ECe through the soil profile from the sites irrigated for 18 and 28 years. The data show that while salinity can be slightly high at the soil surface of the site irrigated for 28 years a very suitable salinity is maintained in the root zone owing to leaching. In comparison the site irrigated for 18 years has a much higher salinity throughout the soil profile.

These soil samples highlight that irrigation with reclaimed water does not have to lead to a rise in soil salinity. While salinity may rise from the long-term application of reclaimed water the addition of a leaching fraction coupled with adequate drainage means the irrigation water can be managed for sustainable crop cultivation over the long term.

The role of leaching in maintaining low soil salinity has been investigated at the research sites by comparing the salinity of the soils irrigated with 100% and 120% of the crop water requirement. It was expected that the soils irrigated with 120% of the water demand would have lower soil salinity than the soil irrigated to 100%.

Table 25.2 shows the average salinity of the soil saturation extract of the root zone of the soil irrigated to 100% and 120% of the crop water requirement following two years of irrigation at each locality. These data show that the application of a 20% leaching fraction reduces the average root zone salinity at Ramtha by just over 20%. However, leaching does not appear to reduce the salinity at Khirbet as Samra and only slightly reduces it at Deir Alla. These observations are interesting as it suggests that at Ramtha leaching is effective in transferring solute through the soil profile, but at the other sites additional factors are preventing this transfer.

At Khirbet as Samra it is possible that the application of excessive quantities of irrigation water is leading to greater upward mobility of solute from deeper in the soil profile resulting in a higher soil solute concentration in the leached soil. The crop water uptake may also exceed that calculated resulting in reduced downward percolation of water. Further data on soil moisture through the profile would be needed to determine the extent to which these factors are responsible for the lack of leaching at Khirbet as Samra. At Deir Alla the solute concentration of the irrigation water is generally fairly low, which is reflected in the low soil salinity in both the leached and unleached soils, and this masks the effects of leaching.

25.3.4 On-farm management methods to prevent soil salinisation

During the interviews the farmers were asked about their leaching schedules and methods in order to identify whether they experienced any salinity problems and to record how they attempted to overcome these problems. A clear difference in perception and management of the water was seen between the direct and indirect water users. Those directly using reclaimed water did not describe any negative observations with regards to the water and its effect on soil. No direct-reuse farmers spoke of soil salinity problems, and this is probably because flood irrigation is used around the treatment plants which results in excessive quantities of water being applied on a regular basis leading to leaching and preventing the accumulation of salt in the soil.

In contrast, soil salinity was mentioned by 51% of the indirect-reuse farmers. Awareness of salinity clearly exists in the Jordan Valley where, even before treated wastewater became part of the water resource, salinity awareness and management were required because of inherent soil salinity and the use of water from saline springs. As a result, all of the Jordan Valley farmers explained how they aim to leach the soil regularly to prevent salt accumulation.

Leaching methods and timings for the Jordan Valley farmers are given in Figure 25.7, which shows that 40% of farmers prefer to use flood irrigation for leaching if there is sufficient water but replace this method by drip irrigation because of water scarcity. Sixty per cent of farmers said that leaching was performed with drip irrigation at the start of the season (September), though two farmers mentioned that they were planning on changing their schedule to leach at the end of the season when they can make use of the residual moisture in the soil. Six farmers (10% of the sample) also said they irrigate by applying excessive quantities of water through the cropping season, as more water is available at this time.

The HYDRUS model has been used to explore how the timing of leaching affects the soil solute concentration during the growing season. Figure 25.8 shows how alterations in the timing of the irrigation water change the soil solute concentration in the root zone (25 cm depth) over a period of two years. The modelled outputs suggest that soil to which leaching water has been applied has a higher solute concentration than soil to which no leaching water has been applied. This is probably because the application of excess reclaimed water for leaching raises the total amount of solute added to soil. However, leaching at the start of the season appears to transfer the solute through the profile to a lower depth, resulting in a reduced concentration of Cl at 25 cm depth during the latter part of the season in both the first and second years of irrigation. Further work into leaching timings based on the experiences of farmers would be of immense value for verifying the modelled solute concentrations.

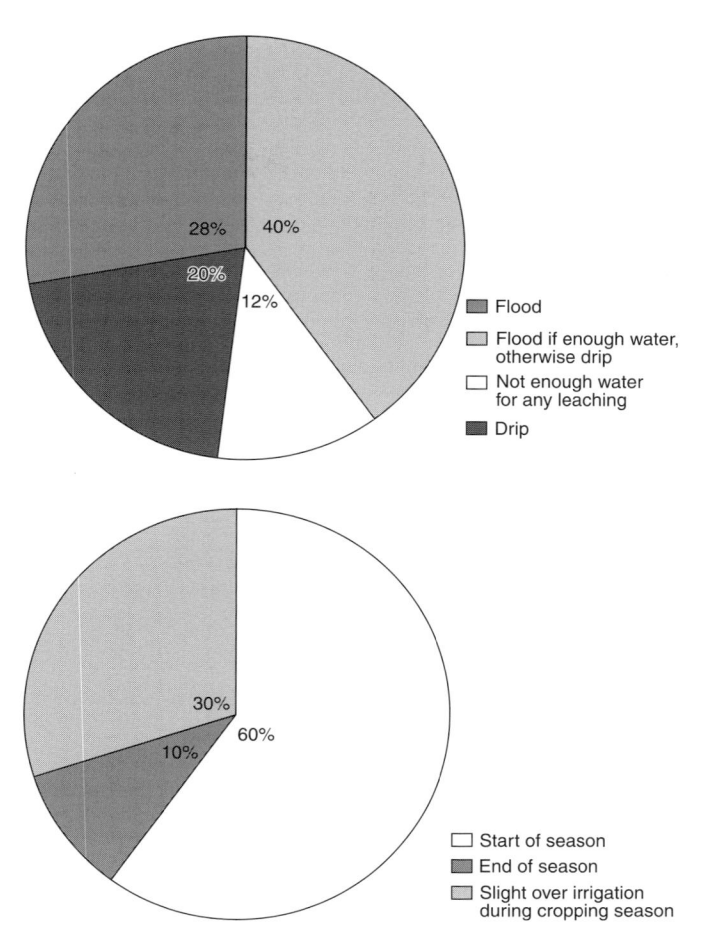

Figure 25.7. Leaching methods and timings of leaching described by farmers irrigating with reclaimed water in the Jordan Valley.

Although leaching was recognised as important by the indirect-reuse farmers, 35% of these farmers said that their leaching schedules were limited by water shortage. The Jordan Valley Authority (JVA) is the government department responsible for water allocations in the Jordan Valley. This institution has an important role in ensuring adequate water is provided to farmers for leaching management but this is not an easy task owing to Jordan's delicate water supply situation. If too much water is sent to farmers for leaching during the non-cropping period there is a risk that the available resources may be insufficient to meet the demands of the cropping period, a particular hazard if rainfall is low or comes late. The ability of the farmer to manage the water on the farm in a manner that allows the sustainability of the soil to be maintained is therefore heavily connected to the management of water off the farm.

25.3.5 Water reuse management: organisations and farmer interactions

During the interviews, an attempt was made to explore the factors which the farmers associated with the water management

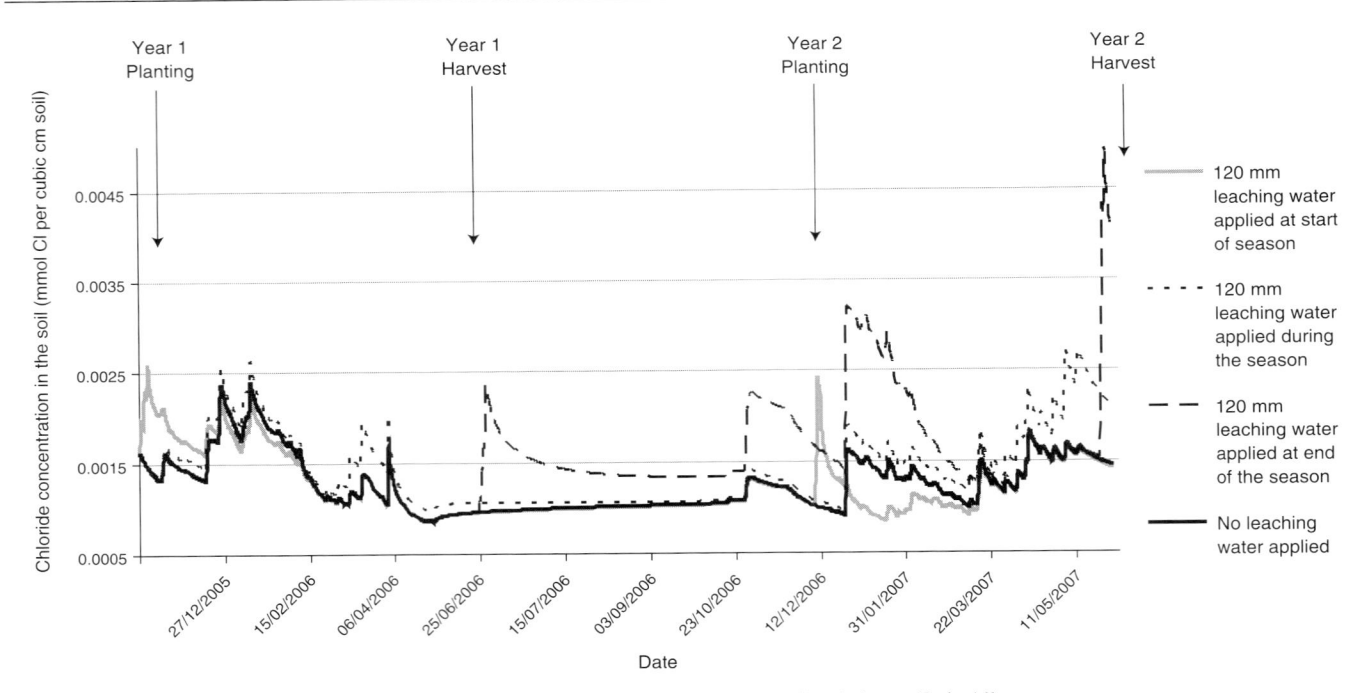

Figure 25.8. Modelled scenario of leaching timings on the Cl concentration in the soil solution at Deir Alla.

challenges they faced and to identify means by which they would be empowered to improve their soil management. Some key observations were made which highlighted the effect of decision-making at the institutional level on soil-water management at the farm level.

While most farmers were aware of the nutrients in the water, those who were not spoke of not having been given information on the nutrient content of the water or that even though they suspected the water had nutrients they continued to apply mineral fertilisers to ensure maximum productivity. The provision of information and training to farmers on the nutrient content of the irrigation water has been recognised by some organisations as being essential to enable farmers to manage their fertilisation schedules and maximise profits. During the interviews with representatives of organisations it became clear that data access and management was one of the challenges which restricted organisations from passing on data on water quality to farmers:

> There is a lot [of data] available which is measured by universities and research institutes, some even have a contract with the JVA to monitor the [water] quality. For example, reports are written and given to JVA and as far as I know nothing happens afterwards. All this information would be interesting for farmers. . . . Every organisation is collecting [data] for themselves so I think there's a lot of duplication. And the information is not passed onto the next organisation who might have a use for it and they have to collect the data again and so there's inefficiency. But there is also a problem because the JVA don't want to publish this data. The quality of water in the King Talal [Reservoir] is a very sensitive issue.
>
> An international organisation representative

The sensitivity of water reuse as an issue in Jordan must be fully appreciated as this adds a further challenge to the sustainable management of water reuse in Jordan. As one Jordan Valley farmer explained when asked if he had any problems selling his produce:

> I don't have problems selling my produce, but in 1994 a foolish journalist wrote that the vegetables from the Jordan Valley are irrigated with water from Kirbet as Samra wastewater treatment plant. Since then Saudi Arabia has banned all Jordanian products. We are still suffering from this article. Though now it is slowly getting better.

In the early 1990s Saudi Arabia placed a ban on the import of Jordan Valley vegetables because of concerns about the use of reclaimed water for irrigation (Haddadin, 2006). As a result, both farmers and the government are extremely cautious in discussing water reuse for irrigation in the Jordan Valley. Comments from government representatives demonstrate this caution:

> No vegetables are irrigated with treated wastewater, by our standards we can irrigate with this water but we do not use it in Jordan.
>
> A government representative

So in Jordan treated wastewater is only for the restricted agriculture, it's not for unrestricted crops. So we irrigate trees, industrial trees, we irrigate wood plants, we irrigate forages. But we do not allow under the Jordanian Standard [for water reuse] to irrigate vegetables or edible crops or plants.

A government representative

For water reuse you have to be sure of a very high quality of treated water to be sure that no problems will be encountered. If pure wastewater would be used then exports would be limited, this makes water reuse a very sensitive topic.

A representative from a non-governmental organisation

These comments show that before farmers can be provided with open and reliable data on the nutrient content of the irrigation water the barrier of sensitivity to water reuse needs to be overcome. Communication from the research and scientific community to society on the risks and benefits of reclaimed water based on empirical data is essential to help overcome these challenges.

The interviews with farmers in the Jordan Valley highlighted concerns about water quality due to its salinity and effect of clogging irrigation pipes. The interviews further suggested that farmers in the Jordan Valley place the responsibility for water quality on the government who have the power to make and enforce regulations to prevent industrial waste and poorly treated sewage entering the irrigation water supply. As one farmer commented: 'The water would be 100 per cent better if there were controls to stop industry dumping waste into the water.'

Through interviews with governmental representatives it became clear that while legislation is in existence to prevent industrial waste from entering the domestic sewage network illegal dumping by industries is considered to be a problem for which action is required:

> Industrial wastewater is legally disposed of to the sewage system from only four or five factories but there is a big problem of illegal disposal by many factories.

A government representative

> Some areas, like Khirbet as Samra wastewater treatment plant, have huge industries linked to their sewage systems. I don't know if I have to say this, but everywhere you can have illegal uses [of the sewage system]. Sometimes [the industries] just open the manholes and drain in their waste. So you cannot control waste like heavy metals. Now the Ministry of Environment has a law for the environment. They restrict for all industries [from putting waste into the sewage system]. Industries have to use their own treatment plants inside the industry and the output of their own treatment plants should conform to the Jordanian standard specifications [on effluent quality]. Otherwise, as it happens in Jordan, the industries will be closed and we already closed around fifty industries. So this issue is really important.

A government representative

The problem of highly saline brine from desalination plants being added to the sewage network or released directly to the wadis was also mentioned by a number of organisations:

> The salinity problem is not only from the domestic wastewater but they also have to find solutions for the brine of the desalination plants which they have in industries and until now, as far as we know, this waste is just drained into the sewer system or released to wadis. This is a management issue and very difficult to solve.

An international organisation representative

Clearly greater enforcement of existing legislation is required with consideration both for the illegal dumping of waste into the sewage system and for waste which is released untreated to the wadis. Control over the waste released by industries is needed in order that the water chain approach is applied which considers the water quality needs of the farmers and provides them with suitable water.

Farmers identified issues of water scarcity with the limited availability of water for leaching. However, the interviews showed that the farmers of the Jordan Valley believed that water shortage was not just a result of environmental conditions. While 38% of indirect-reuse farmers did relate water shortage to environmental scarcity, 40% of these farmers then went on to say how the management of the available water exaggerated the water shortage problems.

Water management issues identified by farmers could be divided into three groups: the allocation of water, which includes the geographical and temporal distribution of water to farmers; the infrastructure for water provision and the competition for water resources from other water use sectors such as tourism and domestic supply.

Many farmers claimed that the distribution of water was unfair; as one farmer said: 'The JVA is biased. They cut 30 per cent of my water and pump it to another person of their favour.' Other farmers said the JVA gave water to those whom it favoured and turned a blind eye to some farmers who stole water (by illegally opening the pipes) while others received fines. However, several farmers maintained that they did not experience problems with the water allocation and this may support claims that water distribution is inequitable. As another farmer explained:

> There isn't mismanagement, but the resource is limited. So what can the JVA do? The JVA tries to fulfil the needs of all the farmers. The problem is in the limitation of water resources.

A second aspect of water distribution involves the seasonal allocation of water by the JVA. As one farmer explained:

> There is water shortage and low water availability – but there is mismanagement. For example, lots of water is provided in winter when it is not needed but less water is given in March when the plant needs it. Why are they sending water when it is not needed – they can keep it in the dam? It is a simple problem of arranging the water balance – like financial accounts.

Water in the Jordan Valley is allocated by the JVA according to land area, crop type and season (Molle et al., 2008) and none of the interviews with organisations raised any issues regarding water distribution. In fact, no institutional agency recognised the importance of the timing of water supply in enabling farmers who irrigate with reclaimed water to leach their land.

Scarcity of water was mentioned by 55% of the organisations interviewed and solving the water shortage problems of Jordan is high on the political agenda. Several of the farmers felt it was the responsibility of the government to install and maintain appropriate infrastructure to meet the agricultural water demands:

> There are three items the government are responsible for: (1) The military, and you will see that they pay military salaries for 50 years. (2) To protect the basic supply of bread for the people through strategic sources. (3) Water, the government is responsible for this. If they can fund the military they can fund the water.
>
> Jordan Valley farmer

Despite the high awareness of water scarcity, very few organisations noted that infrastructure developments offered a way by which greater quantities of wastewater could be collected and transferred to irrigation. This suggests that infrastructure developments are not closely associated with the expansion of water reuse.

Within Jordan there is competition for the available water resources, and an increasing amount of the Yarmouk River water is being transferred from the Jordan Valley to Amman to meet domestic demands. This results in increased scarcity for irrigation water in the Jordan Valley. Some farmers expressed bitterness over this transfer, feeling that the water is put to better use in the Valley:

> The Yarmouk water which is sent to Amman is not acceptable for drinking. They pump it just to be used for car washing and irrigating the gardens!

The transfer of water for tourism was the subject of even more resentment, with comments emphasising the farmers' feelings that agriculture is the sector to which the country should be aligned, not tourism: 'the government is interested in tourism, and I don't know why, we're not even adapted for this sector.'

The organisations were encouraged to speak about the distribution of reclaimed water between the different water use sectors within Jordan to identify whether there were likely to be any competing demands for reclaimed water. Three institutions emphasised that there was no conflict for reclaimed water between agriculture and industry, and during one interview it was clearly shown that the farmers would always be given priority to the reclaimed water:

> Some industries are using reclaimed water and if the industry does want it then they will be given it. If there is any competition between the farmers and the industry then the farmers are given the priority. They are poor, have nothing and are suffering from the treatment plants and so it is important to try to solve their social problems.
>
> A government representative

However, during one organisational interview a situation of conflict for reclaimed water was described:

> That's happening in Aqaba now, there is a conflict now. All of the people want the water. The industry, they are asking for the water, the agriculture, they are asking for the water, the local authority, they are asking for the water for the landscape irrigation.
>
> A representative from the private sector

Of greater potential concern is the allocation of reclaimed water within agriculture. One institutional representative explained:

> But then there's another challenge coming up. Because the [reclaimed] water quality is better the farmers upstream are more interested in it and so it could happen that the water doesn't even reach the Valley. . . . It's a bad thing for the livelihoods of the [Jordan Valley] farmers, and also the question is why invest in upland farming when there's no farming there at the moment, threaten groundwater resources [owing to possible nutrient leaching to groundwater – author's addition] and lose the market advantage of the Jordan Valley in the wintertime. Because upland farming is summer farming. They don't get as much money as the farmers in the Valley as there's a market gap there. It's always an economic decision what or where the water is used. And if they come up with these ideas of growing economic crops, it would be a disaster.
>
> An international organisation representative

> The sustainability of agriculture in the Jordan Valley is more or less connected to water reuse. You cannot imagine how the situation would be if the government decided not to [allow water reuse]. To stop, for example, [because of] health aspects. The water that comes from Khirbet as Samra means that agriculture can continue in the Middle Jordan Valley. And if this stops the situation will be very bad, I cannot

imagine, because there is no water for agriculture. And from a social and economical point of view [reclaimed water] is the only guarantee for the sustainability of agriculture.

An international organisation representative

Farmers were asked about whether they felt supported and to whom they go when they need help, in order to determine the extent to which the institutions went beyond just issuing water and followed through with an aftercare programme that encouraged, trained and supported farmers working with reclaimed water. Of all the issues discussed these questions prompted the most passionate responses, which ranged from bitter disappointment in the lack of support to convinced enthusiasm that there was plenty of support.

The majority of indirect-reuse farmers spoke negatively of institutional support (64%). However, 44% also spoke positively about the support they had received (the total is greater than 100% as some farmers spoke both positively and negatively). The issue of support required further questioning and analysis to determine from whom farmers were and were not receiving support. Most negative comments were directed at state-run institutions such as the JVA and Ministry of Agriculture, with 41% of indirect users speaking about a lack of support supplied by these institutions, compared with 36% of direct water users. The support of projects was viewed positively by 23% of indirect water users and negatively by 8%, while no direct-reuse farmers spoke of project support.

During one interview the frustrations of the JVA were described:

> It's difficult because by law the responsibility of the JVA ends at the entrance to the farm. Any advice that they give to farmers is more than they need to do. But actually they are interested in giving information to the farmers and at the moment they have an information sheet. But it is not fully promoted or given to the farmers because they don't have the staff to do it. Or the skills. They don't have extension workers. They are the water distributors.
>
> An international organisation representative

The importance of communication between the wastewater treatment plant and the end user of the water (the farmer) was demonstrated at Ramtha. Here, several farmers explained how they had a good relationship with the manager of the treatment plant. They gave an example of when the added chlorine (to sterilise the treated effluent) was too high and causing crop damage, and so the manager agreed to reduce the quantity of chlorine being added to the water. This involvement in the treatment process is very valuable and undoubtedly leads to an improved perception of the water and greater ability to manage the potential problems at the farm level.

25.4 CONCLUSIONS

This research has demonstrated the complex and multi-faceted aspects of water reuse and shown that a key approach to improved understanding is that of inter-disciplinarity, which incorporates both the natural and social sciences. The available data on water quality have shown that careful nutrient management and leaching to prevent the accumulation of potentially plant-toxic ions are essential to ensure maximum productivity is obtained when using reclaimed water. The interviews with farmers have shown that awareness of nutrients in the irrigation water is generally high but consideration for the nature of agriculture is required to understand how the farmer responds to this knowledge. The direct-reuse farmers, irrigating fodder crops with low outlay costs appear to appreciate the 'free' nutrients more than the indirect-reuse farmers who irrigate high added value fruit and vegetable crops with high outlay costs. These farmers seem to place greater emphasis on the negative elements of reclaimed water which they cannot control (such as salinity and high suspended load) than the positive elements (nutrients) which they are able to control on the farm through mineral fertiliser additions.

The soil sampling and analysis from sites irrigated with reclaimed water has shown that solute accumulation does appear to occur in soils irrigated with reclaimed water but the application of an adequate leaching fraction reduces the soil salinity. A mathematical model (HYDRUS) has been of value for comparing the timings of leaching and suggests that leaching at the start of the growing season is more effective at maintaining low salinity in the soil root zone than leaching during or after the season. More data would be needed to confirm this result and explore how knowledge of the optimum timing of leaching can be applied to soil management on the farm.

Water quality control and water resource management at the institutional level have been shown to affect the ability of farmers to manage their soil resources successfully. The interviews with both farmers and organisations suggest that greater institutional awareness to the needs of the farmer would be beneficial in overcoming the challenges of water reuse. Consideration for the water quality and quantity requirements of the farmer would help shape an appropriate management strategy for water reuse. However, the sensitivity surrounding water reuse in Jordan raises an additional challenge. Awareness of the needs of the farmer (as the end user of reclaimed water) is made more difficult when the farmer is not recognised as a user of reclaimed water, as in the Jordan Valley.

Water reuse for irrigation will increase in the future, as urban populations rise, *per capita* water consumption goes up and sewerage networks are extended to connect more households with the wastewater treatment plants. The formation of an effective strategy which permits the long-term use of reclaimed water without detrimental effects on natural soil resources is essential. It is also achievable through recognising and building on the knowledge of all stakeholders.

REFERENCES

Abu-Madi, M. O. R. (2004) *Incentive systems for wastewater treatment and reuse in irrigated agriculture in the MENA region: Evidence from Jordan and Tunisia*. Unpublished PhD thesis: Delft University of Technology.

Allan, J. A. (2001) *The Middle East Water Question: Hydro-Politics and the Global Economy*. London: I. B. Tauris.

Allen, R. G., L. S. Pereira, D. Raes and M. Smith (1998) *Crop evapotranspiration – Guidelines for computing crop water requirements*. FAO Irrigation and drainage papers **56**. Rome: Food and Agriculture Organisation of the United Nations.

Al-Zu'bi, Y. (2007) Effect of irrigation water on agricultural soil in Jordan valley: an example from arid area conditions. *Journal of Arid Environments* **70**: 63–79.

Ammary, B. Y. (2007) Wastewater reuse in Jordan: present status and future plans. *Desalination* **211**: 164–176.

Ayers, R. S. and D. W. Westcott (1985) *Water Quality for Agriculture*. Irrigation and Drainage Paper 29 Rome: Food and Agriculture Organisation of the United Nations.

Barth, S. (1998) Application of boron isotopes for tracing sources of anthropogenic contamination in groundwater. *Water Research* **32**: 685–690.

Bashabsheh, I. M. (2007) *Impact of treated wastewater on soil properties and cut flower production under drip irrigation systems*. Unpublished MSc thesis: Jordan University of Science and Technology.

Carr, G., S. Nortcliff and R. Potter (2008) Water reuse for irrigated agriculture in Jordan: what's waste about wastewater? *Bulletin of the Council for British Research in the Levant* **3**: 35–40.

Duran-Ros, M., J. Puig-Barques, G. Arbat, J. Barragan and F. R. de Cartagena (2009) Effect of filter, emitter and location on clogging when using effluents. *Agricultural Water Management* **96**: 67–79.

Falkenmark, M., J Lundquist and C. Widstrand (1989) Macro-scale water scarcity requires micro-scale approaches: aspects of vulnerability in semi-arid development. National Resources Forum 13: 258–267.

FAO (2008) *Aquastats Database of the Food and Agriculture Organisation, Rome, Italy*. Retrieved August, 2008. http://www.fao.org/nr/water/aquastat/dbase/index.stm.

Feign, A., I. Ravina and J. Shalhevet (1991) *Irrigation with Treated Sewage Effluent*. Israel: Springer-Verlag.

GTZ (2005) Long-term groundwater and soil monitoring concept – Final Report 2005. Amman: Reclaimed Water Project of the Deutsche Gesellschaft für Technische Zusammenarbeit.

Haddadin, M. (2006) *Water Resources in Jordan. Evolving Policies for Development, the Environment, and Conflict Resolution*. Washington: RFF Press.

Harmanto, V. M. Salokhe, M. S. Babel, and H. J. Tantau, (2005) Water requirement of drip irrigated tomatoes grown in greenhouse in tropical environment. *Agricultural Water Management* **71**: 225–242.

Hajrasuliha, S., D. K. Cassel and Y. Rezainejad (1991) Estimation of chloride ion concentration in saline soils from measurement of electrical conductivity of saturated soil extracts. *Geoderma* **49**: 117–127.

Huibers, F. P. and J. B. Van Lier (2005) Use of wastewater in agriculture: the water chain approach. *Irrigation and Drainage* **54**: S3–S9.

Hussain, G. and A. J. Al-Saati (1999) Wastewater quality and its reuse in agriculture in Saudi Arabia. *Desalination* **123**: 241–251.

Jimenez, B. (2008) Wastewater use in agriculture: public health considerations. In *Encyclopaedia of Water Science*, ed. S. W. Trimble. Boca Raton, FL: CRC Press pp. 1303–1306.

Kitchin, R. and N. J. Tate (2000) *Conducting Research in Human Geography: Theory, Methodology and Practice*. Harlow: Prentice Hall.

Lazarova, V., H. Bouwer and A. Bahri (2005) Water quality considerations. In *Water Reuse for Irrigation: Agriculture, Landscapes and Turf Grass*, ed. V. Lazarova and A. Bahri. Boca Raton, FL: CRC Press pp. 31–63.

Molle, F. and J. Berkoff (2006) *Cities Versus Agriculture: Revisiting Intersectoral Water Transfers, Political Gains and Conflicts. Comprehensive Assessment Research Report 10*. Colombo, Sri Lanka: IWMI.

Molle, F., J. P. Venot and Y. Hassan (2008) Irrigation in the Jordan Valley: are water pricing policies overly optimistic? *Agricultural Water Management* **95**: 427–438.

MWI (Ministry of Water and Irrigation) (2004) *National Water Master Plan of Jordan*. Amman: GTZ and Ministry of Water and Irrigation.

MWI (2009) *Water for Life. Jordan's Water Strategy 2008–2022*. Jordan: Ministry of Water and Irrigation.

Oron, G., C. Campos, L. Gillerman and M. Salgot (1999) Wastewater treatment, renovation and reuse for agricultural irrigation in small communities. *Agricultural Water Management* **38**: 223–234.

Patterson, R. A. (2001) Wastewater quality relationships with reuse options. *Water Science and Technology* **43**: 147–154.

Postel, S. (1992) *The Last Oasis: Facing Water Scarcity*. London: Earthscan.

Raschid-Sally, L., R. Carr and S. Buechler (2005) Managing wastewater agriculture to improve livelihoods and environmental quality in poor countries. *Irrigation and Drainage* **54**: S11–S22.

Rowell, D. L. (1994) *Soil Science: Methods and Applications*. Harlow: Prentice Hall.

Shainberg, I. and J. D. Oster (1978) *Quality of Irrigation Water*. International Irrigation Information Centre Publication no. 2. Bet Dagan, Israel: Volcani Centre.

Shelef, G. and Y. Azov (1996) The coming era of intensive wastewater reuse in the Mediterranean region. *Water Science and Technology* **33**: 115–125.

Šimůnek, J., M. T. van Genuchten and M. Šejna (2005) *The Hydrus-1D software package for simulating the one-dimensional movement of water, heat, and multiple solutes in variably-saturated media. Version 3.0 HYDRUS Software Series 1*. California: Department of Environmental Sciences, University of California Riverside.

Testi, L., F. J. Villalobos, F. Orgaz and E. Fereres (2006) Water requirements of olive orchards: I simulation of daily evapotranspiration for scenario analysis. *Irrigation Science* **24**: 69–76.

Vallentin, A. (2006) *Agricultural Use of Reclaimed Water – Experiences in Jordan*. Water Practice & Technology. IWA Publishing. DOI: 10.2166/WPT.2006040.

Vogeler, I. (2009) Effect of long-term wastewater application on physical soil properties. *Water Air and Soil Pollution* **196**: 385–392.

Water Authority of Jordan (2007) *Frequently asked questions about the Water Authority of Jordan*. Retrieved March, 2008. http://www.waj.gov.jo/English/left/faq.htm.

WHO (2006) *Wastewater Use in Agriculture*. WHO Guidelines for the safe use of wastewater, excreta and greywater. Geneva: World Health Organisation.

26 Social equity issues and water supply under conditions of 'water stress': a study of low- and high-income households in Greater Amman, Jordan

Khadija Darmame and Robert Potter

ABSTRACT

One of the distinctive characteristics of the water supply regime of Greater Amman is that it has been based on a system of rationing since 1987 with households receiving water once a week for various durations. This reflects the fact that while Amman's recent growth has been phenomenal, Jordan is one of the 10 most water-scarce nations on earth. Amman is highly polarised socio-economically, and by means of household surveys conducted in both high- and low-income divisions of the city, the aim of this chapter has been to provide detailed empirical evidence concerning the storage and use of water, the strategies used by households to manage water, and overall satisfactions with water supply issues, looking specifically at issues of social equity. The analysis demonstrates the social costs of water rationing and consequent management to be high, as well as emphasising that issues of water quality are of central importance to all consumers.

26.1 WATER AND DEVELOPMENT AND THE CASE OF GREATER AMMAN, JORDAN

Access to adequate supplies of water is a universal component and indicator of human well-being and development. It plays a fundamental role in helping to resolve some of the manifold problems associated with poverty, disadvantage and exclusion. The United Nation's second *World Water Development Report* (WWAP (World Water Assessment Programme), 2006) serves to emphasise that human development is inextricably linked with issues of water availability and management, in terms of proximity, quantity and quality.

In order to increase globally the number of households connected to both the water and sanitation networks, several programmes and initiatives related to the water sector have been launched since the 1970s. However, despite these, according to the World Health Organisation and UNICEF, almost one-sixth of the world population of 6.1 billion remains without access to improved water supplies, and two-fifths do not have access to adequate sanitation (WHO and UNICEF, 2000). The United Nations proclaimed the 10-year period from 2005 to 2015 as the international decade for action in respect of Water for Life. At much the same time, the United Nations (2000) announced the Millennium Development Goals (MDGs), which include clear targets for the improvement of access to water and sanitation (see, for example, Potter *et al.*, 2008; Rigg, 2008). One of the principal targets of the MDGs is to reduce by half the proportion of people without sustainable access to safe drinking water by the year 2015 (UNDEP, 2004).

It is noticeable that the principal objective of all these imperatives has been to improve the quantity of water available, while the supply of water, like all urban services, has an equally important qualitative aspect that affects the daily lives of all householders. Reflecting this, several recent research studies have affirmed that having a connection to the network is not always synonymous with adequate access to water, in terms of quantity, quality and regularity. Good examples of this principle have been reported in recent studies by Chikher (1995) in the context of Algiers, Allain-El Mansouri (1996) in Rabat-Sale, Zérah (1999) in Delhi, and Darmame (2004) in Amman. These studies illustrate the impact that an intermittent supply of domestic water can have on households and the costs that are involved in strategies devised to cope with such problems. The lack of a continuous supply of water, whether under a prevailing system of public or private management, often serves to exacerbate socio-spatial inequalities in overall access to water. Thus, although the poor may be well-connected in terms of the network, the quality

Water, Life and Civilisation: Climate, Environment and Society in the Jordan Valley, ed. Steven Mithen and Emily Black. Published by Cambridge University Press.
© Steven Mithen and Emily Black 2011.

of their supply may be irregular and unpredictable (see Zérah, 1998; Graham and Marvin, 2001; Jaglin, 2001; Darmame and Potter, 2008; Mitlin, 2008).

Over the past two decades, Jordan has suffered from a chronic water crisis, as manifest by a deficit in the balance between the demand for water and available water resources and financial investment in the water sector. However, despite such difficulties, Jordan has achieved one of the objectives of the MDGs. As recorded in the *Household Expenditure and Income Survey of 2006* (Hashemite Kingdom of Jordan, 2007) some 98% of households in the country are connected to the public supply network, so universal water access has effectively been achieved for rich and poor alike (Darmame, 2006). However, as in many developing countries, the efficacy of water supply is affected by several dysfunctions. One of these is the intermittent basis of the supply of water, following a weekly rationing programme, which constrains the day-to-day lives of individuals and households (see Iskandarani, 1999; Darmame, 2004; Potter *et al.*, 2007b). It is these key issues of non-continuous supply that are specifically addressed in this chapter in terms of the social equity issues that arise and their potential relevance in the policy arena.

Greater Amman's water supply has been rationed since 1987, with households receiving water once a week for various durations (see also this volume, Chapter 28). This reflects the fact that Jordan is one of the 10 most water-scarce nations on Earth and has long suffered from a structural crisis in the water sector, amounting to what may be referred to as 'water stress'. Further, in 1999 the water supply system of the capital was privatised. The privatised company LEMA oversaw the reduction in 'unaccounted for' or 'lost' water, and upgraded the network as well as making improvements in billing and debt collection (see this volume, Chapter 28). In January 2007, the water supply system of Greater Amman was effectively 'deprivatised' and placed in the hands of a local company, Meyahona ('Our Water'), which is owned by the Water Authority of Jordan, although its remit has remained avowedly commercial.

Water, both for domestic and commercial purposes, is metered and charged for in Jordan, although there is a marked subsidy to the poor, as can be discerned from Figure 26.1. For example, domestic water is charged at just two or three Jordanian Dinars (JD) or euros for quarterly levels of consumption less than 20 cubic metres. In the Water Strategy for Jordan, produced in 1997 (MWI (Ministry of Water and Irrigation), 1997), the Jordanian Government stated its first priority as being to meet the basic water needs of the urban populace, an intention that was confirmed in the National Water Master Plan produced in 2004 (Deutsche Gesellschaft für Technische Zusammenarbeit (GTZ) and MWI, 2004). Indeed, over recent years there has been talk of moving to a comprehensive system of continuous supply, and as an experimental run up to this, in the winter of 2006 continuous supply was introduced to 15.8% of customers

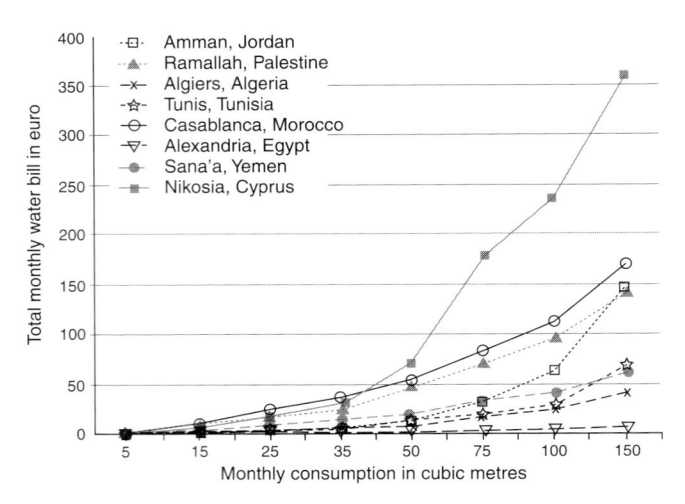

Figure 26.1. Water charges in Amman, Jordan and a selection of other cities in the MENA region. Exchange rates in 2004 (from Magiera *et al.*, 2006).

within Greater Amman (see also this volume, Chapter 28, for further details).

Given the realities of water rationing, water charging, privatisation and deprivatisation, and the possibility of establishing continuous supply, the present chapter reviews how different groups of urban consumers in Greater Amman manage and use water as a household resource, on a daily basis. It also considers the important topic of householders' attitudes to and perceptions of the urban water supply system, and what these reveal about the likely impacts of future developments in the water sector. But before turning to the results of this first-hand research, it is necessary to briefly introduce the Amman region, which is one of the fastest-growing cities in the developing world.

26.2 THE GROWTH AND STRUCTURE OF GREATER AMMAN

The growth of Amman in the second half of the twentieth century through to the start of the twenty-first century has been phenomenal, in terms of its total population, physical extent and regional geopolitical salience (see Figure 26.2). What was in the early 1920s a small town consisting of little more than 2,000–3,000 people is today a major regional city, and had an estimated population of 2.17 million at the start of 2006 (Hashemite Kingdom of Jordan, 2007). The growth of Amman over the past 100 years has principally reflected wider political and geopolitical factors. The city has throughout this period attracted migrants, and its essential political stability has led to successive waves of in-migration from Syria, Palestine, Lebanon and Iraq. Further religious and ethnic minorities, such as the Armenians and Kurds, have also migrated to the city. But it has been those Palestinians displaced as a result of the foundation of the State

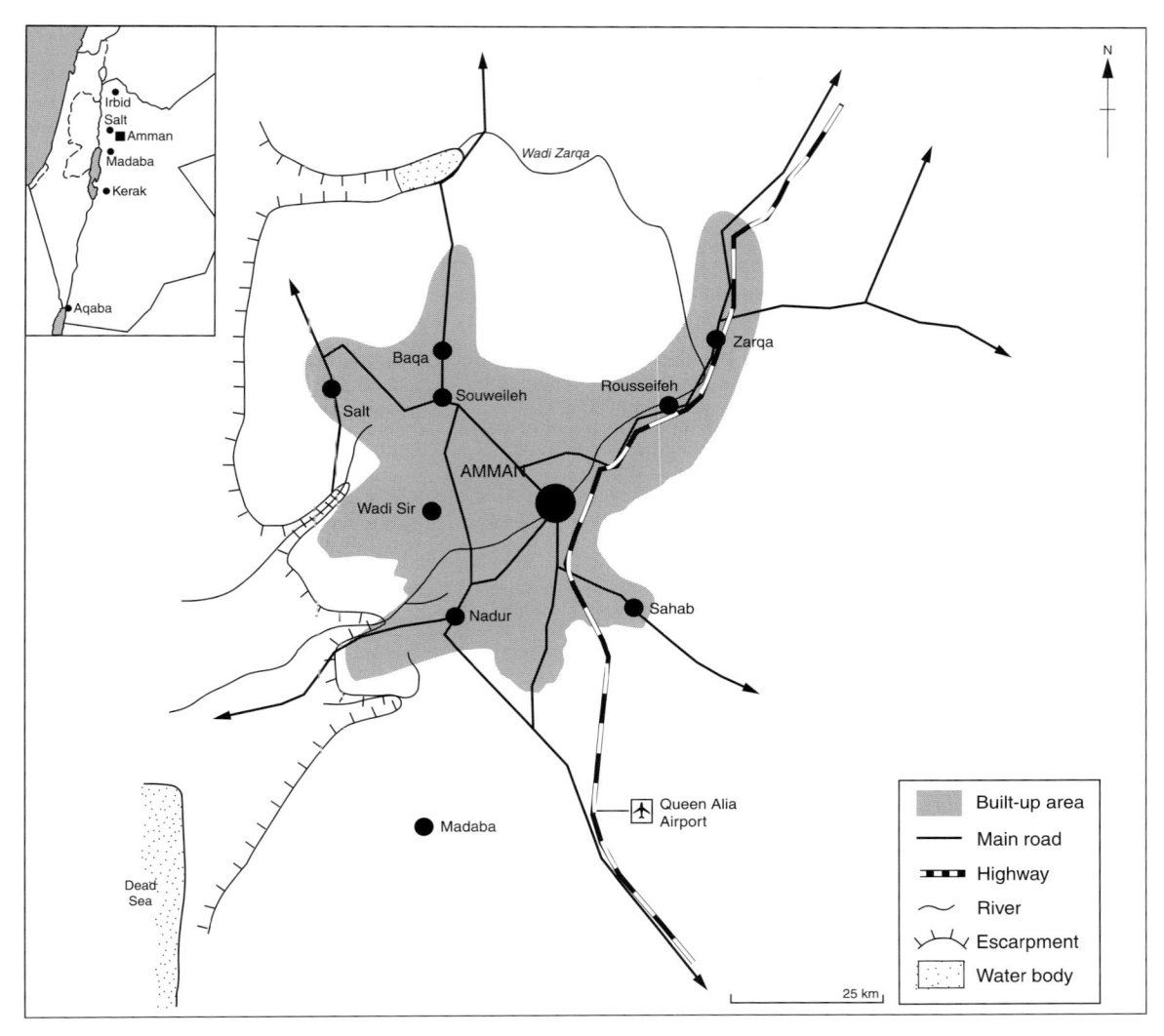

Figure 26.2. The Greater Amman urban region: general location map (revised and adapted from Lavergne, 2004).

of Israel in 1948 that have constituted the main wave of migrants to Amman in particular and Jordan in general.

As a result, the social transformations in the city over this period have been no less spectacular. Abu Dayyah (2004) reports that at the time of the 1952 Census, when the population of the city stood at around one-quarter of a million, as much as 29% of the population was living in tents and a further 8% in natural caves. But by 2006 there were few inhabitants of 'temporary' or 'makeshift' settlements.

Physically, Amman lies on the undulating plateau that makes up the northwest of Jordan (Figure 26.2). The original site of the city occupied seven hills or 'jabals' around the Wadi 'Ras el Ain, which flows northeast from the plateau toward the River Zarqa basin. The original central part of the city was located at an altitude of between 724 and 800 m, and the expansion of the city over the past 25 years has resulted in the occupation of some 19 distinct hills, with an altitudinal extension to 875 m and above, so that the topography of the contemporary city consists of a series of steep hills and deep and sometimes narrow valleys. While initially urban development principally occurred on the upper slopes and crests and on the lower slopes of this hill–valley system, the upsurge in urban growth over the past 60 years has seen extensive urban development on the generally steeper mid-slope locations.

As documented in Potter *et al.* (2009) a correlate of this history of rapid physical growth has been the greater social divide that has come to characterise the residential quarters of present-day Amman. The eastern portions of the city, which include the downtown area, house the urbanised poor living at high overall densities, and the zone is characterised by large Palestinian refugee camps located on its periphery (Ham and Greenway, 2003; Potter *et al.*, 2009). These eastern tracts of the city are characterised by their Islamic wordview and generally more conservative nature. In contrast, the western and northern portions of the city are associated with relatively affluent socio-economic groups within society. Many of these areas are associated with essentially 'modern' lifestyles. For example, Ham and Greenway (2003) note that the upmarket central district of Shmeisani is often referred to by local denizens as 'Shiny Amman', while the trendy western residential area of Abdoun is frequently, and with more than a hint of irony, dubbed 'Paris'. The overall pattern of residential social stratification reflects the early growth of Jabal Amman to the immediate west of the city centre as the initial wealthy quarter. As the city grew in response to successive waves of in-migration, the high-status residential zone has progressively extended westward out from the city centre. Notably, this direction of movement has been towards the relatively high land above sea level (Hannoyer and Shami, 1996; Potter et al., 2009). It is in this context of a fast-growing city in the developing world that we now turn to examine issues concerning water as an essential service for the urban population of what has been called 'ever-growing Amman' (Potter et al., 2009).

26.3 HOUSEHOLD WATER SURVEYS: RESEARCH DESIGN

As part of the Development Studies sub-component of the Water, Life and Civilisation project, social variations in household demand for, management of, and satisfaction with the water supply system in Greater Amman have been examined (Potter et al., 2007b). Specifically, the aim has been to provide detailed empirical evidence concerning the storage and use of water within different social areas of the city, the strategies that are used by households to manage water under conditions of water stress and rationing, and the general degree of satisfaction with contemporary water supply conditions and related issues.

As exemplified in the previous section, Amman shows a marked socio-economic divide (see Potter et al., 2007a; Darmame and Potter, 2008; Potter et al., 2009); so, how do different social and income groups within the city use water and, specifically, how do they react to the realities of rationing, the daily management of water and the prospects of continuous supply? At the societal level, of course, it must be recognised that throughout time, the control and use of water has been closely related to social power and wider forms of societal organisation (see Potter and Lloyd-Evans, 1998).

The study design was based on examining potential social equity dimensions in the use of water within the city, so housing areas were selected in both low- and high-income districts of Greater Amman. As noted, Amman can easily be divided into its relatively high-income western and northwestern tracts (see Potter et al., 2007a; Potter et al., 2009) and its relatively low-income eastern portions, as denoted in Figure 26.3. Twenty-five households were selected in low-income eastern Amman. The specific households were selected from existing contacts, and snowball sampling was employed thereafter. The interviews were carried out in five residential sub-areas of eastern Amman. These were: (i) Wihdat, the second largest Palestinian camp in Jordan; (ii) Wadi Haddada, an area of informal settlement; (iii) Al-Nasser, a Palestinian settlement area; (iv) Nazal; and (v) Quisma, both low-income popular housing areas within the city. The location of these five residential study areas in the central eastern and southeastern areas of the city is depicted in Figure 26.3.

Additionally, 25 households were selected in five areas of high-income western Amman (see Figure 26.3). These were: (i) Abdoun, one of the wealthiest areas of the city, consisting of luxury houses; (ii) Swifieh, an area served by an up-market commercial district; (iii) Deir Ghbar, an area of luxury villas and houses; (iv) Jabal Amman, one of the earliest high-status residential zones of the city; and (v) Jbeha, located close to the University of Jordan in the northwestern neighbourhood (for all residential locations see Figure 26.3).

For both the high- and low-income residential areas, structured interviews collating quantitative data were conducted using a printed *pro forma*. Specifically, issues such as access to the public water supply, the means and extent of household water storage, the daily use of water and the management strategies employed, along with wider perceptions, attitudes and satisfactions, were fully explored with respondent households.

26.4 SOCIO-ECONOMIC AND DEMOGRAPHIC CHARACTERISTICS OF RESPONDENT HOUSEHOLDS

The effectiveness of the surveys would naturally depend on the appropriateness of the household samples selected. As would be expected, the main difference between the low- and high-income households surveyed was in their average income. The high-income households interviewed recorded an average monthly income of JD 1,932, as opposed to JD 235 for the low-income households (see Table 26.1). This eight-fold disparity in household income levels directly reflects the degree of social polarity that characterises contemporary Amman (see Potter et al., 2009). The distribution of households by income is particularly instructive, with 100% of low-income sampled households earning less than JD 500 per month and 64% of high-income households earning in excess of JD 1,001 per month, as shown in Table 26.2.

Looking at the broad occupational categories of the heads of households included in the survey, the principal feature is the noticeably higher level of self-employment recorded among the high-income households, around 44%, as opposed to 28% for the low-income households (Table 26.3). At the time of the survey, unemployment stood at some 4% for the high-income households, but was as high as 24% in the case of the low-income Ammani households we interviewed (Table 26.1).

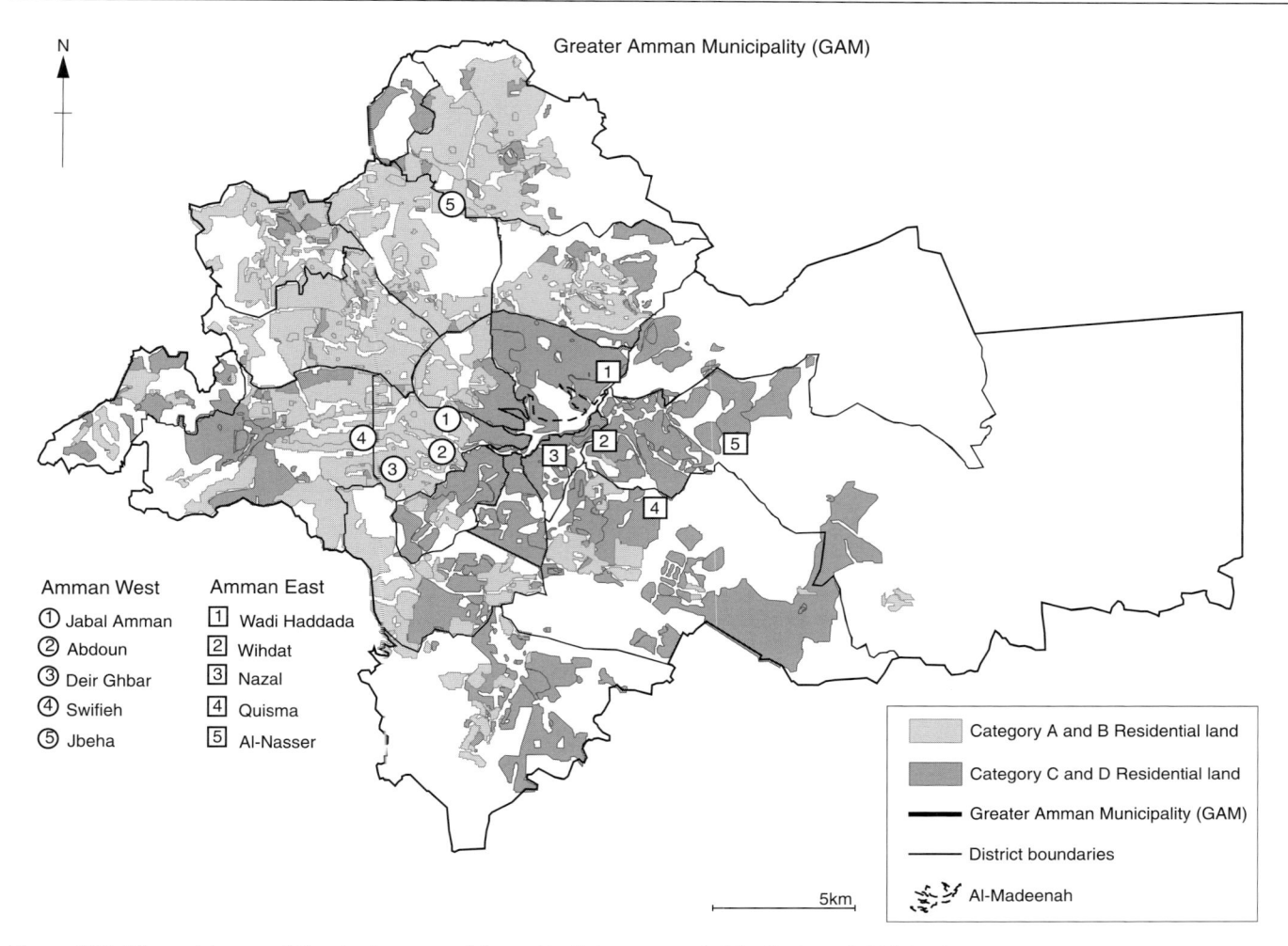

SOCIAL EQUITY AND WATER SUPPLY IN GREATER AMMAN

Figure 26.3. The social areas of Greater Amman and the residential areas sampled for the household interviews.

Table 26.1 *Socio-demographic profile of the respondent households*

Socio-demographic variable	High-income households	Low-income households	Entire sample
Average age (years)	49	43	45
Female (%)	32	40	36
Male (%)	68	60	64
Married (%)	84	84	84
Employed (%)	72	48	60
Unemployed (%)	4	24	14
Retired (%)	16	16	16
Average income (monthly net JD)	1,932	235	1,029

Usefully, however, in wider demographic terms, the sample households showed broad similarities. Thus, both income groups consisted of relatively youthful households, with the average age of those interviewed being in their 40s (Table 26.1). Identical levels of marriage (84%) and retirement (16%) were recorded in

Table 26.2 *Distribution of respondent households by income group*

Income range (net monthly income, JD)	Percentage of households by category		
	High-income households	Low-income households	Entire sample
≥500	4	100	52
501–1,000	32	0	16
≥1,001	64	0	32

both income groups. For the sample as a whole, more males than females were interviewed, although as shown in Table 26.1, this proportion was marginally higher in the case of the respondents from the high-income households, standing at 68%. Families tend to be large in Jordan, and this is evident for the sample households, with an average family size of 5.82 persons for all respondent households, and a somewhat higher figure of 6.44 for the low-income households (see Table 26.4).

Table 26.3 *Occupational categories of the heads of households included in the sample*

Sector	Percentage of household heads		
	High-income households	Low-income households	Entire sample
Government	24	16	20
Private	24	28	26
Self-employed	44	28	36
Military/police	0	4	2
Farmer	8	0	4

Table 26.4 *Residential profiles of the respondent households*

Residential variable	Households by category		
	High-income households	Low-income households	Entire sample
Owner-occupiers (percentage)	92	68	80
Average house size (sq. m)	345	82	213
Average size of family	5.2	6.44	5.82
Sharing with other occupants (percentage)	8	20	14

Table 26.5 *Distribution of respondent households by size of dwelling*

Size of dwelling (m^2)	Percentage of households by category		
	High-income households	Low-income households	Entire sample
≤ 100	0	76	38
101–200	48	24	36
201–300	20	0	10
301–400	12	0	6
≥ 401	20	0	10

Table 26.6 *Household water supply from the public network*

Water supply characteristic	Percentage of households		
	High-income households	Low-income households	Entire sample
Connected to the supply network	100	96	98
Sharing a water meter	12	56	34
With their own private cistern	40	8	24
Using a pump	40	28	34
Using a filter	20	16	18

An examination of the wider socio-economic data for the respondent households shows distinct differences between the two samples. Average house size is 345 square metres for the high-income households interviewed and considerably lower at 82 square metres for the low-income households (Table 26.3). This disparity becomes even more evident if the frequency distribution of households by size of dwelling is examined in detail (Table 26.5). Thus, while 76% of low-income households live in houses comprising 100 square metres or less, some 20% of high-income households live in houses with 401 or more square metres of living space. Levels of owner-occupation of housing are significantly higher at 92% for the high-income households than for the low-income households at 68% (Table 26.4). Once again, the social polarity of everyday life in Amman is clearly revealed by these data.

26.5 HOUSEHOLD ACCESS TO THE PUBLIC WATER SUPPLY

We now turn to the results of the household surveys that we conducted in these two different residential area types. As already noted, Jordan has high levels of overall access to the water supply network. This was confirmed for the sample households with an aggregate level of 98% connection for the entire sample (Table 26.6). The survey data also showed how similar the high- and low-income households are with regard to connection to the public water network, standing at 100% and 96% of households, respectively. Mains water is thus effectively ubiquitous in the Greater Amman urban area, regardless of income level and geographical area of residence within the city.

However, as water is provided only once a week for various durations, the daily continuity of the supply of water depends on the ability of households to store water in roof-top tanks and underground cisterns, something that involves substantial infrastructure costs. Some 56% of the low-income households surveyed shared a water meter, reflecting the fact that several families or two generations of the same family were sharing both the supply and storage of mains water (Table 26.6). Such sharing was much lower at only 12% in the case of the high-income families and stood at 34% for the entire sample of households.

To augment storage capacity and to increase water availability, households have developed strategies for daily water use. For example, households store water in near-ubiquitous roof-top tanks (Figure 26.4) with a typical capacity of 2 cubic metres.

Table 26.7 *Average household water storage capacity by income group*

Income group	Average storage (m^3)
High income	16.24
Low income	3.12
Entire sample	9.72

Figure 26.4. Typical roof-top storage tanks on a property in the northwestern suburbs of Amman.

Table 26.8 *Respondent households by water storage capacity*

Capacity (m^3)	Percentage of households		
	High-income households	Low-income households	Entire sample
1	–	20	10
2	–	40	20
3	4	4	4
4	36	8	22
5	–	12	6
6	–	8	4
7	4	4	4
8	12	–	6
9	–	–	–
≥10	44	4	24

In addition, households can invest in the construction of underground water cisterns as well as using a water pump to aid supply when the water pressure is low. To improve the quality of the water piped into the house, consumers may install and use a filter. As each of these responses potentially involve infrastructural, installation and operating costs, it seems highly likely that their use may closely match, and thereby amplify, existing social inequalities. Thus while some 40% of the high-income households had their own private cistern (Table 26.6), this was much lower at 8% for the sample of low-income households. Pumps were actively used by 40% of the financially well-off consumer households and 28% of those on relatively low incomes. Using a filter stands at less than one-fifth of the entire sample, but is slightly higher for the high-income householders at 20%, as opposed to 16% of low-income families (see Table 26.6).

One of the most pronounced contrasts shown in our survey was in the total water storage capacity that characterised the two social income groups surveyed. The average maximum storage capacity of the entire sample of households surveyed was 9.72 cubic metres (Table 26.7). However, when the high-income households were compared directly with the low-income households, the difference was revealed to be over five-fold, with an average water storage capacity of 16.24 cubic metres for the high-income and 3.12 cubic metres for the low-income

households. The extent to which household water storage capacity is skewed is shown if the more detailed frequency distribution of households by storage capacity is considered, as shown in Table 26.8. This shows, for example, that while 60% of high-income households have 7 cubic metres of storage or more, 72% of low-income households have 4 cubic metres of water storage or less. This socio-economic contrast in water storage is clearly shown in Figure 26.5.

26.6 WATER USE BY HOUSEHOLDS AND HOUSEHOLD WATER MANAGEMENT STRATEGIES

This research also confirmed that, although not quite so unequal and socially polarised, levels of water consumption and household expenditure on water closely reflect socio-economic and income variations in Amman. On average, the high-income households we interviewed consumed 70.24 cubic metres of water per quarter, while the low-income households consumed around half this total at 32.68 cubic metres (Table 26.9). Reflecting the payment subsidy given at relatively low levels of water supply, this two-fold difference in consumption was paralleled by low-income consumers paying, on average, 3.76 times less for their water per quarter. Thus, the average water bill for the high-income households was 55.80 JD per quarter, against 14.84 JD for the low-income households (Table 26.9). However, it is important to note that the low-income households are thereby devoting a higher proportion of their income to the purchase of water. This is reflected in an eight-fold disparity in income versus less than a four-fold difference for water cost between the two sets of households.

The ability of wealthier households to access water more easily was, of course, shown by their greater propensity to buy water from water tankers (Figure 26.6): 24% of high-income

Table 26.9 *Household water consumption and cost levels*

Aspect of consumption	High-income households	Low-income households	Entire sample
Average consumption (m^3 per quarter)	70.24	32.68	51.46
Average water bill per quarter (JD)	55.80	14.84	36
Percentage of households buying bottled water	44	20	32
Average spent on bottled water (JD per week)	10.45	8.2	9.75
Percentage of households buying water from private water tankers	24	4	14
Average spent on water from tankers (JD per summer period)	20	17	19.57

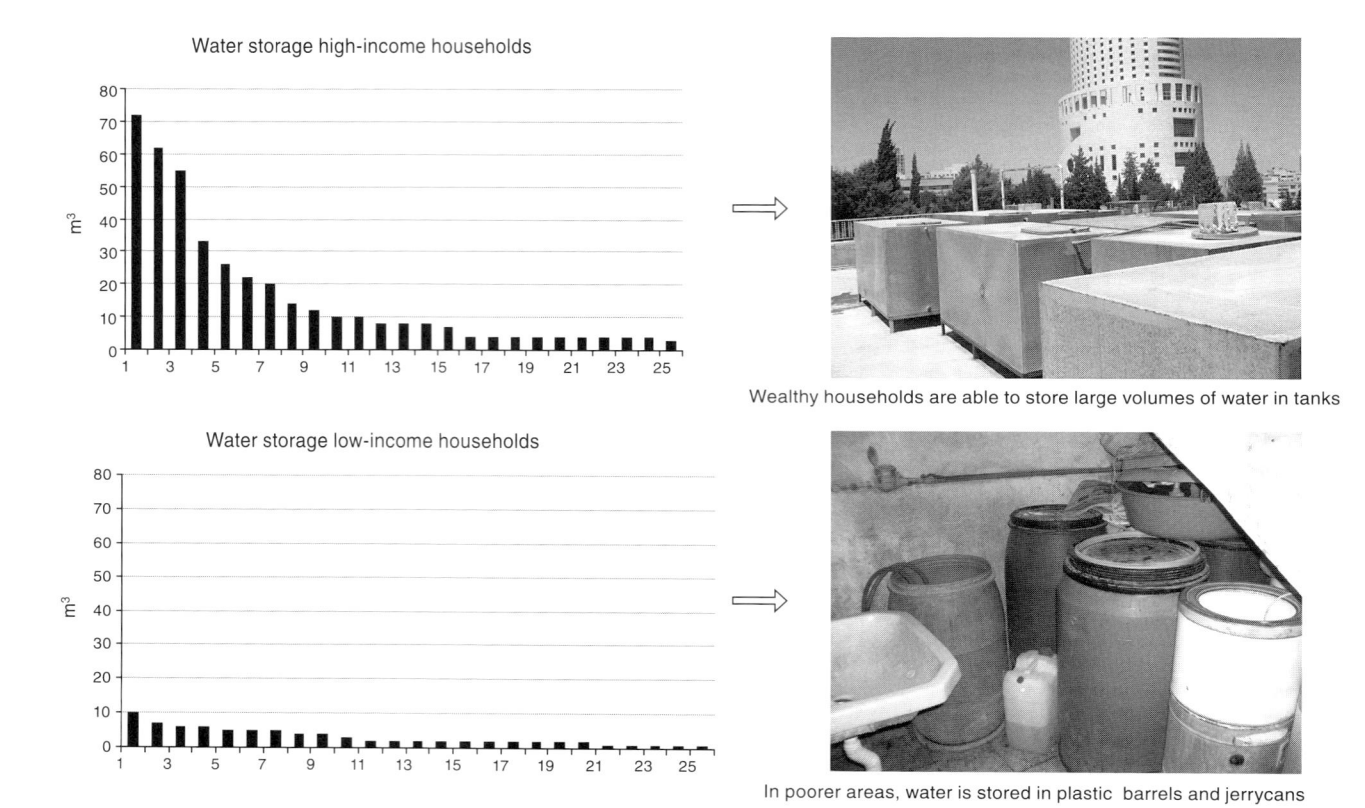

Figure 26.5. Income-related variations in water storage capacity among the interview respondents. Each of the 25 high-income and low-income respondents is represented by a vertical bar.

households report that they regularly buy water in this way, as opposed to only 4% of the poorer households. In purchasing water from tankers, the wealthier households spent on average 20 JD during the summer months, as opposed to 17 JD spent by the lower-income households. Differential access to water resources was also shown by the fact that 44% of high-income households stated that they regularly purchased bottled water, spending on average 10.45 JD per week, against 20% for low-income households at an average cost of 8.2 JD per week.

Respondent households were also asked about how they paid their water bills. Table 26.10 shows that the majority of consumers, some 52%, paid their bills at the post office; but as expected, this was noticeably higher at 62% in the case of the low-income households, whilst accounting for 40% of high-income consumers. The question revealed that just under a third of all households made their payments directly at the offices of the Water Authority of Jordan, and this proportion was identical for the two income groups. The main difference shown in Table 26.10 is that over a quarter of high-income households pay their water bill via their bank accounts, a process that is not possible for the consumers from low-income households who rarely hold bank accounts.

The rationing of water has both financial and time costs for consumers. To cope with rationing, households have developed

Table 26.10 *Method of payment of water bill*

Method	Percentage of households		
	High-income households	Low-income households	Entire sample
Post office	40	64	52
Bank	28	0	14
WAJ office	32	32	32
Other	0	4	2

Table 26.11 *Details of the use of networked water by households*

Aspects of water use	Percentage of households		
	High-income households	Low-income households	Entire sample
Special organisation of tasks on the day of water supply	64	84	74
Using networked water for drinking	12	52	32
Using networked water for cooking	76	96	86
Using networked water for laundry	100	100	100
Using networked water for bathing	100	100	100
Using networked water for cleaning	100	100	100
Using networked water for gardening	40	0	20

Figure 26.6. Typical water delivery tanker in Greater Amman, Jordan.

water management strategies. Some 74% of households stated that they adopt specific strategies to carry out their household tasks on what is commonly referred to as 'the day of water', whereby they organise bathing, housework, laundry, cleaning and gardening at a specific time or in a specific way, as can be seen from Table 26.11. As might be expected, the proportion of households stating that they organise tasks specifically on the day water is supplied was higher among the low-income households, standing at 84%, as opposed to 64% for high-income households.

Further questioning about water use within the household demonstrated all too clearly that water quality is a major issue. Thus, just less than a third of households (32%) stated they used networked water for drinking (Table 26.11). The proportion using mains for drinking was as low as 12% for the high-income households, but over half (52%) of the low-income households. All respondent households recounted past incidents related to water pollution and ill health. Similarly, nearly all low-income consumers (96%) reported that they use networked water for cooking; this figure is lower for the entire sample at 86% because a lower proportion (76%) of high-income consumers report that they use mains water for such purposes. All households interviewed, regardless of income level, reported that they used networked water for doing the laundry, bathing and cleaning.

When asked about the use of networked water for gardening, as expected, a clear socio-economic divide re-emerged, with 40% of high-income households stating that they did so, but none of the low-income households (Table 26.11).

The household surveys showed that 'the day of water' required the specific management of all household tasks, including personal hygiene, in a limited time. In this respect, it was apparent that gender was a salient issue. Revealingly, when asked, 84% of low-income households reported that women take overall responsibility for the day-to-day management and control of the use of water in the home (Table 26.12). The involvement of women in the daily use of water was somewhat lower for high-income households, but remained as high as 68% of households. On the other hand, when it comes to paying water bills and the maintenance of water-related equipment, the household surveys showed that the responsibility was primarily vested with men. Once again there was some variation by income group, with men's primary involvement with maintenance and bill paying being recorded as noticeably higher for the high-income families interviewed (88%) than the low-income households (64%), as shown in Table 26.12.

In conditions of water scarcity, the education of children – and indeed, all members of the family – in the careful use and conservation of water is vital (Middlestadt *et al.*, 2001; Academy for Educational Development, 2005). Prior to the household surveys it was conjectured that women might be expected to be more involved in instructing children in how to use water wisely and to conserve supplies on a daily basis. However, somewhat to our surprise, it was reported that men were generally more

Table 26.12 *Gendered aspects of the management of water within households*

Aspect of water use	High-income households	Low-income households	Entire sample
Managing the daily use of water	68% women	84% women	76% women
Responsible for paying bills and maintaining equipment	88% men	64% men	76% men
Raising children's awareness of the need to conserve water	24% women	36% women	30% women
	40% men	48% men	44% men
	8% both	12% both	10% both
	28% nobody	4% nobody	16% nobody

involved than women in such instructional activities (Table 26.12). For the entire sample of households, of those who stated that they were principally responsible for overseeing such education and monitoring within the home, 44% were men and 30% were women. In 10% of households it was reported that the task was shared between men and women. And as Table 26.12 also shows, in 16% of households, it was reported that nobody undertook such instruction (Table 26.12).

In fact, the proportion of men reported as being responsible for day-to-day guidance in the domestic use of water was higher, standing at 48%, for the low-income households. For the high-income households the proportion was 40%. Perhaps the most striking feature of Table 26.12 is that in well over a quarter of the high-income households in the sample, 28%, it was reported that nobody was concerned with ensuring that children were aware of the need to conserve water, presumably because householders could easily pay their bills and use water without worrying too much about the conservation of supplies (Table 26.12).

Looking finally at the day-to-day realities of water use, personal observations indicate that few urban residents in Greater Amman harvest and make use of rainwater from roof-tops and other surfaces, unlike in some rural areas where rain harvesting has been practised over the years. To substantiate this and to check for any variations, householders were specifically asked whether they were involved in such practices. Table 26.13 shows that none of the high-income households reported the collection and use of rainwater. Indeed, it was only a very small proportion of the low-income households in the sample, some 8%, that reported that they collected rainwater, with an equal number using this for drinking and for non-drinking purposes. Four per cent reported that they treated this water by boiling before use.

26.7 SATISFACTION AND ATTITUDES IN RELATION TO WATER ISSUES

The interviews sought to investigate how aware consumers were of pricing in the water sector and the details of how their bills are derived. The findings indicate that just over a third of all

Table 26.13 *The use made of collected rainwater by households*

Use made of collected rainwater	Percentage of households		
	High-income households	Low-income households	Entire sample
Collect rainwater	0	8	4
Use rainwater for drinking	0	4	2
Use rainwater for non-drinking purposes	0	4	2
Treat water before use	0	4	2

households appear to be fully aware of the intricacies involved in the water tariff system (Table 26.14). The results show that awareness varies markedly by socio-economic group, standing at 60% for high-income households, but only 12% for low-income households (Table 26.14). The formula that relates water consumption to the overall charge made per quarter is complex and it seems that most consumers are not fully aware of the intricacies involved in this.

When asked, however, nearly all respondent households, some 94%, stated that they were aware that prices had increased over the last 10-year period. Socio-economic variations crept in again with the question of whether such price increases were warranted, with 64% of high-income and only 24% of low-income households stating that they felt such increases were justified (Table 26.14). Naturally, low-income consumers are shown to be more price-sensitive than their high-income counterparts. Nearly all respondent households attributed such price increases to the involvement of the private sector since 1999.

In order to examine how satisfied households felt with the water sector in overall terms, households were asked to employ a five-point scale, ranging from 5 representing 'total satisfaction' to 1 denoting 'not satisfied at all'. The frequency distribution of households by stated level of satisfaction is shown in Table 26.15, both for the entire sample and its low- and high-income constituent household divisions. More low-income households (12%) than high-income households stated that they were not at

Table 26.14 *Awareness of selected water tariff issues*

Awareness	Percentage of households		
	High-income households	Low-income households	Entire sample
Aware of water tariff system	60	12	36
Aware of price increases over last 10 years	96	92	94
Felt that price increases were acceptable	64	24	44

Table 26.15 *Variations in household satisfaction levels with the water sector as a whole*

Level of satisfaction	Percentage of households		
	High-income households	Low-income households	Entire sample
1 Not satisfied	0	12	6
2 Moderately satisfied	40	28	34
3 Fairly satisfied	40	28	34
4 Very satisfied	16	32	24
5 Highly satisfied	4	0	2

Table 26.16 *Overall household satisfaction score with the water sector by income group*

Income group	Average satisfaction score
High-income	2.84
Low-income	2.80
Entire sample	2.82

all satisfied, while more high-income households stated they were either moderately or fairly satisfied. However, as shown by Table 26.15, the distribution for low-income households is differentially skewed, with 32 households stating that they were well-satisfied, against 16 high-income households. If these data are summarised by an overall points scoring system, as shown in Table 26.16, then despite these slightly variant distributions, little difference exists between the high- and low-income groups. For the high-income households, the aggregate satisfaction score was recorded at 2.84, and for low-income households, it was fractionally lower at 2.80; in both cases it is noticeable that the scores approximate to the level of being 'fairly satisfied'.

However, when the households were asked to assess their satisfaction with different aspects of the water supply system of Greater Amman, the outcome was highly revealing, as shown by Table 26.17. This is quite a complex table, and so in this account what seem to be the most salient aspects are highlighted. Looking at all households, an overwhelming majority, 92%, stated that they were satisfied with the reliability of the city's water supply system, and 80% stated their general satisfaction with the standard of maintenance of the network. Notably, despite the clear constraints that rationing places on their daily lives, 96% of low-income consumers expressed satisfaction with the reliability of overall supply, although this was marginally lower at 88% of the high-income householders. Similarly, some 52% of all households stated they were satisfied with general management standards in the water sector, and this was identical for both income groups (Table 26.17).

The aggregate level of stated satisfaction with the price paid for water was, however, somewhat lower, standing at 40% for the entire sample of households (Table 26.17). As expected, more high-income households than low-income households stated that they were generally satisfied with the price of water, standing at 52% as opposed to 28% of surveyed households, respectively.

The satisfaction data clearly demonstrate that by far the greatest concern expressed by consumers relates to the quality of the water, with 82% of all households stating that they were not satisfied with the existing quality of water supplied. It is noticeable that when disaggregated by socio-economic group, as shown in Table 26.17, levels of dissatisfaction do not vary much, standing at 84% for the high-income households and 80% of low-income households (Table 26.17). In short, of the two issues causing general dissatisfaction, while price is more important to the relatively poor, water quality clearly represents the chief issue for all households irrespective of income level. This general concern about water quality seems to reflect the appearance and taste of water provided via the network, as well as fears about its impacts on health and general well-being.

A further question then asked the respondent households whether they felt that the water supply system of Greater Amman had improved since 1999, when water was privatised under the control of LEMA. The results are summarised in Table 26.18. Just over half the total sample of respondent households stated that they felt that the water system had improved (52%). Interestingly, the proportion expressing this positive evaluation was higher among the low-income households (60%) than high-income households (44%).

Finally, an overall impression was derived as to just how important water supply issues are to households as part of their day-to-day lives, in relation to the other issues they face. The results are listed in Table 26.19. Just over half of all households surveyed, some 54%, reported that they regarded water as an issue of the utmost importance. This statistic also tends to suggest that while water supply issues are of considerable importance for most consumers, many households have found ways to deal with the situation. Further, as might be anticipated, the data

Table 26.17 *Satisfaction with different aspects of the water supply system*

Aspect of water supply	Percentage of households								
	High-income			Low-income			Entire sample		
	Satisfied	Not satisfied	Don't know	Satisfied	Not satisfied	Don't know	Satisfied	Not satisfied	Don't know
Price	52	48	0	28	72	0	40	60	0
Water quality	16	84	0	20	80	0	18	82	0
Reliability of supply	88	12	0	96	0	4	92	6	2
Standard of management	52	40	8	52	28	20	52	34	14
Standard of maintenance	84	12	4	76	8	16	80	10	10

Table 26.18 *Perceived improvement in the water supply system since privatisation in 1999*

	Percentage of households		
	High-income households	Low-income households	Entire sample
Perceived that water system has improved	44	60	52
Perceived that water system has not improved	56	40	48

Table 26.19 *Priority accorded to water supply issues by households*

	Percentage of households		
	High-income households	Low-income households	Entire sample
Water seen as an issue of the utmost importance	52	56	54
Water seen as an issue of secondary importance	48	44	46

show that a slightly higher proportion of relatively low-income households (56%) regard water supply issues as being of the utmost importance, with the corresponding figure being 52% for the high-income households surveyed.

26.8 CONCLUSIONS

The household survey data derived among the two income groups within Greater Amman show that there are considerable socio-economic differences in the storage and day-to-day use of water, household systems for the management of water, and the satisfactions and attitudes of consumers. However, the survey also shows that urban consumers in Greater Amman have, in overall terms, developed well-articulated strategies to guide their use of water to try to minimise the everyday challenges that water scarcity imposes. The responses of households to the quantitative survey show that they are generally content with the reliability of the urban water supply, its standard of maintenance and the standards of overall management.

Quite simply, by means of carefully developed strategies for household water management, householders have accommodated to rationing, and it is not the major problem they currently perceive as affecting their lives. This is not to say that such realities do not have a daily impact on their lives – far from it. This research shows that the rationing of water constrains poorer households in particular to carry out their daily activities in highly prescribed ways, and that this burden falls disproportionately on the female members of households.

For obvious reasons, more low-income than high-income householders are concerned about the overall quarterly cost of the water they use. But the research shows clearly that the quality of water is the major issue that concerns the vast majority of consumers, regardless of their income levels, socio-economic status or the area of the city in which they live. Consumers generally avoid drinking mains water and report low levels of satisfaction with water quality in general. This is a major challenge that is facing the water supply system of Greater Amman in the coming years, and some might argue far more so than the establishment of continuous water supplies in the city in the near future.

REFERENCES

Abu Dayyah, A. S. (2004) Persistent vision: plans for a modern Arab capital, Amman, 1955–2002. *Planning Perspectives* **19**: 79–110.

Academy for Educational Development (2005) *Water Efficiency and Public Information for Action Project (WEPIA), Final Report 2000–2005*. Amman: Ministry of Water and Irrigation and United States Agency for International Development.

Allain-El Mansouri, B. (1996) L'eau et la Ville, le cas de la Wilaya de Rabat–Salé (Maroc). Unpublished PhD thesis: Université de Poitiers.

Chikher, F. (1995) L'eau à Alger: Ressources, Distribution, Consommation. Unpublished PhD thesis: Université de Toulouse le Mirail.

Darmame, K. (2004) *Gestion de la rareté: le service d'eau potable à Amman entre la Gestion Publique et Privée*. http://www.iwmi.cgiar.org/Assessment/FILES/word/ProjectDocuments/Jordan/RapportDarmame(1).pdf.

Darmame, K. (2006) Enjeux de la Gestion du Service d'eau Potable à Amman (Jordanie) à l'épreuve du partenariat public-privé. Unpublished PhD thesis: Université de Paris X Nanterre.

Darmame, K. and R. Potter (2008) Social variations in the contemporary household use of water in Greater Amman: a preliminary research note. *Bulletin of the Council for British Research in the Levant* 3: 57–60.

Deutsche Gesellschaft für Technische Zusammenarbeit (GTZ) and Ministry of Water and Irrigation (MWI) (2004) *National Water Master Plan for Jordan*. Amman: GTZ and MWI.

Graham, S. and Marvin, S. (2001) *Splintering Urbanism: Networked Infrastructures, Technological Motilities and the Urban Condition*. London and New York: Routledge.

Ham, A. and P. Greenway (2003) *Jordan*, 5th Edition. Victoria, Australia: Lonely Planet.

Hannoyer, J. and S. Shami, eds. (1996) *Amman: the City and its Society* Beirut: CERMOC.

Hashemite Kingdom of Jordan (2007) *Household Expenditure and Income Survey 2006: Main Findings (Preliminary)*. Amman, Jordan: Government of Jordan, Department of Statistics.

Iskandarani, M. (1999) *Analysis of Demand, Access and Usage of Water by Poor Households in Jordan*. Bonn: Centre for Development Research, Bonn University.

Jaglin, S. (2001) L'eau potable dans les villes en développement: les modèles marchands face à la pauvreté. *Revue Tiers Monde* **166**: April–June, 275–303.

Lavergne, M. (2004) Face a l'Extraversion d'Amman, en reseau urbain en quite de sens. *Les Cahiers de L'Orient* **75**: 139–151.

Magiera, P., S. Taha and L. Nolte (2006) Water demand management in the Middle East and North Africa. *Management of Environmental Quality: An International Journal* **17**: 289–298.

Middlestadt, S., M. Grieser, O. Hernandez *et al.* (2001) Turning minds on and faucets off: water conservation education in Jordanian schools. *Journal of Environmental Education* **32**: 37–45.

Mitlin, D. (2008) GATS and water services: the implications for low-income households in the South. *Progress in Development Studies* **8**: 31–44.

MWI (Ministry of Water and Irrigation) (1997) *Water Strategy in Jordan* Jordan: MWI.

Potter, R., N. Barnham and K. Darmame (2007a) The polarised social and residential structure of the city of Amman: a contemporary view using GIS. *Bulletin of the Council for British Research in the Levant* **2**: 48–52.

Potter, R., T. Binns, J. Elliott and J. Smith, eds. (2008) *Geographies of Development: an Introduction to Development Studies*, 3rd Edition. London and New York: Pearson–Prentice Hall.

Potter, R., K. Darmame and S. Nortcliff (2007b) The provision of urban water under conditions of 'water stress', privatization and deprivatization in Amman. *Bulletin of the Council for British Research in the Levant* **2**: 52–54.

Potter, R. and S. Lloyd-Evans (1998) *The City in the Developing World*. London and New York: Pearson-Prentice Hall.

Potter, R. B., K. Darmame, N. Barham and S. Nortcliff (2009) 'Ever-growing Amman', Jordan: Urban expansion, social polarisation and contemporary urban planning issues. *Habitat International* **33**: 81–92.

Rigg, J. (2008) The Millennium Development Goals. In *The Companion to Development Studies*, 2nd Edition, ed. V. Desai and R. Potter. London and New York: Hodder Education and Oxford University Press, Chapter 1.

UNDEP (2004) *Water Governance for Poverty Reduction Key Issues and the UNDP Response to the Millennium Development Goals* N10017. New York: United Nations.

United Nations (2000) *Millennium Declaration* A/RES/55/2 New York. http://www.un.org/millennium/declaration/ares552e.pdf

WHO and UNICEF (2000) *Global Water Supply and Sanitation Assessment, 2000 Report*. New York and Geneva: United Nations Children's Fund and World Health Organization.

WWAP (World Water Assessment Programme) (2006) *Water: A Shared Responsibility*. The United Nations World Water Development Report Paris: UNESCO.

Zérah, M.-H. (1998) How to assess the quality dimension of urban infrastructure: the case of water supply in Delhi. *Cities* **15**: 285–290.

Zérah, M.-H. (1999) *L'accès à l'eau dans les Villes Indiennes*. Paris: Anthropos.

27 The role of water and land management policies in contemporary socio-economic development in Wadi Faynan

Khadija Darmame, Stephen Nortcliff and Robert Potter

ABSTRACT

The role of water and land management practices in promoting current development in the Wadi Faynan area, located in southern Jordan, forms the focus of this chapter. The area represents one of the most arid landscapes in Jordan and, until recently, was primarily used by semi-nomadic Bedouin tribes for winter grazing. Since 1921, the Government of Jordan has encouraged settlement and this transition has led to considerable changes in the day-to-day lives of the Bedouin, including the recent development of irrigated agriculture making use of shallow groundwater sources. The account considers the physical and social contexts of Wadi Faynan before turning to the cooperative model being used in the area to promote this irrigated agricultural activity. The characteristics of the irrigated agricultural system being used to produce tomatoes and watermelons are then outlined and reviewed, including recourse to specific farmer-based case studies. In conclusion, the impacts of the cooperative-led irrigated agricultural activity on reducing poverty in the areas are assessed, along with the need for wider imperatives at the sub-regional scale of southern Jordan, as a part of decentralisation from the national core centred on Greater Amman.

27.1 REDUCING POVERTY THROUGH PARTICIPATORY MANAGEMENT IN ARID AREAS

Located in southern Jordan, the Faynan area is one of the most arid landscapes in Jordan. Until recently, the area was used by the semi-nomadic Bedouin tribes Al-Rashaida and Al-Azazma for winter grazing. With the establishment of the Emirate of Transjordan in 1921 the Government encouraged settlement, which was also stimulated by the limited availability of grazing as a result of droughts (Al-Simadi, 1995; Bocco and Tell, 1995). The passing from nomadic to sedentary activity produced considerable changes in the lifestyle of the Bedouin. While many have joined the army for their livelihoods, others have turned to irrigated agriculture, making use of shallow groundwater (Lewis, 1987; Bocco, 2000). The transition from a nomadic to sedentary lifestyle was encouraged by the government, but in fact, little investment in socio-economic projects and infrastructure has been made to support this transition (Bocco, 2000; World Bank, 2004). A report published in 2004 by the World Bank (WB) on poverty in Jordan recorded that the areas of deprivation and severe poverty were mostly associated with Bedouin villages, particularly in eastern and southern Jordan. This is well illustrated in the case of the Wadi Faynan area. Since the 1990s, a village has been established for the Mal'ab, a branch of the Al-Rashaida using the local Wadi Ghuwayr system for water. In order to develop their livelihoods and improve their quality of life, the tribe members set up cooperatives in agriculture and tourism. These local initiatives are part of national as well as regional measures to aid development and to reduce poverty.

In the terms of Carney (1999), livelihoods comprise the capabilities, assets (including both material and social resources) and activities required to provide a means of living. In the case of Jordan, 18.7% of the approximately 1 million rural people were considered to be living below the income poverty line of 392 Jordanian Dinar (JD) per annum in 2002 (The Household Expenditure and Income Survey, 2002–03). (In November 2009, one Jordanian Dinar was equal to 0.83 Pounds Sterling.) Many factors affect the livelihoods of Jordan's rural poor: limited access to water and land, the lack of alternative sources of income, few opportunities to diversify their agricultural production and the generally poor quality of the soils. Additionally, they

Water, Life and Civilisation: Climate, Environment and Society in the Jordan Valley, ed. Steven Mithen and Emily Black. Published by Cambridge University Press.
© Steven Mithen and Emily Black 2011.

do not own the land, and as tenant farmers they rarely make long-term investments or request loans for development (Ministry of Agriculture (MoA), 2009).

Nevertheless, there is evidence that some of the factors mentioned above – particularly when water is scarce and the infrastructure is lacking – are not always an obstacle to the reduction of poverty. In this regard, good examples are found in Southern Africa, where Maluleke *et al.* (2005) developed a model called Securing Water to Enhance Local Livelihoods (SWELL). Their approach is based on principles of participatory management and community-based planning in poor communities that aim to improve the water resource allocation, the land use, and therefore the social and economic conditions. In the same respect, van Koppen *et al.* (2006) developed and used a Multiple Use Water Services Approach in South Africa, which is defined as '[a] participatory, integrated and poverty-reduction focused approach in poor rural and peri-urban areas, which takes people's multiple water needs as a starting point for providing integrated services'. This approach, framed by three key institutional levels (community, intermediate, national), was adopted by Smits *et al.* (2006) for the case of Zimbabwe. These authors suggest that this approach to scarce water resources management for a range of productive uses has the potential to make significant contributions to the improvement of household food security and nutrition, and potentially to generate more income. In this way and following the settlement of the Al-Rashaida tribe, there has been socio-economic development based on community water management, aimed at enhancing irrigated agriculture (namely tomatoes and watermelons) and ecotourism as part of the strategy to reduce poverty. What assets do the population of Faynan have at their disposal? How does the adopted strategy of managing water and land in Faynan relate to the approaches mentioned above? To what extent are the present strategies of water and land management in (Wadi) Faynan likely to achieve outcomes achieved elsewhere, for example in southern Africa?

This chapter will consider the experience of the Al-Rashaida tribe in Wadi Faynan during the past decade in terms of adaptive strategies to deal with environmental, socio-economic and political changes, as well as their development of crop production practices to improve their livelihoods. The impact of ongoing programmes of development led by different institutions involved in the area is also emphasised in the chapter. The analysis and discussion is primarily based on qualitative and quantitative information collected in the field between March and May 2008 and in March 2009, using participatory and in-depth interviews with the members of the village. In addition a number of interviews were conducted during 2008–09 with representatives from the Ministry of Agriculture (MoA), the Aqaba Special Economic Zone Authority (ASEZA) and the Jordanian Cooperative Corporation (JCC), the official body which authorises and monitors the cooperative associations in rural areas.

In addition, we investigated the available data about the area provided by the Ministry of Planning and International Cooperation (MoPIC), the Department of Statistics (DOS) and the Ministry of Social Development (MoSD) (Department of Statistics, 2007–09). However, most of the information provided from these sources related more generally to the main villages of the Aqaba Governorate, and it proved difficult to obtain specific information on the current socio-economic development in the Wadi Faynan area from the national and regional level institutions.

27.2 PHYSICAL AND SOCIAL CONTEXT OF WADI FAYNAN

27.2.1 Physical and environmental constraint

The dominant climatic influence on Wadi Faynan is the hot, dry climate of the Saharo-Arabian desert belt, although locally there is altitudinal variation (Palmer *et al.*, 2007). Wadi Faynan has an average minimum January temperature of 12 °C and in the summer the temperature frequently exceeds 40 °C. Most of the rain falls in the winter between December and March on westerly winds, with no rain falling between June and September (see Figure 27.1).

Rainfall amounts are extremely variable, but annual totals are in the region of 60 mm, although annual totals in excess of 150 mm have been recorded. Locally the water supply in the Wadi Ghuwayr is maintained throughout the year by a series of perennial springs. The area currently developed for intensive agricultural production lies principally on Pleistocene-age alluvial fans and Holocene-age alluvium and colluviums. The soils are skeletal and sandy loam in texture, showing very few signs of the development of distinctive profile features and organic matter accumulation. Recent analysis, carried out by the authors, of total nitrogen and available phosphorus in these soils showed exceedingly low levels, indicating that any form of crop production would be impossible without an intensive fertiliser management programme particularly for nitrogen and phosphorus. Levels for available potassium were low to moderate, but if crops with a high potassium demand, such as tomatoes, were to be grown there would also be a potassium fertiliser requirement. According to Crook (2009), a contemporary measure of water quality shows that the pH and alkalinity of these waters is very high as a result of their source in the high limestone plateau. If this water were to be used for irrigation, there would be a need for careful management to prevent problems of structural damage to the soil and reduced infiltration capacities in the fields.

27.2.2 From nomadism to sedentary life

Customarily, the Al-Rashaida tribe moved to the highlands around Shawbak for summer grazing and to Wadi Faynan in the winter (see Figure 27.2). Whilst this pastoral activity provided an economic

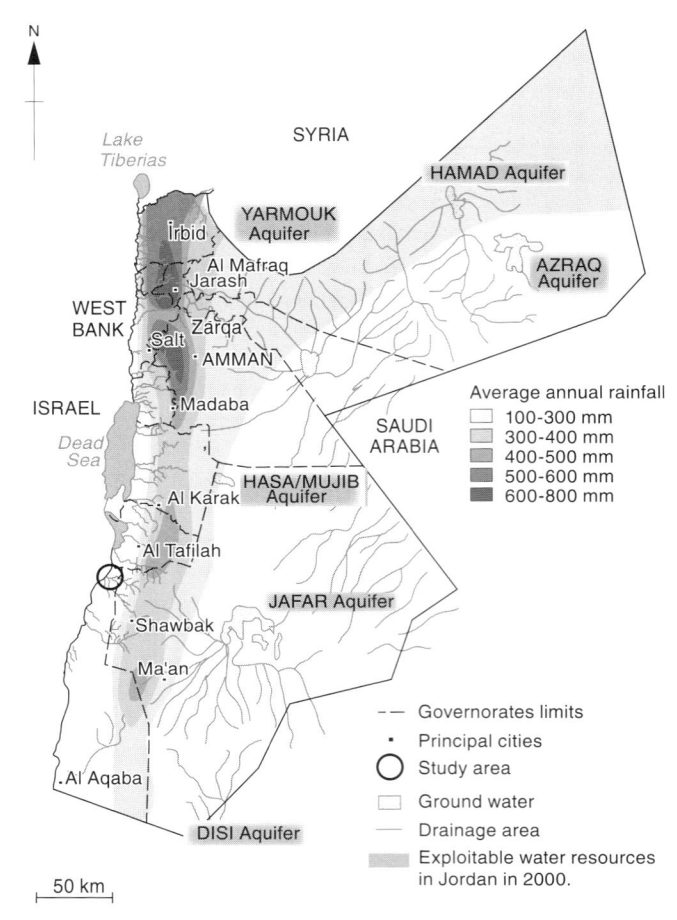

Figure 27.1. Rainfall distribution and water resources in Jordan and location of the study area. Adapted From Darmame (2006).

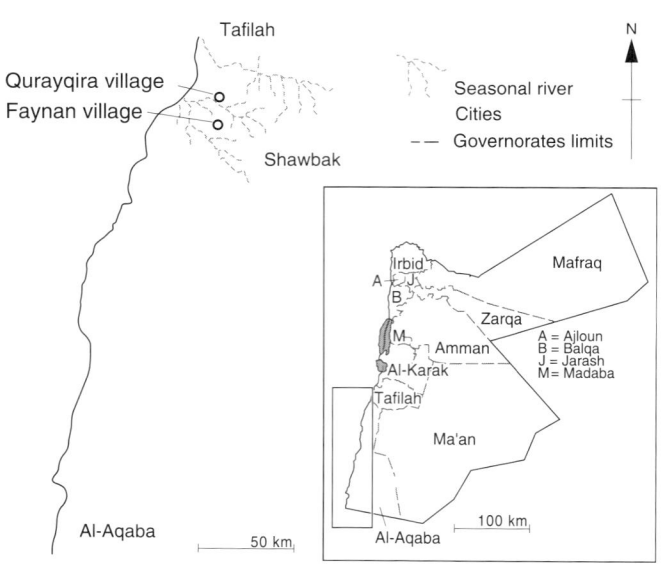

Figure 27.2. The regional and national context of the Faynan and Qurayqira villages.

Figure 27.3. The typical small dwelling provided by the government in Faynan village.

base for their livelihood, it should be recognised that it also represents a key element of what it means to be a Bedouin. During the 1990s, the move towards the development of a settlement began in Wadi Faynan. The Al-Rashaida tribe have settled in a village called *Faynan Al-Jadida* (New Faynan). Although there were 514 inhabitants in 2007, a few families continue to move to the uplands in the summer months (Wadi Araba Development Project, 2007).

In 2004, each family was allowed to use an area of 1 dunum (equivalent to 1,000 m^2) as part of the programme to encourage settlement and reduce poverty in this area. In addition, the government provided small basic dwellings as a royal donation, but it should be recognised they were not in line with the living habits and tradition of Bedouins. For the Bedouin the tent provides both a living space and a space for hospitality while the small dwellings that were provided only cover the needs in terms of living space. A consequence is that whilst some families have used part of the land granted by the government for ornamental plants and an occasional olive tree, many have pitched tents on the land to provide the traditional hospitality space as illustrated in Figures 27.3 and 27.4.

The second tribe, the Al-Azazma, has a population in the order of 200 inhabitants. The Al-Azazma were originally from Bir Saba'a in southern Palestine, but had historically moved between Palestine and Jordan. They moved to the Wadi Faynan area in 1948 following the creation of the State of Israel. They live close to many of the archaeological sites that are to be found locally. About 70 families live in tents and their water supply is from Wadi Ghuwayr or local springs, as shown in Figures 27.5 and 27.6. This tribe consists of nomadic goat herders who generate a limited income from this activity, but who also rely upon support from the Ministry of Social Development in Amman. The tribe is economically very poor and, in part because of their Palestinian origins, they have no rights to use land and water resources in the

Figure 27.4. The tent: place of traditional hospitality and shrubs on the 1 dunum plot.

Figure 27.5. The everyday challenges of water access for the Al-Azazma tribe.

Figure 27.6. A young girl bringing back water from Wadi Ghuwayr for cooking.

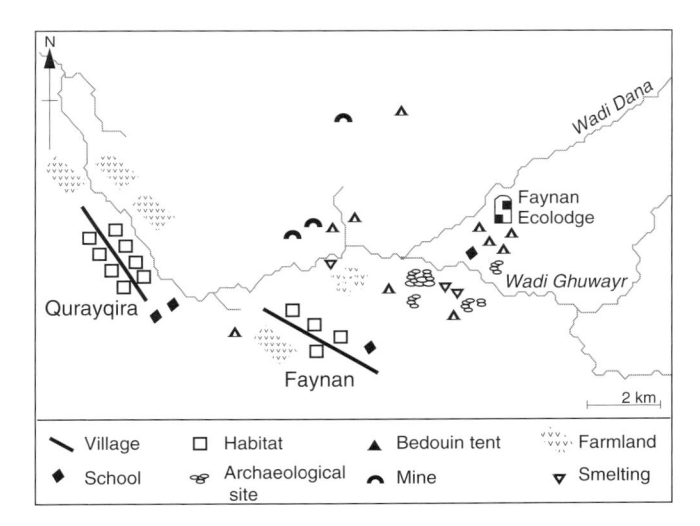

Figure 27.7. Land use in Faynan and Qurayqira villages.

area, although they are recognised as Jordanian nationals. They appear to be excluded from all socio-economic projects in Faynan.

27.2.3 Insufficiency of the socio-economic services in Faynan area

The Qurayqira–Faynan municipality (Figures 27.2 and 27.7), created in July 1999, was identified in 2004 as the second most pressing pocket of poverty in Jordan: about 53% of the population were living at or below the poverty line and the overall rate of illiteracy was 35% (Municipality of Qurayqira-Faynan, 2007). The average size of family is 5.47 persons and the percentage of families with 6 and more is 42.6%. Between 80% and 90% do not have items such as refrigerators, televisions, an oven or washing machine (MoPIC, 2004–06). The level of socio-economic services is, however, relatively acceptable. Since 2004, electricity and water have been provided. The electricity is ensured by the Aqaba Company. The water supply is from an artesian well with a capacity of 76,153 m^3 per year (MoA, 2009) and the distribution, maintenance and billing are assured by the

Aqaba Water Company. The water is stored on the roof of each house in tanks of 2 m^3. The interviewees described the water as quite salty and the storage capacity as insufficient for the size of their families.

Meanwhile, the Al-Azazma tribe, in old Faynan, does not have water or electricity. People have to walk long distances to bring water from the Wadi, which requires time and energy, and the quality of water is not always good. Alternatively, they have to get supplies from a spring managed by the Ecolodge run by the Royal Society for the Conservation of Nature (RSCN), located near to the archaeological sites (see Figure 27.7).

Public transportation is for the most part absent. In winter, the area is completely inaccessible during the flood period. A basic road connects the whole area to the principal Aqaba–Amman road and a bus provides public transport from Qurayqira toward Aqaba at 6.00 am daily. There is only one school and the lack of staff was mentioned as an issue by the population during interviews. The level of education is low and rarely exceeds the secondary level (Municipality of Qurayqira-Faynan, 2007).

There is a health centre, located in Qurayqira, which provides basic facilities for both villages and the Al-Azazma. In short, the status of services provided is basic in the area and Faynan village is very dependent on Qurayqira village and Aqaba. The current situation shows that an improvement has been made by the government in comparison with the situation a few years ago when families were living under traditional tents, but more effort is required to provide the area with infrastructure of sufficient quality.

27.3 THE COOPERATIVE MANAGEMENT MODEL IN SOUTHERN JORDAN: THE CASE OF THE FAYNAN AREA

A cooperative is defined as an autonomous association of persons united voluntarily to meet their common economic, social and cultural needs and aspirations through a jointly owned and democratically controlled venture (International Co-operative Alliance, 2004). For the Bedouin in eastern and southern Jordan, the agro-cooperative is a participatory process model, based exclusively on irrigation from groundwater. It is essentially a coping strategy adopted to respond to the prevailing socio-economic conditions and to ensure the accumulation and management of financial assets. Culturally, many Bedouin have considered that agriculture is degrading work and that, as they are of a higher social status, they should not be involved in such activities (Bin Mohammad, 1999; Van Aken, 2004). Some of the Bedouin tribes in Jordan, however, have a history of engagement in agriculture, as in Salt and in the immediate vicinity of Amman, and there is no evidence that as a consequence of this involvement they have lost their identity (Shryock, 1997). The first agro-cooperative in the rural desert areas began in the 1970s. In 1976 with the royal support of Sharif Nasir, authorisation was given for well drilling to provide irrigation water for agriculture to the Al-Ammarin and Saidiyin tribes in Qurayqira village. In the early 1990s, a similar scheme was implemented in Wadi Faynan but with surface water transferred by pipe from Wadi Ghuwayr. But before studying the cooperative and its activity, it is relevant to highlight the complexity of the property status of the land used by the Al-Rashaida for agriculture, which deprives them of the right to be owners of the land and to have access to groundwater in the area.

27.3.1 The issue of the legal status of land and its use by tribes

Tribal land is defined as the communal land that the tribes have been using for herding, as their primary source of livelihood, and in growing barley for agriculture. The tribal land is described by some authors as a social field (Razzaz, 1994) or a social landscape (Lancaster and Lancaster, 1999) in which the ownership, the rules to get access to the land, and its boundaries were established according to principles of Islamic law, local customs and Ottoman law (Metral, 1996). Since the Emirate of Transjordan was established in 1921 under the British Mandate, a widespread policy of control and centralisation of land was implemented by the government through the Land Registry Law. However, when the government declared sedentarisation as its official policy, the settled tribes had the right of ownership of their pastoral land and all undeveloped land became state property (Lancaster and Lancaster, 1999). The state has also offered the pastoral tribes open-access to grazing and the expansion of barley cultivation (Al-Jaloudy, 2001), but this access does not give them the right to actually own the land or the right to drill wells.

The procedure is that, in general, using the state land for public interest should be requested by a group of people within the framework of a cooperative. The JCC cannot authorise the registration of a cooperative without the provision of an official document by governors delegating the land to the tribe. This has provoked disagreements between tribes about the legitimacy of land use especially in the Aqaba governorate, where most land is owned by the Jordan Valley Authority (JVA), ASEZA or the State Treasury (*khazinat al-dawla*) (Jordanian Cooperative Corporation, 2009). The south bank of Faynan and Wadi Ghuwayr is an example of this kind of conflict between the tribes. The Rashaida have a history of disagreement with the Shawabaka tribe (residents of Shawbak) concerning the ownership of tribal land in Faynan and access to water from Wadi Ghuwayr. Lancaster and Lancaster (1999) described one such incident:

> Al-Rashaida affirmed that they made an agreement with the Shawabaka that they would farm half themselves,

while the Shawabaka had the right to sharecrop with the Rashaida on the other half of the land. In 1992–3, the Rashaida were cultivating two areas, using gravity fed water piped from the Wadi Hammam (the downstream section of Wadi Guwayr [referred to as Wadi Ghuwayr elsewhere in this chapter]). In the autumn of 1993, the pipes were extended west to a new area where water catchment basins were built and the land ploughed and planted. This aroused the Shawabaka who had done nothing about their option. They took their complaint to Government offices at Aqaba, who contacted the army to send a bulldozer from the Jordan Valley Authority to clear the just ripening tomato crops. The Rashaida were furious at the prospect of their investment in pipe and the associated efforts being lost. They stopped the bulldozer from ripping up the crops and pipes by standing in front of it, trying to set the bulldozer on fire, and puncturing its tyres, while the sheikh of the Rashaida arrived from Karak by pickup with supporters, all bearing rifles.

Lancaster and Lancaster (1999, p. 122)

In 1994, the dispute was settled temporarily thanks to then Crown Prince Hassan, and the Rashaida position was upheld.

In all our meetings between 2008 and 2009, the Al-Rashaida members have evoked their tribal history in the Wadi since the 1930s. They claim to have a *hujja* – a traditional document based on Islamic principles and Ottoman law – asserting their presence for decades in Faynan, herding and practising agriculture intermittently. However, if they considered the *hujja* as proof to legitimise the exploitation of land, they failed to register it or to get any official recognition of land ownership. In Jordan, the *hujja* is not recognised by courts of law; it is not considered an official document and is not recorded in Land Registry details, since the tribal members do not own the state land, as confirmed by Razzaz (1994).

The Al-Rashaida tribe believe that they have been in *de facto* possession of the disputed land through a long process of settlement in the area (farming the land and participating actively in the management of the municipality), whilst the Shawabaka refer to the tribal land law and their pastoral life, which gives them rights to a tribal bank in Faynan. The request of the Shawabaka to create their own cooperative has been rejected as the problem of the land partition is not yet resolved. In the context of the tribal structure, the disagreement becomes not only an issue of interest and of securing a livelihood, but also a matter of pride and honour that they must defend (Lancaster and Lancaster, 1999). After more than 14 years, the process of adjustment between the state and the two tribes is still ongoing and the issues have yet to be resolved.

27.3.2 The performance of Sail Faynan cooperative

The Al-Rashaida members have implemented a series of cooperatives according to the laws established by the JCC, which

Figure 27.8. Rudimentary dam in Wadi Ghuwayr to collect water prior to transfer through pipes for irrigation.

also provides technical and administrative assistance. The initial cooperative, known as the Faynan Association, established in 1997, appears to have had only limited activity. In 2002 the Sail Faynan cooperative was established for agriculture and the Hamza Association for tourism in 2004.

The initial idea for undertaking irrigated agricultural activity arose from the Al-Rashaida Tribal Deputy in the mid-1990s, who suggested the transfer of water in pipes from nearby Wadi Ghuwayr as they are not allowed by JVA to drill wells. In order to improve the collection of water and to use it efficiently, the member of the cooperative surveyed the Wadi and introduced a series of small rudimentary stone dams (Figure 27.8) built to channel water to pipes, an effective transfer of water which also affords some protection from pollution due to the presence of heavy metals in the soil. This approach to manage the flows is basic, and the dams are regularly destroyed by the seasonal flood events. The Bedouin in Faynan consider a bad year to be when there are fewer than five floods, as this constrains the growth of wild plants for goats to eat and severely reduces available water for the irrigated land (Lancaster and Lancaster, 1999). To facilitate the transfer of water along the pipeline, the initial diameter of the pipe at the Wadi dam is 4 inches (10 cm), reducing to 3 inches (7.5 cm) approximately mid-distance between the wadi and a small basin close to the cultivated areas where water is stored prior to its use for irrigation (see Figure 27.9).

The first land used for agriculture was land locally identified as a Roman archaeological site, but the soil is considered to be contaminated by metals from smelting. The first pipe established was used to irrigate between 80 and 100 dunums. Following the intervention of the Department of Antiquities, the cultivation site was moved to land closer to the village, incurring additional costs since the pipeline is now in excess of 6 km in length.

Figure 27.9. Water storage basin near the farm for irrigation.

27.3.3 The cooperative functioning

The Sail Faynan cooperative is managed by a council, elected by the male members of the tribe, which is responsible for ensuring that the activities are coordinated and that the needs of the village are met. Women are excluded from participation in the activities of the cooperative, as are members of the Al-Azazma tribe. As the cooperative has developed it has become important in providing a degree of social unity and has a key role to play in the socio-economic development of the tribe. Prior to 2004 there were only 33 members of the cooperative; in 2004 this had increased to 80 members, with a further increase to 180 in 2008, and the projection is to reach 250 members by the end of 2009.

The cooperative has managed the use of water and land allocation among the tribe's members since 2004, and negotiates with entrepreneurs from outside the village (principally from Amman and Qurayqira) who have sought to develop commercial production on the lands managed by the cooperative. The members each have access to one dunum in turn, according to a system of rotation. In order to manage resources more efficiently and to gain greater benefit from the system, the members have often consolidated plots from the family and allocated it for the use of one individual. In order to maintain a sustainable activity and to control the land use, the cooperative charges a rent for each season based on the land area used, and this includes the cost of water supply. There is a rent differential in favour of villagers; in 2008, for example, farmers from outside the village paid an annual rent of 80 JD per dunum, while farmers from the cooperative paid 50 JD per dunum. The external investors are allowed to import labour, principally from Syria and Egypt. The use of the land by non-villagers has resulted in considerable financial benefit to the cooperative. The initial capital of the cooperative was 5,000 JD in 2004, increasing to 8,100 JD in 2006 and 9,645 JD in 2007. The net profit of the cooperative was 14,449 JD in 2006, rising to 27,198 JD in 2007 and almost 40,000 JD in 2008.

27.4 THE ECONOMIC AND ENVIRONMENTAL DIMENSIONS OF THE IRRIGATED AGRICULTURAL SYSTEM

The entire exploitable land area is around 500 dunums, according to the cooperative leader. All farms managed by the local farmers are small-scale systems of 20 dunums, owing to the lack of financial assets and the limited management capability, whilst the private entrepreneur, with his local partnerships, has the largest-scale farm, an area of 200 to 250 dunums, managed by experienced foreign labourers. The land managed by the cooperative is almost exclusively under two crops, namely tomatoes and watermelons. This choice of crops is based on the suitability of local conditions, the timing of production and the availability of markets and a seasonal strategy to optimise production from the land. Currently there are three strategies of land management being practised:

1. The cooperative member leases the land to a third person, often from outside the village. In 2009, for example, a member of the Al-Azazma tribe who wished to cultivate crops on the land was involved. The cooperative member supplies all the necessary planting materials, fertilisers and pesticides. The lessee has the right to 25% of the profit when the harvest is sold. This arrangement is more frequently practised during the tomato growing period.
2. The cooperative member agrees with an individual from outside the village who will invest capital in the activity. Currently there are two cases of this arrangement with involvement of a farmer from nearby Qurayqira and an entrepreneur from Amman.
3. Members of the cooperative work and manage the land themselves.

Each year the local farmers plant tomatoes but do not necessarily plant watermelons, owing to the high costs associated with production, the ever-present risk of production loss as a result of nematode infection and the high demand for labour. In order to understand the strategies developed by farmers for both crops, we will explore the example of a local farmer with limited resources and knowledge, who opted for planting tomatoes, whilst the example of a private entrepreneur will be developed for the case of the watermelons.

27.4.1 The process for tomato farming

Abdurrahman is a young man from Al-Rashaida in his late twenties. He left school at 15 and, as he was unable to find work in the village, decided in 2004 to work in agriculture together with his brother. He was able to bring together 20 dunums from among his family who are members of the cooperative. As with most of the Al-Rashaida, he had limited experience in agriculture and sought guidance and information from a variety of different sources including other farmers in the Qurayqira village who have experience of this form of work, and the merchants selling fertilisers, seeds and other agricultural supplies. With time he gained more experience and knowledge.

Tomatoes are grown from the middle of August to November. After removing stone and boulders from the land, Abdurrahman ploughs and levels it. He connects the 3-inch water pipes from the main pipe from Wadi Ghuwayr and then he fertilises the land in solute form through the water pipes. The land is flood-irrigated for three consecutive days. He lays drip irrigation pipelines with 2 m spacing between irrigation lines, covers the land with black plastic and makes holes in the position of each drip point where the plant is sown. During the establishment phase, the plant is irrigated first for 30 minutes every 2 days until it is established, and then for 60 minutes every 2 days until harvest. The fertilisation process is done over many stages as following: 2 kg of phosphorus is given per dunum 7 days after planting the seedling, nets (Gro-Nets 5×30) are placed over the crop after 30 days to protect against white fly, 2.5 kg of urea is given after 40 days, and finally 3 kg of potassium after 60 days. In 2008, the total cost for tomatoes, including labour, cooperative fees, fertilisers, pesticides etc., reached 4,000 JD for 20 dunums, whilst the harvest from each dunum was between 2–3 tonnes, sold at 200 to 250 JD per tonne. The net income over 20 dunums in 2008 was of the order of 10,000 JD.

27.4.2 The process for watermelon farming

Issa is a private entrepreneur with considerable farming experience in a rural area close to Amman where the costs of pumping water proved prohibitively high. Following a period growing in the Jordan Valley he moved to Wadi Faynan after being contacted and encouraged by local cooperative members to invest in Faynan. The land he occupies is watered by three pipes that he bought with his associates, and he pays 30,000 JD per year to the cooperative for both watermelon and tomato seasons. Watermelon needs a high level of management with respect to land treatment, fertilisers, water and pesticides. For this reason and regarding the size of the farm (an average of 200 dunums in recent years), Issa and his partners brought in qualified foreign workers. Such labour is provided by Egyptians from the Al-Fayoum region who were recruited through agencies in Amman. They have an annual contract with work in Wadi Faynan occurring from July to May. Many return for more than one annual cycle. The salary per month is 200 JD, three-quarters of which is sent home to their families.

The cultivation of watermelons begins between the end of December and the beginning of January. During the early stages of crop establishment and growth the watermelons are cultivated under plastic, which is removed in late March as the temperature rises. The initial land preparation is the same as for tomatoes. However, as the watermelon grows laterally rather than vertically, the planting density for watermelons is 30–50% of that of tomatoes. Also, because of the susceptibility of watermelon to root-knot nematode infection in their roots, which can drastically reduce both the volume and quality of yields, some farmers pass gas (1,3-dichloropropene) through the root zone in a set of pipes prior to planting. Because of the increasing costs associated with this treatment, it is used only infrequently. The alternative approach to dealing with nematode infestation is to move the zone of cultivation to an adjacent area and allow natural sanitation of the soils through 'solarisation', which is effective at the prevailing temperatures but requires a reasonably moist soil, a requirement that may be a constraint in this area of water scarcity.

The production of watermelon is highly dependent on the provision of sufficient water at key stages of the growth cycle. After the land has been ploughed and levelled, it is irrigated for consecutive days – three to five – until it is saturated. Following the initial planting of seedlings, the water is supplied for the first 7 days by drip irrigation across the site for 5 minutes. From this point onward the period of irrigation is incrementally increased to 20 minutes per day. This rate must be achieved before the full growth stage is reached. Once the fruits begin to ripen, the crop is irrigated all day and this normally takes place around mid-April.

The watermelon needs more fertilisers than tomatoes do as it grows. The farmers are not restricted to a fixed time to use fertilisers and it is done according to the strategy of each farmer and to their personal sources of information. The types of fertilisers used are urea, potassium, ammoniac, and a standard 20:20:20 mix of N:P:K. Some farmers use phosphate from the beginning if the quality of the soil is poor, but this is not always necessary. Urea is the most heavily used fertiliser; it is applied 15 days after the planting on a weekly or two-weekly basis depending on the crop needs (around 2.5 kg of urea per dunum). In addition, potassium as well as the standard 20:20:20 mix of N:P:K is applied 50 days after planting to increase the size of the crop if the rate of growth is slow, or during cooler periods to maintain normal growth.

In the case of Issa's farm, the Egyptian workers use the following quantity of fertilisers per dunum: 2 kg of phosphorus after 14 days of planting, 2.5 kg of urea after 21 days, 3 kg of ammoniac after 35 days, 3 kg of 20:20:20 mix of N:P:K after 60 days, and finally 3 kg of potassium after 70 days according to the need of the crop which is usually just 15 days before the first harvesting around 20 April. After the plastic cover has been removed, the plants are treated, if needed, with pesticides such as Leneth or Zeneb or sulphur (liquid). The use of gas through pipes at an early stage, as described above, also provides a degree of sanitation against other microbiological pests such as Fusarium wilt as well as addressing nematode infection. In general, the cropping strategies developed by many farmers in Faynan reflect that the fertiliser needs of plants or pesticides are relatively poorly understood by the farmers and many of the merchants have limited knowledge of the specific demands for both crops in an intensive system such as this. Thus, the strategy for fertiliser applications is often a result of trial and error by farmers.

Watermelon has higher production costs than tomato but it is more profitable. The costs for planting watermelon in terms of rent, labour, fertilisers and pesticides are approximately 6,000 JD for 20 dunums. The returns on harvested watermelons vary during the harvest period. Each dunum produces between 2 to 3 tonnes of saleable product. The price per tonne may reach 500 JD between the end of April and mid-May.

In the past two years Faynan has seen a substantial expansion in the area cultivated both for tomatoes and watermelons, but additionally there is an area where the sole crop is watermelon. There has been an increase to 15 in the number of farmers working together in small groups. This expansion has given rise to considerable pressure on the water resources, with rationing being introduced in respect of the frequency and duration of irrigation. This restriction could seriously affect yields, particularly of watermelon. Furthermore, more pipes will be needed to transfer water to the fields, at more financial cost. Any investment by the cooperative in this area means the increase of rent will constitute a further cost for the local farmers who are dealing with a small budget, an average of 5,000 JD per 20 dunums.

27.4.3 The market for tomatoes and watermelon

The timing of watermelon production in Wadi Faynan is early relative to other production areas in the region, providing farmers with a competitive advantage of at least 2 weeks compared with other farms in south Jordan and 2 to 3 months compared with the international markets. Because of the lack of any local storage facilities, the watermelon crop is harvested and loaded immediately on to large trucks for trans-shipment and sale in Amman and beyond to Syria, Turkey and parts of Eastern Europe (see Figures 27.10 and 27.11). Much of the crop is sold prior to harvesting. Because of the short period over which the watermelon ripens this is the only economic option.

In the case of tomatoes, which ripen over a longer time span, the local farmers often harvest and deliver in smaller quantities to markets in Amman. The economic benefit from tomato and watermelon cultivation has good potential but the activity is very dependent on the intermediary merchants from

Figure 27.10. Watermelons immediately prior to harvest in the Faynan area.

Figure 27.11. Watermelons being packed on to trucks in Faynan for transport north to Amman and for export.

Amman or Aqaba and is exposed to fluctuations in the national and international markets.

27.5 THE FAYNAN COOPERATIVE: IS IT CONTRIBUTING TO POVERTY ALLEVIATION?

During the interviews in 2009 it became apparent that the cooperative has had less impact on poverty than was indicated during the first series of meetings in 2008. The current situation of the village has changed little, and any improvements appear to have been more linked to royal donations and government projects. The tribe members affirm that their condition is still hard and challenging:

Working in agriculture in the Wadi is a hard job, without help or assistance from the Ministry of Agriculture or any entities. No other opportunities to find jobs are available in this area, except to work seasonally in archaeological excavations or as drivers for occasional tourists.

According to the survey carried out by DOS and MoPIC (Department of Statistics (DOS), 2007), approximately 99% of households in the Faynan area do not own livestock, and 95.7% do not have a project that generates income. The lack of funding was the main reason accounting for 83.4% of answers, while 13.3% related it to lack of skills and competences, 1.1% to lack of time, and finally 2.2% to other reasons. Furthermore, it proved to be very difficult to determine the income that households were generating from the cooperative or other activities (for example, some of the sons are employed by the military and public sectors). But when the monthly expenditure for 93 families was considered (Ministry of Planning and Cooperation (MoPIC) and Department of Statistics (DOS), 2007), then it seemed that 85.5% of households spent less than 199 JD, which means they cannot cover all their needs since more than 42% of families have six or more members.

27.5.1 The weak impact of cooperative activity on poverty reduction

Although locally agricultural production geared towards the national and international market has increased, there is still little food production for local consumption and moreover there have been very limited improvements in income for many of the local inhabitants. The families buy their food and other provisions from Qurayqira or from markets in Shawbak, while the backyard plots around the houses are generally not used to plant vegetables or fruits. During interviews the heads of families argued that the land close to their homes is arid. If they irrigate it with fresh water from their tanks, it would increase the water bills, whilst the water from the wadi is exclusively transferred by the cooperative to the farm land. Since 2004, the benefits generated by the cooperative have not been distributed, in order to increase the capital of the cooperative and to enable investment in services and equipment lacking in Faynan area. One of the shareholders explained that in 2009, for the first time, a part of the benefits was distributed but this did not exceed 300 JD per member. As a result, if the capital of the cooperative increases each year, there is no essential revenue from it to all households. Meanwhile, although the growing of commercial crops gives the impression of a commercially thriving community, in fact the income generated from farming is limited to 10 to 12 farmers who have the financial capacity to pursue agriculture.

Despite the apparent lack of work opportunities, most of the villagers do not manifest a real wish to be involved in agriculture; instead, workers from the Al-Azazma tribe, Amman, Egypt and Syria are employed in the different stages of the work. The deficiency of knowledge by tribe members about agriculture was put forward as an argument for non-involvement, but cannot be a convincing one. It appears that some people are still affected by social stereotypes: they consider working the land as labour, like the Egyptians or the Palestinians, to be a low status activity. But it is also because the families see more benefit from being registered as unemployed rather than as agricultural workers, and for this reason the families were discreet about their income in order to keep benefiting from the social aid provided by MoSD. The message that the people are poor and need the assistance of the state may be more a question of 'identity' than a true reflection of the reality. People still consider governmental assistance as a resource and a main form of income. In this context we have the example of Qurayqira village. Since 1970, they have had wells and agro-cooperatives, but this village is still considered to be a poverty pocket.

Therefore, at this early stage of the cooperative's progress, the prime objective of its leaders is not to reduce poverty or to enhance the social welfare of the population but to obtain the right to land and water in preference to the claims of the Shawabaka tribe by *de facto* possession with an ongoing activity. The leader of the Rashaida tribe considers that the main obstacles to improve the performance of their cooperative are at the political and financial levels. This relates to the legal status of land and the use of groundwater resources, which are so vital to improve the life of the villagers. This is the situation of most cooperatives in Wadi Araba, but according to Jaradat (2009), Director of the Wadi Araba Development Project, the lack of skills and a real motivation to get involved in agriculture are the main constraints for the agro-cooperatives.

It is clear that the government seeks to keep the land in state hands and has adopted an attitude of passivity with regard to this problem as well as to the illegal use of land and groundwater in southern Jordan. Many field systems or their extensions are illegal, and many wells have been drilled on state land without authorisation (Jaradat, 2009; Ministry of Agriculture (MoA), 2009), although exact numbers are not available. Actually, the aim of the government is to finance development projects to reduce poverty and unemployment by enhancing the capacity of cooperatives rather than to distribute the land amongst tribes. In this regard, many initiatives of development have been implemented by government and non-government institutions in southern Jordan.

27.6 NATIONAL MEASURES OF POVERTY ALLEVIATION AND THE WADI FAYNAN AREA

Involvement in the management of Faynan takes place at multiple levels and with a variety of actors, the principal being the

RSCN, the Department of Antiquity, JVA, the Municipality of Faynan-Qurayqira, ASEZA and the European Commission (EC) Delegation. How does the involvement of these actors affect the socio-economic development of the area and the initiatives of the local population?

27.6.1 Institutional involvement in socio-economic development

One of the causes of relative poverty in Faynan – as well as other rural poor areas – is the high level of centralisation and the extensive bureaucracy which has resulted in failure to deliver the social and economic infrastructure that is required to enhance the livelihoods of households (UNDP, 2004). For example, Faynan is under the responsibility of the Qurayqira-Faynan Municipality which develops and maintains the infrastructure and services, but which itself is completely dependent on the Ministry of Municipal Affairs (MoMA). The latter is also financially dependent on MoPIC, which supervises many programmes to reduce poverty financed by the donors such as the European Commission Delegation (EC) programme. The EU cooperates with MoMA and MoPIC in implementing socio-economic projects in the municipal area and in enhancing the municipality's capacity-building. Although the programme started in 2004, by 2008 it had not achieved its objectives owing to the high degree of management, overall centralisation and bureaucracy.

27.6.2 Wadi Araba Project

The ASEZA model has been used for the management of the city of Aqaba since 2004 and there has been successful development in the Aqaba Governorate in the past three years as a result of a relative process of decentralisation. Since 2007, ASEZA has managed the Wadi Araba Project, of which Faynan is part. Funded by the Arab Bank and the national government, almost 60% of the investment is designated for agricultural development of eight cooperatives including the Al-Rashaida cooperative, the drilling of 21 wells, developing 2,000 dunums of land, implementing services, and creating a centre for agriculture to provide support in the form of technical assistance or feasibility studies. The other part of the investment is dedicated to infrastructure. The project is still in its first phase and some constraints continue to impede progress, such as the legal status of land and water use, in addition to the tensions between tribes seeking to benefit from the project. As discussed earlier, a concrete resolution has not yet been achieved by the central government, like the issue of obtaining authorisation for water extraction from JVA, which controls the wadi flow and the groundwater, and the authorisation to drill wells (200 to 250 m depth), which is a key to the development of areas like Faynan.

27.6.3 Promoting ecotourism

Wadi Faynan has a long archaeological history of human settlement from the Neolithic through to the Roman and Byzantine periods. Located in south-central Jordan, the Dana Nature Reserve, managed by RSCN in cooperation with local Bedouins, includes the easternmost part of Faynan and offers opportunities to develop ecotourism. With an area of approximately 300 square kilometres, the Dana Nature Reserve is the most important nature reserve in the country, and was proclaimed as a UNESCO Biosphere in 1998. Faynan Ecolodge offers tourists the opportunity to explore the archaeological heritage sites from Neolithic to the Islamic period, the evidence of copper smelting from locally derived materials and the spectacular desert landscape as well as encountering semi-nomadic families in their environment (see Figure 27.6). The RSCN has also established a small centre for the marketing of local products, for example handicrafts and natural soap. The RSCN has an agreement with the Hamza Association for Tourism and Al-Azazma tribes about using their pick-ups for transporting tourists. The visitors are transported to and from the Ecolodge for a sum of 10 JD. While RSCN suggests that it is concerned with supporting local communities by creating jobs and investing in the reserves, locally the impact of its initiative is very limited and controversial. The staff of the Ecolodge point out that they depend strongly on international tourism as the prices are very high (65 JD per night) and are not affordable to the local population.

To date the income generated from the Ecolodge and the Reserve is just sufficient to cover the operating expenses, and there is little investment into supporting local people who are not necessarily involved in the activity of the reserve. The Ecolodge has created only four jobs primarily for those from Qurayqira and Tafila. Only four women from the Al-Azazma tribe are involved in the handicraft centre, generating very limited income. No women from the Rashaida tribe are represented, owing to the cultural stereotypes about women's work (personal interviews, 2009). Additionally, the Al-Azazma and Al-Rashaida herders constantly emphasise the reciprocal relationship between humanity and nature, between the trees and plants and their livelihood as herders (Lancaster and Lancaster, 1999). They consider that the establishment of Dana Nature Reserve was the ending of their local autonomy, restricting their movements and constraining livelihood.

Meanwhile, a project proposing the promotion of eco-tourism in the area was submitted for funding in 2007 by the University of California and the Department of Antiquities to MoPIC. The proposal focuses on the infrastructure of ecotourism in Faynan as well in Qurayqira. The aim is to create an archaeological park to include the Faynan District archaeology, the Faynan Museum with Bedouin ethnographic exhibits, construction of a camp site near to Fidan spring, a freshwater oasis pool in the springs of Wadi Ghuwayr and Ain Fidan, establishment of a Bedouin Tribal Digital Village (building a small wireless tower in the region to connect all the local Bedouin schools), and a centre of training for hospitality. It is a very ambitious project which would bring job opportunities and improve the livelihoods of the population. Unfortunately, by mid-July 2009 only the museum was established and the other components of the proposals have not progressed beyond the planning stage.

27.7 CONCLUSION

This chapter has focused on the role of water and land management practices in promoting current development in the Wadi Faynan area, with particular reference to the recent development of irrigated agriculture making use of shallow groundwater sources and more recently perennial wadi flow. It is clear that the local establishment of cooperative irrigation-based agriculture has brought much-needed capital into the area. It has also been associated with the establishment of a possible means for wider community-based socio-economic development. However, there are real environmental issues to be faced, especially in relation to the long-term viability and sustainability of such economic activity. At this stage, the activity has had less of an impact on poverty reduction than might have been hoped for. Indeed, our enquiries show that only a very small number of residents, some 10–12 farmers, appear to be deriving net benefit at present. The scheme has had negligible influence on patterns of land ownership and very limited impacts on the enhancement of basic infrastructure.

It is hard not to reach the conclusion that wider forms of integrated development are urgently needed if the area is to benefit. In this connection, in the final sections of the chapter we briefly examined the need for wider development schemes set in the context of decentralisation at the sub-regional scale of southern Jordan. At this juncture, a lack of concerted collaboration among the various stakeholders has contributed to delays in the socio-economic development of the Wadi Faynan area. The outcomes of initiatives during the past 10 years, whether implemented by government institutions or by organisations like the RSCN, have served to emphasise two things: firstly, a lack of well-rehearsed and implemented coordination at the institutional level between different entities; and secondly, the generally low levels of cooperation that have been established with the local population to assist in getting them more involved in such local initiatives. These, together with the environmental challenges presented by the irrigation-based agricultural activity, represent the major challenges facing the area.

REFERENCES

Al-Jaloudy, M. (2001) *Country Pasture: Forage Resource Profile for Jordan*. Amman: FAO www.fao.org/ag/AGP/AGPC/doc/Counprof/Jordan.htm

Al-Simadi, F. (1995) The influence of social development and demographic factors on modern orientations and tribal loyalty in Jordan. Unpublished PhD thesis: Brigham Young University.

Bin Mohammad, G. (1999) *The Tribe of Jordan at the Beginning of the 21st Century*. Louisville, USA: Fons Vitae.

Bocco, R. (2000) International organizations and the settlement of nomads in the Arab Middle East, 1950–1990. In *The Transformation of Nomadic Society in the Arab East*, ed. M. Mundy and B. Musallam. Cambridge: Cambridge University Press pp. 197–217.

Bocco, R. and T. Tell (1995) Frontières, tribus et Etats en Jordanie orientale à l'époque du Mandat. In *Tribus, tribalismes et Etats au Moyen-Orient*, ed. R. Bocco and C. Vélud. Paris: La Documentation Française pp. 26–47.

Carney, D. (1999) *Approaches to Sustainable Livelihoods for the Rural Poor*. London: Overseas Development Institute www.odi.org.uk/publications/briefing/pov2.html.

Crook, D. (2009) Hydrology of the combination irrigation system in the Wadi Faynan, Jordan. *Journal of Archaeological Science* **36**: 2427–2436.

Department of Statistics (2007–09) *Annual Statistical Bulletins*. Amman, Jordan: Department of Statistics.

Hashemite Kingdom of Jordan (2002–03) Household Income and Expenditures Survey 2002/03. Amman, Jordan: Department of Statistics.

International Co-operative Alliance (2004) *Annual Report*. Geneva, Switzerland: International Co-operative Alliance.

Jaradat, I. (2009) *Personal Interviews, February 2009*. Aqaba, Jordan: Wadi Araba Technical Development Unit.

Jordanian Cooperative Corporation (2009) *Statistical Database on Cooperatives in Aqaba Governorate, Internal Report*. Amman, Jordan: Directorate of Cooperation.

Lancaster, W. and F. Lancaster (1999) *People, Land and Water in the Arab Middle East. Environments and Landscapes in the Bilad ash-Sham*. Amsterdam: Harwood Academic Publishers.

Lewis, N. (1987) *Nomads and Settlers in Syria and Jordan, 1800–1980*. Cambridge: Cambridge University Press.

Maluleke, T., V. Thomas, T. Cousins, S. Smits and P. Moriarty (2005) *Securing Water to Enhance Local Livelihoods: Community-based Planning of Multiple Uses of Water in Partnership with Service Providers*. South Africa: AWARD www.musproject.net.

Metral, F. (1996) Biens tribaux dans la steppe syrienne entre Coutume et droit écrit. *Revue du Monde Musulman et de la Méditerranée* **79**: 89–112.

Ministry of Agriculture (MoA) (2009) *Pastoral Resources Information Data Base, Monitoring and Evaluation Unit Amman*, Jordan: Ministry of Agriculture.

Ministry of Planning and Cooperation (MoPIC) (2004–2006) *Socio-Economic Development Plan 2004–2006*. Amman: Ministry of Planning and Cooperation.

Ministry of Planning and Cooperation (MoPIC) and Department of Statistics (DOS) (2007) *Socio-Economic Households Indicators in Wadi Araba, household survey, July 2007*. Amman: Ministry of Planning and Cooperation (MoPIC) and Department of Statistics (DOS).

Municipality of Qurayqira-Faynan (2007) *Study on Qurayqira and villages Faynan*: Municipality of Qurayqira-Faynan (unpublished report – in Arabic).

Palmer, C., D. Gilbertson, H. el-Rishi et al. (2007) The Wadi Faynan today: landscape, environment, people. In *Archaeology and Desertification: The Wadi Faynan Landscape Survey, Southern Jordan*, ed. G. Barker, D. Gilbertson and D. Mattingly. Oxford, UK and Amman, Jordan: Council for British Research in the Levant in Association with Oxbow Books pp. 22–57.

Razzaz, O. (1994) Contestation and mutual adjustment: the process of controlling land. *Law and Society Review* **28**: 7–39.

Shryock, A. (1997) Bedouin in suburbia: redrawing the boundaries of urbanity and tribalism in Amman, Jordan. *Arab Studies Journal* **1**: 40–56.

Smits, S., T. Cousins, V. Dlamini *et al.* (2006) *Water, Households and Rural Livelihoods*. The WARFSA conference, 2 November 2006, Lilongwe, Malawi.

UNDP (2004) *Human Development Report 2004*. New York: United Nations Development Programme.

Van Aken, M. (2004) Du Fellah a l'agriculteur: significations symboliques a travers les champs de la vallée du Jordan. *Cahiers de L'Orient* **75**: 101–124.

van Koppen, B., P. Moriarty and E. Boelee (2006) *Multiple-Use Water Services to Advance the Millennium Development Goals Research Report 98*. Colombo, Sri Lanka: International Water Management Institute.

Wadi Araba Development Project (2007) *Progress Report*. Unpublished document, Aqaba, Jordan.

World Bank (2004) *Jordan Poverty Assessment* (2 volumes). Amman: World Bank.

28 Political discourses and public narratives on water supply issues in Amman, Jordan

Khadija Darmame and Robert Potter

ABSTRACT

In its National Water Master Plan 2004, the Jordanian Government stressed that the first priority is to meet the basic needs of the people. On the other hand, water is pressingly needed for agricultural activities as well as in the industrial and tourist sectors. Unlike many cities in the developing world, 98% of households in Amman are connected to the water supply network. However, since 1987, the supply of water has been rationed. For most areas, water is now supplied once a week, although the duration of supply varies considerably. From the first, the rationing of urban water reflected not just scarcity but also the dilapidated physical state of the network. In respect of management, from February 1999 to December 2006, the water supply system of Amman was placed in the hands of the private sector company known as LEMA. During this period, the network was comprehensively upgraded and in the winter of 2006, continuous supply was introduced to some 15.8% of LEMA's customers. Following this, some technical experts argue that the entire system should move to continuous supply, both for technical and supply reasons. However, the political will for this does not appear to exist. Based on critiques of extant water policies, and on semi-structured interviews with key informants including different socio-economic groups of urban consumers in Amman, this chapter analyses political discourses and public narratives on urban water supply issues.

28.1 THE UNRELIABILITY OF WATER SUPPLIES IN DEVELOPING CITIES

28.1.1 Introduction

As noted in Chapter 26 of this volume, one of the principal issues affecting the water sector in many cities of the developing world is the unreliability of supply. Cities such as Algiers, Delhi and Amman are characterised by rationing programmes of a few hours' supply each day or once per week. As a policy, such rationing relates to the perception of cities in 'crisis' or of 'thirsty cities' under a restrictive supply (see Molle and Berkoff (2006) for a review of the arguments behind water transfers from agriculture to the cities used by governments in different countries). Often the reason presented by governments for such rationing relates to water scarcity and to the high level of water demand in all sectors. But the rationing of supply is not always a direct consequence of physical factors, as stressed by Zérah (1998, 2000) or of the increase in domestic demand. It is also linked to structural crises in the management of the water sector. In this chapter we seek to show that this is the case in the instance of Amman in Jordan. Thus, 'unaccounted for water' (UFW) forms a very substantial proportion of the total water budget and includes leakage, illegal connection and unpaid bills, along with errors in the reading of meters. In this account, we focus on the rationing programmes designed and established by the Water Authority of Jordan (WAJ) as a solution to the problems of water shortage, which were designed to limit the substantial level of losses and to ensure equitable access to water by all households.

What are the specificities of the rationing policy in the Jordanian context? What can be learned from an evaluation of the policies implemented by public and private operators in respect of this issue? What are the views of households and how do they influence the use of water? The account addresses the issue of rationing policy and its socio-political dimensions under private management in Amman. Our approach tackles the rationing supply by considering all stakeholders' points of view: private and public operators, the WAJ and the end users or consumers. The study covers the experience of the private company involved in water management in resolving the irregularity of water supply from 1999 to June 2007. The analysis is based mainly on semi-structured interviews undertaken by the authors with the different stakeholders involved, which were carried out during fieldwork in 2007 and 2008.

Water, Life and Civilisation: Climate, Environment and Society in the Jordan Valley, ed. Steven Mithen and Emily Black. Published by Cambridge University Press.
© Steven Mithen and Emily Black 2011.

28.1.2 The notion of the unreliability of water supplies

The first studies to have assessed the impact of water supply unreliability on households were launched by the World Bank (WB) during the 1990s in India (Sethi, 1992), Pakistan (Altaf, 1994) and Dehradun (Choe et al., 1996). The research of Zérah (1997, 1998, 2000) on unreliability linked to household water strategies provides the principal reference for this study. Based on the definitions by Zérah (2000), three attributes characterise water unreliability: (i) intermittence of the supply (number of hours and timing of the supply); (ii) issues of water pressure; and (iii) predictability (sudden breakdowns, regularity of the supply and seasonal variations). Until now, studies have examined the topic as a technical issue or in terms of social consequences. To the best of our knowledge, little or no academic analysis has focused on the reasons for such unreliability or on the political discourses and public narratives that underlie such a state of affairs. Our hypothesis is that the rationing issue in most cities is mainly related to the scarcity of water resources, while in Amman it is closely linked to the mismanagement of the available water within the city and to the deterioration of the public supply network. We assume that irregularity of supply is not only a technical issue, but that exogenous political, economic and social factors also have a crucial role in resolving urban water issues.

28.2 AMMAN: A 'THIRSTY CITY'?

28.2.1 Water resources for Amman: more transfers versus more leakages

Jordan is an arid to semi-arid country and is considered to be one of the ten most water-scarce countries in the world. The kingdom suffers from a deficit between available water resources and water demand. It has a heavily centralised system of water management by the Ministry of Water and Irrigation (MWI). This oversees all policies in the water sector and regulates the WAJ, which is responsible for drinking water. The Jordan Valley Authority (JVA) is in charge of the development in the Jordan Valley. According to the MWI, the renewable water resources in 2004 were estimated at around 817 million cubic metres (MCM), while demand was estimated at 1,297 MCM. Jordan thereby faces a water deficit and it is predicted that this will reach 437 MCM by 2020 (Abu-Shams and Rabadi, 2003).

In 2004, Jordan's population exceeded 5 million, of which 39% live in the capital city region, the Governorate of Amman (Hashemite Kingdom of Jordan, 2006). The metropolitan area is a space of concentrated economic, political and social activities (Potter et al., 2009; see also Chapter 26). Amman and its suburban areas have witnessed successive in-flows of immigrants: Palestinian refugees in 1948 and 1967, the 'returnees' from Kuwait and other Gulf States in 1991 (mostly of Palestinian origin), and approximately 700,000 Iraqi refugees since the first Gulf War (United Nations High Commissioner for Refugees (UNHCR), 2007).

The high natural rate of population growth, together with successive waves of immigration and economic development, has placed heavy burdens on the water resources available in Jordan. According to MWI data in 2004, the average resident in Amman receives less than 100 litres per capita per day, generally nearer to 70 litres. In the Water Strategy Policy of 1997 and the National Water Master Plan produced in 2004, the Jordanian government stated its first priority as meeting the basic water needs of the urban populace (Nortcliff et al., 2008). A large imbalance is shown in water allocation: the water used by all sectors in 2004 did not exceed 866 MCM (MWI, 2004). Irrigated agriculture was the largest consumer, constituting around 64% of the overall usage compared with only 25% in the case of domestic demand and tourism, which stood at around 290 MCM (MWI, 2004).

In order to satisfy the water demands of the capital, in 1975 the Government of Jordan started to transfer water to Amman from the surrounding territories. Of the total water volume allocated to the Governorates, Amman monopolises around 40% of available resources, of which half comes from the Abdellah Canal in the Jordan Valley. The water production allocated to the capital in 2004 was 105 MCM. During the recent dry years from 2000 to 2003, restrictions were placed on irrigated agriculture in the Jordan Valley. During this period, summer crops were forbidden and over 1,000 hectares on farms have been left uncultivated. Priority was given to supplying the urban areas, causing various reactions among farmers and agribusiness.

But regrettably, a large proportion of the water reserved for the cities is, in the final event, 'lost', in the sense of being 'unaccounted for' (UFW). The rate of UFW is high, essentially representing the rate of technical leakage in the network. It is one of the main aspects of the water management failure in Jordan. In 1999, the rate of UFW was 54% of the total water resources distributed within the water network of Amman, of which around 30% represented technical leakages amplified by illegal connections and wasteful practices, which are widespread in Amman, especially in the affluent areas. These facts lead us to question the stereotype of the necessity of weekly rationed water supplies within the city. As a result of water scarcity there is clearly a pressing need to focus on the mismanagement of available resources in the urban areas, principally as the outcome of systemic leakages and losses.

28.2.2 Lack of reservoirs at the top of the hills and inadequate design of the water network in Amman

The water supply network in Amman, as in many other developing world cities, is in a deplorable state. Amman was built on a large number of hills, and rapid urban expansion has taken place from the old downtown towards the hills (Potter *et al.*, 2009), without water reservoirs being established on the higher ground. Indeed, the water network has had to follow the new urban areas and to connect households using cheap, small-diameter extensions. Thus, a poor network design using small-diameter pipes was implemented and extended without the re-design of the system.

The demands of new network users have through time been satisfied by increasing the flows transported, which has in turn led to increasing losses. Most of the sub-networks thereby operate at too high a pressure (up to 15–20 bars). Decker (2004) has observed how the water system of Amman has largely developed in an ad hoc, as opposed to a planned manner. As a result, most distribution is pumped, and generally occurs from the low-lying areas up into the hills, which results in very high pressures in some sections of the distribution system. Amman's network is extensive, and in 2004 comprised around 5,100 km in all. Consequently, the proportion of small-diameter (100 mm and less) galvanised iron pipes, as part of the system, is very high (Decker, 2004; and interviews with M. Bobillier, Director of UFW Service of LEMA, 2007). Furthermore the irregularity or on/off nature of the water supply damages the small pipes that are connected to the dwellings. These are generally made from steel and ductile iron, a material that does not exceed two to three years of use owing to rusting and the impact of rationing (Decker, 2004; J. Naouri (technical manager of management contract programme), personal communication, 2004, 2007). Thus, the water quality is doubly affected because of the deteriorated canalisations; and additionally, the water is stored in tanks on the roof, and is thereby exposed to climatic and other ambient hazards.

But the leakage experienced is not due to technical failures within the system alone. Many illegal connections in rural and peri-urban areas add substantially to the scale of UFW. In some instances, farmers and shepherds break the pipes and leave water running for days, as illustrated in Figures 28.1 and 28.2. The photographs depict substantial losses occurring in the southern areas of the Governorate of Amman.

Figure 28.1. In a country area of the southern Amman Governorate, people use water directly from the piped network.

Figure 28.2. An illegal connection from the principal pipe of the network, southern Amman Governorate.

28.3 MANAGING THE IRREGULARITY IN THE SUPPLY OF WATER

28.3.1 Preamble

Since 1990, the Government of Jordan has adopted specific strategies in an effort to resolve the water supply issue, in the context of structural adjustment reforms, assisted by donors, and with the participation of the private sector. Donors are strongly involved in different levels and aspects of water management by giving grants, donations or providing technical assistance. But all these stakeholders have their own goals, which relate to their own agendas and approaches to water problems. What has been the interaction between these stakeholders? What was the solution proposed to resolve the irregularity of water supply in the capital? Different measures were adopted as key solutions to reduce UFW. Overall, three main steps were implemented successively: (i) rationing water distribution from 1987; (ii) renovating part of the water network; and (iii) implementing public–private participation in Amman from 1999.

Table 28.1 *Selected performance indicators concerning urban water and sanitation utilities in selected cities of the MENA region, 2004*

	Algiers	Amman	Casablanca	Sana'a	Tunis	Good practice
Unaccounted for water (UFW)	51%	46%	34%	~50%	13%	15–25%
Water coverage[a]	100%	100%	100%	65%	100%	100%
Continuous supply	No	No	Yes	No	Yes	Yes
Per capita water use (litres per day)[b]	~70	~70	110	50	~80	120
Operations and management cost recovery	No	Yes	No	(Yes)	Yes	Yes

Adapted from Arce (2004)
[a]Including coverage from standpipes, but excluding private piped systems (in Sana'a).
[b]Estimated amount from public network actually used by the consumer, net of physical losses.

28.3.2 Rationing water distribution from 1987

Because of the lack of financial resource, and as a response to the critical status of supply, the rationing policy previously outlined was adopted in 1987, avowedly to alleviate the impact of water scarcity and to stop increasing leakages. In Jordanian cities, as well as in some other developing urban areas of the MENA region (Table 28.1), water distribution has been assured only once per week, for different durations. The most populated areas, such as the Palestinian refugee camps located in the east and southeast of Amman, were supplied two to three times a week at one day intervals. There were clear socio-economic reasons that guided this decision, namely the small storage capacity characteristic of such areas, their high population densities and low-income levels (M. Balbisi, Ex-General secretary of WAJ from 1987 to 1988, interview carried out in June 2007). In addition to the intermittence of supply, water pressure was generally very low in these areas, and households at the bottom of the hills had to use electric pumps to drain water to different floors and to storage tanks, often depriving those houses situated at the top of the inclines.

28.3.3 Rehabilitating the water network

A big project to rehabilitate part of the water supply network was initiated and financed by the World Bank and other donors in 1999 as a major component of the wider water sector reforms that occurred in Jordan at that time. A study was completed by the international company DORSCH Consultants between 1996 and 1997 (Dorsch Consultants & Assoc., 1997) on how to deal with the problems posed by the technical loss of water and high water pressures. The study recommended the construction of large reservoirs at the tops of the hills, which would then distribute water to dwellings at lower altitudes by gravity. In the associated framework, the city was divided into 44 supply zones, with each zone having a reservoir linked to the principal network. The zones were divided into districts, each one consisting of 1,000 to 2,000 households and comprising approximately the same area and carrying the equivalent water pressure levels (70 m between the users at the top and at the bottom). Furthermore, secondary pipes made from polythene were installed, and such pipes are well known for their good quality, resistance and sustainability. The project was divided into 15 contracts, each one financed by different donors, and the cost of the entire project amounted to some US$ 250,000 (13 of the 15 projects were finished in 2008, two projects in western Amman remain to be completed).

28.3.4 The implementation of the public–private partnership

Meanwhile, the donors laid down the participation of a private operator in the management of Amman's water service as a precondition for funding. The aim of this measure was to tackle the administrative losses, to repair small leakages in the network and to manage the service during the rehabilitation project (Naouri, Technical Manager of Management Contract Programme, interviews carried out in 2004 and 2007). The first Management Contract stemming from this public–private partnership (PPP) was established in Amman in February 1999, and was signed with the multinational company Suez. The contract was managed by a company known as LEMA under a 'Performance Based Management Contract for the Provision of Water and Wastewater Services for Amman Governorate' for a period of 4 years, that is until 2003. In the event, the contract was extended to December 2006. LEMA (Lyonnaise des Eaux – Montgomery Watson – Arabtech Jardaneh) was 75% owned by Suez Environmental, with the other 25% held by MWHAJ (Montgomery Watson and Arabtech Jardaneh).

The service area of LEMA consisted of the whole of Amman and its contiguous suburbs – a total area of some 2,700 sq km (Figure 28.3). The population served under the contract was approximately 2 million, with around 330,000 commercial subscribers, the latter thereby amounting to some 15% of the total (Griffen, 2004).

year and by 25% by contract end in year 4; meanwhile, the performance target relating to hours of supply was set to increase from 48 hours to 72 hours during the peak season (15 May–15 October). In the off-peak period, from the end of the first year, the supply of water was scheduled to be continuous. Therefore, the intended progress of both indicators reflects the close association between UFW reduction and overall improvements in water supply. However, a problem was that even though LEMA was managing water distribution within Greater Amman, the company was not involved directly in the management of resource at the national level or in any decision-making relating to the overall volume of water allocated to Amman.

28.4.2 Continuous supply: unrealistic target without the appropriate framework for management?

The water supply network of Amman is divided into three major zones, as shown on the map in Figure 28.4. These comprise Amman East, Amman West and Amman South, according to the location of the main supplying reservoir (see this volume, Chapter 26, for a description of the economic differences between these regions). Since 1999, a two-fold seasonal schedule comprising winter/summer has been implemented, related to water resource availability and demand. The precise details of the hours of supply to different areas have been changed on a yearly basis. The map in Figure 28.4 shows the hours of supply during the summer of 2006. Data from 1999 to 2002 giving the hours of supply to different areas were not available and so our statements are based on piecing things together from the discussions we have had in interviews with various stakeholders.

In the first year of the contract, as an imperative measure, LEMA reduced the duration of supply to between 24 and 36 hours per week for all households in all areas. The interviewees emphasised how poor households, whose storage capacities were insufficient, were principally affected. The WAJ interpreted this measure as LEMA attempting to reduce the rate of leakage in the system by decreasing the overall quantity of water distributed, in order to achieve the yearly performance indicator for UFW and avoid contractual penalties. Indeed, in the event, LEMA was penalised at the end of the first and the second years for its modest achievements in respect of reductions in UFW. In general, the rationing programmes from 2003 to 2006, as shown in Figure 28.5, involved the Amman East and Amman West areas being favoured in turn with increased durations, while Amman South was always relatively disadvantaged. In fact, from the data shown in the graph we can observe three successive phases in the evolution of water supply in the three areas. Firstly, from May 2003 to April 2004, the distribution did not exceed 40 hours for users in all areas. East and West Amman in turn benefited from marginal enhancements to the supply, while Amman South lagged well behind.

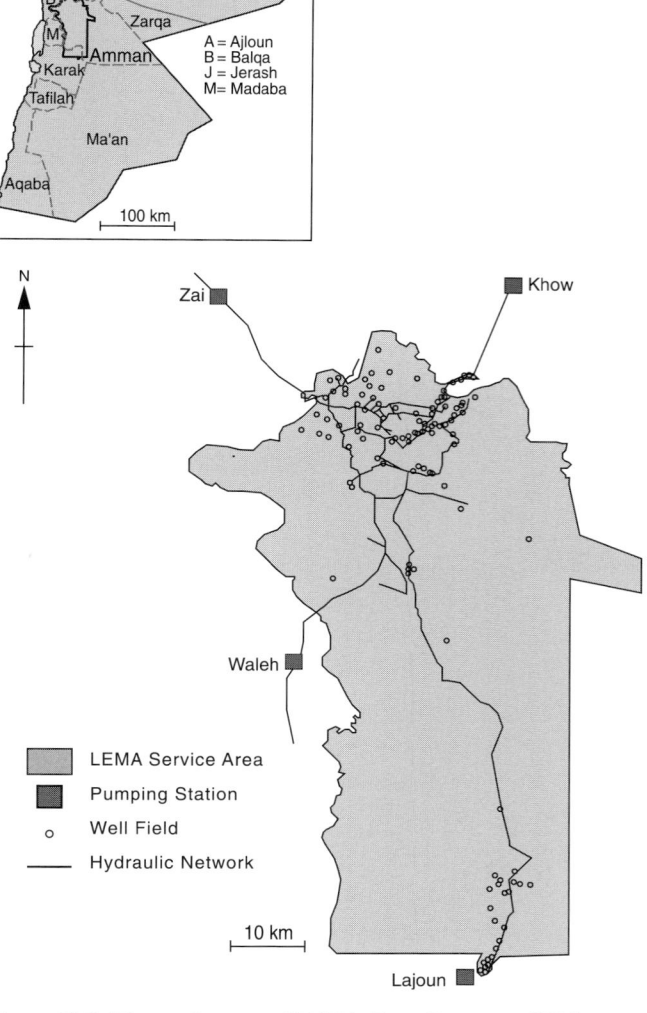

Figure 28.3. The service area of LEMA. From Darmame (2006).

28.4 PRIVATE DISCOURSE VERSUS PUBLIC POLICY DIRECTIVES

28.4.1 Relation between unaccounted for water reduction and improvements in water supply

According to the performance indicators of the Amman Management Contract, reducing the magnitude of UFW was one of the principal contractual targets with which LEMA was charged. The targets set for years 1 to 4 of the contract are summarised in Table 28.2. The table also summarises how, over the same period that UFW was to be reduced gradually, the duration of supply was set to increase progressively.

Thus, according to Table 28.2, the performance target for UFW was envisaged to decrease by 10% at the end of the first

Table 28.2 *Targets set in the LEMA contract for the reduction of unaccounted for water (UFW) and water supply improvements*

	Performance target by year			
	Year 1	Year 2	Year 3	Year 4
UFW reduction target (percentage of total)	10	17	22	25
Hours of supply:				
Peak season (15 May to 15 October)	2 × 24	2 × 30	2 × 36	2 × 36
Off-peak (rest of the year)	2 × 48	continuous	continuous	continuous

Source: Management Contract, Department of Planning, PMU 200

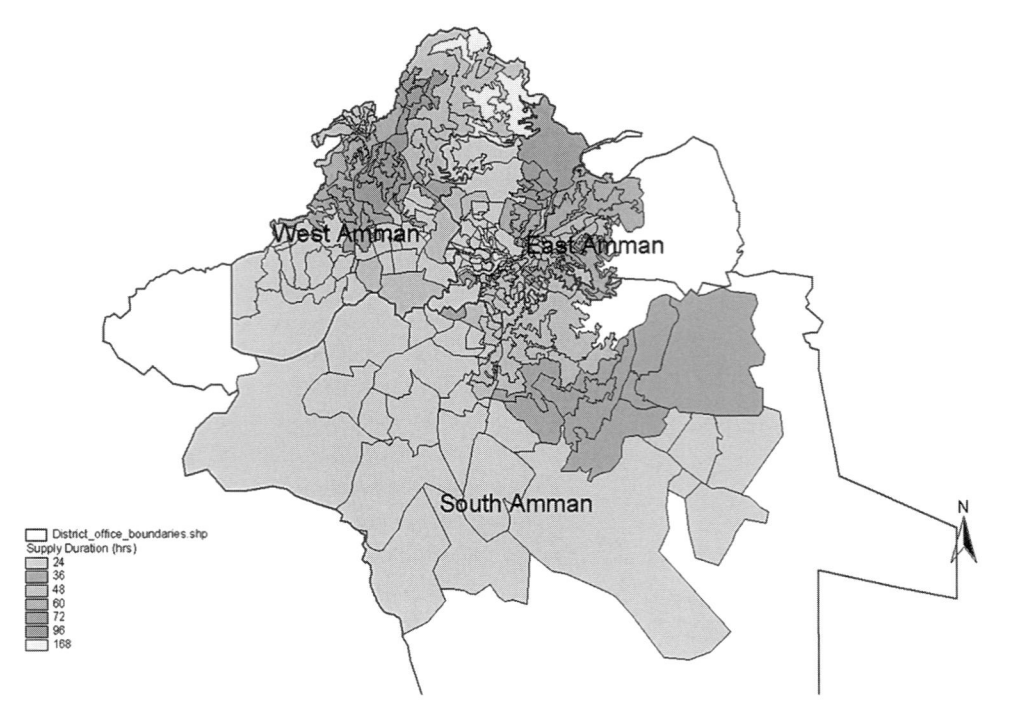

Figure 28.4. The total supply duration (hours per week) in Greater Amman, summer 2006. Source: LEMA Company, personal communication, Amman 2007. See colour plate section.

Secondly, from April 2004 to May 2005, a clear overall improvement was witnessed for users in all areas, but with twin peaks in favour of the more affluent western areas. The supply for South Amman was enhanced but was still disadvantaged compared with Amman East and Amman West. There was then a peak between the winters of 2004 and 2005 when LEMA designated 27% of households as pilot areas to be provided with continuous water during the winter months. In order to compensate for this, additional water was allocated to Amman from Zara-Maïn, the desalinisation station managed by SUEZ. The objective of putting these pilot zones under continuous supply was to detect degraded pipes and illegal connections. A further objective was to test the reaction of households in terms of the water they would consume under continuous supply, and how much this would increase overall consumption.

The outcome was that during this test period, continuous supply did not increase consumption significantly. This was illustrated by the fact that the aggregate water bill for the city as a whole in 2004 increased by only 8% (Darmame, 2006). Thirdly and finally, from May 2005 until 2006, the overall period of supply decreased and was stabilised at around 48 hours for all users (see Figure 28.5). From this point onward, the eastern zone of the city has in overall terms received more hours of supply than Amman West and Amman South. In an interview in 2007, the Director of UFW Service of LEMA affirmed that the decision criteria governing the supply of water to different areas of the city were purely technical, that is relating to hydraulic and network conditions, and not socio-economic ones (M. Bobillier, Director of UFW Service of LEMA, interviews completed in April and May 2007). It is clear that during this period, rationing

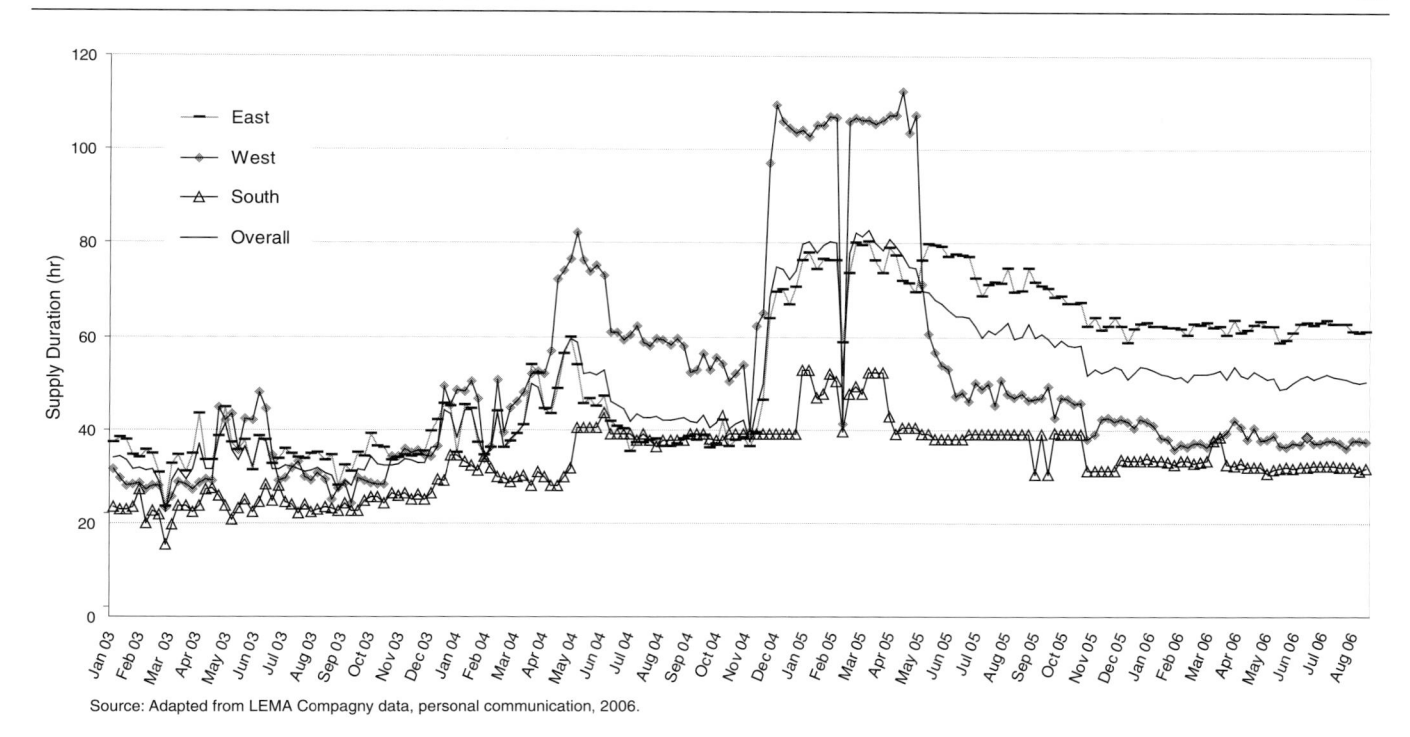

Figure 28.5. Supply duration (hours per week) in Greater Amman by geographical sector, January 2003 to August 2006.

did not deliberately favour the wealthy areas of the city. Rather, the data show that increased supply durations alternated between the low-income popular and wealthier areas of Amman (Bobillier, 2007)[see above]. Furthermore, continuous supply was provided in the popular zone of Amman East, the area with the most recently renewed network as at 2005.

However, LEMA could not maintain this level of supply, a fact that was clearly summarised by Bobillier in our interview with him: 'It is impossible to increase the duration of supply for all subscribers, because the rate of water loss will increase too, and water resources are insufficient to compensate the losses caused by the continuous distribution.' Thus, in 2006, the last year of LEMA's contract, only just under 12% of subscribers received water continuously during the summer months while 66.5% were supplied for 36 hours or less (Table 28.3).

At the end of its mandate in 2006, LEMA submitted a technical study about the necessity of supplying water continually within the network. Decker (2004) and the 2007 interviews with the Director of UFW Service referred to above affirmed that the case studies from other countries and the pilot study carried out in 2004–5 in Amman confirmed that intermittence of supply serves to stress the network about 30 times more than expected in networks of a similar size under conditions of continuous use. Therefore, despite the renovation of the network, water supply irregularity will progressively damage the pipe network if it remains under rationing. The study also demonstrates that by putting zones under continuous supply, the operator can detect leakages and illegal connections that will lead to additional

Table 28.3 *Proportion of water subscribers in Greater Amman by total duration of water supply per week during summer 2006*

Number of hours of supply per week	Number of subscribers	Percentage of subscribers
168	47,055	11.9
120	733	0.2
96	463	0.1
72	9,692	2.5
60	5,201	1.3
48	43,732	11.1
42	29,749	7.5
36	86,726	22.0
30	30,963	7.8
24	140,383	35.6

Source: LEMA database, 2007

quantities of water in the mid to long terms. The progress to a continuous supply should proceed gradually, since there will be a high rate of leakage involved initially as pipes that are damaged become obvious and have to be repaired. In order to achieve the transitional phase, WAJ has had to provide more water resources to Amman to compensate for the outflow.

LEMA's contract finished in 2006 and by that stage the MWI had not responded to the study proposals. According to WAJ, the measures taken by LEMA to solve the water rationing issues were not sufficient and the goal of moving from an irregular offer to a continuous supply at the end of 2006 was not achieved. The private operator attributes this situation to several factors, not least the delay in the programme to rehabilitate the network, illegal connections, successive droughts and the increase in demand due to the war in Iraq in 2003. Also, LEMA reproached MWI for excluding the company from any involvement in wider water management in the country, while more than one-half of all water resources allocated to the Governorate of Amman comes from outside its administrative boundaries. What could be the reaction of donors who have invested in the renovation of part of the network, which included the provision of pipes designed for continuous supply? The main objective for donors would have been to maintain the privatisation processes at the end of the LEMA contract in 2006 and to move to a higher level of PPP as a concession. But, in January 2007, the Government of Jordan convinced them of the utility of its own local model of PPP. Amman's water management is now being placed in the hands of a new 'public company' named Meyahona ('Our Water'). This is owned by the WAJ, but will be run on the lines of a private company. This is exactly the model that has been in operation in the second city of Jordan, Aqaba, for several years. It is being presented by the MWI as a crucial alternative to foreign private sector involvement in the water sector in Jordan (Potter et al., 2007). In this sense, the water system of Amman may no longer be privatised, but it will most definitely remain commercialised. Since then, the Jordanian political discourse on the rationing issue has changed. In the words of I. Dhyat (2007), the Director of the Planning Department at WAJ, 'Progress towards a continuous supply is excluded and rationing meets the needs of 90% of the population of Amman.' Thus, it seems that rationing has been used as an instrument of policy depending on who is managing the water service. The monitoring of the achievements of LEMA was very rigorous and unrealistic regarding some targets such as the rate of UFW and the cultural practices of illegal connections.

28.5 AMBIVALENCE OF USERS' REACTIONS TO CONTINUOUS SUPPLY

Although rationing was basically designed as a solution to provide water access to all households, as shown in Chapter 26 of this volume, it has inevitably become associated with inequalities in water use between poor and rich families. As was also shown in Chapter 26, not all families can afford the cost of alternative strategies in order to access adequate water supplies, and they have also had to face the negative effects of progressive water tariffs over the past few years. Given these circumstances, what are the views of families on improvements to the length and duration of supply and what scenarios do they prefer for adequate water access?

28.5.1 Households' perceptions of overall water supply issues including the policy of rationing

The results of the interviews we conducted with consumers, investigating perceptions of overall water supply issues, including the rationing programme, served initially to highlight a lack of general knowledge among households concerning some detailed aspects of the water crisis at the technical and political levels (see this volume, Chapter 26). This is exemplified in the following statements:

> I think that we don't ask for water from other countries.
> I can't give solutions for water problems in Amman because I really can't see what the problem is.

Most households recognised the national lack of water resources, the impact of the arid climate and the increase in water demand stemming from several phases of refugee influx, plus return migration. But in general, feelings of apathy and a degree of suspicion toward MWI and WAJ's water policies and their efficiency were revealed in our surveys. Most consumers appeared to agree on mismanagement and the failure of WAJ to monitor the water sector competently:

> The dilemma is that we do not have national efficiency in water storage or management. So the water policy is not effective.

Interestingly, this particular remark was illustrated by mention of what is regarded as the weak accountability of WAJ in respect of water flowing in the streets during the 'day of water' (Chapter 26), and WAJ's perceived tolerance toward wasteful practices, mainly in the wealthier areas of the city:

> The main problem is that WAJ doesn't do enough concerning awareness about water use and its value [. . .] They are only looking to see if poor people are stealing water or not.
> Penalties should be established by WAJ for wasting water in gardens and on cars.

Most of the issues raised by families related to the perceived existence of a serious gap in communications between WAJ and public–private operators and users. The concrete example was when we asked about Amman's service contract, progress with water rationing and tariff increases. The respondents were vague and showed a clear lack of information. For example, most of the people we interviewed in East Amman believed that WAJ has been treating relatively wealthy areas favourably by supplying more water:

Water is not distributed with equality and some areas have water every day due to their social connections. Also, how come all the hotels and some people have water that they use with ease, while we don't even have enough water to have a daily shower, especially in summer?

In contrast, of course, our analysis (and Figure 28.5 in particular) shows that East Amman has, in fact, experienced water supply improvements since 2005. Interestingly, the same sorts of comments were made by some of the wealthy households in West Amman, who also expressed the view that the inhabitants of the wealthier areas, where the political and economic elites reside, receive more water:

> The rationing supply goes according to the level of consumption. In rich areas we have more requirements, so they deliver water more than in poor areas. It is according to way of life and income. I think it should be equal.

In all instances, when we asked about the source of information for such views, the interviewees said 'we have been told by others' or 'we are sure that this is the case.'

28.5.2 Overall views of households on the possibility of continuous supplies

From the interviews we completed in Amman in the summer of 2007, focusing on water distribution issues during the past 10 years (1998–2007), it is clear that some amelioration is perceived in the overall water supply situation since LEMA's contract started in 1999, especially in respect of maintenance of the piped network, the regularity of the day of supply, and the overall management of the water service. But for some households moving to continuous supply was regarded as 'a dream – impossible … difficult – or very costly'.

However, as far as households are concerned, is having a continuous supply a priority in comparison with other aspects of water supply (quality, regularity, price, pressure, etc.)? It became clear from our interviews that the issue of water quality is the main concern for the majority of households, despite constant reassurances by WAJ and LEMA. There had indeed been several cases of water pollution during the previous 10 years, the worst being in 1998 at the Zaï water treatment plant which badly affected West Amman.

Only after quality were the factors of duration of water supply, regularity of timing, frequency and pressure cited. This does not mean, of course, that the needs of consumers are satisfied, rather that they have learned to cope with the situation by developing household strategies and by using alternative sources, as elaborated in Chapter 26.

In answering our question 'Would you like to see continuous water supply introduced or would you prefer WAJ to keep to the policy of rationing?' the consensus was that water should be supplied at least twice per week or perhaps daily, with regularity, equality and enough pressure in all areas. However, contrasting attitudes were revealed. These seemed in part to relate to variables such as the income and socio-economic status of consumers and also to general levels of awareness of the issues affecting the water sector in Jordan. We can categorise the various positions taken up as (i) favourable to continuous supply, (ii) ambivalence and opposition, and (iii) indifference. Each of these views is examined in detail in the sections that follow.

28.5.3 The argument for continuous water supplies

Several households in the sample suggested that they felt it was likely they would waste water initially if rationing were to end, but that they would expect to exercise caution after an initial period of adjustment. Many reasons were presented to support such a view. A selection of low- and high-income families alike agreed that such a policy would reduce stress and possibly water consumption and lessen the cost of alternative sources of water:

> It would resolve most of the problems and I would be free to do whatever I needed to at any time.
> I think the water consumption would be less.

Several low-income families argued strongly the case for more water for their area, as a human right and as a mechanism to ensure social justice and social cohesion:

> You can have a war on water because you would leave the poor without water. It's like having the right for air and education.
> We need to have enough water supplied by the system because people don't have money to buy bottled or filtered water. They would only use it for basic needs.

Meanwhile, we note that some wealthy households in West Amman expressed the view that their high standard of living should be seen as the reason for them receiving more water. Typical of such views are the statements that:

> […] in rich areas we have more requirements, so they deliver water more than in the poorer areas. It's according to income and the way of life.
> The supply should be constant or for most of the week. Wealthy areas use more water and the frequency has to be more than once per week.

28.5.4 Ambivalence and firm opposition to the introduction of continuous water supplies

Although most poor families are more affected by the rationing programme, some were still ambivalent regarding an eventual increase in water supplies. Such households seemed to be anxious about changing their habits and therefore having to pay higher water bills. In this vein, a resident of the popular housing area of Wihdat commented:

> I'm not in favor of continuous supply; two days per week is enough for me. I am afraid to use more water and consequently, to pay more [...] I am happy with what I have now.

For this reason, many heads of households expressed real caution and said they preferred to restrain their demands. However, such a position sometimes served to annoy the younger members of households, who clearly wanted to see improvements to their general standards of living. Several alluded to their embarrassment and discomfort at not being able to take daily showers in the summer months. For some affluent consumers, given their knowledge of environmental issues and water shortages in the country, they noted that rationing has advantages as well as disadvantages. Such concerns for moving to a continuous supply stressed an eventual change in customers' attitudes and therefore the risk of putting more pressure on water resources in the middle to long term:

> If I had no restrictions, I would not be as stressed as I am now. I would take a shower whenever I wanted [...] so rationing obliges us to be more careful with water.

But we think that statements such as these, made by wealthy and well-educated consumers, need to be linked to the high storage capacities such consumers can control, and the fact that they are less affected by rationing. Some households, mainly lower-income ones, were firmly opposed to increasing supplies to the wealthy areas because of what they regarded as the wastefulness and lack of responsibility exhibited in these areas:

> Some people don't deserve to have water continually, because they waste water on cleaning cars, gardens and the street with hosepipes.

But it was notable that an essentially similar sentiment was expressed by one resident of a high-income district:

> I am a rich man but I don't need to over-wash my cars or to over-plant my lawn to show that I am wealthy, because then I will deprive others from their right to have enough water.

28.5.5 Indifference regarding the regularity of supply

Some high-income consumers who could clearly afford their water bills, storage capacities and alternative water supply sources tended to exhibit indifference to the issue:

> I am indifferent towards a continuous supply. I have good storage capacity and enough money to buy more.

A few statements were a little more alarming and suggested an overall lack of awareness and responsibility. Thus, an educated professional argued:

> The situation is good for me now and meets my water needs, but I would possibly irrigate my garden every day if I had a continuous supply [...] I would love to have a green lawn like in England!

When in 2007 we interviewed a representative of UNESCO she suggested strongly that attention should not be limited to water scarcity, emphasising that 'matters of culture and ideas regarding water usage' are just as important. She cited the common instance where in an area that is suffering from a lack of water, water is seen being lost over days in the streets as a result of leakages. Yet even when this situation is visible, people do not generally feel directly responsible. This shortfall in awareness and a lack of personal responsibility is seen as also being linked to widespread suspicion on the part of consumers concerning the motivation of government for maintaining a fatalist discourse on the water crisis:

> I believe that even if the rainfall is good, the government discourse doesn't change and every year they say that the water deficiency in summer will be huge.

It should be noted in this context that the term 'government' was used several times rather than reference to the WAJ or the MWI. Such misunderstandings stem from a stance on the management of water that excludes consumers, civil society and even municipalities from taking a full part in the decision-making process. As a consequence, many consumers do not feel directly responsible for water supply issues, and consider that it is entirely the task of the government to supply water and to manage the crisis.

28.6 CONCLUSIONS

The chapter has looked in detail at the policy of water rationing implemented in Amman, including the public–private management of rationing and the impacts on households (on the latter, please also see Chapter 26). The government discourse, which describes the cities as 'thirsty', primarily attributes the need for rationing to water scarcity allied to rapid demographic growth. But the case study of the involvement of the private sector in Amman's water services from 1999 to 2006 shows that rationing in Jordan should mainly be interpreted in the light of technical deficiencies in the water network, exacerbated by illegal connections and wasteful practices. As we have also seen in this chapter, in the light of improvements made to the water supply network since 1999, some technical experts are now arguing strongly that the entire system should move to continuous supply, for both technical and supply reasons. However, at present, the political will on the part of government for the introduction of continuous water supplies does not appear to exist.

REFERENCES

Abu-Shams, I. and A. Rabadi (2003) The strategy of restructuring and rehabilitating the Greater Amman Water Network. *International Journal of Water Resources Development* **19**: 173–183.

Altaf, M. A. (1994) The economics of household response to inadequate water supplies: evidence from Pakistan. *Town and WPR* **16**: 41–53.

Arce, C. (2004) General conditions for private sector participation (PSP) in the water sector in the Middle East: opportunities and constraints. International Water Demand Management Conference, May 30–June 3, Dead Sea, Jordan.

Choe, K., R. C. G. Varley and H. U. Bijlani (1996) *Coping with Intermittent Supply: Problems and Prospects. Dehradun, UttarPradesh, India.* Washington, DC: USAID.

Darmame, K. (2006) *Enjeux de la Gestion du Service d'eau Potable à Amman (Jordanie) à l'épreuve du partenariat public-privé.* Unpublished PhD thesis: Université de Paris X Nanterre.

Decker, C. (2004) *Managing Water Losses in Amman's Renovated Network: A Case Study.* International Water Demand Management Conference, Dead Sea, Jordan.

Dorsch Consultants & Assoc. (1997) *Population and Water Demand Assessment.* Amman: Ministry of Planning, Jordan.

Griffen, R. (2004) Management of water losses due to commercial reasons: a study in Amman. Paper presented at the International Water Demand Management Conference Dead Sea, Jordan, 2 June 2004. Abstract available at http://www.wdm2004.org/conferenceprogram/technical/wednesday/wedx7/roger.php (last accessed August 2010).

Hashemite Kingdom of Jordan (2006) *Jordan in Figures.* Amman, Jordan: Department of Statistics.

Molle, F. and J. Berkoff (2006) *Cities Versus Agriculture: Revisiting Intersectoral Water Transfers, Political Gains and Conflicts.* Comprehensive Assessment Research Report 10. Colombo, Sri Lanka: IWMI.

MWI (Ministry of Water and Irrigation) (2004) *National Water Master Plan of Jordan.* Amman: GTZ and Ministry of Water and Irrigation.

Nortcliff, S., G. Carr, R. Potter and K. Darmame (2008) *Jordan's Water Resources: Challenges for the Future.* Geographical Paper: University of Reading.

Potter, R., K. Darmane and S. Nortcliff (2007) The provision of urban water under conditions of 'water stress', privatization and deprivatization in Amman. *Bulletin of the Council for British Research in the Levant* **2**: 52–54.

Potter, R. B., K. Darmame, N. Barham and S. Nortcliff (2009) 'Ever-growing Amman', Jordan: urban expansion, social polarisation and contemporary urban planning issues. *Habitat International* **33**: 81–92.

Sethi, K. (1992) *Households' Responses to Unreliable Water Supply in Jamshedpur India.* Washington, DC: Transport, Water and Urban Development Department, World Bank.

United Nations High Commissioner for Refugees (UNHCR) (2007) *UNHCR Global Report.* Geneva: United Nations High Commissioner for Refugees. http://www.unhcr.org/gr07/index.html.

Zérah, M.-H. (1997) Contribution à l'analyse des infrastructures urbaines: la réponse des ménages à l'inconstance de l'offre d'eau à Delhi, 2 vol. Unpublished PhD thesis, Institute of Urbanism of Paris, University of Paris XII.

Zérah, M.-H. (1998) How to assess the quality dimension of urban infrastructure: the case of water supply in Delhi. *Cities* **15**: 285–290.

Zérah, M.-H. (2000) *Water: Unreliable Supply in Delhi.* New Dehli: Manohar and Centre de Sciences Humaines.

Part VII
Conclusions

29 Overview and reflections: 20,000 years of water and human settlement in the southern Levant

Steven Mithen and Emily Black

As noted at the start of this volume, the Water, Life and Civilisation (WLC) project aimed:

> to assess the changes in the hydrological climate of the Middle East and North Africa (MENA) region and their impact on human communities between 20,000 BP and AD 2100, with a case study of the Jordan Valley.

The 28 chapters within this volume have certainly made a contribution towards that aim, although inevitably falling short of providing a comprehensive coverage of the vast array of topics that would need to be addressed to meet that aim in full. In this overview, we will highlight selected conclusions made as part of the project, and those areas of research that have now been revealed to require further attention. We will reflect on the challenges involved in developing interdisciplinary research within our study region, the overall achievements and the shortcomings of our project.

29.1 MODELS FOR PRESENT, PAST AND FUTURE CLIMATE CHANGE

The analysis of present-day rainfall and the modelling of past, present and future climates within Chapters 2, 3 and 4 have provided a series of studies of the Middle East and Mediterranean region climate throughout the Holocene and forward to the end of the current century, with sufficient resolution to assess regional climate variability. The use of similar (although not identical) model set-ups to simulate the past and future climate provided an unprecedented opportunity to compare the past, present and future (Chapter 8). Chapter 2 established that the inter-annual variability of present-day rainfall is primarily affected by atmospheric circulation over the Mediterranean. This, in turn, is strongly influenced by larger-scale processes, notably the North Atlantic Oscillation and the East Atlantic/West Russian mode. Any changes in the future location and timing of the Mediterranean storm tracks will have a significant impact on rainfall in the Middle East; such changes are likely to have been the immediate cause for the variation in rainfall detected in the palaeoenvironmental record (Chapters 6 and 7). A key finding of the present-day rainfall analysis was its degree of coherence at a scale of less than 200 km, particularly within physiographic zones. The lack of coherence at a finer scale indicates that the standard Global Circulation Models (GCMs), which have a spatial resolution of $c.$ 250 km^2, are of limited value but that Regional Circulation Models (RCMs) with their $c.$ 50 km^2 resolution have considerable potential for providing useful information for hydrological models. Climate models with an even higher resolution are desirable, but their development remains some way off. In their absence, the use of statistical methods to downscale and correct bias within the RCMs provides the most appropriate way to make progress in hydrological modelling (Chapter 5).

The understanding gained regarding the present-day causes and patterns of precipitation provided the basis for the modelling of past Middle Eastern and North Africa climates within Chapter 3. The authors of that study undertook a series of 'time-slice' experiments between 12,000 years ago and the present day at 2,000-year intervals, using estimates of solar radiation, greenhouse gas concentrations and ice-sheet extents to 'force' the climate into its particular configurations for each time slice. Having demonstrated that the chosen RCM provides a sufficiently accurate representation of present-day rainfall, these time slices provided a suite of projections regarding climate patterns and their change during the Holocene. It confirmed previous proposals that increased greenhouse gas concentrations have led to a gradual rise in mean surface-air temperatures throughout the Holocene, with a particularly rapid rise since the pre-industrial period. The modelling indicated that there has been an overall decrease in rainfall, appearing to reduce relatively swiftly in the period of 10–6 ka BP.

Water, Life and Civilisation: Climate, Environment and Society in the Jordan Valley, ed. Steven Mithen and Emily Black. Published by Cambridge University Press.
© Steven Mithen and Emily Black 2011.

The palaeoenvironmental evidence also indicates a reduction in rainfall during this period. The evidence reviewed in Chapters 6 and 7 suggest that a drying trend started c. 7.5 ka to 5 ka BP, with some variation between the particular proxies being examined. The data from Lake Lisan/Dead Sea (Chapter 9) places drier conditions occurring after 8.5 ka, a date also indicated from the study of Beidha in Chapter 16. Overall, the palaeoenvironmental evidence suggests that a period between 8 and 6 ka was the most critical for a decrease in rainfall, a period compatible with the projections of the climate model.

The climate model found considerable spatial variation in the changing rainfall patterns throughout Europe and the Middle East: eastern areas have experienced a gradual increase in rainfall, while the northern region (Anatolia) experienced a gradual reduction during the course of the Holocene; the eastern coastline of the Mediterranean, extending into the Levant, experienced high precipitation in the early Holocene which then fell off rapidly to low levels in the mid- to late Holocene; the southeast coastline appears to have behaved in a similar manner, although the signal from the simulations was less clear. We should also note that there was no indication from the models that there could have been summer rainfall within the Jordan Valley, despite claims to the contrary (e.g. Rossignol-Strick, 1999).

When the climate modelling explored the future (2070–2100) it found that under a 'business as usual' (A2) scenario in which greenhouse gas emissions continue at the present rate, the whole of the Middle East and Mediterranean is likely to experience a significant increase in temperature and decrease in rainfall, the latter amounting to a 40% reduction of present-day precipitation during the peak of the rainy season (December–January) in the eastern Mediterranean (Chapter 4). That reduction arises from fewer rainy events, as opposed to a decrease in the intensity of rain during those events. Under scenarios of stabilised or reduced greenhouse gas emissions (B1 and B2), the decreases in rainfall become small and statistically insignificant. Consequently, if climate change can be mitigated, the potential social and economic challenges that are likely to arise from reduced rainfall, as considered in Chapters 13, 25, 26 and 27, can be contained – although increasing population pressure is likely to provide ongoing pressure on the water supply.

While the climate modelling within Chapters 3 and 4 has provided valuable projections for both the past and the future, the authors have emphasised that any results must be used cautiously. The projections for the past focus on the mean state of the climate and can account for neither inter-annual variation at the specified time slices nor extreme hydrological events, such as floods and droughts. The extent of that variation and the impacts of such events may have been of greater significance for past human communities than the mean temperature and rainfall values. Given the uncertainty in the projections of the past climate, the hydrological past was explored using a model-based assessment of the flow response to step changes in flow and infiltration. Some aspects of rainfall appear to be captured more accurately than others in current models, notably the frequency and duration of rainy events as opposed to their intensity. It was also noted that the RCM for the past simulations did not include a fully dynamical ocean: feedback processes between atmospheric and oceanic circulations were not accounted for and these could potentially have influenced the projections for temperature and rainfall across the study region.

29.2 THE IMPACT OF FUTURE CLIMATE CHANGE

The potential impact of reduced rainfall and increased temperature from future climate change has been explored in several chapters of this volume. Chapter 24 provided an overview of current water sources, their usage and projected future demands in Jordan. It noted that current demand for water exceeds the available renewable resources, with the shortfall being met by non-renewable reserves and water rationing. The demand for water is predicted to increase because of population growth, urbanisation, tourism and overall economic development. The chapter described how neither the Water Strategy for Jordan (1997) nor the National Water Master Plan for Jordan (2004) takes account of potential changes in the water supply arising from climate change (MWI (Ministry of Water and Irrigation), 1997, 2004). One of the reasons for that is simply that the typical model projections for climate change relate to the end of the twenty-first century whereas the timeframe for economic planning is 20 or at most 40 years – this, of course, being a generic problem throughout the world. A striking statistic within that chapter, and one reiterated in Chapter 28, is that more than 50% of the water within the supply network of Amman is currently lost by leakages, illegal connection and wasteful practices. As such Amman is not so much a 'thirsty' city because of an overall shortage of water but because of the technical and managerial deficiencies of its supply system. Attention to these deficiencies would provide a substantial contribution towards mitigating potential reductions in supply from climate change.

With regard to exploring the impact of future climate change on river flow and runoff, Chapter 10 introduced the methodology of linking an RCM to a weather generator and then a hydrological model, in this case as a means to project the likely rainfall-runoff response of the Upper River Jordan. This was a demanding, and to some extent speculative, approach, not only because of the uncertainties at each of the model steps but also because of a basic shortage of data to calibrate the hydrological models, such as daily flow rates for the Upper Jordan itself and the tributary wadis that make up the drainage network. Furthermore, the need to determine future flood magnitudes requires the consideration of highly uncertain extreme rainfall events projected by the climate models. Lack of accurate information about the volume of abstractions from different reaches provided further complications. The overall conclusion of the hydrological modelling of the Upper River Jordan was that although climate change will reduce the mean annual flow of the River Jordan, the base flow of the Upper Jordan will not change significantly because any decrease in precipitation will be mitigated by the contribution of groundwater. That in itself raises a critical question: how dependable are the groundwater reserves?

Chapter 13 also linked an RCM to a hydrological model via a weather generator, in this case to assess the effects of projected climate change on water security in the rural west of Jordan. The selected study area was Wadi Hasa, which extends from the Jordanian plateau to Ghor Safi, an area of intensive vegetable production, where the wadi drains into the Dead Sea. By utilising a newly developed hydrological model with a monthly time-step, this study found that under the A2 'business as usual' scenario, runoff in Wadi Hasa is projected to decrease by the end of the century and that rainfall is unlikely to be sufficient to replenish the aquifers draining to the Safi plain. The projected increase in temperature, decrease in winter rainfall, shift in seasonality and reduced base-flow exploited for irrigation will all be detrimental for crop growth. To maintain levels of agricultural production it may be necessary to change the cultivated crops to those with lower water requirements, improve irrigation efficiency or simply find additional sources of water, possibly from desalination or water transfers, such as from the Red Sea. The study also highlighted that flood magnitude can increase even as mean annual rainfall decreases because the rainfall amounts falling during the most extreme events may still increase, a result also reported by Pinhas Alpert using an alternative modelling approach at the discussion meeting on Water and Society hosted jointly by the Royal Society and British Academy in November 2009. These results confirm the necessity of considering the hydrological response of surface and groundwaters as well as climate projections of rainfall, since the hydrological response can be non-intuitive.

29.3 CONVERGENCE BETWEEN GEOLOGICAL DATA, CLIMATE MODELS AND HYDROLOGICAL MODELLING FOR PATTERNS OF PAST CLIMATE CHANGE

One of the aims of the WLC project was to develop interdisciplinary approaches to the past by finding means to integrate different sources of data and approaches, exploring whether these converge on particular scenarios for past climates and environments or provide conflicting projections.

Chapter 6 provided a comprehensive review of the currently available evidence for palaeoclimates and palaeoenvironments in the Levant and eastern Mediterranean from 25,000 to 5,000 years ago, hence partially overlapping with the time period of the climate simulations of Chapter 3. The key finding was that the major climatic events that have been primarily defined in the northern hemisphere, such as the Late Glacial Maximum, Younger Dryas and Bølling/Allerød warm interval, can be detected in the geological record of the eastern Mediterranean and Levant and that there is general agreement with the output from GCMs when applied to specific events, notably the Last Glacial Maximum. The availability and quality of the geological data were variable as was the extent to which the inferences from different types of proxies converged for particular periods of study. For the Younger Dryas, for instance, the terrestrial records suggest a greater degree of aridity than is evident from the eastern Mediterranean marine record, which also fails to provide a comparable signal for the Bølling/Allerød warm interval to that coming from the terrestrial records. The early Holocene was identified as the wettest period of the past 25,000 years, confirming the projections from the RCM described in Chapter 3.

The relatively high precipitation of that period was also supported by the new suite of uranium-series dates provided for the Lake Lisan deposits that are exposed on the eastern shore of the Dead Sea, as described in Chapter 9. These new dates and their respective elevations add to the gradually accumulating database for Lake Lisan as described in the chapter, further demonstrating that it provides a sensitive measure for changing levels of precipitation during the Pleistocene and Holocene, while also adding new information about its particular ecological characteristics. The new data provides further evidence for marked changes in the levels of Lake Lisan between 25,000 and 8,000 years ago, confirming the early Holocene as being a period of relatively high precipitation.

The review of palaeoenvironments in the southern Levant from 5,000 years ago to the present day contained within Chapter 7 was in some ways more difficult than that for the earlier periods. This was partly because separating the impact of climate- and human-induced environmental change becomes increasingly problematic as the scale of human activity in the landscape increases and partly because the finer-grained chronological resolutions that are required from the geological evidence are beyond the capabilities of current dating methods. Nevertheless, an overall trend towards increasingly arid conditions was evident. Several fluctuations were imposed on to this trend, notably relatively wet periods that broadly correlate with the Early Bronze Age and Byzantine periods, both of which appear to have had a significant impact on settlement patterns and economic activity, as was later explored for the Early Bronze Age within Chapter 17.

The trend to aridity fits with the projection from the RCM described in Chapter 3. To explore this further, an explicit comparison was undertaken in Chapter 8 between the projections from the RCM for changes in precipitation during the Holocene across the Middle East and Europe and the evidence from palaeoenvironmental data and historical rainfall records. Both sets of information provided evidence for a transition towards a wetter northern Europe and a drier Middle East: the RCM does a sufficiently accurate job of projecting the climate changes that are recorded in the past. As such, we can be confident that these changes in precipitation during the Holocene arose because of a weakening and a poleward shift of the Mediterranean storm track. We can also be encouraged with regard to the utility of such RCMs for projecting future climate changes. Chapter 8 also noted that although the underlying mechanism for future climate change (an increase in greenhouse gas concentrations) is different from that of the past, the projected changes in rainfall patterns across the Middle East and Europe for the end of the twenty-first century show a striking similarity to those that have already occurred during the Holocene.

A further degree of convergence from alternative approaches to the past arose from the development of a hydrological model for the Dead Sea in Chapter 11. While the changes in the Lake Lisan/Dead Sea levels considered in Chapters 6 and 9 arose from changes in precipitation, those during the past 50 years have arisen from abstractions, mainly from the Jordan and Yarmouk Rivers and from Lake Kinneret. The Dead Sea model developed in Chapter 11 predicts that if the current abstractions continue, the level will fall a further 55 m from its current low point of 423 m below sea level by 2050, with serious impacts on the incipient Dead Sea tourist industry, health (arising from increased levels of toxic dust) and the potash and salt mining industries. It seems unlikely that the abstractions will be reduced, and therefore the only way to avoid such impacts appears to be a replenishment of the Dead Sea using water from either the Red Sea or the Mediterranean (Glausiusz, 2010).

Within Chapter 11 the Dead Sea model was used both to explore future levels of the Dead Sea under climate change projections and water transfers, and to estimate levels of past rainfall by drawing on the past levels of the Dead Sea as described within Chapters 6 and 9. The latter required taking changing surface areas, salinity and land use or vegetation conditions into consideration, the latter having a major impact on the extent of runoff. In spite of inevitable uncertainties, this model was able to propose varying levels of rainfall during the Holocene that were compatible with the proxy evidence described in Chapter 6 and the rainfall estimates from the RCM developed in Chapter 3. In other words, these three quite different approaches to the past have converged on similar rainfall scenarios.

29.4 THE IMPACT OF CHANGING WATER AVAILABILITY ON PREHISTORIC SETTLEMENT AND ECONOMY

The climatic modelling and palaeoenvironmental studies of the WLC project have demonstrated that there has been significant variation in water availability within the study region during the Late Pleistocene and Holocene arising from changes in precipitation and evaporation. What impact did this have on human communities? Previous research made coarse-grained comparisons between patterns of human settlement and climatic conditions, sometimes introducing speculations about past climate, such as the presence of summer rainfall that the modelling in Chapter 3 was unable to generate.

The archaeological and palaeoenvironmental study of Beidha in Chapter 16 has provided a particularly detailed comparison between the chronology of a single settlement and the availability of water in its immediate vicinity. It is perhaps surprising that the extensive carbonate deposits, notably the continuous series created by a former spring adjacent to the site, as discovered and analysed by the WLC project, were largely overlooked by previous researchers. That had led to the assumption, first proposed by Kirkbride in the 1960s, that the nearest source of water was at a distance of 3 km from the settlement.

The analysis of those carbonate deposits, along with the derivation of new radiocarbon dates and an evaluation of the existing chronology of settlement, indicated a very close match between human occupation and the fluctuations in environmental conditions that had been recorded in those carbonate deposits, in particular through their isotopic composition and deposition rate. It appears that when water supply fell below a critical threshold during the Younger Dryas and again after 8,500 BP, the settlement was abandoned. While those periods of aridity match with those indicated by the regional-scale palaeoenvironmental evidence (Chapter 6) and the palaeoclimatic modelling (Chapter 3), we must remain cautious about their interpretation. For example, the cessation in the formation of those particular spring carbonate deposits at Beidha may have arisen from a localised event, such as minor tectonic movement interfering with the access of the spring to the aquifer. This may have left other springs still flowing or even generated new springs in the immediate vicinity of Beidha. However, the similar evolution of the carbonate isotopic composition from both the spring carbonates and pedogenic carbonates sampled at the archaeological site give confidence in the fact that the site abandonment was indeed related to changes in the local environmental conditions.

In this regard, the ongoing analysis of spring-formed carbonate deposits in Wadi Faynan is critically important. If this study identifies a similar chronological pattern of isotopic variations and changes in sedimentation rates to that found at Beidha, then this will suggest that a regional signal of climate change has indeed been recognised. A further note of caution is that even if this does prove to be the case, we must not leap to the conclusion that this also caused the abandonment of other Pre-Pottery Neolithic B (PPNB) settlements in the southern Levant: several of these will need to be examined at the same level of detail that we have completed for Beidha, involving the derivation of a precise chronology, before one can be confident of drawing this conclusion.

The study of Bronze Age Jawa (Chapter 18) also pointed towards critical thresholds of water supply below which occupation was not sustainable. This study was the first to utilise Monte Carlo simulation techniques as a means to cope with the inevitable uncertainty within archaeological studies that require estimations of past rainfall, evaporation, water losses and other such factors. In some regards, Jawa was an ideal case study for the application of this approach: it is the earliest known site in the southern Levant to have evidence for a complex water management system, its ponds and channels have been documented in detail by Helms (1981) and the absence of groundwater sources makes the reconstruction of the prehistoric water supply less challenging than at other localities where multiple water sources were simultaneously exploited – such as at Humayma (Chapter 19). On the other hand, Jawa lacks a detailed chronology: we do not know whether it was continuously or intermittently occupied; nor do we know whether the ponds and channels were constructed as one single design or in incremental stages perhaps as the population of the settlement grew or the quantity of runoff declined. Neither do we have adequate information about the Jawa economy: as the modelling in Chapter 18 made clear, the level of sustainable population was highly sensitive to the extent of water being used for irrigation rather than for drinking. In spite of these caveats, the modelling has enhanced our understanding of Jawa: it converged on a similar maximum population estimate (of 6,000 people) to that derived by Helms from a consideration of the archaeological remains alone, explored the interactions between rainfall, runoff, evaporation, storage and demand for drinking and irrigation, and demonstrated the sensitivity of the population to climatic change of the mid-Holocene. The stage has now been set for further archaeological survey and excavation at Jawa to establish a chronology for occupation, a chronology for the development of the hydraulic structures, and the extent of past irrigation. Such research will make an improved level of parameterisation possible for the hydrological models, and hence allow the potential of both the hydrological and climate models for informing about the past to be fully realised.

A Monte Carlo modelling approach to capture the uncertainties of rainfall, water storage and demand was also utilised to develop a water-balance model for Humayma. This was a more advanced model than that used for Jawa because it encompassed crop production and differences in evaporation from covered and uncovered cisterns and reservoirs. It also had the challenge of evaluating the impact of erosion not only on the extent of the watersheds and the flow of surface water, but also on the survival of artificial wadi barriers that may have been constructed to control the past water supply. For this study, the WLC project was fortunate to have been able to draw on the extensive hydraulic surveys undertaken by John Oleson and had his population estimate of 4,007 persons, 3,008 ovicaprids and 300 camels/donkeys to use as a baseline for comparisons (e.g. Oleson, 1992). The project was also fortunate to have his advice on the development and interpretation of the model in light of the complexity of the archaeology. Indeed it is primarily through the similarities and differences between Oleson's estimates and interpretations regarding water usage and those adopted and derived from the WLC model, that an enhanced understanding of the Humayma settlement has been gained. Some key differences of interpretation exist, notably the extent to which the cisterns may have been replenished within and between years and the relative significance of runoff water transmission loss, which the WLC model evaluated to be as critical as the rainfall itself. Overall, the WLC model indicated that the human and animal populations are likely to have been somewhat lower than the levels that Oleson estimated.

As is the case with our Jawa study, now that this initial modelling for Humayma has been completed it is evident what future work needs to be undertaken: further hydrological fieldwork to measure water flows, infiltration and transmission losses, archaeological fieldwork to establish the extent of cultivated fields and chronology of hydraulic structures, and the development of a more sophisticated hydrological model. A further requirement is to explore alternative settlement scenarios, such as assuming that the economy had a higher degree of pastoralism and hence that the resident population at Humayma fluctuated throughout the year, and assuming the importation rather than cultivation of food stuffs, perhaps acquired from caravans in exchange for water.

29.5 THE INTERACTION BETWEEN CLIMATE- AND HUMAN-INDUCED CHANGES IN THE WATER SUPPLY

Whereas the studies of Neolithic Beidha, Bronze Age Jawa and Nabataean Humayma focused on individual settlements, the study in Chapter 17 attempted a landscape approach to the Chalcolithic and Early Bronze Age using a Geographical Information System (GIS)-based analysis to explore how settlement patterns may have been influenced by changes in the availability of, or need for, a water supply. This study had to grapple with the immense methodological challenges that arise when attempting to work with survey data from different sources, in this case that from JADIS (Jordan Archaeological Data Information System) and that from the IAA (Israeli Antiquities Authority). These have their own strengths and weaknesses, and share a lack of consistency or even standardisation of the period divisions into which survey data are classified. Moreover, by their very nature, survey data are often difficult to classify as to period, frequently consisting of rolled or degraded surface material, while the diagnostic criteria for the various sub-periods of the Early Bronze Age are far from adequate. Wrestling with such problems is, unfortunately, the nature of landscape archaeology; they were tackled as effectively as possible within the case study of Chapter 17 making an innovative use of cost-distance analysis for evaluating the significance of potential factors influencing settlement, such as springs, other sites and major route ways.

The study found that Early Bronze Age settlements were located closer to springs than those in the Chalcolithic, perhaps simply as a result of the 'wet event' at $c.$ 5 ka BP that may have caused a higher water table; at the same time, there appeared to be a more formalised settlement pattern with sites showing greater alignment with major route ways and to large, permanent settlements. Overall there was a complex interaction between environmental change and socio-economic developments, neither of which appears to have been more influential than the other, that led to an adjustment of settlement patterns to allow more extensive horticultural practices in the 300–800 metres asl altitude belt.

The interaction between socio-economic development, climate change and the water supply was most explicitly explored in the long-term study of human settlement in Wadi Faynan from the Neolithic to the modern day of Chapter 15. This study was able to build upon the extensive archaeological and palaeoenvironmental research in Wadi Faynan during the past three decades (Barker et al., 2007; Finlayson and Mithen, 2007; Hauptmann, 2007). It provided the most complete interdisciplinary study of the WLC project, drawing on the RCMs for the 2,000-year time slices developed in Chapter 3, the weather generator described in Chapter 5, the hydrological model for Wadi Faynan developed in Chapter 12 and the archaeological and palaeoenvironmental expertise within the project.

The most striking overall result of that modelling, which once again made use of the RCM weather generator – hydrological model chain, was that alterations in ground cover are likely to have been more significant than changes in climate alone for water availability within the wadi. Indeed, in the absence of any changes in vegetation, the fluctuations in rainfall that occurred between 12 and 2 ka would have had limited impact on the hydrology of the wadi. It was found that ground cover – vegetation and soil – significantly enhanced the infiltration and percolation rates, which reduced the number and severity of flood events while increasing the extent and regularity of groundwater flow. Chapter 15 explored the implications of these hydrological relationships between rainfall, vegetation and water availability for the history of Wadi Faynan, reaching a suite of proposals for key phases and types of human settlement.

Chapter 15 suggested that the climate amelioration of the early Holocene had led to the spread of vegetation throughout much of the Wadi and the creation of amenable hydrological circumstances for hunter-gatherers to create semi-sedentary settlements and ultimately the development of early farming communities. The existence of the Pre-Pottery Neolithic A site of WF16 and the PPNB site of Ghuwayr 1 was known at the start of the WLC project, but ongoing excavation has found WF16 to be considerably larger and more complex than previously believed (compare Finlayson and Mithen, 2007 and Mithen et al., 2010), implying a substantial and reliable source of water.

Following the establishment of Neolithic farming, the long-term history of the wadi was of ever-increasing aridity during the mid-Holocene, the impact of which was exacerbated by vegetation clearance. Whether such clearance happened gradually or in short-term bursts remains unclear. One should expect that the foraging of goats in the PPNB, Pottery Neolithic and Bronze Age would have had a significant impact, along with the clearance of woodland to create fields for cultivation and to provide fuel. Demand for fuel would have increased as copper working began during the Chalcolithic and then intensified in the Bronze Age. As described in Chapter 15, the palynological evidence from Wadi Faynan provides evidence for such loss of vegetation. If this was of a significant level, the reduced infiltration and increased runoff may have opened up the possibility, or created the need, for the development of the floodwater farming system suggested by the Bronze Age field system within the wadi. However extensive was the vegetation loss during the Neolithic and Bronze Age, this would have been slight compared with that within historic times in light of the massive scale of the Roman/Byzantine copper industry, and the timber demands for the construction of the Hijaz railway met from the woodland within the water catchment of Wadi Faynan. These removals would have shifted the hydrology of Wadi Faynan to its present-day character of limited infiltration, extensive runoff and often dramatic winter floods. Quite how the evidently large Roman/Byzantine population exploited this water supply remains unclear, in spite of the previous documentation of the aqueduct, reservoir and field system. Mattingly et al. (2007) argued that the field system relied on trapping runoff from the tributary wadis to Wadi Faynan, but the WLC modelling suggested that these would have provided insufficient water, indicating that the field system must have also drawn upon groundwater flows or floods.

However that field system operated, the combination of the climatic and hydrological modelling from the WLC project described in Chapters 12 and 15, and the recent archaeological surveys and excavations in Wadi Faynan, has made this locality one of the most thoroughly explored case studies for the long-term relationship between water and society in the Near East.

This will be further enhanced by the results of the ongoing excavations at WF16 and the analysis of carbonate deposits within Wadi Ghuwayr, which will provide a direct test for the extent of groundwater flows predicted by the WLC model. What the study has already demonstrated is the complex interplay between climate change, environmental change and human activity.

This was brought up to date for Wadi Faynan in Chapter 27, which considered the role of water and land management in the wadi's contemporary socio-economic development. The chapter stressed the complexity of land status within Wadi Faynan, especially with regard to the long-term dispute between the Al-Rashaida, one of the tribes located in Wadi Faynan, and the Shawabaka, residents of the Shawbak region on the Jordanian Plateau, regarding tribal ownership of land in Wadi Faynan and access to the groundwater flows in Wadi Ghuwayr. It is evident that such long-term tribal disputes relate to matters of pride and honour as much as to access to resources, and continue to pervade decision-making concerning land use in the wadi. Within this context, Chapter 27 provided an account of the establishment of the Al-Rashaida cooperative that since the mid-1990s has developed tomato and watermelon farming in the vicinity of the new-found village of Faynan Al-Jadida, where many of the Al-Rashaida have now permanently settled, although some still continue to move to the uplands around Shawbak in the summer months.

The means to develop the cooperative was the use of water pipes to bring groundwater from Wadi Ghuwayr for irrigation, those pipes initially following precisely the same route as that of the Roman aqueduct. Perhaps the key message of this chapter is that even this apparently straightforward programme of development in Wadi Faynan, requiring no more than irrigating plots with groundwater from a nearby source, involves issues about tribal history, ownership and identity that are as important to understand – or at least appreciate – as are the environmental impacts of future climate change and the long-term ecological viability or otherwise of the soil supporting the recently developed tomato and watermelon farming. We are confident that such issues would have been pervasive throughout the past.

The same message of needing to address simultaneously the science and the social science of farming practices was evident in Chapter 25, which considered the use of wastewater for irrigation in the Jordan Valley. Analytical studies of soil samples from sites irrigated with wastewater demonstrated the risk that plant-toxic ions will accumulate within the soil, ultimately reducing productivity. This can be mitigated by a leaching programme to reduce soil salinity. Interviews with farmers in the Jordan Valley who utilise wastewater indicated a high level of understanding and willingness to manage the application of fertilisers and make use of leaching to maintain their soil fertility. Their capacity to do so, however, was limited by the information about the nutrient content of the water supplied to them by the Jordan Valley Authority (JVA) and the timing of that supply. One cause of this appeared to be a general unwillingness for the JVA and other bodies to be transparent about the extent of wastewater usage because of fears about the loss of markets to Islamic countries such as Saudi Arabia, which had banned the import of vegetables from Jordan in the 1990s for this reason. Such problems were exacerbated by what appeared to be a limited understanding of the information that the farmers required, perceptions of the farmers that water was being diverted to support tourism, and the authorities' struggle against illegal dumping of industrial wastewater into the system. Overall, whatever level of scientific understanding is achieved about how to maintain soil fertility in a situation of prolonged use of wastewater, the ability to do so will depend upon developing enhanced understanding and communication between the wastewater providers and the end users.

29.6 DEVELOPING ARCHAEOLOGICAL METHODOLOGIES

One of the challenges that all of the archaeological studies within the WLC project faced was the sparse evidence for past irrigation, in spite of claims that irrigation had a key role in Chalcolithic and later economies. Finding structural evidence for irrigation is always going to be problematic, especially when used on a small scale, because of the ephemeral nature of temporary channels and dams constructed from no more than rocks and sandbanks. Consequently, from the start of its research programme, the WLC project recognised a need to enhance the use of plant remains alone when drawing inferences about past water availability. Claims had previously been made about the impact of irrigation on phytolith formation by Rosen and Weiner (1994) and on the isotopic composition of cereal grain by Araus et al. (1997). The experimental basis for those claims is limited and consequently the WLC project established a plant growing experiment to seek whether these and any further signatures for irrigation from plant remains could be validated.

The plant growing experiment established at the Deir 'Alla, Ramtha and Khirbet as Samra agricultural stations, as described in Chapter 21, was a significant undertaking for the WLC project and continues to provide data, with plots of sorghum now being grown to complement those of wheat and barley during the WLC project. With regard to phytolith formation, this experiment confirmed that the availability of water influences the uptake of silicon and phytolith formation, but found no means to differentiate between water from irrigation or increased rainfall. While large, composite phytoliths of the type described by Rosen and Weiner (1994) did form under high levels of irrigation, these were found to be fragile and unable to resist fragmentation during post-depositional processes and laboratory processing of sediments for their extraction (Jenkins, 2009). More encouraging, however, was the discovery that higher levels of irrigation resulted in a higher percentage of long dendritic cells within the husks of wheat and six-row barley, this finding also being independently discovered and published by Madella *et al.* (2009) while the WLC research was being undertaken. Such long dendritic cells are relatively robust and are likely to survive intact within archaeological deposits, although this requires formal testing. As such, their frequency was designated as a 'water availability index' and applied to archaeological samples, as described in Chapter 23.

The analysis of the isotopic composition of cereal grain from the experimental plots (Chapter 22) also confirmed the patterned response to enhanced irrigation proposed by Araus *et al.* (1997), although again there appeared to be no means to differentiate this from a response to enhanced rainfall. As with the phytolith study, this found high levels of variation in the results coming from each of the three growing plots and in each of the three years of the study – indeed the inter-plot variation for both phytolith formation and the isotopic composition of grain was greater than that caused by different levels of irrigation. Both studies are continuing by making detailed statistical analyses of the influence of the soil chemistry, altitude, transpiration rates and rainfall patterns on phytolith formation and isotopic composition, seeking to understand how these interact with the levels of irrigation. The ongoing research on isotopic composition, being undertaken as PhD research by Pascal Flohr at the University of Reading, involves charring and burial of grain to evaluate whether the water-induced variation in the isotopic composition of grain during growth would still be manifest in archaeological samples.

Chapter 23 involved a test of the 'water availability index' within the context of a review of plant use during the prehistoric and historic periods in Jordan. This analysed 96 sediment samples for phytolith analysis from seven archaeological sites, involving small-scale fieldwork in Wadi Faynan and at Humayma to gain samples from Nabataean/Roman/Byzantine contexts to complement those already available from prehistoric sites. By classifying the samples as to their functional context, it was found that those coming from water channels in field systems had the highest frequencies of long dendritic cells, demonstrating the veracity of the water availability index. The study was also able to demonstrate differences in phytolith density within sediment samples through time. This may reflect increasing intensity of plant use, although the confounding effects of differential preservation and context should not be discounted. Also, some useful insights were gained into the cultivation of particular species, notably date palm at the Pottery Neolithic site of Tell Wadi Feinan and tree crops rather than cereals within the Wadi Faynan Nabataean/Roman/Byzantine field system. Although this does not remove the possibility that cereals had been grown, as argued by Mattingly *et al.* (2007), because only a limited number of samples were studied, it adds to the result of our hydrological modelling (Chapter 15) that there remains considerable uncertainty about the use of that field system.

In addition to enhancing our knowledge about how plant remains can inform about past water availability and usage, the WLC project sought to contribute to our capacity to use bone chemistry to infer the past diet of both humans and domesticated animals. The use of carbon and nitrogen isotopes for dietary analysis has been one of the major growth areas of archaeological science during the past decade, but the number of applications in arid regions has been limited, partly because of assumptions about poor collagen preservation. A considerable amount of effort was involved in the WLC project in gaining access to samples; disappointingly it failed to secure permission to analyse the substantial collection of Roman/Byzantine human skeletal remains from Wadi Faynan stored at Irbid University that would have been an ideal complement to the hydrological, archaeological, phytolith and contemporary farming studies of that wadi. Human and sheep/goat skeletal remains were acquired from Jerash, Ya'amūn, Yajūz, Sa'ad and Pella ranging in date from Middle Bronze Age to the early Islamic period.

Approximately 50% of the samples provided sufficient levels of collagen for isotopic analysis. The most valuable results with regard to water availability came from analysis of sheep/goat bones, with the indication of arid-adapted C_4 plants in the Middle and Late Bronze Age sheep/goat diet along with relatively high levels of $\delta^{15}N$ indicating feeding in low rainfall environments. The number of samples analysed remains too small for any substantive conclusions, but the value of the methodology has been demonstrated and a long-term study of sheep/goat herding strategies within the study area in the context of the climate and hydrological models is now a research priority.

29.7 THE DEVELOPMENT OF WATER MANAGEMENT STRATEGIES

The need to apply the methodologies regarding the analysis of faunal and plant remains with respect to past water availability and exploitation, whether that water is from rainfall or irrigation, appears more pressing at the end of the WLC project than it did at the start. It was always anticipated that identifying the earliest structural evidence for water management in the archaeological record would be problematic because small-scale dams using natural materials and irrigation channels are unlikely to leave any archaeological trace. Nevertheless Chapter 14 found surprisingly sparse structural evidence for water management in the

Neolithic and Chalcolithic archaeological records, periods when substantial sedentary settlements had appeared, some with populations that are claimed to have been greater than 2,000 people. With the impressive architectural achievements of the PPNB settlements, especially those of the size of 'Ain Ghazal, Beidha and Basta, along with the rich and diverse cultural life evident from cultural material and human burial, the absence of any stone-cut cisterns, aqueducts from springs and other hydraulic structures is surprising. The only substantial evidence within the WLC study region is the claimed barrages and cisterns from the Jafr Basin; while these are certainly both impressive and persuasive, they make the absence of similar structures elsewhere all the more striking. The wells from Mylouthkia in Cyprus, and Atlit-Yam in Israel, make the situation even more confusing because these appear to be sophisticated structures, within a cultural interaction zone that suggests that knowledge of such technology would have been widespread. It may, of course, be the case that water management structures are pervasive in the PPNB of the southern Levant: archaeologists are prone to find only what they are looking for, and numerous PPNB settlements may have nearby terraces and wadi barriers that have simply been assumed to be of a much later date. But on present evidence, it appears that up until at least the Pottery Neolithic, and with the exception of what was probably the extreme environment of the Jafr Basin, people in the southern Levant are simply responsive to the natural distributions of water, leaving them unprotected against environmental change, as appears to have been the case at Beidha (Chapter 16).

It is possible that the climatic conditions and population levels were such that water management was not a requirement. But when traces begin to appear within the southern Levant during the Pottery Neolithic, notably the well at Sha'ar Hagolan and terraces at Dhra, they do so at a time when there is no evident increase of demand or reduction of supply arising from a period of aridity. Indeed the well at Sha'ar Hagolan is constructed adjacent to a source of fresh water, the Yarmouk River, suggesting that social factors were at least as important a motivation for its construction as was the need for water.

The Chalcolithic is as surprising as the PPNB for the absence of water management structures; this is made more problematic by the claims for irrigated agriculture within the period, which ultimately have very little – if any – basis for their support. This period appears to be the most in need of the application of the 'water availability index' and the isotopic analysis of sheep/goat bone as a means to verify those claims. It is only when we get into the Bronze Age that structural evidence for water management becomes more widespread – although the evidence remains sparse and often ambiguous. Intriguingly this begins within the Early Bronze Age, identified as a period of relatively high rainfall and consequently the incentive for investing in hydraulic engineering appears to lie within socio-economic forces rather than being the consequence of arid conditions. We must note, however, that a shift towards aridity is recorded towards the end of the Bronze Age, a period which includes a great deal of climatic instability (Chapter 7). Further progress on understanding the relationship between the construction of hydraulic devices and climate change will ultimately depend on securing a finer-grained chronological resolution for both elements of this relationship.

With the relatively patchy archaeological information available and the difficulty of securing absolute dates on water management structures, all that we can conclude at present is that there was a gradual development of water management technology through the Bronze and Iron Ages to reach the diverse and sophisticated systems used in the Nabataean. They were in turn supplemented by the introduction of new techniques, such as arched covers over cisterns, to result in the impressive hydraulic systems that are found at Humayma and Petra.

29.8 POLITICS AND PERCEPTIONS OF WATER

The scale and complexity of the water management systems at Humayma and Petra draws us to a question that is relevant to all of the communities with water management systems that have been referred to in this volume: who controls its supply and distribution? The character of the hydraulic engineering works in the large Nabataean towns suggest that royal patronage had been required for their construction and that a bureaucracy had been necessary for their management – as hinted at by the sparse documentary evidence referred to in Chapter 19. What about the PPNB wadi barriers and cisterns in the Jafr Basin or the well at Sha'ar Hagolan? Who organised their construction and controlled access to the resulting water supply? The nature of the archaeological record suggests that answers will never be forthcoming to such questions and such matters can only be substantially addressed within historically documented or contemporary communities.

We have already referred to Chapter 27, which exposed the political and social complexities involved in the development and running of the cooperative in Wadi Faynan drawing on piped water from Wadi Ghuwayr to irrigate tomatoes and watermelons. Chapter 28 does much the same, but at the vastly larger scale of the city of Amman. This chapter recounted the decisions by the Jordanian government to introduce water rationing to the city in 1987, guaranteeing water for only one day per week, and to prioritise the urban water supply between 2000–03, restricting the amount of irrigation in the Jordan Valley and forbidding summer crops. It then described how control partly moved to the World Bank and other donors who financed a rehabilitation of part of the Amman water system on the condition that a private operator would be brought in for its management. That contract went to a company known as LEMA, which attempted to manage the Amman water supply up until the end of its contract in 2006, but was constrained by not having any involvement in the management of water at a national level and in decisions about the overall amount of water allocated to Amman. So in this case we can see three major tiers of control regarding Amman's water during the past decade: a private company, the national government, and supra-national financing organisations.

Did this matter to the consumers? Chapter 26 undertook a survey of consumer attitudes to the water supply in Amman, exploring how storage and water usage varied between the relatively high- and relatively low-income areas of the city. The study documented the contrasting strategies adopted between the two income groups to adapt to the water rationing – wealthier households could simply buy additional storage cisterns or purchase supplies from tankers while poor households undertook meticulous management of day-to-day tasks to maximise the usage of water when available. The most striking finding was that both groups of households were broadly content with the rationed supply of water, some recognising that a more continuous supply would lead to wastage. Of greater concern to all consumers was the quality of water: there was a widespread avoidance of drinking mains water and overall low level of satisfaction with its quality. Improving the quality rather than quantity of water appears to be the highest priority for the city, irrespective of whether management is provided by public ownership, private company or some combination of the two.

29.9 WATER, LIFE AND CIVILISATION: THE VIRTUES AND CHALLENGE OF INTERDISCIPLINARITY

Archaeologists are prone to look at those who study extant communities with a degree of envy: those working within the WLC project would so much like to know about the governance of the water supply at Humayma and Jawa, or the tribal politics within the Neolithic settlements such as WF16 and Beidha, in the manner that the WLC human geographers know about these for Amman and the Al-Rashaida cooperative in Wadi Faynan. Archaeologists would relish knowing about at least some of the people who lived at their sites of interest, in terms of identity, personality and life history, rather than just having to document and interpret material remains in the effective absence of a social and political context. But such is the discipline of archaeology, and they have the benefit over their colleagues in knowing the long-term history of the communities they study. The origin, development and eventual abandonment of Neolithic towns such as Beidha over a period of several millennia can be documented and studied; what will have happened to the city of Amman or the village of Faynan Al-Jadida by the end of the twenty-first century? What will be the extent and impact of climate change on the Middle East? None of the participants of the WLC project will know.

The questions posed by the WLC archaeologists, human geographers, geologists and hydrologists were often difficult for the climate modellers. Even high-resolution climate models do not produce reliable data at the fine scales of relevance to individual settlements. Moreover, the modelled precipitation was found to be severely biased, to the extent that it was not possible to use model output directly to drive hydrological models. The instinct of a climate modeller faced with these challenges may be to turn away from interdisciplinary collaboration, and return to the model. Perhaps a different model formulation, a hitherto neglected forcing or an increase in resolution will rectify the model precipitation bias and refine the scale at which the model can reliably operate. However, it became apparent during the first years of WLC that model improvements of this magnitude were not feasible during the timescale of the project. This raised two questions: 'How can we use our current models to drive hydrological models for the past and future?' and, crucially, 'With all these problems, can current models say anything meaningful about the future?' Revisiting the model output and comparing it with both palaeoenvironmental and historical data enabled us to address both these questions, and develop statistical techniques to make maximum use of the model's output while compensating for its limitations. Collaboration with physical geographers, hydrologists, human geographers and geologists has thus clarified the questions that climate models can answer, demonstrated their strengths and weaknesses, and motivated the WLC climate modellers to develop methods of working with the available tools.

Becoming more sensitive to, or perhaps simply reminded about, the particular strengths and weaknesses of one's own discipline is one of the benefits of working within an interdisciplinary project such as Water, Life and Civilisation. Another is simply becoming more aware about the aims and methods of other disciplines, so that one can make a more informed critical judgement when needing to draw upon their claims in one's own work, or simply when reading literature for its own sake. Appreciating the complexity of disciplines other than one's own is of immense value. Going significantly beyond these benefits, however, is the simple fact that interdisciplinary work allows one to tackle questions about the past, present and future that disciplines are unable to do when working in isolation.

That was the presumption behind the Water, Life and Civilisation project; whether it has achieved its overall aims is ultimately for others, rather than the participants of the project itself, to judge. But we are confident that the interdisciplinarity has gained insights into the past, present and future that would simply not have otherwise been achieved. Perhaps of most value has been the evaluation of those Regional Climate Models that are being used to predict the future by testing their ability to predict the past against the palaeoenvironmental evidence, the analysis of archaeological site distribution by GIS cost-distance analysis, and the use of hydrological models as the bridge between climate change and human communities. The use of hydrological models is especially important because the complex interactions, feedbacks and thresholds of the various elements within the hydrological system result in a non-intuitive response to climate change. Without such hydrological models, archaeologists are unable to draw on the substantial advances being made in understanding the climate system for their studies of human communities.

It is perhaps the advances in methodology represented by the work within this volume that will have longer-term value than any particular conclusions that have been drawn about past communities and the potential impact of climate change. Such methodological advances do not only lie with the interdisciplinarity but are also found within the core disciplinary areas. In archaeological science, for instance, the formulation of a robust 'water availability index' using phytoliths and the progress regarding the interpretation of isotopes in faunal material, cereal grain and carbonate deposits will have long-term value for this discipline.

By stressing the methodological advances, we do not intend to undervalue the substantive gains in knowledge that have been achieved by the Water, Life and Civilisation project: as a consequence of its research we have acquired a wealth of new understanding about the changes in the hydrological climate and its impact on communities in the southern Levant, if not the more extensive region of the Middle East and North Africa that the project had identified in its original mission.

One of the most effective measures for gains in knowledge is by our growth of ignorance. This has certainly been the case and numerous areas for further study have been highlighted, perhaps most notably the need to improve our chronologies for archaeological settlements and for the construction of hydraulic installations, to enhance our understanding of hydrological processes such as infiltration and transmission rates to refine the existing models, and to develop the climate models further. A great deal of the work required is fairly 'routine' in terms of simply collecting more data from the field and undertaking more laboratory analyses. Such work is nevertheless costly, and ultimately the greatest weakness in the Water, Life and Civilisation project was the amount of funding it had available – in spite of the huge generosity of the Leverhulme Trust in terms of its £1.25m award. This is indeed a mark of its success: the range of topics, case studies and analyses that became both desirable and feasible to study as the research progressed and the interdisciplinarity emerged became both daunting and ultimately frustrating as the available resource came to an end – although several of these are being continued via other sources of funding and are now the subject of new research applications.

29.10 THE IMPACT OF THE HYDROLOGICAL CLIMATE ON HUMAN COMMUNITIES

What was the impact of changes in the hydrological climate on past human communities? Unquestionably this has been substantial: the WLC project has demonstrated that the course of human history in the southern Levant from 20,000 years ago to the present day has been intimately inter-woven with changing patterns of rainfall and water availability. That history is far more complex than people simply having been either the victims or the beneficiaries of climate change. Past communities created their own hydrological environments, sometimes unintentionally by clearing vegetation that changed the dynamics of their environments, sometimes quite deliberately by hydraulic engineering, ranging from simple wadi barriers in the Neolithic to the extensive aqueducts, dams and cisterns of the Nabataean.

What will be the impact of changes in the hydrological climate on future human communities? That is, of course, not only more unknown than the past but is effectively unknowable. While the 'business as usual' climatic scenario does indicate significant reductions in rainfall and increase in temperature, the Water, Life and Civilisation research has perhaps reached rather more optimistic conclusions than might have been expected. The study of Amman has shown that technical and managerial improvements to the water supply system could mitigate against reductions in water supply; the modelling in Wadi Faynan has shown the positive hydrological impacts of encouraging vegetation coverage; the B1 and B2 climate scenarios have shown that reduction in greenhouse gas emissions can leave the predicted rainfall patterns unchanged from those of today – although there remains continued pressure on the water supply from physical factors, such as increased evaporation and changes in land use as well as socio-economic factors, such as population growth and economic development.

Perhaps the most encouraging evidence to have come from the Water, Life and Civilisation project is simply the fact of long-term human settlement in the southern Levant, one of the most arid regions on planet Earth. The archaeological record is one of human ingenuity in managing the water supply, coming from introducing new technology such as wells and dams, from changing settlement patterns, and, no doubt, from new forms of social and community relations. Ongoing adaptation can be seen, whether by the residents of Amman who organise their daily activities around an intermittent water supply, by the Al-Rashaida farmers of Wadi Faynan developing that tomato and watermelon growing cooperative or even by the politicians, engineers and financiers discussing the creation of the Red Sea/Dead Sea canal. Overall, the long-term history of our study region leaves us impressed with the extent to which human communities have always been engaged in the use of social relations and technology as means to mitigate against and adapt to changes in their water supply, whether arising from climate change or human action itself.

REFERENCES

Araus, J. L., A. Febrero, R. Buxo *et al.* (1997) Identification of ancient irrigation practices based on the carbon isotope discrimination of plant seeds: a case study from the south-east Iberian Peninsula. *Journal of Archaeological Science* **24**: 729–740.

Barker, G., D. Gilbertson and D. Mattingly, eds. (2007) *Archaeology and Desertification: The Wadi Faynan Landscape Survey, Southern Jordan.* Wadi Faynan Series Volume 2, Levant Supplementary Series. Oxford, UK and Amman, Jordan: Council for British Research in the Levant in Association with Oxbow Books.

Finlayson, B. L. and S. Mithen (2007) The early prehistory of Wadi Faynan, Southern Jordan: archaeological survey of Wadis Faynan, Ghuwayr and Al Bustan and evaluation of the Pre-Pottery Neolithic A site of WF16. In *Wadi Faynan Series 1, Levant Supplementary Series 4*, ed. B. Finlayson and S. Mithen. Oxford and Amman, Jordan: Oxbow Books and the Council for British Research in the Levant.

Glausiusz, J. (2010) New life for the Dead Sea? *Nature* **464**: 1118–1120.

Hauptmann, A. (2007) *The Archaeometallurgy of Copper: Evidence from Faynan.* New York: Springer.

Helms, S. (1981) *Jawa: Lost City of the Black Desert.* New York: Cornell University Press.

Jenkins, E. (2009) Phytolith taphonomy: a comparison of dry ashing and acid extraction on the breakdown of conjoined phytoliths formed in *Triticum durum*. *Journal of Archaeological Science* **36**: 2402–2407.

Madella, M., M. K. Jones, P. Echlin, A. Powers-Jones and M. Moore (2009) Plant water availability and analytical microscopy of phytoliths: implications for ancient irrigation in arid zones. *Quaternary International* **193**: 32–40.

Mattingly, D., P. Newson, O. Creighton *et al.* (2007) A landscape of Imperial Power: Roman and Byzantine Phaino. In *Archaeology and Desertification. The Wadi Faynan Landscape Survey, Southern Jordan*, ed. G. Barker, D. Gilbertson and D. Mattingly. Oxford: Oxbow Books pp. 305–348.

Mithen, S., Finlayson, B., Najjar, M. *et al.* (2010) Excavations at the PPNA site of WF16: a report on the 2008 Season. *Annual of the Department of Antiquities of Jordan* **53**: 115–126.

MWI (Ministry of Water and Irrigation) (1997) *Water Strategy for Jordan.* Jordan: Ministry of Water and Irrigation.

MWI (Ministry of Water and Irrigation) (2004) *National Water Master Plan of Jordan.* Amman: GTZ and Ministry of Water and Irrigation.

Oleson, J. P. (1992) The water-supply system of ancient Auara: preliminary results of the Humeima Hydraulic Survey. In *Studies in the History and Archaeology of Jordan IV*, ed. M. Zaghoul. Amman: Department of Antiquities pp. 269–276.

Rosen, A. M. and S. Weiner (1994) Identifying ancient irrigation: a new method using opaline phytoliths from emmer wheat. *Journal of Archaeological Science* **21**: 125–132.

Rossignol-Strick, M. (1999) The Holocene climatic optimum and pollen records of sapropel 1 in the eastern Mediterranean, 9000–6000 BP. *Quaternary Science Reviews* **18**: 515–530.

Index

'Ain Abu Nukayla, 200
4.2 ka event, 98–99
8.2 ka cold event, 105
Acacia, 160, 221
agriculture/farming
   altitude and, 285–286
   Bedouin tribes, 192, 240, 247, 291
   Beidha area, 247–248
   Bronze Age, 238–239, 299, 385, 396
   Byzantine period, 386, 397
   Chalcolithic period, 385, 396–397
   Classical period, 386, 396
   crop water requirement threshold, 324
   current and future water demands, 410–411
   current urban water supply issues and, 456, 457
   Humayma, 305, 306–308
   Iron Age, 210
   irrigation, *See* irrigation
   Jawa, 299
   Nabataean period, 385
   Neolithic period, 196, 238, 383–385, 396–397
   origins of, 2, 4
   past Jordanian plant use overview, 383–386
   southern Ghors (current), 177–178, 187
   tomato cultivation, 448–449, 450
   water allocation strategy (current), 409–410
   water reuse issues, *See* water reuse study
   water/land management and development, *See* socio-economic development and water/land management study
   watermelon cultivation, 449–450
Ain el Sultan, 197
Ain Ghazal, 196
Akkadian civilisation, 94
alkenone unsaturation ratios, 82
altitude
   carbon stable isotope analysis and, 376, 378
   cost distance and, 284
   human settlement patterns and, 277, 284–286
Amman
   current water supply issues, *See* water supply (political discourses and public narratives)
   Iron Age water system, 210
   social equity of water supply study, *See* social equity of water supply study (Greater Amman)
Amora/Samra Formation, 115
amplifier lakes, 114
analogue matching, 389
anhydrite deposition, 115
animal population
   Bronze Age, 285, 292
   human population estimation and, 326
   Humayma study, 330
   stable isotope analysis of remains, *See* diet and environment reconstruction (Ancient Jordan)
aqueducts
   Humayma, 212–213, 305, 308, 315
   Petra, 212–213
   Wadi Faynan, 223
aragonite deposition
   dating of, 117–118
   Lake Lisan assessment using geochemical proxies, 120–121
   Lake Lisan findings, 113, 115–116
   Lake Lisan levels reconstruction and, 119
   Lisan Formation, 115–116
   mode of, 81
archaeology methodology, development of, 475–476
archaeology sub-project, 5, 7
aridity
   Holocene aridification investigation, 105–111
   trend towards (5,000 BP–present), 94, 95, 102
arsenic, water concentrations, 167, 172
Artemisia, 89
Arundo donax, 164
ASEZA, 452
ash-Shalaf, 384
Aşıklı Höyük, 339
Atlantic storm track, 47
Atlit-Yam, 199

Ba'ja, 197, 198, 201, 210
Bab edh-Dhra (also known as Bab-adh-Dhra), 98, 208, 385
Babylonian period, 99–100
barley, *See also* cereals
   nutrient requirements, 420
   phytolith formation and irrigation, *See* phytolith formation and irrigation study
baroclinicity (σ), tropospheric, 30, 43–44, 47, 48
barrages, 203–204, 205–206, 212
Basta, 197
Bedouin tribes
   agriculture/farming, 192, 240, 247, 291
   Bedouin dam, 219
   desalination methods, 192
   legal status of land used by, 446–447
   modern tent site description, 387–388
   modern tent site phytolith analysis, *See* phytolith analysis (past plant use study)
   socio-economic development, *See* socio-economic development and water/land management study
Beidha
   buildings at, 196
   groundwater resources, 197
   map of surroundings, 246
   palaeoenvironmental reconstruction, *See* palaeoenvironmental reconstruction (Beidha)
   previous studies, 246, 253
BIOME study, 107
Bir Abu Roga, 253
Birkat Ram Lake, 79, 96, 99
Bølling-Allerød warm interval, 72, 87, 88–89, 252, 471

bone collagen, stable isotope analysis, *See* diet and environment reconstruction (Ancient Jordan)
BRIDGE project, 297
brine evolution studies, 115
Bronze Age
  agriculture/farming, 238–239, 299, 385, 396
  Dead Sea levels, 155, 273, 285
  diet and environment reconstruction, *See* diet and environment reconstruction (Ancient Jordan)
  environmental/climatic review, 95–99
  hydrological context of human settlement study, 238–239, 241
  Pella occupation, 340
  Wadi Faynan occupation, 159, 222–223
  water management archaeology, 194, 206–207, 207–210, 211, 222
  water resource and climate change modelling, *See* water resources and climate change modelling study (Jawa)
  water's influence on settlement patterns, *See* water and settlement pattern investigation (Chalcolithic to Early Bronze Age)
Byblos, 208
Byzantine period
  agriculture/farming, 386, 397
  diet and environment reconstruction, *See* diet and environment reconstruction (Ancient Jordan)
  Khirbet Yajūz occupation, 341
  Sa'ad occupation, 341
  Tell Ya'amūn occupation, 341
  Wadi Faynan occupation, 159, 164, 223

calcretes, 82
carbon dioxide, *See also* greenhouse gas (GHG) concentrations
  carbon stable isotope analysis and atmospheric concentrations of, 376
carbon stable isotope analysis
  archaeological applications, 348, 374–375
  bone collagen, *See* diet and environment reconstruction (Ancient Jordan)
  calcretes, 82
  charred cereal remains, *See* carbon stable isotope analysis and environmental variables study
  effect of environmental variables on, *See* carbon stable isotope analysis and environmental variables study
  lacustrine sediments, 81, 107
  molluscs, *See* molluscs/snails
  palaeoenvironmental reconstruction, *See* palaeoenvironmental reconstruction (Beidha)
  principles of, 373–374
  speleothems, *See* speleothem records
carbon stable isotope analysis and environmental variables study, 378–379, 476
  archaeological applications of carbon stable isotope analysis, 374–375
  identification of environmental variables, 375–376
  method principles, 373–374
  modern reference dataset development, 376–377
carbonate (tufa) deposits
  Dead Sea margin distribution, 122
  sequence analysis, 255–256, 255–256, 259–261, 261–264
Çatalhöyük, 339
ceramics, water management archaeology, 207
cereals
  carbon stable isotope analysis of remains, *See* carbon stable isotope analysis and environmental variables study
  palynological records, limitations of, 382
  past plant use, *See* phytolith analysis (past plant use study)
Chalcolithic period
  agriculture/farming, 385, 396–397
  environmental/climatic review, 95–98
  Wadi Faynan occupation, 159, 222
  water management archaeology, 206–207, 210
  water's influence on settlement patterns, *See* water and settlement pattern investigation (Chalcolithic to Early Bronze Age)
Chenopodiaceae, 78, 89, 228
Choga Mami, 200–201
cisterns
  Bronze Age, 208, 209
  covered, 212–213, 326
  environmental conditions and, 284
  in Humayma, *See* Humayma hydrological modelling study (2,050–1,050 BP)
  Iron Age, 210
  Nabataean, 211, 212–213
  Neolithic, 204–205
  present-day use, 247
civilisation, *See also* human settlement
  attributes of, 2
  evolution in the Levant region, 72
  meaning of, 1
  summary of relationships with water, 2
Classical period
  agriculture/farming, 386, 396
  hydrological context of human settlement study, 239–241, 242
  Wadi Faynan occupation, 159, 223–224
  water management archaeology, 239–241, 242
clastic deposition
  Lake Lisan findings, 113, 115
  Lake Lisan levels reconstruction, 119
climate change
  assessing the impact of, 2–3
  Bronze Age modelling, *See* water resources and climate change modelling study (Jawa)
  carbon stable isotope analysis for reconstructing, 374
  current and future water demand and, 412–413, *See also* water demands and future strategies study (present-day Jordan)
  Dead Sea levels and future projections, 148
  impacts on rainfall-runoff, *See* rainfall-runoff study (River Jordan)
  projected impact on water availability, *See* water availability projection study (Wadi Hasa)
  sea level pressure (SLP) and projections of, 114–115
  summary of current water crisis and, 1
  testing credibility of future projections, 106, 110–111
  WLC project conclusions, 470–471, 471–472, 479–480
climate modelling
  climate change impact assessment use, 2–3
  connecting hydrological models for impacts studies, 63–67
  data use in hydrological models, 132
  future climate, *See* climate modelling study (future climate)
  general circulation models (GCMs), *See* general circulation models (GCMs)
  Holocene aridification investigation, 105–111
  inter-disciplinary collaboration and, 478
  past climates, *See* climate modelling study (past climates)
  precipitation variability and, 23
  regional climate models (RCMs), *See* regional climate models (RCMs)
  spatial resolution refinement, 2–3, 469
  statistical rainfall model, *See* weather generator
  WLC project conclusions on, 469–470
  WLC project model, 5
climate modelling study (future climate), 51, 61, 470
  accuracy of regional model present-day climate representation, 53–55
  data and modelling methodology, 52–53
  key sources of uncertainty, 57–61
  previous studies, 51–52
  projected changes, 55–57
climate modelling study (past climates) 25, 469–470
  experimental configuration and model validation, 26–30
  ocean heat flux error, 29–30, 35, 38, 40, 47
  rainfall, 37–47
  recent past to present day, 30–33
  surface air temperature (SAT), 37
climate variability study (present-day), 13
  daily to inter-annual variability, 19–22
  data and methodology, 14–16
  mean climate, 16–17
  previous study findings, 13–14
  spatial precipitation pattern, 17–19
  teleconnection patterns, 14, 21–22, 23
COHMAP (Cooperative Holocene Mapping Project) model, 105
cooperative management, 446, 447–451
coral records, 85, 90

cost distance analysis
    Chalcolithic to Early Bronze Age, *See* water and settlement pattern investigation (Chalcolithic to Early Bronze Age)
    cost surface calculations, 273
Council for British Research in the Levant (CBRL), 4, 6
CROPWAT model, 144
Cupressus, 221
cyclones
    future Middle East climate projections and, 51
    present-day Middle East climate and, 17, 19
Cypriot Neolithic wells, 198–199

Dana Nature Reserve, 452
Darcy's law, 139
date palm, 383, 396–397
Dead Sea
    future water transfer to, 147, 154, 156, 178, 187
    GIS and shorelines, 148–149
    maps of, 96, 113
    mean rainfall (1860–1960), 150
    rapid climate change response study and, 114
    setting and early history of, 113–114, 114–116
Dead Sea levels
    Bronze Age, 155, 273, 285
    estimates over past 25,000 years, 148
    mean rainfall and levels (1860–1960), 150
    past, present and future levels and rainfall modelling, *See* Dead Sea levels and rainfall modelling study
    recent decline in, 73, 133, 148
    regression of level change against Jerusalem rainfall (1860–1960), 152
    review (25,000–5,000 BP), *See* palaeoenvironmental/palaeoclimatic review (25,000–5,000 BP)
    review (5,000 BP–present), *See* palaeoenvironmental/palaeoclimatic review (5,000 BP–present)
Dead Sea levels and rainfall modelling study
    future levels modelling, 152–154
    GIS and shorelines, 148–149
    modelling levels, 149–152
    rainfall estimation from elevation data, 154–155
Decapolis, the, 101, 340
deep-sea cores, 82–85
Deir 'Alla
    carbon stable isotope analysis, *See* carbon stable isotope analysis and environmental variables study
    phytolith formation, *See* phytolith formation and irrigation study
    water reuse, *See* water reuse study
desalination, 192, 411
development studies
    social equity of water supply, *See* social equity of water supply study (Greater Amman)
    socio-economic development and water/land management, *See* socio-economic development and water/land management study
    water reuse, *See* water reuse study
    WLC sub-project development, 7
    WLC sub-project members, 5
Dhra', 205–206, 385
Dhuweila, 384
diatoms, 113, 121, 122
Dibadiba, 266
diet
    current food security situation, 134
    phytolith analysis of past plant use, *See* phytolith analysis (past plant use study)
    reconstruction of, *See* diet and environment reconstruction (Ancient Jordan)
diet and environment reconstruction (Ancient Jordan), 344–345, 476
    animal and human isotope data, 341–344
    archaeological sites, 339–341
    isotopic approach, 337–339
    location map of sites, 338
    previous isotopic studies in MENA region, 339
Disi aquifer, 411

dolomite deposition, 115
donkey transport, 285
dune formation, 100
DYRESM model, 149–150

Eady growth rate (σ), 30, 43–44, 47, 48
East Atlantic (EA) pattern, 21, 22
East Atlantic/West Russia (EAWR) pattern, 14, 21
ECHAM5/MPI-OM1 coupled model, 60
economy
    impact of past water availability, 472–473
    Humayma taxation, 305
    limitations of palynological records as indicator, 382
    past plant use and changes in, *See* phytolith analysis (past plant use study)
    water/land management role in contemporary development, *See* socio-economic development and water/land management study
Ein Feshkha, 100
el Khawarij, 207
El Niño southern oscillation (ENSO), 21, 22
ENSEMBLES project, 145, 296
Epipalaeolithic period
    environmental/climatic review, *See* palaeoenvironmental/palaeoclimatic review (25,000–5,000 BP)
    Wadi Faynan, 158
    water management archaeology, 193–195, 214
es-Sela, 210
evaporation
    carbonate precipitate composition and, 261, 264
    Humayma hydrological model, 324
    salinity and sea surface, 155
    Wadi Faynan hydrology study, 164–165
evapotranspiration
    carbon stable isotope analysis and, 375–376
    future study directions, 186
    Humayma hydrological model, 324
    potential (PET), 180, 185–186
    southern Ghors, 177–178

farming, *See* agriculture/farming
figs (*Ficus carica*), 164, 383–386
flood water, *See also* runoff
    Bronze Age exploitation, 208
    Neolithic exploitation, 206
    present-day Wadi Faynan, 221
    Wadi Faynan peak flood estimation, 171
fluvial sediments, 77, 79, 95, *See also* sedimentation
forests, *See also* specific tree species
    deforestation, 99, 229
    reforestation, 99, 100, 101
    Wadi Faynan records, 229

Galilee caves, 80, 99, 101
gastropods, 81–82, *See also* molluscs/snails
general circulation models (GCMs)
    associated uncertainty, 132, 293, 297, 469
    HadAM3, 5, 73
    HadAM3P, 52, 107, 153
    HadCM3, 26, 52, 107
    HadSM3, 25, 26–27, 153, 296
    Last Glacial Maximum (LGM) simulation, 73–75
    output and geological proxy data convergence, 90
    spatial resolution of, 2
    stochastic simulation and, *See* water resources and climate change modelling study (Jawa)
    UK Meteorological Office (UKMO) coupled atmosphere-ocean GCM, 73
General Linear Model, 377
GenStat program, 377
GeoB5844–2 core, 72
Gerasa (also known as Jerash)
    archaeological records, 340
    diet and environment reconstruction, *See* diet and environment reconstruction (Ancient Jordan)
Gezer, 209
Ghab Valley, 72, 78–79, 88, 89

Ghassul, 348
Ghor Safi, 176–179, 186–187
Ghuwayr, 1, 221, 222, 238
   phytolith analysis study, *See* phytolith analysis (past plant use study)
GIS (Geographical Information System) analysis, 271–273
GISP 2 ice core, 72, 86
Globaler Wandes des Wasserkreislaufs – Jordan River project (GLOWA–JR), 131–132
grape cultivation, 385
greenhouse gas (GHG) concentrations
   changes during Holocene, 27, 28, 30–33
   emissions scenario uncertainty, 59–61, 412
   future Middle East climate and, 51
   Mediterranean storm track and, 110, 111
   past Middle East climates and, 29, 33, 35–36, 47
Grosswetterlagen (GWL) regimes, 16, 19–21
gypsum deposition
   Dead Sea findings, 88, 98
   Lake Lisan assessment using geochemical proxies, 120–121
   Lake Lisan findings, 81, 113, 115–116
   Lake Lisan level reconstruction and, 118, 119

HadAM3 model, 5, 73
HadAM3P model, 52, 107, 153
HadCM3 model, 26, 52, 107
HadRM3 model, 5, 26, 27, 107
HadRM3P model
   Jordan surface air temperatures simulation, 143
   River Jordan rainfall-runoff study use, 136–137
   water availability projection use, 175–176, 179
HadSM3 model, 25, 26–27, 107, 153, 297
halite deposition
   Lake Lisan findings, 115
   Lake Lisan levels reconstruction and, 119
Hammam Adethni, 163–164, 172
Hamrat al Fidan, 160
   Hamrat al Fidan spring, 167, 172
Harifian culture, 195
Hazeva Formation, 115
Hazor, 209, 210
Heinrich Event 1 (H1), 88
Heinrich Event 2 (H2), 88
Herodotus Basin, 85, 90
Hilazon Tachtit Cave, 194
Holocene, *See also* specific time period subdivisions
   aridification investigation, 111
   mid-Holocene wet event, 90
   review of palaeoenvironmental/palaeoclimatic records for early, 87, 90
   review of palaeoenvironmental/palaeoclimatic records for middle/late, *See* palaeoenvironmental/palaeoclimatic review (5000 BP–present)
Hula Basin/Valley
   current water levels, 133
   Hula Basin location, 72
   palaeoenvironmental/palaeoclimatic review (25,000–5,000 BP), 78–79, 89
   palaeoenvironmental/palaeoclimatic review (5,000 BP–present), 98, 99
human settlement
   chronological framework of Beidha occupation, 248–251
   current understanding of palaeoclimatic events' impact on, 3
   diet and environment reconstruction, *See* diet and environment reconstruction (Ancient Jordan)
   Humayma study, *See* Humayma hydrological modelling study (2050–1050 BP)
   impact of environmental and climatic change 5,000 BP–present, *See* palaeoenvironmental/palaeoclimatic review (5,000 BP–present)
   Jawa study, *See* water resources and climate change modelling study (Jawa)
   Northern Hemisphere climatic events and evolution of, 71–72
   patterns in Neolithic period, 196–197
   Wadi Faynan, 158–159, 172, *See also* Wadi Faynan human settlement study
   water's influence on, *See* water and settlement pattern investigation (Chalcolithic to Early Bronze Age)
   water management archaeology and, *See* water management archaeology

   water needs, 191–192
   WLC project conclusions on impact of hydrological climate change, 479–480
   WLC project conclusions on impact of water availability on prehistoric, 472–473
Humayma
   hydrological modelling, *See* Humayma hydrological modelling study (2050–1050 BP)
   phytolith analysis, *See* phytolith analysis (past plant use study)
   water management archaeology, 212, 213, 214
Humayma Hydraulic Survey, 305, 306–312, 328–330
Humayma hydrological modelling study (2,050–1,050 BP), 302, 473
   hydrogeological remote sensing and digital elevation modelling, 313–320
   modelling methodology, 320–326
   results of modelling, 326–330
   settlement evidence review, 303–313
humidity, 164
hunter-gatherer development, 72
hydraulic hypothesis, 2
Hydrological Model for the Karst Environment (HYMKE), 145
hydrological modelling
   climate change impacts on rainfall-runoff, *See* rainfall-runoff study (River Jordan)
   climate model data use, 132, 153
   connecting climate models for impacts studies, 63–67
   Dead Sea levels/rainfall modelling, *See* Dead Sea levels and rainfall modelling study
   Humayma, *See* Humayma hydrological modelling study (2,050–1,050 BP)
   HYMKE model, 145
   HYSIM model, 180
   HYSIMM model, 175–176, 180–182
   importance of, 479
   INCA model, 131, 137–139
   KINEROS model, 292
   Monte Carlo analysis and, *See* water resources and climate change modelling study (Jawa)
   Pitman model, *See* Pitman model
   Wadi Faynan, *See* Wadi Faynan hydrology study
   water availability projection, *See* water availability projection study (Wadi Hasa)
hydrologically effective rainfall (HER), 137, 293
hydrology sub-project, 5, 6
HYDRUS model, 418–419, 423, 427
HYMKE model, 145
HYSIM model, 180
HYSIMM model, 175–176, 180–182

ice sheet changes, 35
INCA model, 131, 137–139, 145
industry
   current and future water demand, 410
   deforestation and, 229, 239
   Wadi Faynan, 223
   waste management and water quality, 425
infiltration
   hydrological context of human settlement, 224, 228–230, 231–237, 239, 241–242
   palaeo-infiltration simulation, 228–230
   potential infiltration rate determination, 181
   soil moisture relationship with, 139, 141
   storm conditions and, 163
   vegetation and, 157, 164, 170, 172–173
   Wadi Faynan hydrology, 169–170, 172–173
insolation
   Mediterranean storm track and, 108
   past Middle East climates, 27, 28–29, 35–37, 37–38, 40–41, 47
Intergovernmental Panel on Climate Change (IPCC)
   climate change predictions, 2
   future Middle East climate data limitations, 51
intertropical convergence zone (ITCZ), 37–40, 48
Iraq-ed-Dubb, 384
Iron Age
   environmental/climatic review, 99

Wadi Faynan occupation, 159, 223
water management archaeology, 194, 210–211
irrigation, *See also* water management, water management archaeology
   carbon stable isotope analysis, 375, *See also* carbon stable isotope analysis and environmental variables study
   civilisation and, 2
   crop water requirement threshold, 324
   current systems in Wadi Faynan, 159, 162
   demand projections, 144
   irrigated field systems, 224, 241
   Jawa, 299
   phytolith formation and, *See* phytolith formation and irrigation study
   southern Ghors, 178, 187
   water allocation strategy (current), 409
   water reuse, *See* water reuse study
Islamic period
   diet and environment reconstruction, *See* diet and environment reconstruction (Ancient Jordan)
   environmental/climatic review of Early, 100–101
   Pella occupation, 340
   Wadi Faynan occupation, 159, 223
isotope analysis, *See* stable isotope analysis
Israel, climate variability, *See* climate variability study (present day)

Jabal Amman, 432
Jafr Basin Prehistoric Project, 201
Jammam-Amghar, *See* Humayma hydrological modelling study (2,050–1,050 BP)
Jawa
   past plant use overview, 385
   water resources and climate change modelling study, *See* water resources and climate change modelling study (Jawa)
Jebel Druze, 298
Jebel Hamrat al Fidan, 158, 172
Jerash, *See* Gerasa
Jericho, 197, 200
Jerusalem, rainfall data (1846–1996), 152
Jerusalem West Cave, 72, 79, 81, 89
Jordan
   climate variability, *See* climate variability study (present day)
   current environmental and climatic conditions, 133–134
   current water demands and future strategies, *See* water demands and future strategies study (present-day Jordan)
   current water resources, 406–407
   diet and environment reconstruction, *See* diet and environment reconstruction (Ancient Jordan)
   drainage pattern, 404–405
   field work summary, 7–8
   irrigation demand projections, 144
   Jordan Valley water management archaeology, *See* water management archaeology
   past water resources, 192–193
   population size projections, 175
   population statistics (current), 405
   present-day climate summary, 404–405
   Water, Life and Civilisation (WLC) project area, 6
Jordan Archaeological Data Information System (JADIS), 270
Jordan River
   Globaler Wandes des Wasserkreislaufs – Jordan River project (GLOWA–JR), 131–132
   rainfall-runoff study, *See* rainfall-runoff study (River Jordan)
Jordan Valley Authority (JVA), 423, 424, 446, 456
Juniperus, 228

Kharaneh IV, 194
Khirbet an-Nahas, 159
Khirbet as Samra (also known as Kherbet as-Samra)
   carbon stable isotope analysis, *See* carbon stable isotope analysis and environmental variables study
   phytolith formation, *See* phytolith formation and irrigation study
   water reuse, *See* water reuse study
Khirbet Faynan, 159, 223

Khirbet Iskander, 98
Khirbet Sawwan, 201
Khirbet Yajūz
   archaeological records, 341
   diet and environment reconstruction, *See* diet and environment reconstruction (Ancient Jordan)
Khirbet Zeraqoun, 207–208
KINEROS model, 292
Kirkbride, D., 246

lacustrine sediments, 75–76, 81, 107, *See also* sedimentation
Lake Kinneret (also known as Lake Tiberias)
   5000 BP–present, 98, 99, 100, 101, 102
   lake level reconstruction, 120
   previous modelling study, 150
lake levels
   5000 BP–present, *See* palaeoenvironmental/palaeoclimatic review (5000 BP–present)
   Dead Sea, *See* Dead Sea levels
   Lake Lisan, *See* Lake Lisan
   review (25,000–5,000 BP), 75–76, 87, 88, 89, 90
Lake Lisan
   chronology, 121–122
   Dead Sea evolution, 113–114
   evolution and palaeoclimate assessment using geochemical proxies, 120–121
   lake levels reconstruction, 118–120
   levels of, 75–76, 88, 89, 125–126
   map of, 113
   new east side data, 125–126
   palaeoclimatic data limitations, 122
   previous study limitations, 113, 121–122
   rainfall estimation from elevation data, 154–155
   rapid climate change response study and, 114
   sedimentation, 75–76, 81, 115–118
   stratigraphy, 115–116, 121–122
   tectonics, 114–115
   uranium-series dating, 116–118, 122–126, 471
Lake Tiberias, *See* Lake Kinneret
Last Glacial Maximum (LGM)
   geological record and GCM output convergence, 471
   Lake Lisan level reconstruction, 118
   palaeoclimatic records, 87, 88, 252
   palaeotemperature calculation, 82
   simulation of, 73–75
latitude, carbon stable isotope analysis and, 376
LEMA, 430, 439, 455, 458–462, 478
Levant
   palaeoenvironment/palaeoclimate, *See* palaeoenvironmental/palaeoclimatic review (25,000–5,000 BP), palaeoenvironmental/palaeoclimatic review (5,000 BP–present)
Leverhulme Trust, 1, 4
Lisan Formation, 115–116, 117–118, 121–122

M44–1-KL83 core, 72
Ma'ale Efrayim Cave, 72, 79, 80–81, 88, 95
Madaba-Dhiban plateau, 96, 98, 101
magnesium/calcium ratios, 81
Markov models, first-order, 64, 67
Masada, 100
maximum likelihood method, 65
MD84–461 core, 84
Mediterranean Sea, water transfer to Dead Sea, 147, 154, 156
Mediterranean storm track
   climate change projections and, 51, 56–57, 61, 469
   greenhouse gas (GHG) concentrations and, 110
   modelling accuracy, 108–109, 110
   past climates and, 44, 46–47, 48, 469
   present-day, 16–17
MEGA (Middle Eastern Geo-Database for Antiquities), 271
meridional overturning circulation (MOC), 31
Mesopotamia, 2, 200–201
meteorology sub-project, 5, 6

Mg/Ca ratios, 81
Middle Range Theory, 389
Millennium Development Goals (MDGs), 429, 430
mineralogy analyses, 256
Ministry of Municipal Affairs (MoMA), 452
Ministry of Water and Irrigation (MWI), 407
    current urban water supply issues, 456, 461–462, 464
    National Water Master Plan (NWMP), 408–410, 430, 455, 456, 470
MM5 model, 55, 59
molluscs/snails
    25,000–5,000 BP, 81–82, 90, 125
    5,000 BP–present, 95, 96, 99
Monte Carlo analysis
    Humayma hydrological modelling, 303, 320–326
    water resources and climate change modelling, *See* water resources and climate change modelling study (Jawa)
Moringa Cave, 95, 100
Mount Sedom Cave, 98
Mylouthkia, Neolithic wells, 198–199

Nabataean period
    agriculture/farming, 385
    Beidha human occupation, 251
    Humayma occupation, 305–306, 312–313, 315
    Wadi Faynan occupation, 159, 223
    water management archaeology, 194, 211–214
Nahal Qanah Cave, 80, 96, 98, 99
National Water Master Plan (NWMP), 408–410, 430, 455, 456, 470
Natufian period
    Beidha human occupation, 249–250, 253–254, 258, 260–261, 264, 265
    water management archaeology, 194–195
Negev Desert
    25,000–5,000 BP, 82, 88, 89, 252
    5,000 BP–present, 95, 96, 99, 100, 101–102
    location of, 96
    water management archaeology, 195
Neolithic period
    agriculture/farming, 383–385, 396–397
    Beidha, human occupation of, 250–251, 254–255, 258, 264–265, 265–266
    chronological framework, 193
    human settlement patterns, 196–197
    hydrological context of human settlement, 237–238
    site locations, 193
    Wadi Faynan, 158, 164, 222–223
    water management archaeology, *See* Neolithic water management archaeology
Neolithic water management archaeology
    cultural developments, 195–196
    flood water/runoff exploitation, 200–206
    groundwater exploitation, 197–200
    human settlement patterns, 196–197
    water demands, 196, 206
Nerium oleander, 164
Nile (River)
    turbidites, 85
    water levels, 101, 102, 339
Nile Valley, diet studies, 339
nitrogen isotope analysis, bone collagen, *See* diet and environment reconstruction (Ancient Jordan)
North Atlantic Oscillation (NAO), 14, 21–22, 37
North Atlantic storm track, 41–44, 48
North Caspian Pattern (NCP), 14
North Qalkha, *See* Humayma hydrological modelling study (2,050–1,050 BP)
North Rift Basin (NRB) settlement, *See* water and settlement patterns investigation (Chalcolithic to Early Bronze Age)
Northern Dead Sea Basin (NDSB) settlement, *See* water and settlement patterns investigation (Chalcolithic to Early Bronze Age)
Nubia, 339

oak (*Quercus*), 88, 89, 221, 228
oasis theory, 2
Ocean Drilling Program (ODP) Site 967, 84
Ohalo II, 194, 195

Oleson, J., 305
olive cultivation
    5,000 BP–present, 99, 100, 305, 307, 385
    altitude and, 285, 286
    socio-economic development and, 285
orbital parameters, past Middle East climates and, 28–29
orography
    past Middle East climates and, 37, 44
    precipitation and, 17–19, 164, 176
Ottoman period, Wadi Faynan occupation, 159, 164, 223
oxygen isotope analysis, 82–85
    molluscs, *See* molluscs/snails
    palaeoenvironmental reconstruction, *See* palaeoenvironmental reconstruction (Beidha)
    speleothems, *See* speleothem records

Pacific storm track, 47
palaeoclimatic records
    See palaeoenvironmental/palaeoclimatic review (25,000–5,000 BP), palaeoenvironmental/palaeoclimatic review (5,000 BP–present)
palaeoenvironment
    Beidha reconstruction, *See* palaeoenvironmental reconstruction (Beidha)
    palaeoenvironmental studies sub-project, 5, 6–7
    review of records (5,000 BP–present), *See* palaeoenvironmental/palaeoclimatic review (5,000 BP–present)
    review of records (25,000–5,000 BP), *See* palaeoenvironmental/palaeoclimatic review (25,000–5,000 BP)
    Southern Levant (*c.* 20,000–7,000 cal. BP), 253
palaeoenvironmental reconstruction (Beidha), 245, 265–266, 472–473
    chronological framework of human settlement, 248–251
    materials and methods, 255–258
    palaeoenvironmental context, 251–255
    present day environment, 246–248
palaeoenvironmental/palaeoclimatic review (25,000–5,000 BP), 71, 90, 471
    climate models, 73–75
    dating, 73
    generalised climatic evolution of past 25,000 years, 71
    major climatic events, 85–90
    marine records, 82–85
    non-marine sedimentary records, 75–77
    palaeoclimate record summaries, 85
    terrestrial geochemical records, 79–82
    terrestrial palaeobotanical records, 77–79
palaeoenvironmental/palaeoclimatic review (5,000 BP–present), 94, 102, 471–472
    4.2 ka event, 98–99
    aridity trend, 95, 102
    Babylonian period, 99–100
    Bronze Age, 95–99
    Byzantine period, 99–100
    Chalcolithic period, 95–98
    climate change at *c.* 1,400 BP, 100
    climate instability at *c.* 5,200 BP, 98
    climate instability *c.* 1,000 BP and Decapolis society demise, 101
    Early Islamic period, 100–101
    Iron Age, 99
    past 850 years, 101–102
Palaeolithic period, Wadi Faynan, 158
palaeosols
    25,000–5,000 BP, 77, 89, 90
    5,000 BP–present, 95, 98, 99, 101
palynological records
    Beidha, 254–255
    limitations as palaeoeconomic indicator, 382
    previous study use, 107–108
    review of, *See* palaeoenvironmental/palaeoclimatic review (5,000 BP–present); palaeoenvironmental/palaeoclimatic review (25,000–5,000 BP)
    Wadi Faynan, 228–229
Peace Treaty (1994), 411
pedogenesis, 82

INDEX

Pella
   archaeological records, 339–340, 385, 386
   diet and environment reconstruction, See diet and environment reconstruction (Ancient Jordan)
Peqiin Cave, 72, 80–81
Petra
   vegetation, 385–386
   water management system, 211, 212, 213–214
Phoenix, 164, 221
Phragmites australis, 164
phytolith analysis (past plant use study), 381, 397, 476
   archaeological applications of phytolith analysis, 382–383
   methods, 389–391
   overview of Jordanian past plant use, 383–386
   previous ethnoarchaeological studies, 383
   southwest Asian archaeological plant proxies, 382
   study sites, 382, 386–389
phytolith formation and irrigation study, 347, 370–371, 475–476
   aims of, 351
   archaeological background, 347–348
   materials and methods, 351–356
   phytolith formation background, 348–350
   phytoliths as water availability proxy, 350–351
phytoliths
   archaeological applications of analysis, 382–383
   formation process, 348–350
   Humayma study, 307
   irrigation and formation of, See phytolith formation and irrigation study
   modern plant extraction methodology, 353–355
   past plant use study, See phytolith analysis (past plant use study)
   Phytolith Difference Index (PDI), 383
   water availability proxy role, 350–351
Pinus, 228
Pistacia, 79, 89
Pitman model
   rainfall-runoff study, 131, 137, 138
   Wadi Faynan human settlement study, 224
   Wadi Faynan hydrology study, 157, 168–170
pollen data, See palynological records
Populus euphratica, 164
pore size distribution index, 181
post-Lisan Damya Formation, 115
poverty
   national reduction strategies, 451–453
   participatory management and reduction of, 442–443, 450–451
precipitation/rainfall
   Beidha present-day statistics, 247
   current Middle East variability, See climate variability study (present day)
   Dead Sea rainfall modelling, See Dead Sea levels and rainfall modelling study
   early Holocene rainfall signal, 108
   future climate modelling, See climate modelling study (future)
   GWL regimes' relationship summary, 20
   HadSM3 GCM data, 296
   Holocene aridification investigation, 105–111
   hydrological context of human settlement, See Wadi Faynan human settlement study
   Jerusalem rainfall data (1846–1996), 152
   Jordan present-day statistics, 134, 404–405
   LGM and present-day climate model outputs, 73
   mean rainfall and Dead Sea levels data (1860–1960), 150
   modelling past climates, See Wadi Faynan human settlement study, See climate modelling study (past climates)
   predicted rainfall changes for 2070–2100, 412
   probabilities and River Jordan flow, 139–140
   quintiles for NAO positive and negative years, 22
   simulation in Jawa study, 295–296
   RCM representation, 53–55, 111
   simulated and proxy data comparison, 226–228, 241
   Southern Levant present-day statistics, 245
   statistical rainfall model, See weather generator
   Tafilah synthetic rainfall statistics, 225, 226
   unreliability as water source, 192
   Wadi Faynan hydrology study, 172–173
   Wadi Faynan present-day statistics, 164, 219–221
Qa' el Jinz, 177
Quercus (oak), 89, 221, 228
Qurayqira villages, 444

radiocarbon dating, Beidha human occupation, 248
rainfall, See precipitation/rainfall
rainfall-runoff study (River Jordan), 131, 143–145, 470–471
   climate change impact on river flow, 141–144
   future study directions, 145
   modelling framework, 136–139
   sensitivity studies on weather statistics' impact on river flow, 139–141
   study area and data resource, 132
   uncertainties in, 144–145
Ramtha
   carbon stable isotope analysis, See carbon stable isotope analysis and environmental variables study
   phytolith formation, See phytolith formation and irrigation study
   water reuse, See water reuse study
Ran aquifer, 411
Red Sea
   coral records, 85
   palaeoclimatic records compilation, 83–84
   Red Sea Trough, 101, 225
   temperature and salinity, 83–84, 88, 89, 90
   water transfer to Dead Sea, 147, 154, 156, 178, 187
RegCM2/3, 53–55, 59
regional climate models (RCMs)
   climate change predictions, 412–413, 472
   future Middle East climate prediction, See climate modelling study (future climate)
   HadRM3, 5, 26, 27, 107
   HadRM3P, 136–137, 143, 175–176, 179
   Holocene aridification investigation, 105–111
   hydrological context of human settlement, 218, 224–226
   hydrological model input from, 469
   monthly rainfall data disaggregation in conjunction with weather generator, 65–66
   palaeorainfall simulation, 224–226
   past climate modelling issues, 3
   precipitation representation, 111
   RegCM2/3, 53–55, 59
   spatial resolution of, 2
reservoirs
   Humayma, See Humayma hydrological modelling study (2,050–1,050 BP)
   modern Bedouin, 220
Roman period
   Beidha human occupation, 251
   diet and environment reconstruction, See diet and environment reconstruction (Ancient Jordan)
   Humayma occupation, 306, 313
   Khirbet Yajūz occupation, 341
   Pella occupation, 340
   Sa'ad occupation, 341
   Wadi Faynan occupation, 159, 164, 172, 223
runoff
   advantages and disadvantages as water resource, 192
   climate change impacts, See rainfall-runoff study (River Jordan)
   Humayma study coefficients, 324–325
   Neolithic exploitation of, 200–206
   palaeoenvironmental/palaeoclimatic records, 85, 90, 101
   vegetation patterns and, 155
   Wadi Faynan human settlement, 218, 219, 224
   Wadi Faynan hydrology study coefficients, 164
   water availability projection and assessment of, See water availability projection study (Wadi Hasa)

Sa'ad
   archaeological records, 341
   diet and environment reconstruction, See diet and environment reconstruction (Ancient Jordan)
Safi, 183–184

Sail Faynan cooperative, 450–451
salinity
    brackish groundwater desalination, 192, 411
    current trend in Jordan, 134, 144
    evaporation rate and, 155
    Lake Lisan palaeosalinity, 120
    sea surface (SSS) 25,000–5,000 BP, 82–85, 87, 88, 89–90
    seawater desalination, 411
    water reuse for irrigation and soil, 421–423
Salix acmophylla, 164
sapropels, 85, 89–90, 107
sea level pressure (SLP), climate change projections and, 56–57
seawater desalination, 411
Securing Water to Enhance Local Livelihoods (SWELL), 443
sedimentation
    carbonate (tufa) deposits, *See* carbonate (tufa) deposits
    granulometry analysis, 256, 258
    Lake Lisan assessment using geochemical proxies, 120–121
    Lake Lisan findings, 115–116
    Lake Lisan knowledge gaps, 121–122
    Lake Lisan level reconstruction and, 118–120
    palaeoenvironmental reconstruction from observations, *See* palaeoenvironmental reconstruction (Beidha)
    review of records, *See* palaeoenvironmental/palaeoclimatic review (25,000–5,000 BP); palaeoenvironmental/palaeoclimatic review (5,000 BP–present)
Sedom Formation, 115
seed grain size analysis, 348
Sha'ar Hagolan, 199
Shiqmim, 207
Shuttle Radar Topography Mission (SRTM) Digital Elevation Model (DEM), 272, 315–319
silicon
    methodology for analysis of extractable, 357
    phytolith formation and, 348–350, 358–362, 361–362, 366–370
Siloam Tunnel, 209
Smilax aspera, 164
snails, *See* molluscs/snails
social equity of water supply study (Greater Amman), 429, 440
    growth and structure of Greater Amman, 430–432
    household water supply, 434–435
    household water management strategies, 435–438
    interviewees' satisfaction and attitudes, 438–440
    interviewees' socio-economic and demographic characteristics, 432–434
    research design, 432
    water and development issues, 429–430
socio-economic development and water/land management study, 442, 453, 475
    cooperative management model, 446–448
    Faynan cooperative and poverty reduction, 450–451
    irrigated agriculture system dimensions, 448–450
    National poverty reduction strategies, 451–453
    participatory management and poverty reduction background, 442–443
    physical and social context of study area, 443–446
soil
    analysis in phytolith formation and irrigation study, 353
    Humayma sampling, 307
    infiltration and soil moisture relationship, 139, 141
    leaching methods, 423, 427
    Wadi Faynan hydrology, 163–164, 169
    water reuse and soil sustainability, *See* water reuse study
Soreq Cave
    Beidha human occupation studies, 252, 263–264
    differences in speleothem and GCM records, 296–297
    location of, 72
    palaeorainfall estimation evaluation, 226–228
    review of records (25,000–5,000 BP), 79, 80–81, 88, 89
    review of records (5,000 BP–present), 96, 98, 102
South Qalkha, *See* Humayma hydrological modelling study (2,050–1,050 BP)
southern Ghors region, 176–177
southern Levant
    palaeoenvironment (*c.* 20,000–7,000 cal. BP), 253
    present-day rainfall statistics, 245

speleothem records
    25,000–5,000 BP, 79–81, 87, 88, 89
    5,000 BP–present, 95, 98
    Beidha human occupation studies, 252
    GCM records and, 297–296
    palaeorainfall estimation evaluation, 226–228
    Soreq record, *See* Soreq Cave
springs, settlement patterns and, 277–281, 284, 285, 286
stable isotope analysis
    archaeological applications of carbon, 348, 374–375
    Beidha palaeoenvironmental reconstruction use, *See* palaeoenvironmental reconstruction (Beidha)
    bone collagen, *See* diet and environment reconstruction (Ancient Jordan)
    brine evolution investigation, 115
    calcretes, 82
    carbon, principles of, 373–374
    environmental variables and, *See* carbon stable isotope analysis and environmental variables study
    gypsum, 121
    lacustrine sediments, 81, 107
    marine records, 82–85
    molluscs, *See* molluscs/snails
    speleothems, *See* speleothem records
stochastic simulation, *See* water resources and climate change modelling study (Jawa)
stromatolites, 113, 125
strontium isotope analysis, 80, 81, 88, 115
strontium/calcium ratios, 81, 120–121
sulphur isotope analysis, 115, 121
synoptic regimes, present-day Middle East climate and, 16, 19–21

Tafilah (also known as Tafileh), 171, 225, 226
Tamarix, 164, 221
Tannur
    flow records, 178, 183–184
    future flow projections, 186–187
Tannur Dam, 177, 185
taxation, 305
tectonics, 159
Tel Bet Yerach, 98
Tel Rakan I, 384
teleconnection patterns, 14, 21–22, 23, *See also* specific patterns
Teleilat Ghassul, 385
Tell Abu an-Ni'aj, 385
Tell al-Hayyat, 385
Tell Deir Allah, 210
Tell el-Handaquq, 385
Tell esh-Shuna North, phytolith analysis, *See* phytolith analysis (past plant use study)
Tell es-Sa'idiyeh, 209
Tell Handaquq, 208
Tell Seker al-Aheimar, 199
Tell Um Hammad, 291
Tell Wadi Faynan, 159
Tell Wadi Feinan, 221, 222
    phytolith analysis, *See* phytolith analysis (past plant use study)
Tell Ya'amūn
    archaeological records, 340–341
    diet and environment reconstruction, *See* diet and environment reconstruction (Ancient Jordan)
temperature
    Last Glacial Maximum (LGM) calculation, 82
    past climate, surface air (SAT), 30–33, 33–37, 48
    present-day Jordan, 143, 404
    sea surface (SST) 25,000–5,000 BP, 84, 87, 88, 89–90
    simulation of future, 143
    Wadi Faynan (present-day), 164
terracing, 205–206, 214, 307–308, 312
thermohaline circulation (THC), 105
tomato farming, 448–449, 450
tombs, 341
tourism
    ecotourism promotion, 452–453

water demands, 410, 426
TRACK software, 16
TRAIN model, 144
transport, animal, 285
treated wastewater, See water reuse
tropospheric baroclinicity (σ), 30, 43–44, 47, 48
tufa (spring carbonate) deposits
    Dead Sea margin distribution, 122
    sequence analysis, 255–256, 259–261, 261–264
turbidites, 85, 90

UK Meteorological Office (UKMO) coupled atmosphere–ocean GCM, 73
$U_{37}^{k'}$ index, 82
Ulmus, 228
unaccounted for water (UFW), 455, 456–457, 459
United Nations
    Millennium Development Goals (MDGs), 429
    World Water Development Report, 429
University of Reading (UK), 1, 4
uranium-series dating
    Lake Lisan, 116, 117–118, 121–122, 122–126
    palaeoenvironmental reconstruction at Beidha, 255–256, 259
    theory, 116–117

vegetation
    Beidha area, 247
    infiltration and, 157, 164, 170, 172–173
    vegetative feedback systems and monsoon extension, 38–40
    Wadi Faynan human settlement, 221, 228–230, 231–233, 241, 242
    Wadi Faynan hydrology, 163–164, 170, 172–173
Via Nova Triana, 306

Wadi Abu Tulayha, 201–205
Wadi Afra, 177
Wadi al Bustan, 230
Wadi al-Wala, 96
Wadi Amghar, 308, 318–319
Wadi Araba, 159–160
    Wadi Araba Project, 452
Wadi Badda, 205
wadi barriers, 307–308, 312
Wadi Dana, 159–163, 166, 171, 172
Wadi el Farah, 277
Wadi Faynan
    evaporation, 164–165
    fluvial sediments, 77
    Holocene archaeology overview, 221–224
    human settlement, See Wadi Faynan human settlement study
    hydrology, See Wadi Faynan hydrology study
    maps of, 72, 158
    phytolith analysis, See phytolith analysis (past plant use study)
    precipitation (present-day), 164, 219–221
    review of palaeoenvironmental/palaeoclimatic records (25,000–5,000 BP), 77, 79, 253
    review of palaeoenvironmental/palaeoclimatic records (5,000 BP–present), 98, 99, 100, 101–102
    socio-economic development, See socio-economic and water/land management study
    summary of current physical and social conditions, 443–446
    temperature (present-day), 164
    topography, geology and hydrogeology, 159–163
    WF100, 222, 238
    WF16, See WF16
    WF4, See WF4
Wadi Faynan human settlement study, 218, 474–475
    developing integrative methodologies, 224–230
    Holocene archaeology overview, 221–224
    impact of ground cover changes, 231–233
    impact of rainfall changes, 230–231
    palaeo-hydrology scenarios, 233–237
    potential impacts of climatic variation, 237–241
    present-day vegetation, 221
    present-day water availability, 219–221
    scale and uncertainty in human–climate interactions, 218–219

Wadi Faynan hydrology study
    conceptual model, 167–168
    contemporary hydrology and hydrochemistry, 165–167
    evaporation, 164–165
    hydrological simulation results, 170–172
    numerical model, 168–170
    precipitation, 164, 172–173
    soils and vegetation, 163–164, 169, 170, 172–173
    study area and data resource, 158–159
    temperature, 164
    topography, geology and hydrogeology, 159–163
Wadi Fidan, 158, 163
Wadi Fifa, 176
Wadi Ghuwayr, 159–163, 166–167, 171, 172, 179
Wadi Halfa, 339
Wadi Hammeh, 194
Wadi Hasa, 166
    future projections of water availability, See water availability projection study (Wadi Hasa)
Wadi Kafrein, 179
Wadi Kerak, 176
Wadi Khanzeira, 176
Wadi Muqat, 96
Wadi Numeria, 176
Wadi Rajil, 291, 293, 295
Wadi Rayyan, 207
Wadi Ruweishid, 204
Wadi Sarar, 208
Wadi Shagyr, 160
Wadi Yitm, 314
Warren's shaft, 209
water
    settlement patterns and exploitation of, See water and settlement patterns investigation (Chalcolithic to Early Bronze Age)
    unaccounted for (UFW), 455, 456–457, 459
    WLC project conclusions on politics and perceptions of, 477–478
water and settlement pattern investigation (Chalcolithic to Early Bronze Age), 269, 473–474
Water Authority of Jordan (WAJ), current urban water supply issues, 456, 459, 461–462, 462–463, 464
water availability, See also water resources
    carbon stable isotope analysis application, See carbon stable isotope analysis and environmental variables study
    current global crisis, 1
    future projections study, See water availability projection study (Wadi Hasa)
    impact of past changes conclusions, 473
    water availability index, 391, 393, 397, 476
water availability projection study (Wadi Hasa), 175, 187, 471
    model set-up and calibration, 182–184
    modelling frameworks, 179–182
    study area and data resource, 176–179
water demand
    current increases in, 403
    current uses, 405–407, 422, 437
    future climate change impact, 470
    and future strategies, See water demand and future strategies study (present-day Jordan)
    projected, 470
water demands and future strategies study (present-day Jordan), 403, 413
    alternative water sources, 411–412
    climate change issues, 412–413
    current water resources, 406–407
    current water uses, 405–407
    future strategies, 407–410
    introduction, 403–404
    study area, 404–405
    water use planning and prioritisation, 410–411, 413
water management, See also irrigation
    archaeology of, See water management archaeology
    geology and, 284
    Humayma, See Humayma hydrological modelling study (2,050–1,050 BP)
    Jawa system, 295

water management, (cont.)
  social complexity emergence and, 2
  supply (allocation/distribution), *See* water supply (allocation/distribution)
  water quality issues, *See* water quality
water management archaeology, 191, 214
  Bronze Age, 206–207, 207–210, 211, 222
  Chalcolithic period, 206–207, 210
  chronological framework, 193
  Classical period, 239–241, 242
  Epipalaeolithic period, 193–195, 214
  Humayma, 305–306
  Iron Age, 210–211
  Jordan Valley water sources, 192–193
  Nabataean, 211–214
  Neolithic period, 195–206
  strategy development conclusions, 476–477
  Wadi Faynan, 222, 223–224
  water demands, 191–192
water quality
  current control strategies, 410
  current urban issues, 463
  social equity of water supply, 437, 439
  water reuse, 419, 425, 427
water resources, *See also* water availability, *specific resources*
  Beidha, 264
  climate change modelling study, *See* water resources and climate change modelling study (Jawa)
  current alternatives, 411
  current decline in, 404
  current Jordanian, 407
  Jordanian use by sector in 2000, 406
  National Water Master Plan (NWMP) strategy, 409
  past Jordanian, 193
water resources and climate change modelling study (Jawa), 289, 299–300, 473
  methodology, 292–296
  study area and data resource, 289–290, 290–292
water reuse
  background to, 403, 411
  study of, *See* water reuse study
water reuse study, 415, 427–428, 475
  aims of, 417
  Jordanian water reuse background, 415–417
Water Strategy for Jordan (1997), 430, 456, 470
water supply (allocation/distribution)
  current charges, 430, 438
  planning and prioritisation, 409–410, 410–411, 413
  political discourses and public narratives, *See* water supply (political discourses and public narratives)
  social equity of, *See* social equity of water supply study (Greater Amman)
  WLC project conclusions, 470, 473–475, 478
water supply (political discourses and public narratives), 455, 465, 478
  current situation in Amman, 456–457
  management of irregularity, 457–459
  private discourse versus public policy directives, 459–462
  selected performance indicators, 458
  unreliability in developing cities, 455–456
  users' opinions on continuous supply, 462–464
Water, Life and Civilisation (WLC) project
  inter-disciplinarity, 1, 6–8, 478–479
  project development, 6–8
  scope of, 4–6
  scope of this volume, 8
water-balance model, *See* Humayma hydrological modelling study (2,050–1,050 BP)
watermelon farming, 449–450
weather generator
  accuracy of observed seasonal cycle reproduction, 65
  bias issues, 67
  design and development of, 64–65
  hydrological context of human settlement study, 218, 219, 224–226
  modelling future Dead Sea levels use, 153
  monthly rainfall data disaggregation use in conjunction with regional model, 65–66
  palaeorainfall simulation, 224–226
  rainfall-runoff study use, 132, 137, 138, 144
  uses of, 63–64
  water availability projection use, 175–176, 179–180
wells, 197–200, 210, 212, 214
West African monsoon, 38–40, 48
WF100, 222, 238
WF16, 4, 196, 221, 223, 238
  phytolith analysis, *See* phytolith analysis (past plant use study)
WF4, 224, 240–241
  phytolith analysis, *See* phytolith analysis (past plant use study)
wheat, *See also* cereals
  carbon stable isotope analysis, *See* carbon stable isotope analysis and environmental variables study
  phytolith formation and irrigation, *See* phytolith formation and irrigation study
wine production, 340, 341, 386
Wisad pools, 200
woody material index, past plant use study, 391
World Water Assessment Programme (WWAP), 429

Yarmouk River, 406, 411, 426
Younger Dryas
  climate and human settlement, 72, 252
  lake levels, 76
  Lake Lisan level reconstruction, 119–120
  review of palaeoenvironmental/palaeoclimatic records, 87, 89, 471
  tufa analysis, 263–264

Zahrat adh-Dhra (also known as Zahrat edh-Dhra') 2 (ZAD 2), 384
Ze'elim, 100